Acoustic and electromagnetic waves

Acoustic and electromagnetic waves

D. S. Jones

Department of Mathematical Sciences
University of Dundee

CLARENDON PRESS · OXFORD
1986

Oxford University Press, Walton Street, Oxford OX2 6DP

Oxford New York Toronto
Delhi Bombay Calcutta Madras Karachi
Kuala Lumpur Singapore Hong Kong Tokyo
Nairobi Dar es Salaam Cape Town
Melbourne Auckland

and associated companies in
Beirut Berlin Ibadan Nicosia

Oxford is a trade mark of Oxford University Press

Published in the United States
by Oxford University Press, New York

British Library Cataloguing in Publication Data

Jones, D. S.
Acoustic and electromagnetic waves.
1. Sound 2. Electromagnetism
I. Title
534 QC225.15
ISBN 0-19-853365-9

Library of Congress Cataloging in Publication Data

Jones, D. S. (Douglas Samuel)
Acoustic and electromagnetic waves.
Bibliography: p.
Includes index.
1. Electromagnetic waves. 2. Sound-waves. I. Title.
QC661.J66 1985 530.1'41 84–27226
ISBN 0-19-853365-9

Set and printed in Northern Ireland by The Universities Press (Belfast) Ltd.

To

IVY

for the strength of her love and the generosity
of her spirit

Preface

Much research has been devoted to problems in electromagnetism and acoustics in the past two decades. Often the investigations run along parallel lines, the main differences being due to the vector character of the electromagnetic field and the scalar nature of sound waves. Sometimes these differences are significant because of boundary conditions or polarization effects but often the analysis is very similar especially in those cases when the model in electromagnetics leads to a discussion of scalar fields. It was therefore thought that a book which traced the common strands and brought out any appropriate modifications in going from one topic to the other would be helpful to workers in both areas.

One aim of the book is to enable the reader to come within striking distance of modern research. It would therefore have been desirable to include numerous references to recent papers to assist the reader in becoming acquainted with the many contributors to various subjects over the past twenty years. However, the implementation of such a policy has proved to be impracticable. The last twenty years has constituted a period of intense activity and there would not be space for both a text and a complete list of relevant papers. For instance, in microstrips, a topic hardly touched upon, there have been 10 000 papers and there are now specialized journals appearing frequently which deal with areas forming a small fraction of the book. The sheer pace of publication renders any collation on a reasonable scale unmanageable. Accordingly, all efforts to maintain an up-to-date comprehensive list of references have been abandoned. Nevertheless, the hope is that the treatment of fundamentals in this book will place the student in a firm position to undertake research and a few references have been added for those with specialized interests.

As far as electromagnetism is concerned the starting point of this book is Maxwell's equations. The reader is assumed to have reached this point through the customary courses on static electricity, magnetostatics, and current electricity. The book follows logically on these courses in that its viewpoint is macroscopic. No endeavour is made to relate the macroscopic laws to any theory, such as wave mechanics, of microscopic structure. Similarly, fluid motion is regarded as a macroscopic phenomenon whose governing equations are assumed to be familiar to the reader though a brief derivation of the equations pertinent to the propagation of sound is given.

The first chapter is devoted to the general properties of solutions of the equations controlling acoustics and electromagnetism in matter which has certain macroscopic properties. The chief objective, therefore, is a delineation of the methods of construction whether by separation of variables, potentials, or integral representations. Energy considerations and the boundary conditions to be met in traversing the interface between two media are also examined. Chapter 2 is general in nature as well, being occupied with relativistic effects in mechanics, the transmission of light, and moving media.

After Chapter 2 specific problems are considered in more detail. Thus, in Chapter 3, the radiation is determined from a moving source of constant strength and from a fixed source of varying magnitude. The patterns to be expected when the sources are distributed on slender bodies are also studied. Next, the attributes when fields are confined to the interior of a hollow finite container are discussed in Chapter 4. The propagation in tubes and waveguides, together with the influence of intervening obstacles, is the burden of Chapter 5.

The theme of Chapter 6 is the transmission of waves in an infinite domain in the presence of a plane boundary across which the material properties change discontinuously. There are various possibilities depending on whether the media are homogeneous, isotropic, anisotropic, or inhomogeneous and particular attention is paid to the problem of a stratified atmosphere over a plane earth. The guiding of waves outside cylindrical structures is the subject matter of Chapter 7.

Chapters 8 and 9 attend to the scattering of waves by objects, smooth in the one chapter and possessing edges in the other. Both chapters begin with an examination of some exact solutions and then employ these as a basis to develop approximate techniques for general shapes.

The last chapter studies waves of arbitrary time variations, in contrast to Chapters 4–9 where harmonic waves form the main substance. Here will be found short descriptions of SEM (singularity expansion method), nonlinear waves, and features not normally encountered with harmonic waves.

A variety of mathematical techniques and special functions, with their own peculiar traits, is deployed. In one or two places, explanation accompanies the main text but those used most often have been collected together for convenience in seven appendixes. Five of these deal with special functions; two cover tensors and the asymptotic evaluation of integrals respectively. When equations in the appendixes are referred to in the principal text they are prefixed by the identifying letter of the appendix. Similarly, a cross-reference to an equation in another chapter will carry the chapter number first in the absence of any other indicator.

Each chapter has a set of exercises at its end. Some of these are to test how well the reader has grasped the content. There are also some which extend theory begun in the chapter or give details of notions which were excluded for one reason or another from the bulk of the discourse.

Naturally, the parts dealing with electromagnetism have something in common with the author's earlier book The Theory of Electromagnetism published by Pergamon in 1964, which has been out of print for some years. But many topics are covered in this book which were not included in the earlier one and, even where the subject matter overlaps, there are considerable changes in the treatment, emphasis, and approach.

Finally, the author wishes to express his gratitude to Mrs E. D. Ross whose forbearance, good humour, perseverance, and skill were responsible for the production of a readable typescript from a very ragged manuscript. The author's obligation to his wife Ivy is too great to ever be expunged—without her unwavering backing and inspiration nothing would have been achieved.

Dundee 1984 D. S. J.

Contents

1. The representation of acoustic and electromagnetic fields

MUCH of the analysis of the wave motion in acoustics and electromagnetism has common features but the relevant equations can be derived only when the reader has a suitable background knowledge. In this book it will be assumed that the reader has available certain mathematical information and is familiar with, for example, vectors, the divergence theorem, and Cauchy's theorem from previous study. More advanced techniques will also be called on but it is hoped that various explanatory appendixes will aid the understanding of these.

For the sections on electromagnetism it will be supposed that the reader is acquainted with the theory of experiments which lead us to believe that the behaviour of macroscopic electromagnetic phenomena is governed by Maxwell's equations (this knowledge is not required in order to read the material on sound waves). In most undergraduate courses this is achieved by discussing electrostatics and magnetostatics based on Coulomb's law of force between charges, by considering the magnetic forces of steady currents using the laws based on Ampere's experiments and by employing the fundamental results about varying fields that were obtained by Faraday. Accordingly, we shall assume that the reader has this basic information and shall develop from that point our investigation of the predictions ensuing from Maxwell's equations, making no attempt to deduce the principal properties of electrostatics, magnetostatics, and steady currents except in so far as they are required for some particular purpose.

This first chapter will be devoted to deriving general formulae without any relation to specific physical problems; the application to special problems will be undertaken in later chapters.

The equations of sound waves

1.1 The basic equations

Acoustics might be described as the theory of the propagation of small disturbances in fluids. The term fluid includes both liquids and gases. The main difference between liquids and solids is that no force is required to alter the shape of a liquid, provided that sufficient time is available, whereas considerable forces must usually be deployed to deform a solid. Rapid deformations are resisted by liquids—a display of *viscosity*. Liquids of all types, from the mobile to the highly viscous, exist but we shall generally be concerned with freely flowing substances such as water. Another property of liquids is their great resistance to change of volume so that it is very difficult to compress them or to persuade them to expand.

Gases, on the other hand, are distinguished from liquids by their ability to change volume comparatively easily. If they are given more space than they ordinarily occupy they fill the space uniformly. Also they can be compressed into a very small volume by the application of suitable forces. A gas exhibits viscosity so that liquids and gases may be roughly categorized as incompressible and compressible fluids respectively.

In the macroscopic theory of fluids, a fluid is regarded as a continuous distribution. The connection between the microscopic point of view, which considers all substances to be composed of molecules, and the macroscopic will not be discussed here though it is often convenient to think of a fluid as a conglomeration of particles. Such a particle should, however, be visualized as a little bit of fluid rather than as a molecule.

When the position of every particle of the fluid is known at every instant a complete picture of the flow is available. The problem of finding the positions consists in discovering expressions for the location (x, y, z) of a particle at time t given that it was at (X, Y, Z) at the time τ earlier. Then the equations

$$x = x(X, Y, Z, t), \qquad y = y(X, Y, Z, t), \qquad z = z(X, Y, Z, t) \tag{1.1}$$

furnish a curve as t varies which is the *path* of the particle that started from (X, Y, Z). The velocity of this particle has three components (u, v, w) at time t given by

$$u = \partial x/\partial t, \qquad v = \partial y/\partial t, \qquad w = \partial z/\partial t \tag{1.2}$$

when x, y, z are regarded as functions of X, Y, Z, and t.

If a particle initially at (X, Y, Z) arrives at (x, y, z) at time t one commencing from $(X+\xi, Y, Z)$ where ξ is small will be at $(x+\xi\,\partial x/\partial X,\ y+\xi\,\partial y/\partial X,\ z+\xi\,\partial z/\partial X)$ at the same time. Therefore the fluid which originally occupied a small rectangular parallelepiped with sides ξ, η, ζ parallel to the coordinate axes will fill a volume given by the determinant

$$\begin{vmatrix} \dfrac{\partial x}{\partial X} & \dfrac{\partial y}{\partial X} & \dfrac{\partial z}{\partial X} \\[2mm] \dfrac{\partial x}{\partial Y} & \dfrac{\partial y}{\partial Y} & \dfrac{\partial z}{\partial Y} \\[2mm] \dfrac{\partial x}{\partial Z} & \dfrac{\partial y}{\partial Z} & \dfrac{\partial z}{\partial Z} \end{vmatrix} \xi\eta\zeta$$

or $\dfrac{\partial(x, y, z)}{\partial(X, Y, Z)}\xi\eta\zeta$ in abbreviated form. Assuming that there is no mechanism for creating or destroying matter the mass of the two volumes must be the same, i.e.

$$\rho\frac{\partial(x, y, z)}{\partial(X, Y, Z)} = \rho_0 \tag{1.3}$$

where ρ is the density at (x, y, z) and ρ_0 is the density which occurred at (X, Y, Z). Thus, (1.3) is an expression of the conservation of mass. The specification of the flow via (1.1) and (1.2) is known as the *Lagrangian representation* and so (1.3) is the Lagrangian form of the conservation of mass.

An alternative approach is to determine the velocity of the fluid at a point in space, i.e. to find expressions

$$u = u(x, y, z, t), \qquad v = v(x, y, z, t), \qquad w = w(x, y, z, t); \tag{1.4}$$

this is the *Eulerian representation.* The path of a particle may be obtained by substitution from (1.4) in (1.2) and integration, leading back to the representation (1.1).

The *stream-lines* are curves parallel to the velocity and so satisfy the differential equations

$$\frac{dx}{ds} = u, \qquad \frac{dy}{ds} = v, \qquad \frac{dz}{ds} = w.$$

When the motion is *steady* (i.e. independent of the time t) the stream-lines coincide with the paths of the particles. In unsteady motion, however, the stream-lines show the direction of velocity throughout space at a given instant but the path of a particle displays the direction of the velocity of a particle at successive instants of time; in general, the two will differ.

The equation of conservation of mass in the Eulerian approach is derived as follows. The total mass in a volume T (see Fig. 1.1) at time t is $\int_T \rho \, d\boldsymbol{x}$ where $d\boldsymbol{x}$ is the element of volume. Keep this volume *fixed in*

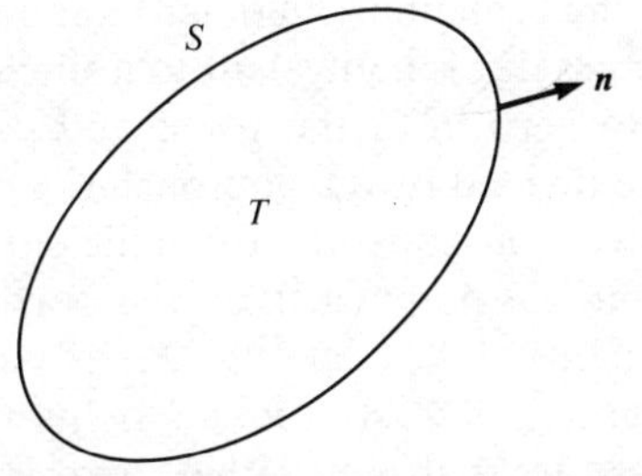

Fig. 1.1 Volume of integration for the equation of continuity

space. Then the increase in mass in a small time δt is

$$\delta t \int_T \frac{\partial \rho}{\partial t} \, \mathrm{d}\boldsymbol{x}.$$

Since mass is conserved this increase must be due to flow across the boundary S of T. Now fluid crosses S only on account of the velocity component along the normal to S. If $\boldsymbol{n}$ is a unit normal vector directed out of T the mass which is transferred across the small element $\mathrm{d}S$ is that contained in a cylinder of volume $\boldsymbol{n} \cdot \boldsymbol{v} \, \delta t \, \mathrm{d}S$ where the vector $\boldsymbol{v}$ is the velocity of the fluid, i.e. $\boldsymbol{v} \equiv (u, v, w)$. Hence

$$\int_T \frac{\partial \rho}{\partial t} \, \mathrm{d}\boldsymbol{x} = -\int_S \rho \boldsymbol{n} \cdot \boldsymbol{v} \, \mathrm{d}S.$$

This equation can be transformed by the *divergence theorem* which will now be stated. Let $\boldsymbol{F}$ be a vector function which, together with its first partial derivatives, is continuous at all points of T and S. Then, if S is a regular surface (for definition see Kellogg (1929)), it can be shown that

$$\int_S \boldsymbol{F} \cdot \boldsymbol{n} \, \mathrm{d}S = \int_T \operatorname{div} \boldsymbol{F} \, \mathrm{d}\boldsymbol{x}. \tag{1.5}$$

Applying (1.5) to our equation we have

$$\int_T \frac{\partial \rho}{\partial t} \, \mathrm{d}\boldsymbol{x} = -\int_T \operatorname{div}(\rho \boldsymbol{v}) \, \mathrm{d}\boldsymbol{x}.$$

Because this holds for an arbitrary volume T, we conclude that

$$\frac{\partial \rho}{\partial t} + \operatorname{div}(\rho \boldsymbol{v}) = 0 \tag{1.6}$$

which is the Eulerian form for the conservation of mass.

Another equation of motion comes from a consideration of the forces in a fluid. The notion of stress in general is that of balancing internal forces which two parts of a fluid exert on one another. Let $\mathrm{d}\Sigma$ be a small area of a given plane containing a point O of the fluid and let $\boldsymbol{n}$ be the normal to the plane. Imagine that the fluid on the side towards which $\boldsymbol{n}$ points exerts force on the remaining fluid across the plane. The system of forces, thus applied across $\mathrm{d}\Sigma$, is equivalent to a single force and couple at O. As $\mathrm{d}\Sigma$ contracts to zero both the force and couple vanish but we postulate that the force divided by $\mathrm{d}\Sigma$ approaches a definite limit whereas couple/$\mathrm{d}\Sigma$ tends to zero. Consequently, for sufficiently small $\mathrm{d}\Sigma$, the fluid on the side towards which $\boldsymbol{n}$ points acts on the portion on the other side with a force of components $(-X_n \, \mathrm{d}\Sigma, -Y_n \, \mathrm{d}\Sigma, -Z_n \, \mathrm{d}\Sigma)$. If the projection on the normal $n_x X_n + n_y Y_n + n_z Z_n$, (n_x, n_y, n_z) being the components of $\boldsymbol{n}$, is positive there is *pressure*; if it is negative there is *tension*.

To show that these forces can be derived from a *stress tensor* consider a tetrahedron having one vertex at O and three edges parallel to the coordinate axes which meet at O. The remaining vertices are the intersections of these edges with a plane near O. Let $\boldsymbol{n}$ be the normal to this plane drawn away from the interior of the tetrahedron and $\mathrm{d}\Sigma$ the area of this face. Then the areas of the other faces are $n_x\,\mathrm{d}\Sigma$, $n_y\,\mathrm{d}\Sigma$, and $n_z\,\mathrm{d}\Sigma$ respectively. Hence

$$-X_n\,\mathrm{d}\Sigma + X_x n_x\,\mathrm{d}\Sigma + X_y n_y\,\mathrm{d}\Sigma + X_z n_z\,\mathrm{d}\Sigma = \rho\,\mathrm{d}\boldsymbol{x} \times \text{external force}.$$

As $d\boldsymbol{x} \to 0$, $\mathrm{d}\boldsymbol{x}/\mathrm{d}\Sigma \to 0$ while the external force remains bounded; so

$$X_n = X_x n_x + X_y n_y + X_z n_z.$$

Thus $X_x n_x + X_y n_y + X_z n_z$ is the component of a vector for all $\boldsymbol{n}$ and the same is true if X is replaced by Y or Z. Now, if the force per unit area on an element of area perpendicular to the x-axis is written (p_{xx}, p_{yx}, p_{zx}) with a similar notation for the other axes, we see that

$$p_{xx}n_x + p_{xy}n_y + p_{xz}n_z,\quad p_{yx}n_x + p_{yy}n_y + p_{yz}n_z,\quad \text{and} \quad p_{zx}n_x + p_{zy}n_y + p_{zz}n_z$$

form the components of a vector for all $\boldsymbol{n}$. Hence the nine quantities of the type p_{xy} constitute a tensor (§F.3).

The rate of change of linear momentum involves the calculation of the rate of change of $\boldsymbol{v}$ with t. Here we must recognize that $\boldsymbol{v}$ carries t not only explicitly but also implicitly via x, y, and z so that

$$\begin{aligned}\frac{\mathrm{d}\boldsymbol{v}}{\mathrm{d}t} &= \frac{\partial \boldsymbol{v}}{\partial x}\frac{\mathrm{d}x}{\mathrm{d}t} + \frac{\partial \boldsymbol{v}}{\partial y}\frac{\mathrm{d}y}{\mathrm{d}t} + \frac{\partial \boldsymbol{v}}{\partial z}\frac{\mathrm{d}z}{\mathrm{d}t} + \frac{\partial \boldsymbol{v}}{\partial t}\\ &= u\frac{\partial \boldsymbol{v}}{\partial x} + v\frac{\partial \boldsymbol{v}}{\partial y} + w\frac{\partial \boldsymbol{v}}{\partial z} + \frac{\partial \boldsymbol{v}}{\partial t}.\end{aligned}$$

To emphasize what this operation entails we shall introduce the notation

$$\frac{\mathrm{D}}{\mathrm{D}t} \equiv u\frac{\partial}{\partial x} + v\frac{\partial}{\partial y} + w\frac{\partial}{\partial z} + \frac{\partial}{\partial t}.$$

The product of the mass and acceleration of the fluid occupying the fixed volume T (Fig. 1.1) is equal to the force acting on the volume. In addition to the forces across S from the stress tensor there may be other forces such as gravity acting on the volume. These are called *body forces* and their resultant at a point will be taken as $\boldsymbol{F}$ per unit mass. Then

$$\int_T \rho\frac{\mathrm{D}u}{\mathrm{D}t}\,\mathrm{d}\boldsymbol{x} = \int_T \rho F_x\,\mathrm{d}\boldsymbol{x} - \int_S (p_{xx}n_x + p_{xy}n_y + p_{xz}n_z)\,\mathrm{d}S.$$

By means of the divergence theorem and noting that T is arbitrary we deduce that

$$\frac{\mathrm{D}u}{\mathrm{D}t}=F_x-\frac{1}{\rho}\left(\frac{\partial p_{xx}}{\partial x}+\frac{\partial p_{xy}}{\partial y}+\frac{\partial p_{xz}}{\partial z}\right). \tag{1.7}$$

Similarly

$$\frac{\mathrm{D}v}{\mathrm{D}t}=F_y-\frac{1}{\rho}\left(\frac{\partial p_{yx}}{\partial x}+\frac{\partial p_{yy}}{\partial y}+\frac{\partial p_{yz}}{\partial z}\right), \tag{1.8}$$

$$\frac{\mathrm{D}w}{\mathrm{D}t}=F_z-\frac{1}{\rho}\left(\frac{\partial p_{zx}}{\partial x}+\frac{\partial p_{zy}}{\partial y}+\frac{\partial p_{zz}}{\partial z}\right). \tag{1.9}$$

Equations (1.7)–(1.9) are sometimes called the *equations of momentum.*

If we ask that the rate of change of angular momentum is equal to the moment of the force we glean the additional fact that the *stress tensor must be symmetric* e.g. $p_{xy}=p_{yx}$.

It remains to discuss the energy balance, according to which the rate of increase of energy is equal to the rate at which work is done by the forces together with the rate at which heat is created. Let e be the internal energy per unit mass so that the total energy per unit mass is $e+\frac{1}{2}(u^2+v^2+w^2)$. Let $\mathscr{H}$ be the rate at which heat is created per unit mass by sources within the fluid. Let $\boldsymbol{h}$ be the thermal flux vector so that h_x is the rate of heat flow through a unit area perpendicular to the x-axis. Following the same route as that employed in deriving (1.7) we arrive at

$$\begin{aligned}\frac{\mathrm{D}}{\mathrm{D}t}\{e+\tfrac{1}{2}(u^2+v^2+w^2)\}=\mathscr{H}+\boldsymbol{F}\cdot\boldsymbol{v}-\frac{1}{\rho}\operatorname{div}\boldsymbol{h}\\ -\frac{1}{\rho}\frac{\partial}{\partial x}(up_{xx}+vp_{yx}+wp_{zx})\\ -\frac{1}{\rho}\frac{\partial}{\partial y}(up_{xy}+vp_{yy}+wp_{zy})\\ -\frac{1}{\rho}\frac{\partial}{\partial z}(up_{xz}+vp_{yz}+wp_{zz}).\end{aligned} \tag{1.10}$$

Replacing $\mathrm{D}u/\mathrm{D}t$ by (1.7) we obtain

$$\begin{aligned}\frac{\mathrm{D}e}{\mathrm{D}t}=\mathscr{H}-\frac{1}{\rho}\operatorname{div}\boldsymbol{h}-\frac{1}{\rho}\Big(p_{xx}\frac{\partial u}{\partial x}+p_{xy}\frac{\partial u}{\partial y}+p_{xz}\frac{\partial u}{\partial z}\\ +p_{yx}\frac{\partial v}{\partial x}+p_{yy}\frac{\partial v}{\partial y}+p_{yz}\frac{\partial v}{\partial z}\\ +p_{zx}\frac{\partial w}{\partial x}+p_{zy}\frac{\partial w}{\partial y}+p_{zz}\frac{\partial w}{\partial z}\Big)\end{aligned} \tag{1.11}$$

which may be regarded as the *equation of energy.* If $\mathscr{H}=0$, (1.11) can be expressed as

$$\frac{\mathrm{D}}{\mathrm{D}t}(\rho e) = -\rho e \operatorname{div} \boldsymbol{v} - \operatorname{div} \boldsymbol{h} - p_{xx}\frac{\partial u}{\partial x} - p_{xy}\frac{\partial u}{\partial y} - p_{xz}\frac{\partial u}{\partial z}$$

$$-p_{yx}\frac{\partial v}{\partial x} - p_{yy}\frac{\partial v}{\partial y} - p_{yz}\frac{\partial v}{\partial z} - p_{zx}\frac{\partial w}{\partial x} - p_{zy}\frac{\partial w}{\partial y} - p_{zz}\frac{\partial w}{\partial z} \tag{1.12}$$

when advantage is taken of (1.6). The left-hand side of (1.12) gives the rate of increase of thermal energy per unit volume moving with the fluid. The right-hand side tells us that it is composed of three parts: (i) the diminution in volume due to the motion, (ii) the heat brought across the boundary, and (iii) the work done due to change of shape and volume.

Equations (1.6)–(1.9) and (1.11) constitute all that can be deduced from simple mechanical principles. They form five equations for the fourteen quantities u, v, w, $\boldsymbol{h}$, ρ, e, p_{xx}, ... Consequently, they must be supplemented by other relations before solution is possible.

1.2 Thermodynamic considerations

In simple systems the pressure p, volume V, and temperature τ are connected by an *equation of state*. For an *ideal gas* the relation is

$$pV/\tau = \text{constant} \tag{2.1}$$

according to the laws of Boyle and Charles. For real gases the law of van der Waals is more appropriate.

The internal energy of a system may be increased by supplying energy either in the form of heat or by doing work. Let the system receive the small quantity of heat $\delta\mathscr{H}$. It is denoted by $\delta\mathscr{H}$ rather than $\mathrm{d}\mathscr{H}$ because it is not a perfect differential since it is not necessary that the total heat of a system be supplied as heat; it may come from mechanical energy in compression. If the work done is accomplished by a change of volume against the pressure p we have

$$\mathrm{d}e = \delta\mathscr{H} - p\,\mathrm{d}(1/\rho). \tag{2.2}$$

Temperature is measured on the *absolute scale* defined so that $\delta\mathscr{H}/\tau$ is a perfect differential $\mathrm{d}\bar{S}$; $\bar{S}$ is known as the *entropy per unit mass*. Thus

$$\delta\mathscr{H} = \tau\,\mathrm{d}\bar{S} = \mathrm{d}e + p\,\mathrm{d}(1/\rho). \tag{2.3}$$

According to the second law of thermodynamics the total entropy of a *closed* system does not decrease.

Provided that thermodynamic relations are valid for motion as well as equilibrium, it follows from (2.3) that

$$\tau\frac{\mathrm{D}\bar{S}}{\mathrm{D}t} = \frac{\mathrm{D}e}{\mathrm{D}t} + p\frac{\mathrm{D}}{\mathrm{D}t}\left(\frac{1}{\rho}\right) \tag{2.4}$$

which enables us to remove e from (1.11). The quantity e can also be eliminated by introducing the *enthalpy* I defined by

$$I = e + p/\rho \tag{2.5}$$

so that, from (2.3),

$$\mathrm{d}I = \tau\, \mathrm{d}\bar{S} + (1/\rho)\, \mathrm{d}p. \tag{2.6}$$

The heat necessary to raise the temperature of a unit mass by one degree is known as *specific heat.* Let c_p and c_v be the *specific heats at constant pressure* and at *constant volume* respectively. In a *polytropic* ideal gas c_p and c_v are independent of τ. This independence will be assumed from now on.

For a polytropic gas

$$p/\rho\tau = R \tag{2.7}$$

where R is a constant. If small changes of pressure and volume are made (2.7) reveals that

$$\delta\mathcal{H} = c_p(p/R)\, \mathrm{d}(1/\rho) + (c_v/\rho R)\, \mathrm{d}p. \tag{2.8}$$

It follows from (2.3) that

$$\mathrm{d}e = \{(c_p/R) - 1\}p\, \mathrm{d}(1/\rho) + (c_v/\rho R)\, \mathrm{d}p.$$

This implies, since $\mathrm{d}e$ is a perfect differential, that

$$c_p - c_v = R \tag{2.9}$$

and then

$$e = c_v p/\rho R = c_v\tau. \tag{2.10}$$

The combination of (2.3), (2.7), and (2.8) gives

$$\mathrm{d}\bar{S} = c_p\rho\, \mathrm{d}(1/\rho) + c_v\, \mathrm{d}p/p$$

so

$$\bar{S} = c_p \ln(1/\rho) + c_v \ln p. \tag{2.11}$$

From (2.5), (2.9), and (2.10)

$$I = c_p\tau. \tag{2.12}$$

Different ways of writing these results stem from the introduction of $\gamma = c_p/c_v$ so that $\gamma - 1 = R/c_v$. Then, from (2.10)–(2.12),

$$e = p/\rho(\gamma - 1), \tag{2.13}$$

$$\bar{S} = c_v \ln(p/\rho^\gamma), \tag{2.14}$$

$$I = \gamma p/\rho(\gamma - 1). \tag{2.15}$$

In *adiabatic* changes no heat is supplied or abstracted and so $\bar{S}$ remains constant according to (2.3). Hence, from (2.14),

$$p/\rho^{\gamma}=\text{constant}. \tag{2.16}$$

The value of γ is 5/3 for monatomic gases and 7/5 for diatomic gases ($\gamma=7/5$ is reasonable for air).

1.3 Thermal flux and stress

The next assumption that will be made is that the rate of flow of heat across a unit area is proportional to the temperature gradient. Then

$$\boldsymbol{h}=-\lambda \text{ grad } \tau. \tag{3.1}$$

According to the kinetic theory of gases the relation between the pressure p and the stress tensor is, for a gas at rest,

$$p=\tfrac{1}{3}(p_{xx}+p_{yy}+p_{zz}). \tag{3.2}$$

The connection between the stress tensor and the velocity depends upon the substance under consideration but a formula of wide applicability is

$$p_{xx}=p+\tfrac{2}{3}\bar{\mu} \text{ div } \boldsymbol{v}-2\bar{\mu}\frac{\partial u}{\partial x}, \tag{3.3}$$

$$p_{xy}=-\bar{\mu}\left(\frac{\partial u}{\partial y}+\frac{\partial v}{\partial x}\right) \tag{3.4}$$

with a similar structure for the other components. These formulae assume that there are effects from internal relaxation; otherwise, the factor $\tfrac{2}{3}\bar{\mu}$ in (3.3) should be replaced by $-\sigma$ where $3\sigma+2\bar{\mu}\geqslant 0$, $\bar{\mu}\geqslant 0$.

The quantity $\bar{\mu}$ is known as the *coefficient of viscosity*. For, when the velocity is a function of y only with $u=u(y)$, $v=0$, $w=0$ so that there is a simple shear flow parallel to the x-axis, (3.3) gives $p_{xx}=p=p_{yy}=p_{zz}$ and the only component which survives is $p_{xy}=-\bar{\mu}\,\partial u/\partial y$. This is the form usually assumed for viscous forces.

In the absence of viscosity, as when a field is at rest or when $\bar{\mu}=0$, (3.3) and (3.4) show that the internal forces across a small surface are always normal to that surface and always have the same magnitude p whatever the direction of the normal.

Bringing together (1.6)–(1.9), (1.11), (3.1), (3.3), and (3.4) we obtain

$$\frac{\mathrm{D}\rho}{\mathrm{D}t}+\rho \text{ div } \boldsymbol{v}=0, \tag{3.5}$$

$$\begin{aligned}\frac{\mathrm{D}u}{\mathrm{D}t}=F_x&-\frac{1}{\rho}\frac{\partial}{\partial x}(p+\tfrac{2}{3}\bar{\mu} \text{ div } \boldsymbol{v})+\frac{2}{\rho}\frac{\partial}{\partial x}\left(\bar{\mu}\frac{\partial u}{\partial x}\right)\\&+\frac{1}{\rho}\frac{\partial}{\partial y}\left\{\bar{\mu}\left(\frac{\partial u}{\partial y}+\frac{\partial v}{\partial x}\right)\right\}+\frac{1}{\rho}\frac{\partial}{\partial z}\left\{\bar{\mu}\left(\frac{\partial u}{\partial z}+\frac{\partial w}{\partial x}\right)\right\},\end{aligned} \tag{3.6}$$

$$\tau\frac{\mathrm{D}\bar{S}}{\mathrm{D}t}=\mathcal{H}+\frac{1}{\rho}\operatorname{div}(\lambda\operatorname{grad}\tau)-\tfrac{2}{3}\nu(\operatorname{div}\boldsymbol{v})^2$$
$$+\nu\left\{2\left(\frac{\partial u}{\partial x}\right)^2+2\left(\frac{\partial v}{\partial y}\right)^2+2\left(\frac{\partial w}{\partial z}\right)^2+\left(\frac{\partial u}{\partial y}+\frac{\partial v}{\partial x}\right)^2\right.$$
$$\left.+\left(\frac{\partial u}{\partial z}+\frac{\partial w}{\partial x}\right)^2+\left(\frac{\partial v}{\partial z}+\frac{\partial w}{\partial y}\right)^2\right\} \tag{3.7}$$

where $\nu=\bar{\mu}/\rho$ is the *kinematic viscosity* and there are two further equations obtained from (3.6) by cyclic permutation. The equation of state permits the expression of all thermodynamic variables in terms of ρ and τ. If the equation for $\bar{S}$ is solved to give τ in terms of ρ and $\bar{S}$, e.g.

$$\tau=g(\rho,\bar{S}), \tag{3.8}$$

the equation of state can be written

$$p=f(\rho,\bar{S}). \tag{3.9}$$

Equations (3.5)–(3.9) supply seven equations for the seven quantities ρ, u, v, w, p, τ, $\bar{S}$.

Equation (2.14) shows that in a polytropic gas

$$f(\rho,\bar{S})=\rho^{\gamma}\exp(\bar{S}/c_v).$$

In many liquids, e.g. water, f is a function of ρ only. Equations (3.9), (3.5), (3.6) and its companions are then sufficient to determine the pressure, density, and velocity of the flow provided that $\bar{\mu}$ is independent of the temperature.

The kinetic theory of gases predicts that $\bar{\mu}$ is independent of density, a law which has been verified experimentally. The predicted variation with temperature differs according to the molecular model adopted but there are discrepancies from the experimental observations in any case. The value of λ for a gas, on the kinetic theory, is such that $\bar{\mu}c_p/\lambda$ is virtually constant and independent of temperature. Its value is about 0.7 for many gases and it is often known as the *Prandtl number.*

1.4 Sound waves

Consider a fluid in which viscosity and thermal conductivity can be neglected. Suppose that the flow is *isentropic* i.e. the entropy is constant throughout the medium. Then the pressure is a function of the density only and the equations of motion (3.5)–(3.9) reduce to

$$\frac{\mathrm{D}\rho}{\mathrm{D}t}+\rho\operatorname{div}\boldsymbol{v}=0, \tag{4.1}$$

$$\frac{\mathrm{D}\boldsymbol{v}}{\mathrm{D}t}=-\frac{1}{\rho}\operatorname{grad}p=-\frac{a^2}{\rho}\operatorname{grad}\rho \tag{4.2}$$

when there is no external stimulus. In (4.2) $a^2 = \mathrm{d}p/\mathrm{d}\rho$ and a is called the *speed of sound;* it may vary throughout the medium.

Assume firstly that the fluid is at rest with the density taking the constant value ρ_0 everywhere. Now allow a disturbance to occur in which the velocity stays small. There will be a corresponding change in the density so that $\rho = \rho_0 + \rho_1$ where ρ_1 is small. Neglect any terms of (4.1) and (4.2) in which two of ρ_1, u, v, w are multiplied together. Then the governing equations become

$$\frac{\partial \rho_1}{\partial t} + \rho_0 \operatorname{div} \boldsymbol{v} = 0, \tag{4.3}$$

$$\frac{\partial \boldsymbol{v}}{\partial t} = -\frac{a_0^2}{\rho_0} \operatorname{grad} \rho_1 \tag{4.4}$$

where $a_0^2 = (\mathrm{d}p/\mathrm{d}\rho)_{\rho=\rho_0}$. Now the speed of sound a_0 is the same at all points of the medium. Eliminate $\boldsymbol{v}$ from (4.3) and (4.4) and write

$$\operatorname{div} \operatorname{grad} f = \frac{\partial^2 f}{\partial x^2} + \frac{\partial^2 f}{\partial y^2} + \frac{\partial^2 f}{\partial z^2} = \nabla^2 f. \tag{4.5}$$

Then

$$\nabla^2 \rho_1 = \frac{1}{a_0^2} \frac{\partial^2 \rho_1}{\partial t^2}. \tag{4.6}$$

In ideal gases the adiabatic law (2.16)

$$p/\rho^\gamma = \text{constant} = p_0/\rho_0^\gamma$$

where p_0 is the pressure corresponding to the density ρ_0 leads to the formula $a_0 = (\gamma p_0/\rho_0)^{\frac{1}{2}}$ for the sound speed. It is appropriate when the sound fluctuations occur at a moderate rate and leads to a speed of sound in air of 332 m s^{-1}. For ideal liquids a small change in local pressure is accompanied instantaneously by a proportional alteration in local density. Thus $a_0 = (\kappa/\rho_0)^{\frac{1}{2}}$ where κ is the bulk modulus in units of pressure. The speed of sound in water at 15 °C is roughly 1430 m s^{-1}. The enormous difference between the speeds of sound in water and air is responsible in part for some of the differences in acoustic phenomena in water and air.

It is also possible to tackle the matter via the Lagrangian equations. Suppose that a particle of fluid at $\boldsymbol{X}$ in the static state moves to $\boldsymbol{X} + \boldsymbol{\xi}$ at time t under the influence of an acoustic disturbance. Then, from (1.2),

$$\boldsymbol{v} = \partial \boldsymbol{\xi}/\partial t. \tag{4.7}$$

Also the conservation of mass (1.3) becomes, when we approximate as above,

$$\rho_1 = -\rho_0 \left(\frac{\partial \xi}{\partial X} + \frac{\partial \eta}{\partial Y} + \frac{\partial \zeta}{\partial Z} \right) \tag{4.8}$$

when $\boldsymbol{\xi} \equiv (\xi, \eta, \zeta)$. Similarly the equations of momentum reduce to

$$\rho_0 \frac{\partial u}{\partial t} = -a_0^2 \frac{\partial \rho_1}{\partial X}, \qquad \rho_0 \frac{\partial v}{\partial t} = -a_0^2 \frac{\partial \rho_1}{\partial Y}, \qquad \rho_0 \frac{\partial w}{\partial t} = -a_0^2 \frac{\partial \rho_1}{\partial Z}. \tag{4.9}$$

The elimination of $\boldsymbol{\xi}$ and $\boldsymbol{v}$ from (4.7)–(4.9) leads to

$$\frac{\partial^2 \rho_1}{\partial X^2} + \frac{\partial^2 \rho_1}{\partial Y^2} + \frac{\partial^2 \rho_1}{\partial Z^2} = \frac{1}{a_0^2} \frac{\partial^2 \rho_1}{\partial t^2} \tag{4.10}$$

on remembering that ρ_0 is constant everywhere.

A more general initial state can be allowed for by dropping the requirement that the flow be isentropic. Then we have the equations

$$\frac{\mathrm{D}\rho}{\mathrm{D}t} + \rho \operatorname{div} \boldsymbol{v} = 0, \tag{4.11}$$

$$\frac{\mathrm{D}\boldsymbol{v}}{\mathrm{D}t} = -\frac{1}{\rho} \operatorname{grad} p, \tag{4.12}$$

$$\frac{\mathrm{D}\bar{S}}{\mathrm{D}t} = 0, \tag{4.13}$$

$$p = f(\rho, \bar{S}). \tag{4.14}$$

Suppose that there is a steady flow in which $\boldsymbol{v} = \boldsymbol{U}$, $p = p_0$, $\rho = \rho_0$, and $\bar{S} = \bar{S}_0$. From (4.11)–(4.14),

$$\boldsymbol{U} \cdot \operatorname{grad} \rho_0 + \rho_0 \operatorname{div} \boldsymbol{U} = 0, \tag{4.15}$$

$$\rho_0(\boldsymbol{U} \cdot \operatorname{grad})\boldsymbol{U} = -\operatorname{grad} p_0, \tag{4.16}$$

$$\boldsymbol{U} \cdot \operatorname{grad} \bar{S}_0 = 0, \tag{4.17}$$

$$\operatorname{grad} p_0 = \frac{\partial f}{\partial \rho} \operatorname{grad} \rho_0 + \frac{\partial f}{\partial \bar{S}} \operatorname{grad} \bar{S}_0 \tag{4.18}$$

where $\partial f/\partial \rho$ and $\partial f/\partial \bar{S}$ are calculated with $\rho = \rho_0$, $\bar{S} = \bar{S}_0$. Now the speed of sound a is given by $a^2 = \partial f/\partial \rho$. If we write $h = \partial f/\partial S$, then a and h will be known at every point once (4.15)–(4.18) have been solved for ρ_0 and $\bar{S}_0$ as functions of position. Therefore, subsequently, a and h can be regarded as known functions of position.

Let sound waves make small perturbations so that $\rho = \rho_0 + \rho_1$, $\boldsymbol{v} = \boldsymbol{U} + \boldsymbol{u}$, $\bar{S} = \bar{S}_0 + \bar{S}_1$. If the products of small quantities are again neglected, the following equations result:

$$\frac{\partial \rho_1}{\partial t} + \boldsymbol{U} \cdot \operatorname{grad} \rho_1 + \boldsymbol{u} \cdot \operatorname{grad} \rho_0 + \rho_0 \operatorname{div} \boldsymbol{u} + \rho_1 \operatorname{div} \boldsymbol{U} = 0, \tag{4.19}$$

$$\rho_0\frac{\partial \boldsymbol{u}}{\partial t}+\rho_0(\boldsymbol{U}\cdot\text{grad})\boldsymbol{u}+\rho_0(\boldsymbol{u}\cdot\text{grad})\boldsymbol{U}+\rho_1(\boldsymbol{U}\cdot\text{grad})\boldsymbol{U}$$

$$=-a^2\,\text{grad}\,\rho_1-h\,\text{grad}\,\bar{S}_1-\left(\rho_1\frac{\partial^2 f}{\partial\rho^2}+\bar{S}_1\frac{\partial^2 f}{\partial\rho\,\partial\bar{S}}\right)\text{grad}\,\rho_0$$

$$-\left(\rho_1\frac{\partial^2 f}{\partial\rho\,\partial\bar{S}}+\bar{S}_1\frac{\partial^2 f}{\partial\bar{S}^2}\right)\text{grad}\,\bar{S}_0, \qquad (4.20)$$

$$\frac{\partial\bar{S}_1}{\partial t}+\boldsymbol{U}\cdot\text{grad}\,\bar{S}_1+\boldsymbol{u}\cdot\text{grad}\,\bar{S}_0=0 \qquad (4.21)$$

where the second partial derivatives $\partial^2 f/\partial\rho^2$, etc. are evaluated at $\rho=\rho_0$, $\bar{S}=\bar{S}_0$.

It is immediately evident that a background flow complicates considerably the analysis which has to be undertaken in order to determine the acoustic disturbance. Some simplifications can be achieved in particular cases. For example, suppose that the basic flow consists of a steady velocity U parallel to the x-axis so that $\boldsymbol{U}=U\boldsymbol{i}$ where $\boldsymbol{i}$ is a unit vector along the x-axis. Assume further that U, ρ_0 are constant and that the flow is isentropic. Then (4.15)–(4.18) are certainly satisfied with $a=a_0$ (a constant) and $h=0$. Now (4.19), (4.20) become

$$\frac{\partial\rho_1}{\partial t}+U\frac{\partial\rho_1}{\partial x}+\rho_0\,\text{div}\,\boldsymbol{u}=0, \qquad (4.22)$$

$$\rho_0\frac{\partial\boldsymbol{u}}{\partial t}+\rho_0 U\frac{\partial\boldsymbol{u}}{\partial x}=-a_0^2\,\text{grad}\,\rho_1 \qquad (4.23)$$

and (4.21) drops out. Eliminate $\boldsymbol{u}$ from (4.22) and (4.23) to obtain

$$\nabla^2\rho_1=\frac{1}{a_0^2}\left(\frac{\partial^2\rho_1}{\partial t^2}+2U\frac{\partial^2\rho_1}{\partial x\,\partial t}+U^2\frac{\partial^2\rho_1}{\partial x^2}\right). \qquad (4.24)$$

When $U=0$ this reduces to (4.6) as it should. The ratio U/a_0 is known as the *Mach number M*. If $M<1$ the flow is said to be *subsonic* whereas if $M>1$ it is *supersonic*.

The substitution $x=\alpha+Ut$ with other variables unchanged converts (4.24) to

$$\frac{\partial^2\rho_1}{\partial\alpha^2}+\frac{\partial^2\rho_1}{\partial y^2}+\frac{\partial^2\rho_1}{\partial z^2}=\frac{1}{a_0^2}\frac{\partial^2\rho_1}{\partial t^2}. \qquad (4.25)$$

Equation (4.25) is of the same type as (4.6) but it must be recognized that its origin is no longer fixed in space. When ρ_1 is independent of t and the flow is subsonic (4.24) may be transformed to

$$\frac{\partial^2\rho_1}{\partial\beta^2}+\frac{\partial^2\rho_1}{\partial y^2}+\frac{\partial^2\rho_1}{\partial z^2}=0 \qquad (4.26)$$

by means of $x = \beta(1 - M^2)^{\frac{1}{2}}$. Here there is no moving origin, merely a change of scale along the x-axis.

The effects of viscosity on acoustic waves can be allowed for by considering small perturbations in (3.5)–(3.9) but the equations will not be written down here (see §6.5). They are, of course, more complex than (4.19)–(4.21).

Having seen the sort of equation which has to be solved we now turn to the corresponding matter for electromagnetic waves, where an equation similar to (4.6) arises. Later sections will discuss solutions of (4.6) but consideration of the effect of background flow will be deferred to another chapter.

Maxwell's equations

1.5 The field equations

The electromagnetic field is produced by a distribution of electric current and charge. It is generally accepted that matter is not continuously divisible but is composed of small discrete particles. The motion of such a charged particle is equivalent to a current. However, we are not concerned with the individual microscopic particles—the theory of their motion belongs to quantum theory—but only the average behaviour of large numbers of them. Naturally, no strict dividing line can be drawn between the macroscopic and quantum theories, but where the processes about to be described are valid should be the domain of macroscopic theory. The final justification for the adoption of the macroscopic point of view can only be provided by predicting results which are in agreement with experiment.

The straightforward definition of the charge density as the limit of the charge per unit volume when the enclosing volume shrinks to zero leads to difficulty when we deal with discrete particles. For, as the volume contracts, the point is eventually reached where the volume contains either one or no particle and the limit is then either infinite or zero. The same kind of trouble arises in the kinetic theory of gases and we may overcome our difficulty in the same way as in that theory. Let δv be the volume of a small volume element, which is big enough to contain a large number of particles although its dimensions are tiny compared with the scale of variation of such macroscopic quantities as the electric and magnetic forces. Let the charge contained in δv be averaged over the time interval from t to $t + \delta t$ which is long compared with the average time that would be taken by a particle to cross δv, yet short compared with the timescale of variations of macroscopic quantities. Then the averaged value of the charge in δv will be proportional only to δv, and will not depend on its shape. Denote it by $\rho\,\delta v$. Then ρ is called the *charge density* at any point within δv at time t. To fix ideas ρ may always be regarded as evaluated at some particular point of δv, the centroid (say), and, with that convention, ρ can be said to be the charge density at a point.

The use of ρ for both the charge density and density of a fluid in §1.1 should cause no confusion since the context should distinguish between them. Many of the letters customarily employed in fluid mechanics are employed in electromagnetic theory for quite different purposes. In any place where they are likely to become entangled a special notation will be introduced.

It will be assumed that, in any physical situation, ρ is a function of the coordinates and time which is continuous and possesses continuous partial derivatives at ordinary points. By an ordinary point is meant one in whose neighbourhood the physical properties of the medium are continuous; this excludes, for example, the points on the bounding surface of a material body. However, it is not always mathematically convenient to restrict ρ in this way. Often idealized models in which ρ may not have derivatives in some direction have to be employed in order that the mathematical problem can be solved. Such a process is justified provided that the physical interpretation is reasonable; it is one which occurs frequently throughout applied mathematics.

The motion of charge constitutes a current which is determined by the magnitude of the charge and its velocity. The average velocity at a point may be defined along lines similar to those used in specifying ρ and will be denoted by the vector $\boldsymbol{v}$. The *current density* at a point is defined as the vector $\boldsymbol{J}$ given by

$$\boldsymbol{J} = \rho\boldsymbol{v}. \tag{5.1}$$

From this it follows that, in metals and electronic devices where the electricity is carried by electrons which are negatively charged, the direction of the current density vector is opposite to that of the moving electrons.

The *current* across a piece of surface S is defined to be the rate at which charge crosses that portion of surface. Since charge can pass through S only by virtue of its velocity normal to S, we have

$$I = \int_S \boldsymbol{J} \cdot \boldsymbol{n}\, \mathrm{d}S \tag{5.2}$$

where $\boldsymbol{n}$ is a unit vector normal to S (Fig. 1.2). It should be noted that the sign of I depends upon the choice of $\boldsymbol{n}$ and so the additional comment is required that I is the current crossing S from the negative to the positive side (the side towards which $\boldsymbol{n}$ points will always be designated positive). When S is closed $\boldsymbol{n}$ is taken to be the *outward normal* and then the

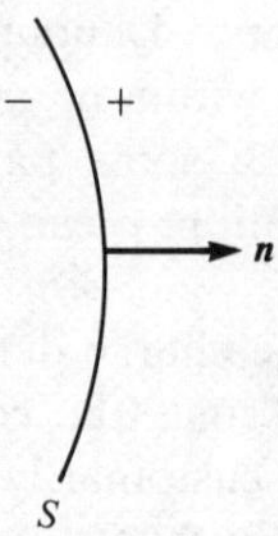

Fig. 1.2 Direction of positive normal

surface integral in (5.2) gives the rate at which charge is lost from the region inside S.

Consider now two isolated charges Q and Q_1 moving in free space. The charge Q is acted on by certain forces due to the presence of Q_1. If Q is at rest the electrical force on it is $Q\boldsymbol{E}$ and the vector $\boldsymbol{E}$ is called the *electric intensity*. When Q is in motion with velocity $\boldsymbol{v}$ there is an additional force $Q\boldsymbol{v}\wedge\boldsymbol{B}$ where $\wedge$ is the symbol for a vector product. The vector $\boldsymbol{B}$ is known as the *magnetic flux density*. Two other vectors play a role in specifying the electromagnetic field and they are related to the lines of force which emanate from charges and currents. The vector $\boldsymbol{D}$, which is called the *electric flux density*, effectively measures the number of lines of force which originate from a charge. The vector $\boldsymbol{H}$, known as the *magnetic intensity*, is such that its value on a closed curve measures in essence the current which passes through the curve.

It will be assumed that the vectors $\boldsymbol{E}$, $\boldsymbol{B}$, $\boldsymbol{D}$, and $\boldsymbol{H}$ are continuous and possess continuous partial derivatives at ordinary points. Their properties at the transition surfaces between various kinds of media will have to be investigated. It will now be postulated that at ordinary points the vectors are connected by *Maxwell's equations*, namely

$$\text{curl}\,\boldsymbol{E}+\frac{\partial\boldsymbol{B}}{\partial t}=\boldsymbol{0}, \tag{5.3}$$

$$\text{curl}\,\boldsymbol{H}-\frac{\partial\boldsymbol{D}}{\partial t}=\boldsymbol{J}, \tag{5.4}$$

$$\text{div}\,\boldsymbol{D}=\rho, \tag{5.5}$$

$$\text{div}\,\boldsymbol{B}=0. \tag{5.6}$$

The substantiation of this postulate lies firstly in the fact that, in the particular cases of electrostatics, magnetostatics, the magnetic effects of steady currents, and slowly varying currents (when certain terms in the equations can be neglected), the phenomena predicted are in agreement with experiment. Furthermore, for rapidly varying currents and for the propagation of disturbances when all terms in the equations have to be retained, the predictions are in harmony with experiment. It is the second aspect which is discussed in this book.

1.6 The equation of continuity

Since the divergence of the curl of any vector vanishes identically we obtain, by taking the divergence of (5.4),

$$\operatorname{div} \boldsymbol{J} = -\operatorname{div} \frac{\partial \boldsymbol{D}}{\partial t} = -\frac{\partial}{\partial t} \operatorname{div} \boldsymbol{D}.$$

The interchange of the operators div and $\partial/\partial t$ is permissible because of the assumed continuity of $\boldsymbol{D}$ and its derivatives at ordinary points. Substitution from (5.5) gives

$$\operatorname{div} \boldsymbol{J} + \frac{\partial \rho}{\partial t} = 0. \tag{6.1}$$

By analogy with (1.6), this is termed the *equation of continuity* or *conservation of charge.*

The application of the divergence theorem (1.5) to (6.1) leads to

$$\int_S \boldsymbol{J} \cdot \boldsymbol{n} \, \mathrm{d}S = -\frac{\partial}{\partial t} \int_T \rho \, \mathrm{d}\boldsymbol{x},$$

i.e. the rate at which charge is crossing the closed surface S is equal to the rate of decrease of charge inside S. Thus, according to Maxwell's equations, charge cannot be created or destroyed in macroscopic amounts; this is confirmed by experimental observation.

In *steady current flow* where there is no variation with time

$$\operatorname{div} \boldsymbol{J} = 0$$

so that the vector $\boldsymbol{J}$ is solenoidal. Therefore, all the streamlines of the current are closed.

One further point must be made. The divergence of (5.3) implies that $\operatorname{div} \boldsymbol{B}$ is independent of time. Therefore $\operatorname{div} \boldsymbol{B}$ will be zero if it was at some earlier time. Thus (5.6) could be dispensed with if it can be asserted that the field was zero at some time in the past and has varied continuously since.

Macroscopic properties of matter

1.7 Dielectric constant and permeability

Assume that the charge and current densities are given. Then, in (5.3)–(5.6), there are twelve unknown scalar quantities to be determined from (at most) eight scalar equations. Other relations between the vectors must be supplied if the system of equations is to become determinate. Experimental data concerning the nature of material media provide the missing links. Certain cases where the constitutive equations take a simple form will now be described.

(a) In *free space*

$$\boldsymbol{D} = \varepsilon_0 \boldsymbol{E}, \qquad \boldsymbol{B} = \mu_0 \boldsymbol{H} \tag{7.1}$$

where μ_0 and ε_0 are constants. The values and dimensions of these constants depend upon the system of units adopted. We shall take

$$\mu_0 = 4\pi \times 10^{-7}\ \mathrm{H\,m^{-1}}.$$

This also determines ε_0 because it is found that

$$1/(\mu_0 \varepsilon_0)^{\frac{1}{2}} = c$$

where c is the speed of light in free space. Many measurements have been made of c and it is known that

$$c = 3 \times 10^8\ \mathrm{m\,sec^{-1}}$$

to a high degree of approximation; this value is sufficiently accurate for the purposes of this book. Hence

$$\varepsilon_0 = 1/\mu_0 c^2 = \frac{1}{36\pi} \times 10^{-9}\ \mathrm{F\,m^{-1}}.$$

(b) *Isotropic bodies* are those in which the physical properties in the neighbourhood of an interior point are the same in all directions. For such bodies we shall put

$$\boldsymbol{D} = \varepsilon \boldsymbol{E}, \qquad \boldsymbol{B} = \mu \boldsymbol{H} \tag{7.2}$$

when they are not ferromagnetic. We call ε and μ the *permittivity* and *permeability* respectively. The dimensionless ratios

$$\kappa = \varepsilon/\varepsilon_0, \qquad \kappa_m = \mu/\mu_0 \tag{7.3}$$

are independent of the choice of units and are usually known as the *specific inductive capacities*. If the physical properties are constant from point to point the medium is said to be *homogeneous*. In a homogeneous medium κ is often termed the *dielectric constant* though this terminology is frequently carried over to a medium which is not homogeneous.

It is found experimentally that (7.2) applies to most gases and liquids provided that the electric field is not too great. Solids, in general, have a crystal structure and are not isotropic on an atomic scale. However, larger portions of these bodies are composed of randomly oriented crystalline aggregates and can be regarded as sensibly isotropic on a macroscopic scale. With large electric fields a dielectric can break down with the occurrence of an electrical discharge.

The dielectric constant is never less than 1. In gases it differs from 1 by a small quantity, being 1.0006 for air. Solids have dielectric constants in the range 2–10 usually, e.g. for mica and porcelain κ is about 6. There is considerable variation in liquids, κ varying from 2 for petroleum to 81 for water. In contrast, κ_m may be either greater or less than 1. If $\kappa_m > 1$, the medium is said to be *paramagnetic* and, if $\kappa_m < 1$, *diamagnetic*. In most paramagnetic and diamagnetic substances κ_m is approximately 1, being 1.00002 for aluminium and 0.999991 for copper.

For many gases and liquids ε is just a function of position and density. In general, κ and κ_m will vary when quantities such as the temperature and frequency are altered. Such changes will normally be neglected subsequently on the grounds that an average value over the frequency range involved, for example, will suffice in many instances.

(c) *Anisotropic* matter has properties which differ according to the direction taken from a point. Indeed, certain anisotropic solids such as oxide films behave as dielectrics in one direction only; they are useful for constructing capacitors of large capacity but small dimensions. For bodies such as crystalline dielectrics there are the relations

$$\left.\begin{aligned} D_x &= \varepsilon_{11}E_x + \varepsilon_{12}E_y + \varepsilon_{13}E_z, \\ D_y &= \varepsilon_{21}E_x + \varepsilon_{22}E_y + \varepsilon_{23}E_z, \\ D_z &= \varepsilon_{31}E_x + \varepsilon_{32}E_y + \varepsilon_{33}E_z \end{aligned}\right\} \tag{7.4}$$

where D_x, D_y, and D_z are the Cartesian components of $\boldsymbol{D}$. The coefficients ε_{ij} are components of a symmetric tensor $\boldsymbol{\varepsilon}$ and (7.4) may be written conveniently as

$$\boldsymbol{D} = \boldsymbol{\varepsilon} \cdot \boldsymbol{E}. \tag{7.5}$$

The corresponding relation for magnetic fields

$$\boldsymbol{B} = \boldsymbol{\mu} \cdot \boldsymbol{H} \tag{7.6}$$

can occur, especially in connection with ferrites (see §5.22).

1.8 Permanent magnetism

One way of accounting for (b) and (c) in §1.7 is by means of the phenomenon of *polarization*. Imposing an electric field on a dielectric tends to separate the negative portions of charge arising from the electrons and the positive portions of charge of the atomic nuclei (which form a lattice structure in a solid) with effects both electrical and mechanical. In fact, a dielectric is inclined to increase its linear dimensions in the direction of the field; an effect known as *electrostriction*. Mechanical stress applied to certain crystals such as tourmaline and Rochelle salt produces a separation of the positive and negative charges; this is called the *piezoelectric* effect. The converse phenomenon, in which the crystal expands or contracts according to the direction of the field is also observed. Piezoelectricity is a useful means of converting mechanical energy (as in sound waves) into electrical energy and vice versa; it therefore forms a basis for certain types of microphone and loudspeaker as well as for the production of ultrasonic waves.

Polarization of a medium can be described theoretically by $\boldsymbol{P}$ and $\boldsymbol{M}$ the *electric* and *magnetic polarizations* defined by

$$\boldsymbol{P} = \boldsymbol{D} - \varepsilon_0 \boldsymbol{E}, \qquad \boldsymbol{M} = \boldsymbol{B} - \mu_0 \boldsymbol{H}. \tag{8.1}$$

In free space the polarizations vectors vanish.

If (8.1) is introduced into (5.3)–(5.6), the equations

$$\operatorname{curl} \boldsymbol{E} + \partial \boldsymbol{B}/\partial t = \boldsymbol{0},$$

$$\operatorname{curl} \boldsymbol{B} - \mu_0 \varepsilon_0 \frac{\partial \boldsymbol{E}}{\partial t} = \mu_0 \left(\boldsymbol{J} + \frac{\partial \boldsymbol{P}}{\partial t} + \frac{1}{\mu_0} \operatorname{curl} \boldsymbol{M} \right),$$

$$\operatorname{div} \boldsymbol{E} = (\rho - \operatorname{div} \boldsymbol{P})/\varepsilon_0,$$

$$\operatorname{div} \boldsymbol{B} = 0$$

are obtained. Thus the presence of material bodies can be accounted for by equivalent distributions of charge density $-\operatorname{div} \boldsymbol{P}/\varepsilon_0$ and current density $\partial \boldsymbol{P}/\partial t + \operatorname{curl} \boldsymbol{M}/\mu_0$ in free space.

In isotropic media it follows from (7.2) and (8.1) that

$$\boldsymbol{P} = (\kappa - 1)\varepsilon_0 \boldsymbol{E}, \qquad \boldsymbol{M} = (\kappa_{\mathrm{m}} - 1)\mu_0 \boldsymbol{H}.$$

The quantities $\chi = \kappa - 1$ and $\chi_{\mathrm{m}} = \kappa_{\mathrm{m}} - 1$ are called the *electric* and *magnetic susceptibilities.* In anisotropic media, where (7.5) and (7.6) hold, the susceptibilities are symmetric tensors.

The polarizations as so defined depend upon the applied field and may therefore be regarded as *induced polarizations.* In certain ferromagnetic substances it is not true that there is a linear relation between $\boldsymbol{B}$ and $\boldsymbol{H}$ (in fact, for given $\boldsymbol{H}$, $\boldsymbol{B}$ depends upon the previous mechanical, thermal, and magnetic history) and, moreover, the material may possess a magnetic field in the absence of an applied field. The dimensions of the body also change on magnetization; this feature of *magnetostriction* has been taken advantage of in the design of oscillators. A body which has a residual magnetic field when the applied field is removed is said to be *permanently magnetized.* The polarization $\boldsymbol{M}$ may still be defined by (8.1) but now it contains two essentially different parts. One which is the polarization which would be present in the absence of the applied field and one which is the induced polarization depending on $\boldsymbol{H}$. For the most part ferromagnets will be excluded from our discussion except in so far as their behaviour can be represented reasonably by (7.2) with constant μ, i.e. where $\boldsymbol{B}$ and $\boldsymbol{H}$ do not change too much.

There are also dielectrics, of which Carnuba wax is one, that exhibit similar electrical properties. If Carnuba wax is melted and allowed to solidify in a strong electric field, it is found that the electric polarization remains after the applied field is removed. Such substances, which are called *electrets*, will be ignored in the considerations which follow.

1.9 Physical properties of conductors

Conduction is achieved by the transfer of charge from one region to another, good conductors being those in which the transfer is accomplished by very small electric fields. The best conductors are metals, which possess a crystalline structure and conduction electrons able to move comparatively freely through the crystal lattice. The motion of an electron under the influence of an applied electric field is retarded by the forces due to the ionic lattice. The retardation, current density and applied field are connected by

$$\boldsymbol{J} = \sigma \mathbf{E} \tag{9.1}$$

in isotropic metals. The current may also involve the applied magnetic field (the *Hall effect*) but, in general, the effect is so small that it may be neglected in the contexts we shall be considering.

Equation (9.1) is known as *Ohm's law* and the quantity σ is called the *conductivity*. For most metals σ is of the order of 10^7 mho m^{-1} at room temperature; consequently, it is often reasonable in theoretical investigations to approximate a metal by a fictitious *perfect conductor* in which σ is taken to be infinite.

The magnitude of σ depends on the temperature and, in general, varies linearly with it for a range of a few hundred degrees. However, there are some metals such as lead, mercury and tin which exhibit a sharp rise in conductivity at temperatures of the order of 5 °K and currents started in these metals persist for many hours. They are displaying the phenomenon of *superconductivity*.

It should be remarked that the process of conduction entails the irreversible transformation of electrical energy into heat. The heat produced raises the temperature of the metal and thereby alters the conductivity of the metal. Thus, when a constant electric field is applied to a metal, the current changes steadily until thermal equilibrium is attained. Until thermal equilibrium is established the relation between the voltage drop across a conductor and the current flowing through it is *nonlinear*. For example, the current in a tungsten filament lamp depends, to a first approximation, upon the cube of the voltage drop as well as the first power. In our applications we shall assume that the temperature is sufficiently stable for the conductivity to be treated as constant.

Some liquids, such as the paraffin oils, have conductivities of the order of 10^{-11} mho m^{-1}. They are mainly used as *insulators* and may be regarded theoretically as dielectrics with zero current density. An electrolytic solution, i.e. an aqueous solution of an acid, base, or salt, possesses a conductivity of about 10^2 mho m^{-1} and the current through it satisfies (9.1). Pure liquids such as alcohol and water have conductivities of about 10^{-3} mho m^{-1} and cannot be deemed to be either insulators or good conductors.

Conduction in gases differs fundamentally from that in solids and liquids, primarily because the current itself can generate ions in the conducting medium. The phenomena are extremely complicated and will be omitted from our discussion.

Certain elements such as silicon, germanium, and selenium are insulators at low temperatures and conductors at high temperatures; they are called *semiconductors.* They have important applications in computers and transistors but they will not be discussed in detail in this book.

Since (9.1) holds for many substances under a wide variety of conditions we shall take it as one of our fundamental equations unless we state otherwise. If σ is nonzero we shall usually refer to the medium as a conductor whereas if $\sigma = 0$ we shall often call the medium a dielectric. When (7.2) and (9.1) are valid with ε, μ, and σ constant the medium may be referred to as a homogeneous isotropic conductor. We shall now show that *there can be no permanent distribution of free charge in a homogeneous isotropic conductor with positive conductivity.*

In a conductor (5.5) and (7.2) show that

$$\rho = \text{div}\,\mathbf{D} = \text{div}\,\varepsilon\mathbf{E} \tag{9.2}$$

while the equation of continuity (6.1) and (9.1) give

$$\text{div}\,\sigma\mathbf{E} + \frac{\partial \rho}{\partial t} = \text{div}\,\mathbf{J} + \frac{\partial \rho}{\partial t} = 0. \tag{9.3}$$

Since div $\varepsilon\mathbf{E} = \varepsilon$ div $\mathbf{E} + \mathbf{E} \cdot$ grad ε (see §F.7 for various vector formulae) we can eliminate div $\mathbf{E}$ from (9.2) and (9.3) to obtain

$$\varepsilon \frac{\partial \rho}{\partial t} + \sigma\rho = \mathbf{E} \cdot (\sigma\,\text{grad}\,\varepsilon - \varepsilon\,\text{grad}\,\sigma)$$

or

$$\varepsilon \frac{\partial \rho}{\partial t} + \sigma\rho = \sigma\mathbf{J} \cdot \text{grad}(\varepsilon/\sigma). \tag{9.4}$$

If the conductor is homogeneous the right-hand side of (9.4) vanishes and

$$\rho = \rho_0 e^{-\sigma t/\varepsilon} \tag{9.5}$$

where ρ_0 is the value of ρ at $t = 0$. Thus, when the conductivity is positive, the *charge density decays exponentially and is independent of the applied field.* If the charge density is zero initially, it remains zero at all subsequent times. The time ε/σ which elapses while the charge density at any point falls to 1/e of its original value is called the *relaxation time.* Except in bad conductors the relaxation time is very small, e.g. in water it is about 10^{-6} sec. For an insulator like petroleum oil, however, the relaxation time is of the order of seconds and for a substance such as sulphur it is of the order of days.

If the conductor is not homogeneous the right-hand side of (9.4) is nonzero (unless ε/σ happens to be constant) and the equation shows that, if a steady state is eventually attained, there is a charge density given by

$$\rho = \boldsymbol{J} \cdot \operatorname{grad}(\varepsilon/\sigma).$$

No conduction current is produced by an applied electric field in a perfect dielectric but in an imperfect dielectric σ is small and the volume charges that arise from the current due to an applied field are appreciable. The phenomenon of the occurrence of free charges throughout a dielectric is known as *dielectric absorption* and is of importance in the construction of high voltage capacitors.

The electromagnetic potentials

1.10 The scalar and vector potentials

The main problem in the theory of electromagnetism is to solve Maxwell's equations subject to certain conditions. Since eight scalar partial differential equations are involved (apart from the constitutive equations) this is usually a problem of some magnitude. One way to try to reduce the difficulty is to adopt a representation for the electromagnetic field which automatically satisfies some of the equations.

Since div $\boldsymbol{B} = 0$ there is a vector $\mathbf{A}$ such that

$$\boldsymbol{B} = \operatorname{curl} \mathbf{A}. \tag{10.1}$$

So, in order that

$$\operatorname{curl} \boldsymbol{E} + \partial \boldsymbol{B}/\partial t = \mathbf{0}$$

it is necessary that

$$\operatorname{curl} \boldsymbol{E} + \operatorname{curl} \partial \mathbf{A}/\partial t = \mathbf{0}$$

which implies that

$$\boldsymbol{E} = -\operatorname{grad} V - \partial \mathbf{A}/\partial t \tag{10.2}$$

where V is a scalar function. The vector $\mathbf{A}$ and function V are known as the *vector* and *scalar potentials* respectively.

Let $\mathbf{A}_0$ be any particular vector such that $\boldsymbol{B} = \text{curl}\,\mathbf{A}_0$ and V_0 the corresponding scalar potential which gives $\boldsymbol{E}$ correctly. Then

$$\mathbf{A} = \mathbf{A}_0 - \text{grad}\,\psi, \tag{10.3}$$

where ψ is an arbitrary scalar, gives the same $\boldsymbol{B}$ and

$$V = V_0 + \partial\psi/\partial t \tag{10.4}$$

supplies the same $\boldsymbol{E}$. Thus the representation (10.1) and (10.2) is unaltered by the transformation (10.3) and (10.4). The representation is said to be *invariant under a gauge transformation.*

If the current and charge densities are zero, (5.4) and (5.5) may be tackled in the same way. $\boldsymbol{D}$ will be expressed in terms of a vector potential and $\boldsymbol{H}$ in terms of both a scalar and vector potential. The connection of these potentials to those of (10.1) and (10.2) cannot be determined until constitutive relations have been imposed. However, it is more convenient to consider the consequences of (5.4) and (5.5) for the representation (10.1) and (10.2) under the constitutive equations.

Suppose that the medium is homogeneous and isotropic so that

$$\boldsymbol{D} = \varepsilon\boldsymbol{E}, \qquad \boldsymbol{B} = \mu\boldsymbol{H}.$$

Then (10.1) and (10.2) satisfy (5.4) and (5.5) provided that

$$\text{curl curl}\,\mathbf{A} + \mu\varepsilon\left(\text{grad}\frac{\partial V}{\partial t} + \frac{\partial^2\mathbf{A}}{\partial t^2}\right) = \mu\boldsymbol{J}, \tag{10.5}$$

$$-\nabla^2 V - \text{div}\frac{\partial\mathbf{A}}{\partial t} = \rho/\varepsilon. \tag{10.6}$$

On account of the gauge invariance, (10.5) and (10.6) are satisfied by $\mathbf{A}$ and V defined by (10.3) and (10.4) if they are satisfied by $\mathbf{A}_0$ and V_0.

In equation (10.5) make use of the identity (F.7.2)

$$\text{curl curl}\,\mathbf{A} = \text{grad div}\,\mathbf{A} - \nabla^2\mathbf{A} \tag{10.7}$$

where the interpretation of $\nabla^2\mathbf{A}$ is given by

$$\nabla^2\mathbf{A} = \boldsymbol{i}\nabla^2 A_x + \boldsymbol{j}\nabla^2 A_y + \boldsymbol{k}\nabla^2 A_z \tag{10.8}$$

with $\boldsymbol{i}$, $\boldsymbol{j}$, $\boldsymbol{k}$ unit vectors parallel to the Cartesian coordinate axes and A_x, A_y, A_z the components of $\mathbf{A}$ parallel to those axes. (Note that in other coordinate systems the expression cannot be split into such a simple component form.) In view of (10.7), (10.5) becomes

$$-\nabla^2\mathbf{A} + \mu\varepsilon\frac{\partial^2\mathbf{A}}{\partial t^2} + \text{grad}\left(\text{div}\,\mathbf{A} + \mu\varepsilon\frac{\partial V}{\partial t}\right) = \mu\boldsymbol{J}.$$

As has been pointed out there is a good deal of arbitrariness in $\mathbf{A}$ and V. The choice can be limited by imposing the condition

$$\text{div}\,\mathbf{A} + \mu\varepsilon\,\partial V/\partial t = 0. \tag{10.9}$$

This can be accomplished by choosing ψ in (10.3) and (10.4) so that

$$\nabla^2\psi - \mu\varepsilon\, \partial^2\psi/\partial t^2 = \operatorname{div} \mathbf{A}_0 + \mu\varepsilon\, \partial V_0/\partial t.$$

Even so any solution of

$$\nabla^2\varphi - \mu\varepsilon\, \partial^2\varphi/\partial t^2 = 0$$

can be added to ψ without affecting the validity of the representation.

Subject to (10.9) the potentials satisfy

$$\nabla^2\mathbf{A} - \mu\varepsilon\, \partial^2\mathbf{A}/\partial t^2 = -\mu\boldsymbol{J}, \tag{10.10}$$

$$\nabla^2 V - \mu\varepsilon\, \partial^2 V/\partial t^2 = -\rho/\varepsilon. \tag{10.11}$$

The close resemblance of (10.11) to the equation (4.6) of sound waves will be noticed.

The method of construction shows that any given electromagnetic field can be represented in the form (10.1) and (10.2) with (10.9) satisfied. Equations (10.10) and (10.11) then give the current and charge densities necessary to maintain this field. Alternatively, when the charge and current densities are given and the resulting field has to be found, the method shows that the field can be written as (10.1) and (10.2) with (10.9) satisfied but now (10.10) and (10.11) have to be solved subject to appropriate boundary conditions.

1.11 The potentials in a homogeneous conductor

Assume that there are no sources of charge and current in the conductor. Then, since the free charge decays independently of the applied field and the relaxation time is usually short (§1.9), the charge density in the interior of the conductor can be neglected. Equations (5.3) and (5.6) are unchanged so that (10.1) and (10.2) still hold. By virtue of (9.1), equations (5.4) and (5.5) become

$$\operatorname{curl} \boldsymbol{H} - \partial\boldsymbol{D}/\partial t = \sigma\boldsymbol{E}, \qquad \operatorname{div} \boldsymbol{D} = 0.$$

There is no difficulty in seeing that, in a similar way to that of §1.10, $\mathbf{A}$ and V can be chosen so that

$$\operatorname{div} \mathbf{A} + \mu\varepsilon\frac{\partial V}{\partial t} + \mu\sigma V = 0$$

with

$$\nabla^2\mathbf{A} - \mu\varepsilon\frac{\partial^2\mathbf{A}}{\partial t^2} - \mu\sigma\frac{\partial\mathbf{A}}{\partial t} = 0,$$

$$\nabla^2 V - \mu\varepsilon\frac{\partial^2 V}{\partial t^2} - \mu\sigma\frac{\partial V}{\partial t} = 0.$$

An alternative representation could be obtained in this case since $\operatorname{div} \boldsymbol{D} = 0$ and so we could take $\boldsymbol{D} = \operatorname{curl} \mathbf{A}'$. This may be convenient in certain circumstances but the analysis, which is similar to that already given, will not be reproduced. The representation must, in any case, be contained in that already set out.

1.12 The Hertz vector

The preceding sections have shown that the electromagnetic field can be represented in terms of the scalar and vector potentials. It will now be demonstrated that, because the potentials are connected by (10.9), there is a representation involving a single vector function. Only the homogeneous isotropic medium will be considered.

The first step is to show that there is a vector $\boldsymbol{p}$ such that

$$\partial \boldsymbol{p}/\partial t = \boldsymbol{J}, \qquad \operatorname{div} \boldsymbol{p} = -\rho. \tag{12.1}$$

Let $\rho = \rho_0$ when $t = 0$ and let $\boldsymbol{p}_0$ be a vector, independent of t, such that $\operatorname{div} \boldsymbol{p}_0 = -\rho_0$. Now choose

$$\boldsymbol{p} = \boldsymbol{p}_0 + \int_0^t \boldsymbol{J}\,\mathrm{d}t.$$

This obviously satisfies the first of (12.1) and as for the second,

$$\begin{aligned}\operatorname{div} \boldsymbol{p} &= \operatorname{div} \boldsymbol{p}_0 + \int_0^t \operatorname{div} \boldsymbol{J}\,\mathrm{d}t\\ &= -\rho_0 - \int_0^t \frac{\partial \rho}{\partial t}\,\mathrm{d}t = -\rho\end{aligned}$$

on employing the equation of continuity (6.1). Thus the existence of $\boldsymbol{p}$ is verified; $\boldsymbol{p}$ is known as the *electric moment per cubic metre*.

By means of (12.1), the equations (10.10) and (10.11) can be written

$$\nabla^2 \mathbf{A} - \mu\varepsilon \frac{\partial^2 \mathbf{A}}{\partial t^2} = -\mu \frac{\partial \boldsymbol{p}}{\partial t}, \tag{12.2}$$

$$\nabla^2 V - \mu\varepsilon \frac{\partial^2 V}{\partial t^2} = \frac{1}{\varepsilon} \operatorname{div} \boldsymbol{p}. \tag{12.3}$$

We can deduce from (12.1) that there is a vector $\boldsymbol{\Pi}'$ such that

$$\mathbf{A} = \mu\varepsilon\, \partial \boldsymbol{\Pi}'/\partial t, \qquad \operatorname{div} \boldsymbol{\Pi}' = -V \tag{12.4}$$

since (10.9) has the same structure as (6.1). On account of (12.2) and (12.3)

$$\frac{\partial}{\partial t}\left(\nabla^2 \boldsymbol{\Pi}' - \mu\varepsilon \frac{\partial^2 \boldsymbol{\Pi}'}{\partial t^2}\right) = -\frac{1}{\varepsilon}\frac{\partial \boldsymbol{p}}{\partial t}, \tag{12.5}$$

$$\operatorname{div}\left(\nabla^2 \boldsymbol{\Pi}' - \mu\varepsilon \frac{\partial^2 \boldsymbol{\Pi}'}{\partial t^2}\right) = -\frac{1}{\varepsilon} \operatorname{div} \boldsymbol{p}. \tag{12.6}$$

Equation (12.6) implies that

$$\nabla^2 \boldsymbol{\Pi}' - \mu\varepsilon \frac{\partial^2 \boldsymbol{\Pi}'}{\partial t^2} + \frac{1}{\varepsilon}\mathbf{p} = \text{curl}\, \boldsymbol{f}$$

and then curl $\partial \boldsymbol{f}/\partial t = 0$ from (12.5) so that $\boldsymbol{f}$ can be treated as independent of t. Put

$$\boldsymbol{\Pi}' = \boldsymbol{\Pi} + \text{curl}\, \mathbf{g}$$

where $\mathbf{g}$ is independent of t and $\nabla^2 \mathbf{g} = \boldsymbol{f}$. Then

$$\nabla^2 \boldsymbol{\Pi} - \mu\varepsilon\, \partial^2 \boldsymbol{\Pi}/\partial t^2 = -\mathbf{p}/\varepsilon \tag{12.7}$$

and, from (12.4),

$$\mathbf{A} = \mu\varepsilon\, \partial \boldsymbol{\Pi}/\partial t, \qquad V = -\text{div}\, \boldsymbol{\Pi}. \tag{12.8}$$

It follows that

$$\boldsymbol{E} = \text{grad div}\, \boldsymbol{\Pi} - \mu\varepsilon\, \partial^2 \boldsymbol{\Pi}/\partial t^2, \tag{12.9}$$

$$\boldsymbol{B} = \mu\varepsilon\, \text{curl}\, \partial \boldsymbol{\Pi}/\partial t. \tag{12.10}$$

Thus it has been shown that there is a vector $\boldsymbol{\Pi}$ satisfying (12.7) such that the electromagnetic field can be expressed in the form (12.9), (12.10). The vector $\boldsymbol{\Pi}$ is called the *Hertz vector;* sometimes the adjective electric is attached. Of course, given the Hertz vector it is easy to check that (12.9) and (12.10) provide a solution of Maxwell's equations but what the above proof reveals is that a given electromagnetic field can be represented in terms of a Hertz vector.

It will be remarked that the representation is gauge invariant in the sense that, if $\boldsymbol{\Pi}$ is replaced by $\boldsymbol{\Pi} + \text{grad}\, \chi$ where

$$\nabla^2 \chi - \mu\varepsilon\, \partial^2 \chi/\partial t^2 = 0,$$

(12.7) continues to hold while $\boldsymbol{E}$ and $\boldsymbol{B}$ are unaltered.

The extension of these results to a homogeneous conductor is straightforward when there are no sources of current or charge. It can be verified readily that

$$\boldsymbol{E} = \text{grad div}\, \boldsymbol{\Pi} - \mu\varepsilon \frac{\partial^2 \boldsymbol{\Pi}}{\partial t^2} - \mu\sigma \frac{\partial \boldsymbol{\Pi}}{\partial t} = \text{curl curl}\, \boldsymbol{\Pi},$$

$$\boldsymbol{B} = \mu\, \text{curl}\left(\varepsilon \frac{\partial \boldsymbol{\Pi}}{\partial t} + \sigma \boldsymbol{\Pi}\right)$$

where

$$\nabla^2 \boldsymbol{\Pi} - \mu\varepsilon \frac{\partial^2 \boldsymbol{\Pi}}{\partial t^2} - \mu\sigma \frac{\partial \boldsymbol{\Pi}}{\partial t} = 0.$$

1.13 Representation by two scalars when there are no sources

The gauge invariance of the Hertz vector makes it possible to find a representation in terms of only two scalars when there are no sources of charge or current present. Suppose that a Hertz vector $\boldsymbol{\Pi}_0$ has been determined. Let χ_0 be a solution of

$$\frac{\partial^2\chi_0}{\partial z^2}-\mu\varepsilon\frac{\partial^2\chi_0}{\partial t^2}=-\frac{\partial\Pi_{0x}}{\partial x}-\frac{\partial\Pi_{0y}}{\partial y}. \tag{13.1}$$

Define

$$\chi=\chi_0+B(x, y)f\{t-z(\mu\varepsilon)^{\frac{1}{2}}\}+C(x, y)g\{t+z(\mu\varepsilon)^{\frac{1}{2}}\} \tag{13.2}$$

where the functions B, C, f, and g will be specified in a moment. It is immediately evident from (13.1) that

$$\frac{\partial^2\chi}{\partial z^2}-\mu\varepsilon\frac{\partial^2\chi}{\partial t^2}=-\frac{\partial\Pi_{0x}}{\partial x}-\frac{\partial\Pi_{0y}}{\partial y}. \tag{13.3}$$

Further

$$\frac{\partial^2\chi}{\partial x^2}+\frac{\partial^2\chi}{\partial y^2}=\frac{\partial^2\chi_0}{\partial x^2}+\frac{\partial^2\chi_0}{\partial y^2}+\left(\frac{\partial^2B}{\partial x^2}+\frac{\partial^2B}{\partial y^2}\right)f+\left(\frac{\partial^2C}{\partial x^2}+\frac{\partial^2C}{\partial y^2}\right)g.$$

Now, from (13.1),

$$\left(\frac{\partial^2}{\partial z^2}-\mu\varepsilon\frac{\partial^2}{\partial t^2}\right)\left(\frac{\partial^2\chi_0}{\partial x^2}+\frac{\partial^2\chi_0}{\partial y^2}\right)=\left(\frac{\partial^2}{\partial z^2}-\mu\varepsilon\frac{\partial^2}{\partial t^2}\right)\left(\frac{\partial\Pi_{0x}}{\partial x}+\frac{\partial\Pi_{0y}}{\partial y}\right)$$

because $\boldsymbol{\Pi}_0$ satisfies (12.7) with the right-hand side zero. Hence (cf. (14.2)).

$$\frac{\partial^2\chi_0}{\partial x^2}+\frac{\partial^2\chi_0}{\partial y^2}=\frac{\partial\Pi_{0x}}{\partial x}+\frac{\partial\Pi_{0y}}{\partial y}+B_1(x, y)f_1\{t-z(\mu\varepsilon)^{\frac{1}{2}}\}$$
$$+C_1(x, y)g_1\{t+z(\mu\varepsilon)^{\frac{1}{2}}\}.$$

Make the following selection of the functions at our disposal:

$$f=f_1, \qquad g=g_1,$$
$$\frac{\partial^2B}{\partial x^2}+\frac{\partial^2B}{\partial y^2}=-B_1, \qquad \frac{\partial^2C}{\partial x^2}+\frac{\partial^2C}{\partial y^2}=-C_1.$$

Then

$$\frac{\partial^2\chi}{\partial x^2}+\frac{\partial^2\chi}{\partial y^2}=\frac{\partial\Pi_{0x}}{\partial x}+\frac{\partial\Pi_{0y}}{\partial y}. \tag{13.4}$$

The combination of (13.3) and (13.4) gives

$$\nabla^2\chi-\mu\varepsilon\,\partial^2\chi/\partial t^2=0. \tag{13.5}$$

Choose now as Hertz vector $\boldsymbol{\Pi}=\boldsymbol{\Pi}_0-\operatorname{grad}\chi$, a permissible choice in view of (13.5). This Hertz vector has the property

$$\frac{\partial\Pi_x}{\partial x}+\frac{\partial\Pi_y}{\partial y}=0$$

on account of (13.4). Hence there is a scalar M such that $\Pi_x = \partial M/\partial y$, $\Pi_y = -\partial M/\partial x$ i.e.

$$\Pi_x \boldsymbol{i} + \Pi_y \boldsymbol{j} = \mathrm{curl}(M\boldsymbol{k}).$$

Replacing $\Pi_z \boldsymbol{k}$ by $\Pi \boldsymbol{k}$ and substituting in (12.9), (12.10) we see that there is a representation

$$\boldsymbol{E} = \mathrm{grad\,div}(\Pi\boldsymbol{k}) - \mu\varepsilon \boldsymbol{k}\, \partial^2\Pi/\partial t^2 - \mu\varepsilon\, \mathrm{curl}(\boldsymbol{k}\, \partial^2 M/\partial t^2), \quad (13.6)$$

$$\boldsymbol{B} = \mu\varepsilon\, \mathrm{curl}(\boldsymbol{k}\, \partial\Pi/\partial t) + \mu\varepsilon\, \mathrm{curl\,curl}(\boldsymbol{k}\, \partial M/\partial t). \quad (13.7)$$

Thus, *in the absence of charges and currents, any electromagnetic field has a representation in terms of the two scalars Π and M.*

Both Π and M are solutions of the partial differential equation (13.5) and so the representation can be regarded as constructed from an electric Hertz vector $\Pi\boldsymbol{k}$ and a *magnetic Hertz vector* $M\boldsymbol{k}$, the adjective magnetic indicating that the roles of $\boldsymbol{E}$ and $\boldsymbol{B}$ have been reversed as far as it is concerned. The two vectors are directed along the z-axis. Since we are at liberty to take the z-axis in any suitable direction we conclude that any electromagnetic field, in a homogeneous isotropic medium free of charges and currents, can be expressed in terms of an electric Hertz vector and a magnetic Hertz vector both along a direction which can be chosen at will.

Integral representations

1.14 The potential of a point source

The partial differential equation for sound waves (4.6) is of the same type as those for the scalar potential, vector potential, and Hertz vector. Therefore, whether we are discussing acoustics or electromagnetism, we need to be able to solve a partial differential equation of the type

$$\nabla^2 p - \frac{1}{v^2}\frac{\partial^2 p}{\partial t^2} = -s$$

where v is a speed (corresponding to a_0 in acoustics and $1/(\mu\varepsilon)^{\frac{1}{2}}$ for electromagnetic waves in a homogeneous isotropic medium) and s stands for the density of the sources.

The simplest source density which can be considered is one which is zero everywhere except for a small neighbourhood of the origin. This leads to the idealized model of a source concentrated at the origin—such a concentrated source is called a *point source.* The density of a point source is zero everywhere except at the source, where it is infinite. The equation

$$\nabla^2 p - \frac{1}{v^2}\frac{\partial^2 p}{\partial t^2} = 0 \quad (14.1)$$

must therefore hold at all points other than the origin. Clearly, p depends only on the radial distance from the origin and not upon its direction, i.e. p is independent of θ and ϕ where R, θ, ϕ are spherical polar coordinates (see §1.39). Hence (14.1) becomes (see (39.7))

$$\frac{1}{R^2}\frac{\partial}{\partial R}\left(R^2\frac{\partial p}{\partial R}\right)-\frac{1}{v^2}\frac{\partial^2 p}{\partial t^2}=0.$$

On putting $p=U/R$ we obtain

$$\frac{\partial^2 U}{\partial R^2}-\frac{1}{v^2}\frac{\partial^2 U}{\partial t^2}=0. \tag{14.2}$$

Make the change of coordinates

$$\xi=t-R/v, \qquad \eta=t+R/v$$

so that

$$\frac{\partial^2 U}{\partial R^2}=\frac{\partial}{\partial R}\left(-\frac{1}{v}\frac{\partial U}{\partial \xi}+\frac{1}{v}\frac{\partial U}{\partial \eta}\right)=\frac{1}{v^2}\left(\frac{\partial^2 U}{\partial \xi^2}-2\frac{\partial^2 U}{\partial \xi\,\partial \eta}+\frac{\partial^2 U}{\partial \eta^2}\right)$$

and

$$\frac{\partial^2 U}{\partial t^2}=\frac{\partial^2 U}{\partial \xi^2}+2\frac{\partial^2 U}{\partial \xi\,\partial \eta}+\frac{\partial^2 U}{\partial \eta^2}.$$

Then (14.2) leads to

$$\frac{\partial^2 U}{\partial \xi\,\partial \eta}=0$$

whence

$$U=f(\xi)+g(\eta)$$

where f and g are arbitrary. Accordingly,

$$p=\frac{f(t-R/v)}{R}+\frac{g(t+R/v)}{R}. \tag{14.3}$$

Consider firstly the function $f(t-R/v)$. If t is increased to $t+T$ and, at the same time, R is altered to $R+vT$ the argument of f is unchanged. In other words, if f has a certain value at time t and distance R it has the same value at the distance vT further out after an interval T has elapsed. Thus there is a wave travelling *out* from the origin with speed v. Consequently, the first term of (14.3) represents an *outgoing* wave whose amplitude is being steadily diminished by the factor $1/R$ as it moves out. Similarly, the second term of p furnishes an *incoming* wave (i.e. travelling towards the origin with speed v) whose amplitude steadily increases as it moves in.

The precise determination of the functions f and g requires further information. If it is assumed that placing a source suddenly at the origin will cause a wave to spread outwards, as occurs with the disturbance of the surface of a pond when a stone is thrown in, then g must be taken as identically zero because it represents a wave travelling inwards. In general, g provides a wave which can be observed *before* it is generated by its source at the origin. Since such behaviour is contrary to what is usually seen it will be assumed, in general, that g is zero. The function f has now to be chosen to give the correct source strength at the origin. Now, a source which has unit strength for all time produces a static potential of $1/4\pi R$. Hence the potential of a source at the origin and of strength $f(t)$ at time t is

$$p = \frac{f(t - R/v)}{4\pi R} \tag{14.4}$$

The same formula is applicable to a point source which is not at the origin provided that R is interpreted as the distance from the source.

1.15 Generalized functions

It has already been noted that the source density of a point source is zero everywhere except at the source where it is infinite. To allow an analytical description of the charge density we introduce the symbol $\delta(x-y)$ to mean something which is zero everywhere except at y where it is infinite and such that

$$\int f(y)\delta(x-y)\,dy = f(x) \tag{15.1}$$

if the interval of integration includes x and equals zero if the interval does not include x. The entity $\delta(x-y)$ is known as the *Dirac delta function;* it is not really a function but a *distribution,* a *generalized function,* or a *weak function.* (For more details on these notions see Jones (1982).) When we say that

$$g(y) = \delta(x-y) \tag{15.2}$$

what we mean is that

$$\int_a^b h(y)g(y)\,dy = \begin{cases} h(x) & (x \text{ in } (a, b)) \\ 0 & (x \text{ outside } (a, b)). \end{cases} \tag{15.3}$$

No meaning is attached to (15.2) other than in terms of integration.

A generalized function always possesses a derivative which is itself a generalized function. So, if a prime denotes a derivative with respect to the argument, $\delta'(x-y)$, $\delta''(x-y), \ldots$ exist and are such that

$$\int_a^b h(y)\delta'(x-y)\,dy = h'(x), \tag{15.3}$$

$$\int_a^b h(y)\delta''(x-y)\,dy = h''(x) \tag{15.4}$$

when x is in (a, b).

It can be shown that (15.3) is still valid when h itself is a generalized function. Further, if

$$Dg(x, y) = \delta(x-y)$$

where D is some differential operator (d^2/dx^2, for example)

$$\begin{aligned} D\int_a^b h(y)g(x, y)\,dy &= \int_a^b h(y)Dg(x, y)\,dy \\ &= \int_a^b h(y)\delta(x-y)\,dy \\ &= h(x) \end{aligned} \tag{15.5}$$

when x is in (a, b) and is zero when x is outside (a, b).

The theory can be extended to two or more dimensions by introducing the generalized function $\delta(\boldsymbol{x}-\mathbf{y})$ such that

$$\begin{aligned} \int_T h(\mathbf{y})\delta(\boldsymbol{x}-\mathbf{y})\,d\mathbf{y} &= h(\boldsymbol{x}) \quad (\boldsymbol{x} \text{ in } T) \\ &= 0 \qquad\quad (\boldsymbol{x} \text{ outside } T) \end{aligned}$$

where $h(\mathbf{y})$ is a generalized function of the position vector $\mathbf{y}$ and $d\mathbf{y}$ is an element of volume of T. It is possible to express $\delta(\boldsymbol{x}-\mathbf{y})$ in terms of the generalized function of (15.1). Thus, if $\mathbf{y}$ is expressed in Cartesian coordinates so that $d\mathbf{y} = dy_1\,dy_2\,dy_3$ we have $\delta(\boldsymbol{x}-\mathbf{y}) = \delta(x_1-y_1)\delta(x_2-y_2)\delta(x_3-y_3)$. Similarly, if $\mathbf{y}$ is put into spherical polar coordinates R', θ', ϕ' so that $d\mathbf{y} = R'^2 \sin\theta'\,dR'\,d\theta'\,d\phi'$ (see (37.2) and §1.39(ii)),

$$\delta(\boldsymbol{x}-\mathbf{y}) = \delta(R-R')\delta(\theta-\theta')\delta(\phi-\phi')/R'^2 \sin\theta'$$

when $R' \neq 0$ and $\sin\theta' \neq 0$.

Analogous to (15.5), when

$$Dg(\boldsymbol{x}, \mathbf{y}) = \delta(\boldsymbol{x}-\mathbf{y}) \tag{15.6}$$

D being a differential operator such as ∇^2, is

$$D\int_T h(\mathbf{y})g(\boldsymbol{x}, \mathbf{y})\,d\mathbf{y} = \begin{cases} h(\boldsymbol{x}) & (\boldsymbol{x} \text{ in } T) \\ 0 & (\boldsymbol{x} \text{ outside } T). \end{cases} \tag{15.7}$$

This result enables the construction of a solution of a differential equation of the form

$$Df(\boldsymbol{x}) = h(\boldsymbol{x}) \tag{15.8}$$

if (15.6) can be solved. For, choose T so that $h(\boldsymbol{x})$ is zero outside T and take

$$f(\boldsymbol{x}) = \int_T h(\mathbf{y})g(\boldsymbol{x}, \mathbf{y})\,d\mathbf{y}. \tag{15.9}$$

Equation (15.7) then shows that (15.8) is satisfied.

It is important to realize that this method of generating a solution is applicable whatever the number of independent variables so long as the right-hand side of (15.6) is modified appropriately and T involves all the variables. Thus, if D contains the Cartesian coordinates x_1, x_2, x_3 and the time t, the appropriate right-hand side of (15.6) is $\delta(x_1-y_1)\delta(x_2-y_2)\delta(x_3-y_3)\delta(t-\tau)$ or some equivalent form and, in (15.9), $\mathrm{d}\mathbf{y}=\mathrm{d}y_1\,\mathrm{d}y_2\,\mathrm{d}y_3\,\mathrm{d}\tau$ the integration being over space and time.

The use of generalized functions therefore permits one to derive a solution of a differential equation with a general right-hand side by solving one (15.6) with a special right-hand side. In general, the solution (15.9) will be a generalized function. Whether this is acceptable or not will depend on the physical interpretation of f. Broadly speaking, one would expect that, when f is physically observable, it should be a conventional function and the right-hand side of (15.9) has to be shown to be such a function. (This is only necessary when discussing general theory; in any particular problem it would be a matter of verification.) But if f is involved only in integration then its existence as a generalized function rather than as a conventional function may well be adequate.

1.16 Retarded potentials

It is now convenient to present the potential of (14.4) as a solution of a differential equation involving a generalized function. Consider the source density

$$s(\boldsymbol{x}, t)=f(t)\delta(\boldsymbol{x}-\mathbf{y}).$$

Then

$$\int_T s(\boldsymbol{x}, t)\,\mathrm{d}\boldsymbol{x}=f(t)\quad(\mathbf{y}\text{ in }T)$$

and the integral is zero where $\mathbf{y}$ is outside T. So the charge density is that due to a point source of strength $f(t)$ placed at the point $\mathbf{y}$. Hence, from (14.4), a solution of

$$\nabla^2 p-\frac{1}{v^2}\frac{\partial^2 p}{\partial t^2}=-f(t)\delta(\boldsymbol{x}-\mathbf{y}) \tag{16.1}$$

which represents an outgoing wave is

$$p(\boldsymbol{x})=\frac{f(t-|\boldsymbol{x}-\mathbf{y}|/v)}{4\pi\,|\boldsymbol{x}-\mathbf{y}|}. \tag{16.2}$$

By putting $f(t)=\delta(t-\tau)$ we see that

$$p_1(\boldsymbol{x}, t)=\frac{\delta(t-\tau-|\boldsymbol{x}-\mathbf{y}|/v)}{4\pi\,|\boldsymbol{x}-\mathbf{y}|} \tag{16.3}$$

is a solution of

$$\nabla^2 p_1 - \frac{1}{v^2}\frac{\partial^2 p_1}{\partial t^2} = -\delta(\mathbf{x}-\mathbf{y})\delta(t-\tau). \tag{16.4}$$

Equation (16.4) corresponds to (15.6) with $D \equiv \nabla^2 - (1/v^2)(\partial^2/\partial t^2)$ when it is remembered that there are space and time variables to be allowed for. We can now deduce a solution of

$$\nabla^2 p - \frac{1}{v^2}\frac{\partial^2 p}{\partial t^2} = -s(\mathbf{x}, t) \tag{16.5}$$

by means of (15.9). It is

$$p(\mathbf{x}, t) = \int_{-\infty}^{\infty}\int_T s(\mathbf{y}, \tau)\frac{\delta(t-\tau-|\mathbf{x}-\mathbf{y}|/v)}{4\pi|\mathbf{x}-\mathbf{y}|}\,\mathrm{d}\mathbf{y}\,\mathrm{d}\tau$$

when s vanishes outside the space volume T. Carrying out the integration with respect to τ we obtain

$$p(\mathbf{x}, t) = \int_T \frac{s(\mathbf{y}, t-|\mathbf{x}-\mathbf{y}|/v)}{4\pi|\mathbf{x}-\mathbf{y}|}\,\mathrm{d}\mathbf{y}. \tag{16.6}$$

Sometimes the briefer notation $[s]$ for $s(\mathbf{y}, t-|\mathbf{x}-\mathbf{y}|/v)$ will be used and sometimes it will be written as $s(\mathbf{y}, T_0)$ with $T_0 = t-|\mathbf{x}-\mathbf{y}|/v$.

The formula (16.6) shows that to calculate the potential at a point $\mathbf{x}$ at time t due to a source at $\mathbf{y}$ the source density at time $t-|\mathbf{x}-\mathbf{y}|/v$ must be taken. This indicates that the effect of a source at one point requires a finite time to reach another point. For this reason, the solution (16.6) is known as a *retarded potential.*

Although the integral in (16.6) has a singularity when $\mathbf{x}$ is in T it can be shown that p is continuous when s is and that p has continuous derivatives when s has.

Equation (16.6) is only one solution of (16.5). Another could be constructed by replacing $t-|\mathbf{x}-\mathbf{y}|/v$ in the argument of s by $t+|\mathbf{x}-\mathbf{y}|/v$. In this case the field would be observed before it was generated by the source; it constitutes an *advanced potential.* Such a potential is rejected on the grounds that it conflicts with experimental evidence that acoustic and electromagnetic waves are detected after the source starts up. However, there is sometimes a theoretical advantage to be gained from an advanced potential (for instance, in one formulation of inverse scattering) and so the fact that it does provide a solution of (16.5) should not be forgotten. Even if the advanced potential is discarded (16.6) is still not the general solution of (16.5) because any solution of the wave equation

$$\nabla^2 p - \frac{1}{v^2}\frac{\partial^2 p}{\partial t^2} = 0$$

could be added to it without destroying it as a solution of (16.5). Further conditions must therefore be imposed before it can be asserted that (16.6) is *the* solution of (16.5). Further consideration of the uniqueness of (16.6) will be deferred until Kirchhoff's solution has been obtained; in the meantime the retarded potential will be accepted as the proper solution.

By means of the potential found, a solution of a vector equation such as (10.10) can be derived through splitting the equation into its three Cartesian components. Accordingly, a formula for the vector potential is

$$\mathbf{A}(\boldsymbol{x}, t) = \frac{\mu}{4\pi} \int_T \frac{\boldsymbol{J}(\mathbf{y}, T_0)}{|\boldsymbol{x}-\mathbf{y}|} \, \mathrm{d}\mathbf{y} \tag{16.7}$$

on the assumption that $\boldsymbol{J}$ is zero outside T.

It is necessary to check whether (16.7) fulfils the condition (10.9) on the scalar and vector potentials when

$$V(\boldsymbol{x}, t) = \int_T \frac{\rho(\mathbf{y}, T_0)}{4\pi\,|\boldsymbol{x}-\mathbf{y}|} \, \mathrm{d}\mathbf{y}. \tag{16.8}$$

Now

$$\begin{aligned} \operatorname{div} \mathbf{A} &= \frac{\mu}{4\pi} \int_T \operatorname{div} \frac{\boldsymbol{J}(\mathbf{y}, T_0)}{|\boldsymbol{x}-\mathbf{y}|} \, \mathrm{d}\mathbf{y} \\ &= \frac{\mu}{4\pi} \int_T \left(\boldsymbol{J} \cdot \operatorname{grad} \frac{1}{|\boldsymbol{x}-\mathbf{y}|} + \frac{1}{|\boldsymbol{x}-\mathbf{y}|} \operatorname{div} \boldsymbol{J} \right) \mathrm{d}\mathbf{y}. \end{aligned} \tag{16.9}$$

Next, note that $\operatorname{grad} |\boldsymbol{x}-\mathbf{y}| = -\operatorname{grad}_y |\boldsymbol{x}-\mathbf{y}|$ where the suffix y indicates that the operations are with respect to the variable $\mathbf{y}$. Also $\boldsymbol{x}$ occurs in $\boldsymbol{J}$ only through T_0 and so

$$\begin{aligned} \operatorname{div} \boldsymbol{J}(\mathbf{y}, T_0) &= \frac{\partial \boldsymbol{J}}{\partial T_0} \cdot \operatorname{grad} T_0 = -\frac{1}{v} \frac{\partial \boldsymbol{J}}{\partial T_0} \cdot \operatorname{grad} |\boldsymbol{x}-\mathbf{y}| \\ &= \frac{1}{v} \frac{\partial \boldsymbol{J}}{\partial T_0} \cdot \operatorname{grad}_y |\boldsymbol{x}-\mathbf{y}|. \end{aligned}$$

But

$$\operatorname{div}_y \boldsymbol{J}(\mathbf{y}, T_0) = (\operatorname{div}_y \boldsymbol{J})_{T_0} - \frac{1}{v} \frac{\partial \boldsymbol{J}}{\partial T_0} \cdot \operatorname{grad}_y |\boldsymbol{x}-\mathbf{y}|$$

where $(\ )_{T_0}$ signifies that the operations with respect to $\mathbf{y}$ are to be implemented with T_0 regarded as constant. Thus (16.9) becomes

$$\operatorname{div} \mathbf{A} = -\frac{\mu}{4\pi} \int_T \operatorname{div}_y \frac{\boldsymbol{J}}{|\boldsymbol{x}-\mathbf{y}|} \, \mathrm{d}\mathbf{y} + \frac{\mu}{4\pi} \int_T \frac{(\operatorname{div}_y \boldsymbol{J})_{T_0}}{|\boldsymbol{x}-\mathbf{y}|} \, \mathrm{d}\mathbf{y}.$$

By the divergence theorem

$$\int_T \operatorname{div}_y \frac{\boldsymbol{J}}{|\boldsymbol{x}-\mathbf{y}|} \, \mathrm{d}\mathbf{y} = \int_S \frac{\boldsymbol{J} \cdot \boldsymbol{n}}{|\boldsymbol{x}-\mathbf{y}|} \, \mathrm{d}S$$

where S is the surface bounding T. Since $\boldsymbol{J}$ and ρ are zero outside T, S can be made sufficiently distant for $\boldsymbol{J}$ to vanish on it. Hence

$$\operatorname{div} \mathbf{A} + \frac{1}{v^2}\frac{\partial V}{\partial t} = \frac{\mu}{4\pi}\int_T \frac{1}{|\boldsymbol{x}-\mathbf{y}|}\left\{(\operatorname{div}_y \boldsymbol{J})_{T_0} + \frac{\partial \rho}{\partial T_0}\right\} \mathrm{d}\mathbf{y} = 0$$

since the equation of continuity takes the form

$$(\operatorname{div}_y \boldsymbol{J}(\mathbf{y}, T_0))_{T_0} + \frac{\partial}{\partial T_0}\rho(\mathbf{y}, T_0) = 0$$

at any point whose local time is T_0. Consequently, the potentials (16.7) and (16.8) do comply with (10.9) when the current and charge densities are confined to a finite volume.

Finally, it should be remarked that because the Hertz vector satisfies (12.7)

$$\boldsymbol{\Pi}(\boldsymbol{x}, t) = \frac{1}{4\pi\varepsilon}\int_T \frac{\mathbf{p}(\mathbf{y}, T_0)}{|\boldsymbol{x}-\mathbf{y}|}\,\mathrm{d}\mathbf{y}. \tag{16.10}$$

1.17 Kirchhoff's solution of the wave equation

The general solution of the wave equation (16.5) was obtained by Kirchhoff by a method which is related to one for solving Poisson's equation. Let T be a bounded domain enclosed by the surface S and consider

$$I = \int_{-\infty}^{\infty}\int_S \left\{p_1(\boldsymbol{x}, t)\frac{\partial}{\partial n_y}p(\mathbf{y}, \tau) - p(\mathbf{y}, \tau)\frac{\partial}{\partial n_y}p_1(\boldsymbol{x}, t)\right\}\mathrm{d}S_y\,\mathrm{d}\tau$$

where again the suffix y indicates what variable is involved in the operation and p_1 is defined in (16.3). The integral for I will be evaluated in two different ways. Firstly, by the divergence theorem,

$$\begin{aligned} I &= \int_{-\infty}^{\infty}\int_T \{p_1(\boldsymbol{x}, t)\nabla_y^2 p(\mathbf{y}, \tau) - p(\mathbf{y}, \tau)\nabla_y^2 p_1(\boldsymbol{x}, t)\}\,\mathrm{d}\mathbf{y}\,\mathrm{d}\tau \\ &= \int_{-\infty}^{\infty}\int_T \left[p_1\left\{\frac{1}{v^2}\frac{\partial^2 p}{\partial \tau^2} - s(\mathbf{y}, \tau)\right\}\right. \\ &\qquad \left. - p\left\{\frac{1}{v^2}\frac{\partial^2 p_1}{\partial \tau^2} - \delta(\boldsymbol{x}-\mathbf{y})\delta(t-\tau)\right\}\right]\mathrm{d}\mathbf{y}\,\mathrm{d}\tau \end{aligned}$$

from (16.5) and (16.4). Since

$$\frac{\partial^2 p_1}{\partial \tau^2} = \frac{\delta''(t-\tau-|\boldsymbol{x}-\mathbf{y}|/v)}{4\pi|\boldsymbol{x}-\mathbf{y}|}$$

a use of (15.4) gives

$$I = p(\boldsymbol{x}, t) - \int_T \frac{s(\mathbf{y}, T_0)}{4\pi\,|\boldsymbol{x}-\mathbf{y}|}\,\mathrm{d}\mathbf{y} \tag{17.1}$$

when $\boldsymbol{x}$ is in T, the first term being absent when $\boldsymbol{x}$ is outside T.

Secondly,

$$\int_{-\infty}^{\infty}\int_S p_1(\boldsymbol{x}, t)\frac{\partial}{\partial n_y}p(\mathbf{y}, \tau)\,\mathrm{d}S_y\,\mathrm{d}\tau = \int_S \frac{1}{4\pi\,|\boldsymbol{x}-\mathbf{y}|}\left(\frac{\partial}{\partial n_y}p(\mathbf{y}, T_0)\right)_{T_0}\mathrm{d}S_y.$$

Also

$$\begin{aligned}\int_{-\infty}^{\infty}\int_S & p(\mathbf{y}, \tau)\frac{\partial}{\partial n_y}p_1(\boldsymbol{x}, t)\,\mathrm{d}S_y\,\mathrm{d}\tau \\ &= \int_{-\infty}^{\infty}\int_S \frac{p(\mathbf{y}, \tau)}{4\pi}\left\{\delta\left(t-\tau-\frac{1}{v}|\boldsymbol{x}-\mathbf{y}|\right)\frac{\partial}{\partial n_y}\frac{1}{|\boldsymbol{x}-\mathbf{y}|}\right. \\ &\quad \left. -\frac{\delta'(t-\tau-|\boldsymbol{x}-\mathbf{y}|/v)}{v\,|\boldsymbol{x}-\mathbf{y}|}\frac{\partial}{\partial n_y}|\boldsymbol{x}-\mathbf{y}|\right\}\mathrm{d}S_y\,\mathrm{d}\tau \\ &= \frac{1}{4\pi}\int_S\left\{p(\mathbf{y}, T_0)\frac{\partial}{\partial n_y}\frac{1}{|\boldsymbol{x}-\mathbf{y}|} - \frac{1}{v\,|\boldsymbol{x}-\mathbf{y}|}\frac{\partial}{\partial T_0}p(\mathbf{y}, T_0)\frac{\partial}{\partial n_y}|\boldsymbol{x}-\mathbf{y}|\right\}\mathrm{d}S_y.\end{aligned}$$

Therefore

$$\begin{aligned}I = \frac{1}{4\pi}\int_S & \left\{\frac{1}{|\boldsymbol{x}-\mathbf{y}|}\left(\frac{\partial}{\partial n_y}p(\mathbf{y}, T_0)\right)_{T_0} - p(\mathbf{y}, T_0)\frac{\partial}{\partial n_y}\frac{1}{|\boldsymbol{x}-\mathbf{y}|}\right. \\ & \left. + \frac{1}{v\,|\boldsymbol{x}-\mathbf{y}|}\frac{\partial}{\partial T_0}p(\mathbf{y}, T_0)\frac{\partial}{\partial n_y}|\boldsymbol{x}-\mathbf{y}|\right\}\mathrm{d}S_y.\end{aligned} \tag{17.2}$$

The combination of (17.1) and (17.2) results in

$$\begin{aligned}p(\boldsymbol{x}, t) = \int_T \frac{s(\mathbf{y}, T_0)}{4\pi\,|\boldsymbol{x}-\mathbf{y}|}\,\mathrm{d}\mathbf{y} + \frac{1}{4\pi}\int_S & \left\{\frac{1}{|\boldsymbol{x}-\mathbf{y}|}\left(\frac{\partial}{\partial n_y}p(\mathbf{y}, T_0)\right)_{T_0}\right. \\ & \left. - p(\mathbf{y}, T_0)\frac{\partial}{\partial n_y}\frac{1}{|\boldsymbol{x}-\mathbf{y}|} + \frac{1}{v\,|\boldsymbol{x}-\mathbf{y}|}\frac{\partial}{\partial T_0}p(\mathbf{y}, T_0)\frac{\partial}{\partial n_y}|\boldsymbol{x}-\mathbf{y}|\right\}\mathrm{d}S_y\end{aligned} \tag{17.3}$$

when $\boldsymbol{x}$ is in T. If $\boldsymbol{x}$ is outside T, p on the left-hand side is replaced by zero.

The volume integral in (17.3) represents the contribution to the potential in T of the sources inside T. The surface integral provides the part due to the sources outside T. If p and its first derivatives are known all over S, p is completely determined at all interior points. However, it is not permissible to assign arbitrary surface values to these quantities. For, suppose that p was known for all time on S then, in principle, all terms in the surface integral could be evaluated except the one involving $\partial p/\partial n_y$ and a certain expression would be obtained for p. In order that this expression reproduce the given values of p on S it would be necessary for $\partial p/\partial n_y$ to satisfy a certain integral equation, which, in general, would not have an arbitrary function as solution.

1.18 Volterra's solution in two dimensions

When the field is independent of the z-coordinate the wave equation reduces to

$$\frac{\partial^2 p}{\partial x^2}+\frac{\partial^2 p}{\partial y^2}-\frac{1}{v^2}\frac{\partial^2 p}{\partial t^2}=-s(x, y, t)$$

and it might be thought that (17.3) would be correspondingly simplified but this is far from true. In fact, the analysis is considerably more complicated and all details of the derivation will be omitted. Regard x, y, and t as Cartesian coordinates in three-dimensional space and take S as a surface in this space. Then the formula, due to Volterra, is

$$\begin{aligned} p(x, y, t)=&\frac{1}{2\pi}\int_D vs(\xi, \eta, \tau)\{v^2(\tau-t)^2-r^2\}^{-\frac{1}{2}}\,d\xi\,d\eta\,d\tau \\ &+\frac{1}{2\pi}\int_{S_1}\frac{v}{\{v^2(\tau-t)^2-r^2\}^{\frac{1}{2}}}\left\{\frac{\partial p}{\partial \xi}\frac{\partial \xi}{\partial n}+\frac{\partial p}{\partial \eta}\frac{\partial \eta}{\partial n}-\frac{1}{v^2}\frac{\partial p}{\partial \tau}\frac{\partial \tau}{\partial n}\right\}dS \\ &-\frac{1}{2\pi}\frac{\partial}{\partial t}\int_{S_1}\frac{vp}{\{v^2(\tau-t)^2-r^2\}^{\frac{1}{2}}}\left\{\frac{\tau-t}{r}\frac{\partial r}{\partial n}+\frac{1}{v^2}\frac{\partial \tau}{\partial n}\right\}dS \qquad (18.1) \end{aligned}$$

where $r^2=(\xi-x)^2+(\eta-y)^2$, S_1 is the area cut out of S by the cone $r=v(\tau-t)$, D is the volume bounded by the cone and the plane $\tau=0$, and $\partial/\partial n$ denotes a derivative along the normal to S. If, in particular, $s=0$ while $p=g(x, y)$ and $\partial p/\partial t=f(x, y)$ at $t=0$ (18.1) simplifies to

$$\begin{aligned} p(x, y, t)=&\frac{1}{2\pi v}\int_0^{vt}\int_0^{2\pi} f(x+u\cos\phi, y+u\sin\phi)\frac{u\,d\phi\,du}{(v^2t^2-u^2)^{\frac{1}{2}}} \\ &+\frac{1}{2\pi v}\frac{\partial}{\partial t}\int_0^{vt}\int_0^{2\pi} g(x+u\cos\phi, y+u\sin\phi)\frac{u\,d\phi\,du}{(v^2t^2-u^2)^{\frac{1}{2}}}. \end{aligned} \qquad (18.2)$$

1.19 An integral formula for the electromagnetic field

The determination of a formula for the electromagnetic field analogous to Kirchhoff's for a scalar field requires a certain amount of analysis. If there were no surface integrals to be considered a straight forward application of (16.7) and (16.8) to (10.1) and (10.2) would be sufficient but when the surface integral is present it is better to proceed in a different way rather than through (17.3) and the scalar and vector potentials.

From (5.3)

$$\text{curl curl}\, \boldsymbol{E} = -\text{curl}\frac{\partial \boldsymbol{B}}{\partial t} = -\mu\frac{\partial \boldsymbol{J}}{\partial t} - \frac{1}{v^2}\frac{\partial^2 \boldsymbol{E}}{\partial t^2}$$

from (5.4). After a use of (10.7) we deduce that

$$\begin{aligned}\nabla^2 \boldsymbol{E} - \frac{1}{v^2}\frac{\partial^2 \boldsymbol{E}}{\partial t^2} &= \mu\frac{\partial \boldsymbol{J}}{\partial t} + \text{grad div}\, \boldsymbol{E} \\ &= (1/\varepsilon)\text{grad}\, \rho + \mu\, \partial \boldsymbol{J}/\partial t\end{aligned}$$

from (5.5). When this vector equation is expressed in Cartesian coordinates three equations of the type (16.5) are obtained and each has a solution of the form (17.3). Hence, when $\boldsymbol{x}$ is in T,

$$\begin{aligned}\boldsymbol{E}(\boldsymbol{x}, t) = &-\frac{1}{4\pi}\int_T \left(\mu\frac{\partial \boldsymbol{J}}{\partial t} + \frac{1}{\varepsilon}\text{grad}_y\, \rho\right)\frac{\mathrm{d}\boldsymbol{y}}{\Delta} \\ &+ \frac{1}{4\pi}\int_S \left\{\frac{1}{\Delta}((\boldsymbol{n}_y \cdot \text{grad}_y)\boldsymbol{E})_{T_0} - \boldsymbol{E}\frac{\partial}{\partial n_y}\frac{1}{\Delta} + \frac{1}{v\Delta}\frac{\partial \boldsymbol{E}}{\partial T_0}\frac{\partial \Delta}{\partial n_y}\right\}\mathrm{d}S_y\end{aligned} \tag{19.1}$$

where, for brevity, Δ has been written for the distance $|\boldsymbol{x}-\boldsymbol{y}|$ and the argument $(\boldsymbol{y}, T_0)$ has been omitted from the functions on the right-hand side. If $\boldsymbol{x}$ is outside T, the left-hand side of (19.1) is changed to zero.

The object of the next manipulation is to remove the space derivatives on $\boldsymbol{E}$ and to rewrite the surface integral in terms of the tangential components of the electromagnetic field as far as possible. Note firstly that

$$(\boldsymbol{n}_y \cdot \text{grad}_y)\boldsymbol{E}(\boldsymbol{y}, T_0) = ((\boldsymbol{n}_y \cdot \text{grad}_y)\boldsymbol{E})_{T_0} - \frac{1}{v}\frac{\partial \boldsymbol{E}}{\partial T_0}\boldsymbol{n}_y \cdot \text{grad}_y\, \Delta$$

and, from the identity (F.8.14),

$$\int_S \{(\boldsymbol{n}_y \cdot \text{grad}_y)\boldsymbol{a} + \boldsymbol{n}_y \wedge \text{curl}_y\, \boldsymbol{a} - \boldsymbol{n}_y\, \text{div}_y\, \boldsymbol{a}\}\, \mathrm{d}S_y = \boldsymbol{0}$$

for a closed surface S,

$$\begin{aligned}\int_S \Bigg\{&\frac{1}{\Delta}(\boldsymbol{n}_y \cdot \text{grad}_y)\boldsymbol{E} + \left(\boldsymbol{n}_y \wedge \text{grad}_y\frac{1}{\Delta}\right)\wedge \boldsymbol{E} \\ &+ \frac{1}{\Delta}\boldsymbol{n}_y \wedge \text{curl}_y\, \boldsymbol{E} - \frac{\boldsymbol{n}_y}{\Delta}\text{div}_y\, \boldsymbol{E}\Bigg\}\mathrm{d}S_y = \boldsymbol{0}\end{aligned}$$

on taking $\boldsymbol{a} = \mathbf{E}/\Delta$. Now

$$\mathrm{div}_y\, \boldsymbol{E}(\mathbf{y}, T_0) = (\mathrm{div}_y\, \boldsymbol{E})_{T_0} - \frac{1}{v}\frac{\partial \boldsymbol{E}}{\partial T_0} \cdot \mathrm{grad}_y\, \Delta,$$

$$\mathrm{curl}_y\, \boldsymbol{E}(\mathbf{y}, T_0) = (\mathrm{curl}_y\, \boldsymbol{E})_{T_0} - \frac{1}{v}\, \mathrm{grad}_y\, \Delta \wedge \frac{\partial \boldsymbol{E}}{\partial T_0}$$

and hence

$$\int_S \frac{1}{\Delta}((\boldsymbol{n}_y \cdot \mathrm{grad}_y)\boldsymbol{E})_{T_0}\, \mathrm{d}S_y$$
$$= \int_S \left\{ \frac{1}{\Delta v}\left(\boldsymbol{n}_y \wedge \frac{\partial \boldsymbol{E}}{\partial T_0}\right) \wedge \mathrm{grad}_y\, \Delta + \frac{1}{\Delta}\, \boldsymbol{n}_y \wedge \frac{\partial \boldsymbol{B}}{\partial T_0} \right.$$
$$+ \frac{1}{\Delta v}\, \boldsymbol{n}_y \wedge \left(\mathrm{grad}_y\, \Delta \wedge \frac{\partial \boldsymbol{E}}{\partial T_0}\right) - \left(\boldsymbol{n}_y \wedge \mathrm{grad}_y \frac{1}{\Delta}\right) \wedge \boldsymbol{E}$$
$$\left. + \frac{\boldsymbol{n}_y}{\Delta}(\mathrm{div}_y\, \boldsymbol{E})_{T_0} \right\} \mathrm{d}S_y$$

since $(\mathrm{curl}\, \boldsymbol{E})_{T_0} = -\partial \boldsymbol{B}/\partial T_0$ from Maxwell's equations. Thus the surface integral in (19.1) can be written as

$$\frac{1}{4\pi}\int_S \left\{ \frac{\boldsymbol{n}_y}{\Delta}(\mathrm{div}_y\, \boldsymbol{E})_{T_0} - (\boldsymbol{n}_y \wedge \boldsymbol{E}) \wedge \mathrm{grad}_y \frac{1}{\Delta} + \frac{1}{v\Delta}\left(\boldsymbol{n}_y \wedge \frac{\partial \boldsymbol{E}}{\partial T_0}\right) \wedge \mathrm{grad}_y\, \Delta \right.$$
$$\left. - \boldsymbol{n}_y \cdot \boldsymbol{E}\, \mathrm{grad}_y \frac{1}{\Delta} + \frac{1}{v\Delta}\, \boldsymbol{n}_y \cdot \frac{\partial \boldsymbol{E}}{\partial T_0}\, \mathrm{grad}_y\, \Delta + \frac{1}{\Delta}\, \boldsymbol{n}_y \wedge \frac{\partial \boldsymbol{B}}{\partial T_0} \right\} \mathrm{d}S_y.$$

Moreover,

$$\mathrm{curl} \int_S \boldsymbol{n}_y \wedge \boldsymbol{E} \frac{\mathrm{d}S_y}{\Delta} = \int_S \left\{ (\boldsymbol{n}_y \wedge \boldsymbol{E}) \wedge \mathrm{grad}_y \frac{1}{\Delta} - \frac{1}{v\Delta}\left(\boldsymbol{n}_y \wedge \frac{\partial \boldsymbol{E}}{\partial T_0}\right) \wedge \mathrm{grad}_y\, \Delta \right\} \mathrm{d}S_y,$$

$$\mathrm{grad} \int_S \boldsymbol{n}_y \cdot \boldsymbol{E} \frac{\mathrm{d}S_y}{\Delta} = \int_S \left\{ \frac{1}{v\Delta}\, \boldsymbol{n}_y \cdot \frac{\partial \boldsymbol{E}}{\partial T_0}\, \mathrm{grad}_y\, \Delta - \boldsymbol{n}_y \cdot \boldsymbol{E}\, \mathrm{grad}_y \frac{1}{\Delta} \right\} \mathrm{d}S_y$$

and

$$(\mathrm{div}_y\, \boldsymbol{E}(\mathbf{y}, T_0))_{T_0} = \rho(\mathbf{y}, T_0)/\varepsilon$$

so that

$$\int_S \frac{\boldsymbol{n}_y}{\Delta}(\mathrm{div}_y\, \boldsymbol{E})_{T_0}\, \mathrm{d}S_y = \frac{1}{\varepsilon}\int_T \mathrm{grad}_y \frac{\rho}{\Delta}\, \mathrm{d}\mathbf{y}.$$

Hence (19.1) is finally transformed to

$$\boldsymbol{E}(\boldsymbol{x}, t) = -\frac{\partial}{\partial t}\frac{\mu}{4\pi}\int_T \frac{\boldsymbol{J}(\mathbf{y}, T_0)}{|\boldsymbol{x} - \mathbf{y}|}\, \mathrm{d}\mathbf{y} - \mathrm{grad} \frac{1}{4\pi\varepsilon}\int_T \frac{\rho(\mathbf{y}, T_0)}{|\boldsymbol{x} - \mathbf{y}|}\, \mathrm{d}\mathbf{y}$$
$$- \mathrm{curl} \frac{1}{4\pi}\int_S \frac{\boldsymbol{n}_y \wedge \boldsymbol{E}(\mathbf{y}, T_0)}{|\boldsymbol{x} - \mathbf{y}|}\, \mathrm{d}S_y + \mathrm{grad} \frac{1}{4\pi}\int_S \frac{\boldsymbol{n}_y \cdot \boldsymbol{E}(\mathbf{y}, T_0)}{|\boldsymbol{x} - \mathbf{y}|}\, \mathrm{d}S_y$$
$$+ \frac{\partial}{\partial t}\frac{1}{4\pi}\int_S \frac{\boldsymbol{n}_y \wedge \boldsymbol{B}(\mathbf{y}, T_0)}{|\boldsymbol{x} - \mathbf{y}|}\, \mathrm{d}S_y. \qquad (19.2)$$

It may be shown in a similar way that

$$\begin{aligned}\boldsymbol{H}(\boldsymbol{x}, t) = {} & \operatorname{curl}\frac{1}{4\pi}\int_T \frac{\boldsymbol{J}(\mathbf{y}, T_0)}{|\boldsymbol{x}-\mathbf{y}|}\,\mathrm{d}\mathbf{y} - \operatorname{curl}\frac{1}{4\pi}\int_S \frac{\boldsymbol{n}_y \wedge \boldsymbol{H}(\mathbf{y}, T_0)}{|\boldsymbol{x}-\mathbf{y}|}\,\mathrm{d}S_y \\ & + \operatorname{grad}\frac{1}{4\pi}\int_S \frac{\boldsymbol{n}_y \cdot \boldsymbol{H}(\mathbf{y}, T_0)}{|\boldsymbol{x}-\mathbf{y}|}\,\mathrm{d}S_y - \frac{\partial}{\partial t}\frac{1}{4\pi}\int_S \frac{\boldsymbol{n}_y \wedge \boldsymbol{D}(\mathbf{y}, T_0)}{|\boldsymbol{x}-\mathbf{y}|}\,\mathrm{d}S_y.\end{aligned} \tag{19.3}$$

The formulae (19.2) and (19.3) are not completely expressed in terms of the tangential components of the electromagnetic field on S. However, the normal components are related to the tangential by virtue of

$$\frac{\partial}{\partial t}\int_S \frac{\boldsymbol{n}_y \cdot \boldsymbol{B}(\mathbf{y}, T_0)}{|\boldsymbol{x}-\mathbf{y}|}\,\mathrm{d}S_y = \operatorname{div}\int_S \frac{\boldsymbol{n}_y \wedge \boldsymbol{E}(\mathbf{y}, T_0)}{|\boldsymbol{x}-\mathbf{y}|}\,\mathrm{d}S_y, \tag{19.4}$$

$$\frac{\partial}{\partial t}\int_S \frac{\boldsymbol{n}_y \cdot \boldsymbol{D}(\mathbf{y}, T_0)}{|\boldsymbol{x}-\mathbf{y}|}\,\mathrm{d}S_y = -\operatorname{div}\int_S \frac{\boldsymbol{n}_y \wedge \boldsymbol{H}(\mathbf{y}, T_0)}{|\boldsymbol{x}-\mathbf{y}|}\,\mathrm{d}S_y - \int_S \frac{\boldsymbol{n}_y \cdot \boldsymbol{J}(\mathbf{y}, T_0)}{|\boldsymbol{x}-\mathbf{y}|}\,\mathrm{d}S_y \tag{19.5}$$

when S is a closed surface.

The above representation assumes certain continuity on the part of the fields involved. It may happen that the charges and currents are confined to the interior of the surface Σ, which itself is entirely within S, but the normal component of $\boldsymbol{J}$ does not vanish on Σ. In that case, there must be added to the right-hand side of (19.2) the term

$$-\operatorname{grad}\int_\Sigma \frac{\boldsymbol{n}_y \cdot \boldsymbol{p}(\mathbf{y}, T_0)}{4\pi\,|\boldsymbol{x}-\mathbf{y}|}\,\mathrm{d}S_y$$

where $\boldsymbol{p}$ is the polarization defined in §1.12.

For some purposes it is convenient to have available solutions of

$$\operatorname{curl}\mathbf{E} + \mu\,\partial\boldsymbol{H}/\partial t = \boldsymbol{J}', \qquad \operatorname{div}\mathbf{E} = 0, \tag{19.6}$$

$$\operatorname{curl}\boldsymbol{H} - \varepsilon\,\partial\mathbf{E}/\partial t = \mathbf{0}, \qquad \operatorname{div}\boldsymbol{H} = -\rho'/\mu \tag{19.7}$$

where now the equation of continuity is

$$\operatorname{div}\boldsymbol{J}' + \partial\rho'/\partial t = 0. \tag{19.8}$$

Put $\mathbf{E} = \boldsymbol{H}'$ and $\boldsymbol{H} = -\varepsilon\boldsymbol{E}'/\mu$. Then equations for the primed quantities are the same as Maxwell's equations. Accordingly, a representation for solutions of (19.6) and (19.7) can be deduced from (19.2) and (19.3). However, the surface integrals satisfy Maxwell's equations with zero right-hand side when $\boldsymbol{J}$ and ρ disappear. Therefore, once $\boldsymbol{J}'$ and ρ' have been accounted for, the same surface integral occurs. In other words, solutions of (19.6) and (19.7) are obtained from (19.2) and (19.3) by replacing the two volume integrals in (19.2) by

$$\text{curl} \int_T \frac{\boldsymbol{J}'(\mathbf{y}, T_0)}{4\pi |\boldsymbol{x}-\mathbf{y}|} \mathrm{d}\mathbf{y}$$

for $\mathbf{E}$ and by dropping the volume integral in (19.3) in favour of

$$\frac{\partial}{\partial t} \frac{\varepsilon}{4\pi} \int_T \frac{\boldsymbol{J}'(\mathbf{y}, T_0)}{|\boldsymbol{x}-\mathbf{y}|} \mathrm{d}\mathbf{y} + \text{grad} \frac{1}{4\pi\mu} \int_T \frac{\rho'(\mathbf{y}, T_0)}{|\boldsymbol{x}-\mathbf{y}|} \mathrm{d}\mathbf{y}$$

for $\boldsymbol{H}$. The additional term

$$\text{grad} \int_\Sigma \frac{\boldsymbol{n}_y \cdot \mathbf{p}'(\mathbf{y}, T_0)}{4\pi |\boldsymbol{x}-\mathbf{y}|} \mathrm{d}S_y$$

would be required in $\boldsymbol{H}$ if the normal component of $\boldsymbol{J}'$ did not vanish on Σ, $\mathbf{p}'$ satisfying $\partial \mathbf{p}'/\partial t = \boldsymbol{J}'$, $\text{div}\, \mathbf{p}' = -\rho'$.

Of course, if we are dealing with equations in which all of $\boldsymbol{J}$, ρ, $\boldsymbol{J}'$, and ρ' are all present, all of the volume integrals that have been mentioned will be in the representations but there will be no change to the surface integrals. In fact, one can take the field $\boldsymbol{E}_1$ due to $\boldsymbol{J}$ and the field $\boldsymbol{E}_2$ due to $\boldsymbol{J}'$ and place $\mathbf{E} = \mathbf{E}_1 + \boldsymbol{E}_2$.

1.20 Uniqueness

The question of whether our equations can have more than one solution in some circumstances must now be considered. When the equations have several solutions it may be that each corresponds to a physical situation. However, it may be that only one kind of behaviour is expected. It then becomes necessary to select one of the several solutions available. This is usually done by applying some additional restriction deduced from the expected physical performance; one must then show that this restriction ensures a *unique* solution. Sometimes one is also concerned as to whether a solution exists at all but since we shall be dealing mainly with explicit solutions that aspect will not be discussed further.

Consider now Kirchhoff's formula (17.3). It provides a general solution at points inside S. Suppose that s is always zero outside a sphere S_1 of radius R_1 and also that ρ is zero until $t = 0$. Then, choose S to be a sphere whose centre coincides with that of S_1 and whose radius at time t is greater than $R_1 + vt$. Any disturbance travels with speed v and cannot arise before $t = 0$ so that it cannot escape from the interior of the sphere of radius $R_1 + vt$. Consequently, the values of p and its derivatives must be zero on S and there is no contribution from the surface integral. Thus, only the volume integral survives in (17.3) with T any volume containing S_1. So we have proved the theorem:

If $s = 0$ for $t < 0$, and is zero for all time outside a finite sphere S_1 of radius

R_1, the solution of

$$\nabla^2 p - \frac{1}{v^2}\frac{\partial^2 p}{\partial t^2} = -s$$

at any finite time which provides an outgoing disturbance outside S_1 is

$$p(\boldsymbol{x}, t) = \int_T \frac{s(\mathbf{y}, T_0)}{4\pi |\boldsymbol{x} - \mathbf{y}|} \, \mathrm{d}\mathbf{y},$$

T being any volume which includes S_1.

Similar considerations apply to the electromagnetic field but observe that the theorem is valid only if there are no obstacles present (or until a disturbance first strikes an obstacle). A theorem is available which takes account of obstacles but this will be dealt with in a later section (§9.1).

Boundary conditions

1.21 The acoustic field

So far the discussion of the wave equation for the acoustic field has concentrated primarily on the case of a medium with a constant sound speed. At an interface separating two different media the speed of sound may change sharply with the consequence that the field may no longer be continuous.

For instance, when a moving fluid impinges on an impenetrable object no fluid can enter the object. Hence, if there is no cavitation of the fluid (as will be assumed), *the normal velocity of the object must be the same as the normal component of the particle velocity of the fluid.* In particular, for a fixed rigid obstacle $\boldsymbol{n} \cdot \boldsymbol{v} = 0$ on the surface. In view of (4.4) this may also be expressed as $\partial\bar{\rho}/\partial n = 0$ where $\bar{\rho}$ has been written in place of ρ_1. For this reason, $\partial\bar{\rho}/\partial n = 0$ is often referred to as the *hard boundary condition.*

There is no condition on the tangential components of the velocity in the absence of viscosity. Therefore, they can be expected to change discontinuously from nonzero just outside the fixed object to zero inside. If the frictional forces of viscosity are present, the tangential components drop rapidly to zero in a thin boundary layer but viscous effects will be excluded from our consideration.

At the other extreme from the hard boundary condition is a fixed surface which is a site of pressure release so that the acoustic pressure vanishes there. Since it has been seen in §1.4 that the acoustic pressure is proportional to $\bar{\rho}$ when there is no background flow this is equivalent to the requirement that $\bar{\rho} = 0$ on the surface. This is known as the *soft boundary condition.*

For a fixed interface where there can be motion on both sides we shall require firstly that the normal components are continuous, i.e. $\boldsymbol{n} \cdot \boldsymbol{v}_1 = \boldsymbol{n} \cdot \boldsymbol{v}_2$ where the suffix 1 indicates the medium on one side of the interface and the suffix 2 that on the other. By virtue of (4.4) this translates to

$$\frac{a_1^2}{\rho_{01}} \frac{\partial \bar{\rho}_1}{\partial n} = \frac{a_2^2}{\rho_{02}} \frac{\partial \bar{\rho}_2}{\partial n}. \tag{21.1}$$

Also the continuity of the pressure will be enforced so that

$$a_1^2 \bar{\rho}_1 = a_2^2 \bar{\rho}_2. \tag{21.2}$$

1.22 Discontinuities in the electromagnetic field

To investigate the discontinuities stemming from Maxwell's equations replace them by equivalent integral formulae. Let S be any closed surface and T the volume enclosed by it. Then the application of the divergence theorem to (5.5) and (5.6) gives

$$\int_S \boldsymbol{n} \cdot \boldsymbol{B} \, \mathrm{d}S = 0, \tag{22.1}$$

$$\int_S \boldsymbol{n} \cdot \boldsymbol{D} \, \mathrm{d}S = Q \tag{22.2}$$

where $Q = \int_T \rho \, \mathrm{d}\boldsymbol{x}$ is the total charge within T.

Let S_0 be an *unclosed* surface with the closed curve C as rim. Then, Stokes' theorem (§F.8) applied to (5.3) and (5.4) gives

$$\int_C \boldsymbol{E} \cdot \mathrm{d}\boldsymbol{s} = -\int_{S_0} \boldsymbol{n} \cdot \frac{\partial \boldsymbol{B}}{\partial t} \mathrm{d}S, \tag{22.3}$$

$$\int_C \boldsymbol{H} \cdot \mathrm{d}\boldsymbol{s} = I + \int_{S_0} \boldsymbol{n} \cdot \frac{\partial \boldsymbol{D}}{\partial t} \mathrm{d}S \tag{22.4}$$

where $\mathrm{d}\boldsymbol{s}$ is the vector element of arc on C and I is the total current across S_0 as defined by (5.2).

Equations (22.1)–(22.4) have been established at ordinary points where Maxwell's equations hold and are equivalent since they are valid for any S, S_0, and C. Now *assume* that (22.1)–(22.4) are still the governing equations where the material parameters are discontinuous. This assumption will imply certain boundary conditions on the field vectors.

Choose S to be a small cylinder whose generators, of length δl, are normal to the boundary between two media (Fig. 1.3). If the cross-section is made sufficiently small, it is reasonable to suppose that $\boldsymbol{B}$ has a constant value over each end. Denote by $\boldsymbol{B}_1$ the value of $\boldsymbol{B}$ in medium 1 and by $\boldsymbol{B}_2$ the value in medium 2. Then, from (22.1)

$$\boldsymbol{B}_1 \cdot \boldsymbol{n}_1 \, \mathrm{d}S + \boldsymbol{B}_2 \cdot \boldsymbol{n}_2 \, \mathrm{d}S + \text{contribution from sides of cylinder} = 0.$$

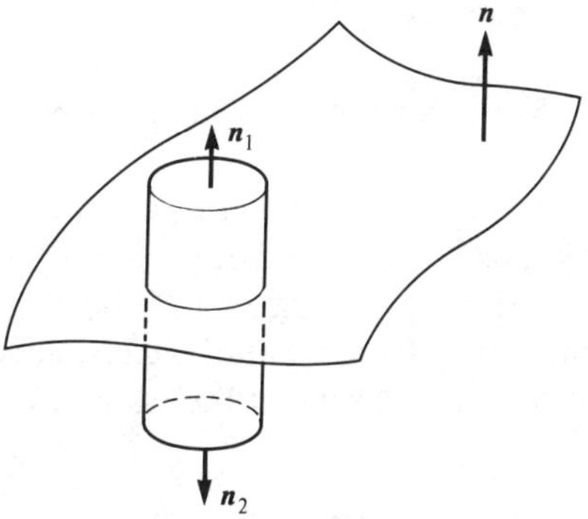

Fig. 1.3 The normal boundary condition

In the limit, as $\delta l \to 0$, the sides supply no contribution. Moreover, if $\boldsymbol{n}$ is the unit normal drawn from medium 2 to medium 1, $\boldsymbol{n}_1 = \boldsymbol{n}$ and $\boldsymbol{n}_2 = -\boldsymbol{n}$. Hence, in the limit,

$$(\boldsymbol{B}_1 - \boldsymbol{B}_2) \cdot \boldsymbol{n} = 0. \tag{22.5}$$

The normal component of $\boldsymbol{B}$ *is continuous across the interface between two media.*

The equation (22.2) may be dealt with in the same way but the possibility that Q may not vanish has to be allowed for. When δl is small Q is approximately $\rho \, \mathrm{d}S \, \delta l$. Replace $\rho \, \delta l$ by a *surface density* ρ_S, defined as the charge per unit area on the boundary. Then, in the limit,

$$(\boldsymbol{D}_1 - \boldsymbol{D}_2) \cdot \boldsymbol{n} = \rho_S. \tag{22.6}$$

The normal component of $\boldsymbol{D}$ *changes discontinuously across an interface by an amount equal to the surface density on the boundary.* Naturally, if the surface density is zero the normal component of $\boldsymbol{D}$ is continuous.

With regard to (22.3) and (22.4) take C to be a rectangle whose sides of length δl are nearly normal to the interface and whose sides of length δs are nearly parallel to it (Fig. 1.4). Let $\boldsymbol{t}$ be a unit vector parallel to the direction of integration on the upper path and $\boldsymbol{n}_0$ the positive normal to the rectangle. Then, when C is sufficiently small and S_0 occupies the

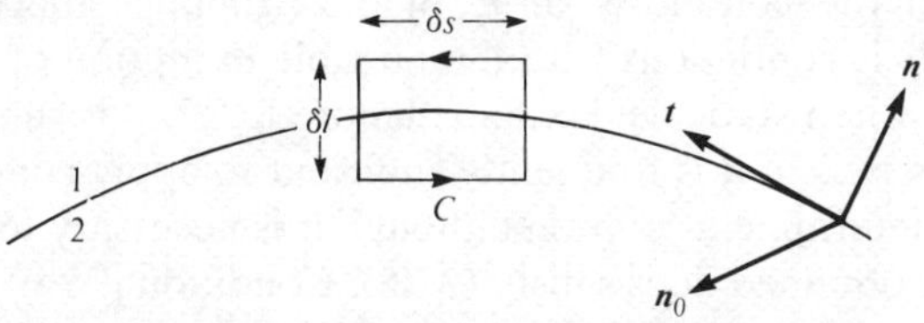

Fig. 1.4 The contour for tangential components

plane interior of the rectangle, (22.3) gives

$$(\mathbf{E}_1 \cdot \boldsymbol{t} - \mathbf{E}_2 \cdot \boldsymbol{t})\,\delta s + \text{contribution from sides } \delta l = -\frac{\partial \boldsymbol{B}}{\partial t} \cdot \boldsymbol{n}_0\,\delta s\,\delta l.$$

In the limit as $\delta l \to 0$, the sides of length δl furnish no contribution and the right-hand side tends to zero because the field is assumed to remain bounded. Hence $(\mathbf{E}_1 - \mathbf{E}_2) \cdot \boldsymbol{t} = 0$. Since $\boldsymbol{t}$ may be selected in any direction tangential to the boundary, *the tangential components of* **E** *are continuous across a change in medium.* In terms of the normal $\boldsymbol{n}$ to the boundary this can be written as

$$\boldsymbol{n} \wedge (\mathbf{E}_1 - \mathbf{E}_2) = \mathbf{0}. \tag{22.7}$$

It may be deduced in a similar manner from (22.4) that

$$\boldsymbol{n} \wedge (\boldsymbol{H}_1 - \boldsymbol{H}_2) = \lim_{\delta l \to 0} \boldsymbol{J}\,\delta l. \tag{22.8}$$

When the conductivities of both are finite and Ohm's law is satisfied ($\boldsymbol{J} = \sigma \mathbf{E}$) $\boldsymbol{J}$ is bounded because the field is. Therefore, in the case of finite conductivity,

$$\boldsymbol{n} \wedge (\boldsymbol{H}_1 - \boldsymbol{H}_2) = \mathbf{0}, \tag{22.9}$$

i.e. *the tangential components of* **H** *are continuous across an interface where the conductivity is bounded.* However, if one medium is a perfect conductor and so possesses infinite conductivity, the right-hand side of (22.8) may have a nonzero value $\boldsymbol{J}_s$, a *surface current*, defined as the current per unit length. Then

$$\boldsymbol{n} \wedge (\boldsymbol{H}_1 - \boldsymbol{H}_2) = \boldsymbol{J}_s. \tag{22.10}$$

There is a connection between (22.5) and (22.7). Take Cartesian axes with the origin at the point of the boundary where the transition is being considered and with the z-axis parallel to $\boldsymbol{n}$. Then (22.5) states that B_z is continuous and (22.7) that E_x and E_y are continuous. But, from (5.3),

$$-\frac{\partial B_z}{\partial t} = \frac{\partial E_y}{\partial x} - \frac{\partial E_x}{\partial y}.$$

The derivatives with respect to x and y are essentially tangential to the surface and so the continuity of E_x and E_y implies the continuity of $\partial B_z/\partial t$. Thus B_z is continuous with the possible exception of a static field. Consequently, when static fields can be ignored, (22.5) is subsumed under (22.7). For this reason it is frequently sufficient to apply conditions (22.7) and (22.9) to determine a solution although it is necessary to confirm that the solution so obtained does satisfy (22.5). Condition (22.6) then informs us what surface density is produced. These observations will prove to be important in §1.31.

Some condition at infinity will also have to be imposed to avoid difficulties with non-uniqueness (§1.20); our assumption will be that the wave at infinity is travelling outwards with finite speed.

Stress and energy

1.23 The energy of a sound wave

The kinetic energy in a fluid involves the square of the fluid speed. Therefore, the energy density of a sound wave would be zero unless second orders were retained. Keeping them but discarding third and higher orders we obtain $\frac{1}{2}\rho_0\boldsymbol{v}^2$ for the kinetic energy density, $\boldsymbol{v}^2$ standing for $\boldsymbol{v}\cdot\boldsymbol{v}$.

There is also potential energy due to the compression of the fluid. Let a small element of fluid of volume V_0 be compressed to $V_0(1-s)$; s is known as the *condensation*. Then the work done in compressing the volume is

$$V_0\int_0^s a_0^2\bar{\rho}\,\mathrm{d}s$$

when the work done by the background pressure is omitted. Thus the density of potential energy in a sound wave is $\int_0^s a_0^2\bar{\rho}\,\mathrm{d}s$. The condensation can be expressed in terms of the particle displacement of the Lagrangian description via

$$s=-\operatorname{div}\boldsymbol{\xi}=\bar{\rho}/\rho_0 \tag{23.1}$$

from (4.8). Consequently, the potential energy density is $\bar{\rho}^2a_0^2/2\rho_0$.

The sum of the densities of the acoustic kinetic energy and potential energy is called the *acoustic energy density* $\mathscr{E}$ i.e.

$$\mathscr{E}=\tfrac{1}{2}(\rho_0\boldsymbol{v}^2+a_0^2\bar{\rho}^2/\rho_0). \tag{23.2}$$

Related to the energy is the *intensity*, which is the rate of flux of energy per unit area across a surface. Let A be an element of area whose normal is parallel to $\boldsymbol{v}$. In a short time interval δt a particle of fluid is displaced by $\boldsymbol{v}\,\delta t$ from its original position (4.7). It is acted on by a force $a_0^2\bar{\rho}A$ if the background pressure is ignored as above and so the work done in the time interval δt is $a_0^2\bar{\rho}\boldsymbol{v}A\,\delta t$. The rate of change is therefore $a_0^2\bar{\rho}\boldsymbol{v}$ per unit area. This can be regarded as the intensity of the sound wave, i.e. the *acoustic intensity* $\boldsymbol{I}$ is defined by

$$\boldsymbol{I}=a_0^2\bar{\rho}\boldsymbol{v}. \tag{23.3}$$

It is a directed quantity parallel to the particle velocity. Its component perpendicular to $\boldsymbol{v}$ is zero corresponding to there being no energy flux across a surface to which $\boldsymbol{v}$ is tangential.

These formulae are applicable when the fluid is stationary in the absence of a sound wave. If the acoustic disturbance is superimposed on a fluid flow then different formulae must be employed (see §6.19).

Observe that, when S is a closed surface with interior T,

$$\int_S \boldsymbol{n} \cdot \boldsymbol{I} \, \mathrm{d}S = \int_T a_0^2(\boldsymbol{v} \cdot \operatorname{grad} \bar{\rho} + \bar{\rho} \operatorname{div} \boldsymbol{v}) \, \mathrm{d}\boldsymbol{x}$$

$$= -\frac{\partial}{\partial t} \int_T \tfrac{1}{2}(\rho_0 \boldsymbol{v}^2 + a_0^2 \bar{\rho}^2/\rho_0) \, \mathrm{d}\boldsymbol{x} \tag{23.4}$$

from (4.3) and (4.4). On account of (23.2) the right-hand side represents the rate at which acoustic energy is decreasing in the volume. This loss must be accompanied by a flow of energy through the bounding surface S. Therefore the surface integral on the left of (23.4) represents the rate at which acoustic energy is flowing across S. This is consistent with calling $\boldsymbol{I}$ the acoustic intensity.

It should be remarked that the interpretation is unaltered by the presence of boundaries separating different media inside T. For, if the boundary is hard or soft, the conditions of §21 ensure that the surface integral over the boundary vanishes. On the other hand, when the boundary is not impenetrable, the continuity of $\boldsymbol{n} \cdot \boldsymbol{v}$ and $a_0^2 \rho$ by (21.1) and (21.2) enables the surface integral to be continued through the boundary and (23.4) still holds with the values of ρ_0 and a_0 appropriate to the various media.

1.24 The electromagnetic stress tensor in free space

Consider a bounded portion of free space within which is a distribution of charge and current. It has already been explained in §1.5 (see also §2.10) that the force on this distribution is $\rho \boldsymbol{E} + \boldsymbol{J} \wedge \boldsymbol{B}$. This fact will now be transformed to an alternative form.

In §2.12 it is shown that, if the tensor $\boldsymbol{S}$ defined by

$$S_{ij} = -\tfrac{1}{2}(\boldsymbol{E} \cdot \boldsymbol{D} + \boldsymbol{H} \cdot \boldsymbol{B})\delta_{ij} + E_i D_j + H_i B_j$$

(using Cartesian tensors with the suffixes referring to three space coordinates) is introduced,

$$\int_S \boldsymbol{S} \cdot \boldsymbol{n} \, \mathrm{d}S = \int_T \left\{ \rho \boldsymbol{E} + \boldsymbol{J} \wedge \boldsymbol{B} + \frac{\partial}{\partial t}(\boldsymbol{D} \wedge \boldsymbol{B}) \right\} \mathrm{d}\boldsymbol{x} \tag{24.1}$$

in free space.

Suppose, firstly, that fields independent of time are considered. Then (24.1) states that the force on the charge and current inside T can be expressed as a surface integral. A similar relation connecting the force on a volume and a surface integral occurs with the stress tensor of §1.1. For this reason **S** is called the *electromagnetic stress tensor.* The tensor was introduced by Maxwell and Faraday to avoid dealing with forces acting at a distance. They postulated the existence of an *ether* which filled the whole of space and was in a state of stress determined by **S**. This state of stress was responsible for the transmission of force from one charge to another in a way similar to that in which force is transmitted through an elastic medium. Subsequent research has not supported their opinion (see Chapter 2) and nowadays the components of **S** are not regarded as having any physical reality. The most that can be said is that forces can be calculated on the basis that there exists a fictitious state of stress given by **S**.

1.25 Electromagnetic force in a general medium

Up to the moment the force on a charge in a medium other than free space has not been specified. Suppose that (24.1) is modified to be relevant to a linear isotropic medium; the appropriate form is

$$\int_S \mathbf{S}\cdot\boldsymbol{n}\,\mathrm{d}S=\int_T\left\{\rho\boldsymbol{E}+\boldsymbol{J}\wedge\boldsymbol{B}-\tfrac{1}{2}|\boldsymbol{E}|^2\,\mathrm{grad}\,\varepsilon-\tfrac{1}{2}|\boldsymbol{H}|^2\,\mathrm{grad}\,\mu+\frac{\partial}{\partial t}(\boldsymbol{D}\wedge\boldsymbol{B})\right\}\mathrm{d}\boldsymbol{x}. \tag{25.1}$$

Minkowski suggested that, by analogy with (24.1), the force at a point of T is

$$\rho\boldsymbol{E}+\boldsymbol{J}\wedge\boldsymbol{B}-\tfrac{1}{2}|\boldsymbol{E}|^2\,\mathrm{grad}\,\varepsilon-\tfrac{1}{2}|\boldsymbol{H}|^2\,\mathrm{grad}\,\mu. \tag{25.2}$$

However, this formula can be correct only if electrostriction and magnetostriction are neglected because it does not take account of the changes in ε and μ due to the deformation produced by the electromagnetic forces. Indeed, in the theory of the electrostriction of a liquid (see, for example, Stratton (1941)) the third term of (25.2) is small compared with the force associated with deformation. At best (25.2) will be applicable to a rigid medium and asserts that the force in a rigid homogeneous linear isotropic medium is $\rho\boldsymbol{E}+\boldsymbol{J}\wedge\boldsymbol{B}$.

1.26 Electromagnetic momentum

The force on a distribution of charge and current in free space in a volume T is

$$\boldsymbol{F}=\int_T(\rho\boldsymbol{E}+\boldsymbol{J}\wedge\boldsymbol{B})\,\mathrm{d}\boldsymbol{x}.$$

The force acting on a mechanical system is related to the linear momentum $\boldsymbol{p}_{\mathrm{m}}$ by $\boldsymbol{F}=\mathrm{d}\boldsymbol{p}_{\mathrm{m}}/\mathrm{d}t$. Hence (24.1) can be written as

$$\int_S \mathbf{S}\cdot\boldsymbol{n}\,\mathrm{d}S=\frac{\mathrm{d}\boldsymbol{p}_{\mathrm{m}}}{\mathrm{d}t}+\int_T\frac{\partial}{\partial t}(\boldsymbol{D}\wedge\boldsymbol{B})\,\mathrm{d}\boldsymbol{x}. \tag{26.1}$$

If the field is outgoing at infinity, T can be chosen sufficiently large for all components of $\mathbf{S}$ to vanish on S; the surface integral then disappears. T can be taken as a fixed volume (all space, say) and the time derivative removed to outside the integral sign so that (26.1) becomes

$$\frac{\mathrm{d}}{\mathrm{d}t}\left(\boldsymbol{p}_{\mathrm{m}}+\int \boldsymbol{D}\wedge\boldsymbol{B}\,\mathrm{d}\boldsymbol{x}\right)=\mathbf{0}.$$

In order that the total momentum of the system remain constant it is necessary to postulate an *electromagnetic momentum* $\int \boldsymbol{D}\wedge\boldsymbol{B}\,\mathrm{d}\boldsymbol{x}$ besides the mechanical momentum. The vector $\boldsymbol{D}\wedge\boldsymbol{B}=\boldsymbol{E}\wedge\boldsymbol{H}/c^2$ is known as the *electromagnetic momentum density*.

The definition of the electromagnetic momentum in a general medium has been the subject of controversy but there is now some support for the view that $\boldsymbol{D}\wedge\boldsymbol{B}$ is the momentum density in a rigid medium. This matter will not, however, be pursued further here.

1.27 Poynting's theorem

The disentangling of the strands which make up electrical energy is no small feat because of the many phenomena. However, if electrostriction and magnetostriction are neglected, if the *dielectric loss* of electrical energy through the generation of heat by internal frictional effects when the molecules are distorted by the field is ignored and the properties of the medium are kept constant some progress is possible. Books of electrostatic and magnetostatic phenomena show that small changes in the field cause changes in the electric and magnetic energies of

$$\delta U_{\mathrm{e}}=\int_T \boldsymbol{E}\cdot\delta\boldsymbol{D}\,\mathrm{d}\boldsymbol{x},\qquad \delta U_{\mathrm{m}}=\int_T \boldsymbol{H}\cdot\delta\boldsymbol{B}\,\mathrm{d}\boldsymbol{x}. \tag{27.1}$$

It will be *assumed* that the same formulae can be adopted for time varying fields. Our object is to find the energy flow across the boundary necessary to maintain this energy change.

The rate at which electric and magnetic energy is increasing in T is, according to the assumption just made,

$$\int_T\left(\boldsymbol{E}\cdot\frac{\partial\boldsymbol{D}}{\partial t}+\boldsymbol{H}\cdot\frac{\partial\boldsymbol{B}}{\partial t}\right)\mathrm{d}\boldsymbol{x}=\int_T(\boldsymbol{E}\cdot\operatorname{curl}\boldsymbol{H}-\boldsymbol{H}\cdot\operatorname{curl}\boldsymbol{E}-\boldsymbol{E}\cdot\boldsymbol{J})\,\mathrm{d}\boldsymbol{x}$$

from (5.3) and (5.4). Since (F.7.6)

$$\operatorname{div}(\boldsymbol{a}\wedge\boldsymbol{b})=\boldsymbol{b}\cdot\operatorname{curl}\boldsymbol{a}-\boldsymbol{a}\cdot\operatorname{curl}\boldsymbol{b}$$

the divergence theorem gives

$$-\int_T\left(\boldsymbol{E}\cdot\frac{\partial\boldsymbol{D}}{\partial t}+\boldsymbol{H}\cdot\frac{\partial\boldsymbol{B}}{\partial t}\right)\mathrm{d}\boldsymbol{x}=\int_S \boldsymbol{E}\wedge\boldsymbol{H}\cdot\boldsymbol{n}\,\mathrm{d}S+\int_T \boldsymbol{E}\cdot\boldsymbol{J}\,\mathrm{d}\boldsymbol{x} \tag{27.2}$$

which is known as *Poynting's theorem*.

The left-hand side of (27.2) is the rate at which the electric and magnetic energy is decreasing the volume. Also $\boldsymbol{E} \cdot \boldsymbol{J}$ is the rate at which heat is produced per unit volume by conduction and so the second term on the right-hand side gives the rate at which energy is being converted into heat in the volume. Since it has been assumed in (27.1) that the media are rigid no electromagnetic energy can be transformed into elastic energy and, consequently, any loss of energy which is not accounted for by heat must be occurring by a flow through S. Therefore, the surface integral in (27.2) represents the rate at which electromagnetic energy is crossing S. In other words, the rate of flow of electromagnetic energy across a closed surface S is

$$\int_S \boldsymbol{E} \wedge \boldsymbol{H} \cdot \boldsymbol{n} \, \mathrm{d}S. \tag{27.3}$$

As in §1.23 our conclusions are unaffected by the presence of boundaries inside T provided that they are perfectly conducting or places where the tangential components of $\boldsymbol{E}$ and $\boldsymbol{H}$ are continuous.

The expression (27.3) suggests that the *Poynting vector* $\boldsymbol{S}$, defined by

$$\boldsymbol{S} = \boldsymbol{E} \wedge \boldsymbol{H} \tag{27.4}$$

should be regarded as the *intensity* of energy flow at a point. This procedure is open to criticism since any vector whose divergence is zero could be added to $\boldsymbol{S}$ without affecting (27.2). However, that is of no consequence so long as only flow through a closed surface is considered.

The energy of an electromagnetic field has been a controversial subject and other forms of the energy balance have been proposed. It has been suggested that the change of energy is not given correctly by (27.1) but there are strong arguments in favour of it. Firstly, it is assuredly right in steady fields. Secondly, its use in quantum mechanics, where it appears as a Hamiltonian, undoubtedly leads to sound results. Thirdly, the electromagnetic field is invariant under a gauge transformation and so the energy should be similarly invariant. This is true of (27.1) but not of some of the alternatives which have been put forward. Nevertheless, even if (27.1) is accepted, it does not follow that (27.4) is the proper formula for the intensity. Two examples will indicate why modification may be deemed necessary. The first is the mass spectrograph in which the fields are steady but not parallel so that the Poynting vector gives a flow of energy in a steady field although, of course, the integral (27.3) shows that there is no change of energy in any volume. The second example concerns steady current in a wire. Here, according to (27.4), the energy being dissipated in the wire arrives from the surrounding medium. The integral (27.3), however, would merely reveal that the heat produced in the wire is supplied by sources outside S. In both examples the energy flow at a point conflicts with what is expected from physical intuition. Notwithstanding, the integrated energy flow is in agreement with expectation; moreover, the Poynting vector indicates that in an electromagnetic disturbance in free space the energy flow is normal to the wavefront (§6.3) which is in harmony with intuition. Therefore, any modification of the Poynting vector must, to be acceptable, leave the rate of change of energy unaltered, remove the seeming anomalies just described, give an energy flow normal to the wavefront in free space, and be invariant under a gauge transformation. It does not seem likely that a formula satisfying all these conditions will be at all simple and so we shall stay with the Poynting vector. Fortunately, the energy flow at a point is rarely of concern; in most applications it is the energy crossing a closed surface which is important.

Harmonic waves

1.28 Harmonic sound waves

When fields vary harmonically in time some simplification takes place in earlier formulae. Suppose that the period of oscillation is $2\pi/\omega$ and that in (16.5)

$$s(\boldsymbol{x}, t) = f(\boldsymbol{x})\cos(\omega t + \alpha).$$

From (16.6)

$$p(\boldsymbol{x}, t) = \int_T \frac{f(\mathbf{y})\cos\{\omega t - (\omega/v)\,|\boldsymbol{x} - \mathbf{y}| + \alpha\}}{4\pi\,|\boldsymbol{x} - \mathbf{y}|}\,\mathrm{d}\mathbf{y}.$$

These may be written alternatively in complex form via ($\mathrm{i} = \sqrt{-1}$)

$$s(\boldsymbol{x}, t) = \mathrm{Re}\{f(\boldsymbol{x})\mathrm{e}^{\mathrm{i}(\omega t + \alpha)}\},$$

$$p(\boldsymbol{x}, t) = \mathrm{Re}\left(\frac{\mathrm{e}^{\mathrm{i}(\omega t+\alpha)}}{4\pi}\int_T \frac{f(\mathbf{y})}{|\boldsymbol{x} - \mathbf{y}|}\,\mathrm{e}^{-\mathrm{i}\omega|\boldsymbol{x}-\mathbf{y}|/v}\,\mathrm{d}\mathbf{y}\right)$$

where Re means the real part. It is advantageous, as in the treatment of forced oscillations in the theory of mechanics, to omit the symbol Re during analysis and to supply it at the end. Thus we write $p(\boldsymbol{x}, t)$ as $p_h(\boldsymbol{x})\mathrm{e}^{\mathrm{i}\omega t}$ with the understanding that the real part is to be taken. Here p_h may be a complex function of $\boldsymbol{x}$ because of the phase α and the complex exponential in the integrand. Such a representation is permissible so long as only *linear operations* are involved, i.e. operations in which it is immaterial whether taking the real part is done before or after the operation. Typical examples of such linear operations are addition, subtraction, integration, and taking a derivative. Clearly, the operations entailed in forming (4.3) and (4.4) are linear.

In a harmonic field put $\bar{\rho} = \bar{\rho}_{\mathrm{h}}\mathrm{e}^{\mathrm{i}\omega t}$, $\boldsymbol{v} = \boldsymbol{v}_{\mathrm{h}}\mathrm{e}^{\mathrm{i}\omega t}$. Then the governing equations become

$$i\omega\bar{\rho}_h + \rho_0 \operatorname{div} \boldsymbol{v}_h = 0, \tag{28.1}$$

$$i\omega\rho_0\boldsymbol{v}_h + a_0^2 \operatorname{grad} \bar{\rho}_h = \mathbf{0} \tag{28.2}$$

after deletion of the common factor $e^{i\omega t}$. Accordingly, $\bar{\rho}_h$ satisfies the equation

$$\nabla^2\bar{\rho}_h + k^2\bar{\rho}_h = 0 \tag{28.3}$$

where $k^2 = \omega^2/a_0^2$. Equation (28.3) is called *Helmholtz's equation.*

A solution of

$$\nabla^2\bar{\rho}_h + k^2\bar{\rho}_h = -s_h \tag{28.4}$$

which is outgoing at infinity is, according to the foregoing analysis,

$$\bar{\rho}_h(\boldsymbol{x}) = \int_T s_h(\mathbf{y})\psi(\boldsymbol{x}, \mathbf{y})\, d\mathbf{y} \tag{28.5}$$

where

$$\psi(\boldsymbol{x}, \mathbf{y}) = \frac{e^{-ik|\boldsymbol{x}-\mathbf{y}|}}{4\pi\,|\boldsymbol{x}-\mathbf{y}|} \tag{28.6}$$

A general solution may be obtained as in (29.6).

The calculation of energy flow is more complicated because it involves the multiplication of two field components—a nonlinear process. Thus the acoustic intensity is given as (23.3)

$$\boldsymbol{I} = a_0^2 \operatorname{Re}(\bar{\rho}_h e^{i\omega t}) \operatorname{Re}(\boldsymbol{v}_h e^{i\omega t}).$$

For most applications, however, it is not the energy flow at an instant which is of interest in a harmonic field but the flow which takes place on average. A convenient average comes from evaluating the energy change during a cycle of the field and dividing by the length of the cycle. Such an average demands the value of integrals like

$$\frac{\omega}{2\pi}\int_0^{2\pi/\omega} \operatorname{Re}(f e^{i\omega t}) \operatorname{Re}(g e^{i\omega t})\, dt$$

where f and g are complex functions of position but independent of t. Suppose $f = Ae^{i\alpha}$, $g = Be^{i\beta}$ where A, B, α, and β are all real. Then the integral is

$$\frac{\omega AB}{2\pi}\int_0^{2\pi/\omega} \cos(\omega t + \alpha)\cos(\omega t + \beta)\, dt = \tfrac{1}{2}AB\cos(\alpha - \beta) = \tfrac{1}{2}\operatorname{Re}(fg^*)$$

where $g^* = Be^{-i\beta}$ is the complex conjugate of g.

Applying this to each component of the acoustic intensity we find that its average is $\frac{1}{2}a_0^2 \operatorname{Re} \bar{\rho}_h \boldsymbol{v}_h^*$. The vector $\frac{1}{2}a_0^2\rho_h\boldsymbol{v}_h^*$ is called the *complex acoustic intensity.*

Let us examine the integral of the complex acoustic intensity over a closed surface. We have

$$\int_S a_0^2\bar{\rho}_h \boldsymbol{v}_h^* \cdot \boldsymbol{n}\, \mathrm{d}S = \int_T a_0^2(\boldsymbol{v}_h^* \cdot \operatorname{grad} \bar{\rho}_h + \bar{\rho}_h \operatorname{div} \boldsymbol{v}_h^*)\, \mathrm{d}\boldsymbol{x}$$

$$= -\mathrm{i}\omega \int_T (\rho_0 \boldsymbol{v}_h \cdot \boldsymbol{v}_h^* - a_0^2 \bar{\rho}_h \bar{\rho}_h^* / \rho_0)\, \mathrm{d}\boldsymbol{x} \qquad (28.7)$$

from (28.2) and the complex conjugate of (28.1). The integral on the right is purely imaginary and, by the remarks of §23 is $4\mathrm{i}\omega$ times the difference between the average potential energy and the average kinetic energy. In fact,

$$\operatorname{Re}\left(\int_S \tfrac{1}{2} a_0^2 \bar{\rho}_h \boldsymbol{v}_h^* \cdot \boldsymbol{n}\, \mathrm{d}S\right) = 0,$$

$$\operatorname{Im}\left(\int_S \tfrac{1}{2} a_0^2 \bar{\rho}_h \boldsymbol{v}_h^* \cdot \boldsymbol{n}\, \mathrm{d}S\right) = 2\omega\,(\text{average potential energy} - \text{average kinetic energy})$$

where Im is the imaginary part.

1.29 Helmholtz's theorem

The same technique may be employed for harmonic electromagnetic waves. Thus the equation of continuity becomes

$$\operatorname{div} \boldsymbol{J}_h + \mathrm{i}\omega\rho_h = 0 \qquad (29.1)$$

whereas the potentials satisfy

$$\nabla^2 \mathbf{A}_h + k^2 \mathbf{A}_h = -\mu \boldsymbol{J}_h, \qquad (29.2)$$

$$\nabla^2 V_h + k^2 V_h = -\rho_h/\varepsilon \qquad (29.3)$$

where now $k^2 = \omega^2\mu\varepsilon$. The field is given by

$$\boldsymbol{E}_h = -\operatorname{grad} V_h - \mathrm{i}\omega \mathbf{A}_h,$$

$$\boldsymbol{B}_h = \operatorname{curl} \mathbf{A}_h.$$

Particular solutions of (29.2) and (29.3) are

$$V_h = (1/\varepsilon) \int_T \rho_h(\mathbf{y}) \psi(\boldsymbol{x}, \mathbf{y})\, \mathrm{d}y, \qquad (29.4)$$

$$\mathbf{A}_h = \mu \int_T \boldsymbol{J}_h(\mathbf{y}) \psi(\boldsymbol{x}, \mathbf{y})\, \mathrm{d}\mathbf{y} \qquad (29.5)$$

from (16.7) and (16.8).

Alternatively, the Hertz vector may be used by taking $\mathbf{p}_h = \boldsymbol{J}_h/\mathrm{i}\omega$ which complies with (12.1) and (29.1). Then

$$\boldsymbol{E}_{\mathrm{h}} = \operatorname{grad}\operatorname{div}\boldsymbol{\Pi}_{\mathrm{h}} + k^2\boldsymbol{\Pi}_{\mathrm{h}}, \qquad \boldsymbol{B}_{\mathrm{h}} = \mathrm{i}k^2\operatorname{curl}\boldsymbol{\Pi}_{\mathrm{h}}/\omega$$

where

$$\nabla^2\boldsymbol{\Pi}_{\mathrm{h}} + k^2\boldsymbol{\Pi}_{\mathrm{h}} = -\boldsymbol{J}_{\mathrm{h}}/\mathrm{i}\omega\varepsilon$$

a particular solution (16.10) being

$$\boldsymbol{\Pi}_{\mathrm{h}} = (1/\mathrm{i}\omega\varepsilon)\int_T \boldsymbol{J}_{\mathrm{h}}(\mathbf{y})\psi(\boldsymbol{x}, \mathbf{y})\,\mathrm{d}\mathbf{y}.$$

A general solution of (29.3) is supplied by (17.3) with the harmonic time dependence incorporated. It is

$$V_{\mathrm{h}}(\boldsymbol{x}) = \frac{1}{\varepsilon}\int_T \rho_{\mathrm{h}}(\mathbf{y})\psi(\boldsymbol{x}, \mathbf{y})\,\mathrm{d}\mathbf{y} + \int_S \left\{\frac{\partial V_{\mathrm{h}}}{\partial n_y}\psi(\boldsymbol{x}, \mathbf{y}) - V_{\mathrm{h}}\frac{\partial}{\partial n_y}\psi(\boldsymbol{x}, \mathbf{y})\right\}\mathrm{d}S_y \tag{29.6}$$

when $\boldsymbol{x}$ is inside S. The left-hand side is replaced by zero when $\boldsymbol{x}$ is outside S. Often formula (29.6) is known as *Helmholtz's representation.*

The corresponding formulae for the electromagnetic field which emanate from (19.2) and (19.3) are

$$\begin{aligned}\boldsymbol{E}_{\mathrm{h}}(\boldsymbol{x}) = {}& -\mathrm{i}\omega\mu\int_T \boldsymbol{J}_{\mathrm{h}}\psi(\boldsymbol{x}, \mathbf{y})\,\mathrm{d}\mathbf{y} - \operatorname{grad}\int_T \rho_{\mathrm{h}}\psi(\boldsymbol{x}, \mathbf{y})\,\mathrm{d}\mathbf{y}/\varepsilon \\ & - \operatorname{curl}\int_S \boldsymbol{n}_y \wedge \boldsymbol{E}_{\mathrm{h}}\psi(\boldsymbol{x}, \mathbf{y})\,\mathrm{d}S_y + \operatorname{grad}\int_S \boldsymbol{n}_y \cdot \boldsymbol{E}\psi(\boldsymbol{x}, \mathbf{y})\,\mathrm{d}S_y \\ & + \mathrm{i}\omega\int_S \boldsymbol{n}_y \wedge \boldsymbol{B}_{\mathrm{h}}\psi(\boldsymbol{x}, \mathbf{y})\,\mathrm{d}S_y,\end{aligned} \tag{29.7}$$

$$\begin{aligned}\boldsymbol{H}_{\mathrm{h}}(\boldsymbol{x}) = {}& \operatorname{curl}\int_T \boldsymbol{J}_{\mathrm{h}}\psi(\boldsymbol{x}, \mathbf{y})\,\mathrm{d}\mathbf{y} - \operatorname{curl}\int_S \boldsymbol{n}_y \wedge \boldsymbol{H}_{\mathrm{h}}\psi(\boldsymbol{x}, \mathbf{y})\,\mathrm{d}S_y \\ & + \operatorname{grad}\int_S \boldsymbol{n}_y \cdot \boldsymbol{H}_{\mathrm{h}}\psi(\boldsymbol{x}, \mathbf{y})\,\mathrm{d}S_y - \mathrm{i}\omega\int_S \boldsymbol{n}_y \wedge \boldsymbol{D}_{\mathrm{h}}\psi(\boldsymbol{x}, \mathbf{y})\,\mathrm{d}S_y.\end{aligned} \tag{29.8}$$

They may also be derived directly from the equations satisfied by $\boldsymbol{E}_{\mathrm{h}}$ and $\boldsymbol{H}_{\mathrm{h}}$ by means of a vector analogue of Green's theorem. In the event that $\boldsymbol{J}_{\mathrm{h}}$ vanishes outside Σ, which is enclosed by S, but $\boldsymbol{n} \cdot \boldsymbol{J}_{\mathrm{h}}$ is not zero on Σ, there must be added to (29.7)

$$-\operatorname{grad}\int_\Sigma \boldsymbol{n}_y \cdot \boldsymbol{J}_{\mathrm{h}}\psi(\boldsymbol{x}, \mathbf{y})\,\mathrm{d}\mathbf{y}/\mathrm{i}\omega.$$

The normal components of the field in (29.7) and (29.8) can be eliminated by means of (19.4) and (19.5). The result is, when (29.1) is remembered,

$$\boldsymbol{E}_{\rm h}(\boldsymbol{x}) = (1/i\omega\varepsilon)(\text{grad div} + k^2)\left(\int_T \boldsymbol{J}_{\rm h}\psi \, d\boldsymbol{y} - \int_S \boldsymbol{n}_y \wedge \boldsymbol{H}_{\rm h}\psi \, dS_y\right)$$

$$- \text{curl} \int_S \boldsymbol{n}_y \wedge \boldsymbol{E}_{\rm h}\psi \, dS_y, \qquad (29.9)$$

$$\boldsymbol{H}_{\rm h}(\boldsymbol{x}) = \text{curl}\left(\int_T \boldsymbol{J}_{\rm h}\psi \, d\boldsymbol{y} - \int_S \boldsymbol{n}_y \wedge \boldsymbol{H}_{\rm h}\psi \, dS_y\right)$$

$$+ (1/i\omega\mu)(\text{grad div} + k^2) \int_S \boldsymbol{n}_y \wedge \boldsymbol{E}_{\rm h}\psi \, dS_y. \qquad (29.10)$$

Although the pairs (29.7), (29.8), and (29.9), (29.10) are equivalent when S is closed they are not when S is unclosed (T then being taken as the volume occupied by the sources). If S is not closed (29.9) and (29.10) still satisfy Maxwell's equations but (29.7) and (29.8) do not. When $\boldsymbol{J}$ and ρ are zero (29.7) and (29.8) can be made to satisfy Maxwell's equations (assuming that $\boldsymbol{E}$ and $\boldsymbol{H}$ do so on S) by adding $-\text{grad} \int_C \psi \boldsymbol{H}_{\rm h} \cdot d\boldsymbol{s}/i\omega\varepsilon$ to (29.7) and $\text{grad} \int_C \psi \boldsymbol{H}_{\rm h} \cdot d\boldsymbol{s}/i\omega\mu$ to (29.8), C being the rim of the unclosed S. These additional fields represent the influence of charges distributed along the rim C of S.

For closed S the formulae for $V_{\rm h}$, $\boldsymbol{E}_{\rm h}$, and $\boldsymbol{H}_{\rm h}$ all reduce to surface integrals only when $\boldsymbol{J}_{\rm h} = \boldsymbol{0}$ and $\rho_{\rm h} = 0$. This representation has been demonstrated under the assumption that the field and its derivatives are continuous. But, when $\boldsymbol{x}$ is not a point of S, ψ is a regular function since $\boldsymbol{x}$ and $\boldsymbol{y}$ never coincide. Therefore, *$V_{\rm h}$, $\boldsymbol{E}_{\rm h}$, and $\boldsymbol{H}_{\rm h}$ are regular functions of $\boldsymbol{x}$ inside S in the absence of interior sources.* This is a general property of elliptic linear partial differential equations.

1.30 The complex Poynting vector

As in §1.28 the average over a cycle of the Poynting vector is $\frac{1}{2}\text{Re}(\boldsymbol{E}_{\rm h} \wedge \boldsymbol{H}_{\rm h}^*)$ and $\frac{1}{2}\boldsymbol{E}_{\rm h} \wedge \boldsymbol{H}_{\rm h}^*$ is known as the *complex Poynting vector*. Since two of Maxwell's equations are

$$\text{curl}\, \boldsymbol{E}_{\rm h} + i\omega \boldsymbol{B}_{\rm h} = \boldsymbol{0}, \qquad (30.1)$$

$$\text{curl}\, \boldsymbol{H}_{\rm h} - i\omega \boldsymbol{D}_{\rm h} = \boldsymbol{J}_{\rm h} \qquad (30.2)$$

the equation of energy balance is

$$\int_S \boldsymbol{E}_{\rm h} \wedge \boldsymbol{H}_{\rm h}^* \cdot \boldsymbol{n} \, dS = -\int_T \{i\omega(\boldsymbol{B}_{\rm h} \cdot \boldsymbol{H}_{\rm h}^* - \boldsymbol{E}_{\rm h} \cdot \boldsymbol{D}_{\rm h}^*) + \boldsymbol{E}_{\rm h} \cdot \boldsymbol{J}_{\rm h}^*\} \, d\boldsymbol{x}.$$

In particular, for a linear isotropic medium in which $\boldsymbol{J} = \sigma \boldsymbol{E}$,

$$\int_S \boldsymbol{E}_{\rm h} \wedge \boldsymbol{H}_{\rm h}^* \cdot \boldsymbol{n} \, dS = \int_T \{i\omega(\varepsilon \boldsymbol{E}_{\rm h} \cdot \boldsymbol{E}_{\rm h}^* - \mu \boldsymbol{H}_{\rm h} \cdot \boldsymbol{H}_{\rm h}^*) - \sigma \boldsymbol{E}_{\rm h} \cdot \boldsymbol{E}_{\rm h}^*\} \, d\boldsymbol{x}.$$

If the electric energy in T is taken as $\int_T \frac{1}{2}\varepsilon \boldsymbol{E}^2\,\mathrm{d}\boldsymbol{x}$ and the magnetic energy as $\int_T \frac{1}{2}\mu \boldsymbol{H}^2\,\mathrm{d}\boldsymbol{x}$ the integrand with the factor $\mathrm{i}\omega$ gives $4\mathrm{i}\omega$ times the difference in the average electric and magnetic energies. The remaining integrand on the right is twice the average heat produced. Hence

$$-\mathrm{Re}\left(\int_S \tfrac{1}{2}\boldsymbol{E}_\mathrm{h}\wedge\boldsymbol{H}_\mathrm{h}^*\cdot\boldsymbol{n}\,\mathrm{d}S\right)=\text{average heat produced}$$

$$\mathrm{Im}\left(\int_S \tfrac{1}{2}\boldsymbol{E}_\mathrm{h}\wedge\boldsymbol{H}_\mathrm{h}^*\cdot\boldsymbol{n}\,\mathrm{d}S\right)=2\omega\text{ average (electric energy}-\text{magnetic energy).}$$

1.31 Boundary conditions and uniqueness

The boundary conditions for harmonic waves may be obtained from the analysis of §§1.21, 1.22 by substituting $\mathrm{i}\omega$ for the time derivative $\partial/\partial t$. Since this derivative does not occur in §1.21, the boundary conditions for acoustic waves are unaffected by the switch to harmonic time variation.

Likewise, in the electromagnetic field, the normal component of $\boldsymbol{B}_\mathrm{h}$ and the tangential components of $\boldsymbol{E}_\mathrm{h}$ are continuous across a change in medium. If the conductivity is bounded the tangential components of $\boldsymbol{H}_\mathrm{h}$ are also continuous. It was pointed out at the end of §1.22 that the continuity of the tangential components of $\boldsymbol{E}$ implies the continuity of the time derivative of the normal component of $\boldsymbol{B}$; this means, for harmonic waves, that the continuity of the normal component $\mathrm{i}\omega\boldsymbol{B}_\mathrm{h}$ and hence of $\boldsymbol{B}_\mathrm{h}$ is implied. Therefore, in harmonic waves, it is sufficient to impose at the boundary the continuity of the tangential components of $\boldsymbol{E}_\mathrm{h}$ and $\boldsymbol{H}_\mathrm{h}$.

The behaviour at infinity has also to be discussed because, as seen in §1.20, this is essential to guaranteeing uniqueness. The condition at infinity requires the field to behave like an outgoing wave. One infers that, in harmonic fields, the field at infinity must behave like $\psi \mathrm{e}^{\mathrm{i}\omega t}$ rather than like $\psi^* \mathrm{e}^{\mathrm{i}\omega t}$. This somewhat vague prescription can be translated to more precise mathematical terms. Let p_h satisfy

$$\nabla^2 p_\mathrm{h}+k^2 p_\mathrm{h}=0$$

outside some finite sphere. Then it is required that, for some finite constant K,

$$|Rp_\mathrm{h}|<K, \tag{31.1}$$

$$R\left(\frac{\partial p_\mathrm{h}}{\partial R}+\mathrm{i}kp_\mathrm{h}\right)\to 0 \tag{31.2}$$

uniformly with respect to direction as $R\to\infty$, R being the distance from the origin. These are known as *Sommerfeld's radiation conditions.* Less restrictive forms are available but for many purposes (31.1) and (31.2) will be adequate. Clearly, ψ satisfies the radiation conditions but ψ^* does not.

Assume that V_h in (29.6) satisfies the radiation conditions and that ρ_h is nonzero only in a finite domain. Choose S to be a large sphere of radius R_0. Then, as $R_0 \to \infty$, $\partial/\partial n_y \sim \partial/\partial R_0$ and $|\mathbf{x}-\mathbf{y}| \sim R_0 - \hat{\mathbf{y}} \cdot \mathbf{x}$ where $\hat{\mathbf{y}}$ is a unit vector in the direction of $\mathbf{y}$. Hence

$$\frac{\partial V_h}{\partial n_y}\psi - V_h\frac{\partial \psi}{\partial n_y} \sim \left(\frac{\partial V_h}{\partial R_0} + ikV_h\right)\frac{e^{-ikR_0}}{R_0} + V_h\frac{e^{-ikR_0}}{R_0^2}.$$

Since the area of S is of the order of R_0^2, (31.1) shows that, as $R_0 \to \infty$,

$$\int_S V_h(e^{-ikR_0}/R_0^2)\,dS \to 0$$

while

$$\int_S \left(\frac{\partial V_h}{\partial R_0} + ikV_h\right)\frac{e^{-ikR_0}}{R_0}\,dS \to 0$$

follows from (31.2). Hence, *when* V_h *satisfies the radiation conditions it is given by* (29.4) *everywhere, T being the whole domain occupied by* ρ_h.

The radiation conditions for the electromagnetic field can be expressed in terms of the scalar and vector potentials but it is usually more convenient to involve the field components directly. Suitable conditions are

$$|R\boldsymbol{E}_h| < K, \qquad |R\boldsymbol{H}_h| < K, \tag{31.3}$$

$$R(\boldsymbol{E}_h + Z_0\hat{\boldsymbol{x}} \wedge \boldsymbol{H}_h) \to 0, \tag{31.4}$$

$$R(\boldsymbol{H}_h - \hat{\boldsymbol{x}} \wedge \boldsymbol{E}_h/Z_0) \to 0 \tag{31.5}$$

uniformly with respect to direction as $R = |\mathbf{x}| \to \infty$. Here $Z_0 = (\mu/\varepsilon)^{\frac{1}{2}}$ is often known as the *impedance of the medium.* Once again weaker versions of the conditions can be constructed. The surface integrals in (29.7) and (29.8) are removed when S goes off at infinity if the field complies with the radiation conditions just stated.

It should be emphasized that in this examination of uniqueness the presence of obstacles is forbidden (see §9.1).

1.32 Green's functions

The general representation of a solution of Helmholtz' equation in (29.6) asks for a knowledge of the function and its normal derivative on the surface S. A somewhat different form will now be described which needs only the function or its normal derivative. Let

$$\nabla^2 p_h + k^2 p_h = 0 \tag{32.1}$$

in T and suppose that φ_h is any solution of

$$\nabla^2 \varphi_h + k^2 \varphi_h = 0 \tag{32.2}$$

which has no singularities in T. Because of (32.1) and (32.2)

$$\int_S \left(\frac{\partial p_h}{\partial n_y}\varphi_h - p_h\frac{\partial \varphi_h}{\partial n_y}\right) dS_y = 0.$$

But, by Helmholtz's representation,

$$p_h(\boldsymbol{x}) = \int_S \left\{\frac{\partial p_h}{\partial n_y}\psi(\boldsymbol{x}, \mathbf{y}) - p_h\frac{\partial}{\partial n_y}\psi(\boldsymbol{x}, \mathbf{y})\right\} dS_y.$$

Subtraction gives

$$p_h(\boldsymbol{x}) = \int_S \left[\frac{\partial p_h}{\partial n_y}\{\psi(\boldsymbol{x}, \mathbf{y}) - \varphi_h\} - p_h\frac{\partial}{\partial n_y}\{\psi(\boldsymbol{x}, \mathbf{y}) - \varphi_h\}\right] dS_y. \quad (32.3)$$

Since the only restriction on φ_h is that it be a singularity-free solution of (32.2) this is a very general representation for a solution of Helmholtz's equation. A nonzero right-hand side of (32.1) can be coped with by adding a volume integral such as (28.5) to (32.3).

Now choose φ_h so that $\varphi_h = \psi(\boldsymbol{x}, \mathbf{y})$ at all points $\mathbf{y}$ of S. Then

$$p_h(\boldsymbol{x}) = -\int_S p_h\frac{\partial}{\partial n_y}G_1(\boldsymbol{x}, \mathbf{y})\, dS_y \quad (32.4)$$

where

$$G_1(\boldsymbol{x}, \mathbf{y}) = \psi(\boldsymbol{x}, \mathbf{y}) - \varphi_h.$$

This particular representation involves only the values of p_h on S. Contrariwise, φ_h might be selected so that $\partial\varphi_h/\partial n_y = \partial\psi(\boldsymbol{x}, \mathbf{y})/\partial n_y$ at every point of S. In that case

$$p_h(\boldsymbol{x}) = \int_S \frac{\partial p_h}{\partial n_y}G_2(\boldsymbol{x}, \mathbf{y})\, dS_y \quad (32.5)$$

where

$$G_2(\boldsymbol{x}, \mathbf{y}) = \psi(\boldsymbol{x}, \mathbf{y}) - \varphi_h;$$

this representation requires only the normal derivative of p_h.

G_1 and G_2 are known as *Green's functions*. Both functions satisfy

$$\nabla^2 G + k^2 G = -\delta(\boldsymbol{x} - \mathbf{y}) \quad (32.6)$$

but G_1 vanishes on S whereas $\partial G_2/\partial n_y = 0$ on S. Thus the price to be paid for the removal of either p_h or its normal derivative from the representation is the determination of a specific kind of solution of Helmholtz's equation. Finding a Green's function in a particular problem may not be any easier than solving the problem directly but neither its determination nor the question of whether it exists will be considered here. The matter of representation by Green's functions outside S will be taken up in §1.34.

One important property of Green's functions is that $G_1(\boldsymbol{x}, \boldsymbol{y}) = G_1(\boldsymbol{y}, \boldsymbol{x})$ and $G_2(\boldsymbol{x}, \boldsymbol{y}) = G_2(\boldsymbol{y}, \boldsymbol{x})$ (see §1.35).

1.33 Green's tensor

For the electromagnetic field it may be desirable to eliminate certain components of the field rather than the potentials or their normal derivatives. This can be achieved by introducing a *Green's tensor.*

In the notation of Cartesian tensors (§F.6) the tensor $\boldsymbol{\Gamma}$ can be written $\boldsymbol{\Gamma}_1\boldsymbol{i}+\boldsymbol{\Gamma}_2\boldsymbol{j}+\boldsymbol{\Gamma}_3\boldsymbol{k}$ where $\boldsymbol{\Gamma}_1$, $\boldsymbol{\Gamma}_2$, and $\boldsymbol{\Gamma}_3$ are vectors. The convention to be adopted in working with $\boldsymbol{\Gamma}$ is that

$$\boldsymbol{u}\wedge\boldsymbol{\Gamma} = (\boldsymbol{u}\wedge\boldsymbol{\Gamma}_1)\boldsymbol{i}+(\boldsymbol{u}\wedge\boldsymbol{\Gamma}_2)\boldsymbol{j}+(\boldsymbol{u}\wedge\boldsymbol{\Gamma}_3)\boldsymbol{k}, \tag{33.1}$$

$$\boldsymbol{\Gamma}\cdot\boldsymbol{u} = \boldsymbol{\Gamma}_1(\boldsymbol{i}\cdot\boldsymbol{u})+\boldsymbol{\Gamma}_2(\boldsymbol{j}\cdot\boldsymbol{u})+\boldsymbol{\Gamma}_3(\boldsymbol{k}\cdot\boldsymbol{u}), \tag{33.2}$$

$$\operatorname{curl}\boldsymbol{\Gamma} = (\operatorname{curl}\boldsymbol{\Gamma}_1)\boldsymbol{i}+(\operatorname{curl}\boldsymbol{\Gamma}_2)\boldsymbol{j}+(\operatorname{curl}\boldsymbol{\Gamma}_3)\boldsymbol{k}, \tag{33.3}$$

For a linear homogeneous isotropic medium we see from (30.1) and (30.2) that

$$\operatorname{curl}\operatorname{curl}\boldsymbol{E}_\mathrm{h} = -\mathrm{i}\omega\mu\operatorname{curl}\boldsymbol{H}_\mathrm{h} = k^2\boldsymbol{E}_\mathrm{h}, \tag{33.4}$$

$$\operatorname{curl}\operatorname{curl}\boldsymbol{H}_\mathrm{h} = \mathrm{i}\omega\varepsilon\operatorname{curl}\boldsymbol{E}_\mathrm{h} = k^2\boldsymbol{H}_\mathrm{h} \tag{33.5}$$

when there is no current density. Introduce the tensor $\boldsymbol{\Gamma}(\boldsymbol{x}, \boldsymbol{y})$ which satisfies, throughout T,

$$\operatorname{curl}\operatorname{curl}\boldsymbol{\Gamma}(\boldsymbol{x}, \boldsymbol{y}) - k^2\boldsymbol{\Gamma}(\boldsymbol{x}, \boldsymbol{y}) = -\boldsymbol{I}\delta(\boldsymbol{x}-\boldsymbol{y}) \tag{33.6}$$

where $\boldsymbol{I}$ is the unit tensor. Now

$$\int_S \{(\boldsymbol{n}\wedge\operatorname{curl}\boldsymbol{E}_\mathrm{h})\cdot\boldsymbol{\Gamma}(\boldsymbol{x}, \boldsymbol{y}) + (\boldsymbol{n}\wedge\boldsymbol{E}_\mathrm{h})\cdot\operatorname{curl}\boldsymbol{\Gamma}(\boldsymbol{x}, \boldsymbol{y})\}\,\mathrm{d}S$$

$$= \int_S \boldsymbol{n}\cdot\{(\operatorname{curl}\boldsymbol{E}_\mathrm{h})\wedge\boldsymbol{\Gamma} + \boldsymbol{E}_\mathrm{h}\wedge\operatorname{curl}\boldsymbol{\Gamma}\}\,\mathrm{d}S$$

$$= \int_T \{(\operatorname{curl}\operatorname{curl}\boldsymbol{E}_\mathrm{h})\cdot\boldsymbol{\Gamma} - \boldsymbol{E}_\mathrm{h}\cdot\operatorname{curl}\operatorname{curl}\boldsymbol{\Gamma}\}\,\mathrm{d}\boldsymbol{x}$$

by the divergence theorem (see §F.8). By virtue of (33.4) and (33.6), an interchange of $\boldsymbol{x}$ and $\boldsymbol{y}$ gives

$$\boldsymbol{E}_\mathrm{h}(\boldsymbol{x}) = \int_S \{(\boldsymbol{n}_y\wedge\operatorname{curl}_y\boldsymbol{E}_\mathrm{h})\cdot\boldsymbol{\Gamma}(\boldsymbol{y}, \boldsymbol{x}) + (\boldsymbol{n}_y\wedge\boldsymbol{E}_\mathrm{h})\cdot\operatorname{curl}_y\boldsymbol{\Gamma}(\boldsymbol{y}, \boldsymbol{x})\}\,\mathrm{d}S_y \tag{33.7}$$

when $\boldsymbol{x}$ is in T, the left-hand side being zero when $\boldsymbol{x}$ is outside T. The same formula holds with $\boldsymbol{H}_\mathrm{h}$ substituted throughout for $\boldsymbol{E}_\mathrm{h}$ on account of (33.5). These general formulae are the vector analogues of (32.3).

Suppose that there is a $\boldsymbol{\Gamma}_1$ satisfying (33.6) and such that $\boldsymbol{n}_y\wedge\boldsymbol{\Gamma}_1(\boldsymbol{y}, \boldsymbol{x})$ is zero for $\mathbf{y}$ on S. Then (33.7) becomes

$$\boldsymbol{E}_{\mathrm{h}}(\boldsymbol{x}) = \int_S (\boldsymbol{n}_{\mathrm{y}} \wedge \boldsymbol{E}_{\mathrm{h}}) \cdot \operatorname{curl}_{\mathrm{y}} \boldsymbol{\Gamma}_1(\mathbf{y}, \boldsymbol{x}) \, \mathrm{d}S_{\mathrm{y}} \tag{33.8}$$

and there is the same formula for $\boldsymbol{H}_{\mathrm{h}}$. On the other hand, if $\boldsymbol{n}_{\mathrm{y}} \wedge \operatorname{curl}_{\mathrm{y}} \boldsymbol{\Gamma}_2(\mathbf{y}, \boldsymbol{x})$ vanishes on S,

$$\begin{aligned} \boldsymbol{E}_{\mathrm{h}}(\boldsymbol{x}) &= \int_S (\boldsymbol{n}_{\mathrm{y}} \wedge \operatorname{curl}_{\mathrm{y}} \boldsymbol{E}_{\mathrm{h}}) \cdot \boldsymbol{\Gamma}_2(\mathbf{y}, \boldsymbol{x}) \, \mathrm{d}S_{\mathrm{y}} \\ &= -\mathrm{i}\omega\mu \int_S (\boldsymbol{n}_{\mathrm{y}} \wedge \boldsymbol{H}_{\mathrm{h}}) \cdot \boldsymbol{\Gamma}_2(\mathbf{y}, \boldsymbol{x}) \, \mathrm{d}S_{\mathrm{y}} \end{aligned} \tag{33.9}$$

from (30.1). There is a similar result for $\boldsymbol{H}_{\mathrm{h}}$.

These Green's tensors enable the electromagnetic field to be expressed exclusively in terms of the tangential components of $\boldsymbol{E}_{\mathrm{h}}$ or those of $\boldsymbol{H}_{\mathrm{h}}$.

Sources of charge and current can be dealt with by considering

$$\operatorname{curl} \boldsymbol{H}_{\mathrm{h}} - \mathrm{i}\omega\varepsilon \boldsymbol{E}_{\mathrm{h}} = \boldsymbol{a}\delta(\boldsymbol{x} - \boldsymbol{x}_1). \tag{33.10}$$

If $\boldsymbol{x}_1$ is outside S there is no alteration but, if $\boldsymbol{x}_1$ is inside S, $\mathrm{i}\omega\mu\boldsymbol{a} \cdot \boldsymbol{\Gamma}(\boldsymbol{x}_1, \boldsymbol{x})$ must be added to the right-hand side of (33.7) and $-\boldsymbol{a} \cdot \operatorname{curl}_1 \boldsymbol{\Gamma}(\boldsymbol{x}_1, \boldsymbol{x})$ to the corresponding expression for $\boldsymbol{H}_{\mathrm{h}}$. As a consequence, for (30.2), (33.7) must be supplemented by $\mathrm{i}\omega\mu \int \boldsymbol{J}_{\mathrm{h}}(\boldsymbol{x}_1) \cdot \boldsymbol{\Gamma}(\boldsymbol{x}_1, \boldsymbol{x}) \, \mathrm{d}\boldsymbol{x}_1$ and the analogous formula for $\boldsymbol{H}_{\mathrm{h}}$ by $-\int \boldsymbol{J}_{\mathrm{h}}(\boldsymbol{x}_1) \cdot \operatorname{curl}_1 \boldsymbol{\Gamma}(\boldsymbol{x}_1, \boldsymbol{x}) \, \mathrm{d}\boldsymbol{x}_1$.

1.34 The exterior problem

Attention has been concentrated so far on representations for fields inside a closed surface S. The next matter is to consider integrals for fields outside S.

Suppose that, at all points outside S,

$$\nabla^2 p_{\mathrm{h}} + k^2 p_{\mathrm{h}} = 0.$$

Since the region outside S includes infinity the field is subjected to the radiation conditions (31.1) and (31.2). Let Ω be the surface of a large sphere of radius R_0 which completely encloses S (Fig. 1.5). Then $S \cup \Omega$ forms a closed surface whose interior is the volume T between S and Ω. Let $\boldsymbol{n}'$ be the outward normal to this volume. Then, when $\boldsymbol{x}$ is between S and Ω, Helmholtz's representation gives

$$p_{\mathrm{h}}(\boldsymbol{x}) = \int_{S \cup \Omega} \left\{ \frac{\partial p_{\mathrm{h}}}{\partial n'_{\mathrm{y}}} \psi(\boldsymbol{x}, \mathbf{y}) - p_{\mathrm{h}} \frac{\partial}{\partial n'_{\mathrm{y}}} \psi(\boldsymbol{x}, \mathbf{y}) \right\} \mathrm{d}S_{\mathrm{y}}.$$

The radiation conditions make the integral over Ω tend to zero as $R_0 \to \infty$ in the same way as in §1.31. Therefore, if $\boldsymbol{n}'$ is replaced by $-\boldsymbol{n}$ (the same

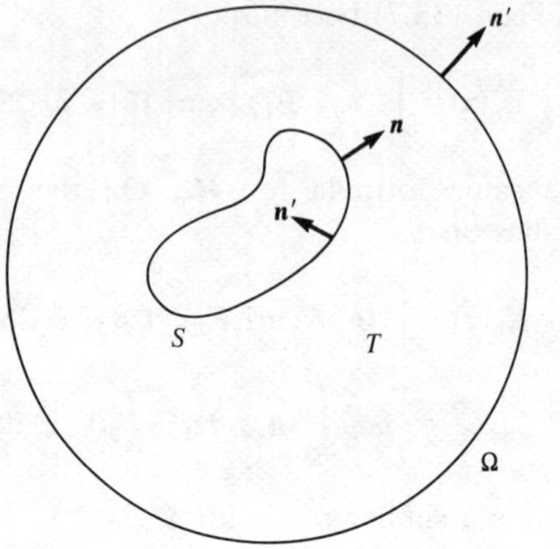

Fig. 1.5 Domain of integration for the exterior problem

normal as used hitherto),

$$p_h(\boldsymbol{x}) = -\int_S \left\{ \frac{\partial p_h}{\partial n_y} \psi(\boldsymbol{x}, \boldsymbol{y}) - p_h \frac{\partial}{\partial n_y} \psi(\boldsymbol{x}, \boldsymbol{y}) \right\} \mathrm{d}S_y \tag{34.1}$$

for any $\boldsymbol{x}$ outside S. In other words, the surface representation for exterior points is the same as that for interior points but with the sign reversed.

It may be shown in a similar manner that, when the electromagnetic field satisfies the radiation conditions (31.3)–(31.5) outside S, the same surface integrals may be used for exterior points as for interior points provided that their sign is reversed.

Similar remarks apply to the representations through Green's functions in (32.4) and (32.5) with an appropriate redefinition of these functions. Thus G_1 and G_2 satisfy (32.6) outside S (with $\boldsymbol{y}$ outside S) subject to $G_1 = 0$ and $\partial G_2/\partial n_y = 0$ on S. In addition, both must satisfy the radiation conditions at infinity.

The formulae (33.7)–(33.9) remain valid for exterior points, again with a change of sign, provided that (33.6), together with the boundary conditions, are satisfied and the Green's tensor meets the radiation conditions

$$|R\boldsymbol{\Gamma}| < K,$$

$$R(\hat{\boldsymbol{x}} \wedge \operatorname{curl} \boldsymbol{\Gamma} - \mathrm{i}k\boldsymbol{\Gamma}) \to 0$$

uniformly with respect to direction as $R \to \infty$. In particular, if (33.10) holds and S can be contracted down to a point

$$\boldsymbol{E}_h(\boldsymbol{x}) = \mathrm{i}\omega\mu\boldsymbol{a} \cdot \boldsymbol{\Gamma}(\boldsymbol{x}_1, \boldsymbol{x}), \tag{34.2}$$

$$\boldsymbol{H}_h(\boldsymbol{x}) = -\boldsymbol{a} \cdot \operatorname{curl} \boldsymbol{\Gamma}(\boldsymbol{x}_1, \boldsymbol{x}). \tag{34.3}$$

1.35 Reciprocity theorems

Certain relations exist between two possible potential and charge distributions on a given system of conductors in electrostatics. Such connections come under the general heading of reciprocity theorems and there are analogous theorems for acoustic and electromagnetic waves.

Suppose that $f_1(\boldsymbol{x})$ and $f_2(\boldsymbol{x})$ satisfy inside T

$$\operatorname{div}(a \operatorname{grad} f_1) + k^2 f_1 = -\delta(\boldsymbol{x} - \boldsymbol{x}_1), \tag{35.1}$$

$$\operatorname{div}(a \operatorname{grad} f_2) + k^2 f_2 = -\delta(\boldsymbol{x} - \boldsymbol{x}_2) \tag{35.2}$$

where $\boldsymbol{x}_1, \boldsymbol{x}_2$ are points of T and $\boldsymbol{x}_1 \neq \boldsymbol{x}_2$. The equation is more general than Helmholtz's but reduces to it when $a = 1$. It is assumed the boundary conditions for f_1 and f_2 are the same.

If S is the boundary of T

$$\int_S a\left(f_1 \frac{\partial f_2}{\partial n} - f_2 \frac{\partial f_1}{\partial n}\right) \mathrm{d}S = f_2(\boldsymbol{x}_1) - f_1(\boldsymbol{x}_2). \tag{35.3}$$

Boundaries within T are permitted so long as the surface integral can be taken continuously through them or vanishes on them. Examples of permitted boundary conditions are (i) f and $a\,\partial f/\partial n$ continuous, (ii) af and $\partial f/\partial n$ continuous, (iii) $a(\alpha f + \partial f/\partial n)$ and $f + \beta\,\partial f/\partial n$ continuous with $\alpha\beta \neq 1$, (iv) $f = 0$, (v) $\partial f/\partial n = 0$ and (vi) $\sigma f + \partial f/\partial n = 0$. If S itself can be chosen so that the surface integral vanishes, say

$$\sigma f_1 + \partial f_1/\partial n = 0, \qquad \sigma f_2 + \partial f_2/\partial n = 0$$

on S, then

$$f_1(\boldsymbol{x}_2) = f_2(\boldsymbol{x}_1). \tag{35.4}$$

In these circumstances the reciprocity theorem asserts that *the field produced by the first source at the second source is the same as that produced by the second source at the first.* One particular application of this theorem proves the statement made about the reciprocity of Green's functions in the last paragraph of §1.32.

It may happen that (35.1) and (35.2) are valid not inside S but outside. In this case (35.3) and (35.4) are still valid provided that f_1 and f_2 satisfy suitable radiation conditions at infinity.

Maxwell's equations can be handled in a similar way. Simplify the notation by dropping the suffix h and suppose there are two fields which satisfy

$$\operatorname{curl} \boldsymbol{E}_1 + \mathrm{i}\omega \boldsymbol{B}_1 = \boldsymbol{0}, \qquad \operatorname{curl} \boldsymbol{H}_1 - \mathrm{i}\omega \boldsymbol{D}_1 = -\boldsymbol{a}\delta(\boldsymbol{x} - \boldsymbol{x}_1), \tag{35.5}$$

$$\operatorname{curl} \boldsymbol{E}_2 + \mathrm{i}\omega \boldsymbol{B}_2 = \boldsymbol{0}, \qquad \operatorname{curl} \boldsymbol{H}_2 - \mathrm{i}\omega \boldsymbol{D}_2 = -\boldsymbol{b}\delta(\boldsymbol{x} - \boldsymbol{x}_2) \tag{35.6}$$

inside S with $\boldsymbol{a}$ and $\boldsymbol{b}$ unit vectors. The fields are assumed to be subject to the same boundary conditions inside and on S. Then

$$\int_S (\mathbf{E}_1 \wedge \boldsymbol{H}_2 - \mathbf{E}_2 \wedge \boldsymbol{H}_1) \cdot \boldsymbol{n} \, \mathrm{d}S$$

$$= \boldsymbol{b} \cdot \mathbf{E}_1(\boldsymbol{x}_2) - \boldsymbol{a} \cdot \mathbf{E}_2(\boldsymbol{x}_1) + \mathrm{i}\omega \int_T (\boldsymbol{H}_1 \cdot \boldsymbol{B}_2 - \boldsymbol{H}_2 \cdot \boldsymbol{B}_1 + \mathbf{E}_2 \cdot \boldsymbol{D}_1 - \mathbf{E}_1 \cdot \boldsymbol{D}_2) \, \mathrm{d}\boldsymbol{x}.$$

Changes of medium inside S are allowed so long as the boundary conditions are of the right type, e.g. (i) the tangential components of $\mathbf{E}$ and $\boldsymbol{H}$ are continuous, (ii) the tangential components of $\mathbf{E}$ or $\boldsymbol{H}$ are zero. Take for the constitutive equations

$$\boldsymbol{D} = \boldsymbol{\varepsilon} \cdot \mathbf{E}, \qquad \boldsymbol{B} = \boldsymbol{\mu} \cdot \boldsymbol{H} \tag{35.7}$$

where $\boldsymbol{\varepsilon}$ and $\boldsymbol{\mu}$ are symmetric tensors. The volume integral is then disposed of and

$$\int_S (\mathbf{E}_1 \wedge \boldsymbol{H}_2 - \mathbf{E}_2 \wedge \boldsymbol{H}_1) \cdot \boldsymbol{n} \, \mathrm{d}S = \boldsymbol{b} \cdot \mathbf{E}_1(\boldsymbol{x}_2) - \boldsymbol{a} \cdot \mathbf{E}_2(\boldsymbol{x}_1). \tag{35.8}$$

If, furthermore, the integral over S vanishes, e.g. if the tangential components of $\mathbf{E}$ or $\boldsymbol{H}$ are zero,

$$\boldsymbol{b} \cdot \mathbf{E}_1(\boldsymbol{x}_2) = \boldsymbol{a} \cdot \mathbf{E}_2(\boldsymbol{x}_1). \tag{35.9}$$

This reciprocity theorem states that, under the specified conditions, *the electric intensity produced by the first source at the second source has a component parallel to the second source which is equal to the component, parallel to the first source, of the electric intensity produced by the second source at the first.*

There may be special fields for which (35.8) and (35.9) are true even when (35.7) fails. For an example see Exercise 25 of Chapter 5.

Once again the radiation conditions at infinity will enable the reciprocity theorem to be transferred from points inside S to points outside.

It may also be deduced from (34.2) and (35.9) that $(\boldsymbol{\Gamma}(\boldsymbol{x}, \boldsymbol{x}_1))_{\alpha\beta} = (\boldsymbol{\Gamma}(\boldsymbol{x}_1, \boldsymbol{x}))_{\beta\alpha}$.

1.36 Two-dimensional fields

When the field is independent of the Cartesian coordinate z it satisfies the simpler partial differential equation

$$\nabla_2^2 p_\mathrm{h} + k^2 p_\mathrm{h} = -s \tag{36.1}$$

where ∇_2^2 stands for $\partial^2/\partial x^2 + \partial^2/\partial y^2$. A particular integral can be found in the same way as in three dimensions (§§1.16 and 1.28) when the cylindrically symmetric potential due to a concentrated source on $x = 0$,

$y=0$ is available. Now, a solution of

$$\nabla_2^2 p_h + k^2 p_h = 0$$

which is cylindrically symmetric and has a singularity at the origin is the Hankel function $H_0^{(2)}(kr)$ where $r=(x^2+y^2)^{\frac{1}{2}}$, (see §1.41(i)). Also

$$H_0^{(2)}(kr)e^{i\omega t} \sim \left(\frac{2}{\pi kr}\right)^{\frac{1}{2}} e^{i(\omega t - kr)} \tag{36.2}$$

as $r \to \infty$ from (A.2.3) so that the pertinent outgoing behaviour is reproduced. Consequently, since for small r

$$H_0^{(2)}(kr) \approx -(2i/\pi)\ln kr$$

from (A.1.13), a solution of

$$\nabla_2^2 p_h + k^2 p_h = -\delta(\mathbf{x}-\mathbf{y}) \tag{36.3}$$

which gives an outgoing wave is

$$p_h = -\tfrac{1}{4} i H_0^{(2)}(k\,|\mathbf{x}-\mathbf{y}|). \tag{36.4}$$

Hence a particular integral of (36.1) is

$$p_h(\mathbf{x}) = -\tfrac{1}{4} i \int_A s(\mathbf{y}) H_0^{(2)}(k\,|\mathbf{x}-\mathbf{y}|)\, dA_y \tag{36.5}$$

where A is the area occupied by s.

Weber's representation, which is the two-dimensional analogue of Helmholtz's, is

$$p_h(\mathbf{x}) = \int_C \left\{ \frac{\partial p_h}{\partial n_y} \chi(\mathbf{x}, \mathbf{y}) - p_h \frac{\partial}{\partial n_y} \chi(\mathbf{x}, \mathbf{y}) \right\} dS_y \tag{36.6}$$

where $\mathbf{x}$ is a point inside the closed curve C and

$$\chi(\mathbf{x}, \mathbf{y}) = -\tfrac{1}{4} i H_0^{(2)}(k\,|\mathbf{x}-\mathbf{y}|).$$

The reader will observe that the only difference between (36.5), (36.6) and the corresponding ones in three dimensions is that T, S, ψ are replaced by A, C, χ respectively. This observation is applicable to all the formulae of §§1.28, 1.29 and enables two-dimensional results to be deduced from those for three. The Green's function formulae of §1.32 will go over to the two-dimensional ones on a similar replacement, with ∇_2^2 for ∇^2 in (32.6). The same is true for the exterior representations of §1.34 except that the radiation conditions must be modified. Here, on account of (36.2),

$$|r^{\frac{1}{2}} p_h| < K, \tag{36.7}$$

$$r^{\frac{1}{2}}\left(\frac{\partial p_h}{\partial r} + ikp_h\right) \to 0 \tag{36.8}$$

as $r \to \infty$.

The radiation conditions for the two-dimensional electromagnetic field are

$$|r^{\frac{1}{2}}\mathbf{E}_{\mathrm{h}}|<K, \qquad |r^{\frac{1}{2}}\mathbf{H}_{\mathrm{h}}|<K, \tag{36.9}$$

$$r^{\frac{1}{2}}(\mathbf{E}_{\mathrm{h}}+Z_0\hat{\mathbf{x}}\wedge\mathbf{H}_{\mathrm{h}})\to 0, \tag{36.10}$$

$$r^{\frac{1}{2}}(\mathbf{H}_{\mathrm{h}}-\hat{\mathbf{x}}\wedge\mathbf{E}_{\mathrm{h}}/Z_0)\to 0 \tag{36.11}$$

as $r\to\infty$.

However, there is one special feature of two-dimensional electromagnetic fields that needs to be commented on. When the field equations are independent of z, the scalar components of Maxwell's equations are, for general time variation,

$$\left.\begin{aligned}\frac{\partial E_z}{\partial y}+\frac{\partial B_x}{\partial t}&=0,\\ -\frac{\partial E_z}{\partial x}+\frac{\partial B_y}{\partial t}&=0,\end{aligned}\right\}\text{(I)} \qquad \left.\begin{aligned}\frac{\partial H_z}{\partial y}-\frac{\partial D_x}{\partial t}&=0,\\ -\frac{\partial H_z}{\partial x}-\frac{\partial D_y}{\partial t}&=0,\end{aligned}\right\}\text{(II)}$$

$$\left.\begin{aligned}\frac{\partial E_y}{\partial x}-\frac{\partial E_x}{\partial y}+\frac{\partial B_z}{\partial t}&=0,\\ \frac{\partial E_x}{\partial x}+\frac{\partial E_y}{\partial y}&=0,\end{aligned}\right\}\text{(II)} \qquad \left.\begin{aligned}\frac{\partial H_y}{\partial x}-\frac{\partial H_x}{\partial y}-\frac{\partial D_z}{\partial t}&=0,\\ \frac{\partial H_x}{\partial x}+\frac{\partial H_y}{\partial y}&=0\end{aligned}\right\}\text{(I)}$$

in the absence of current or charge densities. It will be noted that, in a linear medium, the group (I) involves only E_z, H_x, H_y whereas group (II) concerns only E_x, E_y, H_z. Therefore, *there are two independent fields provided that they are not coupled by the boundary conditions.* Moreover, H_x and H_y can be deduced from E_z while E_x and E_y can be derived from H_z. Hence the determination of the two-dimensional electromagnetic field reduces essentially to finding the two quantities E_z and H_z. In particular, for harmonic fields, there are two possibilities

$$E_x=0, \qquad E_y=0, \qquad E_z=E_z, \qquad \mathrm{i}\omega\mu\mathbf{H}=\mathbf{k}\wedge\operatorname{grad}E_z, \tag{36.12}$$

$$H_x=0, \qquad H_y=0, \qquad H_z=H_z, \qquad \mathrm{i}\omega\varepsilon\mathbf{E}=-\mathbf{k}\wedge\operatorname{grad}H_z \tag{36.13}$$

with $\mathbf{k}$ a unit vector along the z-axis.

Orthogonal curvilinear coordinates

1.37 Curvilinear coordinates

Finding analytical solutions of the partial differential equations under consideration is a non-trivial matter when they are tackled directly instead of via integral representations. Sometimes the task is eased by working in a special coordinate system and so we now consider how the equations will be transformed by a change of coordinates.

Let x, y, z be Cartesian coordinates and let ξ, η, ζ be other coordinates given as independent, differentiable and single-valued functions of x, y, z. In general x, y, z are independent, continuously differentiable single-valued functions of ξ, η, ζ although there may be some singular points where this is not true. To each point (x, y, z) there corresponds a (ξ, η, ζ) and, conversely, with each (ξ, η, ζ) can be associated one (x, y, z) provided that the functions are single-valued. The functions ξ, η, ζ are called *curvilinear coordinates.* A coordinate surface is a surface on which one coordinate is constant and the other two variable, e.g. $\xi = \text{constant}$. Two coordinate surfaces intersect in a curve, called a coordinate curve, which will be designated by the variable coordinate on it. Thus $\xi = 0$ and $\eta = 0$ intersect in a ζ-curve. The tangents to the coordinate curves at a point may be regarded as coordinate axes but the directions of these axes can vary from point to point.

The element of distance $\mathrm{d}s$ between two nearby points (x, y, z) and $(x+\mathrm{d}x, y+\mathrm{d}y, z+\mathrm{d}z)$ is given by

$$\begin{aligned}\mathrm{d}s^2 &= \mathrm{d}x^2+\mathrm{d}y^2+\mathrm{d}z^2\\ &= \left(\frac{\partial x}{\partial \xi}\,\mathrm{d}\xi+\frac{\partial x}{\partial \eta}\,\mathrm{d}\eta+\frac{\partial x}{\partial \zeta}\,\mathrm{d}\zeta\right)^2+\left(\frac{\partial y}{\partial \xi}\,\mathrm{d}\xi+\frac{\partial y}{\partial \eta}\,\mathrm{d}\eta+\frac{\partial y}{\partial \zeta}\,\mathrm{d}\zeta\right)^2\\ &\quad+\left(\frac{\partial z}{\partial \xi}\,\mathrm{d}\xi+\frac{\partial z}{\partial \eta}\,\mathrm{d}\eta+\frac{\partial z}{\partial \zeta}\,\mathrm{d}\zeta\right)^2\end{aligned}$$

whence

$$\mathrm{d}s^2 = h_1^2\,\mathrm{d}\xi^2+h_2^2\,\mathrm{d}\eta^2+h_3^2\,\mathrm{d}\zeta^2+2f\,\mathrm{d}\eta\,\mathrm{d}\zeta+2g\,\mathrm{d}\zeta\,\mathrm{d}\xi+2h\,\mathrm{d}\xi\,\mathrm{d}\eta \tag{37.1}$$

where

$$h_1^2=\left(\frac{\partial x}{\partial \xi}\right)^2+\left(\frac{\partial y}{\partial \xi}\right)^2+\left(\frac{\partial z}{\partial \xi}\right)^2,\qquad h_2^2=\left(\frac{\partial x}{\partial \eta}\right)^2+\left(\frac{\partial y}{\partial \eta}\right)^2+\left(\frac{\partial z}{\partial \eta}\right)^2,$$

$$h_3^2=\left(\frac{\partial x}{\partial \zeta}\right)^2+\left(\frac{\partial y}{\partial \zeta}\right)^2+\left(\frac{\partial z}{\partial \zeta}\right)^2,\qquad f=\frac{\partial x}{\partial \eta}\frac{\partial x}{\partial \zeta}+\frac{\partial y}{\partial \eta}\frac{\partial y}{\partial \zeta}+\frac{\partial z}{\partial \eta}\frac{\partial z}{\partial \zeta},$$

$$g=\frac{\partial x}{\partial \zeta}\frac{\partial x}{\partial \xi}+\frac{\partial y}{\partial \zeta}\frac{\partial y}{\partial \xi}+\frac{\partial z}{\partial \zeta}\frac{\partial z}{\partial \xi},\qquad h=\frac{\partial x}{\partial \xi}\frac{\partial x}{\partial \eta}+\frac{\partial y}{\partial \xi}\frac{\partial y}{\partial \eta}+\frac{\partial z}{\partial \xi}\frac{\partial z}{\partial \eta}.$$

From (37.1) can be deduced that the element of length of a ξ-curve $(\mathrm{d}\eta = 0, \mathrm{d}\zeta = 0)$ is $h_1\,\mathrm{d}\xi$ and, similarly, the elements of the η- and ζ-curves are $h_2\,\mathrm{d}\eta$ and $h_3\,\mathrm{d}\zeta$ respectively. Since the direction cosines of a line joining (x, y, z) to $(x+\mathrm{d}x, y+\mathrm{d}y, z+\mathrm{d}z)$ are $\mathrm{d}x/\mathrm{d}s$, $\mathrm{d}y/\mathrm{d}s$, $\mathrm{d}z/\mathrm{d}s$, the direction cosines of a ξ-curve are

$$\frac{1}{h_1}\frac{\partial x}{\partial \xi},\ \frac{1}{h_1}\frac{\partial y}{\partial \xi},\ \frac{1}{h_1}\frac{\partial z}{\partial \xi}$$

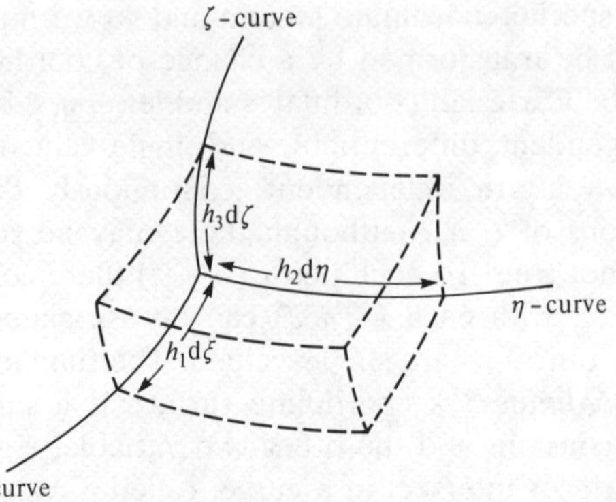

Fig. 1.6 Element of volume in orthogonal curvilinear coordinates

and those of the η- and ζ-curves are

$$\frac{1}{h_2}\frac{\partial x}{\partial \eta},\frac{1}{h_2}\frac{\partial y}{\partial \eta},\frac{1}{h_2}\frac{\partial z}{\partial \eta} \quad \text{and} \quad \frac{1}{h_3}\frac{\partial x}{\partial \zeta},\frac{1}{h_3}\frac{\partial y}{\partial \zeta},\frac{1}{h_3}\frac{\partial z}{\partial \zeta}$$

respectively. Consequently, the angle between the η- and ζ-curves is

$$\cos^{-1}\frac{1}{h_2h_3}\left(\frac{\partial x}{\partial \eta}\frac{\partial x}{\partial \zeta}+\frac{\partial y}{\partial \eta}\frac{\partial y}{\partial \zeta}+\frac{\partial z}{\partial \eta}\frac{\partial z}{\partial \zeta}\right)=\cos^{-1}\frac{f}{h_2h_3}.$$

Similarly $\cos^{-1}(g/h_3h_1)$ is the angle between the ζ- and ξ-curves whereas $\cos^{-1}(h/h_1h_2)$ is that between the ξ- and η-curves. Thus, if f, g, h are all zero at a point the ξ-, η-, ζ-curves are mutually perpendicular at that point. If this is true at all points the coordinates ξ, η, ζ are said to form an *orthogonal curvilinear system.* The positive directions of the curvilinear axes are those which form a right-handed system (Fig. 1.6).

Henceforth only orthogonal curvilinear coordinates will be treated. With such coordinates

$$\mathrm{d}s^2=h_1^2\,\mathrm{d}\xi^2+h_2^2\,\mathrm{d}\eta^2+h_3^2\,\mathrm{d}\zeta^2$$

and the element of volume $\mathrm{d}\boldsymbol{\xi}$ is given by

$$\mathrm{d}\boldsymbol{\xi}=h_1h_2h_3\,\mathrm{d}\xi\,\mathrm{d}\eta\,\mathrm{d}\zeta \tag{37.2}$$

since the lengths of the edges of the elementary parallelepiped in Fig. 1.6 are $h_1\,\mathrm{d}\xi$, $h_2\,\mathrm{d}\eta$, and $h_3\,\mathrm{d}\zeta$. Also the element of surface on $\zeta=$ constant is $h_1h_2\,\mathrm{d}\xi\,\mathrm{d}\eta$ while those on $\xi=$ constant and $\eta=$ constant are $h_2h_3\,\mathrm{d}\eta\,\mathrm{d}\zeta$ and $h_3h_1\,\mathrm{d}\zeta\,\mathrm{d}\xi$ respectively.

1.38 Differential operators

The aim of this section is to discover what happens to the operators grad, div, and curl when orthogonal curvilinear coordinates are introduced. Let

$\boldsymbol{i}_1$, $\boldsymbol{i}_2$, and $\boldsymbol{i}_3$ be unit vectors at a point in the positive directions of the ξ-, η-, and ζ-curves respectively.

The gradient of the scalar p is the vector which gives the direction and magnitude of the maximum rate of change of p with respect to the coordinates. Therefore, in a displacement from $\boldsymbol{x}$ to $\boldsymbol{x}+\mathrm{d}\boldsymbol{x}$ the change effected in p is

$$\mathrm{d}p = \mathrm{d}\boldsymbol{x} \cdot \operatorname{grad} p.$$

Now

$$\mathrm{d}p = \frac{\partial p}{\partial \xi}\,\mathrm{d}\xi + \frac{\partial p}{\partial \eta}\,\mathrm{d}\eta + \frac{\partial p}{\partial \zeta}\,\mathrm{d}\zeta.$$

Make a displacement in which η and ζ are kept constant so that $\mathrm{d}\boldsymbol{x} = h_1\,\mathrm{d}\xi\,\boldsymbol{i}_1$. Then

$$\frac{\partial p}{\partial \xi} = h_1 \boldsymbol{i}_1 \cdot \operatorname{grad} p.$$

The other components can be found in a similar fashion and hence

$$\operatorname{grad} p = \frac{1}{h_1}\frac{\partial p}{\partial \xi}\boldsymbol{i}_1 + \frac{1}{h_2}\frac{\partial p}{\partial \eta}\boldsymbol{i}_2 + \frac{1}{h_3}\frac{\partial p}{\partial \zeta}\boldsymbol{i}_3 \qquad (38.1)$$

on account of the orthogonal nature of the coordinates.

The divergence of a vector $\boldsymbol{F}$ is deduced most easily by *Kelvin's method.* In the divergence theorem

$$\int_S \boldsymbol{F} \cdot \boldsymbol{n}\,\mathrm{d}S = \int_T \operatorname{div} \boldsymbol{F}\,\mathrm{d}\boldsymbol{x}$$

allow S to shrink to a point P of T. When S is sufficiently small $\operatorname{div} \boldsymbol{F}$ does not differ appreciably from its value at P and the volume integral may be approximated by $T \operatorname{div} \boldsymbol{F}$. As a consequence

$$\operatorname{div} \boldsymbol{F} = \lim_{T \to 0} \frac{1}{T} \int_S \boldsymbol{F} \cdot \boldsymbol{n}\,\mathrm{d}S. \qquad (38.2)$$

Choose for T the elementary volume of Fig. 1.6 so that

$$T = h_1 h_2 h_3\,\mathrm{d}\xi\,\mathrm{d}\eta\,\mathrm{d}\zeta.$$

For the part of S on $\zeta = \text{constant}$, $\boldsymbol{n} = -\boldsymbol{i}_3$ and so

$$\boldsymbol{F} \cdot \boldsymbol{n}\,\mathrm{d}S = -\boldsymbol{F} \cdot \boldsymbol{i}_3 h_1 h_2\,\mathrm{d}\xi\,\mathrm{d}\eta = -F_3 h_1 h_2\,\mathrm{d}\xi\,\mathrm{d}\eta$$

where F_1, F_2, and F_3 are the components of $\boldsymbol{F}$ in the positive directions of the ξ-, η-, and ζ-curves respectively. On the portion of S where $\zeta + \mathrm{d}\zeta =$ constant, $\boldsymbol{n} = \boldsymbol{i}_3$ and $\boldsymbol{F} \cdot \boldsymbol{n}\,\mathrm{d}S = [F_3\,\mathrm{d}S]_{\zeta+\mathrm{d}\zeta}$ where the bracket signifies that the enclosed quantity is to be evaluated at $\zeta + \mathrm{d}\zeta$. A Taylor expansion gives to a first approximation

$$f(\zeta+\mathrm{d}\zeta)=f(\zeta)+\mathrm{d}\zeta\ \partial f/\partial\zeta$$

and so

$$[F_3\,\mathrm{d}S]_{\zeta+\mathrm{d}\zeta}=F_3h_1h_2\,\mathrm{d}\xi\,\mathrm{d}\eta+\frac{\partial}{\partial\zeta}(h_1h_2F_3)\,\mathrm{d}\xi\,\mathrm{d}\eta\,\mathrm{d}\zeta.$$

Consequently, the contribution of these two faces to the surface integral is

$$\frac{\partial}{\partial\zeta}(h_1h_2F_3)\,\mathrm{d}\xi\,\mathrm{d}\eta\,\mathrm{d}\zeta.$$

The contributions from the other pairs of faces may be determined similarly. Divide by T and proceed to the limit $\mathrm{d}\xi\to 0$, $\mathrm{d}\eta\to 0$, $\mathrm{d}\zeta\to 0$ so that (38.2) is relevant. The passage to the limit disposes of any contributions from higher terms in the Taylor expansion. Hence

$$\operatorname{div}\boldsymbol{F}=\frac{1}{h_1h_2h_3}\left\{\frac{\partial}{\partial\xi}(h_2h_3F_1)+\frac{\partial}{\partial\eta}(h_3h_1F_2)+\frac{\partial}{\partial\zeta}(h_1h_2F_3)\right\}. \quad (38.3)$$

The combination of (38.1) and (38.3) supplies

$$\nabla^2 p=\frac{1}{h_1h_2h_3}\left\{\frac{\partial}{\partial\xi}\left(\frac{h_2h_3}{h_1}\frac{\partial p}{\partial\xi}\right)+\frac{\partial}{\partial\eta}\left(\frac{h_3h_1}{h_2}\frac{\partial p}{\partial\eta}\right)+\frac{\partial}{\partial\zeta}\left(\frac{h_1h_2}{h_3}\frac{\partial p}{\partial\zeta}\right)\right\} \quad (38.4)$$

when p is an invariant scalar.

The vector curl $\boldsymbol{F}$ may be obtained by a similar technique based on Stokes's theorem. According to Stokes's theorem, if S is a regular two-sided surface with the regular curve C as rim,

$$\int_C \boldsymbol{F}\cdot\mathrm{d}\boldsymbol{s}=\int_S \boldsymbol{n}\cdot\operatorname{curl}\boldsymbol{F}\,\mathrm{d}S \quad (38.5)$$

where $\mathrm{d}\boldsymbol{s}$ is an element of arc along C and $\boldsymbol{n}$ is a unit vector normal to S. The convention relating $\mathrm{d}\boldsymbol{s}$ and $\boldsymbol{n}$ is conveniently described by visualizing S as a hill and C as a path round its base along which the observer is travelling in the direction of $\mathrm{d}\boldsymbol{s}$. Then, if the observer has the hill on his left, the position direction of $\boldsymbol{n}$ is upwards, pointing towards the sky at the top.

Allow C to contract to the point P. When C is small enough select S to be plane and then the surface integral can be replaced by $\boldsymbol{n}\cdot\operatorname{curl}\boldsymbol{F}\,\mathrm{d}S$ approximately. Thus the component of curl $\boldsymbol{F}$ in the direction of $\boldsymbol{n}$ is given by

$$\boldsymbol{n}\cdot\operatorname{curl}\boldsymbol{F}=\lim_{S\to 0}\frac{1}{S}\int_C \boldsymbol{F}\cdot\mathrm{d}\boldsymbol{s}. \quad (38.6)$$

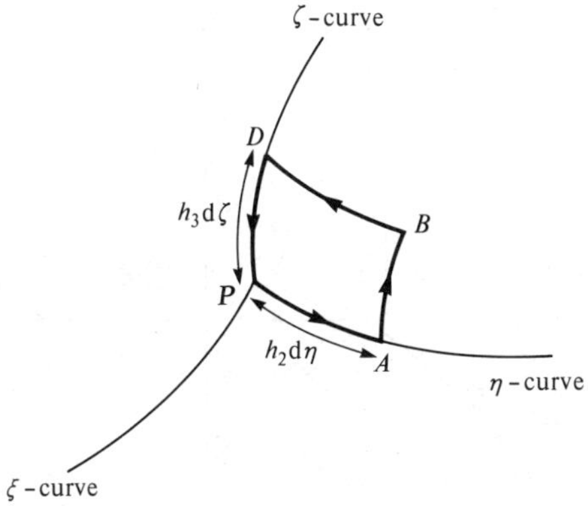

Fig. 1.7 Contour for the calculation of curl

Choose for C the rectangular element of area in the surface $\xi = \text{constant}$ (Fig. 1.7). Then, with the direction of integration on C as shown in Fig. 1.7, $\boldsymbol{n} = \boldsymbol{i}_1$ and $S = h_2 h_3 \, \mathrm{d}\eta \, \mathrm{d}\zeta$. On PA, $\mathrm{d}\boldsymbol{s} = h_2 \, \mathrm{d}\eta \, \boldsymbol{i}_2$ and $\boldsymbol{F} \cdot \mathrm{d}\boldsymbol{s} = h_2 F_2 \, \mathrm{d}\eta$. Likewise, on BD

$$\boldsymbol{F} \cdot \mathrm{d}\boldsymbol{s} = -[h_2 F_2 \, \mathrm{d}\eta]_{\zeta + \mathrm{d}\zeta} = -h_2 F_2 \, \mathrm{d}\eta - \frac{\partial}{\partial \zeta}(h_2 F_2) \, \mathrm{d}\eta \, \mathrm{d}\zeta.$$

Thus these two sides combine to give $-\dfrac{\partial}{\partial \zeta}(h_2 F_2) \, \mathrm{d}\eta \, \mathrm{d}\zeta$. The other two sides may be tackled similarly and then (38.6) implies that

$$\boldsymbol{i}_1 \cdot \operatorname{curl} \boldsymbol{F} = \frac{1}{h_2 h_3} \left\{ \frac{\partial}{\partial \eta}(h_3 F_3) - \frac{\partial}{\partial \zeta}(h_2 F_2) \right\}.$$

The components parallel to $\boldsymbol{i}_2$ and $\boldsymbol{i}_3$ may also be obtained in this way and so, finally,

$$\operatorname{curl} \boldsymbol{F} = \frac{1}{h_2 h_3} \left\{ \frac{\partial}{\partial \eta}(h_3 F_3) - \frac{\partial}{\partial \zeta}(h_2 F_2) \right\} \boldsymbol{i}_1 + \frac{1}{h_3 h_1} \left\{ \frac{\partial}{\partial \zeta}(h_1 F_1) - \frac{\partial}{\partial \xi}(h_3 F_3) \right\} \boldsymbol{i}_2$$
$$+ \frac{1}{h_1 h_2} \left\{ \frac{\partial}{\partial \xi}(h_2 F_2) - \frac{\partial}{\partial \eta}(h_1 F_1) \right\} \boldsymbol{i}_3. \tag{38.7}$$

1.39 Particular coordinate systems

The explicit expressions for the differential operators for a selection of orthogonal coordinates will now be given.

(i) *Cylindrical polar coordinates*

In cylindrical polar coordinates (see Fig. 1.8) $x = r \cos \phi$, $y = r \sin \phi$, $z = z$. Taking $\xi = r$, $\eta = \phi$, $\zeta = z$ we find without difficulty $h_1 = 1$, $h_2 = r$,

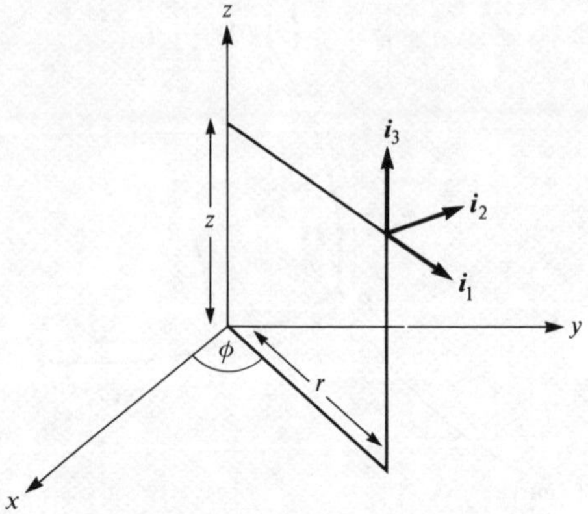

Fig. 1.8 Cylindrical polar coordinates

$h_3 = 1$, and

$$\mathrm{d}s^2 = \mathrm{d}r^2 + r^2\,\mathrm{d}\phi^2 + \mathrm{d}z^2.$$

The differential operators are

$$\operatorname{grad} p = \frac{\partial p}{\partial r}\boldsymbol{i}_1 + \frac{1}{r}\frac{\partial p}{\partial \phi}\boldsymbol{i}_2 + \frac{\partial p}{\partial z}\boldsymbol{i}_3, \tag{39.1}$$

$$\operatorname{div} \boldsymbol{F} = \frac{1}{r}\frac{\partial}{\partial r}(rF_1) + \frac{1}{r}\frac{\partial F_2}{\partial \phi} + \frac{\partial F_3}{\partial z}, \tag{39.2}$$

$$\nabla^2 p = \frac{1}{r}\frac{\partial}{\partial r}\left(r\frac{\partial p}{\partial r}\right) + \frac{1}{r^2}\frac{\partial^2 p}{\partial \phi^2} + \frac{\partial^2 p}{\partial z^2}, \tag{39.3}$$

$$\operatorname{curl} \boldsymbol{F} = \left(\frac{1}{r}\frac{\partial F_3}{\partial \phi} - \frac{\partial F_2}{\partial z}\right)\boldsymbol{i}_1 + \left(\frac{\partial F_1}{\partial z} - \frac{\partial F_3}{\partial r}\right)\boldsymbol{i}_2 + \left\{\frac{1}{r}\frac{\partial}{\partial r}(rF_2) - \frac{1}{r}\frac{\partial F_1}{\partial \phi}\right\}\boldsymbol{i}_3. \tag{39.4}$$

(ii) *Spherical polar coordinates*

In spherical polar coordinates (see Fig. 1.9) $x = R\sin\theta\cos\phi$, $y = R\sin\theta\sin\phi$, $z = R\cos\theta$. Put $\xi = R$, $\eta = \theta$, $\zeta = \phi$ and then $h_1 = 1$, $h_2 = R$, $h_3 = R\sin\theta$ with

$$\mathrm{d}s^2 = \mathrm{d}R^2 + R^2\,\mathrm{d}\theta^2 + R^2\sin^2\theta\,\mathrm{d}\phi^2.$$

The differential operators are

$$\operatorname{grad} p = \frac{\partial p}{\partial R}\boldsymbol{i}_1 + \frac{1}{R}\frac{\partial p}{\partial \theta}\boldsymbol{i}_2 + \frac{1}{R\sin\theta}\frac{\partial p}{\partial \phi}\boldsymbol{i}_3, \tag{39.5}$$

$$\operatorname{div} \boldsymbol{F} = \frac{1}{R^2}\frac{\partial}{\partial R}(R^2F_1) + \frac{1}{R\sin\theta}\frac{\partial}{\partial \theta}(\sin\theta F_2) + \frac{1}{R\sin\theta}\frac{\partial F_3}{\partial \phi}, \tag{39.6}$$

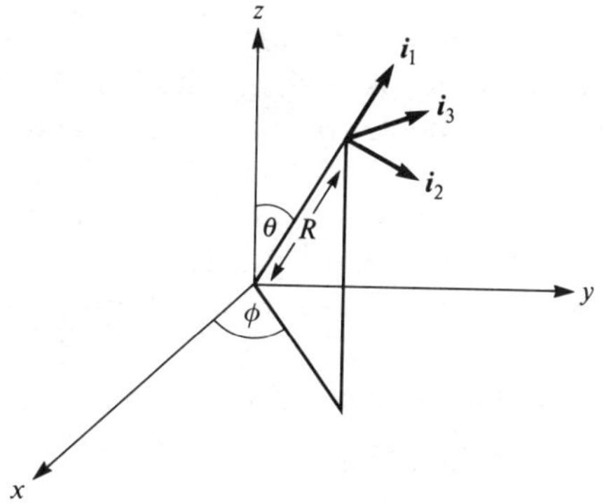

Fig. 1.9 Spherical polar coordinates

$$\nabla^2 p = \frac{1}{R^2}\frac{\partial}{\partial R}\left(R^2\frac{\partial p}{\partial R}\right) + \frac{1}{R^2 \sin\theta}\frac{\partial}{\partial\theta}\left(\sin\theta\frac{\partial p}{\partial\theta}\right) + \frac{1}{R^2\sin^2\theta}\frac{\partial^2 p}{\partial\phi^2}, \tag{39.7}$$

$$\operatorname{curl}\boldsymbol{F} = \frac{1}{R\sin\theta}\left\{\frac{\partial}{\partial\theta}(\sin\theta F_3) - \frac{\partial F_2}{\partial\phi}\right\}\boldsymbol{i}_1 + \frac{1}{R}\left\{\frac{1}{\sin\theta}\frac{\partial F_1}{\partial\phi} - \frac{\partial}{\partial R}(RF_3)\right\}\boldsymbol{i}_2 + \frac{1}{R}\left\{\frac{\partial}{\partial R}(RF_2) - \frac{\partial F_1}{\partial\theta}\right\}\boldsymbol{i}_3. \tag{39.8}$$

(iii) *Elliptic cylinder coordinates*

The transformation $x = l\cosh u\cos v$, $y = l\sinh u\sin v$, $z = z$ with l a positive constant furnishes elliptic cylinder coordinates (Fig. 1.10). The

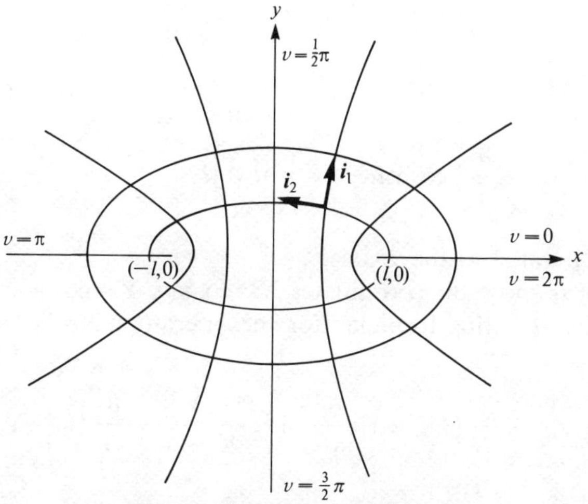

Fig. 1.10 Elliptic cylinder coordinates

surface $u=$constant is a cylinder of elliptic cross-section whose foci are $(\pm l, 0)$ and whose eccentricity is sech u. The surface $v=$constant is a hyperbolic cylinder of two sheets with foci $(\pm l, 0)$ and eccentricity sec v. To make sure that u and v are single-valued functions of x and y except on the x-axis the restrictions $u \geqslant 0$, $0 \leqslant v \leqslant 2\pi$ are imposed. The positive y-axis is then $v=\frac{1}{2}\pi$ and the negative y-axis is $v=\frac{3}{2}\pi$. Alternatively, one can have the limitation $0 \leqslant v \leqslant \pi$ and allow u both positive and negative values.

The fundamental parameters are given by

$$h_1 = h_2 = l(\cosh^2 u - \cos^2 v)^{\frac{1}{2}}, \qquad h_3 = 1,$$

$$ds^2 = l^2(\cosh^2 u - \cos^2 v)(du^2 + dv^2) + dz^2.$$

Therefore

$$\operatorname{grad} p = \frac{1}{l(\cosh^2 u - \cos^2 v)^{\frac{1}{2}}}\left(\frac{\partial p}{\partial u}\boldsymbol{i}_1 + \frac{\partial p}{\partial v}\boldsymbol{i}_2\right) + \frac{\partial p}{\partial z}\boldsymbol{i}_3, \tag{39.9}$$

$$\operatorname{div} \boldsymbol{F} = \frac{1}{l(\cosh^2 u - \cos^2 v)}\left[\frac{\partial}{\partial u}\{(\cosh^2 u - \cos^2 v)^{\frac{1}{2}} F_1\}\right.$$
$$\left. + \frac{\partial}{\partial v}\{(\cosh^2 u - \cos^2 v)^{\frac{1}{2}} F_2\}\right] + \frac{\partial F_3}{\partial z}, \tag{39.10}$$

$$\nabla^2 p = \frac{1}{l^2(\cosh^2 u - \cos^2 v)}\left(\frac{\partial^2 p}{\partial u^2} + \frac{\partial^2 p}{\partial v^2}\right) + \frac{\partial^2 p}{\partial z^2}, \tag{39.11}$$

$$\operatorname{curl} \boldsymbol{F} = \left\{\frac{1}{l}(\cosh^2 u - \cos^2 v)^{-\frac{1}{2}}\frac{\partial F_3}{\partial v} - \frac{\partial F_2}{\partial z}\right\}\boldsymbol{i}_1$$
$$+ \left\{\frac{\partial F_1}{\partial z} - \frac{1}{l}(\cosh^2 u - \cos^2 v)^{-\frac{1}{2}}\frac{\partial F_3}{\partial u}\right\}\boldsymbol{i}_2$$
$$+ \frac{1}{l}(\cosh^2 u - \cos^2 v)^{-1}\left[\frac{\partial}{\partial u}\{(\cosh^2 u - \cos^2 v)^{\frac{1}{2}} F_2\}\right.$$
$$\left. - \frac{\partial}{\partial v}\{(\cosh^2 u - \cos^2 v)^{\frac{1}{2}} F_1\}\right]\boldsymbol{i}_3 \tag{39.12}$$

where $\boldsymbol{i}_3$ is parallel to the z-axis.

Instead of u and v the coordinates $X=\cosh u$, $Y=\cos v$ are sometimes adopted. The resulting formulae for the operators can be written down but only

$$\nabla^2 p = \frac{(X^2-1)^{\frac{1}{2}}}{l^2(X^2-Y^2)}\frac{\partial}{\partial X}\left\{(X^2-1)^{\frac{1}{2}}\frac{\partial p}{\partial X}\right\} + \frac{(1-Y^2)^{\frac{1}{2}}}{l^2(X^2-Y^2)}\frac{\partial}{\partial Y}\left\{(1-Y^2)^{\frac{1}{2}}\frac{\partial p}{\partial Y}\right\} + \frac{\partial^2 p}{\partial z^2} \tag{39.13}$$

will be quoted.

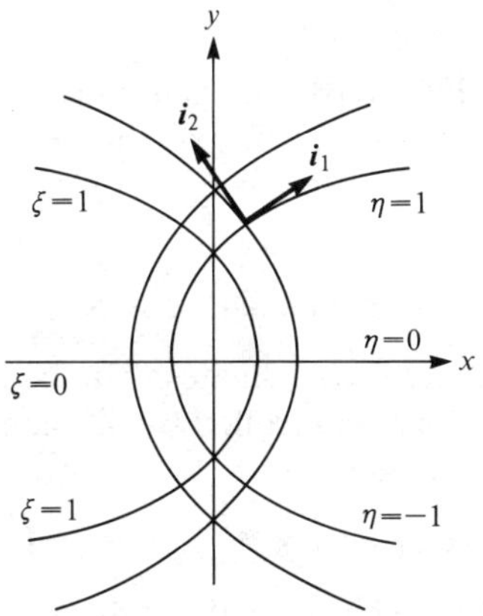

Fig. 1.11 Parabolic cylinder coordinates

(iv) *Parabolic cylinder coordinates*

From the cylindrical polar coordinates r, ϕ, z can be derived new coordinates by the transformation $\xi=(2r)^{\frac{1}{2}}\cos\phi$, $\eta=(2r)^{\frac{1}{2}}\sin\phi$. The new coordinates are related to x and y by

$$x=\tfrac{1}{2}(\xi^2-\eta^2), \qquad y=\xi\eta.$$

The variables ξ, η, z are called parabolic cylinder coordinates because the surfaces $\xi=$constant and $\eta=$constant are orthogonal parabolic cylinders. (Fig. 1.11) One choice for them is $\xi\geqslant 0$ with η unrestricted.

The basic parameters are

$$h_1=h_2=(\xi^2+\eta^2)^{\frac{1}{2}}, \qquad h_3=1, \tag{39.14}$$
$$ds^2=(\xi^2+\eta^2)(d\xi^2+d\eta^2)+dz^2.$$

The formulae for the differential operators are

$$\text{grad } p=\frac{1}{h_1}\left(\frac{\partial p}{\partial\xi}\boldsymbol{i}_1+\frac{\partial p}{\partial\eta}\boldsymbol{i}_2\right)+\frac{\partial p}{\partial z}\boldsymbol{i}_3, \tag{39.15}$$

$$\text{div } \boldsymbol{F}=\frac{1}{h_1^2}\left\{\frac{\partial}{\partial\xi}(h_1F_1)+\frac{\partial}{\partial\eta}(h_1F_2)\right\}+\frac{\partial F_3}{\partial z}, \tag{39.16}$$

$$\nabla^2 p=\frac{1}{\xi^2+\eta^2}\left(\frac{\partial^2 p}{\partial\xi^2}+\frac{\partial^2 p}{\partial\eta^2}\right)+\frac{\partial^2 p}{\partial z^2}, \tag{39.17}$$

$$\text{curl } \boldsymbol{F}=\left(\frac{1}{h_1}\frac{\partial F_3}{\partial\eta}-\frac{\partial F_2}{\partial z}\right)\boldsymbol{i}_1+\left(\frac{\partial F_1}{\partial z}-\frac{1}{h_1}\frac{\partial F_3}{\partial\xi}\right)\boldsymbol{i}_2+\frac{1}{h_1^2}\left\{\frac{\partial}{\partial\xi}(h_1F_2)-\frac{\partial}{\partial\eta}(h_1F_1)\right\}\boldsymbol{i}_3 \tag{39.18}$$

where h_1 is given by (39.14).

(v) *Prolate spheroidal coordinates*

The curves shown in Fig. 1.10 will generate an orthogonal system of surfaces if they are rotated about the x-axis, i.e. about the major axis of the ellipses. The spheroids so produced are prolate and so the coordinates u, v, ϕ such that

$$x = l \cosh u \cos v, \qquad y = l \sinh u \sin v \cos \phi, \qquad z = l \sinh u \sin v \sin \phi$$

are called prolate spheroidal coordinates. The coordinate ϕ is subject to $0 \leqslant \phi \leqslant 2\pi$ while u and v are controlled by the same restrictions as in (iii).

The fundamental parameters are

$$h_1 = h_2 = l(\cosh^2 u - \cos^2 v)^{\frac{1}{2}}, \qquad h_3 = l \sinh u \sin v.$$

Only the Laplacian will be given of the differential operators, namely

$$\nabla^2 p = \frac{1}{l^2 \sinh u \sin v (\cosh^2 u - \cos^2 v)} \left\{ \frac{\partial}{\partial u} \left(\sinh u \sin v \frac{\partial p}{\partial u} \right) + \frac{\partial}{\partial v} \left(\sinh u \sin v \frac{\partial p}{\partial v} \right) \right\} + \frac{1}{l^2 \sinh^2 u \sin^2 v} \frac{\partial^2 p}{\partial \phi^2}. \tag{39.19}$$

The alternative coordinates $X = \cosh u$, $Y = \cos v$ give

$$\nabla^2 p = \frac{1}{l^2(X^2 - Y^2)} \left[\frac{\partial}{\partial X} \left\{ (X^2 - 1) \frac{\partial p}{\partial X} \right\} + \frac{\partial}{\partial Y} \left\{ (1 - Y^2) \frac{\partial p}{\partial Y} \right\} \right] + \frac{1}{l^2 (X^2 - 1)(1 - Y^2)} \frac{\partial^2 p}{\partial \phi^2}. \tag{39.20}$$

(vi) *Oblate spheroidal coordinates*

Another system of coordinates can be created by rotating the curves of Fig. 1.10 about the y-axis, i.e. the minor axis of the ellipses. The resulting oblate spheroidal coordinates are

$$x = l \cosh u \cos v \sin \phi, \qquad y = l \sinh u \sin v, \qquad z = l \cosh u \cos v \cos \phi$$

where $0 \leqslant \phi \leqslant 2\pi$ and u, v are limited as in (iii). Note that $u = 0$ is the circular disc $y = 0$, $x^2 + z^2 \leqslant l^2$.

The fundamental parameters are

$$h_1 = h_2 = l(\cosh^2 u - \cos^2 v)^{\frac{1}{2}}, \qquad h_3 = l \cosh u \cos v,$$

and

$$\nabla^2 p = \frac{1}{l^2 \cosh u \cos v (\cosh^2 u - \cos^2 v)} \left\{ \frac{\partial}{\partial u} \left(\cosh u \cos v \frac{\partial p}{\partial u} \right) + \frac{\partial}{\partial v} \left(\cosh u \cos v \frac{\partial p}{\partial v} \right) \right\} + \frac{1}{l^2 \cosh^2 u \cos^2 v} \frac{\partial^2 p}{\partial \phi^2}. \tag{39.21}$$

In terms of X and Y

$$\nabla^2 p = \frac{(X^2-1)^{\frac{1}{2}}}{l^2 X(X^2-Y^2)} \frac{\partial}{\partial X}\left\{X(X^2-1)^{\frac{1}{2}} \frac{\partial p}{\partial X}\right\} + \frac{(1-Y^2)^{\frac{1}{2}}}{l^2 Y(X^2-Y^2)} \frac{\partial}{\partial Y}\left\{Y(1-Y^2)^{\frac{1}{2}} \frac{\partial p}{\partial Y}\right\} + \frac{1}{l^2 X^2 Y^2} \frac{\partial^2 p}{\partial \phi^2}. \tag{39.22}$$

Sometimes the coordinates $\xi = \sinh u$, $\eta = \sin v$ are employed. For them

$$h_1 = l\left(\frac{\xi^2+\eta^2}{1+\xi^2}\right)^{\frac{1}{2}}, \qquad h_2 = l\left(\frac{\xi^2+\eta^2}{1-\eta^2}\right)^{\frac{1}{2}}, \qquad h_3 = l(1+\xi^2)^{\frac{1}{2}}(1-\eta^2)^{\frac{1}{2}},$$

and

$$\nabla^2 p = \frac{1}{l^2(\xi^2+\eta^2)}\left[\frac{\partial}{\partial \xi}\left\{(1+\xi^2)\frac{\partial p}{\partial \xi}\right\} + \frac{\partial}{\partial \eta}\left\{(1-\eta^2)\frac{\partial p}{\partial \eta}\right\}\right] + \frac{1}{l^2(1+\xi^2)(1-\eta^2)}\frac{\partial^2 p}{\partial \phi^2}. \tag{39.23}$$

Both prolate and oblate spheroidal coordinates go over into spherical polar coordinates (the role of the polar z-axis being played by the x- and y-axes respectively) in the limit as $l \to 0$, $u \to \infty$ in such a way that $le^u = 2R$.

(vii) *Paraboloidal coordinates*

The curves of Fig. 1.11 can be rotated about the x-axis to provide paraboloids. The pertinent paraboloidal coordinates ξ, η, ϕ are given by

$$x = \tfrac{1}{2}(\xi^2-\eta^2), \qquad y = \xi\eta \cos \phi, \qquad z = \xi\eta \sin \phi.$$

In this case

$$h_1 = h_2 = (\xi^2+\eta^2)^{\frac{1}{2}}, \qquad h_3 = \xi\eta,$$

and

$$\nabla^2 p = \frac{1}{\xi\eta(\xi^2+\eta^2)}\left\{\frac{\partial}{\partial \xi}\left(\xi\eta\frac{\partial p}{\partial \xi}\right) + \frac{\partial}{\partial \eta}\left(\xi\eta\frac{\partial p}{\partial \eta}\right)\right\} + \frac{1}{\xi^2\eta^2}\frac{\partial^2 p}{\partial \phi^2}. \tag{39.24}$$

Solutions in special functions

1.40 Separation of variables

The general representation of harmonic fields by means of potentials and surface integrals has been described in earlier sections. There is another method which consists of finding special solutions of the partial differential equations and constructing combinations of these solutions sufficiently general to represent an arbitrary function. The validity of such a process will not be examined; attention will be devoted to discovering the special functions.

The basic partial differential equation to be tackled is Helmholtz's. Since it has been indicated in §§1.28, 1.29 how to find a particular integral only

$$\nabla^2 p + k^2 p = 0$$

needs to be considered.

The method of determining solutions can be illustrated most simply by means of the Cartesian coordinates x, y, z. The equation is then

$$\frac{\partial^2 p}{\partial x^2} + \frac{\partial^2 p}{\partial y^2} + \frac{\partial^2 p}{\partial z^2} + k^2 p = 0. \tag{40.1}$$

The first objective is to see if there is a solution of the form

$$p = X(x)Y(y)Z(z) \tag{40.2}$$

where X, Y, Z are functions of x, y, z respectively. Substitute this trial form in (40.1) and divide by XYZ. Then

$$\frac{X''}{X} + \frac{Y''}{Y} + \frac{Z''}{Z} + k^2 = 0 \tag{40.3}$$

where a prime stands for a derivative with respect to the argument. The only term containing x is the first one so that variations in x alone will not alter the other terms. Consequently, (40.3) can be satisfied only if

$$X''/X = \text{constant} = -l^2 \text{ (say).}$$

The reasoning can be repeated for the second and third terms so that

$$Y''/Y = -m^2, \qquad Z''/Z = -n^2$$

where

$$l^2 + m^2 + n^2 = k^2 \tag{40.4}$$

on account of (40.3). Hence $X = Ae^{ilx} + Be^{-ilx}$, $Y = Ce^{imy} + De^{-imy}$, $Z = Ee^{inz} + Fe^{-inz}$ where A, B, C, D, E, and F are arbitrary constants. Thus the only solutions of (40.1) of the type (40.2) are of the form

$$p = (Ae^{ilx} + Be^{-ilx})(Ce^{imy} + De^{-imy})(Ee^{inz} + Fe^{-inz})$$

subject to (40.4).

A more general solution may be put together by taking a sum of such terms with different values of l and m or by multiplying by a function of l and m and integrating with respect to l and m. The value and relevance of such representations depend entirely upon the problem under consideration.

The above method of finding a solution is known as *the separation of variables.* Its value lies in the fact that the solution of a partial differential equation is converted to solving ordinary differential equations. Obviously, the same technique can be attempted in other coordinate systems. In orthogonal curvilinear coordinates ξ, η, ζ a solution of the form $p = \Xi(\xi)H(\eta)Z(\zeta)$ is tried in Helmholtz's equation as given by (38.4). The method leads to ordinary differential equations for Ξ, H, and Z only in a small number of coordinate systems. Some of these and the special functions which arise from them will be described in the next section.

1.41 Special functions

The special functions which occur in connection with the particular coordinate systems of §1.39 will now be considered.

(i) *Bessel functions*

In cylindrical polar coordinates r, ϕ, z put

$$p = R(r)\Phi(\phi)Z(z)$$

and then

$$Z'' = -\lambda^2 Z, \tag{41.1}$$

$$\Phi'' = -\nu^2 \Phi, \tag{41.2}$$

$$r^2R'' + rR' + (\mu^2r^2 - \nu^2)R = 0 \tag{41.3}$$

where λ and ν are the separation constants corresponding to l and m in the Cartesian case. Also $\mu^2 = k^2 - \lambda^2$. Equations (41.1) and (41.2) are immediately soluble and

$$Z = A\mathrm{e}^{\mathrm{i}\lambda z} + B\mathrm{e}^{-\mathrm{i}\lambda z}, \tag{41.4}$$

$$\Phi = C\mathrm{e}^{\mathrm{i}\nu\phi} + D\mathrm{e}^{-\mathrm{i}\nu\phi}. \tag{41.5}$$

Equation (41.3) is *Bessel's equation.* It has two independent solutions $\mathrm{J}_\nu(\mu r)$ and $\mathrm{J}_{-\nu}(\mu r)$ when ν is not an integer, where

$$\mathrm{J}_\nu(z) = \sum_{m=0}^{\infty} \frac{(-)^m(\frac{1}{2}z)^{\nu+2m}}{m!(\nu+m)!}. \tag{41.6}$$

Further properties of Bessel functions will be found in Appendix A. Here we note from (41.6) that, unless ν is an integer or purely imaginary, one of J_ν, $\mathrm{J}_{-\nu}$ is finite at the origin and the other is unbounded. J_ν is known as *the Bessel function of the first kind and* ν*th order.*

Definition (41.6) also reveals that, when ν is the integer n, $\mathrm{J}_{-n}(z) = (-)^n\mathrm{J}_n(z)$ so that a second solution of Bessel's equation is required. It is customary to define a second solution $\mathrm{Y}_\nu(z)$ for all ν which will fill the gap when ν is an integer. Y_ν is called *the Bessel function of the second kind and* ν*th order.*

Thus the solutions of Helmholtz's equation constructed in this way are

$$p=(A\mathrm{e}^{\mathrm{i}\lambda z}+B\mathrm{e}^{-\mathrm{i}\lambda z})(C\mathrm{e}^{\mathrm{i}\nu\phi}+D\mathrm{e}^{-\mathrm{i}\nu\phi})\{E\mathrm{J}_\nu(\mu r)+F\mathrm{Y}_\nu(\mu r)\} \quad (41.7)$$

where $\mu=(k^2-\lambda^2)^{\frac{1}{2}}$. When Re $\nu>0$, J_ν is finite at $r=0$ but Y_ν is singular. Therefore, if a solution is forced to be bounded at the origin, Y_ν can be excluded from consideration.

Sometimes, when $\lambda^2>k^2$, it is desirable to avoid imaginary arguments. J_ν and Y_ν are then replaced by the *modified Bessel functions* I_ν and K_ν. Of these K_ν is singular at the origin but I_ν is not.

Although J_ν may be finite at the origin it does not provide an outgoing wave at infinity because (A.2.1)

$$\mathrm{J}_\nu(z)\sim(2/\pi z)^{\frac{1}{2}}\cos(z-\tfrac{1}{2}\nu\pi-\tfrac{1}{4}\pi)$$

as $|z|\to\infty$ with $|\mathrm{ph}\, z|<\pi$. Neither does Y_ν supply an outgoing wave; however, it is possible to form a linear combination of them which is satisfactory. These are solutions of Bessel's equation defined by

$$\mathrm{H}_\nu^{(1)}(z)=\mathrm{J}_\nu(z)+\mathrm{i}\mathrm{Y}_\nu(z), \qquad \mathrm{H}_\nu^{(2)}(z)=\mathrm{J}_\nu(z)-i\mathrm{Y}_\nu(z).$$

$\mathrm{H}_\nu^{(1)}$ and $\mathrm{H}_\nu^{(2)}$ are called the *Hankel functions of the first and second kind respectively*. Their asymptotic behaviour is given by

$$\mathrm{H}_\nu^{(1)}(z)\sim(2/\pi z)^{\frac{1}{2}}\mathrm{e}^{\mathrm{i}(z-\frac{1}{2}\nu\pi-\frac{1}{4}\pi)}$$

as $|z|\to\infty$ with $-\pi<\mathrm{ph}\, z<2\pi$ and

$$\mathrm{H}_\nu^{(2)}(z)\sim(2/\pi z)^{\frac{1}{2}}\mathrm{e}^{-\mathrm{i}(z-\frac{1}{2}\nu\pi-\frac{1}{4}\pi)}$$

with $-2\pi<\mathrm{ph}\, z<\pi$. Thus, when z is real, $\mathrm{H}_\nu^{(2)}$ is appropriate for outgoing waves and $\mathrm{H}_\nu^{(1)}$ for incoming. Consequently, if an outgoing solution is wanted the Bessel functions in (41.7) should be replaced by $\mathrm{H}_\nu^{(2)}(\mu r)$; this solution will, however, not be finite at $r=0$.

(ii) *Legendre functions*

Separable solutions in spherical polar coordinate R, θ, ϕ are sought in the form

$$p=\bar{R}(R)\Theta(\theta)\Phi(\phi).$$

After substitution in (39.7) we deduce that

$$\Phi''=-\mu^2\Phi,$$

$$\sin\theta\frac{\mathrm{d}}{\mathrm{d}\theta}\left(\sin\theta\frac{\mathrm{d}\Theta}{\mathrm{d}\theta}\right)+\{\nu(\nu+1)\sin^2\theta-\mu^2\}\Theta=0, \quad (41.8)$$

$$R^2\bar{R}''+2R\bar{R}'+\{k^2R^2-\nu(\nu+1)\}\bar{R}=0 \quad (41.9)$$

where μ^2 and $\nu(\nu+1)$ are separation constants. The choice of $\nu(\nu+1)$ as the constant is made in order to cast (41.8) into a particular form.

With regard to equation (41.9) put $\bar{R}(R) = y(R)/R^{\frac{1}{2}}$ and then

$$R^2y'' + Ry' + \{k^2R^2 - (\nu + \tfrac{1}{2})^2\}y = 0.$$

This is of the same type as (41.3) so that solutions in Bessel functions are available. One form for $\bar{R}$ is $J_{\nu+\frac{1}{2}}(kR)/R^{\frac{1}{2}}$. A convenient notation is

$$j_\nu(z) = (\pi/2z)^{\frac{1}{2}} J_{\nu+\frac{1}{2}}(z)$$

with similar definitions for y_ν, $h_\nu^{(1)}$, and $h_\nu^{(2)}$.

Equation (41.8) is known as *Legendre's associated equation* (*Legendre's equation* when $\mu = 0$) and its solutions are linear combinations of two chosen solutions called the *associated Legendre functions* (see Appendix B). They are denoted by $P_\nu^\mu(\cos\theta)$ and $Q_\nu^\mu(\cos\theta)$. Thus solutions in spherical polars are

$$p = (Ae^{i\mu\phi} + Be^{-i\mu\phi})\{Cj_\nu(kR) + Dy_\nu(kR)\}\{EP_\nu^\mu(\cos\theta) + FQ_\nu^\mu(\cos\theta)\}. \tag{41.10}$$

For solutions finite at the origin drop y_ν whereas for outgoing waves use the combination $h_\nu^{(2)}$.

If ν is the non-negative integer n, $P_n^\mu(x)$ is a polynomial of degree n unless μ is a non-negative integer m when it is a polynomial of degree $n-m$ which vanishes identically when $m > n$. Functions in which $\mu = 0$ are P_ν and, in particular, P_n is called a *Legendre polynomial.* The function $P_\nu^m(x)$ is finite at $x = 1$ and $P_n^m(x)$ is also bounded at $x = -1$. On the other hand, Q_n^m exhibits singularities. Therefore, if one is looking for a field finite at $\theta = 0$ and $\theta = \pi$, the function $Q_\nu^\mu(\cos\theta)$ will be discarded when ν and μ are integers.

(iii) *Mathieu functions*

The use of elliptic cylinder coordinates u, v, z with $u \geqslant 0$, $0 \leqslant v \leqslant 2\pi$ leads, through

$$p = U(u)V(v)Z(z),$$

to

$$Z'' = -\mu^2 Z,$$

$$U'' - (\lambda - 2h\cosh 2u)U = 0, \tag{41.11}$$

$$V'' + (\lambda - 2h\cos 2v)V = 0 \tag{41.12}$$

where λ, μ^2 are separation constants and

$$h = \tfrac{1}{4}l^2(k^2 - \mu^2).$$

Equation (41.12) is known as *Mathieu's equation* (see Appendix C). Equation (41.11) which differs from (41.12) only in that v is exchanged for iu is called the *modified Mathieu equation.*

Mathieu's equation has solutions for all values of λ but for the applications in this book solutions which are periodic in v will be needed.

Solutions which are unaltered by an increase of 2π in v occur only for certain *eigenvalues* of λ; these eigenvalues depend on h. It is found that the periodic solutions are either even or odd in v. The even ones are denoted by $\mathrm{ce}_n(v, h)$ and the odd ones by $\mathrm{se}_n(v, h)$. The functions ce_n and se_n are known as Mathieu functions of the first kind; both are finite.

It is possible to define second solutions of Mathieu's equation; they display singularities. However, solutions of the modified equation are more important in our context. Four even solutions $\mathrm{Mc}_n^{(j)}(u, h)$ $(j=1,2,3,4)$ and four odd solutions $\mathrm{Ms}_n^{(j)}(u, h)$ can be developed for (41.11). They play the same part for the modified Mathieu equation as do J_ν, Y_ν, $\mathrm{H}_\nu^{(1)}$, $\mathrm{H}_\nu^{(2)}$. Thus $\mathrm{Mc}_n^{(1)}$ and $\mathrm{Ms}_n^{(1)}$ are finite, $\mathrm{Mc}_n^{(4)}$ and $\mathrm{Ms}_n^{(4)}$ give outgoing waves whereas $\mathrm{Mc}_n^{(3)}$ and $\mathrm{Ms}_n^{(3)}$ correspond to incoming waves.

The form of solutions in this case is

$$p=(A\mathrm{e}^{\mathrm{i}\mu z}+B\mathrm{e}^{-\mathrm{i}\mu z})\{C\mathrm{ce}_n(v, h)+D\mathrm{se}_n(v, h)\}\{E\mathrm{Mc}_n^{(j)}(u, h)+F\mathrm{Ms}_n^{(j)}(u, h)\} \tag{41.13}$$

where j is taken as 1 for finiteness and as 4 for an outgoing wave.

(iv) *Parabolic cylinder functions*

When the parabolic cylinder coordinates ξ, η, z of §1.39(iv) are employed the substitution

$$p=\Xi(\xi)H(\eta)Z(z)$$

in Helmholtz's equation leads to

$$Z''=\mu^2 Z,$$

$$\Xi''+\{(\mu^2+k^2)\xi^2-\kappa\}\Xi=0, \tag{41.14}$$

$$H''+\{(\mu^2+k^2)\eta^2+\kappa\}H=0 \tag{41.15}$$

where μ and κ are separation constants. Equations (41.14) and (41.15) are essentially the same and can be treated together. Change the independent variable ξ to $\{4(\mu^2+k^2)\}^{-\frac{1}{4}}\mathrm{e}^{\frac{1}{4}\pi\mathrm{i}}x$ and then (41.14) becomes

$$\Xi''+(\nu+\tfrac{1}{2}-\tfrac{1}{4}x^2)\Xi=0 \tag{41.16}$$

where $\mathrm{i}(\nu+\frac{1}{2})=\{4(\mu^2+k^2)\}^{-\frac{1}{2}}\kappa$.

Equation (41.16) is known as *Weber's equation* and its solutions are called *parabolic cylinder functions* or *Weber–Hermite functions* which can be defined in terms of a basis function $\mathrm{D}_\nu(x)$ (see Appendix D). Since changing x to $-x$ does not affect (41.16) another solution of it is $\mathrm{D}_\nu(-x)$. On the other hand, the simultaneous replacement of ν and x by $-\nu-1$ and $\pm\mathrm{i}x$ respectively does not alter Weber's equation so that $\mathrm{D}_{-\nu-1}(\mathrm{i}x)$ and $\mathrm{D}_{-\nu-1}(-\mathrm{i}x)$ are also solutions. From this it is evident that a form satisfying Helmholtz's equation is, if $h=\mathrm{e}^{\frac{1}{4}\pi\mathrm{i}}\{4(\mu^2+k^2)\}^{\frac{1}{4}}$,

$$p=(A\mathrm{e}^{\mu z}+B\mathrm{e}^{-\mu z})\{C\mathrm{D}_{-\nu-1}(h\xi)+D\mathrm{D}_{-\nu-1}(-h\xi)\}\{E\mathrm{D}_\nu(h\eta)+F\mathrm{D}_\nu(-h\eta)\}. \tag{41.17}$$

Since

$$D_\nu(z) \sim z^\nu e^{-\frac{1}{4}z^2}$$

as $|z| \to \infty$ with $|\mathrm{ph}\, z| < \frac{3}{4}\pi$, we see that

$$D_{-\nu-1}(h\xi)D_\nu(-h\eta) \sim (|\eta|^\nu / h\xi^{\nu+1})e^{-\frac{1}{2}ik(\xi^2+\eta^2)} \tag{41.18}$$

when $\mu = 0$, $h\xi \gg 1$, $h\eta \ll -1$ so that this is a combination associated with an outgoing wave in two dimensions.

(v) *Spheroidal functions*

In terms of the prolate spheroidal functions X, Y, ϕ of §1.39(v) the insertion of

$$p = \Xi(X)H(Y)\Phi(\phi)$$

into Helmholtz's equation brings one to

$$\Phi'' = -\mu^2\Phi,$$

$$\frac{d}{dX}\left\{(1-X^2)\frac{d\Xi}{dX}\right\} + \left\{\lambda + \hbar^2(1-X^2) - \frac{\mu^2}{1-X^2}\right\}\Xi = 0, \tag{41.19}$$

$$\frac{d}{dY}\left\{(1-Y^2)\frac{dH}{dY}\right\} + \left\{\lambda + \hbar^2(1-Y^2) - \frac{\mu^2}{1-Y^2}\right\}H = 0 \tag{41.20}$$

where λ, μ^2 are separation constants and $\hbar = kl$.

Equations (41.19) and (41.20) differ only in notation and either can be regarded as the standard differential equation for *spheroidal functions* (Appendix E). Four solutions $S_\nu^{\mu(j)}(X, \hbar)$ $(j = 1, 2, 3, 4)$ of (41.19) can be constructed which enjoy properties similar to the four types of Bessel function (compare the analogous situation in Mathieu functions) when ν is a suitable function of λ. There are also solutions $\mathrm{ps}_\nu^\mu(X, \hbar^2)$, $\mathrm{qs}_\nu^\mu(X, \hbar^2)$ which are related to the Legendre functions. In this case

$$p = (Ae^{i\mu\phi} + Be^{-i\mu\phi})S_\nu^{\mu(j)}(X, \hbar)\{C\mathrm{ps}_\nu^\mu(Y, \hbar^2) + D\mathrm{qs}_\nu^\mu(Y, \hbar^2)\} \tag{41.21}$$

where $j = 4$ is appropriate to a radiating wave as $X \to \infty$ and $j = 1$ is used for bounded solutions.

Solutions in terms of oblate spheroidal coordinates may also be inferred. If

$$p = \Xi(\xi)H(\eta)\Phi(\phi)$$

(see §1.39(vi)) then

$$\Phi'' = -\mu^2\Phi,$$

$$\frac{d}{d\xi}\left\{(1+\xi^2)\frac{d\Xi}{d\xi}\right\} + \left\{-\lambda + \mathcal{H}^2(1+\xi^2) + \frac{\mu^2}{1+\xi^2}\right\}\Xi = 0, \tag{41.22}$$

$$\frac{d}{d\eta}\left\{(1-\eta^2)\frac{dH}{d\eta}\right\} + \left\{\lambda - \mathcal{H}^2(1-\eta^2) - \frac{\mu^2}{1-\eta^2}\right\}H = 0 \tag{41.23}$$

where $\mathscr{H} = kl$. Equation (41.23) is the same as (41.20) apart from notation and having $-\mathscr{H}^2$ for $\hbar^2$. Therefore, the substitution $\hbar = i\mathscr{H}$ converts the solution of one into that of the other. If, in addition $\xi = iX$ (41.22) goes over to (41.19). Consequently, the replacement of X, Y, $\hbar$ in (41.21) by $-i\xi, \eta, i\mathscr{H}$ transforms (41.21) into a solution in oblate spheroidal coordinates.

Exercises on Chapter 1

1. Verify that, when b is a constant, $p = \dfrac{1}{R}\sin\dfrac{R}{b}\cos\dfrac{vt}{b}$ satisfies

$$\frac{\partial^2 p}{\partial R^2} + \frac{2}{R}\frac{\partial p}{\partial R} = \frac{1}{v^2}\frac{\partial^2 p}{\partial t^2}.$$

2. Show that

$$p = \int_{-\infty}^{\infty} F\left(t + \frac{ir}{v}\sinh\alpha, \alpha + i\phi\right) d\alpha$$

is a solution of

$$\frac{1}{r}\frac{\partial}{\partial r}\left(r\frac{\partial p}{\partial r}\right) + \frac{1}{r^2}\frac{\partial^2 p}{\partial \phi^2} - \frac{1}{v^2}\frac{\partial^2 p}{\partial t^2} = 0$$

provided that $\dfrac{\partial F}{\partial \alpha} - \dfrac{2ir}{v}\cosh\alpha\dfrac{\partial F}{\partial t} \to 0$ as $|\alpha| \to \infty$.

3. Prove that

$$p = \frac{1}{2\pi}\int_0^{2\pi} F(z + ir\cos\alpha, t - (r/v)\sin\alpha)\, d\alpha$$

satisfies

$$\frac{1}{r}\frac{\partial}{\partial r}\left(r\frac{\partial p}{\partial r}\right) + \frac{\partial^2 p}{\partial z^2} - \frac{1}{v^2}\frac{\partial^2 p}{\partial t^2} = 0$$

when $\partial F/\partial\alpha$ is single-valued in α.

Show that

$$p = \int_{-\infty}^{\infty} F(z - r\sinh\alpha, t - (r/v)\cosh\alpha)\, d\alpha$$

is also a solution, given that $\partial F/\partial\alpha \to 0$ as $|\alpha| \to \infty$.

4. Show that

$$p = \int_{-\pi}^{\pi}\int_{-\pi}^{\pi} F(x\sin\alpha\cos\beta + y\sin\alpha\sin\beta + z\cos\alpha + vt, \alpha, \beta)\, d\alpha\, d\beta$$

and

$$p=\int_{-\pi}^{\pi} F(x\cos\alpha+y\sin\alpha+iz,\ y+iz\sin\alpha+vt\cos\alpha,\ \alpha)\,d\alpha$$

are solutions of

$$\frac{\partial^2 p}{\partial x^2}+\frac{\partial^2 p}{\partial y^2}+\frac{\partial^2 p}{\partial z^2}-\frac{1}{v^2}\frac{\partial^2 p}{\partial t^2}=0.$$

5. Prove that

$$p=e^{i\omega(t-y/v)}\int_0^{(r+y)^{\frac{1}{2}}} e^{i\omega u^2/v}\,du,$$

with $r=(x^2+y^2)^{\frac{1}{2}}$, satisfies

$$\frac{\partial^2 p}{\partial x^2}+\frac{\partial^2 p}{\partial y^2}-\frac{1}{v^2}\frac{\partial^2 p}{\partial t^2}=0.$$

6. Show that

$$\frac{\partial^2 p}{\partial x^2}+\frac{\partial^2 p}{\partial y^2}+k^2 p=0$$

is satisfied by

$$p=\int_{r+r_0}^{\infty}\frac{\sin k(u+\alpha)\,du}{\{u^2-(x-x_0)^2-(y-y_0)^2\}^{\frac{1}{2}}}$$

where $r=(x^2+y^2)^{\frac{1}{2}}$ and $r_0=(x_0^2+y_0^2)^{\frac{1}{2}}$.

7. Prove that, inside the closed surface S,

$$p_h(\boldsymbol{x})=-\int_S p_h\frac{\partial}{\partial n_y}G(\boldsymbol{x},\boldsymbol{y})\,dS_y=-\int_S\frac{\partial p_h}{\partial n_y}G(\boldsymbol{x},\boldsymbol{y})\,dS_y$$

where $G=G_1-G_2$ (G_1, G_2 defined as in §1.32) so that $\nabla^2 G+k^2 G=0$.

8. Prove that, in a linear homogeneous isotropic medium, Maxwell's equations can be written

$$\text{curl}\,\boldsymbol{Q}+i(\mu\varepsilon)^{\frac{1}{2}}\frac{\partial \boldsymbol{Q}}{\partial t}=\mu\boldsymbol{J},\qquad \text{div}\,\boldsymbol{Q}=i\rho(\mu/\varepsilon)^{\frac{1}{2}}$$

where $\boldsymbol{Q}$ is the complex vector defined by $\boldsymbol{Q}=\boldsymbol{B}+i(\mu\varepsilon)^{\frac{1}{2}}\boldsymbol{E}$.

If $\boldsymbol{Q}$ is expressed in terms of potentials by means

$$\boldsymbol{Q}=\text{curl}\,\mathbf{A}-i(\mu\varepsilon)^{\frac{1}{2}}\frac{\partial\mathbf{A}}{\partial t}-i(\mu\varepsilon)^{\frac{1}{2}}\,\text{grad}\,V$$

where $\text{div}\,\mathbf{A}+\mu\varepsilon\,\partial V/\partial t=0$ show that $\mathbf{A}$ and V satisfy (10.10) and (10.11) respectively.

Show also that, in the absence of currents and charges, there is a representation in terms of a Hertz vector through

$$\boldsymbol{Q} = \mu\varepsilon \operatorname{curl} \frac{\partial \boldsymbol{\Pi}}{\partial t} + \mathrm{i}(\mu\varepsilon)^{\frac{1}{2}} \operatorname{curl} \operatorname{curl} \boldsymbol{\Pi}.$$

9. Verify that a possible electromagnetic field in a region where both ρ and $\boldsymbol{J}$ are zero is given by: $\mathbf{A} = \operatorname{curl}(\boldsymbol{x} W_1 + \boldsymbol{x} \wedge \operatorname{grad} W_2)$ and $V = 0$, where both W_1 and W_2 satisfy

$$\nabla^2 W - \mu\varepsilon\, \partial^2 W/\partial t^2 = 0.$$

10. The field outside a perfectly conducting sphere of radius a decays exponentially with time and has the form of Exercise 9 with, in spherical polar coordinates, $W_1 = 0$ and $W_2 = f(1/R)\mathrm{e}^{\kappa(ct-R)} \cos\theta$ where $f(u)$ is a polynomial in u. Prove that the field falls to $1/\mathrm{e}$ of its original value in time $2a/c$. ($c = 1/(\mu\varepsilon)^{\frac{1}{2}}$).

11. Show that $\boldsymbol{E} = \operatorname{grad}(\boldsymbol{k} \cdot \operatorname{grad} W) - \dfrac{1}{c^2}\dfrac{\partial^2 W}{\partial t^2}\boldsymbol{k}$, $\boldsymbol{B} = \dfrac{1}{c^2} \operatorname{curl}\left(\dfrac{\partial W}{\partial t}\boldsymbol{k}\right)$ is a possible electromagnetic field in free space when $\boldsymbol{k}$ is a constant vector, $c = (\mu_0\varepsilon_0)^{-\frac{1}{2}}$, and $\nabla^2 W = \dfrac{1}{c^2}\dfrac{\partial^2 W}{\partial t^2}$.

12. If $A_x = A_y = 0$ and $A_z = f(r)\mathrm{e}^{\mathrm{i}(\omega t - kz)}$ with $r = (x^2 + y^2)^{\frac{1}{2}}$ find $\boldsymbol{E}$ and $\boldsymbol{H}$ in cylindrical polar coordinates.

13. A harmonic electromagnetic field in a conductor is such that $\sigma/\varepsilon\omega \gg 1$. Show that

$$\boldsymbol{E} = -\mathrm{i}\omega \mathbf{A}, \qquad \boldsymbol{B} = \operatorname{curl} \mathbf{A}, \qquad \boldsymbol{J} = -\mathrm{i}\omega\sigma\mathbf{A},$$
$$\nabla^2 \mathbf{A} - \mathrm{i}\omega\mu\sigma\mathbf{A} = \mathbf{0}, \qquad \operatorname{div} \mathbf{A} = 0$$

approximately.

14. If the right-hand sides of (35.1) and (35.2) are replaced by $\rho_1(\boldsymbol{x})$ and $\rho_2(\boldsymbol{x})$ respectively deduce from (35.4) the reciprocity theorem

$$\int_{T_1} \rho_1(\boldsymbol{x}_1) f_2(\boldsymbol{x}_1)\, \mathrm{d}\boldsymbol{x}_1 = \int_{T_2} \rho_2(\boldsymbol{x}_2) f_1(\boldsymbol{x}_2)\, \mathrm{d}\boldsymbol{x}_2$$

by writing $\rho_1(\boldsymbol{x}) = \int \rho_1(\boldsymbol{x}_1)\delta(\boldsymbol{x} - \boldsymbol{x}_1)\, \mathrm{d}\boldsymbol{x}_1$. Here ρ_1, ρ_2 are zero outside the volumes T_1, T_2 respectively and T_1, T_2 have no points in common.

15. Show that, if

$$\operatorname{curl} \boldsymbol{E}_1 + \mathrm{i}\omega\boldsymbol{B}_1 = \boldsymbol{J}_1', \qquad \operatorname{curl} \boldsymbol{H}_1 - \mathrm{i}\omega\boldsymbol{D}_1 = \boldsymbol{J}_1,$$
$$\operatorname{curl} \boldsymbol{E}_2 + \mathrm{i}\omega\boldsymbol{B}_2 = \boldsymbol{J}_2', \qquad \operatorname{curl} \boldsymbol{H}_2 - \mathrm{i}\omega\boldsymbol{D}_2 = \boldsymbol{J}_2$$

and the conditions are similar to those under which (35.9) is valid, a reciprocity theorem holds, namely

$$\int_{T_1} \{\boldsymbol{J}_1'(\boldsymbol{x}_1)\cdot\boldsymbol{H}_2(\boldsymbol{x}_1)+\boldsymbol{J}_1(\boldsymbol{x}_1)\cdot\boldsymbol{E}_2(\boldsymbol{x}_1)\}\,\mathrm{d}\boldsymbol{x}_1$$

$$=\int_{T_2} \{\boldsymbol{J}_2'(\boldsymbol{x}_2)\cdot\boldsymbol{H}_1(\boldsymbol{x}_2)+\boldsymbol{J}_2(\boldsymbol{x}_2)\cdot\boldsymbol{E}_1(\boldsymbol{x}_2)\}\,\mathrm{d}\boldsymbol{x}_2$$

where $\boldsymbol{J}_1'$, $\boldsymbol{J}_1$ and $\boldsymbol{J}_2'$, $\boldsymbol{J}_2$ vanish outside T_1 and T_2 respectively.

16. Consider formulating reciprocity theorems for (a) the scalar and vector potentials and (b) the Hertz vector.

17. If

$$\operatorname{div}(a \operatorname{grad} f_1)-\frac{\partial^2 f_1}{\partial t^2}=\rho_1(\boldsymbol{x}, t), \qquad \operatorname{div}(a \operatorname{grad} f_2)-\frac{\partial^2 f_2}{\partial t^2}=\rho_2(\boldsymbol{x}, t)$$

and f_1, f_2 die off sufficiently rapidly as $t\to\pm\infty$ show that

$$\int_{-\infty}^{\infty}\int_{T_1} \rho_1(\boldsymbol{x}_1, t)f_2(\boldsymbol{x}_1, t)\,\mathrm{d}\boldsymbol{x}_1\,\mathrm{d}t=\int_{-\infty}^{\infty}\int_{T_2} \rho_2(\boldsymbol{x}_2, t)f_1(\boldsymbol{x}_2, t)\,\mathrm{d}\boldsymbol{x}_2\,\mathrm{d}t_1$$

Form the analogous reciprocity theorem for the electromagnetic field.

18. Verify the formulae for $\nabla^2 p$ for the various coordinate systems of §1.39.

19. The *toroidal coordinates* (ξ, η, ζ) are related to the Cartesian coordinates (x, y, z) by

$$x=\frac{l \sinh \xi \cos \zeta}{\cosh \xi-\cos \eta}, \qquad y=\frac{l \sinh \xi \sin \zeta}{\cosh \xi-\cos \eta}, \qquad z=\frac{l \sin \eta}{\cosh \xi-\cos \eta}.$$

Decide what the surfaces $\xi=$ constant, $\eta=$ constant, and $\zeta=$ constant are, and prove that $h_1=h_2=l(\cosh \xi-\cos \eta)^{-1}$ and $h_3=l \sinh \xi(\cosh \xi-\cos \eta)^{-1}$. Deduce that

$$\nabla^2 p=\frac{(\cosh \xi-\cos \eta)^3}{l^2 \sinh \xi}\left[\frac{\partial}{\partial \xi}\left(\frac{\sinh \xi}{\cosh \xi-\cos \eta}\frac{\partial p}{\partial \xi}\right)+\frac{\partial}{\partial \eta}\left(\frac{\sinh \xi}{\cosh \xi-\cos \eta}\frac{\partial p}{\partial \eta}\right)\right]$$
$$+\frac{(\cosh \xi-\cos \eta)^2}{l^2 \sinh^2 \xi}\frac{\partial^2 p}{\partial \zeta^2}.$$

20. *Bipolar coordinates* are defined by

$$x=\frac{l \sin \eta \cos \zeta}{\cosh \xi-\cos \eta}, \qquad y=\frac{l \sin \eta \sin \zeta}{\cosh \xi-\cos \eta}, \qquad z=\frac{l \sinh \xi}{\cosh \xi-\cos \eta}.$$

Prove that $h_1=h_2=l(\cosh \xi-\cos \eta)^{-1}$ and $h_3=l \sin \eta(\cosh \xi-\cos \eta)^{-1}$.

21. Prove that orthogonal curvilinear coordinates can be defined via

$$x+\mathrm{i}y=f(\xi+\mathrm{i}\eta), \qquad z=\zeta$$

and check that

$$ds^2 = |f'(\xi + i\eta)|^2 (d\xi^2 + d\eta^2) + d\zeta^2.$$

Verify that elliptic cylinder coordinates are obtained by taking $f(W) = \cosh W$. What coordinates does $f(W) = \frac{1}{2}W^2$ give?

22. If ξ, η, z are parabolic cylinder coordinates find all electromagnetic fields in a linear homogeneous isotropic medium of the form $E_x = E_y = 0$, $E_z = f(\xi)e^{i\omega t - \frac{1}{2}ik\eta^2}$.

23. Two-dimensional orthogonal coordinates ξ, ζ in the (x, z)-plane are rotated about the z-axis so as to produce the three-dimensional system ξ, ϕ, ζ with axial symmetry in which

$$ds^2 = h_1^2\, d\xi^2 + r^2\, d\phi^2 + h_3^2\, d\zeta^2$$

where r is the perpendicular distance to the z-axis. A harmonic electromagnetic field in a medium with constant μ, ε, σ is independent of ϕ. Show that the field can be split into two independent parts

$$H_\phi = V_1/r, \qquad H_\xi = H_\zeta = 0, \qquad \mathbf{E} = \{i/(\omega\varepsilon - i\sigma)r\}\mathbf{i}_2 \wedge \text{grad } V_1,$$
$$E_\phi = V_2/r, \qquad E_\xi = E_\zeta = 0, \qquad \mathbf{H} = (1/i\omega\mu r)\mathbf{i}_2 \wedge \text{grad } V_2$$

where $\mathbf{i}_2$ has the same significance as in §1.39 and both V_1, V_2 satisfy

$$\frac{\partial}{\partial \xi}\left(\frac{h_3}{rh_1}\frac{\partial V}{\partial \xi}\right) + \frac{\partial}{\partial \zeta}\left(\frac{h_1}{rh_3}\frac{\partial V}{\partial \zeta}\right) + \omega\mu(\omega\varepsilon - i\sigma)h_1 h_3 \frac{V}{r} = 0.$$

24. Prove that

(i) $d\{z^2 J_{\nu+1}(z) J_{\nu-1}(z)\}/dz = 2z^2 J_\nu(z) J_\nu'(z)$,

(ii) $\dfrac{d}{dz}\dfrac{J_\nu(z^{\frac{1}{2}})}{z^{\frac{1}{2}\nu}} = -\dfrac{1}{2}\dfrac{J_{\nu+1}(z^{\frac{1}{2}})}{z^{\frac{1}{2}\nu+\frac{1}{2}}}$,

(iii) $\displaystyle\int_0^1 u^3 J_0(u)\, du = J_1(1) - 2J_2(1)$.

25. The equation $\nabla^2 p + k^2 p = 0$ is expressed in terms of paraboloidal coordinates ξ, η, ϕ and a solution of the form $p = U(\xi)V(\eta)\Phi(\phi)$ is sought. Show that U and V must satisfy

$$\frac{1}{\xi}\frac{d}{d\xi}\left(\xi\frac{dU}{d\xi}\right) + \left(k^2\xi^2 - ik\nu - \frac{\mu^2}{\xi^2}\right)U = 0,$$

$$\frac{1}{\eta}\frac{d}{d\eta}\left(\eta\frac{dV}{d\eta}\right) + \left(k^2\eta^2 - ik\nu - \frac{\mu^2}{\eta^2}\right)V = 0$$

where μ and ν are separate constants. By putting $k\xi^2 = iz$, $U =$

$(iz)^{\frac{1}{2}\mu}e^{-\frac{1}{2}z}w$ show that w satisfies *the confluent hypergeometric equation*

$$z\frac{d^2w}{dz^2}+(c-z)\frac{dw}{dz}-aw=0$$

where $c=1+\mu$, $a=\frac{1}{2}+\frac{1}{2}\mu-\frac{1}{4}\nu$.

Verify that a solution of this differential equations is $w={}_1F_1(a;c;z)$ where

$${}_1F_1(a;c;z)=1+\frac{a}{c}\frac{z}{1!}+\frac{a(a+1)}{c(c+1)}\frac{z^2}{2!}+\ldots,$$

the series converging for all finite z so long as c is not zero or a negative integer. By considering the substitution $z=-v$, $w=e^zW$ prove *Kummer's transformation*

$${}_1F_1(a;c;z)=e^z\,{}_1F_1(c-a;c;-z).$$

26. The *Laguerre polynomial* $L_n^{\mu}(z)$ (written $L_n(z)$ when $\mu=0$) is defined by

$$L_n^{\mu}(z)=\frac{(\mu+n)!}{n!\mu!}\,{}_1F_1(-n;\mu+1;z)$$

where ${}_1F_1$ is specified in Exercise 25. Verify that

$$L_n^{\mu}(z)=\frac{e^z}{n!z^{\mu}}\frac{d^n}{dz^n}(e^{-z}z^{\mu+n})$$

and that

$$z\frac{d}{dz}L_n^{\mu}(z)=-zL_{n-1}^{\mu+1}(z)=nL_n^{\mu}(z)-(n+\mu)L_{n-1}^{\mu}(z).$$

27. Verify that

$$\psi(a,c;z)=\frac{1}{(a-1)!}\int_0^{\infty e^{i\alpha}}e^{-zt}t^{a-1}(1+t)^{c-a-1}\,dt$$

is a solution of the confluent hypergeometric equation in Exercise 25 if $\operatorname{Re} a>0$, $|\alpha|<\pi$, $|\alpha+\operatorname{ph} z|<\frac{1}{2}\pi$, and t^{a-1}, $(1+t)^{c-a-1}$ have their principal values. The restriction on a can be removed by means of the contour integral representation

$$\psi(a,c;z)=\frac{(-a)!e^{-a\pi i}}{2\pi i}\int_{\infty e^{i\alpha}}^{(0+)}e^{-zt}t^{a-1}(1+t)^{c-a-1}\,dt$$

with $\operatorname{ph} t=\alpha$ at the beginning of the loop. If $\psi_1=\psi(a,c;z)$ and $\psi_2=e^z\psi(c-a,c;-z)$ the Wronskian

$$\psi_1\frac{d\psi_2}{dz}-\psi_2\frac{d\psi_1}{dz}=z^{-c}e^{\pm i\pi(c-a)+z},$$

the upper or lower sign being used according as Im $z>0$ or <0, shows that ψ_1 and ψ_2 are always linearly independent solutions.

Prove that

$$\psi(a, c; z)=\frac{(-c)!}{(a-c)!}\,{}_1F_1(a; c; z)+\frac{(c-2)!}{(a-1)!}z^{1-c}{}_1F_1(a-c+1; 2-c; z)$$
$$=z^{1-c}\psi(a-c+1, 2-c; z).$$

28. Prove that

$$\psi(1+n, 1; z)=\frac{1}{n!}\int_0^\infty \frac{L_n(t)}{t+z}e^{-t}\,dt$$

for $|\text{ph } z|<\pi$, ψ and L_n being defined in Exercises 27 and 26 respectively.

29. Show, from Exercises 25, 27, and Appendixes A, D, that

$$D_\nu(z)=2^{\frac{1}{2}\nu}e^{-\frac{1}{4}z^2}\psi(-\tfrac{1}{2}\nu, \tfrac{1}{2}; \tfrac{1}{2}z^2),$$

$$\lim_{a\to\infty} {}_1F_1(a; c; -z/a)=(c-1)!z^{\frac{1}{2}-\frac{1}{2}c}J_{c-1}(2z^{\frac{1}{2}}),$$

$$\lim_{a\to\infty}(a-c)!\psi(a, c; -z/a)=i\pi e^{-i\pi c}z^{\frac{1}{2}-\frac{1}{2}c}H^{(2)}_{c-1}(2z^{\frac{1}{2}}) \quad (\text{Im } z<0)$$
$$=-i\pi e^{i\pi c}z^{\frac{1}{2}-\frac{1}{2}c}H^{(1)}_{c-1}(2z^{\frac{1}{2}}) \quad (\text{Im } z>0).$$

30. By comparing coefficients of z^n on both sides show that

$$\sum_{n=0}^{\infty}\frac{L_n^\mu(x)z^n}{(\mu+n)!}=\frac{e^zJ_\mu\{2(xz)^{\frac{1}{2}}\}}{(xz)^{\frac{1}{2}\mu}},$$

the Laguerre polynomial being defined in Exercise 26.

31. By deforming the contour over the poles $\nu=2\mu-2+4l+4p$ show that

$$\frac{1}{2\pi i}\int_{-i\infty}^{i\infty}(\tfrac{1}{2}\mu-\tfrac{1}{2}+\tfrac{1}{4}\nu+k)!(\tfrac{1}{2}\mu-\tfrac{1}{2}-\tfrac{1}{4}\nu+l)!(\tan\tfrac{1}{2}\theta)^{\frac{1}{2}\nu}\,d\nu$$
$$=(\mu+k+l)!4(\sin\tfrac{1}{2}\theta)^{\mu+2l+1}(\cos\tfrac{1}{2}\theta)^{\mu+2k+1}$$

where k and l are non-negative integers, Re $\mu>-1$ and $0<\theta\leqslant\frac{1}{2}\pi$. Deduce that, in fact, the result holds for $0<\theta<\pi$. Hence show that

$$\frac{1}{2\pi i}\int_{-i\infty}^{i\infty}(\tan\tfrac{1}{2}\theta)^{\frac{1}{2}\nu}(uv)^{\frac{1}{2}m}e^{-\frac{1}{2}i(u+v)}(\tfrac{1}{2}m-\tfrac{1}{2}+\tfrac{1}{4}\nu)!(\tfrac{1}{2}m-\tfrac{1}{2}-\tfrac{1}{4}\nu)!$$
$$\times{}_1F_1(\tfrac{1}{2}m+\tfrac{1}{2}-\tfrac{1}{4}\nu; m+1; iu){}_1F_1(\tfrac{1}{2}m+\tfrac{1}{2}+\tfrac{1}{4}\nu; m+1; iv)\,d\nu/m!^2$$
$$=4\sum_{k=0}^{\infty}\frac{u^{\frac{1}{2}m}i^k}{(m+k)!}v^{\frac{1}{2}m+k}e^{-\frac{1}{2}i(v+u\cos\theta)}(\cos\tfrac{1}{2}\theta)^{m+2k+1}$$
$$\times(\sin\tfrac{1}{2}\theta)^{m+1}L_k^m(-iu\sin^2\tfrac{1}{2}\theta)$$
$$=2\sin\theta\, e^{-\frac{1}{2}i(u-v)\cos\theta}J_m\{(uv)^{\frac{1}{2}}\sin\theta\}$$

from Exercise 30.

Deduce that, if x, z are expressed in paraboloidal coordinates so that $x=\frac{1}{2}(\xi^2-\eta^2)$, $z=\xi\eta\sin\phi$ and $u=\kappa\xi^2$, $v=\kappa\eta^2$

$$
\begin{aligned}
e^{-i\kappa(x\cos\theta+z\sin\theta)} &= \frac{1}{4\pi i\sin\theta}\int_{-i\infty}^{i\infty}(\tan\tfrac{1}{2}\theta)^{\frac{1}{2}\nu}e^{-\frac{1}{2}i(u+v)} \\
&\quad\times\{(\tfrac{1}{4}\nu-\tfrac{1}{2})!(-\tfrac{1}{2}-\tfrac{1}{4}\nu)!\,{}_1F_1(\tfrac{1}{2}-\tfrac{1}{4}\nu;1;iu)\,{}_1F_1(\tfrac{1}{2}+\tfrac{1}{4}\nu;1;iv) \\
&\quad+2\sum_{m=1}^{\infty}(-i)^m(\tfrac{1}{2}m-\tfrac{1}{2}+\tfrac{1}{4}\nu)!(\tfrac{1}{2}m-\tfrac{1}{2}-\tfrac{1}{4}\nu)!(uv)^{\frac{1}{2}m} \\
&\quad\times{}_1F_1(\tfrac{1}{2}m+\tfrac{1}{2}-\tfrac{1}{4}\nu;m+1;iu) \\
&\quad\times{}_1F_1(\tfrac{1}{2}m+\tfrac{1}{2}+\tfrac{1}{4}\nu;m+1;iv)\cos m(\tfrac{1}{2}\pi-\phi)/m!^2\}\,d\nu
\end{aligned}
$$

by using (A.1.23).

A related formula is

$$
\begin{aligned}
\frac{e^{i\kappa R}}{\kappa R}=\frac{e^{\frac{1}{2}i(u+v)}}{\frac{1}{2}(u+v)} &= -\int_{-i\infty}^{i\infty}e^{\frac{1}{2}i(u+v)}\psi(\tfrac{1}{2}+\nu,1;-iu)\psi(\tfrac{1}{2}-\nu,1;-iv)\frac{d\nu}{\cos\nu\pi} \\
&= 2i\sum_{m=0}^{\infty}m!\,e^{\frac{1}{2}i(u+v)}L_m(-iu)\psi(1+m,1;-iv)
\end{aligned}
$$

for $u<v$.

2. The special theory of relativity

No attempt was made in Chapter 1 to specify the frame of reference within which electromagnetic phenomena are governed by Maxwell's equations. Strong efforts were made at the end of the nineteenth century to resolve this matter; the investigations culminated in the theory of relativity which will be described in this chapter. The theory, which rests on experiments about to be delineated, requires results from the tensor calculus. An account of this calculus giving the principal properties needed will be found in Appendix F.

The Lorentz transformation

2.1 The Michelson–Morley experiment

The origins of the theory of relativity are old and lie in the theory of mechanics. One of the fundamental laws of mechanics, as stated by Newton, may be expressed for a particle essentially as

$$\boldsymbol{F} = m\,\mathrm{d}^2\boldsymbol{x}/\mathrm{d}t^2$$

where $\boldsymbol{F}$ is the force acting on the particle, m its mass, and $\boldsymbol{x}$ its position vector with respect to a fixed origin at time t. In other words, measurements are made in a fixed system of coordinates. This system, the *absolute system*, was regarded by Newton and his followers as having a permanent position in space by reference to fixed stars, for example. However, the law involves the acceleration but not the velocity of the particle and so it would hold equally well in any system moving with uniform velocity relative to the absolute system. Consequently, it is impossible to detect uniform motion relative to the absolute system by the observation of purely mechanical phenomena. Thoughts therefore turned to the question of whether uniform motion in the absolute system could be determined by electrical experiments.

Maxwell and his contemporaries assumed that Maxwell's equations were valid only in the absolute system, which could be regarded as a medium (called the ether) pervading all matter and space thus serving as

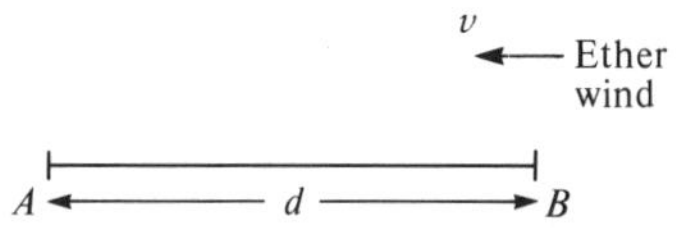

Fig. 2.1 Transmission parallel to the ether wind

the carrier of all electromagnetic and optical phenomena. On this basis experiments can be devised which should determine the velocity of the earth relative to the absolute system.

None of these experiments produced results in conflict with existing theory but neither was any of them sufficiently accurate to evaluate the velocity of the earth until 1881. In that year an experiment was carried out by Michelson which was capable of measuring the velocity with a very small error.

Basically, the idea behind the experiment was to find the difference between light which travelled parallel to the direction of motion of the earth and light which had gone in a perpendicular direction. For calculation of the difference to be expected it is easier to visualize the earth as fixed with the ether flowing over it like a wind than to consider the earth as moving with speed v in a fixed ether. Suppose now that light goes from A to B, two points fixed on the earth (Fig. 2.1). The speed of light in the ether is c and hence, the earth being regarded as fixed, light travels with speed $c-v$ from A to B and with speed $c+v$ from B to A. Therefore, if the distance AB is d, the time taken by light going from A to B and back again is

$$\frac{d}{c-v}+\frac{d}{c+v}=\frac{2cd}{c^2-v^2}$$

Similarly, if AC is perpendicular to the ether wind and of length d (Fig. 2.2), the vector diagram of velocities shows that the time for light to

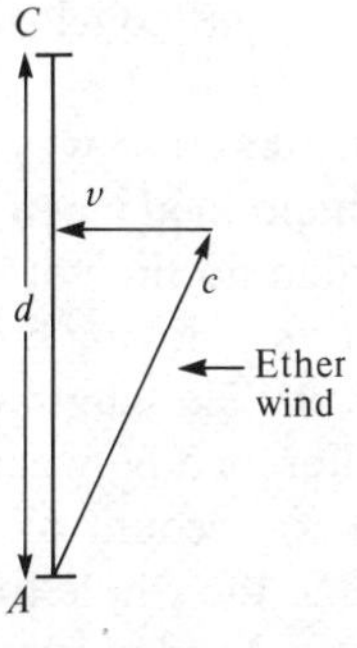

Fig. 2.2 Perpendicular propagation

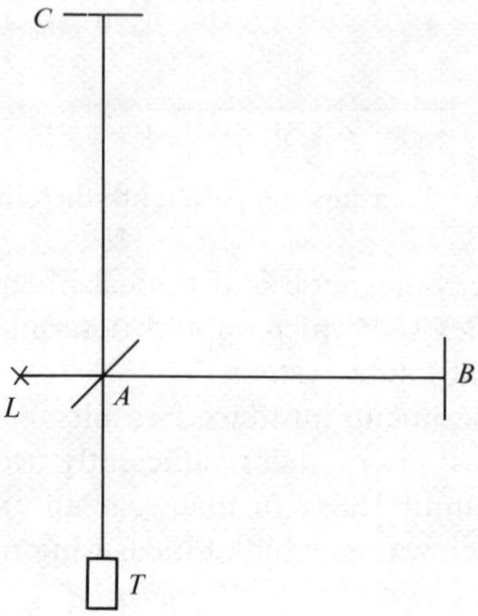

Fig. 2.3 Essence of Michelson–Morley apparatus

travel from A to C and back again is

$$2d/(c^2-v^2)^{\frac{1}{2}}.$$

The difference in time between these two journeys is

$$2d(\beta-1)/(c^2-v^2)^{\frac{1}{2}}$$

where $\beta=(1-v^2/c^2)^{-\frac{1}{2}}$. The time difference was measured by Michelson in the apparatus shown in Fig. 2.3. A light beam from a source L is divided by a semi-reflecting glass plate A into two beams perpendicular to each other. One of these is reflected by a mirror at B back to A, where it is again reflected to the telescope T. The second beam is reflected by a mirror at C and passes through A to T.

In spite of the foregoing theory no time difference between the two paths could be detected in the experiment. On rejecting the conjecture that $v=0$, which would lead to considerable complication in the theory of mechanics, we are forced to the conclusion that the hypothesis of the ether is unsound.

Michelson suggested that the ether was carried round by the earth but this suggestion had to be abandoned when a repetition of the experiments on top of a mountain revealed no time difference. Lorentz and Fitzgerald put forward the notion that any moving body is contracted in its direction of motion by an amount $1/\beta$. The length of AB would then be d/β and there would be no time difference between the two paths. However, this hypothesis was insufficient to account for other experiments and it became necessary to re-examine the whole foundation of the ether theory.

A version of the Michelson–Morley experiment occurs daily now in the provision of International Atomic Time by 7 atomic clocks distributed around the globe. The synchronization of these clocks is checked by a continual exchange of radio signals. If these signals were subjected to an ether wind of the magnitude foreseen the effect would be detected by the clocks.

2.2 The Lorentz transformation

The failure to find any influence in mechanical and electromagnetic phenomena due to the motion of the earth in the ether suggested to Poincaré the hypothesis that physical events are independent of this motion. He also conjectured that no velocity could exceed that of light. Building on this base Einstein, after a radical re-examination of concepts, put forward what now constitutes *the special theory of relativity*. According to this theory (i) the laws of physics, when properly formulated, are the same in all systems which move with uniform velocity relative to one another and, (ii) the speed of light in free space is c whatever the relative uniform motion of source and observer.

The consequences of these postulates are surprising. Consider a source of light moving with uniform velocity which emits a light signal as it passes the point E and suppose that the source is at E' some time later (Fig. 2.4). An observer who stays at E will say that, according to hypothesis (ii), the light signal travels in all directions with speed c being at any time on a sphere with E as centre and, in particular, that it will reach the points A and D at the same time. But an observer, moving with the source, will say that (i) and (ii) imply that the light signal travels with speed c being at any time on a sphere with the source as centre; in particular, he will assert that the signal reaches B and D at the same time. These statements, which refer to the same signal, can be reconciled only if the two observers mean something different by the words 'at the same time'. Thus it becomes necessary to study and specify how time is to be measured by an observer.

Imagine that there is a large number of standard clocks which show the same rate when placed at rest at the same place. These clocks are distributed throughout a given coordinate system S and are synchronized

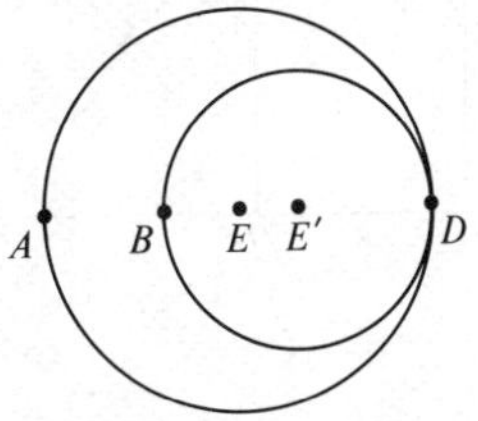

Fig. 2.4 The signals seen by different observers

by sending light signals from one clock O as regulating centre. O sends out a signal at time t_0. On receipt of this signal the clock at the point P is set at $t_0 + l/c$ where l is the distance from O to P measured by *measuring sticks at rest in S.* Events at P and Q are then said to be simultaneous if the clocks at P and Q show the same time when the events occur.

Let S' be another coordinate system moving relative to S. Distribute standard clocks at rest in S' and select a regulating centre O' in S'. The clocks in S' are synchronized from O' in the same fashion as those in S, but distances are measured by *measuring sticks at rest in S'*. The rule for comparing time by different observers is now: an event occurring at a given point is said to happen at time t relative to S and a time t' relative to S' if the clocks in S and S' show the times t and t' respectively when the event takes place.

The relation which connects the times in S and S' has now to be determined. Let $Oxyz$, $O'x'y'z'$ be Cartesian coordinate systems in S and S' respectively, chosen so that S' is moving with uniform speed v relative to S along both Ox and $O'x'$ (Fig. 2.5). An event can be specified as to the place and time that it occurs by the space-time coordinates (x, y, z, t) in S or (x', y', z', t') in S'. Since any uniform motion relative to S along the x-axis is also uniform with respect to S', the coordinates (x', y', z', t') must be linear functions of (x, y, z, t).

Assume that $t=0$ and $t'=0$ when O and O' coincide; then subsequently the position of O' in S is given by $x = vt$ whereas O occupies $x' = -vt'$ in S'.

All the points of S' on a plane $y' = d'$ must form a plane $y = d$ in S. Since distances are measured in S and S' by sticks which are in motion relative to one another the ratio $\alpha = d'/d$ may not be 1. However, if we change the signs of the x-, z-, x'-, and z'-axes neither d nor d' is altered but S is now moving with speed v relative to S' along the positive x-axis. Hence $\alpha = d/d'$ and the two formulae for α imply that $\alpha^2 = 1$. For d and d' to have the same sign we require $\alpha = 1$ and consequently

$$y = y'. \tag{2.1}$$

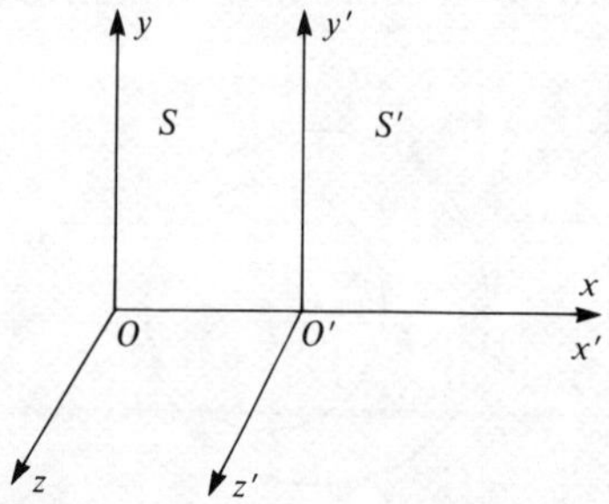

Fig. 2.5 Observers in relative motion

Similarly, it can be proved that

$$z = z'. \tag{2.2}$$

The connection between the other coordinates is fixed by the postulate (ii). If a light signal is emitted from the origin when O and O' coincide, then an observer in S says that its subsequent position lies on

$$x^2 + y^2 + z^2 - c^2t^2 = 0 \tag{2.3}$$

whereas an observer at rest in S' says it is on

$$x'^2 + y'^2 + z'^2 - c^2t'^2 = 0. \tag{2.4}$$

Hence when the left-hand side of (2.3) vanishes so does the left-hand side of (2.4). Since the relation between the coordinates has already been established as linear this is possible only if

$$x'^2 + y'^2 + z'^2 - c^2t'^2 = f(v)(x^2 + y^2 + z^2 - c^2t^2)$$

where $f(v)$ depends solely on v. An argument on the same lines as that used to demonstrate (2.1) shows that $f(v) = 1$. Hence, after invocation of (2.1) and (2.2),

$$x'^2 - c^2t'^2 = x^2 - c^2t^2. \tag{2.5}$$

The fact that x' and t' are linear functions of x and t enables us to write

$$x' = ax + bt, \tag{2.6}$$

$$t' = dx + et \tag{2.7}$$

where a, b, d, and e are constants. Now O' is the point $x' = 0$ or, equally well, $x = vt$ and therefore

$$b/a = -v. \tag{2.8}$$

Also O is $x = 0$ or $x' = -vt'$ and so

$$b/e = -v \tag{2.9}$$

The replacement of x' and t' in (2.5) by means of (2.6)–(2.9) leads to

$$a^2(x - vt)^2 - c^2(dx + at)^2 = x^2 - c^2t^2$$

for all x and t. This forces

$$a^2 - c^2d^2 = 1,$$
$$-2a^2v - 2c^2da = 0,$$
$$a^2v^2 - c^2a^2 = -c^2.$$

The third equation supplies

$$a = (1 - v^2/c^2)^{-\frac{1}{2}} = \beta \tag{2.10}$$

on specifying a to be positive or, alternatively, $x' = x$ when $v = 0$. From the second equation it follows that $d = -\beta v/c^2$ and then the first equation is satisfied automatically.

Incorporating these constants in (2.6) and (2.7) we obtain

$$x' = \beta(x - vt), \qquad y' = y, \tag{2.11}$$

$$t' = \beta(t - vx/c^2), \qquad z' = z. \tag{2.12}$$

These equations may be solved for x and t to furnish

$$x = \beta(x' + vt'), \qquad y = y', \tag{2.13}$$

$$t = \beta(t' + vx'/c^2), \qquad z = z'. \tag{2.14}$$

They may be derived from (2.11), (2.12) by replacing v by $-v$ and by interchanging primed and unprimed coordinates. This is merely recognition of the fact that S is moving relative to S' with speed v along the negative x'-axis.

The transformation described by (2.11), (2.12) was first introduced by Lorentz and so it is designated the *Lorentz transformation.*

When $v \ll c$, $\beta \approx 1$ and the Lorentz transformation degenerates to

$$x' = x - vt, \qquad y' = y,$$

$$t' = t, \qquad z' = z$$

approximately. This is known as a *Galilean transformation* because it corresponds to the relativity theory of Newtonian mechanics in which time is regarded as absolute.

2.3 The Lorentz–Fitzgerald contraction

The hypothesis formulated by Lorentz and Fitzgerald to explain the Michelson–Morley experiment has already been mentioned at the end of §2.1. It will now be examined from the point of view of the theory of relativity.

Let a rod at rest in S' lie along the x'-axis with its ends at x_1' and x_2'. Then its length l' in S' is given by $l' = x_2' - x_1'$. The observer in S must measure with sticks at rest in S, i.e. he notes the ends of the rods at a particular time t. If he finds the ends to be at x_1 and x_2 at this time he takes the length to be $l = x_2 - x_1$. Now, from (2.11),

$$x_2' = \beta(x_2 - vt), \qquad x_1' = \beta(x_1 - vt)$$

and hence, by subtraction,

$$l' = \beta l.$$

Thus to an observer a rod which has length l' when at rest relative to him appears, when moving with uniform speed v along its length, to have contracted to the length l'/β. On the other hand, a rod moving perpendicular to its length would be unchanged by virtue of (2.11), (2.12). Consequently, the Lorentz–Fitzgerald hypothesis is an automatic concomitant of the Lorentz transformation.

As a result of the Lorentz–Fitzgerald contraction a sphere becomes an oblate spheroid when moving and this phenomenon led Lorentz to his theory of the deformable electron.

2.4 The clock paradox

Consider a standard clock which is at rest in S' at a point on the x'-axis with x'-coordinate x'_1. At time $t' = t'_1$ the corresponding time measured in S is t_1 where, from (2.14),

$$t_1 = \beta(t'_1 + vx'_1/c^2)$$

and likewise, to $t' = t'_2$ corresponds t_2 where

$$t_2 = \beta(t'_2 + vx'_1/c^2).$$

Thus, if an event at a fixed point of S' lasts for a time interval T' and the corresponding time which is observed to elapse in S is T,

$$T = t_2 - t_1 = \beta(t'_2 - t'_1) = \beta T'. \tag{4.1}$$

Hence a clock which is travelling with speed v relative to S will be slow compared with the clocks in S.

One application of this result is in radioactive processes. The mean life of a moving radioactive substance will be longer than when it is at rest. For example, suppose that a particle in a cosmic ray has a mean life of about 2×10^{-6} sec when at rest. In the cosmic ray it travels with a speed close to c and has a range of about 20 km; so the mean life in motion is $20 \times 10^3/3 \times 10^8$ sec $\approx 7 \times 10^{-5}$ sec. Putting the two lifetimes in (4.1) we obtain $\beta \approx 35$ and infer that the speed of the particle is about $(1 - 1/1200)c$.

That a moving clock goes slow compared with a stationary clock seems to give rise to a paradox which will now be enunciated. Two clocks C and C' are at the origin at time $t = 0$ at which moment C' is set in uniform motion with speed v along the x-axis. As soon as it arrives at the point $x = x_0$ it is returned to the origin at the same speed v and it is assumed that the reversal of motion of C' has no effect on the clock. (This assumption can be avoided by imagining a clock C'' travelling with speed v in the opposite direction to C' which passes C' at x_0 showing the same time.) According to (4.1) C' (or C'' in the alternative model) will be slow compared with C when it passes the origin. However, for an observer who stays with C', the motion of C relative to C' seems to be analogous to that of C' relative to C and therefore he would expect C to be slow compared with C'. These two assertions make up the paradox which has to be explained.

The resolution of the paradox comes from the realization that the motion of C relative to C' is not analogous to that of C' relative to C. For in the motion of C' relative to C, an observer at C remarks that C' will arrive back after a time $t = 2x_0/v$ but an observer with C' says 'I must travel the contracted distance $2x_0/\beta$ so that I shall return after a time $t' = 2x_0/\beta v$', i.e. C' will be slow by $(2x_0/v)(1-1/\beta)$. In the motion of C relative to C', C says 'I must travel a distance $2x_0$ because that is the distance measured in my system of coordinates' and C' notes that C covers a distance $2x_0/\beta$. The times measured are, therefore, exactly the same either way and the paradox disappears.

Relativistic mechanics

2.5 The transformation of velocity

The space-time coordinates of a particle depend upon the frame of reference chosen and so the velocity of a particle will also vary with the choice. The velocity $\boldsymbol{u}$ in S is defined by

$$\boldsymbol{u} \equiv (u_x, u_y, u_z) = (\mathrm{d}x/\mathrm{d}t, \mathrm{d}y/\mathrm{d}t, \mathrm{d}z/\mathrm{d}t)$$

and the velocity $\boldsymbol{u}'$ in S' by

$$\boldsymbol{u}' \equiv (u'_x, u'_y, u'_z) = (\mathrm{d}x'/\mathrm{d}t', \mathrm{d}y'/\mathrm{d}t', \mathrm{d}z'/\mathrm{d}t').$$

On account of the Lorentz transformation relating S' to S

$$\mathrm{d}x' = \beta(\mathrm{d}x - v\,\mathrm{d}t), \qquad \mathrm{d}y' = \mathrm{d}y,$$
$$\mathrm{d}t' = \beta(\mathrm{d}t - v\,\mathrm{d}x/c^2), \qquad \mathrm{d}z' = \mathrm{d}z.$$

Hence

$$u'_x = \frac{u_x - v}{1 - vu_x/c^2}, \qquad u'_y = \frac{u_y}{\beta(1 - vu_x/c^2)}, \qquad u'_z = \frac{u_z}{\beta(1 - vu_x/c^2)}. \tag{5.1}$$

The equations (5.1) supply the conversion of the velocity of a particle from one coordinate systems to another. If the particle is moving along the x-axis $u_y = u_z = u'_y = u'_z = 0$ and, if the suffix x is dropped from u_x,

$$u' = \frac{u - v}{1 - vu/c^2}, \qquad u = \frac{u' + v}{1 + vu'/c^2} \tag{5.2}$$

which is the relativistic addition theorem for velocities in a straight line.

Sometimes (5.2) is expressed in terms of the *rapidity* $w = \tanh^{-1}(v/c)$. If $q = \tanh^{-1}(u/c)$, (5.2) states that

$$q' = q + w \tag{5.3}$$

which has a much simpler appearance than (5.2). If the rapidity in S'', which is travelling with speed v_1 relative to S' along the x-axis, is q'' we see that

$$q'' = q' + w_1 = q + w + w_1$$

indicating the additivity of relative rapidity for three observers traversing along a line parallel to the velocity of the particle.

The derivation of the Lorentz transformation in §2.2 makes it clear that no real coordinate system S' can exist for which $v > c$ because then the transformation becomes imaginary and the Lorentz–Fitzgerald contraction does as well. In fact, no particle can move with a speed greater than c relative to S. To see this, take a particle whose speed is c and subject a second particle, moving along the same line with the smaller speed v, to continual acceleration in the hope of making it travel faster than the first. The speed of the first relative to the second is provided by the formula (5.2) for u' with $u = c$ and hence is c. Since this is independent of v, the second particle never attains a higher speed than the first.

That the speed of a particle is always less than c may seem to be contrary to experience in some cases. For example, when a radioactive atom emits two electrons moving at almost the speed of light in opposite directions their relative speed seen from the stationary atom seems to be almost $2c$. However, if S is placed at the atom and S' at one of the electrons, $v = c - \varepsilon$ where ε is a small positive quantity. The second electron as observed in S has $u = -(c - \varepsilon)$ and hence, according to (5.2), u' is approximately $-c$, i.e. the approximate speed of one electron relative to the other is c and not $2c$ as appeared as first sight.

It is not only particles which cannot move at a faster speed than c. *Any signal* which carries information is obliged to travel at a speed which is not greater than c. Processes which do not carry information or energy can exceed the speed c. There are examples in §5.5.

2.6 The variation of mass with speed

The method of §2.5 can be extended so that the accelerations in S and S' can be related. Now acceleration and mass are connected via Newton's laws for a particle whose speed is much lower than the speed of light. Therefore, if acceleration depends on the frame of reference, it is a distinct possibility that mass will also vary with the space–time coordinates in the theory of relativity.

To investigate this possibility start from the customary definition of the linear momentum $\boldsymbol{p}$ of a particle, namely $\boldsymbol{p} = m\boldsymbol{u}$. By virtue of postulate (i) of the theory of relativity, physical laws are to be the same in systems in uniform motion relative to one another. Therefore, if linear momentum is conserved in S, we expect that it will be preserved in any system S'. If the particle's direction of motion is the x-axis this can be ensured by demanding that

$$\frac{\mathrm{d}}{\mathrm{d}t}(mu) = \frac{\mathrm{d}}{\mathrm{d}t'}(m'u') \tag{6.1}$$

in the notation of (5.2). Now introduce the proposition that the mass is a function only of the speed of the particle, i.e.

$$mu = f(u), \qquad m'u' = f(u').$$

Then, (6.1) can be written

$$\frac{\mathrm{d}f(u)}{\mathrm{d}t} = \frac{\mathrm{d}f(u')}{\mathrm{d}u'} \cdot \frac{\mathrm{d}u'}{\mathrm{d}t} \cdot \frac{\mathrm{d}t}{\mathrm{d}t'}$$

whence

$$\frac{\mathrm{d}f(u)}{\mathrm{d}u} = \beta^{-3}\left(1 - \frac{vu}{c^2}\right)^{-3} \frac{\mathrm{d}f(u')}{\mathrm{d}u'} \tag{6.2}$$

since

$$\mathrm{d}u'/\mathrm{d}u = \beta^{-2}(1 - vu/c^2)^{-2}, \qquad \mathrm{d}t'/\mathrm{d}t = (1 - vu/c^2)\beta$$

by (5.2). Equation (6.2) must hold for arbitrary v since any S' may be chosen. Indeed, the left-hand side is independent of v and, therefore, so must be the right-hand side. Pick $v = u$. Then, bearing in mind (5.2), we see that

$$\mathrm{d}f(u)/\mathrm{d}u = (1 - u^2/c^2)^{-\frac{3}{2}} f_0$$

where f_0 is the value of $\mathrm{d}f(u)/\mathrm{d}u$ when $u = 0$. Therefore

$$f(u) = f_0 u (1 - u^2/c^2)^{-\frac{1}{2}}$$

since $f(0) = 0$. Hence

$$m = m_0(1 - u^2/c^2)^{-\frac{1}{2}} \tag{6.3}$$

where m_0 is a constant which is equal to the mass when $u = 0$. For this reason m_0 is called the *rest mass*. The variation of mass with speed in (6.3) does not affect Newtonian mechanics where the speeds of particles are much smaller than that of light. Nevertheless, the variation of the mass of an electron with speed was, in fact, observed before the theory of relativity was propounded; subsequent experiments have confirmed (6.3).

In view of (6.3) the linear momentum of a particle in general motion is given by

$$\boldsymbol{p}=m_0(1-u^2/c^2)^{-\frac{1}{2}}\boldsymbol{u} \tag{6.4}$$

where now $u^2=u_x^2+u_y^2+u_z^2$. It can be shown in Newtonian mechanics that the force $\boldsymbol{F}$ acting on a particle satisfies

$$\boldsymbol{F}=\mathrm{d}\boldsymbol{p}/\mathrm{d}t. \tag{6.5}$$

Equation (6.5) is taken over as the relativistic definition of force. It is assumed to hold in S and, with the symbols primed, in S'. The rule for the connection between force in S and force in S' can be deduced from earlier results but it will not be given here except in the particular case when the particle is at rest in S' when

$$\boldsymbol{F}=\boldsymbol{F}'/\beta+\boldsymbol{v}(1-1/\beta)\boldsymbol{v}\cdot\boldsymbol{F}'/v^2 \tag{6.6}$$

where $\boldsymbol{v}$ is the velocity of the particle in S.

In relativistic mechanics the rate at which work is done by $\boldsymbol{F}$ is defined to be $\boldsymbol{F}\cdot\boldsymbol{u}$ just as in Newtonian mechanics. Also the kinetic energy T is defined so that its rate of change is equal to the rate at which work is done. Hence

$$\begin{aligned}\frac{\mathrm{d}T}{\mathrm{d}t}&=\boldsymbol{F}\cdot\boldsymbol{u}\\ &=m_0(1-u^2/c^2)^{-\frac{1}{2}}\boldsymbol{u}\cdot\mathrm{d}\boldsymbol{u}/\mathrm{d}t+m_0(u^3/c^2)(1-u^2/c^2)^{-\frac{3}{2}}\mathrm{d}u/\mathrm{d}t\end{aligned} \tag{6.7}$$

from (6.4) and (6.5). Since

$$\boldsymbol{u}\cdot\frac{\mathrm{d}\boldsymbol{u}}{\mathrm{d}t}=u_x\frac{\mathrm{d}u_x}{\mathrm{d}t}+u_y\frac{\mathrm{d}u_y}{\mathrm{d}t}+u_z\frac{\mathrm{d}u_z}{\mathrm{d}t}=u\frac{\mathrm{d}u}{\mathrm{d}t}$$

we conclude that

$$\mathrm{d}T/\mathrm{d}t=\mathrm{d}\{m_0c^2(1-u^2/c^2)^{-\frac{1}{2}}\}/\mathrm{d}t$$

whence

$$T=m_0c^2(1-u^2/c^2)^{-\frac{1}{2}}+\text{constant.}$$

In order that the kinetic energy vanish with u the constant must be $-m_0c^2$ and

$$T=(m-m_0)c^2. \tag{6.8}$$

If u is small compared with c an expansion in powers of u/c leads to the approximation $T=\frac{1}{2}m_0u^2$ which is in harmony with the kinetic energy of Newtonian mechanics.

Related to T is the *energy of the particle, E,* defined by $E=T+m_0c^2$ so that $E=mc^2$ from (6.8).

A straightforward calculation shows that, if $\boldsymbol{p}\equiv(p_x, p_y, p_z)$, the four quantities p_x, p_y, p_z, and E/c^2 are related their corresponding values in S'

by the same laws as connect the space–time coordinates x, y, z, t to x', y', z', t'. It follows that

$$p_x^2+p_y^2+p_z^2-E^2/c^2=p_x'^2+p_y'^2+p_z'^2-E'^2/c^2. \tag{6.9}$$

Equation (6.7) can be written as

$$\mathrm{d}E/\mathrm{d}t=\boldsymbol{F}\cdot\boldsymbol{u} \tag{6.10}$$

and in this form it holds in S' with primed letters.

2.7 Conservation of momentum and energy

In the preceding section it was assumed that the conservation of linear momentum is a physical law which takes the same form in S and S'. It will now be assumed that the conservation of energy is also such a physical law. In other words, if after a process such as a collision the changes in linear momentum and energy ($\Delta\boldsymbol{p}$, ΔE) in S are zero then the corresponding changes ($\Delta\boldsymbol{p}'$, $\Delta E'$) in S' are zero as well. Confirmation of these hypotheses is forthcoming from experiments in which electrons collide; these experiments also verify the relativistic formula for the variation of mass.

The equation $E=mc^2$ is a statement of the *inertia of energy* which says that the mass increases by $\Delta E/c^2$ when the increment in energy is ΔE and that an increase of mass Δm is accompanied by a growth in energy of $c^2\Delta m$. Atomic power and the bombardment of lithium by protons provide supporting experimental evidence. When a proton strikes a lithium nucleus two alpha-particles are produced. The masses of a proton, lithium nucleus, and alpha-particle are 1.0076, 7.0166, and 4.0028 respectively on a scale in which the mass of an oxygen atom is 16. Hence the loss of mass in the collision is

$$7.0166+1.0076-2\times4.0028=0.0186$$

which represents an energy of about 27.7×10^{-13} Joules. Experiments reveal that this energy does indeed appear as the kinetic energy of the alpha-particle.

2.8 Proper time

When a particle is moving with speed u in S the *proper time* τ is defined by

$$\mathrm{d}\tau/\mathrm{d}t=(1-u^2/c^2)^{\frac{1}{2}}. \tag{8.1}$$

Similarly, if τ' is the proper time in S'

$$\mathrm{d}\tau'/\mathrm{d}t'=(1-u'^2/c^2)^{\frac{1}{2}}$$

where the components of $\boldsymbol{u}'$ are related to those of $\boldsymbol{u}$ by (5.1). From (5.1)

$$(1-u'^2/c^2)^{\frac{1}{2}} = (1-u^2/c^2)^{\frac{1}{2}}/\beta(1-vu_x/c^2) \tag{8.2}$$

and hence

$$\mathrm{d}\tau'/\mathrm{d}t = \mathrm{d}\tau/\mathrm{d}t$$

i.e. the proper time is the same in S and S'.

If we write $x^1 = x$, $x^2 = y$, $x^3 = z$, $x^4 = ct$ a 4-*velocity* U^μ can be defined by

$$U^\mu = \mathrm{d}x^\mu/\mathrm{d}\tau. \tag{8.3}$$

Clearly

$$(U^1, U^2, U^3) = (u_x, u_y, u_z)(1-u^2/c^2)^{-\frac{1}{2}}, \qquad U^4 = c(1-u^2/c^2)^{-\frac{1}{2}}$$

and, from (5.1) and (8.2),

$$U'^1 = \beta(U^1 - vU^4/c), \qquad U'^2 = U^2,$$
$$U'^4 = \beta(U^4 - vU^1/c), \qquad U'^3 = U^3$$

showing that the components U^μ transform according to the same rules as x^μ in the Lorentz transformation (2.11), (2.12).

A *4-momentum* P^μ can be specified from the 4-velocity by $P^\mu = m_0 U^\mu$ where m_0 is the rest mass. It follows from (6.4) that $(P^1, P^2, P^3) \equiv (p_x, p_y, p_z)$ and $P^4 = E/c$. The assumptions of §§2.6, 2.7 that linear momentum and energy are conserved now imply that *4-momentum is preserved* in an isolated system.

If the 4-momentum of a particle changes it can be ascribed to a *4-force* F^μ via $F^\mu = \mathrm{d}P^\mu/\mathrm{d}\tau$. The relation between the 4-force and the force of (6.5) can be deduced without difficulty.

Electrodynamics in free space

2.9 The invariant form of Maxwell's equations

According to the special theory of relativity the laws of physics are the same in all space–time systems which move with uniform velocity relative to one another. The coordinates in two such systems are connected by a Lorentz transformation and, therefore, any physical law must be put in a form which is unaltered by a transformation of this type. Tensor equations are invariant to changes of coordinates and so, if the laws of physics are written as tensor equations, the special theory of relativity will be complied with. Although it was thought at one time that all physical laws could be expressed in tensor form it was subsequently found necessary, in the quantum theory of the electron, to introduce quantities called spinors which do not satisfy the transformation laws of tensors.

Basic to the application of tensors is the introduction of a suitable set of coordinates and a metric tensor (Appendix F.4). As in §2.8 we work in a four-dimensional space with $x^1 = x$, $x^2 = y$, $x^3 = z$, $x^4 = ct$. In these coordinates the Lorentz transformation can be written as

$$x'^{\alpha} = a^{\alpha}_{\mu} x^{\mu}$$

where $a^1_1 = \beta$, $a^4_1 = -\beta v/c = a^1_4$, $a^2_2 = 1 = a^3_3$, $a^4_4 = \beta$, and all other a^{α}_{μ} are zero, and we use the summation convention (see §F.1 of the appendix).

For the metric tensor $g_{\mu\nu}$ the distance between two neighbouring points $\mathrm{d}s$ is given by

$$\mathrm{d}s^2 = g_{\mu\nu}\,\mathrm{d}x^{\mu}\,\mathrm{d}x^{\nu}.$$

It will now be required that when distances are measured at constant time so that $\mathrm{d}x^4 = 0$ they must be the same as in conventional Cartesian space; this forces

$$g_{11} = 1, \quad g_{22} = 1, \quad g_{33} = 1, \quad g_{12} = 0, \quad g_{13} = 0, \quad g_{23} = 0$$

so that

$$\mathrm{d}s^2 = \mathrm{d}x^2 + \mathrm{d}y^2 + \mathrm{d}z^2 + 2g_{14}c\,\mathrm{d}x\,\mathrm{d}t + 2g_{24}c\,\mathrm{d}y\,\mathrm{d}t + 2g_{34}c\,\mathrm{d}z\,\mathrm{d}t + g_{44}c^2\,\mathrm{d}t^2.$$

After a Lorentz transformation to the coordinates x', y', z', ct' the elementary distance $\mathrm{d}s$ must take the usual Cartesian form when $\mathrm{d}t' = 0$ i.e. there must be no terms containing $\mathrm{d}x'\,\mathrm{d}y'$ and $\mathrm{d}x'\,\mathrm{d}z'$. This entails $g_{24} = 0$, $g_{34} = 0$.

The theory of §2.2 indicates that $(\mathrm{d}x)^2 + (\mathrm{d}y)^2 + (\mathrm{d}z)^2 - c^2(\mathrm{d}t)^2$ is invariant under a Lorentz transformation and hence, by subtraction of this quantity from $\mathrm{d}s^2$, we deduce that

$$2g_{14}c\,\mathrm{d}x\,\mathrm{d}t + (g_{44} + 1)c^2\,\mathrm{d}t^2$$

must be invariant. A Lorentz transformation of this will generate a term in $(\mathrm{d}x')^2$ unless

$$2g_{14} + (g_{44} + 1)v/c = 0.$$

Since this must be true for any Lorentz transformation it must hold for arbitrary v and so $g_{14} = 0$, $g_{44} = -1$ because g_{14} and g_{44} are independent of v. Consequently, the metric has been fully determined and

$$\mathrm{d}s^2 = \mathrm{d}x^2 + \mathrm{d}y^2 + \mathrm{d}z^2 - c^2\,\mathrm{d}t^2 = (\mathrm{d}x^1)^2 + (\mathrm{d}x^2)^2 + (\mathrm{d}x^3)^2 - (\mathrm{d}x^4)^2. \tag{9.1}$$

The metric tensor which has been arrived at for our four-dimensional space is such that the square of the distance between neighbouring points can be negative. The geometry of the system must therefore be different from that of a Euclidean space where the square of the distance is always positive. This more general type of geometry was first studied by Riemann and it is therefore called *Riemannian geometry*.

Consider now Maxwell's equations in free space though the presence of charges and convective currents will be permitted. Two of the equations

$$\operatorname{curl} \boldsymbol{E} + \partial \boldsymbol{B}/\partial t = \mathbf{0}, \qquad \operatorname{div} \boldsymbol{B} = 0 \tag{9.2}$$

can be satisfied by the introduction of the vector potential **A** and the scalar potential V (§1.10). Thus

$$\boldsymbol{B} = \operatorname{curl} \mathbf{A}, \qquad \boldsymbol{E} = -\partial \mathbf{A}/\partial t - \operatorname{grad} V.$$

It will now be *assumed* that **A** and V/c form the four components of a contravariant vector A^{μ}, where for brevity we denote a vector by its typical element. Later it will be seen that this assumption leads to the conclusion that electric charge is invariant under a Lorentz transformation. Conversely, it can be shown that the assumption that electric charge is invariant to a Lorentz transformation implies that (**A**, V/c) is a contravariant vector.

The associated covariant vector A_{μ} can be derived by lowering suffixes (Appendix F.4) and, since $g_{11} = g_{22} = g_{33} = 1$, $g_{44} = -1$,

$$A_1 = A^1, \qquad A_2 = A^2, \qquad A_3 = A^3, \qquad A_4 = -V/c.$$

From A_{μ} an antisymmetric covariant tensor $F_{\mu\nu}$ can be defined (Appendix F.5) by

$$F_{\mu\nu} = \frac{\partial A_{\nu}}{\partial x^{\mu}} - \frac{\partial A_{\mu}}{\partial x^{\nu}}.$$

When the derivatives are converted to the electromagnetic field

$$F_{\mu\nu} = \begin{pmatrix} 0 & B_z & -B_y & E_x/c \\ -B_z & 0 & B_x & E_y/c \\ B_y & -B_x & 0 & E_z/c \\ -E_x/c & -E_y/c & -E_z/c & 0 \end{pmatrix} \tag{9.3}$$

in the standard matrix notation.

The identification of the electromagnetic field with the tensor $F_{\mu\nu}$ via (9.3) means that the two equations (9.2) of Maxwell are automatically satisfied. In fact, they may be written as

$$\frac{\partial F_{\mu\nu}}{\partial x^{\sigma}} + \frac{\partial F_{\nu\sigma}}{\partial x^{\mu}} + \frac{\partial F_{\sigma\mu}}{\partial x^{\nu}} = 0 \tag{9.4}$$

where μ, ν, σ are any three of the numbers 1, 2, 3, 4.

The associated contravariant tensor $F^{\mu\nu}$ can be obtained by raising suffixes, an easy operation because $g^{\mu\nu} = g_{\mu\nu}$ for our metric. Then the contravariant tensor $G^{\mu\nu} = F^{\mu\nu}/\mu_0$ has components

$$G^{\mu\nu} = \begin{pmatrix} 0 & H_z & -H_y & -cD_x \\ -H_z & 0 & H_x & -cD_y \\ H_y & -H_x & 0 & -cD_z \\ cD_x & cD_y & cD_z & 0 \end{pmatrix}. \tag{9.5}$$

This tensor is suitable for handling the remaining pair of Maxwell's equations namely

$$\operatorname{curl} \boldsymbol{H} - \partial \boldsymbol{D}/\partial t = \boldsymbol{J}, \qquad \operatorname{div} \boldsymbol{D} = \rho. \tag{9.6}$$

Firstly, the divergence of $G^{\mu\nu}$ can be defined by (5.3) of Appendix F; since the determinant $|g|$ of $g_{\mu\nu}$ is a constant the divergence reduces to $\partial G^{\mu\nu}/\partial x^\nu$. Next, introduce the contravariant vector J^μ given by

$$J^1 = J_x, \qquad J^2 = J_y, \qquad J^3 = J_z, \qquad J^4 = c\rho. \tag{9.7}$$

The equations (9.6) can be transformed to

$$\partial G^{\mu\nu}/\partial x^\nu = J^\mu. \tag{9.8}$$

Equations (9.4) and (9.8) constitute the tensor form of Maxwell's equations when they are supplemented by the identifications (9.3), (9.5), and (9.7).

Since (9.4) and (9.8) are tensor in character they are invariant under a Lorentz transformation. Therefore, if valid in S, they are also valid in S' with F, G, x replaced by F', G', x'. Taking the field, current and charge density in S' to be given by (9.3), (9.5) and (9.7) except that all quantities are primed, the equations of the electromagnetic field in S' are

$$\operatorname{curl}' \boldsymbol{E}' + \partial \boldsymbol{B}'/\partial t' = \boldsymbol{0}, \qquad \operatorname{div}' \boldsymbol{B}' = 0,$$

$$\operatorname{curl}' \boldsymbol{H}' - \partial \boldsymbol{D}'/\partial t' = \boldsymbol{J}', \qquad \operatorname{div}' \boldsymbol{D}' = \rho'.$$

In other words, Maxwell's equations are unchanged by a Lorentz transformation except for the insertion of primes everywhere.

If the tensor transformation law of Appendix F.2 is applied to $F_{\mu\nu}$ and (9.3) invoked in S', there results

$$E'_x = E_x,$$
$$E'_y = \beta(E_y - vB_z),$$
$$E'_z = \beta(E_z + vB_y)$$

since the only nonzero components of b^α_μ are $b^1_1 = \beta$, $b^1_4 = \beta v/c = b^4_1$, $b^2_2 = 1 = b^3_3$, $b^4_4 = \beta$. These equations can be put in a form which is independent of a special choice of x-axis by writing them as

$$\boldsymbol{E}' = \beta \boldsymbol{E} + (1-\beta)(\boldsymbol{E} \cdot \boldsymbol{v})\boldsymbol{v}/v^2 + \beta \boldsymbol{v} \wedge \boldsymbol{B}. \tag{9.9}$$

The corresponding relation for the magnetic flux density which may also be inferred from the transformation of $F_{\mu\nu}$ is

$$\boldsymbol{B}' = \beta \boldsymbol{B} + (1-\beta)(\boldsymbol{B} \cdot \boldsymbol{v})\boldsymbol{v}/v^2 - \beta \boldsymbol{v} \wedge \boldsymbol{E}/c^2. \tag{9.10}$$

Similarly, from the transformation of $G^{\mu\nu}$ and (9.5),

$$\boldsymbol{D}' = \beta \boldsymbol{D} + (1-\beta)(\boldsymbol{D} \cdot \boldsymbol{v})\boldsymbol{v}/v^2 + \beta \boldsymbol{v} \wedge \boldsymbol{H}/c^2, \tag{9.11}$$

$$\boldsymbol{H}' = \beta \boldsymbol{H} + (1-\beta)(\boldsymbol{H} \cdot \boldsymbol{v})\boldsymbol{v}/v^2 - \beta \boldsymbol{v} \wedge \boldsymbol{D}. \tag{9.12}$$

The formulae (9.9)–(9.12) demonstrate that the electric and magnetic fields $\boldsymbol{E}$, $\boldsymbol{B}$ do not exist as separate entities; the resolution of the electromagnetic field into electric and magnetic components is wholly dependent on the motion of the observer. Thus an observer moving with a charge sees a purely electrostatic field but for an observer, relative to whom the charge is in motion, there is both an electric and magnetic field. The generation of a magnetic field by a moving charge was an experimental observation of Rowland.

From the transformation $J'^{\mu} = a^{\mu}_{\nu} J^{\nu}$ comes

$$J'_x = \beta(J_x - \rho v), \qquad J'_y = J_y, \qquad J'_z = J_z, \qquad \rho' = \beta(\rho - vJ_x/c^2) \tag{9.13}$$

or

$$\boldsymbol{J}' = \boldsymbol{J} - (1-\beta)(\boldsymbol{J}\cdot\boldsymbol{v})\boldsymbol{v}/v^2 - \beta\rho\boldsymbol{v}, \qquad \rho' = \beta(\rho - \boldsymbol{J}\cdot\boldsymbol{v}/c^2). \tag{9.14}$$

If the charge is at rest in S', $\boldsymbol{J}' = \boldsymbol{0}$ and then

$$J_x = \rho v, \qquad J_y = 0, \qquad J_z = 0, \qquad \rho = \beta\rho'.$$

Consequently, to an observer in S, the charge density is in motion with speed v and appears as a charge density ρ $(=\beta\rho')$ attended by a convection current ρv. The charge of a volume $\mathrm{d}\boldsymbol{x}'$ of S' is $\rho'\,\mathrm{d}\boldsymbol{x}'$ which is $\rho'\beta\,\mathrm{d}\boldsymbol{x}$ in terms of the corresponding volume of S in view of the contraction of length by the motion. Since $\rho'\beta = \rho$ the charge measured in S is the same as that measured in in S' i.e. *electric charge is invariant under a Lorentz transformation.* Neither mass nor charge density obeys this rule.

Finally, note that (9.8) gives

$$\partial J^{\mu}/\partial x^{\mu} = 0 \tag{9.15}$$

since $G^{\mu\nu}$ is antisymmetric. In S the left-hand side of (9.15) is $\operatorname{div}\boldsymbol{J} + \partial\rho/\partial t$ so that (9.15) is the tensor form of the equation of continuity.

2.10 The Lorentz force

The theory of the preceding section enables one to calculate the force of a given electromagnetic field on charged matter. Suppose a point charge Q is travelling with velocity $\boldsymbol{v}$ relative to the system S. Let S' be a system in which the charge is momentarily at rest. Then the force on the charge as measured in S' is $Q\boldsymbol{E}'$. From (6.6) the force measured in S is

$$Q\{\boldsymbol{E}'/\beta + \boldsymbol{v}(1-1/\beta)\boldsymbol{v}\cdot\boldsymbol{E}'/v^2\}.$$

Substitute for $\boldsymbol{E}'$ from (9.9). Then the force on the charge is

$$\boldsymbol{F} = Q(\boldsymbol{E} + \boldsymbol{v}\wedge\boldsymbol{B}) \tag{10.1}$$

as far as S is concerned. This force is known as the *Lorentz force*; it acts on a charge moving with velocity $\boldsymbol{v}$ relative to S.

When the charge has rest mass m_0 the equations governing its motion are

$$\frac{\mathrm{d}}{\mathrm{d}t}(m_0\beta\boldsymbol{v}) = Q(\boldsymbol{E}+\boldsymbol{v}\wedge\boldsymbol{B}), \tag{10.2}$$

$$\frac{\mathrm{d}}{\mathrm{d}t}(m_0\beta c^2) = Q\boldsymbol{v}\cdot\boldsymbol{E} \tag{10.3}$$

from (6.5) and (6.7).

2.11 The Doppler effect

It is well known that the frequency of a signal received from a source depends upon the velocity of the sender relative to the recipient. The phenomenon is known as the *Doppler effect* and the relevant relativistic formula will now be derived.

Let S' be a system in which the source is at rest and suppose that the electromagnetic signal produced is a plane wave travelling in the direction with direction cosines (l', m', n'). Its speed is c and all field components vary according to the factor

$$\exp[\mathrm{i}\omega'\{t' - (l'x' + m'y' + n'z')/c\}]. \tag{11.1}$$

If S is a system in which the observer is at rest, it follows from the preceding sections that all field components in S have a variation

$$\exp[\mathrm{i}\omega\{t - (lx + my + nz)/c\}].$$

After insertion of (2.11) and (2.12), (11.1) becomes

$$\exp[\mathrm{i}\omega'\{\beta(1 + vl'/c)t - x(l' + v)/c)\beta/c - (m'y + n'z)/c\}].$$

Therefore

$$\omega = \omega'\beta(1 + vl'/c) \tag{11.2}$$

which gives the change in frequency noted by a stationary observer. It depends upon both the speed and direction of the source. A special choice of axes can be avoided by writing it as

$$\omega = \omega'\beta(1 + \boldsymbol{v}\cdot\boldsymbol{n}'/c) \tag{11.3}$$

where $\boldsymbol{n}'$ is a unit vector in the direction of propagation as measured in S'.

There are also the relations

$$l : m : n = \beta(l' + v/c) : m' : n' \tag{11.4}$$

which indicate that the wave appears to be going in different directions for observers in S and S', i.e. *aberration* occurs. If axes are chosen so that $n=0$, $l=\cos\alpha$, $m=\sin\alpha$, (11.4) *gives*

$$\tan\alpha=\frac{\sin\alpha'}{\beta(\cos\alpha'+v/c)} \tag{11.5}$$

which is the relativistic formula for the aberration that takes place for observers in relative motion.

The relativistic red shift predicted by (11.2) has been checked in extremely accurate experiments with rapidly moving sources.

2.12 Electromagnetic stress and momentum

Let f_μ be the covariant vector given by $f_\mu=F_{\mu\nu}J^\nu$. Its first three components coincide with those of $\rho\boldsymbol{E}+\boldsymbol{J}\wedge\boldsymbol{B}$ whereas

$$f_4=-\boldsymbol{J}\cdot\boldsymbol{E}/c. \tag{12.1}$$

Since $(\rho\boldsymbol{E}+\boldsymbol{J}\wedge\boldsymbol{B})\,\mathrm{d}\boldsymbol{x}$ is the Lorentz force on the charge occupying the volume $\mathrm{d}\boldsymbol{x}$ the vector f_μ is called the *force density*. Note that cf_4 represents the power expended by the field on the moving charge.

Another formula for the force density is obtained by substituting for J^μ from (9.8). Thus

$$f_\mu=F_{\mu\nu}\frac{\partial G^{\nu\sigma}}{\partial x^\sigma}=\frac{\partial}{\partial x^\sigma}(F_{\mu\nu}G^{\nu\sigma})-G^{\nu\sigma}\frac{\partial F_{\mu\nu}}{\partial x^\sigma}.$$

Now $F_{\mu\nu}$ and $G^{\mu\nu}$ are antisymmetric so that

$$G^{\nu\sigma}\,\partial F_{\mu\nu}/\partial x^\sigma=G^{\sigma\nu}\,\partial F_{\nu\mu}/\partial x^\sigma$$

or, on interchanging dummy suffixes,

$$G^{\nu\sigma}\,\partial F_{\mu\nu}/\partial x^\sigma=G^{\nu\sigma}\,\partial F_{\sigma\mu}/\partial x^\nu. \tag{12.2}$$

Both sides of (12.2) must equal half their sum i.e.

$$\begin{aligned}G^{\nu\sigma}\,\partial F_{\mu\nu}/\partial x^\sigma&=\tfrac{1}{2}G^{\nu\sigma}(\partial F_{\mu\nu}/\partial x^\sigma+\partial F_{\sigma\mu}/\partial x^\nu)\\&=-\tfrac{1}{2}G^{\nu\sigma}\,\partial F_{\nu\sigma}/\partial x^\mu\end{aligned}$$

from (9.4). However, $G^{\mu\nu}$ is a constant multiple of $F^{\mu\nu}$ and so, since the metric tensor has constant coefficients,

$$G^{\nu\sigma}\,\partial F_{\mu\nu}/\partial x^\sigma=-\tfrac{1}{4}\partial(G^{\nu\sigma}F_{\nu\sigma})/\partial x^\mu.$$

Hence

$$f_\mu=\frac{\partial}{\partial x^\sigma}(F_{\mu\nu}G^{\nu\sigma})+\tfrac{1}{4}\frac{\partial}{\partial x^\mu}(G^{\nu\sigma}F_{\nu\sigma})=\frac{\partial S^\sigma_\mu}{\partial x^\sigma} \tag{12.3}$$

where

$$S^\sigma_\mu=F_{\mu\nu}G^{\nu\sigma}+\tfrac{1}{4}\delta^\sigma_\mu F_{\nu\tau}G^{\nu\tau}. \tag{12.4}$$

Formula (12.3) expresses the force density as the divergence of a mixed tensor. This tensor has, from (12.4), (9.3) and (9.5), components given by

$$S_\mu^\sigma = H_\mu B_\sigma + E_\mu D_\sigma - \tfrac{1}{2}\delta_\mu^\sigma(\boldsymbol{B}\cdot\boldsymbol{H}+\boldsymbol{E}\cdot\boldsymbol{D}) \quad (\mu\neq 4, \sigma\neq 4), \tag{12.5}$$

$$S_\mu^4 = -c(\boldsymbol{D}\wedge\boldsymbol{B})_\mu, \quad S_4^\mu = (\boldsymbol{E}\wedge\boldsymbol{H})_\mu/c \quad (\mu\neq 4), \tag{12.6}$$

$$S_4^4 = \tfrac{1}{2}(\boldsymbol{B}\cdot\boldsymbol{H}+\boldsymbol{E}\cdot\boldsymbol{D}) \tag{12.7}$$

where the suffixes 1, 2, 3 on the field components are understood to mean x, y, z respectively. In free space $\boldsymbol{B} = \mu_0\boldsymbol{H}$, $\boldsymbol{D} = \varepsilon_0\boldsymbol{E}$ and an alternative expression for (12.6) is

$$-S_\mu^4 = S_4^\mu = (\boldsymbol{S})_\mu/c \quad (\mu\neq 4)$$

where $\boldsymbol{S} = c^2\boldsymbol{D}\wedge\boldsymbol{B}$.

Placing $\mu = 4$ in (12.3) and combining it with (12.1) and (12.6) we obtain

$$\boldsymbol{J}\cdot\boldsymbol{E} + \operatorname{div}(\boldsymbol{E}\wedge\boldsymbol{H}) + \partial S_4^4/\partial t = 0. \tag{12.8}$$

The energy balance of the electromagnetic field is covered by (12.8) and leads to the notion of Poynting's vector (§1.27).

When neither μ nor σ is equal to 4 write t_μ^σ for S_μ^σ. Then, from (12.3),

$$f_\mu = \frac{\partial t_\mu^\sigma}{\partial x^\sigma} - \frac{1}{c^2}\frac{\partial(\boldsymbol{S})_\mu}{\partial t} \quad (\mu\neq 4, \sigma\neq 4). \tag{12.9}$$

In stationary fields the time derivative disappears and (12.9) is of the same type as that connecting body forces and stresses in static elasticity. For this reason t_μ^σ is called the *electromagnetic stress tensor.* The equation (12.9) says that in stationary fields the Lorentz force density can be replaced by stresses. It does not state that the Lorentz force is maintained in equilibrium by the stress; equilibrium has to be established by mechanical forces of another kind because a charge distribution cannot be kept in equilibrium by electrical forces alone.

Since the first term on the right-hand side of (12.9) is a divergence its integral over the whole of space will be zero provided that t_μ^σ dies off sufficiently rapidly at infinity. Then

$$\int f_\mu \,\mathrm{d}\boldsymbol{x} = -\frac{1}{c^2}\frac{\partial}{\partial t}\int(\boldsymbol{S})_\mu \,\mathrm{d}\boldsymbol{x}. \tag{12.10}$$

The left-hand side of (12.10) supplies the total force exerted on matter and therefore stems from the rate of change of mechanical momentum $\mathrm{d}\boldsymbol{p}_m/\mathrm{d}t$. Hence, we infer from (12.10)

$$\frac{\mathrm{d}}{\mathrm{d}t}\left(\boldsymbol{p}_m + \frac{1}{c^2}\int\boldsymbol{S}\,\mathrm{d}\boldsymbol{x}\right) = \boldsymbol{0}.$$

In order to assert that the total momentum of the system remains constant it is necessary to assume that there is an *electromagnetic momentum* $\int \boldsymbol{S}\, d\boldsymbol{x}/c^2$ as well as the mechanical momentum. Often $\boldsymbol{S}/c^2$ is called the *electromagnetic momentum density*.

These ideas suggest that a light wave carries momentum and that, accordingly, it will impose a pressure, called the *radiation pressure*, on a body which absorbs it. This pressure has been observed experimentally, as has the torque generated when elliptically polarized light is absorbed.

In a plane wave the rate of transfer of energy is given by the Poynting vector and so the energy density is $|\boldsymbol{E} \wedge \boldsymbol{H}|/c$. Thus the magnitude p of the momentum is related to the energy by $p = E/c$. Since a photon has energy $h\nu$ where h is Planck's constant and ν the frequency, the momentum of a photon is $h\nu/c$. Radiation pressure is regarded in quantum theory as due to bombardment by photons.

There is one hypothetical experiment which should be mentioned. Consider a closed box which contains two exactly similar instruments A and B, each of which can send out a brief light signal to the other or completely absorb an incoming signal. Let A emit a signal at a certain instant. On account of the radiation pressure the box recoils and remains in motion until the signal is absorbed by B. The radiation pressure then brings the box to rest. The centre of mass of the box will then have moved in the absence of an external force, an event which is contrary to the principle of the conservation of momentum. The paradox can be resolved if a momentum (equal and opposite to that of the box) is assigned to the light wave during its lifetime. However, since the box has moved, there will still be an inconsistency unless we assume that the mass of A is reduced by the emission of the light signal and the mass of B is correspondingly increased by its absorption. This is another illustration of the equivalence of mass and energy (§2.7).

Electrodynamics in moving media

2.13 The field equations

The theory of electromagnetism in arbitrary moving media is a complicated subject and only one simple case will be treated here. In essence we find the equations in bodies moving with uniform velocity by making a Lorentz transformation from a system in which the bodies are at rest so that Maxwell's equations can be implemented in conventional form.

In a system of coordinates in which the medium is at rest take the governing equations of the electromagnetic field to be (9.2) and (9.6). If the matter is isotropic and linear

$$\boldsymbol{D} = \varepsilon \boldsymbol{E}, \qquad \boldsymbol{B} = \mu \boldsymbol{H}. \tag{13.1}$$

If, in addition, Ohm's law is satisfied $\boldsymbol{J} = \sigma \boldsymbol{E}$.

Define the tensors $F_{\mu\nu}$, $G^{\mu\nu}$, and vector J^{μ} by (9.3), (9.5), and (9.7) respectively. The field equations in tensor form are then (9.4) and (9.8) because they coincide with Maxwell's equations in a system at rest. Since (9.4) and (9.8) are unaltered by a Lorentz transformation they will continue to hold in a system in uniform motion and can be used to describe the behaviour of the electromagnetic field in a moving medium.

In particular, the equation of continuity (9.15) and the equations (9.14) connecting current and charge densities in the two systems remain valid. Thus, if in the rest system there is a nonzero current flow J_x but the charge density vanishes, the charge density when the body is moving relative to an observer will be nonzero.

The relations between the field components in the two systems will continue to be given by (9.9)–(9.12). An observer at rest in S' will ascribe the primed field to a moving medium. Now, if $\boldsymbol{v}'$ is the velocity of S relative to S', (9.9) gives

$$\boldsymbol{E} = \beta \boldsymbol{E}' + (1-\beta)(\boldsymbol{E}' \cdot \boldsymbol{v}')\boldsymbol{v}'/v'^2 + \beta \boldsymbol{v}' \wedge \boldsymbol{B}'$$

and there are similar equations corresponding to (9.10)–(9.12). Substitute these expressions in (13.1) and take a scalar product with $\boldsymbol{v}'$; then $\boldsymbol{D}' \cdot \boldsymbol{v}' = \varepsilon \boldsymbol{E}' \cdot \boldsymbol{v}'$ and $\boldsymbol{B}' \cdot \boldsymbol{v}' = \mu \boldsymbol{H}' \cdot \boldsymbol{v}'$. Hence

$$\boldsymbol{D}' + \boldsymbol{v}' \wedge \boldsymbol{H}'/c^2 = \varepsilon(\boldsymbol{E}' + \boldsymbol{v}' \wedge \boldsymbol{B}'), \tag{13.2}$$

$$\boldsymbol{B}' - \boldsymbol{v}' \wedge \boldsymbol{E}'/c^2 = \mu(\boldsymbol{H}' - \boldsymbol{v}' \wedge \boldsymbol{D}') \tag{13.3}$$

are the constitutive equations in a moving medium. The analogous form of Ohm's law is

$$\boldsymbol{J}' - \rho' \boldsymbol{v}' = \sigma\beta\{\boldsymbol{E}' + \boldsymbol{v}' \wedge \boldsymbol{B}' - (\boldsymbol{E}' \cdot \boldsymbol{v}')\boldsymbol{v}'/c^2\}. \tag{13.4}$$

In a moving medium the quantities

$$\tilde{\boldsymbol{D}}' = \boldsymbol{D}' + \boldsymbol{v}' \wedge \boldsymbol{H}'/c^2, \qquad \tilde{\boldsymbol{E}}' = \boldsymbol{E}' + \boldsymbol{v}' \wedge \boldsymbol{B}',$$

$$\tilde{\boldsymbol{B}}' = \boldsymbol{B}' - \boldsymbol{v}' \wedge \boldsymbol{E}'/c^2, \qquad \tilde{\boldsymbol{H}}' = \boldsymbol{H}' - \boldsymbol{v}' \wedge \boldsymbol{D}'$$

represent the forces on suitable test bodies at rest in the medium. From (7.7) of Appendix F

$$\text{curl}'\, \tilde{\boldsymbol{E}}' = \text{curl}'\, \boldsymbol{E}' + \boldsymbol{v}' \,\text{div}'\, \boldsymbol{B}' - (\boldsymbol{v}' \cdot \text{grad}')\boldsymbol{B}',$$

$$\text{curl}'\, \tilde{\boldsymbol{H}}' = \text{curl}'\, \boldsymbol{H}' - \boldsymbol{v}' \,\text{div}'\, \boldsymbol{D}' + (\boldsymbol{v}' \cdot \text{grad}')\boldsymbol{D}'.$$

Hence the field equations can also be written as

$$\text{curl}'\, \tilde{\boldsymbol{E}}' + \mathrm{D}\boldsymbol{B}'/\mathrm{D}t' = \boldsymbol{0}, \qquad \text{div}'\, \boldsymbol{B}' = 0, \tag{13.5}$$

$$\text{curl}'\, \tilde{\boldsymbol{H}}' - \mathrm{D}\boldsymbol{D}'/\mathrm{D}t' = \boldsymbol{C}', \qquad \text{div}'\, \boldsymbol{D}' = \rho' \tag{13.6}$$

where $\mathrm{D}/\mathrm{D}t'$ is the total time derivative (§1.1), namely

$$\mathrm{D}/\mathrm{D}t' \equiv \partial/\partial t' + \boldsymbol{v}' \cdot \text{grad}',$$

and the conduction current

$$\boldsymbol{C}' = \boldsymbol{J}' - \rho' \boldsymbol{v}'.$$

The left-hand equation of (13.5) can be viewed as a generalization of Faraday's law of induction since the total time derivative is a derivative following the motion. For, if the equation is integrated over a moving surface the time derivative can be placed outside the integral and then the rate of change of magnetic flux through a moving circuit is equal to the induced electromotive force, *as calculated by an observer moving with the circuit.*

The theory given so far is applicable only to one body moving with constant velocity. However, the linearity of the equations ensures that they hold for several bodies, moving with different uniform velocities, separated by free space. The electromagnetic forces on a moving body will, in general, cause it to accelerate and so the field equations will be valid only in so far as these accelerations can be regarded as small.

2.14 Boundary conditions

The boundary conditions at an interface between two immobile media have been discussed in §1.22. From them can be deduced by means of (9.9)–(9.12) the boundary conditions between moving media. It is found that the tangential components of $\boldsymbol{E}' + \boldsymbol{v}' \wedge \boldsymbol{B}'$ and $\boldsymbol{H}' - \boldsymbol{v}' \wedge \boldsymbol{D}'$ are continuous; the normal component of $\boldsymbol{B}'$ is also continuous but it must be remembered that in these statements the directions are those observed in S'. For instance, when the relative motion is along the x-axis, a direction in S, fixed in the medium, with direction cosines (l, m, n) must be taken to have direction cosines proportional to $(\beta l, m, n)$ in S'.

It should also be noted that $\boldsymbol{v}'$ is a constant of the Lorentz transformation so that, for a boundary between free space and a medium moving with velocity $\boldsymbol{v}'$ relative to S', the tangential components of $\boldsymbol{E}' + \boldsymbol{v}' \wedge \boldsymbol{B}'$ and $\boldsymbol{H}' - \boldsymbol{v}' \wedge \boldsymbol{D}'$ are continuous there. In other words a velocity $\boldsymbol{v}'$ is ascribed to the points of free space just *outside* the medium.

A dielectric has the additional property of possessing no volume or surface distributions of charge when it is stationary. Hence the normal component of $\boldsymbol{D}$ is continuous. It follows that at the surface of a travelling dielectric the normal component of $\boldsymbol{D}'$ is continuous, the direction of the normal being again that in S'.

2.15 The convection of light

In this section the effect of a moving medium on the propagation of light will be examined with the aim of discovering the extent to which the medium convects the light.

Let S be fixed in the medium and moving with it while S' is the system of the fixed laboratory observer. Let S be in motion parallel to the x-axis and let the signal in S be of frequency $\omega/2\pi$ travelling along the x-axis with speed u. It will be observed in the laboratory as a signal of frequency $\omega'/2\pi$ moving along the x'-axis with speed u'. From (5.2)

$$u' = \frac{u+v'}{1+uv'/c^2}$$

where v' is the speed of S relative to S'. The modified form of (11.2) is $\omega = \beta\omega'(1-v'/u')$ or, equivalently.

$$\omega' = \beta\omega(1+v'/u).$$

Usually the speed of the medium will be much lower than that of light so that $v'/c \ll 1$ and $v'/u \ll 1$. If squares of these quantities are neglected ω'/ω will be only slightly different from unity. Also if $N(\omega)$ is the refractive index at frequency $\omega/2\pi$ in the medium at rest $u = c/N(\omega)$. Therefore, to a first approximation,

$$\omega/\omega' \approx 1 - v'N(\omega')/c$$

and

$$N(\omega) \approx N(\omega') - \frac{\omega' v'}{c} N(\omega') \frac{d}{d\omega'} N(\omega').$$

Consequently,

$$u' \approx \frac{c}{N(\omega')} + v'\left\{1 - \frac{1}{N^2(\omega')}\right\} + \frac{\omega' v'}{N(\omega')} \frac{d}{d\omega'} N(\omega') \tag{15.1}$$

to the same order of approximation.

When the medium is not dispersive N does not vary with frequency and only the first two terms of (15.1) survive. Since the coefficient of v' is less than unity the light is only *partially convected* by the medium. The factor $1-1/N^2$ is known as the *Fresnel convection coefficient* because Fresnel suggested it on the basis of an elastic ether theory. The formula (15.1) has been verified by experiments on the speed of light in running water. In these experiments the third term is of no importance but it is necessary to secure agreement with Zeeman's experiments on the speed of light in a moving quartz rod.

The reflection of light by a mirror moving perpendicular to itself may also be treated by the theory. Assume that the mirror is a perfect conductor with velocity $\boldsymbol{v}'$ relative to the laboratory. Then the tangential component of $\boldsymbol{E}' + \boldsymbol{v}' \wedge \boldsymbol{B}'$ is zero on the mirror. Take the mirror to lie in the plane $x = 0$. Let the wave

$$B'_{zi} = \exp[i\omega'\{t' - (x' \cos\theta + y' \sin\theta)/c\}],$$

$$E'_{xi} = -c \sin\theta B'_{zi}, \qquad E'_{yi} = c \cos\theta B'_{zi}$$

be incident on the mirror and excite a reflected wave

$$B'_{zr} = A \exp[i\omega'_r\{t' + (x' \cos\phi - y' \sin\phi)/c\}],$$
$$E'_{xr} = -c \sin\phi B'_{zr}, \qquad E'_{yr} = -c \cos\phi B'_{zr}.$$

The boundary condition is

$$E'_{yi} + E'_{yr} - v'(B'_{xi} + B'_{zr}) = 0$$

on $x' = v't'$ for all t' and y'. Hence

$$\omega'(1 - v' \cos\theta/c) = \omega'_r(1 + v' \cos\phi/c),$$
$$\omega' \sin\theta = \omega'_r \sin\phi,$$
$$c \cos\theta - v' = A(c \cos\phi + v').$$

It is clear that the angle of reflection is not, in general, equal to the angle of incidence. Also, if the mirror is moving away from the observer, both the frequency and amplitude of the reflected wave are reduced by the motion. At normal incidence ($\theta = 0$) they are smaller, by the ratio

$$(1 - v'/c)/(1 + v'/c) \approx 1 - 2v'/c$$

if $v'/c \ll 1$, than their values if the mirror were fixed. There are corresponding increases in the frequency and amplitude of the reflected wave when the mirror moves towards the observer. Experiments with moving mirrors are consistent with the theoretical predictions.

Here is another manifestation of the Doppler effect, the frequency change at normal incidence being twice that of the simpler case in §2.11. The phenomenon has been put to considerable practical use and is employed to determine the speeds of motor cars, aircraft, rockets, and satellites to mention a few examples.

2.16 The convection of charge by a moving medium

The magnetic effect of a moving charge was first noted by Rowland (1876) and that due to a dielectric in motion by Röntgen (1888, 1890). A quantitative basis was provided by Eichenwald (1903) in a beautiful series of experiments.

Eichenwald's basic apparatus consists of a parallel plate capacitor with a slab of dielectric R between the plates P and Q (Fig. 2.6). P, Q, and R are all capable of translation parallel to themselves, i.e. in the direction of the x'-axis in Fig. 2.6, the y'-axis being perpendicular to the plates. The potential difference between P and Q is maintained at a constant value V.

In the first experiment R is absent, P is fixed, and Q is translated with uniform speed v'. Since stationary fields are involved all time derivatives disappear from Maxwell's equations. Therefore, the electric field is de-

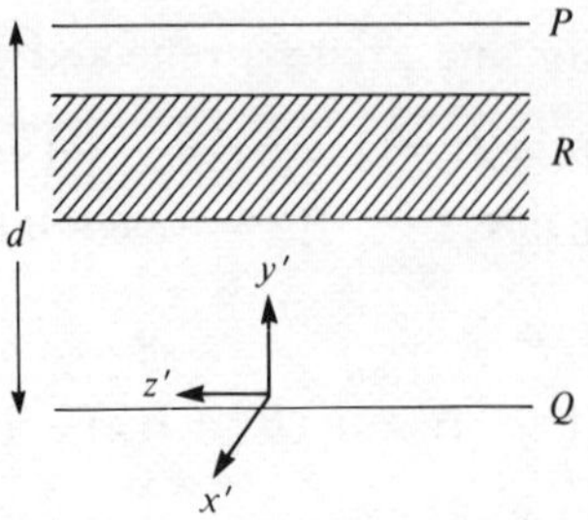

Fig. 2.6 Eichenwald's apparatus

rived from a potential and the non-zero components are given by

$$E'_y = -V/d, \qquad D'_y = \varepsilon_0 E'_y.$$

With regard to the magnetic field the tangential components of $\boldsymbol{H}' - \boldsymbol{v}' \wedge \boldsymbol{D}'$ must be continuous across Q. Since $\boldsymbol{D}'$ is zero as soon as the charged layer in Q is crossed we have

$$(H'_z)_{y'=+0} - (H'_z)_{y'=-0} = -v'\varepsilon_0 V/d.$$

Elsewhere the magnetic field is continuous since P is fixed and hence it is the same as that produced by a surface current $J_x = -v'\varepsilon_0 V/d$. This coincides with the magnetic field arising from all the charge induced on Q, by raising the capacitor to potential V, moving with Q like a rigid body, i.e. there is a convection current. It might be thought that the charge induced on P would move in sympathy. If it did there would be a conduction current in disagreement with the preceding analysis and Eichenwald's observations.

In a second experiment P and Q were fixed whereas R was translated with uniform speed v'. The magnetic field outside R was found to be the same as if the fictitious charge on a dielectric surface were considered to be generating a convection current (the so-called Röntgen current). It was therefore inferred that the Röntgen current existed physically. However, that interpretation is incorrect; in fact, the magnetic field originates from the motion of certain real charges though it is equal, outside R, to that produced by the Röntgen current. To demonstrate this assume that the permeability of R is that of free space and that v'/c is sufficiently small for its square to be neglected. Since the normal component of $\boldsymbol{D}'$ is continuous at the boundaries of R (§2.14) and the tangential components of $\boldsymbol{H}' - \boldsymbol{v}' \wedge \boldsymbol{D}'$ are continuous, it follows that the tangential components of $\boldsymbol{H}'$ are continuous ($\boldsymbol{v}'$ is parallel to the boundary). Hence the Röntgen current does not exist and $\boldsymbol{H}'$ is zero.

From (13.2) and (13.3), with the square of v'/c ignored,

$$\boldsymbol{D}' = \varepsilon \boldsymbol{E}'$$

so that $\boldsymbol{D}'$, $\boldsymbol{E}'$ are exactly the same as if R were stationary. From (13.3)

$$B'_z = -\mu_0 v' D'_y + v' E'_y / c^2 = \mu_0(\varepsilon_0 - \varepsilon) v' E'_y$$

at points inside R and is zero at points outside. Therefore, according to magnetostatics, the moving dielectric behaves as a *permanent magnet*, the difference between $\boldsymbol{B}'$ and $\mu_0\boldsymbol{H}'$ $(=\mathbf{0})$ being the magnetic moment. Actually, when the dielectric is polarized the positive and negative charges of the atoms become slightly displaced and, when they are translated, form two equal and opposite nearby electric currents i.e. a (two-dimensional) magnetic dipole as regards magnetic field. It is well known that the magnetic field of a uniformly magnetized permanent magnet is equivalent to that of a certain surface current (in this case, the Röntgen current). Thus, what Eichenwald's experiments really confirm is that the charge distribution in the dielectric is carried along with it and that the magnetic field of a permanent magnet equals that of a suitable surface current distribution at exterior points.

As a final check on convection theory the space between P and Q was completely filled by R and all three moved together as a rigid body. As before, the dielectric behaves like a permanent magnet and there are convection currents on P and Q. The net *equivalent* current on Q is

$$J'_x = -v'\varepsilon V/d - v'(\varepsilon_0 - \varepsilon)V/d = -\varepsilon_0 Vv'/d$$

which is independent of the dielectric. Experimentally, it was found that the magnetic field was indeed independent of the dielectric.

Further experimental verification was provided by Wilson (1905, 1913). In this experiment P and Q were insulated but uncharged initially. The dielectric R, which filled the space between P and Q, was moving with uniform speed v' $(\ll c)$ and a uniform magnetic field was applied in the direction of the z'-axis. It was discovered that the capacitor became charged.

In the motion the magnetic intensity $\boldsymbol{H}'$ is unaltered and the only non-zero components of the electric field are given by

$$D'_y = \varepsilon E'_y - \mu_0(\varepsilon - \varepsilon_0)v'H'_z = \varepsilon V/d - \mu_0(\varepsilon - \varepsilon_0)v'H'_z$$

from (13.2). Since D'_y is the charge per unit area on the plates this is a relation between the charge and potential; it agrees with the experimental results. The Lorentz force $\boldsymbol{v}' \wedge \boldsymbol{B}'$ causes the positive and negative charges of the dielectric to become separated with a resulting polarization, i.e. the moving dielectric behaves as an electret.

Exercises on Chapter 2

In the following exercises the results and notation of Appendix F are pertinent.

1. If $X(\lambda, \mu, \nu)$ is such that $X(\lambda, \mu, \nu)T^{\mu\alpha}_{\nu} = 0$ for every tensor $T^{\mu\alpha}_{\nu}$ prove that X vanishes identically.

If $Y(\lambda, \mu, \nu)$ is such that, for every tensor $T^{\mu\alpha}_{\nu}$, $Y(\lambda, \mu, \nu)T^{\mu\alpha}_{\nu}$ forms a tensor S^{α}_{λ} show that $Y(\lambda, \mu, \nu)$ is a tensor.

2. $T_{\alpha\beta}$ is an anti-symmetric tensor. Prove that $T_{\alpha\lambda}T_{\mu\beta}g^{\lambda\mu}$ is symmetric.

3. Show that the scalar invariant $c_{\mu\nu}A^{\mu}A^{\nu}$ can always be written $d_{\mu\nu}A^{\mu}A^{\nu}$ where $d_{\mu\nu}$ is symmetric.

4. The *Christoffel symbols* $[\lambda\mu, \nu]$, $\begin{pmatrix}\lambda\\ \mu\nu\end{pmatrix}$ are defined by

$$[\lambda\mu, \nu] = \frac{1}{2}\left(\frac{\partial g_{\lambda\nu}}{\partial x^{\mu}} + \frac{\partial g_{\mu\nu}}{\partial x^{\lambda}} - \frac{\partial g_{\lambda\mu}}{\partial x^{\nu}}\right),$$

$$\begin{pmatrix}\lambda\\ \mu\nu\end{pmatrix} = g^{\lambda\sigma}[\mu\nu, \sigma].$$

Prove that

$$[\alpha\beta, \gamma]' = [\lambda\mu, \nu]b^{\lambda}_{\alpha}b^{\mu}_{\beta}b^{\nu}_{\gamma} + g_{\lambda\mu}b^{\lambda}_{\gamma}\frac{\partial^2 x^{\mu}}{\partial x'^{\alpha}\,\partial x'^{\beta}},$$

$$\begin{pmatrix}\alpha\\ \beta\gamma\end{pmatrix}' = \begin{pmatrix}\lambda\\ \mu\nu\end{pmatrix}a^{\alpha}_{\lambda}b^{\mu}_{\beta}b^{\nu}_{\gamma} + a^{\alpha}_{\lambda}\frac{\partial^2 x^{\lambda}}{\partial x'^{\beta}\,\partial x'^{\gamma}}$$

so that the Christoffel symbols are not tensors in general.

5. Prove that $\partial A_{\nu}/\partial x^{\mu} - \partial A_{\mu}/\partial x^{\nu}$ is a covariant tensor but that $\partial A_{\nu}/\partial x^{\mu}$ is not a tensor. Show, however, that

$$\frac{\partial A_{\nu}}{\partial x^{\mu}} - \begin{pmatrix}\lambda\\ \nu\mu\end{pmatrix}A_{\lambda},$$

with the Christoffel symbol defined as in Exercise 4, is a covariant tensor called the *covariant derivative* of A_{ν} and written $A_{\nu,\mu}$. Note that

$$\partial A_{\nu}/\partial x^{\mu} - \partial A_{\mu}/\partial x^{\nu} = A_{\nu,\mu} - A_{\mu,\nu}.$$

6. A vector $\mathbf{A}$ is given in a Cartesian system and x'^{α} is an orthogonal curvilinear system. Show that $\mathbf{A} = A'_{\alpha}\,\text{grad}\,x'^{\alpha}$. Deduce that the physical components $A_{(1)}$, $A_{(2)}$, $A_{(3)}$ of $\mathbf{A}$, i.e. the projections on tangents to the coordinate curves, are given by relations of the form

$$A_{(1)} = A'_1/(g'_{11})^{\frac{1}{2}} = A'^1(g'_{11})^{\frac{1}{2}}.$$

Similarly, show that the physical components $T_{(\mu\nu)}$ of a tensor are given by equations of the form

$$T_{(11)} = T'_{11}/g'_{11} = g'_{11}T'^{11}, \qquad T_{(12)} = T'_{12}/(g'_{11}g'_{22})^{\frac{1}{2}} = T'^{12}(g'_{11}g'_{22})^{\frac{1}{2}}.$$

7. Prove the following results

$$\text{div}(\phi\mathbf{A}) = (\text{grad}\,\phi)\cdot\mathbf{A} + \phi\,\text{div}\,\mathbf{A},$$

$$\text{curl}(\phi\mathbf{A}) = (\text{grad}\,\phi)\wedge\mathbf{A} + \phi\,\text{curl}\,\mathbf{A},$$

$$\text{div}(\mathbf{A}\wedge\mathbf{B}) = \mathbf{B}\cdot\text{curl}\,\mathbf{A} - \mathbf{A}\cdot\text{curl}\,\mathbf{B},$$

$$\text{curl}(\mathbf{A}\wedge\mathbf{B}) = \mathbf{A}\,\text{div}\,\mathbf{B} - \mathbf{B}\,\text{div}\,\mathbf{A} + (\mathbf{B}\cdot\text{grad})\mathbf{A} - (\mathbf{A}\cdot\text{grad})\mathbf{B},$$

$$\text{grad}(\mathbf{A}\cdot\mathbf{B}) = (\mathbf{A}\cdot\text{grad})\mathbf{B} + (\mathbf{B}\cdot\text{grad})\mathbf{A} + \mathbf{A}\wedge\text{curl}\,\mathbf{B} + \mathbf{B}\wedge\text{curl}\,\mathbf{A}.$$

8. Show that a way of writing the Lorentz transformation without a special choice of axes is

$$\boldsymbol{x}' = \boldsymbol{x} - \beta t\boldsymbol{v} + (\beta - 1)(\boldsymbol{x} \cdot \boldsymbol{v})\boldsymbol{v}/v^2,$$
$$t' = \beta(t - \boldsymbol{v} \cdot \boldsymbol{x}/c^2).$$

9. A particle of rest mass m_0 is moving with velocity $\boldsymbol{u}$ relative to S. If S' and S are connected by a Lorentz transformation prove that $\gamma m_0 u_x$, $\gamma m_0 u_y$, $\gamma m_0 u_z$, and γm_0 where $\gamma = (1 - u^2/c^2)^{-\frac{1}{2}}$ are related to the corresponding quantities in S' in the same way as the space–time coordinates.

10. A particle of rest mass m_0 has a lifetime of t_0 seconds when at rest. It is observed in the system S to be in uniform motion with total energy E_0. Show that it will travel a distance $t_0(E_0^2 - m_0^2c^4)^{\frac{1}{2}}/m_0c$ in S during its lifetime.

11. At the end of its life the particle of Exercise 10 splits into two equal particles, each of rest mass M_0 ($<\frac{1}{2}m_0$) and with kinetic energy due to the difference in mass. Show that the speed of the secondary particles is $c(1 - 4M_0^2/m_0^2)^{\frac{1}{2}}$ in a system in which the original particle is at rest. If, in the system S, one of the secondary particles has total energy E show that its path makes an angle θ with the path of the original particle where

$$\cos\theta = \frac{2EE_0 - m_0^2c^4}{2(E^2 - M_0^2c^4)^{\frac{1}{2}}(E_0^2 - m_0^2c^4)^{\frac{1}{2}}}.$$

12. A particle of charge Q and rest mass m_0 is moving under the influence of a magnetic flux density $\boldsymbol{B}$ parallel to the z-axis and a perpendicular electric intensity $\boldsymbol{E}$. If at some time the particle is in the plane $z = 0$ and moving parallel to it show that the particle remains in the plane $z = 0$.

13. A particle of charge Q and rest mass m_0 is repelled by the Coulomb force due to a charge q fixed at the origin. Initially, Q is in $z = 0$ and is projected with speed u in $z = 0$ so as to pass q at a distance b in the absence of the repulsive force. If the position of Q is specified by the cylindrical polar coordinates r, ϕ, z when its speed is v show that $\beta r^2\dot{\phi} = \beta_0 bu$, $\beta_0 - \beta = Qq/4\pi\varepsilon_0 c^2 m_0 r$ and that

$$\mathrm{d}(\beta\dot{r})/\mathrm{d}t - \beta r\dot{\phi}^2 = Qq/4\pi\varepsilon_0 m_0 r^2$$

where $\beta = (1 - v^2/c^2)^{-\frac{1}{2}}$, $\beta_0 = (1 - u^2/c^2)^{-\frac{1}{2}}$.

With $w = 1/r$ show that

$$\mathrm{d}^2w/\mathrm{d}\theta^2 + \omega^2 w = -a$$

where

$$a = Qq/4\pi\varepsilon_0 m_0 \beta_0 b^2 u^2, \qquad \omega^2 = 1 - a^2b^2u^2/c^2$$

and hence show that the orbit of Q is

$$\frac{1}{r}=\frac{a}{\omega^2}(\cos\omega\theta-1)+\frac{1}{b\omega}\sin\omega\theta.$$

Deduce that Q is deflected through an angle $\pi-(2/\omega)\cot^{-1}(ab/\omega)$ and that its closest distance of approach to q is $ab^2+b(a^2b^2+\omega^2)^{\frac{1}{2}}$.

14. If the particle in Exercise 12 is initially at rest at the origin and $\boldsymbol{E}=\boldsymbol{i}\ a\ \cos\omega(t-y/c)$, $c\boldsymbol{B}=-\boldsymbol{k}a\cos\omega(t-y/c)$ prove that $\beta(c-\dot{y})=c$ where $\beta=(1-v^2/c^2)^{-\frac{1}{2}}$. Prove that the equation of the path can be written

$$x=b(1-\cos\theta),\qquad y=(b^2\omega/8c)(2\theta-\sin 2\theta)$$

where $b=Qa/m_0\omega^2$, $\theta=\omega(t-y/c)$.

15. The particle in Exercise 12 is projected with velocity $(u_0, v_0, 0)$ from the origin and $\boldsymbol{E}=E\boldsymbol{j}$, $\boldsymbol{B}=B\boldsymbol{k}$ where E and B are positive constants such that $cB>E$. Prove that the path of the particle is

$$x(1-E^2/c^2B^2)^{\frac{1}{2}}=b\psi+a\sin\psi_0-a\sin(\psi+\psi_0),$$

$$y=a\{\cos\psi_0-\cos(\psi+\psi_0)\}$$

where $a\cos\psi_0=m_0\beta_0c^2(E-u_0B)/Q(c^2B^2-E^2)$, $a\sin\psi_0=m_0\beta_0cv_0/Q(c^2B^2-E^2)^{\frac{1}{2}}$,

$$b=\frac{m_0\beta_0E(c^2B-u_0E)}{QB(c^2B^2-E^2)},\qquad \psi=\frac{E(c^2Bt-Ex)}{bcB(c^2B^2-E^2)^{\frac{1}{2}}},\qquad \beta_0=\left(1-\frac{u_0^2+v_0^2}{c^2}\right)^{-\frac{1}{2}}.$$

Deduce that, if the initial speed is much smaller than c, the particle strikes $y=0$ at a point which depends only on m_0/Q and the field strengths.

16. Show that (11.5) can be expressed as

$$(c+v)^{\frac{1}{2}}\tan\tfrac{1}{2}\alpha=(c-v)^{\frac{1}{2}}\tan\tfrac{1}{2}\alpha'.$$

17. If, in §2.11, the wave is travelling with speed w' instead of speed c in S' show that $\omega=\omega'\beta(1+vl'/w')$ and that the analogue of (11.5) is

$$\tan\alpha=\frac{\sin\alpha'}{\beta(\cos\alpha'+vw'/c^2)}.$$

Prove also that the speed w of the wave in S satisfies

$$1-c^2/w^2=(1-c^2/w'^2)(1-v^2/c^2)/(1+vl'/w')^2.$$

18. Show that, in the reflection by a mirror in §2.15,

$$\frac{c-v\cos\theta}{c+v\cos\phi}=\frac{c\cos\theta-v}{c\cos\phi+v}.$$

19. Prove that, if S and S' are connected by a Lorentz transformation,

$$|\boldsymbol{E}|^2-c^2|\boldsymbol{B}|^2=|\boldsymbol{E}'|^2-c^2|\boldsymbol{B}'|^2,\qquad \boldsymbol{E}\cdot\boldsymbol{B}=\boldsymbol{E}'\cdot\boldsymbol{B}'.$$

20. A particle of charge Q at rest at the origin of S' produces an electric intensity of $Q\mathbf{x}'/4\pi\varepsilon_0\,|\mathbf{x}'|^3$ at the point $\mathbf{x}'$. Show that the electric intensity observed in S of a charge Q moving with uniform velocity $\boldsymbol{v}$ is

$$\frac{Q(1-v^2/c^2)\boldsymbol{R}}{4\pi\varepsilon_0\,|\boldsymbol{R}|^3\{1-(v^2/c^2)\sin^2\theta\}^{\frac{3}{2}}}$$

where $\boldsymbol{R}$ is the position vector (as observed in S) of the point of observation with respect to Q and θ is the angle between $\boldsymbol{v}$ and $\boldsymbol{R}$.

21. If $\boldsymbol{E}$, $\boldsymbol{H}$, and ρ are zero outside a finite volume T show that the total force exerted on the charge by the field is $-\int_T \frac{\partial}{\partial t}\left(\frac{1}{c^2}\boldsymbol{E}\wedge\boldsymbol{H}\right)\mathrm{d}\mathbf{x}$.

3. Radiation

The field of a moving point source

3.1 Radiation from a moving acoustic source

IT has been seen in Chapter 1 that the acoustic field satisfies the partial differential equation

$$\nabla^2 p - \frac{1}{v^2}\frac{\partial^2 p}{\partial t^2} = -s(\boldsymbol{x}, t). \tag{1.1}$$

The particular problem that will be examined now is the field from a point source moving in a given manner. In order that the source move in a prescribed way forces will have to be applied but their magnitude and direction are not the subject of this enquiry. Assume that the point occupies the position $\mathbf{X}(t)$ at time t. Then we shall take

$$s(\boldsymbol{x}, t) = \delta\{\boldsymbol{x} - \mathbf{X}(t)\}. \tag{1.2}$$

The solution of (1.1) as a retarded potential has been derived in §1.16. Inserting (1.2) in (1.16.6) we obtain

$$p(\boldsymbol{x}, t) = \int_T \frac{\delta\{\mathbf{y} - \mathbf{X}(t - |\boldsymbol{x} - \mathbf{y}|/v)\}}{4\pi |\boldsymbol{x} - \mathbf{y}|}\, \mathrm{d}\mathbf{y} \tag{1.3}$$

where T is the whole of space.

Introduce a new variable of integration into (1.3) by means of

$$\boldsymbol{x}_1 = \mathbf{y} - \mathbf{X}(t - |\boldsymbol{x} - \mathbf{y}|/v).$$

Then, if $\mathrm{d}\boldsymbol{x}_1 = J\, \mathrm{d}\mathbf{y}$,

$$p(\boldsymbol{x}, t) = \frac{1}{4\pi}\int_T \frac{\delta(\boldsymbol{x}_1)\, \mathrm{d}\boldsymbol{x}_1}{|\boldsymbol{x} - \mathbf{y}|\; J} = \frac{1}{4\pi}\left[\frac{1}{|\boldsymbol{x} - \mathbf{y}|\, J}\right]_0$$

where $[\ ]_0$ signifies the value at $\boldsymbol{x}_1 = \mathbf{0}$ (§1.15). Now, if y_1, y_2, and y_3 are the components of $\mathbf{y}$,

$$\partial\, |\boldsymbol{x} - \mathbf{y}|/\partial y_j = (y_j - x_j)/|\boldsymbol{x} - \mathbf{y}|.$$

Hence, if $\mathbf{X}'$ denotes a derivative of $\mathbf{X}$ with respect to its argument and

$z_j=(y_j-x_j)/v\,|\boldsymbol{x}-\mathbf{y}|,$

$$J=\begin{vmatrix}1+z_1X_1' & z_2X_1' & z_3X_1'\\ z_1X_2' & 1+z_2X_1' & z_3X_2'\\ z_1X_3' & z_2X_3' & 1+z_3X_3'\end{vmatrix}$$
$$=1+(\mathbf{y}-\boldsymbol{x})\cdot\mathbf{X}'(t-|\boldsymbol{x}-\mathbf{y}|/v)/v\,|\boldsymbol{x}-\mathbf{y}|.$$

The time derivative of $\mathbf{X}(t)$ is the velocity $\boldsymbol{u}(t)$ of the particle. Accordingly,

$$p(\boldsymbol{x},t)=\frac{1}{4\pi[|\boldsymbol{x}-\mathbf{y}|-(\boldsymbol{x}-\mathbf{y})\cdot\boldsymbol{u}(t-|\boldsymbol{x}-\mathbf{y}|/v)/v\,]_0}, \tag{1.4}$$

where $[\;]_0$ means that $\mathbf{y}$ satisfies

$$\mathbf{y}-\mathbf{X}(t-|\boldsymbol{x}-\mathbf{y}|/v)=0, \tag{1.5}$$

is the desired solution.

The interpretation of (1.4) and (1.5) is that, to determine the field at the point P at time t (Fig. 3.1), the position P_1 of the particle at an earlier time must be found. This position must be such that a wave from it travelling at speed v just reaches P at time t. Then, if the source was at P_1 at time t_1

$$P_1P=v(t-t_1). \tag{1.6}$$

Thus, in (1.4), $\mathbf{y}$ is the position vector of P_1, $\boldsymbol{x}-\mathbf{y}$ is the vector $\overrightarrow{P_1P}$ and $\boldsymbol{u}$ is the velocity of the source at time t_1, i.e. at the point P_1.

Another way of stating this result is that, at each instant, the source emits a spherical wave and this wave travels outwards with speed v. At time t the wavefront emitted from P_1 lies on a sphere of radius $v(t-t_1)$ and centre P_1 on account of (1.6).

In calculating the gradient of p it has to be remembered that $\mathbf{y}$ is a function of $\boldsymbol{x}$ and t given implicitly by (1.5). Consequently,

$$\partial\mathbf{y}/\partial x_j+\mathbf{X}'\,\partial\,|\boldsymbol{x}-\mathbf{y}|/\partial x_j v=0$$

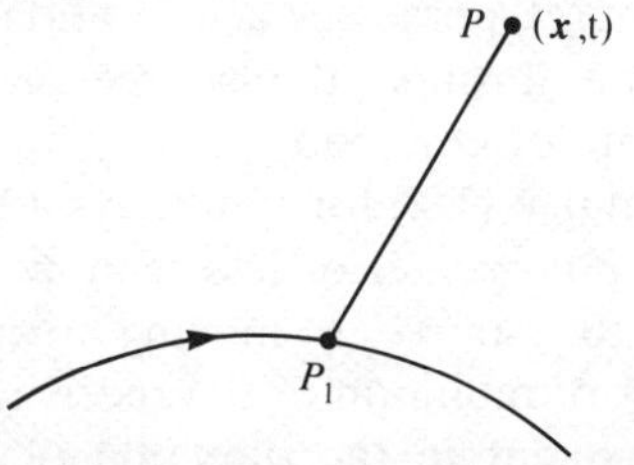

Fig. 3.1 Radiation from a moving source

and

$$\partial\,|\boldsymbol{x}-\mathbf{y}|/\partial x_j = \{x_j - y_j - (\boldsymbol{x}-\mathbf{y})\cdot\partial\mathbf{y}/\partial x_j\}/|\boldsymbol{x}-\mathbf{y}|.$$

Combining these two equations we see that

$$\frac{\partial}{\partial x_j}|\boldsymbol{x}-\mathbf{y}| = \frac{x_j - y_j}{|\boldsymbol{x}-\mathbf{y}| - (\boldsymbol{x}-\mathbf{y})\cdot\boldsymbol{u}/v}$$

or

$$\text{grad}\,|\boldsymbol{x}-\mathbf{y}| = (\boldsymbol{x}-\mathbf{y})4\pi p. \tag{1.7}$$

Hence

$$\partial\mathbf{y}/\partial x_j = -\boldsymbol{u}(x_j - y_j)4\pi p/v. \tag{1.8}$$

It follows from (1.4) that

$$\text{grad}\,p = -(4\pi)^2 p^3\left[(\boldsymbol{x}-\mathbf{y})\left(1-\frac{\boldsymbol{u}\cdot\boldsymbol{u}}{v^2}\right)-\frac{\boldsymbol{u}}{v}\left\{|\boldsymbol{x}-\mathbf{y}|-(\boldsymbol{x}-\mathbf{y})\cdot\frac{\boldsymbol{u}}{v}\right\}\right.$$
$$\left.+(\boldsymbol{x}-\mathbf{y})\{(\boldsymbol{x}-\mathbf{y})\cdot\boldsymbol{u}'\}/v^2\right]_0 \tag{1.9}$$

where $\boldsymbol{u}'$ is the derivative of $\boldsymbol{u}$ with respect to its argument.

With regard to the radiation of energy a measure is provided by an intensity $\boldsymbol{I}$ defined by

$$\boldsymbol{I} = p\,\text{grad}\,p$$

(compare §1.23). When $|\boldsymbol{x}-\mathbf{y}|$ is large the dominant term in this is, from (1.9),

$$\boldsymbol{I} \sim -(4\pi)^2 p^4[(\boldsymbol{x}-\mathbf{y})\{(\boldsymbol{x}-\mathbf{y})\cdot\boldsymbol{u}'\}/v^2]_0. \tag{1.10}$$

To this order of approximation the energy at large distances is directed along the radius vector from the point of emission to the point of observation. In fact, the terms neglected in (1.9) in arriving at (1.10) supply no net contribution to the energy flow through a large sphere at infinity because they are $O(|\boldsymbol{x}-\mathbf{y}|^{-3})$. We conclude that there is no radiation of energy if $\boldsymbol{u}'=\mathbf{0}$, i.e. *only an accelerating source can radiate energy*. Even if the source possesses acceleration there is no appreciable radiation in the plane through the source and perpendicular to the acceleration at the time of emission.

Notice that the formulae (1.4) for p displays a Doppler effect (§2.11) in that the amplitude of p is greater or less than it would be for a stationary source according as the source is moving towards or away from the observer at the instant of radiation of the received wave. Of course, if the velocity has no component in the direction of the observer the signal received is the same as if the source were immobile.

3.2 The Liénard–Wiechert potentials

The equations satisfied by the electromagnetic scalar and vector potentials are similar to (1.1). Indeed, for a moving point charge Q in free space take the charge density to be $Q\delta\{\boldsymbol{x}-\mathbf{X}(t)\}$ and then

$$\nabla^2 V - \frac{1}{c^2}\frac{\partial^2 V}{\partial t^2} = -Q\delta\{\boldsymbol{x}-\mathbf{X}(t)\}/\varepsilon_0, \tag{2.1}$$

$$\nabla^2 \mathbf{A} - \frac{1}{c^2}\frac{\partial^2 \mathbf{A}}{\partial t^2} = -\mu_0 Q\boldsymbol{u}(t)\delta\{\boldsymbol{x}-\mathbf{X}(t)\} \tag{2.2}$$

from §1.10. It is an immediate deduction from (1.4) that

$$V(\boldsymbol{x}, t) = Q/4\pi\varepsilon_0 q, \tag{2.3}$$

$$\mathbf{A}(\boldsymbol{x}, t) = \mu_0 Q[\boldsymbol{u}]_0/4\pi\varepsilon_0 q \tag{2.4}$$

where

$$q = [|\boldsymbol{x}-\mathbf{y}| - (\boldsymbol{x}-\mathbf{y})\cdot \boldsymbol{u}(t-|\boldsymbol{x}-\mathbf{y}|/c)/c]_0,$$
$$\mathbf{y} = \mathbf{X}(t-|\boldsymbol{x}-\mathbf{y}|/c).$$

The potentials (2.3) and (2.4) are known as the *Liénard-Wiechert potentials.*

For the electromagnetic field derivatives of V and $\mathbf{A}$ are involved because

$$\boldsymbol{E} = -\text{grad } V - \frac{\partial \mathbf{A}}{\partial t}, \qquad \mu_0 \boldsymbol{H} = \text{curl } \mathbf{A}.$$

Here (1.7) and (1.8) are pertinent. Also required are

$$\partial |\boldsymbol{x}-\mathbf{y}|/\partial t = -(\boldsymbol{x}-\mathbf{y})\cdot \boldsymbol{u}/q,$$
$$\partial(\boldsymbol{x}-\mathbf{y})/\partial t = -|\boldsymbol{x}-\mathbf{y}|\, \boldsymbol{u}/q.$$

When these formulae are used and the terms rearranged there results

$$\boldsymbol{E} = \frac{Q}{4\pi\varepsilon_0 q^3}\left[\left(1-\frac{\boldsymbol{u}\cdot\boldsymbol{u}}{c^2}\right)\left(\boldsymbol{x}-\mathbf{y}-\frac{\boldsymbol{u}}{c}|\boldsymbol{x}-\mathbf{y}|\right)\right.$$
$$\left. + \frac{1}{c^2}(\boldsymbol{x}-\mathbf{y})\wedge\left\{\left(\boldsymbol{x}-\mathbf{y}-\frac{\boldsymbol{u}}{c}|\boldsymbol{x}-\mathbf{y}|\right)\wedge \boldsymbol{u}'\right\}\right]_0, \tag{2.5}$$

$$\boldsymbol{H} = c\varepsilon_0(\boldsymbol{x}-\mathbf{y})\wedge \boldsymbol{E}/|\boldsymbol{x}-\mathbf{y}|. \tag{2.6}$$

The magnetic intensity is always perpendicular to the electric intensity and to the radius vector from the position of the charge at the moment of emission. The electric intensity is not perpendicular to the radius vector although it is nearly so at large distances where the last term in the square bracket of (2.5) predominates.

As regards energy radiation the relevant entity is the Poynting vector (§1.27)

$$\boldsymbol{S}=\boldsymbol{E}\wedge\boldsymbol{H}=c\varepsilon_0\boldsymbol{E}\wedge\{(\boldsymbol{x}-\mathbf{y})\wedge\boldsymbol{E}\}/|\boldsymbol{x}-\mathbf{y}|$$

from (2.6). At large distances where only the last term in (2.5) is retained

$$\boldsymbol{S}\sim c\varepsilon_0\boldsymbol{E}\cdot\boldsymbol{E}(\boldsymbol{x}-\mathbf{y})/|\boldsymbol{x}-\mathbf{y}|.$$

Once again *there is no radiation unless the charge has acceleration.*

If the acceleration of the charge is in the same direction as its velocity i.e. $\boldsymbol{u}$ and $\boldsymbol{u}'$ are parallel

$$\frac{\boldsymbol{S}\cdot(\boldsymbol{x}-\mathbf{y})}{|\boldsymbol{x}-\mathbf{y}|}=\frac{Q^2\boldsymbol{u}'\cdot\boldsymbol{u}'}{16\pi^2\varepsilon_0c^3\,|\boldsymbol{x}-\mathbf{y}|^2}\frac{\sin^2\theta}{\{1-(u/c)\cos\theta\}^6} \tag{2.7}$$

where θ is the angle between $\boldsymbol{u}$ and $\boldsymbol{x}-\mathbf{y}$, and u is the magnitude of $\boldsymbol{u}$. For small enough speed the ratio u/c can be neglected and the maximum radiation occurs when $\theta=\frac{1}{2}\pi$ i.e. in a direction transverse to that of the velocity. As u increases from small values the angle of maximum radiation changes from $\theta=\frac{1}{2}\pi$ and approaches $\theta=0$, being given by

$$\cos\theta\approx1-(1-u^2/c^2)/10$$

when $u/c\approx1$. This phenomenon has been observed in the X-rays which are produced when rapidly moving electrons are suddenly brought to rest as in an X-ray tube or television screen, although the direction of maximum radiation can only be observed with very thin anodes because the deceleration takes place along a zigzag path in solid objects.

Integration of (2.7) over a large sphere with centre $\mathbf{y}$ reveals that the rate of transmission of energy though the sphere is

$$\frac{Q^2\boldsymbol{u}'\cdot\boldsymbol{u}'}{6\pi\varepsilon_0c^3}\left(1+\frac{1}{5}\frac{u^2}{c^2}\right). \tag{2.8}$$

At low speeds when $u/c\ll1$ the part of the electromagnetic field in (2.5), (2.6) due to the acceleration of the charge is

$$\boldsymbol{E}=\frac{Q}{4\pi\varepsilon_0c^2\,|\boldsymbol{x}-\mathbf{y}|^3}(\boldsymbol{x}-\mathbf{y})\wedge\{(\boldsymbol{x}-\mathbf{y})\wedge\boldsymbol{u}'\},$$

$$\boldsymbol{H}=-\frac{Q}{4\pi c\,|\boldsymbol{x}-\mathbf{y}|^2}(\boldsymbol{x}-\mathbf{y})\wedge\boldsymbol{u}'.$$

Later on this field will be related to that of a particular source.

3.3 The self-force of an electron

When a charge moves in a given field it will, in general, accelerate and therefore radiate energy by the theory of the preceding section. This energy loss, which is not accounted for by the given field, must cause the energy of the charge to be less than its value if it moved in the given field without radiating. One possible explanation was offered by Larmor who suggested in 1912 that the proper mass of the charge did not remain constant, i.e. mass was converted into energy (see §2.7). If this explanation is not accepted the energy balance can be restored by introducing a resistive force, the *radiation reaction*, which dissipates energy at the same rate as it is radiated. The radiation reaction affects the motion of the charge and hence the exact trajectory of the charge can be found only by assuming at the outset that it is caused by the given field together with the radiation reaction. In practice, the radiation reaction is small enough not to effect the emission of radiation so that a good approximation to the path of the charge can be obtained by neglecting it. Apart from these difficulties there are others (some to be mentioned below) and so there would seem to be a good deal of merit in adopting Larmor's point of view.

In considering the radiation reaction assume, as a first approximation, that the motion of the particle is unaltered by radiation and that its speed is small. Then the rate at which energy is radiated is given by (2.8) with the terms in u^2/c^2 missing. Let the acceleration last only during the interval from t_1 to t_2. Then

$$\int_{t_1}^{t_2} \boldsymbol{u}' \cdot \boldsymbol{u}' \, \mathrm{d}t = [\boldsymbol{u}' \cdot \boldsymbol{u}]_{t_1}^{t_2} - \int_{t_1}^{t_2} \boldsymbol{u} \cdot \boldsymbol{u}'' \, \mathrm{d}t = -\int_{t_1}^{t_2} \boldsymbol{u} \cdot \boldsymbol{u}'' \, \mathrm{d}t.$$

Hence the radiation reaction or *self-force* $\boldsymbol{F}_s$ complies with

$$\int_{t_1}^{t_2} \boldsymbol{F}_s \cdot \boldsymbol{u} \, \mathrm{d}t = \int_{t_1}^{t_2} \frac{Q^2}{6\pi\varepsilon_0 c^3} \boldsymbol{u} \cdot \boldsymbol{u}'' \, \mathrm{d}t$$

which can be satisfied by putting

$$\boldsymbol{F}_s = Q^2 \boldsymbol{u}''/6\pi\varepsilon_0 c^3.$$

It can be proved that this force is consistent with the conservation of linear momentum and of angular momentum.

Another way of calculating the self-force is to assume that the charge is a small sphere carrying a specific charge distribution. An element of charge $\rho_1 \, \mathrm{d}\boldsymbol{x}_1$ produces on the element $\rho \, \mathrm{d}\boldsymbol{x}$ a force (§2.10) $\rho \, \mathrm{d}\boldsymbol{x}(\boldsymbol{E} + \boldsymbol{v} \wedge \mu_0 \boldsymbol{H})$ where $\boldsymbol{E}$ and $\boldsymbol{H}$ are calculated as in §3.2. The force on the whole charge due to $\rho_1 \, \mathrm{d}\boldsymbol{x}_1$ is then obtained by integration over the sphere; a further integration over the sphere (with respect to $\boldsymbol{x}_1$) supplies the total force due to all elements. The force so found is

$$\frac{Q^2}{6\pi\varepsilon_0 c^2}\left(\frac{\boldsymbol{u}''}{c} - \frac{\boldsymbol{u}'}{a}\right) + O(a) \tag{3.1}$$

where a is the radius of the sphere.

The first term of (3.1) is the radiation reaction already postulated. The second term represents an *inertial force* since it is proportional to the acceleration. Thus $Q^2/6\pi\varepsilon_0 c^2 a$ is regarded as a mass, the *electromagnetic mass*. This mass must be combined with the inertial mass so it would seem more logical to follow Larmor's hypothesis than to introduce a change of mass in this roundabout manner.

It will be noted that the second term of (3.1) tends to infinity as $a \rightarrow 0$, thereby implying that the self-energy also tends to infinity. Avoiding an infinite self-energy for a point charge is obviously a desirable goal.

In any case the radiation reaction has been discussed on the basis that the velocity and acceleration are small. Roughly speaking, this means that in the interval in which a signal traverses the charge distribution the time derivative of a quantity has to change by a smaller amount than the quantity itself i.e. $a\,|\boldsymbol{u}''| \ll c\,|\boldsymbol{u}'|$, $a\,|\boldsymbol{u}'''| \ll c\,|\boldsymbol{u}''|, \ldots$. For periodic processes with frequency $\omega/2\pi$ and wavelength λ this will be true if $\omega \ll c/a$ or $\lambda \gg a$. For an electron a is of the order of 10^{-13} cm so that the radiation reaction is negligible whenever $\lambda \gg 10^{-13}$ cm.

The accelerations in the betatron and synchrotron are large so that the above theory will not be applicable to them. Indeed, the determination of the radiation reaction along the lines described is extremely difficult.

3.4 Cerenkov radiation

The derivation of (1.4) assumed that the speed u of the source was less than the speed v of the propagation of signals. Likewise, (2.5) and (2.6) are based on the speed u being less than the speed of light c. This is enforced by the theory of relativity (§2.5) which demands that the speed of a particle in free space should not exceed the speed of light. There is no corresponding restriction in acoustics where $v \ll c$ so that it is feasible to have particles travelling faster than the waves. Moreover, in a dielectric where (2.5) and (2.6) are still relevant provided that μ_0, ε_0, and c are replaced by μ, ε, and v $(=1/(\mu\varepsilon)^{\frac{1}{2}})$ respectively it is possible for the speed of a particle to be less than c but greater than v since $v < c$.

What happens when the speed of the particle is greater than the speed of wave propagation will now be discussed and the acoustic case will be concentrated on firstly. The source now moves ahead of the waves which it emits. Some of the wavefronts can then intersect each other as shown in Fig. 3.2 and the spheres have an envelope, part of which is displayed as PQR; there is a cusp at Q. In this case every point to the left of PQR is on the intersection of at least two waves and at every point of PQR two waves touch. To the right of PQR there are no waves at all. Thus (1.5) will have at least two solutions when $\boldsymbol{x}$ is to the left of PQR and none when $\boldsymbol{x}$ is to the right. It is to be expected that the field is much larger on PQR than elsewhere so that the linear theory of acoustics may no longer

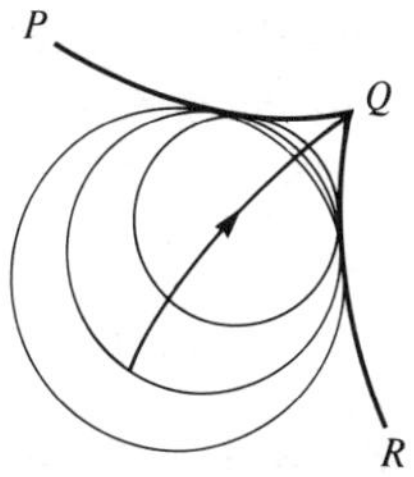

Fig. 3.2 Waves emitted by a source moving faster than the waves propagate

be adequate so that nonlinear terms may have to be kept and the presence of a shock wave allowed for.

To study a case in detail suppose that the source is moving with uniform speed u along the x-axis; an application would be to the production of bangs by a supersonic aircraft in steady flight. Let the source be at the origin at time $t=0$ so that at time t_1 its position is $(ut_1, 0, 0)$. For the point of observation (x, y, z) at time t (1.6) then becomes

$$(x-ut_1)^2+y^2+z^2=v^2(t-t_1)^2$$

or

$$(t-t_1)^2(v^2-u^2)-2u(x-ut)(t-t_1)-(x-ut)^2-y^2-z^2=0.$$

Hence

$$(v^2-u^2)(t-t_1)=(x-ut)u\pm\{(x-ut)^2v^2+(y^2+z^2)(v^2-u^2)\}^{\frac{1}{2}}. \quad (4.1)$$

In order that a signal be not received before it is transmitted it is necessary to insist that $t_1 \leqslant t$. There are two possibilities to be considered.

(a) *Speed of source less than the speed of waves*

When the source travels slower than the rate at which signals are propagated $u<v$. The square root in (4.1) then never vanishes, except in the special case when the point of observation coincides with the source, a case which will not be considered henceforth. The square root is never less than $|(x-ut)v|$ and so is always greater than $|(x-ut)u|$. Thus, only the upper sign in (4.1) leads to $t \geqslant t_1$ and, since $\mathbf{y}$ in (1.4), (1.5) is identified with $(ut_1, 0, 0)$,

$$|\boldsymbol{x}-\mathbf{y}|-(\boldsymbol{x}-\mathbf{y})\cdot\mathbf{u}/v=\{(x-ut)^2+(y^2+z^2)(1-u^2/v^2)\}^{\frac{1}{2}},$$

Hence

$$p=1/4\pi\{(x-ut)^2+(y^2+z^2)(1-u^2/v^2)\}^{\frac{1}{2}}. \quad (4.2)$$

At every point of observation a disturbance exists for all time.

(b) *Speed of source greater than that of waves*

For a source moving faster than the speed of signals $u > v$ and the square root in (4.1) can be real, zero, or imaginary. The square root vanishes at points which satisfy

$$(x-ut)^2v^2+(y^2+z^2)(v^2-u^2)=0. \tag{4.3}$$

This is a cone PQR whose vertex Q is at the source and whose semi-vertical angle is $\sin^{-1}(v/u)$ (see Fig. 3.3).

When the square root is imaginary no real value of t_1 can be obtained from (4.1) and so no disturbance can reach points where the left-hand side of (4.3) is negative. All such points are to the right of PQR in Fig. 3.3.

When the square root is real it cannot exceed $|(x-ut)v|$ and therefore cannot be greater than $|(x-ut)u|$. Therefore, $t-t_1$ can be positive only for $ut > x$ i.e. for points to the left of Q in Fig. 3.3. To the left of PQR the left-hand side of (4.3) is positive and becomes negative on crossing PQR. Hence $t-t_1$ is real and positive only for points to the left of PQR. Thus the semi-cone PQR divides the space into two regions: (i) that to the right where there is no disturbance and (ii) that to the left where the field is due to two waves. For example, the disturbance at A arises from waves radiated from the two earlier positions S_1 and S_2 of the source. Because of its importance in separating the two regions PQR is called the *Mach cone.*

For the two waves which arrive simultaneously at a point of observation the values of $|\mathbf{x}-\mathbf{y}|-(\mathbf{x}-\mathbf{y})\cdot\mathbf{u}/v$ are $\{(x-ut)^2-(y^2+z^2)(u^2/v^2-1)\}^{\frac{1}{2}}$ and $-\{(x-ut)^2-(y^2+z^2)(u^2/v^2-1)\}^{\frac{1}{2}}$. However, in the derivation of (1.4) the quantity J must always be chosen positive. Therefore, the sign of the

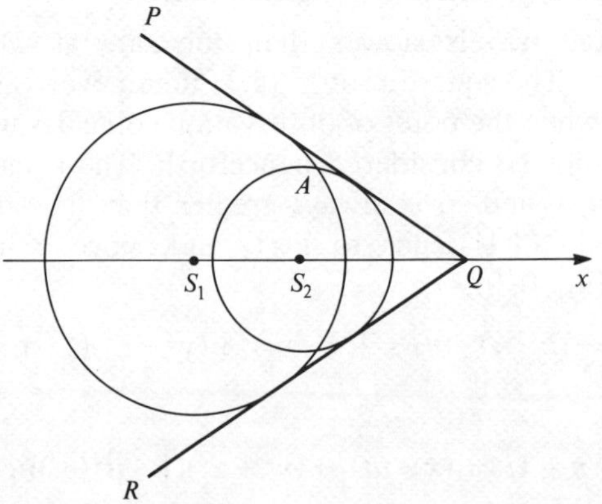

Fig. 3.3 The radiation of a supersonic source

second value must be reversed and

$$p = \begin{cases} 1/2\pi\{(x-ut)^2-(y^2+z^2)(u^2/v^2-1)\}^{\frac{1}{2}} & \text{(to the left of } PQR\text{)} \\ 0 & \text{(to the right of } PQR\text{)} \end{cases}. \tag{4.4}$$

It will be noted that p becomes infinite on the Mach cone PQR, though this infinity can disappear when extended distributions are under consideration instead of a single point source.

For the electromagnetic radiation from a point charge moving with uniform velocity in a dielectric the scalar potential can be deduced immediately from (4.2) and (4.4). If $u<v$

$$V = Q/4\pi\varepsilon\{(x-ut)^2+(y^2+z^2)(1-u^2/v^2)\}^{\frac{1}{2}}, \tag{4.5}$$

If $u>v$, V is zero to the right of PQR whereas it is given by

$$V = Q/2\pi\varepsilon\{(x-ut)^2+(y^2+z^2)(1-u^2/v^2)\}^{\frac{1}{2}} \tag{4.6}$$

to the left of PQR. The vector potential $\mathbf{A}$ is easily inferred and its only nonzero component is parallel to the direction of motion, being given by

$$A_x = \mu\varepsilon u V = uV/v^2. \tag{4.7}$$

It follows that the electric intensity at (x, y, z) to the left of PQR is

$$E_x = (x-ut)\Phi, \qquad E_y = y\Phi, \qquad E_z = z\Phi \tag{4.8}$$

where

$$\Phi = Q\mu(v^2-u^2)/2\pi\{(x-ut)^2+(y^2+z^2)(1-u^2/v^2)\}^{\frac{3}{2}}.$$

Thus, the direction cosines of the electric intensity are proportional to $x-ut$, y, z. In particular, *the electric intensity on the Mach cone lies along a generator of the cone.* It may also be shown that the magnetic intensity on the Mach cone is tangential to the circular cross-section of the cone and that *the Poynting vector is normal to the cone.* At a fixed point (x, y, z) the term ut dominates the others as the time becomes large and then the radiation is essentially polarized with the electric vector parallel to the direction of motion of the charge.

The radiation can be analysed into harmonic components by means of a Fourier transform with respect to θ where $\theta = ut-x$, namely

$$\int_{-\infty}^{\infty} V e^{-i\alpha\theta}\,d\theta = (-iQ/4\varepsilon)H_0^{(2)}\{\alpha(y^2+z^2)^{\frac{1}{2}}\tan\beta\} \tag{4.9}$$

where $H_0^{(2)}$ is the Hankel function (Appendix A) and $\cos\beta = v/u$. Formula (4.9) holds when $\alpha>0$; if $\alpha<0$ replace $-iH_0^{(2)}$ by $iH_0^{(1)}\{-\alpha(y^2+z^2)^{\frac{1}{2}}\tan\beta\}$. The spectrum is continuous and, in the electromagnetic case, stretches from the infra-red to the ultraviolet. Whether or not such radiation is observable depends upon its intensity. As $u\to v$ the intensity tends to zero but as $u \to \infty$ the intensity becomes of the order of unity. Therefore, a rapidly moving source is more likely to produce detectable radiation than one of low speed.

The radiation which occurs when $u > v$ is known as *Cerenkov radiation* since Cerenkov was the first person to carry out careful experiments into the phenomenon in the visible spectrum. The experimental results bear out completely the theoretical predictions and the fact that the intensity increases with the speed makes it a useful mechanism for detecting high energy sources.

The field of a source of variable strength

3.5 Simple sources and sinks

Hitherto the radiation from a moving source of constant strength has been under consideration. Now the field from a source fixed in space but varying in strength will be discussed. The solution of

$$\nabla^2 p - \frac{1}{a_0^2}\frac{\partial^2 p}{\partial t^2} = -f(t)\delta(\boldsymbol{x}-\mathbf{y}) \tag{5.1}$$

has been examined in §1.16. An outgoing wave is provided by

$$p(\boldsymbol{x}) = \frac{f(t-|\boldsymbol{x}-\mathbf{y}|/a_0)}{4\pi\,|\boldsymbol{x}-\mathbf{y}|} \tag{5.2}$$

and, since it causes energy to leave the point $\mathbf{y}$, is said to be due to a *simple source* at $\mathbf{y}$, its strength at time t being $f(t)$. Similarly, the field

$$p(\boldsymbol{x}) = \frac{g(t+|\boldsymbol{x}-\mathbf{y}|/a_0)}{4\pi\,|\boldsymbol{x}-\mathbf{y}|}, \tag{5.3}$$

which makes energy flow to the point $\mathbf{y}$, is attributed to a *simple sink* of strength $g(t)$ at $\mathbf{y}$.

It will be observed that, if a simple source is active for only a finite time, the acoustic wave is subsequently confined to the space between two spherical surfaces progressing outwards with speed a_0.

The power radiated by a simple source can be calculated from the sound field at a large distance from the source. No energy is dissipated as the sound propagates and so what leaves the source must eventually pass through any closed surface surrounding the source. Allowance has to be made for the time taken by the energy to reach the surface so that computation on the sphere Ω of radius R, centred on the source, refers to what was happening at the source at a time R/a_0 earlier. Now the velocity $\boldsymbol{v}$ is related to p by $\rho_0\, \partial \boldsymbol{v}/\partial t = -\text{grad}\, p$ with ρ_0 the undisturbed density. Place the origin at the source and then (5.2) reveals that $\boldsymbol{v}$ has only a radial component v_R satisfying

$$\rho_0\,\partial v_R/\partial t = \{Rf'(t-R/a_0) + a_0 f(t-R/a_0)\}/4\pi a_0 R^2. \tag{5.4}$$

At large distances the second term can be neglected and $v_R \sim f(t-R/a_0)/4\pi a_0\rho_0 R$. Thus, on Ω, the acoustic intensity (§1.23) is given by

$$\boldsymbol{I} = p\boldsymbol{v} = \hat{\boldsymbol{R}} f^2(t-R/a_0)/(4\pi R)^2 a_0\rho_0 \tag{5.5}$$

where $\hat{\boldsymbol{R}}$ is a unit vector along the radius vector from the origin. The power crossing the sphere Ω is therefore

$$P = \int_\Omega \boldsymbol{I}\cdot\hat{\boldsymbol{R}}\,\mathrm{d}\Omega = f^2(t-R/a_0)/4\pi a_0\rho_0 \tag{5.6}$$

at time t. This implies that the power radiated in the immediate vicinity of the simple source at time t_0 is $f^2(t_0)/4\pi a_0\rho_0$.

When the time variation is harmonic take $f(t) = A\mathrm{e}^{\mathrm{i}\omega t}$ and then the field of a simple source at the origin is

$$p = A\mathrm{e}^{\mathrm{i}(\omega t - kR)}/4\pi R \tag{5.7}$$

where $k = \omega/a_0$. This is the outgoing solution of

$$\nabla^2 p + k^2 p = -A\delta(\boldsymbol{x}). \tag{5.8}$$

The velocity is obtained from $\boldsymbol{v} = -\mathrm{grad}\, p/\mathrm{i}\omega\rho_0$ and has only the radial component

$$v_R = \frac{A}{4\pi a_0\rho_0 R}\left(1 - \frac{\mathrm{i}}{kR}\right)\mathrm{e}^{\mathrm{i}(\omega t - kR)}. \tag{5.9}$$

Suppose that A is chosen so that

$$A = 4\pi\rho_0 a_0 b u(1-\mathrm{i}/kb)^{-1}\mathrm{e}^{ikb}.$$

Then, from (5.9), $v_R = u\mathrm{e}^{\mathrm{i}\omega t}$ on $R = b$. Such a velocity could be regarded as being generated by a pulsating sphere of radius b making minute harmonic radial oscillations in which the velocity of the surface was $u\mathrm{e}^{\mathrm{i}\omega t}$ (it is assumed that there is no cavitation to the fluid). The total pressure on the sphere due to p is

$$4\pi\rho_0 a_0 b^2 u(1-\mathrm{i}/kb)^{-1}.$$

The ratio of this to u is known as the *radiation impedance* Z_r i.e.

$$Z_r = 4\pi\rho_0 a_0 b^2(1-\mathrm{i}/kb)^{-1}. \tag{5.10}$$

The average power radiated by the vibrating sphere can be calculated by averaging (5.6) or by using the complex acoustic intensity $\frac{1}{2}p\boldsymbol{v}^*$ (§1.28). Either way

$$P = |A|^2/8\pi\rho_0 a_0 = 2\pi\rho_0 a_0 b^2\,|u|^2\,(1 + 1/k^2b^2)^{-1}. \tag{5.11}$$

Write $Z_r = R_r + iX_r$ where R_r and X_r are real. Then

$$R_r = 4\pi\rho_0 a_0 k^2 b^4/(1+k^2b^2), \tag{5.12}$$

$$P = \tfrac{1}{2}R_r\,|u|^2. \tag{5.13}$$

The power input necessary to sustain the oscillation of the sphere in the fluid is given by (5.13). It depends linearly on R_r which may be called the *radiation resistance.*

Dividing the radiation resistance by $\rho_0 a_0$ and the area of the sphere supplies the *radiation efficiency* R_e, i.e.

$$R_e = k^2b^2/(1+k^2b^2). \tag{5.14}$$

Since $R_e \to 1$ as $kb \to \infty$, the radiation efficiency is a measure of how well the sphere radiates as compared with a flat plate of the same area as the sphere. The variation of R_e with kb is shown on a logarithmic scale in Fig. 3.4.

By spreading point sources along a line parallel to the z-axis a field which is independent of z can be produced. So the prime mover of a two-dimensional disturbance may be termed a *line source.* A harmonic line source which satisfies

$$\frac{\partial^2 p}{\partial x^2} + \frac{\partial^2 p}{\partial y^2} + k^2 p = -\delta(\boldsymbol{x} - \boldsymbol{x}_0) \tag{5.15}$$

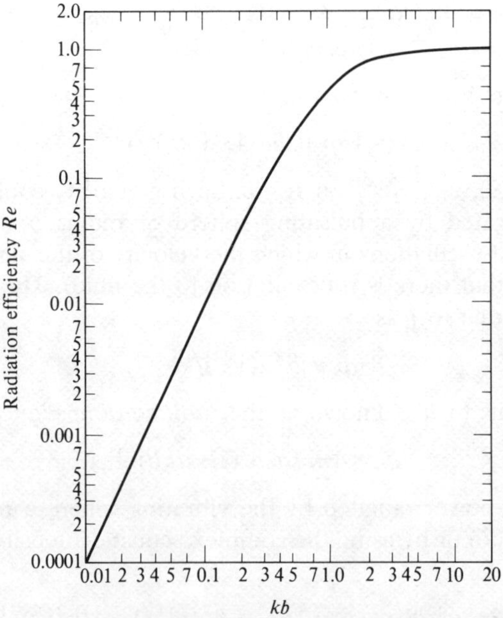

Fig. 3.4 The radiation efficiency of a pulsating sphere of radius b.

where $\boldsymbol{x}$ represents the point (x, y) is

$$p = -\tfrac{1}{4}\mathrm{i}\mathrm{H}_0^{(2)}(k\,|\boldsymbol{x}-\boldsymbol{x}_0|) \tag{5.16}$$

if the time dependence is suppressed.

3.6 The electric dipole

The electromagnetic field in free space due to a varying source may be derived from the Hertz vector (§1.12) satisfying

$$\nabla^2 \boldsymbol{\Pi} - \partial^2\boldsymbol{\Pi}/c^2\,\partial t^2 = -f(t)\mathbf{p}_0\delta(\boldsymbol{x})/\varepsilon_0 \tag{6.1}$$

when the source is at the origin, $\mathbf{p}_0$ being a constant vector. Since each component of (6.1) is of the form (5.1)

$$\boldsymbol{\Pi} = \mathbf{p}_0 f(t-R/c)/4\pi\varepsilon_0 R \tag{6.2}$$

where R is the distance from the source. Hence

$$\begin{aligned}\mathbf{E} &= \operatorname{grad}\operatorname{div}\boldsymbol{\Pi} - \partial^2\boldsymbol{\Pi}/c^2\,\partial t^2\\ &= \frac{1}{4\pi\varepsilon_0 R^2}\left(\frac{f'}{c}+\frac{f}{R}\right)\{3\hat{\boldsymbol{R}}(\hat{\boldsymbol{R}}\cdot\mathbf{p}_0)-\mathbf{p}_0\} + \frac{f''}{4\pi\varepsilon_0 c^2 R}\hat{\boldsymbol{R}}\wedge(\hat{\boldsymbol{R}}\wedge\mathbf{p}_0),\end{aligned} \tag{6.3}$$

$$\mathbf{H} = \varepsilon_0 \operatorname{curl}\left(\frac{\partial\boldsymbol{\Pi}}{\partial t}\right) = -\frac{\hat{\boldsymbol{R}}\wedge\mathbf{p}_0}{4\pi R}\left(\frac{f''}{c}+\frac{f'}{R}\right) \tag{6.4}$$

where the arguments of f, f', f'' are $t-R/c$ and the primes indicate derivatives with respect to the argument. This electromagnetic field is an outgoing solution of

$$\operatorname{curl}\mathbf{E} + \mu_0\,\partial\mathbf{H}/\partial t = \mathbf{0}, \qquad \operatorname{div}\mathbf{E} = -f(t)\operatorname{div}\mathbf{p}_0\delta(\boldsymbol{x})/\varepsilon_0, \tag{6.5}$$

$$\operatorname{curl}\mathbf{H} - \varepsilon_0\,\partial\mathbf{E}/\partial t = f'(t)\mathbf{p}_0\delta(\boldsymbol{x}), \qquad \operatorname{div}\mathbf{H} = 0, \tag{6.6}$$

When R is small the electric intensity is effectively given by

$$\mathbf{E} = f(t)\{3\hat{\boldsymbol{R}}(\hat{\boldsymbol{R}}\cdot\mathbf{p}_0)-\mathbf{p}_0\}/4\pi\varepsilon_0 R^3$$

which is the same as the electrostatic field of a dipole of moment $f(t)\mathbf{p}_0$. Also the dominant term in the magnetic intensity is $-\hat{\boldsymbol{R}}\wedge\mathbf{p}_0 f'(t)/4\pi R^2$ which coincides with the field, obtained by the Biot–Savart law, of a current element is the direction of $\mathbf{p}_0$ of length $\mathrm{d}s$ and current I where $I\,\mathrm{d}s = f'(t)$. Therefore, (6.2)–(6.4) are called the Hertz vector and electromagnetic field of *an electric dipole of moment* $f(t)\mathbf{p}_0$.

It will be observed that *the magnetic intensity is always perpendicular to the plane containing the dipole and the radius vector.* On the other hand, the electric field always *lies in that plane* (Fig. 3.5).

At large distances from the dipole the dominant field is

$$\mathbf{E} \sim f''\hat{\boldsymbol{R}}\wedge(\hat{\boldsymbol{R}}\wedge\mathbf{p}_0)/4\pi\varepsilon_0 c^2 R, \tag{6.7}$$

$$\mathbf{H} \sim -\hat{\boldsymbol{R}}\wedge\mathbf{p}_0 f''/4\pi c R. \tag{6.8}$$

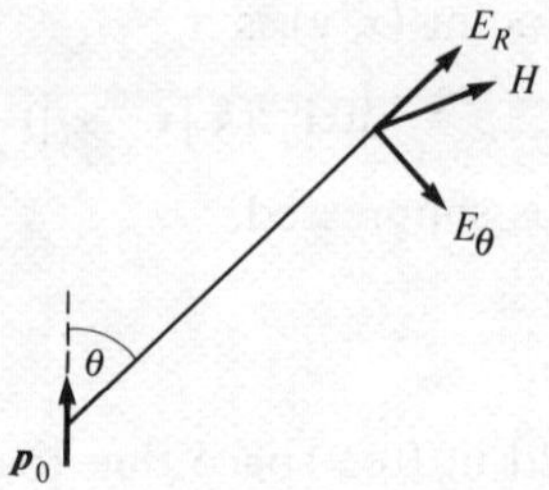

Fig. 3.5 The field of an electric dipole

Thus, *at large distances from the dipole,* the field is *transverse to the direction of propagation.* The rate at which energy is transmitted across the large sphere Ω is supplied by integrating the Poynting vector and for this (6.7), (6.8) are pertinent. If the z-axis is selected to lie along $\boldsymbol{p}_0$ and the spherical polar coordinates R, θ, ϕ are introduced (see Fig. 3.5) the rate at which energy is radiated is

$$\int_\Omega \mathbf{E}\wedge\boldsymbol{H}\cdot\boldsymbol{n}\,\mathrm{d}\Omega=(f'')^2\int_\Omega|\hat{\boldsymbol{R}}\wedge\boldsymbol{p}_0|^2\,\mathrm{d}\Omega/16\pi^2\varepsilon_0R^2c^3$$
$$=(f'')^2\int_0^\pi\int_0^{2\pi}\sin^3\theta\,\mathrm{d}\phi\,\mathrm{d}\theta/16\pi^2\varepsilon_0c^3$$
$$=\{f''(t-R/c)\}^2/6\pi\varepsilon_0c^3=\mu_0\{f''(t-R/c)\}^2/6\pi c. \quad (6.9)$$

In order to maintain the dipole energy has to be supplied at this rate. Once again it will be remarked that there is no radiation of energy unless the second derivative of the dipole moment is nonzero.

3.7 The magnetic dipole

At nearby points the electric dipole looks like a linear current element. In contrast, a steady current in a small circular wire is equivalent to a magnetostatic dipole. It may be expected, therefore, that there is an electromagnetic field in which the magnetic field is the counterpart of the electric intensity of the electric dipole. This suggests that if the formulae for $\boldsymbol{E}$ and $\boldsymbol{H}$ of the electric dipole are interchanged the field of a magnetic dipole will be obtained. One way of accomplishing this is to make the change

$$\boldsymbol{E}_1=-(\mu_0/\varepsilon_0)^{\frac{1}{2}}\boldsymbol{H},\qquad \boldsymbol{H}_1=(\varepsilon_0/\mu_0)^{\frac{1}{2}}\boldsymbol{E}$$

but to drop the suffix 1 after the substitution has been achieved. At the same time put $\boldsymbol{p}_0=(\varepsilon_0/\mu_0)^{\frac{1}{2}}\boldsymbol{m}_0$. Then a solution of (c.f. §1.19)

$$\operatorname{curl}\mathbf{E}+\mu_0\,\partial\boldsymbol{H}/\partial t=-f'(t)\boldsymbol{m}_0\delta(\boldsymbol{x}),\qquad \operatorname{div}\mathbf{E}=0, \quad (7.1)$$
$$\operatorname{curl}\boldsymbol{H}-\varepsilon_0\,\partial\mathbf{E}/\partial t=\mathbf{0},\qquad \operatorname{div}\boldsymbol{H}=-f(t)\operatorname{div}\boldsymbol{m}_0\delta(\boldsymbol{x})/\mu_0 \quad (7.2)$$

is

$$\boldsymbol{E} = \frac{\hat{\boldsymbol{R}} \wedge \boldsymbol{m}_0}{4\pi R}\left(\frac{f''}{c} + \frac{f'}{R}\right), \tag{7.3}$$

$$\boldsymbol{H} = \frac{1}{4\pi\mu_0 R^2}\left(\frac{f'}{c} + \frac{f}{R}\right)\{3\hat{\boldsymbol{R}}(\hat{\boldsymbol{R}} \cdot \boldsymbol{m}_0) - \boldsymbol{m}_0\} + \frac{f''}{4\pi\mu_0 c^2 R}\hat{\boldsymbol{R}} \wedge (\hat{\boldsymbol{R}} \wedge \boldsymbol{m}_0). \tag{7.4}$$

This electromagnetic field can be derived from a Hertz vector for which

$$\frac{\partial \boldsymbol{\Pi}}{\partial t} = \frac{\boldsymbol{m}_0 \wedge \hat{\boldsymbol{R}}}{4\pi\varepsilon_0\mu_0 R}\left(\frac{f'}{c} + \frac{f}{R}\right). \tag{7.5}$$

As $R \to 0$ the magnetic intensity in (7.4) behaves like that of a magnetostatic dipole and so (7.3), (7.4) furnish the electromagnetic field of a magnetic dipole of moment $\boldsymbol{m}_0 f(t)$. When $\boldsymbol{m}_0$ is a unit vector the local magnetic field is the same as that of a current I flowing in a small circuit of area S in a plane perpendicular to $\boldsymbol{m}_0$ where $\mu_0 IS = f(t)$.

It is straightforward to deduce from §3.6 that the rate at which energy is radiated by the magnetic dipole is

$$\{f''(t - R/c)\}^2/6\pi c^3 \mu_0.$$

3.8 The harmonic electric and magnetic dipoles

When the time variation of the dipoles is harmonic the electromagnetic field can be deduced from §§3.6, 3.7 by replacing $f(t)$ by $Ae^{i\omega t}$ where A is a complex constant. The field that arises physically can be obtained by taking the real part of the resulting expressions. Actually, we shall suppress the factor $e^{i\omega t}$ but the reader must remember to insert it before taking the real part when a real field is desired.

With this convention, the electromagnetic field of a harmonic electric dipole satisfies

$$\operatorname{curl} \boldsymbol{E} + i\omega\mu_0 \boldsymbol{H} = \boldsymbol{0}, \qquad \operatorname{div} \boldsymbol{E} = -A \operatorname{div} \boldsymbol{p}_0 \delta(\boldsymbol{x})/\varepsilon_0, \tag{8.1}$$

$$\operatorname{curl} \boldsymbol{H} - i\omega\varepsilon_0 \boldsymbol{E} = i\omega A \boldsymbol{p}_0 \delta(\boldsymbol{x}), \qquad \operatorname{div} \boldsymbol{H} = 0 \tag{8.2}$$

and is given by

$$\boldsymbol{E} = \frac{Ae^{-ikR}}{4\pi\varepsilon_0 R}\left[\frac{1}{R}\left(ik + \frac{1}{R}\right)\{3\hat{\boldsymbol{R}}(\hat{\boldsymbol{R}} \cdot \boldsymbol{p}_0) - \boldsymbol{p}_0\} - k^2 \hat{\boldsymbol{R}} \wedge (\hat{\boldsymbol{R}} \wedge \boldsymbol{p}_0)\right], \tag{8.3}$$

$$\boldsymbol{H} = A\omega \hat{\boldsymbol{R}} \wedge \boldsymbol{p}_0 e^{-ikR}(kR - i)/4\pi R^2 \tag{8.4}$$

where $k = \omega/c$. The corresponding Hertz vector is

$$\boldsymbol{\Pi} = A\boldsymbol{p}_0 e^{-ikR}/4\pi\varepsilon_0 R. \tag{8.5}$$

The components in (8.3), (8.4) may be conveniently expressed in terms of spherical polar coordinates R, θ, ϕ with $\theta=0$ along $\boldsymbol{p}_0$. The only components which are not zero are

$$E_R = A\mathrm{e}^{-\mathrm{i}kR}(\mathrm{i}kR+1)\cos\theta/2\pi\varepsilon_0 R^3, \tag{8.6}$$

$$E_\theta = A\mathrm{e}^{-\mathrm{i}kR}(1+\mathrm{i}kR-k^2R^2)\sin\theta/4\pi\varepsilon_0 R^3, \tag{8.7}$$

$$H_\phi = \omega A\mathrm{e}^{-\mathrm{i}kR}(\mathrm{i}-kR)\sin\theta/4\pi R^2. \tag{8.8}$$

At large distances from the dipole where $kR \gg 1$, E_R is much smaller than the other two components and

$$E_\theta \sim -Ak^2\mathrm{e}^{-\mathrm{i}kR}\sin\theta/4\pi\varepsilon_0 R, \tag{8.9}$$

$$H_\phi \sim -A\omega k\mathrm{e}^{-\mathrm{i}kR}\sin\theta/4\pi R. \tag{8.10}$$

The average rate of energy radiation can be determined either by averaging (6.9) or from the complex Poynting vector (§1.30) and (8.9), (8.10); it is

$$\mu_0\omega^4\,|A|^2/12\pi c = \omega^4\,|A|^2\,10^{-15}/9 \text{ watt}.$$

The equivalent current element of length ds and carrying current $I_0\mathrm{e}^{\mathrm{i}\omega t}$ has $I_0\,\mathrm{d}s = \mathrm{i}\omega A$ so that the average rate at which energy is radiated can be written as

$$\tfrac{1}{3}\mu_0\pi c(\mathrm{d}s/\lambda)^2\,|I_0|^2$$

where $\lambda = 2\pi/k$ is the wavelength. A current I_0 maintained in a resistance R_0 dissipates energy at the rate $\frac{1}{2}R_0\,|I_0|^2$. So the *radiation resistance* R_r of a harmonic electric dipole is

$$R_r = \tfrac{2}{3}\mu_0\pi c\left(\frac{\mathrm{d}s}{\lambda}\right)^2 = 790(\mathrm{d}s/\lambda)^2 \text{ ohm}.$$

This theory is relevant to the radiation from the current in a wire whose length is much smaller than a wavelength and is found to be in tolerable agreement with experiment.

For a harmonic magnetic dipole at the origin, the Hertz vector is

$$\boldsymbol{\Pi} = B\boldsymbol{m}_0 \wedge \hat{\boldsymbol{R}}\mathrm{e}^{-\mathrm{i}kR}(kR-\mathrm{i})/4\pi\varepsilon_0\mu_0\omega R^2$$

and, when the dipole is oriented along the z-axis, the non-vanishing components are

$$E_\phi = -B\omega\mathrm{e}^{-\mathrm{i}kR}(\mathrm{i}-kR)\sin\theta/4\pi R^2, \tag{8.11}$$

$$H_R = B\mathrm{e}^{-\mathrm{i}kR}(1+\mathrm{i}kR)\cos\theta/2\pi\mu_0 R^3, \tag{8.12}$$

$$H_\theta = B\mathrm{e}^{-\mathrm{i}kR}(1+\mathrm{i}kR-k^2R^2)\sin\theta/4\pi\mu_0 R^3. \tag{8.13}$$

When $R \gg \lambda$ the dominant part of the field is

$$E_\phi \sim Bk\omega\mathrm{e}^{-\mathrm{i}kR}\sin\theta/4\pi R, \tag{8.14}$$

$$H_\phi \sim -Bk^2\mathrm{e}^{-\mathrm{i}kR}\sin\theta/4\pi\mu_0 R \tag{8.15}$$

and the rate of which energy is radiated is

$$\omega^4 |B|^2/12\pi c^3\mu_0 = 10k^4 |B|^2/\mu_0^2 \text{ watt}.$$

A model of a harmonic magnetic dipole is provided by the alternating current $I_0 e^{i\omega t}$ flowing in a small circular loop of radius a. Then $\mu_0 I_0 \pi a^2 = B$ and the radiation resistance is given by

$$R_r = \tfrac{8}{3}\pi^5 \mu_0 c(a/\lambda)^4 = 3.075 \times 10^5 (a/\lambda)^4 \text{ ohm}.$$

The radiation resistance is about 31 ohm when $\lambda = 10a$ and decreases rapidly as λ increases. This property can be used to design resistive coils which do not radiate.

A feature of (8.14) is that the electric intensity is a maximum when $\theta = \frac{1}{2}\pi$, i.e. at points in the plane containing the loop. By the reciprocity theorem (§1.35) the signal received by a loop will be a maximum when the plane of the loop passes through the transmitter. This forms the basis of a navigational aid in which a loop antenna is fitted to a vehicle to take bearings on radio beacons at known places.

3.9 Two-dimensional dipoles

It has been pointed out in §1.36 that for two-dimensional fields it is necessary to find only the components E_z and H_z when there is no current or charge density present. A slight modification permits a similar splitting in the presence of dipoles.

Consider first an electric line dipole of moment $Ae^{i\omega t}\boldsymbol{k}\delta(\boldsymbol{x})$ where $\boldsymbol{x}$ stands for (x, y) and $\boldsymbol{k}$ is a unit vector parallel to the z-axis. Then

$$\begin{aligned} \partial E_z/\partial y + i\omega\mu_0 H_x &= 0, & \partial H_z/\partial y - i\omega\varepsilon_0 E_x &= 0,\\ -\partial E_z/\partial x + i\omega\mu_0 H_y &= 0, & -\partial H_z/\partial x - i\omega\varepsilon_0 E_y &= 0,\\ \partial E_y/\partial x - \partial E_x/\partial y + i\omega\mu_0 H_z &= 0, & \partial H_y/\partial x - \partial H_x/\partial y - i\omega\varepsilon_0 E_z &= i\omega A\delta(\boldsymbol{x}),\\ \partial E_x/\partial x + \partial E_y/\partial y &= 0, & \partial H_x/\partial x + \partial H_y/\partial y &= 0. \end{aligned}$$

From these equations it is clear that H_z, E_x, and E_y can be taken to be zero. Also

$$\partial^2 E_z/\partial x^2 + \partial^2 E_z/\partial y^2 + k^2 E_z = -\omega^2 \mu_0 A\delta(\boldsymbol{x})$$

so that (5.15), (5.16) furnish

$$E_z = -\tfrac{1}{4} i\omega^2 \mu_0 A H_0^{(2)}(kr), \qquad i\omega\mu_0 \boldsymbol{H} = \boldsymbol{k} \wedge \operatorname{grad} E_z \tag{9.1}$$

where $r = (x^2 + y^2)^{\frac{1}{2}}$.

For the electric line dipole of moment $Ae^{i\omega t}\boldsymbol{i}\delta(\boldsymbol{x})$ where $\boldsymbol{i}$ is a unit vector along the x-axis it can be seen that E_z, H_x, and H_y are zero while

$$H_z = -\tfrac{1}{4}\omega A \frac{\partial}{\partial y} \mathrm{H}_0^{(2)}(kr), \qquad \mathrm{i}\omega\varepsilon_0 \boldsymbol{E} = -\boldsymbol{k}\wedge \operatorname{grad} H_z + \mathrm{i}\omega A \boldsymbol{i}\delta(\boldsymbol{x}), \quad (9.2)$$

the last term vanishing away from the origin.

Similarly, for the magnetic line dipole $B\mathrm{e}^{\mathrm{i}\omega t}\boldsymbol{k}\delta(\boldsymbol{x})$

$$H_z = -\tfrac{1}{4}\mathrm{i}\omega^2\varepsilon_0 B \mathrm{H}_0^{(2)}(kr), \qquad \mathrm{i}\omega\varepsilon_0 \boldsymbol{E} = -\boldsymbol{k}\wedge \operatorname{grad} H_z \quad (9.3)$$

and for the magnetic line dipole $B\mathrm{e}^{\mathrm{i}\omega t}\boldsymbol{i}\delta(\boldsymbol{x})$

$$E_z = \tfrac{1}{4}\omega B \frac{\partial}{\partial y} \mathrm{H}_0^{(2)}(kr), \qquad \mathrm{i}\omega\mu_0 \boldsymbol{H} = \boldsymbol{k}\wedge \operatorname{grad} E_z - \mathrm{i}\omega B \boldsymbol{i}\delta(\boldsymbol{x}), \quad (9.4)$$

The characteristics of linear antenna systems

3.10 Radiation from a thin wire

From a practical point of view, electric and magnetic dipoles are radiators whose dimensions are small compared with the wavelength. The next simplest radiator is one in which a single dimension is not restricted to being short. A typical example is a thin wire in which the size of the cross-section is small compared with the wavelength. Such a wire is a particular instance of an *antenna* which is the name applied to any device designed to transmit or receive radio waves (sometimes the term *aerial* is employed in Britain).

In the first place it will be assumed that the current in the wire is specified; the more difficult problem of determining the current under given excitation will be deferred to later sections. In reality radiation from an antenna is affected by the earth and the presence of objects. Such effects will be ignored in this chapter and considered in later chapters. Only an isolated straight wire will be discussed for the moment.

The first case to be examined is that in which the field is derived from a Hertz vector

$$\boldsymbol{\Pi} = \frac{\boldsymbol{k}}{4\pi\varepsilon_0}\int_{-l}^{l} \frac{f(t-\mathcal{R}/c, \zeta)}{\mathcal{R}}\,\mathrm{d}\zeta \quad (10.1)$$

where

$$\mathcal{R}^2 = x^2 + y^2 + (z-\zeta)^2.$$

This corresponds to a distribution of electric dipoles of moment $f(t, z)\boldsymbol{k}$ along the z-axis over the interval $-l \leqslant z \leqslant l$. The electromagnetic field can be calculated directly from the Hertz vector or by integrating (6.3), (6.4). At large distances (6.7) and (6.8) are relevant.

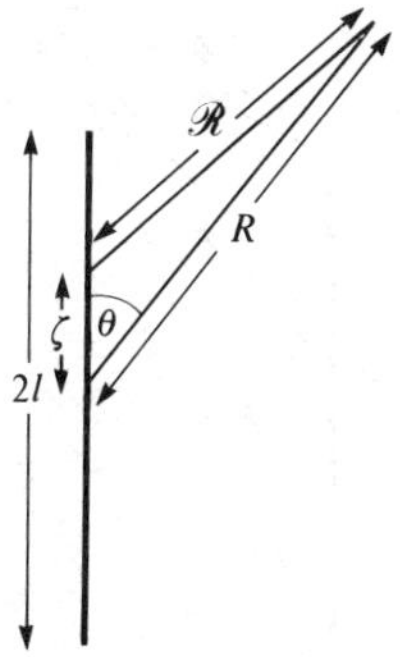

Fig. 3.6 The thin wire antenna

When the point of observation is distant let R be its distance from the origin (Fig. 3.6). Then

$$\mathscr{R} = R - z\zeta/R + O(R^{-1}) = R - \zeta \cos\theta + O(R^{-1})$$

where θ is the angle between $\boldsymbol{R}$ and the z-axis. The field (6.7) and (6.8) is transverse to the radius vector from the dipole at ζ but, since terms which are $o(R^{-1})$ can be neglected, it will be sufficient to treat the field as transverse to $\boldsymbol{R}$. Accordingly,

$$E_\theta = \left(\frac{\mu_0}{\varepsilon_0}\right)^{\frac{1}{2}} H_\phi = \frac{\sin\theta}{4\pi\varepsilon_0 c^2 R}\frac{\partial^2}{\partial t^2}\int_{-l}^{l} f\{t-(\mathscr{R}-\zeta\cos\theta)/c, \zeta\}\,\mathrm{d}\zeta + o(R^{-1}) \tag{10.2}$$

at large distances.

Further evaluation is impossible without an explicit form for f. There is one instance in which the integrals for the electric intensity can be calculated completely and that is when $f(t, \zeta) = g(t - \zeta/c)$. In this case

$$\begin{aligned} E_z &= \left(\frac{\partial^2}{\partial z^2} - \frac{1}{c^2}\frac{\partial^2}{\partial t^2}\right)\frac{1}{4\pi\varepsilon_0}\int_{-l}^{l}\frac{g\{t-(\mathscr{R}+\zeta)/c\}}{\mathscr{R}}\,\mathrm{d}\zeta \\ &= -\frac{\partial}{\partial z}\frac{1}{4\pi\varepsilon_0}\int_{-l}^{l}\left\{\frac{\partial}{\partial\zeta}\left(\frac{g}{\mathscr{R}}\right) + \frac{g'}{c\mathscr{R}}\right\}\mathrm{d}\zeta - \frac{1}{4\pi\varepsilon_0 c^2}\int_{-l}^{l}\frac{g''}{\mathscr{R}}\,\mathrm{d}\zeta \\ &= -\frac{\partial}{\partial z}\frac{1}{4\pi\varepsilon_0}\int_{-l}^{l}\frac{\partial}{\partial\zeta}\left(\frac{g}{\mathscr{R}}\right)\mathrm{d}\zeta + \frac{1}{4\pi\varepsilon_0 c}\int_{-l}^{l}\frac{\partial}{\partial\zeta}\left(\frac{g'}{\mathscr{R}}\right)\mathrm{d}\zeta \\ &= \frac{1}{4\pi\varepsilon_0}\left(\frac{\partial}{\partial z} - \frac{1}{c}\frac{\partial}{\partial t}\right)\left[\frac{g\{t-(\mathscr{R}_1-l)/c\}}{\mathscr{R}_1} - \frac{g\{t-(\mathscr{R}_2+l)/c\}}{\mathscr{R}_2}\right] \end{aligned} \tag{10.3}$$

where $\mathscr{R}_1^2 = x^2 + y^2 + (z+l)^2$, $\mathscr{R}_2^2 = x^2 + y^2 + (z-l)^2$. The parameters $\mathscr{R}_1$ and $\mathscr{R}_2$ are the distances to the point of observation from the ends of the

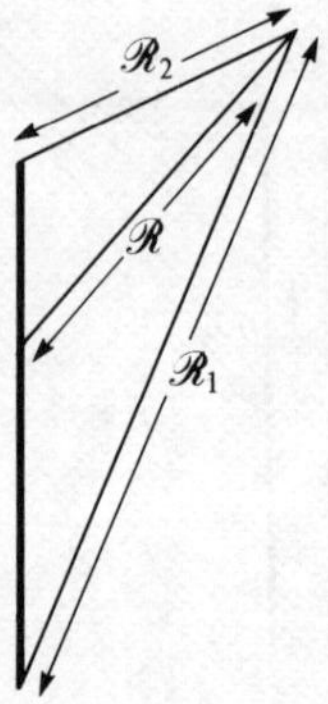

Fig. 3.7 Parameters for field calculation due to $g(t-\zeta/c)$

wire (Fig. 3.7). Moreover,

$$E_x = \frac{\partial^2}{\partial x\,\partial z}\int_{-l}^{l}\frac{g\{t-(\mathcal{R}+\zeta)/c\}}{4\pi\varepsilon_0\mathcal{R}}\,d\zeta$$

$$= -\frac{\partial}{\partial x}\frac{1}{4\pi\varepsilon_0}\int_{-l}^{l}\left\{\frac{\partial}{\partial\zeta}\left(\frac{g}{\mathcal{R}}\right)+\frac{g'}{c\mathcal{R}}\right\}d\zeta.$$

In the second integrand make the substitution $u=\mathcal{R}+\zeta-z$ so that $du/u=d\zeta/\mathcal{R}$. Then

$$\int_{-l}^{l}\frac{g'}{\mathcal{R}}\,d\zeta=\int_{u_1}^{u_2}g'\{t-(z+u)/c\}\frac{du}{u}$$

where $u_1=\mathcal{R}_1-l-z$ and $u_2=\mathcal{R}_2+l-z$. Since the integrand does not involve x we deduce that

$$E_x = \frac{1}{4\pi\varepsilon_0}\frac{\partial}{\partial x}\left[\frac{g\{t-(\mathcal{R}_1-l)/c\}}{\mathcal{R}_1}-\frac{g\{t-(\mathcal{R}_2+l)/c\}}{\mathcal{R}_2}\right]$$
$$+\frac{x}{4\pi\varepsilon_0 c}\left[\frac{g'\{t-(\mathcal{R}_1-l)/c\}}{u_1\mathcal{R}_1}-\frac{g'\{t-(\mathcal{R}_2+l)/c\}}{u_2\mathcal{R}_2}\right]. \tag{10.4}$$

For E_y replace $\partial/\partial x$ and x by $\partial/\partial y$ and y respectively.

3.11 Harmonic radiation from a wire

For harmonic time variation replace $f(t,\zeta)$ in (10.1) by $I(\zeta)e^{i\omega t}/i\omega$ and regard I as being the current in the wire. Then

$$\boldsymbol{\Pi}=\frac{\boldsymbol{k}}{4\pi i\omega\varepsilon_0}\int_{-l}^{l}I(\zeta)\frac{e^{-ik\mathcal{R}}}{\mathcal{R}}\,d\zeta \tag{11.1}$$

and the far field is, from (10.2),

$$E_\theta = \left(\frac{\mu_0}{\varepsilon_0}\right)^{\frac{1}{2}} H_\phi = \frac{\mathrm{i}\omega\mu_0}{4\pi R}\,\mathrm{e}^{-\mathrm{i}kR}\sin\theta\int_{-l}^{l} I(\zeta)\mathrm{e}^{\mathrm{i}k\zeta\cos\theta}\,\mathrm{d}\zeta + o(R^{-1}). \tag{11.2}$$

In the particular case when $I(\zeta) = I_0\mathrm{e}^{-\mathrm{i}k\zeta}$ where I_0 is constant the change of variable to u, used to evaluate E_x in the preceding section, leads to

$$\boldsymbol{\Pi} = \frac{I_0\mathrm{e}^{-\mathrm{i}kz}}{4\pi\mathrm{i}\omega\varepsilon_0}\{\mathrm{Ei}(-\mathrm{i}ku_2) - \mathrm{Ei}(-\mathrm{i}ku_1)\}\boldsymbol{k} \tag{11.3}$$

where

$$\mathrm{Ei}(\pm\mathrm{i}x) = -\int_x^\infty \frac{\mathrm{e}^{\pm\mathrm{i}u}}{u}\,\mathrm{d}u = \mathrm{Ci}\,x \pm \mathrm{i}\,\mathrm{si}\,x, \tag{11.4}$$

$$\mathrm{Ci}\,x = -\int_x^\infty \frac{\cos u}{u}\,\mathrm{d}u, \qquad \mathrm{si}\,x = -\int_x^\infty \frac{\sin u}{u}\,\mathrm{d}u \tag{11.5}$$

with x positive. Furthermore (10.3) and (10.4) supply

$$E_z = \frac{I_0}{4\pi\mathrm{i}\omega\varepsilon_0}\left(\frac{\partial}{\partial z} - \mathrm{i}k\right)\left\{\frac{\mathrm{e}^{-\mathrm{i}k(\mathcal{R}_1-l)}}{\mathcal{R}_1} - \frac{\mathrm{e}^{-\mathrm{i}k(\mathcal{R}_2+l)}}{\mathcal{R}_2}\right\}, \tag{11.6}$$

$$E_x = \frac{I_0 x}{4\pi\mathrm{i}\omega\varepsilon_0}\left[\left\{\mathrm{i}k\frac{\mathcal{R}_1}{u_1}(z+l) - 1\right\}\frac{\mathrm{e}^{-\mathrm{i}k(\mathcal{R}_1-l)}}{\mathcal{R}_1^3} - \left\{\mathrm{i}k\frac{\mathcal{R}_2}{u_2}(z-l) - 1\right\}\frac{\mathrm{e}^{-\mathrm{i}k(\mathcal{R}_2+l)}}{\mathcal{R}_2^3}\right]. \tag{11.7}$$

Similarly, if $I(\zeta) = I_0\mathrm{e}^{\mathrm{i}k\zeta}$,

$$E_z = \frac{I_0}{4\pi\mathrm{i}\omega\varepsilon_0}\left(\frac{\partial}{\partial z} + \mathrm{i}k\right)\left\{\frac{\mathrm{e}^{-\mathrm{i}k(\mathcal{R}_2-l)}}{\mathcal{R}_2} - \frac{\mathrm{e}^{-\mathrm{i}k(\mathcal{R}_1+l)}}{\mathcal{R}_1}\right\}, \tag{11.8}$$

$$E_x = \frac{I_0 x}{4\pi\mathrm{i}\omega\varepsilon_0}\left[\left\{\mathrm{i}k\frac{\mathcal{R}_2}{v_2}(l-z) - 1\right\}\frac{\mathrm{e}^{-\mathrm{i}k(\mathcal{R}_2-l)}}{\mathcal{R}_2^3} + \left\{\mathrm{i}k\frac{\mathcal{R}_1}{v_1}(z+l) + 1\right\}\frac{\mathrm{e}^{-\mathrm{i}k(\mathcal{R}_1+l)}}{\mathcal{R}_1^3}\right] \tag{11.9}$$

where $v_1 = \mathcal{R}_1 + l + z$, $v_2 = \mathcal{R}_2 - l + z$.

As far as the distant field is concerned a convenient starting point is $I(\zeta) = I_0 \sin k(\zeta + l)$. The integral in (11.2) can be evaluated in a straightforward manner and

$$E_\theta = \left(\frac{\mu_0}{\varepsilon_0}\right)^{\frac{1}{2}}\frac{I_0\mathrm{e}^{-\mathrm{i}kR}}{4\pi R}\left\{\mathrm{e}^{\mathrm{i}kl}\frac{\sin kl(1+\cos\theta)}{1+\cos\theta} - \mathrm{e}^{-\mathrm{i}kl}\frac{\sin kl(1-\cos\theta)}{1-\cos\theta}\right\}\sin\theta.$$

In practice, the current is often sinusoidally distributed on a wire to a good approximation and the wavelength of the sinusoidal wave is virtually the same as that of the excitation. Moreover, the current usually vanishes at the ends of the wire. All these aspects can be covered by choosing $kl = \frac{1}{2}m\pi$ or $l = m\lambda/4$ where m is a positive integer. With this choice

$$E_\theta = \mathrm{i}I_0\left(\frac{\mu_0}{\varepsilon_0}\right)^{\frac{1}{2}} \frac{\cos(\frac{1}{2}m\pi\cos\theta)}{2\pi R\sin\theta}\,\mathrm{e}^{-ikR} \quad (m \text{ odd}),$$

$$E_\theta = I_0\left(\frac{\mu_0}{\varepsilon_0}\right)^{\frac{1}{2}} \frac{\sin(\frac{1}{2}m\pi\cos\theta)}{2\pi R\sin\theta}\,\mathrm{e}^{-ikR} \quad (m \text{ even}).$$

The factor $(\mu_0/\varepsilon_0)^{\frac{1}{2}}/2\pi$ is 59.92 or 60 approximately.

The *power* radiated per unit solid angle $P(\theta, \phi)$ or the average rate at which energy is radiated per unit solid angle is

$$P(\theta, \phi) = \left(\frac{\mu_0}{\varepsilon_0}\right)^{\frac{1}{2}} \frac{I_0^2}{8\pi^2}\left\{\frac{\cos(\frac{1}{2}m\pi\cos\theta)}{\sin\theta}\right\}^2 \quad (m \text{ odd}),$$

$$P(\theta, \phi) = \left(\frac{\mu_0}{\varepsilon_0}\right)^{\frac{1}{2}} \frac{I_0^2}{8\pi^2}\left\{\frac{\sin(\frac{1}{2}m\pi\cos\theta)}{\sin\theta}\right\}^2 \quad (m \text{ even}).$$

An important case is that in which the wire is a half-wavelength long, i.e. $m = 1$. The power patterns of such a radiator and of a harmonic electric dipole which has the same field in the equatorial plane $\theta = \frac{1}{2}\pi$ are shown in Fig. 3.8. The patterns do no differ much; both are uniform in azimuth and have a single maximum in the equatorial plane. There is no radiation in the direction in which the wire is pointing.

In general, the patterns exhibit *lobes*, i.e. regions in which power is transmitted separated by nulls. For example, there is no radiation in the directions where

$$\cos\theta = 1/m, 3/m, \ldots, 1 \quad (m \text{ odd}),$$

$$\cos\theta = 0, 2/m, 4/m, \ldots, 1 \quad (m \text{ even})$$

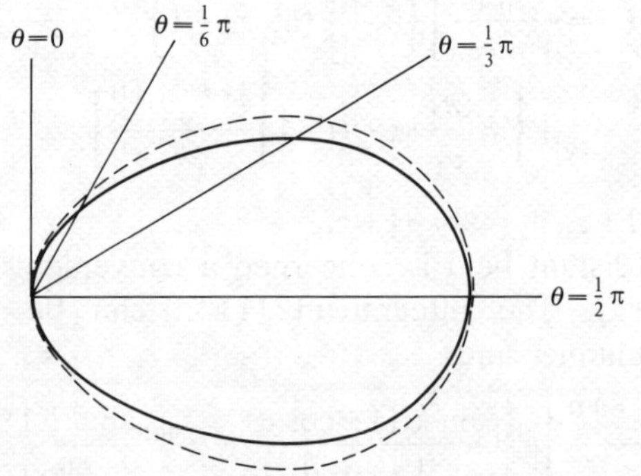

Fig. 3.8 Polar diagram of the half-wavelength antenna. —— $\frac{1}{2}\lambda$ wire; ------ dipole

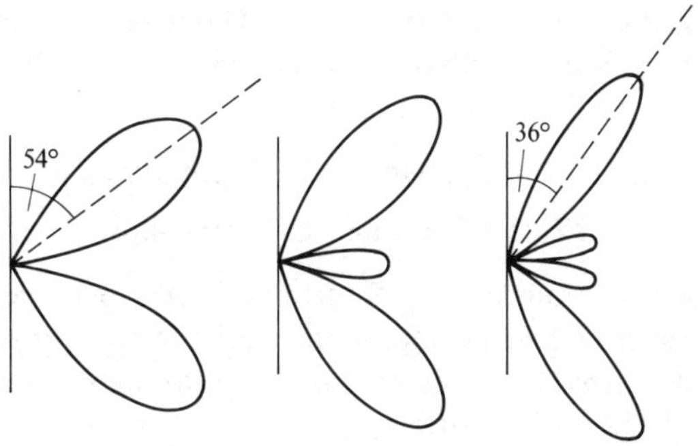

Fig. 3.9 Polar diagrams for various current distributions

so that there are m lobes. Typical patterns are displayed in Fig. 3.9. The directions of the maxima of the lobes are provided by

$$m\pi \tan\theta \sin\theta - 2\cot(\tfrac{1}{2}m\pi\cos\theta) = 0 \quad (m \text{ odd}),$$

$$m\pi \tan\theta \sin\theta + 2\tan(\tfrac{1}{2}m\pi\cos\theta) = 0 \quad (m \text{ even}).$$

A measure of *directivity* of an antenna is yielded by comparison with a hypothetical antenna which would radiate the same total power uniformly in all directions. The power per unit solid angle radiated by the hypothetical antenna is $P_t/4\pi$ when P_t is the total power radiated by the antenna. The increase of directivity of the actual antenna in a specific direction may then be estimated from $4\pi P(\theta, \phi)/P_t$. If P_M is the maximum of $P(\theta, \phi)$ which occurs the *power gain* G can be defined by

$$G = 4\pi P_M/P_t.$$

The value of P_t is given, when m is odd, by

$$P_t = \left(\frac{\mu_0}{\varepsilon_0}\right)^{\frac{1}{2}} \frac{I_0^2}{4\pi} \int_0^{\pi} \frac{\{\cos(\frac{1}{2}m\pi\cos\theta)\}^2}{\sin\theta}\, d\theta$$

$$= \left(\frac{\mu_0}{\varepsilon_0}\right)^{\frac{1}{2}} \frac{I_0^2}{4\pi} \int_{-1}^{1} \frac{\cos^2(\frac{1}{2}m\pi u)}{1-u^2}\, du$$

on making the substitution $\cos\theta = u$. Split $(1-u^2)^{-1}$ into partial fractions and, in the integral containing $(1-u)^{-1}$, change the sign of u. Then

$$P_t = \left(\frac{\mu_0}{\varepsilon_0}\right)^{\frac{1}{2}} \frac{I_0^2}{8\pi} \int_{-1}^{1} \frac{1+\cos m\pi u}{1+u}\, du$$

and, after the transformation $1+u = v/m\pi$,

$$P_t = \left(\frac{\mu_0}{\varepsilon_0}\right)^{\frac{1}{2}} \frac{I_0^2}{8\pi} (\gamma + \ln 2m\pi - \mathrm{Ci}\, 2m\pi) \qquad (11.10)$$

where γ is Euler's constant 0.5772.... The same technique may be employed when m is even; there is no change to (11.10).

The consequent radiation resistance is

$$R_r = (\mu_0/\varepsilon_0)^{\frac{1}{2}}(\gamma + \ln 2m\pi - \text{Ci}\, 2m\pi)/4\pi$$
$$\approx 72.45 + 30(\ln m - \text{Ci}\, 2m\pi)\ \text{ohm}.$$

With P_t known the gain can be determined. For instance, when $m = 1$, $G = 1.65$. The ratio of two powers is, however, often expressed in terms of the *decibel*, which is *ten times the logarithm to the base* 10 *of the ratio.* Thus, when $m = 1$, $G = 2.17$ dB.

The theoretical patterns are modified to some extent in practice by the damping due to heat and radiation losses. The current amplitude tends to diminish from the theoretical as one proceeds from the centre of the antenna to the end. The lobes near the equatorial plane tend to be strengthened at the expense of the others and the effect becomes more pronounced as the resistance of the wire is increased.

3.12 Linear arrays

The half-wavelength wire antenna is directive in the sense that most of its radiation is in directions for which $120° > \theta > 60°$ (see Fig. 3.8) but it offers no discrimination in azimuth since the pattern does not vary with ϕ. To produce a preferred direction in the equatorial plane it is necessary to replace the single antenna by an *array* of antennas arranged suitably in space and appropriately excited. Let the wires be all equal and parallel with their centres located on a rectangular lattice (Fig. 3.10). The lattice may be specified by the unit vectors $\boldsymbol{i}$, $\boldsymbol{j}$, $\boldsymbol{k}$ along the axes and the

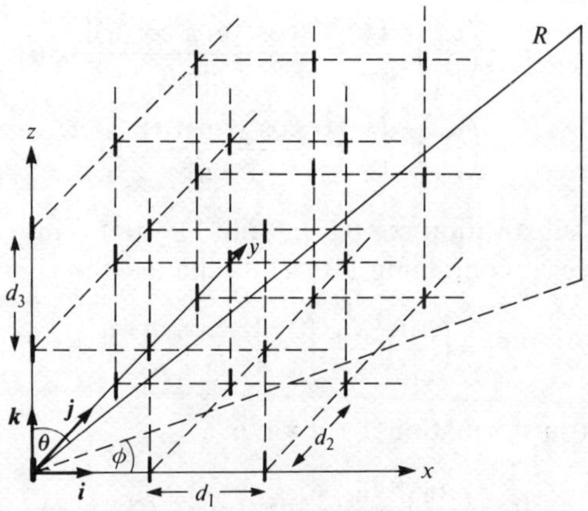

Fig. 3.10 An array of antennas arranged in a lattice

distances d_1, d_2, d_3 between consecutive centres in these three directions. A typical centre then has position vector $n_1 d_1 \boldsymbol{i} + n_2 d_2 \boldsymbol{j} + n_3 d_3 \boldsymbol{k}$ where n_1, n_2, and n_3 are non-negative integers. The array is kept finite by prescribing

$$n_1 \leqslant N_1, \qquad n_2 \leqslant N_2, \qquad n_3 \leqslant N_3$$

where N_1, N_2, and N_3 are given non-negative integers.

Take the current distribution in each antenna to be the same except for amplitude and phase which are to be $A_{n_1 n_2 n_3}$ and $n_1\alpha_1 + n_2\alpha_2 + n_3\alpha_3$ respectively. Suppose that a current of unit amplitude and zero phase in the antenna with centre $\boldsymbol{x}_n$ produces a field

$$E_\theta = \left(\frac{\mu_0}{\varepsilon_0}\right)^{\frac{1}{2}} \frac{\mathrm{e}^{-ik|\boldsymbol{x}-\boldsymbol{x}_n|}}{2\pi|\boldsymbol{x}-\boldsymbol{x}_n|} F(\theta, \phi) \tag{12.1}$$

by the analysis of §3.11. Then the field of a typical antenna in the array is

$$E_\theta = \left(\frac{\mu_0}{\varepsilon_0}\right)^{\frac{1}{2}} A_{n_1 n_2 n_3} \frac{F(\theta, \phi)}{2\pi|\boldsymbol{x}-\boldsymbol{x}_n|} \exp\left(-ik\,|\boldsymbol{x}-\boldsymbol{x}_n| - \sum_{j=1}^{3} n_j\alpha_j\right).$$

Here θ and ϕ are the angles as measured at the antenna but, in view of the large distance to the point of observation, may be identified with the corresponding angles at the origin. Also

$$|\boldsymbol{x}-\boldsymbol{x}_n| = R - \sum_{j=1}^{3} n_j\beta_j + O(1/R)$$

where $\beta_1 = d_1 \sin\theta\cos\phi$, $\beta_2 = d_2 \sin\theta \sin\phi$ and $\beta_3 = d_3 \cos\theta$. Thus, at a large distance from every element of the array, the total field of the array is

$$E_\theta = \left(\frac{\mu_0}{\varepsilon_0}\right)^{\frac{1}{2}} \frac{F(\theta, \phi)}{2\pi R} \mathrm{e}^{-ikR} \sum_{n_1=0}^{N_1} \sum_{n_2=0}^{N_2} \sum_{n_3=0}^{N_3} A_{n_1 n_2 n_3} \exp\left\{\mathrm{i}\sum_{j=1}^{3} (k\beta_j - \alpha_j)\right\}. \tag{12.2}$$

It is consequently the product of the field due to a single antenna (as exemplified by (12.1)) and a factor which depends upon the lattice.

One particularly simple case of (12.2) is when all the current amplitudes are unity. The resulting geometric series can then be summed and the power radiated per unit solid angle $P_\mathrm{A}(\theta, \phi)$ is given by

$$P_\mathrm{A}(\theta, \phi) = P(\theta, \phi)(F_1 F_2 F_3)^2 \tag{12.3}$$

where $P(\theta, \phi)$ is defined in §3.11 and

$$F_j = \frac{\sin\frac{1}{2}(N_j+1)(k\beta_j - \alpha_j)}{\sin\frac{1}{2}(k\beta_j - \alpha_j)}.$$

In (12.2) the currents in the antennas are assumed to be known. The technical problem of supplying specified currents will not be considered here. In fact, the currents are not independent because the presence of any antenna can affect the current in another. At long wavelengths it is possible to feed the antennas so that the amplitude and phase are independent of the space, except for effects from external field coupling between the radiators. If, however, the array has to be fed from a transmission line, as in some microwave antennas, there can be a close relationship between the spacing and phase. When this is taken into account and the mutual interactions between the elements of the array are allowed for the analysis can become extremely complicated.

As an application of (12.2) suppose that the array consists of two antennas, carrying currents of equal amplitude I_0, on the x-axis. Then, from (12.3)

$$P_A(\theta, \phi) = 4P(\theta, \phi)\cos^2\tfrac{1}{2}(kd_1 \sin\theta\cos\phi - \alpha_1). \qquad (12.4)$$

If the currents are in phase $\alpha_1 = 0$ and if the antennas are separated by half a wavelength $d_1 = \frac{1}{2}\lambda$; then, from §3.11,

$$P_A(\theta, \phi) = \left(\frac{\mu_0}{\varepsilon_0}\right)^{\frac{1}{2}} \frac{I_0^2}{2\pi^2} \left[\frac{\cos(\frac{1}{2}\pi\cos\theta)}{\sin\theta}\cos(\tfrac{1}{2}\pi\sin\theta\cos\phi)\right]^2 \qquad (12.5)$$

when the antennas are half a wavelength long.

The pattern resulting from (12.5) is depicted in Fig. 3.11 for the plane $\theta = \frac{1}{2}\pi$. It is evident that there are two principal directions of radiation in the equatorial plane and that both are perpendicular to the line of the array. This is one example of a *broadside array*. If the number of elements is increased the main beam is sharpened and small side lobes appear. Fig. 3.12 displays the power pattern of four antennas spaced a half wavelength apart and carrying equal currents in phase.

If the currents in the two antennas have a phase difference $\alpha_1 = \frac{1}{2}\pi$ and the antennas are separated by a quarter wavelength so that $d_1 = \frac{1}{4}\lambda$, (12.4)

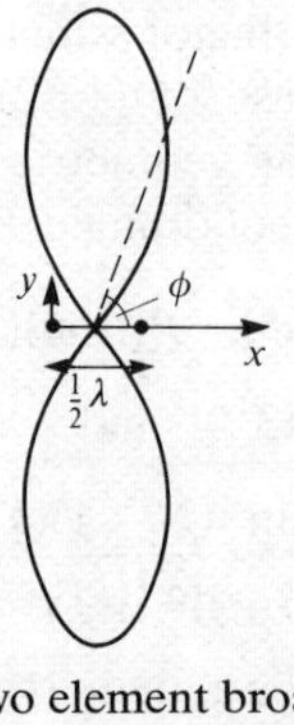

Fig. 3.11 Two element broadside array

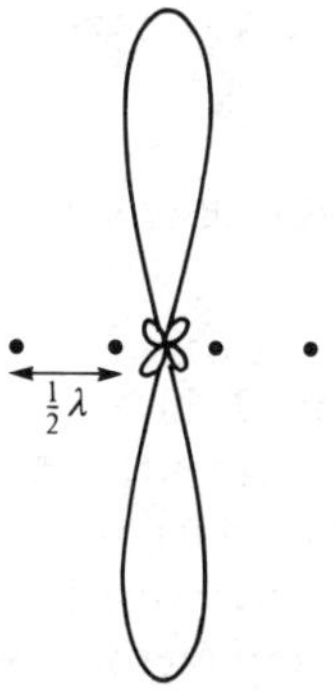

Fig. 3.12 Four element broadside array

leads to

$$P_A(\theta, \phi) = \left(\frac{\mu_0}{\varepsilon_0}\right)^{\frac{1}{2}} \frac{I_0^2}{2\pi^2} \left[\frac{\cos(\frac{1}{2}\pi \cos \theta)}{\sin \theta} \cos\{\tfrac{1}{4}\pi(\sin \theta \cos \phi - 1)\}\right]^2. \tag{12.6}$$

This pattern is symmetrical about the (x, z)-plane, sometimes called the *E-plane*, and about the (x, y)-plane or *H-plane*. The power patterns in these two planes are shown in Fig. 3.13. Clearly, only a small portion of the energy radiated goes into the regions $x < 0$; the main power is in a broad beam in the direction of the positive x-axis, i.e. the line of the array. This system is an illustration of an *end-fire array*.

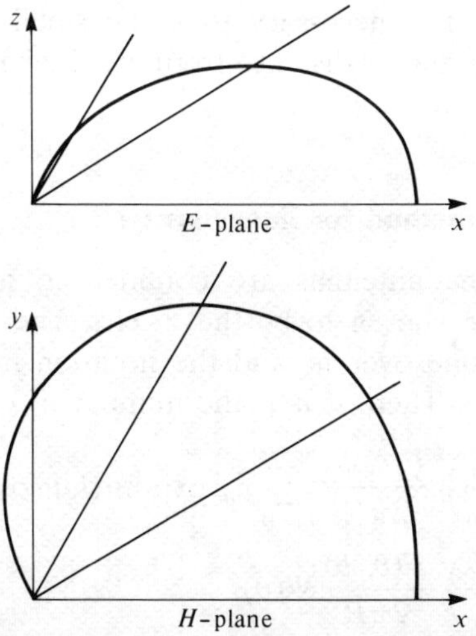

Fig. 3.13 Two element end-fire array

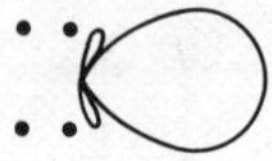

Fig. 3.14 Combination of broadside and end-fire arrays

The factor $\cos^2(\frac{1}{2}\pi\cos\theta)/\sin^2\theta$ in (12.5) and (12.6) represents the pattern due to a single element. If it is ignored the factors $\cos^2(\frac{1}{2}\pi\sin\theta\cos\phi)$ and $\cos^2\frac{1}{4}\pi(\sin\theta\cos\phi-1)$ are left; they characterize the broadside and end-fire arrays respectively and may be designated *array factors.* The broadside array factor is something like a toroid with a directive pattern in any plane through the x-axis. In contrast, the end-fire array factor has a pattern rather like a pencil beam possessing directivity in planes through the y-axis as well as those through the x-axis. The same is true for arrays containing more than two elements and end-fire arrays usually have a higher gain than broadside arrays of the same size.

More highly directive systems can be obtained by combining broadside and end-fire arrays. The resulting pattern is just the product of the two patterns on account of (12.3). Thus Fig. 3.14 shows the pattern for a two element broadside array backed by a similar array $\frac{1}{4}\lambda$ behind with the currents differing in phase by $\frac{1}{2}\pi$ from those in front. The pattern can be sharpened by increasing the number of antennas. These theoretical predictions are substantially borne out by experiment.

In some end-fire arrays the rear antenna is not connected directly to the current supply. It is then *parasitic* and the current in it is excited by the field from the forward one. In other words it acts as a reflector and the radiation from the induced current is much the same as if it were fed directly. In practice it is necessary to make small adjustments to the length of the reflector and its distance from the driven antenna in order to achieve the best results.

3.13 Schelkunoff's method for linear arrays

In a *linear array* the antennas are confined to a single line which, conveniently, may be chosen to be the z-axis. The triple sum in (12.2) reduces to a single one over n_3 and the notation may be simplified by putting $d_3=d$, $\alpha_3=\alpha$. Then, if n is the number of elements,

$$E_\theta=\left(\frac{\mu_0}{\varepsilon_0}\right)^{\frac{1}{2}}\frac{F(\theta,\phi)}{2\pi R}\sum_{m=0}^{n-1}a_m\exp\{im(kd\cos\theta-\alpha)\}$$

$$=\left(\frac{\mu_0}{\varepsilon_0}\right)^{\frac{1}{2}}\frac{F(\theta,\phi)}{2\pi R}\Psi(\theta) \tag{13.1}$$

say, after writing a_m for the current amplitude in the mth radiator. The factor of Ψ in (13.1) varies only with the pattern of an individual wire whereas Ψ depends upon the arrangement of antennas in the array. Therefore the characteristics peculiar to the array are contained in Ψ alone and so, consistent with §3.12,

$$|\Psi(\theta)|^2 = \left|\sum_{m=0}^{n-1} a_m \exp\{im(kd\cos\theta - \alpha)\}\right|^2 \tag{13.2}$$

may be deemed the array factor.

Consider now the case in which the phase difference between currents in adjacent elements is entirely accounted for by α. Then a_m is of the form $|a_m|\,e^{i\gamma}$ and γ disappears from (13.2). There is, therefore, no loss of generality in taking a_m to be real, as will be done from now on.

For simple arrays, the series in (13.2) can be evaluated in the Argand diagram by regarding it as the resultant of n complex vectors. General arrays are handled more easily by Schelkunoff's device of putting

$$z = e^{i(kd\cos\theta - \alpha)}. \tag{13.3}$$

Then, if $f(z)$ is the polynomial specified by

$$f(z) = \sum_{m=0}^{n-1} a_m z^m, \tag{13.4}$$

the array factor is given by

$$|\Psi(\theta)|^2 = |f(z)|^2. \tag{13.5}$$

As θ varies from 0 to π, z moves on the unit circle in the complex z-plane from $e^{i(kd-\alpha)}$ to $e^{-i(kd+\alpha)}$; in doing so z may traverse only part of the unit circle or it may make several complete circuits, depending on the value of kd. The path that is traced on the unit circle will be called the *scope* of z.

Any polynomial can be expressed as a product of linear factors. Thus

$$f(z) = a_{n-1}(z - z_1)(z - z_2)\ldots(z - z_{n-1})$$

assuming that $a_{n-1} \neq 0$ in (13.4). The complex numbers $z_1, \ldots, z_{n-1}$ are known as the *zeros* of the polynomial. If z_0 is a zero so is the complex conjugate z_0^* because the coefficients a_m have been chosen to be real. It must be emphasized, however, that the zeros of f do not necessarily lie on the unit circle in the z-plane and so may not be within the scope of z.

Obviously $\Psi(\theta)$ vanishes if, and only if, a zero of f occurs in the scope of z. Thus, zeros in the scope of z determine the nulls of the array pattern. The maxima of $|\Psi(\theta)|^2$ will be round about the midpoints of arcs joining consecutive zeros. As an illustration take $a_m = 1$, $d = \frac{1}{4}\lambda$, $\alpha = 0$, $n = 6$. The five zeros of this six element array are shown in Fig. 3.15 but only two of them correspond to nulls of the array. A principal maximum occurs at $z = 1$ or $\theta = \frac{1}{2}\pi$.

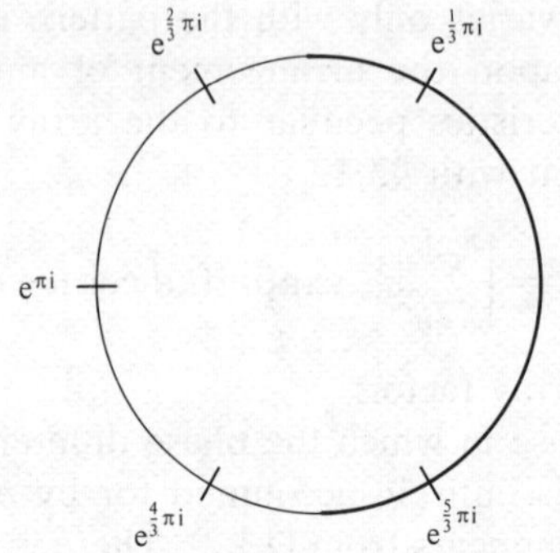

Fig. 3.15 The zeros of $1+z+z^2+z^3+z^4+z^5$. The scope of z is shown thickened

This example is a special case of the array in which all currents have the same amplitude. Then $a_m = a_{n-1}$ and

$$f(z) = a_{n-1} \sum_{m=0}^{n-1} z^m = a_{n-1}(z^n - 1)/(z-1) = a_{n-1} z^{\frac{1}{2}(n-1)}(z^{\frac{1}{2}n} - z^{-\frac{1}{2}n})/(z^{\frac{1}{2}} - z^{-\frac{1}{2}})$$

whence

$$|\Psi(\theta)|^2 = a_{n-1}^2 \sin^2(\tfrac{1}{2}n\chi)/\sin^2(\tfrac{1}{2}\chi) \tag{13.6}$$

with $\chi = kd \cos\theta - \alpha$. Nulls of the array occur where

$$\cos\theta = \frac{1}{kd}\left(\frac{2p\pi}{n} + \alpha\right), \tag{13.7}$$

p being any nonzero integer, which is not an integral multiple of n, such that θ is real. There are maxima of $|\Psi|^2$ at $\chi = 2M\pi$ for any integer M; the relevant values of θ are given by

$$\cos\theta = (2M\pi + \alpha)/kd \tag{13.8}$$

which has real solutions in θ only if $|2M\pi + \alpha| \leqslant kd$. Other maxima can exist and are obtained by solving a transcendental equation.

The array factor of a broadside array has an absolute maximum in directions perpendicular to the line of the array, i.e. for $\theta = \frac{1}{2}\pi$. If $\alpha = 0$, there is a solution $\theta = \frac{1}{2}\pi$ of (13.8) and it is the only solution if $d < \lambda$. In particular, if $d = \frac{1}{2}\lambda$, the explicit broadside pattern is given by (13.6) with $\chi = \pi \cos\theta$ and then the *half-power width*, i.e. the full angle in which the power radiated in a direction is not less than half the maximum value, is approximately $102°/n$, the approximation improving as n grows. Also, when $d = \frac{1}{2}\lambda$ and $\alpha = 0$, there are solutions of (13.7) when $n > 2$ so that side lobes with their associated secondary maxima appear for $n > 0$.

In an effort to combat the appearance of side lobes consider arrays in which the currents are not the same in all elements. One possibility is

$$\begin{aligned} f(z) &= (1+z+z^2+\dots+z^r)^2 \\ &= 1+2z+\dots+(r+1)z^r+\dots+2z^{2r-1}+z^{2r}. \end{aligned}$$

Now the amplitudes of the currents decrease uniformly from the central element to the ends of the array, forming a *gabled array*. The array factor is the square of that for a uniformly illuminated array. Therefore, if a is the ratio of a side lobe maximum to the main maximum for uniform currents, that ratio will be a^2 for a gabled array. Hence the side lobes of a gabled array are at a lower level than those of a uniformly illuminated array. However, the main beam is broader and has a half-power width of about $146°/n$ for n elements. Higher powers of the polynomials lead to still smaller side lobes but there is a corresponding increase in the width of the main beam.

Side lobes can be totally absent. For instance,

$$f(z) = (1+z)^r$$

has an array factor

$$|\Psi(\theta)|^2 = \cos^{2r}(\tfrac{1}{2}\pi \cos\theta)$$

when $d = \frac{1}{2}\lambda$ and $\alpha = 0$.

It can be proved that, of all arrays in which $d = \frac{1}{2}\lambda$ and $\alpha = 0$, the one in which the current amplitudes are uniform has the maximum gain. The uniform array in which $d = \frac{1}{2}\lambda$ also has the largest gain of all uniform arrays in which d is such that the distance between the first and last elements is $\frac{1}{2}\lambda$ but there are non-uniform arrays with greater gain.

The end-fire array possesses a maximum along the line of the array; it may be selected to be $\theta = 0$. There is a solution of (13.8) at $\theta = 0$ when $\alpha = kd$. If $kd = \pi$ the scope of z is the whole unit circle going from $z = 1$ to $z = e^{-2i\pi}$ and the maximum is repeated at $\theta = \pi$. To avoid this the scope must be reduced which can be achieved by $kd < \pi$ or $d < \frac{1}{2}\lambda$. With $d = \frac{1}{4}\lambda$, the phase difference between currents is $\frac{1}{2}\pi$ and the gain is n, the same as that of the (longer) broadside array. The gain can be increased to about $1.8n$ by making the phase difference $\frac{1}{2}\pi + 2.94/n$ when n is large; the width of the main beam is then about halved although the maxima of the side lobes are almost doubled.

Schelkunoff has suggested another way of improving the gain. When $\alpha = kd = \frac{1}{2}\pi$

$$f(z) = \sum_{m=0}^{n-1} z^m = \prod_{m=1}^{n-1} (z - \omega^m)$$

where ω is an nth root of unity. The scope is the semicircle from $z = 1$ to $z = e^{-i\pi}$ (Fig. 3.16). The zeros outside the scope of z have little effect on

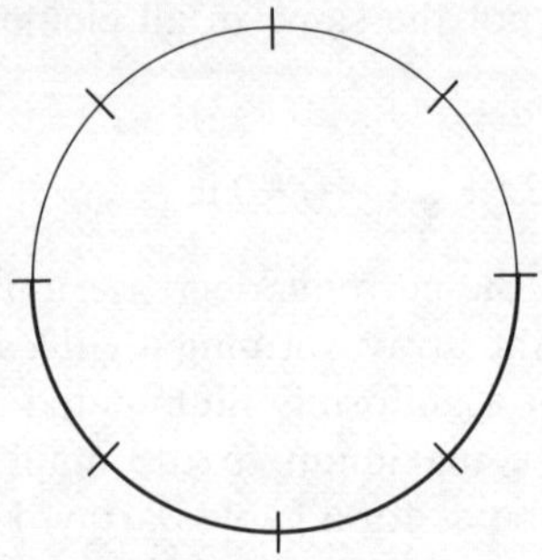

Fig. 3.16 The zeros for $a_m = 1$, $\alpha = kd = \frac{1}{2}\pi$, $n = 8$

the directivity. Therefore retain only those zeros in the scope and take

$$f(z) = \prod_{m=[\frac{1}{2}(n+1)]}^{n-1} (z - \omega^m)$$

where $[p]$ is the largest integer which does not exceed p. The nulls of the pattern are unaltered so that gain is about the same as before but the number of elements in the array has been almost halved.

3.14 Beam synthesis for linear arrays

The determination of the pattern of a given array has been explained in the preceding sections. A more difficult problem is to find the array which produces a prescribed pattern (the analogous problem for continuous current distributions is discussed in Jones (1979)). Obviously, if the only requirement is to have nulls in specified places a simple choice of zeros in $f(z)$ will fulfil it. However, our desires are rarely so simple.

It should be recognized from the beginning that, in general, an arbitrarily chosen pattern is not realizable; only patterns which satisfy certain conditions can be radiated. Of those patterns which are capable of radiation only a restricted number can be derived from linear arrays. The appropriate linear array will depend upon how the pattern is specified whether by as little as nominating the beam width or by a detailed specification such as the angular variation in some plane. Usually, there will not be a unique solution to the problem.

Under suitable constraints an exact solution is feasible. An illustrative example is provided by the attempt to create a broadside array in which the side lobes are to be smallest possible consistent with a given beam width, the beam width being defined as the angle between the two nulls which separate the main beam from the side lobes. A solution is sought from arrays in which $\alpha = 0$.

With these restrictions the array factor given by (13.5) is

$$|\Psi(\theta)|^2 = \sum_{m=0}^{n-1} A_m \cos(mkd \cos\theta) \tag{14.1}$$

where $A_m = 2\sum_{r=m}^{n-1} a_r a_{r-m}$. Any positive-definite expression of the form (14.1) is an array factor since any positive trigonometric polynomial can be written as the square of some trigonometric polynomial, e.g. $8+4\cos(kd\cos\theta) = |\sqrt{3}+1+(\sqrt{3}-1)z|^2$.

It is often convenient to rewrite (14.1) in terms of the *Chebyshev polynomial* $\mathrm{T}_m(y)$ defined by

$$\mathrm{T}_m(y) = \cos(m \cos^{-1} y).$$

This polynomial lies between -1 and 1 in the interval $|y| \leqslant 1$; outside this interval its magnitude steadily increases as $|y|$ departs further from 1. When m is even the polynomial is symmetrical about $y=0$.

By virtue of the definition of the Chebyshev polynomial (14.1) can be expressed as

$$|\Psi(\theta)|^2 = \sum_{m=0}^{n-1} A_m \mathrm{T}_{2m}(\zeta) \tag{14.2}$$

where $\zeta = \cos(\frac{1}{2}kd\cos\theta)$ so that the array factor is a polynomial of degree $n-1$ in ζ^2.

A special case of (14.2) is

$$|\Psi_0(\theta)|^2 = \tfrac{1}{2}\{1+\mathrm{T}_{2n-2}(C\zeta)\} \tag{14.3}$$

where $C>1$. For it is non-negative since $\mathrm{T}_{2n-2} \geqslant -1$ and it is a polynomial of degree $n-1$ in ζ^2. Therefore it can be expressed in the form (14.2) and so (14.3) is a possible array factor.

The direction $\theta = \frac{1}{2}\pi$ corresponds to $\zeta = 1$ and hence the main beam maximum of (14.3) is $B = \frac{1}{2}\{1+\mathrm{T}_{2n-2}(C)\}$. Because of the growth of the Chebyshev polynomial outside the interval $(-1, 1)$ no side lobe can have a maximum greater than 1 provided that $C\zeta \geqslant -1$ for $0 \leqslant \theta \leqslant \frac{1}{2}\pi$ (there is obvious symmetry in the pattern about $\theta = \frac{1}{2}\pi$). This requires $C\cos\frac{1}{2}kd \geqslant -1$, for which $d<\lambda$ is necessary but not sufficient. If $d \geqslant \frac{1}{2}\pi$ there are certainly side lobes whose maxima are 1. Therefore, if $\frac{1}{2}\lambda \leqslant d \leqslant (\lambda/\pi)\cos^{-1}(-1/C) < \lambda$, (14.3) supplies an array factor in which the ratio of the side lobe maximum to the main beam maximum is $1/B$ and the beam width is determined by the zero of (14.3) nearest $\zeta = 1$, i.e.

$$\zeta_1 = \frac{1}{C}\cos\frac{\pi}{2n-2}.$$

It will now be shown that, with the above restrictions on *d*, *no other array factor can have the same beam width and the same or smaller strength of side lobe.* Any array factor, symmetrical about $\theta = \frac{1}{2}\pi$, is a polynomial $P_{2n-2}(\zeta)$ of degree $n-1$ in ζ^2. Suppose it takes the value B at $\zeta=1$, vanishes at $\zeta=\zeta_1$ and that its maxima do not exceed those of (14.3) in $|\zeta|<\zeta_1$. Then

$$\tfrac{1}{2}\{1+T_{2n-2}(C\zeta_2)\}-P_{2n-2}(\zeta_2)\geqslant 0, \qquad \tfrac{1}{2}\{1+T_{2n-2}(C\zeta_3)\}-P_{2n-2}(\zeta_3)\leqslant 0, \ldots$$

where $C\zeta_k=\cos\{k\pi/(2n-2)\}$. Thus $\frac{1}{2}\{1+T_{2n-2}(C\zeta)\}-P_{2n-2}(\zeta)$ has $2n-6$ zeros between $-\zeta_1$ and ζ_1. It also has zeros at $\zeta=\pm1$ and double zeros at $\zeta=\pm\zeta_1$ because both polynomials have minima there. Hence the polynomial has $2n$ zeros and it must be identically zero because it is of degree $2n-2$. The statement at the commencement of the paragraph has been verified.

A further application of the Chebyshev polynomial is in the design of *super-gain arrays*, which are theoretically able to attain high gain with small dimensions. For an example, consider a broadside array containing five antennas with $d<\frac{1}{2}\lambda$. As θ varies from 0 to $\frac{1}{2}\pi$, $y=\cos(kd\cos\theta)$ moves from $\cos kd$ to 1 and then swings back again as θ increases from $\frac{1}{2}\pi$ to π. Choose a and b so that

$$a\cos kd+b=-1, \qquad a+b>1.$$

Then $T_2(ay+b)$ will be symmetrical about $\theta=\frac{1}{2}\pi$, starting from 1 at $\theta=0$, passing through two zeros as θ increases to the broadside direction where it is greater than 1. Hence $|\Psi(\theta)|^2=T_2^2(ay+b)$ has a pattern in which the side lobes do not exceed 1 whereas the main beam can be large.

Numerical values shed light on the phenomenon. Suppose that $d=\lambda/16$ (the length of the array being $\frac{1}{4}\lambda$) and that the main beam maximum is to be thirty-one times that of the side lobes. Then $a=65.8$ and $b=-61.8$. The nulls occur at approximately $\theta=18°, 37°, 143°, 162°$ so that the main beam width is 106°. The corresponding $f(z)$ is, when a constant factor is omitted,

$$f(z)=z^4-3.8z^3+5.61z^2-3.8z+1.$$

Consequently, the current ratios in the antennas necessary to generate the pattern are

$$1:-3.8:5.61:-3.8:1.$$

In the direction of the bore sight of the main beam $z=1$ and the field strength is proportional to the sum 0.01 of the currents. Therefore the effective current radiating towards the maximum is only 0.2% of the current in the central antenna. The low value of the effective current is because super-gain is achieved by the addition of radiation from antennas carrying large currents which are nearly equal and opposite. In practice, the theoretical predictions are unlikely to be accomplished partly because a small error in current destroys the super-gain pattern and partly because the interaction of one antenna on another has been neglected in the theory.

3.15 The helical antenna

The radiation from straight wires is particularly simple to handle but that from curved wires can be calculated; the helix furnishes an example. Let the axis of the helix be the z-axis and suppose that it is wound on a circular cylinder of radius a with pitch angle β (Fig. 3.17). Then the equation of the antenna is

$$x = a\cos\phi, \qquad y = a\sin\phi, \qquad z = a\phi\tan\beta$$

and the distance d between adjacent turns is $2\pi a\tan\beta$. Assume that the helix consists of M complete turns so that ϕ runs from 0 to $2M\pi$. The total length of the wire is then $2M\pi a\sec\beta$.

Let the current $A\sin q(\phi+2M\pi)$ flow in the wire. If $q = m/4M$, m being an integer, the current vanishes at the end $\phi = 0$ whereas if m is odd this end can be regarded as a driving point. The Hertz vector of the radiated field is

$$\boldsymbol{\Pi} = \frac{-\mathrm{i}Aa}{4\pi\omega\varepsilon_0}\int_0^{2M\pi}(-\boldsymbol{i}\sin\alpha+\boldsymbol{j}\cos\alpha+\boldsymbol{k}\tan\beta)\frac{\mathrm{e}^{-\mathrm{i}k|\boldsymbol{x}-\boldsymbol{y}|}}{|\boldsymbol{x}-\boldsymbol{y}|}\sin q(\alpha+2M\pi)\,\mathrm{d}\alpha$$

where

$$|\boldsymbol{x}-\boldsymbol{y}|^2 = (x-a\cos\alpha)^2+(y-a\sin\alpha)^2+(z-a\alpha\tan\beta)^2.$$

At a large distance R from the origin in the direction specified by the spherical polar angles θ, ϕ this simplifies to

$$\boldsymbol{\Pi} \sim \frac{-\mathrm{i}Aa\mathrm{e}^{-\mathrm{i}kR}}{4\pi\omega\varepsilon_0 R}\int_0^{2M\pi}(-\boldsymbol{i}\sin\alpha+\boldsymbol{j}\cos\alpha+\boldsymbol{k}\tan\beta)\sin q(\alpha+2M\pi)$$
$$\times\exp[\mathrm{i}ka\{\sin\theta\cos(\phi-\alpha)+\alpha\cos\theta\tan\beta\}]\,\mathrm{d}\alpha. \qquad (15.1)$$

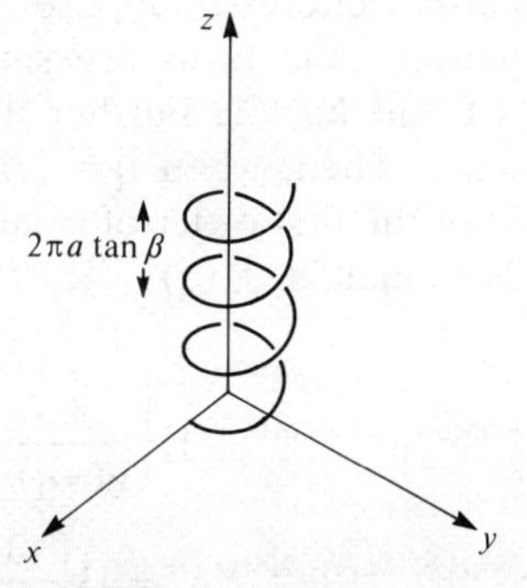

Fig. 3.17 The helical antenna

The integration in (15.1) can be performed by introducing the formula (A.1.23)

$$e^{iz\cos(\phi-\alpha)} = J_0(z) + 2\sum_{s=1}^{\infty} (-)^s J_{2s}(z)\cos 2s(\phi-\alpha)$$
$$+ 2i\sum_{s=0}^{\infty} (-)^s J_{2s+1}(z)\cos(2s+1)(\phi-\alpha).$$

Then

$$\boldsymbol{\Pi} \sim \frac{-Aa e^{-ikR}}{16\pi\omega\varepsilon_0 R}\left[\boldsymbol{a}_0 J_0(ka\sin\theta) + 2\sum_{s=1}^{\infty}(-)^s \boldsymbol{a}_{2s} J_{2s}(ka\sin\theta)\right.$$
$$\left. + 2i\sum_{s=0}^{\infty}(-)^s \boldsymbol{a}_{2s+1} J_{2s+1}(ka\sin\theta)\right] \quad (15.2)$$

where

$$\boldsymbol{a}_r = \{a_r(-q) - a_r(q)\}\boldsymbol{i} + \{b_r(q) - b_r(-q)\}\boldsymbol{j} + \{c_r(q) - c_r(-q)\}\boldsymbol{k}\tan\beta,$$

$$a_r(q) = -e^{-2iMq\pi}(e^{2iM\chi\pi} - 1)\left\{\frac{e^{ir\phi}}{(\chi-r)^2-1} + \frac{e^{-ir\phi}}{(\chi+r)^2-1}\right\}, \quad (15.3)$$

$$b_r(q) = ie^{-2iMq\pi}(e^{2iM\chi\pi} - 1)\left\{\frac{(\chi-r)e^{ir\phi}}{(\chi-r)^2-1} + \frac{(\chi+r)e^{-ir\phi}}{(\chi+r)^2-1}\right\}, \quad (15.4)$$

$$c_r(q) = ie^{-2iMq\pi}(e^{2iM\chi\pi} - 1)\left\{\frac{e^{ir\phi}}{\chi-r} + \frac{e^{-ir\phi}}{\chi+r}\right\}, \quad (15.5)$$

$$\chi(q) = ka\cos\theta\tan\beta - q = (kd/2\pi)\cos\theta - q.$$

In any case where the denominator would be zero the pertinent value is obtained by a limiting process.

The electric field can be deduced from (15.2) by means of

$$E_\theta \sim k^2\{(\Pi_x\cos\phi + \Pi_y\sin\phi)\cos\theta - \Pi_z\sin\theta\},$$
$$E_\phi \sim k^2(\Pi_y\cos\phi - \Pi_x\sin\phi).$$

Some particular cases are of interest. Suppose that the diameter of the helix and the spacing between the turns are small compared with the wavelength so that $ka \ll 1$ and $kd \ll 1$. Further, if $q \ll 1$, $\chi \ll 1$ although $M\chi$ is not necessarily small. Then, when $q = m/2M$, $a_0(q) - a_0(-q)$ and $b_0(q) - b_0(-q)$ are at least of the first order of small quantities whereas c_0, a_1 and b_1 are $O(1/\chi)$. For small z, $J_n(z) \approx (\frac{1}{2}z)^n/n!$ (A.1.12) and so the dominant terms are

$$E_\theta \sim \frac{\omega\mu_0 Aa}{8\pi i R} e^{-ikR+2iMq\pi}\{e^{2iM\chi(q)\pi} - 1\}\left\{\frac{1}{\chi(-q)} - \frac{1}{\chi(q)}\right\}\sin\theta\tan\beta,$$

$$E_\phi \sim \frac{\omega\mu_0 Aa^2 k}{16\pi R} e^{-ikR+2iMq\pi}\{e^{2iM\chi(q)\pi} - 1\}\left\{\frac{1}{\chi(q)} - \frac{1}{\chi(-q)}\right\}\sin\theta.$$

These fields bear some resemblance to those which would be produced by the superposition of an electric dipole and magnetic dipole parallel to the axis of the helix. The main radiation can occur perpendicular to the axis of the helix; this is often referred to as the *normal mode*.

At all angles $E_\theta/E_\phi = (2\mathrm{i} \tan \beta)/ka$ so that the radiated wave is elliptically polarized in general. If, however, $\tan \beta = \frac{1}{2}ka$ or $d = \pi ka^2$ the radiation is circularly polarized in all directions.

In general, each of the coefficients in (15.3)–(15.5) can be viewed as the product of $\mathrm{e}^{-2\mathrm{i}M(q-1)\pi}(\mathrm{e}^{2\mathrm{i}M\chi\pi} - 1)/(\mathrm{e}^{2\mathrm{i}\chi\pi} - 1)$ and the coefficient generated by a single loop. Therefore the helix can be regarded as an array of loops with an array factor $|(\mathrm{e}^{2\mathrm{i}M\chi\pi} - 1)/(\mathrm{e}^{2\mathrm{i}\chi\pi} - 1)|^2$. There will be no radiation in a direction in which the array factor vanishes. The radiation will also tend to be a maximum in the directions where the array factor is greatest. A detailed discussion would follow the lines of preceding sections.

Strong radiation along the axis $\theta = 0$ is to be expected when $q - kd/2\pi$ is an integer, one important case being

$$kd/2\pi = 1 + q. \tag{15.6}$$

The existence of such an *axial mode*, with the radiation concentrated about the axis of the helix, has been verified experimentally. Its practical value lies in its frequency insensitivity, a given helix producing an axial beam for all wavelengths from about one-tenth to one-fifth of the length of the helix when ka is about 1. For larger or smaller wavelengths the maximum radiation takes place off the axis.

On the axis $\theta = 0$ it is clear that, at any wavelength,

$$\frac{E_x}{E_y} = \frac{a_0(-q) - a_0(q)}{b_0(q) - b_0(-q)} = \frac{-\mathrm{i}kd/\pi}{1 - q^2 + (kd/2\pi)^2}.$$

Thus the axial radiation is generally elliptically polarized. However, if $q - kd/2\pi \approx \pm 1$, the ratio is of nearly unit magnitude and the polarization is virtually circular. Since (15.6) is included in these conditions, it follows that the axial mode produces a beam which is almost circularly polarized.

The antenna boundary value problem

3.16 The straight tube

The preceding sections have indicated that the radiation from a wire can be evaluated in principle once the current is known though the actual calculation may be laborious and involve numerical quadrature. When only the source which stimulates the antenna is known the problem becomes much more difficult because the current which is induced in the antenna has first to be determined. This is essentially a boundary value problem and two basic methods have been developed for tackling it. One is exact and the other is approximate but both have the underlying aim of formulating an integral equation for the unknown current.

In this section the exact approach is considered, the approximate model being deferred till later. To fix ideas the specific problem of finding the current induced in a perfectly conducting straight cylindrical tube of radius a when it is irradiated by a known electromagnetic field in which the electric intensity is $\boldsymbol{E}^{\mathrm{i}}$ is examined. The time variation is assumed to be harmonic and the factor $\mathrm{e}^{\mathrm{i}\omega t}$ is suppressed. Surround the walls of the tube by a toroid-like surface S (see Fig. 3.18). Then, by §1.34 and (1.29.7), the total electric intensity outside S can be represented by

$$\boldsymbol{E}(\boldsymbol{x}) = \boldsymbol{E}^{\mathrm{i}}(\boldsymbol{x}) + \operatorname{curl}\int_S \boldsymbol{n}_y \wedge \boldsymbol{E}\psi(\boldsymbol{x}, \mathbf{y})\,\mathrm{d}S_y - \operatorname{grad}\int_S \boldsymbol{n}_y \cdot \boldsymbol{E}\psi(\boldsymbol{x}, \mathbf{y})\,\mathrm{d}S_y$$
$$-\mathrm{i}\omega\mu_0\int_S \boldsymbol{n}_y \wedge \boldsymbol{H}\psi(\boldsymbol{x}, \mathbf{y})\,\mathrm{d}S_y$$

where $\psi(\boldsymbol{x}, \mathbf{y}) = \mathrm{e}^{-\mathrm{i}k|\boldsymbol{x}-\mathbf{y}|}/4\pi\,|\boldsymbol{x}-\mathbf{y}|$. Now allow S to collapse so that it coincides with the two sides of the wall of the tube. The tube is perfectly conducting and so the tangential components of $\boldsymbol{E}$ must vanish on it; accordingly the curl disappears from the representation. Also the tangential components of $\boldsymbol{H}$ are discontinuous (1.22.10); let us write $k^2\boldsymbol{I}_s/2\pi a$ for the discontinuity of $-\mathrm{i}\omega\mu_0\boldsymbol{n} \wedge \boldsymbol{H}$ in going from the outside of the tube to the inside. Take the z-axis as the axis of the tube. Then, if r, ϕ, z are cylindrical polar coordinates,

$$\mathrm{i}\omega\varepsilon_0 E_r = \frac{1}{r}\frac{\partial H_z}{\partial \phi} - \frac{\partial H_\phi}{\partial z}$$

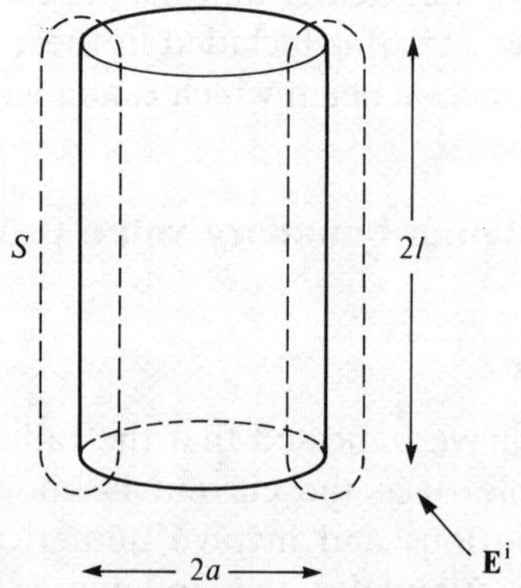

Fig. 3.18 The straight tube as an antenna

from Maxwell's equations and §1.39(i). This enables us to express the discontinuity of $\boldsymbol{n}\cdot\boldsymbol{E}$ in terms of $\boldsymbol{I}_s$ and

$$\boldsymbol{E}(\boldsymbol{x})=\boldsymbol{E}^{\mathrm{i}}(\boldsymbol{x})+\operatorname{grad}\int_{-l}^{l}\int_{0}^{2\pi}\left(\frac{1}{a}\frac{\partial I_{s\phi}}{\partial\phi'}+\frac{\partial I_{sz}}{\partial\zeta}\right)\psi(\boldsymbol{x},\mathbf{y})\,\mathrm{d}\phi'\,\mathrm{d}\zeta/2\pi$$
$$+k^2\int_{-l}^{l}\int_{0}^{2\pi}\boldsymbol{I}_s\psi(\boldsymbol{x},\mathbf{y})\,\mathrm{d}\phi'\,\mathrm{d}\zeta/2\pi \tag{16.1}$$

where $\mathbf{y}$ has cylindrical polar coordinates a, ϕ', ζ and the tube extends from $z=-l$ to $z=l$.

An integral equation for $\boldsymbol{I}_s$ is obtained from (16.1) by requiring that $\boldsymbol{n}\wedge\boldsymbol{E}$ vanish on the tube. To simplify the presentation it will be assumed that $\boldsymbol{E}^{\mathrm{i}}$ has the single component E_z^{i} which will be taken to be independent of ϕ. Then $I_{s\phi}$ can be dropped and only the z-component of $\boldsymbol{E}$ has to be made zero on the tube. If I_{sz} is replaced by I the boundary condition gives

$$0=E_z^{\mathrm{i}}+\frac{\partial}{\partial z}\int_{-l}^{l}\frac{\partial}{\partial\zeta}I(\zeta)K(z-\zeta)\,\mathrm{d}\zeta+k^2\int_{-l}^{l}I(\zeta)K(z-\zeta)\,\mathrm{d}\zeta \tag{16.2}$$

for $-l<z<l$. Here

$$K(z)=\frac{1}{8\pi^2}\int_{-\pi}^{\pi}\frac{\exp\{-\mathrm{i}k(z^2+4a^2\sin^2\frac{1}{2}\phi)^{\frac{1}{2}}\}}{(z^2+4a^2\sin^2\frac{1}{2}\phi)^{\frac{1}{2}}}\,\mathrm{d}\phi. \tag{16.3}$$

The kernel $K(z)$ possesses a singularity at $z=0$. It is the same as that of

$$\frac{1}{8\pi^2}\int_{-\pi}^{\pi}\frac{\mathrm{d}\phi}{(z^2+a^2\phi^2)^{\frac{1}{2}}}=\frac{1}{8\pi a^2}\ln\frac{a\pi+(z^2+a^2\pi^2)^{\frac{1}{2}}}{-a\pi+(z^2+a^2\pi^2)^{\frac{1}{2}}}.$$

In other words $K(z)-(1/4\pi a)\ln(2a\pi/|z|)$ is non-singular.

There are other ways of expressing (16.2). For instance,

$$\int_{-l}^{l}\frac{\partial}{\partial\zeta}I(\zeta)K(z-\zeta)\,\mathrm{d}\zeta=[I(\zeta)K(z-\zeta)]_{-l}^{l}-\int_{-l}^{l}I(\zeta)\frac{\partial}{\partial\zeta}K(z-\zeta)\,\mathrm{d}\zeta$$
$$=[I(\zeta)K(z-\zeta)]_{-l}^{l}+\frac{\partial}{\partial z}\int_{-l}^{l}I(\zeta)K(z-\zeta)\,\mathrm{d}\zeta,$$

the processes being legitimate, despite the singularity of $K(z)$, so long as the principal value of the integral is used. Hence (16.2) becomes

$$0=E_z^{\mathrm{i}}+I(l)K(z-l)-I(-l)K(z+l)+\left(\frac{\partial^2}{\partial z^2}+k^2\right)\int_{-l}^{l}I(\zeta)K(z-\zeta)\,\mathrm{d}\zeta. \tag{16.4}$$

Again, integration of (16.4) gives

$$\int_{-l}^{l} I(\zeta)K(z-\zeta)\,\mathrm{d}\zeta = A\cos kz + B\sin kz$$

$$-\int_{-l}^{z} \{E_z^{\mathrm{i}}(t) + I(l)K(t-l) - I(-l)K(t+l)\}\sin k(z-t)\,\mathrm{d}t \quad (16.5)$$

where A and B are constants.

It is clear from (16.4) that none of the integral equations can be solved completely unless additional information is available. One possibility, when the tube is a receiving antenna, is to demand that the current vanishes at the ends in compliance with the edge conditions (§9.2). It can be shown (Jones 1981) that, when $I(-l) = I(l) = 0$, the solution of (16.4) exists and is unique. The same is true of (16.5) so long as A and B are properly chosen. In this connection it is important to observe that when the right-hand side of (16.5) consists solely of $\cos kz$ the resulting $I(\zeta)$ must possess a singularity which behaves like $(l^2 - \zeta^2)^{-\frac{1}{2}}$ at the ends; this statement also holds if $\cos kz$ is replaced by $\sin kz$.

From the transmitting antenna the integral equation depends upon the feeding arrangements. In a symmetrical feed insert a gap in the tube in $|z| < d$. Then, since there is no incoming wave and the current vanishes at $z = \pm l$, the integral equation

$$0 = K(z+d) - K(z-d) + \left(\frac{\partial^2}{\partial z^2} + k^2\right)\left(\int_{-l}^{-d} + \int_{d}^{l}\right) I(\zeta)K(z-\zeta)\,\mathrm{d}\zeta \quad (16.6)$$

is solved in $d < |z| < l$ subject to $I(d) = I(-d) = 1$. With the resulting I the right-hand side of (16.6) takes certain values, which will be denoted by $E(z)$, in $|z| < d$. The *input impedance* Z and the *input admittance* Y of the antenna are then defined by

$$Z = 1/Y = \mathrm{i}(\mu_0/\varepsilon_0)^{\frac{1}{2}} \int_{-d}^{d} E(z)\,\mathrm{d}z. \quad (16.7)$$

3.17 Murray–Pidduck theory

One way of tackling (16.2) analytically was suggested by Murray (1931, 1933) and elaborated by Pidduck (1946). It is really an early example of the *method of moments* or *Galerkin's method* but it may conveniently be called Murray–Pidduck theory. A modified version of the theory will be presented here.

Some formulae are slightly less cumbersome if a minor change of notation is made. One end of the tube will be placed at the origin and the other at $z = h$. Then the solution of

$$0 = E_z^{\mathrm{i}} + \frac{\partial}{\partial z}\int_0^h \frac{\partial I(\zeta)}{\partial \zeta} K(z-\zeta)\,\mathrm{d}\zeta + k^2\int_0^h I(\zeta)K(z-\zeta)\,\mathrm{d}\zeta \qquad (17.1)$$

with zero current at the ends is required. Such a solution can be represented by a Fourier sine series, i.e. $I(\zeta) = \sum_{n=1}^{\infty} I_n \sin(n\pi\zeta/h)$. Murray and Pidduck made a similar substitution for E_z^{i} but since E_z^{i} is not forced to be zero at the ends it seems that better results should be achieved with a cosine series. Therefore put

$$\tfrac{1}{2}hE_z^{\mathrm{i}} = \tfrac{1}{2}E_0 + \sum_{m=1}^{\infty} E_m \cos(m\pi z/h)$$

where

$$E_m = \int_0^h E_z^{\mathrm{i}}(\zeta)\cos(m\pi\zeta/h)\,\mathrm{d}\zeta.$$

Multiply (17.1) by $\cos(m\pi z/h)$ and integrate from $z = 0$ to $z = h$. Then

$$-4\pi E_m = \sum_{n=1}^{\infty} Z_{mn} I_n \quad (m = 0, 1, 2, \ldots) \qquad (17.2)$$

where

$$Z_{mn} = k_n\int_0^h \cos k_n\zeta\{(-)^m K_1(h-\zeta) - K_1(-\zeta)\}\,\mathrm{d}\zeta$$
$$+ \int_0^h\int_0^h (k_m k_n \sin k_m z \cos k_n\zeta + k^2 \cos k_m z \sin k_n\zeta) K_1(z-\zeta)\,\mathrm{d}z\,\mathrm{d}\zeta,$$

$K_1(z) = 4\pi K(z), \qquad k_m = m\pi/h.$

Define

$$\zeta_{m,n} = \int_0^h \mathrm{e}^{\mathrm{i}k_n\zeta}\{(-)^m K_1(h-\zeta) - K_1(-\zeta)\}\,\mathrm{d}\zeta,$$

$$z_{m,n} = \int_0^h\int_0^h \mathrm{e}^{\mathrm{i}(k_m z + k_n\zeta)} K_1(z-\zeta)\,\mathrm{d}z\,\mathrm{d}\zeta$$

and then

$$Z_{mn} = \tfrac{1}{2}k_n(\zeta_{m,n} + \zeta_{m,-n}) + \frac{1}{4\mathrm{i}}\{(k^2 + k_m k_n)(z_{m,n} - z_{-m,-n})$$
$$-(k^2 - k_m k_n)(z_{m,-n} - z_{-m,n})\}.$$

Note that because of (16.3)

$$\zeta_{m,n} = \int_0^h \{(-)^{m+n}\mathrm{e}^{-\mathrm{i}k_n\zeta} - \mathrm{e}^{\mathrm{i}k_n\zeta}\}K_1(\zeta)\,\mathrm{d}\zeta,$$

from which it is evident that $\zeta_{m,-n} = -\zeta_{m,n}$ when $m+n$ is even and $\zeta_{m,-n} = \zeta_{m,n}$ when $m+n$ is odd.

The substitution $z = \zeta + t$ in $z_{m,n}$ gives

$$z_{m,n} = \int_0^h e^{i(k_m+k_n)\zeta} \int_{-\zeta}^{h-\zeta} e^{ik_m t} K_1(t)\, dt\, d\zeta.$$

An integration by parts with respect to ζ supplies, if $k_m + k_n \neq 0$,

$$z_{m,n} = (\zeta_{m,n} + \zeta_{n,m})/i(k_m + k_n). \tag{17.3}$$

It is now clear that $z_{m,n} = z_{-m,-n}$ when $m+n$ is even and $z_{m,n} = -z_{-m,-n}$ when $m+n$ is odd provided that $k_m + k_n \neq 0$. On the other hand, when $k_m + k_n = 0$ it is obvious from the definition of $z_{m,n}$ that $z_{m,n} = z_{-m,-n}$. It thus follows that $Z_{mn} = 0$ whenever $m+n$ is even.

The evaluation of Z_{mn} when $m+n$ is odd needs a certain amount of manipulation. In carrying it out the approximation that the tube is thin will be introduced. Then a can be placed equal to zero in $K_1(z)$ except for small $|z|$.

Let η be such that $k\eta$ is a small positive quantity but $\eta \gg a$. Then write

$$\int_0^h e^{ik_n\zeta} K_1(\zeta)\, d\zeta = \left(\int_{-\infty}^{\infty} - \int_{-\eta}^{0} - \int_{-\infty}^{-\eta} - \int_h^{\infty}\right) e^{ik_n\zeta} K_1(\zeta)\, d\zeta. \tag{17.4}$$

In the last integral a^2 may be neglected in K_1 and

$$\int_h^{\infty} = \int_h^{\infty} e^{i(k_n - k)\zeta} \frac{d\zeta}{\zeta} = -\text{Ei}\{i(k_n - k)h\}$$

from (11.4). The third integral in (17.4) can be treated similarly and

$$\int_{-\infty}^{-\eta} = -\text{Ei}\{-i(k_n + k)\eta\}.$$

For the second integral the range of ζ is so small that the exponential in K_1 can be replaced by unity and so can $e^{ik_n\zeta}$ if n is not too large. This leads to

$$\int_{-\eta}^{0} = \int_0^{\eta} \int_{-\pi}^{\pi} \frac{d\phi\, d\zeta}{2\pi(\zeta^2 + 4a^2 \sin^2 \tfrac{1}{2}\phi)^{\frac{1}{2}}} \approx \frac{1}{\pi} \int_0^{\pi} \ln \frac{\eta}{a \sin \phi}\, d\phi \approx \ln \frac{2\eta}{a}.$$

When x is small

$$\text{Ei}(ix) \approx \gamma + \ln |x| - \tfrac{1}{2}\pi i \operatorname{sgn} x$$

where $\operatorname{sgn} x$ is 1 when $x > 0$ and -1 when $x < 0$. Hence

$$\begin{aligned}\int_0^h e^{ik_n\zeta} K_1(\zeta)\, d\zeta = \int_{-\infty}^{\infty} & e^{ik_n\zeta} K_1(\zeta)\, d\zeta + \gamma + \tfrac{1}{2}\pi i \operatorname{sgn}(k_n + k) \\ & + \ln \tfrac{1}{2}a\, |k_n + k| + \text{Ei}\{i(k_n - k)h\}.\end{aligned}$$

Now, from (A.1.40) and (A.1.35)

$$\int_{-\infty}^{\infty} e^{ik_n\zeta} K_1(\zeta)\, d\zeta = -\tfrac{1}{2}i \int_0^{2\pi} H_0^{(2)}\{2a(k^2-k_n^2)^{\frac{1}{2}} \sin \tfrac{1}{2}\phi\}\, d\phi$$

$$= -\pi i J_0\{a(k^2-k_n^2)^{\frac{1}{2}}\} H_0^{(2)}\{a(k^2-k_n^2)^{\frac{1}{2}}\} \qquad (17.5)$$

where $(k^2-k_n^2)^{\frac{1}{2}}$ is positive when $k^2 > k_n^2$ and negative imaginary for $k^2 < k_n^2$. Hence, so long as n is not too large

$$\int_0^h e^{ik_n\zeta} K_1(\zeta)\, d\zeta = \mathrm{Ei}\{i(k_n-k)h\} - \gamma - \ln \tfrac{1}{2}a\,|k_n-k| + \tfrac{1}{2}\pi i\, \mathrm{sgn}(k_n-k)$$

and, when $m+n$ is odd,

$$\zeta_{m,n} = 2\gamma + \ln \tfrac{1}{4}a^2\,|k_n^2-k^2| - \tfrac{1}{2}\pi i\, \mathrm{sgn}(k_n-k) + \tfrac{1}{2}\pi i\, \mathrm{sgn}(k_n+k)$$
$$- \mathrm{Ei}\{i(k_n-k)h\} - \mathrm{Ei}\{-i(k_n+k)h\}.$$

Therefore, for odd $m+n$,

$$Z_{mn} = \frac{k_n(k^2-k_n^2)}{k_m^2-k_n^2}[2\gamma + \ln \tfrac{1}{4}a^2\,|k_n^2-k^2| + \pi i H(k^2-k_n^2)$$
$$- \mathrm{Ei}\{i(k_n-k)h\} - \mathrm{Ei}\{-i(k_n+k)h\}]$$
$$+ \frac{k_n(k^2-k_m^2)}{k_m^2-k_n^2}[2\gamma + \ln \tfrac{1}{4}a^2\,|k_m^2-k^2| + \pi i H(k^2-k_m^2)$$
$$- \mathrm{Ei}\{i(k_m-k)h\} - \mathrm{Ei}\{-i(k_m+k)h\}] \qquad (17.6)$$

where $H(x)$ is the Heaviside unit function which is 1 for $x > 0$ and 0 for $x < 0$. If $k = k_n$, (17.6) reduces to

$$Z_{mn} = -k_n[2\gamma + \ln \tfrac{1}{4}a^2\,|k_m^2-k_n^2| + \pi i H(n^2-m^2)$$
$$- \mathrm{Ei}\{i(k_m-k_n)h\} - \mathrm{Ei}\{-i(k_m+k_n)h\}]. \qquad (17.7)$$

The formulae (17.6) and (17.7) can be expected to be valid for those values of m and n such that $k_m a$ and $k_n a$ are small, the error being roughly of the order of the greater of these two quantities. Given the incident field an approximate solution of (17.2) can be derived on the assumption that I_n is zero after the first few values of n.

3.18 The axisymmetric antenna

An integral equation bearing some resemblance to that for the tube can be found for an axisymmetric antenna under suitable conditions. Take the axis of symmetry to be the z-axis and let the antenna be broken by a gap which may contain a generator (Fig. 3.19). The electromagnetic fields to be considered will be such that at any point the electric intensity lies in the meridian plane through the point and z-axis whereas the magnetic

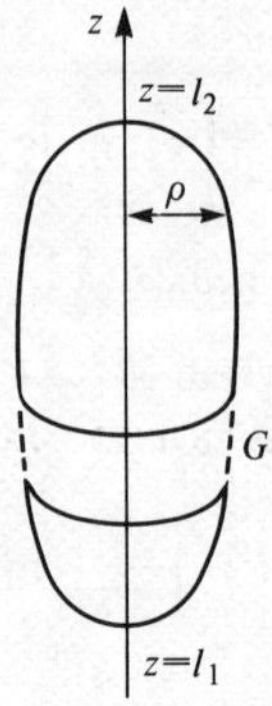

Fig. 3.19 The axisymmetric antenna with gap

intensity is perpendicular to the meridian plane. The field $\boldsymbol{E}_2$, $\boldsymbol{H}_2$ is due to an electric dipole parallel to the z-axis placed on the axis of the antenna at the point $(0, 0, z_2)$ or $\boldsymbol{x}_2$ between the ends. Then, from (8.3) and (8.4), the nonzero components in cylindrical polar coordinates are

$$E_{2r} = -\frac{1}{4\pi i\omega\varepsilon_0}\frac{\partial^2}{\partial r\,\partial z}\psi(\boldsymbol{x}, \boldsymbol{x}_2),\ E_{2z} = -\frac{1}{4\pi i\omega\varepsilon_0}\left(\frac{\partial^2}{\partial z^2}+k^2\right)\psi(\boldsymbol{x}, \boldsymbol{x}_2),$$

$$H_{2\phi} = \frac{1}{4\pi i\omega}\frac{\partial}{\partial r}\psi(\boldsymbol{x}, \boldsymbol{x}_2).$$

Let $\boldsymbol{E}_1$, $\boldsymbol{H}_1$ be a similar field produced by a source $\boldsymbol{a}$ at $\boldsymbol{x}_1$ outside the surface S which consists of the boundary of the antenna together with the surface G covering the gap.

Both electromagnetic fields are radiating at infinity and so by the reciprocity theorem of §1.35

$$\int_S (\boldsymbol{E}_1 \wedge \boldsymbol{H}_2 - \boldsymbol{E}_2 \wedge \boldsymbol{H}_1)\cdot \boldsymbol{n}\,dS = \boldsymbol{a}\cdot \boldsymbol{E}_2(\boldsymbol{x}_1),$$

Insert the expressions for $\boldsymbol{E}_2$, $\boldsymbol{H}_2$ and use the fact that

$$dS = \rho(z)\left\{1+\left(\frac{d\rho}{dz}\right)^2\right\}^{\frac{1}{2}} d\phi\, dz$$

where $\rho(z)$ is the radius of the cross-section of the antenna at height z. Then, if E_{1t} is the tangential component of $\boldsymbol{E}_1$ on S

$$-\int_{l_1}^{l_2}\int_0^{2\pi} H_\phi \rho(z)\left(k^2 - \frac{\partial^2}{\partial z\,\partial z_2}\right)\psi\, d\phi\, dz$$

$$= \int_{l_1}^{l_2}\int_0^{2\pi} \varepsilon_0 E_{1t}\rho^2(z)\left\{1+\left(\frac{d\rho}{dz}\right)^2\right\}^{\frac{1}{2}}\frac{1}{D}\frac{\partial\psi}{\partial D}\, d\phi\, dz + 4\pi i\omega\varepsilon_0 \boldsymbol{a}\cdot \boldsymbol{E}_2(\boldsymbol{x}_1)$$

where now $\psi = e^{-ikD}/D$, $D^2 = \rho^2(z) + (z - z_2)^2$ and the antenna extends from $z = l_1$ to $z = l_2$. The quantity $-2\pi\rho(z)H_\phi$ may be taken to be the current $I(z)$.

Suppose now that the source at $\boldsymbol{x}_1$ is absent but that the field $\boldsymbol{E}_1$, $\boldsymbol{H}_1$ is produced by a generator in the gap giving a known tangential electric field on G. Then

$$\int_{l_1}^{l_2} I(z)\left(k^2 - \frac{\partial^2}{\partial z\, \partial z_2}\right)\psi\, \mathrm{d}z = 2\pi\varepsilon_0 \int_{l_1}^{l_2} E_{1t}\rho^2(z)\left\{1 + \left(\frac{d\rho}{dz}\right)^2\right\}^{\frac{1}{2}} \frac{1}{D}\frac{\partial\psi}{\partial D}\, \mathrm{d}z \qquad (18.1)$$

where the integration on the right can be limited to G since $E_{1t} = 0$ on the perfectly conducting boundary of the antenna. This constitutes an integral equation on $l_1 < z_2 < l_2$ for I, in the case when the antenna is acting as a transmitter.

Next assume that there is no gap but there is a dipole at $\boldsymbol{x}_1$; this is the reception problem. Now

$$\int_{l_1}^{l_2} I(z)\left(k^2 - \frac{\partial^2}{\partial z\, \partial z_2}\right)\psi\, \mathrm{d}z = 4\pi i\omega\varepsilon_0 \boldsymbol{a} \cdot \mathbf{E}_2(\boldsymbol{x}_1).$$

Let $\mathbf{E}^{\mathrm{i}}$ be the field generated from $\boldsymbol{x}_1$ when there is no antenna. Applying the reciprocity theorem to $\mathbf{E}^{\mathrm{i}}$ and $\boldsymbol{E}_2$ we obtain

$$\boldsymbol{k} \cdot \mathbf{E}^{\mathrm{i}}(\boldsymbol{x}_2) = \boldsymbol{a} \cdot \mathbf{E}_2(\boldsymbol{x}_1).$$

Since $\boldsymbol{x}_2$ is $(0, 0, z_2)$ the integral equation can be written

$$\int_{l_1}^{l_2} I(z)\left(k^2 - \frac{\partial^2}{\partial z\, \partial z_2}\right)\psi\, \mathrm{d}z = 4\pi i\omega\varepsilon_0 E^{\mathrm{i}}_z(z_2) \qquad (18.2)$$

and holds for $l_1 < z_2 < l_2$.

If it is assumed that $I(l_1) = I(l_2) = 0$ this integral equation can be cast into the same form as that for the straight tube but it holds on the axis of the antenna and the kernel is different. Moreover, there is no obligation for ρ to be constant provided that the initial assumptions on the current and field are met. An arbitrary incident field will probably vary round the cross-section and, in such circumstances, the assumptions will not be fulfilled unless ρ is small. However, when the antenna is excited by a generator in the gap it may well be that the field has the desired symmetry even when ρ is not small.

3.19 The thin-wire approximation

One-dimensional integral equations can also be derived for other shapes so long as they have the structure of a *thin wire*. In a thin wire the dimensions of the cross-section are small in comparison with both the length of the wire and the wavelength of operation. The surface of the

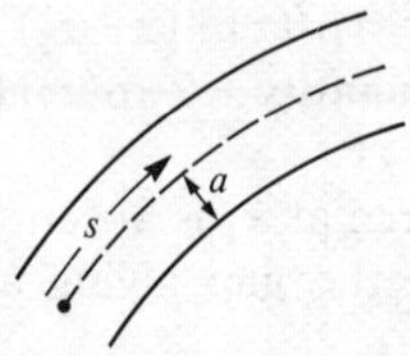

Fig. 3.20 The thin wire antenna

wire is generated by moving a circle of radius a so that its centre lies on a curve, called the axis of the wire, and its plane is always perpendicular to the axis. Any point on the axis can be specified by its arc-length s from some fixed point (Fig. 3.20).

At a point on the perimeter of the cross-section associated with the point on the axis identified by s the incident electric intensity can be replaced by its value at s with an error of $O(a)$. The field due to the induced current can be split into two parts—one due to cross-sections in $(s-\delta, s+\delta)$ where δ is small but $\delta \gg a$ and the other due to the more distant circumference. The second part provides a finite field on the periphery which may be replaced by its value on the axis with an error of $O(a)$. A rough estimate of the field due to the nearby current can be obtained by imagining that a circular cylinder is involved. From the theory of scattering by an infinite circular cylinder (§8.1), a finite electric intensity along the axis is produced by a current density of $O\{1/(a \ln ka)\}$ whereas an intensity which varies with the azimuthal angle of the cross-section needs a current density of $O(1)$ with a similar angular variation. A field perpendicular to the axis must vary with azimuthal angle and requires a lateral current density of $O(a)$. Therefore, if terms of $O(a)$ are rejected, it is only necessary to consider the component of $\mathbf{E}^{\mathrm{i}}$ parallel to the axis and an induced current flow, parallel to the axis, which is uniform round the cross-section.

Let $I(s)$ be the axial current at s, i.e. $2\pi a$ times the surface current density. Also let $\psi(\boldsymbol{x}, \mathbf{y})$ when it is averaged for $\mathbf{y}$ on the circular cross-section of the point ζ on the axis be denoted by $\psi_1(s, \zeta)$ when $\boldsymbol{x}$ is at the point s on the axis. Then the analogue of (16.2) is

$$-\mathrm{i}\omega\varepsilon_0 E_s^{\mathrm{i}}(s) = \int_0^l \left\{ I(\zeta)k^2 \cos\alpha K(s, \zeta) + \frac{\partial I}{\partial \zeta}\frac{\partial}{\partial s} K(s, \zeta) \right\} \mathrm{d}\zeta \qquad (19.1)$$

where l is the arc-length of the axis, the initial point being taken at one end, $K(s, \zeta) = \psi_1(s, \zeta)$, E_s^{i} is the component of the incident field parallel to the axis and α is the angle between the tangents to the axis at s and ζ. It has to be accepted that (19.1) may not hold at a point where there is a sharp change in the curvature of the axis.

Equation (19.1) is supplemented by the conditions $I(0) = I(l) = 0$.

While these are exact for a hollow tube they are only approximate for a solid wire because small currents flow over the end caps. The neglect of these currents may entail an error of order a which is no worse than quantities which have already been ignored.

In the absence of an external field Pocklington derived an approximate solution of (19.1) as follows. The dominant terms come from a neighbourhood $(s-\delta, s+\delta)$ of s; δ may be chosen small enough for the wire to be regarded as straight and the exponential in K to be dropped in this neighbourhood. Hence, bearing in mind the vanishing of the current at the ends of the wire, the dominant part of (19.1) is

$$\begin{aligned} 0 &= \left(\frac{\partial^2}{\partial s^2}+k^2\right)\frac{I(s)}{2\pi}\int_{s-\delta}^{s+\delta}\int_0^{2\pi}\frac{\mathrm{d}\phi\,\mathrm{d}\zeta}{\{(s-\zeta)^2+4a^2\sin^2\frac{1}{2}\phi\}^{\frac{1}{2}}} \\ &= 2\ln(2\delta/a)\left(\frac{\partial^2}{\partial s^2}+k^2\right)I(s) \end{aligned}$$

when only the largest terms are retained. Hence $I(s)=A\mathrm{e}^{iks}+B\mathrm{e}^{-iks}$ showing that the current is sinusoidal, propagates with the speed of light, and maintains its amplitude when there is no incident field. It is this result of Pocklington which forms the basis for the theory of radiation from wires (§3.11).

There is another way of arriving at an approximate integral equation and that is to assume that the current is concentrated on the axis of the wire. Applying the boundary condition on the surface of the wire leads to (19.1) again but the kernel is different. If it is denoted by $K_1(s, \zeta)$ it takes the form

$$K_1(s,\zeta)=\exp[-ik\{(s-\zeta)^2+a^2\}^{\frac{1}{2}}]/4\pi\{(s-\zeta)^2+a^2\}^{\frac{1}{2}}$$

for a straight wire. This is always finite though $K(s, \zeta)$ is singular at $s=\zeta$. Nevertheless, K_1 and K differ only to order ka when s is not near ζ and so the greater simplicity of K_1 has made it preferable to K in the eyes of some. The physical notion on which this model is based has led to developments by Gear (1975) for antennas of slender shape.

3.20 Numerical procedure

Most numerical methods for solving the integral equation of a thin wire are analogous to the Murray–Pidduck theory of §3.17 with some modifications. To simplify the presentation only the straight wire or tube will be treated but the principles are clearly more widely applicable. From (16.4) the integral equation to be solved is

$$\left(\frac{\partial^2}{\partial z^2}+k^2\right)\int_{-l}^{l} I(\zeta)K_1(z-\zeta)\,\mathrm{d}\zeta=f(z) \tag{20.1}$$

subject to $I(\pm l)=0$. Here $K_1(z)=4\pi K(z)$ with K defined in (16.3).

Expand I in a set of known *basis functions* u_n by means of

$$I(\zeta) = \sum_{n=1}^{N} I_n u_n(\zeta). \tag{20.2}$$

One would expect the representation in (20.2) to be exact when N is infinite but only a finite number of terms is kept in the numerical approximation. After substituting (20.2) in (20.1) multiply by a weight function $w_m(z)$ and integrate from $z=-l$ to l. The resulting algebraic equations for the coefficients I_n are

$$\sum_{n=1}^{N} Z_{mn} I_n = \int_{-l}^{l} w_m(z) f(z)\,\mathrm{d}z \quad (m=1,\ldots,N) \tag{20.3}$$

where

$$Z_{mn} = \int_{-l}^{l} w_m(z)\left(\frac{\partial^2}{\partial z^2}+k^2\right)\int_{-l}^{l} u_n(\zeta) K_1(z-\zeta)\,\mathrm{d}\zeta\,\mathrm{d}z. \tag{20.4}$$

Considerable latitude is available for the choice of weight and basis functions but it is usual in numerical work to select them to be zero outside a segment of wire. Let $d=2l/(N+1)$ and put $z_p=-l+pd$ $(p=0,1,\ldots,N+1)$. Possibilities are the *pulse functions*

$$p_m(z) = \begin{cases} 1 & \text{for } |z-z_m| < \tfrac{1}{2}d \\ 0 & \text{for other } z, \end{cases} \tag{20.5}$$

the *triangle functions*

$$t_m(z) = \begin{cases} (d-|z-z_m|)/d & \text{for } z_{m-1} \leqslant z \leqslant z_{m+1} \\ 0 & \text{for other } z, \end{cases} \tag{20.6}$$

and the *piecewise sinusoidal functions*

$$s_m(z) = \begin{cases} \sin k(d-|z-z_m|)/\sin kd & \text{for } z_{m-1} \leqslant z \leqslant z_{m+1} \\ 0 & \text{for other } z. \end{cases} \tag{20.7}$$

Other functions have been considered (see, for example, Jones (1979) where there is a lengthier discussion of numerical processes in electromagnetism). To ensure that the current is zero at the ends (20.5) is often modified there by taking $p_0(z)=0$ and $p_{N+1}(z)=0$ everywhere when I is expanded in pulse functions.

Suppose that w_m is identified with t_m and u_m with s_n, which may be signified by writing Z_{mn}^{ts} for Z_{mn}. Then

$$\begin{aligned} Z_{mn}^{ts} &= \int_{z_{m-1}}^{z_{m+1}} \frac{d-|z-z_m|}{d}\left(\frac{\partial^2}{\partial z^2}+k^2\right)\int_{z_{n-1}}^{z_{n+1}} \frac{\sin k(d-|\zeta-z_n|)}{\sin kd} K_1(z-\zeta)\,\mathrm{d}\zeta\,\mathrm{d}z \\ &= \int_{-d}^{d} \frac{d-|z|}{d}\left(\frac{\partial^2}{\partial z^2}+k^2\right)\int_{-d}^{d} \frac{\sin k(d-|\zeta|)}{\sin kd} K_1(z-\zeta+z_m-z_n)\,\mathrm{d}\zeta\,\mathrm{d}z. \end{aligned} \tag{20.8}$$

If the signs of both z and ζ are changed the result is the same as interchanging m and n, by virtue of the properties of K_1. Hence $Z^{ts}_{mn} = Z^{ts}_{nm}$, i.e. the matrix in (20.3) is symmetric. In like manner it can be shown that Z^{tt}_{mn}, Z^{tp}_{mn}, Z^{st}_{mn}, Z^{ss}_{mn}, Z^{sp}_{mn} are symmetric.

On the other hand, integration by parts in (20.8) furnishes

$$Z^{ts}_{mn} = \int_{-d}^{d}\int_{-d}^{d}\left\{(d-|z|)\frac{\sin k(d-|\zeta|)}{d\sin kd} - k \operatorname{sgn} z \operatorname{sgn}\zeta \frac{\cos k(d-|\zeta|)}{d\sin kd}\right\}$$
$$\times K_1(z-\zeta+z_m-z_n)\,\mathrm{d}\zeta\,\mathrm{d}z$$

from which it is evident that $Z^{ts}_{mn} = Z^{st}_{mn}$, i.e. the algebraic system on the left of (20.3) is unaltered if the basis and weight functions are interchanged.

Likewise it can be shown that $Z^{tp}_{mn} = Z^{pt}_{mn}$ and $Z^{sp}_{mn} = Z^{ps}_{mn}$. Thus the same numerical matrices can occur for different choices of basis and weight functions.

One may also note the following approximations, if $f(z)$ is sufficiently well-behaved,

$$\int_{-l}^{l} p_m(z)f(z)\,\mathrm{d}z \approx df(z_m),$$

$$\int_{-l}^{l} t_m(z)f(z)\,\mathrm{d}z \approx f(z_m)d,$$

$$\int_{-l}^{l} s_m(z)f(z)\,\mathrm{d}z \approx 2(1-\cos kd)f(z_m)/k\sin kd.$$

For small kd the last approximant is almost $df(z_m)$ so that the right-hand sides of (20.3) do not differ by much with any of the three weight functions so long as the segmentation is fine enough.

The only remaining numerical point is the logarithmic singularity of K_1. This can be dealt with by writing

$$K_1(z) = \frac{1}{2\pi}\int_{-\pi}^{\pi}\frac{\exp\{-ik(z^2+4a^2\sin^2\tfrac{1}{2}\phi)^{\frac{1}{2}}\}-1}{(z^2+4a^2\sin^2\tfrac{1}{2}\phi)^{\frac{1}{2}}}\,\mathrm{d}\phi$$
$$+\frac{1}{2\pi}\int_{-\pi}^{\pi}(z^2+4a^2\sin^2\tfrac{1}{2}\phi)^{-\frac{1}{2}}\,\mathrm{d}\phi.$$

The first integral presents no problems and the second can be expressed as a complete elliptic integral. The behaviour of the elliptic integral is well known, consisting of a slow variation on top of a logarithm, and so it can be handled in a straightforward manner.

As for N it should be chosen so that d does not exceed $\lambda/5$ and larger values of N may well be necessary in some circumstances. One method of checking the accuracy of the approximate current is to observe how well it reproduces $f(z)$ over the whole length of the wire. Wide discrepancies in some interval will indicate a poor approximation to the true current either because N is inadequately large or through a bad choice of basis and weight functions or both.

3.21 Wires in the time domain

By starting from the formulae of §1.16 and following the lines of §3.16 we may obtain

$$E_z^i(z,t) = \frac{\mu_0}{2\pi}\int_{-\pi}^{\pi}\int_{-l}^{l}\frac{I(\zeta,T)}{|\mathbf{z}-\boldsymbol{\zeta}|}\,d\zeta\,d\phi + \frac{1}{2\pi\varepsilon_0}\frac{\partial}{\partial z}\int_{-l}^{l}\int_{-\pi}^{\pi}\frac{1}{|\mathbf{z}-\boldsymbol{\zeta}|}\int_{-\infty}^{T}\frac{\partial}{\partial\zeta}I(\zeta,u)\,du\,d\phi\,d\zeta \quad (21.1)$$

as the integral equation for a tube in the time domain. Here $T = t - |\mathbf{z}-\boldsymbol{\zeta}|/c$ and $|\mathbf{z}-\boldsymbol{\zeta}| = \{(z-\zeta)^2 + 4a^2\sin^2\frac{1}{2}\phi\}^{\frac{1}{2}}$. Clearly corresponding integral equations for curved wires can also be derived.

One method of tackling (21.1) would be to take a Fourier transform in time of (21.1), solve the consequent integral equation in the frequency domain as in previous sections and then take an inverse transform. In this section, however, we shall be concerned with a direct numerical attack on (21.1).

Clearly, an algebraic system can be formulated in terms of basis and weight functions as in the preceding section. In order to bring out the vital fact that (21.1) can be solved step-by-step in time only the case in which the basis functions are pulse functions will be discussed; in practice rather smoother basis functions may be desirable. Also $\delta(z-z_m)$ will be employed as the weight function. This corresponds to *point matching* in which the integral equation is imposed at discrete points.

The discussion is easier, if instead of working with (21.1), the integral equation

$$g(z,t) = \int_{-l}^{l} f(\zeta, T_0)K_2(z,\zeta)\,d\zeta \quad (21.2)$$

is examined where $T_0 = t - |z-\zeta|/c$. The function f is required given g and K_2.

The time intervals will be related to the space intervals in a special way to take account of the speed of propagation of waves. Thus the pulse function $q_n(t)$ is defined to be 1 for $|t - nd/c| < d/2c$ and to be zero elsewhere. Let

$$f(z, T_0) = \sum_{n=1}^{N}\sum_{m=1}^{M} I_{n,m}p_n(\zeta)q_m(T_0).$$

Inserting this in (21.2) and applying point matching at $z = z_r$, $t =$

$t_s(=sd/c)$, we obtain

$$g(z_r, t_s) = \int_{-l}^{l} \sum_{n=1}^{N} \sum_{m=1}^{M} I_{n,m} p_n(\zeta) q_m(T_{0s}) K_2(z_r, \zeta)\, d\zeta$$

$$= \sum_{n=1}^{N} \int_{z_n - \frac{1}{2}d}^{z_n + \frac{1}{2}d} \sum_{m=1}^{M} I_{n,m} q_m(T_{0s}) K_2(z_r, \zeta)\, d\zeta$$

where $T_{0s} = t_s - |z_r - \zeta|/c$. Since q_m vanishes unless T_{0s} is within $d/2c$ of t_m this simplifies to

$$g(z_r, t_s) = \sum_{n=1}^{N} Z_{rn} I_{n, s-|r-n|} \tag{21.3}$$

where

$$Z_{rn} = \int_{-\frac{1}{2}d}^{\frac{1}{2}d} K_2(z_r, \zeta + z_n)\, d\zeta.$$

From (21.3)

$$I_{r,s} = \left\{ g(z_r, t_s) - \sum_{n=1}^{N}{}' Z_{rn} I_{n, s-|r-n|} \right\} \Big/ Z_{rr} \tag{21.4}$$

where $\sum'$ means omit $n = r$ from the summation. The second suffix of I_{nm} on the right of (21.4) is always less than s. Therefore if $I_{n,m}$ is known for $n = 1, \ldots, N$ and $m = 1, \ldots, s-1$ the value of I_{rs} can be calculated immediately. In other words, knowledge of I_{nm} up to a certain time permits the determination at once of values one time step later. Moreover, the explicit form of (21.4) reveals that no matrix inversion is involved. If there is no current flow until excitation arrives the procedure is particularly straightforward since, at the first time step, there is no summation on the right-hand side of (21.4). Consequently, calculations in the time domain are less exacting on the computer than those in the frequency domain unless the response over a long period is needed when instabilities may appear. Results in the frequency domain can be obtained from the time domain by a Fourier transform but the net effort may not be all that different from a direct approach in the frequency domain.

Exercises on Chapter 3

1. The pressure p satisfies

$$\frac{\partial^2 p}{\partial x'^2} + \frac{\partial^2 p}{\partial y'^2} + \frac{\partial^2 p}{\partial z'^2} - \frac{1}{v^2}\frac{\partial^2 p}{\partial t'^2} = -\delta(\boldsymbol{x}'),$$

Show that, after a Lorentz transformation $x' = \beta(x - ut)$, $y' = y$, $z' = z$, $t' = \beta(t - ux/v^2)$, p satisfies

$$\nabla^2 p - \partial^2 p / v^2\, \partial t^2 = -\delta(x - ut)\delta(y)\delta(z)/\beta$$

where $\beta=(1-u^2/v^2)^{-\frac{1}{2}}$. Use this result to verify (4.2) for a moving subsonic source.

2. An acoustic point source describes a circle which lies in $z=0$ and has the origin as centre. Show that on the z-axis both p and $\partial p/\partial z$ are independent of the motion of the source.

3. If an acoustic simple source be regarded as a *monopole*, then *dipole* radiation comes from

$$\nabla^2 p+k^2 p=-\delta'(x)\delta(y)\delta(z).$$

Prove that, for a dipole,

$$p=\frac{\partial}{\partial x}\frac{e^{i(\omega t-kR)}}{4\pi R}=-\frac{x}{4\pi R^3}(1+ikR)e^{i(\omega t-kR)}.$$

Find p for a *quadrupole* where

$$\nabla^2 p+k^2 p=-\frac{\partial^2}{\partial x_i\,\partial x_j}\delta(\boldsymbol{x}).$$

4. A point source of magnitude 1 oscillates harmonically along the x-axis with very small amplitude A about a fixed point source of magnitude -1. Show that p obtained from (1.4) gives the real part of a dipole of magnitude A as calculated from Exercise 3.

5. Find the power radiated from an acoustic quadrupole.

6. Verify that the electric intensity of a point charge moving with uniform velocity in free space as determined by §3.4 agrees with that given in Exercise 20 of Chapter 2.

7. A point charge travels on a circle which has the origin as centre and is in $z=0$. Show that the value of E_z on the z-axis is independent of the motion of the charge.

8. A point charge Q oscillates harmonically with very small amplitude in a straight line about a fixed point charge $-Q$. Verify that the field supplied by (2.5), (2.6) can be regarded as due to a certain harmonic electric dipole.

9. Point charges Q and $-Q$ describe a circle of radius a, centre the origin, in $z=0$ with angular speed ω. The charges $\pm Q$ are at $(\pm a, 0, 0)$ at $t=0$. If $a\to 0$ and $Q\to\infty$ in such a way that $2Qa$ stays equal to p_0 show that, in spherical polar coordinates,

$$H_\phi \sim \frac{\omega^2 p_0 \cos\theta}{4\pi cR}\cos\{\omega(t-R/c)-\phi\}$$

at a large distance R from the origin.

10. Harmonic electric dipoles of moments $p_0\boldsymbol{k}e^{i\omega t}$ and $-p_0\boldsymbol{k}e^{i\omega t}$ are placed at $(0,0,0)$ and $(-a,0,0)$ respectively. If $a\to 0$ and $p_0\to\infty$ in such a way that $p_0 a$ is always equal to q_{31} show that the Hertz vector of the

field is

$$\boldsymbol{\Pi} = \frac{q_{31}x}{4\pi\varepsilon_0 R^3}(1+\mathrm{i}kR)\boldsymbol{k}\mathrm{e}^{-\mathrm{i}kR}.$$

Harmonic dipoles $p_0\boldsymbol{i}\mathrm{e}^{\mathrm{i}\omega t}$ at (0, 0, 0) and (0, 0, $-a$) are similarly combined to give

$$\boldsymbol{\Pi} = \frac{q_{13}z}{4\pi\varepsilon_0 R^3}(1+\mathrm{i}kR)\boldsymbol{i}\mathrm{e}^{-\mathrm{i}kR}.$$

A *plane quadrupole* is derived by taking the average of these two fields with $q_{13} = q_{31}$, so that

$$\boldsymbol{\Pi} = \frac{q_{13}}{8\pi\varepsilon_0 R^3}(1+\mathrm{i}kR)(x\boldsymbol{k}+z\boldsymbol{i})\mathrm{e}^{-\mathrm{i}kR}.$$

A *linear quadrupole* is derived by taking half the field due to $p_0\boldsymbol{i}\mathrm{e}^{\mathrm{i}\omega t}$ and $-p_0\boldsymbol{i}\mathrm{e}^{\mathrm{i}\omega t}$ at (0, 0, 0) and ($-a$, 0, 0) respectively; then

$$\boldsymbol{\Pi} = \frac{q_{11}x}{8\pi\varepsilon_0 R^3}(1+\mathrm{i}kR)\boldsymbol{i}\mathrm{e}^{-\mathrm{i}kR}.$$

In general the Hertz vector of the quadrupole $\boldsymbol{q}$ at the origin is given by

$$\boldsymbol{\Pi} = -\frac{1}{8\pi\varepsilon}(\boldsymbol{q}\cdot\mathrm{grad})\frac{\mathrm{e}^{-\mathrm{i}kR}}{R}.$$

Show that at large distances from the plane quadrupole above

$$E_\theta \sim \frac{\mathrm{i}k^3 q_{13}}{4\pi\varepsilon_0 R}\cos\phi\cos 2\theta\mathrm{e}^{-\mathrm{i}kR}, \qquad E_\phi \sim -\frac{\mathrm{i}k^3 q_{13}}{4\pi\varepsilon_0 R}\cos\theta\sin\phi\mathrm{e}^{-\mathrm{i}kR},$$

$$H_\theta \sim -(\varepsilon_0/\mu_0)^{\frac{1}{2}}E_\phi, \qquad H_\phi \sim (\varepsilon_0/\mu_0)^{\frac{1}{2}}E_\theta$$

whereas the radial components are $O(1/R^2)$. Hence show that the average rate at which energy is radiated is $k^6 c q_{13}^2/160\pi\varepsilon_0$.

11. Show that at a large distance from a harmonic electric dipole the r.m.s. field intensity $|\boldsymbol{E}|/\sqrt{2}$ is $6.7\sqrt{P_t}(\sin\theta)/R$. The corresponding quantity for a straight wire which is half a wavelength long is $7\sqrt{P_t}\cos(\frac{1}{2}\pi\cos\theta)/R\sin\theta$ when Ci 2π is taken to be -0.02.

12. The current $I_0\mathrm{e}^{\mathrm{i}\omega t}$, where I_0 is constant, flows in a circular wire of radius a which lies in the plane $z=0$. If the origin is at the centre of the circle and $a \ll \lambda$, show that at a large distance from the origin

$$E_\phi \sim \frac{k^3 a^2 I_0}{4\omega\varepsilon_0 R}\sin\theta\mathrm{e}^{\mathrm{i}(\omega t - kR)}$$

and deduce that the loop produces the same distant field as a magnetic dipole of moment $\mu_0 I_0 \pi a^2 \boldsymbol{k}\mathrm{e}^{\mathrm{i}\omega t}$.

13. If $f(t, \zeta) = g(t+\zeta/c)$ in (10.1) prove that

$$E_z = \frac{1}{4\pi\varepsilon_0}\left(\frac{\partial}{\partial z} + \frac{1}{c}\frac{\partial}{\partial t}\right)\left[\frac{g\{t-(\mathcal{R}_1+l)/c\}}{\mathcal{R}_1} - \frac{g\{t-(\mathcal{R}_2-l)/c\}}{\mathcal{R}_2}\right],$$

$$E_x = \frac{1}{4\pi\varepsilon_0}\frac{\partial}{\partial x}\left[\frac{g\{t-(\mathcal{R}_1+l)/c\}}{\mathcal{R}_1} - \frac{g\{t-(\mathcal{R}_2-l)/c\}}{\mathcal{R}_2}\right]$$

$$+\frac{x}{4\pi\varepsilon_0 c}\left[\frac{g'\{t-(\mathcal{R}_1+l)/c\}}{v_1\mathcal{R}_1} - \frac{g'\{t-(\mathcal{R}_2-l)/c\}}{v_2\mathcal{R}_2}\right]$$

where $v_1 = \mathcal{R}_1 + l + z$, $v_2 = \mathcal{R}_2 - l + z$.

14. The current $I_0 e^{i\omega t - ikz}$ with I_0 constant flows in a thin wire along the entire positive z-axis. Show that the Hertz vector has

$$\Pi_z = -\frac{I_0 e^{i(\omega t - kz)}}{4\pi i\omega\varepsilon_0}\,\mathrm{Ei}\{ik(z-R)\}.$$

15. A travelling wave of current $I_0 e^{i(\omega t - kz)}$ flows along the thin wire extending from $z = -\frac{1}{2}l$ to $z = \frac{1}{2}l$. Show that the radiation resistance is given by

$$R_r = 60(\ln 2kl - \mathrm{Ci}\, 2kl + \gamma - 1 + \sin 2kl/2kl).$$

If $kl \gg 1$, show that the power gain is

$$G = 5.97 - 10\log_{10}[\lambda\{0.915 + \log_{10}(l/\lambda)\}/l] \quad \mathrm{dB}.$$

[The lowest root of $\tan x = 2x$ may be taken to be 1.16.]

16. Show that, in Schelkunoff's method, the gain of a linear array can be expressed as

$$G = 2\,|\Psi|_m^2 \Big/ \int_0^\pi |f(z)|^2 \sin\theta\, d\theta$$

where $|\Psi|_m^2$ is the maximum of $|\Psi(\theta)|^2$. The integral can always be evaluated by the substitution $kd\cos\theta - \alpha = u$.

Show that the gain of a broadside array of n elements spaced a half-wavelength apart is n. For an end-fire array of n elements spaced a quarter-wavelength apart the gain is n.

17. The currents in a linear array of n elements are all in phase and the spacing between elements is half a wavelength. Show that the gain in the direction $\theta = \frac{1}{2}\pi$ is $(\sum_{m=0}^{n-1} a_m)^2/\sum_{m=0}^{n-1} a_m^2$ with a_m as in (13.4) and deduce that it is a maximum when the amplitudes of the currents are all equal.

18. The spacing of a three element array is a quarter wavelength and

$$f(z) = (a + bz + cz^2)/(a+b+c).$$

Show that the gain has a maximum of 2.4 when $a = c$ and $(\pi-2)b = (\pi-4)a$. [This demonstrates that a non-uniform array $\frac{1}{2}\lambda$ long can have a greater gain than the two element uniform array of the same length.]

19. If the antenna of §3.19 forms a closed curve and $I(\zeta) = \sum_{n=-\infty}^{\infty} I_n e^{2in\pi\zeta/l}$ show that

$$-i\omega\varepsilon_0 E_m = \sum_{n=-\infty}^{\infty} Z_{mn} I_n$$

where

$$E_m = \int_0^l E_s^i(s) e^{-2im\pi s/l}\, ds,$$

$$Z_{mn} = \int_0^l \int_0^l (k^2 \cos\alpha - 4mn\pi^2/l^2) e^{2\pi i(n\zeta - ms)/l} K(s, \zeta)\, d\zeta\, ds.$$

If the curve is a circle of radius b show that $Z_{mn} = 0$ $(m \neq n)$ and that

$$Z_{00} = -2ik\{-ikb + \tfrac{1}{2}ikb \ln(8b/a) - k^2b^2 S_1(-2kb)\},$$
$$Z_{11} = -2ik[\{\tfrac{4}{3}k^2b^2 - 2 + (1 - k^2b^2)\ln(8b/a)\}/2ikb$$
$$-\tfrac{1}{2}k^2b^2 S_0(-2kb) + S_1(-2kb) - \tfrac{1}{2}k^2b^2 S_2(-2kb)]$$

where

$$S_m(x) = \int_0^{\frac{1}{2}\pi} \frac{1 - e^{ix\sin\psi}}{ix \sin\psi} \cos 2m\psi\, d\psi.$$

If $E_s^i(\phi) = E_0(\cos\phi)e^{-ikb\cos\phi\cos\beta}$ and $kb = 1 - \xi$ where $\xi \ll 1$ show that $I_1 = I_{-1}$ approximately and

$$\left(\frac{2I_1}{\lambda E_0}\right)^2 = \frac{\{J_0(\cos\beta) - J_2(\cos\beta)\}^2}{(264 - 332\xi)^2 + \{749\xi - 202 - 1735\xi \ln(b/a)\}^2}.$$

[$S_0(-2) = -1.12 + 0.80i$, $S_1(-2) = -0.22 - 0.22i$, $S_2(-2) = -0.01 - 0.03i$, $S_0'(-2) = -0.38 - 0.22i$, $S_1'(-2) = 0.17 + 0.00i$, $S_2'(-2) = 0.02 + 0.05i$.]

20. Generalize the theory of the integral equation for a single thin wire to a system of thin wires.

4. Resonators

THIS chapter deals with the electromagnetic and sound fields in closed regions like boxes. The representation of such fields rests upon the theory of eigenfunctions and it will be necessary to make a few remarks about this theory in suitable places.

One-dimensional eigenfunctions

4.1 Fourier series

As a starting point consider harmonic fields which have no y- or z-dependence. Then, whether acoustic or electromagnetic fields are under consideration, the equation to be solved has the form

$$\mathrm{d}^2u/\mathrm{d}x^2 + k^2u = 0. \tag{1.1}$$

This equation is satisfied between the walls of the box at $x = 0$ and $x = l$ (Fig. 4.1). A solution is required which meets appropriate boundary conditions on the walls. For instance, if u stands for E_z and the walls are perfectly conducting, u must vanish at $x = 0$ and $x = l$. On the other hand, if u represents sound pressure and the walls are rigid, $\mathrm{d}u/\mathrm{d}x$ has to be zero at $x = 0$, $x = l$. There are several other possibilities but, to fix ideas, we shall settle on

$$u(0) = u(l) = 0. \tag{1.2}$$

The equation (1.1) does not have a solution which complies with (1.2) unless $k = k_n = n\pi/l$ where n is a positive integer; then

$$u_n = A_n \sin k_n x$$

satisfies (1.1) and (1.2) for any choice of the constant A_n.

The question arises as to whether more general fields can be constructed from these elementary solutions by writing

$$f(x) = \sum_{n=1}^{\infty} A_n \sin k_n x \tag{1.3}$$

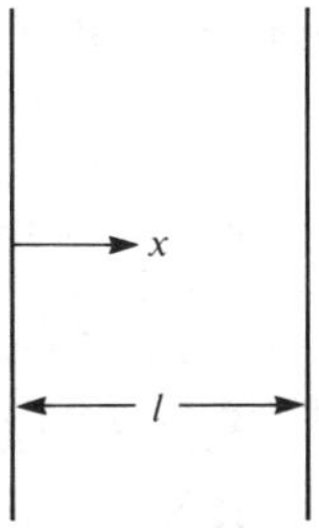

Fig. 4.1 One-dimensional boundary value problem

assuming that the A_n are such that the infinite series converges. If (1.3) holds there is a relation between A_n and f. For, multiply (1.3) by $\sin k_m x$ and integrate with respect to x from $x = 0$ and $x = l$. Since

$$\int_0^l \sin k_n x \sin k_m x \, dx = \begin{cases} 0 & (m \neq n) \\ \frac{1}{2}l & (m = n) \end{cases} \tag{1.4}$$

we have

$$\int_0^l f(x) \sin k_m x \, dx = \tfrac{1}{2} l A_m. \tag{1.5}$$

When A_m is determined by (1.5) the series on the right of (1.3) is called a *half-range Fourier series*. It is not by any means obvious that such a series converges and even if it does converge its sum may be different from $f(x)$.

Actually, (1.3) is a particular case of a more general expansion of a function $f(x)$ defined in $(-l, l)$, namely

$$f(x) = \tfrac{1}{2}a_0 + \sum_{n=1}^{\infty} (a_n \cos k_n x + b_n \sin k_n (x). \tag{1.6}$$

The series in (1.6) is known as a *Fourier series*. Often it is more convenient to express the trigonometric functions in terms of $e^{\pm ik_n x}$ and put

$$f(x) = \sum_{m=-\infty}^{\infty} A_m e^{ik_m x}$$

where the constant A_m can be complex. The analogue of (1.5) is

$$\int_{-l}^{l} f(x) e^{-ik_n x} \, dx = 2lA_n. \tag{1.7}$$

If f is *piecewise continuous* on $(-l, l)$ (i.e. it is continuous except at a finite number of discontinuities where the jumps are finite) and its derivative f', calculated at the points where f is continuous, is piecewise continuous it can be shown that the Fourier series is *pointwise convergent.* It has a sum for every x; if $-l<x<l$ its sum is $f(x)$ where f is continuous and its sum is $\frac{1}{2}\{f(x+0)+f(x-0)\}$ at a point of discontinuity of f; at $x=\pm l$ its sum is $\frac{1}{2}\{f(-l+0)+f(l-0)\}$.

If one is prepared to ask that (1.3) holds only on average instead of pointwise the conditions on f can be relaxed. For example,

$$\int_{-l}^{l} |f(x)-\sum_{m=-\infty}^{\infty} A_m e^{ik_m x}|^2 \, dx = 0 \tag{1.8}$$

provided that $\int_{-l}^{l} |f(x)|^2 \, dx$ is finite i.e. f is a member of L_2.

More generally still, (1.6) is valid as a relation between generalized functions for any periodic generalized function f with period $2l$.

Some care is necessary in applications where pointwise convergence is involved because the Fourier series is *not uniformly convergent* at a point of discontinuity. In fact, if $f(x)$ switches from -1 to 1 as x increases through zero, it can be arranged that $n\to\infty$ and $x\to 0$ in such a way that $\sum_{m=-n}^{n} A_m e^{ik_m x}$ takes any value from -1.18 to 1.18. This behaviour is known as *Gibbs' phenomenon.*

A question which might be asked is whether the coefficients in (1.8) need to be chosen as in (1.7) in order that (1.8) remains true. The answer is 'yes' and is part of the general result that the Fourier series minimizes the *mean square error* between it and f. To see this let $S_n = \sum_{m=-n}^{n} B_m e^{ik_m x}$. For the mean square error between S_n and f to be least, the coefficients B_m must be picked so that

$$y = \int_{-l}^{l} |f(x)-S_n|^2 \, dx$$

is a minimum. Treating B_m and B_m^* as distinct we require

$$0 = \partial y/\partial B_m = -\int_{-l}^{l} f^*(x) e^{ik_m x} \, dx + 2lB_m^*$$

where the asterisk indicates a complex conjugate. It is then clear that $B_m = A_m$ and the assertion is verified.

When f' can also be represented by a Fourier series

$$f'(x) = \sum_{m=-\infty}^{\infty} C_m e^{ik_m x}$$

where

$$2lC_m = \int_{-l}^{l} f'(x) e^{-ik_m x} \, dx = [f(x)e^{-ik_m x}]_{-l}^{l} + ik_m \int_{-l}^{l} f(x) e^{-ik_m x} \, dx$$

after an integration by parts. The first term on the right-hand side is

$(-1)^m\{f(l)-f(-l)\}$ so that $C_m = ik_m A_m$ whenever $f(l)=f(-l)$. In other words, f' can be obtained from the Fourier series for f by taking derivatives term-by-term provided that $f(l)=f(-l)$ and that f' can be represented as a Fourier series.

4.2 The Sturm–Liouville equation

In some problems the differential equation is more complicated than (1.1). A typical possibility is the *Sturm–Liouville equation*

$$(py')' + (\lambda r - q)y = 0 \tag{2.1}$$

where p, q, and r are given, continuous functions of x in $a \leqslant x \leqslant b$. The constant λ corresponds to k^2 of (1.1). The additional assumption that p and r are positive in $a \leqslant x \leqslant b$, with p possessing a continuous first derivative, will also be made.

The appropriate boundary conditions vary with the physical problem under consideration. A selection of possibilities is

(i) $y(a)=0,\ y(b)=0;$
(ii) $y'(a)=0,\ y'(b)=0;$
(iii) $y'(a)-\sigma_0 y(a)=0,\ y'(b)+\sigma_1 y(b)=0;$
(iv) $y(a)=y(b),\ p(a)y'(a)=p(b)y'(b).$

Here σ_0 and σ_1 are positive. Other combinations such as $y(a)=0$, $y'(b)=0$ can be handled by the techniques which follow.

When the boundary conditions are imposed there will be solutions of (2.1), in general, only for certain values of λ, say $\lambda_1, \lambda_2, \ldots$. These values are called *eigenvalues* and the corresponding solutions $y_1, y_2, \ldots$ are known as *eigenfunctions*. For example, the eigenvalues of (1.1) are $\lambda_m = k_m^2$ and the eigenfunctions are $y_m = A_m \sin k_m x$.

Since

$$(py_m')' + (\lambda_m r - q)y_m = 0 \tag{2.2}$$

it follows that

$$\int_a^b (\lambda_m - \lambda_n) r y_m y_n \, dx = \int_a^b \{y_m (py_n')' - y_n (py_m')'\} \, dx$$
$$= \int_a^b \{p(y_n' y_m - y_m' y_n)\}' \, dx$$

and the right-hand side vanishes for any one of the sets (i)–(iv) of boundary conditions. Hence, if $\lambda_m \neq \lambda_n$,

$$\int_a^b r y_m y_n \, dx = 0. \tag{2.3}$$

This is known as the *orthogonal property* of eigenfunctions.

If a function can be expanded as a series of eigenfunctions the orthogonal property leads to the coefficients. Thus, multiplying

$$f(x) = \sum_{m=1}^{\infty} A_m y_m$$

by ry_n and invoking (2.3), we obtain

$$A_n \int_a^b r y_n^2 \, dx = \int_a^b f r y_n \, dx$$

so long as all the eigenvalues are distinct.

When q is real all the eigenvalues are real. For suppose $\lambda_m \neq \lambda_m^*$. Choose $\lambda_n = \lambda_m^*$ and then (2.3) implies that

$$\int_a^b r \, |y_m|^2 \, dx = 0$$

which is impossible since $r > 0$. Therefore $\lambda_m = \lambda_m^*$ and the eigenvalues are real. The eigenfunctions y_m may therefore be chosen to be real.

Another property when q is real is, from (2.2),

$$\begin{aligned} \lambda_m \int_a^b r y_m^2 \, dx &= \int_a^b \{q y_m^2 - y_m (p y_m')'\} \, dx \\ &= \int_a^b (q y_m^2 + p y_m'^2) \, dx - [p y_m y_m']_a^b \end{aligned} \tag{2.4}$$

after integration by parts. The last term in (2.4) is zero for boundary conditions (i), (ii), (iv) and makes a positive contribution to the equation for (iii). Hence the right-hand side is positive if $q \geqslant 0$. Therefore no eigenvalue can be negative if $q \geqslant 0$. More can be said. For $\lambda_m = 0$ if, and only if, $y_m' = 0$ for all x, i.e. y_m is constant. This is impossible unless $q \equiv 0$ and, when $q \equiv 0$, is excluded by the boundary conditions (i), (iii). Hence, if $q \geqslant 0$, all the eigenvalues are positive except when $q \equiv 0$ and the boundary conditions (ii) or (iv) apply.

For future reference it should be observed that (2.4) is a particular case of

$$\lambda_m \int_a^b r y_m y_n \, dx = \int_a^b (q y_m y_n + p y_m' y_n') \, dx - [p y_n y_m']_a^b. \tag{2.5}$$

When the last term on the right is absent and $\lambda_m \neq \lambda_n$ this implies that

$$\int_a^b (q y_m y_n + p y_m' y_n') \, dx = 0.$$

Broadly speaking, the convergence properties of an expansion in eigenfunctions of a function with the coefficients derived as above are similar to those for Fourier series when the function satisfies similar conditions.

4.3 The variational method

Calculation of an eigenvalue and eigenfunction can often be extremely complex, if not an impossible, analytical task. Therefore, approximate methods have been devised. An important one is the variational method which will now be described for the case when q is real so that the eigenfunctions and eigenvalues are real. It will be assumed that the eigenvalues have been arranged in increasing order with λ_1 the lowest. Only the boundary condition (i) will be discussed though others can be treated the same way.

Define $D(y)$ and $H(y)$ by

$$D(y)=\int_a^b (py'^2+qy^2)\,dx, \qquad H(y)=\int_a^b ry^2\,dx.$$

Because of (2.4) and the boundary conditions (i)

$$D(y_m)=\lambda_m H(y_m). \tag{3.1}$$

It will now be demonstrated that, *if* y *satisfies* (*i*) *and is such that* $H(y)=1$,

$$D(y)\geqslant\lambda_1. \tag{3.2}$$

In other words, $D(y)$ always provides an upper bound for the lowest eigenvalue λ_1. By allowing y to contain adjustable parameters one can vary them to make $D(y)$ as small as possible and as close to λ_1 as feasible. Equality is achieved in (3.2) when $y=y_1$ so that the nearer $D(y)$ is to λ_1 the more one would hope that the trial function y is approaching y_1.

Let y be expanded in a series of eigenfunctions so that

$$y=\sum_{m=1}^{\infty} c_m y_m. \tag{3.3}$$

To meet $H(y)=1$ it is necessary that $\sum_{m=1}^{\infty} c_m^2 H(y_m)=1$ on account of the orthogonality (2.3) of the eigenfunctions. By virtue of (2.5), (2.3), (3.1), and (i)

$$D(y)=\sum_{m=1}^{\infty}\lambda_m c_m^2 H(y_m) \tag{3.4}$$

$$\geqslant\lambda_1\sum_{m=1}^{\infty} c_m^2 H(y_m)\geqslant\lambda_1.$$

Thus the earlier statement is verified. Note that there can be equality only when $y=y_1$ if $\lambda_1>0$.

Upper bounds for higher eigenvalues can be found if y_1 has been determined. Then y is further restricted by enforcing $\int_a^b ryy_1\,dx = 0$. As a result the sums in (3.3) and (3.4) start at $m = 2$ instead of $m = 1$ and $D(y) \geqslant \lambda_2$. Obviously, the process can be generalized and *if y satisfies (i), $H(y) = 1$ and is such that*

$$\int_a^b ryy_m\,dx = 0 \quad (m = 1, 2, \ldots, n)$$

then $D(y) \geqslant \lambda_{n+1}$.

4.4 The inhomogeneous differential equation

Consider the differential equation

$$(py')' + (\lambda_0 - q)y = -\delta(x - z) \tag{4.1}$$

and suppose that a solution is desired which satisfies one of the boundary conditions (i)–(iv). Let $y_1, y_2, \ldots$ be the eigenfunctions of (2.1) subject to the same boundary conditions when $r \equiv 1$. A solution is sought in terms of these eigenfunctions when they are normalised so that $\int_a^b y_m^2\,dx = 1$. To put it another way we seek a solution in which

$$y = \sum_{m=1}^{\infty} c_m y_m$$

with $c_m = \int_a^b yy_m\,dx$. Multiply (4.1) by $y_m(x)$ and integrate from a to b. Then, as in the derivation of (2.5),

$$(\lambda_0 - \lambda_m)c_m = -y_m(z).$$

This procedure gives as the solution of (4.1) which complies with the boundary conditions

$$y(x) = \sum_{m=1}^{\infty} \frac{y_m(x)y_m(z)}{\lambda_m - \lambda_0} \tag{4.2}$$

provided that $\lambda_0 \neq \lambda_m$ for any m. If $\lambda_0 = \lambda_n$ say an arbitrary multiple of y_n can be added to any solution of (4.1) and it will still remain a solution.

For the more general equation

$$(py')' + (\lambda_0 - q)y = -f(x) \tag{4.3}$$

we write $f(x) = \int f(z)\delta(x - z)\,dz$ and deduce from (4.2) that

$$y = \sum_{m=1}^{\infty} \frac{y_m(x)}{\lambda_m - \lambda_0} \int_a^b f(z)y_m(z)\,dz \tag{4.4}$$

so long as $\lambda_0 \neq \lambda_m$ for any m. It is this kind of approach that enables one to find the field generated by sources within a one-dimensional box.

Higher dimensions

4.5 The boundary condition u=0

Most of the ideas for eigenfunctions in one dimension can be generalized to higher dimensions and it will be sufficient to discuss two-dimensional space to see what is involved. Consider therefore

$$(pu_x)_x + (pu_y)_y + (\lambda r - q)u = 0 \tag{5.1}$$

where the notation u_x stands for $\partial u/\partial x$. The partial differential equation (5.1) holds on a bounded domain D whose perimeter is C. Various boundary conditions may be imposed on C; in this section the discussion will be limited to $u = 0$ on C. It will be assumed that p, q, r are real and continuous with continuous derivatives on $D + C$. Furthermore, it will be supposed that $p > 0$ and $r > 0$.

An eigenfunction φ_m and its associated eigenvalue λ_m are now such that

$$(p(\varphi_m)_x)_x + (p(\varphi_m)_y)_y + (\lambda_m r - q)\varphi_m = 0 \tag{5.2}$$

with $\varphi_m = 0$ on C. Hence

$$\begin{aligned}\int_D (\lambda_m - \lambda_n) r\varphi_m\varphi_n \,\mathrm{d}x\,\mathrm{d}y &= \int_D [\varphi_m\{(p(\varphi_n)_x)_x + (p(\varphi_n)_y)_y \\ &\quad - \varphi_n\{(p(\varphi_m)_x)_x + (p(\varphi_m)_y)_y\}]\,\mathrm{d}x\,\mathrm{d}y \\ &= \int_D [(p\{\varphi_m(\varphi_n)_x - \varphi_n(\varphi_m)_x\})_x \\ &\quad + (p\{\varphi_m(\varphi_n)_y - \varphi_n(\varphi_m)_y\})_y]\,\mathrm{d}x\,\mathrm{d}y \\ &= \int_C p(\varphi_m \operatorname{grad} \varphi_n - \varphi_n \operatorname{grad} \varphi_m) \cdot \boldsymbol{n}\,\mathrm{d}s \end{aligned} \tag{5.3}$$

by the divergence theorem, $\boldsymbol{n}$ being the unit outward normal to C. On account of the boundary condition on φ_m the right-hand side of (5.3) vanishes. Therefore, if $\lambda_m \neq \lambda_n$,

$$\int_D r\varphi_m\varphi_n \,\mathrm{d}x\,\mathrm{d}y = 0 \tag{5.4}$$

which is the analogue of (2.3) and shows that the eigenfunctions are orthogonal. It may also be inferred, as in §4.2, that the eigenvalues are real. In an expansion of the form

$$f(x, y) = \sum_{m=1}^{\infty} A_m\varphi_m \tag{5.5}$$

the coefficients A_m are given by

$$A_n \int_D r\varphi_n^2 \, dx \, dy = \int_D fr\varphi_n \, dx \, dy. \tag{5.6}$$

If (5.2) is multiplied by φ_n we obtain

$$\lambda_m \int_D r\varphi_m\varphi_n \, dx \, dy = \int_D (q\varphi_m\varphi_n + p \operatorname{grad} \varphi_m \cdot \operatorname{grad} \varphi_n) \, dx \, dy - \int_C p\varphi_n \boldsymbol{n} \cdot \operatorname{grad} \varphi_m \, ds \tag{5.7}$$

and the integral over C disappears because $\varphi_n = 0$ on C. In particular

$$\lambda_m \int_D r\varphi_m^2 \, dx \, dy = \int_D (q\varphi_m^2 + p \operatorname{grad}^2 \varphi_m) \, dx \, dy. \tag{5.8}$$

It follows that no eigenvalue can be negative if $q \geqslant 0$. Indeed $\lambda_m = 0$ requires $q \equiv 0$ and grad $\varphi_m = 0$. This forces φ_m to be constant which must be zero because of the boundary condition on C. Hence, if $q \geqslant 0$, all the eigenvalues are positive.

For a variational method define

$$D(\varphi) = \int_D (p \operatorname{grad}^2 \varphi + q\varphi^2) \, dx \, dy, \qquad H(\varphi) = \int_D r\varphi^2 \, dx \, dy.$$

Then, for trial functions φ which are zero on C and such that $H(\varphi) = 1$, $D(\varphi) \geqslant \lambda_1$. The proof runs parallel to that of §4.3 but based on (5.7) and (5.8). Similarly, trial functions such that $H(\varphi) = 1$ and $\int_D r\varphi\varphi_m \, dx \, dy = 0$ $(m = 1, 2, \ldots, n)$ give $D(\varphi) \geqslant \lambda_{n+1}$.

For the solution of

$$(pu_x)_x + (pu_y)_y + (\lambda_0 - q)u = -\delta(x - x_0)\delta(y - y_0) \tag{5.9}$$

with (x_0, y_0) in D proceed as in §4.4 with φ_m the eigenfunction for $r \equiv 1$. Then, if $\lambda_0 \neq \lambda_m$ for any m,

$$u(x, y) = \sum_{m=1}^{\infty} \varphi_m(x, y)\varphi_m(x_0, y_0)/(\lambda_m - \lambda_0). \tag{5.10}$$

The corresponding formula for a general right-hand side in (5.9) is similar to (4.4).

The expansion (5.5) is valid whenever f is in L_2 provided that convergence in mean is employed, i.e. when A_m is given by (5.6)

$$\lim_{M \to \infty} \int_D r\left(f - \sum_{m=1}^{M} A_m\varphi_m\right)^2 dx \, dy = 0. \tag{5.11}$$

The φ_m can be normalized by choosing them so that

$$\int_D r\varphi_m^2 \, dx \, dy = 1.$$

They are then said to constitute an *orthonormal set.* An orthornormal set such that $\int_D rf\varphi_m \, dx \, dy = 0$ for $m = 1, 2, \ldots$ implies that $f \equiv 0$ is said to be *complete.*

When φ_m is orthonormal

$$\begin{aligned}\int_D r\left\{f - \sum_{m=1}^{M} c_m\varphi_m\right\}^2 dx\,dy &= \int_D r\left(f^2 - 2\sum_{m=1}^{M} c_m f\varphi_m\right) dx\,dy + \sum_{m=1}^{M} c_m^2 \\ &= \int_D rf^2 \, dx\,dy - \sum_{m=1}^{M}\left(\int_D rf\varphi_m \, dx\,dy\right)^2 \\ &\quad + \sum_{m=1}^{M}\left(\int_D rf\varphi_m \, dx\,dy - c_m\right)^2. \qquad (5.12)\end{aligned}$$

On the right-hand side only the third term contains the coefficients c_m and so the mean-square error is minimized when $c_m = \int_D rf\varphi_m \, dx \, dy$. In other words, the mean-square error between f and a finite expansion in terms of an orthonormal set is least when c_m agrees with A_m of (5.6). Since the left-hand side of (5.12) cannot be negative this choice of c_m tells us that

$$\sum_{m=1}^{M}\left(\int_D rf\varphi_m \, dx\,dy\right)^2 \leq \int_D rf^2 \, dx\,dy \qquad (5.13)$$

which is known as *Bessel's inequality*. Moreover by letting $M \to \infty$ and invoking (5.11), we deduce that

$$\sum_{m=1}^{\infty}\left(\int_D rf\varphi_m \, dx\,dy\right)^2 = \int_D rf^2 \, dx\,dy. \qquad (5.14)$$

Apply this result to $f + g$ and $f - g$; after subtraction there follows

$$\int_D rfg \, dx\,dy = \sum_{m=1}^{\infty} a_m b_m \qquad (5.15)$$

when $f = \sum_{m=1}^{\infty} a_m\varphi_m$ and $g = \sum_{m=1}^{\infty} b_m\varphi_m$; this is *Parseval's formula.*

The question of the expansion of a derivative will be considered only in the case when $q \equiv 0$. If φ_m is orthonormal

$$f = \sum_{m=1}^{\infty} a_m\varphi_m, \qquad a_m = \int_D rf\varphi_m \, dx\,dy,$$

and if $\operatorname{grad} f = \sum_{m=1}^{\infty} b_m \operatorname{grad} \varphi_m$ then

$$\lambda_n b_n = \int_D p \operatorname{grad} f \cdot \operatorname{grad} \varphi_n \, dx\,dy$$

from (5.7) and (5.4). Note that b_n is determined by this formula because it has already been proved that λ_n is positive when $q \equiv 0$. From the divergence theorem and (5.2)

$$\lambda_n b_n = \int_C p f \boldsymbol{n} \cdot \operatorname{grad} \varphi_n \, \mathrm{d}s + \lambda_n \int_D r f \varphi_n \, \mathrm{d}x \, \mathrm{d}y. \tag{5.16}$$

Thus, when $f = 0$ on C, $b_n = a_n$ and the series for grad f is the term-by-term derivative of that for f.

If the set φ_m is complete the question arises as to whether the set grad φ_m which, from the preceding paragraph, can be made orthonormal is also complete i.e. does $b_n = 0$ for $n = 1, 2, \ldots$ imply that grad $f \equiv 0$? Suppose $b_n = 0$; then, if $f = 0$ on C, $a_n = 0$ and $f \equiv 0$. Therefore, grad $f \equiv 0$ in this case. Generally $b_m = 0$ entails, via the divergence theorem,

$$0 = \int_C \varphi_m \boldsymbol{n} \cdot p \operatorname{grad} f \, \mathrm{d}s - \int_D \varphi_m \{(pf_x)_x + (pf_y)_y\} \, \mathrm{d}x \, \mathrm{d}y.$$

The first term vanishes because $\varphi_m = 0$ on C and so $(pf_x)_x + (pf_y)_y = 0$ from the completeness of the set φ_m. Consequently, if there is a vector $\mathbf{A}$ such that p grad $f = \operatorname{curl} \mathbf{A}$ the coefficients b_m are zero but grad f is not. The conclusion is that grad φ_m is not a complete set though it is effectively so for those functions f which vanish on the boundary C.

4.6 The boundary conditions with normal derivative

The exchange of the boundary condition $u = 0$ for the vanishing of the normal derivative $\partial u/\partial n$ makes some small differences to the foregoing analysis. The curvilinear integrals in (5.3) and (5.7) still disappear. Therefore the eigenfunctions are orthogonal and (5.8) holds. It continues to be true that the eigenvalues are positive when $q \geqslant 0$ unless $q \equiv 0$; if $q \equiv 0$, the least eigenvalue is zero with the corresponding eigenfunction a constant.

The variational method takes the same form as in §4.5 but there is now no necessity for the trial function to satisfy a boundary condition on C because $\partial u/\partial n = 0$ constitutes what is called a natural boundary condition. The functions which are free of a boundary condition must include those which vanish on C. Therefore the minimum of $D(\varphi)$ when φ is unrestricted cannot be greater than that when φ is forced to be zero on C. Hence, the lowest eigenvalue for $\partial u/\partial n = 0$ on C cannot exceed the smallest eigenvalue when $u = 0$ on C.

The remarks about orthonormal sets, completeness, Bessel's inequality, and Parseval's formula also carry over. With regard to the completeness of grad ϕ_m when $q \equiv 0$ note that $\lambda_1 = 0$ and φ_1 is constant so that φ_1 does not figure in the expansion of grad f. Also (5.16) reveals that $b_n = a_n$ $(n \neq 1)$ and so a complete set is provided by grad ϕ_m. Hence, *when* φ_m *is* *an eigenfunction (for $q \equiv 0$) such that $\partial\varphi_m/\partial n = 0$ on C and f is in L_2*

$$f = a_1 + \sum_{m=2}^{\infty} a_m \varphi_m, \qquad \operatorname{grad} f = \sum_{m=2}^{\infty} a_m \operatorname{grad} \varphi_m.$$

The theory of this section can be extended to the equation

$$(pu_x)_x + (pu_y)_y + (\lambda r - q + a_x + b_y)u = 0$$

subject to the boundary condition $p\,\partial u/\partial n + \sigma u = 0$, where $\sigma = a\,\partial x/\partial n + b\,\partial y/\partial n$, provided that there is a positive constant K such that

$$p(\xi^2 + \eta^2) + 2a\xi\zeta + 2b\eta\zeta + q\zeta^2 \geq K(\xi^2 + \eta^2 + \zeta^2)$$

for all real values of ξ, η, ζ. The choice $\xi = -a\zeta/p$, $\eta = -b\zeta/p$ shows that $q > (a^2 + b^2)/p$ is necessary for the validity of the inequality.

4.7 The electromagnetic cavity resonator

When electromagnetic waves are considered the above notions have to be transferred to vector equations. In free space, Maxwell's equations for periodic phenomena in the absence of sources of charge and current are

$$\operatorname{curl} \boldsymbol{E} + \mathrm{i}\omega\mu_0 \boldsymbol{H} = \boldsymbol{0}, \qquad \operatorname{div} \boldsymbol{E} = 0,$$
$$\operatorname{curl} \boldsymbol{H} - \mathrm{i}\omega\varepsilon_0 \boldsymbol{E} = \boldsymbol{0}, \qquad \operatorname{div} \boldsymbol{H} = 0.$$

From the equations of the first column it can be deduced that

$$\operatorname{curl}\operatorname{curl} \boldsymbol{E} - \omega^2\mu_0\varepsilon_0 \boldsymbol{E} = \boldsymbol{0}. \tag{7.1}$$

If $\omega \neq 0$, once $\boldsymbol{E}$ is known from (7.1), $\boldsymbol{H}$ can be found from

$$\boldsymbol{H} = -\operatorname{curl} \boldsymbol{E}/\mathrm{i}\omega\mu_0. \tag{7.2}$$

However, it cannot be assumed *a priori* that ω, which plays the role of an eigenvalue, is nonzero for fields in a box.

Let the field occupy the *simply connected* volume T bounded by the perfectly conducting surface S. Then the boundary condition on S is

$$\boldsymbol{n} \wedge \boldsymbol{E} = \boldsymbol{0}. \tag{7.3}$$

When $\omega = 0$ a solution of (7.1) is $\boldsymbol{E} = \operatorname{grad} \varphi$ which complies with (7.3) if φ is constant on S. There are thus many solutions of (7.1) subject to (7.3) when $\omega = 0$ and it is not a straightforward eigenvalue problem. Let the equation

$$\operatorname{div} \boldsymbol{E} = 0 \tag{7.4}$$

be added to (7.1). Then $\nabla^2\varphi = 0$ and so φ must be constant in T since it is constant on S and T is simply connected. Consequently, $\operatorname{grad} \varphi = \boldsymbol{0}$ and the troublesome arbitrary solution at $\omega = 0$ is eliminated.

Therefore the eigenvalue problem will be taken as (7.1), (7.4) subject to (7.3). Observe that (7.4) is an automatic consequence of (7.1) when $\omega \neq 0$.

Let $\boldsymbol{E}_m$ be a typical eigenfunction and ω_m the corresponding value of ω. Then

$$(\omega_m^2 - \omega_n^2)\mu_0\varepsilon_0 \int_T \boldsymbol{E}_m \cdot \boldsymbol{E}_n \,\mathrm{d}\boldsymbol{x} = \int_T (\boldsymbol{E}_n \cdot \operatorname{curl} \boldsymbol{E}_m - \boldsymbol{E}_m \cdot \operatorname{curl} \boldsymbol{E}_n)\,\mathrm{d}\boldsymbol{x}$$
$$= \int_S \boldsymbol{E}_m \wedge \boldsymbol{E}_n \cdot \boldsymbol{n}\,\mathrm{d}S \tag{7.5}$$

by the divergence theorem (see (F. 7.6)). By virtue of (7.3) the right-hand side vanishes and we have the *orthogonal property*

$$\int_T \boldsymbol{E}_m \cdot \boldsymbol{E}_n \,\mathrm{d}\boldsymbol{x} = 0 \tag{7.6}$$

when $\omega_m \neq \omega_n$. As in §4.2 it may be concluded that ω_m^2 is real and so $\boldsymbol{E}_m$ may be chosen to be real.

Furthermore

$$\omega_m^2\mu_0\varepsilon_0 \int_T \boldsymbol{E}_m \cdot \boldsymbol{E}_n \,\mathrm{d}x = \int_T \operatorname{curl} \boldsymbol{E}_n \cdot \operatorname{curl} \boldsymbol{E}_m \,\mathrm{d}\boldsymbol{x} + \int_S (\operatorname{curl} \boldsymbol{E}_m) \wedge \boldsymbol{E}_n \cdot \boldsymbol{n}\,\mathrm{d}S. \tag{7.7}$$

The last term vanishes because of the boundary condition on S and so

$$\omega_m^2\mu_0\varepsilon_0 \int_T \boldsymbol{E}_m \cdot \boldsymbol{E}_m \,\mathrm{d}\boldsymbol{x} = \int_T \operatorname{curl} \boldsymbol{E}_m \cdot \operatorname{curl} \boldsymbol{E}_m \,\mathrm{d}\boldsymbol{x}. \tag{7.8}$$

Since $\boldsymbol{E}_m$ is real it follows that ω_m^2 cannot be negative and consequently ω_m is real. Equation (7.2) then implies that $\boldsymbol{H}_m$ is pure imaginary. If $\omega_m^2 = 0$, (7.8) shows that $\operatorname{curl} \boldsymbol{E}_m = \boldsymbol{0}$, i.e. $\boldsymbol{E}_m = \operatorname{grad} \varphi$. On account of the argument just after (7.4) $\boldsymbol{E}_m$ is then identically zero. Hence ω_1^2 must be positive and the possibility of a zero eigenvalue is excluded.

A useful inference from (7.7) and (7.6) is

$$\int_T \operatorname{curl} \boldsymbol{E}_n \cdot \operatorname{curl} \boldsymbol{E}_m \,\mathrm{d}\boldsymbol{x} = 0. \tag{7.9}$$

For a variational method take

$$\mathscr{D}(\boldsymbol{E}) = \int_T \operatorname{curl} \boldsymbol{E} \cdot \operatorname{curl} \boldsymbol{E}\,\mathrm{d}\boldsymbol{x}, \qquad \mathscr{H}(\boldsymbol{E}) = \int_T \boldsymbol{E} \cdot \boldsymbol{E}\,\mathrm{d}\boldsymbol{x}.$$

Then, for trial functions which satisfy (7.3), (7.4), and $\mathscr{H}(\boldsymbol{E}) = 1$, $\mathscr{D}(\boldsymbol{E}) \geqslant$

$\omega_1^2\mu_0\varepsilon_0$. If, in addition,

$$\int_T \boldsymbol{E}\cdot\boldsymbol{E}_m\,\mathrm{d}\boldsymbol{x}=0 \quad (m=1,\ldots,n)$$

then $\mathscr{D}(\boldsymbol{E})\geqslant\omega_{n+1}^2\mu_0\varepsilon_0$.

An arbitrary $\boldsymbol{E}$ may be expanded as

$$\boldsymbol{E}=\sum_{m=1}^{\infty} A_m\boldsymbol{E}_m \tag{7.10}$$

where

$$A_m\int_T \boldsymbol{E}_m\cdot\boldsymbol{E}_m\,\mathrm{d}\boldsymbol{x}=\int_T \boldsymbol{E}\cdot\boldsymbol{E}_m\,\mathrm{d}\boldsymbol{x} \tag{7.11}$$

and A_m may be complex if $\boldsymbol{E}$ is. The specification of A_m by (7.11) minimizes the mean-square error between $\boldsymbol{E}$ and finite expansion in the $\boldsymbol{E}_m$.

A solution of

$$\operatorname{curl}\operatorname{curl}\boldsymbol{E}-\omega_0^2\mu_0\varepsilon_0\boldsymbol{E}=-\boldsymbol{a}\delta(\boldsymbol{x}-\boldsymbol{x}_0), \tag{7.12}$$

where $\boldsymbol{a}$ is a constant vector and $\boldsymbol{E}$ satisfies (7.3), is provided by

$$\boldsymbol{E}(\boldsymbol{x})=-\sum_{m=1}^{\infty}\boldsymbol{a}\cdot\boldsymbol{E}_m(\boldsymbol{x}_0)\boldsymbol{E}_m(\boldsymbol{x})/(\omega_m^2-\omega_0^2)\mu_0\varepsilon_0 \tag{7.13}$$

if $\omega_0^2\neq\omega_m^2$ for any m.

Clearly, $\boldsymbol{E}_m$ can be made into an orthonormal set and it is, in fact, complete. Bessel's inequality takes the form

$$\sum_{m=1}^{M}\left|\int_T \boldsymbol{E}\cdot\boldsymbol{E}_m\,\mathrm{d}\boldsymbol{x}\right|^2\leqslant\int_T |\boldsymbol{E}|^2\,\mathrm{d}\boldsymbol{x} \tag{7.14}$$

with equality as $M\to\infty$. Parseval's formula is

$$\int_T \boldsymbol{E}^{(1)}\cdot\boldsymbol{E}^{(2)*}\,\mathrm{d}\boldsymbol{x}=\sum_{m=1}^{\infty}a_m b_m^*$$

if $\boldsymbol{E}^{(1)}=\sum_{m=1}^{\infty}a_m\boldsymbol{E}_m$ and $\boldsymbol{E}^{(2)}=\sum_{m=1}^{\infty}b_m\boldsymbol{E}_m$.

With $\boldsymbol{E}_m$ orthonormal let

$$\boldsymbol{E}=\sum_{m=1}^{\infty}a_m\boldsymbol{E}_m,\qquad a_m=\int_T \boldsymbol{E}\cdot\boldsymbol{E}_m\,\mathrm{d}\boldsymbol{x}.$$

Then, if $\operatorname{curl}\boldsymbol{E}=\sum_{m=1}^{\infty}b_m\operatorname{curl}\boldsymbol{E}_m$, (7.8) and (7.9) show that

$$\omega_m^2\mu_0\varepsilon_0 b_m=\int_T \operatorname{curl}\boldsymbol{E}\cdot\operatorname{curl}\boldsymbol{E}_m\,\mathrm{d}\boldsymbol{x} \tag{7.15}$$

and b_m is well-defined since no ω_m is zero. Also, by the divergence theorem,

$$\omega_m^2\mu_0\varepsilon_0 b_m = \omega_m^2\mu_0\varepsilon_0 a_m + \int_S \boldsymbol{n}\cdot\boldsymbol{E}\wedge\text{curl}\,\boldsymbol{E}_m\,\text{d}S.$$

Consequently, when $\boldsymbol{n}\wedge\boldsymbol{E}=\boldsymbol{0}$ on S, $b_m = a_m$ and the series for curl $\boldsymbol{E}$ can be obtained by taking term-by-term derivatives of the expansion for $\boldsymbol{E}$.

Next, suppose that $b_m = 0$ for $m = 1, 2, \ldots$. If $\boldsymbol{n}\wedge\boldsymbol{E}=\boldsymbol{0}$ on S, $a_m = 0$ by what has just been proved and $\boldsymbol{E}\equiv\boldsymbol{0}$ which enforces curl $\boldsymbol{E}\equiv\boldsymbol{0}$. If, however, $\boldsymbol{E}$ does not fulfil this boundary condition

$$0=\int_S \boldsymbol{n}\cdot\boldsymbol{E}_m\wedge\text{curl}\,\boldsymbol{E}\,\text{d}S+\int_T \mathbf{E}_m\cdot\text{curl curl}\,\boldsymbol{E}\,\text{d}\boldsymbol{x}.$$

The surface integral is zero and so curl curl $\boldsymbol{E}=\boldsymbol{0}$, the set $\boldsymbol{E}_m$ being complete. Thus curl $\boldsymbol{E}=$ grad φ and the divergence of this gives $\nabla^2\varphi=0$. So any vector of the form curl $\boldsymbol{E}$ has the expansion

$$\text{curl}\,\boldsymbol{E}=\text{grad}\,\varphi+\sum_{m=1}^{\infty} b_m\,\text{curl}\,\boldsymbol{E}_m \tag{7.16}$$

with b_m specified in (7.15). The set curl $\boldsymbol{E}_m$ is not complete. It is already known that the term grad φ is missing when $\boldsymbol{n}\wedge\boldsymbol{E}=\boldsymbol{0}$ on S and it is also absent when $\boldsymbol{n}\cdot$ curl $\boldsymbol{E}=0$ on S. For grad φ is orthogonal to curl $\boldsymbol{E}_m$ and hence

$$\int_T |\text{grad}\,\varphi|^2\,\text{d}\boldsymbol{x}=\int_T \text{curl}\,\boldsymbol{E}\cdot\text{grad}\,\varphi^*\,\text{d}\boldsymbol{x}=\int_S \varphi^*\boldsymbol{n}\cdot\text{curl}\,\boldsymbol{E}\,\text{d}S.$$

A particular vector which is a curl is $\boldsymbol{H}$ and so we can contemplate the expansion

$$\boldsymbol{H}=\text{grad}\,\varphi+\sum_{m=1}^{\infty} c_m\,\text{curl}\,\boldsymbol{E}_m \tag{7.17}$$

where $\nabla^2\varphi=0$ and

$$\omega_m^2\mu_0\varepsilon_0 c_m=\int_T \boldsymbol{H}\cdot\text{curl}\,\boldsymbol{E}_m\,\text{d}\boldsymbol{x}. \tag{7.18}$$

With $\boldsymbol{H}$ as a solution of Maxwell's equations the integral can be transformed with the result

$$\omega_m^2\mu_0\varepsilon_0 c_m=\text{i}\omega\varepsilon_0\int_T \boldsymbol{E}\cdot\boldsymbol{E}_m\,\text{d}\boldsymbol{x}. \tag{7.19}$$

It has thereby been demonstrated that *a solution of Maxwell's harmonic equations in the simply connected volume T can be expanded in the orthonormal set* $\boldsymbol{E}_m$ *via*

$$\boldsymbol{E} = \sum_{m=1}^{\infty} a_m \boldsymbol{E}_m, \tag{7.20}$$

$$\boldsymbol{H} = \text{grad } \varphi + \text{i}\omega \sum_{m=1}^{\infty} a_m \text{ curl } \boldsymbol{E}_m/\omega_m^2 \mu_0 \tag{7.21}$$

where $\nabla^2 \varphi = 0$. *If* $\boldsymbol{n} \cdot \boldsymbol{H} = 0$ *on* S *or if* $\boldsymbol{n} \wedge \boldsymbol{E} = \boldsymbol{0}$ *and* $\omega \neq 0$ *the term* grad φ *can be removed.*

It may be noted that

$$\int_S \boldsymbol{n} \cdot \boldsymbol{E} \wedge \text{curl } \boldsymbol{E}_m \, \text{d}S = \int_T (\text{curl } \boldsymbol{E}_m \cdot \text{curl } \boldsymbol{E} - \boldsymbol{E} \cdot \text{curl curl } \boldsymbol{E}_m) \, \text{d}\boldsymbol{x}$$

$$= (\omega^2 - \omega_m^2)\mu_0 \varepsilon_0 a_m \tag{7.22}$$

from (7.18) and (7.19).

4.8 Typical eigenfunctions

The construction of explicit eigenfunctions involves the solution of a partial differential equation and is therefore difficult in general. In practice the eigenfunctions whose analytical structure can be derived are mostly limited to those which can be formed by the combination of solutions of ordinary differential equations. Usually ordinary differential equations are somewhat easier to investigate. Some examples to indicate the process will now be given.

(i) *The rectangle*

Eigenfunctions for the rectangle $0 \leqslant x \leqslant a$, $0 \leqslant y \leqslant b$ can be derived from

$$u_{xx} + u_{yy} + k^2 u = 0. \tag{8.1}$$

One orthonormal set can be obtained by imposing the boundary condition $u = 0$ on the perimeter of the rectangle. Firstly, u is a function of x for fixed y which, by §4.1, can be expanded in the sine series

$$u = \sum_{m=1}^{\infty} \alpha_m(y) \sin(m\pi x/a)$$

where

$$\alpha_m(y) = (2/a) \int_0^a u(x, y) \sin(m\pi x/a) \, \text{d}x.$$

Multiply (8.1) by $\sin(m\pi x/a)$ and integrate with respect to x from 0 to a.

Since

$$\int_0^a u_{xx}\sin(m\pi x/a)\,dx = -(m\pi/a)\int_0^a u_x\cos(m\pi x/a)\,dx$$

$$= -(m\pi/a)^2\int_0^a u\sin(m\pi x/a)\,dx$$

on account of $u=0$ at $x=0$, a we have

$$\frac{d^2\alpha_m}{dy^2}+\left\{k^2-\left(\frac{m\pi}{a}\right)^2\right\}\alpha_m=0.$$

A solution such that $\alpha_m=0$ at $y=0$, b is $\sin(m\pi y/b)$ provided that

$$k^2=(m\pi/a)^2+(n\pi/b)^2 \qquad (m=1,2,\ldots;n=1,2,\ldots). \tag{8.2}$$

Thus the eigenfunctions are $\sin(m\pi x/a)\sin(n\pi y/b)$ and the eigenvalues are determined by (8.2). Any u which belongs to L_2 has the expansion

$$u=\sum_{m=1}^{\infty}\sum_{n=1}^{\infty} a_{mn}\sin(m\pi x/a)\sin(n\pi y/b)$$

where

$$a_{mn}=(4/ab)\int_0^b\int_0^a u\sin(m\pi x/a)\sin(n\pi y/b)\,dx\,dy.$$

It should be noted that the eigenvalues are not necessarily simple, i.e. there can be more than one eigenfunction per eigenvalue. If there are integers m, n, m_1, n_1 such that

$$(m/a)^2+(n/b)^2=(m_1/a)^2+(n_1/b)^2$$

there will be at least two eigenfunctions corresponding to this eigenvalue. For example, in the square in which $a=b=\pi$ the eigenfunctions $\sin mx\sin ny$ and $\sin nx\sin my$ both correspond to $k^2=m^2+n^2$.

When the eigenvalues are simple the *nodal curves*, i.e. the curves where an eigenfunction vanishes, are straight lines parallel to the coordinate axes. For eigenvalues of higher multiplicity eigenfunctions can be constructed which have quite different nodal curves (see Fig. 4.2).

Equation (8.1) serves as a basis for other complete orthonormal sets. For instance, the boundary condition $\partial u/\partial n=0$ on the perimeter of the rectangle leads to the expansion

$$u=\tfrac{1}{4}b_{00}+\sum_{m=1}^{\infty}\{b_{m0}\cos(m\pi x/a)+b_{0m}\cos(m\pi y/b)\}$$

$$+\sum_{m=1}^{\infty}\sum_{n=1}^{\infty} b_{mn}\cos(m\pi x/a)\cos(n\pi y/b)$$

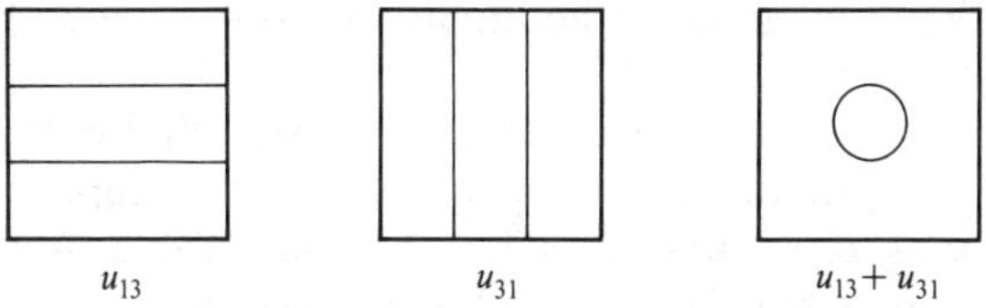

Fig. 4.2 Nodal curves for a square

where

$$b_{mn} = (4/ab)\int_0^b \int_0^a u \cos(m\pi x/a)\cos(n\pi y/b)\, dx\, dy.$$

Further orthonormal sets can be devised by combining boundary conditions e.g.

$$u = 0 \quad \text{on} \quad x = 0, a; \qquad \partial u/\partial y = 0 \quad \text{on} \quad y = 0, b$$

or

$$\partial u/\partial x = 0 \quad \text{on} \quad x = 0, a; \qquad u = 0 \quad \text{on} \quad y = 0, b.$$

(ii) *The circle and associated domains*

The introduction of polar coordinates r, ϕ converts (8.1) to

$$r(ru_r)_r + u_{\phi\phi} + k^2r^2u = 0. \tag{8.3}$$

If this holds in a circle with centre the origin and radius a, u can be regarded as a function of ϕ defined on the interval $(0, 2\pi)$ when r is fixed. By §4.1 u can be expanded in a Fourier series

$$u(r, \phi) = \sum_{m=-\infty}^{\infty} \alpha_m(r)e^{im\phi}$$

where

$$\alpha_m(r) = (1/2\pi)\int_0^{2\pi} u(r, \phi)e^{-im\phi}\, d\phi.$$

On multiplying (8.3) by $e^{-im\phi}$ and integrating with respect to ϕ from 0 to 2π we obtain

$$\frac{1}{r}\frac{d}{dr}\left(r\frac{d\alpha_m}{dr}\right) + \left(k^2 - \frac{m^2}{r^2}\right)\alpha_m = 0$$

since $u(r, 2\pi) = u(r, 0)$. This is Bessel's equation (Appendix A and §1.41(i)) and

$$\alpha_m = A_m J_m(kr) + B_m Y_m(kr).$$

The function Y_m is singular at the origin and so α_m can be finite at $r=0$ only if $B_m=0$.

For the boundary condition $u=0$ on $r=a$ we require $J_m(ka)=0$. Let $j_{\nu n}$ $(n=1,2,\dots)$ be the zeros of $J_\nu(x)$, i.e. the numbers such that $J_\nu(j_{\nu n})=0$, arranged so that $j_{\nu 1}<j_{\nu 2}<\dots$ (see Table A.1). Then the eigenfunctions are $e^{im\phi}J_m(j_{mn}r/a)$ and, if u belongs to L_2,

$$u=\sum_{m=-\infty}^{\infty}\sum_{n=1}^{\infty}a_{mn}e^{im\phi}J_m(j_{mn}r/a) \tag{8.4}$$

where, from (A.1.44),

$$a_{mn}=\{a^2\pi J_{m+1}^2(j_{mn})\}^{-1}\int_0^{2\pi}\int_0^a ue^{-im\phi}J_m(j_{mn}r/a)r\,dr\,d\phi.$$

A series of the type $\sum_{m=1}^{\infty}a_mJ_\nu(j_{\nu m}/a)$ where

$$a_m=\{\tfrac{1}{2}a^2\,J_{\nu+1}^2(j_{\nu m})\}^{-1}\int_0^a rf(r)J_\nu(j_{\nu m}r/a)\,dr$$

is called a *Fourier–Bessel series* for $f(r)$. Since (8.4) is valid if $\int_0^{2\pi}\int_0^a|u|^2\,r\,dr\,d\phi$ exists the selection of $f(r)e^{im\phi}$ for u shows that the Fourier–Bessel expansion of $f(r)$ holds when ν is an integer if $\int_0^a|f|^2\,r\,dr$ exists.

An alternative boundary condition on the circumference of the circle is $\partial u/\partial r=0$. Now, it is required that $J_m'(ka)=0$. One solution is $k=0$ but then $J_m(ka)=0$ unless $m=0$ when the value 1 is obtained. Therefore the only non-trivial solution provided by $k=0$ is $u=$constant. Let $j_{\nu n}'$ $(n=1,2,\dots)$ be the positive zeros of $J_\nu'(x)$. Then

$$u=b_{00}+\sum_{m=-\infty}^{\infty}\sum_{n=1}^{\infty}b_{mn}e^{im\phi}J_m(j_{mn}'r/a) \tag{8.5}$$

where

$$b_{mn}=\{\pi a^2J_m^2(j_{mn}')\}^{-1}\int_0^{2\pi}\int_0^a ue^{-im\phi}J_m(j_{mn}'r/a)r\,dr\,d\phi.$$

A series of the form $\sum_{n=1}^{\infty}b_mJ_\nu(j_{\nu m}'r/a)$ with

$$b_m=\{\tfrac{1}{2}a^2J_\nu^2(j_{\nu m}')\}^{-1}\int_0^a rf(r)J_\nu(j_{\nu m}'r/a)\,dr$$

is known as a *Dini series* for f. From (8.5) the Dini expansion of f is valid if $\int_0^a r\,|f|^2\,dr$ exists when ν is an integer.

The sector of a circle can be considered instead of the full circle. If the sector is $0\leqslant r\leqslant a$, $0\leqslant\phi\leqslant\Phi\leqslant 2\pi$ a half-range Fourier series is used instead of the whole Fourier series. For example, if u is zero round the boundary of the sector the eigenfunctions are $J_\nu(j_{\nu n}r/a)\sin\nu\phi$ with $\nu = m\pi/\Phi$ where m is a positive integer. If $\partial u/\partial\phi = 0$ on $\phi = 0$, Φ replace $\sin\nu\phi$ by $\cos\nu\phi$ and include $\nu = 0$. There are similar modifications to the eigenfunctions when $\partial u/\partial r = 0$ on $r = a$.

The selection of different values of Φ permits one to confirm that the Fourier–Bessel and Dini expansions of f are valid when $\nu = 0$ and when $\nu \geqslant \frac{1}{2}$ (they hold, in fact, for $\nu \geqslant -\frac{1}{2}$) provided that $\int_0^a |f|^2 r\,dr$ exists.

When the domain is the annulus $b \leqslant r \leqslant a$, $0 \leqslant \phi \leqslant 2\pi$ the eigenfunctions are

$$e^{im\phi}\{J_m(\chi_{mn}r/b)Y_m(\chi_{mn}) - Y_m(\chi_{mn}r/b)J_m(\chi_{mn})\}$$

where

$$J_m(\chi_{mn})Y_m(\chi_{mn}a/b) - Y_m(\chi_{mn})J_m(\chi_{mn}a/b) = 0$$

when u vanishes on the perimeter (c.f. §5.7(e)). There are obvious modifications for other boundary conditions and when the domain is $b \leqslant r \leqslant a$, $0 \leqslant \phi \leqslant \Phi$.

(iii) *The elliptical domain*

Let the ellipse have semi-axes, a, b and put $a = l\cosh u_0$, $b = l\sinh u_0$ so that the eccentricity is $\operatorname{sech} u_0$. The substitution $x = l\cosh u\cos v$, $y = l\sinh u\sin v$ (c.f. §1.39(iii)) transforms

$$f_{xx} + f_{yy} + k^2 f = 0 \tag{8.6}$$

to a form which can be separated into a Mathieu equation and a modified Mathieu equation (see §C.1 and §1.41(iii)). Thus possible solutions of (8.6) are

$$\{A\,\mathrm{Ce}_m(u, h) + B\,\mathrm{Fe}_m(u, h)\}\{C\,\mathrm{ce}_m(v, h) + D\,\mathrm{fe}_m(v, h)\},$$
$$\{A'\,\mathrm{Se}_m(u, h) + B'\,\mathrm{Ge}_m(u, h)\}\{C'\,\mathrm{se}_m(v, h) + D'\,\mathrm{ge}_m(v, h)\}$$

where $h = \frac{1}{4}k^2 l^2$. The function f must be periodic in v and so $D = D' = 0$ because fe_m and ge_m are not periodic. When $u = 0$, $\mathrm{Se}_m = 0$ but $\mathrm{Ge}_m \neq 0$; however a change in sign of v does not alter x and should therefore leave f unchanged. Since se_m is odd this is possible only if $B' = 0$. Similarly, a consideration of the derivative of f demonstrates that $B = 0$ because $\mathrm{Fe}_m'(0, h) \neq 0$. Hence the basic solutions are $\mathrm{Ce}_m(u, h)\mathrm{ce}_m(v, h)$ and $\mathrm{Se}_m(u, h)\mathrm{se}_m(v, h)$.

The vanishing on the perimeter of the ellipse leads to the expansion

$$f = \sum_m \sum_n \{a_{mn}\,\mathrm{Ce}_m(u, c_{mn})\mathrm{ce}_m(v, c_{mn}) + b_{mn}\,\mathrm{Se}_m(u, s_{mn})\mathrm{se}_m(v, s_{mn})\}$$

where

$$\mathrm{Ce}_m(u_0, c_{mn}) = 0, \qquad \mathrm{Se}_m(u_0, s_{mn}) = 0. \tag{8.7}$$

Here a_{mn} is given by

$$A_{mn}a_{mn}=\int_0^{2\pi}\int_0^{u_0} f\,\mathrm{Ce}_m(u,c_{mn})\mathrm{ce}_m(v,c_{mn})(\cosh 2u-\cos 2v)\,du\,dv$$

where

$$A_{mn}=\int_0^{2\pi}\int_0^{u_0}\mathrm{Ce}_m^2(u,c_{mn})\mathrm{ce}_m^2(v,c_{mn})(\cosh 2u-\cos 2v)\,du\,dv.$$

There is a similar formula for b_{mn}.

If the normal derivative vanishes on the elliptical boundary there is a similar set of eigenfunctions with c_{mn}, s_{mn} replaced c'_{mn}, s'_{mn} respectively where

$$\mathrm{Ce}'_m(u_0,c'_{mn})=0,\qquad \mathrm{Se}'_m(u_0,s'_{mn})=0.$$

There are also appropriate expansions for the region between two confocal ellipses and for the domain between an ellipse and a confocal hyperbola.

(iv) *The rectangular parallelepiped*

The eigenfunctions of a rectangular parallelepiped bear a close resemblance to those in (i) for the rectangle. For instance, the eigenfunction which vanishes on the boundary is $\sin(m\pi x/a)\sin(n\pi y/b)\sin(p\pi z/c)$ with $k^2=(m^2/a^2+n^2/b^2+p^2/c^2)\pi^2$. Therefore the eigenfunctions of a rectangular parallelepiped will not be discussed further nor will those for a cylinder whose cross-section is one of (i)–(iii).

(v) *The sphere*

The equation

$$u_{xx}+u_{yy}+u_{zz}+k^2u=0 \tag{8.8}$$

when transformed to spherical polar coordinates R, θ, ϕ can be separated into three ordinary differential equations (§1.41(ii)). That in terms of ϕ originates Fourier series while that involving R leads to $\mathrm{j}_m(kR)$ where $\mathrm{j}_m(x)=(\pi/2x)^{\frac{1}{2}}\mathrm{J}_m(x)$. For the θ-dependence

$$\sin\theta\frac{\mathrm{d}}{\mathrm{d}\theta}\left(\sin\theta\frac{\mathrm{d}\Theta}{\mathrm{d}\theta}\right)+\{m(m+1)\sin^2\theta-p^2\}\Theta=0$$

which is Legendre's equation (see Appendix B). Thus eigenfunctions which vanish on the sphere $R=a$ are

$$\mathrm{j}_m(j_{m+\frac{1}{2},n}R/a)\mathrm{P}_m^p(\cos\theta)\mathrm{e}^{\mathrm{i}p\phi} \tag{8.9}$$

where m, n, and p are integers. The coefficients in an expansion of these eigenfunctions can be deduced from earlier results together with (B.3.20).

It can be deduced that a function of θ, ϕ only can be expanded in a series of products of Legendre functions and $e^{ip\phi}$ whereas a function of θ only, with θ limited to $(0, \pi)$, has an expansion in Legendre functions of integral order.

There are many extensions, similar to those for the circle, to domains of the type $b \leqslant R \leqslant a$, $0 \leqslant \beta \leqslant \theta \leqslant \alpha \leqslant \pi$, $0 \leqslant \phi \leqslant \gamma \leqslant 2\pi$. One of interest is the cone with spherical end $0 \leqslant R \leqslant a$, $0 \leqslant \theta \leqslant \alpha \leqslant \pi$. The eigenfunctions which vanish on the boundary are still of the form (8.9) but m is no longer an integer. Instead it must satisfy (c.f. §9.20) $P_m(\cos\alpha) = 0$.

(vi) *The spheroid*

In prolate spheroidal coordinates X, Y, ϕ the harmonic wave equation (8.8) separates (§1.41(v)) with solutions $S_n^{m(1)}(X, \hbar)\mathrm{ps}_n^m(Y, \hbar^2)e^{im\phi}$ where m and n are integers. Here the semi-axes a, b $(a > b)$ of the spheroid $X = X_0$ are given by $a = lX_0$, $b = l(X_0^2 - 1)^{\frac{1}{2}}$ and $\hbar = kl$. The eigenfunctions which vanish on the spheroid are supplied by making $\hbar$ satisfy $S_n^{m(1)}(X_0, \hbar) = 0$.

For the oblate spheroid with coordinates ξ, η, ϕ the eigenfunctions are $S_n^{m(1)}(-i\xi, i\mathcal{H})\mathrm{ps}_n^m(\eta, -\mathcal{H}^2)e^{im\phi}$ with $S_n^{m(1)}(-i\xi_0, i\mathcal{H}) = 0$.

(vii) *Maxwell's equations*

The fields of several electromagnetic oscillations in regions bounded by perfectly conducting walls can be deduced from the eigenfunctions already found. Two-dimensional fields which are independent of the coordinate z require only the determination of E_z and H_z according to §1.36. Since E_z and the normal derivative of H_z are zero on the boundary appropriate eigenfunctions are available from the foregoing. If the domain is cylindrical it is a particular case of a waveguide and relevant expansions stem from §5.7 Therefore attention will be confined to the sphere and spheroid.

Only fields which possess axial symmetry and are independent of the coordinate ϕ will be considered. It is a straightforward matter to show that, in spherical polar and spheroidal coordinates, the component E_ϕ (when it is independent of ϕ) satisfies

$$\nabla^2(E_\phi e^{i\phi}) + k^2 E_\phi e^{i\phi} = 0.$$

Therefore $E_\phi e^{i\phi}$ can be determined from earlier results. So can $H_\phi e^{i\phi}$ since it satisfies the same equation.

(a) *The sphere.* The field in which $H_\phi \equiv 0$ is

$$E_\phi = \mathrm{j}_m(j_{m+\frac{1}{2},n}R/a)P_m^1(\cos\theta),$$
$$H_R = -(1/i\omega\mu_0 R\sin\theta)\,\partial(\sin\theta E_\phi)/\partial\theta,$$
$$H_\theta = (1/i\omega\mu_0 R)\,\partial(RE_\phi)/\partial\phi.$$

If $E_\phi \equiv 0$ the boundary condition on $R = a$ is $\partial(RH_\phi)/\partial R = 0$ and

$$H_\phi = \mathrm{j}_m(\mu_{mn}R/a)\mathrm{P}_m^1(\cos\theta),$$
$$E_R = (1/\mathrm{i}\omega\varepsilon_0 R\sin\theta)\,\partial(\sin\theta H_\phi)/\partial\theta,$$
$$E_\theta = -(1/\mathrm{i}\omega\varepsilon_0 R)\,\partial(RH_\phi)/\partial R$$

where

$$\mu_{mn}\mathrm{j}'_m(\mu_{mn}) + \mathrm{j}_m(\mu_{mn}) = 0.$$

(b) *The spheroid.* When $H_\phi \equiv 0$ the field in a prolate spheroid is

$$E_\phi = \mathrm{S}_n^{1(1)}(X, \hbar)\mathrm{ps}_n^1(Y, \hbar^2),$$
$$H_X = -\{1/\mathrm{i}\omega\varepsilon_0 l(X^2 - Y^2)^{\frac{1}{2}}\}\,\partial\{(1-Y^2)^{\frac{1}{2}}E_\phi\}/\partial Y,$$
$$H_Y = \{1/\mathrm{i}\omega\varepsilon_0 l(X^2 - Y^2)^{\frac{1}{2}}\}\,\partial\{(X^2-1)^{\frac{1}{2}}E_\phi\}/\partial X$$

where $\mathrm{S}_n^{1(1)}(X_0, \hbar) = 0$. The standard modification gives the field in oblate spheroidal coordinates ξ, η, ϕ.

For the prolate spheroid with $E_\phi \equiv 0$

$$H_\phi = \mathrm{S}_n^{1(1)}(X, \hbar)\mathrm{ps}_n^1(Y, \hbar^2),$$
$$E_X = \{1/\mathrm{i}\omega\varepsilon_0 l(X^2 - Y^2)^{\frac{1}{2}}\}\,\partial\{(1-Y^2)^{\frac{1}{2}}H_\phi\}/\partial Y,$$
$$E_Y = -\{1/\mathrm{i}\omega\varepsilon_0 l(X^2 - Y^2)^{\frac{1}{2}}\}\,\partial\{(X^2-1)^{\frac{1}{2}}H_\phi\}/\partial X$$

where

$$\partial\{(X^2-1)^{\frac{1}{2}}\mathrm{S}_n^{1(1)}(X, \hbar)\}/\partial X = 0$$

when $X = X_0$.

4.9 Time dependent fields

Eigenfunctions can also be employed when the time dependence is not harmonic. As an illustration consider the equation

$$a^2\nabla^2 p - \partial^2 p/\partial t^2 = 0. \tag{9.1}$$

Suppose that the eigenfunctions φ_m and eigenvalues λ_m satisfy

$$\nabla^2\varphi_m + \lambda_m\varphi_m = 0. \tag{9.2}$$

It will be assumed that the eigenfunctions have been normalized so that the φ_m form an orthonormal set. In the expansion $p = \sum_{m=1}^{\infty} a_m\varphi_m$, a_m is then given by

$$a_m = \int_T p\varphi_m\,\mathrm{d}\boldsymbol{x}.$$

By the divergence theorem and (9.2)

$$\int_T \varphi_m \nabla^2 p \, \mathrm{d}\boldsymbol{x} = \int_S \left(\frac{\partial p}{\partial n} \varphi_m - p \frac{\partial \varphi_m}{\partial n} \right) \mathrm{d}S - \lambda_m a_m.$$

Hence (9.1) provides

$$\ddot{a}_m + \lambda_m a^2 a_m = a^2 \int_S \left(\frac{\partial p}{\partial n} \varphi_m - p \frac{\partial \varphi_m}{\partial n} \right) \mathrm{d}S \tag{9.3}$$

where a dot indicates a derivative with respect to t.

Suppose $p = 0$ on S. Select the orthonormal set so that $\varphi_m = 0$ on S. The right-hand side of (9.3) vanishes and a_m is a linear combination of $\exp(\mathrm{i}\lambda_m^{\frac{1}{2}} at)$ and $\exp(-\mathrm{i}\lambda_m^{\frac{1}{2}} at)$ (the precise form depending on the initial conditions). Thus any free oscillation consists of a linear superposition of modes each oscillating at its resonant frequency. Clearly, the same conclusion can be drawn when $\partial p/\partial n = 0$ on S by picking $\partial \varphi_m/\partial n = 0$.

The electromagnetic field satisfies

$$\operatorname{curl} \boldsymbol{E} + \mu_0 \, \partial \boldsymbol{H}/\partial t = \boldsymbol{0}, \qquad \operatorname{curl} \boldsymbol{H} - \varepsilon_0 \, \partial \boldsymbol{E}/\partial t = \boldsymbol{0}. \tag{9.4}$$

On account of the theory of §4.7 take

$$\boldsymbol{E} = \sum_{m=1}^{\infty} a_m \boldsymbol{E}_m, \tag{9.5}$$

$$\boldsymbol{H} = \operatorname{grad} \varphi + \sum_{m=1}^{\infty} b_m \operatorname{curl} \boldsymbol{E}_m \tag{9.6}$$

where $\nabla^2 \varphi = 0$ in the simply connected domain T. Now

$$\begin{aligned} \omega_m^2 \mu_0 \varepsilon_0 b_m &= \int_T \boldsymbol{H} \cdot \operatorname{curl} \boldsymbol{E}_m \, \mathrm{d}\boldsymbol{x} \\ &= \int_T \boldsymbol{E}_m \cdot \operatorname{curl} \boldsymbol{H} \, \mathrm{d}\boldsymbol{x} \end{aligned}$$

since $\boldsymbol{n} \wedge \boldsymbol{E}_m = \boldsymbol{0}$ on S. Invoking (9.4) we deduce that

$$b_m = \dot{a}_m / \omega_m^2 \mu_0. \tag{9.7}$$

Integration of the scalar product of the first of (9.4) with $\operatorname{curl} \boldsymbol{E}_m$ now leads to

$$\ddot{a}_m + \omega_m^2 a_m = -c^2 \int_S \boldsymbol{n} \cdot \boldsymbol{E} \wedge \operatorname{curl} \boldsymbol{E}_m \, \mathrm{d}S \tag{9.8}$$

where $c^2 = 1/\mu_0 \varepsilon_0$.

If $\boldsymbol{n} \wedge \boldsymbol{E} = \boldsymbol{0}$ on S the right-hand side of (9.8) vanishes and a_m is a linear combination of $\mathrm{e}^{\mathrm{i}\omega_m t}$, $\mathrm{e}^{-\mathrm{i}\omega_m t}$. Also, if static fields are ignored, the theory of §1.22 tells us that $\boldsymbol{n} \cdot \boldsymbol{H} = 0$ on S; the consequence is that φ is a constant and grad $\varphi \equiv \mathbf{0}$. Therefore, again the free oscillation is a linear combination of resonant modes.

Perturbation theory

The preceding theory gives information about the oscillations which can occur in boxes when there are no losses present. In practice, however, it will be desirable to have some mechanism for transferring energy to and from the cavity. There may also be some loss of energy in the walls. The existence of such energy losses influences the oscillations. The effect will now be examined on the assumption that it is small.

4.10 Dissipation in the walls

In an acoustic cavity when the walls are perfectly rigid $\partial p/\partial n = 0$ on S. It will now be assumed that the boundary is not quite rigid and that

$$\partial p/\partial n = Z_m p \tag{10.1}$$

where $|Z_m|$ does not differ by much from zero. The formula (9.3) is still valid and the selection for φ_m is that $\partial\varphi_m/\partial n = 0$ on S. Now, since a small effect is being considered, we investigate how p differs from the oscillation based on φ_m. The normal derivative $\partial p/\partial n$ will be correct to the first order if, in (10.1) p is taken to be the same as when $Z_m = 0$, i.e. $\partial p/\partial n = Z_m a_m \varphi_m$. Assuming that the oscillation is proportional to $e^{i\omega t}$ where $\omega \approx a\lambda_m^{\frac{1}{2}}$, the deduction from (9.3) is that

$$\lambda_m a^2 - \omega^2 = a^2 Z_m \int_S \varphi_m^2 \, dS.$$

Hence, the first approximation to ω is

$$\omega = a\lambda_m^{\frac{1}{2}} - \frac{aZ_m}{2\lambda_m^{\frac{1}{2}}} \int_S \varphi_m^2 \, dS.$$

Now, if $Z_m = \alpha - i\beta$ with α and β positive, it is clear that dissipation in the wall causes both a change of frequency and damping of the oscillation. The energy, being proportional to the square of the amplitude, decreases exponentially according to the factor $\exp(-a\lambda_m^{\frac{1}{2}} t/Q_u)$ where

$$\frac{1}{Q_u} = \frac{\beta}{\lambda_m} \int_S \varphi_m^2 \, dS.$$

Q_u can be deemed to be the ratio of the average energy stored to the average energy absorbed by the wall. The suffix u indicates that the cavity is *unloaded* i.e. no devices have been inserted for supplying energy or abstracting it.

Similar evidence to that to follow for the electromagnetic case suggests that Q_u is largest, for a given surface area, when S is a sphere, i.e. the least damping occurs for a given surface area when the surface is spherical.

A parallel line may be pursued for the electromagnetic case but founded on (9.8). When the walls are highly conducting the theory of the skin effect (§6.8) says that near the oscillation in which $\omega = \omega_m$

$$\boldsymbol{n} \wedge \boldsymbol{E} = \tfrac{1}{2}\delta\mu_0\omega_m(1+\mathrm{i})\boldsymbol{H}$$

where $\delta = (2/\omega_m\mu_0\sigma)^{\frac{1}{2}}$ is the skin depth. Taking $\boldsymbol{H}$ to have its unperturbed value of $(\mathrm{i}a_m/\omega_m\mu_0)\mathrm{curl}\,\boldsymbol{E}_m$ we find from (9.8) that

$$\omega = \omega_m\left\{1+\frac{1}{4}\frac{\delta(1+\mathrm{i})}{\omega_m^2\mu_0\varepsilon_0}\int_S (\mathrm{curl}\,\boldsymbol{E}_m)^2\,\mathrm{d}S\right\}.$$

The energy decays as $\mathrm{e}^{-\omega_m t/Q_u}$ where

$$\frac{1}{Q_\mathrm{u}} = \frac{\delta}{2\omega_m^2\mu_0\varepsilon_0}\int_S (\mathrm{curl}\,\boldsymbol{E}_m)^2\,\mathrm{d}S.$$

For a rectangular parallelepiped of sides a, b, l

$$\frac{1}{Q_\mathrm{u}} = 2\delta\left\{\frac{p^2}{a^2}\left(\frac{1}{a}+\frac{1}{l}\right)+\frac{q^2}{b^2}\left(\frac{1}{b}+\frac{1}{l}\right)\right\}\Big/\left(\frac{p^2}{a^2}+\frac{q^2}{b^2}\right)$$

when $\omega_m^2\mu_0\varepsilon_0 = (p^2/a^2+q^2/b^2+n^2/l^2)\pi^2$. This formula may be written roughly as $1/Q_u = S\delta/T$ where S is the surface area and T is the volume of the cavity. Other resonators also comply with this crude estimate. Since the sphere has the largest volume of all shapes with the same surface area the largest Q_u and the smallest damping would be predicted for the sphere when the surface area is fixed.

4.11 Boundary perturbation

A small change of frequency can also be effected by making a slight alteration to S. Suppose that the walls have no dissipation but the part S_0 of the boundary is pushed out a small distance S' (Fig. 4.3). Inside T, (9.3) remains valid but the right-hand side is no longer zero because, although $\partial\varphi_m/\partial n$ vanishes on S_0, $\partial p/\partial n$ does not. However, $\partial p/\partial n$ is zero on S' and the surface integral over S_0 can be transformed into one over T' by the divergence theorem provided that φ_m can be continued suitably into that region. As a first approximation take $p = a_m\varphi_m$ in T' and then

$$\omega^2 = \lambda_m a^2 - a^2\int_{T'} (\mathrm{grad}^2\varphi_m - \lambda_m\varphi_m^2)\,\mathrm{d}\boldsymbol{x}.$$

Hence, when the surface is pushed out, the frequency decreases if the

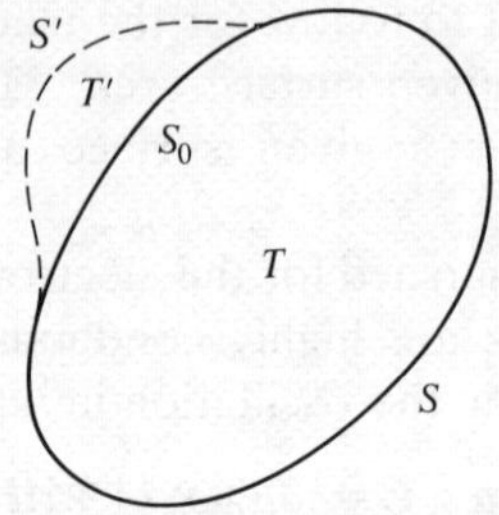

Fig. 4.3 Boundary perturbation

perturbation is in a region where $|\text{grad}\ \varphi_m|$ dominates $|\varphi_m|$ and increases if $|\varphi_m|$ is sufficiently larger than $|\text{grad}\ \varphi_m|$. The opposite effect on the frequency occurs if the surface is pushed in.

The analogous formula for electromagnetic oscillations is

$$\omega^2 = \omega_m^2 - \int_{T'} \{(c\ \text{curl}\ \boldsymbol{E}_m)^2 - \omega_m^2 \boldsymbol{E}_m^2\}\,\mathrm{d}\boldsymbol{x}.$$

When the surface is pushed out the frequency decreases when the modal magnetic field is strong and increases if the modal electric field is the stronger partner.

4.12 The effect of an aperture

Another situation of interest is when the walls are loss-free but there is an aperture S' in them through which energy can be supplied or extracted. The discussion will centre on the electromagnetic problem although a like study can be undertaken for sound waves. It is also convenient to take the aperture as the cross-section of a waveguide; much of the analysis can be generalized to more arbitrary apertures.

Formulae (9.5)–(9.8) are still applicable but it cannot be assumed that the grad φ is zero. To obtain a formula for φ let $G(\boldsymbol{x}, \mathbf{y})$ be the Green's function which satisfies

$$\nabla^2 G = -\delta(\boldsymbol{x} - \mathbf{y})$$

in T and is such that $\partial G/\partial n = 0$ on S. Then

$$\int_S \left(\frac{\partial \varphi}{\partial n} G - \varphi \frac{\partial G}{\partial n}\right) \mathrm{d}S_y = \int_T (G\nabla^2\varphi - \varphi\nabla^2 G)\,\mathrm{d}\mathbf{y} = \varphi(\boldsymbol{x})$$

since φ satisfies Laplace's equation in T. Now, $\boldsymbol{n} \cdot \text{grad}\ \varphi = \boldsymbol{n} \cdot \boldsymbol{H}$ on S since the normal component of curl $\boldsymbol{E}_m$ is zero there. Further, $\boldsymbol{n} \cdot \boldsymbol{H} = 0$

on $S-S'$ and hence

$$\varphi(\boldsymbol{x})=\int_{S'}\boldsymbol{n}\cdot\boldsymbol{H}G\,\mathrm{d}S_y$$

which determines φ in terms of the normal component of $\boldsymbol{H}$ on S'.

The formula (9.8) may be simplified by the observation that the tangential component of $\boldsymbol{E}$ vanishes on $S-S'$. Thus, for periodic phenomena,

$$a_m(\omega^2-\omega_m^2)=c^2\int_{S'}\boldsymbol{n}\cdot\boldsymbol{E}\wedge\operatorname{curl}\boldsymbol{E}_m\,\mathrm{d}S.$$

A suitable way of expressing the interrelation of the fields in the aperture is to expand them in terms of the modes of the waveguide connected to the aperture. Denote such modes by $\boldsymbol{H}_{\mathrm{tp}}$ for the component tangential to S' and by $\boldsymbol{H}_{\mathrm{np}}$ for the normal component. Then, for points in S', we can write (see §5.4)

$$(\boldsymbol{E}_m)_{\mathrm{t}}=\sum_{\mathrm{p}}V_{m\mathrm{p}}\boldsymbol{E}_{\mathrm{tp}},\qquad(\boldsymbol{H}_m)_{\mathrm{t}}=\sum_{\mathrm{p}}I_{m\mathrm{p}}\boldsymbol{H}_{\mathrm{tp}},$$

$$\boldsymbol{H}=\sum_{\mathrm{p}}(I_{\mathrm{p}}\boldsymbol{H}_{\mathrm{tp}}+V_{\mathrm{p}}\boldsymbol{H}_{\mathrm{np}}/Z_{\mathrm{p}}),\qquad(\boldsymbol{E})_{\mathrm{t}}=\sum_{\mathrm{p}}V_{\mathrm{p}}\boldsymbol{E}_{\mathrm{tp}}$$

and we seek the relation between I_{p} and V_q. The waveguide modes are such that $\boldsymbol{H}_{\mathrm{tp}}=\boldsymbol{E}_{\mathrm{tp}}\wedge\boldsymbol{n}$ and

$$\int_{S'}\boldsymbol{H}_{\mathrm{tp}}\cdot\boldsymbol{H}_{\mathrm{tq}}\,\mathrm{d}S=\int_{S'}\boldsymbol{E}_{\mathrm{tp}}\cdot\boldsymbol{H}_{\mathrm{tq}}\,\mathrm{d}S=\delta_{\mathrm{pq}}=\int_{S'}\boldsymbol{E}_{\mathrm{tp}}\wedge\boldsymbol{H}_{\mathrm{tq}}\cdot\boldsymbol{n}\,\mathrm{d}S.$$

Therefore

$$a_m(\omega^2-\omega_m^2)=\sum_{\mathrm{p}}V_{\mathrm{p}}v_{m\mathrm{p}}\tag{12.1}$$

where

$$v_{m\mathrm{p}}=c^2\int_{S'}\boldsymbol{n}\cdot\boldsymbol{E}_{\mathrm{tp}}\wedge\operatorname{curl}\boldsymbol{E}_m\,\mathrm{d}S.$$

Hence

$$I_q=\int_{S'}\boldsymbol{H}\cdot\boldsymbol{H}_{\mathrm{tq}}\,\mathrm{d}S=\mathrm{i}\omega\sum_m\frac{a_m}{\mu_0\omega_m^2}\int_{S'}(\operatorname{curl}\boldsymbol{E}_m)\cdot\boldsymbol{E}_{\mathrm{tq}}\wedge\boldsymbol{n}\,\mathrm{d}S$$

$$+\int_{S'}\boldsymbol{H}_{\mathrm{tq}}\cdot\operatorname{grad}\varphi\,\mathrm{d}S,$$

i.e.

$$I_q = \sum_p Y_{qp} V_p \tag{12.2}$$

where

$$Y_{qp} = -i\omega\varepsilon_0 \sum_m \frac{v_{mq} v_{mp}}{\omega_m^2(\omega^2 - \omega_m^2)} + \int_{S'} \boldsymbol{H}_{tq} \cdot \operatorname{grad} \int_{S'} \boldsymbol{H}_{np}(G/Z_p)\, dS_y\, dS. \tag{12.3}$$

The matrix with entries Y_{qp} is known as the *admittance matrix.*

If S' can be placed so that only one waveguide mode is relevant (notice that moving S' along the guide alters $\boldsymbol{E}_m$) (12.2) becomes much simpler. Such a procedure may be feasible for a waveguide in which only one mode can propagate and so the case when only the lowest mode is pertinent will be considered. Then

$$I_1 = Y_{11} V_1.$$

At frequencies near ω_m the denominator of Y_{11} ensures that one term is dominant and

$$Y_{11} = \frac{-i\omega\varepsilon_0 v_{m1}^2}{\omega_m^2(\omega^2 - \omega_m^2)} + Y'_{11}$$

where Y'_{11} is a correction term representing the remainder of the series and integral in (12.3).

If the coupling is such that there is no reflected wave in the guide $Y_{11} = 1/Z_1 = Y_1$ and

$$(\omega^2 - \omega_m^2)(Y_1 - Y'_{11}) = -i\omega\varepsilon_0 v_{m1}^2/\omega_m^2.$$

This equation determines the change in frequency and gives an energy loss with

$$\frac{1}{Q} = \mathscr{R} \frac{\varepsilon_0 v_{m1}^2}{\omega_m^3(Y'_{11} - Y_1)} = \frac{g}{Q_{me}} \tag{12.4}$$

where

$$1/Q_{me} = \varepsilon_0 v_{m1}^2/\omega_m^3 Y_1, \qquad g + ib = Y_1/(Y'_{11} - Y_1).$$

If there are surface losses in the walls (12.4) has to be modified. For ω near ω_m the field in the cavity will still be near $\boldsymbol{E}_m$ as in §4.10. There will be an additional term in (12.1) and

$$\frac{Y_{11}}{Y_1} = \sum_n \frac{-i\omega\omega_n/Q_{ne}}{\omega^2 - \omega_n^2 + (1+i)\omega\omega_n/Q_u}.$$

The resulting energy loss is now determined by

$$1/Q_L = 1/Q_u + g/Q_{me},$$

Q_L being called the *loaded* Q.

Exercises on Chapter 4

1. The function $f(x)$ is defined to be zero for $0<x<l_1$ and for $l_1+l_2<x<l$, whereas it is unity for $l_1<x<l_1+l_2$. Show that the following are possible expansions:

$$f(x)=(l_2/l)+\sum_{m=1}^{\infty}(2/m\pi)\sin(m\pi l_2/l)\cos 2m\pi(x-l_1-\tfrac{1}{2}l_2)/l.$$

$$f(x)=(l_2/l_1)+\sum_{m=1}^{\infty}(4/m\pi)\sin(m\pi l_2/2l)\cos m\pi(l_1+\tfrac{1}{2}l_2)/l\cos(m\pi x/l),$$

$$f(x)=\sum_{m=1}^{\infty}(4/m\pi)\sin(m\pi l_2/2l)\sin m\pi(l_1+\tfrac{1}{2}l_2)/l\sin(m\pi x/l).$$

Deduce that, if $f(x)=1\,(0<x<a)$, $f(x)=0\ (a<x<2a)$,

$$f(x)=\tfrac{1}{2}+\sum_{m=0}^{\infty}\{2/(2m+1)\pi\}\sin\{(2m+1)\pi x/a\}.$$

2. Show that, in $(-\pi,\pi)$,

$$\mathrm{e}^x=\frac{1}{\pi}\sinh\pi\sum_{m=-\infty}^{\infty}\frac{(-)^m\mathrm{e}^{imx}}{1-im}.$$

Deduce that

$$\sum_{m=0}^{\infty}\frac{(-)^m}{1+m^2}=\tfrac{1}{2}(\pi\operatorname{cosech}\pi+1).$$

Is it true that

$$\mathrm{e}^x=\frac{1}{\pi}\sinh\pi\sum_{m=-\infty}^{\infty}\frac{(-)^m im\mathrm{e}^{imx}}{1-im}?$$

3. The functions u_1 and u_2 are linearly independent solutions of $u''-qu=0$ such that $u_1(0)=0$ and $u_2(l)=0$. Show that

$$G(x,y)=\begin{cases}Au_1(x)u_2(y) & (0\leqslant x\leqslant y)\\ Au_1(y)u_2(x) & (y\leqslant x\leqslant l)\end{cases}$$

is the solution of

$$\mathrm{d}^2G(x,y)/\mathrm{d}x^2-qG(x,y)=-\delta(x-y)$$

such that $G(0,y)=0$, $G(l,y)=0$ for a suitable choice of the constant A. Hence show that an integral equation equivalent to $u''+(\lambda r-q)u=0$ subject to $u=0$ at $x=0,l$ is

$$u(x)=\lambda\int_0^l rG(y,x)u(y)\,\mathrm{d}y.$$

More generally, if u_1 and u_2 both satisfy $(pu')' - qu = 0$ and $u_1(0) = \beta$, $u_1'(0) = -\alpha$, $u_2(l) = \delta$, $u_2'(l) = -\gamma$ show that G defined above satisfies $(pG')' - qG = -\delta(x-y)$ and $\alpha G(0, y) + \beta G'(0, y) = 0$, $\gamma G(l, y) + \delta G'(l, y) = 0$.

4. The function u satisfies $u_{tt} + 2au_t = c^2 u_{xx}$ and $c^2\pi^2 > 4a^2$. At $x = \pm 1$, $u = 0$ for all t. At $t = 0$, $u = (1-x^2)^{\frac{1}{2}}$ and $u_t = 0$ for $-1 \leqslant x \leqslant 1$. Show that

$$u = \sum_{n=1}^{\infty} \frac{J_1\{(n-\frac{1}{2})\pi\}}{n-\frac{1}{2}} \left(\cos q_n t + \frac{a}{q_n} \sin q_n t\right) e^{-at} \cos(n-\tfrac{1}{2})\pi x$$

where $q_n^2 = (n-\frac{1}{2})^2 c^2 \pi^2 - a^2$.

5. Show that a solution of $u_{xx} + u_{yy} = 0$ such that (i) $u = 0$ on $x = 0$ and $x = a$, (ii) $u = x$ on $0 < x < \frac{1}{2}a$, $y = 0$ and $u = a - x$ on $\frac{1}{2}a < x < a$, $y = 0$, (iii) $u \to 0$ as $y \to \infty$ is

$$u = (4a/\pi^2) \sum_{n=1}^{\infty} (1/n^2) e^{-n\pi y/a} \sin \tfrac{1}{2} n\pi \sin(n\pi x/a).$$

6. Show that

$$E_z = A J_p(j_{pn} r/a) \cos(s\pi z/l) \cos p\phi$$

can represent the z-component of an eigenfunction $\boldsymbol{E}_m$ for a cylindrical resonator of radius a and height l with $\omega_m^2 \mu_0 \varepsilon_0 = j_{pm}^2/a^2 + s^2\pi^2/l^2$.

If the cylinder is not perfectly conducting show that

$$1/Q_u = \begin{cases} \delta(1/l + 1/a) & (s = 0) \\ \delta(2/l + 1/a) & (s \neq 0) \end{cases}.$$

7. A resonator consists of the space between $r = b$ and $r = a$, with $a > b$, closed by the planes $z = 0$ and $z = l$. Show that there is an eigenfunction with $E_r = \sin(s\pi z/l)/r\{l\pi \ln(a/b)\}^{\frac{1}{2}}$.

If the surfaces are not perfectly conducting show that

$$\frac{1}{Q_u} = \delta\left\{\frac{2}{l} + \frac{1}{2\ln(a/b)}\left(\frac{1}{a} + \frac{1}{b}\right)\right\}.$$

8. Show that $Q_u = 0{\cdot}725 a/\delta$ for the mode μ_{11} $(= 2.75)$ of §4.8(vii)(a).

9. The boundaries in Exercises 6 and 8 are made from copper instead of being perfectly conducting. Calculate the change in frequency when the radius is 20 cm and the height 5 cm in Exercise 6 and when the radius is 20 cm in Exercise 8.

10. If regions surrounding $\theta = 0$ and $\theta = \pi$ are excluded show that there are fields of the form §4.8(vii)(b) with $H_\phi = j_m(kR) Q_m^1(\cos\theta)$. Using (B.3.1), (B.3.8), and $j_0(x) = (\sin x)/x$ show that there is a field in $\alpha \leqslant \theta \leqslant \pi - \alpha$, $0 \leqslant R \leqslant a$ whose only nonzero components are

$$H_\phi = \frac{\sin(\frac{1}{2}\pi R/a)}{R \sin\theta}, \qquad E_\theta = \frac{\pi \cos(\frac{1}{2}\pi R/a)}{2i\omega\varepsilon_0 R a \sin\theta}.$$

Show that, for such a resonant mode,

$$1/Q_u = (\delta/a)\{1 + b/\sin\alpha \ln\cot\tfrac{1}{2}\alpha\}$$

where $b = \frac{1}{2}(\gamma + \ln\pi - \text{Ci}\,\pi) = 0.82$. Show that Q_u has a maximum of $0.45a/\delta$ at $\alpha = 33\frac{1}{2}°$.

11. Compare the values of Q_u in Exercises 6, 8, 10 at a wavelength of 40 cm when the walls are copper and the dimensions are (i) cylinder height = diameter = 30.6 cm, (ii) sphere radius = 17.5 cm, (iii) sphere radius = 10 cm and angle of cone $\alpha = 34°$. [They are in the ratio 53:49:18].

12. Consider the eigenvalue problem for

$$\mathrm{i}\omega\rho_0\boldsymbol{u} + \text{grad}\, p = 0, \qquad \text{div}\,\boldsymbol{u} + \mathrm{i}\omega p/\rho_0 a_0^2 = 0$$

when ρ_0 varies with position.

5. The theory of waveguides

ONE mechanism for transporting energy from one point to another is a hollow cylindrical tube. The method is of some antiquity in acoustics having been used for many years in ships. It is also extremely valuable in the transmission of microwaves between transmitter or receiver and an antenna. The theory of waves in such tubes will be developed in this chapter.

5.1 The modal expansion in acoustics

The tube that will be considered is a cylinder with axis parallel to the z-axis and attention will be restricted to harmonic waves. The cross-section in $z = \text{constant}$ will be denoted by $\mathscr{D}$ and its perimeter by C (Fig. 5.1). The medium inside the waveguide will be taken to be homogeneous and there is no loss of generality in making it free space. Then, for waves with time dependence $e^{i\omega t}$ the equation to be satisfied is

$$\nabla^2 p + k^2 p = 0 \tag{1.1}$$

where $k = \omega/a$, a being the speed of sound in free space.

A rigid tube is likely to be employed in practice so that a solution of (1.1) is sought such that

$$\partial p/\partial n = 0 \tag{1.2}$$

on C. Designating this as the *hard boundary condition* we may also consider in conjunction with it the *soft boundary condition*

$$p = 0 \tag{1.3}$$

although this is less likely to be physically realizable. Once p has been determined the velocity $\boldsymbol{v}$ is furnished by

$$\boldsymbol{v} = (i/\omega\rho_0)\text{grad } p \tag{1.4}$$

where ρ_0 is the density of the medium when there is no sound wave.

Turn firstly to the hard boundary condition. The square of the absolute value of p is integrable over $\mathscr{D}$. Therefore p, regarded as a function of x and y, can be expanded in a series of eigenfunctions of $\mathscr{D}$ (§4.6). Let ψ_m

be a typical eigenfunction of the orthonormal set which solves

$$\frac{\partial^2\psi_m}{\partial x^2}+\frac{\partial^2\psi_m}{\partial y^2}+\mu_m^2\psi_m=0 \tag{1.5}$$

in $\mathscr{D}$ and is such that $\partial\psi_m/\partial n=0$ on C. Then

$$\int_{\mathscr{D}} \psi_p\psi_m \, dx\, dy=\delta_{pm}$$

and also (c.f. (5.7), (5.8) of Chapter 4)

$$\int_{\mathscr{D}} \text{grad}_t\,\psi_m\cdot\text{grad}_t\,\psi_p\, dx\, dy=\mu_p^2\,\delta_{pm}$$

where grad_t signifies the two-dimensional gradient depending on x and y, i.e.

$$\text{grad}_t\,\varphi\equiv \boldsymbol{i}\,\partial\varphi/\partial x+\boldsymbol{j}\,\partial\varphi/\partial y.$$

with $\boldsymbol{i}, \boldsymbol{j}, \boldsymbol{k}$ unit vectors along the Cartesian axes.

One thing pointed out in §4.6 is that the lowest eigenvalue is zero with the associated eigenfunction a constant. In order that the smallest positive eigenvalue may correspond to $m=1$, the zero eigenvalue will be identified by $m=0$. It follows that $\psi_0=A$ where A^2 is the reciprocal of the area of $\mathscr{D}$.

From §4.6

$$p=\sum_{m=0}^{\infty} a_m(z)\psi_m, \qquad \text{grad}_t\,p=\sum_{m=1}^{\infty} a_m(z)\text{grad}_t\,\psi_m$$

where $a_m(z)=\int_{\mathscr{D}} p\psi_m\, dx\, dy$. Multiply (1.1) by ψ_m and integrate over $\mathscr{D}$. Let

$$\nabla_t^2\equiv\partial^2/\partial x^2+\partial^2/\partial y^2.$$

Then

$$\begin{aligned}\int_{\mathscr{D}} \psi_m\nabla_t^2 p\, dx\, dy &= \int_{\mathscr{D}} (\psi_m\nabla_t^2 p - p\nabla_t^2\psi_m)\, dx\, dy+\int_{\mathscr{D}} p\nabla_t^2\psi_m\, dx\, dy\\ &= \int_C (\psi_m\,\partial p/\partial n - p\,\partial\psi_m/\partial n)\, ds-\mu_m^2 a_m\\ &= -\mu_m^2 a_m\end{aligned}$$

on account of (1.5) and both p, ψ_m satisfying the hard boundary condition on C. Hence (1.1) gives

$$d^2a_m/dz^2+(k^2-\mu_m^2)a_m=0.$$

Therefore

$$a_m(z)=A_m e^{i\kappa_m z}+B_m e^{-i\kappa_m z}$$

where A_m and B_m are constants. The quantity κ_m is such that $\kappa_m^2 = k^2 - \mu_m^2$ and κ_m is defined precisely by

$$\kappa_m = \begin{cases} (k^2 - \mu_m^2)^{\frac{1}{2}} & (k^2 > \mu_m^2) \\ -\mathrm{i}(\mu_m^2 - k^2)^{\frac{1}{2}} & (k^2 < \mu_m^2) \end{cases}$$

Thus the expansion of the field is

$$p = \sum_{m=0}^{\infty} (A_m \mathrm{e}^{\mathrm{i}\kappa_m z} + B_m \mathrm{e}^{-\mathrm{i}\kappa_m z})\psi_m, \tag{1.6}$$

$$v_z = (\mathrm{i}/\omega\rho_0) \sum_{m=0}^{\infty} \mathrm{i}\kappa_m (A_m \mathrm{e}^{\mathrm{i}\kappa_m z} - B_m \mathrm{e}^{-\mathrm{i}\kappa_m z})\psi_m, \tag{1.7}$$

$$\boldsymbol{v}_\mathrm{t} = (\mathrm{i}/\omega\rho_0) \sum_{m=0}^{\infty} (A_m \mathrm{e}^{\mathrm{i}\kappa_m z} + B_m \mathrm{e}^{-\mathrm{i}\kappa_m z}) \operatorname{grad}_\mathrm{t} \psi_m. \tag{1.8}$$

If all the terms involving $\mathrm{e}^{\mathrm{i}\kappa_m z}$ in (1.6)–(1.8) are suppressed a wave travelling in the direction of the positive z-axis is obtained. The speed of propagation in a mode ω/κ_m is known as the *phase speed* of the mode. The terms with $\mathrm{e}^{\mathrm{i}\kappa_m z}$ are responsible for waves moving in the direction of the negative z-axis; they have the same phase speed. The general field consists of waves travelling in both directions.

It will be observed that propagation along the positive z-axis takes place without attenuation only if κ_m is real, i.e. $k^2 > \mu_m^2$. If $k^2 < \mu_m^2$, κ_m is negative imaginary and the wave is exponentially damped. Since $\mu_0 = 0$ propagation in this mode is always possible and the phase speed is the same as the speed of sound in free space. If $k^2 < \mu_1^2$ no other mode can propagate. Faithful reproduction of a signal is more likely if the energy can be kept in the lowest mode without any transfer to higher modes. This suggests that speaking tubes should be designed so that $k^2 < \mu_1^2$. Since $1/\mu_1^2$ is of the order of the area of the cross-section this means that the wavelength needs to be somewhat larger than the maximum chord of the cross-section. To put it another way, it is the shorter wavelengths or higher frequencies that tend to be distorted by a speaking tube.

Expansions in eigenfunctions can also be achieved for the soft boundary condition. Here the orthonormal set is built from

$$\nabla_\mathrm{t}^2 \varphi_m + \nu_m^2 \varphi_m = 0$$

in $\mathscr{D}$ with $\varphi_m = 0$ on C. Also

$$\int_{\mathscr{D}} \varphi_p \varphi_m \, \mathrm{d}x \, \mathrm{d}y = \delta_{pm},$$

$$\int_{\mathscr{D}} \operatorname{grad} \varphi_p \cdot \operatorname{grad} \varphi_m \, \mathrm{d}x \, \mathrm{d}y = \nu_p^2 \delta_{pm}.$$

In this case the lowest eigenvalue is nonzero and

$$p = \sum_{m=1}^{\infty} (C_m e^{i\lambda_m z} + D_m e^{-i\lambda_m z})\varphi_m \tag{1.9}$$

where

$$\lambda_m = \begin{cases} (k^2 - \nu_m^2)^{\frac{1}{2}} & (k^2 > \nu_m^2) \\ -i(\nu_m^2 - k^2)^{\frac{1}{2}} & (k^2 < \nu_m^2). \end{cases}$$

The main difference from the hard boundary condition is that there may be no propagating mode if the frequency is low enough. In fact, if $k^2 < \nu_1^2$ no mode propagates. Since $k = 2\pi/\lambda$ where λ is the wavelength there is no propagation if $\lambda > 2\pi/\nu_1$. For this reason $2\pi/\nu_1$ is known as the *cut-off wavelength* and the first propagating mode is called the *fundamental* or *dominant mode*. Tubes with soft walls can only be useful at higher frequencies of operation.

5.2 The electromagnetic boundary conditions

When the walls of the waveguide in Fig. 5.1 are perfectly conducting the electromagnetic field must be such that the tangential component of the electric intensity $\boldsymbol{E}$ is zero on C. If s denotes the arc-length measured on C from some fixed point (Fig. 5.2), this condition may be expressed as

$$E_z = 0, \qquad E_s = 0 \tag{2.1}$$

on C. For subsequent purposes it is helpful to replace these conditions by equivalent ones. With the time variation $e^{i\omega t}$

$$\text{curl}\, \boldsymbol{E} + i\omega\mu_0 \boldsymbol{H} = \boldsymbol{0}. \tag{2.2}$$

Let H_n be the component of $\boldsymbol{H}$ in the direction of the unit normal $\boldsymbol{n}$ to C

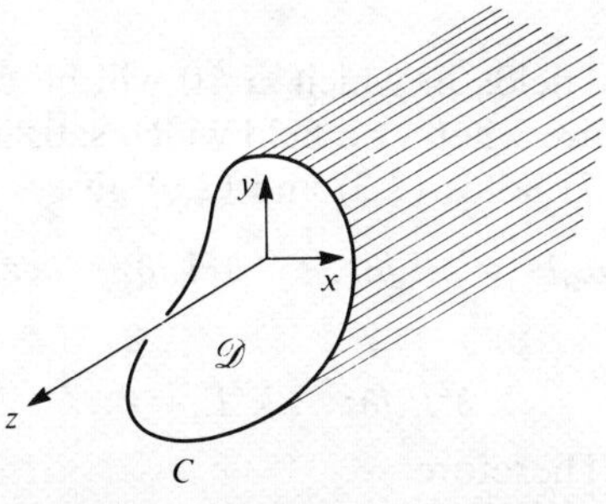

Fig. 5.1 Waveguide geometry

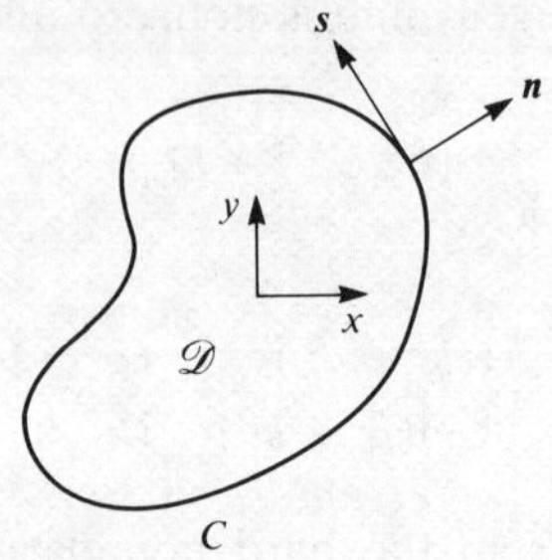

Fig. 5.2 The tangential and normal vectors to C

and let the direction cosines of $\boldsymbol{n}$ be $(l, m, 0)$. Then

$$-\mathrm{i}\omega\mu_0 H_n = -\mathrm{i}\omega\mu_0(lH_x + mH_y)$$

$$= l\left(\frac{\partial E_z}{\partial y} - \frac{\partial E_y}{\partial z}\right) + m\left(\frac{\partial E_x}{\partial z} - \frac{\partial E_z}{\partial x}\right) = \frac{\partial E_z}{\partial s} - \frac{\partial E_s}{\partial z} \tag{2.3}$$

since l and m are independent of z. Hence (2.1) implies that

$$H_n = 0 \tag{2.4}$$

on C if $\omega \neq 0$.

An alternative condition stems from

$$\operatorname{curl} \boldsymbol{H} = \mathrm{i}\omega\varepsilon_0 \boldsymbol{E}. \tag{2.5}$$

This implies that

$$l(\partial H_x/\partial z - \partial H_z/\partial x) - m(\partial H_z/\partial y - \partial H_y/\partial z) = \mathrm{i}\omega\varepsilon_0 E_s$$

or

$$\partial H_n/\partial z - \partial H_z/\partial n = \mathrm{i}\omega\varepsilon_0 E_s. \tag{2.6}$$

Then (2.4) and (2.1) make it necessary that $\partial H_z/\partial n = 0$ on C if $\omega \neq 0$. Thus every electromagnetic field which complies with (2.1) also satisfies

$$E_z = 0, \qquad \partial H_z/\partial n = 0 \tag{2.7}$$

on C if $\omega \neq 0$.

From now on only fields in which $\omega \neq 0$ will be considered.

We shall now enquire whether a field which satisfies (2.7) also complies with (2.1). When (2.7) holds (2.3) and (2.6) give

$$\mathrm{i}\omega\mu_0 H_n = \partial E_s/\partial z, \qquad \partial H_n/\partial z = \mathrm{i}\omega\varepsilon_0 E_s$$

on C. Hence

$$\partial^2 E_s/\partial z^2 + k^2 E_s = 0$$

where $k^2 = \omega^2\mu_0\varepsilon_0$. Therefore

$$E_s = A\mathrm{e}^{\mathrm{i}kz} + B\mathrm{e}^{-\mathrm{i}kz}, \qquad H_n = (\varepsilon_0/\mu_0)^{\frac{1}{2}}(A\mathrm{e}^{\mathrm{i}kz} - B\mathrm{e}^{-\mathrm{i}kz})$$

on C. We conclude that, unless the boundary behaviour is consistent with $e^{\pm ikz}$, (2.7) ensures that (2.1) holds. If the boundary variation is of the type $e^{\pm ikz}$ a separate investigation is necessary to secure $E_s = 0$ on C.

5.3 The modal structure of the electromagnetic field

One consequence of (2.2) and (2.5) is, since $\operatorname{div} \boldsymbol{E} = 0$,

$$\nabla^2 \mathbf{E} + k^2 \mathbf{E} = \mathbf{0}. \tag{3.1}$$

Since $E_z = 0$ on C the determination of E_z is no different from finding p in §5.1 under the soft boundary condition. Hence

$$E_z = \sum_{m=1}^{\infty} (A_m e^{i\lambda_m z} + B_m e^{-i\lambda_m z})\varphi_m. \tag{3.2}$$

$\boldsymbol{H}$ also satisfies (3.1) and, if (2.7) is applied, the problem is that of the acoustic hard boundary condition. Therefore

$$H_z = \sum_{m=0}^{\infty} (C_m e^{i\kappa_m z} + D_m e^{-i\kappa_m z})\psi_m. \tag{3.3}$$

Now, according to the last paragraph of the preceding section, the field generated by (3.3) will make $E_s = 0$ on C except when $\kappa_m^2 = k^2$. However, $\kappa_m^2 = k^2$ only when $\mu_m^2 = 0$, i.e. $m = 0$ and then $\psi_0 = \text{constant}$. But, from (2.2),

$$-i\omega\mu_0 \int_{\mathscr{D}} H_z \, dx \, dy = \int_{\mathscr{D}} (\partial E_y/\partial x - \partial E_x/\partial y) \, dx \, dy$$
$$= \int_C E_s \, ds = 0$$

from (2.1). Thus if the H_z is constant on a cross-section it must be zero. Therefore, the term $m = 0$ must be excluded from (3.3) and there is then no trouble with the boundary condition.

Next, consider the field $\boldsymbol{E}_m^{M}$, $\boldsymbol{H}_m^{M}$ in which the z-components are $e^{-i\lambda_m z}\varphi_m$ and 0 respectively. Assuming the same z-dependence for the whole field it is easily confirmed from (2.2) and (2.5) that

$$(\boldsymbol{E}_m^{M})_z = \varphi_m e^{-i\lambda_m z}, \qquad (\boldsymbol{E}_m^{M})_t = -i\lambda_m e^{-i\lambda_m z} \operatorname{grad}_t \varphi_m/\nu_m^2, \tag{3.4}$$

$$(\boldsymbol{H}_m^{M})_z = 0, \qquad (\boldsymbol{H}_m^{M})_t = -i\omega\varepsilon_0 e^{-i\lambda_m z} \boldsymbol{k} \wedge \operatorname{grad} \varphi_m/\nu_m^2 \tag{3.5}$$

where $(\boldsymbol{E})_t = E_x \boldsymbol{i} + E_y \boldsymbol{j}$. Similarly, there is another field

$$(\boldsymbol{E}_m^{E})_z = 0, \qquad (\boldsymbol{E}_m^{E})_t = i\omega\mu_0 e^{-i\kappa_m z} \boldsymbol{k} \wedge \operatorname{grad} \psi_m/\mu_m^2, \tag{3.6}$$

$$(\boldsymbol{H}_m^{E})_z = \psi_m e^{-i\kappa_m z}, \qquad (\boldsymbol{H}_m^{E})_t = -i\kappa_m e^{-i\kappa_m z} \operatorname{grad}_t \psi_m/\mu_m^2. \tag{3.7}$$

All terms in (3.2) can be taken care of by adding together fields of the type (3.4); let the resulting field be $\mathbf{E}^M, \mathbf{H}^M$. Also all of (3.3) can be disposed of since $m \neq 0$ by a sum of (3.7); denote this total field by $\mathbf{E}^E, \mathbf{H}^E$. Let $\mathbf{E}_0 = \mathbf{E} - \mathbf{E}^M - \mathbf{E}^E$, $\mathbf{H}_0 = \mathbf{H} - \mathbf{H}^M - \mathbf{H}^E$. By construction $(\mathbf{E}_0)_z = 0$, $(\mathbf{H}_0)_z = 0$. From (2.2) and (2.5), both $(\mathbf{H}_0)_x$ and $(\mathbf{H}_0)_y$ satisfy

$$\partial^2 H/\partial z^2 + k^2 H = 0$$

and $\partial(\mathbf{H}_0)_x/\partial x + \partial(\mathbf{H}_0)_y/\partial y = 0$. This requires

$$(\mathbf{H}_0)_x = \mathrm{e}^{-ikz}\, \partial\chi_1/\partial y + \mathrm{e}^{ikz}\, \partial\chi_2/\partial y,$$

$$(\mathbf{H}_0)_y = -\mathrm{e}^{-ikz}\, \partial\chi_1/\partial x - \mathrm{e}^{ikz}\, \partial\chi_2/\partial x$$

and then (2.5) gives

$$(\mathbf{E}_0)_x = (\mu_0/\varepsilon_0)^{\frac{1}{2}}(\mathrm{e}^{ikz}\, \partial\chi_2/\partial x - \mathrm{e}^{-ikz}\, \partial\chi_1/\partial x),$$

$$(\mathbf{E}_0)_y = (\mu_0/\varepsilon_0)^{\frac{1}{2}}(\mathrm{e}^{ikz}\, \partial\chi_2/\partial y - \mathrm{e}^{-ikz}\, \partial\chi_1/\partial y).$$

But $\partial(\mathbf{E}_0)_x/\partial x + \partial(\mathbf{E}_0)_y/\partial y = 0$ and so

$$\nabla_t^2 \chi_1 = 0, \qquad \nabla_t^2 \chi_2 = 0. \tag{3.8}$$

In order that $(\mathbf{E}_0)_s = 0$ it is necessary that $\chi_1 =$ constant and $\chi_2 =$ constant on C. If $\mathscr{D}$ is simply connected this is possible only if χ_1 and χ_2 are constant throughout $\mathscr{D}$; but then $\mathbf{E}_0$ and $\mathbf{H}_0$ disappear. On the other hand when $\mathscr{D}$ is multiply connected (as in a coaxial cable) χ_1 and χ_2 are not forced to be constant everywhere in $\mathscr{D}$ and there may be a field of this type.

To sum up then every field inside a waveguide can be expressed in the form

$$E_z = \sum_{m=1}^{\infty} (A_m \mathrm{e}^{i\lambda_m z} + B_m \mathrm{e}^{-i\lambda_m z})\varphi_m, \qquad H_z = \sum_{m=1}^{\infty} (C_m \mathrm{e}^{i\kappa_m z} + D_m \mathrm{e}^{-i\kappa_m z})\psi_m, \tag{3.9}$$

$$\begin{aligned}(\mathbf{E})_t = &\sum_{m=1}^{\infty} i\lambda_m (A_m \mathrm{e}^{i\lambda_m z} - B_m \mathrm{e}^{-i\lambda_m z})\,\mathrm{grad}_t\, \varphi_m/\nu_m^2 \\ &+ i\omega\mu_0 \sum_{m=1}^{\infty} (C_m \mathrm{e}^{i\kappa_m z} + D_m \mathrm{e}^{-i\kappa_m z})\mathbf{k} \wedge \mathrm{grad}\, \psi_m/\mu_m^2 \\ &+ (\mu_0/\varepsilon_0)^{\frac{1}{2}}(\mathrm{e}^{ikz}\, \mathrm{grad}_t\, \chi_2 - \mathrm{e}^{-ikz}\, \mathrm{grad}_t\, \chi_1),\end{aligned} \tag{3.10}$$

$$\begin{aligned}(\mathbf{H})_t = &-i\omega\varepsilon_0 \sum_{m=1}^{\infty} (A_m \mathrm{e}^{i\lambda_m z} + B_m \mathrm{e}^{-i\lambda_m z})\mathbf{k} \wedge \mathrm{grad}\, \varphi_m/\nu_m^2 \\ &+ \sum_{m=1}^{\infty} i\kappa_m (C \mathrm{e}^{i\kappa_m z} - D_m \mathrm{e}^{-i\kappa_m z})\,\mathrm{grad}_t\, \psi_m/\mu_m^2 \\ &- \mathbf{k} \wedge (\mathrm{e}^{ikz}\, \mathrm{grad}_t\, \chi_2 + \mathrm{e}^{-ikz}\, \mathrm{grad}_t\, \chi_1)\end{aligned} \tag{3.11}$$

where χ_1 and χ_2 are absent if $\mathscr{D}$ is simply connected.

The wave $\boldsymbol{E}_m^{\mathrm{M}}, \boldsymbol{H}_m^{\mathrm{M}}$ of (3.4), (3.5) is travelling in the direction of the positive z-axis with phase speed ω/λ_m. It is characterized by having no component of magnetic field in the direction of transmission. It is therefore known as a *transverse magnetic* or *TM-mode.* Similarly, the wave $\boldsymbol{E}_m^{\mathrm{E}}, \boldsymbol{H}_m^{\mathrm{E}}$ of (3.6), (3.7), whose phase speed is ω/κ_m is called a *transverse electric* or *TE-mode.* The wave $\boldsymbol{E}_0, \boldsymbol{H}_0$, whose phase speed is that of free space, has neither the electric nor magnetic vector with a component in the direction of propagation and is termed a *TEM-mode.*

Every TE-mode suffers attenuation during propagation if $k^2 < \mu_1^2$ and every TM-mode is attenuated if $k^2 < \nu_1^2$. It turns out that μ_1^2 can never exceed ν_1^2 and so the cut-off wavelength is $2\pi/\mu_1$. Thus the first propagating or fundamental mode is a TE-mode. This statement needs qualification when a TEM-mode exists for it has no cut-off wavelength and so can occur for any frequency of excitation. However, for a simply connected cross-section there is no TEM-mode and the TE-mode is dominant. For this reason hollow metallic tubes are useful for guiding waves only for free-space wavelengths of 10 cm or less.

Before leaving the subject it should perhaps be stressed that χ_1 and χ_2 in (3.10) are the most general solutions of Laplace's equation which take constant values on C. They may contain several independent elements.

5.4 Lumped circuit equations

In this section it will be supposed that there is no TEM-mode. Then (3.9)–(3.11) can be written as

$$E_z = \mathrm{i} \sum_{m=1}^{\infty} \nu_m^2 I_m^{\mathrm{M}}(z)(\boldsymbol{e}_m^{\mathrm{M}})_z, \qquad H_z = \sum_{m=1}^{\infty} \mu_m^2 V_m^{\mathrm{E}}(z)(\boldsymbol{h}_m^{\mathrm{E}})_z/\mathrm{i}\omega\mu_0, \tag{4.1}$$

$$(\boldsymbol{E})_t = \sum_{m=1}^{\infty} \{V_m^{\mathrm{E}}(z)(\boldsymbol{e}_m^{\mathrm{E}})_t + V_m^{\mathrm{M}}(z)(\boldsymbol{e}_m^{\mathrm{M}})_t\}, \tag{4.2}$$

$$(\boldsymbol{H})_t = \sum_{m=1}^{\infty} \{I_m^{\mathrm{E}}(z)(\boldsymbol{h}_m^{\mathrm{E}})_t + I_m^{\mathrm{M}}(z)(\boldsymbol{h}_m^{\mathrm{M}})_t\} \tag{4.3}$$

where

$$\begin{aligned} V_m^{\mathrm{M}}(z) &= \mathrm{i}\lambda_m (A_m \mathrm{e}^{\mathrm{i}\lambda_m z} - B_m \mathrm{e}^{-\mathrm{i}\lambda_m z})/\nu_m, \\ I_m^{\mathrm{M}}(z) &= -\mathrm{i}\omega\varepsilon_0 (A_m \mathrm{e}^{\mathrm{i}\lambda_m z} + B_m \mathrm{e}^{-\mathrm{i}\lambda_m z})/\nu_m, \end{aligned} \tag{4.4}$$

$$\begin{aligned} V_m^{\mathrm{E}}(z) &= \mathrm{i}\omega\mu_0 (C_m \mathrm{e}^{\mathrm{i}\kappa_m z} + D_m \mathrm{e}^{-\mathrm{i}\kappa_m z})/\mu_m, \\ I_m^{\mathrm{E}}(z) &= -\mathrm{i}\kappa_m (C_m \mathrm{e}^{\mathrm{i}\kappa_m z} - D_m \mathrm{e}^{-\mathrm{i}\kappa_m z})/\mu_m, \end{aligned} \tag{4.5}$$

and

$$\boldsymbol{e}_m^{\mathrm{E}} = \boldsymbol{k} \wedge \operatorname{grad} \psi_m/\mu_m, \qquad \boldsymbol{h}_m^{\mathrm{E}} = (\boldsymbol{k} - \operatorname{grad})\psi_m/\mu_m \tag{4.6}$$

while

$$\boldsymbol{e}_m^{\mathrm{M}} = (\boldsymbol{k} + \operatorname{grad})\varphi_m/\nu_m, \qquad \boldsymbol{h}_m^{\mathrm{M}} = \boldsymbol{k} \wedge \operatorname{grad} \varphi_m/\nu_m. \tag{4.7}$$

It will be remarked that (4.4) and (4.5) furnish

$$\mathrm{d}V_m^{\mathrm{M}}/\mathrm{d}z = -\mathrm{i}\lambda_m Z_m^{\mathrm{M}} I_m^{\mathrm{M}}, \qquad \mathrm{d}I_m^{\mathrm{M}}/\mathrm{d}z = -\mathrm{i}\lambda_m Y_m^{\mathrm{M}} V_m^{\mathrm{M}}, \tag{4.8}$$

$$\mathrm{d}V_m^{\mathrm{E}}/\mathrm{d}z = -\mathrm{i}\kappa_m Z_m^{\mathrm{E}} I_m^{\mathrm{E}}, \qquad \mathrm{d}I_m^{\mathrm{E}}/\mathrm{d}z = -\mathrm{i}\kappa_m Y_m^{\mathrm{E}} V_m^{\mathrm{E}} \tag{4.9}$$

where $Z_m^{\mathrm{M}} = 1/Y_m^{\mathrm{M}} = \lambda_m/\omega\varepsilon_0$ and $Z_m^{\mathrm{E}} = 1/Y_m^{\mathrm{E}} = \omega\mu_0/\kappa_m$. These equations have the same structure as those which arise in standard transmission line theory. For this reason V_m and I_m are known as the *mode voltage* and *mode current* of the relevant mode whereas Z_m and Y_m are called the *mode impedance* and *mode admittance* respectively.

Next, observe that, on account of the formulae of §5.1,

$$\int_{\mathscr{D}} (\boldsymbol{e}_m^{\mathrm{E}})_{\mathrm{t}} \cdot (\boldsymbol{e}_p^{\mathrm{E}})_{\mathrm{t}} \,\mathrm{d}x\,\mathrm{d}y = \int_{\mathscr{D}} (\boldsymbol{e}_m^{\mathrm{M}})_{\mathrm{t}} \cdot (\boldsymbol{e}_p^{\mathrm{M}})_{\mathrm{t}} \,\mathrm{d}x\,\mathrm{d}y = \delta_{pm},$$

$$\int_{\mathscr{D}} (\boldsymbol{e}_m^{\mathrm{E}})_{\mathrm{t}} \cdot (\boldsymbol{e}_p^{\mathrm{M}})_{\mathrm{t}} \,\mathrm{d}x\,\mathrm{d}y = 0$$

and similarly for $\boldsymbol{h}$ so that

$$V_m^{\mathrm{E}} = \int_{\mathscr{D}} (\boldsymbol{E})_{\mathrm{t}} \cdot (\boldsymbol{e}_m^{\mathrm{E}})_{\mathrm{t}} \,\mathrm{d}x\,\mathrm{d}y, \qquad V_m^{\mathrm{M}} = \int_{\mathscr{D}} (\boldsymbol{E})_{\mathrm{t}} \cdot (\boldsymbol{e}_m^{\mathrm{M}})_{\mathrm{t}} \,\mathrm{d}x\,\mathrm{d}y,$$

$$I_m^{\mathrm{E}} = \int_{\mathscr{D}} (\boldsymbol{H})_{\mathrm{t}} \cdot (\boldsymbol{h}_m^{\mathrm{E}})_{\mathrm{t}} \,\mathrm{d}x\,\mathrm{d}y, \qquad I_m^{\mathrm{M}} = \int_{\mathscr{D}} (\boldsymbol{H})_{\mathrm{t}} \cdot (\boldsymbol{h}_m^{\mathrm{M}})_{\mathrm{t}} \,\mathrm{d}x\,\mathrm{d}y.$$

5.5 Energy flow

Equations similar to (4.9) can be obtained for the acoustic field by rewriting (1.6) and (1.7) as

$$p = \sum_{m=0}^{\infty} V_m^{\mathrm{H}}(z)\psi_m, \qquad v_z = \sum_{m=0}^{\infty} I_m^{\mathrm{H}}(z)\psi_m. \tag{5.1}$$

In this case the hard impedance $Z_m^{\mathrm{H}} = \omega\rho_0/\kappa_m$. Similarly for the soft boundary condition

$$p = \sum_{m=1}^{\infty} V_m^{\mathrm{S}}\varphi_m, \qquad v_z = \sum_{m=1}^{\infty} I_m^{\mathrm{S}}\varphi_m \tag{5.2}$$

with $Z_m^{\mathrm{S}} = \omega\rho_0/\lambda_m$.

The average energy crossing a plane $z = \text{constant}$ per unit time is

$$P_{\mathrm{a}} = \tfrac{1}{2}\,\mathrm{Re}\int_{\mathscr{D}} p\boldsymbol{v}^* \cdot \boldsymbol{k} \,\mathrm{d}x\,\mathrm{d}y$$

where the star indicates a complex conjugate. Inserting (5.1) and taking notice of the orthogonal properties of the ψ_m we see that

$$P_a = \tfrac{1}{2}\,\mathrm{Re} \sum_{m=0}^{\infty} V_m^H I_m^{H*} \tag{5.3}$$

for the hard boundary condition. The soft boundary produces a similar expression which will not be written down explicitly. If all the modes are travelling along the positive z-axis then

$$V_m^H = Z_m^H I_m^H \tag{5.4}$$

and the average power is

$$P_a = \tfrac{1}{2}\,\mathrm{Re} \sum_{m=0}^{\infty} Z_m^H\, |I_m^H|^2. \tag{5.5}$$

When a mode can propagate without attenuation Z_m^H is real and there is a contribution to the power. On the other hand, if the mode is attenuated Z_m^H is purely imaginary and that mode falls out of the power calculation. It will be remarked that the mode $m=0$ always provides a power flow and that when all other modes are cut off, $P_a = \frac{1}{2} Z_0^H\, |I_0^H|^2$. The main difference for the soft boundary condition is that all modes may be cut off.

Let v_g be the speed of average energy flow. Then, in the short time δt, the energy crossing the plane $z=$ constant fills a cylinder whose base is the cross-section and whose height is $v_g\,\delta t$. The average energy in this volume when only a single mode is propagating along the positive z-axis is

$$\tfrac{1}{4}\,\mathrm{Re}\int_0^{v_g \delta t}\int_{\mathscr{D}} (\rho_0 \boldsymbol{v}\cdot\boldsymbol{v}^* + pp^*/\rho_0 a^2)\,\mathrm{d}x\,\mathrm{d}y\,\mathrm{d}z$$

$$= \tfrac{1}{4}\,\mathrm{Re}\; v_g\,\delta t\{\rho_0\,|I_m|^2 + (1 + a^2\mu_m^2/\omega^2)\,|V_m|^2/\rho_0 a^2\}.$$

The substitution of (5.4) leads to

$$\tfrac{1}{2} v_g\,\delta t\,|I_m|^2\,\rho_0 k^2/\kappa_m^2.$$

This must equal $P_a\,\delta t$ or $\frac{1}{2} Z_m\,|I_m|^2\,\delta t$. Consequently

$$v_g = \kappa_m^2 Z_m/\rho_0 k^2 = \omega\kappa_m/k^2.$$

Since $\kappa_m \leqslant k$ this shows that v_g never exceeds the speed of sound. In fact, it equals the speed of sound only for the mode $m=0$. If v_p is written for the phase speed ω/κ_m the relation may be expressed as

$$v_p v_g = a^2. \tag{5.6}$$

The speed v_g is known as the *group speed*.

The considerations for electromagnetic waves are very similar. If a possible TEM-mode is omitted (its contribution to $m=0$ is as above) the average power P crossing $z=$ constant is

$$P=\tfrac{1}{2}\operatorname{Re}\int_{\mathscr{D}} \boldsymbol{E}\wedge\boldsymbol{H}^*\cdot\boldsymbol{k}\,\mathrm{d}x\,\mathrm{d}y=\tfrac{1}{2}\operatorname{Re}\sum_{m=1}(V_m^{\mathrm{E}}I_m^{\mathrm{E}*}+V_m^{\mathrm{M}}I_m^{\mathrm{M}*}) \qquad (5.7)$$

and the formula analogous to (5.5) can be deduced at once. As far as the group speed of a mode is concerned, there is the different formula

$$\tfrac{1}{4}\operatorname{Re}\int_0^{v_{\mathrm{g}}\delta t}\int_{\mathscr{D}}(\varepsilon_0\boldsymbol{E}\cdot\boldsymbol{E}^*+\mu_0\boldsymbol{H}\cdot\boldsymbol{H}^*)\,\mathrm{d}x\,\mathrm{d}y\,\mathrm{d}z$$

for the average energy in the cylindrical volume, but the net result is that

$$v_{\mathrm{p}}v_{\mathrm{g}}=c^2 \qquad (5.8)$$

for both TE- and TM-modes as well as TEM-modes. The relation (5.8) is important because, although the phase speed of a mode is never less than the speed of light, the group speed is never greater. Thus, despite the phase speed apparently being in conflict with the theory of relativity, the resolution of the seeming paradox is achieved by the energy travelling slower than the speed of light.

5.6 The effect of wall losses

The walls of practical guides are rarely perfect and so some part of the travelling wave will penetrate them. A full treatment allowing for this requires the determination of fields in both the guide and the walls which are matched at the interface by relevant boundary conditions. Frequently, however, this boundary value problem can be evaded because the imperfections in the walls are slight enough for the approximate method of §4.10 to be applicable. For example, when the boundary is almost acoustically hard the appropriate condition on C is

$$\partial p/\partial n = Yp \qquad (6.1)$$

where Y is small in magnitude while, in the electromagnetic case, the fact that metals are nearly perfectly conducting leads, when the frequency of operation is high enough, to

$$\boldsymbol{n}\wedge\boldsymbol{E}=\tfrac{1}{2}\,\delta\mu_0\omega(1+\mathrm{i})\boldsymbol{H} \qquad (6.2)$$

where $\delta=(2/\omega\mu_0\sigma)^{\frac{1}{2}}$ is the skin depth (§6.8).

The average power $P_{\mathrm{a}}(z_2)$ transmitted through $z=z_2$ differs from $P_{\mathrm{a}}(z_1)$ only in the loss through the intervening wall. Hence, for sound waves,

$$\begin{aligned}P_{\mathrm{a}}(z_2)-P_{\mathrm{a}}(z_1)&=-\tfrac{1}{2}\operatorname{Re}\int_{z_1}^{z_2}\int_C p\boldsymbol{v}^*\cdot\boldsymbol{n}\,\mathrm{d}s\,\mathrm{d}z\\&=-\tfrac{1}{2}\operatorname{Re}(-\mathrm{i}Y^*/\omega\rho_0)\int_{z_1}^{z_2}\int_C |p|^2\,\mathrm{d}s\,\mathrm{d}z\end{aligned}$$

from (1.4) and (6.1). Therefore

$$\mathrm{d}P_a/\mathrm{d}z = \mathrm{Re}(\mathrm{i}Y^*/2\omega\rho_0)\int_C |p|^2\,\mathrm{d}s. \tag{6.3}$$

Clearly, there is power loss unless Y is purely real. For a mode propagating in the positive z-direction the right-hand side of (6.3) is proportional to the square of the amplitude of the field. So is P_a and therefore the right-hand side of (6.3) may be regarded as proportional to P_a i.e.

$$\mathrm{d}P_a/\mathrm{d}z = -2\alpha P_a. \tag{6.4}$$

The power decays exponentially according to $P_a(z) = P_0 e^{-2\alpha z}$ and the field attenuates as $e^{-\alpha z}$. The *attenuation constant* α is given by

$$\mathrm{Re}(\mathrm{i}Y^*/2\omega\rho_0)\int_C |p|^2\,\mathrm{d}s = -\alpha\,\mathrm{Re}\int_{\mathscr{D}} p\boldsymbol{v}^*\cdot\boldsymbol{k}\,\mathrm{d}x\,\mathrm{d}y \tag{6.5}$$

from (6.3), (6.4), and §5.5. When a single mode is propagating α will be correct to the first order if p and $\boldsymbol{v}$ are given the same values as when the tube is rigid. Thus

$$\alpha = -\mathrm{Re}(\mathrm{i}Y^*/2\omega\rho_0)Z_m^{\mathrm{H}}\int_C |\psi_m|^2\,\mathrm{d}s. \tag{6.6}$$

In the electromagnetic case

$$P(z_2) - P(z_1) = -\tfrac{1}{2}\,\mathrm{Re}\int_{z_1}^{z_2}\int_C \boldsymbol{E}\wedge\boldsymbol{H}^*\cdot\boldsymbol{n}\,\mathrm{d}s\,\mathrm{d}z$$

and, by virtue of (6.2), the analogues of (6.5) and (6.6) are

$$\tfrac{1}{4}\delta\mu_0\omega\int_C \boldsymbol{H}\cdot\boldsymbol{H}^*\,\mathrm{d}s = \alpha\,\mathrm{Re}\int_{\mathscr{D}} \boldsymbol{E}\wedge\boldsymbol{H}^*\cdot\boldsymbol{k}\,\mathrm{d}x\,\mathrm{d}y, \tag{6.7}$$

$$\alpha = (\delta\mu_0\omega/4Z_m\,|I_m|^2)\int_C \boldsymbol{H}_m\cdot\boldsymbol{H}_m^*\,\mathrm{d}s. \tag{6.8}$$

The detailed evaluation of (6.6) and (6.8) depends upon the mode under consideration and the shape of the cross-section. Results for some typical cases will be found in the following section.

5.7 Typical waveguides

It will be sufficient, in displaying formulae for various waveguides, to set down only φ_m and ψ_m since the consequent fields can be deduced immediately. However, it will be convenient to replace the single suffix m indicating the order of a mode by a double suffix mp in order to improve the identification of the mode.

(a) *Parallel plate waveguide*

The first region to be considered is that between the parallel planes $y=0$ and $y=b$ (Fig. 5.3). This is not included in the general theory because of the possibility of propagation in the direction of the x-axis. However, if propagation is restricted to the z-direction alone with no dependence on x, the modes take a particularly simple form. The modal functions are

$$\varphi_m=\sin(m\pi y/b) \quad (m=1,2,\ldots),$$
$$\psi_m=\cos(m\pi y/b) \quad (m=0,1,2,\ldots).$$

In a typical TE-mode moving in the direction of the positive z-axis the only nonzero components are

$$E_x=(\mathrm{i}\omega\mu_0 b/m\pi)\mathrm{e}^{-\mathrm{i}\kappa_m z}\sin(m\pi y/b), \tag{7.1}$$

$$H_z=\mathrm{e}^{-\mathrm{i}\kappa_m z}\cos(m\pi y/b), \qquad H_y=(\mathrm{i}\kappa_m b/m\pi)\mathrm{e}^{-\mathrm{i}\kappa_m z}\sin(m\pi y/b) \tag{7.2}$$

for $m\geqslant 1$, while in a typical TM-mode

$$E_z=\mathrm{e}^{-\mathrm{i}\lambda_m z}\sin(m\pi y/b), \qquad E_y=-(\mathrm{i}\lambda_m b/m\pi)\mathrm{e}^{-\mathrm{i}\lambda_m z}\cos(m\pi y/b), \tag{7.3}$$

$$H_x=(\mathrm{i}\omega\varepsilon_0 b/m\pi)\mathrm{e}^{-\mathrm{i}\lambda_m z}\cos(m\pi y/b) \tag{7.4}$$

for $m\geqslant 1$. In (7.1)–(7.4)

$$\kappa_m=\lambda_m=(k^2-m^2\pi^2/b^2)^{\frac{1}{2}}.$$

It will be noted that the limit of (7.1) as $m\to 0$ does not satisfy the boundary conditions on E_x at $y=0$, b showing that there is no TE-mode with $m=0$ in harmony with the general theory. There is, however, a TEM-mode corresponding to $m=0$ and it is (taking $\chi_1=-y$)

$$E_y=(\mu_0/\varepsilon_0)^{\frac{1}{2}}\mathrm{e}^{-\mathrm{i}kz}, \qquad H_x=-\mathrm{e}^{-\mathrm{i}kz} \tag{7.5}$$

with all other components zero.

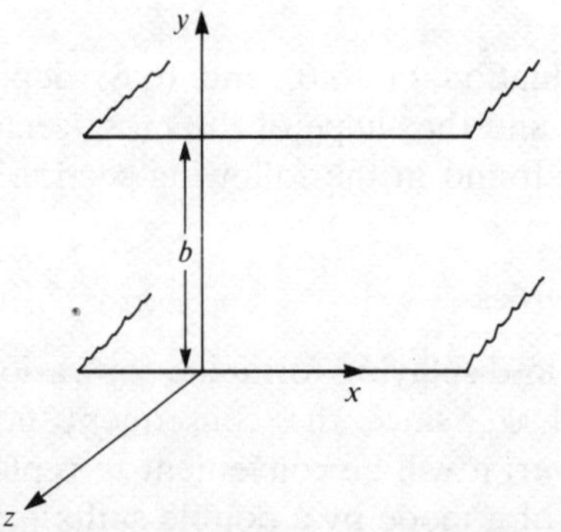

Fig. 5.3 The parallel plate waveguide

(b) *The rectangular waveguide*

Let

$$\eta_m = 2 - \delta_{m0} \tag{7.6}$$

so that η_m is 1 when $m = 0$ and 2 for $m \geqslant 1$. Then for the rectangular waveguide of Fig. 5.4 in which $a > b$

$$\psi_{mp} = \left(\frac{\eta_m \eta_p}{ab}\right)^{\frac{1}{2}} \cos\frac{m\pi x}{a} \cos\frac{p\pi y}{b}$$

for $m, p = 0, 1, 2, \ldots$. Also

$$\kappa_{mp}^2 = k^2 - (m\pi/a)^2 - (p\pi/b)^2, \qquad \mu_{mp}^2 = (m\pi/a)^2 + (p\pi/b)^2.$$

The acoustic attenuation is given by

$$\alpha_{mp}^{\mathrm{H}} = -2\,\mathrm{Re}(\mathrm{i}Y^*)(\eta_p a + \eta_m b)/\kappa_{mp} ab$$

and in the electromagnetic case

$$\alpha_{mp} = \frac{\delta}{2ab\kappa_{mp}\mu_{mp}^2}\left\{\mu_{mp}^4(\eta_p a + \eta_m b) + \kappa_{mp}^2\left(\eta_p \frac{m^2}{a} + \eta_m \frac{p^2}{b}\right)\right\}.$$

Notice that there is no electromagnetic TE_{mp}-mode in which m and p are simultaneously zero.

For the TM_{mp}-modes

$$\varphi_{mp} = \frac{2}{(ab)^{\frac{1}{2}}} \sin\frac{m\pi x}{a} \sin\frac{p\pi y}{b} \quad (m, p = 1, 2, \ldots),$$

$$\lambda_{mp}^2 = k^2 - (m\pi/a)^2 - (p\pi/b)^2, \qquad \nu_{mp}^2 = (m\pi/a)^2 + (p\pi/b)^2,$$

$$\alpha_{mp} = \frac{\delta k^2 \pi^2}{ab\lambda_{mp}\nu_{mp}^2}\left(\frac{m^2 b}{a^2} + \frac{p^2 a}{b^2}\right).$$

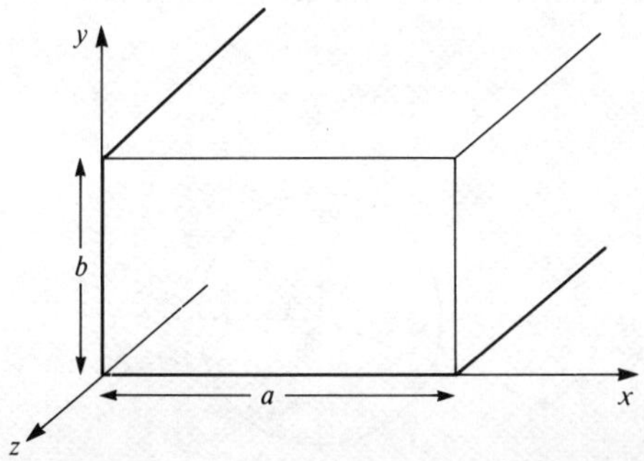

Fig. 5.4 The rectangular waveguide

There is no TEM-mode because the cross-section of the guide is simply connected.

The cut-off wavelength for TE-modes is $2\pi/\mu_{10}$ or $2a$ and this is also the cut-off for the guide. For TM-modes the cut-off wavelength is $2\pi/\nu_{11}$. The fundamental mode is the TE_{10}-mode and its field components in a wave travelling in the positive z-direction are

$$E_x = 0, \qquad E_y = -\frac{i\omega\mu_0 a}{\pi}\left(\frac{2}{ab}\right)^{\frac{1}{2}} e^{-i\kappa_{10}z} \sin\frac{\pi x}{a}, \qquad E_z = 0,$$

$$H_x = \frac{i\kappa_{10}a}{\pi}\left(\frac{2}{ab}\right)^{\frac{1}{2}} e^{-i\kappa_{10}z} \sin\frac{\pi x}{a}, \qquad H_y = 0, \qquad H_z = \left(\frac{2}{ab}\right)^{\frac{1}{2}} e^{-i\kappa_{10}z} \cos\frac{\pi x}{a}.$$

Typical dimensions of a guide designed to transmit the TE_{10}-mode primarily at a free space wavelength of 10 cm are $a = 7.5$ cm, $b = 2.5$ cm.

(c) *The circular waveguide*

Appropriate coordinates for the circular waveguide are cylindrical polar r, ϕ, z in which $x = r\cos\phi$, $y = r\sin\phi$ (Fig. 5.5). The TE_{mp}-modes have sinusoidal dependence on ϕ and the two forms are contained in

$$\psi_{mp} = \{\pi(j'^2_{mp} - m^2)\}^{-\frac{1}{2}} \frac{j'_{mp} J_m(j'_{mp} r/a)}{a J_m(j'_{mp})} \begin{cases} \eta_m^{\frac{1}{2}} \cos m\phi \\ 2^{\frac{1}{2}} \sin m\phi \end{cases}$$

for $m = 0, 1, \ldots$, with j'_{mp} $(p = 1, 2, \ldots)$ the pth nonzero root of $J'_m(z)$. In addition, when $m = 0$, there is an acoustic mode in which j'_{0p} is replaced by zero. Observe that there is no mode with a sine dependence on ϕ when $m = 0$. The relevant parameters are

$$\kappa^2_{mp} = k^2 - j'^2_{mp}/a^2, \qquad \mu_{mp} = j'_{mp}/a,$$

$$\alpha^{H}_{mp} = -2\,\mathrm{Re}(iY^*) j'^2_{mp}/\kappa_{mp} a (j'^2_{mp} - m^2),$$

$$\alpha_{mp} = \frac{\delta}{2a\kappa_{mp}}\left(\frac{j'^2_{mp}}{a^2} + \frac{k^2 m^2}{j'^2_{mp} - m^2}\right).$$

In deriving these results the formulae of Appendix A.1 will be found helpful.

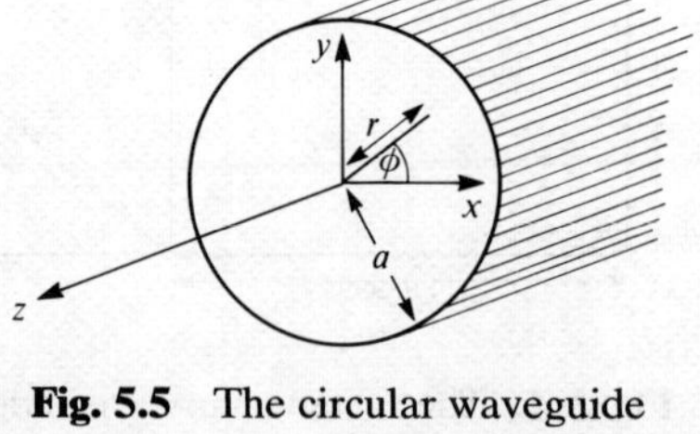

Fig. 5.5 The circular waveguide

The TM_{mp}-modes are given by

$$\varphi_{mp} = \frac{J_m(j_{mp}r/a)}{\pi^{\frac{1}{2}} a J_{m+1}(j_{mp})} \begin{cases} \eta_m^{\frac{1}{2}} \cos m\phi \\ 2^{\frac{1}{2}} \sin m\phi \end{cases}$$

where j_{mp} is the pth zero of $J_m(z)$. The parameters for this mode are

$$\lambda_{mp}^2 = k^2 - j_{mp}^2/a^2, \qquad \nu_{mp} = j_{mp}/a, \qquad \alpha_{mp} = \delta k^2/2a\lambda_{mp}.$$

Because the cross-section is simply connected there is to TEM-mode.

Typical values of j_{mp} and j'_{mp} are set out in Tables A.1 and A.2. A glance at these reveals that the TE_{11}-mode is fundamental and that the cut-off wavelength is $2\pi a/j'_{11} \approx 3.41a$. The next mode to propagate is the TM_{01}-mode with a cut-off wavelength of $2.61a$; it is independent of ϕ and so has circular symmetry. The modes with $m \neq 0$ are degenerate in the sense that they really consist of two modes, one with a cosine dependence on ϕ and the other with a sine dependence. This is one reason for employing a rectangular guide in preference to a circular guide.

The modes when the cross-section is a sector rather than the full circle can be determined (c.f. §4.8(ii)).

(d) *The elliptical waveguide*

The modes for a waveguide of elliptical cross-section can be expressed in terms of Mathieu functions as in §4.8(iii). The first mode to propagate is a TE-mode derived from a longitudinal component proportional to $Ce_1(u, c_{11})ce_1(v, c_{11})$ where c_{11} is the first nonzero root of equation (8.7) of §4.8(iii).

Consideration of elliptical cross-sections of small eccentricity indicates that waves in a circular waveguide with $m \neq 0$ are unstable to small deformations of the cross-section whereas those with $m = 0$ are stable. This was first shown by Chu (1938) and Brillouin (1938) who also gave the results of numerical calculation for the attenuation and power flow.

(e) *The coaxial line*

The coaxial line consists of an inner cylinder of radius b surrounded by a concentric circular cylinder of inner radius a. (Fig. 5.6).

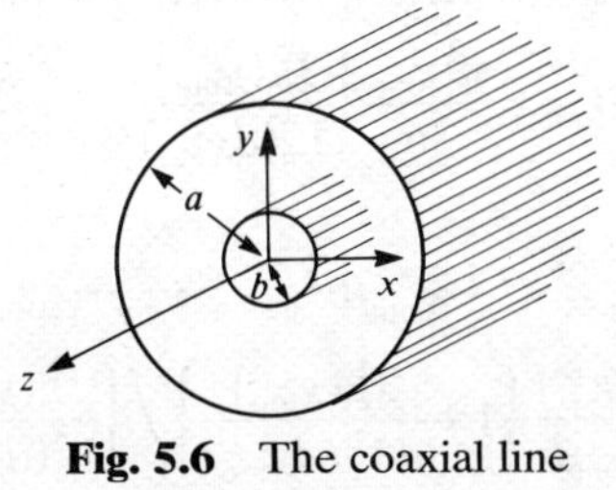

Fig. 5.6 The coaxial line

The modal functions are

$$\psi_{mp} = A_{mp}\{J_m(\chi'_{mp}r/b)Y'_m(\chi'_{mp}) - Y_m(\chi'_{mp}r/b)J'_m(\chi'_{mp})\}\begin{cases}\eta_m^{\frac{1}{2}} \cos m\phi \\ 2^{\frac{1}{2}} \sin m\phi\end{cases}$$

where $m = 0, 1, 2, \ldots, p = 1, 2, \ldots$ and

$$J'_m(d\chi'_{mp})Y'_m(\chi'_{mp}) - Y'_m(d\chi'_{mp})J'_m(\chi'_{mp}) = 0$$

with $d = a/b$. The evaluation of A_{mp} can be achieved by §A.1 and

$$\frac{1}{A_{mp}} = \frac{2b}{\chi'_{mp}\pi^{\frac{1}{2}}}\left[\left\{\frac{J'_m(\chi'_{mp})}{J'_m(d\chi'_{mp})}\right\}^2\left\{1 - \left(\frac{m}{d\chi'_{mp}}\right)^2\right\} - 1 + \left(\frac{m}{\chi'_{mp}}\right)^2\right]^{\frac{1}{2}}.$$

We have

$$\kappa_{mp}^2 = k^2 - \chi'^2_{mp}/b^2, \qquad \mu_{mp} = \chi'_{mp}/b,$$

$$\alpha_{mp}^H = -\frac{\mathrm{Re}(\mathrm{i}Y^*)}{\pi\kappa_{mp}} 2A_{mp}^2\left(\frac{2}{\chi'_{mp}}\right)^2\frac{b}{a}\left[a + b\left\{\frac{J'_m(\chi'_{mp})}{J'_m(d\chi'_{mp})}\right\}^2\right],$$

$$\alpha_{mp} = \frac{2\,\delta A_{mp}^2}{\pi\kappa_{mp}b}\left[1 + \left(\frac{m\kappa_{mp}b}{\chi'^2_{mp}}\right)^2 + \frac{b}{a}\left\{\frac{J'_m(\chi'_{mp})}{J'_m(d\chi'_{mp})}\right\}^2\left\{1 + \left(\frac{m\kappa_{mp}b^2}{\chi'^2_{mp}a}\right)^2\right\}\right]$$

assuming that both the inner and outer cylinders are made from the same material. The case of the mode in which $m = 0$, $p = 0$, and $\chi'_{00} = 0$ will be examined shortly.

Some values of χ'_{mp} are displayed in Table A.4. A reasonably good estimate of χ'_{mp} is

$$\chi'_{m1} = 2m/(1 + d),$$

$$\chi'_{mp} = (p-1)\pi/(d-1) \quad (p \geqslant 2)$$

which is borne out by the tabulated figures.

In the TM_{mp}-mode

$$\varphi_{mp} = B_{mp}\{J_m(\chi_{mp}r/b)Y_m(\chi_{mp}) - Y_m(\chi_{mp}r/b)J_m(\chi_{mp})\}\begin{cases}\eta_m^{\frac{1}{2}} \cos m\phi \\ 2^{\frac{1}{2}} \sin m\phi\end{cases}$$

where $m = 0, 1, \ldots, p = 1, 2, \ldots$ and

$$J_m(d\chi_{mp})Y_m(\chi_{mp}) - Y_m(\chi_{mp}d)J_m(\chi_{mp}) = 0,$$

$$B_{mp} = \frac{\pi^{\frac{1}{2}}\chi_{mp}}{2b}\left\{\frac{J_m^2(\chi_{mp})}{J_m^2(d\chi_{mp})} - 1\right\}^{-\frac{1}{2}}.$$

Also

$$\lambda_{mp}^2 = k^2 - \chi_{mp}^2/b^2, \qquad \nu_{mp} = \chi_{mp}/b,$$

$$\alpha_{mp} = \frac{\delta k^2}{2\lambda_{mp}b}\left\{1 + \frac{J_m^2(\chi_{mp})}{dJ_m^2(d\chi_{mp})}\right\}\Big/\left\{\frac{J_m^2(\chi_{mp})}{J_m^2(d\chi_{mp})} - 1\right\}$$

assuming that both inner and outer cylinders are constructed from the same substance.

Tabular values of χ_{mp} occur in Table A.3 and a tolerably good approximation is

$$\chi_{mp} = p\pi/(d-1).$$

The cross-section of the coaxial line is not simply connected so the possibility of a TEM-mode cannot be excluded. Since χ_1 satisfies $\nabla_t^2\chi_1 = 0$ and is constant on $r = a, b$ it is independent of ϕ. Hence

$$\chi_1 = A + B \ln r$$

where A and B are constants. It follows that the nonzero components of the TEM-mode are

$$(\varepsilon_0/\mu_0)^{\frac{1}{2}} E_r = -e^{-ikz} B/r = H_\phi$$

when propagating in the direction of the positive z-axis. The attenuation, when the two are made of the same material, is given by

$$\alpha = \delta k(a+b)/4ab \ln(a/b). \tag{7.7}$$

The fundamental electromagnetic mode of the coaxial line is the TEM-mode and it propagates at all frequencies. The next mode to propagate is the TE_{11}-mode with a cut-off wavelength of $2\pi b/\chi'_{11}$ or $\pi(a+b)$ approximately. The coaxial line is therefore extremely useful for carrying signals with a wavelength greater than the TE_{11} cut-off and it has many applications in radar and television. As the frequency of the signal increases the dimensions of the line have to be reduced to prevent the TE_{11}-mode being stimulated. There is therefore a greater risk of break-down if the same power is being carried. Also the losses, as indicated by (7.7), become steadily larger as the frequency grows and there will be additional losses in the dielectric supports separating the two conductors. These shortcomings are absent in the circular and rectangular waveguides because, as can be checked, the attenuation is less than that of a coaxial line when the frequency is about 1.2 times the cut-off frequency. Thus, for wavelengths less than a few centimetres, it is more efficient to use a rectangular or circular waveguide than a coaxial line.

Approximate methods of calculating κ_m and λ_m when the cross-section is not simple have been developed. For details see Jones (1979).

5.8 Modes produced by a sound source

Suppose that a point source is placed at the point (x_0, y_0, z_0) inside a rigid tube. Then the equation to be solved for p is

$$\nabla^2 p + k^2 p = -\delta(\mathbf{x} - \mathbf{x}_0). \tag{8.1}$$

In addition the boundary condition $\partial p/\partial n = 0$ must be satisfied on C.

Put

$$p = \sum_{m=0}^{\infty} a_m(z)\psi_m(x, y)$$

where $a_m(z) = \int_{\mathscr{D}} p\psi_m \, dx \, dy$. The terms in the expansion meet the boundary condition on C. Also (8.1) entails

$$d^2a_m/dz^2 + \kappa_m^2 a_m = -\delta(z - z_0)\psi_m(x_0, y_0) \tag{8.2}$$

as in §5.1. Equation (8.2) tells us that $a_m(z)$ is continuous as z passes through z_0 but its derivative is discontinuous by an amount $-\psi_m(x_0, y_0)$. Therefore different forms of a_m are required in $z > z_0$ and $z < z_0$. Since we expect the waves to be going away from the source at infinity we choose

$$a_m = \begin{cases} Ae^{-i\kappa_m z} & (z > z_0) \\ Be^{i\kappa_m z} & (z < z_0). \end{cases}$$

For a_m to be continuous at $z = z_0$ we must have

$$Ae^{-i\kappa_m z_0} = Be^{i\kappa_m z_0}$$

and, to achieve the discontinuity of the derivative,

$$-i\kappa_m Ae^{-i\kappa_m z_0} - i\kappa_m Be^{i\kappa_m z_0} = -\psi_m(x_0, y_0).$$

Thus $a_m = e^{-i\kappa_m|z-z_0|}\psi_m(x_0, y_0)/2i\kappa_m$ solves (8.2) and the desired solution of (8.1) is

$$p = \sum_{m=0}^{\infty} (1/2i\kappa_m)\psi_m(x, y)\psi_m(x_0, y_0)e^{-i\kappa_m|z-z_0|} \tag{8.3}$$

assuming, of course, that k does not coincide with any μ_m.

The formula (8.3) can be employed as a Green's function for tackling more complicated distributions of sources.

Junctions

The distribution of energy to various places may be facilitated by splitting it from one pipe into several. Also tubes can carry energy usefully only if they can be connected to transmitting and receiving mechanisms or radiating devices. Such couplings will influence the fields in guides and the next few sections are concerned with relevant theory.

5.9 General junction

A typical junction is depicted in Fig. 5.7. It consists of a perforated surface S_0 with guides attached at the apertures. The extent of the

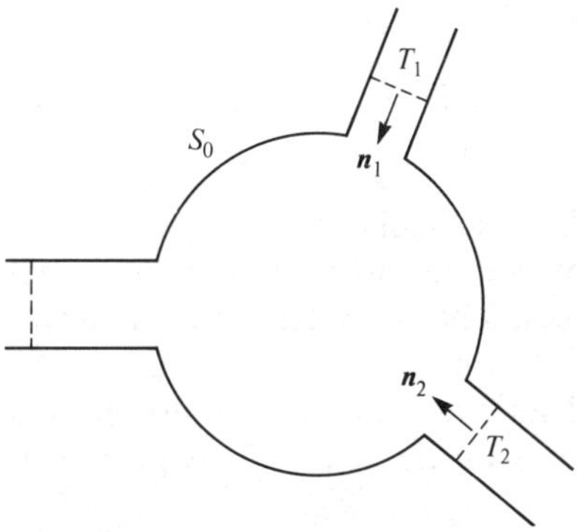

Fig. 5.7 A general junction

junction is a matter of one's convenience since personal preference plays a role in deciding how far a junction penetrates into a particular guide. Select *terminal planes* $T_1, T_2, \ldots, T_L$ so that there is one in each of the L guides. Each terminal plane is perpendicular to the axis of the guide in which it is placed. The terminal planes are sufficiently far from the apertures in S_0 for any non-propagating modes which have been generated to have died away. Evidently, the positions of the terminal planes are, to a large degree, arbitrary since they could always be put further down the guides without violating our criterion. At any rate, the field on a terminal plane can be expressed entirely in terms of propagating modes; for simplicity, only the fundamental will be considered but the theory can be extended in a straightforward manner to the case of several propagating modes. It will also be helpful in the following to regard S_0 as including the walls of the guides as far as the terminal planes. Then $S_0 \cup T_1 \cup \ldots \cup T_L$ is a closed surface. The region within this closed surface may contain other substances, dielectrics, conductors and the like; if there are such materials they will be assumed to be linear and passive.

Let ψ_l be the fundamental mode in the lth guide. Note that the suffix is here used to indicate the tube number and not the modal type. For the sound wave on the terminal T_l

$$p(l) = V_l \psi_l, \qquad v_z(l) = I_l \psi_l$$

in the representation of §5.5. The positive z-axis in the lth tube is in the direction of the unit vector $\boldsymbol{n}_l$ shown in Fig. 5.7. Unless ω corresponds to an eigenfrequency (which may be complex when there are losses present) the field inside $S_0 \cup T_1 \cup \ldots \cup T_L$ is completely determined by a knowledge of p or the normal component of $\boldsymbol{v}$ on $S_0 \cup T_1 \cup \ldots \cup T_L$ (see §9.1). We assume that ω is not such that there are eigenoscillations; then the values of $V_1, \ldots, V_L$ determine those of $I_1, \ldots, I_L$ and vice versa. Hence, by the assumed linearity,

$$V_m = \sum_{l=1}^{L} Z_{ml} I_l \qquad (m = 1, \ldots, L) \tag{9.1}$$

for some suitable Z_{ml}. The similarity of these equations to those occurring for lumped-constant circuits suggests that the coefficients Z_{ml} might be viewed as impedances. That they have the requisite properties of impedances will now be shown.

Let S be a closed surface with outward unit normal $\boldsymbol{n}$. Then, if $p^{(1)}$, $\boldsymbol{v}^{(1)}$ and $p^{(2)}$, $\boldsymbol{v}^{(2)}$ are two acoustic fields satisfying the same boundary conditions,

$$\int_S \{p^{(1)}\boldsymbol{v}^{(2)} - p^{(2)}\boldsymbol{v}^{(1)}\} \cdot \boldsymbol{n} \, \mathrm{d}S = 0 \tag{9.2}$$

as in the reciprocity theorem of §1.35. Take S to be the surface $S_0 \cup T_1 \cup \ldots \cup T_L$ and suppose that the boundary conditions make the integral over S_0 vanish (as they will if the boundaries are hard or soft). Then

$$\sum_{m=1}^{L} V_m^{(1)} I_m^{(2)} = \sum_{m=1}^{L} V_m^{(2)} I_m^{(1)}.$$

Choose $I_m^{(1)} = 0$ $(m \neq 1)$ and $I_m^{(2)} = 0$ $(m \neq 2)$. The equation reduces, via (9.1), to

$$Z_{21} I_1^{(1)} I_2^{(2)} = Z_{12} I_2^{(2)} I_1^{(1)}$$

whence $Z_{21} = Z_{12}$. By carrying out this process for every pair of suffixes we conclude that

$$Z_{ml} = Z_{lm}. \tag{9.3}$$

It has been shown already in §1.28 that, if T is the volume enclosed by S,

$$\begin{aligned} \int_S p\boldsymbol{v}^* \cdot \boldsymbol{n} \, \mathrm{d}S &= -\mathrm{i}\omega \int_T (\rho_0 |\boldsymbol{v}|^2 - |p|^2/\rho_0 a^2) \, \mathrm{d}\boldsymbol{x} \\ &= -4\mathrm{i}\omega (W_\mathrm{K} - W_\mathrm{P}) \end{aligned} \tag{9.4}$$

where W_K and W_P are the average kinetic and potential energies. Apply this result to the region bounded by $S_0 \cup T_1 \cup \ldots \cup T_L$. If S_0 is hard or soft there is no contribution from the integral over it. If it is dissipative there will be some power flow through it. Then

$$\int_{T_1 \cup \ldots \cup T_L} p\boldsymbol{v}^* \cdot \boldsymbol{n} \, \mathrm{d}S = 4\mathrm{i}\omega (W_\mathrm{P} - W_\mathrm{K}) - 2P'$$

where P' is the power loss. Hence

$$\sum_{m=1}^{L}\sum_{l=1}^{L} Z_{ml}I_m^*I_l = \sum_{m=1}^{L} V_m I_m^* = 4\mathrm{i}\omega(W_K - W_P) + 2P' \tag{9.5}$$

because of the difference in directions of $\boldsymbol{n}_l$ and $\boldsymbol{n}$.

Let $Z_{ml} = R_{ml} + \mathrm{i}X_{ml}$ where R_{ml} and X_{ml} are real. Then, because of (9.3), (9.5) implies that

$$\mathrm{Re} \sum_{m=1}^{L}\sum_{l=1}^{L} R_{ml}I_m^*I_l = 2P'.$$

If there are no losses present this forces $R_{ml} = 0$. When there is dissipation the quadratic form on the left must be positive definite for any choice of I_m. This requires

$$R_{11} \geqslant 0, \qquad \begin{vmatrix} R_{11} & R_{12} \\ R_{21} & R_{22} \end{vmatrix} \geqslant 0, \ldots .$$

These properties of the coefficients Z_{mn} are exactly analogous to those of lumped-circuit theory and permit the representation of the junction in terms of a lumped-constant circuit. The equivalent circuit arrived at depends upon the positioning of the terminal planes and a careful choice of these often leads to a simplified circuit. The matrix

$$\boldsymbol{Z} \equiv \begin{pmatrix} Z_{11} & Z_{12} & \ldots & Z_{1L} \\ Z_{21} & Z_{22} & \ldots & Z_{2L} \\ & \cdots\cdots & & \\ Z_{L1} & Z_{L2} & \ldots & Z_{LL} \end{pmatrix}$$

is called the *impedance matrix* of the junction. It has been demonstrated above that every element of $\boldsymbol{Z}$ is purely imaginary when there are no losses.

The discussion of the electromagnetic junction treads a parallel path. On T_l

$$\boldsymbol{E}_t(l) = V_l\boldsymbol{e}_t(l), \qquad \boldsymbol{H}_t(l) = I_l\boldsymbol{h}_t(l), \qquad \boldsymbol{h}_t = \boldsymbol{n}_l \wedge \boldsymbol{e}_t$$

in the notation of §5.4. The tangential component of $\boldsymbol{E}$ or that of $\boldsymbol{H}$ is sufficient to determine the interior field if ω is not an eigenfrequency (§9.1). Thus (9.1) is reached again. The analogue of (9.2) is

$$\int_S (\boldsymbol{E}^{(1)} \wedge \boldsymbol{H}^{(2)} - \boldsymbol{E}^{(2)} \wedge \boldsymbol{H}^{(1)}) \cdot \boldsymbol{n} \, \mathrm{d}S = 0 \tag{9.6}$$

(c.f. §1.35) and of (9.4) is

$$\int_S \boldsymbol{E} \wedge \boldsymbol{H}^* \cdot \boldsymbol{n} \, \mathrm{d}S = 4\mathrm{i}\omega(W_E - W_H) - 2P' \tag{9.7}$$

(§1.30), where W_{E} and W_{H} are the average stored electric and magnetic energies while P' is the average dissipated power. The net result is that there is no change to the properties of the impedance matrix $\boldsymbol{Z}$.

The equations (9.1) could be inverted so that

$$I_m = \sum_{l=1}^{L} Y_{ml} V_l \quad (m = 1, \ldots, L). \tag{9.8}$$

The matrix $\boldsymbol{Y}$ defined by

$$\boldsymbol{Y} \equiv \begin{pmatrix} Y_{11} & Y_{12} & \cdots & Y_{1L} \\ Y_{21} & Y_{22} & \cdots & Y_{2L} \\ & \cdots & & \\ Y_{L1} & Y_{L2} & \cdots & Y_{LL} \end{pmatrix}$$

is known as the *admittance matrix*. Obviously $\boldsymbol{Y} = \boldsymbol{Z}^{-1}$ and $Y_{ml} = Y_{lm}$. Consequently, a junction could be represented in terms of admittances. Whether or not the ensuing circuit is simpler than that for impedances can only be discovered by detailed investigation.

5.10 The scattering matrix

There is an alternative way of characterizing a junction in which the wave reaching T_l from the guide (called *the incident wave*) is distinguished from the one coming to T_l from the junction (designated *the reflected wave*). It will be sufficient to deal with the electromagnetic junction since the acoustic analysis follows a parallel course.

Let a_l be the amplitude of the incident wave and b_l that of the reflected wave. Then, from §5.4,

$$\mathbf{E}_t(l) = (a_l + b_l)\mathbf{e}_t(l), \qquad \mathbf{H}_t(l) = (a_l - b_l)\mathbf{h}_t(l)/Z_l$$

where Z_l is the mode impedance of the sole propagating wave in the lth guide. From the preceding section

$$a_m = \tfrac{1}{2}(V_m + Z_m I_m) = \tfrac{1}{2} \sum_{l=1}^{L} (Z_{ml} + \delta_{ml} Z_m) I_l \quad (m = 1, \ldots, L).$$

These equations can be expressed in matrix form by writing $\mathbf{A}$ and $\boldsymbol{I}$ for the vectors with components $a_1, \ldots, a_L$ and $I_1, \ldots, I_L$ respectively. Thus

$$\mathbf{A} = \tfrac{1}{2}(\boldsymbol{Z} + \boldsymbol{Z}_0)\boldsymbol{I} \tag{10.1}$$

where $\boldsymbol{Z}_0$ is the diagonal matrix with elements $Z_1, \ldots, Z_L$. Similarly, if $\boldsymbol{B}$ has components $b_1, \ldots, b_L$,

$$\begin{aligned} \boldsymbol{B} &= \tfrac{1}{2}(\boldsymbol{Z} - \boldsymbol{Z}_0)\boldsymbol{I} \\ &= (\boldsymbol{Z} - \boldsymbol{Z}_0)(\boldsymbol{Z} + \boldsymbol{Z}_0)^{-1}\mathbf{A} \end{aligned}$$

from (10.1). Hence

$$\boldsymbol{B} = \boldsymbol{SA} \tag{10.2}$$

where

$$\boldsymbol{S} = (\boldsymbol{Z} - \boldsymbol{Z}_0)(\boldsymbol{Z} + \boldsymbol{Z}_0)^{-1}. \tag{10.3}$$

The matrix $\boldsymbol{S}$ is called the *scattering matrix*. It gives the amplitudes of the reflected waves in terms of the incident waves or conversely. The formula (10.3) may also be written in terms of the admittance matrix as

$$\boldsymbol{S} = (\boldsymbol{I} - \boldsymbol{Z}_0\boldsymbol{Y})(\boldsymbol{I} + \boldsymbol{Z}_0\boldsymbol{Y})^{-1}$$

where $\boldsymbol{I}$ is the unit matrix.

The symmetry of $\boldsymbol{Z}$ and $\boldsymbol{Z}_0$ guarantees that $\boldsymbol{S}$ is symmetric, i.e. $\boldsymbol{S}^{\mathrm{T}} = \boldsymbol{S}$ where $\boldsymbol{S}^{\mathrm{T}}$ is the transpose of $\boldsymbol{S}$. Also, the electromagnetic analogue of (9.5) supplies by virtue of the reality of Z_l

$$\sum_{m=1}^{L} (a_m a_m^* - b_m b_m^*)/Z_m = 2P'.$$

Substitution from (10.2) gives

$$\mathbf{A}^{\mathrm{T}*}\boldsymbol{Y}_0^{\frac{1}{2}}(\boldsymbol{I} - \boldsymbol{S}^*\boldsymbol{S}^{\mathrm{T}})\boldsymbol{Y}_0^{\frac{1}{2}}\mathbf{A} = 2P'$$

where $\boldsymbol{Y}_0^{\frac{1}{2}}$ is a diagonal matrix with elements $1/Z_1^{\frac{1}{2}}, \ldots$. When $P' > 0$ we infer from the quadratic form on the left that all the principal minors of $\boldsymbol{I} - \boldsymbol{S}^*\boldsymbol{S}$ (remember $\boldsymbol{S}$ is symmetric) are positive. If $P' = 0$, then

$$\boldsymbol{S}^*\boldsymbol{S} = \boldsymbol{I} \tag{10.4}$$

of necessity. Equation (10.4) is the definition of a symmetric *unitary matrix*. Hence the scattering matrix of a lossless junction is symmetric and unitary.

Description of a junction through any of the impedance, admittance, and scattering matrices is permissible. Depending on the purpose in hand one description may be more suitable than the others but there is no universal rule for always picking one. The impedance and admittance descriptions have the great advantage of schematic representation. The circuit elements are familiar to engineers and the laws for their combination and movement from one reference plane to another are well known. On the other hand, a change of terminal plane for the scattering matrix involves only a simple phase shift; in many experiments the reflection coefficient can be measured virtually directly. Whichever description seems most convenient will be adopted subsequently.

5.11 T-junctions

A junction at which three guides are connected is known as a T-*junction*. Some interesting theorems can be derived about lossless T-junctions. A *short circuit* in an electromagnetic waveguide is a perfectly conducting plane perpendicular to the axis of the guide which completely fills the cross-section. The analogue for sound waves in rigid tubes is a rigid plane perpendicular to the axis.

The first theorem states that *a short circuit can always be placed in one guide of a T-junction such that no power can be transmitted between the other two guides.* Place the short circuit in guide 3 beyond the terminal plane T_3. Its effect is equivalent to a certain impedance Z' at T_3, i.e.

$$V_3 = -Z' I_3$$

(currents and voltages are positive when they represent power flow into the network). By adjusting the position of the short circuit the magnitude of Z', which is purely imaginary, can be varied from $-\infty$ to ∞.

Elimination of V_3 and I_3 from (9.1) leads to relations between the voltages and currents in the other two guides governed by the impedance matrix

$$\begin{pmatrix} Z'_{11} & Z'_{12} \\ Z'_{12} & Z'_{22} \end{pmatrix}$$

where $Z'_{11} = Z_{11} - Z_{13}^2/(Z_{33} + Z')$, $Z'_{12} = Z_{12} - Z_{13}Z_{23}/(Z_{33} + Z')$, $Z'_{22} = Z_{22} - Z_{23}^2/(Z_{33} + Z')$. In view of the possible adjustment of Z' a position of the short circuit can always be found for which $Z'_{12} = 0$. But then there is no coupling between guides 1 and 2, and hence no power flow between them.

If the T-junction is *symmetrical* about guide 3 *the short circuit can be arranged so that an incident wave in guide 1 produces no reflected wave in that guide.* Since there is no incident wave in guide 2 this is feasible if $V_1 = Z_1 I_1$ and $V_2 = -Z_1 I_2$ hold simultaneously. This can be organized by placing the short circuit so that Z_3 makes

$$Z'^2_{11} - Z'^2_{12} = Z_1^2.$$

When there is no short circuit *there is at least one arm in which an incident wave will produce a reflected wave in itself.* Suppose there were no such arm. Then all the diagonal components of the scattering matrix $\mathbf{S}$ would be zero. Calculation of the diagonal elements of $\mathbf{S}^*\mathbf{S}$ and imposition of (10.4) then requires $|S_{12}|^2 = |S_{13}|^2 = |S_{32}|^2 = \frac{1}{2}$. In contrast, the off-diagonal elements of $\mathbf{S}^*\mathbf{S}$ necessitate $S_{13}^* S_{23} = 0$. The evident inconsistency confirms the earlier statement.

5.12 Directional couplers

A junction of four guides such that a wave incident in guide 1 produces only waves leaving in guides 2 and 4 whereas an incident wave in guide 2 generates only waves leaving in guides 1 and 3 is called a *directional*

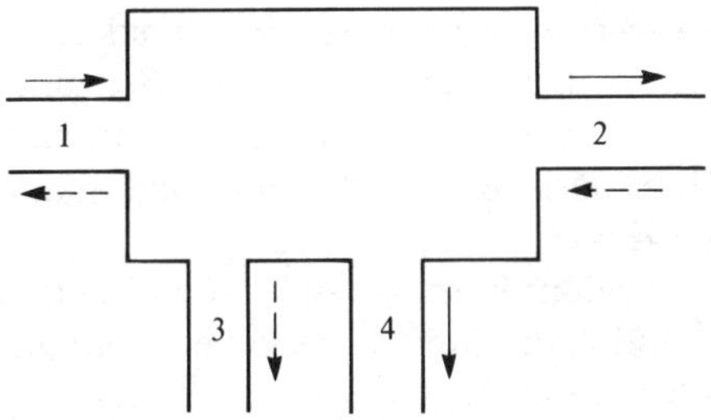

Fig. 5.8 The directional coupler

coupler (Fig. 5.8). The scattering matrix of a directional coupler must be such that

$$S_{11}=0, \qquad S_{22}=0, \qquad S_{13}=0, \qquad S_{24}=0.$$

In conjunction with $\boldsymbol{S}$ being unitary these imply that

$$|S_{12}|^2+|S_{23}|^2=1, \qquad |S_{12}|^2+|S_{14}|^2=1, \tag{12.1}$$

$$S_{23}^*S_{33}=0, \qquad S_{14}^*S_{44}=0. \tag{12.2}$$

If $S_{33}\neq 0$, (12.2) makes $S_{23}=0$ and then $|S_{12}|^2=1$ from the first of (12.1). The second of (12.1) then shows that $S_{14}=0$. In this case the pair of guides 1 and 2 are isolated from the pair 3 and 4. This is not a genuine directional coupler but two junctions, each of two guides, side by side. This possibility is therefore rejected and so is that in which $S_{44}\neq 0$.

Accordingly, a true directional coupler has the property

$$S_{33}=0, \qquad S_{44}=0.$$

Evidently, a wave incident in guide 3 will leave via guides 2 and 4 whereas one originating in 4 will depart through 1 and 3. From (12.1), $|S_{23}|^2=|S_{14}|^2$ so that the power leaving 4 for unit incident power in 1 is the same as that transmitted through 3 when the incident power in 2 is unity.

Equations additional to (12.1) and (12.2) are needed to make $\boldsymbol{S}$ unitary. The missing ones are

$$S_{12}^*S_{23}+S_{14}^*S_{43}=0, \qquad S_{21}^*S_{14}+S_{23}^*S_{34}=0, \tag{12.3}$$

$$|S_{32}|^2+|S_{34}|^2=1. \tag{12.4}$$

By virtue of (12.1) and (12.4), $|S_{12}|^2=|S_{34}|^2$ so that a wave incident in 1 passes on the same fraction of its power to 2 that a wave in 3 transmits to 4.

The solution of (12.1), (12.3), (12.4) is

$$S_{12}=A\mathrm{e}^{\mathrm{i}\alpha}, \qquad S_{23}=(1-A^2)^{\frac{1}{2}}\mathrm{e}^{\mathrm{i}\gamma}$$
$$S_{34}=A\mathrm{e}^{\mathrm{i}\beta}, \qquad S_{14}=-(1-A^2)^{\frac{1}{2}}\mathrm{e}^{\mathrm{i}(\alpha+\beta-\gamma)}$$

where $0<A<1$ to warrant a genuine directional coupler. An alteration of T_1 merely changes all the phases of S_{11}, S_{13}, and S_{14} by the same amount. Therefore the real α, β, γ can be allotted any desired values by manipulation of the positions of the terminal planes; one particular allocation might be $\alpha=0$, $\beta=0$, $\gamma=\frac{1}{2}\pi$.

When the power incident in the guide 1 of a directional coupler divides equally between 2 and 4 the junction is called a *magic T*. For a magic T $A=1/2^{\frac{1}{2}}$ and we elect to make

$$\alpha=\beta=\gamma=\tfrac{1}{2}\pi.$$

The formula for the scattering matrix is

$$\mathbf{S}=\frac{\mathrm{i}}{2^{\frac{1}{2}}}\begin{pmatrix} 0 & 1 & 0 & -1 \\ 1 & 0 & 1 & 0 \\ 0 & 1 & 0 & 1 \\ -1 & 0 & 1 & 0 \end{pmatrix}.$$

Consequently $\mathbf{S}=\mathrm{i}\mathbf{X}$ where $\mathbf{X}$ is real. Invoking (10.4) we have $\mathbf{X}^2=\mathbf{I}$ and hence

$$\mathbf{S}^2=-\mathbf{I}. \tag{12.5}$$

From (10.3)

$$\mathbf{S}(\mathbf{Z}+\mathbf{Z}_0)=\mathbf{Z}-\mathbf{Z}_0$$

whence

$$(\mathbf{S}-\mathbf{I})\mathbf{Z}=-\mathbf{Z}_0-\mathbf{S}\mathbf{Z}_0=(\mathbf{S}-\mathbf{I})\mathbf{S}\mathbf{Z}_0$$

from (12.5). The imaginary nature of $\mathbf{S}$ ensures that $\mathbf{S}-\mathbf{I}$ is non-singular and therefore

$$\mathbf{Z}=\mathbf{S}\mathbf{Z}_0.$$

Similarly, $\mathbf{Y}=-\mathbf{Y}_0\mathbf{S}$. These formulae enable an easy determination of the equivalent circuit for the magic T. The convenable properties of the magic T make it a popular candidate in the design of directional couplers.

Matrix elements

Quantitative results for the matrices can only be arrived at by finding their elements through the solution of appropriate boundary value problems. Some of the analytical methods which have been used will now be illustrated. Usually, it will be assumed that the dimensions are such that the fundamental is the single propagating mode.

5.13 The source method

The first problem to be tackled is the effect of the insertion of a soft post on the fundamental acoustic mode in a rigid tube. The cross-section will be taken to be rectangular and a soft cylindrical post parallel to the shorter side connects the two wider faces (Fig. 5.9). The fundamental mode is independent of y and so its impact on the post may be expected to generate a field which is independent of y. Thus only modes which vary with x need to be considered.

Suppose now that the post is *thin*. The surface field induced on it may, to a first approximation, be regarded as concentrated on the axis of the cylinder and forming a line source. If the axis lies along $x = x_0$, $z = z_0$ the total field may be taken as, in view of (8.3),

$$p = \mathrm{e}^{-\mathrm{i}kz} + A \sum_{m=0}^{\infty} (1/\kappa_{m0})\mathrm{e}^{-\mathrm{i}\kappa_{m0}|z-z_0|} \cos(m\pi x/a)\cos(m\pi x_0/a)$$

where A is a constant to be determined and $\kappa_{m0}^2 = k^2 - (m\pi/a)^2$.

A plan view of the post is displayed in Fig. 5.10. In order to meet the soft boundary condition $p = 0$ on $x = x_0 + \frac{1}{2}d \sin\phi$, $z = z_0 + \frac{1}{2}d \cos\phi$ where d is the diameter of the post. Thus

$$\exp\{-\mathrm{i}k(z_0 + \tfrac{1}{2}d\cos\phi)\} + A \sum_{m=0}^{\infty} \frac{1}{\kappa_{m0}} \mathrm{e}^{-\mathrm{i}\kappa_{m0}\frac{1}{2}d|\cos\phi|}$$
$$\times \cos\frac{m\pi x_0}{a} \cos\frac{m\pi}{a}(x_0 + \tfrac{1}{2}d\sin\phi) = 0$$

for $0 \leqslant \phi \leqslant 2\pi$. Now, as ϕ varies, the left-hand side alters by at most $O(d)$. Since we are concerned with a thin post such corrections can be omitted and a calculation undertaken for some convenient ϕ. An obvious one is $\phi = \frac{1}{2}\pi$ and then

$$\mathrm{e}^{-\mathrm{i}kz_0} + A \sum_{m=0}^{\infty} (1/\kappa_{m0})\cos(m\pi x_0/a)\cos m\pi(x_0 + \tfrac{1}{2}d)/a = 0.$$

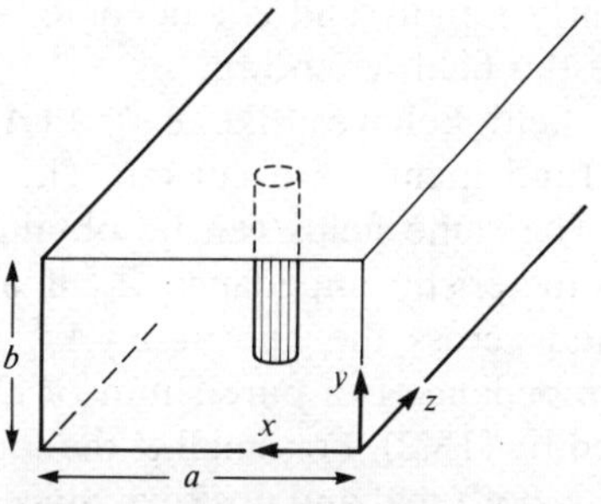

Fig. 5.9 The soft post

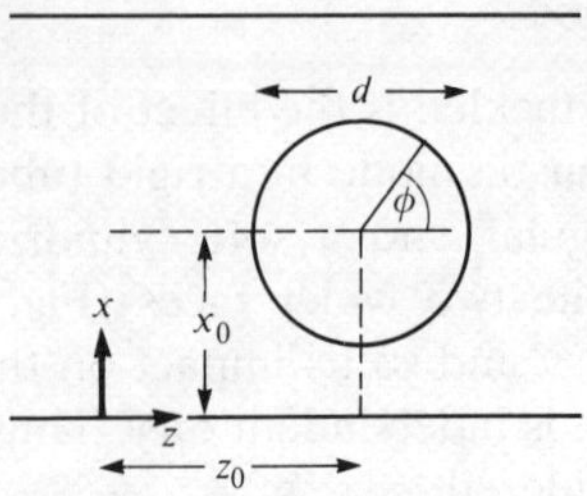

Fig. 5.10 Plan of the thin post

To estimate the sum note that $\kappa_{m0} \to -im\pi/a$ as $m \to \infty$. Also

$$\begin{aligned}\sum_{m=1}^{\infty} (1/m)\cos(m\pi x_0/a)\cos m\pi(x_0+\tfrac{1}{2}d)/a \\ &= \sum_{m=1}^{\infty} \frac{1}{4m}\{e^{im\pi(4x_0+d)/2a}+e^{-im\pi(4x_0+d)/2a}+e^{im\pi d/2a}+e^{-im\pi d/2a}\} \\ &= -\tfrac{1}{4}\ln[\{1-e^{i\pi(4x_0+d)/2a}\}\{1-e^{-i\pi(4x_0+d)/2a}\}] \\ &\quad -\tfrac{1}{4}\ln[(1-e^{i\pi d/2a})(1-e^{-i\pi d/2a})] \\ &= -\tfrac{1}{4}\ln[16\sin^2\{\pi(4x_0+d)/4a\}\sin^2(\pi d/4a)] \\ &= -\tfrac{1}{2}\ln\left(\frac{\pi d}{a}\sin\frac{\pi x_0}{a}\right)\end{aligned}$$

if terms of $O(d)$ are rejected. Hence, ignoring terms of $O(d)$, we see that

$$e^{-ikz_0}+\frac{A}{k}\left(1+\frac{2iX}{Z_0}\right)=0 \tag{13.1}$$

where

$$\frac{X}{Z_0}=-\frac{ka}{4\pi}\ln\left(\frac{\pi d}{a}\sin\frac{\pi x_0}{a}\right)-\tfrac{1}{2}k\sum_{m=1}^{\infty}\left(\frac{i}{\kappa_{m0}}+\frac{a}{m\pi}\right)\cos^2\frac{m\pi x_0}{a}, \tag{13.2}$$

Z_0 being the mode impedance $\omega\rho_0/k$ of the fundamental. The series in (13.2) converges fairly rapidly and is amenable to numerical evaluation. Thus A, and hence the field, is known.

As $z \to -\infty$ the field behaves like $e^{-ikz}+(A/k)e^{ik(z-z_0)}$ because all modes except the fundamental are cut-off. The behaviour as $z \to \infty$ is $(1+Ae^{ikz_0}/k)e^{-ikz}$. The same fields can be obtained in an infinite transmission line of characteristic impedance Z_0 if an impedance $-\tfrac{1}{2}Z_0(1+ke^{-ikz_0}/A)$ is shunted across the line at $z=z_0$. By virtue of (13.1) we conclude that the impedance is a pure shunt of iX placed at $z=z_0$ (Fig. 5.11), X being given by (13.2). For small d the logarithmic term in (13.2) dominates. Thus X is both real and positive. Since an inductance provides a positive impedance the post may be referred to as being *inductive*.

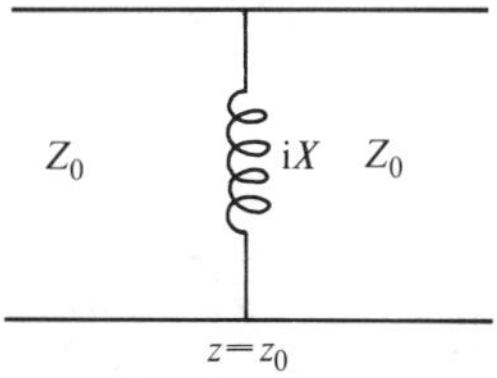

Fig. 5.11 The impedance corresponding to the thin post

The procedure can be adapted to retain higher powers of d by introducing a line doublet, a quadrupole, etc. on the axis of the post but details are omitted. It is also possible to replace the soft boundary condition by others such as $p = Z_\nu\, \partial p/\partial n$.

5.14 Integral equations

Another method of attack is based on integral equations. This will be illustrated by the electromagnetic problem of an *iris*, which is a thin sheet of perfectly conducting metal placed across part of the cross-section of a waveguide. The particular iris to be considered is that formed by positioning strips of metal parallel to the longer side of a rectangular waveguide, the plane of the strips being $z=0$ (Fig. 5.12). The dimensions of the guide permit only the fundamental TE_{10}-mode to propagate. In the TE_{10}-mode $E_x=0$ and E_y varies only as $\sin(\pi x/a)$ (§5.7(b)). On account of the shape of the iris we seek a total field of the same type. Such a field can be constructed by a combination of TE- and TM-modes; it is

$$E_x = 0, \qquad E_y = \left\{ e^{-i\kappa_{10}z} + \sum_{n=0}^{\infty} a_n e^{-i\kappa_{1n}|z|} \cos(n\pi y/b) \right\} \sin(\pi x/a),$$

$$E_z = \sum_{n=1}^{\infty} (in\pi/b\kappa_{1n}) a_n e^{-i\kappa_{1n}|z|} \sin(n\pi y/b) \sin(\pi x/a) \operatorname{sgn} z,$$

$$-i\omega\mu_0 H_x = \left\{ i\kappa_{10} e^{-i\kappa_{10}z} + \sum_{n=0}^{\infty} (i\kappa_{10}^2/\kappa_{1n}) a_n e^{-i\kappa_{1n}|z|} \right.$$

$$\left. \times \cos(n\pi y/b) \operatorname{sgn} z \right\} \sin(\pi x/a),$$

$$-i\omega\mu_0 H_y = -\sum_{n=1}^{\infty} (in\pi^2/ab\kappa_{1n}) a_n e^{-i\kappa_{1n}|z|} \sin(n\pi y/b) \cos(\pi x/a) \operatorname{sgn} z,$$

$$-i\omega\mu_0 H_z = (\pi/a) \left\{ e^{-i\kappa_{10}z} + \sum_{n=0}^{\infty} a_n e^{-i\kappa_{1n}|z|} \cos(n\pi y/b) \right\} \cos(\pi x/a)$$

where $\kappa_{1n}^2 = k^2 - (n\pi/b)^2 - (\pi/a)^2$ and $\operatorname{sgn} z$ is 1 for $z>0$ and -1 for $z<0$.

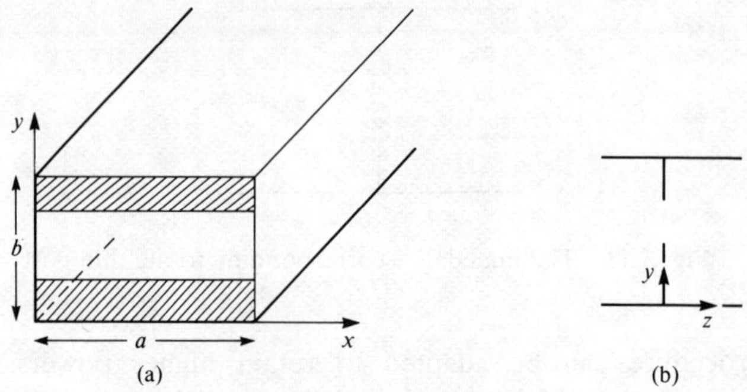

Fig. 5.12 (a) The capacitive iris; (b) view in the direction of the x-axis

In E_y the ratio of the amplitude of the reflected wave to the incident is a_0 and for the transmitted field is $1+a_0$. Thus, as in §5.13, the iris is equivalent to a shunt iX at $z=0$ in a transmission line of characteristic impedance Z_0, where now Z_0 is the impedance of the TE_{10}-mode, with

$$\mathrm{i}X/Z_0 = -\tfrac{1}{2}(1+1/a_0). \tag{14.1}$$

In the derivation of a_0 let $\mathcal{S}$ be the perfectly conducting metal portion of the iris and $\mathcal{A}$ the aperture. On $z=0$ let E_y be written as $E(y)\sin(\pi x/a)$. Then all conditions will be fulfilled if $E(y)$ is zero on $\mathcal{S}$. There are two ways of attaining this via integral equations. One is to formulate an integral equation over $\mathcal{A}$ and the other over $\mathcal{S}$. The former will be considered first and the latter second.

From §4.1

$$1+a_0 = (1/b)\int_0^b E(t)\,\mathrm{d}t = (1/b)\int_{\mathcal{A}} E(t)\,\mathrm{d}t, \tag{14.2}$$

$$a_n = (2/b)\int_{\mathcal{A}} E(t)\cos(n\pi t/b)\,\mathrm{d}t \quad (n \geqslant 1) \tag{14.3}$$

since $E(y)$ vanishes on $\mathcal{S}$. As z crosses the aperture H_x must be continuous and this is possible only if

$$\sum_{n=0}^{\infty} (a_n/\kappa_{1n})\cos(n\pi y/b) = 0 \tag{14.4}$$

for y in $\mathcal{A}$. Equation (14.4), after a derivative with respect to y, also ensures that H_y and E_z are continuous through the aperture.

Substitution of (14.2) and (14.3) in (14.4) leads for $E(y)$ to the integral equation

$$a_0 + \int_{\mathcal{A}} E(t) \sum_{n=1}^{\infty} (2\kappa_{10}/\kappa_{1n}b)\cos(n\pi y/b)\cos(n\pi t/b)\,\mathrm{d}t = 0 \tag{14.5}$$

with y in $\mathcal{A}$.

Normally, a field must be expected to be complex. However, in this special case, note that κ_{10} is real and $\kappa_{1n} = -i\kappa_n$ $(n \geqslant 1)$ where κ_n is real. Therefore the change

$$E(t) = \tfrac{1}{2} b i a_0 g(t) \tag{14.6}$$

results in

$$1 - \int_{\mathscr{A}} g(t) \sum_{n=1}^{\infty} (\kappa_{10}/\kappa_n) \cos(n\pi y/b) \cos(n\pi t/b)\, dt = 0 \tag{14.7}$$

for y in $\mathscr{A}$. All quantities in (14.7) are real so that g is real.

Having determined g from (14.7) we find a_0 through (14.2) and (14.6), i.e.

$$1 + a_0 = \tfrac{1}{2} i a_0 \int_{\mathscr{A}} g(t)\, dt.$$

From (14.1)

$$X/Z_0 = -\tfrac{1}{4} \int_{\mathscr{A}} g(t)\, dt. \tag{14.8}$$

Observe that multiplication of (14.7) by g(y) and integration over $\mathscr{A}$ discloses that $\int_{\mathscr{A}} g(t)\, dy$ is positive. Thus X is negative and the equivalent impedance of the iris is *capacitive*.

The source of the second integral equation is the current in the metal of the iris. A quantity proportional to this current is

$$J(y) \sin(\pi x/a) = i\omega\mu_0\{(H_x)_{z=+0} - (H_x)_{z=-0}\}$$

from which

$$J(y) = 2 \sum_{n=0}^{\infty} (\kappa_{10}^2 / i\kappa_{1n}) a_n \cos(n\pi y/b).$$

Consequently

$$a_0 = (i/2\kappa_{10} b) \int_{\mathscr{S}} J(t)\, dt, \tag{14.9}$$

$$a_n = (i\kappa_{1n}/\kappa_{10}^2 b) \int_{\mathscr{S}} J(t) \cos(n\pi t/b)\, dt \quad (n \geqslant 1)$$

when (14.4) is recalled. To make $E(y)$ vanish on $\mathscr{S}$ requires

$$1 + \sum_{n=0}^{\infty} a_n \cos(n\pi y/b) = 0$$

whence

$$1 + a_0 + \sum_{n=1}^{\infty} (i\kappa_{1n}/\kappa_{10}^2 b) \int_{\mathscr{S}} J(t) \cos(n\pi t/b) \cos(n\pi y/b)\, dt = 0$$

for y in $\mathscr{S}$. Introducing a new unknown by

$$J(t) = (1 + a_0) \kappa_{10}^2 b g_1(t)$$

we obtain

$$1+\sum_{n=1}^{\infty} \kappa_n \int_{\mathscr{S}} g_1(t)\cos(n\pi t/b)\cos(n\pi y/b)\,dt = 0 \tag{14.10}$$

for y in $\mathscr{S}$. Also from (14.9) $a_0 = \frac{1}{2}i\kappa_{10}(1+a_0)\int_{\mathscr{S}} g_1(t)\,dt$ and so

$$Z_0/X = \kappa_{10}\int_{\mathscr{S}} g_1(t)\,dt. \tag{14.11}$$

Notice that (14.10) makes g_1 real and $\int_{\mathscr{S}} g_1(t)\,dt$ negative so there is no conflict with what has already been asserted about X.

The integral equations (14.7) and (14.10) cannot be solved in closed form and recourse to approximate techniques cannot be avoided. The standard method is to expand the unknown in a series of specified functions and generate algebraic equations for the coefficients. An indication of the steps involved is offered by the special case in which $\mathscr{A}$ extends from $y=0$ to $y=d$ whereas $\mathscr{S}$ occupies from $y=d$ to $y=b$. It must be stressed, however, that the process is intended to serve as an example; there is no obligation to select the particular series chosen here and other expansions, depending on one's aim, may be more suitable (see also next section).

The function g can be expanded in a Fourier cosine series appropriate to $\mathscr{A}$, i.e.

$$g(y) = \sum_{m=0}^{\infty} b_m \cos(m\pi y/d).$$

Multiply (14.7) by $\cos(r\pi y/d)$ and integrate over $\mathscr{A}$. Then

$$d - \sum_{m=0}^{\infty} b_m \sum_{n=1}^{\infty} K_{mn}K_{0n}\kappa_{10}/\kappa_n = 0, \tag{14.12}$$

$$\sum_{m=0}^{\infty} b_m \sum_{n=1}^{\infty} K_{mn}K_{rn}/\kappa_n = 0 \quad (r \geqslant 1) \tag{14.13}$$

where

$$K_{mn} = \int_0^d \cos(m\pi t/d)\cos(n\pi t/b)\,dt.$$

Actually we do not seem to be any further forward since (14.12) and (14.13) constitute an infinite set of equations for the coefficients b_m and resolving them exactly is a task of at least the same order of magnitude as solving (14.7). This suggests *truncation* in which the series for g is terminated at b_M and b_m is taken to be zero for $m > M$. At the same time all the equations in (14.13) in which $r > M$ are discarded. Thus we are left with M linear equations for the M unknown coefficients. One hopes that if M is sufficiently large a good enough approximation to g will be obtained. But this is by no means certain and infinite systems are known in which the solution of the truncated set does not tend to the solution of the infinite system. There have been numerous investigations of conditions that will assure a satisfactory approximation but it would take us too far afield to discuss them here (some details will be found in Jones (1979)). Unless M is small, numerical procedures will be vital and then other representations of g may be more profitable.

A flavour of the procedure may be gained by making the rough approximation $M=0$ which means estimating g as the constant b_0. Only (14.12) is retained and now

$$d-b_0\sum_{n=1}^{\infty}K_{0n}^2\kappa_{10}/\kappa_n=0.$$

It follows from (14.8) that in this approximation

$$X/Z_0=-d^2\Big/4\kappa_{10}\sum_{n=1}^{\infty}(K_{0n}^2/\kappa_n)$$
$$=-\pi^2d^2\Big/4\kappa_{10}b^2\sum_{n=1}^{\infty}(1/n^2\kappa_n)\sin^2(n\pi d/b).$$

Some idea of numerical values can be grasped from the special case in which $d=\frac{1}{2}b$ and $\kappa_{10}b\ll\pi$. Then $\kappa_n\approx n\pi/b$ and

$$X/Z_0\approx-\pi^3\Big/16\kappa_{10}b\sum_{m=0}^{\infty}(2m+1)^{-3}\approx-0.59\pi/\kappa_{10}b \tag{14.14}$$

since $\sum_{m=0}^{\infty}(2m+1)^{-3}\approx1.04$. It will be shown in the next section that X/Z_0 cannot be greater than the value on the right-hand side of (14.14).

There is an analogous path for (14.10). Here a one-term approximation might be

$$g_1(y)=C_0\cos\{\pi(b-y)/2(b-d)\}$$

which ensures that the current vanishes at the edge of the metal as it should (§9.2). Then

$$X/Z_0=-(1/\kappa_{10})\sum_{n=1}^{\infty}\kappa_n\cos^2(n\pi d/b)\{1-4n^2(b-d)^2/b^2\}^{-2}.$$

To compare with (14.14) take $d=\frac{1}{2}b$ and $\kappa_{10}b\ll\pi$. A limiting process is necessary for the term $n=1$ with the result that

$$\frac{X}{Z_0}\approx-\frac{\pi}{\kappa_{10}b}\left\{\frac{\pi^2}{16}+\sum_{n=1}^{\infty}\frac{2n}{(4n^2-1)^2}\right\}\approx-0.87\pi/\kappa_{10}b \tag{14.15}$$

since

$$\sum_{n=1}^{\infty}\frac{2n}{(4n^2-1)^2}=\frac{1}{4}\sum_{n=1}^{\infty}\left\{\frac{1}{(2n-1)^2}-\frac{1}{(2n+1)^2}\right\}=\frac{1}{4}.$$

In the next section it will be demonstrated that the true value of X/Z_0 must be greater than that in (14.15).

The mean of (14.14) and (14.15) is $-0.73\pi/\kappa_{10}b$ which should be compared with the correct value of $-0.71\pi/\kappa_{10}b$.

Often $\kappa_{10} \approx \pi/a$ and $b = a/3$ so that

$$\kappa_n = (n\pi/b)(1 - 1/9n^2)^{\frac{1}{2}}.$$

In these circumstances κ_n never differs from $n\pi/b$ by more than 10 per cent and the difference is less than 1 per cent for $n \geqslant 4$.

5.15 General theory and variational principles

This section will be devoted to placing the analysis of the previous section in the context of a general theory so that it will be applicable to other problems. If we write

$$Lg = \int_{\mathscr{A}} g(t) \sum_{n=1}^{\infty} (\kappa_{10}/\kappa_n) \cos(n\pi y/b) \cos(n\pi t/b) \, dt$$

the integral equation (14.7) is a particular example of the equation

$$Lg = f. \tag{15.1}$$

When g has been found from (14.7) an integral of it has to be calculated to determine X from (14.8). The proposal is to put this in other language so as to be capable of generalization. Introduce the *inner product* (h_1, h_2) defined by

$$(h_1, h_2) = \int_{\mathscr{A}} h_1(t) h_2(t) \, dt. \tag{15.2}$$

Then finding the integral of g can be expressed as calculating $(g, 1)$.

Thus the general problem to be considered is the solution of (15.1) when L is a linear operator, followed by the evaluation of (g, h) where h is given. Of course, once the mantle of generality has been donned, it is not obligatory to make the inner product that of (15.2). Any other inner product will be acceptable but it will be insisted that the inner product satisfy

$$(h_1, h_2) = (h_2, h_1). \tag{15.3}$$

The truncated series method of the preceding section starts by assuming an approximation G to g of the form

$$G = \sum_{n=1}^{N} b_n \varphi_n$$

where the φ_n are specified and the coefficients b_n are to be found. It is obviously desirable that the φ_n should be independent to evade unnecessary duplication of effort but there is no necessity for them to be orthogonal. If the approximation were an exact solution of (15.1) then

$$\sum_{n=1}^{N} b_n L\varphi_n = f \tag{15.4}$$

would hold. However, it is unlikely that one will be lucky enough to hit on this happy situation. Now, the next stage in dealing with (14.7) was to multiply the integral equation by one of the expansion functions and integrate over $\mathscr{A}$. Since we have replaced integration by the inner product the analogous operation for (15.4) is to take the inner product with φ_m. Thus we are led to

$$\sum_{n=1}^{N} b_n (L\varphi_n, \varphi_m) = (f, \varphi_m) \quad (m = 1, \ldots, N), \tag{15.5}$$

a system of N linear equations for the N unknowns $b_1, \ldots, b_N$. The determination of b_n by (15.5) is known as *Galerkin's method.* Once the b_n have been found (G, h) is taken as the approximation to the desired quantity (g, h).

The simplest possibility is $N = 1$ and then (15.5) reduces to $b_1(L\varphi_1, \phi_1) = (f, \varphi_1)$ with the consequence that

$$(G, h) = (\varphi_1, h)(f, \varphi_1)/(L\varphi_1, \varphi_1).$$

On many occasions L has the property that

$$(Lh_1, h_2) = (h_1, Lh_2). \tag{15.6}$$

Galerkin's method is then related to a *variational principle.* Let γ be such that

$$L\gamma = h. \tag{15.7}$$

Then

$$(g, h) = (g, L\gamma) = (Lg, \gamma) = (f, \gamma) \tag{15.8}$$

on invoking (15.6) and (15.1). Equation (15.8) constitutes a *reciprocity theorem* for the solutions involving f and h. Since $(f, \gamma) = (g, L\gamma)$ we can write

$$(g, h) = (g, h)(f, \gamma)/(g, L\gamma). \tag{15.9}$$

Suppose now that g is replaced on the right-hand side by $g + \eta g_0$ where η is a scalar parameter and g_0 is fixed. It is the intention to regard η as a small quantity and thus a *variation in* g is being introduced. The right-hand side becomes

$$\frac{(f, \gamma)}{(g, L\gamma)}\left[(g, h) + \eta\left\{(g_0, h) - (g_0, L\gamma)\frac{(g, h)}{(g, L\gamma)}\right\} + O(\eta^2)\right].$$

If γ satisfies (15.7) the coefficient of η is zero for all g_0 i.e. the expression (15.9) does not vary to the first order for small changes in g. Conversely, if the coefficient of η is zero for all g_0 we must have

$$(g, h)L\gamma = (g, L\gamma)h.$$

Now, if γ_0 is a solution of this so is $C\gamma_0$ where C is a constant. Choose C so that

$$(g, h) = (g, CL\gamma_0)$$

and then, with $\gamma = C\gamma_0$, $L\gamma = h$, i.e. (15.7) is satisfied.

Likewise, if small variations of γ are considered it is found that g satisfies (15.1). Hence the following *variational principle* has been shown: *a necessary and sufficient condition for* (15.1) *and* (15.7) *to be satisfied is that*

$$(g, h)(f, \gamma)/(g, L\gamma)$$

be stationary for small independent variations of g *and* γ *about their correct determination.*

The variational principle suggests that if g and γ are replaced by trial functions chosen to make the expression stationary a good approximation to (g, h) should be reached. At first sight, this technique seems to be superior to the simpler Galerkin's method but, in fact, it conveys precisely the same information, no more and no less.

To see this consider the trial functions

$$G = \sum_{m=1}^{N} c_m \varphi_m, \qquad \Gamma = \sum_{m=1}^{N} d_m \varphi_m.$$

The coefficients c_m and d_m are to be chosen so that small variations in them do not alter the variational expression to the first order. Choose a particular d_n and change it slightly to $d_n + \eta_n$. There will be no first order alteration to the variational expression provided that

$$(G, L\Gamma)(f, \varphi_n) = (f, \Gamma)(LG, \varphi_n). \tag{15.10}$$

Similarly, consideration of a variation in c_n leads to

$$(G, L\Gamma)(\varphi_n, h) = (G, h)(\varphi_n, L\Gamma). \tag{15.11}$$

Equations (15.10) and (15.11) constitute $2N$ equations to determine the coefficients. It will be noticed that if G is multiplied by a constant neither (15.10) nor (15.11) is affected. Therefore, we will elect to make

$$(G, L\Gamma) = (f, \Gamma) \tag{15.12}$$

and then (15.10) reduces to

$$(f, \varphi_n) = (LG, \varphi_n). \tag{15.13}$$

Γ may also be multiplied by a constant since that does not influence

(15.11)–(15.13). Choosing it so that $(G, L\Gamma)=(G, h)$ we have

$$(\varphi_n, h)=(\varphi_n, L\Gamma). \tag{15.14}$$

It is evident that (15.13) is the same as (15.5) whereas (15.14) is the corresponding form for (15.7). Because of the way in which the coefficients have been selected the variational expression becomes (G, h) or (f, Γ) and so the results of the variational principle agree with those from Galerkin's method.

One valuable property of these techniques will now be proved for the case in which $h=f$ i.e. the quantity desired from the solution of (15.1) is (g, f). Let G be an approximation determined by Galerkin's method (or the variational principle). Then, from (15.5) (or (15.13)),

$$(G, f)=(G, LG) \tag{15.15}$$

(remember (15.3)). Now

$$(g-G, f)=(g, f)-2(G, f)+(G, f)=(g, Lg)-2(G, Lg)+(G, LG)$$

from (15.1) and (15.15). Hence

$$(g, f)=(G, f)+(g-G, f)=(G, f)+(g-G, Lg-LG).$$

Accordingly, if $(p, Lp)\geqslant 0$ for all p,

$$(g, f)\geqslant(G, f),$$

i.e. *if* $(p, Lp)\geqslant 0$ *for all* p *the approximation to the desired quantity provided by Galerkin's method is never greater than the true value.* On the other hand, *if* $(p, Lp)\leqslant 0$ *for all* p *the approximation is never less than the correct value.*

As an illustration return to (14.7) with $f=1$. For this integral equation $(p, Lp)\geqslant 0$ for all p and so $\int_{\mathscr{A}} G(t)\,\mathrm{d}t$ will never be greater than the correct value. So the estimate of X from (14.8) will not be less than the true quantity. Similarly, for (14.10) with $f=1$, $(p, Lp)\leqslant 0$ for all p so that $\int_{\mathscr{A}} G_1(t)\,\mathrm{d}t$ will never be less than the exact value and the estimate of X from (14.11) can be expected to be below the exact answer. The statements that (14.14) and (14.15) supply upper and lower bounds for X have been justified.

5.16 Approximation to the kernel

Galerkin's method seeks to solve an equation approximately by replacing the unknown g by an approximant. Another attack on the problem is to try to exchange L for a linear operator where inversion can be carried out precisely. For example, suppose that we know how to find the exact solution of

$$L_0 h=f.$$

Let us write it as $h = L_0^{-1}f$. Then (15.1) is reformulated as

$$L_0 g = f - (L - L_0)g$$

whence

$$g = L_0^{-1}f - L_0^{-1}(L - L_0)g.$$

This is actually another integral equation for g. It is of the *second kind* or *Fredholm type.* Its advantage over (15.1) is that there is a considerable theory concerning its solution and for estimating the error made by approximation. It is usually better behaved both analytically and numerically. Here we shall merely note that, if L_0 is a good enough approximation to L to make $L - L_0$ small in some sense, a first stab at a solution is $L_0^{-1}f$ and a better one is

$$L_0^{-1}f - L_0^{-1}(L - L_0)L_0^{-1}f.$$

Although formally simple, the better approximation may involve calculation which is far from trivial.

The technique may be illustrated by (14.7) when $\mathscr{A}$ is the interval of y from 0 to d. The fact that κ_n is not much different from $n\pi/b$ for a fair number of values of n suggests that a suitable approximate kernel is

$$\sum_{n=1}^{\infty} \frac{b}{n\pi} \cos\frac{n\pi y}{b} \cos\frac{n\pi t}{b} = -\frac{b}{2\pi} \ln 2 \left|\cos\frac{\pi y}{b} - \cos\frac{\pi t}{b}\right|. \tag{16.1}$$

In order to solve an integral equation with this kernel it is helpful to make the change of variable

$$\cos(\pi y/b) = \sin^2(\pi d/2b)\cos\theta + \cos^2(\pi d/2b). \tag{16.2}$$

This makes the interval $0 \leqslant y \leqslant d$ correspond to $0 \leqslant \theta \leqslant \pi$. If $\theta = \phi$ when $y = t$ in (16.2) the right-hand side of (16.1) becomes

$$-\frac{b}{2\pi} \ln\left(2 \sin^2\frac{\pi d}{2b} |\cos\theta - \cos\phi|\right)$$
$$= -\frac{b}{2\pi}\left(\ln \sin^2\frac{\pi d}{2b} - 2\sum_{n=1}^{\infty} \frac{1}{n} \cos n\theta \cos n\phi\right).$$

Now, if $G(\phi) = g(t)\,\mathrm{d}t/\mathrm{d}\phi$, the approximation represented by $L_0 g = f$ is

$$1 + (\kappa_{10}b/2\pi)\int_0^{\pi} G(\phi)\left\{\ln \sin^2(\pi d/2b) - 2\sum_{n=1}^{\infty} (1/n)\cos n\theta \cos n\phi\right\} \mathrm{d}\phi = 0.$$

It holds for θ in $(0, \pi)$ and the coefficient of $\cos n\theta$ must vanish, i.e.

$$\int_0^{\pi} G(\phi)\,\mathrm{d}\phi = -\pi/\kappa_{10}b \ln \sin(\pi d/2b),$$

$$\int_0^{\pi} G(\phi) \cos n\phi\,\mathrm{d}\phi = 0 \quad (n = 1, 2, \ldots).$$

According to (14.8), $X/Z_0 = -\frac{1}{4}\int_0^\pi G(\phi)\,d\phi$ and so, in this approximation,

$$X/Z_0 = -\pi/4\kappa_{10}b \ln \sin(\pi d/2b).$$

When $d = \frac{1}{2}b$ this estimate is $X/Z_0 = -0.71\pi/\kappa_{10}b$.

The most serious error in this approximation to the kernel is the replacement of κ_1 by π/b. To improve matters let us change every κ_n but κ_1. Then the approximate integral equation is

$$1 + (\kappa_{10}b/\pi)\int_0^\pi G(\phi)\left\{\ln \sin(\pi d/2b) - \sum_{n=1}^{\infty} (1/n)\cos n\theta \cos n\phi\right\} d\phi$$
$$- (\kappa_{10}/\kappa_1)(1-\kappa_1 b/\pi)\int_0^\pi G(\phi)\cos(\pi y/b)\cos(\pi t/b)\,d\phi = 0$$

with (16.2) to be substituted for the two cosines in the last integral. The constant term vanishes if

$$1 + \{(\kappa_{10}b/\pi)\ln \sin(\pi d/2b) - (\kappa_{10}/\kappa_1)(1-\kappa_1 b/\pi)\cos^4(\pi d/2b)\}\int_0^\pi G(\phi)\,d\phi$$
$$- (\kappa_{10}/4\kappa_1)(1-\kappa_1 b/\pi)\sin^2(\pi d/b)\int_0^\pi G(\phi)\cos\phi\,d\phi = 0$$

while the coefficient of $\cos\theta$ disappears if

$$\{(\kappa_{10}b/\pi) + (\kappa_{10}/\kappa_1)(1-\kappa_1 b/\pi)\sin^4(\pi d/2b)\}\int_0^\pi G(\phi)\cos\phi\,d\phi$$
$$+ (\kappa_{10}/4\kappa_1)(1-\kappa_1 b/\pi)\sin^2(\pi d/b)\int_0^\pi G(\phi)\,d\phi = 0.$$

These constitute two linear equations for the two integrals and hence $\int_0^\pi G(\phi)\,d\phi$ is determined. The result is

$$\frac{Z_0}{X} = \frac{4b\kappa_{10}}{\pi}\left\{\ln \sin\left(\frac{\pi d}{2b}\right) - \frac{C\cos^4(\pi d/2b)}{1 + C\sin^4(\pi d/2b)}\right\}$$

where $C = -1 + \pi/\kappa_1 b$.

5.17 Multi-mode propagation

The advantage of Galerkin's method is its simplicity to implement but it suffers from a lack of knowledge of how many equations are needed for accuracy though the bounds derived in §5.15 can be of assistance. The strength of the method of the last section is its ability to take care of terms neglected by Galerkin's method but it depends heavily on solving the approximate integral equation. This suggests that one should try to combine the best features of the two methods especially as there will be more equations to handle when several modes can propagate. We may also make virtue of necessity by allowing ourselves the freedom to fix at a convenient stage the number of modes that are to be approximated.

The example of a discontinuous change in cross-section in a rectangular guide will now be treated in order to bring out the main ideas. At $z=0$ the longer side changes suddenly from a to a_1 while the shorter side is unaltered (Fig. 5.13). Only TE-modes which are independent of y will be discussed. There is no specification of how many modes can propagate. However, one of them is incident on the discontinuity and our aim is to determine the scattered field. To assist in the solution the coordinate x_1 is introduced in $z>0$; it is measured from the wall of the guide there in the same way that x is measured in $z<0$ (Fig. 5.13). In fact, $x_1=x-d$. Let the incident mode in $z<0$ be $e^{-i\chi_m z}\sin(m\pi x/a)$ where

$$\chi_m=\{k^2-(m\pi/a)^2\}^{\frac{1}{2}}.$$

Then the relevant components of the total field in $z<0$ can be expressed as

$$E_y=e^{-i\chi_m z}\sin(m\pi x/a)+\sum_{n=1}^{\infty}a_n e^{i\chi_n z}\sin(n\pi x/a), \tag{17.1}$$

$$-i\omega\mu_0 H_x=i\chi_m e^{-i\chi_m z}\sin(m\pi x/a)-\sum_{n=1}^{\infty}i\chi_n a_n e^{i\chi_n z}\sin(n\pi x/a). \tag{17.2}$$

The corresponding expansions in $z>0$ are

$$E_y=\sum_{n=1}^{\infty}b_n e^{-i\zeta_n z}\sin(n\pi x_1/a_1), \tag{17.3}$$

$$-i\omega\mu_0 H_x=\sum_{n=1}^{\infty}i\zeta_n b_n e^{-i\zeta_n z}\sin(n\pi x_1/a_1) \tag{17.4}$$

where

$$\zeta_n=\{k^2-(n\pi/a_1)^2\}^{\frac{1}{2}}.$$

On $z=0$, E_y vanishes except on the aperture A where it is continu-

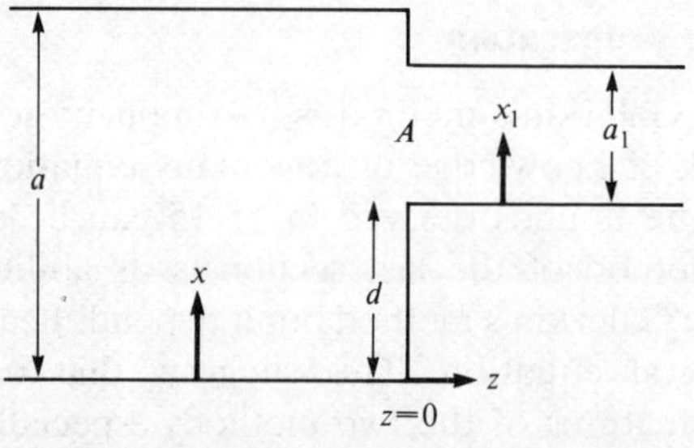

Fig. 5.13 An inductive change of cross-section

ous. Let its common value on A be denoted by $E(x)$ or $E(x_1)$ according to which coordinate system is pertinent. Then multiplication of (17.1) by $\sin(n\pi x/a)$ and integration from 0 to a gives

$$\int_A E(t)\sin(n\pi t/a)\,\mathrm{d}t = \tfrac{1}{2}a(a_n + \delta_{mn}). \tag{17.5}$$

Similarly

$$\int_A E(t_1)\sin(n\pi t_1/a_1)\,\mathrm{d}t_1 = \tfrac{1}{2}a_1 b_n. \tag{17.6}$$

Now H_x must be continuous through the aperture and so

$$\mathrm{i}\chi_m \sin(m\pi x/a) - \sum_{n=1}^{\infty} \mathrm{i}\chi_n a_n \sin(n\pi x/a) - \sum_{n=1}^{\infty} \mathrm{i}\zeta_n b_n \sin(n\pi x_1/a_1) = 0 \tag{17.7}$$

in A. Incorporation of (17.5) and (17.6) into (17.7) leads to

$$\mathrm{i}\chi_m \sin(m\pi x/a) - \sum_{n=1}^{\infty} (\mathrm{i}\chi_n/a)\int_A E(t)\sin(n\pi t/a)\sin(n\pi x/a)\,\mathrm{d}t$$
$$- \sum_{n=1}^{\infty} (\mathrm{i}\zeta_n/a_1)\int_A E(t_1)\sin(n\pi t_1/a_1)\sin(n\pi x_1/a_1)\,\mathrm{d}t_1 = 0 \tag{17.8}$$

as the integral equation to be solved to find E.

We acknowledge that we are prepared to make some approximations in χ_n and ζ_n but we do not wish to decide which ones yet. Therefore, we adopt an idea of Rozzi (1973) and designate M of the modes in $z<0$ and N of the modes in $z>0$ as *accessible.* All other modes will be called *localized.* There is no need to specify the number of accessible modes at this stage except to ensure they contain all propagating modes but there is no reason why they should not include non-propagating modes. The intention is to make approximations to the localized modes so these should be well beyond cut-off. Nevertheless, the precise division into the two groups can be deferred until later.

Now return to (17.7) and substitute for a_n and b_n only in the localized modes. Then

$$\mathrm{i}\chi_m \sin(m\pi x/a) - \sum_{n=1}^{M} \mathrm{i}\chi_n a_n \sin(n\pi x/a) - \sum_{n=1}^{N} \mathrm{i}\zeta_n b_n \sin(n\pi x_1/a_1) - LE = 0 \tag{17.9}$$

where

$$LE \equiv \sum_{n=M+1}^{\infty} (2\mathrm{i}\chi_n/a)\int_A E(t)\sin(n\pi t/a)\sin(n\pi x/a)\,\mathrm{d}t$$
$$+ \sum_{n=N+1}^{\infty} (2\mathrm{i}\zeta_n/a_1)\int_A E(t_1)\sin(n\pi t_1/a_1)\sin(n\pi x_1/a_1)\,\mathrm{d}t_1. \tag{17.10}$$

Suppose that E_n and F_n can be found such that

$$LE_n = \sin(n\pi x/a), \qquad LF_n = \sin(n\pi x_1/a_1). \tag{17.11}$$

Then the solution to (17.9) is

$$E = \mathrm{i}\chi_m E_m - \sum_{n=1}^{M} \mathrm{i}\chi_n a_n E_n - \sum_{n=1}^{N} \mathrm{i}\zeta_n b_n F_n. \tag{17.12}$$

Substitute this formula in (17.5) and (17.6). Then

$$a_r + \delta_{rm} = -\frac{Z_{rm}^{(1)}}{Z_m} + \sum_{n=1}^{M} a_n \frac{Z_{rn}^{(1)}}{Z_n} + \sum_{n=1}^{N} b_n \frac{Z_{rn}^{(2)}}{Z_n'} \quad (r = 1, \ldots, M), \tag{17.13}$$

$$b_r = -\frac{Z_{rm}^{(3)}}{Z_m} + \sum_{n=1}^{M} a_n \frac{Z_{rn}^{(3)}}{Z_n} + \sum_{n=1}^{N} b_n \frac{Z_{rn}^{(4)}}{Z_n'} \quad (r = 1, \ldots, N) \tag{17.14}$$

where $Z_n = \omega\mu_0/\chi_n$, $Z_n' = \omega\mu_0/\zeta_n$,

$$Z_{rn}^{(1)} = -(2\mathrm{i}\omega\mu_0/a)\int_A E_n(t)\sin(r\pi t/a)\,\mathrm{d}t,$$

$$Z_{rn}^{(2)} = -(2\mathrm{i}\omega\mu_0/a)\int_A F_n(t)\sin(r\pi t/a)\,\mathrm{d}t,$$

$$Z_{rn}^{(3)} = -(2\mathrm{i}\omega\mu_0/a_1)\int_A E_n(t_1)\sin(r\pi t_1/a_1)\,\mathrm{d}t_1,$$

$$Z_{rn}^{(4)} = -(2\mathrm{i}\omega\mu_0/a_1)\int_A F_n(t_1)\sin(r\pi t_1/a_1)\,\mathrm{d}t_1.$$

The problem has therefore been split into two. First, the solution of the integral equations (17.11) is obtained and then the solution of the $M+N$ algebraic equations (17.13) and (17.14). The latter is a straight matter of computation so the key to the method lies in solving (17.11) sufficiently accurately to get good approximations to the impedances $Z_{rn}^{(1)}, \ldots, Z_{rn}^{(4)}$ and here adjustment of M and N can be very beneficial.

Equation (17.11) will be solved by Galerkin's method. Select expansion functions $f_1, f_2, \ldots, f_P$ and put $E_n = \sum_{p=1}^{P} \alpha_p f_p$. The coefficients $\alpha_1, \ldots, \alpha_P$ are determined by the Galerkin equations

$$\sum_{p=1}^{P} \alpha_p (Lf_p, f_q) = (\sin(n\pi x/a), f_q) \quad (q = 1, \ldots, P)$$

where the inner product is defined as in (15.2). Define the $P\times P$-matrix $\boldsymbol{B}$ to have components $B_{qp} = (Lf_p, f_q)$ and the P-vectors $\boldsymbol{\alpha}$, $\mathbf{X}_n$ to have components α_p and $(\sin(n\pi x/a), f_p)$ respectively. Then our equations in matrix form are

$$\boldsymbol{B\alpha} = \mathbf{X}_n.$$

In this notation

$$Z_{rn}^{(1)} = -(2i\omega\mu_0/a)\mathbf{X}_r^{\mathrm{T}}\boldsymbol{\alpha} = -(2i\omega\mu_0/a)\mathbf{X}_r^{\mathrm{T}}\mathbf{B}^{-1}\mathbf{X}_n. \tag{17.15}$$

Note that, although this is an approximate result, bounds can be established by the theory of §5.15 because (15.6) is satisfied, $(h, Lh) \geqslant 0$ for all h (every localized mode is cut-off) and the exact formula for $Z_{rr}^{(1)}$ can be expressed as a constant times (E_r, LE_r).

Actually a substantial reduction in the amount of computation on a change of frequency can be achieved by means of an idea proposed by Rozzi. It is first assumed that $f_1, \ldots, f_p$ are independent of frequency; it is highly likely that such a choice would be made anyway. Then define

$$Y_{np} = \int_A f_p(t_1)\sin(n\pi t_1/a_1)\,\mathrm{d}t_1 \quad (p = 1, \ldots, P).$$

Also write the pth component of $\mathbf{X}_n$ as X_{np}. The definition (17.10) then supplies

$$B_{qp} = \sum_{n=M+1}^{\infty} 2i\chi_n X_{nq} X_{np}/a + \sum_{n=N+1}^{\infty} 2i\zeta_n Y_{nq} Y_{np}/a_1.$$

Our purpose is to approximate $i\chi_n$ but, if it is replaced by $n\pi/a$ on the grounds that M is large enough, $\mathbf{B}$ will become independent of frequency. So we need a higher approximation. Now, an obvious one for $(1-c^2/n^2)^{\frac{1}{2}}$ is $1-c^2/2n^2$ but better accuracy can be obtained for a wider range of n by putting

$$(1-c^2/n^2)^{\frac{1}{2}} = c_1^{(n)} - c_2^{(n)}c^2/2n^2 \tag{17.16}$$

and determine $c_1^{(n)}, c_2^{(n)}$ so that the approximation is best in the Chebyshev sense for the interval of c under consideration. For instance, when $1.2 \leqslant c \leqslant 2$, $c_1^{(3)} = 1.014$, $c_2^{(3)} = 1.194$. The precise values of $c_1^{(n)}$ and $c_2^{(n)}$ do not concern us; they are positive and approach 1 rapidly as $n \to \infty$.

With (17.16) incorporated

$$B_{qp} = C_{qp}^{\mathrm{s}} - (ka/\pi)^2 C_{qp}^{\mathrm{d}} \tag{17.17}$$

where

$$C_{qp}^{\mathrm{s}} = (2\pi/a^2) \sum_{n=M+1}^{\infty} nc_1^{(n)} X_{nq} X_{np} + (2\pi/a_1^2) \sum_{n=N+1}^{\infty} nc_1^{(n)} Y_{nq} Y_{np},$$

$$C_{qp}^{\mathrm{d}} = \pi \sum_{n=M+1}^{\infty} c_2^{(n)} X_{nq} X_{np}/na^2 + \pi \sum_{n=N+1}^{\infty} c_2^{(n)} Y_{nq} Y_{np}/na_1^2.$$

The matrix $\mathbf{C}^{\mathrm{s}}$ represents the static behaviour since it is left when $ka = 0$. The matrix $\mathbf{C}^{\mathrm{d}}$ determines the dynamic correction when the frequency is changed. Both matrices are geometric quantities and do not need to be recalculated when the frequency is altered.

Although (17.17) separates off the frequency dependence it is, in fact, the inverse of $\boldsymbol{B}$ which occurs in (17.15). An illuminating formula for the inverse can be derived from (17.17). First, note that if $\mathbf{x}$ is any nonzero P-vector $\mathbf{x}^{\mathrm{T}}\boldsymbol{C}^{\mathrm{s}}\mathbf{x}>0$ (recall that $c_1^{(n)}$ and $c_2^{(n)}$ are positive). Thus $\boldsymbol{C}^{\mathrm{s}}$ is a positive-definite matrix. Similarly, $\boldsymbol{C}^{\mathrm{d}}$ is a positive-definite matrix. Consequently, $\boldsymbol{C}^{\mathrm{s}}$ has a (symmetric) square root $(\boldsymbol{C}^{\mathrm{s}})^{\frac{1}{2}}$. The matrix $(\boldsymbol{C}^{\mathrm{s}})^{-\frac{1}{2}}\boldsymbol{C}^{\mathrm{d}}(\boldsymbol{C}^{\mathrm{s}})^{-\frac{1}{2}}$ is positive-definite so that its eigenvalues $1/\omega_1^2, \ldots, 1/\omega_P^2$ are positive and they can be numbered so that $\omega_1^2 \leqslant \omega_2^2 \leqslant \ldots \leqslant \omega_P^2$. Let $\boldsymbol{u}_m$ be the eigenvector corresponding to $1/\omega_m^2$ so that

$$(\boldsymbol{C}^{\mathrm{s}})^{-\frac{1}{2}}\boldsymbol{C}^{\mathrm{d}}(\boldsymbol{C}^{\mathrm{s}})^{-\frac{1}{2}}\boldsymbol{u}_m = \boldsymbol{u}_m/\omega_m^2.$$

Because the matrix is positive-definite the $\boldsymbol{u}_m$ can be adjusted so that $\boldsymbol{u}_m^{\mathrm{T}}\boldsymbol{u}_m = \delta_{mn}$. Construct the matrix $\boldsymbol{U}$ which has the $\boldsymbol{u}_m$ as columns i.e. $\boldsymbol{U} \equiv (\boldsymbol{u}_1, \boldsymbol{u}_2, \ldots, \boldsymbol{u}_P)$. Then $\boldsymbol{U}^{\mathrm{T}}\boldsymbol{U}$ is the unit matrix so that $\boldsymbol{U}$ is an orthogonal $P \times P$-matrix and

$$\boldsymbol{U}^{\mathrm{T}}(\boldsymbol{C}^{\mathrm{s}})^{-\frac{1}{2}}\boldsymbol{C}^{\mathrm{d}}(\boldsymbol{C}^{\mathrm{s}})^{-\frac{1}{2}}\boldsymbol{U} = \mathrm{diag}[\omega_1^{-2}, \omega_2^{-2}, \ldots, \omega_P^{-2}].$$

Call the diagonal matrix on the right $\boldsymbol{\Omega}$. Then

$$\boldsymbol{C}^{\mathrm{d}} = (\boldsymbol{C}^{\mathrm{s}})^{\frac{1}{2}}\boldsymbol{U}\boldsymbol{\Omega}\boldsymbol{U}^{\mathrm{T}}(\boldsymbol{C}^{\mathrm{s}})^{\frac{1}{2}}$$

since $\boldsymbol{U}\boldsymbol{U}^{\mathrm{T}}$ is the unit matrix. Therefore

$$\begin{aligned}\boldsymbol{B} &= (\boldsymbol{C}^{\mathrm{s}})^{\frac{1}{2}}\boldsymbol{U}\boldsymbol{U}^{\mathrm{T}}(\boldsymbol{C}^{\mathrm{s}})^{\frac{1}{2}} - (ka/\pi)^2(\boldsymbol{C}^{\mathrm{s}})^{\frac{1}{2}}\boldsymbol{U}\boldsymbol{\Omega}\boldsymbol{U}^{\mathrm{T}}(\boldsymbol{C}^{\mathrm{s}})^{\frac{1}{2}}\\ &= \boldsymbol{D}\{\boldsymbol{I} - (ka/\pi)^2\boldsymbol{\Omega}\}\boldsymbol{D}^{\mathrm{T}}\end{aligned}$$

where $\boldsymbol{D} = (\boldsymbol{C}^{\mathrm{s}})^{\frac{1}{2}}\boldsymbol{U}$ and $\boldsymbol{I}$ is the $P \times P$ unit matrix. If $\mathbf{y}_n = \boldsymbol{D}^{-1}\mathbf{X}_n$, (17.15) supplies

$$Z_{rn}^{(1)} = -(2\mathrm{i}\omega\mu_0/a)\mathbf{y}_r^{\mathrm{T}}\{\boldsymbol{I} - (ka/\pi)^2\boldsymbol{\Omega}\}\mathbf{y}_n.$$

Taking advantage of $\boldsymbol{\Omega}$ being diagonal and letting $(\mathbf{y}_n)_{\mathrm{p}}$ be the pth component of $\mathbf{y}_n$ we see that

$$Z_{rn}^{(1)} = -\frac{2\mathrm{i}\omega\mu_0}{a}\sum_{\mathrm{p}=1}^{P}\frac{s_{rn}^{(\mathrm{p})}}{1-(ka/\pi\omega_{\mathrm{p}})^2} \tag{17.18}$$

where $s_{rn}^{(\mathrm{p})} = (\mathbf{y}_r)_{\mathrm{p}}(\mathbf{y}_n)_{\mathrm{p}}$.

The quantity $s_{rn}^{(\mathrm{p})}$ does not involve k, as the method of construction shows nor does ω_{p}. Therefore all the effects of frequency change are collected in the denominators of (17.18). Only these denominators have to be recomputed for a new value of k. There will be no trouble with a denominator becoming small so long as M and N are chosen large enough for ω_1^2 to be much greater than the upper limit of $(ka/\pi)^2$ in the frequency band of interest.

Only the formula for $Z_{rn}^{(1)}$ has been given but there is no difficulty in deriving the corresponding relations for the other impedances by travelling along the same lines.

The first step in the above procedure is picking the f_p. This is a crucial matter. One wants to be able to calculate X_{np} and Y_{np} readily. Also, if it can be organized that either or both of $\sum_{n=1}^{\infty} nX_{np}X_{nq}$ and $\sum_{n=1}^{\infty} nY_{np}Y_{nq}$ can be summed easily, the determination of $\boldsymbol{C}^{\mathrm{s}}$ is much facilitated because $c_1^{(n)}$ tends to 1 rapidly as $n \to \infty$. Once $\boldsymbol{C}^{\mathrm{s}}$ and $\boldsymbol{C}^{\mathrm{d}}$ are known it is necessary to compute a square root of a positive-definite matrix followed by the eigenvalues $1/\omega_m^2$ and eigenvectors of another positive definite matrix. For all of these processes reliable numerical procedures are available so that the evaluation of the impedance from (17.18) can be regarded as a practical proposition.

5.18 Cascades

Several discontinuities in a guide may be necessary for the purpose and these form a cascade (Fig. 5.14). The combined effect can be found from a knowledge of the behaviour of each obstacle separately. To indicate the procedure we consider again the problem of the preceding section when there is an additional aperture A_2 present.

Suppose firstly that the aperture A_2 is absent. Assume that when the problem of the change in cross-section has been solved it is found in $z<0$ that

$$E_y = \sum_{n=1}^{M} V_n(z)\sin(n\pi x/a), \qquad H_x = \sum_{n=1}^{M} I_n(z)\sin(n\pi x/a)$$

where M is the number of modes taken into account and

$$V_n(z) = A_n \mathrm{e}^{-\mathrm{i}\chi_n z} + B_n \mathrm{e}^{\mathrm{i}\chi_n z},$$
$$I_n(z) = (B_n \mathrm{e}^{\mathrm{i}\chi_n z} - A_n \mathrm{e}^{-\mathrm{i}\chi_n z})/Z_n.$$

The impedance matrix $\boldsymbol{Z}$ representing the effect of the change in cross-section is given by

$$\mathbf{V}(0) = \boldsymbol{ZI}(0). \tag{18.1}$$

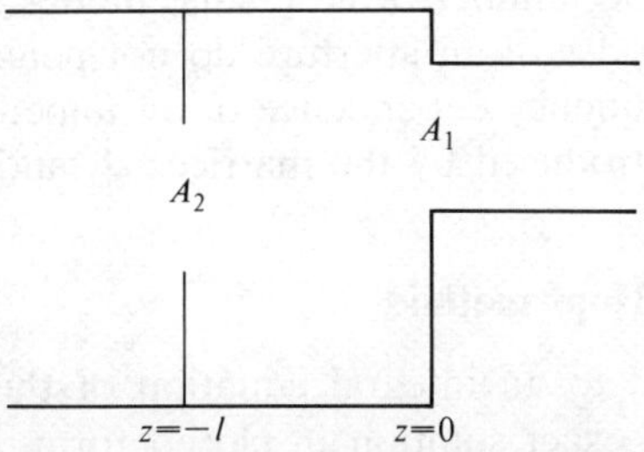

Fig. 5.14 Typical cell of a cascade

To transfer this information to $z=-l$, the location of A_2, the impedance matrix $\boldsymbol{Z}'$ is required where

$$\mathbf{V}(-l)=\boldsymbol{Z}'\boldsymbol{I}(-l). \tag{18.2}$$

Let $\mathbf{A}$, $\boldsymbol{B}$ be M-vectors with components A_n, B_n respectively and let $\boldsymbol{J}$, $\boldsymbol{K}$, $\boldsymbol{Y}_0$ be diagonal matrices whose diagonal elements are $e^{i\chi_n l}$, $e^{-i\chi_n l}$, $1/Z_n$ respectively. Then (18.1) can be written as

$$\mathbf{A}+\boldsymbol{B}=\boldsymbol{ZY}_0(\boldsymbol{B}-\mathbf{A})$$

while (18.2) becomes

$$\boldsymbol{JA}+\boldsymbol{KB}=\boldsymbol{Z}'\boldsymbol{Y}_0(\boldsymbol{KB}-\boldsymbol{JA}).$$

Elimination of $\boldsymbol{B}$ from these two equations provides

$$\{\boldsymbol{J}+\boldsymbol{K}(\boldsymbol{ZY}_0-\boldsymbol{I})^{-1}(\boldsymbol{ZY}_0+\boldsymbol{I})\}\mathbf{A}=\boldsymbol{Z}'\boldsymbol{Y}_0\{\boldsymbol{K}(\boldsymbol{ZY}_0-\boldsymbol{I})^{-1}(\boldsymbol{ZY}_0+\boldsymbol{I})-\boldsymbol{J}\}\mathbf{A}.$$

As $\mathbf{A}$ is arbitrary this implies that

$$\begin{aligned}\boldsymbol{Z}'&=\{\boldsymbol{J}+\boldsymbol{K}(\boldsymbol{ZY}_0-\boldsymbol{I})^{-1}(\boldsymbol{ZY}_0+\boldsymbol{I})\}\{\boldsymbol{K}(\boldsymbol{ZY}_0-\boldsymbol{I})^{-1}(\boldsymbol{ZY}_0+\boldsymbol{I})-\boldsymbol{J}\}^{-1}\boldsymbol{Y}_0^{-1}\\&=\{\boldsymbol{J}(\boldsymbol{ZY}_0-\boldsymbol{I})+\boldsymbol{K}(\boldsymbol{ZY}_0+\boldsymbol{I})\}\{\boldsymbol{K}(\boldsymbol{ZY}_0+\boldsymbol{I})-\boldsymbol{J}(\boldsymbol{ZY}_0-\boldsymbol{I})\}^{-1}\boldsymbol{Y}_0\end{aligned}$$

since $(\boldsymbol{ZY}_0-\boldsymbol{I})^{-1}$ and $\boldsymbol{ZY}_0+\boldsymbol{I}$ commute.

Now $\boldsymbol{J}+\boldsymbol{K}=2\boldsymbol{J}_1$, $\boldsymbol{J}-\boldsymbol{K}=2i\boldsymbol{J}_2$ where $\boldsymbol{J}_1$, $\boldsymbol{J}_2$ are diagonal matrices with elements $\cos\chi_n l$, $\sin\chi_n l$ respectively. Hence

$$\boldsymbol{Z}'=4(\boldsymbol{J}_1\boldsymbol{ZY}_0-i\boldsymbol{J}_2)(\boldsymbol{J}_1-i\boldsymbol{J}_2\boldsymbol{ZY}_0)^{-1}\boldsymbol{Y}_0^{-1}. \tag{18.3}$$

The expression (18.3) gives the equivalent circuit of A_1 as seen at A_2. Now consider the problem of A_2 alone with A_1 removed. With M_1 (say) accessible modes to the left of A_2 and M to the right the impedance matrix $\boldsymbol{Z}_2$ is determined. Both $\boldsymbol{Z}_2$ and $\boldsymbol{Z}'$ are at $z=-l$ so they can be combined by standard network rules to supply the equivalent circuit of the total effect of A_1 and A_2. If there were another aperture A_3 to the left of A_2 we could transfer the combined impedance of A_1 and A_2 to A_3 by another application of (18.3). Thus a cascade of discontinuities can be handled in this way.

As a precaution the number of accessible modes should be chosen so that the localized modes of an aperture do not penetrate to an adjacent one. Usually the frequency dependence of an impedance matrix is small compared to that introduced by the matrices $\boldsymbol{J}_1$ and $\boldsymbol{J}_2$.

5.19 The Wiener–Hopf method

Some problems lead to an integral equation of the Wiener–Hopf type which is capable of exact solution in closed form. For an example the effect of a bifurcation will be discussed. Only the main steps in the argument will be pointed out since a fuller treatment of the technique can be found in §9.7.

Suppose that a sound wave is propagating along the z-axis in the fundamental mode between the two soft planes $x = 0$ and $x = a$. Let a soft half-plane be inserted on $x = \frac{1}{2}a$ from $z = 0$ to $z = \infty$ (Fig. 5.15). A *bifurcation* is then said to occur. To simplify the analysis it will be assumed that the smaller guides I and IV are cut off at the operating wavelength and that propagation of only the dominant mode can be supported by the larger guide. Then, as $z \to \infty$, p will be exponentially damped whereas, when $z \to -\infty$, p will behave like

$$(e^{-i\lambda_{10}z} + Re^{i\lambda_{10}z})\sin(\pi x/a)$$

where

$$\lambda_{m0}^2 = k^2 - (m\pi/a)^2.$$

The reflection coefficient R is the quantity to be determined.

Write

$$p = g(x, z) + e^{-i\lambda_{10}z} \sin(\pi x/a).$$

The z-dependence of g as $z \to -\infty$ will be $e^{i\lambda_{10}z}$; on the other hand, as $z \to \infty$ it must remove the incident wave in order to allow for the exponential attenuation and so behave like $-e^{-i\lambda_{10}z}$. It will now be assumed that k, instead of being real, has a small negative imaginary part. This assumption is made for analytical convenience though it could be regarded as accounting for a small amount of dissipation in the medium. At the end of the analysis the imaginary part will be made zero. The legitimacy of such a procedure will not be considered here. The effect of this change to k is to ensure that λ_{m0} always has a negative imaginary part. The imaginary part of k will be denoted by k_i so that $k = k_r - ik_i$ where k_r and k_i are both positive.

Define

$$G(x, s) = \int_{-\infty}^{\infty} g(x, z)e^{-sz}\, dz$$

where s is the complex variable $\sigma + i\tau$. The behaviour of g at infinity

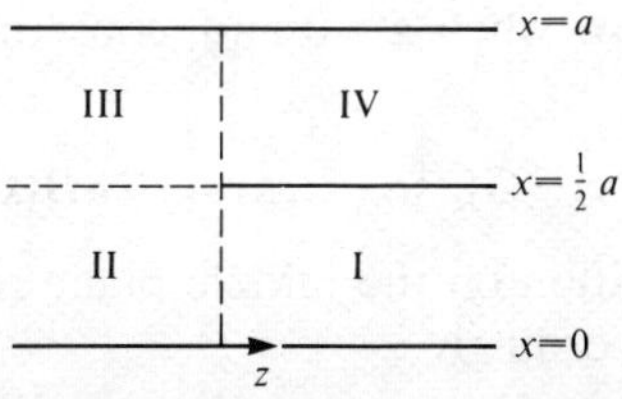

Fig. 5.15 Plan view of a bifurcation

guarantees that G is regular in the strip $\text{Im}\,\lambda_{10} < \sigma < -\text{Im}\,\lambda_{10}$. From

$$\partial^2 p/\partial x^2 + \partial^2 p/\partial z^2 + k^2 p = 0$$

it follows that

$$\mathrm{d}^2 G/\mathrm{d}x^2 + \kappa^2 G = 0 \tag{19.1}$$

where $\kappa^2 = s^2 + k^2$.

In order to specify precisely which root is to be identified with κ we select that branch of $(s^2+k^2)^{\frac{1}{2}}$ which reduces to k when $s=0$. Then κ has a negative imaginary part for $-k_i < \sigma < k_i$. Also the smallness of k_i ensures $\text{Im}\,\lambda_{10} < -k_i$ and hence (19.1) holds in the strip $-k_i < \sigma < k_i$. In this strip

$$G(x, s) = \begin{cases} A_1(s)\cos\kappa x + B_1(s)\sin\kappa x & (x > \frac{1}{2}a) \\ A_2(s)\cos\kappa x + B_2(s)\sin\kappa x & (x < \frac{1}{2}a). \end{cases}$$

Since $g(0, z) = 0$ it follows that $G(0, s) = 0$ with the result that $A_2(s) = 0$. Similarly, from $g(a, z) = 0$,

$$A_1(s)\cos\kappa a + B_1(s)\sin\kappa a = 0.$$

Moreover, $g(x, z)$ is continuous as x crosses $\frac{1}{2}a$ so $G(x, s)$ has the same property. Thus

$$G(x, s) = \begin{cases} -B(s)\sin\kappa(x-a) & (x > \frac{1}{2}a) \\ B(s)\sin\kappa x & (x < \frac{1}{2}a). \end{cases} \tag{19.2}$$

Further progress can be made by splitting the interval of integration into two so that $G(x, s) = G_R(x, s) + G_L(x, s)$ where

$$G_R(x, s) = \int_0^\infty g(x, z)\mathrm{e}^{-sz}\,\mathrm{d}z, \qquad G_L(x, s) = \int_{-\infty}^0 g(x, z)\mathrm{e}^{-sz}\,\mathrm{d}z.$$

Then, from (19.2),

$$G_R(\tfrac{1}{2}a, s) + G_L(\tfrac{1}{2}a, s) = B(s)\sin\tfrac{1}{2}\kappa a, \tag{19.3}$$

$$G_R'(\tfrac{1}{2}a+0, s) + G_L'(\tfrac{1}{2}a+0, s) = -\kappa B(s)\cos\tfrac{1}{2}\kappa a, \tag{19.4}$$

$$G_R'(\tfrac{1}{2}a-0, s) + G_L'(\tfrac{1}{2}a-0, s) = \kappa B(s)\cos\tfrac{1}{2}\kappa a \tag{19.5}$$

where the primes signify derivatives with respect to x. Now $\partial g/\partial x$ is continuous through $x = \frac{1}{2}a$ when $z < 0$ and, consequently, $G_L'(\frac{1}{2}a+0, s) = G_L'(\frac{1}{2}a-0, s)$. Hence

$$G_R'(\tfrac{1}{2}a+0, s) - G_R'(\tfrac{1}{2}a-0, s) = -2\kappa B(s)\cos\tfrac{1}{2}\kappa a. \tag{19.6}$$

The soft boundary condition on the middle plane requires that $g(\frac{1}{2}a, z) + \mathrm{e}^{-\mathrm{i}\lambda_{10}z} = 0$ for $z > 0$. Accordingly

$$G_R(\tfrac{1}{2}a, s) = -1/(s + \mathrm{i}\lambda_{10}).$$

Combining this with (19.3) and (19.6) to eliminate B we obtain

$$G_L(\tfrac{1}{2}a, s) - 1/(s + i\lambda_{10}) = -\{H_R(s)/2\kappa\}\tan\tfrac{1}{2}\kappa a$$

where the left-hand side of (19.6) has been abbreviated to H_R. This equation is in the standard Wiener–Hopf form because G_L is regular in $\sigma < k_i$ and H_R in $\sigma > -k_i$. After application of the usual technique it is discovered that

$$H_R(s) = \frac{2K_-(-i\lambda_{10})}{(s + i\lambda_{10})K_+(s)}$$

where

$$K_-(s) = \frac{2\prod_{n=1}^{\infty}\{(+i\lambda_{n0} - s)(a/n\pi)e^{sa/n\pi}\}e^{-(sa/\pi)\ln 2}}{a[\prod_{n=1}^{\infty}\{(i\lambda_{2n,0} - s)(a/2n\pi)e^{sa/2n\pi}\}]^2}, \tag{19.7}$$

$$K_+(s)K_-(-s) = 2/a, \qquad K_+(s)/K_-(s) = (1/\kappa)\tan(\tfrac{1}{2}\kappa a).$$

The field at any point can be found by means of the Mellin inversion formula

$$g(x, z) = \frac{1}{2\pi i}\int_{c-i\infty}^{c+i\infty} G(x, s)e^{sz}\, ds$$

where $-k_i < c < k_i$. When $x \leqslant \tfrac{1}{2}a$ we infer from (19.2) and (19.6) that

$$g(x, z) = \frac{1}{2\pi i}\int_{c-i\infty}^{c+i\infty} \frac{\sin \kappa x}{2\kappa \cos\tfrac{1}{2}\kappa a} H_R(s)e^{sz}\, ds$$

so that

$$g(\tfrac{1}{2}a, z) = -\frac{1}{2\pi i}\int_{c-i\infty}^{c+i\infty} \frac{K_-(-i\lambda_{10})}{(s + i\lambda_{10})K_-(s)} e^{sz}\, ds.$$

When $z < 0$ the contour can be deformed over the poles of the integrand in the right-half plane due to the zeros of $K_-(s)$. But, as $z \to -\infty$, only the contribution from that at $s = i\lambda_{10}$ survives. As a result

$$R = \frac{K_-(-i\lambda_{10})}{2i\lambda_{10}K'_-(i\lambda_{10})} \tag{19.8}$$

where $K'_-(s) = dK_-(s)/ds$. At this point put $k_i = 0$. It may then be verified easily that $|R| = 1$ so that R is of the form $e^{i\theta}$ where θ is real.

The equivalent circuit is a terminating impedance of $(1 + R)/(1 - R)$ at $z = 0$. Alternatively, it can be regarded as a short circuit (i.e. $p = 0$) placed at $z = d$ in the large guide where $e^{-2i\lambda_{10}d} = -R$ or, on using (19.7) and (19.8),

$$\lambda_{10}d = (\kappa_{10}a/\pi)(1 - \ln 2) - S_2(\kappa_{10}a/\pi; 1) + 2S_1(\kappa_1 a/2\pi; \tfrac{1}{2})$$

with

$$S_N(x; y) = \sum_{n=N}^{\infty}[\sin^{-1}\{x(n^2 - y^2)^{-\frac{1}{2}}\} - x/n].$$

5.20 Equivalence theorems

The direct determination of the reflection coefficient or circuit parameters of an obstacle involves a considerable amount of labour. Any theory which enables the behaviour of one obstacle to be deduced from that of another is therefore valuable. One method of doing this is to reflect known fields in planes and examine what boundary conditions are satisfied by the fields constructed thereby. (For more details see Karp and Williams (1951)). For example, consider the field built from that of the preceding section by putting in regions I–IV of Fig. 5.15.

$$\begin{aligned} \text{I and II:}\quad & g(x, z)+e^{-i\lambda_{10}z}\sin(\pi x/a)+g(x, -z)+e^{i\lambda_{10}z}\sin(\pi x/a), \\ \text{III:}\quad & g(x, z)-g(a-x, -z)+(e^{-i\lambda_{10}z}-e^{i\lambda_{10}z})\sin(\pi x/a), \\ \text{IV:}\quad & -g(a-x, z)+g(x, -z)-(e^{-i\lambda_{10}z}-e^{i\lambda_{10}z})\sin(\pi x/a). \end{aligned}$$

Now, for $z>0$, $g(\frac{1}{2}a, z)+e^{-i\lambda_{10}z}=0$ and $g(x, -z)$ is continuous across $x=\frac{1}{2}a$. Hence the constructed field is continuous through the interface between I and IV, as are also its derivatives. The same argument applies to the boundary between II and III. At the interface between I and II there is also continuity. On the boundary connecting III and IV the field vanishes because $g(x, 0)=g(a-x, 0)$ but the derivative is discontinuous. Hence the field acts as if it were a waveguide which is empty except for a soft wall occupying $z=0$, $x\geqslant\frac{1}{2}a$.

The incident wave in this field has two parts, one of $e^{-i\lambda_{10}z}\sin(\pi x/a)$ starting from $z=-\infty$ and one of $e^{i\lambda_{10}z}\sin(\pi x/a)$ approaching from $z=\infty$. To disentangle the two portions let p_1 be the field constructed and p_2, p_3 those due to the two separate incident waves. Then

$$p_2+p_3=p_1.$$

Moreover, if the two incident waves are subtracted their difference satisfies the boundary condition on $z=0$ and there is no further scattered field i.e.

$$p_2-p_3=(e^{-i\lambda_{10}z}-e^{i\lambda_{10}z})\sin(\pi x/a).$$

It follows that

$$p_2=\tfrac{1}{2}\{p_1+(e^{-i\lambda_{10}z}-e^{i\lambda_{10}z})\sin(\pi x/a)\}$$

and this is the solution for the single incident wave. The reflection coefficient is $\frac{1}{2}(R-1)$ and the inductive iris is equivalent to a shunt impedance iX where

$$X/Z_0=(1+R)/2i(1-R)=\tfrac{1}{2}\tan\kappa_{10}d. \qquad (20.1)$$

This method, which works only when suitable symmetry is present, gives an exact answer only for certain sizes of obstacle. Subject to this qualification it can provide a useful check on results obtained for arbitrary dimensions by approximate methods.

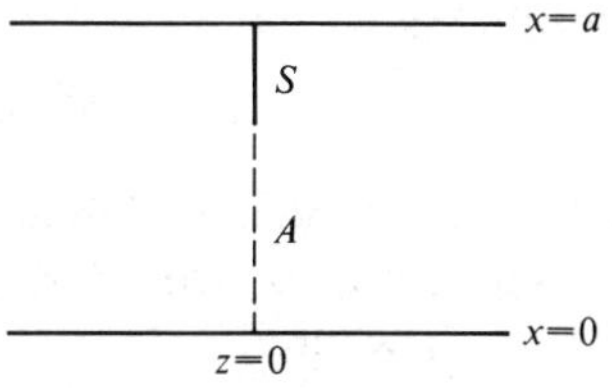

Fig. 5.16 The inductive iris

Another source of equivalent theorems is Babinet's principle (§9.3). Consider, for example, the inductive iris of Fig. 5.16, S and the walls being soft. Let the incident wave $e^{-i\lambda_{10}z}\sin(\pi x/a)$ produce a total field of $p(z)$. The total field for the incident wave $e^{i\lambda_{10}z}\sin(\pi x/a)$ is then $p(-z)$. Because $p(z)-p(-z)$ automatically satisfies the boundary condition on S

$$p(z)-p(-z)=-2i\sin\lambda_{10}z\sin(\pi x/a).$$

On A, $p'(z)=\partial p(z)/\partial z$ is continuous and so

$$p'(0)=-i\lambda_{10}\sin(\pi x/a) \tag{20.2}$$

whereas, on S,

$$p'(+0)+p'(-0)=-2i\lambda_{10}\sin(\pi x/a). \tag{20.3}$$

Construct $\tilde{p}$ by the rule

$$\tilde{p}(z)=\begin{cases} p(z)+e^{i\lambda_{10}z}\sin(\pi x/a) & (z<0)\\ e^{-i\lambda_{10}z}\sin(\pi x/a)-p(z) & (z>0).\end{cases}$$

It is contended that $\tilde{p}$ solves the problem of the complementary iris of Fig. 5.17 in which S' is a hard screen occupying the same region as A while A' is an aperture. For $\tilde{p}(+0)=\tilde{p}(-0)$ on A' because p vanishes on S and $\tilde{p}'$ is continuous by (20.3). On S', $\tilde{p}'=0$ on account of (20.2). Moreover, if

$$p(z)\sim(e^{-i\lambda_{10}z}+R_0e^{i\lambda_{10}z})\sin(\pi x/a)$$

as $z\to-\infty$, then

$$\tilde{p}(z)\sim\{e^{-i\lambda_{10}z}+(1+R_0)e^{i\lambda_{10}z}\}\sin(\pi x/a)$$

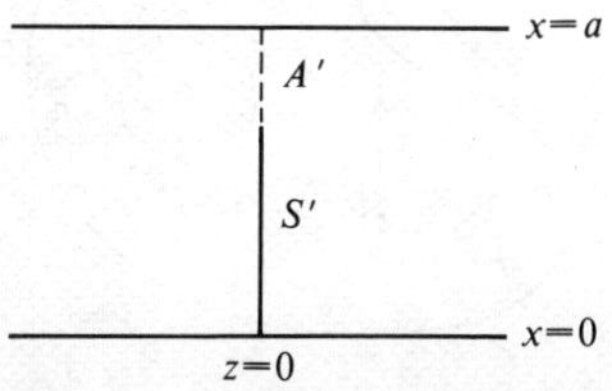

Fig. 5.17 The complementary iris

whereas, in the opposite direction

$$p(z) \sim (1+R_0)e^{-i\lambda_{10}z}\sin(\pi x/a) \quad \text{and} \quad \tilde{p}(z) \sim -R_0 e^{-i\lambda_{10}z}\sin(\pi x/a).$$

Thus, the reflection coefficient for the complementary iris is $1+R_0$. If the inductive iris presents a shunt impedance of Z the complementary iris gives a series impedance of Z_1 where

$$Z_0/Z_1 = -1 - Z/Z_0 = (1-R_0)/2R_0.$$

5.21 Non-uniform cross-section

When there is a change of medium within the cross-section the form of the modes is changed in general and the main problem is the determination of the possible propagating modes. Only the more difficult electromagnetic case will be discussed; the adaptation of the analysis to acoustics is straightforward. Consider modes travelling along the positive z-axis with a z-dependence $e^{-i\kappa z}$. Then, in a medium with permittivity ε and permeability μ, the transverse fields can be verified to be, when $\kappa^2 \neq \omega^2\mu\varepsilon$,

$$\boldsymbol{E}_t = (i\omega\mu\boldsymbol{k} \wedge \operatorname{grad} H_z - i\kappa \operatorname{grad}_t E_z)/(k^2N^2 - \kappa^2), \tag{21.1}$$

$$\boldsymbol{H}_t = (-i\kappa \operatorname{grad}_t H_z - i\omega\varepsilon\boldsymbol{k} \wedge \operatorname{grad} E_z)/(k^2N^2 - \kappa^2) \tag{21.2}$$

where $k^2 = \omega^2\mu_0\varepsilon_0$ as before and $N^2 = \mu\varepsilon/\mu_0\varepsilon_0$. The case when $\kappa^2 = k^2N^2$ will not be examined since, if such a case occurred with the cross-section of Fig. 5.18, it would be necessary to have $N_1 = N_2$ in order to accomplish continuity in the z-dependence across C_1. In general $N_1 \neq N_2$ and there will be no TEM-mode in a waveguide containing two media.

The longitudinal components satisfy

$$(\nabla^2 + k^2N^2 - \kappa^2)E_z = 0, \qquad (\nabla^2 + k^2N^2 - \kappa^2)H_z = 0. \tag{21.3}$$

The process is first to find E_z, H_z from (21.3) with $N = N_1$ and then with $N = N_2$, the corresponding values of the transverse field coming from

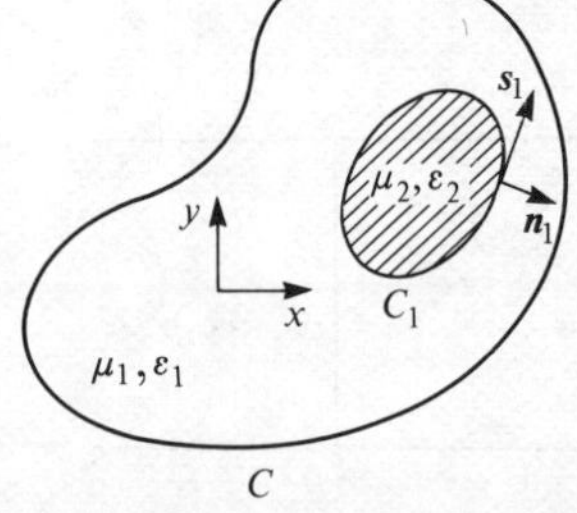

Fig. 5.18 A partly filled waveguide

(21.1) and (21.2) with the ε, μ appropriate to the region. Imposition of the boundary conditions on C and C_1 leads to an equation for κ which gives the possible modes of the type under consideration. The boundary conditions are that E_z and $\partial H_z/\partial n$ vanish on C together with the tangential components of $\boldsymbol{E}$ and $\boldsymbol{H}$ being continuous across C_1: This means that E_z, H_z must be continuous through C_1 and, from (21.1), and (21.2), the values of

$$(\omega\mu\, \partial H_z/\partial n_1 - \kappa\, \partial E_z/\partial s_1)/(k^2N^2 - \kappa^2)$$

and

$$(\kappa\, \partial H_z/\partial s_1 + \omega\varepsilon\, \partial E_z/\partial n_1)/(k^2N^2 - \kappa^2)$$

on C_1 in medium 1 must coincide with those in medium 2.

Suppose that there are TE-modes in which $E_z \equiv 0$. The boundary conditions on C_1 necessitate the continuity of both H_z and $(k^2N^2 - \kappa^2)^{-1}\kappa\, \partial H_z/\partial s_1$. But this is impossible unless $N_1 = N_2$ or $\partial H_z/\partial s_1 = 0$. Therefore, TE-modes do not exist, in general. Similarly, TM-modes must be excluded except in the special circumstances $N_1 = N_2$ or $\partial E_z/\partial s_1 = 0$. The absence of a simple structure for the modes makes the evaluation of κ much more tedious than when the cross-section is uniform. In the worst situation the reduction of a fourth order determinant to a transcendental equation may be involved.

A relatively simple illustration of the detailed procedure is provided when C_1 is a plane parallel to the narrow side of a rectangular waveguide (Fig. 5.19). For brevity, a mode is sought which reduces to the fundamental TE_{10}-mode when the cross-section is uniform. In this mode the field is independent of y so try

$$H_z = \begin{cases} A e^{-i\kappa z} \cos \alpha x & (0 \leqslant x \leqslant d) \\ B e^{-i\kappa z} \cos \beta(x-a) & (d \leqslant x \leqslant a) \end{cases}$$

where $\alpha^2 + \kappa^2 = k^2N_1^2$ and $\beta^2 + \kappa^2 = k^2N_2^2$. This certainly satisfies (21.3) and also meets the boundary conditions on C. Since C_1 is $x = d$ we have $\partial/\partial s_1 = -\partial/\partial y$ and $\partial/\partial n_1 = -\partial/\partial x$. Now $\partial H_z/\partial s_1 = 0$ and so the boundary conditions on C_1 can be complied with when $E_z \equiv 0$ provided that A and B can be chosen so that H_z and $(k^2N^2 - \kappa^2)^{-1}\omega\mu\, \partial H_z/\partial x$ are continuous

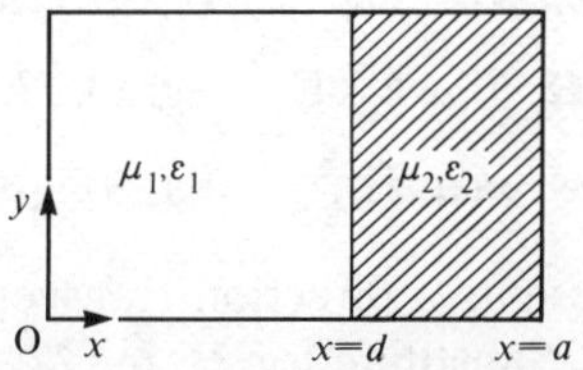

Fig. 5.19 The partially filled rectangular waveguide

across $x = d$. This is possible if

$$A \cos \alpha d = B \cos \beta(d-a),$$
$$(\mu_1 A/\alpha)\sin \alpha d = (\mu_2 B/\beta)\sin \beta(d-a).$$

Elimination of A and B supplies

$$(\mu_1/\alpha)\tan \alpha d = (\mu_2/\beta)\tan \beta(d-a).$$

When the formulae for α and β are substituted there results an equation for κ which can, however, only be solved numerically.

It is important to stress that the occurrence of a TE-mode reducing to the TE_{10}-mode in the uniform guide must be regarded as fortuitous. When the boundary between the dielectrics is parallel to the x-axis there is no TE-mode which degenerates to the TE_{10}-mode of the uniform guide; both E_z and H_z are nonzero.

Ferrites in waveguides

5.22 Waves in a gyromagnetic medium

A *gyromagnetic medium*, such as a ferrite, is one where the scalar permeability is replaced by a tensor. Correspondingly there is a *gyroelectric medium* in which the permittivity is a tensor (some information about such a medium will be found in §6.11). In a gyromagnetic medium the intensity and flux density are connected by $\boldsymbol{D} = \varepsilon \boldsymbol{E}$, $\boldsymbol{B} = \boldsymbol{\mu} \cdot \boldsymbol{H}$. When the medium is a ferrite with the applied magnetic field parallel to the z-axis

$$\boldsymbol{\mu} \equiv \begin{pmatrix} \mu_1 & -\mathrm{i}\mu_2 & 0 \\ \mathrm{i}\mu_2 & \mu_1 & 0 \\ 0 & 0 & \mu_3 \end{pmatrix}$$

when the time variation is harmonic. This is the form which will be adopted from now on. Let the dependence of the waves on z be $\mathrm{e}^{-\mathrm{i}\kappa z}$. Then Maxwell's equations become

$$\mathrm{i}\kappa \boldsymbol{E}_t + \omega\mu_2 \boldsymbol{H}_t + \mathrm{i}\omega\mu_1 \boldsymbol{k} \wedge \boldsymbol{H}_t = -\mathrm{grad}_t\, E_z, \tag{22.1}$$

$$\mathrm{i}\kappa \boldsymbol{H}_t - \mathrm{i}\omega\varepsilon \boldsymbol{k} \wedge \boldsymbol{E}_t = -\mathrm{grad}_t\, H_z, \tag{22.2}$$

$$\partial E_y/\partial x - \partial E_x/\partial y = -\mathrm{i}\omega\mu_3 H_z, \qquad \partial H_y/\partial x - \partial H_x/\partial y = \mathrm{i}\omega\varepsilon E_z \tag{22.3}$$

where, as earlier, t represents a direction transverse to the unit vector $\boldsymbol{k}$ along the positive z-axis. Substitute for $\boldsymbol{H}_t$ in (22.1) from (22.2) and then multiply vectorially by $\boldsymbol{k}$. It is found that

$$(p^2-q^2)\boldsymbol{E}_t=\boldsymbol{k}\wedge\{-i\omega(\mu_1 p+\mu_2 q)\operatorname{grad}_t H_z-\kappa q\operatorname{grad}_t E_z\}$$
$$-\omega\mu_2\kappa^2\operatorname{grad}_t H_z+i\kappa p\operatorname{grad}_t E_z, \qquad (22.4)$$

$$(p^2-q^2)\boldsymbol{H}_t=\boldsymbol{k}\wedge(i\omega\varepsilon p\operatorname{grad}_t E_z-\kappa q\operatorname{grad}_t H_z)$$
$$+i\kappa p\operatorname{grad}_t H_z+\omega\varepsilon q\operatorname{grad}_t E_z \qquad (22.5)$$

where

$$p=\kappa^2-\omega^2\mu_1\varepsilon, \qquad q=\omega^2\mu_2\varepsilon.$$

So long as $p^2\neq q^2$, (22.4) and (22.5) allow the determination of $\boldsymbol{E}_t$ and $\boldsymbol{H}_t$ from E_z and H_z.

Insertion of the transverse components from (22.4), (22.5) in (22.3) results in

$$i\omega(\mu_2 q+\mu_1 p)\nabla_t^2 H_z+\kappa q\nabla_t^2 E_z=i\omega\mu_3(p^2-q^2)H_z,$$
$$i\omega\varepsilon p\nabla_t^2 E_z-\kappa q\nabla_t^2 H_z=i\omega\varepsilon(p^2-q^2)E_z.$$

Two equations are the consequence of eliminating $\nabla_t^2 E_z$ and $\nabla_t^2 H_z$; they are

$$\nabla_t^2 E_z-(\mu_2 q+\mu_1 p)E_z/\mu_1+i\kappa\mu_3 qH_z/\omega\mu_1\varepsilon=0, \qquad (22.6)$$
$$\nabla_t^2 H_z-p\mu_3 H_z/\mu_1-i\kappa qE_z/\omega\mu_1=0. \qquad (22.7)$$

The formula for H_z in (22.6) may be used in (22.7) and then the equation for E_z is

$$\varepsilon(\nabla_t^2-p\mu_3/\mu_1)\{\nabla_t^2-(\mu_2 q+\mu_1 p)/\mu_1\}E_z-\mu_3(\kappa q/\omega\mu_1)^2 E_z=0. \qquad (22.8)$$

The stages in the solution are now to employ (22.8) for E_z, deduce H_z from (22.6) followed by the transverse field from (22.4) and (22.5). This works unless $\kappa q=0$ when the equations for E_z and H_z become independent of one another. When $\kappa q=0$ the case $\kappa=0$ can be discarded of no interest; the remaining alternative is $\mu_2=0$ and then the problem is similar to those discussed earlier in this chapter.

When $\kappa q\neq 0$ and $H_z=0$, (22.7) reveals that $E_z=0$ and the transverse field vanishes. Therefore there are no TM- or TE-modes. A TEM-mode can exist only if $p^2=q^2$. Then (22.1) and (22.2) imply that

$$i\boldsymbol{E}_t\pm\boldsymbol{k}\wedge\boldsymbol{E}_t=\boldsymbol{0} \qquad (22.9)$$

according as $p=\pm q$. Hence a TEM-mode, when it exists, is circularly polarized (§6.4). Furthermore, $\boldsymbol{E}_t=-\operatorname{grad}_t\chi$ where $\nabla_t^2\chi=0$.

Equation (22.8) can be written as

$$(\nabla_t^2+k_1^2)(\nabla_t^2+k_2^2)E_z=0 \qquad (22.10)$$

where $t^2 = k_1^2$, $t^2 = k_2^2$ are the roots of

$$t^4 + \{(\mu_1 + \mu_2)p + \mu_2 q\}t^2/\mu_1 + \mu_3(p^2 - q^2)/\mu_1 = 0. \qquad (22.11)$$

In case of doubt k_1^2 will be selected to be greater than k_2^2. Make the change of variable

$$t^2 + (\mu_1 p + \mu_2 q)/\mu_1 = \kappa^2 \mu_2/\mu_1 u \qquad (22.12)$$

and (22.11) becomes

$$\omega^2 \mu_2 \varepsilon u^2 - u\{\kappa^2(\mu_3 - \mu_1) + \omega^2 \varepsilon(\mu_1^2 - \mu_1\mu_3 - \mu_2^2)\}/\mu_2 - \kappa^2 = 0. \qquad (22.13)$$

Removing κ^2 from (22.13) by means of (22.12) leaves

$$t^2 = \omega^2 \mu_3 \varepsilon(u^2 - 1)/\{1 + u(\mu_3 - \mu_1)/\mu_2\}. \qquad (22.14)$$

An inference from (22.10) is that

$$(\nabla_t^2 + k_2^2)E_z = \varphi$$

where $(\nabla_t^2 + k_1^2)\varphi = 0$. If $k_1^2 \neq k_2^2$ it follows that

$$E_z = \phi_2 + \phi/(k_2^2 - k_1^2) = \varphi_2 + \varphi_1 \qquad (22.15)$$

where $(\nabla_t^2 + k_2^2)\varphi_2 = 0$, $(\nabla_t^2 + k_1^2)\varphi_1 = 0$. This solution is nugatory when $k_2^2 = k_1^2$ but then

$$E_z = \varphi_1 - \frac{1}{2k_1^2}\left(x\frac{\partial\varphi}{\partial x} + y\frac{\partial\varphi}{\partial y}\right) \qquad (22.16)$$

where $(\nabla^2 + k_1^2)\varphi_1 = 0$.

Turning now our attention to particular waves let the space variation of all components be $e^{-i\kappa(lx+my+nz)}$ where $l^2 + m^2 + n^2 = 1$, representing a wave travelling in the direction (l, m, n). The only alteration this causes to (22.8) is to exchange κ for κn. Thus (22.8) implies that

$$\{\kappa^2(l^2 + m^2) + \mu_3(\kappa^2 n^2 - \omega^2 \mu_1 \varepsilon)/\mu_1\}\{\kappa^2(l^2 + m^2)$$
$$+ \kappa^2 n^2 - \omega^2 \mu_1 \varepsilon + \mu_2 q/\mu_1\} - (\kappa n q)^2 \mu_3/\omega^2 \mu_1 = 0. \qquad (22.17)$$

Equation (22.17) is a quadratic in κ^2. The function on the left-hand side tends to $+\infty$ as $\kappa^2 \to \pm\infty$ and has the value $\omega^4 \varepsilon^2 \mu_3(\mu_1^2 - \mu_2^2)/\mu_1$ when $\kappa^2 = 0$. The left-hand side is negative when κ^2 has the positive value $\omega^2 \mu_3 \varepsilon/(l^2 + m^2 + n^2 \mu_3/\mu_1)$ unless $n = 0$. Setting aside $n = 0$ for the moment we see that (22.17) furnishes two real values for κ^2, both of which are positive if $\mu_1^2 > \mu_2^2$; if $\mu_1^2 < \mu_2^2$ they have opposite signs. Consequently, there are two waves associated with a direction both of which can propagate when $\mu_1^2 > \mu_2^2$ (the usual circumstance) but, if $\mu_1^2 < \mu_2^2$, one propagates and the other is attenuated.

The exclusion of $n = 0$ will now be remedied. Here (22.17) is not strictly applicable because (22.8) fails. Nevertheless, (22.6) and (22.7) are still valid; they reveal that either $E_z = 0$ and $\kappa^2 = \omega^2 \mu_3 \varepsilon$ or $H_z = 0$ and

$\kappa^2 = \omega^2\varepsilon(\mu_1^2 - \mu_2^2)/\mu_1$ or both E_z and H_z are zero. The last possibility can be abandoned because it entails a null field. Hence, when $n = 0$, there are two waves, both propagating if $\mu_1^2 > \mu_2^2$, unless $\mu_1^2 - \mu_1\mu_3 - \mu_2^2 = 0$ when only a single value of κ^2 arises.

The interpretation is that a ferrite behaves as a *doubly refracting medium* with characteristics similar to a biaxial crystal supporting optical waves. In contrast to what happens in crystals there are not normally directions for which the two values of κ^2 coalesce. An important point to note is that, as with crystals, the normals to the wavefronts are not the rays which carry energy (§6.11).

Whether n is zero or not, $\operatorname{div} \boldsymbol{E} = 0$ can be expressed as $lE_x + mE_y + nE_z = 0$ so that $\boldsymbol{E}$ is perpendicular to the direction of propagation. $\boldsymbol{B}$ (but not $\boldsymbol{H}$) has the same property because $\operatorname{div} \boldsymbol{B} = 0$.

The tensor $\boldsymbol{\mu}$ is invariant to a rotation of axes about the z-axis. Therefore, the x-axis can be chosen so that $m = 0$ i.e. the direction of propagation lies in the (x, z)-plane. With this choice let E' be the component of $\boldsymbol{E}$ in the (x, z)-plane and perpendicular to the direction of propagation. Then

$$E' = E_x n - E_z l = -E_z/l$$

since now $lE_x + nE_z = 0$ and $l^2 + n^2 = 1$. H_z is an imaginary multiple of E_z by virtue of (22.6) and, in the absence of any y-dependence, (22.4) makes E_y an imaginary multiple of E_z. Therefore E_y/E' is pure imaginary and the wave is elliptically polarized (§6.4). For propagation along the z-axis, $l = 0$ and E_z, H_z both vanish with the consequence $p^2 = q^2$. Therefore, in propagation along the z-axis

$$\kappa^2 = \omega^2\varepsilon(\mu_1 \pm \mu_2) \tag{22.18}$$

and (22.9) can be asserted. The waves are circularly polarized in opposite senses.

In optics the rotation of the plane of polarization of a linearly polarized beam as it traverses a body along the lines of force of a uniform magnetic field is known as the *Faraday effect.* (When the direction is perpendicular to the lines of force the rotation is called the *Cotton–Mouton effect;* the corresponding phenomenon when an electric field is applied is the *Kerr effect.*) Thus a ferrite exhibits the Faraday effect and the size can be large for frequencies well above ferromagnetic resonance when $\varepsilon/\varepsilon_0$ is of the order of 10 and μ_1/μ_0, μ_2/μ_0 about 1 and 0.3 respectively. (Note, however, that the resistivity of a microwave ferrite may be as high as 10^7 ohm cm as compared with the 10^{-5} ohm cm of iron.)

An elliptically polarized wave in an isotropic medium can be broken down into the superposition of two linearly polarized plane waves but there is no such decomposition in a gyromagnetic medium. This has an important bearing on the reflection of a wave at the plane interface between an isotropic medium and a gyromagnetic substance. A linearly polarized plane wave, incident from the isotropic medium, with its electric vector perpendicular to the plane of incidence will normally generate two transmitted waves in the gyromagnetic substance. There will then have to be *two reflected waves* in the isotropic medium and their electric vectors will be perpendicular to and in the plane of incidence respectively.

The permeability tensor of a gyromagnetic medium is not symmetric. So the proof of the reciprocity theorem in §1.35 fails and, in general, a reciprocity theorem is not available for ferrites. There are special circumstances when a reciprocity theorem can be supported (see Exercise 25).

The failure of reciprocity for ferrites makes it possible to design microwave devices having non-reciprocal properties. Let two circularly polarized waves of equal amplitude be travelling in the direction of magnetization and, thereby, governed by (22.18), (22.9). Write $\kappa_+^2 = \omega^2\varepsilon(\mu_1+\mu_2)$, $\kappa_-^2=\omega^2\varepsilon(\mu_1-\mu_2)$ and let $\boldsymbol{E}_{t+}, \boldsymbol{E}_{t-}$ be the corresponding transverse electric intensities, with $|E_{x+}|=|E_{x-}|=1$. It will be discovered that

$$\text{Re}\,(E_{x+}+E_{x-}) = 2\cos\{\omega t - \tfrac{1}{2}(\kappa_+ + \kappa_-)z\}\cos\tfrac{1}{2}(\kappa_+ - \kappa_-)z,$$

$$\text{Re}\,(E_{y+}+E_{y-}) = 2\cos\{\omega t - \tfrac{1}{2}(\kappa_+ + \kappa_-)z\}\sin\tfrac{1}{2}(\kappa_+ - \kappa_-)z.$$

Accordingly, the wave $\boldsymbol{E}_+ + \boldsymbol{E}_-$ is linearly polarized but the plane of polarization turns through the angle $\frac{1}{2}(\kappa_+ - \kappa_-)$ for every unit length covered by the wave. The total angle of rotation in traversing a length of ferrite is often known as the phase shift. A device which produces phase shift for propagation in one direction but none in the opposite direction is called a *directional phase shifter.* The particular directional phase shifter which causes a phase shift of π is sometimes termed a *gyrator.* Such a device can be created from the Faraday rotation in a suitable length of ferrite; a linearly polarized wave entering will emerge linearly polarized but with the plane of polarization rotated through an angle equal to the phase shift, though there can be anomalous effects due to reflection at ferrite faces in waveguides.

Another non-reciprocal device is a *circulator* which is a junction in which energy entering from guide 1 leaves at guide 2, energy arriving from guide 2 departs in guide 3 and so on. A ferrite which produces a Faraday rotation of 45° can play an integral part in such a device.

Isolators which transmit energy without absorption in one direction but attenuate it strongly in the opposite direction can also be manufactured from ferrites.

5.23 Waveguide modes

A waveguide filled with ferrite involves the solution of (22.8) subject to the boundary conditions. The formula (22.15) can be of assistance espe-

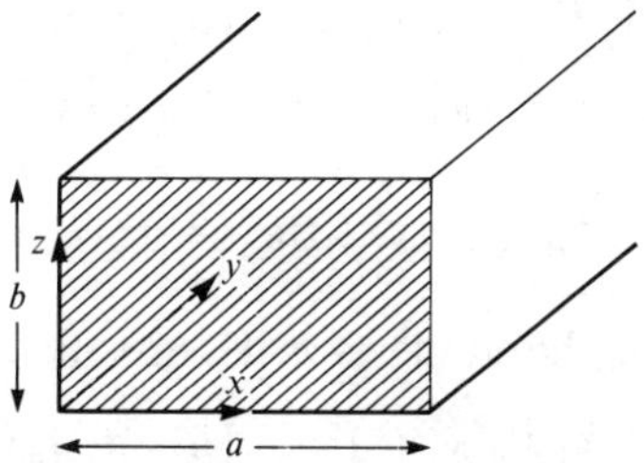

Fig. 5.20 Rectangular waveguide with transversely magnetized ferrite

cially if the ferrite is magnetized longitudinally but it may be more expedient to start at another component of the field if the magnetization is in another direction.

Suppose that the cross-section of a rectangular waveguide is completely occupied by transversely magnetized ferrite (Fig. 5.20). Continuing with the z-axis in the direction of magnetization we take the y-axis along the guide. A field which satisfies the governing equations is

$$E_x = 0, \qquad E_y = 0, \qquad E_z = e^{-i\beta y} \sin \alpha x, \qquad H_z = 0,$$
$$\omega(\mu_1^2 - \mu_2^2)H_x = (\beta\mu_1 \sin \alpha x + \alpha\mu_2 \cos \alpha x)e^{-i\beta y},$$
$$i\omega(\mu_1^2 - \mu_2^2)H_y = (\alpha\mu_1 \cos \alpha x + \beta\mu_2 \sin \alpha x)e^{-i\beta y}$$

provided that $\alpha^2 + \beta^2 = \omega^2\varepsilon(\mu_1^2 - \mu_2^2)/\mu_1$. The boundary conditions on $x = 0$ and $x = a$ where E_z vanishes are met if $\alpha = m\pi/a$ and so there is a mode with

$$\beta^2 - \omega^2\varepsilon(\mu_1^2 - \mu_2^2)/\mu_1 = -(m\pi/a)^2.$$

Clearly, the larger m is the more likely the mode is to be cut-off.

An examination of modes of this type when the cross-section is partially filled with ferrite can be undertaken as in §5.21.

When the magnetization is parallel to the axis of the guide put $E_z = \varphi_1 + \varphi_2$ from (22.15) and then, by invoking (22.6), (22.12),

$$H_z = -i\kappa(u_2\varphi_1 + u_1\varphi_2)/\omega\mu_3 u_1 u_2$$

where u_1, u_2 are the roots of (22.13). To secure E_z and E_s zero on the perimeter C of the general cross-section requires

$$\varphi_1 + \varphi_2 = 0,$$
$$k_2^2\left(\frac{\partial\varphi_1}{\partial s} - \frac{i}{u_1}\frac{\partial\varphi_1}{\partial n}\right) + k_1^2\left(\frac{\partial\varphi_2}{\partial s} - \frac{i}{u_2}\frac{\partial\varphi_2}{\partial n}\right) = 0.$$

A circular waveguide is suited to the introduction of polar coordinates r, ϕ in the cross-section. The field will be finite at the axis if

$$\varphi_1 = AJ_m(k_1 r)e^{im\phi}, \qquad \varphi_2 = BJ_m(k_2 r)e^{im\phi}$$

where m is an integer. From the boundary conditions

$$A\mathrm{J}_m(k_1 a)+B\mathrm{J}_m(k_2 a)=0,$$

$$Ak_2^2\left\{\frac{m}{a}\mathrm{J}_m(k_1 a)-\frac{k_1}{u_1}\mathrm{J}'_m(k_1 a)\right\}+Bk_1^2\left\{\frac{m}{a}\mathrm{J}_m(k_2 a)-\frac{k_2}{u_2}\mathrm{J}'_m(k_2 a)\right\}=0$$

assuming the radius of the guide is a. Hence the equation fixing κ is

$$k_2^2\left\{\frac{m}{a}-\frac{k_1}{u_1}\frac{\mathrm{J}'_m(k_1 a)}{\mathrm{J}_m(k_1 a)}\right\}=k_1^2\left\{\frac{m}{a}-\frac{k_2}{u_2}\frac{\mathrm{J}'_m(k_2 a)}{\mathrm{J}_m(k_2 a)}\right\}.$$

Changing m to $-m$ does not leave this equation unaltered. Thus, the propagation constant associated with $\mathrm{e}^{im\phi}$ is different from that for $\mathrm{e}^{-im\phi}$. Consequently, there are no modes with ϕ-dependence of the form of either $\cos m\phi$ or $\sin m\phi$, if $m=0$ is excepted.

Having a high dielectric constant in a ferrite means that more modes can propagate in a guide of given size than when the medium is air. For instance, a guide which just passes the first mode when empty will permit the passage of about the first three modes when filled with a medium of dielectric constant $10\varepsilon_0$.

The variation of the field across the ferrite also differs from that in an empty guide. Suppose that $z=0$ divides free space $(z<0)$ from ferrite $(z>0)$. A single mode incident from $z<0$ will create an infinite number of transmitted and reflected waves. Undoubtedly, the analysis of such a junction is much more complicated than that for the discontinuities investigated in earlier sections of this chapter.

The problem of a circular waveguide containing a concentric longitudinally magnetized ferrite core of smaller radius can be tackled by picking φ_1, φ_2 as above in the ferrite and a general Bessel function expansion in the annulus. There are now six equations, two on the perimeter of the waveguide and four on the surface of the ferrite, from which constants have to be eliminated to arrive at the transcendental equation for κ. It turns out that this equation is not invariant to a change in sign of m so that again there is no simple sinusoidal ϕ-dependence.

Radiation from waveguides and horns

One way of communicating a signal in a guide to free space is by means of a *horn*. Horns may be crudely classified into two types, the *sectoral* where part of the perimeter is not flared out (Fig. 5.21) and the *conical* where the whole perimeter expands (Fig. 5.23). Sectoral horns were fairly popular for public address systems but conical ones tend to be adopted for hi-fi though, of course, the feed is usually an electric circuit rather than a guide. However, small conical horns are often fitted to the ends of voice pipes.

5.24 The sectoral horn

One example of a sectoral horn occurs when the shorter side of a rectangular guide is flared out but no change is made to the dimension of the other side (Fig. 5.21). The line of intersection of the two slanting sides is taken as the x-axis with $x = 0$ and $x = a$ as the two fixed walls. The y-axis is made parallel to the shorter side so that the two faces, before flaring, are $y = \pm\frac{1}{2}b$. Polar coordinates r, ϕ are introduced in the (y, z)-plane with one sloping face $\phi = 0$ and the other $\phi = \Phi$; thus $z = r\cos(\phi - \frac{1}{2}\Phi)$, $y = r\sin(\phi - \frac{1}{2}\Phi)$. Free space will be assumed to be the medium in the horn and in the guide.

An acoustic wave in the horn must be such that

$$\frac{1}{r}\frac{\partial}{\partial r}\left(r\frac{\partial p}{\partial r}\right) + \frac{1}{r^2}\frac{\partial^2 p}{\partial \phi^2} + \frac{\partial^2 p}{\partial x^2} + k^2 p = 0 \tag{24.1}$$

and subject to the normal derivative of p vanishing on the boundaries when these are rigid, as will be assumed. A Fourier expansion in $\cos(m\pi x/a)\cos(n\pi\phi/\Phi)$ will meet the boundary conditions and the remaining factor of p will be a solution of Bessel's equations. Since $r = 0$ is not within the horn both types of Bessel function must be kept in the first place. Therefore a typical mode in the sectoral horn is

$$p_{mn} = \{A_{mn}\mathrm{H}^{(1)}_{n\pi/\Phi}(\kappa_{m0}r) + B_{mn}\mathrm{H}^{(2)}_{n\pi/\Phi}(\kappa_{m0}r)\}\cos\frac{m\pi x}{a}\cos\frac{n\pi\phi}{\Phi} \tag{24.2}$$

where $\kappa^2_{m0} = k^2 - (m\pi/a)^2$, $(m, n = 0, 1, \ldots)$.

The asymptotic formulae of (A.2.2) and (A.2.3) reveal that the $\mathrm{H}^{(1)}$ will have a dependence $e^{i\kappa_{m0}r}$ for large r while $\mathrm{H}^{(2)}$ offers $e^{-i\kappa_{m0}r}$. Thus, $\mathrm{H}^{(2)}$

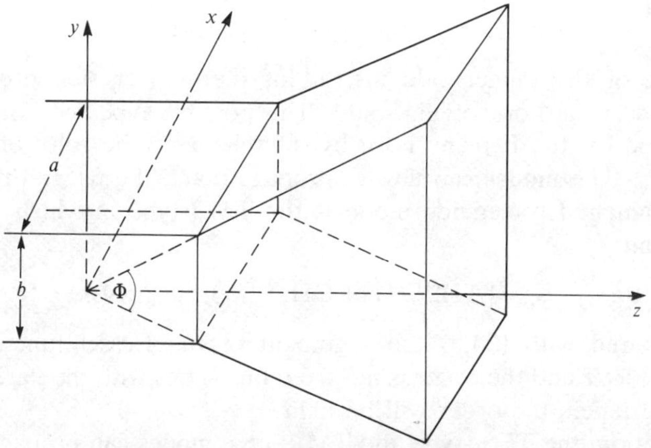

Fig. 5.21 The E-plane sectoral horn

corresponds to waves which are outgoing at infinity (attenuated if κ_{m0} is imaginary) and $\mathrm{H}^{(1)}$ is responsible for incoming waves. Therefore, a decision on whether the energy flow at infinity is outward or inward allows one to dispose of one of the Hankel functions in (24.2).

The fundamental mode ($m = 0$) is always present but the number of other modes which can propagate depends on the size of k. That modes are cut off can account for the distortion which is experienced sometimes with sectoral horns. Notwithstanding, the x-dependence of p_{mn} is the same as that in a waveguide mode and so the two modes may be expected to have a reasonable degree of coupling, particularly where the flare angle Φ is not too large.

For the corresponding electromagnetic problem the horn of Fig. 5.21 is designated as an *E-plane sectoral horn.* The structure of the field can be based on E_x and H_x, all other components being expressed in terms of them. Since E_x, H_x both satisfy (24.1) typical modes are

$$E_x = 0, \qquad H_x = \{A_{mn}\mathrm{H}^{(1)}_{n\pi/\Phi}(\kappa_{m0}r) + B_{mn}\mathrm{H}^{(2)}_{n\pi/\Phi}(\kappa_{m0}r)\}\sin\frac{m\pi x}{a}\cos\frac{n\pi\phi}{\Phi}$$

and

$$H_x = 0, \qquad E_x = \{A_{mn}\mathrm{H}^{(1)}_{n\pi/\Phi}(\kappa_{m0}r) + B_{mn}\mathrm{H}^{(2)}_{n\pi/\Phi}(\kappa_{m0}r)\}\cos\frac{m\pi x}{a}\sin\frac{n\pi\phi}{\Phi}$$

the former being named TE_{mn}*-type* and the latter TM_{mn}*-type.* Again, information about the progression of waves at infinity may permit one Hankel function to be favoured over the other. When the waveguide is carrying the fundamental TE_{10}-mode its E_x is zero and it will not be expected to initiate any TM_{mn}-type modes in the horn. Rather, most of its energy will go into the TE_{10}-type mode, at any rate when the angle of taper Φ is small, in which $E_x = E_r = H_\phi = 0$ and

$$H_x = \{A\mathrm{H}^{(1)}_0(\kappa_{10}r) + B\mathrm{H}^{(2)}_0(\kappa_{10}r)\}\sin(\pi x/a), \qquad (24.3)$$

Flaring of the longer side instead of the shorter side provides the H-plane sectoral horn of Fig. 5.22. The possible types can be deduced from those for the E-plane horn by interchanging the roles of x and y. The TM_{0n}-type modes can always propagate and the natural relation of the fundamental waveguide mode is the TM_{01}-type in which $E_r = E_\phi = H_y = 0$ and

$$E_y = \{A\mathrm{H}^{(1)}_{\pi/\Phi}(kr) + B\mathrm{H}^{(2)}_{\pi/\Phi}(kr)\}\sin(\pi\phi/\Phi).$$

As compared with (24.3) the argument of the Bessel functions is kr instead of $\kappa_{10}r$ and the order is not fixed but varies with the flare angle Φ. For small tapers the order will be high.

In addition, the TE_{1n}-type and TM_{1n}-type modes can propagate in the horn when the fundamental waveguide mode is unattenuated.

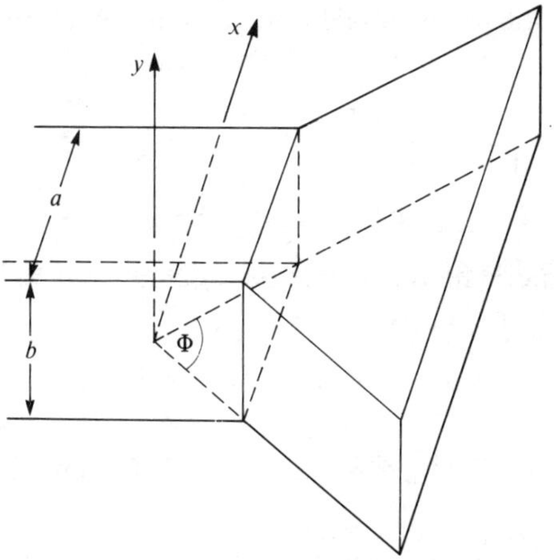

Fig. 5.22 The H-plane sectoral horn

The computation of the excitation of a horn by a waveguide might be tackled by expanding in all possible modes in both regions and matching the two expansions at the throat. The procedure is more awkward than appears at first sight since the surface $r=\text{constant}$ does not coincide with a plane perpendicular to the axis of the guide.

5.25 The conical horn

The conical horn consists of part of a right circular cone (Fig. 5.23). In terms of spherical polar coordinates R, θ, ϕ with the z-axis on the axis of the cone and origin at the cone's apex it is $\theta = \Theta$, R being limited to some interval. A solution of

$$(\nabla^2 + k^2)p = 0 \tag{25.1}$$

in spherical polar coordinates is

$$p_{m\nu}^{(j)} = z_\nu^{(j)}(kR)\mathrm{e}^{\mathrm{i}m\phi}\mathrm{P}_\nu^{|m|}(\cos\theta)$$

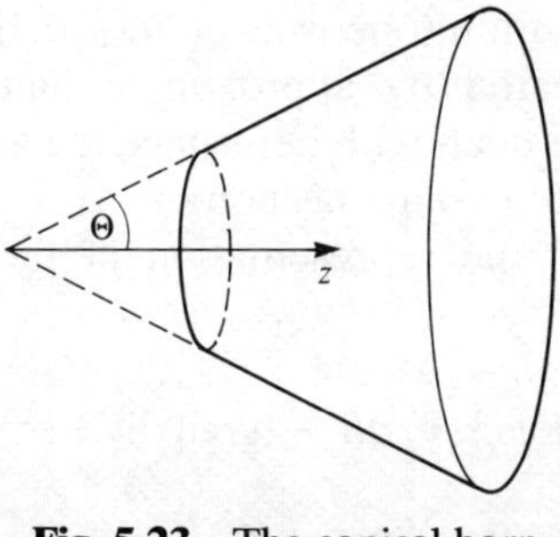

Fig. 5.23 The conical horn

where $z_\nu^{(j)}$ are the spherical Bessel functions of Appendix A.4 and $\mathrm{P}_\nu^{|m|}$ is a Legendre function (Appendix B). If the horn is soft ν must satisfy

$$\mathrm{P}_\nu^{|m|}(\cos\Theta)=0 \tag{25.2}$$

whereas if it is hard

$$\mathrm{dP}_\nu^{|m|}(\cos\Theta)/\mathrm{d}\Theta=0. \tag{25.3}$$

The electromagnetic field can also be erected from the soft solution $P_{m\nu}$ and the hard solution $M_{m\nu}$ via

$$E_R=\partial^2(RP_{m\nu})/\partial R^2+k^2RP_{m\nu}$$

and

$$E_\theta=\partial^2(RP_{m\nu})/R\,\partial R\,\partial\theta-\mathrm{i}\omega\,\partial M_{m\nu}/\sin\theta\,\partial\phi$$

for example (§9.20).

When Θ is small, (25.2) and (25.3) lead to a definite value of ν only when ν is large in which case

$$\mathrm{P}_\nu^\mu(\cos\theta)\sim\{(\nu+\tfrac{1}{2})\cos\tfrac{1}{2}\theta\}^\mu\,\mathrm{J}_{-\mu}\{(2\nu+1)\sin\tfrac{1}{2}\theta\}$$

according to (B.2.4). Thus, in a soft mode, $(\nu+\frac{1}{2})\Theta$ is approximately j_{mn} but j'_{mn} in a hard mode. The electromagnetic mode in which $(\nu+\frac{1}{2})\Theta\approx j'_{11}$ latches on naturally to the fundamental TE_{11}-mode of a circular waveguide.

The solutions $p_{m\nu}^{(3)}$ and $p_{m\nu}^{(4)}$ represent incoming and outgoing waves respectively due to the asymptotic behaviour of spherical Bessel functions. Furthermore, all modes propagate when R is large enough. However, if R is not too large, the modes with large orders of Bessel function are damped. There is therefore a tendency for a mode in a circular waveguide to excite only the associated mode in the horn at some distance from the throat; there is filtering similar to that in the sectoral horn.

5.26 Radiation properties

Few problems concerning the radiation from the open end of a waveguide are capable of exact solution (one is given in §§9.14 and 9.16). Therefore they are generally treated by approximate methods. Only the electromagnetic case will be dealt with here since the analysis may be adapted to acoustics in a straightforward fashion.

The starting point is the representation of the electromagnetic field (§1.29) in which

$$\boldsymbol{E}(\boldsymbol{x})=-\mathrm{curl}\int_S \boldsymbol{n}\wedge\boldsymbol{E}\psi(\boldsymbol{x},\boldsymbol{y})\,\mathrm{d}S_y-(\mathrm{grad\,div}+k^2)\int_S \boldsymbol{n}\wedge\boldsymbol{H}\psi\,\mathrm{d}S_y/\mathrm{i}\omega\varepsilon_0$$

for $\boldsymbol{x}$ inside the closed surface S. Let S consist of the mouth and exterior of the guide together with the sphere at infinity. Since there is an outgoing wave at infinity there is no contribution from the large sphere. On the exterior of the guide the tangential component of $\boldsymbol{E}$ disappears but the tangential component of $\boldsymbol{H}$ is nonzero. Nevertheless, the currents induced on the outside of the guide can be expected to be small to a first approximation so the surface integral over the guide's exterior can be dropped. There is left an integral over the aperture A formed by the mouth. The final approximation is to assume that the field on A is comprised of a known incident and reflected mode.

The method has much in common with Kirchhoff's theory of diffraction by an aperture (§9.26) and one might feel that the two methods would have errors of the same order of magnitude. On the other hand, Kirchhoff's theory applies to apertures which are large compared with the wavelength while the dimensions of a guide are often of the order of a wavelength. In spite of this discrepancy the approximation does seem to be tolerably accurate at points whose angular deviation from the waveguide axis is small.

Temporarily the reflected wave will be omitted. In a mode travelling along the positive z-axis

$$\boldsymbol{H}_t = Y\boldsymbol{k}\wedge\boldsymbol{E}_t$$

where $Y=\kappa_m/\omega\mu_0$ in a TE-mode and $Y=\omega\varepsilon_0/\lambda_m$ in a TM-mode. Choose the origin in the aperture A (Fig. 5.24) and let $\boldsymbol{x}=R\hat{\boldsymbol{x}}$ where $\hat{\boldsymbol{x}}$ is a unit vector in the direction of $\boldsymbol{x}$. Then, when $kR\gg 1$.

$$\boldsymbol{E}(\boldsymbol{x})\sim-\frac{ik\mathrm{e}^{-ikR}}{4\pi R}\hat{\boldsymbol{x}}\wedge\{\boldsymbol{k}\wedge\boldsymbol{P}-(\mu_0/\varepsilon_0)^{\frac{1}{2}}Y\boldsymbol{P}\wedge\hat{\boldsymbol{x}}\} \tag{26.1}$$

where

$$\boldsymbol{P}=\int_A \boldsymbol{E}_t\mathrm{e}^{ik\boldsymbol{y}\cdot\hat{\boldsymbol{x}}}\,\mathrm{d}S_y. \tag{26.2}$$

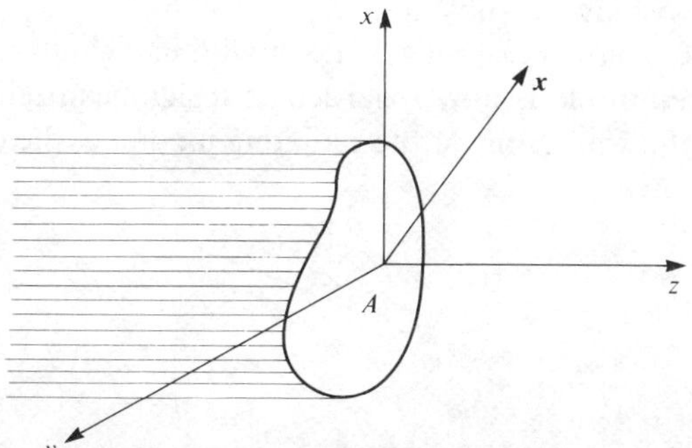

Fig. 5.24 Radiation from the open end of a waveguide

Advantage has been taken of the fact that A is plane and that, on it, $\boldsymbol{n}=-\boldsymbol{k}$.

The presence of a reflected mode with reflection coefficient R_m makes the total electric intensity in the aperture $(1+R_m)\boldsymbol{E}_t$. The effect on (26.1) is to multiply it by $(1+R_m)$ and, at the same time, replace Y by $Y(1-R_m)/(1+R_m)$; (26.2) is unchanged.

The TE_{mp}-mode in a rectangular guide has

$$\boldsymbol{E}_t=\frac{\mathrm{i}\omega\mu_0\pi}{\mu_{mp}^2}\left(\frac{\eta_m\eta_p}{ab}\right)^{\frac{1}{2}}\left(\frac{p}{b}\cos\frac{m\pi x}{a}\sin\frac{p\pi y}{b}\,\boldsymbol{i}-\frac{m}{a}\sin\frac{m\pi x}{a}\cos\frac{p\pi y}{b}\,\boldsymbol{j}\right)$$

where $\mu_{mp}^2=(m\pi/a)^2+(p\pi/b)^2$. Keeping the origin at the corner we obtain

$$\begin{aligned}\boldsymbol{P}=&-\frac{4\omega\mu_0\pi^2}{\mu_{mp}^2}\left(\frac{\eta_m\eta_p}{ab}\right)^{\frac{1}{2}}\\&\times\frac{k\sin\theta\sin\frac{1}{2}(m\pi+ka\sin\theta\cos\phi)\sin\frac{1}{2}(p\pi+kb\sin\theta\sin\phi)}{\{k^2\sin^2\theta\cos^2\phi-(m\pi/a)^2\}\{k^2\sin^2\theta\sin^2\phi-(p\pi/b)^2\}}\\&\times\left(\frac{p^2}{b^2}\cos\phi\,\boldsymbol{i}-\frac{m^2}{a^2}\sin\phi\,\boldsymbol{j}\right)\exp\{\tfrac{1}{2}\pi\mathrm{i}(m+p)+\tfrac{1}{2}\mathrm{i}k(a\cos\phi+b\sin\phi)\sin\theta\},\end{aligned}$$

θ and ϕ being the usual spherical polar angles.

The field radiated by a TM_{mp}-mode may be inferred from

$$(\boldsymbol{P}_x)_{TM}=-\lambda_m mb(\boldsymbol{P}_x)_{TE}/\omega\mu_0 pa,$$

$$(\boldsymbol{P}_y)_{TM}=\lambda_m ap(\boldsymbol{P}_y)_{TE}/\omega\mu_0 bm.$$

One interesting feature to note is that $E_\phi=0$.

The peak of the beam occurs on the z-axis. The first zero of the radiation pattern from the TE_{10}-mode occurs in the plane $x=0$ at an angle $2\sin^{-1}(2\pi/kb)$ and in the plane $y=0$ at an angle $2\sin^{-1}(3\pi/ka)$. The agreement between predicted and experimental patterns is surprisingly good for small values of θ. At large values of θ the deviations between the two may be substantial.

The radiation from a horn may be dealt with in a similar manner except that the reflected mode is now regarded as insignificant and the aperture coincides with the wavefront of the radiating mode i.e. the dash–dot curve

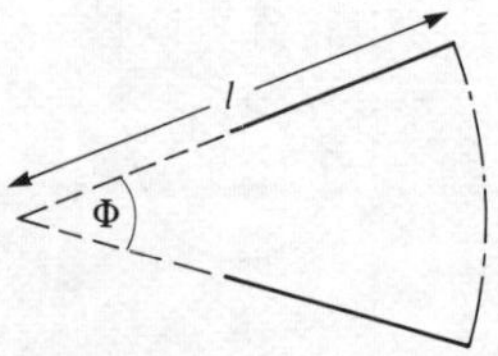

Fig. 5.25 Arrangement for the sectoral horn

for the E-plane sectoral horn in Fig. 5.25. With the origin at the apex of the horn

$$\boldsymbol{E}(\boldsymbol{x}) = \frac{ikl\mathrm{e}^{-ikR}}{4\pi R}\int_0^{\Phi}\int_0^a$$

$$\times\left[\hat{\boldsymbol{x}}\wedge i\boldsymbol{E}_{\phi} + \left(\frac{\mu_0}{\varepsilon_0}\right)^{\frac{1}{2}} H_x\hat{\boldsymbol{x}}\wedge(\{\cos(\phi'-\tfrac{1}{2}\Phi)\boldsymbol{j} - \sin(\phi'-\tfrac{1}{2}\Phi)\boldsymbol{k}\}\wedge\hat{\boldsymbol{x}})\right]$$

$$\times\exp[ik\{x\sin\theta\cos\phi + l\sin(\phi'-\tfrac{1}{2}\Phi)\sin\theta\sin\phi$$

$$+ l\cos(\phi'-\tfrac{1}{2}\Phi)\cos\theta\}]\,\mathrm{d}x\,\mathrm{d}\phi'.$$

In the TE_{10}-type mode the representation of (24.3) is employed with $r = l$ and $A = 0$. The integrals have to be evaluated numerically unless $\Phi \ll 1$. If $\Phi \ll 1$ the change of variable $l\sin(\phi'-\frac{1}{2}\Phi) = t$ gives for the field on the z-axis ($\hat{\boldsymbol{x}} = \boldsymbol{k}$, $\theta = 0$),

$$\boldsymbol{E}(R\boldsymbol{k}) = \frac{\mathrm{i}\omega\mu_0 a}{\pi^2 R}\left(\frac{2k}{\kappa_{10}}\right)^{\frac{1}{2}}\left(\frac{1}{\kappa_{10}} + \frac{1}{k}\right)$$

$$\times\left[\mathrm{C}\left\{\tfrac{1}{2}\Phi\left(\frac{kl}{\pi}\right)^{\frac{1}{2}}\right\} - \mathrm{iS}\left\{\tfrac{1}{2}\Phi\left(\frac{kl}{\pi}\right)^{\frac{1}{2}}\right\}\right]$$

$$\times\boldsymbol{j}\exp[-\mathrm{i}\{kR + l(k-\kappa_{10}) + \tfrac{1}{4}\pi\}]$$

for the TE_{10}-type mode when the Hankel functions are replaced by their asymptotic expansions. Here

$$\mathrm{C}(u) - \mathrm{iS}(u) = \int_0^u \mathrm{e}^{-\frac{1}{2}\pi\mathrm{i}t^2}\,\mathrm{d}t.$$

The quantity $\mathrm{C}^2(u) + \mathrm{S}^2(u)$ has a maximum when $u \approx 1.3$. Since the intensity involves this square it is a maximum when $kl\Phi^2 = 5.2\pi$. This is a criterion of optimal design for a sharp beam and high gain.

Horns of small flare angle have wavefronts which are virtually plane and so the fields at the aperture resemble those at the mouth of a uniform waveguide. To a first approximation, therefore, the pattern from the horn should agree with that from a rectangular waveguide of sides a and $l\Phi$. Indeed, experiment confirms that errors in beam width are negligible if neither a nor $l\Phi$ is less than the wavelength.

The H-plane horn can be investigated in a similar way. The conical horn also yields to the same technique but the analysis is more complex; it indicates that the optimal dimensions for maximum intensity at small flare angle are given by $kl(\sec\Theta - 1) = 0.6\pi$.

Exercises on Chapter 5

In the following questions a soft(hard)-mode is one which exists in an acoustic tube with soft(hard) boundaries.

1. The cross-section of a waveguide consists of the semicircle $x^2+y^2 \leqslant a^2$, $y \geqslant 0$. Show that there are soft- and TM-modes of a waveguide in which $J_n\{a(k^2-\lambda_{mp}^2)^{\frac{1}{2}}\}=0$.

2. The cross-section of a waveguide has the form of the quadrant of a circle $x^2+y^2 \leqslant a^2$, $x \geqslant 0$, $y \geqslant 0$. Show that there are hard- and TE-modes in which

$$J'_{2n}\{a(k^2-\kappa_{2n,p}^2)^{\frac{1}{2}}\}=0.$$

3. The boundary of the cross-section of a waveguide is specified by $r=b$, $r=a$ $(>b)$, $\phi=0$ and $\phi=\Phi$ in the usual polar coordinates. Show that the cut-off wavelength of TE-modes is obtained from $2\pi b/\chi'_{\nu 1}$ (see §5.7(e)) where $\nu=n\pi/\Phi$ $(n=0, 1, 2, \ldots)$ and of TM-modes is $2\pi b/\chi_{\mu 1}$ where $\mu=(n+1)\pi/\Phi$.

By considering the case $b=0$, $\Phi=2\pi$ show that the insertion of a radial baffle in a circular waveguide increases the cut-off frequency of the TM-wave and reduces it for the TE-wave. [The smallest positive root of $\tan x=2x$ is $x=1.16$.]

4. The perimeter of the cross-section of a waveguide is an equilateral triangle of side a. The centre of the triangle is taken as origin, the x-axis being parallel to one side. Put $u=\frac{1}{2}(3^{\frac{1}{2}}x+y)+2b$, $v=\frac{1}{2}(y-3^{\frac{1}{2}}x)$, $w=-y$. Verify that hard- and TE-modes are obtained from

$$A\psi_{mn}=\cos\frac{l\pi u}{3b}\cos\frac{\pi}{9b}(m-n)(v-w)+\cos\frac{m\pi u}{3b}\cos\frac{\pi}{9b}(n-l)(v-w)$$
$$+\cos\frac{n\pi u}{3b}\cos\frac{\pi}{9b}(l-m)(v-w)$$

where $\kappa_{mn}^2=k^2-(4\pi/3a)^2(m^2+mn+n^2)$, $b=a/3^{\frac{1}{2}}2$, $l=-m-n$. Confirm also that soft- and TM-modes are derived from

$$A\varphi_{mn}=\sin\frac{l\pi u}{3b}\cos\frac{\pi}{9b}(m-n)(v-w)+\sin\frac{m\pi u}{3b}\cos\frac{\pi}{9b}(n-l)(v-w)$$
$$+\sin\frac{n\pi u}{3b}\cos\frac{\pi}{9b}(l-m)(v-w)$$

with $\lambda_{mn}^2=\kappa_{mn}^2$.

Prove that the electromagnetic cut-off wavelength is $1.5a$.

5. The cross-section of a waveguide is nearly circular, the equation of the boundary being $r=a\{1+\eta f(\phi)\}$ where $\eta \ll 1$. Show that the boundary conditions on φ_m and ψ_m may be approximated by

$$\varphi_m+a\eta f(\phi)\,\partial\varphi_m/\partial r=0,$$
$$\partial\psi_m/\partial r+a\eta f(\phi)\,\partial^2\psi_m/\partial r^2-\eta f'(\phi)\,\partial\psi_m/\partial\phi=0$$

on $r=a$.

If $f(\phi)=\cos 2\phi$, prove that possible eigenfunctions are

$$\varphi_{0n}=\mathrm{J}_0(j_{0n}r/a)+\tfrac{1}{2}\eta j_{0n}^2\mathrm{J}_2(j_{0n}r/a)\cos 2\phi,$$

$$\psi_{0n}=\mathrm{J}_0(j'_{0n}r/a)+\tfrac{1}{2}\eta j'^2_{0n}\mathrm{J}_2(j'_{0n}r/a)\cos 2\phi.$$

Can you find corresponding results for the modes which are not symmetric when $\eta=0$?

6. A *bent pipe* has the boundaries $z=0$, $z=b$, $r=r_1$, $r=r_2$ $(>r_1)$ in cylindrical polar coordinates. Show that there are modes which travel along the pipe with

$$\varphi_m=\{\mathrm{J}_\kappa(\chi r)\mathrm{Y}_\kappa(\chi r_1)-\mathrm{J}_\kappa(\chi r_1)\mathrm{Y}_\kappa(\chi r)\}\mathrm{e}^{-\mathrm{i}\kappa\phi}\cos(m\pi z/b)$$

where $\chi^2=k^2-(m\pi/b)^2$ and κ satisfies

$$\mathrm{J}_\kappa(\chi r_2)\mathrm{Y}_\kappa(\chi r_1)=\mathrm{J}_\kappa(\chi r_1)\mathrm{Y}_\kappa(\chi r_2).$$

Prove also that there are modes

$$\psi_m=\{\mathrm{J}_\kappa(\chi r)\mathrm{Y}'_\kappa(\chi r_1)-\mathrm{J}'_\kappa(\chi r_1)\mathrm{Y}_\kappa(\chi r)\}\mathrm{e}^{-\mathrm{i}\kappa\phi}\sin(m\pi z/b)$$

when κ satisfies

$$\mathrm{J}'_\kappa(\chi r_2)\mathrm{Y}'_\kappa(\chi r_1)=\mathrm{J}'_\kappa(\chi r_1)\mathrm{Y}'_\kappa(\chi r_2).$$

These eigenfunctions do not lead to true electromagnetic TM- and TE waves unless $m=0$ when the first group give TE-waves. Hence, in a rectangular waveguide, a wave with a single component of electric intensity parallel to one pair of sides can be transmitted without modification round a bend in the plane perpendicular to the electric intensity.

7. Find the formulae analogous to (6.5) and (6.6) for the attenuation constant α^{s} when the boundary condition on a soft-walled pipe is changed to $p=Z\,\partial p/\partial n$ with Z small.

8. An electric dipole of moment $p_0\boldsymbol{k}\mathrm{e}^{\mathrm{i}\omega t}$ is placed at (x_0, y_0, z_0) inside a waveguide parallel to the z-axis. Prove that

$$(\nabla^2+k^2)E_z=-(p_0/\varepsilon_0)\{\delta(x-x_0)\delta(y-y_0)\delta''(z-z_0)+k^2\delta(\boldsymbol{x}-\boldsymbol{x}_0)\}$$

and hence show that

$$E_z=\sum_{m=1}^{\infty}(\nu_m^2p_0/2\mathrm{i}\varepsilon_0\lambda_m)\varphi_m(x_0, y_0)\varphi_m(x, y)\mathrm{e}^{-\mathrm{i}\lambda_m|z-z_0|}.$$

Deduce that, if an antenna carries a current $I_0\sin k(l-|z|)$ along $-l\leqslant z\leqslant l$, $r=b$, $\phi=0$ inside the circular waveguide $r=a$, the resulting field can be found in $|z|>l$ from

$$E_z=\sum_{m=0}^{\infty}\sum_{p=1}^{\infty}\left\{\frac{k(\cos kl-\cos\lambda_{mp}l)\eta_m}{\varepsilon_0\omega\pi a(k^2a^2-j_{mp}^2)^{\frac{1}{2}}}\right.$$
$$\left.\times\frac{\mathrm{J}_m(j_{mp}r/a)\mathrm{J}_m(j_{mp}b/a)}{\mathrm{J}_{m+1}^2(j_{mp})}\mathrm{e}^{-\mathrm{i}\lambda_{mp}|z|}\cos m\phi\right\}$$

where $\lambda_{mp}^2=k^2-(j_{mp}/a)^2$.

9. A junction of four waveguides is such that an incident wave in guide 1 produces only waves leaving in guides 2 and 3. An incident wave in guide 4 produces only waves leaving in guides 2 and 3. Prove that an incident wave in either guide 2 or guide 3 produces only waves leaving in guides 1 and 4.

Prove that the elements of the scattering matrix have the representation

$$S_{12}=A\mathrm{e}^{\mathrm{i}\alpha},\qquad S_{13}=(1-A^2)^{\frac{1}{2}}\mathrm{e}^{\mathrm{i}\beta},\qquad S_{24}=(1-A^2)^{\frac{1}{2}}\mathrm{e}^{\mathrm{i}\gamma},\qquad S_{34}=A\mathrm{e}^{\mathrm{i}(\beta+\gamma-\alpha)}$$

where $0<A<1$ and α, β, γ are arbitrary real numbers.

10. The soft post of §5.13 is replaced by a thin hard pillar. Examine the effect on the analysis.

11. The post of §5.13 is made perfectly conducting and is subjected to the incident TE_{10}-mode in which $E_y=\mathrm{e}^{-\mathrm{i}\kappa_{10}z}\sin(\pi x/a)$. Show that the post is inductive, presenting a shunt impedance $\mathrm{i}X$ at the post where

$$\frac{X}{Z_0}=\frac{\kappa_{10}a}{4\pi}\operatorname{cosec}^2\frac{\pi x_0}{a}\left[\ln\left(\frac{4a}{\pi d}\sin\frac{\pi x_0}{a}\right)-2\sin^2\frac{\pi x_0}{a}\right.$$
$$\left.-2\sum_{m=2}^{\infty}\left(\frac{\mathrm{i}\pi}{a\kappa_{m0}}+\frac{1}{m}\right)\sin^2\frac{m\pi x_0}{a}\right].$$

12. A capacitive iris in a rectangular waveguide (in which only the fundamental mode can propagate) consists of a thin sheet of metal occupying $0\leqslant y\leqslant\frac{1}{2}d_1$ and $d+\frac{1}{2}d_1\leqslant y\leqslant b$. Show that X/Z_0 cannot exceed

$$-\pi^2d^2/4\kappa_{10}b^2\sum_{n=1}^{\infty}\frac{1}{n^2\kappa_n}\left\{\sin\frac{n\pi}{b}(d+\tfrac{1}{2}d_1)-\sin\frac{n\pi d_1}{2b}\right\}^2.$$

Demonstrate that the method of §5.16 gives

$$\frac{Z_0}{X}=\frac{2b\kappa_{10}}{\pi}\left[\ln\left\{\sin\frac{\pi d}{2b}\sin\frac{\pi}{2b}(d+d_1)\right\}-\frac{2C\cos^2\dfrac{\pi d}{2b}\cos^2\dfrac{\pi}{2b}(d+d_1)}{1+C\sin^2\dfrac{\pi d}{2b}\sin^2\dfrac{\pi}{2b}(d+d_1)}\right]$$

by using the substitution

$$\cos\frac{\pi y}{b}=\sin\frac{\pi d}{2b}\sin\frac{\pi}{2b}(d+d_1)\cos\theta+\cos\frac{\pi d}{2b}\cos\frac{\pi}{2b}(d+d_1).$$

13. Prove that $(f,g)^2/(g,Lg)$ is stationary to small variations in g, if, and only if, $Lg=f$. If the trial function $\sum_{m=1}^{N}c_m\varphi_m$ is employed for g compare the results from the variational formula with those for Galerkin's method for $Lg=f$.

14. Show that the integral equation

$$\sin\frac{\pi x}{a}+\sum_{m=2}^{\infty}\frac{a\kappa_{10}}{m\pi}\int_0^d g_1(t)\sin\frac{m\pi t}{a}\sin\frac{m\pi x}{a}\,\mathrm{d}t=0\quad(0\leqslant x\leqslant d)$$

can be converted to the form §5.16 by taking a derivative with respect to x and then integrating by parts.

15. An inductive iris in a rectangular waveguide consists of a thin sheet of metal occupying $0 \leqslant x \leqslant d < a$, $0 \leqslant y \leqslant b$ at $z = 0$. If $2 > ka/\pi > 1$ prove that the equivalent circuit is a pure shunt of impedance iX at $z = 0$ where

$$X/Z_0 \leqslant (\kappa_{10}/dK_{11}) \sum_{m=2}^{\infty} K_{1m}^2/\kappa_m$$

with $\kappa_m^2 = (m\pi/a)^2 - k^2$, $K_{1m} = \int_0^d \sin(\pi x/d)\sin(m\pi x/a)\,dx$.

Use the theory of Exercise 14 to obtain

$$X/Z_0 = \kappa_{10} a\beta^2/2\pi\alpha(1+\beta)$$

as a first approximation, where $\alpha = \sin^2(\pi d/2a)$, $\beta = \cos^2(\pi d/2a)$, and

$$\frac{X}{Z_0} = \frac{\kappa_{10} a\beta^2}{2\pi\alpha(1+\beta)}\left\{1 + \frac{8\alpha\beta^2 C_1}{1+\beta+\alpha^3(\alpha^2+6\beta)C_1}\right\}$$

with $C_1 = -1 + 2\pi/a\kappa_2$ as a second approximation.

Hint: $\int_0^\pi G_1(\phi)\phi\,d\phi$ can be evaluated by Parseval's formula (§4.5) as $\sum_{m=1}^{\infty} \{(-)^{m-1} 2/m\} \int_0^\pi G_1(\phi)\sin m\phi\,d\phi$.

16. A circular guide of radius a contains a concentric circular guide of radius b but the inner guide extends only from $z = -\infty$ to $z = 0$. The fields are axially symmetric and $2.61ka < 2\pi$. Show, by the Wiener–Hopf method, that for the fundamental mode in the coaxial guide the equivalent impedance is an open circuit located at $z = d$ where

$$\begin{aligned} kd = {} & (ka/\pi)\{(\alpha-1)\ln(1-\alpha) - \alpha\ln\alpha\} \\ & + S(ka/\pi) - S(kb/\pi) - S^{(1)}\{(1-\alpha)ka/\pi\}, \end{aligned}$$

$$S(x) = \sum_{n=1}^{\infty} \{\sin^{-1}(\pi x/j_{0n}) - x/n\},$$

$$S^{(1)}(x) = \sum_{n=1}^{\infty} [\sin^{-1}\{\alpha\pi x/(1-\alpha)\chi_{0n}\} - x/n]$$

where $\alpha = b/a$ and χ_{0n} is defined §5.7(e).

17. A rectangular waveguide is bifurcated by placing a strip from $z = 0$ to $z = \infty$ in the plane $x = a_2$. If $1 < ka/\pi < 2$ and $ka_2 > \pi$ show that for the fundamental mode in the largest guide the equivalent network is the junction of two uniform lines, the ratio of their characteristic impedances being κ_{10}/κ'_{10}, situated at $z = -d$ where

$$\begin{aligned} \kappa_{10} d = {} & \frac{\kappa_{10} a}{\pi}\left(\frac{a_2}{a}\ln\frac{a}{a_2} + \frac{a_3}{a}\ln\frac{a}{a_3}\right) - \sin^{-1}\frac{\kappa_{10} a_3 a}{\pi(a^2 - a_3^2)^{\frac{1}{2}}} \\ & + S_2(\kappa_{10} a/\pi; 1) - S_2(\kappa_{10} a_2/\pi; a_2/a) - S_2(\kappa_{10} a_3/\pi; a_3/a) \end{aligned}$$

where $\kappa'_{10} = (k^2 - \pi^2/a_2^2)^{\frac{1}{2}}$, $a_3 = a - a_2$ and S_N is defined in §5.19.

18. A rectangular waveguide is bifurcated by placing a strip from $z=0$ to $z=\infty$ in the plane $y=b_1$. If $\kappa_{10}b<\pi$ show that for the fundamental mode in each guide the equivalent circuit consists of three lines in series, the ratio of the characteristic impedances being $Z_1/Z_0=b_1/b$ and $Z_2/Z_0=b_2/b$ (with $b_2=b-b_1$), located at $z=-d$ where

$$\kappa_{10}d=\frac{\kappa_{10}b}{\pi}\left(\frac{b_1}{b}\ln\frac{b}{b_1}+\frac{b_2}{b}\ln\frac{b}{b_2}\right)$$
$$+S_1\left(\frac{\kappa_{10}b}{\pi};0\right)-S_1\left(\frac{\kappa_{10}b_1}{\pi};0\right)-S_1(\kappa_{10}b_2/\pi;0).$$

19. By Babinet's principle show how the impedance of a symmetric inductive iris in a rectangular waveguide is related to that of a capacitive iris in a parallel plate guide supporting only the first TM-mode.

20. The region $0\leqslant y\leqslant b-d$ of a rectangular waveguide consists of dielectric, with parameters ε_1 and μ_0, whereas $b-d<y\leqslant b$ is dielectric with constants ε_2 and μ_0. Show that there are TE-modes with $H_z=e^{-i\kappa z}f(y)$ provided that κ satisfies $\beta\tan\alpha(b-d)+\alpha\tan\beta d=0$ where $\alpha^2=k^2N_1^2-\kappa^2$, $\beta^2=k^2N_2^2-\kappa^2$.

Show also that there are modes, which are neither TE or TM, in which $H_z=e^{-i\kappa z}f(y)\cos(\pi x/a)$, $E_z=e^{-i\kappa z}g(y)\sin(\pi x/a)$ so long as either $\varepsilon_1\beta_1\tan\beta_1 d+\varepsilon_2\alpha_1\tan\alpha_1(b-d)=0$ or $\alpha_1\tan\beta_1 d+\beta_1\tan\alpha_1(b-d)=0$ where $\alpha_1^2=k^2N_1^2-\kappa^2-(\pi/a)^2$, $\beta_1^2=k^2N_2^2-\kappa^2-(\pi/a)^2$. The first of these groups supplies the fundamental mode.

21. The cross-section of a circular waveguide has parameters ε_1, μ_0 for $r\leqslant b$ and ε_2, μ_0 for $b<r\leqslant a$. Show that there are modes with no angular dependence provided that κ satisfies either

$$\varepsilon_1\beta J_1(\alpha b)\{J_0(\beta b)Y_0(\beta a)-Y_0(\beta b)J_0(\beta a)\}$$
$$=\varepsilon_2\alpha J_0(\alpha b)\{J_1(\beta b)Y_0(\beta a)-Y_1(\beta b)J_0(\beta a)\}$$

or

$$\beta J_1(\alpha b)\{J_0(\beta b)Y_1(\beta a)-Y_0(\beta b)J_1(\beta a)\}$$
$$=\alpha J_0(\alpha\beta)\{J_1(\beta b)Y_1(\beta a)-Y_1(\beta b)J_1(\beta a)\}$$

where α and β are defined in Exercise 20.

22. A wave in which all the components are constant multiples of $\exp\{i(\omega t-k_0N\boldsymbol{x}\cdot\boldsymbol{n}_0)\}$ is moving in a gyromagnetic medium. Here $k_0=\omega(\mu_0\varepsilon_0)^{\frac{1}{2}}$ and N is the refractive index associated with the direction of the (constant) unit vector $\boldsymbol{n}_0$. Prove that $\boldsymbol{E}$ is perpendicular to $\boldsymbol{n}_0$, $\boldsymbol{H}$, and $\boldsymbol{B}$. Show also that $\boldsymbol{B}$ is perpendicular to $\boldsymbol{n}_0$ and that

$$\boldsymbol{B}\cdot\boldsymbol{H}=k_0^2N^2\{\boldsymbol{H}\cdot\boldsymbol{H}-(\boldsymbol{H}\cdot\boldsymbol{n}_0)^2\}/\omega^2\varepsilon.$$

Prove that Poynting's vector $\boldsymbol{S}$ lies in the plane of $\boldsymbol{n}_0$, $\boldsymbol{H}$, and $\boldsymbol{B}$ being given by

$$\boldsymbol{S}=k_0N\{\boldsymbol{H}\cdot\boldsymbol{H}\boldsymbol{n}_0-(\boldsymbol{H}\cdot\boldsymbol{n}_0)\boldsymbol{H}\}/\omega\varepsilon.$$

23. A rectangular waveguide contains traversely magnetized ferrite in $0 \leqslant x \leqslant d$ and free space in $d < x \leqslant a$. Prove that there are TE-modes with $E_z = \mathrm{e}^{-i\beta y} f(x)$ (the y-axis being along the guide) provided that β satisfies

$$\mu_1(\alpha\mu_1 + \beta\mu_2 \tan \alpha d)\tan \gamma(d-a) = \gamma(\mu_1^2 - \mu_2^2)\tan \alpha d$$

where $\alpha^2 + \beta^2 = \omega^2\varepsilon(\mu_1^2 - \mu_2^2)/\mu_1$, $\gamma^2 + \beta^2 = \omega^2\mu_0\varepsilon_0$.

24. A circular waveguide of radius a contains a concentric longitudinally magnetized ferrite core of radius b. A wave is propagating that would be the TE_{11}-mode if there were no ferrite. Show that in this wave $\kappa^2 = k^2 - (j'_{11}/a)^2 + O(b^2)$ when b is very small.

25. A gyromagnetic medium is magnetized parallel to the z-axis and occupies $x > 0$. Free space lies in $x < 0$. $\boldsymbol{E}_1$, $\boldsymbol{H}_1$ is created by a line source of moment $\boldsymbol{a}_1$ at $\boldsymbol{x}_1$ $(x_1 < 0)$ and $\boldsymbol{E}_2$, $\boldsymbol{H}_2$ by $\boldsymbol{a}_2$ at $\boldsymbol{x}_2$ $(x_2 < 0)$; both fields are independent of y. By integrating $E_{2y}H_{1z} + E_{2z}H_{1y} - E_{1y}H_{2z} - E_{1z}H_{2y}$ with respect to z from $-\infty$ to ∞ and with respect to x from $-\infty$ to ∞ prove the reciprocity theorem $a_{2y}E_{1y}(\boldsymbol{x}_2) = a_{1y}E_{2y}(\boldsymbol{x}_1)$ when $\boldsymbol{a}_1$ and $\boldsymbol{a}_2$ are both parallel to the y-axis. Show also that there is reciprocity when both $\boldsymbol{a}_1$ and $\boldsymbol{a}_2$ are perpendicular to the y-axis.

There is reciprocity in amplitude but a phase difference of π when $a_{1x} = a_{1z} = a_{2y} = 0$.

Does the introduction of perfectly conducting plane(s) $x =$ constant affect the reciprocity?

Is there a reciprocity theorem when the gyromagnetic medium is magnetized parallel to the x-axis?

26. A conical waveguide is made of the boundaries $\theta = \Theta_1$, $\theta = \Theta_2$ $(>\Theta_1)$ in spherical polar coordinates and the space in between. Show that there is a mode

$$P = (\mathrm{e}^{-ikR}/2R)\ln\{(1+\cos\theta)/(1-\cos\theta)\}$$

which, although not satisfying the soft boundary condition, does give an acceptable electromagnetic field with $E_\theta = (ik/R\sin\theta)\mathrm{e}^{-ikR}$, $B_\phi = (ik^2/\omega R\sin\theta)\mathrm{e}^{-ikR}$. This is the fundamental mode which is the natural analogue of the TEM-mode in the coaxial cylindrical guide. In addition, there are other modes arising from the values of ν satisfying either

$$\mathrm{P}_\nu^m(\cos\Theta_2)\mathrm{Q}_\nu^m(\cos\Theta_1) = \mathrm{P}_\nu^m(\cos\Theta_1)\mathrm{Q}_\nu^m(\cos\Theta_2)$$

or

$$\mathrm{P}_\nu^{m\prime}(\cos\Theta_2)\mathrm{Q}_\nu^{m\prime}(\cos\Theta_1) = \mathrm{P}_\nu^{m\prime}(\cos\Theta_1)\mathrm{Q}_\nu^{m\prime}(\cos\Theta_2).$$

27. Show that the average rate at which energy is radiated by the rectangular waveguide of §5.26 is $\omega\mu_0 a^2\kappa_{10}/2\pi^2$ in the TE_{10}-mode when $R_m = 0$. If $R_m \neq 0$ multiply by $1 - |R_m|^2$. [Hint: transform the integral of Poynting's vector over a large sphere into one over the mouth of the guide.]

28. With the theory of §5.26 show that the radiation from a circular waveguide, operating in the TE-mode in which

$$H_z = (j'_{mp}/a)^2 \mathrm{J}_m(j'_{mp}r/a)\mathrm{e}^{-\mathrm{i}\kappa_{mp}z} \cos m\phi,$$

has

$$E_\theta \sim \frac{\mathrm{i}^{m+1} m\omega\mu_0}{2kR \sin\theta}(k+\kappa_{mp}\cos\theta)\mathrm{e}^{-ikR}\mathrm{J}_m(j'_{mp})\mathrm{J}_m(ka\sin\theta)\sin m\phi,$$

$$E_\phi \sim \frac{\mathrm{i}^{m+1} a\omega\mu_0 \mathrm{e}^{-ikR}}{2R\{1-(ka\sin\theta/j'_{mp})^2\}}(\kappa_{mp}+k\cos\theta)\mathrm{J}_m(j'_{mp})\mathrm{J}'_m(ka\sin\theta)\cos m\phi$$

when $R_m = 0$. The average power radiated in the TE_{11}-mode is $\frac{1}{4}\pi\omega\mu_0\kappa_{11}(j'^2_{11}-1)\mathrm{J}_1^2(j'_{11})$. For large a the gain is approximately $10\pi(a/\lambda)^2$.

The TM-mode in which $E_z = (j_{mp}/a)^2\mathrm{J}_m(j_{mp}r/a)\mathrm{e}^{-\mathrm{i}\lambda_{mp}z}\cos m\phi$ generates $E_\phi \sim 0$ and

$$E_\theta \sim \frac{\mathrm{i}^{m-1} j_{mp}\mathrm{e}^{-ikR}\cos m\phi}{2R\sin\theta\{1-(j_{mp}/ka\sin\theta)^2\}}(\lambda_{mp}+k\cos\theta)\mathrm{J}_m(ka\sin\theta)\mathrm{J}'_m(j_{mp}).$$

29. Undertake a study of the radiation of sound waves from the open end of a pipe along the same lines as that for electromagnetic waves.

30. An acoustic wave e^{-ikz} propagates along a semi-infinite rigid tube with circular cross-section of radius b. The tube lies in $z \leq 0$ and a semi-infinite rigid conical horn of small angle θ_0 is attached at $z=0$. If the tube can support only the fundamental mode show that, at a large distance from the junction, the reflected wave is approximately $R\mathrm{e}^{ikz}$ where

$$R = -\exp\{-2ik\alpha(A/\zeta)^{\frac{1}{2}} - 2k^2A/\zeta\},$$

$$\alpha = \frac{\zeta}{4\pi}\left\{1-c\left(\frac{\zeta}{4\pi}\right)^{\frac{1}{2}}\right\}+\left(1-\frac{\zeta}{4\pi}\right)^{\frac{1}{2}},$$

$$A = \pi b^2, \qquad \zeta = 2\pi(1-\cos\theta_0), \qquad c = 16\sum_{n=1}^{\infty}(1/j'_{0n})^3 \approx 0.364.$$

31. In a *gyrotropic* medium Maxwell's equations are

$$\mathrm{curl}\,\boldsymbol{E} + \mathrm{i}\omega\boldsymbol{\mu}\cdot\boldsymbol{H} = \boldsymbol{0}, \qquad \mathrm{curl}\,\boldsymbol{H} - \mathrm{i}\omega\boldsymbol{\varepsilon}\cdot\boldsymbol{E} = \boldsymbol{0}$$

where the tensors $\boldsymbol{\mu}$ and $\boldsymbol{\varepsilon}$ are given by

$$\boldsymbol{\mu} = \begin{bmatrix} \mu_1 & -\mathrm{i}\mu_2 & 0 \\ \mathrm{i}\mu_2 & \mu_1 & 0 \\ 0 & 0 & \mu_3 \end{bmatrix}, \qquad \boldsymbol{\varepsilon} = \begin{bmatrix} \varepsilon_1 & -\mathrm{i}\varepsilon_2 & 0 \\ \mathrm{i}\varepsilon_2 & \varepsilon_1 & 0 \\ 0 & 0 & \varepsilon_2 \end{bmatrix}.$$

If $\boldsymbol{\mu}^{\mathrm{T}}$ and $\boldsymbol{\varepsilon}^{\mathrm{T}}$ are the transposes of $\boldsymbol{\mu}$ and $\boldsymbol{\varepsilon}$ respectively, verify that a solution is

$$\boldsymbol{E} = \boldsymbol{\varepsilon}^{-1} \cdot \operatorname{curl}\{\boldsymbol{\varepsilon} \cdot \operatorname{curl}(u\boldsymbol{k})\} - \mathrm{i}\omega\mu_1 \boldsymbol{\varepsilon}^{\mathrm{T}} \cdot \operatorname{curl}(v\boldsymbol{k})/\varepsilon_1,$$
$$\boldsymbol{H} = \boldsymbol{\mu}^{-1} \cdot \operatorname{curl}\{\boldsymbol{\mu} \cdot \operatorname{curl}(v\boldsymbol{k})\} + \mathrm{i}\omega\varepsilon_1 \boldsymbol{\mu}^{\mathrm{T}} \cdot \operatorname{curl}(u\boldsymbol{k})/\mu_1$$

where

$$\nabla_{\mathrm{t}}^2 u + \frac{\varepsilon_3}{\varepsilon_1}\frac{\partial^2 u}{\partial z^2} + \frac{\omega^2 \varepsilon_3}{\mu_1}(\mu_1^2 - \mu_2^2)u = \omega\mu_1\left(\frac{\varepsilon_2}{\varepsilon_1} + \frac{\mu_2}{\mu_1}\right)\frac{\varepsilon_3}{\varepsilon_1}\frac{\partial v}{\partial z},$$

$$\nabla_{\mathrm{t}}^2 v + \frac{\mu_3}{\mu_1}\frac{\partial^2 v}{\partial z^2} + \frac{\omega^2 \mu_3}{\varepsilon_1}(\varepsilon_1^2 - \varepsilon_2^2)v = -\omega\varepsilon_1\left(\frac{\varepsilon_2}{\varepsilon_1} + \frac{\mu_2}{\mu_1}\right)\frac{\mu_3}{\mu_1}\frac{\partial u}{\partial z}.$$

See if you can prove that any electromagnetic field in an infinite gyrotropic medium can be expressed in this form. [This is a *representation in two scalars* for a gyrotropic medium analogous to that of §1.13 for an isotropic one.]

6. Refraction

Two basic solutions play an important role in a uniform medium. One is the field produced by a point source, which is significant because it is a fundamental radiating device. Its properties were elaborated in Chapter 3. It is also important in representing fields by surface integrals as enumerated in Chapter 1. The second basic solution is the plane wave which is valuable for at least two reasons. Firstly, it can be generated experimentally over a band of frequencies to an acceptable level of accuracy. Secondly, it is a highly important analytical tool because of its simplicity and because of the facility of manufacturing more highly structured fields from it by means of Fourier's theorem.

This chapter starts by examining the plane wave in a homogeneous medium. However, in reality, the parameters of a medium often change from point to point. It is therefore desirable to be able to estimate the effect of a discontinuous jump in material properties as well as gradual continuous variations. Some account of these effects is also contained in this chapter when the changes occur over a plane. Discussion of other boundaries is deferred to later chapters.

The homogeneous isotropic medium

6.1 The acoustic plane wave

A solution of

$$a^2\nabla^2 p - \partial^2 p/\partial t^2 = 0 \tag{1.1}$$

which depends only on the time t and the single space coordinate x must satisfy

$$a^2\,\partial^2 p/\partial x^2 = \partial^2 p/\partial t^2. \tag{1.2}$$

Make the transformation $\xi = x - at$, $\eta = x + at$. Since a is a constant (1.2) reduces to

$$\partial^2 p/\partial\xi\,\partial\eta = 0.$$

Therefore $\partial p/\partial\eta$ is a function of η only and so p is of the form

$F(\xi)+G(\eta)$ where F and G are functions of ξ only and of η only respectively. On restoring the original variables we have

$$p = F(x-at)+G(x+at) \tag{1.3}$$

as the general solution of (1.2).

At a given time, the field is constant on any plane perpendicular to the x-axis. Moreover, suppose that $x = x_0$ at $t = 0$; then $x - at$ has the same value at the later time t_0 if $x = x_0 + at_0$. Thus, as time increases, the field distribution due to $F(x-at)$ moves in the direction of the positive x-axis with speed a and without change of shape. Hence $F(x-at)$ is known as a *plane wave* travelling in the direction of the positive x-axis. Similarly, $G(x+at)$ is a plane wave moving in the direction of the negative x-axis.

A plane wave going in the direction of the unit vector $\boldsymbol{\nu}$ can be found by temporarily selecting that direction as the x-axis. There results

$$p = F(\boldsymbol{\nu}\cdot\mathbf{x} - at). \tag{1.4}$$

The velocity $\boldsymbol{v}$ in a sound wave satisfies

$$\rho_0\, \partial\boldsymbol{v}/\partial t = -\operatorname{grad} p$$

so that, in a plane wave moving along the x-axis only v_x survives and

$$\rho_0\, \partial v_x/\partial t = -F'(x-at) - G'(x+at)$$

where the primes indicate derivatives with respect to the argument. After integration

$$\rho_0 a v_x = F(x-at) - G(x+at) \tag{1.5}$$

if v_x is to have no part which is time-independent.

The kinetic energy density of a plane wave travelling in the direction of the positive x-axis is $F^2(x-at)/2\rho_0 a^2$ and the potential energy density is exactly the same. Thus the kinetic and potential energy densities of a plane wave are equal everywhere and the acoustic energy density

$$\mathscr{E} = F^2(x-at)/\rho_0 a^2. \tag{1.6}$$

Likewise the acoustic intensity is given by

$$\boldsymbol{I} = \boldsymbol{i}F^2(x-at)/\rho_0 a = \boldsymbol{i}a\mathscr{E} \tag{1.7}$$

where $\boldsymbol{i}$ is a unit vector along the x-axis.

The acoustic intensity is the rate of flux of energy density so that (1.7) states that the energy density of a plane wave is moving in the same direction as the wave with energy speed equal to the speed of sound a.

It must not be imagined that the energy of a general wave travels either along the normal to the wavefront or with the speed of sound. In some special cases this is true but often the velocity of pressure fluctuation has to be distinguished from the velocity of energy flow.

6.2 The electromagnetic plane wave

For an electromagnetic field depending only on x and t in a homogeneous isotropic dielectric Maxwell's equations become

$$\partial H_x/\partial t = 0, \qquad \partial E_x/\partial t = 0, \tag{2.1}$$

$$-\partial E_z/\partial x + \mu\ \partial H_y/\partial t = 0, \qquad -\partial H_z/\partial x - \varepsilon\ \partial E_y/\partial t = 0, \tag{2.2}$$

$$\partial E_y/\partial x + \mu\ \partial H_z/\partial t = 0, \qquad \partial H_y/\partial x - \varepsilon\ \partial E_z/\partial t = 0, \tag{2.3}$$

$$\partial H_x/\partial x = 0, \qquad \partial E_x/\partial x = 0. \tag{2.4}$$

Equations (2.1) and (2.4) reveal that both E_x and H_x are independent of the two variables x and t. Therefore, if they are zero initially, they are zero at all subsequent times. Since the primary interest is in variable fields it is convenient to set both E_x and H_x zero.

From (2.2) and (2.3)

$$\partial^2 E_z/\partial x^2 = \mu\ \partial^2 H_y/\partial x\ \partial t = \mu\varepsilon\ \partial^2 E_z/\partial t^2$$

and it may be checked that E_y, H_y, and H_z satisfy the same equation. With $\mu\varepsilon = 1/u^2$, the equation is of the same type as (1.1). Therefore the solution (1.3) can be quoted and

$$E_z = F(x - ut) + G(x + ut) \tag{2.5}$$

where F and G are arbitrary. Similarly

$$E_y = F_1(x - ut) + G_1(x + ut). \tag{2.6}$$

No further arbitrary functions are introduced because, in order to be consistent with (2.2) and (2.3),

$$\mu u H_y = G(x + ut) - F(x - ut), \tag{2.7}$$

$$\mu u H_z = F_1(x - ut) - G_1(x + ut). \tag{2.8}$$

Invoking the interpretation of the preceding section the field with argument $x - ut$ constitutes a *plane wave* travelling along the positive x-axis with speed u. That with argument $x + ut$ is moving in the direction of the negative x-axis.

Since $E_x = H_x = 0$, *the electric and magnetic vectors of a plane wave are perpendicular to the direction of propagation.* Furthermore, for a plane wave along the positive x-axis

$$\mu u \mathbf{E} \cdot \mathbf{H} = -FF_1 + FF_1 = 0$$

from (2.5)–(2.8) and there is a like result for the wave in the opposite direction. In other words, *the electric and magnetic vectors of a plane wave are orthogonal to one another.*

The plane wave in the direction of the unit vector $\boldsymbol{\nu}$ has

$$\mathbf{E} = \mathbf{F}(\boldsymbol{\nu} \cdot \mathbf{x} - ut), \tag{2.9}$$

$$(\mu/\varepsilon)^{\frac{1}{2}}\mathbf{H} = \boldsymbol{\nu} \wedge \mathbf{F}(\boldsymbol{\nu} \cdot \mathbf{x} - ut) = \boldsymbol{\nu} \wedge \mathbf{E} \tag{2.10}$$

where $\boldsymbol{\nu} \cdot \boldsymbol{F} = 0$ in order to make the electric intensity perpendicular to the direction of propagation.

The Poynting vector $\boldsymbol{S}$ (§1.27) for a plane wave along the positive x-axis is

$$\boldsymbol{S} = \boldsymbol{E} \wedge \boldsymbol{H} = (\varepsilon/\mu)^{\frac{1}{2}}\{F_1^2(x-ut) + F^2(x-ut)\}\boldsymbol{i} \tag{2.11}$$

so that, as with the acoustic plane wave, the energy is flowing in the direction in which the wave is going.

6.3 Harmonic plane waves

When the variations with time are simple harmonic the field of a plane wave will contain a factor such as $\cos(\omega t - \omega x/u)$. The quantity $\omega/2\pi$ is called the *frequency* and $\omega/u = k$ is known as the *wavenumber.* The *wavelength* λ is defined by $\lambda = 2\pi/k$. Because only a single frequency is involved the disturbance is sometimes said to be *monochromatic.* The entity u is often designated the *phase speed,* especially when one wishes to draw a distinction with the speed of energy flow as in Cerenkov radiation (§3.4,.

As in §1.28 the time variation may be taken as $e^{i\omega t}$, with the real part of the field understood, and then the factor $e^{i\omega t}$ suppressed. For a plane sound wave moving along the x-axis (1.3) and (1.5) give

$$p = Ae^{-ikx} + Be^{ikx},$$
$$\boldsymbol{v} = (Ae^{-ikx} - Be^{ikx})\boldsymbol{i}/\rho_0 a$$

with $k = \omega/a$. In fact, for the plane wave in the direction of the unit vector $\boldsymbol{\nu}$,

$$p = Ae^{-ik\boldsymbol{\nu}\cdot\boldsymbol{x}}, \qquad \boldsymbol{v} = A\boldsymbol{\nu}e^{-ik\boldsymbol{\nu}\cdot\boldsymbol{x}}/\rho_0 a \tag{3.1}$$

from (1.4). Comparison with corresponding formulae for transmission lines invites the notion that $\rho_0 a$ may be referred to as the *characteristic impedance* of the fluid.

The complex acoustic intensity for (3.1) is

$$\boldsymbol{I} = \tfrac{1}{2}p\boldsymbol{v}^* = \tfrac{1}{2}|A|^2\boldsymbol{\nu}/\rho_0 a, \tag{3.2}$$

the asterisk signifying a complex conjugate.

In the electromagnetic harmonic plane wave we see from (2.9)–(2.11) that

$$\boldsymbol{E} = \boldsymbol{A}e^{-ik\boldsymbol{\nu}\cdot\boldsymbol{x}}, \qquad (\mu/\varepsilon)^{\frac{1}{2}}\boldsymbol{H} = \boldsymbol{\nu} \wedge \boldsymbol{A}e^{-ik\boldsymbol{\nu}\cdot\boldsymbol{x}} \tag{3.3}$$

where now $k = \omega/u$ and the constant complex vector $\boldsymbol{A}$ is such that $\boldsymbol{A} \cdot \boldsymbol{\nu} = 0$. The complex Poynting vector $\boldsymbol{S}$ (§1.30) is given by

$$\boldsymbol{S} = \tfrac{1}{2}\operatorname{Re}(\boldsymbol{E} \wedge \boldsymbol{H}^*) = \tfrac{1}{2}(\varepsilon/\mu)^{\frac{1}{2}}|\boldsymbol{A}|^2\boldsymbol{\nu} \tag{3.4}$$

so that the average flow of energy is in the direction of propagation and depends upon the square of the amplitude of the electric vector.

6.4 Polarization

The vector character of the electric intensity means that its orientation can vary in time in a plane wave, a phenomenon which does not occur in acoustics, and this variation is of some practical importance. Let the wave be going in the direction of the positive z-axis so that in complex form

$$E_x = A_1 e^{-ikz}, \qquad E_y = A_2 e^{-ikz}, \qquad E_z = 0.$$

Our interest lies in the movement of the electric intensity in time so the factor $e^{i\omega t}$ must be reinserted and the real part taken. Let $A_1 = a_1 e^{i\delta_1}$, $A_2 = a_2 e^{i\delta_2}$ where a_1, a_2, δ_1, δ_2 are real and a_1, a_2 non-negative. Then the abstraction of the real part results in

$$E_x = a_1 \cos(\omega t - kz + \delta_1), \qquad E_y = a_2 \cos(\omega t - kz + \delta_2) \tag{4.1}$$

where the real part sign has been omitted from the left-hand side for brevity. To find the curve described by the tip of the electric vector when z is fixed the quantity $\omega t - kz$ is eliminated. One arrives at

$$(E_x/a_1)^2 - 2(E_x E_y/a_1 a_2)\cos\delta + (E_y/a_2)^2 = \sin^2\delta \tag{4.2}$$

where $\delta = \delta_2 - \delta_1$.

Make the substitution

$$E_x = X\cos\chi - Y\sin\chi, \qquad E_y = X\sin\chi + Y\cos\chi$$

where

$$\tan 2\chi = (2a_1 a_2 \cos\delta)/(a_1^2 - a_2^2). \tag{4.3}$$

Then (4.2) goes over to

$$(X/a)^2 + (Y/b)^2 = 1 \tag{4.4}$$

where

$$a^2 = 2(a_1 a_2 \sin\delta)^2/(a_1^2 + a_2^2 - \alpha), \tag{4.5}$$

$$b^2 = 2(a_1 a_2 \sin\delta)^2/(a_1^2 + a_2^2 + \alpha), \tag{4.6}$$

$$\alpha = (a_1^4 + a_2^4 + 2a_1^2 a_2^2 \cos 2\delta)^{\frac{1}{2}}. \tag{4.7}$$

The quantities a and b are clearly non-negative and so (4.4) represents an ellipse with major axis of length $2a$ on the X-axis and minor axis of length $2b$ (Fig. 6.1). The end of the electric vector rotates round this ellipse, whose major axis makes an angle χ with the x-axis, as time increases completing a circuit every time that an interval of $2\pi/\omega$ elapses. The sense of rotation depends on δ. If $\sin\delta < 0$ the rotation is anticlockwise as in Fig. 6.1. The wave is said to be *elliptically polarized.*

When $a_1 = 0$ the ellipse degenerates into the straight line $x = 0$, $-a_2 \leqslant y \leqslant a_2$. If $a_1 \neq 0$ but $\sin\delta = 0$ the ellipse also reduces to a straight line

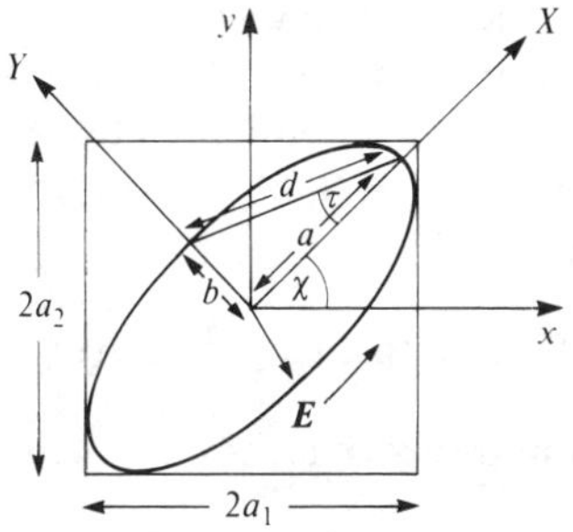

Fig. 6.1 The polarization ellipse

$y/x = \pm a_2/a_1$ depending on δ. In these cases the plane wave is said to be *linearly polarized.*

If $\delta = \pm\frac{1}{2}\pi$ and $a_1 = a_2$, then $\alpha = 0$ from (4.7) and $a = b$. The ellipse is now a circle of radius a_1 and the wave is deemed to be *circularly polarized.* When the wave is approaching and the electric vector rotates in the anticlockwise direction as in Fig. 6.1 the wave is called *right circularly polarized.* Clockwise rotation during an approach gives a *left circularly polarized* wave. In other words, it is right circularly polarized for $\delta = -\frac{1}{2}\pi$ and left circularly polarized when $\delta = \frac{1}{2}\pi$. Following a right-handed screw in the direction of propagation gives a rotation in the same sense as for right circular polarization.

Any plane wave can be represented as the linear superposition of two linearly polarized waves (e.g. choose $a_1 = 0$ in one and $a_2 = 0$ in the other). It can also be expressed as the linear combination of a right circularly polarized wave and a left polarized wave. For, in the complex form, we can take one with $E_x = \frac{1}{2}(A_1 - iA_2)$, $E_y = \frac{1}{2}i(A_1 - iA_2)$ and, in the other, $E_x = \frac{1}{2}(A_1 + iA_2)$, $E_y = -\frac{1}{2}i(A_1 + iA_2)$.

The parameters of the ellipse can also be described by introducing the angle τ where $\tan\tau = b/a$. (Fig. 6.1) Then $a = d\cos\tau$, $b = d\sin\tau$ where

$$d^2 = a^2 + b^2 = a_1^2 + a_2^2, \qquad \sin 2\tau = 2a_1a_2\,|\sin\delta|/(a_1^2 + a_2^2). \tag{4.8}$$

Polarization may also be described in terms of *Stokes parameters* defined by

$$g_0 = a_1^2 + a_2^2 = d^2,$$
$$g_1 = a_1^2 - a_2^2 = d^2\cos 2\tau\cos 2\chi,$$
$$g_2 = 2a_1a_2\cos\delta = d^2\cos 2\tau\sin 2\chi,$$
$$g_3 = 2a_1a_2\sin\delta = \pm d^2\sin 2\tau.$$

Note that $g_0^2 = g_1^2 + g_2^2 + g_3^2$. The component g_0 gives the intensity whereas the other components are the average powers of various entities. Sometimes, instead of g_0 and g_1, the parameters $g_{m0} = a_1^2$, $g_{m1} = a_2^2$ are used; they are then called *modified Stokes parameters.*

Regard (g_1, g_2, g_3) as the Cartesian coordinates of a point in three-dimensional space. This represents a state of polarization as a point on a sphere of radius g_0, known in this context as the *Poincaré sphere*. In terms of spherical polar coordinates (R, Θ, Φ) on the Poincaré sphere the polarization is identified by

$$R = a_1^2 + a_2^2, \qquad \Theta = \cos^{-1}\{(2a_1a_2 \sin\delta)/(a_1^2 + a_2^2)\}, \qquad \Phi = 2\chi.$$

Each of the representations has some virtue in the recording of polarization effects.

6.5 The effect of dissipation

Conductivity in a medium can originate a convection current with current density $\boldsymbol{J} = \sigma\boldsymbol{E}$. Associated with the current there might also be a charge. However, it was demonstrated in §1.9 that any initial charge distribution is dissipated without reference to any electromagnetic disturbance. There is, therefore, no loss in having the charge density zero during the passage of a plane wave. As far as harmonic waves are concerned the only difference thereby is that the equation $\operatorname{curl}\boldsymbol{H} = \mathrm{i}\omega\varepsilon\boldsymbol{E}$ is replaced by $\operatorname{curl}\boldsymbol{H} = (\mathrm{i}\omega\varepsilon + \sigma)\boldsymbol{E}$. Hence the theory is unchanged except for the substitution of $\varepsilon - \mathrm{i}\sigma/\omega$ for ε. The field is given by (3.3) with this alteration and

$$k^2 = \omega^2\mu(\varepsilon - \mathrm{i}\sigma/\omega).$$

To identify k precisely it is defined to be that root of k^2 which has a negative imaginary part, i.e. $k = k_\mathrm{r} - \mathrm{i}k_\mathrm{i}$ where k_r and k_i are real and positive with

$$k_\mathrm{r}^2 - k_\mathrm{i}^2 = \omega^2\mu\varepsilon, \qquad 2k_\mathrm{r}k_\mathrm{i} = \omega\mu\sigma.$$

The field contains the exponential factor $\mathrm{e}^{-\mathrm{i}k\boldsymbol{\nu}\cdot\boldsymbol{x}} = \mathrm{e}^{-(\mathrm{i}k_\mathrm{r}+k_\mathrm{i})\boldsymbol{\nu}\cdot\boldsymbol{x}}$ showing that the wave is steadily attenuated as it progresses. Indeed, the amplitude falls to $1/\mathrm{e}$ of its value after a distance $1/k_\mathrm{i}$ has been traversed.

When $\sigma/\omega\varepsilon \gg 1$, $k_\mathrm{r} \approx k_\mathrm{i} \approx (\frac{1}{2}\omega\mu\sigma)^{\frac{1}{2}}$ and the wave decays very rapidly. At the same time the magnetic intensity lags behind the electric vector by a phase difference of $\frac{1}{4}\pi$, because of the alteration to $\varepsilon^{\frac{1}{2}}$ in (3.3). The phase lag should be compared to the case when conductivity is absent and the two vectors oscillate in phase.

It is sometimes convenient to write $k = N\omega(\mu_0\varepsilon_0)^{\frac{1}{2}}$ so that

$$N = \{(\mu\varepsilon/\mu_0\varepsilon_0)(1 - \mathrm{i}\sigma/\varepsilon\omega)\}^{\frac{1}{2}}.$$

Then N is known as the *refractive index*, being complex when $\sigma \neq 0$.

Power changes are often measured in *decibels*, a decibel being ten times the common logarithm of the change. Hence the power loss due to conductivity is $10\log_{10}\mathrm{e}^{2k_\mathrm{i}} = 8.686k_\mathrm{i}\ \mathrm{dB\,m^{-1}}$.

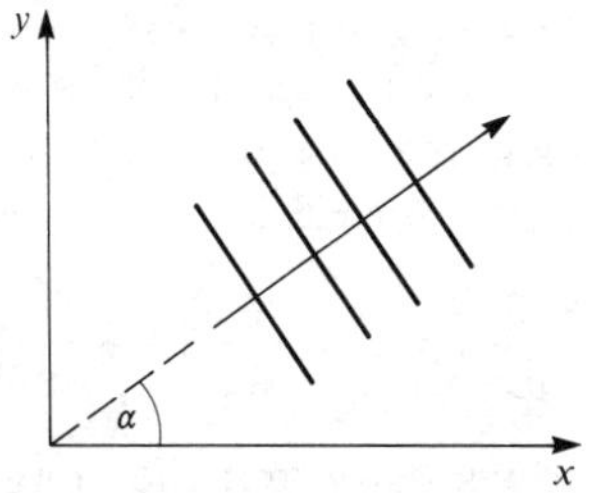

Fig. 6.2 The homogeneous plane wave (α real); the heavy lines indicate both the planes of constant amplitude and of constant phase

There is another plane wave in which the amplitude falls exponentially. Consider

$$e^{-ik(x\cos\alpha+y\sin\alpha)}=e^{-ikr\cos(\phi-\alpha)}$$

in which k is real and r, ϕ are polar coordinates related to x, y by $x=r\cos\phi$, $y=r\sin\phi$. So far, it has been assumed that α is real and then α is the angle between the direction of propagation and the positive x-axis. (Fig. 6.2). The planes of constant phase and constant amplitude coincide. To emphasize this it is referred to as a *homogeneous plane wave.* But it is possible to make α complex; this will not prevent the field satisfying the governing equations. The planes of constant phase and constant amplitude may then differ; it is an *inhomogeneous plane wave.* Let $\alpha=\alpha_r+i\alpha_i$ where α_r, α_i are real and $\alpha_i\neq 0$. Then

$$e^{-ikr\cos(\phi-\alpha)}=e^{-ikr\cosh\alpha_i\cos(\phi-\alpha_r)}\cdot e^{kr\sinh\alpha_i\sin(\phi-\alpha_r)}.$$

It is immediately evident that the planes of constant phase are perpendicular to the planes of constant amplitude (Fig. 6.3). The direction of phase propagation makes an angle α_r with the x-axis and the wavelength is diminished by the factor sech α_i. On the other hand, there is exponential decay or growth, of magnitude determined by $k\sinh\alpha_i$, in the direction orthogonal to that of phase propagation.

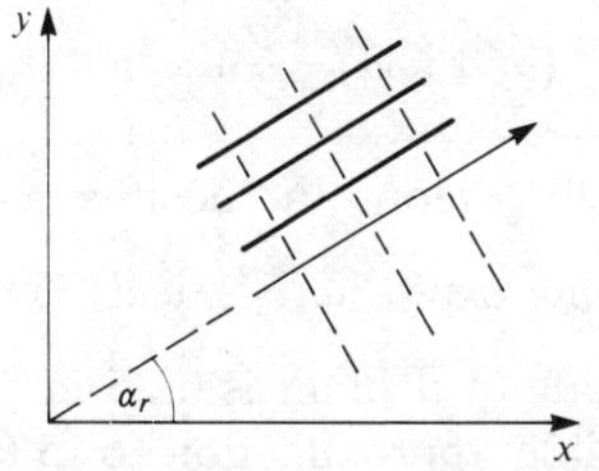

Fig. 6.3 The inhomogeneous plane wave (α not real); the heavy lines are the planes of constant amplitude while the dashed lines denote planes of constant phase

Insight into the influence of dissipation on sound waves can be gained from considering how viscosity affects a disturbance which depends only on x and t. Thermal effects may be discussed in a similar manner. The starting point, instead of $\rho(\partial v_i/\partial t + v_j\,\partial v_i/\partial x_j) = -\partial p/\partial x_i$, is (1.3.6)

$$\frac{\partial v_i}{\partial t} + v_j\frac{\partial v_i}{\partial x_j} = -\frac{1}{\rho}\frac{\partial}{\partial x_i}\left(p + \tfrac{2}{3}\rho\nu\frac{\partial v_j}{\partial x_j}\right) + \frac{1}{\rho}\frac{\partial}{\partial x_j}\left\{\rho\nu\left(\frac{\partial v_i}{\partial x_j} + \frac{\partial v_j}{\partial x_i}\right)\right\} \tag{5.1}$$

where ν is the kinematic viscosity and the summation convention (Appendix F) has been employed. When all quantities depend only on x and t the equation for v_x is

$$\frac{\partial v_x}{\partial t} + v_x\frac{\partial v_x}{\partial x} = -\frac{1}{\rho}\frac{\partial p}{\partial x} + \frac{4}{3\rho}\frac{\partial}{\partial x}\left(\rho\nu\frac{\partial v_x}{\partial x}\right).$$

Neglecting the squares of small quantities and treating ν as constant we obtain

$$\frac{\partial v_x}{\partial t} = -\frac{1}{\rho_0}\frac{\partial p}{\partial x} + \frac{4\nu}{3}\frac{\partial^2 v_x}{\partial x^2}. \tag{5.2}$$

Now, from (1.43),

$$\frac{\partial \rho}{\partial t} + \rho_0\frac{\partial v_x}{\partial x} = 0$$

which becomes, with $p = a^2\rho$,

$$\frac{\partial p}{\partial t} + a^2\rho_0\frac{\partial v_x}{\partial x} = 0. \tag{5.3}$$

A partial derivative of (5.2) with respect to x empowers the removal of v_x from (5.2) and (5.3) with the result

$$\frac{\partial^2 p}{\partial t^2} - a^2\frac{\partial^2 p}{\partial x^2} = d\frac{\partial^3 p}{\partial x^2\,\partial t} \tag{5.4}$$

where $d = 4\nu/3$ is a measure of the dissipation.

For harmonic waves (5.4) becomes

$$(a^2 + \mathrm{i}\omega d)\frac{\mathrm{d}^2 p}{\mathrm{d}x^2} + \omega^2 p = 0 \tag{5.5}$$

so that in a wave travelling along the positive x-axis

$$p = \exp\{-\mathrm{i}\omega(a^2 + \mathrm{i}\omega d)^{-\frac{1}{2}}x\}. \tag{5.6}$$

The order of magnitude of d in air is $0.25\ \mathrm{cm}^2\,\mathrm{s}^{-1}$ and $d/2a^2$ is around 10^{-10} s. Thus a reasonable approximation to (5.6) is

$$p = \exp\left\{-\frac{\mathrm{i}\omega}{a}\left(1 - \frac{\mathrm{i}\omega d}{2a^2}\right)x\right\} \tag{5.7}$$

and the factor controlling the attenuation is $\omega^2 d/2a^3$. The decline in p will take a long distance, with the higher frequencies suffering first, unless the frequency is greater than, say, 20 kHz when the attentuation may be appreciable enough to invalidate (5.7) as an approximation to (5.6) at distances over 30 m.

It will be observed that an equation similar to (5.5) can be obtained by ascribing a positive imaginary part to the sound speed.

The power loss due to viscosity according to (5.6) is

$$10 \log_{10} \exp(\omega^2 d/a^3) = 4.343\omega^2 d/a^3 \text{ dB m}^{-1}.$$

6.6 Refraction and reflection at a plane

(a) *Sound waves*

When the properties of the medium are different on the two sides of a plane the propagation of a plane wave cannot go undisturbed; the governing equation may not even be the same in the two media. Choose axes so that the interface is $x = 0$ and a plane wave impinges on the boundary from $x < 0$ (Fig. 6.4). The y-axis can be selected so that the direction of propagation lies in the (x, y)-plane. Generally, the plane containing the direction of propagation and the normal to the interface is called the *plane of incidence.* Our special choice of axes makes the (x, y)-plane the plane of incidence.

For generality it will be assumed that the rest density and speed of sound are ρ_1, a_1 in $x < 0$ and ρ_2, a_2 in $x > 0$. The boundary conditions depend upon what one is prepared to assert about the interface. One possibility is to ask for the pressure and normal velocity to be continuous; discontinuity in the tangential velocity must then be accepted unless $\rho_1 a_1 = \rho_2 a_2$. On the other hand, one might prefer to have the velocity

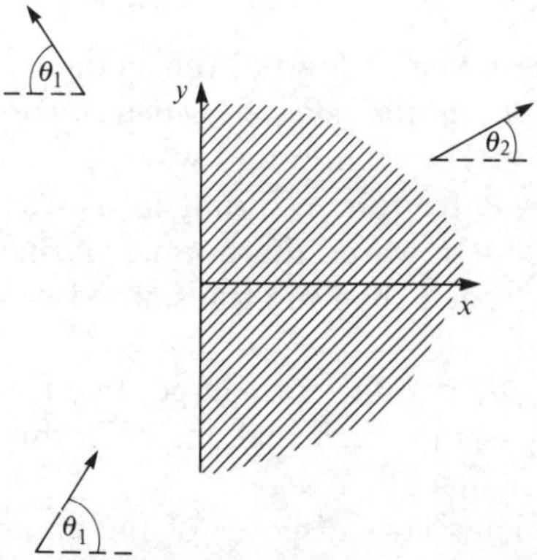

Fig. 6.4 Reflection and refraction at $x = 0$

continuous and agree that the pressure might be discontinuous. Mathematical generality is secured by imposing

$$p_1 = p_2, \qquad v_{1x} = \zeta v_{2x}, \tag{6.1}$$

the constant ζ being adjusted according to the prevailing circumstances. The suffix 1 refers to the region in $x<0$ and 2 to that in $x>0$. Notice that if the apparently more general boundary condition $p_1 = \eta p_2$ were imposed its solution could be deduced from (6.1). For, if p_2 is the solution from (6.1), the field in medium 2 is taken as p_2/η and ζ is replaced by ζ/η.

Let the incident wave p^{i} be travelling at an angle θ to the x-axis so that

$$p^{\mathrm{i}} = \mathrm{e}^{-\mathrm{i}k_1(x\cos\theta + y\sin\theta)}$$

where $k_1 = \omega/a_1$. There is no z-dependence in p^{i} and the boundary conditions do not introduce any so it is a reasonable supposition that any additional fields are independent of z. The incident wave can be expected to transmit a disturbance $p^{(2)}$ in $x>0$, balanced by a reflected wave $p^{(1)}$ in $x<0$. Taking there extra waves to be plane we assume (Fig. 6.4)

$$p^{(1)} = C_1\mathrm{e}^{-\mathrm{i}k_1(-x\cos\theta_1 + y\sin\theta_1)},$$

$$p^{(2)} = C_2\mathrm{e}^{-\mathrm{i}k_2(x\cos\theta_2 + y\sin\theta_2)}$$

where $k_2 = \omega/a_2$. The total field in $x<0$ is $p^{\mathrm{i}} + p^{(1)}$ and that in $x>0$ is $p^{(2)}$.

Identifying $p^{\mathrm{i}} + p^{(1)}$ with p_1 and $p^{(2)}$ with p_2 the first of (6.1) gives on $x=0$

$$\mathrm{e}^{-\mathrm{i}k_1 y\sin\theta} + C_1\mathrm{e}^{-\mathrm{i}k_1 y\sin\theta_1} = C_2\mathrm{e}^{-\mathrm{i}k_2 y\sin\theta_2}. \tag{6.2}$$

This equation cannot be valid for all y unless

$$k_1\sin\theta = k_1\sin\theta_1 = k_2\sin\theta_2.$$

Put another way

$$\theta_1 = \theta, \qquad \sin\theta_2 = (a_2/a_1)\sin\theta. \tag{6.3}$$

Equations (6.3) express *Snell's laws* of reflection and refraction. The first states that *the incident and the reflected waves make equal angles with the normal to the interface.* The transmitted wave is bent towards the normal to the interface or away from it according as $a_2 < a_1$ or $a_2 > a_1$. Thus the incident wave gets bent towards the normal during transmission if it is travelling at a higher speed than the transmitted wave. The phenomenon is known as *refraction.*

Observe that, if a_2/a_1 is sufficiently large, $(a_2/a_1)\sin\theta$ exceeds unity in magnitude so that θ_2 cannot be a real angle. In this case the transmitted field is an inhomogeneous plane wave.

Complying with Snell's laws disposes of the exponential in (6.2) and

$$1 + C_1 = C_2. \tag{6.4}$$

For the satisfaction of the second of (6.1) we require

$$(k_1/\rho_1)(\cos\theta - C_1\cos\theta_1) = (\zeta k_2/\rho_2)C_2\cos\theta_2. \tag{6.5}$$

Consequently

$$C_1 = \frac{k_1\rho_2\cos\theta - k_2\rho_1\zeta\cos\theta_2}{k_1\rho_2\cos\theta_1 + k_2\rho_1\zeta\cos\theta_2}, \tag{6.6}$$

$$C_2 = \frac{k_1\rho_2(\cos\theta + \cos\theta_1)}{k_1\rho_2\cos\theta_1 + k_2\rho_1\zeta\cos\theta_2}. \tag{6.7}$$

By means of Snell's laws (6.3) the formulae (6.6), (6.7) can be expressed in terms of the angle of incidence solely. Thus

$$C_1 = \frac{a_2\rho_2\cos\theta - \rho_1\zeta(a_1^2 - a_2^2\sin^2\theta)^{\frac{1}{2}}}{a_2\rho_2\cos\theta + \rho_1\zeta(a_1^2 - a_2^2\sin^2\theta)^{\frac{1}{2}}}, \tag{6.8}$$

$$C_2 = \frac{2a_2\rho_2\cos\theta}{a_2\rho_2\cos\theta + \rho_1\zeta(a_1^2 - a_2^2\sin^2\theta)^{\frac{1}{2}}}. \tag{6.9}$$

(b) *Electromagnetic waves*

As in the acoustical problem the incident field will be taken as independent of z and the scattered waves likewise. Then, by §1.36, the total field can be viewed as composed of two separate parts, one depending on E_z alone and the other on H_z only. Either the electric vector or the magnetic intensity will be perpendicular to the plane of incidence. The investigation can treat the two problems as divorced from one another.

(i) $\mathbf{E}^i$ *perpendicular to the plane of incidence.* When the electric intensity is normal to the plane of incidence

$$E_z^i = \exp\{-ik_1(x\cos\theta + y\sin\theta)\}, \qquad \omega\mu_1 H_x^i = k_1 E_z^i \sin\theta,$$
$$\omega\mu_1 H_y^i = -k_1 E_z^i \cos\theta$$

where ε_1, μ_1 are the permittivity and permeability of the medium and now $k_1 = \omega(\mu_1\varepsilon_1)^{\frac{1}{2}}$. The reflected wave has

$$E_z^{(1)} = A_1\exp\{-ik_1(-x\cos\theta_1 + y\sin\theta_1)\}, \qquad \omega\mu_1 H_x^{(1)} = k_1 E_z^{(1)}\sin\theta_1,$$
$$\omega\mu_1 H_y^{(1)} = k_1 E_z^{(1)}\cos\theta_1$$

and in the transmitted wave

$$E_z^{(2)} = A_2\exp\{-ik_2(x\cos\theta_2 + y\sin\theta_2)\}, \qquad \omega\mu_2 H_x^{(2)} = k_2 E_z^{(2)}\sin\theta_2,$$
$$\omega\mu_2 H_y^{(2)} = -k_2 E_z^{(2)}\cos\theta_2$$

where ε_2, μ_2 are the permittivity and permeability in $x > 0$ and $k_2 = \omega(\mu_2\varepsilon_2)^{\frac{1}{2}}$.

The tangential components of the electric and magnetic intensities must be continuous across $x=0$. Therefore, on $x=0$,

$$E_z^i + E_z^{(1)} = E_z^{(2)}, \qquad H_y^i + H_y^{(1)} = H_y^{(2)}. \tag{6.10}$$

It is evident at once that Snell's laws hold. They may be put in the form

$$\theta_1 = \theta, \qquad \sin\theta_2 = (N_1/N_2)\sin\theta \tag{6.11}$$

where benefit has been drawn from $k_1/k_2 = N_1/N_2$, N_1 and N_2 being the refractive indices of the two media. The refracted wave is pushed towards or away from the normal to the interface according as the incident field is in the medium of smaller or larger refractive index. Again, the transmitted plane wave will be inhomogeneous if N_2 is sufficiently smaller than N_1.

A point concerning the refractive index in this context needs airing. To avoid undue complication take $\mu_1 = \mu_2 = \mu_0$. Then N might be determined for a particular substance by measuring the value of ε by electrostatic methods. Alternatively, a slab of it might be placed in free space and N calculated from a refraction experiment via (6.11). The two values are are found to agree for many substances, e.g. the monatomic gases, at all wavelengths down to those of the spectrum of visible light. Other materials can exhibit a wide discrepancy between the two values as the frequency goes up. For instance, $(\varepsilon/\varepsilon_0)^{\frac{1}{2}} = 9$ in electrostatic experiments with water but refraction in the visible spectrum offers $N = 1.3$. The difference is explained by the presence of *polar molecules*, i.e. molecules which have a permanent dipole moment. This moment has an important effect at low frequencies and therefore in the electrostatic determination of ε but little influence at high frequencies where the large mass of the molecule impedes its response to the rapid fluctuations of the field. At optical frequencies N will be independent of the permanent molecular moment and disagree with the electrostatic $(\varepsilon/\varepsilon_0)^{\frac{1}{2}}$. Therefore ε cannot be regarded as independent of frequency in general but thought of as defined from N^2. At long wavelengths this is in harmony with the electrostatic definition.

The imposition of (6.10) furnishes

$$1 + A_1 = A_2,$$
$$(k_1/\mu_1)(-\cos\theta + A_1\cos\theta_1) = -(k_2/\mu_2)A_2\cos\theta_2$$

whence

$$A_1 = \frac{N_1\mu_2\cos\theta - N_2\mu_1\cos\theta_2}{N_1\mu_2\cos\theta + N_2\mu_1\cos\theta_2}, \tag{6.12}$$

$$A_2 = \frac{2N_1\mu_2\cos\theta}{N_1\mu_2\cos\theta + N_2\mu_1\cos\theta_2} \tag{6.13}$$

or, in terms of the angle of incidence only,

$$A_1 = \frac{\mu_2 N_1 \cos\theta - \mu_1(N_2^2 - N_1^2 \sin^2\theta)^{\frac{1}{2}}}{\mu_2 N_1 \cos\theta + \mu_1(N_2^2 - N_1^2 \sin^2\theta)^{\frac{1}{2}}}, \tag{6.14}$$

$$A_2 = \frac{2\mu_2 N_1 \cos\theta}{\mu_2 N_1 \cos\theta + \mu_1(N_2^2 - N_1^2 \sin^2\theta)^{\frac{1}{2}}}. \tag{6.15}$$

(ii) $\boldsymbol{H}^{\mathrm{i}}$ *perpendicular to the plane of incidence.* Here the appropriate forms are

$$H_z^{\mathrm{i}} = \exp\{-ik_1(x\cos\theta + y\sin\theta)\},$$
$$H_z^{(1)} = B_1 \exp\{-ik_1(-x\cos\theta_1 + y\sin\theta_1)\},$$
$$H_z^{(2)} = B_2 \exp\{-ik_2(x\cos\theta_2 + y\sin\theta_2)\}.$$

The analysis is very similar to (i) except that continuity of H_z and E_y is demanded. The conclusion is that Snell's laws hold and

$$B_1 = \frac{\mu_1 N_2^2 \cos\theta - \mu_2 N_1(N_2^2 - N_1^2 \sin^2\theta)^{\frac{1}{2}}}{\mu_1 N_2^2 \cos\theta + \mu_2 N_1(N_2^2 - N_1^2 \sin^2\theta)^{\frac{1}{2}}}, \tag{6.16}$$

$$B_2 = \frac{2\mu_1 N_2^2 \cos\theta}{\mu_1 N_2^2 \cos\theta + \mu_2 N_1(N_2^2 - N_1^2 \sin^2\theta)^{\frac{1}{2}}}. \tag{6.17}$$

6.7 Lossless media

The amplitudes of the reflected and refracted waves in the previous section all have a similar structure. They are known as *Fresnel coefficients* because the reflection problem was first tackled successfully by Fresnel on the basis of a dynamical theory.

There is great similarity between (6.8), (6.9) and (6.14), (6.15). In fact, the identification $\zeta = \mu_1\rho_2/\rho_1\mu_2$, $a_1 = 1/N_1$, $a_2 = 1/N_2$ makes them the same. It will therefore be sufficient to discuss only the electromagnetic coefficients.

Assume that the media are lossless and that $\mu_1 = \mu_2$. Then $N_1/N_2 = (\varepsilon_1/\varepsilon_2)^{\frac{1}{2}}$. From (6.11) the angle of refraction is real for angles of incidence such that $(\varepsilon_1/\varepsilon_2)^{\frac{1}{2}} \sin\theta \leqslant 1$; in particular, this is true if $\varepsilon_1 < \varepsilon_2$. Naturally there will be no reflected wave if $\varepsilon_1 = \varepsilon_2$ as well.

The first case that will be discussed is when the angle of refraction is real and $\varepsilon_1 \neq \varepsilon_2$. Then (6.14)–(6.17) can be expressed as

$$A_1 = \sin(\theta_2 - \theta)/\sin(\theta_2 + \theta), \qquad A_2 = 2\sin\theta_2\cos\theta/\sin(\theta_2 + \theta),$$
$$B_1 = \tan(\theta - \theta_2)/\tan(\theta + \theta_2), \qquad B_2 = \sin 2\theta/\sin(\theta + \theta_2)\cos(\theta - \theta_2)$$

and are all real. The coefficients A_2 and B_2 are positive ensuring that the transmitted wave is always in phase with the incident field. The reflected wave is either in phase with the incident or out of phase by π depending upon the relative magnitudes of θ and θ_2. For example, if $\theta > \theta_2$ (i.e. $N_2 > N_1$) and $\theta + \theta_2 < \frac{1}{2}\pi$ both $E_z^{(1)}$ and $E_y^{(1)}$ are π out of phase with E_z^{i} and E_y^{i}. This is important in certain interference phenomena.

When $\theta + \theta_2 = \frac{1}{2}\pi$, $B_1 = 0$ and there is no reflected wave with $\boldsymbol{H}^{\mathrm{i}}$ perpendicular to the plane of incidence. The effect occurs, according to (6.11), at the angle of incidence

$$\tan\theta = (\varepsilon_2/\varepsilon_1)^{\frac{1}{2}}. \tag{7.1}$$

The angle satisfying (7.1) is called the *Brewster angle.* For a general incident wave the polarization of the reflected field differs from that in the incident because the coefficients of the two component reflected waves are not equal. In particular, for incidence at the Brewster angle, the reflected wave is entirely polarized with the electric vector perpendicular to the plane of incidence. Polarized beams can be produced by means of this property. *Natural light* (light whose planes of polarization have random orientations) at the Brewster angle is reflected as linearly polarized light. Typical values of the Brewster angle for water are 53°–85°, depending on the frequency.

Further $A_1/B_1 = -\cos(\theta - \theta_2)/\cos(\theta + \theta_2)$ and $A_2/B_2 = \sin\theta_2 \cos(\theta - \theta_2)/\sin\theta$ so that, when the incident wave contains both polarizations, both the reflected and transmitted waves will have their polarizations in different planes from the incident wave in general. Measurement of the deviation between these planes provides an experimental check on the theory. In addition, $|A_2/B_2| < 1$ so that, if a wave undergoes a number of successive refractions of the type under consideration, the transmitted wave will become more and more polarized with its magnetic intensity normal to the plane of incidence. In this way, linearly polarized light can be created by passing ordinary light through a series of transparent plates.

Turn now to the case when $(\varepsilon_1/\varepsilon_2)^{\frac{1}{2}}\sin\theta > 1$ so that the angle of refraction is complex. Then $\cos\theta_2$ may be positive or negative imaginary. To preclude the transmitted wave from growing exponentially as it departs from the interface only the negative imaginary value can be accepted. Hence

$$\cos\theta_2 = -\mathrm{i}\{(\varepsilon_1/\varepsilon_2)\sin^2\theta - 1\}^{\frac{1}{2}}. \tag{7.2}$$

The refracted wave is now attenuated by the factor $\exp[-k_2 x\{(\varepsilon_1/\varepsilon_2)\sin^2\theta - 1\}^{\frac{1}{2}}]$.

On the basis of (7.2)

$$A_1 = \frac{\cos\theta + \mathrm{i}(\sin^2\theta - \varepsilon_2/\varepsilon_1)^{\frac{1}{2}}}{\cos\theta - \mathrm{i}(\sin^2\theta - \varepsilon_2/\varepsilon_1)^{\frac{1}{2}}}, \qquad A_2 = \frac{2\cos\theta}{\cos\theta - \mathrm{i}(\sin^2\theta - \varepsilon_2/\varepsilon_1)^{\frac{1}{2}}}, \tag{7.3}$$

$$B_1 = \frac{(\varepsilon_2/\varepsilon_1)\cos\theta + \mathrm{i}(\sin^2\theta - \varepsilon_2/\varepsilon_1)^{\frac{1}{2}}}{(\varepsilon_2/\varepsilon_1)\cos\theta - \mathrm{i}(\sin^2\theta - \varepsilon_2/\varepsilon_1)^{\frac{1}{2}}}, \qquad B_2 = \frac{2(\varepsilon_2/\varepsilon_1)\cos\theta}{(\varepsilon_2/\varepsilon_1)\cos\theta - \mathrm{i}(\sin^2\theta - \varepsilon_2/\varepsilon_1)^{\frac{1}{2}}}. \tag{7.4}$$

The complex character of the coefficients excludes any simple phase relationship between the reflected, transmitted and incident waves.

The behaviour of the energy warrants a little thought. Let the incident wave have $\boldsymbol{E}^{\mathrm{i}}$ perpendicular to the plane of incidence. The average energy crossing a unit area perpendicular to the incident wave is $\frac{1}{2}(\varepsilon_1/\mu_1)^{\frac{1}{2}}$ from (3.4). The reflected wave carries average energy across a unit area perpendicular to its direction of propagation of $\frac{1}{2}(\varepsilon_1/\mu_1)^{\frac{1}{2}}|A_1|^2=\frac{1}{2}(\varepsilon_1/\mu_1)^{\frac{1}{2}}$. In other words, all the incident energy is reflected on average. Repetition of the calculation reveals that the same statement is true when $\boldsymbol{H}^{\mathrm{i}}$ is perpendicular to the plane of incidence. The phenomenon is described as *total reflection.* A check on the energy flow in the second medium brings out a surprising feature about the energy balance. Take again the incident wave to have $\boldsymbol{E}^{\mathrm{i}}$ perpendicular to the plane of incidence. The complex Poynting vector gives for the average flow in the second medium

$$\tfrac{1}{2}\,\mathrm{Re}\{E_z^{(2)}(H_x^{(2)*}\boldsymbol{j}-H_y^{(2)*}\boldsymbol{i})\}$$
$$=\tfrac{1}{2}(\varepsilon_1/\mu_1)^{\frac{1}{2}}|A_2|^2\sin\theta\boldsymbol{j}\exp[-2k_2x\{(\varepsilon_1/\varepsilon_2)\sin^2\theta-1\}^{\frac{1}{2}}],$$

$\boldsymbol{j}$ being a unit vector parallel to the y-axis. Thus there is energy flow in the second medium parallel to the interface. Yet all the incident energy is reflected and the media are loss-free so the energy in the second medium could have been expected to be zero. The analysis does not tell us where the energy in the second medium comes from but the explanation of the paradox lies in the plane waves being of infinite extent. In practice the incident wave will be of finite size and will be switched on at some time. When these facts are taken into account it is found that energy does indeed flow along the interface because the reflected beam is displaced laterally from the primary beam on the interface (see Exercise 41). In spite of the paradox the above analysis works well in practical situations.

The reflected amplitudes may be expressed from (7.3) and (7.4) as

$$A_1=\mathrm{e}^{2\mathrm{i}\delta_{\mathrm{A}}},\qquad B_1=\mathrm{e}^{2\mathrm{i}\delta_{\mathrm{B}}}$$

where

$$\tan\delta_{\mathrm{A}}=(\sin^2\theta-\varepsilon_2/\varepsilon_1)^{\frac{1}{2}}/\cos\theta,$$
$$\tan\delta_{\mathrm{B}}=\varepsilon_1(\sin^2\theta-\varepsilon_2/\varepsilon_1)^{\frac{1}{2}}/\varepsilon_2\cos\theta.$$

Separate an arbitrary incident field into its two components in which $\boldsymbol{E}^{\mathrm{i}}$ and $\boldsymbol{H}^{\mathrm{i}}$ are respectively perpendicular to the plane of incidence. The reflected field will consist of the superposition of two linearly polarized waves with their electric vectors perpendicular to one another and differing in phase by $2\delta_{\mathrm{A}}-2\delta_{\mathrm{B}}$. Now

$$\tan(\delta_{\mathrm{A}}-\delta_{\mathrm{B}})=\frac{\tan\delta_{\mathrm{A}}-\tan\delta_{\mathrm{B}}}{1+\tan\delta_{\mathrm{A}}\tan\delta_{\mathrm{B}}}=-\frac{\cos\theta}{\sin^2\theta}(\sin^2\theta-\varepsilon_2/\varepsilon_1)^{\frac{1}{2}} \qquad (7.5)$$

which is not zero in general. Hence, unless the incident wave is polarized perpendicular to the plane of incidence, *the reflected wave is elliptically polarized when total reflection occurs.*

The behaviour of the phase difference for various θ is of interest. As the angle of incidence θ increases from zero there is no total reflection until it reaches the value given by $\sin\theta=(\varepsilon_2/\varepsilon_1)^{\frac{1}{2}}$. At that point (7.5) becomes pertinent and $\tan(\delta_A-\delta_B)=0$. As θ increases $\tan(\delta_A-\delta_B)$ decreases steadily until it attains the value $(\varepsilon_2-\varepsilon_1)/2(\varepsilon_1\varepsilon_2)^{\frac{1}{2}}$ at $\sin\theta=\{2\varepsilon_2/(\varepsilon_1+\varepsilon_2)\}^{\frac{1}{2}}$. Further increase of θ makes $(\delta_A-\delta_B)$ grow and it eventually returns to zero at $\theta=\frac{1}{2}\pi$. Put succinctly, $\tan(\delta_A-\delta_B)$ never exceeds 0 and has a single minimum of $(\varepsilon_2-\varepsilon_1)/2(\varepsilon_1\varepsilon_2)^{\frac{1}{2}}$.

Total reflection has formed the basis of the production of circularly polarized waves. The incident wave is arranged to be polarized in a plane inclined at 45° to the plane of incidence. Its two basic components then have equal magnitudes and hence the reflected waves have as well. Therefore the reflected wave is circularly polarized if $2(\delta_A-\delta_B)=\pm\frac{1}{2}\pi$, i.e. if $\tan(\delta_A-\delta_B)=\pm1$. By the preceding paragraph the upper sign is impossible to achieve and the lower sign can be attained only if $-1\geqslant(\varepsilon_2-\varepsilon_1)/2(\varepsilon_1\varepsilon_2)^{\frac{1}{2}}$, i.e. $(\varepsilon_2/\varepsilon_1)^{\frac{1}{2}}\leqslant0.414$. Subject to this condition there will be two angles of incidence (one on either side of the value giving the minimum of $\tan(\delta_A-\delta_B)$) at which the reflected wave is circularly polarized.

6.8 Dissipative media

When the medium in $x>0$ possesses conductivity the only change to Maxwell's equations is, as has been explained in §6.5, that ε_2 is replaced by $\varepsilon_2-\mathrm{i}\sigma_2/\omega$ where σ_2 is the conductivity. Consequently, (6.14)–(6.17) are still valid provided that

$$N_2^2=\mu_2\varepsilon_2(1-\mathrm{i}\sigma_2/\varepsilon_2\omega)/\mu_0\varepsilon_0. \tag{8.1}$$

It was also pointed out in §6.5 that dissipation can be introduced in a sound wave by making the sound speed a complex number with positive imaginary part. Also, in the conversion between (6.8), (6.9) and (6.14), (6.15), $a_2=1/N_2$ so that a positive imaginary part in a_2 gives a negative imaginary part in N_2, consistent with (8.1). Hence discussion of (8.1) will cover both the electromagnetic and acoustic cases.

Write $N_2=N_r-\mathrm{i}N_i$ where N_r and N_i are both positive. On account of (6.11), $\sin\theta_2$ has a positive imaginary part. Of the two possible choices for $\cos\theta_2$ that ensuring attenuation is required and so $\cos\theta_2=\alpha-\mathrm{i}\beta$ where α and β are positive. Then the exponential dependence of the refracted wave is

$$\begin{aligned}&\exp\{-\mathrm{i}k_2(x\cos\theta_2+y\sin\theta_2)\}\\&=\exp[-\omega(\mu_0\varepsilon_0)^{\frac{1}{2}}(N_i\alpha+N_r\beta)x-\mathrm{i}\omega(\mu_0\varepsilon_0)^{\frac{1}{2}}\{(N_r\alpha-N_i\beta)x+N_1y\sin\theta\}].\end{aligned}$$

The planes of constant amplitude are $x=$ constant, the wave decaying as x increases. The planes of constant phase are

$$(N_r\alpha - N_i\beta)x + N_1 y \sin\theta = \text{constant}$$

which do not coincide with the planes of constant amplitude except at normal incidence $\theta = 0$. Thus the refracted field is an inhomogeneous plane wave, generally speaking.

A direction of propagation might be defined as perpendicular to the planes of constant phase. A real angle of refraction ψ would then be specified by

$$\cos\psi = \frac{N_r\alpha - N_i\beta}{\{(N_r\alpha - N_i\beta)^2 + N_1^2\sin^2\theta\}^{\frac{1}{2}}}, \qquad \sin\psi = \frac{N_1\sin\theta}{\{(N_r\alpha - N_i\beta)^2 + N_1^2\sin^2\theta\}^{\frac{1}{2}}}.$$

This suggests that a real refractive index N be defined by

$$N/N_1 = \sin\theta/\sin\psi = \{\sin^2\theta + (N_r\alpha - N_i\beta)^2/N_1^2\}^{\frac{1}{2}}.$$

Such a refractive index depends upon the angle of incidence—a remarkable state of affairs when the law for lossless media is recalled. Nevertheless experimental confirmation has been forthcoming at optical frequencies.

The plane wave is no longer inhomogeneous at normal incidence $\theta = 0$. Moreover, the magnitude of the reflected wave is $|(N_1 - N_2)/(N_1 + N_2)|$ with either polarization when $\mu_1 = \mu_2$ i.e.

$$\left\{\frac{(N_1 - N_r)^2 + N_i^2}{(N_1 + N_r)^2 + N_i^2}\right\}^{\frac{1}{2}}.$$

The larger N_i the nearer this fraction is to unity. In other words, the bigger the absorption in the dissipative medium the more strongly is the wave reflected. Consequently, the colours observed by transmitted and reflected light are complementary; a thin film of gold, for example, appears blue by transmitted light.

Two important limiting cases must now be discussed.

(a) *Strong dissipation*

When the imaginary part of N_2^2 is very much larger than the real part, as occurs for strong dissipation,

$$N_2 \approx (1 - \mathrm{i})(\mu_2\sigma_2/2\omega\mu_0\varepsilon_0)^{\frac{1}{2}},$$

Hence, $\sin\theta_2 = O(1/\sigma_2^{\frac{1}{2}})$ and $\cos\theta_2 = 1 - O(1/\sigma_2)$. Therefore

$$\sin\psi \approx (2\omega\mu_1\varepsilon_1/\mu_2\sigma_2)^{\frac{1}{2}}\sin\theta$$

so that $\psi \to 0$ as $\sigma_2 \to \infty$. Thus, the planes of constant phase are nearly parallel to the interface and *the refracted wave is essentially travelling along the normal to the interface whatever the angle of incidence.*

The transmitted wave is attenuated and falls to 1/e of its value at the surface in the distance

$$\delta = (2/\omega\mu_2\sigma_2)^{\frac{1}{2}}.$$

A measure of the extent to which the refracted wave penetrates into the dissipative medium is δ, which is therefore termed the *skin depth.* In copper $\sigma_2 \approx 6\times 10^7$, $\varepsilon_2 = \varepsilon_0$ and $\sigma_2/\omega\varepsilon_2 \approx 3\times 10^9\lambda$ where λ is the wavelength in metres and so the approximation should be valid in copper for wavelengths greater than 10^{-3} cm. The skin depth is $3.2\lambda^{\frac{1}{2}}\times 10^{-4}$ cm, which is 0.007 cm and 0.0007 cm at 300 m (1 MHz) and 3 m (100 MHz) respectively. In sea water $\sigma_2 = 3$, $\varepsilon_2 = 81\varepsilon_0$ and the skin depth is $1.7\lambda^{\frac{1}{2}}$ cm; when $\lambda = 300$ the skin depth is 29 cm and $\psi < 0.35°$. The confinement of the transmitted disturbance to a thin layer or *skin* near the interface is known as the *skin effect.*

With regard to the reflected wave

$$A_1 \approx -1 + (1+\mathrm{i})\gamma\cos\theta \approx (\gamma\cos\theta - 1)\mathrm{e}^{-\mathrm{i}\gamma\cos\theta} \tag{8.2}$$

where $\gamma = (2\omega\varepsilon_1\mu_2/\mu_1\sigma_2)^{\frac{1}{2}}$. Also

$$B_1 \approx \{\cos\theta - \tfrac{1}{2}(1+\mathrm{i})\gamma\}/\{\cos\theta + \tfrac{1}{2}(1+\mathrm{i})\gamma\} \tag{8.3}$$

and

$$|B_1|^2 = \frac{2\cos^2\theta - 2\gamma\cos\theta + \gamma^2}{2\cos^2\theta + 2\gamma\cos\theta + \gamma^2}.$$

The second order terms are retained in $|B_1|$ because $\cos\theta$ may be small. The expression for $|B_1|^2$ has a minimum of $3 - 2\sqrt{2}$ when $\cos\theta = \gamma/\sqrt{2}$. Provided that this minimum is sharply defined and θ is not too close to $\frac{1}{2}\pi$ an experimental method of determining γ (and hence μ_2/σ_2) can be based on locating the minimum. Clearly an arbitrary incident wave will usually be reflected as elliptically polarized.

At normal incidence $\theta = 0$ and

$$|A_1| = |B_1| = 1 - \gamma$$

a relation which has been verified experimentally for metals in the infrared region. Generally it is adequate for wavelengths greater than 10^{-3} cm but there is disagreement with experiment for smaller wavelengths.

As θ increases from 0, $|A_1|$ grows steadily from the initial value of $1-\gamma$ to 1 at $\theta = \frac{1}{2}\pi$. In contrast, $|B_1|$ starts at $1-\gamma$, decreases steadily to its minimum and then increases to 1 at $\theta = \frac{1}{2}\pi$. Therefore $|A_1| \geqslant |B_1|$ at all angles of incidence. Near the minimum of $|B_1|$ some 80 per cent of the energy is lost on reflection. For example, a vertically polarized wave striking sea water at close to grazing incidence will lose a substantial portion of its energy so long as the wavelength is large enough for γ to be small.

The notion that the field in a strongly dissipative medium is restricted to a thin layer near the interface leads to an important simplification in the treatment of scattering by curved boundaries. The case of acoustics will be considered first. For strong dissipation $a_2 = ae^{i\alpha}$ where $a \ll 1$ and $0 < \alpha < \pi$. Therefore

$$C_1 \approx -1 + \eta \cos\theta \tag{8.4}$$

with $\eta = 2a_2\rho_2/\zeta\rho_1 a_1$, so long as $\rho_1\zeta a_1/\rho_2$ is not small. Now consider the total field in $x < 0$

$$p = (e^{-ik_1 x\cos\theta} + C_1 e^{ik_1 x\cos\theta})e^{-iky\sin\theta}.$$

On $x = 0$, $p = \eta\cos\theta e^{-iky\sin\theta}$ according to (8.4). In addition,

$$\partial p/\partial x = ik_1\cos\theta(-2 + \eta\cos\theta)e^{-iky\sin\theta}$$

on $x = 0$. Therefore, correct to the first order in η,

$$p = (i\eta a_1/2\omega)\,\partial p/\partial x \tag{8.5}$$

on $x = 0$. Conversely, if (8.5) is imposed as a boundary condition, C_1 is recovered as (8.4) to the first order. Thus, as far as reflection is concerned, all the necessary information about the strongly dissipative medium is carried by (8.5). The relation (8.5) does not involve the angle of incidence and so it is suggested that even *the curved boundary of a strongly dissipative medium can be taken care of, for scattering purposes, by satisfying the boundary condition*

$$p = (i\eta a_1/2\omega)\boldsymbol{\nu}\cdot\operatorname{grad} p \tag{8.6}$$

where $\boldsymbol{\nu}$ is a unit vector normal to the interface directed into the absorbing medium. If $\rho_1\zeta a_1/\rho_2$ is small, then (8.4) must be replaced by

$$C_1 = (a_2\rho_2\cos\theta - \rho_1\zeta a_1)/(a_2\rho_2\cos\theta + \rho_1\zeta a_1).$$

The analysis may be repeated and (8.5) (and hence (8.6)) holds again. A certain degree of universality may therefore be attributed to (8.6).

It will be observed that, in the limit as $a_2 \to 0$, other parameters being fixed, (8.6) goes over to $p = 0$, i.e. the soft boundary condition. In contradistinction, if ζ tends to zero at the same time as a_2 in such a way that $\zeta/a_2 \to 0$, the limit is $\partial p/\partial\nu = 0$, i.e. the hard boundary condition.

For the electromagnetic wave with $\boldsymbol{E}^i$ perpendicular to the plane of incidence

$$E_z = \{e^{-ik_1 x\cos\theta} + A_1 e^{ik_1 x\cos\theta}\}e^{-ik_1 y\sin\theta}$$

so that, from (8.2), on $x = 0$

$$E_z = (1+i)\gamma\cos\theta e^{-ik_1 y\sin\theta}.$$

Also, on $x = 0$

$$H_y = (\varepsilon_1/\mu_1)^{\frac{1}{2}}\{-2 + (1+i)\gamma\cos\theta\}\cos\theta e^{-ik_1 y\sin\theta}$$

whence, correct to the first order in γ,

$$E_z = -\tfrac{1}{2}(1+\mathrm{i})\,\delta\mu_2\omega H_y. \tag{8.7}$$

Likewise, from (8.3), we find for the other polarization that

$$E_y = \tfrac{1}{2}(1+\mathrm{i})\,\delta\mu_2\omega H_z. \tag{8.8}$$

Both (8.7) and (8.8) are encompassed in

$$\boldsymbol{E}_\mathrm{t} = \tfrac{1}{2}(1+\mathrm{i})\,\delta\mu_2\omega\boldsymbol{H}_\mathrm{t}\wedge\boldsymbol{\nu} \tag{8.9}$$

where the suffix t refers to the tangential component and $\boldsymbol{\nu}$ has been defined already.

Both (8.6) and (8.9) are founded on the theory of plane waves. However, each is independent of the angle of incidence and so should be applicable to any field which can be composed of plane waves; the large family with Fourier representations is included in this group. Another point in favour of the boundary conditions is that when a curved boundary has radii of curvature large compared with the wavelength, each portion of the surface will behave locally as a plane and the field in the absorber will be restricted to a thin layer (see also §7.2). Therefore, *for absorbing substances of any shape in which the appropriate parameters are small and the radii of curvature large compared with the wavelength, the boundary condition* (8.6) *or* (8.9) *may be applied on the surface to determine the field outside the absorber to a first approximation.*

As $\sigma_2 \to \infty$, (8.9) transforms to $\boldsymbol{E}_\mathrm{t} = \boldsymbol{0}$, the boundary condition for a perfectly conducting body. The current in the conductor is squeezed into a surface layer and the tangential component of $\boldsymbol{H}$ is no longer continuous (§1.22). That the current does become compressed into a surface layer can be seen from the behaviour at a plane interface. For example, the component

$$J_y = \sigma_2\gamma(1+\mathrm{i})\mathrm{e}^{-(1+\mathrm{i})x/\delta}$$

so that the current flowing in the y-direction per unit width in the z-direction is

$$\int_0^\infty J_y\,\mathrm{d}x = 2(\varepsilon_1/\mu_1)^{\frac{1}{2}}$$

which does not vanish as $\sigma_2 \to \infty$ although J_y becomes zero at every point at which $x > 0$.

(b) *Light damping*

When the imaginary part of N_2^2 is small there is not much absorption and

$$N_2 \approx (\mu_2\varepsilon_2/\mu_0\varepsilon_0)^{\frac{1}{2}}(1 - \mathrm{i}\sigma_2/2\varepsilon_2\omega)$$

with a corresponding formula for the acoustic case. The net effect is that the amplitudes of the waves are the same as if there were no dissipation when terms of the first order are neglected. Therefore the reflected field will look as if the scatterer were lossless, although the transmitted wave will display attenuation. The approximation should be valid for electromagnetic waves of frequency higher than 1 MHz over both water and land. However, most dielectrics exhibit a great increase in absorption at wavelengths of a few centimetres on account of molecular resonance (2.8 cm or about 11 GHz for water). For metals the theory seems to cope down to infrared wavelengths.

6.9 The plane slab

The next problem to be considered is that of three homogeneous media separated by parallel plane boundaries as shown in Fig. 6.5. The media are designated 1, 2, 3 and, for greater generality, the dividing planes are placed at $x = x_1$ and $x = x_2$. A wave incident from $x < x_1$ will excite a reflected wave in region 1 and a refracted wave in region 2. The refracted wave on striking $x = x_2$ will cause a reflected wave in region 2 and a transmitted wave in region 3. The reflected wave will cause further disturbance when it hits $x = x_1$ and the process will continue. Rather than calculating each individual reflection and trying to sum the result we imagine that they have already been combined into bumper reflected and refracted plane waves so that only the five waves of Fig. 6.5 need to be considered. The angle of incidence will be denoted by θ_1 now, consonant with the notation in other regions. Also, much of the analysis will be concerned with the boundary conditions so it will simplify things if measurements are made from the separating planes where feasible.

Turning now to the acoustic case we assume

Region 1: $p = \{C_1 e^{-ik_1(x-x_1)\cos\theta_1} + D_1 e^{ik_1(x-x_1)\cos\theta_1}\} e^{-ik_1 y \sin\theta_1}$,

Region 2: $p = \{C_2 e^{-ik_2(x-x_2)\cos\theta_2} + D_2 e^{ik_2(x-x_2)\cos\theta_2}\} e^{-ik_2 y \sin\theta_2}$,

Region 3: $p = C_3 e^{-ik_3(x-x_2)\cos\theta_3 - ik_3 y \sin\theta_3}$.

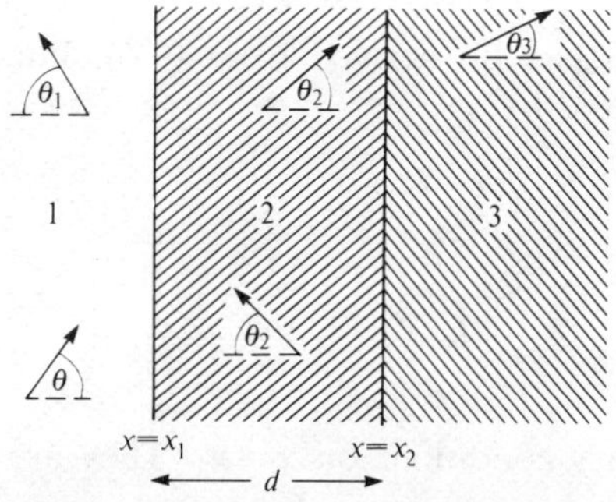

Fig. 6.5 Reflection by a plane slab

The constants C_1 and C_2 are not the same as those in the preceding sections but are to be determined afresh; in fact, C_1 is to be regarded as known and allows for adjustment of the incident wave.

Snell's laws insist that

$$k_1 \sin\theta_1 = k_2 \sin\theta_2 = k_3 \sin\theta_3. \tag{9.1}$$

Bearing in mind (6.1) we have on $x = x_1$ the relations

$$C_1 + D_1 = C_2 e^{ik_2 d\cos\theta_2} + D_2 e^{-ik_2 d\cos\theta_2}, \tag{9.2}$$

$$(C_1 - D_1)\cos\theta_1/\rho_1 a_1 = (C_2 e^{ik_2 d\cos\theta_2} - D_2 e^{-ik_2 d\cos\theta_2})\zeta\cos\theta_2/\rho_2 a_2 \tag{9.3}$$

where $d = x_2 - x_1$ is the thickness of the slab. Equations (9.2), (9.3) will be rewritten by introducing the notation

$$p_1 = C_1 + D_1, \qquad q_1 = (C_1 - D_1)/\eta_1, \qquad \eta_1 = \rho_1 a_1 \sec\theta_1,$$

$$p_2 = C_2 + D_2, \qquad q_2 = (C_2 - D_2)/\eta_2, \qquad \eta_2 = \rho_2 a_2 \sec\theta_2.$$

Thus

$$p_1 = p_2 \cos\xi_2 + i\eta_2 q_2 \sin\xi_2, \tag{9.4}$$

$$q_1 = \zeta\{q_2\cos\xi_2 + (i/\eta_2)p_2\sin\xi_2\} \tag{9.5}$$

where $\xi_2 = k_2 d\cos\theta_2$. In matrix form (9.4) and (9.5) are

$$\mathbf{p}_1 = \mathbf{L}_2\mathbf{p}_2 \tag{9.6}$$

where $\mathbf{p}_j$ is the column vector with components p_j and q_j, and $\mathbf{L}_2$ is the matrix defined by

$$\mathbf{L}_2 \equiv \begin{pmatrix} \cos\xi_2 & i\eta_2\sin\xi_2 \\ (i\zeta/\eta_2)\sin\xi_2 & \zeta\cos\xi_2 \end{pmatrix}. \tag{9.7}$$

Notice that $\det \mathbf{L}_2 = \zeta$. Consequently, if l_{ij} is a typical element of $\mathbf{L}_2$

$$\mathbf{L}_2^{-1} = (1/\zeta)\begin{pmatrix} l_{22} & -l_{12} \\ -l_{21} & l_{11} \end{pmatrix}. \tag{9.8}$$

Various quantities can be derived from (9.6). Put $R_1 = D_1/C_1$, $Z_1 = p_1/q_1$. They are related by

$$Z_1 = \eta_1(1+R_1)/(1-R_1), \qquad R_1 = (Z_1 - \eta_1)/(Z_1 + \eta_1). \tag{9.9}$$

From (9.6)

$$Z_1 = (l_{11}Z_2 + l_{12})/(l_{21}Z_2 + l_{22}) \tag{9.10}$$

where $Z_2 = p_2/q_2$.

Apply the boundary conditions on $x = x_2$. They are

$$p_2 = C_3, \qquad q_2 = \zeta C_3/\eta_3 \tag{9.11}$$

with η_3 defined in analogous way to η_1 and η_2. Hence $p_2/q_2 = \eta_3/\zeta$ and, from (9.10),

$$Z_1 = (l_{11}\eta_3 + l_{12}\zeta)/(l_{21}\eta_3 + l_{22}\zeta). \tag{9.12}$$

All quantities are known on the right-hand side so that R_1 can now be determined from (9.9) and thus the reflected wave in region 1 has been found.

The disturbance caused in region 3 can be obtained by noting that $\mathbf{p}_2 = \mathbf{L}_2^{-1}\mathbf{p}_1$ and so, from (9.11) and (9.8),

$$C_3 = (l_{22}p_1 - l_{12}q_1)/\zeta = (l_{22} - l_{12}/Z_1)p_1/\zeta.$$

Since $p_1/C_1 = 1 + R_1$ we have from (9.9) and (9.12)

$$C_3/C_1 = 2\eta_3/\{l_{11}\eta_3 + l_{12}\zeta + \eta_1(l_{21}\eta_3 + l_{22}\zeta)\}. \tag{9.13}$$

From this point onwards it will be assumed that there is no total reflection at either interface. Then η_1, η_2, η_3, and ξ_2 are all real. There is no wave reflected from the slab if $R_1 = 0$, i.e. $Z_1 = \eta_1$. Inserting the values in (9.12) this occurs when

$$(\eta_3 - \eta_1\zeta^2)\cos\xi_2 = \mathrm{i}(\zeta/\eta_2)(\eta_1\eta_3 - \eta_2^2)\sin\xi_2. \tag{9.14}$$

The left-hand side is real whereas the right-hand side is purely imaginary and so each side has to be zero. One possibility is $\eta_3 = \eta_1\zeta^2$ and $\eta_1\eta_3 = \eta_2^2$. Thus there is no reflected wave if $\eta_3 = \eta_1\zeta^2$, $\eta_1^2\zeta^2 = \eta_2^2$ whatever the thickness of the slab; in particular, this is true for normal incidence if $\rho_3 a_3 = \rho_1 a_1 \zeta^2$, $\rho_1 a_1 \zeta = \pm\rho_2 a_2$.

Another solution of (9.14) is $\eta_3 = \eta_1\zeta^2$, $\sin\xi_2 = 0$. At normal incidence with $\rho_3 a_3 = \rho_1 a_1 \zeta^2$ there is no reflection by slabs whose thickness satisfies $\sin k_2 d = 0$.

The final solution of (9.14) is $\eta_1\eta_3 = \eta_2^2$, $\cos\xi_2 = 0$ so that there is no reflection at normal incidence when $\rho_1 a_1 \rho_3 a_3 = (\rho_2 a_2)^2$ if the dimension of the slab satisfies $\cos k_2 d = 0$.

In general, R_1 contains the oscillatory factors $\cos\xi_2$, $\sin\xi_2$ and so will fluctuate as the width of the slab increases at fixed frequency. This will also happen if total reflection is present or the media are dissipative but there will no longer be any thicknesses at which the reflected wave disappears.

Observe that the same technique will also work if ζ has different values on $x = x_1$ and $x = x_2$.

In the electromagnetic case consider the field when the electric vector is perpendicular to the plane of incidence. Take as the field in

Region 1: $E_z = \{A_1 e^{-\mathrm{i}k_1(x-x_1)\cos\theta_1} + B_1 e^{\mathrm{i}k_1(x-x_1)\cos\theta_1}\}e^{-\mathrm{i}k_1 y\sin\theta_1}$,

$$H_y = -(\varepsilon_1/\mu_1)^{\frac{1}{2}}\cos\theta_1\{A_1 e^{-\mathrm{i}k_1(x-x_1)\cos\theta_1} - B_1 e^{\mathrm{i}k_1(x-x_1)\cos\theta_1}\}e^{-\mathrm{i}k_1 y\sin\theta_1},$$

Region 2: $E_z = \{A_2 e^{-ik_2(x-x_2)\cos\theta_2} + B_2 e^{ik_2(x-x_2)\cos\theta_2}\} e^{-ik_2 y \sin\theta_2}$,

Region 3: $E_z = A_3 e^{-ik_3(x-x_2)\cos\theta_3 - ik_3 y \sin\theta_3}$.

Again A_1 and A_2 must not be identified with those of the previous section.

The continuity of E_z and H_y imply (9.1) and

$$A_1 + B_1 = A_2 e^{i\xi_2} + B_2 e^{-i\xi_2},$$
$$(A_1 - B_1)(\varepsilon_1/\mu_1)^{\frac{1}{2}} \cos\theta_1 = (A_2 e^{i\xi_2} - B_2 e^{-i\xi_2})(\varepsilon_2/\mu_2)^{\frac{1}{2}} \cos\theta_2.$$

Now put

$$E_m = A_m + B_m, \qquad H_m = (B_m - A_m)/\eta_m, \qquad \eta_m = (\mu_m/\varepsilon_m)^{\frac{1}{2}} \sec\theta_m.$$

Then

$$E_1 = E_2 \cos\xi_2 - i\eta_2 H_2 \sin\xi_2,$$
$$H_1 = H_2 \cos\xi_2 - (i/\eta_2) E_2 \sin\xi_2$$

or

$$\boldsymbol{e}_1 = \boldsymbol{K}_2 \boldsymbol{e}_2 \tag{9.15}$$

where $\boldsymbol{e}_m$ has elements E_m, $-H_m$ and

$$\boldsymbol{K}_2 = \begin{pmatrix} \cos\xi_2 & i\eta_2 \sin\xi_2 \\ (i/\eta_2)\sin\xi_2 & \cos\xi_2 \end{pmatrix}. \tag{9.16}$$

Thus $\boldsymbol{K}_2$ is the formally the same as $\boldsymbol{L}_2$ with $\zeta = 1$.

We can now infer that, putting $R_1 = B_1/A_1$ and $Z_1 = -E_1/H_1$, (9.9) still holds and

$$Z_1 = (k_{11} Z_2 + k_{12})/(k_{21} Z_2 + k_{22}) \tag{9.17}$$

where $Z_2 = -E_2/H_2$.

On $x = x_2$

$$E_2 = A_3, \qquad -H_2 = A_3/\eta_3$$

and so

$$Z_1 = (k_{11}\eta_3 + k_{12})/(k_{21}\eta_3 + k_{22}) \tag{9.18}$$

from which R_1 can be found. In a similar way to which (9.13) was derived, the transmitted wave is discovered to be

$$A_3/A_1 = 2\eta_3/\{k_{11}\eta_3 + k_{12} + \eta_1(k_{21}\eta_3 + k_{22})\}. \tag{9.19}$$

When there is no total reflection at an interface the reflected wave is absent from region 1 when (9.14) is satisfied with $\zeta = 1$. Apart from the special case $\eta_1 = \eta_2 = \eta_3$ there are two possibilities. In one $\eta_1\eta_3 = \eta_2^2$ and $\cos\xi_2 = 0$. At normal incidence this requires $(\varepsilon_1\varepsilon_3/\mu_1\mu_3)^{\frac{1}{2}} = \varepsilon_2/\mu_2$ and

$$d = (2m+1)\lambda_2/4$$

where m is any integer and λ_2 is the wavelength in region 2. In particular, if $\mu_1 = \mu_2 = \mu_3$, *the introduction of a quarter wavelength slab whose permittivity is the geometric mean of its surroundings will produce no reflected wave.* This principle can be used to make optically invisible glass and to reduce radar reflections from obstacles.

The alternative possibility is $\eta_3 = \eta_1$, $\sin \xi_2 = 0$; i.e. $\varepsilon_3/\mu_3 = \varepsilon_1/\mu_1$ and $d = \frac{1}{2}m\lambda_2$ at normal incidence. This property has been employed in the design of radomes on occasion.

In general $|R_1|$ is an oscillatory function of the thickness. While it may never become precisely zero in the presence of dissipation or total reflection careful design can arrange that the reflection is low over a band of frequencies—a property which is useful in radar camouflage.

6.10 The sandwich

Another important combination of media is the *sandwich* which, in general form, involves four interfaces as illustrated in Fig. 6.6. For the electric vector perpendicular to the plane of incidence assume that the field in region m is

$$E_z = \{A_m e^{-ik_m(x-x_m)\cos\theta_m} + B_m e^{ik_m(x-x_m)\cos\theta_m}\}e^{-ik_m y \sin\theta_m}$$

except in region 5 where we take

$$E_z = A_5 e^{-ik_5(x-x_4)\cos\theta_5 - ik_5 y \sin\theta_5}.$$

By Snell's law $k_m \sin\theta_m$ has the same value in each region. Now we remark that (9.15) was derived from the boundary condition on $x = x_1$ alone. Therefore its validity continues. Similarly, from the other boundary conditions, we have

$$\boldsymbol{e}_{m-1} = \boldsymbol{K}_m \boldsymbol{e}_m \quad (m = 2, 3, 4) \tag{10.1}$$

in an obvious notation with $\xi_m = k_m(x_m - x_{m-1})\cos\theta_m$. Thus

$$\boldsymbol{e}_1 = \boldsymbol{K}\boldsymbol{e}_4$$

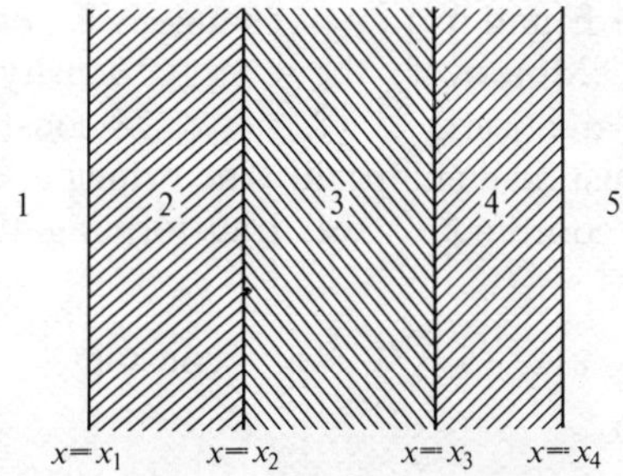

Fig. 6.6 The sandwich

where $\boldsymbol{K}=\boldsymbol{K}_2\boldsymbol{K}_3\boldsymbol{K}_4$. Since $\det \boldsymbol{K}_n=1$ it is transparent that $\det \boldsymbol{K}=1$ and $\boldsymbol{K}$ has similar properties to $\boldsymbol{K}_2$ in the preceding section.

If R_1 and Z_m are defined as before, (9.9) can be quoted and from (9.18)

$$Z_1=(k_{11}\eta_5+k_{12})/(k_{21}\eta_5+k_{22}) \tag{10.2}$$

where k_{ij} is now an element of $\boldsymbol{K}$.

The sandwich causes no reflected wave in region 1 when $Z_1=\eta_1$. However, the resulting equation is nothing like as simple as (9.14) even when total reflection is missing because of the several factors in $\boldsymbol{K}$. One special case is the symmetrical configuration in which regions 1 and 5 have the same parameters as do regions 2 and 4 while $x_4-x_3=x_2-x_1$. Then $\eta_1=\eta_5$, $\eta_2=\eta_4$ and $\xi_2=\xi_4$. There is no reflected wave at normal incidence when

$$\tan \xi_3=\frac{2\eta_2\eta_3(\eta_2^2-\eta_1^2)\sin 2\xi_2}{(\eta_1^2+\eta_2^2)(\eta_2^2-\eta_3^2)+(\eta_1^2-\eta_2^2)(\eta_2^2+\eta_3^2)\cos 2\xi_2}$$

The sandwich has been popular in the construction of radomes because it offers a structure with mechanical strength which is comparatively reflection-free.

The homogeneous anisotropic medium

6.11 The plane wave

The properties of an anisotropic substance depend upon orientation with a profound effect on wave propagation. For simplicity, the conductivity will be ignored and the magnetic properties will be assumed to isotropic. The medium will therefore be *gyroelectric*. (A discussion of gyromagnetic media is to be found in §5.22). The constitutive equations are

$$\boldsymbol{D}=\boldsymbol{\varepsilon}\cdot\boldsymbol{E}, \qquad \boldsymbol{B}=\mu\boldsymbol{H} \tag{11.1}$$

where $\boldsymbol{\varepsilon}$ is a real tensor. If it is assumed, as in §1.27, that the electric energy density is $\frac{1}{2}\boldsymbol{D}\cdot\boldsymbol{E}$ and its rate of change is $\boldsymbol{E}\cdot\partial\boldsymbol{D}/\partial t$ then $\boldsymbol{\varepsilon}$ must be a symmetric tensor. Moreover, the energy density $\boldsymbol{D}\cdot\boldsymbol{E}$ cannot be negative so that the quadric $\sum_{j=1}^{3}\sum_{k=1}^{3}\varepsilon_{jk}x_jx_k=\text{constant}$, x_1, x_2, and x_3 being coordinates, must be an ellipsoid. The principal axes of the ellipsoid may be selected as coordinate axes, transforming the equation of the quadric to

$$\varepsilon_1x^2+\varepsilon_2y^2+\varepsilon_3z^2=\text{constant}.$$

Based on these axes

$$D_x=\varepsilon_1E_x, \qquad D_y=\varepsilon_2E_y, \qquad D_z=\varepsilon_3E_z \tag{11.2}$$

and ε_1, ε_2, and ε_3 are called the *principal dielectric constants of the anisotropic medium* (*or crystal*). The ellipsoid degenerates into a sphere and $\varepsilon_1 = \varepsilon_2 = \varepsilon_3$ when the medium is isotropic. The suffixes x, y, z will always refer to the principal axes in the rest of this section.

In §6.6 it was remarked that the permittivity in an isotropic material depends upon the frequency. For an anisotropic medium not only do ε_1, ε_2, and ε_3 depend upon the frequency but the directions of the principal axes may also be frequency dependent.

Let us examine the possibility of a harmonic plane wave propagating in the direction of the unit vector $\boldsymbol{\nu}$. In such a wave all components of the field are proportional to $\exp\{i(\omega t - k_0 N\boldsymbol{x} \cdot \boldsymbol{\nu})\}$ where $k_0 = \omega(\mu_0\varepsilon_0)^{\frac{1}{2}} = \omega/c$ and N is the refractive index associated with $\boldsymbol{\nu}$. The quantity c/N is called the *phase speed*. Maxwell's equations supply

$$k_0 N\boldsymbol{E} \wedge \boldsymbol{\nu} + \omega\mu\boldsymbol{H} = \boldsymbol{0}, \tag{11.3}$$

$$k_0 N\boldsymbol{H} \wedge \boldsymbol{\nu} - \omega\boldsymbol{D} = \boldsymbol{0}. \tag{11.4}$$

Consequently

$$\boldsymbol{D} = -(k_0^2 N^2/\mu\omega^2)(\boldsymbol{E} \wedge \boldsymbol{\nu}) \wedge \boldsymbol{\nu} = (k_0^2 N^2/\mu\omega^2)\{\boldsymbol{E} - (\boldsymbol{E} \cdot \boldsymbol{\nu})\boldsymbol{\nu}\}. \tag{11.5}$$

Equations (11.3) and (11.4) disclose that $\boldsymbol{H}$ is perpendicular to all of $\boldsymbol{\nu}$, $\boldsymbol{E}$, and $\boldsymbol{D}$; these last three vectors must therefore be coplanar. The vector $\boldsymbol{D}$ is perpendicular to the direction of propagation $\boldsymbol{\nu}$ but $\boldsymbol{E}$ is not. (Fig. 6.7)

The directions along which energy travels are called *rays*. The rays are not, in general, normal to the surface of constant phase. The direction of Poynting's vector $\boldsymbol{S}$ is given by

$$\boldsymbol{E} \wedge \boldsymbol{H} = (k_0 N/\omega\mu)\{\boldsymbol{E} \cdot \boldsymbol{E}\boldsymbol{\nu} - \boldsymbol{E} \cdot \boldsymbol{\nu}\boldsymbol{E}\}$$

a vector which lies in the plane of $\boldsymbol{\nu}$, $\boldsymbol{E}$, and $\boldsymbol{D}$. If α_0 is the angle between $\boldsymbol{D}$ and $\boldsymbol{E}$ then $\boldsymbol{S}$ makes an angle α_0 with $\boldsymbol{\nu}$. An *energy speed* u_E can be

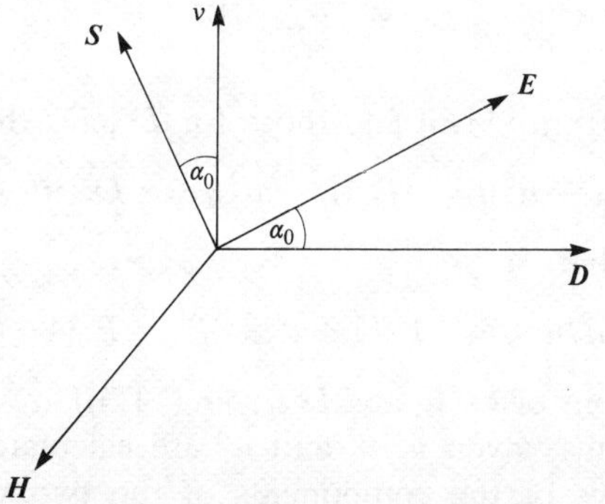

Fig. 6.7 Plane wave in a crystal

defined as $|\boldsymbol{E}\wedge\boldsymbol{H}|/W$ where W is the energy density, i.e. $W=\frac{1}{2}(\boldsymbol{E}\cdot\boldsymbol{D}+\boldsymbol{H}\cdot\boldsymbol{B})$. Hence $u_E=c/N\cos\alpha_0$, i.e. *the phase speed is the projection of the energy speed on the wave normal* $\boldsymbol{\nu}$. It follows that a surface of constant phase is the pedal surface of the surface generated by the ends of rays.

With the form (11.2), (11.5) gives

$$D_x=(k_0^2N^2/\mu\omega^2)\{-\nu_1(\boldsymbol{E}\cdot\boldsymbol{\nu})+D_x/\varepsilon_1\}$$

where ν_1, ν_2, and ν_3 are the components of $\boldsymbol{\nu}$ along the principal axes. On putting $\omega/k_0N=u$ and $u_j=1/(\mu\varepsilon_j)^{\frac{1}{2}}$ we obtain

$$D_x=-\nu_1(\boldsymbol{E}\cdot\boldsymbol{\nu})/\mu(u^2-u_1^2). \tag{11.6}$$

There are two similar equations for the components D_y, D_z. Substitution in $\boldsymbol{D}\cdot\boldsymbol{\nu}=0$ leads to

$$\frac{\nu_1^2}{u^2-u_1^2}+\frac{\nu_2^2}{u^2-u_2^2}+\frac{\nu_3^2}{u^2-u_3^2}=0, \tag{11.7}$$

which is known as *Fresnel's equation.* It is a quadratic in u^2 and shows that *for every direction of the wave normal there are two distinct phase speeds* u' *and* u''. Let $\boldsymbol{D}'$, $\boldsymbol{D}''$ be the corresponding flux densities. Subtract the two equations (11.7) with $u=u'$ and $u=u''$ from one another and invoking (11.6) we see that $\boldsymbol{D}'\cdot\boldsymbol{D}''=0$. Since $\boldsymbol{D}'$, $\boldsymbol{E}'$, and $\boldsymbol{\nu}$ are coplanar it follows that

$$\boldsymbol{D}'\cdot\boldsymbol{E}''=\boldsymbol{D}''\cdot\boldsymbol{E}'=0.$$

There is an alternative formula for u. Let l_1, l_2, and l_3 be the direction cosines of $\boldsymbol{D}$ with respect to the principal axes. Then (11.5) gives

$$\mu(u^2-u_i^2)l_i=-\nu_i\sum_{j=1}^{3}(n_jl_j/\varepsilon_j).$$

Multiplication by l_i and summation over i furnishes

$$u^2=\sum_{j=1}^{3}l_j^2u_j^2 \tag{11.8}$$

since $\boldsymbol{D}\cdot\boldsymbol{\nu}=0$.

It also follows from (11.6) and those for D_y, D_z that

$$D_x(u^2-u_1^2)/\nu_1=D_y(u^2-u_2^2)/\nu_2=D_z(u^2-u_3^2)/\nu_3 \tag{11.9}$$

or, from (11.2), that

$$E_x(1-u^2/u_1^2)/\nu_1=E_y(1-u^2/u_2^2)/\nu_2=E_z(1-u^2/u_3^2)/\nu_3. \tag{11.10}$$

Hence the directions of $\boldsymbol{D}$, $\boldsymbol{E}$, and $\boldsymbol{H}$ (from (11.3)) are known as soon as u is determined. Thus, given $\boldsymbol{\nu}$, u' and u'' are calculated from (11.7) and then the directions of the components of the two waves are specified uniquely. On the other hand, if $\boldsymbol{D}$ is given, $\boldsymbol{E}$ is known at once and the direction of $\boldsymbol{\nu}$ follows since it is in the plane of $\boldsymbol{D}$ and $\boldsymbol{E}$ but perpendicular to $\boldsymbol{D}$; u is derived from (11.8) and $\boldsymbol{H}$ from (11.3).

Assume that the principal axes are chosen so that $u_1 > u_2 > u_3$. Then the left-hand side of Fresnel's equation (11.7) when multiplied by $(u^2 - u_1^2)(u^2 - u_2^2)(u^2 - u_3^2)$ is positive as $u^2 \to \infty$ and when $u^2 = 0$; it cannot be positive when $u^2 = u_2^2$, being zero only if $\nu_2 = 0$. Therefore, *the values of u' and u'' are different unless $\nu_2 = 0$*. If $\nu_2 = 0$, $u' = u'' = u_2$ and there are two possible direction of the wave normal satisfying

$$u_2^2 - u_1^2 - \nu_1^2(u_3^2 - u_1^2) = 0.$$

The two directions in which there is only one phase speed are called *optical axes* and the medium said to be *biaxial*. The optical axes lie in the (x, z)-plane and are symmetrically placed about the x-axis. On an optical axis the direction of $\boldsymbol{D}$ is no longer specified uniquely by (11.9) so that plane waves of arbitrary polarization can propagate along an optical axis.

There is a single optical axis if $u_1^2 = u_2^2$, i.e. $\varepsilon_1 = \varepsilon_2$, and it is the z-axis. When this occurs put $u_2 = u_1 = u_0$ and $u_3 = u_e$. Let θ_0 be the angle between $\boldsymbol{\nu}$ and the z-axis. Then $\nu_3^2 = \cos^2\theta_0$ and $\nu_1^2 + \nu_2^2 = \sin^2\theta_0$ so that (11.7) becomes

$$(u^{\cdot} - u_0^2)\{(u^2 - u_e^2)\sin^2\theta_0 + (u^2 - u_0^2)\cos^2\theta_0\} = 0.$$

The two phase speeds are given by

$$u^2 = u_0^2, \quad u^2 = u_0^2\cos^2\theta_0 + u_e^2\sin^2\theta_0.$$

The wave with phase speed u_0 independent of the direction of propagation is called the *ordinary wave*. The other wave is termed the *extraordinary wave*; its phase speed varies with the direction of propagation. Only for the wave normal along the z-axis ($\theta_0 = 0$) are the two speeds the same. Perpendicular to the z-axis ($\theta_0 = \frac{1}{2}\pi$) the phase speed of the extraordinary wave is u_e. In some substances (e.g. quartz) $u_0 > u_e$ and in others (e.g. calcspar) $u_0 < u_e$.

There is no loss of generality in placing the direction of propagation in the (x, z)-plane in a uniaxial crystal. Then, a reference to (11.6) demonstrates that $\boldsymbol{E} \cdot \boldsymbol{\nu} = 0$ in the ordinary wave provided that $\nu_1 \neq 0$. This forces D_z, E_z to be zero and hence D_x, E_x to secure the necessary orthogonality with $\boldsymbol{\nu}$. Therefore, when $\nu_1 \neq 0$, $\boldsymbol{D}$ is perpendicular to the plane containing the direction of propagation and the z-axis (the *principal section*) in the ordinary wave. $\boldsymbol{D}$ must lie in the principal section in the extraordinary wave.

6.12 Refraction in a crystal

Reflection at the plane face of a biaxial medium is a more elaborate affair than when there is isotropy. Suppose a plane wave is incident on a crystal from free space; let its spatial dependence be $\exp(-i\omega \boldsymbol{x} \cdot \boldsymbol{\nu}/c)$. There may be two waves in the biaxial medium; let their spatial variation be $\exp(-i\omega \boldsymbol{x} \cdot \boldsymbol{\nu}'/u')$ and $\exp(-i\omega \boldsymbol{x} \cdot \boldsymbol{\nu}''/u'')$. Dropping the convention on x, y, z of the last section we let the interface be $z = 0$. The continuity of the exponentials on the boundary demands

$$\nu_1 x + \nu_2 y = c(\nu_1' x + \nu_2' y)/u' = c(\nu_1'' x + \nu_2'' y)/u''$$

for all x and y. Therefore

$$\nu_1 = c\nu_1'/u' = c\nu_1''/u'', \qquad \nu_2 = c\nu_2'/u' = c\nu_2''/u''. \tag{12.1}$$

Hence the two waves in the biaxial medium travel in different directions and each direction has its own refractive index associated with it. The phenomenon is deemed *double refraction.*

To fix ideas let the incident plane wave be

$$H_x = e^{-ik_0(y \sin\theta + z\cos\theta)}, \qquad E_y = -(\mu_0/\varepsilon_0)^{\frac{1}{2}} H_x \cos\theta,$$
$$E_z = (\mu_0/\varepsilon_0)^{\frac{1}{2}} H_x \sin\theta, \qquad E_x = H_y = H_z = 0.$$

The knowledge that there are two refracted waves precludes the assumption that the reflected wave is linearly polarized. Therefore, the proposed form for the reflected wave is

$$E_x = (\mu_0/\varepsilon_0)^{\frac{1}{2}} R'' e^{-ik_0(y\sin\theta - z\cos\theta)}, \qquad E_y = (\mu_0/\varepsilon_0)^{\frac{1}{2}} H_x \cos\theta,$$
$$E_z = (\mu_0/\varepsilon_0)^{\frac{1}{2}} H_x \sin\theta, \qquad H_x = R' e^{-ik_0(y\sin\theta - z\cos\theta)},$$
$$H_y = -(\varepsilon_0/\mu_0)^{\frac{1}{2}} E_x \cos\theta, \qquad H_z = -(\varepsilon_0/\mu_0)^{\frac{1}{2}} E_x \sin\theta.$$

Then the statement of (12.1) is

$$\nu_1' = \nu_1'' = 0, \qquad \sin\theta = N'\nu_2' = N''\nu_2'' \tag{12.2}$$

in terms of the refractive indices N', N'' of the two refracted waves. The first set of equations in (12.2) tells us that the directions of the refracted wave normals lie in the plane of incidence. Hence, if the wave normals make angles θ' and θ'' with the normal to the interface, the second set of (12.2) can be expressed as

$$\sin\theta = N' \sin\theta' = N'' \sin\theta'', \tag{12.3}$$

i.e. *the wave normals obey Snell's law of refraction.*

The determination of θ' and θ'', while straightforward in principle, is involved in practice. First, with θ' supposed known, the direction cosines of the wave normal with respect to the principal axes are determined. With this information and $u = c \sin\theta'/\sin\theta$, (11.7) can be set up and is, in fact, a quartic for $\tan\theta'$. Two of the roots are $\tan\theta'$ and $\tan\theta''$ respectively; the remaining two are for waves travelling in the negative z-direction. As soon as θ' has been evaluated, u' is known and hence the directions of $\boldsymbol{E}'$, $\boldsymbol{H}'$ and $\boldsymbol{D}'$ according to the remark just after (11.10). What has yet to be found is the complex amplitude of the field.

Of the waves travelling in the direction of $\boldsymbol{\nu}'$ choose one whose magnetic intensity is of unit magnitude; denote it by $(\mu_0/\varepsilon_0)^{\frac{1}{2}}\boldsymbol{e}'$, $\boldsymbol{h}'$ where $|\boldsymbol{h}'|=1$. Let $(\mu_0/\varepsilon_0)^{\frac{1}{2}}\boldsymbol{e}''$, $\boldsymbol{h}''$ be a similar choice for the $\boldsymbol{\nu}''$-wave. Then the refracted waves can be taken as

$$\boldsymbol{E}'=(\mu_0/\varepsilon_0)^{\frac{1}{2}}H'\boldsymbol{e}',\qquad \boldsymbol{H}'=H'\boldsymbol{h}',\qquad \boldsymbol{E}''=(\mu_0/\varepsilon_0)^{\frac{1}{2}}H''\boldsymbol{e}'',\qquad \boldsymbol{H}''=H''\boldsymbol{h}''$$

where the only unknowns are the scalars H' and H''.

In order that the tangential components of the total fields on the two sides of $z=0$ be continuous it is necessary that

$$\begin{aligned} R''&=H'e'_x+H''e''_x,\\ (R'-1)\cos\theta&=H'e'_y+H''e''_y,\\ R'+1&=H'h'_x+H''h''_x,\\ -R''\cos\theta&=H'h'_y+H''h''_y \end{aligned}$$

whence

$$H'=-2(h''_y+e''_x\cos\theta)\cos\theta/(A+C), \tag{12.4}$$

$$H''=2(h'_y+e'_x\cos\theta)\cos\theta/(A+C), \tag{12.5}$$

$$R'=(A-C)/(A+C), \tag{12.6}$$

$$R''=2(h'_ye''_x-h''_ye'_x)\cos\theta/(A+C) \tag{12.7}$$

where

$$A\sec\theta=\begin{vmatrix} h'_x & h''_x & 0\\ h'_y & h''_y & \cos\theta\\ e'_x & e''_x & -1 \end{vmatrix},\qquad C=\begin{vmatrix} e'_x & e''_x & 1\\ e'_y & e''_y & 0\\ h'_y & h''_y & -\cos\theta \end{vmatrix}.$$

The reflected and refracted fields are now completely determined; obviously they are extremely intricate functions of θ and the material constants.

Although the wave normals are subject to Snell's laws *the refracted rays do not, in general, comply with Snell's laws.* Energy flow is also relevant to another phenomenon. When the wave normal in a crystal is along an optical axis there is no special direction of $\boldsymbol{D}$ attached to the wave. Therefore there is no particular orientation for $\boldsymbol{S}$ except that it lies in the plane of $\boldsymbol{\nu}$ and $\boldsymbol{D}$ at a certain angle to $\boldsymbol{\nu}$. Accordingly, as $\boldsymbol{D}$ rotates round $\boldsymbol{\nu}$ the rays generate a cone. Let natural light fall on the crystal in such a way that the refracted wave normal coincides with an optical axis. The light rays in the crystal will form the cone just described and when this cone of rays emerges to free space from a parallel interface it will generate an elliptic cylinder with axis parallel to the incident light. The phenomenon is that of *internal conical refraction.* It may also be shown that a ray travelling along the axis of the cone in the crystal, on emerging to free space, is responsible for a cone of rays so giving rise to *external conical refraction.* Experimental confirmation of these conical refraction effects has been forthcoming.

The inhomogeneous isotropic medium

6.13 General considerations

The equation

$$\nabla^2 p + k^2 p = 0 \tag{13.1}$$

is a mite harder to handle when k varies from point to point. Exact solutions are available only for a limited number of forms of k; usually k and p are required to be functions of one variable only. Some of the equations which have found favour are

$$\mathrm{d}^2p/\mathrm{d}z^2 - zp = 0$$

which has the Airy functions (§A.2) $\mathrm{Ai}(z)$, $\mathrm{Ai}(z\mathrm{e}^{\pm\frac{2}{3}\pi\mathrm{i}})$ as solutions and

$$\mathrm{d}^2p/\mathrm{d}z^2 + (\kappa^2 + \tfrac{1}{4})p/z^2 = 0$$

which is solved by $z^{\frac{1}{2}\pm\mathrm{i}\kappa}$. More fearsome looking is

$$\mathrm{d}^2p/\mathrm{d}z^2 + \kappa^2\mathrm{e}^z\{\kappa_1\mathrm{e}^{-z} + (\kappa_2 - \kappa_1)(\mathrm{e}^z+1)^{-1} + \kappa_3(\mathrm{e}^z+1)^{-2}\}p = 0.$$

The substitutions $\mathrm{e}^z = u - 1$, $p = (u-1)^\alpha u^\beta q$ where $\alpha^2 + \kappa^2\kappa_1 = 0$ and $\beta(\beta - 1) = \kappa^2\kappa_3$ lead to

$$u(1-u)\,\mathrm{d}^2q/\mathrm{d}u^2 + \{2\beta - (2\alpha + 2\beta + 1)u\}\,\mathrm{d}q/\mathrm{d}u$$
$$-\{\beta^2 + 2\alpha\beta + (\kappa_2 - \kappa_1)\kappa^2\}q = 0.$$

This is the *hypergeometric equation,* possessing solutions $\mathrm{F}(a, b; 2\beta; u)$ and $u^{1-2\beta}\mathrm{F}(a - 2\beta + 1, b - 2\beta + 1; 2 - 2\beta; u)$ where $a = \alpha + \beta + \mathrm{i}\kappa\kappa_2^{\frac{1}{2}}$, $b = \alpha + \beta - \mathrm{i}\kappa_2^{\frac{1}{2}}$. Other examples are in Exercises.

But generally it has to be accepted that in dealing with an inhomogeneous medium resort to approximate methods is, in many circumstances, inevitable. There is the additional misfortune to be coped with in Maxwell's equations that the governing equations may have a fiercer structure than (13.1) even when the medium is non-conducting. Evidently the equations satisfied by the harmonic electric and magnetic vectors in an isotropic non-conducting medium are

$$\mathrm{curl}\{(1/\mu)\mathrm{curl}\,\boldsymbol{E}\} = \omega^2\varepsilon\boldsymbol{E}, \qquad \mathrm{curl}\{(1/\varepsilon)\mathrm{curl}\,\boldsymbol{H}\} = \omega^2\mu\boldsymbol{H}.$$

The equation for $\boldsymbol{E}$ may be rewritten as (§F.7)

$$(1/\mu)\mathrm{curl}\,\mathrm{curl}\,\boldsymbol{E} + \{\mathrm{grad}(1/\mu)\}\wedge\mathrm{curl}\,\boldsymbol{E} = \omega^2\varepsilon\boldsymbol{E}. \tag{13.2}$$

Since curl curl $\boldsymbol{E} = \text{grad div } \boldsymbol{E} - \nabla^2 \boldsymbol{E}$ and

$$0 = \text{div}(\varepsilon \boldsymbol{E}) = \varepsilon \text{ div } \boldsymbol{E} + \boldsymbol{E} \cdot \text{grad } \varepsilon$$

(13.2) may also be expressed as

$$\nabla^2 \boldsymbol{E} + \omega^2 \mu\varepsilon \boldsymbol{E} = \mu\{\text{grad}(1/\mu)\} \wedge \text{curl } \boldsymbol{E} - \text{grad}\{(1/\varepsilon)\boldsymbol{E} \cdot \text{grad } \varepsilon\}. \tag{13.3}$$

The stumbling block of the right-hand side of (13.3) means that the chances of exact solution are even more remote than for (13.1). Both right-hand terms disappear only if the medium is homogeneous. If μ is constant the improvement to (13.3) is scarcely appreciable and it may, indeed, be preferable to operate with (13.2).

6.14 The Rayleigh–Gans approximation

The essence of the Rayleigh–Gans technique is conversion of the equation so that the knowledge of homogeneous media can be drawn on. Write (13.1) as

$$\nabla^2 p + k_0^2 p = -(k^2 - k_0^2)p \tag{14.1}$$

where k_0 is a constant. Then, from §1.28, a solution of (14.1) is

$$p(\boldsymbol{x}) = p^{\text{i}}(\boldsymbol{x}) + \int_T (k^2 - k_0^2) p(\mathbf{y}) \psi(\boldsymbol{x}, \mathbf{y}) \, \mathrm{d}\mathbf{y} \tag{14.2}$$

where p^{i} is any solution of (14.1) with zero right-hand side and $\psi(\boldsymbol{x}, \mathbf{y}) = \exp(-\mathrm{i}k_0 |\boldsymbol{x} - \mathbf{y}|)/4\pi |\boldsymbol{x} - \mathbf{y}|$. T is the domain over which $k^2 - k_0^2$ is non-zero. All quantities are assumed to have sufficient continuity to justify the representation. Preferably, T is finite to avoid delicate convergence considerations.

Equation (14.2) is an exact solution of (14.1) and, when impressed in T, constitutes a Fredholm integral equation to determine p in T if p^{i} is supposed known (say as an incident field). Once p is known in T, p can be found elsewhere by evaluating the integral on the right of (14.2). One method of tackling the problem is to employ approximate and numerical methods for Fredholm integral equations.

If $|k^2 - k_0^2| \ll 1$, the integral offers a small correction and a first approximation to p is p^{i}. A second is clearly

$$p(\boldsymbol{x}) = p^{\text{i}}(\boldsymbol{x}) + \int_T (k^2 - k_0^2) p^{\text{i}}(\mathbf{y}) \psi(\boldsymbol{x}, \mathbf{y}) \, \mathrm{d}\mathbf{y}. \tag{14.3}$$

A third approximation is obtained by inserting (14.3) in the integral of (14.2) and obviously the *iteration* can be performed as many times as one is willing to evaluate the increasingly cumbersome integrals. The iteration process can be expected to converge, roughly speaking, if $k^2 - k_0^2$ is not too large though a precise statement would involve T and possibly the form of p^{i}. However, even if the infinite series of iterations does not strictly converge, the result of a couple of iterations may be accurate enough for the purpose in hand, i.e. the iteration process is acting as a manufacturer of an asymptotic series.

The formula displayed in (14.3) is known as the *Rayleigh–Gans approximation*; sometimes it is named after *Born* because of his use of it in atomic theory.

The theory for Maxwell's equations runs on parallel lines. First of all they are written as

$$\operatorname{curl} \boldsymbol{E} + \mathrm{i}\omega\mu_0 \boldsymbol{H} = \mathrm{i}\omega(\mu_0 - \mu)\boldsymbol{H},$$
$$\operatorname{curl} \boldsymbol{H} - \mathrm{i}\omega\varepsilon_0 \boldsymbol{E} = \mathrm{i}\omega(\varepsilon - \varepsilon_0)\boldsymbol{E}$$

where the homogeneous medium has been taken to be free space though that is not obligatory. These equations represent the inhomogeneous medium as the source of electric and magnetic current densities in free space. From §§1.19, 1.29

$$\boldsymbol{E} = \boldsymbol{E}^{\mathrm{i}} - (\operatorname{grad}\operatorname{div} + k_0^2)\int_T (1 - \varepsilon/\varepsilon_0)\boldsymbol{E}\psi(\boldsymbol{x}, \boldsymbol{y})\,\mathrm{d}\boldsymbol{y}$$
$$+ \mathrm{i}\omega \operatorname{curl}\int_T (\mu_0 - \mu)\boldsymbol{H}\psi(\boldsymbol{x}, \boldsymbol{y})\,d\boldsymbol{y}, \qquad (14.4)$$

$$\boldsymbol{H} = \boldsymbol{H}^{\mathrm{i}} - (\operatorname{grad}\operatorname{div} + k_0^2)\int_T (1 - \mu/\mu_0)\boldsymbol{H}\psi(\boldsymbol{x}, \boldsymbol{y})\,d\boldsymbol{y}$$
$$+ \mathrm{i}\omega \operatorname{curl}\int_T (\varepsilon - \varepsilon_0)\boldsymbol{E}\psi(\boldsymbol{x}, \boldsymbol{y})\,d\boldsymbol{y} \qquad (14.5)$$

where $k_0^2 = \omega^2\mu_0\varepsilon_0$ and $\boldsymbol{E}^{\mathrm{i}}$, $\boldsymbol{H}^{\mathrm{i}}$ is any solution of Maxwell's equations in free space. Given $\boldsymbol{E}^{\mathrm{i}}$ and $\boldsymbol{H}^{\mathrm{i}}$, (14.4) and (14.5) provide integral equations (six simultaneous scalar equations) for $\boldsymbol{E}$ and $\boldsymbol{H}$ in T. If $\mu = \mu_0$, attention can be confined to (14.4) and similarly, if $\varepsilon = \varepsilon_0$, only (14.5) needs to be solved but the problem of solution, whether approximate or numerical, is still formidable despite the reduction to three scalar integral equations.

The Rayleigh-Gans approximation to (14.4) and (14.5) merely replaces $\boldsymbol{E}$ and $\boldsymbol{H}$ in the integrals by $\boldsymbol{E}^{\mathrm{i}}$ and $\boldsymbol{H}^{\mathrm{i}}$.

Discussion of the convergence of the iterative procedure is similar to that for (14.2) but more elaborate because of the vector character of the integral equations and the extra differential operators.

6.15 The high-frequency approximation

When the wavelength is so small that a significant change in the medium occupies many wavelengths, a reasonable hypothesis is that in local regions the field behaves as if it were in a homogeneous medium. Thus, locally, the field may be expected to look like a plane wave. A first approximation to the solution of

$$\nabla^2 p + k_0^2 N^2 p = 0 \tag{15.1}$$

might be taken to be $e^{-ik_0 L}$. Here, the variation of the medium is in N, k_0 being constant, and the sound speed is $a = \omega / k_0 N$.

A more recondite assumption or *Ansatz* is

$$p = e^{-ik_0 L}(\gamma_0 + \gamma_1/k_0 + \gamma_2/k_0^2 + \dots) \tag{15.2}$$

where L, γ_0, $\gamma_1, \dots$ may depend upon position but are independent of k_0 and L is real. A priori, there is nothing to say that (15.2) is valid either as a convergent or as an asymptotic series. Indeed, it can be extremely difficult to prove its legitimacy even in the best circumstances. Nevertheless, experience is overwhelming in its support. The expansion (15.2) offers the prospect of progress when other methods are intractable. And there are varian.. which sometimes succeed when (15.2) fails.

Substitute (15.2) in (15.1) and take derivatives term by term, grouping together like powers of k_0, to obtain

$$\begin{aligned} &k_0^2(N^2 - \text{grad}^2 L)\gamma_0 \\ &\qquad -ik_0\{\gamma_0 \nabla^2 L + 2\,\text{grad}\,L \cdot \text{grad}\,\gamma_0 + i\gamma_1(N^2 - \text{grad}^2 L)\} \\ &+ \gamma_2(N^2 - \text{grad}^2 L) - i\{\gamma_1 \nabla^2 L + 2\,\text{grad}\,L \cdot \text{grad}\,\gamma_1\} + \nabla^2\gamma_0 + O(1/k) = 0. \end{aligned} \tag{15.3}$$

If L, $\gamma_0, \dots$ and their derivatives are finite and do not alter appreciably over a few wavelengths, the coefficients of the individual powers of k_0 may be equated to zero. Hence

$$\text{grad}^2 L = N^2, \tag{15.4}$$

$$\gamma_0 \nabla^2 L + 2\,\text{grad}\,L \cdot \text{grad}\,\gamma_0 = 0, \tag{15.5}$$

$$\gamma_j \nabla^2 L + 2\,\text{grad}\,L \cdot \text{grad}\,\gamma_j = -i\nabla^2 \gamma_{j-1} \quad (j = 1, 2, \dots). \tag{15.6}$$

Equation (15.4) acts as a partial differential equation for L, the function which defines the surfaces of constant phase, i.e. *the wavefronts.* Often L is called the *eikonal* and (15.4) is known as the *eikonal equation.* The differential equations (15.5) and (15.6) which fix the amplitudes γ_j are termed *transport equations.*

If the approximation is stopped at the first term, i.e. $p = \gamma_0 e^{-ik_0 L}$, it is evident that the velocity is

$$\boldsymbol{v} = (k_0 \gamma_0 / \omega \rho_0) e^{-ik_0 L}\,\text{grad}\,L. \tag{15.7}$$

The complex acoustic intensity is thereby directed parallel to grad L. In other words, the *direction of energy flow is normal to the wavefront.* Consequently, the field has all the properties locally of a plane wave and may be regarded as satisfactory at high frequencies. A convenient terminology for the approximation $p = \gamma_0 e^{-ik_0 L}$ is to describe it as *geometrical acoustics.* In geometrical acoustics the rays, already defined as the carriers of energy, are normal to the wavefronts.

There is an analogous expansion for the electromagnetic field. The starting hypothesis is

$$\boldsymbol{E} = e^{-ik_0 L}(\boldsymbol{e}_0 + \boldsymbol{e}_1/k_0 + \ldots), \qquad \boldsymbol{H} = e^{-ik_0 L}(\boldsymbol{h}_0 + \boldsymbol{h}_1/k_0 + \ldots). \tag{15.8}$$

These series satisfy Maxwell's equations if

$$-ik_0(\mathrm{grad}\, L) \wedge (\boldsymbol{e}_0 + \boldsymbol{e}_1/k_0 + \ldots) + \mathrm{curl}(\boldsymbol{e}_0 + \boldsymbol{e}_1/k_0 + \ldots) + ik_0 c\mu(\boldsymbol{h}_0 + \boldsymbol{h}_1/k_0 + \ldots) = \boldsymbol{0}, \tag{15.9}$$

$$-ik_0(\mathrm{grad}\, L) \wedge (\boldsymbol{h}_0 + \boldsymbol{h}_1/k_0 + \ldots) + \mathrm{curl}(\boldsymbol{h}_0 + \boldsymbol{h}_1/k_0 + \ldots) - ik_0 c\varepsilon(\boldsymbol{e}_0 + \boldsymbol{e}_1/k_0 + \ldots) = \boldsymbol{0} \tag{15.10}$$

since $\omega = ck_0$. The coefficient of the highest power of k_0 is erased by making

$$c\mu\boldsymbol{h}_0 = \mathrm{grad}\, L \wedge \boldsymbol{e}_0, \qquad c\varepsilon\boldsymbol{e}_0 + \mathrm{grad}\, L \wedge \boldsymbol{h}_0 = \boldsymbol{0}. \tag{15.11}$$

It is evident from (15.11) that $\boldsymbol{h}_0$ is perpendicular to $\boldsymbol{e}_0$ and the wave normal while $\boldsymbol{e}_0$ is also orthogonal to the wave normal. The Poynting vector is normal to the wavefront and of magnitude $\frac{1}{2}(\varepsilon/\mu)^{\frac{1}{2}}|\boldsymbol{e}_0|^2$. The approximation therefore exhibits all the characteristics of a plane wave locally; it is called *geometrical optics.* The elimination of $\boldsymbol{h}_0$ from (15.11) results in

$$(\mathrm{grad}^2 L - N^2)\boldsymbol{e}_0 = \boldsymbol{0}$$

since $\boldsymbol{e}_0 \cdot \mathrm{grad}\, L = 0$ and $N^2 = c^2\mu\varepsilon$. For a nonzero field it is necessary that L satisfy the same eikonal equation (15.4) as in acoustics with appropriate parameters.

The derivation of transport equations entails a certain amount of manoeuvring since the rate of change normal to a wavefront is needed. Here we shall be content with finding that for $\boldsymbol{e}_0$. The vanishing of the coefficient of the next lower power of k_0 brings

$$i\boldsymbol{e}_1 \wedge \mathrm{grad}\, L + \mathrm{curl}\, \boldsymbol{e}_0 + ic\mu\boldsymbol{h}_1 = \boldsymbol{0}, \tag{15.12}$$

$$i\boldsymbol{h}_1 \wedge \mathrm{grad}\, L + \mathrm{curl}\, \boldsymbol{h}_0 - ic\varepsilon\boldsymbol{e}_1 = \boldsymbol{0}. \tag{15.13}$$

Take the vector product of (15.12) and grad L to remove $\boldsymbol{h}_1$ via (15.13) with the result

$$-i(\boldsymbol{e}_1 \cdot \mathrm{grad}\, L)\mathrm{grad}\, L + \mathrm{grad}\, L \wedge \mathrm{curl}\, \boldsymbol{e}_0 + c\mu\, \mathrm{curl}\, \boldsymbol{h}_0 = \boldsymbol{0}.$$

Now $\mathrm{div}(\varepsilon\boldsymbol{E}) = 0$ enforces $\mathrm{div}(\varepsilon\boldsymbol{e}_0) = i\varepsilon\boldsymbol{e}_1 \cdot \mathrm{grad}\, L$ so that

$$-(1/\varepsilon)\mathrm{div}(\varepsilon\boldsymbol{e}_0)\mathrm{grad}\, L + \mathrm{grad}\, L \wedge \mathrm{curl}\, \boldsymbol{e}_0 + c\mu\, \mathrm{curl}\, \boldsymbol{h}_0 = \boldsymbol{0}. \tag{15.14}$$

From (15.11),

$$\begin{aligned} c \operatorname{curl} \boldsymbol{h}_0 &= \operatorname{grad}(1/\mu) \wedge (\operatorname{grad} L \wedge \boldsymbol{e}_0) + (1/\mu)\operatorname{curl}(\operatorname{grad} L \wedge \boldsymbol{e}_0) \\ &= \operatorname{grad}(1/\mu) \wedge (\operatorname{grad} L \wedge \boldsymbol{e}_0) + (1/\mu)\operatorname{curl}(\operatorname{grad} L \wedge \boldsymbol{e}_0) \\ &\quad + (1/\mu)\operatorname{grad}(\boldsymbol{e}_0 \cdot \operatorname{grad} L) \end{aligned}$$

since $\boldsymbol{e}_0 \cdot \operatorname{grad} L = 0$. The application of (F.7.5) and (F.7.7) now supplies

$$\begin{aligned} c \operatorname{curl} \boldsymbol{h}_0 = {} & \operatorname{grad}(1/\mu) \wedge (\operatorname{grad} L \wedge \boldsymbol{e}_0) - \boldsymbol{e}_0 \nabla^2 L + \operatorname{div} \boldsymbol{e}_0 \operatorname{grad} L \\ & - 2(\operatorname{grad} L \cdot \operatorname{grad})\boldsymbol{e}_0 - \operatorname{grad} L \wedge \operatorname{curl} \boldsymbol{e}_0. \end{aligned}$$

After substitution of this in (15.14) we obtain

$$2(\operatorname{grad} L \cdot \operatorname{grad})\boldsymbol{e}_0 + \mu \boldsymbol{e}_0 \operatorname{div}\{(1/\mu)\operatorname{grad} L\} + (1/\mu\varepsilon)\boldsymbol{e}_0 \cdot \operatorname{grad}(\mu\varepsilon)\operatorname{grad} L = 0 \quad (15.15)$$

the desired transport equations.

The satisfactory features of geometrical acoustics and optics have been remarked on. However, the approximation will merit close scrutiny where the hypotheses seem likely to be violated e.g. at points where N becomes small or undergoes a discontinuous change and at points where γ_0 and $\boldsymbol{e}_0$ are predicted to be unduly large.

6.16 Properties of rays

The rays which carry the energy are normal to the wavefront in both acoustics and electromagnetism (note that this statement would be false for the high-frequency approximation in a crystal). The eikonal equation is the same in both so that it is a matter of indifference whether one talks about geometrical acoustics or geometrical optics.

Since the rays are normal to the wavefront the tangent to a ray must be parallel to grad L. Hence a unit vector $\mathbf{s}$ along the tangent is

$$\mathbf{s} = (1/N)\operatorname{grad} L \quad (16.1)$$

in view of (15.4). Let κ be the curvature at the same point and $\boldsymbol{n}$ a unit vector in the direction of the radius of curvature. Then $\kappa \boldsymbol{n} = \mathrm{d}\mathbf{s}/\mathrm{d}s$ where s is arc length measured along the ray. Now

$$\mathrm{d}\mathbf{s}/\mathrm{d}s = (\mathbf{s} \cdot \operatorname{grad})\mathbf{s} = -\mathbf{s} \wedge \operatorname{curl} \mathbf{s}$$

and so $\kappa = -\boldsymbol{\nu} \cdot \mathbf{s} \wedge \operatorname{curl} \mathbf{s}$. On using (16.1) and replacing grad$(1/N)$ by $-(1/N)\operatorname{grad} \ln N$ we have

$$\kappa = \boldsymbol{n} \cdot \operatorname{grad} \ln N \quad (16.2)$$

because $\boldsymbol{n}$ is perpendicular to grad L.

Equation (16.2) conveys the information that the rate of change of N in the direction of the centre of curvature is positive. Hence *a ray bends towards a region of higher refractive index.*

In a homogeneous medium N is independent of position and the right-hand side of (16.2) disappears. Thus the curvature κ is zero and the rays are straight lines.

As regards the amplitude of the energy on a ray observe that (15.5) implies that

$$\operatorname{div}(|\gamma_0|^2 \operatorname{grad} L) = 0 \tag{16.3}$$

and (15.15) gives

$$\operatorname{div}\{(|\boldsymbol{e}_0|^2/\mu)\operatorname{grad} L\} = 0. \tag{16.4}$$

The aim is to apply these formulae to a tube of rays. Let the tube intersect the wavefronts L_1 and L_2 in the surface elements $\mathrm{d}S_1$ and $\mathrm{d}S_2$ respectively (Fig. 6.8). No power will cross the sides of the tube because the energy moves in the direction of the ray. The flow through a cross-section must be invariable and, in particular, the power crossing $\mathrm{d}S_1$ must be the same as that traversing $\mathrm{d}S_2$. Therefore, if values on L_1 and L_2 are denoted by the suffixes 1 and 2 respectively,

$$|\gamma_0|_1^2 N_1 \,\mathrm{d}S_1 = |\gamma_0|_2^2 N_2 \,\mathrm{d}S_2, \tag{16.5}$$

$$(\varepsilon_1/\mu_1)^{\frac{1}{2}} \,|\boldsymbol{e}_0|_1^2 \,\mathrm{d}S_1 = (\varepsilon_2/\mu_2)^{\frac{1}{2}} \,|\boldsymbol{e}_0|_2^2 \,\mathrm{d}S_2 \tag{16.6}$$

as may also be seen by applying the divergence theorem to the ray tube and invoking (16.3), (16.4).

The implications of (16.5), (16.6) in a homogeneous medium are illuminating. Let the ray, which is a straight line, through the point A of the wavefront L_1 be the z-axis. Take A as origin and choose the (x, z)-, (y, z)-planes to be the planes containing the principal radii of curvature ρ_1 and ρ_2 of the wavefront at A. (Fig. 6.9). A ray through B, a point on the x-axis adjacent to A, will intersect the z-axis at O_1 and $O_1B = \rho_1$. Similarly, a ray through the adjacent point C on the y-axis will intersect the z-axis at O_2 where $O_2C = \rho_2$. Take a radius of curvature as positive

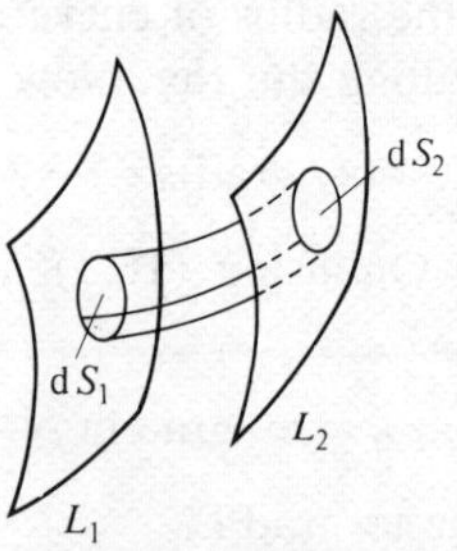

Fig. 6.8 Propagation along a ray tube

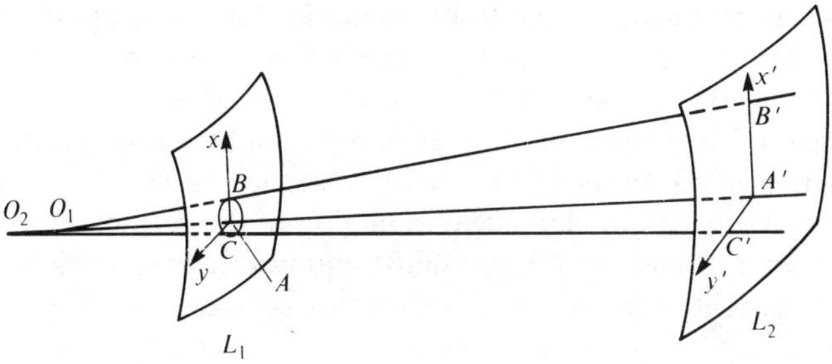

Fig. 6.9 Propagation of energy in a homogeneous medium

when the centre of curvature is on the negative z-axis as in Fig. 6.9; otherwise it is negative. Continue the rays through A, B, and C until they meet the wavefront L_2 at A', B', and C' respectively. The rays are normal to L_2 and so the normals at A' and B' intersect at O_1 i.e. the (x', z')-plane contains a principal radius of curvature. This radius of curvature can be no other than ρ_1+s where s is the constant length of ray cut off by the wavefronts L_1 and L_2. Similarly, the other principal radius of curvature ρ_2+s lies in the (y', z')-plane.

Let dS_1 be an element of area surrounding A and let the rays through dS_1 intersect L_2 in dS_2. Corresponding points in the two areas are related by

$$x' = |(\rho_1+s)/\rho_1|\, x, \qquad y' = |(\rho_2+s)/\rho_2|\, y.$$

Hence

$$dS_2 = |(\rho_1+s)(\rho_2+s)/\rho_1\rho_2|\, dS_1.$$

Bearing in mind the homogeneity of the medium we have, from (16.5), (16.6)

$$|\gamma_0|_2 = |\rho_1\rho_2/(\rho_1+s)(\rho_2+s)|^{\frac{1}{2}}\, |\gamma_0|_1, \tag{16.6}$$

$$|\boldsymbol{e}_0|_2 = |\rho_1\rho_2/(\rho_1+s)(\rho_2+s)|^{\frac{1}{2}}\, |\boldsymbol{e}_0|_1. \tag{16.7}$$

Evidently trouble may arise if ρ_1 or ρ_2 is negative i.e. the approximation becomes suspect when a principal centre of curvature of a wavefront is in the region into which the energy is propagating. In other words, when there are focussing effects some modification of the theory will be necessary.

The bending of rays towards regions of higher refractive index accounts for a number of optical phenomena. Normally, the refractive index of the atmosphere decreases with height. Therefore, light rays from stars are bent towards the surface of the earth so that celestial bodies appear to us slightly higher above the horizon than they really are. For objects on the earth the bending of the rays is usually insignificant compared with the earth's curvature but in a favourable climatic state (a temperature inversion, for example) the gradient of refractive index can be so pronounced that objects normally concealed behind the horizon become visible, i.e. a *mirage* occurs. The source of another mirage can be a thin boundary layer of hot and rarefied air next to a heated surface (such as an asphalt or concrete road on a sunny day). The reduction of the refractive index often produces a sharp bending of rays almost equivalent to reflection (the road then looks as if it were covered with pools of water—reflections of the distant sky).

It remains to mention how rays are calculated. If (x, y, z) is a point on a ray, the direction cosines of the tangent are $(\mathrm{d}x/\mathrm{d}s, \mathrm{d}y/\mathrm{d}s, \mathrm{d}z/\mathrm{d}s)$ with

$$(\mathrm{d}x/\mathrm{d}s)^2+(\mathrm{d}y/\mathrm{d}s)^2+(\mathrm{d}z/\mathrm{d}s)^2=1. \tag{16.8}$$

It follows from (16.1) that

$$N\,\mathrm{d}x/\mathrm{d}s=\partial L/\partial x, \qquad N\,\mathrm{d}y/\mathrm{d}s=\partial L/\partial y, \qquad N\,\mathrm{d}z/\mathrm{d}s=\partial L/\partial z. \tag{16.9}$$

If the wavefronts are known, (16.9) are differential equations which enable the rays to be determined. However, they are not suitable if L is unknown. The elimination of L is accomplished by observing that

$$\frac{\mathrm{d}}{\mathrm{d}s}\left(N\frac{\mathrm{d}x}{\mathrm{d}s}\right)=\frac{\mathrm{d}}{\mathrm{d}s}\left(\frac{\partial L}{\partial x}\right)=\frac{\partial^2 L}{\partial x^2}\frac{\mathrm{d}x}{\mathrm{d}s}+\frac{\partial^2 L}{\partial y^2}\frac{\mathrm{d}y}{\mathrm{d}s}+\frac{\partial^2 L}{\partial z^2}\frac{\mathrm{d}z}{\mathrm{d}s}$$

$$=(1/2N)\frac{\partial}{\partial x}(\mathrm{grad}^2 L).$$

Quoting the eikonal equation we deduce that the differential equations of the rays are

$$\frac{\mathrm{d}}{\mathrm{d}s}\left(N\frac{\mathrm{d}x}{\mathrm{d}s}\right)=\frac{\partial N}{\partial x}, \qquad \frac{\mathrm{d}}{\mathrm{d}s}\left(N\frac{\mathrm{d}y}{\mathrm{d}s}\right)=\frac{\partial N}{\partial y}, \qquad \frac{\mathrm{d}}{\mathrm{d}s}\left(N\frac{\mathrm{d}z}{\mathrm{d}s}\right)=\frac{\partial N}{\partial z}. \tag{16.10}$$

Once the rays have been found the eikonal L can be evaluated by noting the fact that its values at two points on a ray differ by $\int N\,\mathrm{d}s$.

6.17 Fermat's principle

Let C be any curve joining the points P_0 and P. Then the curvilinear integral

$$\int_C N\,\mathrm{d}s$$

is known as the *optical* (*acoustical*) *path length*. In general, the optical path length will depend not only upon the end-points P_0 and P but also on the actual track C traced between them. If C is a ray, the optical path length is $L(P) - L(P_0)$. An immediate inference is that the optical path length is the same for every ray between two wavefronts.

Comparison of (16.10) with the equations of motion of a particle in the theory of mechanics discloses that the optical path length is stationary on a ray. This way of characterizing a ray is embodied in *Fermat's principle: The rays between P_0 and P are those curves along which the optical (acoustical) path length is stationary with respect to infinitesimal variations in the path.* Of course, there may be several rays, following various routes, joining P_0 and P.

It is important to recognize that Fermat's principle does not require the optical path length to be a minimum. It is sufficient that the optical path length of a trajectory be stationary under local deformations of the path. If this point be forgotten significant rays may be lost.

For a homogeneous medium the optical path length is a constant multiple of the geometrical path length. A straight line supplies a minimum for both; therefore Fermat's principle states that in a homogeneous medium the rays are straight lines in accordance with what has already been asserted in the preceding section.

Consider now the interface separating two different homogeneous media with P_0 and P on the same side of the interface (Fig. 6.10). Of the curves joining P_0 and P the straight line connecting P_0 and P is one path where the arc length is stationary. Indeed, it is an absolute minimum of all paths between P_0 and P. It corresponds to the direct ray from P_0 to P and carries the primary radiation as if there were no interface present.

The next possibility is to investigate whether there is a point O on the interface for which $P_0O + OP$ is stationary for small displacements of O. The links P_0O and OP must be straight lines because the medium is homogeneous. Let O' be at a short distance $\delta\sigma$ from O so that its position is specified with respect to O by $\boldsymbol{t}\,\delta\sigma$ where $\boldsymbol{t}$ is a unit vector

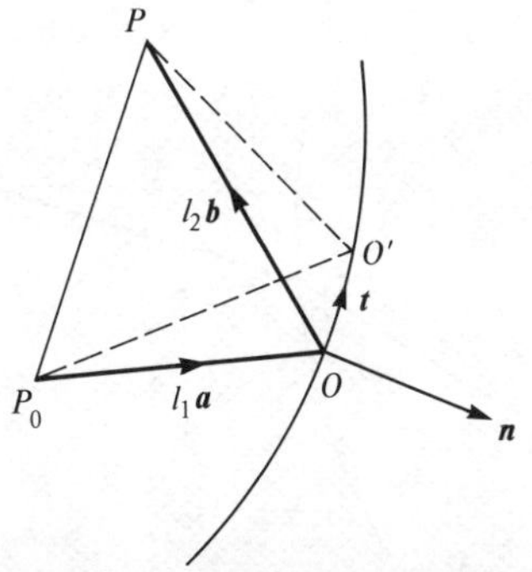

Fig. 6.10 Fermat's principle applied to reflection

tangential to the interface at O. Draw $\boldsymbol{n}$ the unit vector along the normal to the interface at O and let $\boldsymbol{a}$, $\boldsymbol{b}$ be unit vectors along P_0O and OP respectively. Then the vectors P_0O and OP may be expressed as $l_1\boldsymbol{a}$ and $l_2\boldsymbol{b}$ respectively. Let the lengths of P_0O' and $O'P$ be $l_1+\delta l_1$ and $l_2+\delta l_2$ respectively. The difference between the optical path length along $P_0O'P$ and that along P_0OP is $N(\delta l_1+\delta l_2)$. According to Fermat's principle this must vanish, at least to the first order.

If $\boldsymbol{a}+\delta\boldsymbol{a}$ is a unit vector along $\boldsymbol{P}_0O'$

$$(l_1+\delta l_1)(\boldsymbol{a}+\delta\boldsymbol{a})=l_1\boldsymbol{a}+\boldsymbol{t}\,\delta\sigma$$

or, correct to the first order,

$$l_1\,\delta\boldsymbol{a}+\boldsymbol{a}\,\delta l_1=\boldsymbol{t}\,\delta\sigma.$$

But $\boldsymbol{a}\cdot\delta\boldsymbol{a}=0$ because $\boldsymbol{a}$ and $\boldsymbol{a}+\delta\boldsymbol{a}$ are unit vectors. Hence $\delta l_1=\boldsymbol{a}\cdot\boldsymbol{t}\,\delta\sigma$ and, similarly, $\delta l_2=-\boldsymbol{b}\cdot\boldsymbol{t}\,\delta\sigma$. Therefore, Fermat's principle demands that

$$(\boldsymbol{a}-\boldsymbol{b})\cdot\boldsymbol{t}=0$$

for all $\boldsymbol{t}$ in the tangent plane at O. Accordingly, the plane containing $\boldsymbol{a}$ and $\boldsymbol{b}$ is perpendicular to the tangent plane, i.e. the incident ray, reflected ray, and surface normal all lie in the same plane. Moreover, the incident and reflected rays make equal angles with the interface normal. There is thus agreement with Snell's laws of reflection.

The law of refraction may be derived in a similar manner. The quantity to be made zero is now $N_1\,\delta l_1+N_2\,\delta l_3$ where N_1 and N_2 are the refractive indices of the two media (Fig. 6.11). It may be verified that $\delta l_1=\boldsymbol{a}\cdot\boldsymbol{t}\,\delta\sigma$ and $\delta l_3=-\boldsymbol{c}\cdot\boldsymbol{t}\,\delta\sigma$; consequently $(N_1\boldsymbol{a}-N_2\boldsymbol{c})\cdot\boldsymbol{t}=0$. Hence the incident ray, refracted ray, and surface normal are all in the same plane. With $\boldsymbol{t}$ in that plane the cosine of the angle between $\boldsymbol{a}$ and $\boldsymbol{t}$ is the same as the sine

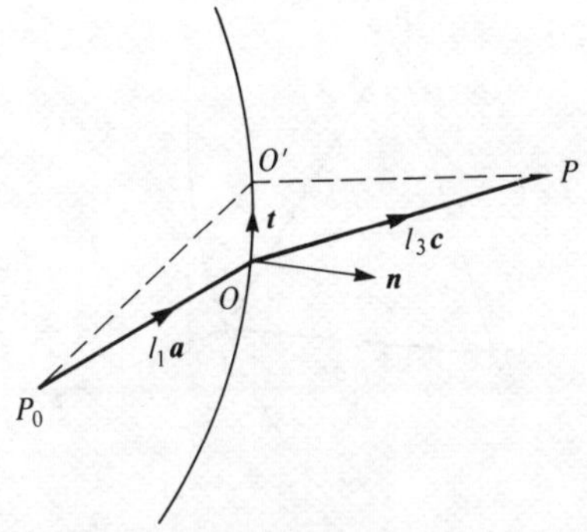

Fig. 6.11 Refraction by Fermat's principle

of the angle between $\boldsymbol{a}$ and $\boldsymbol{n}$. Thus the angle of refraction is predicted correctly and complete consistency with Snell's laws has been secured.

Snell's laws were derived rigorously for a plane wave striking a plane interface. At high frequencies each small portion of an interface can be regarded locally as a piece of tangent plane provided that its principal radii of curvature are large compared with the wavelength. The field of geometrical optics behaves locally as a plane wave and so Snell's laws might be thought of as an unsurprising outcome.

Nevertheless, the satisfaction of Snell's laws by rays is an unexpected and welcome bonus. The point is that the validity of geometrical optics is questionable near an interface because the discontinuity in N conflicts with the hypothesis on which the theory is based—namely, that the material parameters shall be slowly varying. The explanation is that there is no assertion that actual rays are reflected and refracted at the interface according to Snell's laws. What is affirmed is that, away from the interface where the ray representation is valid, the reflected and refracted waves can be visualized as composed of rays, and these rays if continued back to the interface (despite the ray picture being no longer a proper description there) comply with Snell's laws.

Associated with the rays are the wavefronts. In a homogeneous medium the rays are straight lines normal to the wavefronts. The rays thus constitute a *normal congruence*, i.e. a family of straight lines for which there exists a family of orthogonal surfaces. It may be asked whether reflected and refracted rays form a normal congruence when the incident rays do. The answer comes from *Malus' theorem* that *a normal congruence remains a normal congruence after any number of reflections and refractions*. The theorem may be proved from Fermat's principle and the proof also demonstrates that the family of surfaces orthogonal to the reflected (refracted) rays forms the system of wavefronts.

However, this theorem does not preclude the intersection of neighbouring rays. Consider a small pencil of rays which meets a wavefront in the small rectangle $PQRS$ bounded by lines of curvature (Fig. 6.12). The normals at P, Q intersect at X and those at R, S cross at Y. Similarly, the normals at P, S meet at Z and those at Q, R intersect at W. All rays through the rectangle pass through the lines XY and WZ, to the first order of approximation. They are called the *focal lines* of the pencil and the concentration of rays is easily observable in a lens held obliquely in the sun. A small fan-like pencil near PQ is brought to a focus at X and a similar fan along PS is focussed at Z. Fans which do not coincide with either of these directions do not focus at a point at all. Note that the planes PXY and PWZ are perpendicular but XY is not, in general, perpendicular to PX nor WZ to PZ.

Only in the special case of a spherical wavefront do all the rays of a pencil intersect in a common point; the pencil is then said to be

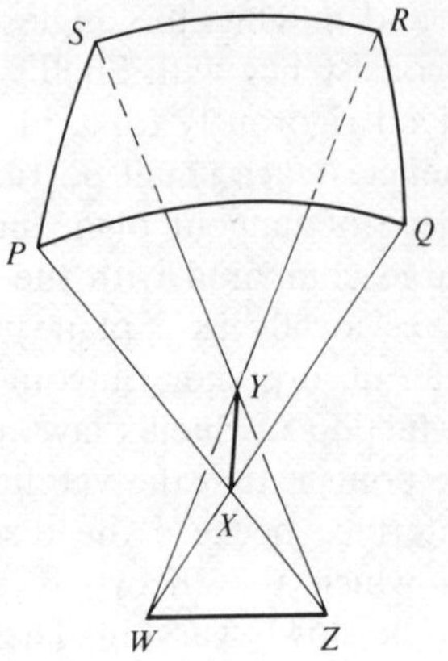

Fig. 6.12 The focal lines of a pencil

homocentric. A point source produces a homocentric pencil but the pencil does not usually remain homocentric after reflection or refraction. Instead, there is a concentration in the focal lines and *astigmatism* occurs. One of the problems in designing equipment to steer rays is to keep astigmatism to a minimum.

For different pencils of the wavefront Z and X will occupy different positions, i.e. each will generate a surface. These two surfaces are known as the *focal surfaces* or *caustics.* Every normal to a wavefront is tangent to both the caustics. The caustics are the loci of the two principal centres of curvature for the wavefront. Therefore, a wavefront which is a surface of revolution possesses caustics which consist of a portion of the axis of rotation and a surface of revolution whose meridian section is the evolute of the meridian section of the wavefront.

As an example of the construction of a caustic consider a point source P in a medium of refractive index N_1 separated by a plane interface from a medium of refractive index N_2 with $N_2 < N_1$. (Fig. 6.13). It will be

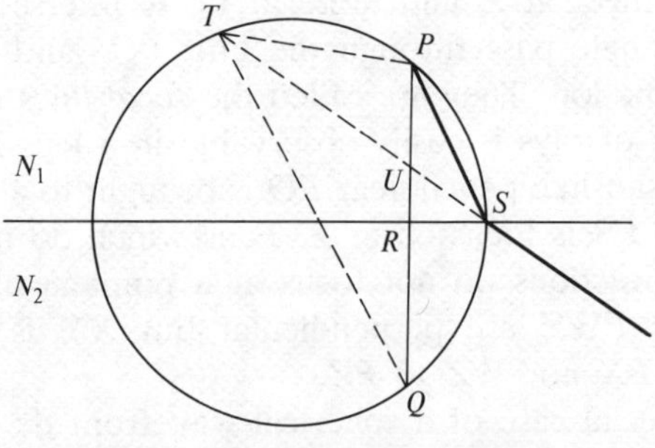

Fig. 6.13 Construction of a caustic

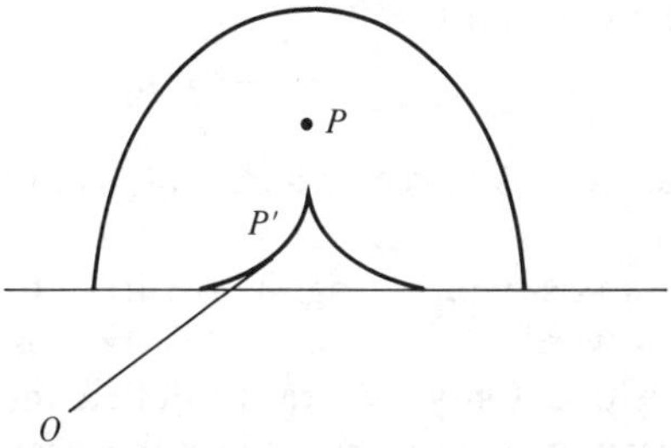

Fig. 6.14 An observer at O sees P at P^1

sufficient to consider the behaviour in a plane containing the normal PRQ to the interface from P. Let $QR = PR$ and let PS be a ray from P meeting the interface at S. Draw the circle through P, Q, S and let the refracted ray continued backwards intersect the circle in T. The angles of incidence and refraction are $\angle SPQ$ and $\angle SUQ$ respectively. Obviously

$$\angle QTS = \angle SPQ = \angle PQS = \angle STP$$

so that $UQ/TQ = PU/PT = N_2/N_1$. Hence

$$N_1 \cdot PQ = N_1(PU + UQ) = N_2(PT + TQ)$$

so that, as S varies, T describes an ellipse with P and Q as foci. Since $\angle QTU = \angle PTU$, ST is normal to this ellipse so that the refracted rays are normal to this ellipse. The caustic of the refracted rays is therefore the evolute of the ellipse and hence an asteroid. (Fig. 6.14). An observer at O therefore sees P at P' where OP' is tangent to the asteroid. An object in water ($N_1/N_2 = 4/3$) when viewed in a direction perpendicular to the interface will appear to be only three quarters of its true distance from the surface and a straight rod will look slightly curved.

6.18 Stratified media

A medium in which N is constant on any one of a set of parallel planes is said to be *stratified*. Let the family of planes be parallel to $z = 0$. There is no loss of generality in taking the z-axis vertically upwards. Then N is a function of z only. Attention can be restricted to the rays which lie in the (x, z)-plane since the rays in other planes through the z-axis can be obtained from the (x, z)-plane by rotation about the z-axis.

The ray equations (16.10) reduce to

$$\frac{\mathrm{d}}{\mathrm{d}s}\left(N\frac{\mathrm{d}x}{\mathrm{d}s}\right) = 0, \qquad \frac{\mathrm{d}}{\mathrm{d}s}\left(N\frac{\mathrm{d}z}{\mathrm{d}s}\right) = \frac{\partial N}{\partial z}.$$

Hence

$$\mathrm{d}x/\mathrm{d}s = C/N \tag{18.1}$$

where C is a constant. From (16.8)

$$dz/ds = \pm(1 - C^2/N^2)^{\frac{1}{2}}. \tag{18.2}$$

The ray will be real only if $C < N$ which will always be assumed.

Suppose that the ray starts from the point $x = 0$, $z = z_0$ and that $N(z_0)$ is positive. Measure s positively along the ray from the initial point. Then, if a ray is to travel towards $x > 0$, dx/ds must be positive initially and so, from (18.1), C is positive. Only such rays will be considered from now on. If the ray goes upward, z must increase and the positive sign is necessary in (18.2); equally well, the negative sign in (18.2) corresponds to a downward ray. From (18.1) and (18.2)

$$dz/dx = \{(N/C)^2 - 1\}^{\frac{1}{2}} \tag{18.3}$$

on an upward ray and

$$dz/dx = -\{(N/C)^2 - 1\}^{\frac{1}{2}} \tag{18.4}$$

on a downward ray. Clearly, the constant C fixes the slope of the ray at its starting point. Evidently, the equation of an upward ray has the form

$$x = \int^z [\{N(w)/C\}^2 - 1]^{-\frac{1}{2}}\, dw; \tag{18.5}$$

a change of sign is pertinent for the downward ray.

Let the tangent to the ray make an angle θ with the z-axis; then $dx/ds = \sin\theta$. Hence (18.1) entails

$$N(z)\sin\theta = C = N(z_0)\sin\theta_0 \tag{18.6}$$

where θ_0 is the value of θ at the beginning. If $N(z)$ increases with z, on the upward ray θ decreases with height and the ray tends to become more vertical but the downward ray tends to the horizontal because θ grows on it. If $N(z)$ decreases as z increases the reverse effect occurs; the upward ray inclines to the horizontal and the downward to the vertical. Where a ray is being forced to the horizontal the possibility arises of the ray being bent so far round that it starts to reverse its motion. In essence the ray acts as if it were reflected and returned to its original level. Such action is feasible only if at some height h the ray is horizontal, i.e. $\theta = \frac{1}{2}\pi$. Thus this phenomenon requires the existence of $N(h)$ such that

$$N(h) = N(z_0)\sin\theta_0. \tag{18.7}$$

No real angle of launching is possible unless $N(h) < N(z_0)$, i.e the refractive index at height h is less than that at the source.

Suppose, in fact, that N decreases as z increases from z_0, reaches a minimum at $z = m$ and then increases. Let θ_m be the angle such that $\sin\theta_m = N(m)/N(z_0)$. If $\theta_0 < \theta_m$, (18.7) can never be attained for any value of h. A ray launched at such an angle will always move upwards and

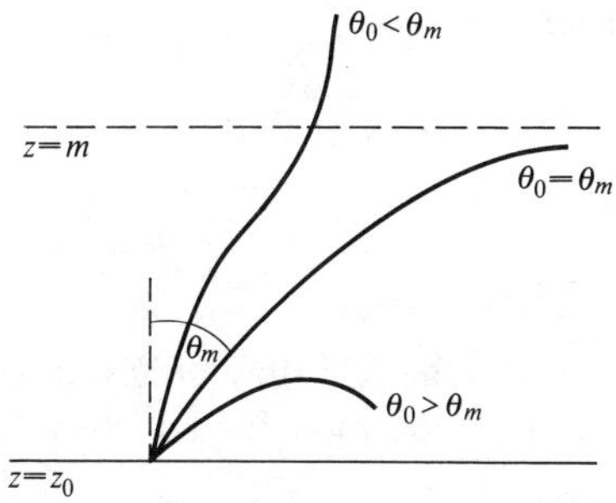

Fig. 6.15 Rays when N has a minimum at $z = m$

steadily depart from the level $z = z_0$ (Fig. 6.15). If $\theta_0 > \theta_m$ the ray will be reflected and turn back at the height h given by (18.7). The horizontal distance x_h to the point of turning is, from (18.5) and (18.6), supplied by

$$x_h = N(z_0)\sin\theta_0 \int_{z_0}^{h} \{N^2(w) - N^2(z_0)\sin^2\theta_0\}^{-\frac{1}{2}}\,dw. \qquad (18.8)$$

When $\theta_0 = \theta_m$ then $h = m$. Also, for w near m,

$$N^2(w) - N^2(m) = O\{(w-m)^2\}$$

because of the minimum of $N(w)$ at $w = m$. The integrand in x_h becomes non-integrable and hence $x_h \to \infty$ as $\theta_0 \to \theta_m$. Thus the ray $\theta_0 = \theta_m$ approaches the level $z = m$ asymptotically.

As θ_0 increases there is no reflection until it reaches θ_m at which point x_h is infinite. Further growth in θ_0 must force x_h to decrease. Yet eventually x_h must go back to infinity because that is the situation when $\theta_0 = \frac{1}{2}\pi$. Hence x_h has a minimum for some value of θ_0. Clearly, *there are at least two rays with different angles of launching which turn round at any value of x_h greater than the minimum.*

Once the ray has been reflected it acquires a new equation because it is now a downward ray and (18.4) is relevant. Hence, for $x \geqslant x_h$,

$$x = x_h - N(h) \int_{h}^{z} \{N^2(w) - N^2(h)\}^{-\frac{1}{2}}\,dw. \qquad (18.9)$$

The eikonal is obtained from the optical path length $\int N\,ds$ calculated along a ray. For an upgoing ray

$$L = \int_{z_0}^{z} N^2(w)\{N^2(w) - C^2\}^{-\frac{1}{2}}\,dw$$

$$= Cx + \int_{z_0}^{z} \{N^2(w) - C^2\}^{\frac{1}{2}}\,dw \qquad (18.10)$$

if $L = 0$ at the initial point. When $\theta_0 > \theta_m$, (18.10) holds until the point of reflection but thereafter

$$L = N(h)x + \int_{z_0}^{h} \{N^2(w) - N^2(h)\}^{\frac{1}{2}}\,\mathrm{d}w - \int_{h}^{z} \{N^2(w) - N^2(h)\}^{\frac{1}{2}}\,\mathrm{d}w \tag{18.11}$$

where x is given by (18.9).

One or two explicit formulae will now be given for the refractive index in which $N^2(z) = 1 + z$. The equation for an upward ray from $z = z_0$ is

$$x = 2C(1 + z - C^2)^{\frac{1}{2}} - 2C(1 + z_0 - C^2)^{\frac{1}{2}} \tag{18.12}$$

and this ray is never reflected. For the downward ray

$$x = 2C(1 + z_0 - C^2)^{\frac{1}{2}} - 2C(1 + z - C^2)^{\frac{1}{2}}. \tag{18.13}$$

Its slope can become zero at $z = C^2 - 1$. On its upward path after reflection its equation is

$$x = 2C(1 + z_0 - C^2)^{\frac{1}{2}} + 2C(1 + z - C^2)^{\frac{1}{2}}. \tag{18.14}$$

As for the eikonal

$$L = Cx + \tfrac{2}{3}(1 + z_0 - C^2)^{\frac{3}{2}} - \tfrac{2}{3}(1 + z - C^2)^{\frac{3}{2}} \tag{18.15}$$

on (18.13) and

$$L = Cx + \tfrac{2}{3}(1 + z_0 - C^2)^{\frac{3}{2}} + \tfrac{2}{3}(1 + z - C^2)^{\frac{3}{2}} \tag{18.16}$$

on (18.14).

6.19 Rays in a moving fluid

Phenomena which have some kinship with the anisotropy of electromagnetism occur when a sound wave is superimposed on a flowing fluid. Let the velocity $\boldsymbol{U}$ of the background flow be steady with density ρ_b and entropy S_b such that, from (1.4.15)–(1.4.17),

$$\boldsymbol{U} \cdot \operatorname{grad} \rho_b + \rho_b \operatorname{div} \boldsymbol{U} = 0, \tag{19.1}$$

$$\rho_b(\boldsymbol{U} \cdot \operatorname{grad})\boldsymbol{U} = -\operatorname{grad} p_b, \tag{19.2}$$

$$\boldsymbol{U} \cdot \operatorname{grad} S_b = 0. \tag{19.3}$$

The equation of state is $p = f(\rho, S)$; thus

$$\operatorname{grad} p_b = (\partial f/\partial \rho)\operatorname{grad} \rho_b + (\partial f/\partial S)\operatorname{grad} S_b.$$

The speed of sound a is given by $a^2 = \partial f/\partial \rho$ (with $\rho = \rho_b$, $S = S_b$). Put $h = \partial f/\partial S$. When (19.1)–(19.3) have been solved for ρ_b and S_b as functions of position, the values of a^2 and h will be known at every point. They can be regarded, therefore, as known functions of position.

Let sound waves make small perturbations so that the velocity, density, and entropy become $\boldsymbol{U}+\boldsymbol{v}$, $\rho_b+\rho$, S_b+S respectively. Then, if squares of small quantities are neglected,

$$\partial\rho/\partial t+\boldsymbol{U}\cdot\text{grad}\,\rho+\boldsymbol{v}\cdot\text{grad}\,\rho_b+\rho_b\,\text{div}\,\boldsymbol{v}+\rho\,\text{div}\,\boldsymbol{U}=0, \tag{19.4}$$

$$\begin{aligned}\rho_b\,\partial\boldsymbol{v}/\partial t+\rho_b(\boldsymbol{U}\cdot\text{grad})\boldsymbol{v}+\rho_b(\boldsymbol{v}\cdot\text{grad})\boldsymbol{U}+\rho(\boldsymbol{U}\cdot\text{grad})\boldsymbol{U}\\ =-a^2\,\text{grad}\,\rho-h\,\text{grad}\,S-(\rho\,\partial^2 f/\partial\rho^2+S\,\partial^2 f/\partial\rho\,\partial S)\text{grad}\,\rho_b\\ -(\rho\,\partial^2 f/\partial\rho\,\partial S+S\,\partial^2 f/\partial S^2)\text{grad}\,S_b,\end{aligned} \tag{19.5}$$

$$\partial S/\partial t+\boldsymbol{U}\cdot\text{grad}\,S+\boldsymbol{v}\cdot\text{grad}\,S_b=0, \tag{19.6}$$

the partial derivatives of f being evaluated at $\rho=\rho_b$, $S=S_b$.

Let a_0 be a constant typical of the speed of sound. It can be identified with a if a does not vary with position but otherwise might be some average of a. Write $k_0=\omega/a_0$ for harmonic vibrations of time factor $e^{i\omega t}$. At high frequencies try the Ansatz

$$\begin{aligned}\rho&=e^{-ik_0L}(r_0+r_1/ik_0+\ldots),\\ \boldsymbol{v}/a_0&=e^{-ik_0L}(\boldsymbol{m}_0+\boldsymbol{m}_1/ik_0+\ldots),\\ S&=e^{-ik_0L}(S_0+S_1/ik_0+\ldots).\end{aligned}$$

Then the first approximation from (19.4)–(19.6) is

$$r_0(1-\boldsymbol{M}\cdot\text{grad}\,L)-\rho_b\boldsymbol{m}_0\cdot\text{grad}\,L=0, \tag{19.7}$$

$$(1-\boldsymbol{M}\cdot\text{grad}\,L)\rho_b\boldsymbol{m}_0=(\alpha^2 r_0+hS_0/a_0^2)\text{grad}\,L, \tag{19.8}$$

$$(1-\boldsymbol{M}\cdot\text{grad}\,L)S_0=0 \tag{19.9}$$

where $\boldsymbol{M}=\boldsymbol{U}/a_0$, $\alpha=a/a_0$. Setting aside the possibility $\boldsymbol{M}\cdot\text{grad}\,L=1$ we have $S_0=0$ from (19.9) and then (19.8) gives

$$\rho_b\boldsymbol{m}_0=(\alpha^2 r_0\,\text{grad}\,L)/(1-\boldsymbol{M}\cdot\text{grad}\,L). \tag{19.10}$$

Using this in (19.7) we obtain for $r_0\neq 0$

$$\alpha^2\,\text{grad}^2 L=(1-\boldsymbol{M}\cdot\text{grad}\,L)^2. \tag{19.11}$$

Equation (19.11) is the eikonal equation analogous to (15.4). It is responsible for the behaviour of the wavefronts of constant phase. One consequence of (19.10) and (19.11) is

$$\rho_b^2\boldsymbol{m}_0\cdot\boldsymbol{m}_0=\alpha^2 r_0^2 \tag{19.12}$$

which expresses a certain relationship between parts of the momentum associated with the sound wave.

The formula for grad p_b and (19.3) show that entropy variations do not enter (19.2). Therefore, in this context, p_b may be treated as if it did not involve S and so, if the argument of §1.23 is repeated, the direction of energy flow is parallel to $\rho a^2\boldsymbol{U}+\rho_b a^2\boldsymbol{v}$ in the acoustic wave. Hence, the equations of the rays are, on account (19.10),

$$\mathrm{d}\boldsymbol{x}/\mathrm{d}\sigma = \alpha^2 \operatorname{grad} L + (1 - \boldsymbol{M} \cdot \operatorname{grad} L)\boldsymbol{M} \tag{19.13}$$

where σ is a parameter on the ray which normally does not coincide with the arc length s. Evidently, the energy does not travel perpendicular to the surfaces of constant phase.

In contrast to (15.4), L cannot be removed from (19.13). Some supplementary equations are needed. They are provided by

$$\mathrm{d}(\operatorname{grad} L)/\mathrm{d}\sigma = \{(\mathrm{d}\boldsymbol{x}/\mathrm{d}\sigma) \cdot \operatorname{grad}\}\operatorname{grad} L.$$

When $\mathrm{d}\boldsymbol{x}/\mathrm{d}\sigma$ is substituted from (19.13) and derivatives of (19.11) are taken, the equation can be written as

$$\begin{aligned}\frac{\mathrm{d}}{\mathrm{d}\sigma}\operatorname{grad} L = {} & \alpha \operatorname{grad}^2 L \operatorname{grad} \alpha \\ & - (1 - \boldsymbol{M} \cdot \operatorname{grad} L) \\ & \times \{\operatorname{grad}(\boldsymbol{M} \cdot \operatorname{grad} L) - (\boldsymbol{M} \cdot \operatorname{grad})\operatorname{grad} L\}\end{aligned} \tag{19.14}$$

which contains no second derivatives of L on the right-hand side. (19.14) is an equation which traces how the normal to the wavefront changes, owing to refraction caused by non-uniformities in the flow velocity $\boldsymbol{U}$ and the sound speed a. Between them, (19.13) and (19.14) form a system of six partial differential equations which have to be solved before the rays and wavefronts can be determined.

By virtue of (19.11) the right-hand side of (19.13) can be written

$$(1 - \boldsymbol{M} \cdot \operatorname{grad} L)\{\boldsymbol{U} + a(\operatorname{grad} L)/|\operatorname{grad} L|\}/a_0.$$

Thus, relative to the local fluid velocity $\boldsymbol{U}$, a packet of energy moves at the local speed a in a direction normal to a phase front. A ray AB (Fig. 6.16) connecting two points on adjacent phase fronts may therefore be drawn according to the usual parallelogram of velocities. To put it

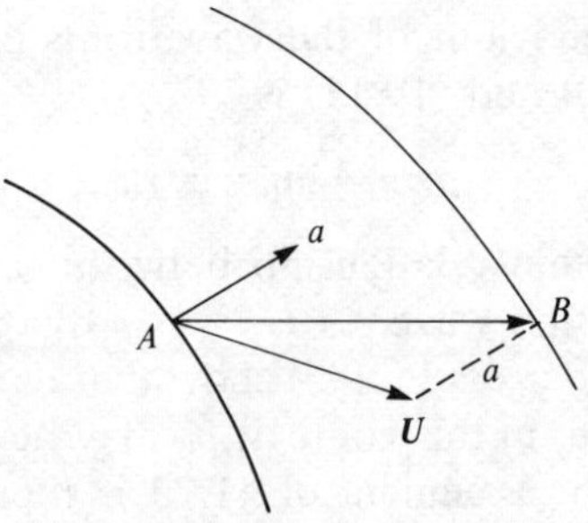

Fig. 6.16 Geometry of a ray between two phase fronts

another way, in a frame of reference in which the fluid is locally at rest the sound propagates as if it were in a medium at rest. Naturally, in this instantaneous frame, the frequency observed is not $\omega/2\pi$ but is modified by the Doppler shift (§2.11) and must be calculated from $\omega(1-\boldsymbol{M}\cdot\text{grad } L)/2\pi$; the local frame frequency thereby varies from point to point. It will be observed that, in the excluded case of $\boldsymbol{M}\cdot\text{grad } L=1$, the sound wave is not fluctuating in time in the local frame which is one reason for omitting this possibility.

By proceeding to the next order in the expansion a transport equation can be derived. It implies that along a ray tube

$$\frac{\alpha^2 r_0^2\,|\alpha^2\,\text{grad } L+(1-\boldsymbol{M}\cdot\text{grad } L)\boldsymbol{M}|}{\rho_b(1-\boldsymbol{M}\cdot\text{grad } L)^2}\,\text{d}S=\text{constant} \tag{19.15}$$

where $\text{d}S$ is the cross-sectional area of the ray tube. Comparison with (16.5) indicates that the background flow affects the movement of energy by pushing it in a direction oblique to the propagating direction of the phase front and by inserting a Doppler factor in the denominator.

6.20 Propagation in shear flow

Suppose now that $\boldsymbol{U}$ and a depend only on the variable z as in a stratified medium. Then (19.14) tells us that $\text{d}(\partial L/\partial x)/\text{d}\sigma$ and $\text{d}(\partial L/\partial y)/\text{d}\sigma$ are both zero, i.e. $\partial L/\partial x$ and $\partial L/\partial y$ are constant on a ray. Let θ_1, θ_2, θ_3 be the angles made by the normal to the wavefront with the x-, y-, z-axes respectively; then

$$\cos^2\theta_1+\cos^2\theta_2+\cos^2\theta_3=1. \tag{20.1}$$

The constancy of $\partial L/\partial x$ and $\partial L/\partial y$ on a ray then implies

$$|\text{grad } L|\cos\theta_1=C_1,\qquad |\text{grad } L|\cos\theta_2=C_2 \tag{20.2}$$

where C_1 and C_2 are constants.

Assume now that the background flow has no vertical velocity, i.e. $M_z=0$. Then, from (20.1) and (20.2),

$$1-\boldsymbol{M}\cdot\text{grad } L=1-M_xC_1-M_yC_2.$$

So long as this quantity is positive (19.11) furnishes

$$\alpha C_1\sec\theta_1+C_1M_x+C_2M_y=1 \tag{20.3}$$

which determines θ_1. The value of θ_2 follows from $\cos\theta_2=(C_2/C_1)\cos\theta_1$, and now θ_3 can be deduced from (20.1).

According to (19.13)

$$\text{d}z/\text{d}\sigma=\alpha^2\,|\text{grad } L|\cos\theta_3$$

so that the equations for the rays can be cast in the form

$$dx/dz = (\cos\theta_1 + M_x/\alpha)\sec\theta_3, \tag{20.4}$$

$$dy/dz = (\cos\theta_2 + M_y/\alpha)\sec\theta_3 \tag{20.5}$$

in which the right-hand sides are known functions.

A ray attains its maximum or minimum height where $\cos\theta_3 = 0$. On account of (20.1) and (20.3) this necessitates

$$\alpha^2(C_1^2 + C_2^2) = (1 - C_1M_x - C_2M_y)^2 \tag{20.6}$$

as the condition for an altitude of reflection.

If the direction of the background stream remains the same at all altitudes, axes can be chosen so that $M_y = 0$. As a result, (20.3) simplifies to

$$\alpha\sec\theta_1 + M_x = \text{constant} \tag{20.7}$$

which reveals how refraction due to inhomogenities in the sound speed and the main stream influences the inclination of the ray to the main stream. In terms of the inclination to the vertical (20.7) can be expressed as

$$(C_1^2 + C_2^2)^{\frac{1}{2}}\alpha\,\text{cosec}\,\theta_3 + C_1M_x = 1 \tag{20.8}$$

for an upgoing ray.

The temperature has an effect on α in still air and a typical temperature lapse-rate of 10 K km^{-1} causes a reduction in a of about 5 m s^{-1} km^{-1}. This makes the rays bend upwards with a curvature the same as that of a circle of radius 66 km roughly. While this is not a great curvature it is sufficient to prevent the rectilinear propagation of sound and, on very hot days, passing vehicles may be inaudible at a range of 50 m.

A temperature inversion can make α increase with height and then the situation is similar to that of Fig. 6.15. Rays can be returned to the ground and longer ranges of propagation are feasible, though there may be intermediate shadow zones in which the sound, while not completely silent, is exponentially small.

A rule of thumb for the effect of wind is that a 1 K change in temperature has the same consequence as a $\frac{1}{2}$ m s^{-1} alteration in wind for propagation in the plane $y = 0$ (sometimes known as *Livermore's Law*). Increases in wind speed far in excess of 5 m s^{-1} km^{-1} may easily occur for strong winds; one concludes that refraction phenomena are more often due to shearing flow than to temperature changes and so temperature variations will be ignored beyond this point. An increase in wind speed corresponds to a reduction in θ_3 if C_1 is negative, i.e. $\theta_1 > \frac{1}{2}\pi$. Thus, in propagation upstream the rays bend towards the vertical, usually with substantially greater curvature than originates from a change in sound speed except on still days (Fig. 6.17). On the other hand, when the ray is

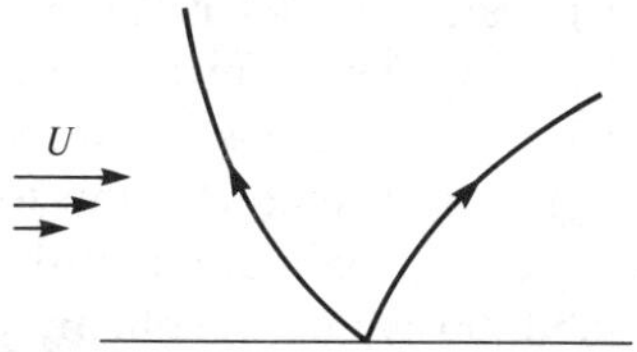

Fig. 6.17 Possible upwind and downwind rays in shear flow

going downwind C_1 is positive and θ_3 increases with M_x. The rays may now bend round and return to the ground as in a temperature inversion.

6.21 The Hertz vector in a stratified medium

The awkward form (13.3) assumed by the equation for the electric vector has already been remarked upon. On occasion the right-hand side can be essentially disposed of in a stratified medium. For example, suppose that $\mu = \mu_0$ and ε is a function of z only. Then there is a field with $H_z = 0$ and the Hertz vector parallel to the z-axis. With $\boldsymbol{k}$ a unit vector parallel to the z-axis, it can be verified that a solution of Maxwell's equations is

$$\boldsymbol{E} = (1/\varepsilon)\text{curl curl}(\varepsilon^{\frac{1}{2}}\Pi\boldsymbol{k}), \tag{21.1}$$

$$\boldsymbol{H} = \text{i}\omega\,\text{curl}(\varepsilon^{\frac{1}{2}}\Pi\boldsymbol{k}) \tag{21.2}$$

provided that

$$\nabla^2\Pi + k_0^2N_1^2\Pi = 0 \tag{21.3}$$

where

$$k_0^2N_1^2 = \omega^2\mu_0\varepsilon - \varepsilon^{\frac{1}{2}}\,\text{d}^2\varepsilon^{-\frac{1}{2}}/\text{d}z^2 = k_0^2N^2 - N\,\text{d}^2N^{-1}/\text{d}z^2.$$

Similarly there is a solution with $E_z = 0$, namely

$$\boldsymbol{E} = -(\text{i}\omega\mu_0/\varepsilon)\text{curl}(\varepsilon M\boldsymbol{k}), \tag{21.4}$$

$$\boldsymbol{H} = \text{curl}\{(1/\varepsilon)\text{curl}(\varepsilon M\boldsymbol{k})\} \tag{21.5}$$

provided that

$$\nabla^2 M + k_0^2N^2M = 0. \tag{21.6}$$

In fact, the right-hand side of (13.3) disappears if $E_z = 0$ under the given conditions on μ and ε.

Both (21.3) and (21.6) are of a type already examined.

6.22 Laminated media

The methods delineated so far are all attempts to find approximate solutions of the exact equations for inhomogeneous media. There is another technique in which the physical model is adjusted so that the resulting equations can be solved exactly. The continuous variations of the material parameters are subdivided into regions of homogeneous media. If the subdivisions are fine enough the exact solution to this problem might be expected to be a reasonable approximation to the answer for continuous variation. A stratified medium which is split by planes perpendicular to the variation into homogeneous blocks is called a *laminated medium.* As well as their value as approximations to continuous media, laminated media have important applications in their own right in electromagnetism and acoustics.

The necessary technique for laminated media is already to hand in §§6.9, 6.10 and the electromagnetic case will be considered. Let the dividing planes be $x = x_1, \ldots x_M$ and designate region m as between $x = x_{m-1}$ and $x = x_m$. Then the validity of (10.1) carries over and

$$\boldsymbol{e}_1 = \boldsymbol{K}\boldsymbol{e}_M$$

where $\boldsymbol{K} = \boldsymbol{K}_2\boldsymbol{K}_3 \ldots \boldsymbol{K}_M$. Then, with $R_m = B_m/A_m$ and $Z_m = -E_m/H_m$,

$$Z_1 = (k_{11}\eta_{M+1} + k_{12})/(k_{21}\eta_{M+1} + k_{22})$$

determines the reflection coefficient R_1 of the lamination and the transmission coefficient A_m/A_1 is

$$2\eta_{M+1}/\{k_{11}\eta_{M+1} + k_{12} + \eta_1(k_{21}\eta_{M+1} + k_{22})\}.$$

A periodic laminated medium is obtained by making $\boldsymbol{K}_{m+2} = \boldsymbol{K}_m$ and having an infinite number of layers. Only two matrices will appear throughout the structure, say $\boldsymbol{K}_2$ and $\boldsymbol{K}_3$. Then

$$\boldsymbol{e}_1 = \boldsymbol{K}_2\boldsymbol{K}_3\boldsymbol{e}_3 = \boldsymbol{P}\boldsymbol{e}_3$$

where

$$\boldsymbol{P} = \begin{pmatrix} \cos\xi_2\cos\xi_3 - (\eta_2/\eta_3)\sin\xi_2\sin\xi_3 & \mathrm{i}\eta_3\cos\xi_2\sin\xi_3 + \mathrm{i}\eta_2\sin\xi_2\cos\xi_3 \\ (\mathrm{i}/\eta_2)\sin\xi_2\cos\xi_3 + (\mathrm{i}/\eta_3)\cos\xi_2\sin\xi_3 & \cos\xi_2\cos\xi_3 - (\eta_3/\eta_2)\sin\xi_2\sin\xi_3 \end{pmatrix}$$

A wave propagating in one direction can be achieved with no reflected waves if $\boldsymbol{e}_1 = \mathrm{e}^{\mathrm{i}\kappa}\boldsymbol{e}_3$. There can be a non-trivial wave only if $\mathrm{e}^{\mathrm{i}\kappa}$ is an eigenvalue of $\boldsymbol{P}$. Since $\det \boldsymbol{P} = 1$ the determinantal equation is $2\cos\kappa = p_{11} + p_{12}$ or

$$2\cos\kappa = 2\cos\xi_2\cos\xi_3 - (1/\eta_2\eta_3)(\eta_2^2 + \eta_3^2)\sin\xi_2\sin\xi_3. \qquad (22.1)$$

Whenever κ is a solution of (22.1) so is $-\kappa$, the two values corresponding to waves in opposite directions. The waves do not decay only if κ is real. Studies of (22.1) indicate that ranges of frequencies for real κ alternate with those where κ is complex so that a periodic laminated medium possesses *the ability to filter.*

By allowing the dividing planes to come together the continuously stratified medium with the field y-dependence $e^{-iky\sin\theta}$ is approached. Now Z is continuous and $Z=-E_z/H_y$. Then $R=(Z-\eta)/(Z+\eta)$ where $\eta=(\mu/\varepsilon)^{\frac{1}{2}}\sec\theta$. Then, from (9.17),

$$dZ/dx = i\omega\mu(1-Z^2/\eta^2) \tag{22.2}$$

and hence

$$dR/dx = (2i\omega\mu/\eta)R - \tfrac{1}{2}(1-R^2)\, d(\ln\eta)/dx. \tag{22.3}$$

These differential equations, which are of *Riccati's generalized type*, have been extensively studied and form the basis for numerical work.

6.23 The WKB method

In the consideration of stratified media the equation

$$d^2w/dz^2 + k_0^2N^2(z)w = 0 \tag{23.1}$$

turns up. At high frequencies k_0 is large and the method of geometrical optics suggests the trial of

$$w = e^{-ik_0L}(w_0 + w_1/k_0 + \ldots).$$

(A full discussion of such expansions in this context can be found in Olver (1974).) After the usual modus operandi, the equations

$$L'^2 = N^2, \tag{23.2}$$

$$2L'w_0' + L''w_0 = 0, \tag{23.3}$$

where primes indicate derivatives with respect to z, hold. Equation (23.3) gives $w_0 = A(L')^{-\frac{1}{2}}$ and (23.2) supplies $L = \pm\int^z N(t)\,dt$. Thus the proposed high-frequency solution of (23.1) is

$$w(z) = (Ae^{-ik_0L} + Be^{ik_0L})/\{N(z)\}^{\frac{1}{2}} \tag{23.4}$$

where, for definiteness, the upper sign has been selected outside the integral for L. The approximation (23.4) is said to have been arrived at by the *WKB method* (sometimes the letters are in a different order and sometimes other initials are added).

For instance, consider

$$d^2v/dz^2 + \{\nu^2(1-1/z^2) + 1/4z^2\}v = 0 \tag{23.5}$$

where ν is large. For $z>1$ and z not near 1, the $1/4z^2$ can be neglected and $N^2(z) = 1-1/z^2$. Hence

$$L = \int^z (1-1/t^2)^{\frac{1}{2}}\,dt = \int \tan^2\beta\, d\beta = \tan\beta - \beta$$

where $z = \sec\beta$. Hence, according to (23.4),

$$v = \{A e^{-i\nu(\tan\beta - \beta)} + B e^{i\nu(\tan\beta - \beta)}\}/(\sin\beta)^{\frac{1}{2}}. \qquad (23.6)$$

But a solution of (23.4) is $z^{\frac{1}{2}} J_\nu(\nu z)$ and, as $z \to \infty$, (A.2.1)

$$J_\nu(\nu z) \sim (2/\pi\nu z)^{\frac{1}{2}} \cos(\nu z - \tfrac{1}{2}\nu\pi - \tfrac{1}{4}\pi). \qquad (23.7)$$

For large z, $\beta \sim \frac{1}{2}\pi - 1/z$ and $\tan\beta \sim z$, so that (23.6) is consistent with (23.7) if $A e^{-\frac{1}{4}\pi i} = B e^{\frac{1}{4}\pi i} = (1/2\pi\nu)^{\frac{1}{2}}$. Now (23.6) can be used for $z^{\frac{1}{2}} J_\nu(\nu z)$ for large ν, i.e.

$$J_\nu(\nu \sec\beta) \sim (2/\pi\nu \tan\beta)^{\frac{1}{2}} \cos\{\nu(\tan\beta - \beta) - \tfrac{1}{4}\pi\}. \qquad (23.8)$$

This is Meissel's formula (A.2.7). Similarly, Carlini's formula (A.2.8) can be derived when $z < 1$, though L is purely imaginary in this event.

6.24 Langer's method

The WKB method is unsatisfactory near a zero of N, as is evident from the denominator of (23.3). The simple device of replacing N by the first term in its Taylor series is unrewardng because the resulting solution will not match the WKB formula away from the zero. It is obviously desirable to overlap the WKB thereby attaining an approximation which is *uniformly valid.* The trick is to make a change of variable which retains the character of N while bringing out the features of the zero.

Let $N^2(z)$ have a simple zero at $z = 0$ and be positive for $z > 0$, negative for $z < 0$. Introduce the new variable $\xi = f(z)$; then (23.1) becomes

$$\frac{d^2 w}{d\xi^2} + \frac{f''(z)}{\{f'(z)\}^2}\frac{dw}{d\xi} + \left\{\frac{k_0 N(z)}{f'(z)}\right\}^2 w = 0$$

where the primes indicate derivatives with respect to z. The coefficient of w will clearly display the zero if the choice

$$\{N(z)/f'(z)\}^2 = f(z) \qquad (24.1)$$

is made. The unwanted first derivative can be eliminated by the substitution $w = v(\xi)/\{f'(z)\}^{\frac{1}{2}} = g(z)v(\xi)$ (say). Since

$$dw/d\xi = g\, dv/d\xi + g'v/f'$$

it is found that

$$d^2 v/d\xi^2 + (k_0^2 \xi + g''/gf'^2)v = 0. \qquad (24.2)$$

A solution of (24.1) is required such that ξ is positive or negative according as N^2 is positive or negative. Therefore

$$\tfrac{2}{3}\xi^{\frac{3}{2}} = \int_0^z N(t)\,\mathrm{d}t \quad (z>0, \xi>0), \tag{24.3}$$

$$\tfrac{2}{3}(-\xi)^{\frac{3}{2}} = -\int_0^z |N(t)|\,\mathrm{d}t \quad (z<0, \xi<0) \tag{24.4}$$

taking N positive for $z>0$ and pure imaginary for $z<0$. Either (24.3) or (24.4) ensures that, for small z,

$$\xi \approx N_0^{\frac{2}{3}} z \tag{24.5}$$

where $N^2(z) \approx N_0^2 z$ with N_0 a positive constant.

As a first approximation the factor of v in (24.2) which does not involve k_0^2 can be dropped with the consequence

$$\mathrm{d}^2 v/\mathrm{d}\xi^2 + k_0^2 \xi v = 0. \tag{24.6}$$

Now, a solution of

$$\mathrm{d}^2 u/\mathrm{d}\eta^2 - \eta u = 0$$

is the Airy function $\mathrm{Ai}(\eta)$ (§A.2). A simple change of variable reveals that $\mathrm{Ai}(\eta \mathrm{e}^{\pm\frac{2}{3}\pi\mathrm{i}})$ are also solutions; they are related by

$$\mathrm{Ai}(\eta) = \mathrm{e}^{\frac{1}{3}\pi\mathrm{i}}\,\mathrm{Ai}(\eta e^{-\frac{2}{3}\pi\mathrm{i}}) + \mathrm{e}^{-\frac{1}{3}\pi\mathrm{i}}\,\mathrm{Ai}(\eta \mathrm{e}^{\frac{2}{3}\pi\mathrm{i}}). \tag{24.7}$$

The Airy function is an entire function and $\mathrm{Ai}(\eta \mathrm{e}^{2\pi\mathrm{i}}) = \mathrm{Ai}(\eta)$. The general solution of (24.6) is

$$v = C\,\mathrm{Ai}(-k_0^{\frac{2}{3}}\xi) + D\,\mathrm{Ai}(k_0^{\frac{2}{3}}\xi \mathrm{e}^{-\frac{1}{3}\pi\mathrm{i}})$$

and so our approximation to (23.1) in the presence of a simple zero of N^2 is

$$w(z) = (\xi/N^2)^{\frac{1}{4}}\{C\,\mathrm{Ai}(-k_0^{\frac{2}{3}}\xi) + D\,\mathrm{Ai}(k_0^{\frac{2}{3}}\xi \mathrm{e}^{-\frac{1}{3}\pi\mathrm{i}})\}. \tag{24.8}$$

Higher order approximations are feasible. The experience of geometrical optics suggests that a better high frequency solution of (24.2) is obtained by taking instead of $\mathrm{Ai}(-k_0^{\frac{2}{3}}\xi)$

$$\mathrm{Ai}(-k_0^{\frac{2}{3}}\xi)(1 + a_1/k_0^2 + a_2/k_0^4 + \ldots) + \mathrm{Ai}'(-k_0^{\frac{2}{3}}\xi)(b_0 + b_1/k_0^2 + \ldots)$$

where $\mathrm{Ai}'(\eta)$ is the derivative of $\mathrm{Ai}(\eta)$. This matter will not, however, be pursued further.

The next item is to compare (24.8) with the WKB approximation when z is not near the zero of N^2. Suppose z is positive. Then $k_0^{\frac{2}{3}}\xi$ is large and positive, permitting the employment of asymptotic expansions for the Airy function. These are (A.2.17), (A.2.19)

$$\mathrm{Ai}(z) \sim (1/2\pi^{\frac{1}{2}} z^{\frac{1}{4}})\exp(-\tfrac{2}{3} z^{\frac{3}{2}}) \quad (|\mathrm{ph}\, z| < \pi), \tag{24.9}$$

$$\mathrm{Ai}(z) \sim (\mathrm{e}^{\frac{1}{4}\pi\mathrm{i}}/\pi^{\frac{1}{2}} z^{\frac{1}{4}})\cos(\tfrac{2}{3}\mathrm{i} z^{\frac{3}{2}} - \tfrac{1}{4}\pi) \quad (\tfrac{1}{3}\pi < \mathrm{ph}\, z < \tfrac{5}{3}\pi). \tag{24.10}$$

L and $\frac{2}{3}\xi^{\frac{3}{2}}$ can be identified so long as $L=0$ at $z=0$. Therefore, applying (24.9) to the factor of D with $z=k_0^{\frac{2}{3}}\xi e^{-\frac{1}{3}\pi i}$ and (24.10) to the factor of C with $z=k_0^{\frac{2}{3}}\xi e^{\pi i}$, we have

$$w \sim (1/2\pi^{\frac{1}{2}}k_0^{\frac{1}{6}}N^{\frac{1}{2}})\{(Ce^{-\frac{1}{4}\pi i}+De^{\frac{1}{12}\pi i})e^{ik_0L}+Ce^{\frac{1}{4}\pi i-ik_0L}\}.$$

This agrees with (23.4) provided that

$$C=2\pi^{\frac{1}{2}}k_0^{\frac{1}{6}}Ae^{-\frac{1}{4}\pi i}, \qquad D=2\pi^{\frac{1}{2}}k_0^{\frac{1}{6}}(B+iA)e^{-\frac{1}{12}\pi i}. \tag{24.11}$$

Consequently, conversion from the WKB to the uniformly valid formula can be accomplished by making the identification (24.11) provided that $L=0$ at $z=0$. In addition, the uniformly valid expression supplies the behaviour near the zero (at a sufficiently close distance (24.5) can be used) and passes through it to latch on to the WKB on the other side.

For example, consider the Bessel function of §6.23. Here the zero is at $z=1$ so that, from (24.3) and (24.4) adjusted appropriately,

$$\tfrac{2}{3}\xi^{\frac{3}{2}}=\int_1^z(1-1/\zeta^2)^{\frac{1}{2}}\,d\zeta=(z^2-1)^{\frac{1}{2}}-\sec^{-1}z \quad (z>1,\ \xi>0), \tag{24.12}$$

$$\tfrac{2}{3}(-\xi)^{\frac{3}{2}}=-\int_1^z\{(1/\zeta^2)-1\}^{\frac{1}{2}}\,d\zeta=\operatorname{sech}^{-1}z-(1-z^2)^{\frac{1}{2}} \quad (0<z<1,\ \xi<0). \tag{24.13}$$

Invoking (24.11) we have $C=(2/\nu)^{\frac{1}{2}}$, $D=0$ and

$$J_\nu(\nu z)\sim 2^{\frac{1}{2}}\nu^{-\frac{1}{3}}\{\xi/(z^2-1)\}^{\frac{1}{4}}\,\mathrm{Ai}(-\nu^{\frac{2}{3}}\xi) \tag{24.14}$$

where ξ is given by (24.12) and (24.13). The expression (24.14) contains both Meissel's and Carlini's formulae as well as showing that

$$J_\nu(\nu)\sim 2^{\frac{1}{3}}/(-\tfrac{1}{3})!\,3^{\frac{2}{3}}\nu^{\frac{1}{3}} \tag{24.15}$$

(A.2.35) for large ν. The formula (24.14) cannot be carried over to $z<0$ (there is another zero at $z=-1$) but results can be deduced from the relation (A.1.3) for $J_\nu(ze^{\pi i})$. Other Bessel functions can be handled in a similar way and the relevant approximations are in §A.2.

Propagation over a plane earth

6.25 The Earth's atmosphere

Predicting the behaviour of signals in a neighbourhood of the Earth is beset with complication. The earth's surface is composed of a multiplicity of substances and its level can undulate rapidly and unpredictably, as well as carrying artefacts which can obstruct the free course of a wave. A first stab is to smooth everything away and make the earth a plane leaving the effects of sphericity and irregular terrain as later problems. The composition of the material below the plane is taken as some average of the real components, perhaps over some depth because radio waves of 1 MHz can penetrate 15 m into the Earth. How far the predictions are likely to be valid depends upon the range of propagation since the dielectric constant may vary from 5 for a city through 14 for a rocky soil to 80 for water and the conductivity from 10^{-2} mho m^{-1} for moist ground to 5 mho m^{-1} for sea water. Indeed, the average for a path mainly over the land is appreciably different from that for propagation entirely over the sea. Nevertheless, we shall assume that such averages exist, that they are unaffected by the weather and can be regarded as unvarying with frequency over a wide frequency band.

The atmosphere can be divided into several regions. That nearest the earth and extending for a height of 10–15 km is called the *troposphere*. Above the troposphere is the *stratosphere*. Further out, starting at a height of about 50 km is the *ionosphere*. Although the troposphere may be dealt with as free space in a first treatment, the presence of fog, hail, rain, cloud, dust, meteorological fronts, turbulence, etc. can cause significant disturbance to a signal. Even in their absence the troposphere is not exactly free space. The dielectric constant of dry air varies from 1.0006 at zero frequency to 1.0003 in the visible spectrum. Moreover, the molecular resonance of oxygen is responsible for attenuations of 14 dB km^{-1} and 2 dB km^{-1} at 0.5 cm and 0.25 cm respectively; there is also attenuation of 0.3 dB km^{-1} at 1.3 cm due to the molecular resonance of water vapour.

All of these variations, as well as horizontal changes, will be neglected but their interference with predictions based on their extinction may be profound. Our oversimplified model will be

$$N^2 = N_1^2(1+qz) \tag{25.1}$$

where z is the height above the surface of the Earth. The constant q is a measure of the rate of change with altitude. For radio waves it is about 0.25×10^{-6} m^{-1} near the earth. Possible values for acoustic waves can be inferred from §6.20.

The deviations of the stratosphere from free space are not of great significance except for communication with rockets and satellites. The ionosphere, on the other hand, plays a vital role. The constituent gases of the atmosphere are ionized by radiation from outside the earth in the ionosphere, which extends from 50 km to 400 km roughly above the earth. It displays great fluctuations over intervals of all lengths from short periods of a few minutes to cycles of several years. In addition, the earth's magnetic field exerts an important influence. An investigation of what the ionosphere does to signals is deferred to §6.34, after some knowledge has been gained of characteristics without it.

6.26 Propagation in a homogeneous atmosphere

The problem to be considered is the radiation from a point source in a homogeneous medium separated from another homogeneous medium by a plane interface and so offers a natural generalization of the reflection of a plane wave studied in §6.6. The technique of solution can be adapted to any excitation but will be limited to the point source here.

Let the dividing plane be $z=0$ and let the source be placed above the plane at $(0, 0, z_0)$ (Fig. 6.18). The properties of medium 1 in $z>0$ will be designated by the suffix 1 and those below $z=0$ by the suffix 2.

The equation to be satisfied by p is

$$\nabla^2 p + k_1^2 p = -\delta(x)\delta(y)\delta(z-z_0) \tag{26.1}$$

in $z>0$ and k_1 is replaced by k_2 in $z<0$. The boundary conditions will be taken to be the same as (6.1) adjusted to the context, namely

$$p_1 = p_2, \qquad v_{1z} = \zeta v_{2z} \tag{26.2}$$

on $z=0$.

Introduce the transform

$$\mathscr{P}(\alpha, \beta, z) = \int_{-\infty}^{\infty}\int_{-\infty}^{\infty} \mathrm{e}^{-\mathrm{i}\alpha x - \mathrm{i}\beta y} p(x, y, z)\,\mathrm{d}x\,\mathrm{d}y. \tag{26.3}$$

Then (26.1) becomes

$$\mathrm{d}^2\mathscr{P}/\mathrm{d}z^2 + (k_1^2 - \alpha^2 - \beta^2)\mathscr{P} = -\delta(z-z_0). \tag{26.4}$$

Write $\alpha^2+\beta^2=\lambda^2$ and $\kappa_1=(k_1^2-\lambda^2)^{\frac{1}{2}}$ with κ_1 to be either positive real or negative imaginary. Similarly $\kappa_2=(k_2^2-\lambda^2)^{\frac{1}{2}}$. The primary wave solution of (26.4) is (c.f. §4.4)

$$\mathscr{P}^{\mathrm{i}} = (1/2\mathrm{i}\kappa_1)\mathrm{e}^{-\mathrm{i}\kappa_1|z-z_0|}. \tag{26.5}$$

Therefore the hypothesis on the field is

$$\mathscr{P}_1 = \{\mathrm{e}^{-\mathrm{i}\kappa_1|z-z_0|} + C_1\mathrm{e}^{-\mathrm{i}\kappa_1(z+z_0)}\}/2\mathrm{i}\kappa_1 \quad (z>0),$$
$$\mathscr{P}_2 = C_2\mathrm{e}^{-\mathrm{i}\kappa_1 z_0 + \mathrm{i}\kappa_2 z}/2\mathrm{i}\kappa_1 \quad (z<0)$$

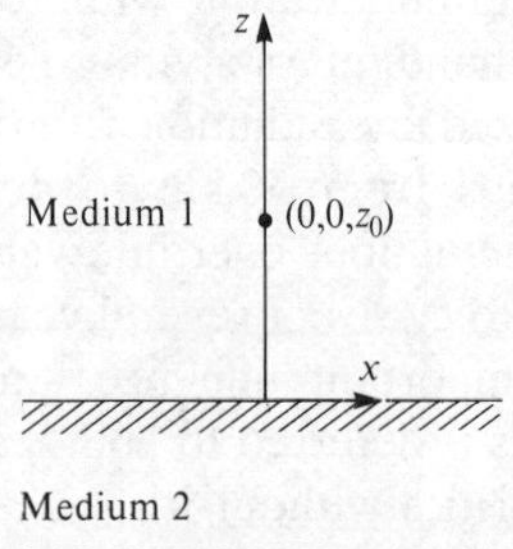

Fig. 6.18 Model for a plane earth

to provide additional fields that are outgoing as $z \to \infty$ and $z \to -\infty$ respectively.

The boundary conditions do not alter their form after the transform and so

$$1 + C_1 = C_2, \qquad (\kappa_1/\rho_1)(1 - C_1) = \zeta(\kappa_2/\rho_2)C_2.$$

Accordingly

$$C_1 = \frac{\kappa_1\rho_2 - \zeta\kappa_2\rho_1}{\kappa_1\rho_2 + \zeta\kappa_2\rho_1}, \qquad C_2 = \frac{2\kappa_1\rho_2}{\kappa_1\rho_2 + \zeta\kappa_2\rho_1}.$$

An inverse transform now supplies

$$p_1 = p^{\mathrm{i}} + \frac{1}{4\pi^2}\int_{-\infty}^{\infty}\int_{-\infty}^{\infty} \frac{1}{2\mathrm{i}\kappa_1}\frac{\kappa_1\rho_2 - \zeta\kappa_2\rho_1}{\kappa_1\rho_2 + \zeta\kappa_2\rho_1} \mathrm{e}^{\mathrm{i}\alpha x + \mathrm{i}\beta y - \mathrm{i}\kappa_1(z+z_0)}\, \mathrm{d}\alpha\, \mathrm{d}\beta, \tag{26.6}$$

$$p_2 = \frac{1}{4\pi^2 \mathrm{i}}\int_{-\infty}^{\infty}\int_{-\infty}^{\infty} \frac{\rho_2}{\kappa_1\rho_2 + \zeta\kappa_2\rho_1} \mathrm{e}^{\mathrm{i}\alpha x + \mathrm{i}\beta y - \mathrm{i}\kappa_1 z_0 + \mathrm{i}\kappa_2 z}\, \mathrm{d}\alpha\, \mathrm{d}\beta. \tag{26.7}$$

One of the integrations can be carried out by putting $x = r\cos\phi$, $y = r\sin\phi$, $\alpha = \lambda\cos\psi$, $\beta = \lambda\sin\psi$. The only part of the integrand which involves ψ is the exponential and $\int_0^{2\pi} \mathrm{e}^{\mathrm{i}r\lambda\cos(\psi-\phi)}\,\mathrm{d}\psi = 2\pi \mathrm{J}_0(\lambda r)$ (A.1.24). Hence

$$p_1 = p^{\mathrm{i}} + \frac{1}{2\pi}\int_0^{\infty} \frac{\lambda}{2\mathrm{i}\kappa_1}\frac{\kappa_1\rho_2 - \zeta\kappa_2\rho_1}{\kappa_1\rho_2 + \zeta\kappa_2\rho_1} \mathrm{J}_0(\lambda r)\mathrm{e}^{-\mathrm{i}\kappa_1(z+z_0)}\, \mathrm{d}\lambda, \tag{26.8}$$

$$p_2 = \frac{1}{2\pi \mathrm{i}}\int_0^{\infty} \frac{\lambda\rho_2}{\kappa_1\rho_2 + \zeta\kappa_2\rho_1} \mathrm{J}_0(\lambda r)\mathrm{e}^{-\mathrm{i}\kappa_1 z_0 + \mathrm{i}\kappa_2 z}\, \mathrm{d}\lambda. \tag{26.9}$$

An interesting observation is that, since $p^{\mathrm{i}} = \mathrm{e}^{-\mathrm{i}k_1 R_s}/4\pi R_s$ where R_s is the distance from the source, (26.5) permits the statement

$$\frac{\mathrm{e}^{-\mathrm{i}k_1 R_s}}{4\pi R_s} = \frac{1}{4\pi^2}\int_{-\infty}^{\infty}\int_{-\infty}^{\infty} \frac{1}{2\mathrm{i}\kappa_1} \mathrm{e}^{\mathrm{i}\alpha x + \mathrm{i}\beta y - \mathrm{i}\kappa_1|z-z_0|}\, \mathrm{d}\alpha\, \mathrm{d}\beta \tag{26.10}$$

$$= \frac{1}{4\pi \mathrm{i}}\int_0^{\infty} \frac{\lambda}{\kappa_1} \mathrm{J}_0(\lambda r)\mathrm{e}^{-\mathrm{i}\kappa_1|z-z_0|}\, \mathrm{d}\lambda \tag{26.11}$$

with the convention on κ_1 already explained.

The same transform is also successful for electromagnetic waves. Suppose that the field due to a vertical electric dipole is desired. The equations to be solved are

$$\operatorname{curl} \boldsymbol{E} + \mathrm{i}\omega\mu_1\boldsymbol{H} = \boldsymbol{0}, \qquad \operatorname{curl} \boldsymbol{H} - \mathrm{i}\omega\varepsilon_1\boldsymbol{E} = -\delta(x)\delta(y)\delta(z - z_0)\boldsymbol{k} \tag{26.12}$$

in $z > 0$ with apposite adjustments in $z < 0$. Before taking the transform it is convenient to make a temporary rotation of axes in the (x, y)-plane by putting $X=(\alpha x+\beta y)/\lambda$, $Y=(\alpha y-\beta x)/\lambda$. Then

$$d\mathscr{E}_Y/dz = i\omega\mu_1\mathscr{H}_X, \qquad d\mathscr{H}_Y/dz = -i\omega\varepsilon_1\mathscr{E}_X,$$
$$d\mathscr{E}_X/dz - i\lambda\mathscr{E}_z = -i\omega\mu_1\mathscr{H}_Y, \qquad d\mathscr{H}_X/dz - i\lambda\mathscr{H}_z = i\omega\varepsilon_1\mathscr{E}_Y,$$
$$\lambda\mathscr{E}_Y = -\omega\mu_1\mathscr{H}_z, \qquad i\lambda\mathscr{H}_Y = i\omega\varepsilon_1\mathscr{E}_z - \delta(z-z_0).$$

It is immediately evident that $\mathscr{H}_X=\mathscr{E}_Y=\mathscr{H}_z$ and that the remaining field can be expressed in terms of $\mathscr{H}_Y$ or $\mathscr{E}_z$. In fact

$$d^2\mathscr{H}_Y/dz^2+\kappa_1^2\mathscr{H}_Y = -i\lambda\delta(z-z_0).$$

Hence we take

$$\mathscr{H}_Y = \lambda\{e^{-i\kappa_1|z-z_0|}+B_1e^{-i\kappa_1(z+z_0)}\}/2\kappa_1 \quad (z>0),$$
$$= \lambda B_2 e^{-i\kappa_1 z_0 + i\kappa_2 z}/2\kappa_1 \quad (z<0).$$

The continuity of $\mathscr{H}_Y$ and $\mathscr{E}_X$ across $z=0$ now furnishes

$$B_1=\frac{\kappa_1\varepsilon_2-\kappa_2\varepsilon_1}{\kappa_1\varepsilon_2+\kappa_2\varepsilon_1}, \qquad B_2=\frac{2\kappa_1\varepsilon_2}{\kappa_1\varepsilon_2+\kappa_2\varepsilon_1}.$$

Rather than list the components we recognize that they can be regarded as constructed from a Hertz vector parallel to the z-axis with $\mathscr{H}_Y/\omega\varepsilon\lambda$ the transform of that vector. Hence the desired solution of (26.12) is

$$\boldsymbol{E} = (\text{grad div} + k_1^2)\Pi\boldsymbol{k}, \qquad \boldsymbol{H} = i\omega\varepsilon_1 \text{ curl}(\Pi\boldsymbol{k}) \tag{26.13}$$

where

$$\omega\varepsilon_1\Pi = \frac{ie^{-ik_1R_s}}{4\pi R_s} + \frac{1}{4\pi}\int_0^\infty \frac{\lambda}{\kappa_1}\frac{\kappa_1\varepsilon_2-\kappa_2\varepsilon_1}{\kappa_1\varepsilon_2+\kappa_2\varepsilon_1} J_0(\lambda r)e^{-i\kappa_1(z+z_0)}\,d\lambda \tag{26.14}$$

in $z>0$ and

$$\boldsymbol{E} = (\text{grad div} + k_2^2)\Pi_2\boldsymbol{k}, \qquad \boldsymbol{H} = i\omega\varepsilon_2 \text{ curl}(\Pi_2\boldsymbol{k}) \tag{26.15}$$

where

$$\omega\Pi_2 = \frac{1}{2\pi}\int_0^\infty \frac{\lambda J_0(\lambda r)}{\kappa_1\varepsilon_2+\kappa_2\varepsilon_1} e^{-i\kappa_1 z_0 + i\kappa_2 z}\,d\lambda \tag{26.16}$$

in $z<0$.

The vertical magnetic dipole can be handled likewise. In this case a solution of

$$\text{curl } \boldsymbol{E} + i\omega\mu_1\boldsymbol{H} = \delta(x)\delta(y)\delta(z-z_0)\boldsymbol{k}, \qquad \text{curl } \boldsymbol{H} = i\omega\varepsilon_1\boldsymbol{E} \tag{26.17}$$

is sought. It is discovered that

$$\boldsymbol{E} = -i\omega\mu_1 \text{ curl}(M\boldsymbol{k}), \qquad \boldsymbol{H} = (\text{grad div} + k_1^2)M\boldsymbol{k} \tag{26.18}$$

where

$$\omega\mu_1 M = \frac{\mathrm{i}e^{-\mathrm{i}k_1R_s}}{4\pi R_s} + \frac{1}{4\pi}\int_0^\infty \frac{\lambda}{\kappa_1}\frac{\kappa_1\mu_2 - \kappa_2\mu_1}{\kappa_1\mu_2 + \kappa_2\mu_1} \mathrm{J}_0(\lambda r)\mathrm{e}^{-\mathrm{i}\kappa_1(z+z_0)}\,\mathrm{d}\lambda, \tag{26.19}$$

$$\omega M_2 = \frac{1}{2\pi}\int_0^\infty \frac{\lambda \mathrm{J}_0(\lambda r)}{\kappa_1\mu_2 + \kappa_2\mu_1}\mathrm{e}^{-\mathrm{i}\kappa_1 z_0 + \mathrm{i}\kappa_2 z}\,\mathrm{d}\lambda. \tag{26.20}$$

6.27 Asymptotic behaviour of the field

The evaluation of the field at many wavelengths from the source has been undertaken by numerous authors and there is an exhaustive treatment in the book by Baños (1966). Only the reflected wave will be discussed here. Also, it is evident that (26.8), (26.14), and (26.19) all have the same structure so that it will be sufficient to consider one of them.

Since

$$(\kappa_1\varepsilon_2 - \kappa_2\varepsilon_1)/(\kappa_1\varepsilon_2 + \kappa_2\varepsilon_1) = -1 + 2\kappa_1\varepsilon_2/(\kappa_1\varepsilon_2 + \kappa_2\varepsilon_1)$$

we can draw on (26.11) to obtain

$$\omega\varepsilon_1 \Pi = \frac{\mathrm{i}e^{-\mathrm{i}k_1R_s}}{4\pi R_s} - \frac{\mathrm{i}e^{-\mathrm{i}k_1R_i}}{4\pi R_i} + \Pi_0 \tag{27.1}$$

where R_i is the distance from the image $(0, 0, -z_0)$ (Fig. 6.19) and

$$\Pi_0 = \frac{\varepsilon_2}{2\pi}\int_0^\infty \frac{\lambda \mathrm{J}_0(\lambda r)}{\kappa_1\varepsilon_2 + \kappa_2\varepsilon_1}\mathrm{e}^{-\mathrm{i}\kappa_1(z+z_0)}\,\mathrm{d}\lambda. \tag{27.2}$$

The second term of (27.1) shows that one effect of the boundary is to produce a field from the image which cancels the primary wave on the interface, i.e. the medium below $z = 0$ appears to be impenetrable and perfectly reflecting. The term Π_0 corrects this viewpoint and takes into account the properties of the earth.

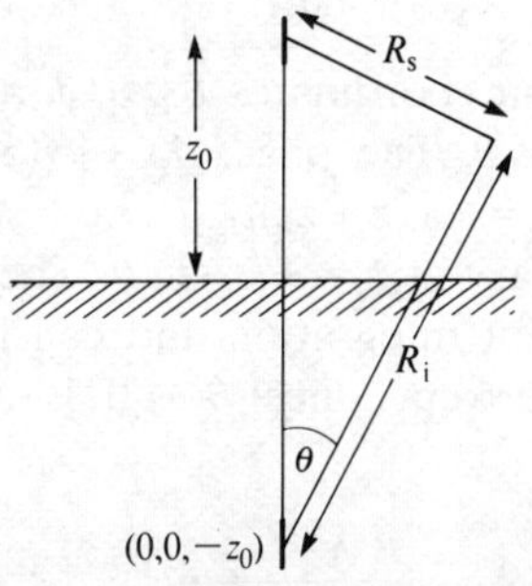

Fig. 6.19 The position relative to the image

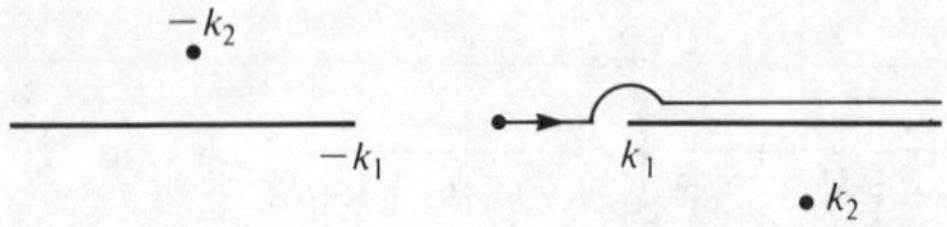

Fig. 6.20 Contour of integration in (27.2)

In (27.2) the path of integration has to be drawn so that κ_1 and κ_2 have the right properties. The singularities of κ_1 are at $\lambda = \pm k_1$ so that, if branch lines go from these points to infinity, the correct path with κ_1 positive at the origin is shown in Fig. 6.20. Furthermore, κ_2 has singularities at $\lambda = \pm k_2$. If k_2 came ahead of k_1 the path would have to surmount it in the same way that it makes an upper detour of k_1 in Fig. 6.20. There is also the possibility that k_2 has a negative imaginary part if there are losses in the lower medium (change ε_2 to $\varepsilon_2 - i\sigma/\omega$). It will be assumed that there is no dissipation in the atmosphere and that the properties of the earth ensure that $|k_2| > k_1$ so that the location of the singularities is typically as in Fig. 6.20.

The integral for Π_0 will be evaluated asymptotically on the assumption that $k_1 r \gg 1$ thereby excluding the region directly over the source. Then λr (λ has the same dimensions as k_1) will be large unless λ is small. Let d be a small quantity but such that $dr \gg 1$. The smallness of λ in the interval up to d enables that part of the integral to be written

$$\frac{\varepsilon_2 \mathrm{e}^{-\mathrm{i}k_1(z+z_0)}}{2\pi(k_1\varepsilon_2 + k_2\varepsilon_1)} \int_0^d \lambda \mathrm{J}_0(\lambda r)\,\mathrm{d}\lambda = \frac{\varepsilon_2 \mathrm{e}^{-\mathrm{i}k_1(z+z_0)}}{2\pi(k_1\varepsilon_2 + k_2\varepsilon_1)r} d\mathrm{J}_1(dr)$$

on account of the recurrence relation (A.1.6). The Bessel function can be approximated asymptotically because of the largeness of dr and so the contribution of this portion of the integral is $O(1/r^{\frac{3}{2}})$. Since only behaviour like $1/r$ is of interest this contribution can be discarded. For $\lambda \geqslant d$ the asymptotic formula for J_0 can be employed (A.2.1) with the result

$$\Pi_0 \sim \frac{\varepsilon_2}{(2\pi^3 r)^{\frac{1}{2}}} \int_d^\infty \frac{\lambda^{\frac{1}{2}}}{\kappa_1\varepsilon_2 + \kappa_2\varepsilon_1} \cos(\lambda r - \tfrac{1}{4}\pi) \mathrm{e}^{-\mathrm{i}\kappa_1(z+z_0)}\,\mathrm{d}\lambda. \qquad (27.3)$$

Introduce spherical polar coordinates R_i, θ, ϕ relative to the image, i.e. $r = R_i \sin\theta$, $z + z_0 = R_i \cos\theta$. The phase $\lambda r - \kappa_1(z + z_0)$ is stationary (see Appendix G) when $r = -\lambda(z + z_0)/\kappa_1$, i.e. $\lambda = -k_1 \sin\theta$. Similarly $-\lambda r - \kappa_1(z + z_0)$ is stationary at $\lambda = +k_1 \sin\theta$. The first stationary point is well outside the interval of integration and can be ignored. The second stationary point must be kept since $\theta \to 0$ is an excluded possibility. Hence

$$\Pi_0 \sim \frac{\varepsilon_2}{(8\pi^3 r)^{\frac{1}{2}}} \int_d^\infty \frac{\lambda^{\frac{1}{2}}}{\kappa_1\varepsilon_2 + \kappa_2\varepsilon_1} \mathrm{e}^{-\mathrm{i}\lambda r - \mathrm{i}\kappa_1(z+z_0) + \frac{1}{4}\pi\mathrm{i}}\,\mathrm{d}\lambda. \qquad (27.4)$$

In the neighbourhood of the stationary point put $\lambda = k_1 \sin\theta + t$ and expand the exponent to the second order. Then

$$\Pi_0 \sim \frac{\varepsilon_2(k_1 \sin\theta)^{\frac{1}{2}} e^{-ik_1R_i + \frac{1}{4}\pi i}}{(8\pi^3 r)^{\frac{1}{2}}(\kappa_1\varepsilon_2 + \kappa_2\varepsilon_1)} \int_{-\infty}^{\infty} e^{\frac{1}{2}iR_i t^2/k_1 \cos^2\theta}\, dt$$

$$\sim \frac{ik_1\varepsilon_2 \cos\theta e^{-ik_1R_i}}{2\pi R_i(\kappa_1\varepsilon_2 + \kappa_2\varepsilon_1)},$$

since $\int_{-\infty}^{\infty} e^{iax^2}\, dx = \pi^{\frac{1}{2}} e^{\frac{1}{4}\pi i}/a^{\frac{1}{2}}$ for positive a, with $\lambda = k_1 \sin\theta$ in κ_1 and κ_2. Combining this result with (27.1) we obtain

$$\omega\varepsilon_1\Pi \sim \frac{ie^{-ik_1R_s}}{4\pi R_s} + R_v(\theta)\frac{ie^{-ik_1R_i}}{4\pi R_i} \tag{27.5}$$

where

$$R_v(\theta) = \frac{\varepsilon_2 k_1 \cos\theta - \varepsilon_1(k_2^2 - k_1^2 \sin^2\theta)^{\frac{1}{2}}}{\varepsilon_2 k_1 \cos\theta + \varepsilon_1(k_2^2 - k_1^2 \sin^2\theta)^{\frac{1}{2}}}. \tag{27.6}$$

Comparison of (27.6) with (6.16) discloses that $R_v(\theta)$ *is the Fresnel reflection coefficient for a plane wave, incident at the angle* θ, *when the magnetic vector is perpendicular to the plane of incidence.* Ray theory would give precisely the same result because it treats wavefronts as locally plane.

The reflection coefficient R_v is -1 when $\theta = \frac{1}{2}\pi$. At nearby values of θ, R_v may be anywhere from close to -1 to nearly 1 depending on the size of $\varepsilon_2/\varepsilon_1$. For radio waves at an earth-air boundary $|k_2^2/k_1^2| \geqslant 20$ (except for very dry soil and even then $|k_2^2/k_1^2| \geqslant 5$) and R_v will fluctuate rapidly when θ is near $\frac{1}{2}\pi$. At the same time the angle θ_B such that $\tan\theta_B = k_2/k_1$, known as the *complex Brewster angle* by analogy with (7.1), is also in the neighbourhood of $\frac{1}{2}\pi$. On occasion the angle at which the phase of R_v is $-\frac{1}{2}\pi$ is known as the *pseudo Brewster angle.*

The sharp variation of R_v means that the condition that the integrand of Π_0 should be slowly varying apart from the phase is not met. Therefore (27.5) will be invalid in such circumstances. Moreover, the point of stationary phase is approaching the branch point $\lambda = k_1$ when $\theta \approx \frac{1}{2}\pi$. Another look at the calculation of Π is therefore called for, particularly since $\theta \approx \frac{1}{2}\pi$ corresponds to propagation near the surface of the earth.

This time write

$$\omega\varepsilon_1\Pi = \frac{ie^{-ik_1R_s}}{4\pi R_s} + \frac{ie^{-ik_1Ri}}{4\pi R_i} - \Pi_1 \tag{27.7}$$

where

$$\Pi_1 = \frac{\varepsilon_1}{2\pi}\int_0^{\infty} \frac{\kappa_2\lambda J_0(\lambda r)}{\kappa_1(\kappa_1\varepsilon_2 + \kappa_2\varepsilon_1)} e^{-i\kappa_1(z+z_0)}\, d\lambda.$$

The argument replacing the Bessel function by its asymptotic expression is unaltered and the stationary point near $\lambda=-k_1$ may be discarded again. Hence

$$\Pi_1 \sim \frac{\varepsilon_1}{(8\pi^3 r)^{\frac{1}{2}}} \int_d^\infty \frac{\kappa_2 \lambda^{\frac{1}{2}}}{\kappa_1(\kappa_1\varepsilon_2+\kappa_2\varepsilon_1)} \mathrm{e}^{-\mathrm{i}\lambda r - \mathrm{i}\kappa_1(z+z_0)+\frac{1}{4}\pi\mathrm{i}}\, \mathrm{d}\lambda.$$

Limiting ourselves to $\theta \approx \frac{1}{2}\pi$ the main contribution comes from the vicinity of the point of stationary phase and the branch point $\lambda = k_1$. For $\lambda \leqslant k_1$ make the substitution $\kappa_1 = \tau$. Expand the exponent, which causes fast variations in phase, as far as τ^2 but keep only powers up to τ in the remainder of the integrand. For $\lambda \geqslant k_1$ make the change of variable $\lambda^2 - k_1^2 = u^2$ and carry out an expansion in u similar to that for τ. The limit of integration may be set at ∞ without impairing the quality of the approximation and so

$$\Pi_1 \sim \frac{\varepsilon_1\kappa_2}{(8\pi^3 k_1 r)^{\frac{1}{2}}} \mathrm{e}^{-\mathrm{i}k_1 r+\frac{1}{4}\pi\mathrm{i}} \int_0^\infty \frac{\mathrm{e}^{-\mathrm{i}\tau(z+z_0)+\mathrm{i}r\tau^2/2k_1}}{\tau\varepsilon_2+\kappa_2\varepsilon_1}\,\mathrm{d}\tau$$
$$+\frac{\mathrm{i}\varepsilon_1\kappa_2}{(8\pi^3 k_1 r)^{\frac{1}{2}}} \mathrm{e}^{-\mathrm{i}k_1 r+\frac{1}{4}\pi\mathrm{i}} \int_0^\infty \frac{\mathrm{e}^{-u(z+z_0)-\mathrm{i}ru^2/2k_1}}{-\mathrm{i}u\varepsilon_2+\kappa_2\varepsilon_1}\,\mathrm{d}u$$

where now κ_2 is an abbreviation for $(k_2^2-k_1^2)^{\frac{1}{2}}$. Deform the contour of integration in the u-integral into the negative imaginary axis and put $u=\mathrm{i}\tau$. The pole does not lie in the fourth quadrant so that

$$\Pi_1 \sim \frac{\varepsilon_1\kappa_2}{(8\pi^3 k_1 r)^{\frac{1}{2}}} \mathrm{e}^{-\mathrm{i}k_1 r+\frac{1}{4}\pi\mathrm{i}} \int_{-\infty}^\infty \frac{\mathrm{e}^{-\mathrm{i}\tau(z+z_0)+\mathrm{i}r\tau^2/2k_1}}{\tau\varepsilon_2+\kappa_2\varepsilon_1}\,\mathrm{d}\tau$$
$$\sim \frac{\varepsilon_1\kappa_2}{(8\pi^3 k_1 r)^{\frac{1}{2}}} \mathrm{e}^{-\mathrm{i}k_1 R_i+\frac{1}{4}\pi\mathrm{i}} \int_{-\infty}^\infty \frac{\mathrm{e}^{\mathrm{i}r\tau^2/2k_1}}{\tau\varepsilon_2+\kappa_2\varepsilon_1+k_1\varepsilon_2(z+z_0)/r}\,\mathrm{d}\tau \tag{27.8}$$

on changing the variable of integration and recognizing that $r+(z+z_0)^2/2r$ is effectively R_i when $\theta\approx\frac{1}{2}\pi$.

Let

$$I(a)=\int_{-\infty}^\infty \frac{\mathrm{e}^{\mathrm{i}a\tau^2}}{\tau+b}\,\mathrm{d}\tau. \tag{27.9}$$

Then

$$\mathrm{d}\{\mathrm{e}^{-\mathrm{i}ab^2}I(a)\}/\mathrm{d}a = -2b\mathrm{i}\int_0^\infty \mathrm{e}^{\mathrm{i}a(\tau^2-b^2)}\,\mathrm{d}\tau = b\pi^{\frac{1}{2}}\mathrm{e}^{-\mathrm{i}ab^2-\frac{1}{4}\pi\mathrm{i}}/a^{\frac{1}{2}}.$$

But $I(a)\to 0$ as $a\to\infty$ and so

$$I(a) = b\pi^{\frac{1}{2}}\mathrm{e}^{\mathrm{i}ab^2-\frac{1}{4}\pi\mathrm{i}}\int_\infty^a t^{-\frac{1}{2}}\mathrm{e}^{-\mathrm{i}tb^2}\,\mathrm{d}t$$
$$= -2\pi^{\frac{1}{2}}\mathrm{e}^{-\frac{1}{4}\pi\mathrm{i}}\mathrm{F}(ba^{\frac{1}{2}}) \tag{27.10}$$

where F(z) is defined by

$$\mathrm{F}(z)=\mathrm{e}^{\mathrm{i}z^2}\int_z^\infty \mathrm{e}^{-\mathrm{i}t^2}\,\mathrm{d}t. \tag{27.11}$$

Some properties of $F(z)$ are set out in Chapter 9 (§9.7).

Bringing together (27.8)–(27.10) we obtain

$$\Pi_1\sim-\frac{\kappa_2\varepsilon_1\mathrm{e}^{-\mathrm{i}k_1R_\mathrm{i}}}{(2k_1r)^{\frac{1}{2}}\pi\varepsilon_2}\mathrm{F}\left\{(\tfrac{1}{2}k_1r)^{\frac{1}{2}}\left(\cot\theta+\frac{\kappa_2\varepsilon_1}{k_1\varepsilon_2}\right)\right\}.$$

To the degree of approximation that has been worked to the replacement of r, $\cot\theta$, and κ_2 by R_i, $\cos\theta$, and $(k_2^2-k_1^2\sin^2\theta)^{\frac{1}{2}}$ is acceptable. Hence

$$\omega\varepsilon_1\Pi=\frac{\mathrm{i}\mathrm{e}^{-\mathrm{i}k_1R_\mathrm{s}}}{4\pi R_\mathrm{s}}+\frac{\mathrm{i}\mathrm{e}^{-\mathrm{i}k_1R_\mathrm{i}}}{4\pi R_\mathrm{i}}+\frac{(k_2^2-k_1^2\sin^2\theta)^{\frac{1}{2}}\varepsilon_1}{(2k_1R_\mathrm{i})^{\frac{1}{2}}\pi\varepsilon_2}\mathrm{e}^{-\mathrm{i}k_1R_\mathrm{i}}\mathrm{F}(\gamma) \tag{27.12}$$

where

$$\gamma=(\tfrac{1}{2}k_1R_\mathrm{i})^{\frac{1}{2}}\{\cos\theta+(k_2^2-k_1^2\sin^2\theta)^{\frac{1}{2}}\varepsilon_1/k_1\varepsilon_2\}. \tag{27.13}$$

The formula (27.12) has been determined specifically for when θ is near $\frac{1}{2}\pi$. However, when θ is not in the vicinity of $\frac{1}{2}\pi$, γ is large and then $\mathrm{F}(\gamma)\sim-\mathrm{i}/2\gamma$; the formula (27.5) is recovered. Thus (27.12) provides a smooth transition from (27.5) down to $\theta=\frac{1}{2}\pi$. Accordingly, (27.12) *can be used for every value of* θ.

At sufficiently distant points R_i and R_s will be indistinguishable as far as the denominator is concerned (but not for the phase) and (27.12) gives

$$1+\{1+4(k_2^2-k_1^2\sin^2\theta)^{\frac{1}{2}}\varepsilon_1\mathrm{F}(\gamma)/(2k_1R_\mathrm{i})^{\frac{1}{2}}\mathrm{i}\varepsilon_2\}\mathrm{e}^{-\mathrm{i}k_1(R_\mathrm{i}-R_\mathrm{s})}$$

as the factor by which the Hertz vector of the primary wave must be multiplied in order to obtain the actual Hertz vector.

The formulae corresponding to (27.5) for acoustics and the magnetic vertical dipole can now be deduced from (26.8) and (26.19). They are

$$p_1=\frac{\mathrm{e}^{-\mathrm{i}k_1R_\mathrm{s}}}{4\pi R_\mathrm{s}}+R_\mathrm{a}(\theta)\frac{\mathrm{e}^{-\mathrm{i}k_1R_\mathrm{i}}}{4\pi R_\mathrm{i}} \tag{27.14}$$

where

$$R_\mathrm{a}(\theta)=\frac{\rho_2k_1\cos\theta-\zeta\rho_1(k_2^2-k_1^2\sin^2\theta)^{\frac{1}{2}}}{\rho_2k_1\cos\theta+\zeta\rho_1(k_2^2-k_1^2\sin^2\theta)^{\frac{1}{2}}}, \tag{27.15}$$

and

$$\omega\mu_1M=\frac{\mathrm{i}\mathrm{e}^{-\mathrm{i}k_1R_\mathrm{s}}}{4\pi R_\mathrm{s}}+R_\mathrm{h}(\theta)\frac{\mathrm{i}\mathrm{e}^{-\mathrm{i}k_1R_\mathrm{i}}}{4\pi R_\mathrm{i}}, \tag{27.16}$$

$$R_\mathrm{h}(\theta)=\frac{\mu_2k_1\cos\theta-\mu_1(k_2^2-k_1^2\sin^2\theta)^{\frac{1}{2}}}{\mu_2k_1\cos\theta+\mu_1(k_2^2-k_1^2\sin^2\theta)^{\frac{1}{2}}}, \tag{27.17}$$

Both R_a and R_h are what would be inferred from the relevant Fresnel coefficients for a plane wave. The more refined formula (27.12) will not be needed for M because μ_2 and μ_1 are always of the same order of magnitude and so R_h is never far from -1 when θ is near $\frac{1}{2}\pi$. The relative size of ρ_2 and $\zeta\rho_1$ will dictate whether it is necessary to go to (27.12) for sound waves but the adaptation is straightforward.

Ground-to-ground transmission occurs when both source and point of observation are on the surface of the earth. Then $z = z_0 = 0$ and $\theta = \frac{1}{2}\pi$ so that, from (27.12),

$$\omega\varepsilon_1\Pi = \{2 - 4i\gamma_0 F(\gamma_0)\}\Pi^i \tag{27.18}$$

where $\gamma_0 = (\frac{1}{2}k_1R_i)^{\frac{1}{2}}(k_2^2 - k_1^2)^{\frac{1}{2}}\varepsilon_1/k_1\varepsilon_2$, $\Pi^i = ie^{-ikR_i}/4\pi R_i$. If $|\gamma_0| \ll 1$, $\omega\varepsilon_1\Pi \approx 2\Pi^i$ while, if $|\gamma_0| \gg 1$, $\omega\varepsilon\Pi \approx -i\Pi^i/\gamma_0^2$ since $F(\gamma) + i/2\gamma \sim 1/4\gamma^3$ as $\gamma \to \infty$. It will be noted in the latter case that the wave is falling off like $1/R^2$ rather than $1/R$ and terms which have been rejected in the earlier analysis might be just as important.

The ground-to-ground field for the magnetic vertical dipole is zero by (27.16). So it is for the acoustic field unless $|\zeta\rho_1/\rho_2| \ll 1$ when the behaviour has the same character as Π.

When θ is not near $\frac{1}{2}\pi$, which occurs whenever the source or point of observation is sufficiently elevated the fields are the same as those predicted by ray theory. When $|k_2^2| \gg |k_1^2|$, as is usually true for an earth-air boundary, R_h never differs by much from -1 and so, *for elevated transmission, the earth behaves as a perfect conductor for the horizontally polarized waves produced by a vertical magnetic dipole.* Equally well, R_v is 1 for angles of incidence up to a moderate level but the size of $|N_2|$ depends on σ_2/ω and this may be reduced by diminishing the wavelengths. *At high conductivities or low frequencies the earth acts essentially as a perfect conductor for vertically polarized waves.* On the other hand, at large angles of incidence and high frequencies we can have $R_v = -1$; *if the angle of incidence is not too small, the earth behaves as a perfect dielectric for high frequency waves.*

Generally speaking $|\rho_2| \gg |\zeta\rho_1|$ in the acoustic case so that *the earth behaves as a hard boundary*; if the inequality were reversed the earth would appear as a soft interface.

When the heights of the source and receiver are low enough for the reflection coefficient to be -1, a typical formula is

$$\omega\mu_1 M = i\{1 - e^{ik_1(R_s - R_i)}\}e^{-ik_1R_s}/4\pi R_s$$

if R_s and R_i are close enough to be equalized except in the phase. Since $R_s \approx r + (z - z_0)^2/2r$ and $R_i \approx r + (z + z_0)^2/2r$

$$|\omega\mu_1 M| = (1/2\pi R_s)\,|\sin(k_1 z z_0/r)|.$$

Thus the field exhibits *interference effects*, going from a maximum of

$1/2\pi R_s$ to a minimum of 0 and back again. At the maxima the field strength is double that of the primary wave. The zeros cause the pattern to be broken up into lobes. The first maximum occurs when $2k_1 z z_0 = \pi r$ so that the low-angle radiation improves, for fixed height of transmission, as the wavelength decreases. Interference can also be observed at all locations where the reflection coefficient can be regarded as invariant with respect to angle but the maxima are not as great nor the minima as small as those in low-angle transmission. When the variation of the reflection coefficient with angle has to be allowed for the qualitative description still holds but the interference at the maxima and minima is no longer complete.

6.28 The impedance boundary condition and wave tilt

It will now be assumed that $|k_2| \gg |k_1|$ so that

$$R_v(\theta) \approx (\varepsilon_2 k_1 \cos\theta - \varepsilon_1 k_2)/(\varepsilon_2 k_1 \cos\theta + \varepsilon_1 k_2), \tag{28.1}$$

$$R_h(\theta) \approx (\mu_2 k_1 \cos\theta - \mu_1 k_2)/(\mu_2 k_1 \cos\theta + \mu_1 k_2), \tag{28.2}$$

$$R_a(\theta) \approx (\rho_2 k_1 \cos\theta - \zeta\rho_1 k_2)/(\rho_2 k_1 \cos\theta + \zeta\rho_1 k_2). \tag{28.3}$$

A check on the analysis reveals that the same reflection coefficients can be obtained by totally disregarding the field in $z < 0$ provided that the boundary conditions

$$\zeta k_2 \rho_1 p = -\mathrm{i}\rho_2\, \partial p/\partial z, \tag{28.4}$$

$$k_2 E_x = -\mu_2 \omega H_y, \qquad k_2 E_y = \mu_2 \omega H_x \tag{28.5}$$

are imposed on $z = 0$. Indeed, (27.12) will be reproduced with the approximation $|k_2| \gg |k_1|$ incorporated. The vector form of these boundary conditions is

$$p = (\mathrm{i}\rho_2/\zeta k_2 \rho_1)\boldsymbol{\nu} \cdot \operatorname{grad} p, \tag{28.6}$$

$$\boldsymbol{E}_t = (\omega\mu_2/k_2)\boldsymbol{H}_t \wedge \boldsymbol{\nu} \tag{28.7}$$

where $\boldsymbol{\nu}$ is a unit normal to the boundary directed into the high wavenumber region and the suffix t means a tangential component.

The proposal is that in calculating the scattering from an object, *in which* $|k_2| \gg |k_1|$, *it is sufficient to satisfy the impedance boundary conditions* (28.6) *and* (28.7). These boundary conditions should be compared with those derived in (8.6) and (8.9) for a wave striking a strongly dissipative medium. The impedance conditions have been found to be very valuable in applications and have formed the basis for discussing the characteristics of transitions such as coastlines.

An electromagnetic wave travelling over a highly conducting earth must have a small horizontal electric vector to sustain the energy dissipated by the earth's conductivity. The horizontal component is not in phase with the vertical component so the wave will be elliptically polarized and the electric vector tilted forward from vertical. An estimate of this *wave tilt* is furnished by consideration of a point on the x-axis. Previous interpretation indicates that $M=0$ and thereby $E_y=0$. Therefore the horizontal component is E_x, satisfying (28.5). H_y and E_z can be assessed as if the earth were perfectly conducting and then $E_z = -(\mu_1/\varepsilon_1)^{\frac{1}{2}}H_y$. Hence

$$E_x/E_z = \omega\mu_2\varepsilon_1^{\frac{1}{2}}/k_2\mu_1^{\frac{1}{2}}, \tag{28.8}$$

If $k_2=|k_2|e^{-i\delta_2}$ it follows from §6.4 that the major axis of the polarization ellipse is inclined at angle χ to the horizontal where $\tan 2\chi = \tan 2\psi \cos \delta_2$ and $\cot \psi = |E_x/E_z|$. Thus the experimental determination of the wave tilt provides a measure of the properties of the earth.

6.29 The quasi-homogeneous atmosphere

Modification of the theory to allow for continuous variation in the material parameters involves appreciable effort. It has already been pointed out in §6.13 that complication ensues for the governing equations though there is some simplification when the medium is stratified (§6.21). The additional terms due to the derivatives of the parameters are a bugbear but quite often their variations within a wavelength are small enough for their neglect to be justified. A medium in which the derivatives of the material parameters can be ignored is called *quasi-homogeneous.* In a stratified quasi-homogeneous medium the representations (26.13) and (26.18) are still valid with the Hertz vectors satisfying the Helmholtz equation with variable refractive index (§6.21). The same equation is also satisfied by p in the acoustic case.

The problem to be examined is that of a quasi-homogeneous stratified atmosphere with a linear vertical variation so that the equation to be solved, whether sound or electromagnetic waves are under consideration, is

$$\nabla^2 p + k_1^2(1+qz)p = -\delta(x)\delta(y)\delta(z-z_0) \tag{29.1}$$

where k_1 and q are positive constants. For the boundary condition, advantage will be taken of the preceding section which shows that the presence of the Earth can be accounted for by an impedance boundary condition as far as the reflected wave is concerned. So the condition

$$\partial p/\partial z = i\nu_0 p \tag{29.2}$$

will be imposed on $z=0$.

The finding of a solution of (29.1) is made slightly more convenient by introducing, as a temporary measure, $X = k_1 x$ and $Y = k_1 y$. Then, if

$$\mathcal{P} = \int_{-\infty}^{\infty}\int_{-\infty}^{\infty} p e^{-i\alpha X - i\beta Y}\, dX\, dY,$$

$$d^2\mathcal{P}/dz^2 + k_1^2(1 + qz - \lambda^2)\mathcal{P} = -k_1^2 \delta(z - z_0). \tag{29.3}$$

Consider first (29.3) with zero on the right-hand side. Put

$$\zeta = qz, \qquad \eta = (k_1/q)^{\frac{2}{3}}. \tag{29.4}$$

The substitution $\zeta + 1 - \lambda^2 = \xi/\eta$ leads to

$$d^2\mathcal{P}/d\xi^2 + \xi\mathcal{P} = 0.$$

As already seen in §6.24 this equation has as solutions the Airy functions $\mathrm{Ai}(-\xi)$, $\mathrm{Ai}(\xi e^{\pm\frac{1}{3}\pi i})$. To shorten the subsequent notation we write

$$f(\zeta) = e^{\frac{1}{3}\pi i}\,\mathrm{Ai}\{\eta(1+\zeta)e^{\frac{1}{3}\pi i}\}, \tag{29.5}$$

$$g(\zeta) = \mathrm{Ai}\{-\eta(1+\zeta)\}, \tag{29.6}$$

$$h(\zeta) = e^{-\frac{1}{3}\pi i}\,\mathrm{Ai}\{\eta(1+\zeta)e^{-\frac{1}{3}\pi i}\}. \tag{29.7}$$

Equation (24.7) may then be expressed as (A.2.14)

$$g(\zeta) = f(\zeta) + h(\zeta). \tag{29.8}$$

The solutions relevant to (29.3) are $f(\zeta-\lambda^2)$, $g(\zeta-\lambda^2)$, $h(\zeta-\lambda^2)$. As $\zeta \to \infty$, (24.9) and (24.10) indicate that f is an outgoing wave, g is oscillatory and h is incoming. Therefore, to cope with the δ which gives rise to the primary radiation, f must be selected for $z > z_0$. As $\zeta \to -\infty$, f and h increase exponentially whereas g falls exponentially; therefore g is the appropriate choice here. Hence, if $\zeta_0 = qz_0$,

$$\mathcal{P}^i = \begin{cases} Af(\zeta-\lambda^2)g(\zeta_0-\lambda^2) & (\zeta > \zeta_0) \\ Af(\zeta_0-\lambda^2)g(\zeta-\lambda^2) & (\zeta < \zeta_0). \end{cases}$$

In order that the discontinuity in $d\mathcal{P}^i/d\zeta$ be $-k_1/q$ it is necessary that

$$A\{f'(\zeta_0-\lambda^2)g(\zeta_0-\lambda^2) - f(\zeta_0-\lambda^2)g'(\zeta_0-\lambda^2)\} = -k_1 q.$$

From the Wronskian relation (A.2.39), $A = -2\pi i \eta^{\frac{1}{2}}$.

The disturbance due to the earth must be outgoing at infinity and so be a multiple of $f(\zeta-\lambda^2)$. Therefore, below ζ_0,

$$\mathcal{P} = Af(\zeta_0-\lambda^2)\{g(\zeta-\lambda^2) + Cf(\zeta-\lambda^2)\}.$$

To meet the boundary condition (29.2) we must have

$$C = -\{g'(-\lambda^2) - i\nu g(-\lambda^2)\}/\{f'(-\lambda^2) - i\nu f(-\lambda^2)\}$$

where $\nu = \nu_0/q$ and the field has been completely determined.

From now on only points of observation such that $\zeta \leq \zeta_0$ will be considered. There is no real loss of generality in this restriction because the reciprocity theorem (§1.35) enables one to assert that the field at ζ due to a source at ζ_0 coincides with the field at ζ_0 caused by a source at ζ. Thus the field in $\zeta > \zeta_0$ can be deduced from that below ζ_0.

The total field in $\zeta \leq \zeta_0$ is

$$p = \frac{1}{2\pi}\int_0^\infty \mathscr{P} \mathrm{J}_0(\lambda k_1 r)\lambda \, \mathrm{d}\lambda$$

$$= -\mathrm{i}\eta^{\frac{1}{2}}\int_0^\infty \lambda \mathrm{J}_0(\lambda k_1 r) f(\zeta_0 - \lambda^2)$$

$$\times \left\{ g(\zeta - \lambda^2) - \frac{g'(-\lambda^2) - \mathrm{i}\nu g(-\lambda^2)}{f'(-\lambda^2) - \mathrm{i}\nu f(-\lambda^2)} f(\zeta - \lambda^2) \right\} \mathrm{d}\lambda. \qquad (29.9)$$

While (29.9) constitutes the exact solution to the posed problem some evaluation of the integral is needed before interpretation is feasible. The Airy functions are entire functions so that the only singularities of the integrand in the complex λ-plane are poles at the zeros where

$$f'(-\lambda^2) - \mathrm{i}\nu f(-\lambda^2) = 0. \qquad (29.10)$$

The location of the poles varies considerably with ν but something can be said about limiting cases. If the magnitude of ν is large there will be poles near the zeros of f; these zeros are related to those α_s such that $\mathrm{Ai}(-\alpha_s) = 0$. The α_s are real and positive; their first few values are listed in Table A.5. Denote the roots of (29.10) by $\pm\lambda_s$, with λ_s reserved for the one with the positive real part. Then a first approximation for large ν is

$$\lambda_s = (1 + \alpha_s \mathrm{e}^{-\frac{1}{3}\pi\mathrm{i}}/\eta)^{\frac{1}{2}}. \qquad (29.11)$$

A second approximation can be found by expanding about the first; this yields

$$\lambda_s^2 = 1 + (\alpha_s \mathrm{e}^{-\frac{1}{3}\pi\mathrm{i}}/\eta) + \mathrm{i}/\nu.$$

When ν is very small the pertinent approximation is

$$\lambda_s = (1 + \beta_s \mathrm{e}^{-\frac{1}{3}\pi\mathrm{i}}/\eta)^{\frac{1}{2}} \qquad (29.12)$$

where $\mathrm{Ai}'(-\beta_s) = 0$; some values of the positive real constants β_s are set out in Table A.6. Expansion in powers of ν leads to higher approximations.

Some help in tracing the path of λ_s as ν varies is provided by taking a derivative of (29.10) and invoking the differential equation

$$f''(\zeta) + \eta^3(1+\zeta)f(\zeta) = 0 \qquad (29.13)$$

which is satisfied because of (29.5). Then

$$2\lambda_s\{\eta^3(1-\lambda_s^2) - \nu^2\} \, \mathrm{d}\lambda_s/\mathrm{d}\nu = \mathrm{i}. \qquad (29.14)$$

When $|\lambda_s|$ is large, comparison of (24.9), (24.10) with

$$\mathrm{Ai}'(z) \sim -(z^{\frac{1}{4}}/2\pi^{\frac{1}{2}})\exp(-\tfrac{2}{3}z^{\frac{3}{2}}) \quad (|\mathrm{ph}\, z|<\pi), \tag{29.15}$$

$$\mathrm{Ai}'(z) \sim (z^{\frac{1}{4}}\mathrm{e}^{-\frac{1}{4}\pi\mathrm{i}}/\pi^{\frac{1}{2}})\sin(\tfrac{2}{3}\mathrm{i}z^{\frac{3}{2}}-\tfrac{1}{4}\pi) \quad (\tfrac{1}{3}\pi<\mathrm{ph}\, z<\tfrac{5}{3}\pi) \tag{29.16}$$

demonstrates that the derivative always dominates. Therefore (29.12) gives the behaviour of the large zeros and shows that in the fourth quadrant they are going off to infinity in the direction of $\mathrm{e}^{-\frac{1}{6}\pi\mathrm{i}}$.

Our objective is now to deform the contour of integration over the poles which will be taken, in view of the foregoing investigation, to lie in the fourth and second quadrants. However, the asymptotic performance of the integrand does not permit us to do this without some manipulation.

First J_0 is replaced by Hankel functions via $2\mathrm{J}_0=\mathrm{H}_0^{(1)}+\mathrm{H}_0^{(2)}$. The integral involving $\mathrm{H}_0^{(1)}$ is deformed into the positive imaginary axis. The formulae (24.9), (24.10), (29.15), (29.16) together with the asymptotic performance of $\mathrm{H}_0^{(1)}$ indicate that this is permissible at infinity. For the integrand with $\mathrm{H}_0^{(2)}$, deformation down to $\mathrm{e}^{-\frac{1}{6}\pi\mathrm{i}}$ can be achieved in the same way. Beyond that, the two terms inside { } in (29.9) are combined and (29.8) used to eliminate f from the numerator in favour of h. Deformation down to the negative imaginary axis is then valid provided that $k_1 r>3^{\frac{1}{2}}\eta^{\frac{3}{2}}(\zeta+\zeta_0)$ or

$$r>3^{\frac{1}{2}}(z+z_0). \tag{29.17}$$

On the positive imaginary axis put $\lambda=\mathrm{i}u$ and on the negative imaginary axis $\lambda=-\mathrm{i}u$. Then the fact that $\mathrm{H}_0^{(2)}(\mathrm{i}u\mathrm{e}^{-\pi\mathrm{i}})=-\mathrm{H}_0^{(1)}(\mathrm{i}u)$ (A.1.20) means that the two integrals over these axes cancel. Hence, subject to (29.17), the result of the deformation is to pick up just the contributions of the poles in the fourth quadrant.

Consequently, when (29.17) holds, i.e. the point of observation is far enough out horizontally,

$$p=-\frac{\mathrm{i}k_1}{4q}\sum_{s=1}^{\infty}\frac{\mathrm{H}_0^{(2)}(\lambda_s k_1 r)f(\zeta_0-\lambda_s^2)f(\zeta-\lambda_s^2)}{\{\eta^3(1-\lambda_s^2)-\nu^2\}f^2(-\lambda_s^2)} \tag{29.18}$$

where the numerator has been simplified by means of the Wronskian relation and (29.10). If r is great enough for $\lambda_s k_1 r$ to be large for all s the Hankel function may be replaced by the first term in its asymptotic expression and then

$$p\sim\frac{\mathrm{e}^{-\frac{1}{4}\pi\mathrm{i}}}{4q}\left(\frac{2k_1}{\pi r}\right)^{\frac{1}{2}}\sum_{s=1}^{\infty}\frac{f(\zeta_0-\lambda_s^2)f(\zeta-\lambda_s^2)\exp(-\mathrm{i}\lambda_s k_1 r)}{\lambda_s^{\frac{1}{2}}\{\eta^3(1-\lambda_s^2)-\nu^2\}f^2(-\lambda_s^2)} \tag{29.19}$$

when (29.17) holds.

It will be noted that change of altitude of the point of observation merely alters one factor. Thus a measure of the effect of elevation is the *height-gain factor* $f(\zeta-\lambda_s^2)/f(-\lambda_s^2)$ which is unity at ground level where

$\zeta=0$. If $\eta\zeta \ll 1$ the first two terms of the Taylor series and (29.10) give for the height-gain factor $1+\mathrm{i}\nu\zeta$ or $1+\mathrm{i}\nu_0 z$. If $\nu_0=|\nu_0|\,\mathrm{e}^{\mathrm{i}\delta}$ the square of the modulus of the height-gain factor is $1-2\,|\nu_0|\,z\sin\delta+|\nu_0|^2 z^2$ which has a minimum when $z=|\nu_0|^{-1}\sin\delta$. That the field does go through a minimum as the point of observation is raised from the ground has been observed experimentally. The height-gain factor for the source exhibits a similar phenomenon.

6.30 The ray approximation

The formula (29.18) has the virtue of being exact but it holds only in a certain region. Also if $\eta(1+\zeta_0-\lambda_s^2)$ and $\eta(1+\zeta-\lambda_s^2)$ are sufficiently large compared with unity the asymptotic expressions for f may be employed in the early terms of (29.19) with the result

$$p\sim\frac{\mathrm{e}^{\frac{1}{4}\pi\mathrm{i}}}{16\pi q}\left(\frac{2k_1}{\pi\eta r}\right)^{\frac{1}{2}}$$

$$\times\sum_s\frac{\exp[-\frac{2}{3}\mathrm{i}\eta^{\frac{3}{2}}\{(1+\zeta_0-\lambda_s^2)^{\frac{3}{2}}+(1+\zeta-\lambda_s^2)^{\frac{3}{2}}\}-\mathrm{i}\lambda_s k_1 r]}{(1+\zeta_0-\lambda_s^2)^{\frac{1}{4}}(1+\zeta-\lambda_s^2)^{\frac{1}{4}}\lambda_s^{\frac{1}{2}}\{\eta^3(1-\lambda_s^2)-\nu^2\}f^2(-\lambda_s^2)}. \tag{30.1}$$

It can be checked that the real part of the exponent becomes more negative as s increases provided that $k_1 r>2\eta^{\frac{3}{2}}(\zeta^{\frac{1}{2}}+\zeta_0^{\frac{1}{2}})$. When this inequality is not satisfied a fair number of terms will be needed in the series to secure accuracy and another form for the solution is desirable.

In devising an alternative expression for p it will be assumed that $\eta \gg 1$ (which is likely to be true in practice) so that the Airy functions in (29.9) can be replaced by asymptotic formulae. It will also be supposed that $k_1 r\gg 1$. Then, as in §6.27, the interval of integration need involve only that portion where J_0 may be approximated asymptotically. Furthermore, only the factor $\mathrm{e}^{-\mathrm{i}\lambda k_1 r}$ will give rise to points of stationary phase in the range of integration.

It is straightforward to confirm that there are no points of stationary phase on the real axis where $\lambda^2>1+\zeta_0$. In the remaining interval the analysis is somewhat easier if (29.8) is used to replace g by h throughout the integrand. The condition for stationary phase for the term containing $h(\zeta-\lambda^2)$ is

$$\frac{\partial}{\partial\lambda}[-\lambda k_1 r-\tfrac{2}{3}\eta^{\frac{3}{2}}\{(1+\zeta_0-\lambda^2)^{\frac{3}{2}}-(1+\zeta-\lambda^2)^{\frac{3}{2}}\}]=0,$$

i.e.

$$-qr+2\lambda\{(1+\zeta_0-\lambda^2)^{\frac{1}{2}}-(1+\zeta-\lambda^2)^{\frac{1}{2}}\}=0 \tag{30.2}$$

from (29.4). Another derivative with respect to λ shows that (30.2) has no real solution unless

$$qr<2(1+\zeta)^{\frac{1}{2}}(\zeta_0-\zeta)^{\frac{1}{2}} \tag{30.3}$$

when it has one real root, say $\lambda = s_1$. Define L_s by

$$L_s = s_1 qr + \tfrac{2}{3}\{(1+\zeta_0 - s_1^2)^{\frac{3}{2}} - (1+\zeta - s_1^2)^{\frac{3}{2}}\} \tag{30.4}$$

when (30.3) is valid.

The term involving $f(\zeta - \lambda^2)$ in (29.9) has a stationary point of phase where

$$\frac{\partial}{\partial\lambda}[-\lambda k_1 r - \tfrac{2}{3}\eta^{\frac{3}{2}}\{(1+\zeta_0-\lambda^2)^{\frac{3}{2}} + (1+\zeta-\lambda^2)^{\frac{3}{2}} - 2(1-\lambda^2)^{\frac{3}{2}}\}] = 0,$$

i.e.

$$-qr + 2\lambda\{(1+\zeta_0-\lambda^2)^{\frac{1}{2}} + (1+\zeta-\lambda^2)^{\frac{1}{2}} - 2(1-\lambda^2)^{\frac{1}{2}}\} = 0 \tag{30.5}$$

provided that $\lambda^2 < 1$. Equation (30.5) has one real root s_2 so long as $\frac{1}{2}qr < \zeta^{\frac{1}{2}} + \zeta_0^{\frac{1}{2}}$ which is the region under consideration. Define

$$L_r = s_2 qr + \tfrac{2}{3}\{(1+\zeta_0 - s_2^2)^{\frac{3}{2}} + (1+\zeta - s_2^2)^{\frac{3}{2}} - 2(1-s_2^2)^{\frac{3}{2}}\}. \tag{30.6}$$

When $\lambda^2 > 1$ the last term in (30.5) is dropped because of the alteration in asymptotic behaviour of the integrand. The equation then has no real root except when (30.3) fails, in which case it has one zero $\lambda = s_3$. Define L_s, when the inequality in (30.3) is reversed, by

$$L_s = s_3 qr + \tfrac{2}{3}\{(1+\zeta_0 - s_3^2)^{\frac{3}{2}} + (1+\zeta - s_3^2)^{\frac{3}{2}}\}. \tag{30.7}$$

When equality prevails in (30.3), $s_1^2 = s_3^2 = 1 + \zeta$ and the definitions of L_s in (30.4) and (30.7) coincide.

The contributions from the points of stationary phase may now be evaluated. For simplicity, only the result when $1+\zeta-\lambda^2$ and $1+\zeta_0-\lambda^2$ are negligible compared with λ^2 will be quoted. It is

$$p \sim \frac{-\mathrm{i}}{4\pi k_1 r}\left\{\mathrm{e}^{-\mathrm{i}k_1 L_s/q} + D\frac{k_1\cos\chi - \nu_0}{k_1\cos\chi + \nu_0}\mathrm{e}^{-\mathrm{i}k_1 L_r/q}\right\} \tag{30.8}$$

where $\sin\chi = s_2$ and

$$D = (1 + 2\xi_1\xi_2/qr\sin 2\chi)^{-\frac{1}{2}}, \tag{30.9}$$

$$\xi_1 = 2(\zeta_0 + \cos^2\chi)^{\frac{1}{2}}\sin\chi - \sin 2\chi, \qquad \xi_2 = 2(\zeta + \cos^2\chi)^{\frac{1}{2}}\sin\chi - \sin 2\chi \tag{30.10}$$

Comparison of (30.2) and (30.4) with (18.13) and (18.15) reveal that the same conclusions would have been drawn from ray theory for the direct downward ray before it turned. After turning there is also agreement because of (30.7) and (18.16). L_r stems from a ray which strikes the ground before turning round and is reflected by it (Fig. 6.21).

The ray strikes at an angle of incidence χ. It is reflected at an equal angle to the vertical but with its amplitude multiplied by the Fresnel reflection coefficient for an angle of incidence of χ. The extra factor D is

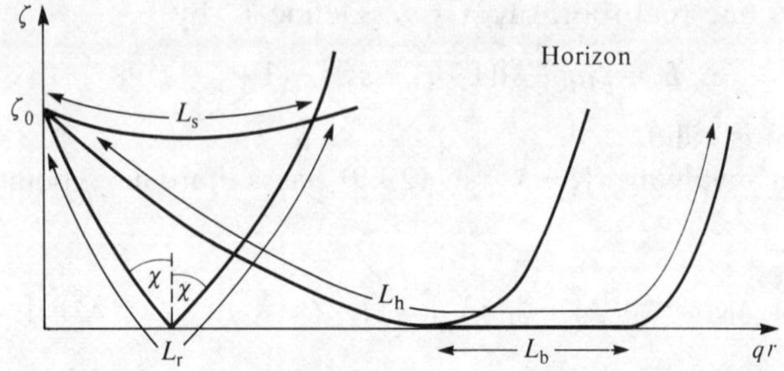

Fig. 6.21 Ray picture for the quasi-homogeneous atmosphere

a *divergence factor* representing the alteration in amplitude caused by the rays being curved, ξ_1 measuring that up to reflection and ξ_2 that after.

The furthest direct ray which can turn round is one which just touches $\zeta = 0$. From (30.2) its equation is

$$\tfrac{1}{2}qr = \zeta^{\frac{1}{2}} + \zeta_0^{\frac{1}{2}}. \tag{30.11}$$

This curve beyond its point of tangency may be called the *horizon.* According to ray theory there is no propagation through the horizon. Up to the horizon there is no conflict between calculating the field from our integral or from rays.

After the horizon ray theory predicts zero field but here (30.1) is applicable. The penetration of the field through the horizon is an example of *diffraction.* In other words, diffraction accounts for propagation past the horizon in a quasi-homogeneous atmosphere.

If only the first term in (30.1) is retained as will be reasonable sufficiently over the horizon for $\tfrac{1}{2}qr - \zeta^{\frac{1}{2}} - \zeta_0^{\frac{1}{2}}$ to be large

$$p \sim \frac{\mathrm{e}^{\frac{1}{4}\pi \mathrm{i}}}{16\pi q}\left(\frac{2k_1}{\pi\eta r}\right)^{\frac{1}{2}} \frac{\exp[-\mathrm{i}k_1\{L_\mathrm{h} + \frac{1}{2}(\lambda_1^2 - 1)L_\mathrm{b}\}]}{(1+\zeta_0-\lambda_1^2)^{\frac{1}{4}}(1+\zeta-\lambda_1^2)^{\frac{1}{4}}\lambda_1^{\frac{1}{2}}\{\eta^3(1-\lambda_1^2)-\nu^2\}f^2(-\lambda_1^2)} \tag{30.12}$$

where

$$L_\mathrm{h} = r + \tfrac{2}{3}(\zeta_0^{\frac{3}{2}} + \zeta^{\frac{3}{2}})/q, \qquad L_\mathrm{b} = r - 2(\zeta_0^{\frac{1}{2}} + \zeta^{\frac{1}{2}})/q,$$

remembering that λ_1 is unity to a first approximation. Formula (30.12) permits an extension of ray theory to cover diffraction. L_h can be regarded as the optical path length of a ray which leaves ζ_0 on the glancing path to the boundary and travels along the boundary until it reaches the point where departure on a tangential ray will take it to the point of observation. The optical path length which it is obliged to follow on the boundary is L_b. The ray path is in accord with Fermat's principle, being the shortest one joining the source and a point in the shadow.

The calculation of the amplitude is, of course, normal off the boundary. Along the boundary there will be a continued loss of energy through the shedding of rays upwards. This causes the exponential decay evident in (30.12). In addition, the amplitude will be diminishing because the wavefront is enlarging as it moves steadily away from the source. Allowance has also to be made for the act of launching the ray along the boundary instantaneously altering the amplitude on arrival.

The theory of rays, extended for diffraction gives the field well into the shadow, i.e. to the right of the dot-dash curve of Fig. 6.22, and also in the fully illuminated zone to the left of the dashed line. But there is a penumbral region about the horizon where neither (30.8) nor (30.12) is effective. In this region (30.12) is deficient because many more terms of the series must be kept while (30.8) is inadequate because the point of stationary phase is close to $\lambda = 1$, thereby invalidating the approximation of using an asymptotic formula for the Airy function.

If we assume that ζ and ζ_0 are big enough for asymptotic approximation of the Airy functions except those of argument $(1-\lambda^2)$ we may employ the technique of §6.27 of considering a neighbourhood of $\lambda = 1$ and taking $\lambda^2 - 1$ as a new variable of integration. The result is that, in the transition zone,

$$\begin{aligned} p \sim \frac{e^{\frac{1}{4}\pi i}}{4\pi\zeta_0^{\frac{1}{4}}\zeta^{\frac{1}{4}}}\left(\frac{1}{2\pi k_1 r}\right)^{\frac{1}{2}} e^{-ik_1L_h}\Bigg[\left(\frac{1}{k_1L_t}\right)^{\frac{1}{2}} \mathrm{F}\left\{\tfrac{1}{2}\left(\frac{k_1}{L_t}\right)^{\frac{1}{2}} L_b\right\} \\ + \frac{e^{-\frac{1}{3}\pi i}}{2\eta}\int_0^\infty \frac{\{\eta\,\mathrm{Ai}'(\mu) + i\nu\,\mathrm{Ai}(\mu)\}\exp(-\frac{1}{2}i\eta^{\frac{1}{2}}L_b\mu)}{\eta e^{\frac{1}{3}\pi i}\,\mathrm{Ai}'(\mu e^{-\frac{2}{3}\pi i}) - i\nu\,\mathrm{Ai}(\mu e^{-\frac{2}{3}\pi i})}\,d\mu \\ + \frac{e^{-\frac{1}{3}\pi i}}{2\eta}\int_0^\infty \frac{\{i\nu\,\mathrm{Ai}(\mu) - \eta e^{-\frac{1}{3}\pi i}\,\mathrm{Ai}'(\mu)\}\exp(-\frac{1}{2}i\eta^{\frac{1}{2}}L_b\mu e^{-\frac{2}{3}\pi i})}{\eta e^{\frac{1}{3}\pi i}\,\mathrm{Ai}'(\mu e^{\frac{2}{3}\pi i}) - i\nu\,\mathrm{Ai}(\mu e^{\frac{2}{3}\pi i})}\,d\mu\Bigg] \end{aligned} \tag{30.13}$$

where

$$qL_t = \zeta_0^{-\frac{1}{2}} + \zeta^{-\frac{1}{2}} - \tfrac{1}{2}qr. \tag{30.14}$$

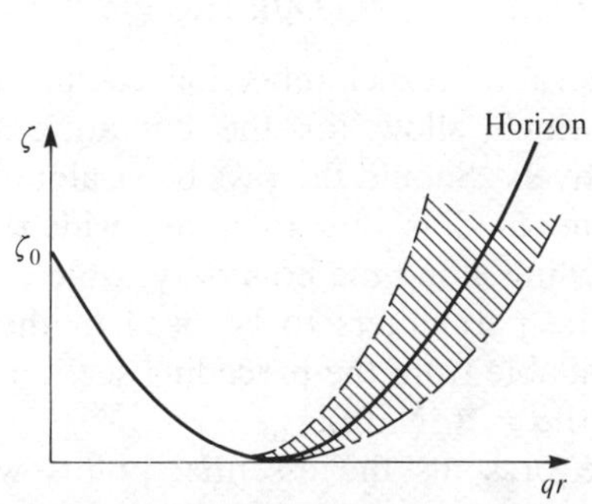

Fig. 6.22 The transition region

The additional approximation of large ν/η reduces (30.13) to

$$p \sim \frac{e^{\frac{1}{4}\pi i}}{4\pi\zeta_0^{\frac{1}{4}}\zeta^{\frac{1}{4}}}\left(\frac{1}{2\pi k_1 r}\right)^{\frac{1}{2}} e^{-ik_1 L_h}\left[\left(\frac{1}{k_1 L_t}\right)^{\frac{1}{2}} F\left\{\tfrac{1}{2}\left(\frac{k_1}{L_t}\right)^{\frac{1}{2}} L_b\right\} + \frac{i}{2\eta}\{f_1(-\tfrac{1}{2}\eta^{\frac{1}{2}}L_b) + f_2(-\tfrac{1}{2}\eta^{\frac{1}{2}}L_b e^{-\frac{2}{3}\pi i})\}\right] \tag{30.15}$$

where

$$f_1(t) = e^{\frac{1}{6}\pi i}\int_0^\infty \{\mathrm{Ai}(\mu)e^{i\mu t}/\mathrm{Ai}(\mu e^{-\frac{2}{3}\pi i})\}\,d\mu, \tag{30.16}$$

$$f_2(t) = e^{\frac{1}{6}\pi i}\int_0^\infty \{\mathrm{Ai}(\mu)e^{i\mu t}/\mathrm{Ai}(\mu e^{\frac{2}{3}\pi i})\}\,d\mu. \tag{30.17}$$

The functions f_1 and f_2 will also be met in Chapter 8 (§8.4).

When $|\eta^{\frac{1}{2}}L_b| \gg 1$, but the point of observation is near the horizon, the formula (30.13) goes over to (30.8) or (30.12) according as L_b is negative or positive and so as $|L_b|$ increases from zero a smooth transition to the fields already obtained is assured.

6.31 The stratified atmosphere

The exact theory of the linear gradient is reassuring in reproducing ray theory in the directly illuminated region and indicating how it should be modified to include diffraction into the shadow. One may therefore apply it confidently to the stratified medium in which the refractive index is $N(z)$ when the free space wavenumber is k_0. The general behaviour will be that depicted in Fig. 6.21. To the left of the horizon primary and reflected waves will be the constituents and to the right extended ray theory will be needed. Let $L = \int N(z)\,ds$ on a ray and let d denote the geometrical length of a ray. Then, in the directly illuminated zone, the field will be a constant multiple of

$$(D_s/d_s)e^{-ik_0 L_s} + (D_r/d_r)R_F(\chi)e^{-ik_0 L_r}$$

where R_F is the relevant Fresnel reflection coefficient and D_s, D_r are divergence factors which allow for the curvature of the direct and reflected rays respectively. Should the rays be straight lines put $D_s = D_r = 1$. In the shadow the field is due to a ray which touches $z = 0$ and originates a ray travelling along the boundary while continually dispatching rays upwards. The parameters to be used in the instigation of the boundary ray are available from the preceding section with the identification $k_1 = k_0 N(0)$, $N(0)q = 2(dN/dz)_{z=0}$.

The above picture presents the essential points when the refractive index increases upwards. A negative gradient entails some modification and in order to bring out the distinctive features any diffraction zones will be ignored. Suppose that N is constant up to a height h_1 and passes through a minimum $N(m)$ between h_1 and h_2 before becoming constant again above h_2. The ray picture will be that of Fig. 6.15. However, any reflected ray will impinge on the earth and be reflected by it. It will then proceed back to the region of variable refractive index where it will be returned to strike the earth again. Obviously the process carries on indefinitely.

When $N(m)>0$ and the source is on the ground no reflected ray can hit the earth closer than a certain distance from the source, known as the *skip distance.* Even if N has a slight positive gradient below h_1 this will still be qualitatively true. However, a wave reflected in the upper altitudes, or *sky wave* as it may be called, can penetrate the shadow zone on its way back to the earth. The sky wave will usually be much stronger than any signal that has got past the horizon by means of diffraction. This is because the sky wave falls off as the reciprocal of the distance it has travelled whereas the amplitude of a diffracted ray is exponentially attenuated. Sky waves can be responsible for the reception of signals well beyond the limitations set by the horizon.

The rays being reflected by the minimum in refractive index produce a caustic because neighbouring rays intersect. A ray which passes through a caustic suffers no loss of amplitude but its phase is advanced by $\frac{1}{2}\pi$ (§8.34). Hence, *a sky wave, immediately after reflection by a minimum in refractive index, has undergone no change in amplitude but its phase has advanced by* $\frac{1}{2}\pi$. The field on a typical ray of a sky wave, launched at an angle θ, is therefore

$$e^{\frac{1}{2}l\pi i}D\{R_F(\theta)\}^m e^{-ik_0L}/d$$

where m is the number of reflections at the earth's surface, l is the number of reflections at the variable region and D, L, d are as defined before for the whole path including reflections.

The total field at a given point is obtained by summing over all rays (the direct as well) which pass through it. Not only the rays emitted upwards by the source but those which are sent out downwards must be taken into account in the summation. If both source and point of observation are on the surface of the earth the resulting formula for the sky wave is

$$\sum D\{R_F(\theta_0)\}^m\{1+R_F(\theta_0)\}^2 \exp\{-ik_0L+\tfrac{1}{2}(m+1)\pi i\}/d \qquad (31.1)$$

where m is the number of earth reflections between the source and receiver. The factor $1+R_F$ is included, once because the receiver is at a point where both a downcoming and upgoing ray occur, and a second time because the source is on the ground. The summation must be carried out for those values of θ_0 which ensure a ray through the point of observation, i.e. which satisfy

$$x = 2(m+1)\sin\theta_0 \int_0^h \{N^2(w) - \sin^2\theta_0\}^{-\frac{1}{2}}\,dw \tag{31.2}$$

where $N(h) = \sin\theta_0$. The values of L and D are given by

$$L = 2(m+1)\int_0^h N^2(w)\{N^2(w) - \sin^2\theta_0\}^{-\frac{1}{2}}\,dw, \tag{31.3}$$

$$D/d = \{(1/x)\tan\theta_0\,|d\theta_0/dx|\}^{\frac{1}{2}}. \tag{31.4}$$

6.32 Ducts

Temperature inversions and other diverse climatological conditions can originate *ducts* in the troposphere. In a duct N decreases upwards from the earth's surface, passes through a minimum and then starts to increase. The theory of the preceding section is applicable to a duct but with the difference that the skip distance is zero. As a result it is possible to transmit over large distances, well beyond the horizon, a phenomenon called *super-refraction.*

When both source and receiver are on the ground (31.1) is relevant. A ray reaches the receiver for every value of m and the series is a sum from $m = 0$ to $m = \infty$. The series is more conveniently handled by applying the *Poisson summation formula* (Jones (1982)) to it. This summation mechanism says that, if $\mathscr{F}(u) = \int_0^\infty f(t)e^{-iut}\,dt$,

$$\tfrac{1}{2}f(0) + \sum_{m=1}^{\infty} f(m) = \sum_{m=-\infty}^{\infty} \mathscr{F}(2\pi m). \tag{32.1}$$

Elect f to be a typical term of (31.1). Then the sum of the series is essentially equal to

$$\sum_{s=-\infty}^{\infty} \int_0^\infty (D/d)\{R_F(\theta_0)\}^t\{1 + R_F(\theta_0)\}^2 \exp\{-ik_0L + \tfrac{1}{2}(t+1)\pi i - 2s\pi it\}\,dt.$$

The angle θ_0 in the integrand is regarded as a function of t, determined from (31.2) with m replaced by t. Also L is given by (31.3) with a similar replacement and D by (31.4)

The integral will be estimated by the method of stationary phase, assuming that R_F is a slowly varying function of t. The phase is stationary when

$$i \ln R_F + 2s\pi - \tfrac{1}{2}\pi + k_0\,\partial L/\partial t = 0,$$

i.e.

$$2k_0 \int_0^h \{N^2(w) - \sin^2\theta_0\}^{\frac{1}{2}}\,dw = \tfrac{1}{2}\pi - 2s\pi - i\ln R_F \tag{32.2}$$

from (31.2) and (31.3). The equation (32.2) determines a value of θ_0, say θ_s; the corresponding value of t follows from (31.2). The series is now

$$\sum_s \left(\frac{\pi \sin \theta_s}{k_0 x}\right)^{\frac{1}{2}} \frac{\{1+R_F(\theta_s)\}^2}{R_F(\theta_s)\cos \theta_s} \exp(-ik_0 x \sin \theta_s - \tfrac{1}{4}\pi i). \qquad (32.3)$$

Generally speaking, θ_s and $\sin \theta_s$ will be complex, causing the terms in (32.3) to be exponentially damped. However, should θ_s be real, a comparatively strong signal occurs and there can be communication over much greater distances than when there is no duct. The duct acts as a waveguide despite the absence of a surrounding metal sheath. In solving (32.2) it should be noticed that s can be negative; when R_F is $e^{-\pi i}$ a real θ_s for negative s is not uncommon.

To assess the chance of real θ_s let $R_F = e^{-\pi i}$ and let the minimum of N differ only slightly from $N(0)$. There is no loss in taking $N(0) = 1$ and the minimum to be $1-\tau$, occurring at $z = h_0$. The left-hand side of (32.2) is positive and does not exceed $2k_0h_0(2\tau)^{\frac{1}{2}}$. The smallest positive value of the right-hand side of (32.2) is $3\pi/2$ for $s = -1$. Hence no θ_s can be real if

$$\lambda_0 > 8h_0(2\tau)^{\frac{1}{2}}/3 \qquad (32.4)$$

where $\lambda_0 = 2\pi/k_0$.

For N decreasing linearly at the rate of $0.12\times 10^{-6}\,\mathrm{m}^{-1}$, (32.4) gives $\lambda_0 > 1.3\times 10^{-3} h_0^{\frac{3}{2}}$ as the condition for no real θ_s. Thus there can be no long-range transmission via the duct if $\lambda_0 > 1.3$ m for $h_0 = 100$ m or if $\lambda_0 > 3.6$ cm if $h_0 = 9$ m. *The presence of the duct is of most assistance at short wavelengths of operation.*

6.33 Rays in a medium with slight absorption

Some assessment of the effect of small losses in a medium on a ray is desirable since there are few real media which are without dissipation. Suppose that $N = N_r - iN_i$ where squares and higher powers of N_i can be neglected. The derivation of §6.15 may still be followed and

$$\mathrm{grad}^2 L = N^2, \qquad \mathrm{div}(|\gamma_0|^2\, \mathrm{grad}\, L) = 0.$$

With $L = L_r - iL_i$ regard L_i as small. The surface L = constant is no longer real. However

$$\mathrm{grad}^2 L_r = N_r^2, \qquad \mathrm{grad}\, L_r \cdot \mathrm{grad}\, L_i = N_r N_i. \qquad (33.1)$$

Thus L_r is the same as if there were no absorption and L_r = constant provides a phase front. The effect of L_i is to produce exponential attenuation as this phase front propagates.

The second of (33.1) states that $\partial L_i/\partial s = N_i$ where $\partial/\partial s$ is a derivative normal to L_r = constant. Hence $L_i = \int N_i\, ds$, the integration being along a ray perpendicular to L_r = constant.

Since grad L_i is of the order of N_i its influence on γ_0 will be of the same order. This will be insignificant compared to the exponential damping introduced into the wave by L_i. Therefore γ_0 can be kept the same as if there were no dissipation.

In summary, the rule is: *when the absorption is small, calculate the rays and fields on them as if the medium were loss free and then multiply the field by* $\exp(-k_0 \int N_i \, ds)$.

6.34 The ionosphere

The ionosphere, being that part of the atmosphere where ionization has taken place, contains particles, like electrons and ions, which are charged. The ions and electrons can be set in motion by a radio wave and will generate a convection current which will affect the wave. From time to time an electron will collide with a gas molecule. During such a collision energy acquired from the radio wave will be partly transferred to the molecule and partly re-radiated. The radio wave will lose energy thereby.

Let $\boldsymbol{u}$ be the velocity of a particle of mass m and charge Q subject to the electric intensity $\boldsymbol{E}$ and the magnetic flux density $\boldsymbol{B}$. The equation of motion is (§2.10)

$$m \, d\boldsymbol{u}/dt + mf\boldsymbol{u} = Q(\boldsymbol{E} + \boldsymbol{u} \wedge \boldsymbol{B}) \tag{34.1}$$

where f is a measure of the frequency of collision. No contribution from Lorentz polarization has been allowed for. The convection current produced by M such charges per unit volume is

$$\boldsymbol{J} = MQ\boldsymbol{u}. \tag{34.2}$$

For the rest of this section it will be assumed that the magnetic field can be neglected. Then (34.1) reduces to the linear equation

$$m \, d\boldsymbol{u}/dt + mf\boldsymbol{u} = Q\boldsymbol{E}. \tag{34.3}$$

When the time variation of $\boldsymbol{E}$ is harmonic, as $e^{i\omega t}$, the forced oscillations of $\boldsymbol{u}$ in (34.3) are

$$\boldsymbol{u} = Q\boldsymbol{E}/m(f + i\omega)$$

and, from (34.2),

$$\boldsymbol{J} = MQ^2\boldsymbol{E}/m(f + i\omega).$$

Thus the ionosphere can be ascribed a complex conductivity $MQ^2/m(f + i\omega)$. Taking the ionosphere to be free space otherwise, we see that it has a complex refractive index given by

$$N^2 = 1 - \omega_c^2/\omega(\omega - if)$$

where $\omega_c^2 = MQ^2/m\varepsilon_0$. The frequency f_c such that $f_c = \omega_c/2\pi$ is called the *critical frequency*. When Q is 1.59×10^{-19} coulomb and m is 9×10^{-31} kg, the critical frequency $f_c = 9M^{\frac{1}{2}}$.

The refractive index of the ionosphere is the same as that which would be found in a medium of dielectric constant $1-\omega_c^2/(\omega^2+f^2)$ and conductivity $\varepsilon_0 f\omega_c^2/(\omega^2+f^2)$. Therefore, *the ionosphere behaves at high frequencies like a dielectric and at low frequencies like a conductor.* Note that, even if there are no collisions, the refractive index can be imaginary if ω_c increases with height to a sufficient extent.

Ignore collisions in the first place. There is then a stratified medium as in §6.18 because of the variation of ω_c with a height. According to (18.7) a ray can be reflected only if there is a height h such that $N(h)=\sin\theta_0$, i.e.

$$\omega=\omega_c(h)\sec\theta_0. \tag{34.4}$$

Transmission beyond the skip distance by means of the sky wave will then be possible. If h_m is the height at which ω_c is a maximum only waves with frequencies satisfying $\omega<\omega_c(h_m)\sec\theta_0$ will be capable of long range communication. The maximum frequency which can be reflected $f_c(h_m)\sec\theta_0$ may be called the *maximum usable frequency*. At vertical incidence the maximum usable frequency is $f_c(h_m)$ or $9M_m^{\frac{1}{2}}$ where M_m is the greatest electron density. Waves with frequencies higher than 30 MHz are rarely reflected. For frequencies from 1.5 MHz to 30 MHz the assumption that $\omega\gg f$ is acceptable for the F-layer and, at night, for the D- and E-layers. (During the day f corresponds to a frequency of about 2 MHz in the D- and E-layers). The sky wave may then be calculated by (31.1). For the total field the ground-to-ground contribution must be added.

One way of visualizing the action of the ionosphere is to picture it as a perfect mirror placed at a suitable height, called the *virtual height*. The time taken by a wave packet to cover a trajectory is $\int \mathrm{d}s/v_g$ where v_g is the group speed (§10.5). The group speed is $\mathrm{d}\omega/\mathrm{d}k$ where $k^2=\omega^2N^2/c^2$ and, since $N^2=1-\omega_c^2/\omega^2$, $v_g=cN$. Hence the time taken by a wave packet going from transmitter to receiver via one reflection at the ionosphere is $(x/c)\operatorname{cosec}\theta_0$. This is the same as a wave packet travelling in free space being reflected by a mirror at height $\frac{1}{2}x\cot\theta_0$ (Fig. 6.23). Such a height is

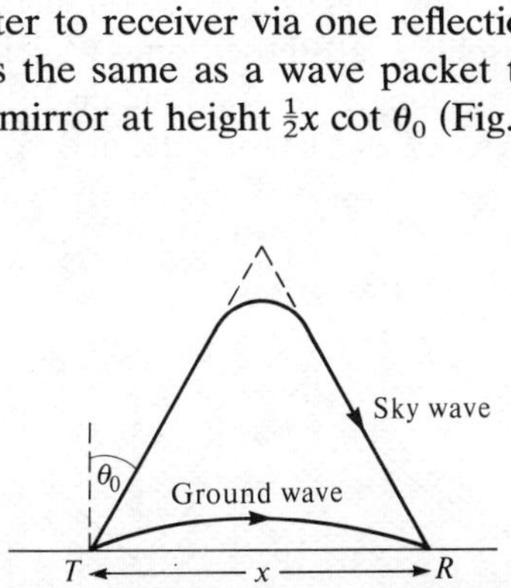

Fig. 6.23 Reflection by the ionosphere

observed in pulse transmission but the relation between the virtual height and the actual position of a layer requires a knowledge of the distribution of electrons with height.

At frequencies just above 2 MHz the absorption is no longer negligible but it can be deemed small. The rule of the preceding section is then available with $N_i = (f\omega_c^2/2\omega^3)(1-\omega_c^2/\omega^2)^{-\frac{1}{2}}$. Consequently, the attenuation in a ray which leaves the earth and returns to it after one reflection is $T(\omega, \theta_0)$ where

$$c\omega \ln T = -\int_0^h f\omega_c^2(\omega^2 \cos^2\theta_0 - \omega_c^2)^{-\frac{1}{2}}\, dz.$$

If ω is replaced by $\omega \cos \theta_0$ and θ_0 by zero, h is unaltered on account of (34.4). Hence there is the *equivalence theorem*

$$\ln T(\omega, \theta_0) = \cos \theta_0 \ln T(\omega \cos \theta_0, 0).$$

The effect of the absorption is to add a factor T^{m+1} to each term of (31.1). At high frequencies

$$c\omega^2 \ln T \approx -\sec \theta_0 \int_0^h f\omega_c^2\, dz$$

if θ_0 is not too close to $\frac{1}{2}\pi$. When $\sec \theta_0$ can be replaced by $\tan \theta_0$

$$T^{m+1} \approx \exp\left\{-(x/2c\omega^2 H)\int_0^h f\omega_c^2\, dz\right\}$$

where H is the virtual height. This should be compared with Eckersley's empirical formula of $e^{-\alpha x/\omega^2}$.

The absorption is so strong in the frequency band 0.5–1.5 MHz during the day that the sky wave practically disappears although it often reappears at night. At low frequencies below 500 KHz, $\omega \ll f$ and $N^2 \approx -i\omega_c^2/\omega f$. The ionosphere now acts as a pure conductor. Ray theory is not applicable in the ionosphere at these long wavelengths because the variation of ω_c over a wavelength is by no means little. However, the D- or E-layer can now be treated as a sharply defined boundary at the virtual

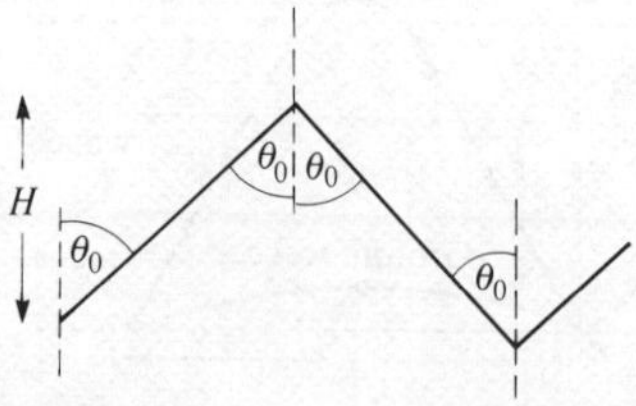

Fig. 6.24 Low frequency reflection

height. The field of the sky wave is then

$$\sum D\{R_F(\theta_0)\}^m\{T_F(\theta_0)\}^{m+1}\{1+R_F(\theta_0)\}^2 e^{-ik_0L}/L$$

where $L=2(m+1)H\sec\theta_0$ and T_F is the appropriate Fresnel reflection coefficient from §6.6 for the ionospheric layer. For a vertically polarized wave the largeness of $\omega_c^2/\omega f$ permits the approximation

$$T_F(\theta_0)=\{\cos\theta_0-(-i\omega_c^2/\omega f)^{-\frac{1}{2}}\}/\{\cos\theta_0+(-i\omega_c^2/\omega f)^{-\frac{1}{2}}\}.$$

If $\cos\theta_0$ is replaced by $2(m+1)H/L$ and $L\gg H$

$$T_F^{m+1}\approx\exp\left\{-\left(-\frac{i\omega_c^2}{\omega f}\right)^{-\frac{1}{2}}\frac{L}{H}\right\}\approx\exp\left\{-\left(-\frac{i\omega_c}{\omega f}\right)^{-\frac{1}{2}}\frac{x}{H}\right\}$$

where x is the horizontal range. Note that this formula is independent of m. Similarly $R_F^{m+1}\approx\exp(-x/N_2H)$. Thus the modulus of the field is proportional to

$$(1/x)\exp[-(\omega^{\frac{1}{2}}x/H)\{(f/2\omega_c)^{\frac{1}{2}}+(\varepsilon_0/2\sigma_2)^{\frac{1}{2}}\}]$$

which can be compared with the empirical *Austin–Cohen formula*: $(1/x)\exp(-0.045\ x/\lambda^{\frac{1}{2}}H)$.

6.35 The influence of the Earth's magnetic field

The term involving $\boldsymbol{B}$ in (34.1) was dropped in the last section. Its effect, in so far as it is due to the Earth's magnetic field, will now be examined. Any magnetic field due to the wave and particles will be omitted. Also $\boldsymbol{B}$ will be allotted the constant form $\boldsymbol{B}_0$. The governing equations (34.1) and (34.2) can be combined, for harmonic waves, into

$$(f+i\omega)\boldsymbol{J}=\varepsilon_0\omega_c^2\mathbf{E}+Q\boldsymbol{J}\wedge\boldsymbol{B}_0/m=\varepsilon_0\omega_c^2\mathbf{E}+\boldsymbol{J}\wedge\boldsymbol{\Omega}_0 \qquad (35.1)$$

where $\boldsymbol{\Omega}_0=Q\boldsymbol{B}_0/m$.

The quantity $\Omega_0/2\pi=|\boldsymbol{\Omega}_0|/2\pi$ is known as the *gyrofrequency*. Since $|\boldsymbol{B}_0|=0.5\times10^{-4}$ weber m^{-2} the gyrofrequency for electrons is about 1.4 MHz (212 m) and that for a hydrogen ion near 800 Hz. Thus only electrons need to be considered in relation to the influence of the Earth's magnetic field on radio propagation. Frequencies in the neighbourhood of the gyrofrequency undergo much severer absorption in the presence of the Earth's magnetic field so that transmission by the sky wave is virtually impossible during the day.

The solution of (35.1) for $\boldsymbol{J}$ may be expressed as

$$i(1-\Omega_2^2)\boldsymbol{J}=\varepsilon_0\omega\Omega_1\{\mathbf{E}+i\Omega_2\mathbf{E}\wedge\hat{\boldsymbol{\Omega}}_0-\Omega_2^2(\mathbf{E}\cdot\hat{\boldsymbol{\Omega}}_0)\hat{\boldsymbol{\Omega}}_0\}$$

where $\hat{\boldsymbol{\Omega}}_0$ is a unit vector parallel to $\boldsymbol{\Omega}_0$ and

$$\Omega_1=\omega_c^2/\omega(\omega-if),\qquad \Omega_2=\Omega_0/(\omega-if).$$

The equation may also be expressed in tensor form as $\boldsymbol{J}=\boldsymbol{\sigma}\cdot\boldsymbol{E}$. Then $\boldsymbol{\sigma}$ is effectively an anisotropic conductivity. Its components, when $\boldsymbol{B}_0$ makes angles α, β, γ with x-, y-, z-axes respectively, are

$$
\begin{aligned}
(1-\Omega_2^2)\sigma_{xx} &= -\mathrm{i}\varepsilon_0\Omega_1\omega(1-\Omega_2^2\cos^2\alpha),\\
(1-\Omega_2^2)\sigma_{yy} &= -\mathrm{i}\varepsilon_0\Omega_1\omega(1-\Omega_2^2\cos^2\beta),\\
(1-\Omega_2^2)\sigma_{xy} &= -\mathrm{i}\varepsilon_0\omega\Omega_1\Omega_2(\mathrm{i}\cos\gamma-\Omega_2\cos\alpha\cos\beta),\\
(1-\Omega_2^2)\sigma_{xz} &= \mathrm{i}\varepsilon_0\omega\Omega_1\Omega_2(\mathrm{i}\cos\beta+\Omega_2\cos\alpha\cos\gamma),\\
(1-\Omega_2^2)\sigma_{yz} &= -\mathrm{i}\varepsilon_0\omega\Omega_1\Omega_2(\mathrm{i}\cos\alpha-\Omega_2\cos\beta\cos\gamma),\\
\sigma_{yx} = -\sigma^*_{xy}, &\qquad \sigma_{zx} = -\sigma^*_{xz}, \qquad \sigma_{zy} = -\sigma^*_{yz}.
\end{aligned}
$$

The lack of symmetry in the tensor precludes normal reciprocity (§1.35) but there can be special circumstances under which fields are reciprocal.

The combination of the convection current $\boldsymbol{J}$ with $\mathrm{i}\omega\varepsilon_0\boldsymbol{E}$ in Maxwell's equations leads to an anisotropic medium whose formal properties are similar to those of §6.11.

In particular, consider a plane wave propagating along the z-axis. Choose the x-axis perpendicular to the plane containing the z-axis and $\boldsymbol{B}_0$ so that $\alpha=\frac{1}{2}\pi$, $\beta=\frac{1}{2}\pi-\gamma$. In the manner of §6.11 it is found that the refractive index N is given by

$$
N^2 = 1-2\Omega_1(1-\Omega_1)[2(1-\Omega_1)-\Omega_2^2\sin^2\gamma \pm\{\Omega_2^4\sin^4\gamma+4\Omega_2^2(1-\Omega_1)^2\cos^2\gamma\}^{\frac{1}{2}}]^{-1}.
$$

There are two waves travelling at an angle γ to the magnetic field. This is similar to a biaxial medium but there are no directions in which the two values of N are equal and so there is no analogue of the optical axes. The ray corresponding to the upper sign in N is called the *ordinary ray*, that to the lower sign the *extraordinary ray*.

The critical frequency can be defined, when there are no collisions, as that for which $N^2=1$ with the z-axis vertical. For the ordinary ray this requires $\Omega_1=1$ or $\omega=\omega_c$. In other words, the critical frequency for the ordinary ray is independent of the magnetic field. In contrast, the condition for the extraordinary ray is $\Omega_1=1\pm\Omega_2$ i.e. $\omega_c^2=\omega(\omega\pm\Omega_0)$ which depends upon the magnitude of the Earth's magnetic field but not its direction.

Let the maximum value of ω_c be ω_M. Reflection of the ordinary ray at vertical incidence can occur only if $\omega<\omega_M$. The extraordinary ray can be reflected at the level $\omega_c^2=\omega(\omega-\Omega_0)$ only if $\omega>\Omega_0$ otherwise the electron density would be negative. Nor can it happen unless $\omega_M^2>\omega(\omega-\Omega_0)$, i.e.

$$
\Omega_0<\omega<\tfrac{1}{2}\Omega_0+(\tfrac{1}{4}\Omega_0^2+\omega_M^2)^{\frac{1}{2}}.
$$

Reflection at this level prevents any higher reflection because the wave is exponentially damped ($N^2<0$). To achieve reflection at the upper level

$\omega_c^2 = \omega(\omega + \Omega_0)$ it is necessary that

$$\omega < -\tfrac{1}{2}\Omega_0 + (\tfrac{1}{4}\Omega_0^2 + \omega_M^2)^{\frac{1}{2}}. \tag{35.2}$$

If $\omega < \Omega_0$ reflection will occur for those values of ω which are smaller than Ω_0 and the right-hand side of (35.2). When $\omega > \Omega_0$ the wave passes through the lower level and is reflected by the upper layer only if the right-hand side of (35.2) is less than Ω_0, i.e. $\Omega_0 > \omega_M/2^{\frac{1}{2}}$. Hence, if $\Omega_0 < \omega_M/2^{\frac{1}{2}}$, *there are two bands of frequency which can be reflected*, one for the ordinary ray and one for the extraordinary. *If, however,* $\Omega_0 > \omega_M/2^{\frac{1}{2}}$ *there are three bands eligible for reflection*, one for the ordinary and one for each of the two levels for the extraordinary. Both situations have been observed, the former more often than the latter.

At long wavelengths the ionosphere can be treated as a sharply defined sheet just as when the magnetic field is missing. The sky wave may be calculated as in the previous section subject to one qualification. There are now two reflection coefficients, essentially R' and R'' of (12.6), (12.7), because an incident vertically polarized wave produces two reflected waves, one vertically polarized and the other horizontally polarized. After each reflection at the ionosphere the number of linearly polarized waves is doubled. The most obvious feature is that the field reaching the Earth is elliptically polarized. If the elliptically polarized wave is split into two circularly polarized waves it is found that the right-handed component is dominant in the northern hemisphere and the left-handed in the southern hemisphere.

The WKB method is an obvious candidate for tackling high frequencies. Two additional complications appear as compared with the isotropic case. Both the ordinary and extraordinary rays have to be catered for and the direction of energy flow is not orthogonal to the phase front (cf. §6.11). Some simplification is attained by neglecting cubes and higher powers of $\Omega_1^{\frac{1}{2}}$ and Ω_2. A phase advance of $\frac{1}{2}\pi$ still takes place on reflection by a caustic.

6.36 Scattering by atmospheric irregularities

Atmospheric irregularities are responsible for the scattering of waves which would otherwise proceed serenely on their way. Sometimes the scattering is unwanted but, on other occasions, it can be used deliberately to deflect energy into a desired direction. Assuming that the cause is a small change of dielectric constant in an air mass in the troposphere the Rayleigh–Gans approximation (§6.14) may be employed to estimate the effect. Thus, if $\varepsilon = \varepsilon_0(1+\varepsilon')$ in a volume T, (14.4) supplies

$$\mathbf{E} = \mathbf{E}^{\mathrm{i}} + (\text{grad div} + k_0^2)\int_T \varepsilon' \mathbf{E}^{\mathrm{i}} \psi(\boldsymbol{x}, \mathbf{y})\, \mathrm{d}\mathbf{y}.$$

When $|\boldsymbol{x}| = R$ is large, $\psi \sim \exp\{-ik_0(R - \hat{\boldsymbol{x}} \cdot \mathbf{y})\}/4\pi R$ and the operators $\partial/\partial x$, $\partial/\partial y$, and $\partial/\partial z$ merely multiply by $-ik_0x/R$, $-ik_0y/R$, and $-ik_0z/R$ respectively. Hence

$$\boldsymbol{E} \sim \boldsymbol{E}^{\mathrm{i}} + (k_0^2 \mathrm{e}^{-\mathrm{i}k_0R}/4\pi R) \int_T \{\boldsymbol{E}^{\mathrm{i}} - (\boldsymbol{E}^{\mathrm{i}} \cdot \hat{\boldsymbol{x}})\hat{\boldsymbol{x}}\}\varepsilon' \exp(\mathrm{i}k_0 \hat{\boldsymbol{x}} \cdot \mathbf{y})\, \mathrm{d}\mathbf{y},$$

$$\boldsymbol{H} \sim \boldsymbol{H}^{\mathrm{i}} + (\varepsilon_0/\mu_0)^{\frac{1}{2}} \hat{\boldsymbol{x}} \wedge (\boldsymbol{E} - \boldsymbol{E}^{\mathrm{i}}).$$

If the incident wave is plane so that $\boldsymbol{E}^{\mathrm{i}} = \boldsymbol{e}^{\mathrm{i}} \exp(-\mathrm{i}k_0 \boldsymbol{n}^{\mathrm{i}} \cdot \boldsymbol{x})$ where $\boldsymbol{n}^{\mathrm{i}} \cdot \boldsymbol{e}^{\mathrm{i}} = 0$ the field becomes

$$\boldsymbol{E} \sim \boldsymbol{E}^{\mathrm{i}} + \frac{k_0^2 \mathrm{e}^{-\mathrm{i}k_0R}}{4\pi R} \{\boldsymbol{e}^{\mathrm{i}} - (\boldsymbol{e}^{\mathrm{i}} \cdot \hat{\boldsymbol{x}})\hat{\boldsymbol{x}}\} \int_T \varepsilon' \exp\{\mathrm{i}k_0 \boldsymbol{y} \cdot (\hat{\boldsymbol{x}} - \boldsymbol{n}^{\mathrm{i}})\}\, \mathrm{d}\boldsymbol{y}.$$

Exercises on Chapter 6

1. A plane sound wave strikes a plane interface where $a_1 = a_2$. Show that the reflection and transmission coefficients are independent of the angle of incidence. If $\zeta = \rho_2/\rho_1$ show that there is no reflected wave and that $p^{(2)}$ is the same as p^{i}.

2. An acoustic plane wave is reflected by a plane boundary where $\zeta = \rho_2/\rho_1$ and $a_2 = a_1(1+\alpha)$ where $|\alpha| \ll 1$. Show that the reflection coefficient is approximately $\frac{1}{2}\alpha \sec^2\theta$ when θ is not near $\frac{1}{2}\pi$. If $\zeta = \rho_2(1-\beta)/\rho_1$ where $|\beta| \ll 1$ show that the reflection coefficient is modified to $\frac{1}{2}\beta + \frac{1}{2}\alpha \sec^2\theta$.

3. A plane sound wave is reflected at normal incidence from a medium in which $a_2 = \frac{1}{2}(1+\mathrm{i})\alpha$ where $\alpha \ll a_1$. Show that the amplitude is reduced on reflection by a factor of $1 - \rho_2\alpha/\rho_1\zeta a_1$. If the incidence is oblique show that the energy is reduced by the factor $1 - 2\rho_2\alpha \cos\theta/\rho_1\zeta a_1$ on reflection.

4. A circularly polarized electromagnetic plane wave is incident normally on a plane perfect conductor. Calculate the instantaneous energy flow at any point.

5. At the troposphere the permittivity ε_0 changes suddenly to $\varepsilon_0(1+2\eta)$ where $|\eta| \ll 1$ and the permeability remains at the same constant value. Show that a plane wave incident at an angle θ (not near $\frac{1}{2}\pi$) is reflected from the troposphere with amplitude $-\frac{1}{2}\eta \sec^2\theta$ or $\frac{1}{2}\eta(1-\tan^2\theta)$ according as $\boldsymbol{E}$ is perpendicular or parallel to the plane of incidence. Total reflection can occur only if $\eta < 0$ and θ is near $\frac{1}{2}\pi$.

6. At the surface $x = 0$ of a conductor in which $\varepsilon = \varepsilon_0$, $\mu = \mu_0$ the tangential magnetic field H_z is maintained at $A \cos \omega t$. Calculate the average rate of production of heat.

7. A plane wave is reflected at normal incidence from an Earth surface in which $\varepsilon = 10\varepsilon_0$. The conductivity of the earth varies from 10^{-4} to 3×10^{-2} mho m^{-1} according to the rainfall. Calculate the amplitude and

phase of the reflected wave as a function of conductivity for a wave of frequency 1000 kHz and one of 100 MHz.

8. A plane wave is reflected by a good conductor. Show that its energy is reduced on reflection by the factor $1-(8\omega\varepsilon_1\mu_2/\mu_1\sigma_2)^{\frac{1}{2}}\cos\theta$ or $1-(8\omega\varepsilon_1\mu_2/\mu_1\sigma_2)^{\frac{1}{2}}\sec\theta$ depending on the polarization if θ is not near $\frac{1}{2}\pi$.

9. A plane wave of intensity I in free space is reflected at normal incidence by a medium of refractive index N. Show that the pressure exerted on the interface by the free space field is $2(1+N^2)I/c(1+N)^2$.

Generalize the result to oblique incidence.

10. A plane electromagnetic wave linearly polarized at 45° to the plane of incidence is totally reflected in a prism which it enters and leaves at normal incidence. Show that the emerging wave is elliptically polarized and has intensity $16N^2/(1+N)^4$ times that in the incident wave, N being the refractive index of the prism.

11. Monochromatic red light of wavelength 600 nm is incident normally on a soap film of refractive index $\frac{4}{3}$. For what thickness of the film is the reflected intensity a minimum (i) if there is air on both sides of the film, (ii) if the film is on a perfect conductor, (iii) if the film is on a glass surface of refractive index $\frac{3}{2}$?

12. Show that, at normal incidence, (9.12) can be expressed as

$$Z_1=\rho_2a_2(1+r_{23}e^{-2ik_2d})/\zeta(1-r_{23}e^{-2ik_2d})$$

where $r_{23}=(\rho_3a_3-\rho_2a_2\zeta)/(\rho_3a_3+\rho_2a_2\zeta)$. Deduce that

$$R_1=(r_{12}+r_{23}e^{-2ik_2d})/(1+r_{12}r_{23}e^{-2ik_2d})$$

where r_{12} is defined in a similar manner to r_{23} with suitable changes of suffix.

13. In Exercise 12 the media 1 and 3 are the same while medium 2 is dissipative with $k_2=\alpha_2-i\beta_2$ and $r_{12}=|r_{12}|\,e^{-i\delta_{12}}$. Show that

$$|R_1|^2=(\sin^2\alpha_2d+\sinh^2\beta_2d)/\{\sin^2(\alpha_2d+\delta_{12})+\sinh^2(\beta_2d-\ln|r_{12}|)\}.$$

14. Show that the formula for R_1 in Exercise 12 remains valid for oblique incidence provided that k_2d is replaced by $k_2d\cos\theta_2$, and in r_{12}, r_{23} the quantity $\rho_ja_j\sec\theta_j$ is substituted for ρ_ja_j.

15. Undertake exercises analogous to 12–14 for both polarizations of electromagnetic waves.

If the medium on either side of the slab is free space calculate the pressure exerted by the wave.

16. A harmonic plane wave is incident at an angle θ on a slab of refractive index N in free space. The frequency is varied and λ_1, λ_2 are two consecutive values of the wavelength for which there is no reflected wave. Show that the thickness of the slab is $\lambda_1\lambda_2/2(\lambda_2-\lambda_1)(N^2-\sin^2\theta)^{\frac{1}{2}}$.

17. The refractive indices of the regions $x<0$, $0\leqslant x<x_2$, $x_2\leqslant x<x_3$, $x_3\leqslant x$ are N_1, N_2, N_3, and N_4 respectively. If $k_2x_2=\frac{1}{2}\pi$ and $k_3(x_3-x_2)=$

$\frac{1}{2}\pi$ show that there is no reflected wave at normal incidence if $N_4N_2^2 = N_1N_3^2$.

If the medium of refractive index N_4 terminates at $x = x_4$ with $k_4(x_4 - x_3) = \frac{1}{2}\pi$ and the refractive index in $x \geqslant x_4$ is N_5 show that there is no reflected wave at normal incidence if $N_1N_5N_3^2 = N_2^2N_4^2$. However, if $k_3(x_3 - x_2) = \frac{1}{2}\pi$ is altered to $k_3(x_3 - x_2) = \pi$ there is no reflected wave if $N_5N_2^2 = N_1N_4^2$.

18. Each of the three systems in Exercise 17 is designed to give no reflected wave at 550 nm In the first system $N_1 = 1$, $N_2 = 1.47$, $N_3 = 1.80$, $N_4 = 1{\cdot}53$ and in the other two $N_1 = 1$, $N_2 = 1{\cdot}47$, $N_3 = 2{\cdot}14$, $N_4 = 1{\cdot}80$, $N_5 = 1{\cdot}53$. Plot the reflection coefficient against wavelength between 400 nm and 700 nm [Optically invisible glass over a broader spectral range is obtained by using 3 coatings instead of 1; there is further improvement in the last system if $N_3 = 2{\cdot}40$.]

19. The energy velocity $\boldsymbol{u}_\mathrm{E}(=\boldsymbol{S}/W)$ of a plane wave in a crystal is written as $u_\mathrm{E}\boldsymbol{\tau}$ where $\boldsymbol{\tau}$ is a unit vector. Show that, in the notation of §6.11, $u_\mathrm{E}\tau_1(u^2 - u_1^2) = u\nu_1(u_\mathrm{E}^2 - u_1^2)$ and that

$$\frac{u_1^2\tau_1^2}{u_\mathrm{E}^2 - u_1^2} + \frac{u_2^2\tau_2^2}{u_\mathrm{E}^2 - u_2^2} + \frac{u_3^2\tau_3^2}{u_\mathrm{E}^2 - u_3^2} = 0.$$

The surface described by $\boldsymbol{u}_\mathrm{E}$ is known as the *ray surface.* The *ray axes,* i.e. the directions of $\boldsymbol{\tau}$ for which u_E has only a single value, are $(\pm u_3n_1/u_2,\ 0,\ \pm u_1n_3/u_2)$ where $(\pm n_1,\ 0,\ \pm n_3)$ are the optical axes when $u_1 > u_2 > u_3$. Show that the tangent plane to the ray surface at $\boldsymbol{u}_\mathrm{E}$ is perpendicular to the wave normal $\boldsymbol{\nu}$.

If the wave normal is along an optical axis show that the corresponding rays lie on the cone

$$(x^2 + y^2 + z^2)u_2^2 = (n_1x + n_3z)(n_1u_3^2x + n_3u_1^2z).$$

Deduce that the plane perpendicular to an optical axis at the point of intersection with the ray surface touches the ray surface along a circle.

20. If $u_1 = u_2 = u_0$, $u_3 = u_\mathrm{e}$ in a crystal show that the ray surface consists of a sphere $x^2 + y^2 + z^2 = u_0^2$ and a spheroid $u_0^2(x^2 + y^2) + u_\mathrm{e}^2z^2 = u_0^2u_\mathrm{e}^2$.

21. Show that, in general, Poynting's vector does not satisfy Snell's laws at the surface of a crystal.

A slab of crystal has $u_1^2 > u_2^2 > u_3^2$ but $u_1^2 \approx u_3^2$. A plane wave falls on such a slab of thickness d. Show that, after passage through the slab, the phase difference of the two emergent waves (multiple reflections being ignored) is

$$2^{\frac{1}{2}}\omega d(u_1^2 - u_3^2)\sin\alpha_1 \sin\alpha_2/(u_2^2 + u_3^2)^{\frac{3}{2}}\cos\alpha$$

where one of the refracted waves makes angles α, and α_1 and α_2 with the slab normal and optical axes respectively.

22. By starting from the hypergeometric equation one can make substitutions to see whether they lead to apposite inhomogeneous media. The equation for $F(a, b; c; u)$ is

$$u(1-u)\,d^2F/du^2+\{c-(a+b+1)u\}\,dF/du-abF=0.$$

The substitution $F=u^{-\frac{1}{2}c}(1-u)^{\frac{1}{2}(c-a-b-1)}(du/dz)^{\frac{1}{2}}p$ gives

$$d^2p/dz^2+k^2p=0$$

where

$$k^2=\tfrac{1}{2}\frac{d^2}{dz^2}\ln\frac{du}{dz}-\tfrac{1}{4}\left(\frac{d}{dz}\ln\frac{du}{dz}\right)^2-\left\{c_1+\frac{c_2u}{1-u}+\frac{c_3u}{(1-u)^2}\right\}\left(\frac{d}{dz}\ln u\right)^2,$$

$$c_1=\tfrac{1}{4}c(c-2),\qquad c_2=\tfrac{1}{4}\{1-(a-b)^2+c(c-2)\},\qquad c_3=\tfrac{1}{4}\{(a+b-c)^2-1\}.$$

Choosing different functions of z for u furnishes various media.

23. Verify that solutions of $d^2p/dz^2+k^2p=0$ can be found in the following cases. (i) $k^2=k_0^2(a^2-b^2/z^2)$ $(p=z^{\frac{1}{2}}H_\nu^{(1)}(k_0az)$ with $\nu^2=k_0^2b^2+\frac{1}{4})$, (ii) $k^2=k_0^2e^{2az}-b^2(p=H_\nu^{(1)}(k_0e^{az}/a)$ with $\nu=b/a)$, (iii) $k^2=k_0^2(1-\sum_{n=2}^{6}c_nz^n)$ (p is given in terms of parabolic cylinder functions).

24. In a medium in which $\mu=\mu_0$, $\varepsilon=\varepsilon_0(1+hz)^n$ the only nonzero component of the electric intensity is E_x which is a function of z only. If $u=2k_0(1+hz)^{\frac{1}{2}(n+2)}/h(n+2)$ show that E_x satisfies

$$2\varepsilon(d^2E_x/du^2+E_x)+(d\varepsilon/du)(dE_x/du)=0.$$

Deduce that

$$E_x=u^{-\nu}\{AH_\nu^{(1)}(u)+BH_\nu^{(2)}(u)\}$$

where $\nu=-(n+2)^{-1}$.

Use this result to infer that reflection at normal incidence from a transition in ε is appreciable only if the thickness of the inhomogeneous zone is less than a wavelength.

25. The plane wave $\boldsymbol{E}^i=\boldsymbol{e}^ie^{-ik_0z}$ where $\boldsymbol{e}^i\cdot\boldsymbol{k}=0$ is initiated in a medium in which $\mu=\mu_0$, $\varepsilon=\varepsilon_0\{1-V(R)/k_0^2\}$ where R is the length of the radius vector from the origin. If $|V(R)/k_0^2|\ll 1$ show that at a large distance from the origin in the direction of $\hat{\boldsymbol{x}}$

$$\boldsymbol{E}\sim\boldsymbol{e}^ie^{-ik_0z}-(e^{-ik_0R}/\kappa R)\{\boldsymbol{e}^i-\hat{\boldsymbol{x}}(\boldsymbol{e}^i\cdot\hat{\boldsymbol{x}})\}\int_0^\infty V(x)x\sin\kappa x\,dx$$

where $\kappa=2k_0\sin\frac{1}{2}\theta$, $\hat{\boldsymbol{x}}\cdot\boldsymbol{k}=\cos\theta$.

26. The plane sound wave $e^{-i\omega z/a_0}$ starts in a medium in which $a^{-2}=a_0^{-2}\{1+V(R)/k_0^2\}$. Show that at a large distance from the origin

$$p=e^{-ik_0z}+(e^{-ik_0R}/\kappa R)\int_0^\infty V(x)x\sin\kappa x\,dx$$

where κ is defined in Exercise 25.

27. In (14.2) suppose that $k^2 \neq k_0^2$ only in a finite domain. Then a Fourier transform of (14.2) with respect to $\boldsymbol{x}$ is

$$P(\boldsymbol{\alpha}) = P^{\mathrm{i}}(\boldsymbol{\alpha}) + \tilde{P}(\boldsymbol{\alpha})\Psi(\boldsymbol{\alpha})$$

where $P(\boldsymbol{\alpha}) = \int_{-\infty}^{\infty} p(\boldsymbol{x})\mathrm{e}^{-\mathrm{i}\boldsymbol{\alpha}\cdot\boldsymbol{x}}\,\mathrm{d}\boldsymbol{x}$, $\tilde{P}$ is the Fourier transform of $(k^2 - k_0^2)p$ and $\Psi(\boldsymbol{\alpha}) = (\boldsymbol{\alpha}\cdot\boldsymbol{\alpha} - k_0^2)^{-1} - 2\pi\mathrm{i}\delta(\boldsymbol{\alpha}\cdot\boldsymbol{\alpha} - k_0^2)$, $(\ )^{-1}$ implying a principal value. This equation can be solved iteratively by the fast Fourier transform F and its inverse F^{-1} through

$$P_n(\boldsymbol{\alpha}) = P^{\mathrm{i}}(\boldsymbol{\alpha}) + \tilde{P}_n(\boldsymbol{\alpha})\Psi(\boldsymbol{\alpha}),\ p_n(\boldsymbol{x}) = \mathsf{F}^{-1}P_n(\boldsymbol{\alpha}),$$
$$\tilde{P}_{n+1}(\boldsymbol{\alpha}) = \mathsf{F}\{(k^2 - k_0^2)p_n(\boldsymbol{x})\}$$

with $\tilde{P}_0(\boldsymbol{\alpha}) \equiv 0$. This procedure, known sometimes as the $\boldsymbol{\alpha}$-*space formulation* or *spectral domain method*, places less demands on the computer than matrix methods when it converges.

Obtain a corresponding formulation for (14.4) and (14.5).

28. Prove for a ray that

$$\mathrm{d}L/\mathrm{d}s = N, \qquad \mathrm{d}(\mathrm{grad}\ L)/\mathrm{d}s = \mathrm{grad}\ N.$$

29. If $\boldsymbol{x}$ is a point of a ray and $\boldsymbol{\kappa} = k_0\,\mathrm{grad}\ L$ prove that

$$\mathrm{d}\boldsymbol{x}/\mathrm{d}s = \boldsymbol{\kappa}/k_0N, \qquad \mathrm{d}\boldsymbol{\kappa}/\mathrm{d}s = k_0\,\mathrm{grad}\ N.$$

30. The coordinates of a point on a surface S are specified by the parameters σ_2 and σ_3. A point on a ray with Cartesian coordinates (x_1, x_2, x_3) is also specified by the parameters of the point on S from which it starts and the arc length s on the ray from S. If J is the Jacobian $\partial(x_1, x_2, x_3)/\partial(s, \sigma_2, \sigma_3)$ show that

$$\frac{\partial J}{\partial s} = \frac{J}{N^2}(N\nabla^2 L - \mathrm{grad}\ L \cdot \mathrm{grad}\ N).$$

Deduce that the transport equation (15.5) can be written

$$\mathrm{d}(NJ\,|\gamma_0|^2)/\mathrm{d}s = 0.$$

[Hint: Obtain a relation between the co-factor of $\partial x_i/\partial\sigma_j$ in J and $J\,\partial\sigma_j/\partial x_i$.]

31. In a region with cylindrical symmetry the rays lie in planes $\phi =$ constant (r, ϕ, z being cylindrical polar coordinates). The equation of a ray which is initially inclined at θ_0 to the z-axis is $f(r, z, \theta_0) = 0$. With J defined as in Exercise 30 and $\sigma_2 = \theta_0$, $\sigma_3 = \phi$ show that

$$J = -r(\partial f/\partial\theta_0)\{(\partial f/\partial r)^2 + (\partial f/\partial z)^2\}^{-\frac{1}{2}}$$

and deduce how the amplitude γ_0 varies.

32. If $\boldsymbol{\kappa} = k_0\,\mathrm{grad}\ L$ for a ray in a moving fluid and $\mathrm{d}\tau =$

$(1-\boldsymbol{M}\cdot \text{grad}\, L)\,\mathrm{d}\sigma$ show that

$$\mathrm{d}\boldsymbol{x}/\mathrm{d}\tau = \boldsymbol{M} + \alpha\boldsymbol{\kappa}/|\boldsymbol{\kappa}|.$$

What is the corresponding formula for $\mathrm{d}\boldsymbol{\kappa}/\mathrm{d}\tau$?

33. In a simple model of a shear layer the only nonzero component of $\boldsymbol{U}$ is $U(z)$ along the x-axis. $U(z)$ is zero for $z<0$, increases linearly to U_0 between $z=0$ and $z=h(>0)$ and has the constant value U_0 for $z>h$. The speed of sound is constant throughout. A ray starts from $(0, 0, -d)$ at an angle θ_1 to the horizontal. Show that it passes through the layer if $\cos\theta_1 < (1+M_0)^{-1}$, and that the difference in abscissae of its points of entry to and exit from the layer is

$$(h/2M_0)\sec\theta_1[\tan\theta_1 - (1+M_0\cos\theta_1)\tan\theta_2 \\ +\cos\theta_1(\cosh^{-1}\sec\theta_1 - \cosh^{-1}\sec\theta_2)]$$

where $\sec\theta_2 = \sec\theta_1 - M_0$.

Show that if $z>h$

$$L = d\,\text{cosec}\,\theta_1 - (h/M_0)(\tan\theta_2 - \tan\theta_1) + (z-h)\text{cosec}\,\theta_2.$$

34. Demonstrate (19.15).

35. In a horizontally stratified medium $N^2 = N_1^2(1+a^2z^2)$. A ray leaves $(0, h_0)$ at an angle $\cos^{-1}A/N_1(1+a^2h_0^2)^{\frac{1}{2}}$ with the x-axis in a downward direction. If $A>N_1$ show that the ray turns at a height $z=h_1=(A^2-N_1^2)^{\frac{1}{2}}/aN_1$ and that its equation thereafter is

$$N_1 ax = A\ln\{z+(z^2-h_1^2)^{\frac{1}{2}}\}\{h_0+(h_0^2-h_1^2)^{\frac{1}{2}}\}/h_1^2.$$

Show that the change of phase along the ray is

$$\tfrac{1}{2}N_1 a\{h_0(h_0^2-h_1^2)^{\frac{1}{2}} - z(z^2-h_1^2)^{\frac{1}{2}}\} \pm x(N_1^2+A^2)/2A$$

the upper or lower sign being adopted according as the rising or falling part of the ray is being considered.

36. A ray is confined to the plane $z=0$ and (r, ϕ) are polar coordinates in this plane. Show that the differential equations for the ray are

$$N(\ddot{r} - r\dot{\phi}^2) + \dot{N}\dot{r} = \partial N/\partial r,$$
$$N(r\ddot{\phi} + 2\dot{r}\dot{\phi}) + \dot{N}r\dot{\phi} = (1/r)\,\partial N/\partial\phi$$

where a dot indicates a derivative with respect to s. Hence show that in a radially stratified medium in which $\partial N/\partial\phi = 0$ the rays have equations of the form

$$\phi = \int^r \frac{A\,\mathrm{d}r}{r(r^2N^2 - A^2)^{\frac{1}{2}}}.$$

[Hint: Use the vector form of the differential equations at the end of §6.16.]

37. Show that, in a periodic laminated medium in which $\boldsymbol{K}_{m+2}=\boldsymbol{K}_m$, there are waves with $\boldsymbol{e}_1=\boldsymbol{e}_3$ provided that either (i) $\eta_2\tan\frac{1}{2}\xi_3+\eta_3\tan\frac{1}{2}\xi_2=0$ or (ii) $\eta_2\tan\frac{1}{2}\xi_2+\eta_3\tan\frac{1}{2}\xi_3=0$. In case (i) show that $E_1\tan\frac{1}{2}\xi_2=\mathrm{i}\eta_2H_1$ and deduce that $E_2=E_1$, $H_2=-H_1$.

If $|\xi_2|\ll 1$ and $|\xi_3|\ll 1$ show that (i) is satisfied by $\mu_3(\varepsilon_3h_3+\varepsilon_2h_2)=\varepsilon_2(\mu_2h_3+\mu_3h_2)\sin^2\theta_2$ where $h_i=x_i-x_{i-1}$ but that (ii) cannot be satisfied.

38. In case (i) of Exercise 37 define an average field by $\bar{E}=\int_{x_1}^{x_3}E\,\mathrm{d}x/(h_2+h_3)$. Show that $\bar{H}_y=0$,

$$\partial\bar{E}_z/\partial y=\mathrm{i}\omega\mu^{\mathrm{e}}H_x,\qquad \partial\bar{H}_x/\partial y=-\mathrm{i}\omega\varepsilon^{\mathrm{e}}\bar{E}_z$$

where $\mu^{\mathrm{e}}=\mu_2\mu_3(\mu_2\eta_3^2-\mu_3\eta_2^2)/(\mu_2^2\eta_3^2-\mu_3^2\eta_2^2)$, $\mu^{\mathrm{e}}\varepsilon^{\mathrm{e}}=\mu_2\varepsilon_2\sin^2\theta_2$. If $|\xi_2|\ll 1$ and $|\xi_3|\ll 1$, show that $\mu^{\mathrm{e}}\approx\mu_2\mu_3(h_2+h_3)/(\mu_2h_3+\mu_3h_2)$, $\varepsilon^{\mathrm{e}}\approx(\varepsilon_3h_3+\varepsilon_2h_2)/(h_3+h_2)$,

39. A continuously stratified medium has constant properties below $z=0$ and above $z=h$. An iterative method of tackling (22.2) is to replace Z on the right-hand side by the value of η at $z=0$; the equation is then solved subject to the condition at $z=h$; the Z so found is inserted into the right-hand side of (22.2) and the process is repeated. Alternatively, one might make R^2 on the right-hand side of (22.3) zero, solve for R (with $R\to 0$ as $z\to\infty$), then use the new R in the R^2 term and continue in this way. Show that the first method is appropriate if $kh\ll 1$ and the second if $kh\gg 1$.

40. In the differential equation

$$(1-x^2)\,\mathrm{d}^2w/\mathrm{d}z^2-2x\,\mathrm{d}w/\mathrm{d}z+\{\nu(\nu+1)-\mu^2/(1-x^2)\}w=0,$$

which is satisfied by $\mathrm{P}_\nu^\mu(x)$, make the substitution $w=(1-x^2)^{-\frac{1}{2}}v$. Show that the WKB method gives for large ν and $0<\theta<\pi$ $(x=\cos\theta)$

$$w\sim\{a\mathrm{e}^{\mathrm{i}(\nu+\frac{1}{2})\theta}+B\mathrm{e}^{-\mathrm{i}(\nu+\frac{1}{2})\theta}\}/\sin^{\frac{1}{2}}\theta.$$

It is known that when $\theta\approx\frac{1}{2}\pi$ and ν is large

$$\mathrm{P}_\nu^\mu(\cos\theta)\sim\nu^{\mu-\frac{1}{2}}(2/\pi)^{\frac{1}{2}}\{\cos(\mu+\nu)\tfrac{1}{2}\pi+\nu\sin(\mu+\nu)\tfrac{1}{2}\pi\cos\theta\}.$$

Deduce that

$$\mathrm{P}_\nu^\mu(\cos\theta)\sim\nu^{\mu-\frac{1}{2}}(2/\pi\sin\theta)^{\frac{1}{2}}\cos\{(\nu+\tfrac{1}{2})\theta-\tfrac{1}{4}\pi+\tfrac{1}{2}\mu\pi\}.$$

41. A primary wave has the form

$$\int_{-\infty}^{\infty}\frac{\sin k(\alpha-\sin\beta)}{\pi(\alpha-\sin\beta)}\exp[-\mathrm{i}k\{\alpha y+(1-\alpha^2)^{\frac{1}{2}}x\}]\,\mathrm{d}\alpha$$

where β is a real angle such that $0<\beta<\frac{1}{2}\pi$ and the contour of integration passes below $\alpha=-1$ and above $\alpha=1$. $(1-\alpha^2)^{\frac{1}{2}}$ is positive for $|\alpha|<1$ and negative imaginary for $|\alpha|>1$. If k is large show, by the method of stationary phase, that integral is small except for $y-1<x\tan\beta<y+1$

where it is $e^{-ik(x\cos\beta+y\sin\beta)}$. The primary wave is thus a *beam of finite width* travelling at an angle β to the x-axis.

The beam impinges on another medium at $x = d$ and the reflected wave is

$$\int_{-\infty}^{\infty} \frac{\sin k(\alpha-\sin\beta)}{\pi(\alpha-\sin\beta)} R(\theta)\exp[-ik\{\alpha y-(1-\alpha^2)^{\frac{1}{2}}(x-2d)\}]\,d\alpha$$

where $\sin\theta=\alpha$ and R is the Fresnel reflection coefficient. If $R(\theta)=|R(\theta)|\,e^{i\phi(\alpha)}$ show that the reflected wave is a beam $R(\beta)\exp[-ik\{y\sin\beta-(x-2d)\cos\beta\}]$ occupying $y-1-h<(2d-x)\tan\beta<y+1-h$ where $kh=[d\phi(\alpha)/d\alpha]_{\alpha=\sin\beta}$. This shows that the beam is displaced laterally by a distance h on reflection. If β is less than the angle of total reflection $\phi=0$ and $h=0$. Determine h when β is greater than the angle of total reflection.

42. A point source is at a distance d from a rigid plane. Find the pressure on the plane.

43. An electric dipole is distant d from, and parallel to, a perfectly conducting plane sheet. Determine the surface current and charge densities on the sheet.

44. Show that, if Π in (26.14) is written $\Pi(r, z, z_0)$,

$$\omega\varepsilon_1\Pi(r,z,z_0)=\frac{ie^{-ik_1R_s}}{4\pi R_s}-\frac{ie^{-ik_1R_i}}{4\pi R_i}+\omega\varepsilon_1\Pi(r,z+z_0,0).$$

45. Show that at large distances an approximation to (22.16) is

$$\Pi_2\sim\frac{ik_1}{2\pi\omega}\frac{(k_2\sin\theta_1)^{\frac{1}{2}}\exp(-ik_1r\sin\theta_1-ik_1z_0\cos\theta_1+ik_2z\cos\theta_2)}{(k_1\varepsilon_2\cos\theta_1+k_2\varepsilon_1\cos\theta_2)r^{\frac{1}{2}}(k_2z_0\sec^3\theta_1-k_1z\sec^3\theta_2)^{\frac{1}{2}}}$$

where $k_1\sin\theta_1=k_2\sin\theta_2$, $r=z_0\tan\theta_1-z\tan\theta_2$.

46. A horizontal circular loop of wire of radius 15 cm carries a current of amplitude 10 amp at 600 Hz and is at a height of 1000 m above ground with permittivity $4\varepsilon_0$ and conductivity 10^{-4}. Calculate the field produced on the ground at a distance of 10 000 m. [Treat the loop as a vertical magnetic dipole.]

47. Two aircraft a distance d apart are at a height h above a lake of permeability the same as the atmosphere and the ratio of the permittivities is κ. The transmitting and receiving antennas are short vertical dipoles. If $d \gg h$ show that the intensity of the signal received by one aircraft from the other by reflection to that received directly is

$$\left[\frac{\{(\kappa-1)d^2+4\kappa h^2\}^{\frac{1}{2}}-2\kappa h}{\{(\kappa-1)d^2+4\kappa h^2\}^{\frac{1}{2}}+2\kappa h}\right]^2.$$

48. Medium 1 extends upwards from $z=h$ and contains a source at z_0. Medium 2 is below $z=h$ and is terminated by a rigid plane at $z=0$.

Show that the field in $z>h$ has the same form as in §6.26 with

$$C_1=(\kappa_1\rho_2-i\zeta\kappa_2\rho_1 \tan \kappa_2 h)e^{2i\kappa_1 h}/(\kappa_1\rho_2+i\zeta\kappa_2\rho_1 \tan \kappa_2 h).$$

49. A vertical electric dipole at $(0, 0, z_0)$ is in a medium with constants ε_1, μ_1 above $z=h$; the medium $h \geqslant z>0$ has constants ε_2, μ_2 and $z=0$ is perfectly conducting. Show that

$$\omega\varepsilon_1 \Pi=\frac{ie^{-ik_1R_s}}{4\pi R_s}+\frac{1}{4\pi}\int_0^\infty \frac{\lambda}{\kappa_1}\frac{\kappa_1\varepsilon_2-i\kappa_2\varepsilon_1 \tan \kappa_2 h}{\kappa_1\varepsilon_2+i\kappa_2\varepsilon_1 \tan \kappa_2 h}J_0(\lambda r)e^{-i\kappa_1(z+z_0-2h)}\,d\lambda.$$

50. In a quasi-homogeneous medium

$$\nabla^2 p+k_1^2(1+q^2z^2)p=-\delta(x)\delta(y)\delta(z-z_0).$$

Show that, for $z>z_0$,

$$p^i=\frac{k_1^2}{(2\pi)^{\frac{3}{2}}h}\int_0^\infty \nu!D_{-\nu-1}(hz)D_{-\nu-1}(-hz_0)J_0(\lambda k_1 r)\lambda\,d\lambda$$

where $h=(2k_1)^{\frac{1}{2}}e^{\frac{1}{4}\pi i}$, $\nu=-\frac{1}{2}-ik_1(1-\lambda^2)/2q$. Interchange z and z_0 for $z<z_0$.

Deduce the corresponding expression when there is an impedance boundary condition at $z=0$.

51. A vertical electric dipole is at a height of 60 m above a perfectly conducting horizontal plane and is transmitting at a wavelength of 3 m in a quasi-homogeneous atmosphere in which $1/q=4\times10^6$ m. Plot the field strength near the plane against distance from the dipole (see Tables 8.2–8.4). Try to estimate the difference if the perfectly conducting plane were replaced by sea water.

52. In a quasi-homogeneous atmosphere $\nabla^2 p+k_1^2N^2p=-\delta(x)\delta(y)\times \delta(z-z_0)$ and N increases monotonically with z from the value 1 at $z=0$. Show that the quantities L_h and L_b which occur in Fig. 6.21 are given by

$$L_h=r+\int_0^{z_0}\{N^2(w)-1\}^{\frac{1}{2}}\,dw+\int_0^{z}\{N^2(w)-1\}^{\frac{1}{2}}\,dw,$$

$$L_b=r-\int_0^{z_0}\{N^2(w)-1\}^{-\frac{1}{2}}\,dw-\int_0^{z}\{N^2(w)-1\}^{-\frac{1}{2}}\,dw.$$

53. A certain stratified atmosphere has $N=1$ for $z<h_1$, passes through a minimum N_m between h_1 and h_2, and equals 1 for $z>h_2$. A plane wave, making an angle θ_0 with the z-axis, is excited in $z<h_1$. If $N_m<\sin\theta_0$ show that the wave leaks through the layer being damped to the extent of the factor

$$\exp\left[\tfrac{1}{4}\pi i-k_0\int_{H_1}^{H_2}\{\sin^2\theta_0-N^2(w)\}^{\frac{1}{2}}\,dw\right]$$

where H_1 and H_2 ($<H_1$) are the heights at which $N=\sin\theta_0$.

54. A quasi-homogeneous atmosphere lies above a plane earth at $z=0$ and $N=N_1$ up to $z=h_1$, but above that height $N=N_1\{1+q(z-h_1)\}^{-1}$. Show that a ray from the origin cannot suffer more than $q^{\frac{1}{2}}r/4h_1^{\frac{1}{2}}$ reflections at the earth by the time it reaches $(r, 0, 0)$. Show that the rays are arcs of circles and that the time to travel along a ray from transmitter to receiver is

$$\frac{m}{qc}\left(\frac{2qh_1}{\cos\theta_0}+\ln\frac{1+\cos\theta_0}{1-\cos\theta_0}\right)$$

where θ_0 is the inclination of the initial ray to the vertical. If $\theta_0=\frac{1}{2}\pi-\eta$ where $\eta \ll 1$ show that this can be written

$$\frac{r}{c}\left\{1+\frac{3qh_1-\eta^2}{6(qh_1+\eta^2)}\eta^2\right\}.$$

If $\eta \leqslant (2qh)^{\frac{1}{2}}$ show that the signal is effectively lengthened by $qr(2h-3h_1)/6c$.

55. The height $h_1=0$ in Exercise 54 and $q/k_0 \ll 1$. Show that in the method of §6.32 for ducts θ_s is determined approximately by

$$\sin\theta_s=1-\{-q\pi(2s+\tfrac{1}{2})/2k_0\}^{\frac{1}{2}}$$

for moderate negative values of s when $R_F=e^{-\pi i}$.

56. Two independent solutions of

$$d^2\mathscr{P}/dz^2+k_1^2(N^2-\lambda^2)\mathscr{P}=-\delta(z-z_0)$$

when the right-hand side is zero are $p_1(z)$ and $p_2(z)$, p_2 representing an outgoing wave as $z\to\infty$. If $d\mathscr{P}/dz=0$ on $z=0$ show that in $z>z_0$

$$\mathscr{P}=\{p_2(z_0)p_1'(0)-p_1(z_0)p_2'(0)\}p_2(z)/p_2'(0)W(z_0)$$

where $W(z_0)=p_1(z_0)p_2'(z_0)-p_1'(z_0)p_2(z_0)$. The poles due to the zeros of $p_2'(0)$ will give a residue series for the field in the atmosphere but the presence of branch points may have to be taken into account.

57. The medium $z<0$ has permittivity ε_2, permeability μ_2, and conductivity σ_2; the parameters in $z>h$ are ε_3, μ_3, σ_3 whereas those in $0\leqslant z\leqslant h$ are ε_1, μ_1, 0. If there is a source at $(0, 0, z_0)$ with $0<z_0<h$ show that, for $z<z_0$,

$$\mathscr{P}=-i\frac{\{e^{i\kappa_1 z}+R_F(\theta_1)e^{-i\kappa_1 z}\}\{e^{-i\kappa_1(z_0-h)}+R_{1F}(\theta_1)e^{i\kappa_1(z_0-h)}\}}{2\kappa_1 e^{i\kappa_1 h}\{1-R_F(\theta_1)R_{1F}(\theta_1)e^{-2i\kappa_1 h}\}}$$

where $\cos\theta_1=\kappa_1/k_1$ and R_F, R_{1F} are the Fresnel reflection coefficients at $z=0$ and $z=h$ respectively.

If $d\mathscr{P}/dz=0$ at $z=h$ and $\mu_1=\mu_2$ show that there are poles where $\tan\kappa_1 h=ik_1^2/\kappa_1 k_2$. If the medium in $z<0$ is also highly conducting show that these poles are located where $\kappa_1^2\approx ik_1^2/k_2 h$ or $\kappa_1=(n\pi/h)+ik_1^2/n\pi k_2$,

n being a nonzero integer. Prove that the contribution of these poles to the field is a multiple of

$$\mathrm{H}_0^{(2)}(k_1 r \cos \theta_0) + 2 \sum_{n=1}^{\infty} \cos(n\pi z/h)\cos(n\pi z_0/h)\mathrm{H}_0^{(2)}(k_1 r \cos \theta_n)$$

where $\lambda = \cos \theta_n$ at a pole.

Compare with the result for low-frequency propagation below the ionosphere at the end of §6.34.

58. The maximum usable frequency is 30 MHz at an angle $\theta_0 = 75°$. Show that the electron density M is about $8 \times 11^{11}\ \mathrm{m}^{-3}$.

If the density is 3×10^{11} electron m^{-3} show that the smallest value of θ_0 for reflection is about $\cos^{-1}(\lambda/70)$.

59. Calculate the attenuation per unit length in an ionosphere where $M = 10^{12}$ electron m^{-3}, $f = 1$ Hz and the frequency of operation is 20 MHz.

60. If $|\mathbf{B}_0| = 0.5 \times 10^{-4}$ weber m^{-2}, explain why marked absorption may be expected at wavelengths of 212 m and 375 km.

61. A plane harmonic electromagnetic wave is propagating in the direction of the unit vector $\boldsymbol{\nu}$ in an ionized atmosphere in the presence of a fixed magnetic field. Obtain the equations connecting the components of the electric intensity.

62. Verify the formula for N in §6.35 for propagation (i) along, (ii) perpendicular to the magnetic field.

63. In an ionized atmosphere with $|\mathbf{B}_0| = 0.5 \times 10^{-4}$ weber m^{-1} the critical frequency for the ordinary ray is 20 MHz. Show that the critical frequency for the extraordinary ray differs by about 700 kHz.

7. Surface waves

WAVES inside tubular structures were investigated in Chapter 5. The reflection of waves by a plane was discussed in Chapter 6. This chapter is concerned with the exterior problem in which waves are supported by a structure, i.e. waves which can exist outside a surface whether metal or not. Such natural waves often do not decay inversely as the distance and so are unlike spherical waves. They are, in effect, guided by the surface and therefore may be called *surface waves*.

Propagation along a cylindrical surface

7.1 Acoustic waves on a circular cylinder

The first problem to be examined is the propagation outside a circular cylinder of radius b. The cylinder will be assumed to run parallel to the z-axis and to be homogeneous. Following the pattern for internal guiding we look for solutions whose z-dependence is $e^{-i\kappa z}$. Then solutions are required of

$$\nabla_t^2 p + (k_1^2 - \kappa^2)p = 0 \quad \text{(outside)}, \qquad \nabla_t^2 p + (k_2^2 - \kappa^2)p = 0 \quad \text{(inside)}$$

subject to the boundary conditions that p is continuous and the ratio of external to internal normal derivative is ζ. The field p must also be outgoing at infinity.

Solutions in terms of Bessel functions are available for the cylindrical polar coordinates r, ϕ, z. To secure a field outgoing at infinity and finite at the cylinder's axis we must choose $\mathrm{H}^{(2)}$ outside and J inside. Also the field must not be affected by an increase of 2π in ϕ. Thus the appropriate choice is

$$p = \begin{cases} \displaystyle\sum_{m=-\infty}^{\infty} A_m \mathrm{J}_m(\kappa_2 r) e^{im\phi - i\kappa z} & (r \leqslant b) \\ \displaystyle\sum_{m=-\infty}^{\infty} B_m \mathrm{H}_m^{(2)}(\kappa_1 r) e^{im\phi - i\kappa z} & (r \geqslant b) \end{cases}$$

where $\kappa_1^2 = k_1^2 - \kappa^2$, $\kappa_2^2 = k_2^2 - \kappa^2$.

The boundary conditions are satisfied if

$$A_m J_m(\kappa_2 b) = B_m H_m^{(2)}(\kappa_1 b), \qquad \kappa_1 B_m H_m^{(2)\prime}(\kappa_1 b) = \zeta\kappa_2 A_m J_m'(\kappa_2 b).$$

Elimination of A_m and B_m leads to the equation

$$\kappa_1 H_m^{(2)\prime}(\kappa_1 b) J_m(\kappa_2 b) = \zeta\kappa_2 H_m^{(2)}(\kappa_1 b) J_m'(\kappa_2 b) \tag{1.1}$$

for the determination of the possible values of κ.

Equation (1.1) takes a much simpler form if the cylinder is soft or hard. If the cylinder is soft, $H_m^{(2)}(\kappa_1 b) = 0$ is required and, if it is hard, $H_m^{(2)\prime}(\kappa_1 b) = 0$. However, it is known that $H_m^{(2)}(z)$ and $H_m^{(2)\prime}(z)$ have no zeros in $-\pi \leqslant \text{ph } z \leqslant 0$. Hence, there is no solution representable in Bessel function when the cylinder is soft or hard. But the governing equation has a different kind of solution when $\kappa = k_1$, because Laplace's equation has to be satisfied. Solutions meeting the boundary condition on the cylinder are $p = Be^{-ik_1 z}\ln(r/b)$ in the soft case and $p = Be^{-ik_1 z}$ in the hard case. The velocity, in the second case, has only a single component parallel to the axis of the cylinder. Neither of these solutions tends to zero at infinity and so will generally have to be rejected. In circumstances for which they are acceptable there are waves travelling along the cylinder at a speed characteristic of the outer medium.

Let us now examine what happens when the cylinder is almost perfectly rigid so that ζ is small. Any waves moving at nearly the speed of sound in the outer medium will have $\kappa_1 b$ small. Now, for small x, (A.1.12–15) gives

$$x H_m^{(2)\prime}(x)/H_m^{(2)}(x) \approx -m + x^2/2(m-1) \quad (m > 1) \tag{1.2}$$

$$\approx -1 - x^2 \ln \tfrac{1}{2}x \quad (m = 1), \tag{1.3}$$

$$H_0^{(2)}(x)/x H_0^{(2)\prime}(x) \approx \ln \tfrac{1}{2}x + \gamma + \tfrac{1}{2}\pi i. \tag{1.4}$$

Consider first $m \geqslant 1$. Then, from (1.1), $\kappa_2 J_m'(\kappa_2 b)/J_m(\kappa_2 b)$ must be large and so $\kappa_2 b \approx j_{mp}$ but $\kappa_2 \approx 0$ is not permitted. The frequencies corresponding to these modes are given by $\omega^2(a_2^{-2} - a_1^{-2})b^2 = j_{mp}^2$ and can exist provided that $a_2 < a_1$. There is consequently a *cut-off frequency* of $j_{m1}/2\pi b(a_2^{-2} - a_1^{-2})^{\frac{1}{2}}$ below which asymmetric modes cannot propagate even when $a_2 < a_1$. The symmetric modes ($m = 0$) exhibit a different behaviour because of (1.4) and are influenced by the magnitude of $\zeta \ln \kappa_1 b$. If this is small the symmetric mode propagation is not dissimilar to the asymmetric but if it is large j_{mp} needs replacing by j_{mp}' and $\kappa_2 b \approx 0$ is a possibility and there is no cut-off.

7.2 The conducting circular cylinder

The next problem to be considered is the propagation of electromagnetic waves outside a good conductor. Absorbing the conductivity, as usual, into a complex permittivity, k_2^2 has a large negative imaginary part and $|k_2| \gg |k_1|$.

It is sufficient to consider expansions for E_z and H_z since we know that other components can be expressed in terms of them when the z-dependence is $e^{-i\kappa z}$. Let

$$E_z = \begin{cases} \sum_{m=-\infty}^{\infty} a_m J_m(\kappa_2 r) e^{im\phi - i\kappa z} & (r \leqslant b) \\ \sum_{m=-\infty}^{\infty} c_m H_m^{(2)}(\kappa_1 r) e^{im\phi - i\kappa z} & (r \geqslant b), \end{cases}$$

$$H_z = \begin{cases} \sum_{m=-\infty}^{\infty} b_m J_m(\kappa_2 r) e^{im\phi - i\kappa z} & (r \leqslant b) \\ \sum_{m=-\infty}^{\infty} d_m H_m^{(2)}(\kappa_1 r) e^{im\phi - i\kappa z} & (r \geqslant b). \end{cases}$$

The boundary conditions are the continuity of E_z, H_z, E_ϕ, and H_ϕ at $r = b$. The elimination of a_m, b_m, c_m, and d_m from these four equations leads to κ being determined by

$$\left\{\frac{\mu_1 H_m^{(2)\prime}(\kappa_1 b)}{\kappa_1 H_m^{(2)}(\kappa_1 b)} - \frac{\mu_2 J_m'(\kappa_2 b)}{\kappa_2 J_m(\kappa_2 b)}\right\}\left\{\frac{k_1^2 H_m^{(2)\prime}(\kappa_1 b)}{\mu_1 \kappa_1 H_m^{(2)}(\kappa_1 b)} - \frac{k_2^2 J_m'(\kappa_2 b)}{\mu_2 \kappa_2 J_m(\kappa_2 b)}\right\} = \kappa^2 m^2 (\kappa_2^2 - \kappa_1^2)^2 / b^2 \kappa_2^4 \kappa_1^4. \quad (2.1)$$

The cylinder may be perfectly conducting. There may then be TM-modes if $H_m^{(2)}(\kappa_1 b) = 0$ and TE-modes if $H_m^{(2)\prime}(\kappa_1 b) = 0$. The argument of the preceding section indicates that neither of these possibilities is feasible. The special case when $\kappa_1 = 0$ has now to be examined. There is a TEM-mode in which the only nonzero components of the field are $(\varepsilon_0/\mu_0)^{\frac{1}{2}} E_r = H_\phi = A/r$. Thus, *if the cylinder is perfectly conducting, the field outside is propagated, as a* TEM*-mode, parallel to the axis of the cylinder at the speed of light in the surrounding medium.*

When the conductivity is not perfect there is no alternative to an examination of (2.1). There is no analytical method of finding the values of κ which satisfy (2.1). A complete coverage will always involve a numerical treatment. However, before any approximations are introduced to (2.1), we note that the right-hand side does not vanish unless $m = 0$. Hence, *except when* $m = 0$, *the boundary conditions on a conducting cylinder cannot be fulfilled by either transverse electric or transverse magnetic waves alone.* For the symmetric mode $m = 0$ there are TM-modes determined by

$$k_1^2 H_1^{(2)}(\kappa_1 b)/\mu_1 \kappa_1 H_0^{(2)}(\kappa_1 b) = k_2^2 J_1(\kappa_2 b)/\mu_2 \kappa_2 J_0(\kappa_2 b) \quad (2.2)$$

and TE-modes for which κ satisfies

$$\mu_1 H_1^{(2)}(\kappa_1 b)/\kappa_1 H_0^{(2)}(\kappa_1 b) = \mu_2 J_1(\kappa_2 b)/\kappa_2 J_0(\kappa_2 b). \tag{2.3}$$

Here $J_0' = -J_1$, $H_0^{(2)\prime} = -H_1^{(2)}$ have been used.

The approximation of high conductivity mentioned in the first paragraph will now be introduced. It is reasonable to suppose that there will be a solution close to that for infinite conductivity. Therefore, look for one in which $\kappa \approx k_1$, i.e. $\kappa_1 b$ is small. From (1.4) the left-hand sides of both (2.2) and (2.3) are large. The right-hand side of (2.3) can be large only if $\kappa_2 b$ is near a zero of J_0. But all the zeros of J_0 are real whereas $\kappa_2 b$ has a large negative imaginary part because $\kappa \approx k_1$. Hence, (2.3) does not provide any solutions of the type sought.

Cognizant that $\kappa_2 b$ has a large negative imaginary part, we can approximate the Bessel functions on the right of (2.2) asymptotically through (A.2.1), whence

$$J_1(\kappa_2 b)/J_0(\kappa_2 b) \sim \tan(\kappa_2 b - \tfrac{1}{4}\pi) \sim -i$$

and (2.2) becomes

$$\kappa_1^2(\tfrac{1}{2}\pi i + \gamma + \ln \tfrac{1}{2}\kappa_1 b) = -i\mu_2 k_1^2/\mu_1 k_2 b \tag{2.4}$$

since κ_2 and k_2 are effectively equal. Equation (2.4) is susceptible to solution by successive approximations in which, if f_n is an estimate of $\kappa_1 b$,

$$f_{n+1}^2(\tfrac{1}{2}\pi i + \gamma + \ln \tfrac{1}{2} f_n) = -i\mu_2 k_1^2 b/\mu_1 k_2. \tag{2.5}$$

Convergence of the iteration is accelerated by selecting f_0 so that the quantity inside ( ) is -10.

A basic assumption in the derivation of (2.4) was that $\kappa_2 b$ was large. If b was very tiny this might be fallacious and replacement of J_0 and J_1 by their asymptotic representations would not be allowable. Apart from this qualification, we can assert that (2.4) supplies a TM-mode which is not far off the TEM-mode which travels along a perfect conductor.

There may be other roots of (2.2). When $\kappa_1 b$ is not small $H_1^{(2)}$ and $H_0^{(2)}$ are of about the same order of magnitude so that to a first approximation $J_1(\kappa_2 b) = 0$ will be needed. Since the roots of J_1 are real it follows that $\operatorname{Im} \kappa^2 = \operatorname{Im} k_2^2$. Consequently, κ has a large negative imaginary part and the wave is rapidly attenuated in its progress down the z-axis. *All waves corresponding to other roots of* (2.2) *are heavily damped.* More will be said about these waves later.

Likewise, it can be shown that all the roots of (2.3) are responsible for waves which decay fast.

The discussion of solutions of (2.1) for asymmetric modes with $m \neq 0$ is more complicated but it turns out that all the waves are strongly attenuated.

Further study of the symmetric TM-mode which is kin to the TEM-mode is justified because it is the only one capable of carrying energy beyond a short distance. The current density within the cylinder has a component parallel to the axis of $J_z = \sigma_2 E_z$. The ratio of the current density at r to that at the boundary is $J_0(\kappa_2 r)/J_0(\kappa_2 b)$, i.e. $J_0(k_2 r)/J_0(k_2 b)$ to the level of approximation adopted. This quantity can be easily tabulated as a function of r. As soon as r increases to a point where $|k_2 r|$ is large asymptotic formulae can be employed. Since $k_2 \approx (1-\mathrm{i})(\frac{1}{2}\mu_2\sigma_2\omega)^{\frac{1}{2}}$ the ratio becomes $(b/r)^{\frac{1}{2}} \exp\{(1+\mathrm{i})(r-b)/\delta\}$ where $\delta = (2/\mu_2\sigma_2\omega)^{\frac{1}{2}}$ is the *skin depth* as defined in §6.8(a). The current density decreases exponentially in moving away from the cylinder boundary and its behaviour is similar to the field in a conductor with a plane face. If $b \gg \delta$ the central portion of the cylinder is virtually free of current, i.e. *the current flows in a thin surface layer.*

Moreover, at the surface of the cylinder,

$$E_z/H_\phi = \kappa_2 \omega \mu_2 J_0(\kappa_2 b)/\mathrm{i}k_2^2 J_1(\kappa_2 b) \approx \tfrac{1}{2}(1+\mathrm{i})\,\delta\omega\mu_2.$$

Since E_z and H_ϕ are perpendicular this is the same relation as occurs at the interface with a plane conductor, thereby offering additional support for the use of impedance boundary conditions for scattering by good conductors. *It can be assumed that the field and current distribution near the surface of an arbitrary good conductor differ negligibly from those near an infinite plane surface provided that the radii of curvature are large compared with the skin depth; the exterior field will satisfy an impedance boundary condition.*

The current I flowing through the wire is $2\pi b$ times the surface value of H_ϕ. Hence $E_z = ZI$ where

$$Z = (1+\mathrm{i})\,\delta\omega\mu_2/4\pi b = \tfrac{1}{2}(1+\mathrm{i})bR_0/\delta$$

and $R_0 = 1/\pi b^2 \sigma_2$ is the direct current resistance per unit length. Combining this result with the previous paragraph we see that the high-frequency resistance of a good conductor is $\frac{1}{2}\,\delta\omega\mu_2/l$ where l is the length of the perimeter of the cross-section. Of course, this result would not hold if the perimeter had a small radius of curvature somewhere, such as at the corner of a conductor of rectangular cross-section.

7.3 The dielectric circular rod

The absence of conductivity in the cylinder does not affect the validity of (2.1)–(2.3) but the fact that $\sigma_2 = 0$ means that it is no longer legitimate to ascribe a large negative imaginary part to k_2. Let us search for solutions in which κ is real. Also if there is to be no transfer of energy to the surrounding medium κ_1 must be negative imaginary in order to ensure exponential decay at infinity. The lowest value of $\mathrm{i}\kappa_1$ which can be permitted then is zero which makes the wave move with speed characteristic of the outer medium. Now, if $\kappa_1 = 0$, the left-hand side of (2.2) is infinite and therefore $J_0(\kappa_2 b) = 0$. Hence there are symmetric TM-modes with $\kappa = \kappa_{0p}$ where $\kappa_{0p}^2 = k_2^2 - j_{0p}^2/b^2$ and $k_1 = \kappa_{0p}$. The frequencies are given by $\omega^2(\mu_2\varepsilon_2 - \mu_1\varepsilon_1) = j_{0p}^2/b^2$ which shows that *there is a cut-off frequency* $j_{01}/2\pi b(\mu_2\varepsilon_2 - \mu_1\varepsilon_1)^{\frac{1}{2}}$ below which no TM-modes propagate. It is found in a similar way that there are TE-modes for $\kappa_1 = 0$ which possess a cut-off frequency.

The asymmetric modes are *hybrid*, i.e. neither E_z nor H_z is zero. For hybrid waves in which $m > 1$ the substitution of (1.2) into (2.1) leads to a term in $1/\kappa_1^2$ (there is automatic cancellation of the terms involving $1/\kappa_1^4$) whose coefficient disappears if

$$(\mu_2\varepsilon_1 + \mu_1\varepsilon_2)\kappa_2 b J_{m-1}(\kappa_2 b)/J_m(\kappa_2 b)$$
$$= m(\varepsilon_2 - \varepsilon_1)(\mu_1 - \mu_2) + \mu_1\varepsilon_1\kappa_2^2 b^2/(m-1)$$

where (A.1.6) has been invoked. This equation determines the permissible values of κ and once again there is a cut-off frequency (since $\kappa_2 = 0$ is not a solution in general). The situation is different, however, if $m = 1$ because (1.3) in (2.1) supplies a term $\kappa_1^{-2} \ln \frac{1}{2}\kappa_1 b$ whose coefficient vanishes only when $J_1(\kappa_2 b) \approx 0$. Possible frequencies are $\omega = 0$ and $\omega^2(\mu_2\varepsilon_2 - \mu_1\varepsilon_1) = j_{1p}^2/b^2$. Thus, *there is no cut-off frequency for hybrid waves with* $m = 1$.

For large values of $k_1 b$, κ will not differ greatly from k_2.

General values of $k_1 b$ entail a numerical treatment but an indication of the location of possible κ can be obtained graphically. The curve of equation (2.1) is drawn in the plane with Cartesian coordinates $i\kappa_1 b$ and $\kappa_2 b$. Then the curve with equation $(\kappa_2 b)^2 - (\kappa_1 b)^2 = (k_2^2 - k_1^2)b^2 = \omega^2 b^2(\mu_2\varepsilon_2 - \mu_1\varepsilon_1)$ is traced. The intersections of the two curves lead at once to the appropriate values of κ.

The case in which $\varepsilon_2 \gg \varepsilon_1$ is of some interest. If $\kappa_1 b$ is not small the left-hand side of (2.2) is of moderate magnitude whereas the right-hand side is large unless $\kappa_2 b$ is near a zero of J_1. Hence

$$\kappa^2 = k_2^2 - j_{1p}^2/b^2, \tag{3.1}$$

the value of $\kappa = k_2$ being omitted because of the factor κ_2 in the denominator of the right-hand side of (2.2). It is convenient to consider (3.1) for a fixed moderate value of p as b decreases. So long as b is not too small $k_2 b \gg j_{1p}$ and $\kappa \approx k_2$ so that $\kappa_1 b$ cannot be near zero and the use of (3.1) is legitimate. Therefore, as b falls, $\kappa \approx k_2$ until $k_2 b$ is of the order of j_{1p}. Now, both $k_1 b$ and κb are small so that $\kappa_1 b$ is also. Accordingly, (3.1) must be discarded. However the case of $\kappa_1 b$ small has already been discussed and requires $k_2 b \approx j_{0m}$. A curve of k_1/κ against $\kappa_1 b$ will consist essentially of the straight line $k_1/\kappa = (\varepsilon_1/\varepsilon_2)^{\frac{1}{2}}$ (if $\mu_1 = \mu_2$) unless $k_1 b$ is small, when it is the straight line $k_1 b = j_{0m}(\varepsilon_1/\varepsilon_2)^{\frac{1}{2}}$, the two lines being joined smoothly around their point of intersection near the origin. Each mode can be dealt with similarly.

Except for one type of hybrid wave, there is a critical radius below which waves of a given frequency cannot propagate. For slightly greater radii κ_1 is little different from zero and the field in the exterior medium dies away very slowly. *Only a small part of the energy flow takes place within the cylinder.* On the other hand, a substantial increase in the radius makes $|\kappa_1|\,b$ large and the Hankel function is quickly attenuated as r grows. *Most of the energy flow is inside the cylinder.*

It may be verified that the component of the Poynting vector normal to the cylindrical interface is purely imaginary when κ_1 is purely imaginary and so there is no energy flow out of the cylinder. Therefore, the modes which have been described are *surface waves,* i.e. waves which travel without any transfer of energy across the dividing surface other than that necessary to make good resistive losses.

Rods of small diameter are suitable for guiding waves in space since most of the energy flow is external to the rod and will not suffer much attenuation from losses in the air. The surface wave may, naturally, be distorted if the cylinder passes too close to other objects. In contrast, for propagation within the rod, as in an *optical fibre,* the radius needs to be large unless only single mode propagation is contemplated. Also the material needs to be one whose dielectric loss is not significant. At large radii where several modes can propagate it may be more fitting to abandon the modal picture and use geometrical optics. The rays are comparatively easy to visualize and the main problem is keeping track of the reflections at the boundary, each of which introduces a Fresnel reflection coefficient. The technique for dealing with this is allied to that of §6.31 for tackling multiple reflections in a stratified atmosphere.

7.4 The modal structure

An important feature of fields in a waveguide is that they can always be expanded in a series of modes. The question arises as to whether it is true that any field associated with the rods of this chapter can be expressed as a series of modes. The short answer to the question is 'no' and the reason can be seen from an example. The illustration will be one from acoustics because it simplifies the analysis without modifying the principles involved; for a fuller investigation of the electromagnetic case see Jones (1979).

A ring source is placed inside the circular cylinder so that a solution of

$$\nabla^2 p + k_2^2 p = -\delta(r - r_0)\delta(z - z_0) \tag{4.1}$$

with $0 < r_0 < b$ is desired. The field is clearly independent of ϕ and so this coordinate may be ignored from now on. Define

$$P=\int_{-\infty}^{\infty} p\mathrm{e}^{-\mathrm{i}\kappa z}\,\mathrm{d}z.$$

Then

$$\frac{1}{r}\frac{\mathrm{d}}{\mathrm{d}r}\left(r\frac{\mathrm{d}P}{\mathrm{d}r}\right)+\kappa_2^2P=-\delta(r-r_0)\mathrm{e}^{-\mathrm{i}\kappa z_0}$$

inside the cylinder and a similar equation with zero right-hand side holds outside. Now assume

$$\begin{aligned}P&=A\,\mathrm{J}_0(\kappa_2 r)\quad(r\leqslant r_0)\\&=B\,\mathrm{J}_0(\kappa_2 r)+C\mathrm{Y}_0(\kappa_2 r)\quad(r_0\leqslant r\leqslant b)\\&=D\mathrm{H}_0^{(2)}(\kappa_1 r)\quad(r\geqslant b).\end{aligned}$$

After applying the boundary conditions it is found that

$$A=-\tfrac{1}{2}\pi r_0\mathrm{e}^{-\mathrm{i}\kappa z_0}\left\{\mathrm{Y}_0(\kappa_2 r_0)+\mathrm{J}_0(\kappa_2 r_0)\right.$$
$$\left.\times\frac{\zeta\kappa_2\mathrm{Y}_0'(\kappa_2 b)\mathrm{H}_0^{(2)}(\kappa_1 b)-\kappa_1\mathrm{Y}_0(\kappa_2 b)\mathrm{H}_0^{(2)\prime}(\kappa_1 b)}{\kappa_1\mathrm{J}_0(\kappa_2 b)\mathrm{H}_0^{(2)\prime}(\kappa_1 b)-\zeta\kappa_2 J_0'(\kappa_2 b)\mathrm{H}_0^{(2)}(\kappa_1 b)}\right\}.$$

Use $\mathrm{J}_0'=-\mathrm{J}_1$, etc., we see that in $r<r_0$

$$p=-\tfrac{1}{4}r_0\int_{-\infty}^{\infty}\left\{\mathrm{Y}_0(\kappa_2 r_0)+\mathrm{J}_0(\kappa_2 r_0)\right.$$
$$\left.\times\frac{\zeta\kappa_2\mathrm{Y}_1(\kappa_2 b)\mathrm{H}_0^{(2)}(\kappa_1 b)-\kappa_1\mathrm{Y}_0(\kappa_2 b)\mathrm{H}_1^{(2)}(\kappa_1 b)}{\kappa_1\mathrm{J}_0(\kappa_2 b)\mathrm{H}_1^{(2)}(\kappa_1 b)-\zeta\kappa_2\mathrm{J}_1(\kappa_2 b)\mathrm{H}_0^{(2)}(\kappa_1 b)}\right\}$$
$$\times\mathrm{J}_0(\kappa_2 r)\mathrm{e}^{\mathrm{i}\kappa(z-z_0)}\,\mathrm{d}\kappa. \tag{4.2}$$

The corresponding result for a rigid waveguide is obtained by allowing $\zeta\to\infty$. Then

$$p=-\tfrac{1}{4}r_0\int_{-\infty}^{\infty}\{\mathrm{Y}_0(\kappa_2 r_0)-\mathrm{J}_0(\kappa_2 r_0)\mathrm{Y}_1(\kappa_2 b)/\mathrm{J}_1(\kappa_2 b)\}\mathrm{J}_0(\kappa_2 r)\mathrm{e}^{\mathrm{i}\kappa(z-z_0)}\,\mathrm{d}\kappa. \tag{4.3}$$

The singularities of the integrand are poles arising from the zeros of J_1 and possible branch points where κ_2 vanishes. In fact, the branch points do not materialize because of (A.1.10) which cancels potential trouble-makers. The contour may be deformed over the poles and a modal series results as in §5.8.

For (4.2) repetition of the same process may be attempted. There are poles at the zeros of the denominator. Because of (1.1) these correspond to the modes. The potential branch points where $\kappa_2=0$ disappear by the same argument as before. There are, however, other branch points where

$\kappa_1 = 0$ and these are not removed. Therefore, when the contour is pushed over the poles, there is also a contribution from a branch line. It is this extra term which spoils a straightforward modal expansion. It stems from the radiation field in the external medium.

The expansion of a general field contains not only the surface wave modes but also a term contributed by the radiation field. Criteria which specify when the radiation term is of no importance are difficult to formulate precisely. However, one can say that the larger $k_2 b$ is, the less likely it is that the radiation term will play any significant role.

7.5 Several conductors

A perfectly conducting circular cylinder supports only one mode of any significance, according to the theory of §7.2, and that is the TEM. The propagation of the TEM-mode along any number of parallel perfect conductors of arbitrary cross-section will now be discussed.

The medium outside the conductors will be attributed parameters ε, μ with $k^2 = \omega^2 \mu \varepsilon$ and solutions depending on z as the factor e^{-ikz} will be the target. Subject to $E_z = H_z = 0$, Maxwell's equations give $\boldsymbol{H}_t = (\varepsilon/\mu)^{\frac{1}{2}} \boldsymbol{k} \wedge \boldsymbol{E}_t$ and $\boldsymbol{E}_t = -\text{grad}_t V$ where, since $\text{div}\, \boldsymbol{E}_t = 0$, $\nabla_t^2 V = 0$. The single unknown V is a two-dimensional solution of Laplace's equation and, as soon as it is known, the whole field is available. Perfect conductivity demands that the tangential component of $\boldsymbol{E}_t$ vanish on the perimeter of a conductor and so V must be a constant there, say $V = V_1$ on C_1, $V = V_2$ on $C_2, \ldots$ (Fig. 7.1). The problem has therefore been transformed to one of electrostatics.

The charge Q_1 per unit length on a conductor is given by

$$Q_1 = -\varepsilon \int_{C_1} (\partial V/\partial n)\, ds. \tag{5.1}$$

The current I_1 on the conductor is related to it by

$$I_1 = \int_{C_1} \boldsymbol{H} \cdot ds = -(\varepsilon/\mu)^{\frac{1}{2}} \int_{C_1} (\partial V/\partial n)\, ds = Q_1/(\mu\varepsilon)^{\frac{1}{2}}. \tag{5.2}$$

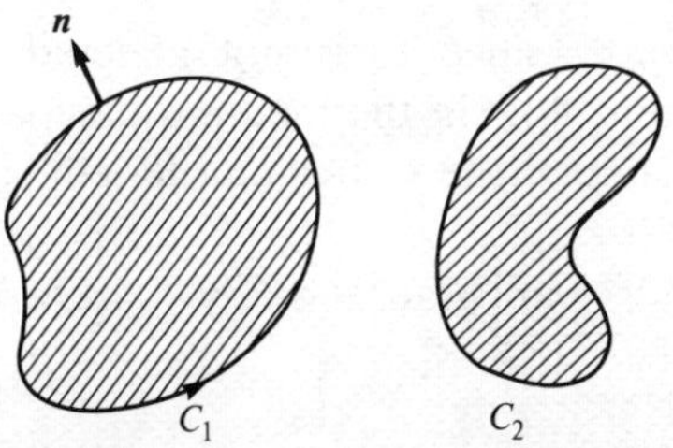

Fig. 7.1 Boundaries of the cross-sections of the conductors

The total charge on all conductors is $-\varepsilon \int_C (\partial V/\partial n)\,ds$ where C is the circle at infinity. If V is not to be logarithmic at infinity, this integral is zero and so the total charge of the system must vanish.

A Lecher arrangement has only two conductors. For the total charge to be zero $Q_1+Q_2=0$ and so, from (5.2), *the current flows along one wire and returns via the other.* The capacitance C per unit length is defined by

$$C(V_1-V_2)=Q_1. \tag{5.3}$$

The inductance L per unit length is derived from integrating the magnetic energy over the area outside the conductors by

$$\tfrac{1}{2}LI_1^2=\tfrac{1}{2}\mu\int \boldsymbol{H}_t\cdot\boldsymbol{H}_t\,dx\,dy=\tfrac{1}{2}\varepsilon\int \text{grad}_t^2 V\,dx\,dy$$

$$=-\tfrac{1}{2}\varepsilon\int_{C_1\cup C_2} V(\partial V/\partial n)\,ds=\tfrac{1}{2}Q_1(V_1-V_2).$$

From (5.2) it follows that

$$LI_1=(V_1-V_2)(\mu\varepsilon)^{\frac{1}{2}}. \tag{5.4}$$

An inference from (5.2)–(5.4) is that $LC=\mu\varepsilon$.

Writing $V_0=(V_1-V_2)e^{-ikz}$, $I_0=I_1e^{ikz}$ we see that (5.2)–(5.4) make V_0 and I_0 satisfy

$$\partial V_0/\partial z=-i\omega LI_0, \qquad \partial I_0/\partial z=-i\omega CV_0.$$

The medium outside the conductors may be allowed to possess conductivity. The determination of V is unaltered but in the representation of the field ε is replaced by $\varepsilon-i\sigma/\omega$. There is no change to the definitions of Q_1 and C so they are the same as when the medium is non-conducting. I_1 is still defined by the curvilinear integral of $\boldsymbol{H}$ but, in view of the changed field representation, (5.2) is altered to

$$I_1=Q_1(1-i\sigma/\omega\varepsilon)^{\frac{1}{2}}/(\mu\varepsilon)^{\frac{1}{2}}. \tag{5.5}$$

Also L continues to be defined by the integral of magnetic energy with the result that (5.4) is modified to

$$LI_1=(V_1-V_2)\{\mu(\varepsilon-i\sigma/\omega)\}^{\frac{1}{2}}. \tag{5.6}$$

It follows that $LC=\mu\varepsilon$; since C is not affected by the conductivity neither is L. The current flow in the medium dissipates heat and may be accounted for by a conductance G per unit length specified by

$$\tfrac{1}{2}G(V_1-V_2)^2=\tfrac{1}{2}\sigma\int \text{grad}_t^2 V\,dx\,dy=\tfrac{1}{2}\sigma Q_1(V_1-V_2)/\varepsilon.$$

Hence $G=\sigma C/\varepsilon$ and

$$L(C+G/i\omega)=\mu(\varepsilon-i\sigma/\omega). \tag{5.7}$$

The alteration to k means that

$$\partial V_0/\partial z = -i\omega L I_0, \qquad \partial I_0/\partial z = -(i\omega C + G)V_0.$$

When the conductivity of the wires is not perfect a TEM-mode is no longer possible. Nevertheless, if the conductivity is high, there is a TM-mode which is not far from TEM and in it E_z is small. Let the dependence on z be $e^{-i\kappa z}$. The fact that $H_z = 0$ means that

$$\boldsymbol{E}_t = -\text{grad } V, \qquad \boldsymbol{H} = (\omega\varepsilon - i\sigma)\boldsymbol{k} \wedge \boldsymbol{E}/\kappa, \qquad \nabla_t^2 V = -i\kappa E_z$$

with ε, σ taking the values ε_i, σ_i in the interior of a conductor. To determine the field in the conductor we follow the idea of §5.6. The field outside is assumed to be the same as for perfectly conducting wires so that its magnetic field is essentially responsible for the interior behaviour. Let

$$W = \int \boldsymbol{E} \wedge \boldsymbol{H}^* \cdot \boldsymbol{k} \, dx \, dy = \int \kappa \boldsymbol{H} \cdot \boldsymbol{H}^* \, dx \, dy/(\omega\varepsilon - i\sigma),$$

the integral being taken over the whole (x, y)-plane. It follows that, if $\omega\varepsilon_i$ is neglected in comparison with σ_i,

$$W = \frac{\kappa}{\mu(\omega\varepsilon - i\sigma)} \tfrac{1}{2} L I_0 I_0^* + \frac{\kappa}{2i\sigma_i\mu_i} L_i I_0 I_0^*$$

where L_i is the internal inductance per unit length of the conductor. It may be shown similarly that

$$\partial W/\partial z = -\tfrac{1}{2}\{R_i + i\omega(L + L_i)\} I_0 I_0^* - \tfrac{1}{2} G V_0 V_0^* + i\kappa\kappa^* L I_0 I_0^*/2\mu(\omega\varepsilon - i\sigma).$$

Since $\partial W/\partial z = -i(\kappa - \kappa^*)W$ an equation for κ is obtained which reduces to

$$\kappa^2 = \omega\mu(\omega\varepsilon - i\sigma)\{(1 + L_i/L) - iR_i/\omega L\}$$

when the second term in W is neglected and σ is assumed small enough to drop G in comparison with L. By virtue of (5.7)

$$\kappa^2 = -(G + i\omega C)\{R_i + i\omega(L + L_i)\}. \tag{5.8}$$

Now

$$\partial V_0/\partial z = -\{R_i + i\omega(L + L_i)\} I_0, \tag{5.9}$$

$$\partial I_0/\partial z = -(G + i\omega C)V_0. \tag{5.10}$$

The behaviour of the system has thus been determined as soon as R_i, L, L_i, G, and C are known. These are available for various standard systems. For example, if the wires are circular of radius b and the distance between the centres is $2d$,

$$\pi\varepsilon/C = \ln\{d + (d^2 - b^2)^{\frac{1}{2}}\}/b$$

from which G and L follow at once. The formulae for the coaxial cable in §5.7(e) are also relevant.

The time and z-dependence have been taken to be $e^{i(\omega t-\kappa z)}$. Very little alteration is necessary to make the dependence $f(\omega t-\kappa z)$; just alter $i\omega$ to $\partial/\partial t$. Then (5.9) and (5.10) become

$$\partial V_0/\partial z+R_i I_0+(L+L_i)\,\partial I_0/\partial t=0, \tag{5.11}$$

$$\partial I_0/\partial z+GV_0+C\,\partial V_0/\partial t=0. \tag{5.12}$$

Elimination of V_0 gives

$$\frac{\partial^2 I_0}{\partial z^2}-C(L+L_i)\frac{\partial^2 I_0}{\partial t^2}-\{CR_i+G(L+L_i)\}\frac{\partial I_0}{\partial t}-R_i G I_0=0 \tag{5.13}$$

which is known as the *telegraph equation.*

7.6 Transmission lines

The system of two conductors discussed in the preceding section constitutes a *transmission line.* The equations (5.9)–(5.12) have been derived on the assumption that the line is infinite in length. Notwithstanding, a finite line which is long compared with the wavelength may be expected to fulfil the same equations. The ends might deserve special attention but since these are usually places where the behaviour is specified the governing equations will be used up to the ends.

For a wave on the transmission line κ^2 is given by (5.8). The attenuation constant $-\mathrm{Im}\,\kappa$ varies from $(R_i G)^{\frac{1}{2}}$ at low frequencies to $\{CR_i+G(L+L_i)\}/2\{C(L+L_i)\}^{\frac{1}{2}}$. However, if $R_i/(L+L_i)=G/C$ the attenuation constant has the value $(R_i G)^{\frac{1}{2}}$ at all frequencies; such a line is said to be *distortionless.*

A *characteristic impedance* Z_c can be defined by

$$Z_c=\{R_i+i\omega(L+L_i)\}^{\frac{1}{2}}/(G+i\omega C)^{\frac{1}{2}}.$$

A wave travelling in the direction of the positive z-axis then has

$$V_0=Ae^{-i\kappa z}, \qquad I_0=Ae^{-i\kappa z}/Z_c$$

while one going the opposite way has

$$V_0=Be^{i\kappa z}, \qquad I_0=-Be^{i\kappa z}/Z_c.$$

Suppose now that the line is terminated at $z=l$ by an impedance Z_l so that $V_0=Z_l I_0$. In general, $V_0=Ae^{-i\kappa z}+Be^{i\kappa z}$ so that

$$B/A=(Z_l-Z_c)e^{-2i\kappa l}/(Z_l+Z_c)$$

which gives the coefficient of the wave reflected by the termination. The effect of the terminating impedance can be represented as an impedance at $z=0$ by calculating V_0/I_0 at $z=0$; the ratio Z_t is

$$(Z_l+iZ_c\tan\kappa l)Z_c/(Z_c+iZ_l\tan\kappa l).$$

Three particular limiting cases warrant mention. They are (i) the *open-circuit line* when Z_l is infinite and the impedance appearing at $z=0$ is $-iZ_c\cot\kappa l$ which can be varied from 0 to $\pm i\infty$ by selecting l suitably when κ is real; (ii) the *short-circuit line* in which $Z_l=0$ and the transmitting impedance is $iZ_c\tan\kappa l$, also capable of wide variation with l; (iii) the *matched line* for which $Z_l=Z_c$; here the transmitting impedance is Z_c whatever the length of the line and, in addition, there is no reflected wave.

The propagation constant κ is real when there are no losses present so that $R_i=0$ and $G=0$; then $\kappa=\omega\{C(L+L_i)\}^{\frac{1}{2}}$. One choice of l is to make $\tan\kappa l=0$ so that $Z_t=Z_l$, i.e. *if the length of the line is a whole number of half wavelengths the transmitting impedance equals the load impedance.* Alternatively, $\tan\kappa l$ might be made infinite so that $Z_t=Z_c^2/Z_l$, i.e. *the characteristic impedance of a line, an odd number of quarter wavelengths long, is the geometric mean of the transmitting and load impedances.* The flexibility in setting Z_t by suitable choices of l and Z_l makes the transmission line a valuable matching device.

7.7 General cylindrical structures

The techniques employed for the homogeneous circular cylinder can clearly be adapted to more complicated problems, e.g. several concentric circular cylinders and when there is cross-sectional inhomogenity in the parameters as in the graded-index optical fibre. The main difficulty, once the tedious algebra of elimination from the boundary conditions has been completed, is the resolution of the roots of the transcendental equation determining κ. This often stretches to the limits analytical and numerical facilities.

Without attempting to enter such detail one can ask the question whether there is a criterion which will indicate if a boundary can support a surface wave. Let us discuss this query in the context of structures with circular symmetry.

Let $r=b$ be the outermost boundary and consider TM-waves which are independent of ϕ. Then, in $r\geqslant b$,

$$E_z=c_0\mathrm{H}_0^{(2)}(\kappa_1 r)e^{-i\kappa z},$$
$$H_\phi=-(i\omega\varepsilon c_0/\kappa_1)\mathrm{H}_0^{(2)\prime}(\kappa_1 r)e^{-i\kappa z},$$

where c_0 is an arbitrary constant. Suppose that on the boundary the impedance condition $E_z=Z_sH_\phi$ is satisfied. Then

$$Z_s=i\kappa_1\mathrm{H}_0^{(2)}(\kappa_1 b)/\omega\varepsilon\mathrm{H}_0^{(2)\prime}(\kappa_1 b).$$

If there are no losses in the system κ_1 must be real or negative imaginary. The former possibility implies a radiating field whereas the latter alternative give a surface wave. For the surface wave use the formula $-\pi i H_0^{(2)}(\kappa_1 b) = 2K_0(i\kappa_1 b)$ so that the argument of the modified Bessel function is real and positive. Then

$$Z_s = -\kappa_1 K_0/\omega\varepsilon_1 K_1$$

since $K_0' = -K_1$. Evidently, *a surface wave is possible only if Z_s is positive imaginary.* The larger $|\kappa_1|$ the more closely concentrated around the guiding structure is the surface wave. Since $K_0/K_1 \sim 1$ as $i\kappa_1 b \to \infty$, large values of $i\kappa_1$ entail a big magnitude of Z_s. In other words, *the larger the reactance of the surface, the more pronounced is the surface wave.*

A dielectric rod has a suitable reactance to support a surface wave, as has been seen earlier. The placing of a perfectly conducting coaxial cylinder within the rod does not invalidate this statement. The position is not altered appreciably if the adjective 'perfectly conducting' is exchanged for 'highly conducting'; a small amount of loss is introduced giving κ_1 a small real part but the level of damping of the surface wave as $r \to \infty$ is not affected. It may therefore be deduced that *a single length of ordinary enamelled wire can be employed to guide high frequency waves without significant radiation.*

The corrugated cylinder is an example of a boundary which can support a surface wave. The structure repeates itself at intervals of l_2 (Fig. 7.2). Choose the origin so that $z = 0$ corresponds to the beginning of a slot. On account of the periodic properties of the corrugation, the longitudinal electric intensity in a TM-mode independent of ϕ is

$$E_z = \sum_{n=-\infty}^{\infty} a_n H_0^{(2)}(\zeta_n r)\exp\{-i(\kappa + 2\pi n/l_2)z\} \quad (r \geqslant b)$$

where $\zeta_n^2 = k_1^2 - (\kappa + 2\pi n/l_2)^2$. Inside the slot the longitudinal component of $\boldsymbol{E}$ must vanish on the base $r = d$ and the radial component must be

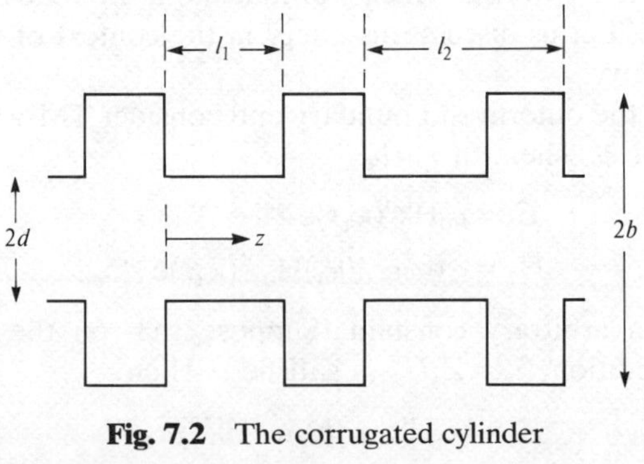

Fig. 7.2 The corrugated cylinder

zero on the sides $z=0$, l_1. Hence, inside the slot,

$$E_z = \sum_{m=0}^{\infty} b_m\{J_0(\eta_m r)Y_0(\eta_m d) - Y_0(\eta_m r)J_0(\eta_m d)\}\cos(m\pi z/l_1)$$

where $\eta_m^2 = k_1^2 - (m\pi/l_1)^2$.

The fields have now to be matched to each other across the opening of the slot and, furthermore, E_z must vanish on the boundary $r=b$, $l_1 \leqslant z \leqslant l_2$. Hence

$$\sum_{n=-\infty}^{\infty} a_n H_0^{(2)}(\zeta_n b)\exp\{-i(\kappa + 2\pi n/l_2)z\}$$
$$= \begin{cases} \sum_{m=0}^{\infty} b_m\{J_0(\eta_m b)Y_0(\eta_m d) - Y_0(\eta_m b)J_0(\eta_m d)\}\cos(m\pi z/l_1) & (0 \leqslant z \leqslant l_1) \\ 0 & (l_1 \leqslant z \leqslant l_2). \end{cases} \tag{7.1}$$

Also H_ϕ must be continuous across the mouth of the slot so that

$$\sum_{n=-\infty}^{\infty} (a_n/\zeta_n)H_1^{(2)}(\zeta_n b)\exp\{-i(\kappa + 2\pi n/l_2)z\}$$
$$= \sum_{m=0}^{\infty} (b_m/\eta_m)\{J_1(\eta_m b)Y_0(\eta_m d) - Y_1(\eta_m b)J_0(\eta_m d)\}\cos(m\pi z/l_1)$$
$$(0 \leqslant z \leqslant l_1). \tag{7.2}$$

Multiply (7.1) by $\exp\{i(\kappa + 2\pi p/l_2)z\}$ and integrate with respect to z from $z=0$ to $z=l_2$. Then

$$l_2 a_p H_0^{(2)}(\zeta_p b) = \sum_{m=0}^{\infty} b_m c_{pm}\{J_0(\eta_m b)Y_0(\eta_m d) - Y_0(\eta_m b)J_0(\eta_m d)\} \tag{7.3}$$

where

$$c_{pm} = \int_0^{l_1} \cos(m\pi z/l_1)\exp\{i(\kappa + 2\pi p/l_2)z\}\,dz.$$

Multiplication of (7.2) by $\cos(p\pi z/l_1)$ and integration from $z=0$ to $z=l_1$ provides

$$\sum_{n=-\infty}^{\infty} (-)^p(a_n/\zeta_n)c_{np}H_1^{(2)}(\zeta_n b)\exp\{-il_1(\kappa + 2\pi n/l_2)\}$$
$$= (b_p l_1/2\eta_p)(1+\delta_{p0})\{J_1(\eta_p b)Y_0(\eta_p d) - Y_1(\eta_p b)J_0(\eta_p d)\}. \tag{7.4}$$

The value of b_p in (7.4) may be substituted into (7.3) leading to an infinite set of linear homogeneous equations for the a_n. A non-trivial solution asks for the determinant of the coefficients to be zero. This determinantal equation fixes κ. It is, however, impossible to solve and the usual procedure is to truncate both series representations so that a determinant of finite size is assured.

Often the period of the structure is much less than a wavelength, i.e. $k_1 l_2 \ll 2\pi$. Then all the exterior modes with the exception of $n=0$ are heavily attenuated. In the slot the fact that $k_1 l_1 < k_1 l_2 \ll 2\pi$ results in all the modes except $m=0$ being heavily damped since η_m is negative imaginary and of large magnitude. It is therefore reasonable to postulate a_0 and b_0 as the only coefficients significantly different from zero. Also $\kappa l_1 \ll 1$ and so the surface impedance $Z_s = E_z/H_\phi$ is

$$Z_s = -i\left(\frac{\mu_1}{\varepsilon_1}\right)^{\frac{1}{2}} \frac{l_1}{l_2} \frac{J_0(k_1 b)Y_0(k_1 d) - Y_0(k_1 b)J_0(k_1 d)}{J_1(k_1 b)Y_0(k_1 d) - Y_1(k_1 b)J_0(k_1 d)}.$$

If $k_1 b \ll 1$ the ratio of the Bessel functions reduces to $k_1 b \ln(d/b)$ so that Z_s is positive imaginary and a surface wave exists. Thus, until b gets too big, the structure can sustain a surface wave.

Propagation along a plane surface

7.8 General remarks

Propagation along a plane interface may be treated in the same manner as that for a cylindrical surface. There is the simplification that the Bessel functions are replaced by exponential functions but often the extra complication that an additional region is involved.

Sound waves, TE- and TM-modes may be treated simultaneously since they differ only in the identification of parameters. An isolated slab of thickness $2l$ occupying the region $-l \leqslant x \leqslant l$ will be discussed. For modes travelling along the z-axis and independent of y

$$\begin{aligned} p &= A_1 \exp(-i\kappa_1 x - i\kappa z) \quad (x \geqslant l) \\ &= (A_2 e^{-i\kappa_2 x} + A_3 e^{i\kappa_2 x})e^{-i\kappa z} \quad (-l \leqslant x \leqslant l) \\ &= A_4 \exp(i\kappa_1 x - i\kappa z) \quad (x \leqslant -l). \end{aligned}$$

The boundary conditions are the continuity of p and the equality of $\partial p/\partial x$ outside with $\zeta\, \partial p/\partial x$ inside. After the constants A_1, A_2, A_3, and A_4 have been disposed of it is found that

$$\tan \kappa_2 l = i\zeta\kappa_2/\kappa_1 \quad \text{or} \quad \tan \kappa_2 l = i\kappa_1/\zeta\kappa_2.$$

On the assumptions that $k_2^2 > k_1^2$ and ζ is positive the second equation always possesses a solution in which κ_2 is real and κ_1 is negative imaginary. Hence a surface wave exists. The first also has a solution of this type if $k_2^2 - k_1^2 > (\pi/2l)^2$.

An electromagnetic TM-surface wave can exist above a plane interface

$x=0$ subject to the impedance boundary condition $E_z = Z_s H_y$ if

$$Z_s = -\kappa_1/\omega\varepsilon_1. \tag{8.1}$$

Hence a pure surface wave with κ_1 negative imaginary requires Z_s to be positive imaginary. When dissipation is present a surface wave occurs if the imaginary part of Z_s is positive. *A surface wave of this type cannot exist above a semi-infinite dielectric.* For, in the dielectric, $Z_s = \kappa_2/\omega\varepsilon_2$ which can never be reconciled with (8.1) because the right-hand sides have opposite signs.

A semi-infinite good conductor is capable of sustaining a surface wave because the skin effect ensures that Z_s has a positive imaginary part. In practice, the surface wave is not readily distinguishable because the reactive component of Z_s is small and approximately equal to the resistive component. The reactance can be enhanced by corrugating the metal surface or by coating it with a dielectric. Suppose the face of the conductor is at $x=-l$ and there is dielectric in $0 \geqslant x \geqslant -l$. In the dielectric,

$$E_x e^{i\kappa z} = A e^{-i\kappa_2 x} + A(\kappa_2 + \omega\varepsilon_2 Z_m) e^{i\kappa_2(2l+x)}/(\omega\varepsilon_2 Z_m - \kappa_2)$$

where $Z_m = \frac{1}{2}(1+i)\,\delta\omega\mu_0$ and δ is the skin depth of the metal; the metal is assumed to have the same permeability as free space. Consequently, at $x=0$,

$$\frac{E_z}{H_y} = \frac{(\omega\varepsilon_2 Z_m + i\kappa_2 \tan \kappa_2 l)\kappa_2}{(\kappa_2 + i\omega\varepsilon_2 Z_m \tan \kappa_2 l)\omega\varepsilon_2}.$$

For a first approximation Z_m may be taken as zero and (8.1) is matched if

$$-\kappa_1/\varepsilon_1 = (i\kappa_2/\varepsilon_2)\tan \kappa_2 l \approx i\kappa_2^2 l/\varepsilon_2$$

if the dielectric layer is thin. Thus $\kappa = k_1 + O(l)$ and

$$\kappa_1 \approx -il\varepsilon_1(k_2^2 - k_1^2)/\varepsilon_2.$$

It is better to retain Z_m when $\kappa_2 l$ is small and then the surface impedance is

$$Z_m + il(k_2^2 - k_1^2)/\omega\varepsilon_2.$$

The relative enhancement of the reactive component over the resistive component, for $\mu_2 = \mu_1 = \mu_0$, is $(2l/\delta)(1-\varepsilon_1/\varepsilon_2)$. This can be large for a thin dielectric coat so long as thickness of the layer is much greater than the skin depth and ε_2 is not too close to ε_1. Materials which make the factor a 1000 or more are not difficult to come by.

7.9 Launching efficiency

The existence of a surface wave does not mean that it is easy to stimulate. An estimate of how much energy goes into it may be derived from the model problem of a point source above a flat surface. It is related to the problem discussed in Chapter 6 but now allows for the presence of surface waves. To benefit from the earlier theory it is convenient to make the boundary $z = 0$ and place the source at $(0, 0, h)$. Then the solution of

$$\nabla^2 p + k^2 p = -\delta(x)\delta(y)\delta(z-h)$$

is desired, the suffix not being required since only one medium is of concern.

As before the transform $\mathscr{P}$ is formed and the boundary condition imposed on $z = 0$ is $\mathscr{P} = Z(\kappa)\,\mathrm{d}\mathscr{P}/\mathrm{d}z$ where λ is the transform variable. The resulting formula is

$$p = \frac{1}{4\pi \mathrm{i}} \int_0^\infty \frac{\lambda}{\kappa} \mathrm{J}_0(\lambda r) \left\{ \mathrm{e}^{-\mathrm{i}\kappa|z-h|} + \frac{\mathrm{i}\kappa Z(\kappa) - 1}{\mathrm{i}\kappa Z(\kappa) + 1} \mathrm{e}^{-\mathrm{i}\kappa(z+h)} \right\} \mathrm{d}\lambda$$

where now $\kappa^2 = k^2 - \lambda^2$.

The first term of the integrand is responsible for the primary spherical wave and so its contribution to p can be written down immediately.

The second term of the integrand can have poles where $\mathrm{i}\kappa Z(\kappa) + 1 = 0$. The ones of interest are those on the real axis with $\lambda > k$. These are the origin of any propagating surface waves. The residue at such a pole λ_s entails a spatial variation of $\mathrm{H}_0^{(2)}(\lambda_s r)\mathrm{e}^{-\kappa_s z}$ in the wave. For fixed z, the surface wave behaves like a cylindrical wave with amplitude falling off like $1/r^{\frac{1}{2}}$ in spite of being generated by a source which produces spherical waves whose decay is $1/(r^2 + z^2)^{\frac{1}{2}}$. However, the surface wave is exponentially damped as z increases and it is not a true cylindrical wave. The power flow through the surface $r = \text{constant}$, $z \geqslant 0$ is infinite in a true cylindrical wave but is finite for our surface waves. Nevertheless near the interface before the attenuation has taken full hold the behaviour looks like that of a cylindrical wave. Fields which have the pseudo-cylindrical character of our surface waves have sometimes been dubbed *Zenneck waves*.

In addition to the surface waves there is a radiation field which can be assessed at large distances by the method of stationary phase. With $r = R\cos\theta$, $z = R\sin\theta$ the point of stationary phase is $\lambda = k\cos\theta$ and, for the radiation field

$$p \sim \frac{\mathrm{e}^{-ikR}}{4\pi R} \left\{ \mathrm{e}^{ikh\sin\theta} + \frac{ik\sin\theta Z(k\sin\theta) - 1}{ik\sin\theta Z(k\sin\theta) + 1} \mathrm{e}^{-ikh\sin\theta} \right\} \tag{9.1}$$

as $R \to \infty$ with $\theta > 0$. The field exhibits the normal decay of a spherical wave. Although it may appear to be dominated by the surface wave this is not so because as $R \to \infty$ with fixed θ one progresses steadily into the region where the surface wave is heavily attenuated. The one possible exceptional direction is $\theta = 0$ where (9.1) indicates that $p \sim 0$.

While (9.1) does signify that the radiation field might be of smaller magnitude than the surface wave close to the interface it is not in fact valid there, because the point of stationary phase has coalesced with a branch point. It is therefore necessary to recalculate the integral when θ is near 0. By the same mechanism as employed in §6.27, the radiation field for $\theta \approx 0$ is

$$p \sim \frac{ik e^{-ikr}}{2\pi r^2}[\{h + Z(0)\}\{z + Z(0)\} + iZ'(0)]. \tag{9.2}$$

When $Z(\kappa)$ is a function of κ^2, as it often is, the term $Z'(0)$ can be omitted and will be from this point onwards.

The expression (9.2) already reveals that near to the interface the surface wave is stronger than the radiation field. The difference can be accentuated by selecting h so that

$$h = -Z(0). \tag{9.3}$$

The radiation field will then be $O(1/r^3)$ and the total field will be almost a pure surface wave. *The height specified by* (9.3) *may be regarded as the most efficient for launching a surface wave by means of the given source.*

Since h is real and positive the value of (9.3) is a feasible proposition only if $Z(0)$ is negative real. If Z is complex and equal to $\rho + i\sigma$ the quantity $|h + Z(0)|$ can be made a minimum by choosing h so that $h = -\rho(0)$.

For example, if the medium below $z = 0$ is a dielectric coating on a perfect conductor, the theory of the preceding section tells us that

$$Z = -(1/\kappa_2)\cot \kappa_2 l$$

where, in the present notation, $\kappa_2^2 = k_2^2 - k_1^2 + \kappa^2$ so that

$$Z(0) = -(k_2^2 - k_1^2)^{-\frac{1}{2}} \cot(k_2^2 - k_1^2)^{\frac{1}{2}} l$$

and $Z'(0) = 0$. Provided that the coating is not too thick $Z(0)$ is negative and the optimal height of launching follows from (9.3).

Of course, one is not obliged to accept this criterion as the best for launching; one may prefer to compare the relative powers in the surface wave and radiation field though the calculations are usually more complicated.

The polyrod antenna

7.10 The radiation pattern

The use of dielectric antennas or *polyrods* is common practice in many applications. The calculation of the radiation pattern in general is a complex affair and here only that of a circular dielectric rod projecting beyond the end of a circular waveguide will be considered.

The technique is that of §5.26. The starting point is the same representation of the electromagnetic field in terms of its tangential components on a closed surface S. In this case S is chosen to be the exterior of the polyrod and guide together with the sphere at infinity (Fig. 7.3). Since the field is radiating from the dielectric antenna the contribution from the sphere at infinity is zero. The fields on the guide and polyrod are not known exactly but they can be estimated in the same way as for electromagnetic horns. On the perfectly conducting guide the induced currents are assumed to be small enough to be negligible. The integral over the end of the rod will be omitted on the grounds that its area is a small fraction of the rest of the rod. Finally, it is assumed that the distribution along the sides of the rod is just as if the cylinder were infinitely long; the energy reflected from the end of the rod is thereby treated as negligible. The evaluation of $\boldsymbol{E}$ has now become a matter of integrating known functions over the sides of the rod.

Taking the environment to be free space we see that at a large distance R from the origin in the direction of the spherical polar angles θ, ϕ

$$E_\theta \sim \frac{\mathrm{i}e^{-\mathrm{i}k_0R}}{4\pi R}\int_0^l\int_0^{2\pi}[k_0E_z\cos(\phi'-\phi)-\omega\mu_0\{H_\phi\sin\theta-H_z\cos\theta\sin(\phi'-\phi)\}]$$
$$\times\exp[\mathrm{i}k_0\{b\sin\theta\cos(\phi'-\phi)+z'\cos\theta\}]b\,\mathrm{d}\phi'\,\mathrm{d}z',$$

$$E_\phi \sim \frac{\mathrm{i}e^{-\mathrm{i}k_0R}}{4\pi R}\int_0^l\int_0^{2\pi}[k_0E_\phi\sin\theta-k_0E_z\cos\theta\sin(\phi'-\phi)-\omega\mu_0H_z\cos(\phi'-\phi)]$$
$$\times\exp[\mathrm{i}k_0\{b\sin\theta\cos(\phi'-\phi)+z'\cos\theta\}]b\,\mathrm{d}\phi'\,\mathrm{d}z'$$

where b is the radius of the polyrod.

For radially symmetric TM-modes the field inside the dielectric cylinder has components (§7.3)

$$E_z=\mathrm{J}_0(\kappa_2 r)e^{-\mathrm{i}\kappa z},\qquad H_\phi=(\mathrm{i}\omega\varepsilon_2/\kappa_2)\mathrm{J}_1(\kappa_2 r)e^{-\mathrm{i}\kappa z},\qquad E_\phi=H_z=0.$$

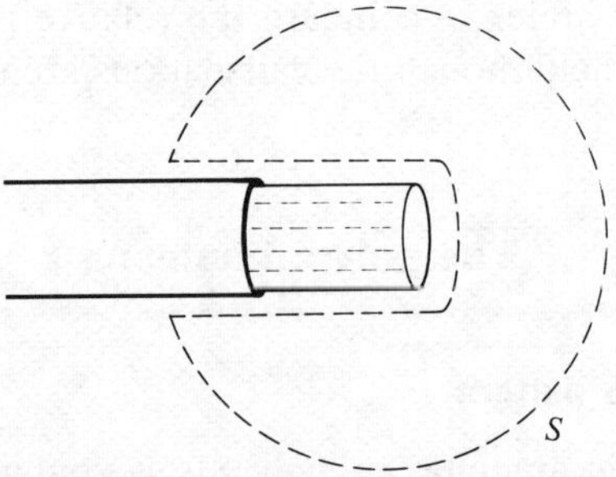

Fig. 7.3 The surface of integration for polyrod radiation

Invoking (A.1.25) we have

$$RE_\theta \sim \{k_0 J_0(\kappa_2 r)J_1(k_2 b \sin\theta) + (k_2^2/\kappa_2)J_1(\kappa_2 r)J_0(k_2 b\sin\theta)\}b e^{-ik_0 R}$$
$$\times \exp\{\tfrac{1}{2}il(k_0\cos\theta-\kappa)\}\sin\tfrac{1}{2}l(k_0\cos\theta-\kappa)/(k_0\cos\theta-\kappa),$$
$$E_\phi \sim 0.$$

Strictly r should be placed equal to b in this expression but sometimes giving r a somewhat smaller value secures a better agreement with observation. It is also useful to incorporate a weighting function which allows the dependence on z to deviate from $e^{-i\kappa z}$.

The function $(\sin\xi)/\xi$ where $\xi = \frac{1}{2}l(k_0\cos\theta-\kappa)$ is the factor produced by the phase differences between the elementary radiators along a generator of the cylinder. It has a maximum at $\xi = 0$ and zeros at $\xi = n\pi$. The values of θ such that $\frac{1}{2}l(k_0\cos\theta-\kappa) = n\pi$ determine the widths of the lobes. As l increases so does the number of lobes. Evidently, there will be a large major lobe only if $\theta = 0$ nearly corresponds to $\xi = 0$, i.e. when κ is not very different from k_0. The polar diagram of the polyrod will not consist merely of the pattern of $(\sin\xi)/\xi$ because of the other factor in E_θ but the number of lobes cannot be less than those of $(\sin\xi)/\xi$.

Exercises on Chapter 7

1. The symmetric TM-mode, which approximates the TEM of perfect conductivity, is propagating in air at a frequency of 1 GHz on a copper wire ($\sigma_2 = 5.8\times10^7$ mho m) of radius 1 mm. In the iteration scheme of (2.5) put $\frac{1}{2}\pi i + \gamma + \ln\frac{1}{2}f_n = \frac{1}{2}\ln(g_n 10^{-8})$. If $\ln(g_0 10^{-8}) = -20$ show that $g_1 = 3.6(1-i)$, $g_2 = 4.1-4.5i$, $g_3 = 4.2-4.5i$. Deduce that the waves are attenuated to 1/e of their initial value in 770 m and that their speed is $(1-6\times10^{-5})c$.

2. Examine the Poynting vector of the symmetric TM-mode, which approximates the TEM of perfect conductivity, on a highly conducting cylinder and show that the propagation of energy takes place mainly in the surrounding medium.

3. Waves of frequency 30 GHz are propagating in air along a circular cylinder of distilled water of radius 2 cm contained in a vessel with very thin glass walls. Show that their speed of propagation is approximately $c/9$. [Neglect the conductivity of the water and take $\varepsilon_2 = 81.1\varepsilon_0$.]

4. A perfectly conducting circular cylinder of radius b is surrounded by a concentric perfectly conducting circular cylinder of radius d. Show that the TEM-mode is determined from $V = A + B\ln r$. Hence show that the capacitance per unit length is given by $C = 2\pi\varepsilon/\ln(d/b)$ and that $2\pi Z_c = (\mu/\varepsilon)^{\frac{1}{2}}\ln(d/b)$. If the conductivity of the intervening medium is σ prove that $G = 2\pi\sigma/\ln(d/b)$. If both cylinders have the same conductivity σ_i and

skin depth δ_i

$$R_i = (b+d)/2\pi bd\sigma_i\,\delta_i, \qquad L_i = R_i/\omega.$$

Set up the exact equations governing this problem and verify that they have an approximate solution that is in accord with the predictions of §7.5.

5. Two parallel wires of radius b have their centres a distance $2d$ ($\gg b$) apart. Verify that

$$V = A\ln[\{x-d)^2+y^2\}/\{(x+d)^2+y^2\}]$$

is a pertinent approximation for the TEM-mode. Hence show that $C = \pi\varepsilon/\ln(2d/b)$ and that $\pi Z_c = (\mu/\varepsilon)^{\frac{1}{2}}\ln(2d/b)$. For this system $R_i = 1/\pi b\sigma_i\delta_i$.

6. Calculate numerical values for Exercise 4 when $b = 1$ mm, $d = 1$ cm at frequencies of 1 kHz, 1 MHz and 1 GHz. Repeat for Exercise 5 with the same parameters and compare.

7. The system of Exercise 5 is surrounded by a circular cylinder of radius D whose axis is midway between the axes of the wires. If $D \gg d$, show that

$$R_i = \frac{1}{\pi b\sigma_i\delta_i}\{1+(1+2p^2)(1-4q^2)/8p^4\} + \frac{4q^2}{\pi b\sigma_i\delta_i}\{1+q^2-(1+4p^2)/8p^4\},$$

$$Z_c = \frac{1}{\pi}\left(\frac{\mu}{\varepsilon}\right)^{\frac{1}{2}}\left\{\ln\left(\frac{1-q^2}{1+q^2}2p\right) - (1+4p^2)(1-4q^2)/16p^4\right\}$$

where $p = d/b$, $q = d/D$.

8. If inductance and leakage can be neglected verify that a solution of the telegraph equation is $t^{-\frac{1}{2}}\exp(-R_iCz^2/4t)$.

9. In a transmission line without leakage $V_0 = 3$, $I_0 = 1$ at $z = 0$. Find the current I_0 for arbitrary z.

10. When there are losses in a transmission line let $\kappa = \kappa_r - i\kappa_i$ and call $2\pi/\kappa_r$ the wavelength. Show that, if $R_i \ll \omega(L+L_i)$ and $G \ll \omega C$, the impedance offered by an open-circuit line, whose length l is a whole number of half wavelengths, is $2(L+L_i)/\{R_iC+G(L+L_i)\}l$.

The Q of a resonant system may be defined as average energy stored divided by average energy absorbed when ω has the resonant value ω_0 (c.f. §4.10); equivalently, the impedance at $\omega = \omega_0(1+\eta)$, where $|\eta| \ll 1$, is $Z(1-2i\eta Q)$ where Z is the impedance at $\omega = \omega_0$. Show that for the above transmission line

$$\omega_0/Q = R_i/(L+L_i) + G/C.$$

11. A plane wave travelling along the x-axis in free space has $\boldsymbol{E}$ parallel to the z-axis and is normally incident on a slab of good conductor between $x = a$ and $x = b$. The parameters of the slab are ε, μ, and σ. By considering the magnitude of the transmitted wave in $x > b$ show that the

power loss in decibels is

$$8.686(\tfrac{1}{2}\omega\mu\sigma)^{\frac{1}{2}}(b-a)+20\log_{10}|(1+\alpha)^2/4\alpha|$$
$$+20\log_{10}\left|1-\left(\frac{1-\alpha}{1+\alpha}\right)^2 e^{-2ik(b-a)}\right|$$

where $k=(1-i)(\omega\mu\sigma/2)^{\frac{1}{2}}$, $\alpha=k_0\mu/\mu_0 k$. The problem may be regarded as a model of a *shield*. The first term represents the loss due to attenuation in the slab, the second the loss due to mismatch at the interfaces and the third similar losses for waves which have undergone more than one reflection.

A less crude model is to place a current line element at the centre of a cylindrical shield occupying $a\leqslant r\leqslant b$. Assume that the primary wave due to the line source has $E_z=H_0^{(2)}(k_0r)$ and that $|ka|\gg 1$. If $k_0b\ll 1$ show that the power loss is

$$8.686(\tfrac{1}{2}\omega\mu\sigma)^{\frac{1}{2}}(b-a)+20\log_{10}|\tfrac{1}{2}(k_0/\alpha)(ab)^{\frac{1}{2}}\ln(\tfrac{1}{2}k_0b)|$$
$$+20\log_{10}|1-e^{-2ik(b-a)}|$$

decibels and, if $k_0a\gg 1$, is

$$8.686(\tfrac{1}{2}\omega\mu\sigma)^{\frac{1}{2}}(b-a)$$
$$+20\log_{10}\tfrac{1}{2}|(1+\alpha)(\alpha-i\cot\beta)+(1-\alpha)(\alpha+i\cot\beta)e^{-2ik(a-b)}|$$
$$-20\log_{10}|\alpha(1-i\cot\beta)|$$

where $\beta=k_0a-\frac{1}{4}\pi$.

12. A circular waveguide of radius a carries an internal coating of dielectric of thickness $a-b$ where $a-b\ll a$. Show that, in a first approximation, the field in $r<b$ is the same as that satisfying impedance boundary conditions on $r=a$ but in matrix form. If $b-a=0.048$ mm and $\varepsilon=2\varepsilon_0$ show that the axial and transverse components of the impedance are $1.7\times 10^{-2}i$ and $3.3\times 10^{-4}i$ at 9 mm.

13. A signal is excited in a perfectly conducting coaxial line by applying unit electric intensity over the length $|z|<l$ of the inner conductor, i.e. on $r=b$, $E_z=\int_{-\infty}^{\infty} e^{i\alpha z}\sin\alpha l\, d\alpha/\alpha\pi$. Construct the electromagnetic field in which the z-component of $\boldsymbol{E}$ is $e^{i\alpha z}$ on $r=b$ and 0 on $r=a$ and use this to express the field produced by the excitation as integrals with respect to α. If $V(z)=\int_b^a E_r\,dr$ show that

$$V(z)=\frac{i}{\pi}\int_{-\infty}^{\infty}\frac{\sin\alpha l}{k_0^2-\alpha^2}e^{i\alpha z}\,d\alpha,$$

the contour passing below $-k_0$ and above k_0. Find $I(z)$ defined as $2\pi bH_\phi(b,z)$. If $k_0a<1$ show that, in $z>l$, $2\pi V(z)=(\mu_0/\varepsilon_0)^{\frac{1}{2}}I(z)\ln(a/b)$ when exponentially damped terms are neglected. This result provides further justification for the use of transmission line theory if the frequency is not too high.

14. A perfectly conducting circular cylinder of radius b is coated with dielectric of thickness d. Examine the possibility of surface waves with $|\kappa_1 b| \gg 1$.

15. In a corrugated waveguide with $k_1 l_2 \ll 2\pi$ and $\kappa l_1 \ll 1$ show that $Z_s = l_1 Z_{sl}/l_2$ where Z_{sl} is the surface impedance of a slot at $r=b$.

16. Examine the possibility of surface waves being produced in the system of Exercise 49 of Chapter 6.

17. A dielectric slab of thickness l is placed in free space a distance h below a vertical electric dipole. Show that the surface waves are determined by

$$i(\varepsilon_2^2\kappa_0^2 + \varepsilon_0^2\kappa_2^2)\tan \kappa_2 l + 2\kappa_0\kappa_2\varepsilon_2\varepsilon_0 = 0.$$

18. Repeat the analysis of §7.9 when the primary wave is a plane wave.

19. The region $r \leq b$ of a circular waveguide is occupied by dielectric with the permeability of free space. Show that, as $\kappa \to 0$, the power flow can be in the opposite direction to the phase velocity if ε is large enough. This constitutes a *backward wave*. For the TE_{11} mode with $k_0 a = 1$ the condition is $\varepsilon > 9.8\varepsilon_0$. There is a similar phenomenon if the dielectric is replaced by a longitudinally magnetized ferrite (c.f. Exercises 21 and 24 of Chapter 5).

20. An approximate version of (2.4) is

$$(\kappa_1 b)^2(\tfrac{1}{2}\pi i + \ln \tfrac{1}{2}\kappa_1 b) = -C$$

where C is a real positive constant. Search for complex roots of this equation of the form $\kappa_1 b = d e^{i\delta}$ where $0 < \delta \leq \frac{1}{2}\pi$. By plotting d against δ from the real part of the equation and from the imaginary part show that there is one root with $0 < \delta < \frac{1}{4}\pi$. Show that, as $C \to 0$, $\delta \to 0$ and $d \to 0$. Obtain numerical estimates of d and δ. (Waves of this type grow as $r \to \infty$ and are known as *leaky waves*).

When $C \ll 1$ show that the leaky wave travels along the z-axis with a phase speed faster than light.

21. An optical fibre consists of circular inner dielectric core of refractive index N_3 surrounded by an annular cladding of smaller refractive index N_2. Around the cladding is a dielectric with refractive index N_1. Set up the equation governing the modes of propagation.

The outermost region is constructed of black glass so as to be highly lossy while the core and cladding are only slightly dissipative. Discuss suitable approximations for simplifying the modal equation and examine to what extent further simplification can be achieved by making the permittivities of the cladding and its environment nearly equal. (The permeabilities of all three media are the same as free space.)

Calculate some values of κ at a wavelength of 0.9 μm when $N_3 = 1.61$ for (a) core radius 20 μm, cladding thickness 5 μm, $N_2 = N_1 = 0.99N_3$ and (b) core radius 40 μm, cladding thickness 5 μm, $N_2 = N_1 = 0.96N_3$.

8. Scattering by smooth objects

To date the behaviour of waves in space as modified by infinite structures has been discussed. In this chapter obstacles of finite size will be examined. Only a limited class of objects is susceptible to analysis but it furnishes valuable clues on the behaviour to be expected generally. Also, solutions make convenient benchmarks against which to test approximate and numerical techniques. Often it is possible to amplify the range of a set of results by an appeal to the quantitative answers and physical principles revealed by an analytical investigation. Only obstacles which do not possess edges will be the subject of this chapter.

Two-dimensional scattering problems

There are sufficient practical situations for which a two-dimensional model is appropriate for the investigation to commence in this area. There is the added advantage that mechanisms can be evolved without the extra complication brought in by three dimensions.

In §1.36 it has been shown that the electromagnetic field in two dimensions is completely specified by the two scalars E_z and H_z when the z-axis is perpendicular to the two-dimensional plane. Both E_z and H_z satisfy the same partial differential equation. The boundary conditions at an interface separating two media are: E_z, $\partial E_z/\mu\,\partial n$ continuous in the one case and H_z, $\partial H_z/\varepsilon\,\partial n$ continuous in the other. If the boundary is perfectly conducting $E_z = 0$ and $\partial H_z/\partial n = 0$ in the two respective types.

Correspondingly, in acoustics, p satisfies the partial differential equation and similar boundary conditions; $p = 0$ is a soft boundary and $\partial p/\partial n = 0$ a hard or rigid one.

It will therefore be convenient to use one letter u to be interpreted as E_z, H_z, or p according to context. For a brief nomenclature $u = 0$ will be referred to as soft and $\partial u/\partial n = 0$ as hard so, for electromagnetic waves, a soft boundary condition has $\boldsymbol{E}$ parallel to the z-axis and a hard one $\boldsymbol{H}$ parallel. The impedance boundary condition $\eta u + \partial u/\partial n = 0$ will also be allowed; it includes both the soft and hard boundaries as special cases.

8.1 The circular cylinder

The first problem is the scattering by a homogeneous isotropic circular cylinder of radius b, irradiated by a plane wave. The partial differential equation to be solved is

$$\partial^2u/\partial x^2+\partial^2u/\partial y^2+k_1^2u=0 \tag{1.1}$$

outside the cylinder; inside k_2 is interchanged with k_1. Let u^{i} be the incident wave, u_1 the field scattered by the cylinder and u_2 that transmitted inside. The boundary conditions on the cylindrical interface are

$$u^{\mathrm{i}}+u_1=u_2, \qquad \partial(u^{\mathrm{i}}+u_1)/\partial n=\zeta\,\partial u_2/\partial n. \tag{1.2}$$

The method of attack is to represent the field by separable solutions of (1.1). Take r and ϕ as polar coordinates in the plane so that the cylindrical boundary is $r=b$. Since u is unaltered if ϕ is increased by 2π it can be expanded in a full-range Fourier series (§4.1) $u=\sum_{m=-\infty}^{\infty}a_m(r)\mathrm{e}^{\mathrm{i}m\phi}$ where $2\pi a_m=\int_0^{2\pi}u\mathrm{e}^{-\mathrm{i}m\phi}\,\mathrm{d}\phi$. It follows from §A.1 that a_m is a Bessel function of order m and argument k_1r.

In particular, the plane wave $u^{\mathrm{i}}=\mathrm{e}^{-\mathrm{i}k_1r\cos\phi}$ has, via (A.1.25), the representation

$$u^{\mathrm{i}}=\sum_{m=-\infty}^{\infty}\mathrm{J}_m(k_1r)\mathrm{e}^{\mathrm{i}m(\phi-\frac{1}{2}\pi)}.$$

Since u_1 must be outgoing at infinity we postulate

$$u_1=\sum_{m=-\infty}^{\infty}b_m\mathrm{H}_m^{(2)}(k_1r)\mathrm{e}^{\mathrm{i}m(\phi-\frac{1}{2}\pi)}.$$

On the other hand u_2 must be bounded at $r=0$ and so a suitable expansion is

$$u_2=\sum_{m=-\infty}^{\infty}c_m\mathrm{J}_m(k_2r)\mathrm{e}^{\mathrm{i}m(\phi-\frac{1}{2}\pi)}.$$

The boundary conditions (1.2) can be met provided that

$$\mathrm{J}_m(k_1b)+b_m\mathrm{H}_m^{(2)}(k_1b)=c_m\mathrm{J}_m(k_2b),$$
$$k_1\{\mathrm{J}_m'(k_1b)+b_m\mathrm{H}_m^{(2)\prime}(k_1b)\}=\zeta k_2c_m\mathrm{J}_m'(k_2b).$$

Hence

$$b_m=\frac{\zeta k_2\mathrm{J}_m'(k_2b)\mathrm{J}_m(k_1b)-k_1\mathrm{J}_m(k_2b)\mathrm{J}_m'(k_1b)}{k_1\mathrm{J}_m(k_2b)\mathrm{H}_m^{(2)\prime}(k_1b)-\zeta k_2\mathrm{J}_m'(k_2b)\mathrm{H}_m^{(2)}(k_1b)}, \tag{1.3}$$

$$c_m=-2\mathrm{i}/\pi b\{k_1\mathrm{J}_m(k_2b)\mathrm{H}_m^{(2)\prime}(k_1b)-\zeta k_2\mathrm{J}_m'(k_2b)\mathrm{H}_m^{(2)}(k_1b)\} \tag{1.4}$$

after use of (A.1.21).

An alternative to (1.2) is the impedance boundary condition

$$\partial(u^{\mathrm{i}}+u_1)/\partial n+\eta(u^{\mathrm{i}}+u_1)=0. \tag{1.5}$$

Here u_2 is not required and to distinguish this case for u_1 we put B_m for b_m. Then

$$B_m=-\frac{\eta \mathrm{J}_m(k_1b)+k_1\mathrm{J}'_m(k_1b)}{\eta \mathrm{H}^{(2)}_m(k_1b)+k_1\mathrm{H}^{(2)\prime}_m(k_1b)}. \tag{1.6}$$

The field is now completely determined as soon as the series has been computed. The computation is relatively straightforward until the frequency becomes high when the slow convergence of the series is a deterrent. The matter of high frequency will be turned to later. In the meantime the low frequency behaviour will be studied.

When k_1b and k_2b are both small the approximations for Bessel functions of small argument (A.1.12–A.1.14) can be used supplemented by $\mathrm{J}_m(z)=(-)^m\mathrm{J}_{-m}(z)$ and $\mathrm{H}^{(2)}_m(z)=(-)^m\mathrm{H}^{(2)}_{-m}(z)$. It is then transparent that b_m is of the order of $(k_1b)^{2m}$ so that, for a first approximation, it is sufficient to retain only the terms $m=0, \pm 1$. Hence

$$u_1\approx b_0\mathrm{H}^{(2)}_0(k_1r)-2\mathrm{i}b_1\mathrm{H}^{(2)}_1(k_1r)\cos\phi, \tag{1.7}$$

$$b_0=\tfrac{1}{4}\pi\mathrm{i}b^2(k_1^2-\zeta k_2^2), \qquad b_1=\tfrac{1}{4}\pi\mathrm{i}k_1^2b^2(\zeta-1)/(\zeta+1). \tag{1.8}$$

At a large distance from the origin (A.2.3) gives

$$u_1\sim\left\{\tfrac{1}{2}(k_1^2-\zeta k_2^2)b^2+\frac{\zeta-1}{\zeta+1}k_1^2b^2\cos\phi\right\}\mathrm{i}\left(\frac{\pi}{2k_1r}\right)^{\frac{1}{2}}\mathrm{e}^{-\mathrm{i}k_1r+\frac{1}{4}\pi\mathrm{i}}. \tag{1.9}$$

The first term in (1.7) and (1.9) represents the field of a simple line source along the axis of the cylinder whereas the second comes from a double line source. Thus, to this degree of approximation, the effect of the cylinder is equivalent to these two line sources. (The omitted terms in which $|m|>1$ are due to line sources of higher order.) The two line sources are not always present; for instance, $\zeta k_2^2=k_1^2$ removes the simple source while $\zeta=1$ disposes of the double source.

The comparable formula for the impedance boundary condition is, from (1.6),

$$u_1\sim\left\{\frac{\frac{1}{4}k_1^2b^2(2+\eta b)-\eta b}{1+b\eta(\frac{1}{2}\pi\mathrm{i}+\gamma+\ln\frac{1}{2}k_1b)}+\frac{\eta b+1}{\eta b-1}k_1^2b^2\cos\phi\right\}$$
$$\times\mathrm{i}\left(\frac{\pi}{2k_1r}\right)^{\frac{1}{2}}\mathrm{e}^{-\mathrm{i}k_1r+\frac{1}{4}\pi\mathrm{i}}. \tag{1.10}$$

Again there are two line sources involved in general. For the soft case $\eta=\infty$

$$u_1\sim-\frac{\mathrm{i}}{\ln\frac{1}{2}k_1b}\left(\frac{\pi}{2k_1r}\right)^{\frac{1}{2}}\mathrm{e}^{-\mathrm{i}k_1r+\frac{1}{4}\pi\mathrm{i}} \tag{1.11}$$

and only the simple source is significant. In the hard case $\eta = 0$

$$u_1 \sim \mathrm{i}k_1^2 b^2(\tfrac{1}{2} - \cos\phi)(\pi/2k_1 r)^{\frac{1}{2}} \mathrm{e}^{-\mathrm{i}k_1 r + \frac{1}{4}\pi\mathrm{i}}, \tag{1.12}$$

both line sources being needed. It will be remarked that the scattered field in (1.12) is much smaller than that in (1.11), i.e. a soft cylinder scatters much more efficiently than a hard one. As regards the electromagnetic field this means that the scattered field is considerably greater when $\boldsymbol{E}$ is parallel to the cylinder than when $\boldsymbol{H}$ is. Thus the current induced in the cylinder is much larger in the former case than in the latter. This property is important in the design of whip antennas (cf. §3.19).

8.2 The scattering coefficient

A measure of the effectiveness of an obstacle is the *scattering coefficient* σ_c defined by

$$\sigma_c = \frac{\text{Total energy scattered}}{\text{Energy incident on obstacle}}. \tag{2.1}$$

In two dimensions the energy is calculated per unit length of cylinder to prevent both numerator and denominator being infinite.

The average energy flow through a curve C is $\mathrm{Im} \int_C u(\partial u^*/\partial n)\,\mathrm{d}s$ apart from a multiplicative real constant for all types of wave we are considering. So long as only ratios of energies are the object in view the multiplying constant is of no concern. It is advantageous to obtain a general formula for σ_c rather than restrict attention to the circular cylinder.

As a temporary expedient assume that the plane wave is travelling along the x-axis and drop the suffix on the wavenumber so that $u^{\mathrm{i}} = \mathrm{e}^{-ikx}$. The energy incident on the obstacle is $\mathrm{Im} \int_D u^{\mathrm{i}}(\partial u^{\mathrm{i}*}/\partial x)\,\mathrm{d}s$ where D is the projection of the obstacle on a line perpendicular to the direction of propagation (Fig. 8.1). In view of the form of u^{i} this is kD.

Let C_r be a circle of large radius r. Then the scattered energy is

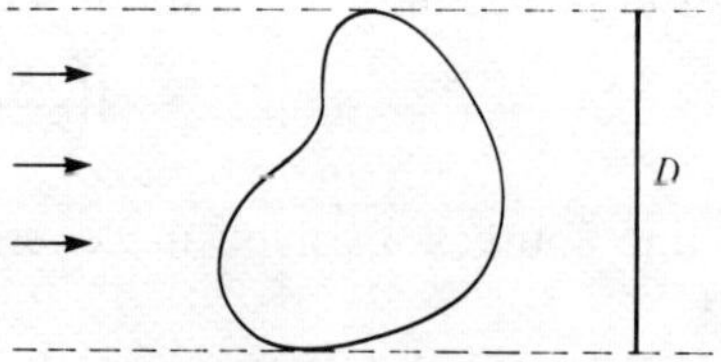

Fig. 8.1 The line cut off by the incident plane wave

$\operatorname{Im} \int_{C_r} u_1(\partial u_1^*/\partial n)\, ds$. Now

$$\operatorname{Im} \int_{C_r} (u^i + u_1)\{\partial(u^i + u_1)^*/\partial n\}\, ds = 0$$

if no energy is absorbed by the target. Hence

$$\sigma_c = -(1/kD)\operatorname{Im} \int_{C_r} \left(u^i \frac{\partial u^{i*}}{\partial n} + u^i \frac{\partial u_1^*}{\partial n} + u_1 \frac{\partial u^{i*}}{\partial n}\right) ds.$$

The first term arises from the energy of the incident wave which crosses a closed curve and this is zero since the source of a plane wave is at infinity. As for the other two terms let

$$u_1 \sim (2/\pi kr)^{\frac{1}{2}} A(\phi) e^{-ikr + \frac{1}{4}\pi i} \tag{2.2}$$

as $r \to \infty$ in the direction ϕ. Then

$$\sigma_c = -(1/kD)\operatorname{Im} \lim_{r\to\infty} \left[(2kr/\pi)^{\frac{1}{2}} e^{\frac{1}{4}\pi i} \int_0^{2\pi} \{A^*(\phi) e^{ikr - ikr\cos\phi} \right.$$

$$\left. + iA(\phi)\cos\phi\, e^{ikr(\cos\phi - 1)}\}\, d\phi\right].$$

The integral can be evaluated by the method of stationary phase (Appendix G) since kr is large. The phase is stationary where $\sin\phi = 0$ but the contribution from $\phi = \pi$ is purely real. Hence

$$\sigma_c = -(2/kD)\operatorname{Im}[i\{A^*(0) + A(0)\}] = -(4/kD)\operatorname{Re} A(0).$$

The result can be adapted immediately to arbitrary incidence. If $u^i = \exp\{-ikr\cos(\phi - \phi^i)\}$ and (2.2) is still assumed,

$$\sigma_c = -(4/kD)\operatorname{Re} A(\phi^i), \tag{2.3}$$

i.e. *the scattering coefficient can be calculated from the amplitude of the scattered wave in the direction in which the incident wave is travelling.*

If the cylinder is dissipative an *absorption coefficient* σ_a can be defined analogously in terms of the energy it absorbs. The only alteration to (2.3) is that

$$\sigma_a + \sigma_c = -(4/kD)\operatorname{Re} A(\phi^i). \tag{2.4}$$

Another measure of interest is the ratio of $2\pi r$ times the intensity scattered back towards the source to the intensity coming in along that line. It is called the *back scattering* or *radar cross-section* and denoted by σ_b. Thus, for the far field behaviour (2.2)

$$\sigma_b = 4\,|A(\phi^i + \pi)|^2/k. \tag{2.5}$$

These general results can be applied to the circular cylinder for which

$D=2b$. The original definition (2.1) can be employed and this gives

$$\sigma_c = 2 \sum_{m=-\infty}^{\infty} |b_m|^2 \Big/ k_1 b. \tag{2.6}$$

In contrast, the suggestion of (2.3) is

$$\sigma_c = -2 \operatorname{Re} \sum_{m=-\infty}^{\infty} b_m \Big/ k_1 b. \tag{2.7}$$

Since the two results must be the same a certain property of the coefficients b_m is implied. The B_m have the same property since (2.6) and (2.7) hold for the impedance boundary condition with B_m in place of b_m.

At low frequencies the information from (1.8) is that

$$\sigma_c \approx \tfrac{1}{8}\pi^2 (k_1 b)^3 \{(1-\zeta k_2^2/k_1^2)^2 + 2(\zeta-1)^2/(\zeta+1)^2\}. \tag{2.8}$$

Observe that (2.7) gives an answer of zero if only b_0, $b_{\pm 1}$ are kept. It is thus necessary to work to higher orders in the coefficients if (2.8) is to be achieved from (2.7). Notwithstanding, there are instances in which an estimate of σ_c is forthcoming from (2.3) but not from (2.1).

The scattering coefficients for the soft and hard boundaries are

$$\sigma_c^S \approx \pi^2/2k_1 b \ln^2 k_1 b, \qquad \sigma_c^H \approx 3\pi^2 (k_1 b)^3/8. \tag{2.9}$$

Computed values of σ_c^S, σ_c^H are shown in Table 8.1. The formulae (2.9) fit well at the lower end of the range.

For the back scattering

$$\sigma_b = 4 \left| \sum_{m=-\infty}^{\infty} (-)^m b_m \right|^2 \Big/ k_1 \tag{2.10}$$

in general and, at low frequencies,

$$\sigma_b^H \approx 9\pi^2 (k_1 b)^4 / 4k_1, \qquad \sigma_b^S \approx \pi^2 / k_1 \ln^2 \tfrac{1}{2} k_1 b. \tag{2.11}$$

Table 8.1 The soft and hard scattering coefficients

$k_1 b$	σ_c^S	σ_c^H	$k_1 b$	σ_c^S	σ_c^H
0.2	4.572	0.029	2.5	2.530	1.447
0.4	3.696	0.206	3.0	2.471	1.517
0.6	3.319	0.533	4.5	2.361	1.639
0.8	3.101	0.834	5.0	2.337	1.665
1.0	2.957	1.000	9.5	2.221	1.789
1.5	2.740	1.200	10.0	2.213	1.797
2.0	2.621	1.359			

8.3 The high frequency field on the cylinder

The series solution is exact but it does not shed much light on what happens at high frequencies because so many terms of the series have to be summed to secure any kind of accuracy. It is therefore desirable to convert the series to a form with an improved rate of convergence. The technique is to discover for the series an integral representation which is more susceptible to manipulation. For purposes of illustration the field on the cylinder will be discussed but for the impedance boundary condition (1.5) only. The high frequency assumption is that $k_1 b \gg 1$.

On $r = b$ the total field $u = u^{\mathrm{i}} + u_1$ is given by, from (1.6),

$$u = -2\mathrm{i} \sum_{m=-\infty}^{\infty} \mathrm{e}^{\mathrm{i}m(\phi - \frac{1}{2}\pi)} \Big/ \pi b\{\eta \mathrm{H}_m^{(2)}(k_1 b) + k_1 \mathrm{H}_m^{(2)\prime}(k_1 b)\} \tag{3.1}$$

by virtue of (A.1.21).

Now consider the rectangular contour C of Fig. 8.2 symmetrically placed about the origin with sides parallel and perpendicular to the real axis of the complex z-plane. If $f(z)$ is regular within and on C, Cauchy's theorem supplies

$$\int_C \frac{f(z)}{\sin \pi z}\, \mathrm{d}z = 2\mathrm{i} \sum_{n=-N}^{N} (-)^n f(n)$$

where N is the largest zero of $\sin \pi z$ inside the contour. If f diminishes sufficiently rapidly the rectangle can be stretched along the real axis to infinity and the vertical sides omitted. Then

$$\int_{C_+ \cup C_-} \frac{f(z)}{\sin \pi z}\, \mathrm{d}z = 2\mathrm{i} \sum_{n=-\infty}^{\infty} (-)^n f(n) \tag{3.2}$$

where C_+ and C_- are the upper and lower horizontals respectively.

The objective is to use (3.2) to transform (3.1). Choose ϕ so that $|\phi| \leqslant \pi$. Instead of working in the z-plane we shall employ the complex ν-plane and put

$$f(\nu) = -\mathrm{e}^{\mathrm{i}\nu(\phi + \frac{1}{2}\pi)}/\pi b\{\eta \mathrm{H}_\nu^{(2)}(k_1 b) + k_1 \mathrm{H}_\nu^{(2)\prime}(k_1 b)\}.$$

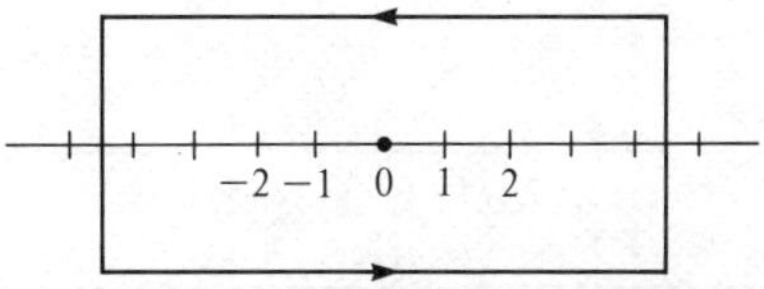

Fig. 8.2 The preliminary contour of integration

The result is that

$$u = -\frac{1}{\pi b}\int_{C_+ \cup C_-} \frac{e^{i\nu(\phi+\frac{1}{2}\pi)}}{\eta H_\nu^{(2)}(k_1 b) + k_1 H_\nu^{(2)\prime}(k_1 b)} \frac{d\nu}{\sin \nu\pi}.$$

However, $H_{-\nu}^{(2)} = e^{-\nu\pi i} H_\nu^{(2)}$ enables the integral over C_+ to be made into one over C_- by the change of variable $\nu = -\mu$. The net effect is that

$$u = -\frac{2}{\pi b}\int_D \frac{e^{\frac{1}{2}\nu\pi i}\cos \nu\phi}{\eta H_\nu^{(2)}(k_1 b) + k_1 H_\nu^{(2)\prime}(k_1 b)} \frac{d\nu}{\sin \nu\pi} \tag{3.3}$$

where D is the straight line running from $-\infty$ to ∞ parallel to and below the real axis but above any singularities of the factors other than cosec $\nu\pi$ in the integrand (Fig. 8.3). The integral of (3.3) is the sought alternative to the series for the field.

The singularities of the terms in the integrand other than cosec $\nu\pi$ must now be settled (a) to check that D can, indeed, meet its criterion and (b) to permit deformation of the contour. It is transparent that the only singularities can be poles at the zeros of the denominator of the Hankel functions. For moderate values of $|\nu|$ the largeness of $k_1 b$ makes the adoption of (A.2.3) permissible so that, as far as ν-dependence is concerned, the denominator behaves like $e^{\frac{1}{2}\nu\pi i}$. Evidently, there are no zeros unless $|\nu| \gg 1$. For $|\nu| \gg 1$ and $k_1 b \gg 1$ the uniform expansion of (A.2.30) is obligatory, i.e.

$$H_\nu^{(2)}(\nu z) \sim 2^{\frac{3}{2}} e^{\frac{1}{3}\pi i} \nu^{-\frac{1}{3}} \{\xi/(z^2-1)\}^{\frac{1}{4}} \operatorname{Ai}(\nu^{\frac{2}{3}} \xi e^{\frac{1}{3}\pi i}), \tag{3.4}$$

$$H_\nu^{(2)\prime}(\nu z) \sim 2^{\frac{3}{2}} \nu^{-\frac{2}{3}} z^{-1} e^{\frac{2}{3}\pi i} \{(z^2-1)/\xi\}^{\frac{1}{4}} \operatorname{Ai}'(\nu^{\frac{2}{3}} \xi e^{\frac{1}{3}\pi i}) \tag{3.5}$$

where

$$\tfrac{2}{3}\xi^{\frac{3}{2}} = (z^2-1)^{\frac{1}{2}} - \sec^{-1} z \tag{3.6}$$

and $(z^2-1)^{\frac{1}{2}} = i$ at $z = 0$. In our case $z = k_1 b/\nu$ and for our purposes $\sec^{-1} z$ can be identified according to

$$\sec^{-1} z = -i \ln[\{1 + i(z^2-1)^{\frac{1}{2}}\}/z]. \tag{3.7}$$

In view of the higher growth in (3.5) than in (3.4) a first approximation to finding the zeros of the denominator is to make the derivative of the Airy function vanish. Those in the neighbourhood of $\nu = k_1 b$ can be estimated by noting that here $\xi \approx -2^{\frac{1}{3}}(1 - k_1 b/\nu)$. Hence, if ν_s denotes a zero,

$$\operatorname{Ai}'\{-2^{\frac{1}{3}} \nu_s^{\frac{2}{3}} e^{\frac{1}{3}\pi i}(1 - k_1 b/\nu_s)\} = 0,$$

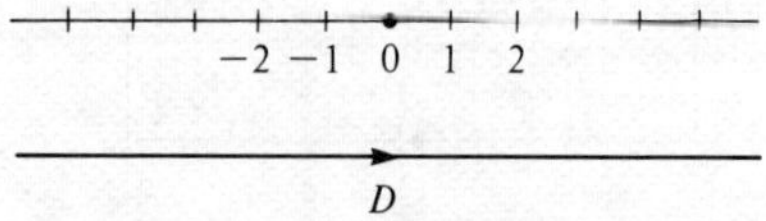

Fig. 8.3 The path of integration in the complex ν-plane

i.e.

$$\nu_s = k_1 b + (\tfrac{1}{2}k_1 b)^{\frac{1}{3}} e^{-\frac{1}{3}\pi i} \beta_s \tag{3.8}$$

where some values pertinent to the zeros $-\beta_s$ of Ai′ are displayed in Table A.6. If ν is not near $k_1 b$ this approximation is not feasible and the requirement is that $\operatorname{ph}(\nu^{\frac{2}{3}} \xi e^{\frac{1}{3}\pi i}) = \pm\pi$. Since $|\nu| \gg k_1 b$, $\frac{2}{3}\xi^{\frac{3}{2}} e^{\frac{3}{2}\pi i} \sim \ln(2\nu/k_1 b)$ (A.2.27) for $|\operatorname{ph}\nu| \leqslant \frac{1}{2}\pi$ and hence the only possibility is $\operatorname{ph}\nu \approx -\frac{1}{2}\pi$. There is a matching set of poles in the second quadrant and so their distribution is somewhat as shown in Fig. 8.4.

It may happen that η is large enough to overthrow the dominance of the derivative at the early poles. In that case, the zeros of Ai will be relevant and

$$\nu_s = k_1 b + (\tfrac{1}{2}k_1 b)^{\frac{1}{3}} e^{-\frac{1}{3}\pi i} \alpha_s \tag{3.9}$$

(see Table A.5) will stand in for (3.8). The more distant poles will continue to originate in the derivative unless η is actually infinite when the zeros of Ai are appropriate; they still lie near the imaginary axis.

Deforming the contour D over the poles in the lower half-plane is now a course of action to be contemplated. With $|\nu| \gg k_1 b$, (A.2.17), (A.2.21), and (A.2.27) show that

$$H_\nu^{(2)}(k_1 b) \sim i(2/\pi\nu)^{\frac{1}{2}} \exp[\nu\{\ln(2\nu/k_1 b) - 1\}] \tag{3.10}$$

for $\frac{1}{2}\pi \geqslant \operatorname{ph}\nu \geqslant -\frac{1}{2}\pi + \delta$ $(\delta > 0)$. If $-\pi \leqslant \operatorname{ph}\nu \leqslant -\frac{1}{2}\pi$, (A.1.16) reveals that the exponential in (3.10) should be replaced by its reciprocal. On the basis of this information it can be confirmed that the proposed deformation is permissible provided that $|\phi| < \frac{1}{2}\pi$.

The calculation of the residue requires the derivative of the denominator with respect to ν. Such a derivative of $H_\nu^{(2)}$ is practically $\{H_{\nu+h}^{(2)} - H_{\nu-h}^{(2)}\}/2h$ for h small. For $|\nu|$ large take $h = 1$ and then (A.1.5) shows that $\partial H_\nu^{(2)}/\partial\nu \sim -H_\nu^{(2)\prime}$. So, if this approximation be adopted, the

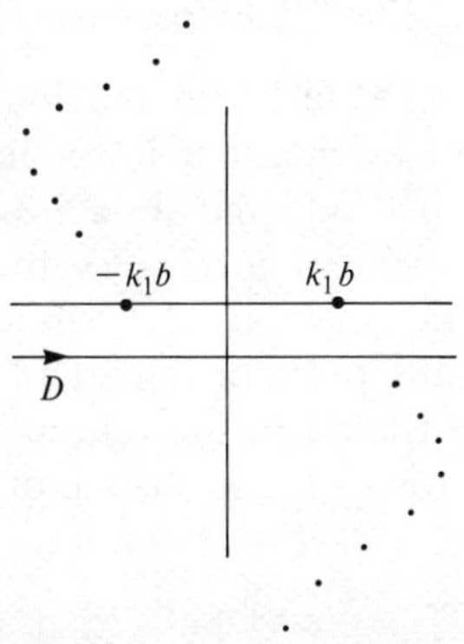

Fig. 8.4 The distribution of poles

second derivative replaced by means of (A.1.1) and then the first derivative expressed in terms of $H_\nu^{(2)}$ through the equation for the zero there results

$$\partial\{\eta H_\nu^{(2)} + k_1 H_\nu^{(2)\prime}\}/\partial\nu \sim -k_1\{b\eta(1-b\eta)-(k_1b)^2+\nu_s^2\}H_{\nu_s}^{(2)}(k_1b)/(k_1b)^2$$

at $\nu = \nu_s$. Since even the smaller zeros have a large negative imaginary part $\sin \nu_s\pi$ can be approximated by $-\frac{1}{2}ie^{i\nu_s\pi}$. Hence

$$u = \sum_{s=1}^{\infty} \frac{8k_1 b e^{-\frac{1}{2}\nu_s\pi i}\cos \nu_s\phi}{\{\nu_s^2-(k_1b)^2+b\eta(1-b\eta)\}H_{\nu_s}^{(2)}(k_1b)}. \tag{3.11}$$

As $s\to\infty$ the behaviour of the terms is essentially $\exp\{-|\text{Im}\,\nu_s|\,(\frac{1}{2}\pi\pm\phi)\}$ so that there is excellent convergence so long as ϕ keeps away from $\pm\frac{1}{2}\pi$. Indeed the exponential decay is preserved down to $s=1$ on account of (3.8) and (3.9) provided that $\frac{1}{2}\pi-|\phi|$ exceeds $(k_1b)^{-\frac{1}{3}}$ sufficiently, say $\frac{1}{2}\pi-|\phi|>(k_1b)^{-\frac{1}{3}+\alpha}$ where α is a small positive number. Under this restriction, each term is exponentially smaller than its predecessor and *the retention of the first term only should be adequate for most purposes.*

The interval $|\phi|<\frac{1}{2}\pi$ corresponds to the back of the cylinder which is not illuminated by the incident wave. The extremes $\phi=\pm\frac{1}{2}\pi$ mark the division between the lit and dark sides. They are points where incident rays are tangent to the perimeter of the cylinder and may be called *points of glancing incidence* or *penumbral points.* What has been established above is that *on the back of the cylinder the field is exponentially small at a short distance of the order of* $(k_1b)^{-\frac{1}{3}}$ *from the penumbral points.* As one moves away from a penumbral point the field falls steadily reaching its lowest value at $\phi=0$ the mid-point of the back. In other words the shadow deepens steeply as one goes round the cylinder from the last point of illumination.

The other side of the cylinder must now be considered. Here (3.11) is of no help and (3.3) must be returned to. Write

$$\operatorname{cosec} \nu\pi = 2ie^{-i\nu\pi} + (\operatorname{cosec} \nu\pi - 2ie^{-i\nu\pi}).$$

The integral with $\operatorname{cosec}\nu\pi - 2ie^{-i\nu\pi}$ in the integrand is now deformed over the poles. The extra attenuation introduced means that the constraint on ϕ can be dropped and the series (3.11) is obtained with the additional factor $e^{-2i\nu_s\pi}$. Even when $\phi=\pm\pi$ the first term is no greater than $e^{-\frac{3}{2}i\nu_1\pi}$ and so this series may be safely discarded without any fear of serious error. The remaining piece of integrand has no poles on the real axis so that D can be translated to the real axis. The negative real axis furnishes a replica apart from a phase change of the positive on changing ν to $-\nu$ and hence

$$u = -\frac{4i}{\pi b}\int_0^\infty \frac{(e^{\frac{3}{2}\nu\pi i}+e^{-\frac{1}{2}\nu\pi i})\cos \nu\phi}{\eta H_\nu^{(2)}(k_1b)+k_1H_\nu^{(2)\prime}(k_1b)}\,d\nu. \tag{3.12}$$

Observe now that (3.4) and (3.5), although derived on the assumption that $|\nu| \gg 1$, are in fact valid on the whole real axis provided that $k_1 b \gg 1$. Moreover, if a small neighbourhood of $\nu = k_1 b$ is excluded, asymptotic expressions can be substituted for the Airy functions. Then the integrand can be inspected for points of stationary phase. It is immediately obvious from (3.10) that there are none in $\nu > k_1 b$. For $\nu < k_1 b$ one exponent under investigation is

$$-\tfrac{1}{2}\nu\pi\mathrm{i} \pm \mathrm{i}\nu\phi + \mathrm{i}(k_1^2 b^2 - \nu^2)^{\frac{1}{2}} - \mathrm{i}\nu \sec^{-1}(k_1 b/\nu).$$

A derivative with respect to ν gives as the equation for stationary phase

$$-\tfrac{1}{2}\pi \pm \phi - \sec^{-1}(k_1 b/\nu) = 0. \tag{3.13}$$

Since $0 < \sec^{-1}(k_1 b/\nu) \leqslant \frac{1}{2}\pi$ there is a stationary point $\nu = k_1 b \sin\phi$ with the upper sign in (3.13) when $\frac{1}{2}\pi < \phi \leqslant \pi$. There is also one at $\nu = -k_1 b \sin\phi$ with the lower sign for $-\pi \leqslant \phi < -\frac{1}{2}\pi$. No stationary points arise from the factor $\mathrm{e}^{\frac{3}{2}\nu\pi\mathrm{i}}$ in the integrand for $|\phi| \leqslant \pi$. In view of the symmetry present it will suffice to concentrate on $\frac{1}{2}\pi < \phi \leqslant \pi$. Then the formula of Appendix G gives

$$u = \frac{2\mathrm{i}k_1 \cos\phi}{\eta + \mathrm{i}k_1 \cos\phi}\,\mathrm{e}^{-\mathrm{i}k_1 b \cos\phi} \tag{3.14}$$

which is precisely the result that ray theory would predict.

The formula (3.14) is invalid as ϕ approaches $\frac{1}{2}\pi$ because then the stationary point is tending to $k_1 b$ and it is illegitimate to employ asymptotic formulae for the Airy functions. The situation at high frequencies then is that on the front of the cylinder the field is that of ray theory and on the back it is not zero as ray theory would suggest but it is exponentially small (Fig. 8.5). In between these regions is a tiny transitional zone of angular dimensions about $(k_1 b)^{-\frac{1}{3}}$ where so far no expression has been derived for the field.

When $\phi \approx \frac{1}{2}\pi$ and the point of stationary phase is near $k_1 b$ experience in previous chapters tells us that an expansion near $\nu = k_1 b$ will be

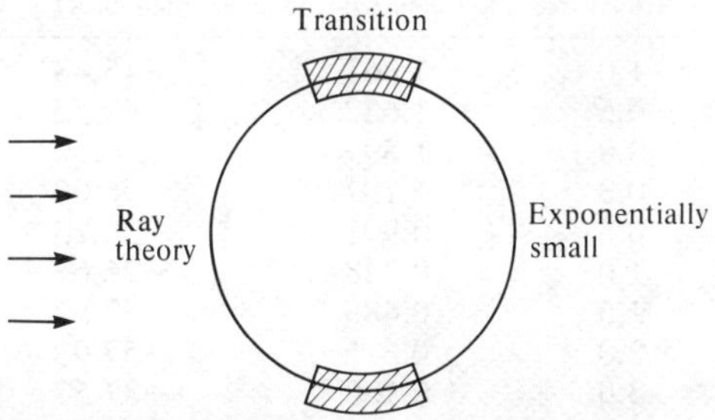

Fig. 8.5 The different regions on the perimeter

effective. Put $\nu^{\frac{2}{3}}\xi = \pm t$ according as $\nu \lessgtr k_1 b$ and then in the exponent of the numerator $\nu \approx k_1 b \pm (\frac{1}{2}k_1 b)^{\frac{1}{3}} t$. Then

$$u = -\frac{\mathrm{i}}{\pi}(\tfrac{1}{2}k_1 b)^{\frac{2}{3}} \mathrm{e}^{\mathrm{i}k_1 b(\phi - \frac{1}{2}\pi)}$$

$$\times \left[\int_0^\infty \frac{\exp\{-\mathrm{i}(\frac{1}{2}k_1 b)^{\frac{1}{3}}(\frac{1}{2}\pi - \phi)\mathrm{e}^{-\frac{2}{3}\pi \mathrm{i}} t\}}{b\eta \operatorname{Ai}(t\mathrm{e}^{\frac{2}{3}\pi \mathrm{i}}) + 2^{\frac{1}{3}}\mathrm{e}^{\frac{1}{3}\pi \mathrm{i}}(k_1 b)^{\frac{2}{3}} \operatorname{Ai}'(t\mathrm{e}^{\frac{2}{3}\pi \mathrm{i}})} \mathrm{d}t \right.$$

$$\left. + \mathrm{e}^{-\frac{1}{3}\pi \mathrm{i}} \int_0^\infty \frac{\exp\{-\mathrm{i}(\frac{1}{2}k_1 b)^{\frac{1}{3}}(\frac{1}{2}\pi - \phi) t\}}{b\eta \operatorname{Ai}(t\mathrm{e}^{-\frac{2}{3}\pi \mathrm{i}}) + 2^{\frac{1}{3}}\mathrm{e}^{\frac{1}{3}\pi \mathrm{i}}(k_1 b)^{\frac{2}{3}} \operatorname{Ai}'(t\mathrm{e}^{-\frac{2}{3}\pi \mathrm{i}})} \mathrm{d}t \right] \qquad (3.15)$$

where, in the first integral, the contour has been turned through $\frac{1}{3}\pi$ to sharpen the convergence.

The hard boundary condition corresponds to $\eta = 0$ and for this

$$u = \mathrm{e}^{\mathrm{i}k_1 b(\phi - \frac{1}{2}\pi)} \mathrm{J}_\mathrm{h}\{(\tfrac{1}{2}k_1 b)^{\frac{1}{3}}(\tfrac{1}{2}\pi - \phi)\} \qquad (3.16)$$

where

$$\mathrm{J}_\mathrm{h}(\zeta) = \frac{\mathrm{i}\mathrm{e}^{\frac{2}{3}\pi \mathrm{i}}}{2\pi} \left\{ \int_0^\infty \frac{\exp(-\mathrm{i}\zeta t \mathrm{e}^{-\frac{2}{3}\pi \mathrm{i}})}{\operatorname{Ai}'(t\mathrm{e}^{\frac{2}{3}\pi \mathrm{i}})} \mathrm{d}t + \mathrm{e}^{-\frac{1}{3}\pi \mathrm{i}} \int_0^\infty \frac{\exp(-\mathrm{i}\zeta t)}{\operatorname{Ai}'(t\mathrm{e}^{-\frac{2}{3}\pi \mathrm{i}})} \mathrm{d}t \right\}. \qquad (3.17)$$

Some values of J_h are listed in Table 8.2.

On the soft boundary u is, of course, zero. But $\partial u/\partial n$ can be obtained by multiplying (3.14) by $-\eta$ and allowing η to tend to infinity. The consequent formula is

$$\partial u/\partial n = (2\mathrm{i}/b)(\tfrac{1}{2}k_1 b)^{\frac{2}{3}} \mathrm{e}^{\mathrm{i}k_1 b(\phi - \frac{1}{2}\pi)} \mathrm{J}_\mathrm{s}\{(\tfrac{1}{2}k_1 b)^{\frac{1}{3}}(\tfrac{1}{2}\pi - \phi)\} \qquad (3.18)$$

where

$$2\pi \mathrm{J}_\mathrm{s}(\zeta) = \int_0^\infty \frac{\exp(-\mathrm{i}\zeta t \mathrm{e}^{-\frac{2}{3}\pi \mathrm{i}})}{\operatorname{Ai}(t\mathrm{e}^{\frac{2}{3}\pi \mathrm{i}})} \mathrm{d}t + \mathrm{e}^{-\frac{1}{3}\pi \mathrm{i}} \int_0^\infty \frac{\exp(-\mathrm{i}\zeta t)}{\operatorname{Ai}(t\mathrm{e}^{-\frac{2}{3}\pi \mathrm{i}})} \mathrm{d}t. \qquad (3.19)$$

A range of values for J_s is displayed in Table 8.3.

Table 8.2 The hard boundary integral

ζ	$\lvert \mathrm{J}_\mathrm{h}(\zeta) \rvert$	ph $\mathrm{J}_\mathrm{h}(\zeta)$
−1.0	1.861	−15.43°
−0.5	1.682	+1.52
0.0	1.399	0.00
0.3	1.197	−6.06
0.6	0.991	−14.23
1.0	0.738	−26.63
1.5	0.488	−42.57
2.0	0.315	−57.98
3.0	0.130	−87.57
4.0	0.054	−116.75

Table 8.3 The normal derivative on the soft boundary

ζ	$\|J_s(\zeta)\|$	ph $J_s(\zeta)$
−1.0	2.16	−25.9°
−0.5	1.38	−16.8
0.0	0.77	−30.0
0.3	0.515	−44.8
0.6	0.327	−62.9
1.0	0.167	−90.1
1.5	0.066	−125.9
2.0	0.025	−161.6
3.0	0.0033	−230.7
4.0	0.00043	−298.0

It can be shown that, when $\zeta > 0$, J_s can be expanded in a series of residues from poles at the zeros of Ai. On account of (3.9) this series matches with (3.11) for a region extending beyond the transition. Contrariwise, when $\zeta < 0$ and $|\zeta|$ is large enough the method of steepest descents (Appendix G) gives $J_s \sim -2\zeta \exp(\frac{1}{3}i\zeta^3)$. This makes $\partial u/\partial n$ agree completely with the ray picture (3.14) over the interval near $\phi = \frac{1}{2}\pi$ where $\cos\phi$ can be represented by the first two terms in its Taylor series about $\frac{1}{2}\pi$.

8.4 The far field at high frequencies

The field at some distance from the cylinder can be calculated from knowledge of u on the perimeter. The three forms of u in the different zones of Fig. 8.5 merit individual consideration.

The field outside the cylinder is given by (§1.36)

$$u = u^i + \tfrac{1}{4}i \int \left\{ \frac{\partial u}{\partial n} H_0^{(2)}(k_1 |\mathbf{x}-\mathbf{y}|) - u \frac{\partial}{\partial n} H_0^{(2)} \right\} ds.$$

At a large distance this becomes, when $\partial u/\partial n$ is inserted from the boundary condition (1.5),

$$u \sim u^i - \tfrac{1}{4}bi\left(\frac{2}{\pi k_1 r}\right)^{\frac{1}{2}} e^{-ik_1 r + \frac{1}{4}\pi i}$$

$$\times \int_{-\pi}^{\pi} u(\alpha)\{\eta + ik_1 \cos(\alpha - \phi)\} e^{ik_1 b \cos(\alpha-\phi)}\, d\alpha. \quad (4.1)$$

For α going from just above $\frac{1}{2}\pi$ to π the formula (3.14) can be quoted. In view of the largeness of $k_1 b$ the method of stationary phase is applicable. The phase is stationary at $\alpha=\frac{1}{2}(\pi+\phi)$ provided that $0<\phi\leqslant\pi$. Hence, from (G.1.4), in $0<\phi\leqslant\pi$

$$u\sim u^{\mathrm{i}}+(R_{\mathrm{F}}D/r^{\frac{1}{2}})\mathrm{e}^{-\mathrm{i}k_1r+2\mathrm{i}k_1b\sin\frac{1}{2}\phi} \tag{4.2}$$

where

$$R_{\mathrm{F}}=-(\eta+\mathrm{i}k_1\sin\tfrac{1}{2}\phi)/(\eta-\mathrm{i}k_1\sin\tfrac{1}{2}\phi), \qquad D=(\tfrac{1}{2}b\sin\tfrac{1}{2}\phi)^{\frac{1}{2}}. \tag{4.3}$$

The second term in (4.2) is the reflected wave as would be calculated by ray theory. And it can be seen from Fig. 8.6 that the point of stationary phase corresponds to the point of reflection of the incoming ray which goes out at the angle ϕ. The factor R_{F} is the Fresnel reflection coefficient for a plane with the specified impedance boundary condition when the angle of incidence is $\frac{1}{2}(\pi-\phi)$, i.e. each portion of the illuminated side of the cylinder behaves locally like a plane tangent to the cylinder. D, on the other hand, is the *divergence factor;* although the rays are initially parallel they diverge from one another after reflection. Since the energy in a ray tube remains constant this divergence causes the amplitude to diminish and the presence of D records this. The final factor contains the amplitude and phase of an ordinary cylindrical wave spreading from the centre but with the phase associated with the path travelled by the incident and reflected rays to the point of observation. The picture loses its validity as $\phi\to 0$ because (3.14) does not hold near $\alpha=\frac{1}{2}\pi$.

The interval from $-\frac{1}{2}\pi$ to $-\pi$ in (4.1) provides, in a similar manner, the ray prediction in $-\pi\leqslant\phi<0$.

Next we turn to the contribution of the back where (3.11) is relevant. Write $\cos\nu_s\alpha=\frac{1}{2}(\mathrm{e}^{\mathrm{i}\nu_s\alpha}+\mathrm{e}^{-\mathrm{i}\nu_s\alpha})$. The term $\mathrm{e}^{\mathrm{i}\nu_s\alpha}$ when in (4.1) has a point of stationary phase near $\alpha=\frac{1}{2}\pi+\phi$ when $-\pi<\phi<0$. However, the second derivative is small and there is another point of stationary phase not far away. So we follow the device of §G.1 of expanding the exponent about

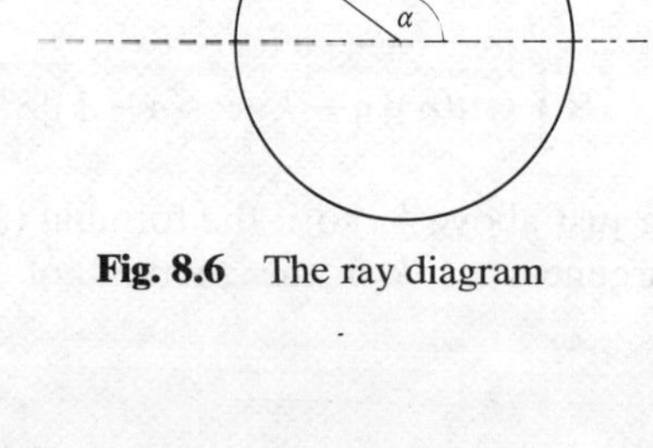

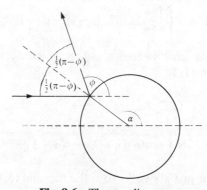

Fig. 8.6 The ray diagram

the point where the second derivative vanishes, i.e. $\alpha = \frac{1}{2}\pi + \phi$. Then, from (G.1.5),

$$\int e^{i\nu_s\alpha + ik_1 b\cos(\alpha-\phi)}\,d\alpha \sim 2\pi\left(\frac{2}{k_1 b}\right)^{\frac{1}{3}} e^{i\nu_s(\phi+\frac{1}{2}\pi)}\,\mathrm{Ai}\left\{\left(\frac{2}{k_1 b}\right)^{\frac{1}{3}}(\nu_s - k_1 b)\right\}.$$

The additional $\cos(\alpha-\phi)$ in the integrand of (4.1) can be accommodated by taking a derivative with respect to this result. The consequent formula can be simplified somewhat by invoking the equation satisfied by ν_s and (A.2.39).

Consequently the contribution of the back in $-\pi<\phi<0$ is

$$u_c = \left(\frac{32}{\pi k_1 r}\right)^{\frac{1}{2}} e^{-ik_1 r - \frac{1}{12}\pi i}$$
$$\times \sum_{s=1} \frac{(\frac{1}{2}k_1 b)^{\frac{4}{3}} e^{i\nu_s\phi}}{\{b\eta(1-b\eta)-(k_1 b)^2+\nu_s^2\}H^{(2)}_{\nu_s}(k_1 b)\,\mathrm{Ai}\left\{(k_1 b-\nu_s)\left(\frac{2}{k_1 b}\right)^{\frac{1}{3}} e^{\frac{1}{3}\pi i}\right\}}. \tag{4.4}$$

The interpretation of a typical term of (4.4) according to ray theory is that a ray arriving at the upper penumbral point initiates a ray which travels round the back of the cylinder. As it moves round it sheds energy in a tangential direction. Its amplitude thereby diminishes at an exponential rate. Once the energy has been dispatched tangentially from the surface the amount in a ray tube remains constant. The diminution in a term of (4.4) is determined by how long the ray stays on the surface before leaving for the point of observation. Other factors represent the change in amplitude which occurs when an incident ray starts creeping round the back. Equation (4.4) is the contribution from the *creeping waves* originating at the upper penumbral; it is not valid as $\phi \to 0$.

Similarly, there are creeping waves from the lower penumbral point (due to the term $e^{-i\nu_s\alpha}$) which shed energy into $\phi>0$.

The contribution from the points of glancing incidence may be calculated in a like manner. There is no point of stationary phase except when

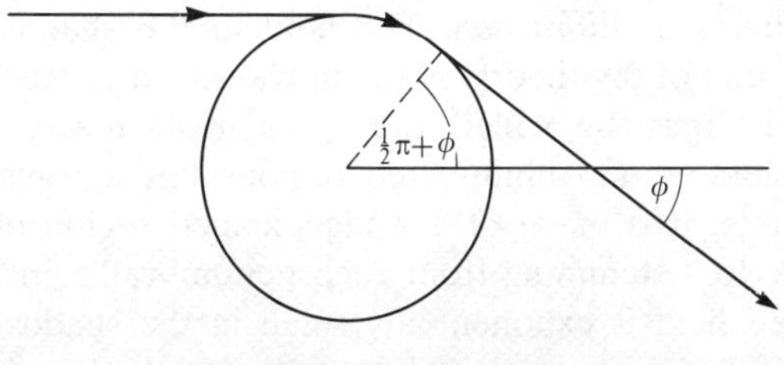

Fig. 8.7 The creeping ray

ϕ is small. So long as $|\phi|$ is not separated from zero by more than about order $(k_1b)^{-\frac{1}{3}}$

$$u_p \sim -\left(\frac{2}{\pi k_1 r}\right)^{\frac{1}{2}} \mathrm{e}^{-\mathrm{i}k_1 r+\frac{1}{4}\pi\mathrm{i}}$$
$$\times[(\sin k_1 b\phi)/\phi - \mathrm{i}(\tfrac{1}{2}k_1 b)^{\frac{1}{3}}\tfrac{1}{2}\{F_1(\phi)+F_1(-\phi)+F_2(\phi)+F_2(-\phi)\}] \tag{4.5}$$

where

$$F_1(\phi)=\mathrm{e}^{\mathrm{i}k_1 b\phi+\frac{1}{6}\pi\mathrm{i}}\int_0^\infty \frac{\{b\eta\,\mathrm{Ai}(t)-2^{\frac{1}{3}}(k_1b)^{\frac{2}{3}}\,\mathrm{Ai}'(t)\}\exp\{\mathrm{i}(\frac{1}{2}k_1b)^{\frac{1}{3}}t\phi\}}{b\eta\,\mathrm{Ai}(t\mathrm{e}^{-\frac{2}{3}\pi\mathrm{i}})+2^{\frac{1}{3}}\mathrm{e}^{\frac{1}{3}\pi\mathrm{i}}(k_1b)^{\frac{2}{3}}\,\mathrm{Ai}'(t\mathrm{e}^{-\frac{2}{3}\pi\mathrm{i}})}\,\mathrm{d}t, \tag{4.6}$$

$$F_2(\phi)=\mathrm{e}^{\mathrm{i}k_1 b\phi+\frac{1}{6}\pi\mathrm{i}}\int_0^\infty \frac{\{b\eta\,\mathrm{Ai}(t)-2^{\frac{1}{3}}(k_1b)^{\frac{2}{3}}\mathrm{e}^{\frac{2}{3}\pi\mathrm{i}}\,\mathrm{Ai}'(t)\}\exp\{\mathrm{i}(\frac{1}{2}k_1b)^{\frac{1}{3}}\mathrm{e}^{-\frac{2}{3}\pi\mathrm{i}}t\phi\}}{b\eta\,\mathrm{Ai}(t\mathrm{e}^{\frac{2}{3}\pi\mathrm{i}})+2^{\frac{1}{3}}\mathrm{e}^{\frac{1}{3}\pi\mathrm{i}}(k_1b)^{\frac{2}{3}}\,\mathrm{Ai}'(t\mathrm{e}^{\frac{2}{3}\pi\mathrm{i}})}\,\mathrm{d}t. \tag{4.7}$$

For the soft boundary condition $F_1(\phi)+F_2(\phi)$ is replaced by $\mathrm{e}^{\mathrm{i}k_1b\phi}[\mathrm{f}_1\{(\frac{1}{2}k_1b)^{\frac{1}{3}}\phi\}+\mathrm{f}_2\{(\frac{1}{2}k_1b)^{\frac{1}{3}}\mathrm{e}^{-\frac{2}{3}\pi\mathrm{i}}\phi\}]$ where

$$\mathrm{f}_1(v)=\mathrm{e}^{\frac{1}{6}\pi\mathrm{i}}\int_0^\infty \frac{\mathrm{Ai}(t)}{\mathrm{Ai}(t\mathrm{e}^{-\frac{2}{3}\pi\mathrm{i}})}\mathrm{e}^{\mathrm{i}tv}\,\mathrm{d}t, \tag{4.8}$$

$$\mathrm{f}_2(v)=\mathrm{e}^{\frac{1}{6}\pi\mathrm{i}}\int_0^\infty \frac{\mathrm{Ai}(t)}{\mathrm{Ai}(t\mathrm{e}^{\frac{2}{3}\pi\mathrm{i}})}\mathrm{e}^{\mathrm{i}tv}\,\mathrm{d}t, \tag{4.9}$$

while for the hard boundary $F_1(\phi)+F_2(\phi)$ is exchanged for $-\mathrm{e}^{\mathrm{i}k_1b\phi}[\mathrm{g}_1\{(\frac{1}{2}k_1b)^{\frac{1}{3}}\phi\}+\mathrm{g}_2\{(\frac{1}{2}k_1b)^{\frac{1}{3}}\mathrm{e}^{-\frac{2}{3}\pi\mathrm{i}}\phi\}]$ where

$$\mathrm{g}_1(v)=\mathrm{e}^{-\frac{1}{6}\pi\mathrm{i}}\int_0^\infty \frac{\mathrm{Ai}'(t)}{\mathrm{Ai}'(t\mathrm{e}^{-\frac{2}{3}\pi\mathrm{i}})}\mathrm{e}^{\mathrm{i}tv}\,\mathrm{d}t, \tag{4.10}$$

$$\mathrm{g}_2(v)=\mathrm{i}\int_0^\infty \frac{\mathrm{Ai}'(t)}{\mathrm{Ai}'(t\mathrm{e}^{\frac{2}{3}\pi\mathrm{i}})}\mathrm{e}^{\mathrm{i}tv}\,\mathrm{d}t. \tag{4.11}$$

Values of $\mathrm{f}_S(v)=\mathrm{f}_1(v)+\mathrm{f}_2(v\mathrm{e}^{-\frac{2}{3}\pi\mathrm{i}})$ and $\mathrm{f}_H(v)=-\mathrm{g}_1(v)-\mathrm{g}_2(v\mathrm{e}^{-\frac{2}{3}\pi\mathrm{i}})$ are set out in Table 8.4.

The position may be summarized as follows. In the illuminated region $\pi \geqslant |\phi| > O\{(k_1b)^{-\frac{1}{3}}\}$ (see Fig. 8.8) the field consists of the primary wave and the reflected wave estimated by conventional ray theory. The remaining field does not come within the scope of conventional ray theory and displays the effects of *diffraction.* The field in the shadow may be put under the umbrella of ray theory by the introduction of creeping rays; this extended theory may be called the *geometrical theory of diffraction.* Between the shadow and illuminated regions lies a transitional region where (4.5) holds; it is made of a wedge-shaped region of angle rather larger than $2(k_1b)^{-\frac{1}{3}}$ stemming from each penumbral point.

The diffracted field is exponentially small in the shadow. One consequence of this is that it is very much less in magnitude than the field behind

Table 8.4 Values of f_S and f_H

v	$\lvert f_S(v)\rvert$	ph $f_S(v)$ (degrees)	$\lvert f_H(v)\rvert$	ph $f_H(v)$ (degrees)
−5.0	0.200	−0.02	0.221	9.71
−4.0	0.250	−0.12	0.264	22.3
−3.0	0.338	+0.12	0.288	45.1
−2.0	0.511	+3.79	0.281	92.0
−1.5	0.648	8.48	0.317	130.6
−1.0	0.823	15.2	0.459	167.8
−0.5	1.030	22.9	0.724	193.7
0	1.254	30.0	1.089	210.0
0.5	1.486	34.2	1.522	218.6
1.0	1.715	32.5	1.985	219.5
1.5	1.945	+21.6	2.44	211.3
2.0	2.21	−1.8	2.80	192.4
3.0	3.13	−93.5	3.16	104.3

a semi-infinite plane (§9.7). Thus, *the shadow behind a large circular cylinder is much more intense than that behind a plane screen of similar dimensions.*

The scattering coefficient from (2.3) and (4.5) is

$$\sigma_c \sim 2+(\tfrac{1}{2}k_1 b)^{-\frac{2}{3}}\,\mathrm{Im}\{F_1(0)+F_2(0)\}. \tag{4.12}$$

Higher order corrections can be calculated and, in particular,

$$\sigma_c^S \sim 2+0.9962(k_1 b)^{-\frac{2}{3}}-0.0224(k_1 b)^{-\frac{4}{3}}, \tag{4.13}$$

$$\sigma_c^H \sim 2-0.8642(k_1 b)^{-\frac{2}{3}}-0.4274(k_1 b)^{-\frac{4}{3}} \tag{4.14}$$

for the soft and hard boundary conditions respectively.

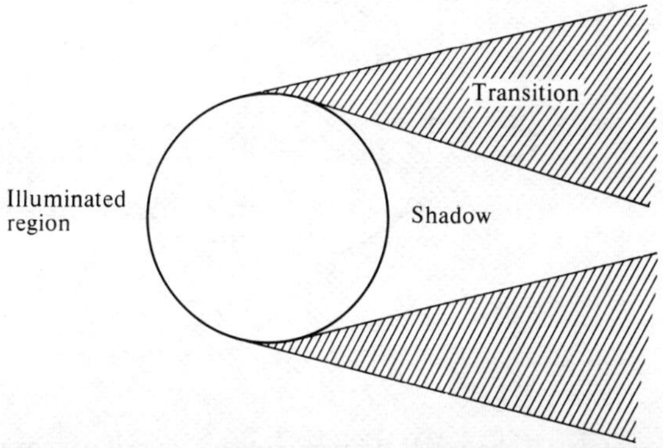

Fig. 8.8 The various domains where different behaviour is exhibited

8.5 The line source

If the excitation of the incident field is a line source at (r_0, π) the governing equation in $r > b$ becomes

$$\frac{1}{r}\frac{\partial}{\partial r}\left(r\frac{\partial u}{\partial r}\right)+\frac{1}{r^2}\frac{\partial^2 u}{\partial \phi^2}+k_1^2 u=-\delta(r-r_0)\delta(\phi-\pi)/r_0. \tag{5.1}$$

A Fourier series expansion of the incident field leads to

$$u^{\mathrm{i}}=-\tfrac{1}{4}\mathrm{i}\sum_{m=-\infty}^{\infty}\mathrm{H}_m^{(2)}(k_1r_0)\mathrm{J}_m(k_1r)\mathrm{e}^{\mathrm{i}m(\phi-\pi)}\quad (r<r_0). \tag{5.2}$$

For $r > r_0$ interchange r and r_0 in the terms of the series.

The only difference from §8.1 is that a typical term has been multiplied by $-\frac{1}{4}\mathrm{i}\mathrm{H}_m^{(2)}(k_1r_0)\mathrm{e}^{-\frac{1}{2}m\pi\mathrm{i}}$. Therefore the scattered field can be determined from the coefficients of §8.1 by a similar multiplication. This can be carried through to the high frequency analysis. For example, (3.11) and (4.4) still hold if each term is multiplied by $-\frac{1}{4}\mathrm{i}\mathrm{H}_{\nu_s}^{(2)}(k_1r_0)\mathrm{e}^{-\frac{1}{2}\nu_s\pi\mathrm{i}}$. The explanation via creeping waves in the shadow (Fig. 8.9) continues to hold good. Conventional ray theory applies in the illuminated region. There remains a transitional zone where the field cannot be split into individual rays. So long as r_0/b differs from 1 by more than $O\{(k_1b)^{-\frac{1}{3}}\}$ the formula (4.5) may be employed provided that an additional multiplying factor of

$$-\tfrac{1}{4}\mathrm{i}(2/\pi k_1r_0)^{\frac{1}{2}}\exp(-\mathrm{i}k_1r_0+\tfrac{1}{4}\pi\mathrm{i})$$

is inserted.

The case when $r_0 = b$ is rather special in that the transitional region

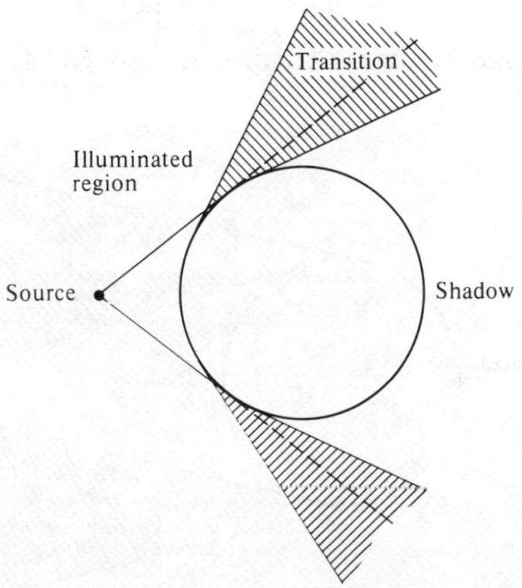

Fig. 8.9 The geometrical theory of diffraction can be used except in the shaded zone

starts from the source. Practically the whole of the circumference is in shadow, only a small interval about the source being excluded. Thus, except in the immediate neighbourhood of the source, the field on the boundary is exponentially small.

8.6 The parabolic cylinder at axial incidence

As a guide to the behaviour of arbitrary obstacles the circular cylinder suffers from two defects—it is symmetrical and has a single radius of curvature. The parabolic cylinder has neither of these deficiencies but is of infinite size. A sort of halfway house is the elliptic cylinder but more progress can be made with its analysis after some familiarity has been gained with the circle and parabola.

The first case to be considered is when a plane wave strikes a parabola when travelling along its axis. The explicit solution then takes a particularly simple form. Parabolic cylinder coordinates ξ, η (§1.39(iv)) will be adopted with

$$x=\tfrac{1}{2}(\xi^2-\eta^2), \qquad y=\xi\eta, \qquad \xi=(2r)^{\frac{1}{2}}\cos\tfrac{1}{2}\phi, \qquad \eta=(2r)^{\frac{1}{2}}\sin\tfrac{1}{2}\phi.$$

The scattering parabola will be specified as $\xi=\xi_0$ so that its equation is

$$-y^2=2\xi_0^2x-\xi_0^4;$$

its axis is the negative x-axis and its nose is at $(\tfrac{1}{2}\xi_0^2, 0)$. The parabola will be assumed to be hard or soft and, since only one medium is involved k will be written for k_1.

On putting $u=e^{ikx}v$, we obtain for the equation for v

$$\partial^2v/\partial x^2+2ik\,\partial v/\partial x+\partial^2v/\partial y^2=0.$$

In terms of ξ and η this becomes

$$\partial^2v/\partial\xi^2+\partial^2v/\partial\eta^2+2ik(\xi\,\partial v/\partial\xi-\eta\,\partial v/\partial\eta)=0.$$

Separable solutions of this equation are feasible. Those which are independent of η satisfy

$$d^2v/d\xi^2+2ik\xi\,dv/d\xi=0.$$

Hence $dv/d\xi=Ae^{-ik\xi^2}$ and

$$v=A\int^{\xi}e^{-ikt^2}\,dt.$$

The incident wave is $u^i=e^{ikx}$ and so $v^i=1$. To satisfy the soft boundary condition a scattered wave v_s is required such that $v_s=-1$ on the cylinder $\xi=\xi_0$. Now, as $\xi\to\infty$,

$$e^{ikx}\int_{\infty}^{\xi}e^{-ikt^2}\,dt\sim-e^{ik(x-\xi^2)}/2ik\xi$$

which is a constant multiple of $e^{-ikr}/r^{\frac{1}{2}}$ and so exhibits the correct outgoing behaviour. Hence $v_s = -A_0 \int_\infty^\xi e^{-ikt^2}\,dt$ where $1/A_0 = \int_\infty^{\xi_0} e^{-ikt^2}\,dt$. Therefore

$$u = e^{ikx}\left(1 - A_0 \int_\infty^\xi e^{-ikt^2}\,dt\right) \tag{6.1}$$

is the solution for the soft boundary.

The hard boundary condition is

$$\partial v/\partial\xi + ik\xi_0 v = 0$$

on $\xi = \xi_0$. The formula (6.1) remains valid but

$$1/A_0 = \int_\infty^{\xi_0} e^{-ikt^2}\,dt + e^{-ik\xi_0^2}/ik\xi_0.$$

It is especially simple to calculate A_0 when $\xi_0 = 0$. The parabolic cylinder then degenerates into the *half-plane* $y = 0$, $x \leqslant 0$. For the soft boundary

$$u = e^{ikx}\left\{1 + 2(k/\pi)^{\frac{1}{2}} e^{\frac{1}{4}\pi i} \int_\infty^\xi e^{-ikt^2}\,dt\right\}$$

while the solution for the hard boundary condition is $u = e^{ikx}$; in the second case the incident wave automatically satisfies the boundary condition and no scattered wave is generated.

8.7 The parabolic cylinder—general incidence

A plane wave falling on the exterior of the parabolic cylinder will now be considered. It is assumed that the direction of propagation makes an angle ϕ^i with the positive x-axis and that $0 < \phi^i < \pi$. The case $\phi^i = \pi$ was discussed in the preceding section and, if $\phi^i = 0$, the incident wave does not impinge on the exterior of the parabola. Only a boundary condition of impedance type will be treated, slightly modified from (1.2) in order to simplify some of the subsequent analysis; the form to be examined is

$$\partial u/\partial n + Zu(\xi_0^2 + \eta^2)^{-\frac{1}{2}} = 0 \tag{7.1}$$

on $\xi = \xi_0$. Transformation to parabolic cylinder coordinates gives

$$\partial u/\partial\xi + Zu = 0. \tag{7.2}$$

The incident wave $u^i = e^{-ikr\cos(\phi - \phi^i)}$ can be expressed in parabolic cylinder coordinates as

$$u^i = \exp[-\tfrac{1}{4}h^2\{(\xi^2 - \eta^2)\cos\phi^i + 2\xi\eta\sin\phi^i\}]$$

where $h = (2k)^{\frac{1}{2}} e^{\frac{1}{4}\pi i}$. The aim is now to represent this by means of the parabolic cylinder function D_ν (see §1.4(iv) and Appendix D). The starting point is *Cherry's formula*

$$-2\mathrm{i}D_0\{h(\xi\cos\tfrac{1}{2}\phi_0+\eta\sin\tfrac{1}{2}\phi_0)\}D_{-1}\{h(\eta\cos\tfrac{1}{2}\phi_0-\xi\sin\tfrac{1}{2}\phi_0)\}$$
$$=\int_{-\frac{1}{2}-\mathrm{i}\infty}^{-\frac{1}{2}+\mathrm{i}\infty}\frac{(\tan\frac{1}{2}\phi_0)^\nu}{\cos\frac{1}{2}\phi_0}D_\nu(-h\xi)D_{-\nu-1}(h\eta)\frac{d\nu}{\sin\nu\pi}. \quad (7.3)$$

From (D.1.9) and (D.1.10)

$$D_0(hx)=e^{-\frac{1}{2}ikx^2},\qquad D_{-1}(hx)=2^{\frac{1}{2}}e^{\frac{1}{4}\pi i-\frac{1}{2}ikx^2}F(k^{\frac{1}{2}}x), \quad (7.4)$$

$$F(x)=e^{ix^2}\int_x^\infty e^{-it^2}\,dt. \quad (7.5)$$

On the basis of (7.3)–(7.5)

$$u^{\mathrm{i}}=\frac{\mathrm{i}}{2(2\pi)^{\frac{1}{2}}}\int_{-\frac{1}{2}-\mathrm{i}\infty}^{-\frac{1}{2}+\mathrm{i}\infty}\left\{\frac{(\tan\frac{1}{2}\phi^{\mathrm{i}})^\nu}{\cos\frac{1}{2}\phi^{\mathrm{i}}}D_\nu(-h\xi)D_{-\nu-1}(h\eta)\right.$$
$$\left.+\frac{(\cot\frac{1}{2}\phi^{\mathrm{i}})^\nu}{\sin\frac{1}{2}\phi^{\mathrm{i}}}D_{-\nu-1}(h\xi)D_\nu(-h\eta)\right\}\frac{d\nu}{\sin\nu\pi}. \quad (7.6)$$

The decomposition in (7.6) is such that the first term in the integrand gives the plane wave plus a scattered wave as $\xi\to\infty$ while the second integrand supplies a cancelling scattered wave. Therefore, the addition of a scattered wave to the first integrand so as to make it comply with the boundary condition (7.2) will provide a representation of the desired field. Accordingly

$$u=\frac{\mathrm{i}}{2(2\pi)^{\frac{1}{2}}}\int_{-\frac{1}{2}-\mathrm{i}\infty}^{-\frac{1}{2}+\mathrm{i}\infty}\frac{(\tan\frac{1}{2}\phi^{\mathrm{i}})^\nu}{\cos\frac{1}{2}\phi^{\mathrm{i}}}$$
$$\times\left\{D_\nu(-h\xi)-\frac{ZD_\nu(-h\xi_0)-hD_\nu'(-h\xi_0)}{ZD_\nu(h\xi_0)+hD_\nu'(h\xi_0)}D_\nu(h\xi)\right\}D_{-\nu-1}(h\eta)\frac{d\nu}{\sin\nu\pi}. \quad (7.7)$$

The parabolic cylinder degenerates into the semi-infinite plane $y=0$, $x\leqslant0$ when $\xi_0=0$. Then (7.7) simplifies and, in the particular case of the hard boundary ($Z=0$), (7.3) and (7.4) give

$$u^{\mathrm{H}}=(1/\pi^{\frac{1}{2}})e^{-ikr+\frac{1}{4}\pi i}[F\{(2kr)^{\frac{1}{2}}\sin\tfrac{1}{2}(\phi-\phi^{\mathrm{i}})\}+F\{(2kr)^{\frac{1}{2}}\sin\tfrac{1}{2}(\phi+\phi^{\mathrm{i}})\}]. \quad (7.8)$$

For the soft boundary condition reverse the sign of the second term. It is noteworthy that for general Z the formula solves the half-plane problem with the impedance condition $x\,\partial u/\partial y=Zu$ on the upper surface and $x\,\partial u/\partial y+Zu=0$ on the lower.

At high frequencies $h\xi_0$ is large in magnitude and it is permissible to employ the uniformly valid asymptotic expansion in (D.1.14) for the parabolic cylinder function. The course of action followed for the circular cylinder may therefore be pursued here.

The field on the cylinder is, from (7.7) and the Wronskian relation (D.1.11),

$$u = \frac{ih}{2\pi}\int_{-\frac{1}{2}-i\infty}^{-\frac{1}{2}+i\infty} \nu!\,\frac{(\tan\frac{1}{2}\phi^{i})^{\nu}}{\cos\frac{1}{2}\phi^{i}}\,\frac{D_{-\nu-1}(h\eta)}{ZD_{\nu}(h\xi_0)+hD'_{\nu}(h\xi_0)}\,d\nu. \tag{7.9}$$

Investigation of the possible saddle-points reveals that when $\eta < -\xi_0 \cot\phi^{i}$ *the field on the boundary agrees with that predicted by ray optics.* This is the part of the cylinder which is illuminated by the incident wave since the point of glancing incidence is $\eta = -\xi_0 \cot\phi^{i}$ (see Fig. 8.10). In the shadow there are creeping waves but where the attenuation for the circle would be $a^{\frac{1}{3}}\phi$, for the parabola it is $\int_A^C ds/\rho^{\frac{2}{3}}$ where A is the penumbral point and C is the point of observation; the integral is taken along the parabola and s is the arc-length from A while ρ is the radius of curvature of the parabola at s $(\rho = (\xi_0^2+\eta^2)^{\frac{3}{2}}/\xi_0)$. The transition between illumination and shadow takes place according to

$$u = \exp\{-ikr\cos(\phi-\phi^{i}) - \tfrac{1}{3}i\zeta^3\}J_h(\zeta) \tag{7.10}$$

for the hard boundary condition; here J_h is defined in (3.17) and

$$\zeta = (k/2\xi_0)^{\frac{1}{3}}(\xi_0\cos\phi^{i} + \eta\sin\phi^{i}).$$

Correspondingly, for the soft boundary,

$$\partial u/\partial\xi = (2k^2\xi_0)^{\frac{1}{3}} i\exp\{-ikr\cos(\phi-\phi^{i}) - \tfrac{1}{3}i\zeta^3\}J_s(\zeta). \tag{7.11}$$

It should be observed that, when η is close to $-\xi_0\cot\phi^{i}$, ζ is interchangeable with $(\frac{1}{2}k)^{\frac{1}{3}}\int_A^{\eta} ds/\rho^{\frac{2}{3}}$ when s and ρ are expressed in terms of η.

The field at large distance may also be estimated in an analogous manner to that for the circle. In the illuminated region

$$u = u^{i} + R_F D\exp\{-ik(x_r\cos\phi^{i} + y_r\sin\phi^{i} + r_F)\}/r_F^{\frac{1}{2}} \tag{7.12}$$

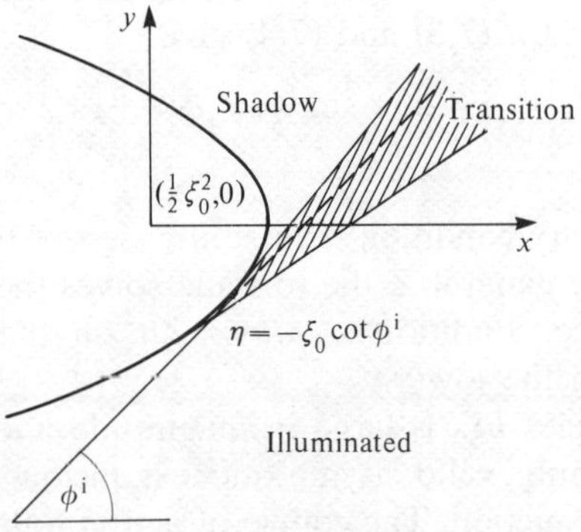

Fig. 8.10 The various zones at high frequencies

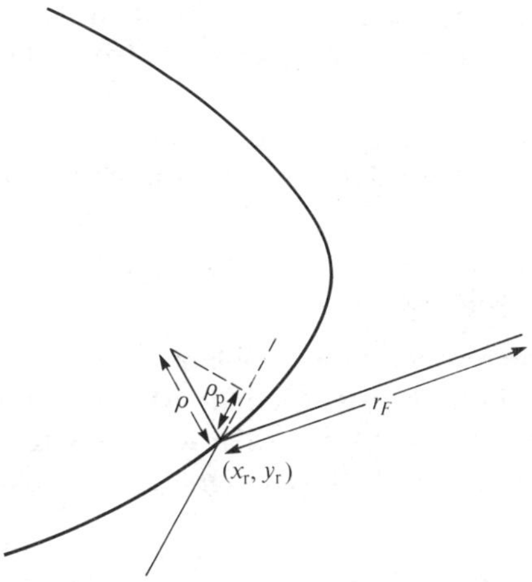

Fig. 8.11 Parameters for reflection

where R_F is the Fresnel coefficient at the point of reflection (x_r, y_r) and r_F the distance of the point of observation from (x_r, y_r) (Fig. 8.11). D is the divergence factor given by

$$D = \{\rho_p r_F/(\rho_p + r_F)\}^{\frac{1}{2}}$$

where ρ_p is the projection on the incident ray of the radius of curvature at the point of reflection. Comparison with (4.2) shows that *reflection from a parabolic cylinder is the same as that from a circular cylinder with cross-section the circle of curvature at the point of reflection.*

The field in the shadow zone can be interpreted in terms of creeping waves with the modifications already indicated for the varying radius of curvature.

There is also a formula for the transition region where $\phi \approx \phi^i$. For the case of the soft boundary

$$\begin{aligned} u \sim {} & (1/2\pi^{\frac{1}{2}})e^{-ikr+\frac{1}{4}\pi i}F\{(2kr)^{\frac{1}{2}}\sin\tfrac{1}{2}(\phi-\phi^i)\} \\ & -(1/2\pi kr)^{\frac{1}{2}}e^{-ikr+\frac{1}{4}\pi i}\{[1-\exp\{-\tfrac{1}{2}ik\xi_0^2(\phi-\phi^i)\operatorname{cosec}\phi^i\}]/i(\phi-\phi^i) \\ & -i(\tfrac{1}{2}k\rho_g)^{\frac{1}{2}}\exp[-\tfrac{1}{2}ik\xi_0^2(\phi-\phi^i)\operatorname{cosec}\phi^i][f_1\{(\tfrac{1}{2}k\rho_g)^{\frac{1}{3}}(\phi^i-\phi)\} \\ & +f_2\{(\tfrac{1}{2}k\rho_g)^{\frac{1}{3}}e^{-\frac{2}{3}\pi i}(\phi^i-\phi)\}] \end{aligned} \qquad (7.13)$$

where $\rho_g = \xi_0^2 \operatorname{cosec}^3\phi^i$ is the radius of curvature of the parabola at the penumbral point.

It should be remarked that *the terms in* (7.13) *are the same as those that would be produced by a circular cylinder whose cross-section coincided with the circle of curvature at the point of glancing incidence provided that only the lower half of the cylinder scattered.* Another point of interest is that the terms excluding f_1 and f_2 are the same as those that are generated by a

Kirchhoff calculation (§9.26) using the incident field over a line, perpendicular to the direction of propagation, from the point of glancing incidence.

The hard boundary condition merely alters f_1 and f_2 to $-g_1$ and $-g_2$ of (4.10), (4.11). The general impedance requires the equivalents of F_1 and F_2.

8.8 The elliptic cylinder

The technique for resolving the problem of the circular cylinder can be adapted to the elliptical cross-section; Mathieu functions are involved instead of Bessel functions. In order to retain the notation of elliptic cylinder coordinates we shall redesignate the unknown as p. Thus a solution of

$$\partial^2 p/\partial x^2 + \partial^2 p/\partial y^2 + k^2 p = -\delta(\boldsymbol{x} - \boldsymbol{x}_0) \tag{8.1}$$

is sought. Only the hard and soft boundaries will be discussed.

Let the major and minor axes of the ellipse be of lengths $2a$ and $2b$ respectively. With these as coordinate axes introduce the elliptic cylinder coordinates u and v where

$$x = l \cosh u \cos v, \qquad y = l \sinh u \sin v, \qquad l^2 = a^2 - b^2.$$

The boundary of the ellipse is then given by $u = U$ where

$$a = l \cosh U, \qquad b = l \sinh U.$$

Let $\boldsymbol{x}_0$ be (u_0, v_0) in the elliptical coordinates.

The primary wave $p^{\mathrm{i}} = -\frac{1}{4}\mathrm{i}H_0^{(2)}(k\,|\boldsymbol{x} - \boldsymbol{x}_0|)$. Hence, from (C.4.18), the expansion in Mathieu functions when $u_0 > u$ is

$$p^{\mathrm{i}} = -\tfrac{1}{2}\mathrm{i} \sum_{m=0}^{\infty} \mathrm{ce}_m(v, h)\mathrm{ce}_m(v_0, h)\mathrm{Mc}_m^{(1)}(u, h)\mathrm{Mc}_m^{(4)}(u_0, h)$$
$$-\tfrac{1}{2}\mathrm{i} \sum_{m=1}^{\infty} \mathrm{se}_m(v, h)\mathrm{se}_m(v_0, h)\mathrm{Ms}_m^{(1)}(u, h)\mathrm{Ms}_m^{(4)}(u_0, h)$$

where $h = \frac{1}{4}k^2 l^2$.

The scattered wave p_{s} can be constructed from outgoing waves. Therefore write

$$p_{\mathrm{s}} = \tfrac{1}{2}\mathrm{i} \sum_{m=0}^{\infty} b_m\, \mathrm{ce}_m(v, h)\mathrm{ce}_m(v_0, h)\mathrm{Mc}_m^{(4)}(u, h)\mathrm{Mc}_m^{(4)}(u_0, h)$$
$$+\tfrac{1}{2}\mathrm{i} \sum_{m=1}^{\infty} c_m\, \mathrm{se}_m(v, h)\mathrm{se}_m(v_0, h)\mathrm{Ms}_m^{(4)}(u, h)\mathrm{Ms}_m^{(4)}(u_0, h).$$

The soft boundary condition $p^{\mathrm{i}} + p_{\mathrm{s}} = 0$ on $u = U$ is satisfied by

$$b_m = \mathrm{Mc}_m^{(1)}(U, h)/\mathrm{Mc}_m^{(4)}(U, h), \qquad c_m = \mathrm{Ms}_m^{(1)}(U, h)/\mathrm{Ms}_m^{(4)}(U, h) \tag{8.2}$$

and the hard boundary condition if

$$b_m = \mathrm{Mc}_m^{(1)\prime}(U, h)/\mathrm{Mc}_m^{(4)\prime}(U, h), \qquad c_m = \mathrm{Ms}_m^{(1)\prime}(U, h)/\mathrm{Ms}_m^{(4)\prime}(U, h). \tag{8.3}$$

Expressions for the incident plane wave $p^{\mathrm{i}} = \exp\{-ik(x \cos \phi^{\mathrm{i}} + y \sin \phi^{\mathrm{i}})\}$ may be deduced directly from these by allowing the source to tend to infinity or from (C.4.17). Thus

$$\begin{aligned} p_{\mathrm{s}} = &-2 \sum_{m=0}^{\infty} (-\mathrm{i})^m b_m \,\mathrm{ce}_m(\phi^{\mathrm{i}}, h)\mathrm{ce}_m(v, h)\mathrm{Mc}_m^{(4)}(u, h) \\ &-2 \sum_{m=1}^{\infty} (-\mathrm{i})^m c_m \,\mathrm{se}_m(\phi^{\mathrm{i}}, h)\mathrm{se}_m(v, h)\mathrm{Ms}_m^{(4)}(u, h) \end{aligned} \tag{8.4}$$

where b_m, c_m are the same as in (8.2) or (8.3) according to which boundary conditions are operative. At a large distance r from the origin $r \sim \frac{1}{2}l\mathrm{e}^u$ so that the asymptotic formulae (C.3.8), (C.3.9) may be quoted; thus

$$\begin{aligned} p_{\mathrm{s}} \sim &-2\left(\frac{2}{\pi kr}\right)^{\frac{1}{2}} \mathrm{e}^{-ikr+\frac{1}{4}\pi\mathrm{i}} \\ &\times \left\{\sum_{m=0}^{\infty} b_m \,\mathrm{ce}_m(\phi^{\mathrm{i}}, h)\mathrm{ce}_m(v, h) + \sum_{m=1}^{\infty} c_m \,\mathrm{se}_m(\phi^{\mathrm{i}}, h)\mathrm{se}_m(v, h)\right\}. \end{aligned} \tag{8.5}$$

When the wavelength is large compared with the ellipse $h \ll 1$. The approximations of §C.4 may then be taken advantage of. For instance, for the soft boundary condition

$$b_0 \approx \tfrac{1}{2}\pi\mathrm{i}/\ln \tfrac{1}{4}k(a+b)$$

and b_m and c_m are $O(h^m)$. Hence

$$\begin{aligned} p_{\mathrm{s}} &\approx -\tfrac{1}{2}\pi\mathrm{i}\,\mathrm{Mc}_0^{(4)}(u, h)/\ln \tfrac{1}{4}k(a+b) \\ &\sim -\mathrm{i}(\pi/2kr)^{\frac{1}{2}}\mathrm{e}^{-ikr+\frac{1}{4}\pi\mathrm{i}}/\ln \tfrac{1}{4}k(a+b) \end{aligned}$$

showing that *the distant scattered field is the same as that produced by a soft circular cylinder of radius* $\frac{1}{2}(a+b)$ (§8.1). The dominant coefficients for the hard boundary are

$$b_0 \approx -\tfrac{1}{4}\mathrm{i}\pi k^2 ab, \qquad b_1 \approx \tfrac{1}{8}\mathrm{i}\pi k^2 b(a+b), \qquad c_1 \approx \tfrac{1}{8}\mathrm{i}\pi k^2 a(a+b).$$

The scattering coefficient for the plane wave incidence can be calculated from the theory of §8.2. By virtue of the orthogonal relations of §C.1 the direct definition gives

$$\sigma_{\mathrm{c}} = \frac{4/kl}{(\cosh^2 U - \cos^2\phi^{\mathrm{i}})^{\frac{1}{2}}}\left[\sum_{m=0}^{\infty} |b_m|^2 \{\mathrm{ce}_m(\phi^{\mathrm{i}}, h)\}^2 + \sum_{m=1}^{\infty} |c_m|^2 \{\mathrm{se}_m(\phi^{\mathrm{i}}, h)\}^2\right] \tag{8.6}$$

while (2.3) provides

$$\sigma_c = \frac{4/kl}{(\cosh^2 U - \cos^2\phi^i)^{\frac{1}{2}}} \operatorname{Re}\left[\sum_{m=0}^{\infty} b_m\{\mathrm{ce}_m(\phi^i, h)\}^2 + \sum_{m=1}^{\infty} c_m\{\mathrm{se}_m(\phi^i, h)\}^2\right]. \tag{8.7}$$

It can be verified from (8.2) and (8.3) that Re $b_m = |b_m|^2$ and Re $c_m = |c_m|^2$ so that (8.6) and (8.7) are consistent. Actually (8.6) and (8.7) are valid whatever boundary conditions are imposed provided that p_s is expressed in the form (8.4).

The coefficients just evaluated enable saying that at low frequencies

$$\sigma_c^S \approx \pi^2(a^2\sin^2\phi^i + b^2\cos^2\phi^i)^{-\frac{1}{2}}/2k \ln^2 \tfrac{1}{4}k(a+b), \tag{8.8}$$

$$\sigma_c^H \approx \tfrac{1}{32}\pi^2 k^3 a^2 b^2 (a^2\sin^2\phi^i + b^2\cos^2\phi^i)^{-\frac{1}{2}} + \tfrac{1}{16}\pi^2 k^3 (a+b)^2 (a^2\sin^2\phi^i + b^2\cos^2\phi^i)^{\frac{1}{2}}. \tag{8.9}$$

Identical results can be achieved by (8.7) but the coefficients have to be expanded to higher order.

As for the circle the series solution is unsuitable for high frequencies. One may attempt to transform it by the Poisson summation formula (§6.32) or try an integral representation along the lines adopted for the circle. In view of the results obtained for the circle and parabola it is plausible to expect illuminated, shadow, and transition zones. The illuminated region will be covered by ray theory. In the shadow the field is exponentially small and caused by the creeping waves of the geometrical theory of diffraction. Near a shadow boundary a penumbral point will produce a transitional field as if it were a circle of radius equal to the radius of curvature of the ellipse at that point. No detailed verification of this behaviour will be undertaken.

8.9 Inhomogeneous cylinders

The complication introduced by the properties of the cylinder varying from point to point is substantial. All the difficulties encountered in §6.13 are present and sometimes they can be overcome by the same devices. The equation to be solved, if the cylinder is circular, is

$$\frac{1}{r}\frac{\partial}{\partial r}\left(r\frac{\partial p}{\partial r}\right) + \frac{1}{r^2}\frac{\partial^2 p}{\partial\phi^2} + k^2 p = 0. \tag{9.1}$$

When k is a function of r only solution of (9.1) may be feasible. For example, if $k^2 = C^2 r^{a+b} - a^2/4r^2$, one possibility is

$$p = \mathscr{L}_\nu(\gamma r^{1+\frac{1}{2}(a+b)})e^{in\phi} \tag{9.2}$$

where $(a+b+2)\nu = \pm(a^2+4n^2)^{\frac{1}{2}}$, $(a+b+2)\gamma = 2C$ and $\mathscr{L}_\nu$ is a Bessel function of order ν.

The electromagnetic field can still be split into components depending on E_z and H_z alone but E_z and H_z no longer satisfy (9.1). For instance

$$\frac{1}{r}\frac{\partial}{\partial r}\left(r\frac{\partial E_z}{\partial r}\right)+\frac{1}{r^2}\frac{\partial^2 E_z}{\partial \phi^2}-\frac{1}{\mu}\,\text{grad}\,\mu\cdot\text{grad}\,E_z+\omega^2\mu\varepsilon E_z=0 \qquad (9.3)$$

while H_z satisfies the same equation with ε in place of μ. If μ is a function of r only, (9.3) can be rewritten as

$$\frac{1}{r}\frac{\partial}{\partial r}\left(r\frac{\partial u}{\partial r}\right)+\frac{1}{r^2}\frac{\partial^2 u}{\partial \phi^2}+\left\{\omega^2\mu\varepsilon+\frac{\mu'}{2\mu r}+\frac{1}{4\mu^2}(2\mu\mu''-3\mu'^2)\right\}u=0 \quad (9.4)$$

where $u=E_z/\mu^{\frac{1}{2}}$. Equation (9.4) has the same structure as (9.1). Thus, if $\mu=Ar^a$ and $\varepsilon=Br^b$, a solution can be derived from (9.2).

Even if μ is a function of both r and ϕ the substitution $u=E_z/\mu^{\frac{1}{2}}$ leads to an equation of the type (9.1) but its usefulness would vary with circumstances. Should the medium be quasi-homogeneous no transformation of (9.3) is necessary because grad μ can then be neglected.

Three-dimensional scattering problems

The transfer from two to three dimensions brings in a feature which distinguishes the electromagnetic field from the acoustic. The fields depending on E_z and H_z can no longer be treated separately, in general. A scattered field will usually involve both elements, even if the incident field does not, because the boundary conditions will normally concern both. Consequently, the scattered field will exhibit polarization effects. Moreover, the shape of the boundary may be inconvenient for a representation in terms of E_z and H_z.

8.10 Scattering by infinitely long cylinders

The simplest three-dimensional scattering problem is really quasi-two-dimensional. It consists of a plane wave falling obliquely on an infinitely long perfectly reflecting cylinder. Whatever the cross-section of the cylinder the solution can be found if that for incidence perpendicular to the axis of the cylinder is known.

Let $u^{\mathrm{S}}(x, y, k)$ be the solution of

$$\partial^2 u/\partial x^2+\partial^2 u/\partial y^2+k^2 u=0,$$

the z-axis being parallel to a generator of the cylindrical surface, which is produced by the plane wave $u^{\mathrm{i}}(x, y, k)=\exp\{-ik(x\cos\phi^{\mathrm{i}}+y\sin\phi^{\mathrm{i}})\}$ under the soft boundary condition $u=0$. Let $u^{\mathrm{H}}(x, y, k)$ be the corresponding total field for the hard boundary where $\partial u/\partial n=0$.

The obliquely incident plane wave is

$$v^{\mathrm{i}} = \exp\{-\mathrm{i}k(x \sin\theta^{\mathrm{i}} \cos\phi^{\mathrm{i}} + y \sin\theta^{\mathrm{i}} \sin\phi^{\mathrm{i}} + z\cos\theta^{\mathrm{i}})\}$$
$$= u^{\mathrm{i}}(x, y, k\sin\theta^{\mathrm{i}})\exp(-\mathrm{i}kz\cos\theta^{\mathrm{i}}).$$

Clearly, the fields v^{S}, v^{H} satisfying the soft and hard boundary conditions on the cylinder are

$$v^{\mathrm{S}} = u^{\mathrm{S}}(x, y, k\sin\theta^{\mathrm{i}})\exp(-\mathrm{i}kz\cos\theta^{\mathrm{i}}), \tag{10.1}$$
$$v^{\mathrm{H}} = u^{\mathrm{H}}(x, y, k\sin\theta^{\mathrm{i}})\exp(-\mathrm{i}kz\cos\theta^{\mathrm{i}}). \tag{10.2}$$

It will now be shown that the electromagnetic field can also be expressed in terms of v^{S} and v^{H} when the cylinder is perfectly conducting. Let the electric intensity in the plane wave be

$$\boldsymbol{E}^{\mathrm{i}} = (l\boldsymbol{i} + m\boldsymbol{j} + n\boldsymbol{k})v^{\mathrm{i}}$$

where $\boldsymbol{i}$, $\boldsymbol{j}$, $\boldsymbol{k}$ are unit vectors parallel to the Cartesian axes. In order that $\boldsymbol{E}^{\mathrm{i}}$ be transverse to the direction of propagation it is necessary that

$$(l\cos\phi^{\mathrm{i}} + m\sin\phi^{\mathrm{i}})\sin\theta^{\mathrm{i}} + n\cos\theta^{\mathrm{i}} = 0. \tag{10.3}$$

Write $\boldsymbol{E}^{\mathrm{i}} = \boldsymbol{E}_{\mathrm{S}} + \boldsymbol{E}_{\mathrm{H}}$ where

$$\boldsymbol{E}_{\mathrm{S}} = n\cot\theta^{\mathrm{i}}(-\boldsymbol{i}\cos\phi^{\mathrm{i}} - \boldsymbol{j}\sin\phi^{\mathrm{i}} + \boldsymbol{k}\tan\theta^{\mathrm{i}})v^{\mathrm{i}},$$
$$\boldsymbol{E}_{\mathrm{H}} = (l\sin\phi^{\mathrm{i}} - m\cos\phi^{\mathrm{i}})(\boldsymbol{i}\sin\phi^{\mathrm{i}} - \boldsymbol{j}\cos\phi^{\mathrm{i}})v^{\mathrm{i}}.$$

Let $\boldsymbol{B}_{\mathrm{S}}$, $\boldsymbol{B}_{\mathrm{H}}$ be the corresponding magnetic flux densities. The incident wave is now composed of the linear combination of two plane waves $\boldsymbol{E}_{\mathrm{S}}$, $\boldsymbol{B}_{\mathrm{S}}$ and $\boldsymbol{E}_{\mathrm{H}}$, $\boldsymbol{B}_{\mathrm{H}}$ each of which may be handled separately.

Consider the electromagnetic field

$$\boldsymbol{E}^{\mathrm{S}} = n\,\mathrm{cosec}^2\theta^{\mathrm{i}}(k^2\boldsymbol{k} - \mathrm{i}k\cos\theta^{\mathrm{i}}\,\mathrm{grad})v^{\mathrm{S}}/k^2, \tag{10.4}$$
$$\boldsymbol{B}^{\mathrm{S}} = -(\mathrm{i}n/\omega)\mathrm{cosec}^2\theta^{\mathrm{i}}\boldsymbol{k}\wedge\mathrm{grad}\,v^{\mathrm{S}}. \tag{10.5}$$

It satisfies Maxwell's equations and, because of the behaviour of v^{S}, consists of the plane wave $\boldsymbol{E}_{\mathrm{S}}$, $\boldsymbol{B}_{\mathrm{S}}$ plus a scattered wave at infinity. The tangential component of $\boldsymbol{E}^{\mathrm{S}}$ is zero on the cylinder because $v^{\mathrm{S}} = 0$ there. Hence $\boldsymbol{E}^{\mathrm{S}}$, $\boldsymbol{B}^{\mathrm{S}}$ is the total field excited by the incident wave $\boldsymbol{E}_{\mathrm{S}}$, $\boldsymbol{B}_{\mathrm{S}}$.

Similarly

$$\boldsymbol{E}^{\mathrm{H}} = \mathrm{i}(l\sin\phi^{\mathrm{i}} - m\cos\phi^{\mathrm{i}})\boldsymbol{k}\wedge\mathrm{grad}\,v^{\mathrm{H}}/k\sin\theta^{\mathrm{i}}, \tag{10.6}$$
$$\boldsymbol{B}^{\mathrm{H}} = (m\cos\phi^{\mathrm{i}} - l\sin\phi^{\mathrm{i}})(k\boldsymbol{k} - \mathrm{i}\cos\theta^{\mathrm{i}}\,\mathrm{grad})v^{\mathrm{H}}/\omega\sin\theta \tag{10.7}$$

is the field initiated by $\boldsymbol{E}_{\mathrm{H}}$, $\boldsymbol{B}_{\mathrm{H}}$.

It now follows that when the wave $\boldsymbol{E}^{\mathrm{i}}$, $\boldsymbol{B}^{\mathrm{i}}$ is incident on the cylinder the total field is $\boldsymbol{E}^{\mathrm{S}} + \boldsymbol{E}^{\mathrm{H}}$, $\boldsymbol{B}^{\mathrm{S}} + \boldsymbol{B}^{\mathrm{H}}$ as set out in (10.4)–(10.7).

Different boundary conditions may entail a more complex procedure because, although the incident wave can again be split into the elements

$\boldsymbol{E}_S$ and $\boldsymbol{E}_H$, the same division of the total field may be prevented by the boundary conditions.

8.11 Spherical waves

Before turning to specific scattering problems it is desirable to say something about solutions of

$$\nabla^2 p + k^2 p = 0 \tag{11.1}$$

in spherical polar coordinates R, θ, ϕ. Assuming that p is continuous on a sphere, we can expand it in a series of Legendre functions (Appendix B) via

$$p = \sum_{n=0}^{\infty} \sum_{m=-n}^{n} a_{mn}(R) e^{im\phi} P_n^{|m|}(\cos\theta) \tag{11.2}$$

where, on account of the orthogonality relations (B.3.20),

$$a_{mn}(R) = \frac{(n-|m|)!}{(n+|m|)!} \frac{2n+1}{4\pi} \int_0^{2\pi} \int_0^{\pi} p e^{-im\phi} P_n^{|m|}(\cos\theta) \sin\theta \, d\theta \, d\phi. \tag{11.3}$$

By multiplying (11.1) by $e^{-im\phi} P_n^{|m|}$ and integrating over the spherical surface we deduce that a_{mn} is a spherical Bessel function $\mathfrak{z}_n^{(j)}(kR)$ in so far as its dependence on R is concerned. The choice of j varies according to the sort of field under consideration.

The plane wave $e^{-ikR\cos\theta}$ satisfies (11.1) and is finite at the origin. Since it is independent of ϕ the appropriate expansion is

$$e^{-ikR\cos\theta} = \sum_{n=0}^{\infty} c_n j_n(kR) P_n(\cos\theta).$$

On account of (11.3)

$$c_n j_n(kR) = \tfrac{1}{2}(2n+1) \int_0^{\pi} e^{-ikR\cos\theta} P_n(\cos\theta) \sin\theta \, d\theta.$$

Take a derivative n times with respect to R, put $R = 0$ and use (B.3.22) together with

$$[d^n j_n(\rho)/d\rho^n]_{\rho=0} = (n!)^2 2^n/(2n+1)!. \tag{11.4}$$

There results $c_n = (2n+1)(-i)^n$. Hence

$$e^{-ikR\cos\theta} = \sum_{n=0}^{\infty} (2n+1)(-i)^n j_n(kR) P_n(\cos\theta) \tag{11.5}$$

and

$$\int_0^{\pi} e^{-ikR\cos\theta} P_n(\cos\theta) \sin\theta \, d\theta = 2(-i)^n j_n(kR). \tag{11.6}$$

Now rotate the axes so that the direction of propagation no longer coincides with the z-axis but instead is in the direction with spherical polar angles θ^{i} and ϕ^{i}. Let Θ be the angle between this direction and the radius vector to the point of observation. Then

$$\cos \Theta = \sin \theta \sin \theta^{\mathrm{i}} \cos(\phi - \phi^{\mathrm{i}}) + \cos \theta \cos \theta^{\mathrm{i}}. \tag{11.7}$$

Putting $\theta = \Theta$ in (11.5) and invoking (B.3.24) we derive

$$\mathrm{e}^{-\mathrm{i}kR\cos\Theta} = \sum_{n=0}^{\infty} (2n+1)(-\mathrm{i})^n \mathrm{j}_n(kR)\Big\{\mathrm{P}_n(\cos\theta)\mathrm{P}_n(\cos\theta^{\mathrm{i}}) + 2\sum_{m=1}^{n} \frac{(n-m)!}{(n+m)!} \mathrm{P}_n^m(\cos\theta)\mathrm{P}_n^m(\cos\theta^{\mathrm{i}})\cos m(\phi-\phi^{\mathrm{i}})\Big\}. \tag{11.8}$$

By virtue of (11.3) and (11.8)

$$\int_0^{2\pi}\int_0^{\pi} \mathrm{e}^{-\mathrm{i}kR\cos\Theta}\mathrm{P}_n^m(\cos\theta)\cos m(\phi-\phi^{\mathrm{i}})\sin\theta\,\mathrm{d}\theta\,\mathrm{d}\phi = 4\pi(-\mathrm{i})^n \mathrm{P}_n^m(\cos\theta^{\mathrm{i}})\mathrm{j}_n(kR). \tag{11.9}$$

Let $R_1 = (R^2 + R_0^2 - 2RR_0\cos\theta)^{\frac{1}{2}}$ where $R < R_0$. Then $\mathrm{e}^{-\mathrm{i}R_1}/R_1$ satisfies (11.1) with $k = 1$ and is finite at $R = 0$. Hence

$$\mathrm{h}_0^{(2)}(R_1) = \mathrm{i}\mathrm{e}^{-\mathrm{i}R_1}/R_1 = \sum_{n=0}^{\infty} d_n \mathrm{j}_n(R)\mathrm{P}_n(\cos\theta)$$

where

$$d_n \mathrm{j}_n(R) = \tfrac{1}{2}(2n+1)\int_0^{\pi} \mathrm{h}_0^{(2)}(R_1)\mathrm{P}_n(\cos\theta)\sin\theta\,\mathrm{d}\theta. \tag{11.10}$$

From (A.4.8) and (A.4.9)

$$\mathrm{h}_0^{(2)}(R_1) = \sum_{m=0}^{\infty} (R^2 - 2RR_0\cos\theta)^m(-)^m \mathrm{h}_m^{(2)}(R_0)/m!\,(2R_0)^m$$

for sufficiently small R. Since P_n is orthogonal to all powers of $\cos\theta$ less than n the first term in this expansion which makes a contribution to the right-hand side of (11.10) is that in which $m = n$. Divide both sides by R^n and put $R = 0$. Then, from (11.4) and (B.3.22), $d_n = (2n+1)\mathrm{h}_n^{(2)}(R_0)$. It follows that

$$\mathrm{h}_0^{(2)}(kR_1) = \mathrm{i}\mathrm{e}^{-\mathrm{i}kR_1}/kR_1 = \sum_{n=0}^{\infty}(2n+1)\mathrm{h}_n^{(2)}(kR_0)\mathrm{j}_n(kR)\mathrm{P}_n(\cos\theta) \quad (R<R_0). \tag{11.11}$$

Since R_1 is symmetrical in R and R_0, it is only ncccssary to reverse the roles of R and R_0 in (11.11) when $R > R_0$.

One conclusion from (11.10) or (11.11) is that

$$\int_0^{\pi} \mathrm{h}_0^{(2)}(kR_1)\mathrm{P}_n(\cos\theta)\sin\theta\,\mathrm{d}\theta = 2\mathrm{h}_n^{(2)}(kR_0)\mathrm{j}_n(kR) \quad (R<R_0). \tag{11.12}$$

By replacing θ by Θ and using (B.3.24) we can modify (11.11) into a form analogous to (11.8). From this can be deduced

$$\int_0^{2\pi}\int_0^{\pi} \mathrm{h}_0^{(2)}(kR_2)\mathrm{P}_n^m(\cos\theta)\sin\theta\, \mathrm{e}^{\pm \mathrm{i}m\phi}\,\mathrm{d}\theta\,\mathrm{d}\phi$$

$$= 4\pi \mathrm{e}^{\pm \mathrm{i}m\phi'}\mathrm{P}_n^m(\cos\theta')\mathrm{h}_n^{(2)}(kR_0)\mathrm{j}_n(kR) \quad (R<R_0,\ m\geqslant 0) \quad (11.13)$$

where $R_2=(R^2+R_0^2-2RR_0\cos\Theta)^{\frac{1}{2}}$.

8.12 Acoustic scattering by a sphere

This section is devoted to finding the Green's function of a sphere. According to §1.32 the field caused when the sphere is subject to the radiation from any distribution of sources can then be written down.

Let the medium outside the sphere be homogeneous with wave number k_1 and let the interior have the constant wave number k_2. The origin will be selected at the centre of the sphere which is of radius b. Then a solution of

$$\nabla^2 G(\boldsymbol{x}, \boldsymbol{x}_0)+k^2 G(\boldsymbol{x}, \boldsymbol{x}_0) = -\delta(\boldsymbol{x}-\boldsymbol{x}_0) \quad (12.1)$$

is needed such that G is continuous through the surface of the sphere while the exterior normal derivative is ζ times the interior (cf. (1.2)). In (12.1) k stands for k_1 outside the sphere and k_2 inside. Furthermore G must be an outgoing wave at infinity. The related problem of an impedance boundary condition can also be tackled by the same procedure. The spherical polar coordinates of $\boldsymbol{x}_0$ are taken as R_0, θ_0, ϕ_0.

It will be supposed that $R_0>b$. Then the primary wave due to the source in (12.1) is $\mathrm{e}^{-\mathrm{i}k_1R_2}/4\pi R_2$ where R_2 is defined at the end of the preceding section. Hence, from (11.11),

$$G^{\mathrm{i}}(\boldsymbol{x}, \boldsymbol{x}_0) = \mathrm{e}^{-\mathrm{i}k_1R_2}/4\pi R_2 = -\mathrm{i}k_1\mathrm{h}_0^{(2)}(k_1R_2)/4\pi$$

$$= \sum_{n=0}^{\infty}\sum_{m=0}^{n} a_{mn}\mathrm{h}_n^{(2)}(k_1R_0)\mathrm{j}_n(k_1R)\mathrm{P}_n^m(\cos\theta)\mathrm{P}_n^m(\cos\theta_0)\cos m(\phi-\phi_0)$$

for $R<R_0$, where

$$a_{mn} = (n-m)!(2n+1)(2-\delta_{m0})k_1/(n+m)!4\pi\mathrm{i}.$$

Now assume an exterior radiating field

$$G_1(\boldsymbol{x}, \boldsymbol{x}_0) = \sum_{n=0}^{\infty}\sum_{m=0}^{n} b_{mn}\mathrm{h}_n^{(2)}(k_1R_0)\mathrm{h}_n^{(2)}(k_1R)\mathrm{P}_n^m(\cos\theta)\mathrm{P}_n^m(\cos\theta_0)\cos m(\phi-\phi_0)$$

and an interior field

$$G_2(\boldsymbol{x}, \boldsymbol{x}_0) = \sum_{n=0}^{\infty}\sum_{m=0}^{n} c_{mn}\mathrm{h}_n^{(2)}(k_1R_0)\mathrm{j}_n(k_2R)\mathrm{P}_n^m(\cos\theta)\mathrm{P}_n^m(\cos\theta_0)\cos m(\phi-\phi_0).$$

The requisite continuity on the boundary demands

$$a_{mn}\mathrm{j}_n(k_1b)+b_{mn}\mathrm{h}_n^{(2)}(k_1b)=c_{mn}\mathrm{j}_n(k_2b)$$

whereas the change in normal derivative gives

$$a_{mn}\mathrm{j}_n'(k_1b)+b_{mn}\mathrm{h}_n^{(2)\prime}(k_1b)=\zeta k_2c_{mn}\mathrm{j}_n'(k_2b)/k_1.$$

Hence

$$b_{mn}=\frac{\zeta k_2\mathrm{j}_n(k_1b)\mathrm{j}_n'(k_2b)-k_1\mathrm{j}_n'(k_1b)\mathrm{j}_n(k_2b)}{k_1\mathrm{h}_n^{(2)\prime}(k_1b)\mathrm{j}_n(k_2b)-\zeta k_2\mathrm{h}_n^{(2)}(k_1b)\mathrm{j}_n'(k_2b)}a_{mn}, \tag{12.2}$$

$$c_{mn}=-\mathrm{i}a_{mn}/k_1b^2\{k_1\mathrm{h}_n^{(2)\prime}(k_1b)\mathrm{j}_n(k_2b)-\zeta k_2\mathrm{h}_n^{(2)}(k_1b)\mathrm{j}_n'(k_2b)\} \tag{12.3}$$

after simplification of the numerator of c_{mn} by the Wronskian (A.4.11).

For the impedance boundary condition $\partial u/\partial n+\eta u=0$, replace b_{mn} by B_{mn} where

$$B_{mn}=-\frac{k_1\mathrm{j}_n'(k_1b)+\eta\mathrm{j}_n(k_1b)}{k_1\mathrm{h}_n^{(2)\prime}(k_1b)+\eta\mathrm{h}_n^{(2)}(k_1b)}a_{mn}. \tag{12.4}$$

The Green's function is now determined by $G^{\mathrm{i}}+G_1$ in $R\geqslant b$ and G_2 in $R\leqslant b$.

The scattering due to a plane wave travelling in the direction of θ^{i}, ϕ^{i} can be deduced by allowing $R_0\to\infty$ or directly from (11.8). Thus

$$p=\sum_{n=0}^{\infty}\sum_{m=0}^{n}\{a_{mn}\mathrm{j}_n(k_1R)+b_{mn}\mathrm{h}_n^{(2)}(k_1R)\}\mathrm{P}_n^m(\cos\theta)\mathrm{P}_n^m(\cos\theta^{\mathrm{i}})\cos m(\phi-\phi^{\mathrm{i}})$$

$$(R\geqslant b), \tag{12.5}$$

$$=\sum_{n=0}^{\infty}\sum_{m=0}^{n}c_{mn}\mathrm{j}_n(k_2R)\mathrm{P}_n^m(\cos\theta)\mathrm{P}_n^m(\cos\theta^{\mathrm{i}})\cos m(\phi-\phi^{\mathrm{i}})\quad(R\leqslant b) \tag{12.6}$$

where now $a_{mn}=(n-m)!(2n+1)(2-\delta_{m0})(-\mathrm{i})^n/(n+m)!$ and b_{mn}, c_{mn} are still given by (12.2) and (12.3). Change b_{mn} to B_{mn} for the impedance boundary condition.

8.13 Spherical electromagnetic waves

It is straightforward to check that

$$\boldsymbol{E}=\operatorname{grad}\frac{\partial}{\partial R}(RP)-\boldsymbol{x}\mu\varepsilon\frac{\partial^2P}{\partial t^2}-\operatorname{curl}\left(\frac{\partial M}{\partial t}\boldsymbol{x}\right), \tag{13.1}$$

$$\boldsymbol{B}=\mu\varepsilon\operatorname{curl}\left(\frac{\partial P}{\partial t}\boldsymbol{x}\right)+\operatorname{curl}\operatorname{curl}(M\boldsymbol{x}) \tag{13.2}$$

satisfy the time dependent form of Maxwell's equations in a homogeneous medium when

$$\nabla^2P-\mu\varepsilon\,\partial^2P/\partial t^2=0,\qquad\nabla^2M-\mu\varepsilon\,\partial^2M/\partial t^2=0. \tag{13.3}$$

Conversely, it can be shown that *every field can be expressed in the form* (13.1), (13.2) *at points unoccupied by charges or currents.* For harmonic waves the components in spherical polar coordinates are

$$\left.\begin{aligned} E_R &= k^2RP+\frac{\partial^2}{\partial R^2}(RP), \qquad E_\theta = \frac{1}{R}\frac{\partial^2}{\partial R\,\partial\theta}(RP)-\frac{\mathrm{i}\omega}{\sin\theta}\frac{\partial M}{\partial\phi}, \\ E_\phi &= \frac{1}{R\sin\theta}\frac{\partial^2}{\partial R\,\partial\phi}(RP)+\mathrm{i}\omega\frac{\partial M}{\partial\theta}, \end{aligned}\right\} \tag{13.4}$$

$$\left.\begin{aligned} B_R &= \frac{\partial^2}{\partial R^2}(RM)+k^2RM, \qquad B_\theta = \frac{1}{R}\frac{\partial^2}{\partial R\,\partial\theta}(RM)+\frac{\mathrm{i}k^2}{\omega\sin\theta}\frac{\partial P}{\partial\phi}, \\ B_\phi &= \frac{1}{R\sin\theta}\frac{\partial^2}{\partial R\,\partial\phi}(RM)-\frac{\mathrm{i}k^2}{\omega}\frac{\partial P}{\partial\theta} \end{aligned}\right\} \tag{13.5}$$

where now

$$\nabla^2P+k^2P=0, \qquad \nabla^2M+k^2M=0. \tag{13.6}$$

It will be observed that waves in which $P=0$ have $E_R=0$ so that the electric intensity has no radial component; they may be called *transverse electric waves.* Similarly, $M=0$ supplies *transverse magnetic waves.*

It must be emphasized that (13.4) and (13.5) are generally true only in regions where there are no sources. Thus the field due to an electric dipole situated away from the origin *cannot usually be expressed in terms of P and M alone in any region which includes the dipole.* An exception to this rule occurs when the dipole is directed along the radius vector from the origin; if the dipole is at $\boldsymbol{x}_0$

$$P=A\exp(-\mathrm{i}k(\boldsymbol{x}-\boldsymbol{x}_0)/4\pi\varepsilon R_0\,|\boldsymbol{x}-\boldsymbol{x}_0|, \qquad M=0.$$

Likewise, for the radial magnetic dipole

$$P=0, \qquad M=B\exp(-\mathrm{i}k\,|\boldsymbol{x}-\boldsymbol{x}_0|)/4\pi R_0\,|\boldsymbol{x}-\boldsymbol{x}_0|.$$

The plane wave, propagating along the z-axis, in which the only nonzero components are $E_x=\mathrm{e}^{-\mathrm{i}kz}$, $B_y=kE_x/\omega$ has

$$P=\mathrm{e}^{-\mathrm{i}kR\cos\theta}\cos\phi/k^2R\sin\theta, \qquad M=\mathrm{e}^{-\mathrm{i}kR\cos\theta}\sin\phi/\omega kR\sin\theta.$$

On account of (13.6) the building blocks in manufacturing electromagnetic waves in spherical coordinates are $p_{mn}^{(j)}$ where

$$p_{mn}^{(j)}=\mathfrak{z}_n^{(j)}(kR)\mathrm{P}_n^{|m|}(\cos\theta)\mathrm{e}^{\mathrm{i}m\phi}.$$

The corresponding vectors which arise are

$$\boldsymbol{p}_{mn}^{(j)}=\operatorname{curl}(p_{mn}^{(j)}\boldsymbol{x}), \qquad k\boldsymbol{q}_{mn}^{(j)}=\operatorname{curl}\operatorname{curl}(p_{mn}^{(j)}\boldsymbol{x}). \tag{13.7}$$

Note that

$$\operatorname{curl}\boldsymbol{p}_{mn}^{(j)}=k\boldsymbol{q}_{mn}^{(j)}, \qquad \operatorname{curl}\boldsymbol{q}_{mn}^{(j)}=k\boldsymbol{p}_{mn}^{(j)} \tag{13.8}$$

since $p^{(j)}_{mn}$ satisfies (13.6). Explicit formulae are

$$\boldsymbol{p}^{(j)}_{mn}=\left\{\frac{im}{\sin\theta}\mathrm{P}^{|m|}_n(\cos\theta)\boldsymbol{i}_2-\frac{\mathrm{d}}{\mathrm{d}\theta}\mathrm{P}^{|m|}_n(\cos\theta)\boldsymbol{i}_3\right\}\mathfrak{z}^{(j)}_n(kR)\mathrm{e}^{im\phi}, \tag{13.9}$$

$$\begin{aligned}k\boldsymbol{q}^{(j)}_{mn}=\Bigg[&\frac{n(n+1)}{R}\mathfrak{z}^{(j)}_n(kR)\mathrm{P}^{|m|}_n(\cos\theta)\boldsymbol{i}_1\\&+\frac{1}{R}\frac{\mathrm{d}}{\mathrm{d}R}\{R\mathfrak{z}^{(j)}_n(kR)\}\frac{\mathrm{d}}{\mathrm{d}\theta}\mathrm{P}^{|m|}_n(\cos\theta)\boldsymbol{i}_2\\&+\frac{im}{R\sin\theta}\frac{\mathrm{d}}{\mathrm{d}R}\{R\mathfrak{z}^{(j)}_n(kR)\}\mathrm{P}^{|m|}_n(\cos\theta)\boldsymbol{i}_3\Bigg]\mathrm{e}^{im\phi}\end{aligned} \tag{13.10}$$

where $\boldsymbol{i}_1$, $\boldsymbol{i}_2$, $\boldsymbol{i}_3$ are unit vectors in the directions R-, θ-, and ϕ-increasing respectively. Notice that $\boldsymbol{q}^{(j)}_{mn}$ has a radial component but $\boldsymbol{p}^{(j)}_{mn}$ does not.

Since P and M can be represented as infinite series of $p^{(j)}_{mn}$ it follows from (13.1)–(13.5) that any electromagnetic field can be written in the form

$$\boldsymbol{E}=\sum_{n=0}^{\infty}\sum_{m=-n}^{n}(a_{mn}\boldsymbol{q}^{(j)}_{mn}+b_{mn}\boldsymbol{p}^{(j)}_{mn}), \tag{13.11}$$

$$\boldsymbol{B}=(\mathrm{i}k/\omega)\sum_{n=0}^{\infty}\sum_{m=-n}^{n}(a_{mn}\boldsymbol{p}^{(j)}_{mn}+b_{mn}\boldsymbol{q}^{(j)}_{mn}) \tag{13.12}$$

subject to the restrictions already mentioned. Naturally, the choice of j varies with the type of field under consideration.

The determination of the coefficients a_{mn} and b_{mn} can be based on vector relations analogous to (11.3). These are derived from (13.9), (13.10), (B.3.20), (B.3.21), (B.3.10), (B.3.13) and are

$$\int_0^{2\pi}\int_0^{\pi}\boldsymbol{p}^{(j)}_{mn}\cdot\boldsymbol{p}^{(j)}_{-s,t}\sin\theta\,\mathrm{d}\theta\,\mathrm{d}\phi=\delta_{ms}\delta_{nt}P^{(j)}_{mn}, \tag{13.13}$$

$$\int_0^{2\pi}\int_0^{\pi}\boldsymbol{q}^{(j)}_{mn}\cdot\boldsymbol{q}^{(j)}_{-s,t}\sin\theta\,\mathrm{d}\theta\,\mathrm{d}\phi=\delta_{ms}\delta_{nt}Q^{(j)}_{mn}, \tag{13.14}$$

$$\int_0^{2\pi}\int_0^{\pi}\boldsymbol{p}^{(j)}_{mn}\cdot\boldsymbol{q}^{(j)}_{st}\sin\theta\,\mathrm{d}\theta\,\mathrm{d}\phi=0 \tag{13.15}$$

where

$$P^{(j)}_{mn}=\frac{(n+|m|)!}{(n-|m|)!}\frac{4\pi n(n+1)}{2n+1}\{\mathfrak{z}^{(j)}_n(kR)\}^2, \tag{13.16}$$

$$Q^{(j)}_{mn}=\frac{(n+|m|)!}{(n-|m|)!}\frac{4\pi n(n+1)}{(2n+1)k^2R^2}\left[n(n+1)\{\mathfrak{z}^{(j)}_n(kR)\}^2+\left\{\frac{\mathrm{d}}{\mathrm{d}R}(R\mathfrak{z}^{(j)}_n)\right\}^2\right]. \tag{13.17}$$

Thus

$$a_{mn} = \frac{1}{Q_{mn}^{(j)}} \int_0^{2\pi} \int_0^{\pi} \boldsymbol{E} \cdot \boldsymbol{q}_{-m,n}^{(j)} \sin\theta \, d\theta \, d\phi$$

$$= \frac{\omega}{ikP_{mn}^{(j)}} \int_0^{2\pi} \int_0^{\pi} \boldsymbol{B} \cdot \boldsymbol{p}_{-m,n}^{(j)} \sin\theta \, d\theta \, d\phi, \tag{13.18}$$

$$b_{mn} = \frac{1}{P_{mn}^{(j)}} \int_0^{2\pi} \int_0^{\pi} \boldsymbol{E} \cdot \boldsymbol{p}_{-m,n}^{(j)} \sin\theta \, d\theta \, d\phi$$

$$= \frac{\omega}{ikQ_{mn}^{(j)}} \int_0^{2\pi} \int_0^{\pi} \boldsymbol{B} \cdot \boldsymbol{q}_{-m,n}^{(j)} \sin\theta \, d\theta \, d\phi. \tag{13.19}$$

8.14 The plane wave

The coefficients in the expansion for the plane wave

$$\boldsymbol{E} = \boldsymbol{i} e^{-ikz}, \qquad \boldsymbol{B} = (k/\omega) \boldsymbol{j} e^{-ikz} \tag{14.1}$$

can be obtained from the preceding theory. According to (13.19) and (13.9)

$$b_{mn} = \frac{1}{P_{mn}^{(j)}} \int_0^{2\pi} \int_0^{\pi} \left\{ -im \cot\theta \cos\phi \, P_n^{|m|}(\cos\theta) + \sin\theta \frac{d}{d\theta} P_n^{|m|}(\cos\theta) \right\}$$
$$\times \mathfrak{z}_n^{(j)}(kR) e^{-im\phi - ikR\cos\theta} \sin\theta \, d\theta \, d\phi.$$

The integral with respect to ϕ disappears unless $m = \pm 1$. Also P_0^1 vanishes identically so that the only surviving coefficients are $b_{\pm 1,n}$ with $n > 0$. When $m = 1$, the derivative of the Legendre function may be transformed by (B.3.15) and then an application of (B.3.17) gives

$$b_{1n} = -\frac{\pi i}{P_{1n}^{(j)}} (n+1) \int_0^{\pi} \{\cos\theta \, P_n^1(\cos\theta) - P_{n-1}^1(\cos\theta)\} \mathfrak{z}_n^{(j)}(kR) e^{-ikR\cos\theta} \, d\theta$$

$$= \frac{\pi i}{P_{1n}^{(j)}} n(n+1) \mathfrak{z}_n^{(j)}(kR) \int_0^{\pi} P_n(\cos\theta) \sin\theta \, e^{-ikR\cos\theta} \, d\theta.$$

Invoking (11.6) and (13.16) we have

$$b_{1n} = -\frac{(2n+1)(-i)^{n+1} j_n(kR)}{2n(n+1) \mathfrak{z}_n^{(j)}(kR)}.$$

In order that b_{1n} shall be independent of R it is necessary to choose $j = 1$ and then

$$b_{1n} = -(2n+1)(-i)^{n+1}/2n(n+1).$$

There is no difficulty in seeing that $b_{-1,n} = -b_{1n}$; further (13.18) furnishes

$a_{1n}=a_{-1,n}=-b_{1n}$. Hence the representation of the plane wave (14.1) is

$$\boldsymbol{E}=\sum_{n=1}^{\infty}\frac{(2n+1)(-\mathrm{i})^{n+1}}{2n(n+1)}\{\boldsymbol{q}_{1n}^{(1)}+\boldsymbol{q}_{-1,n}^{(1)}+\boldsymbol{p}_{-1,n}^{(1)}-\boldsymbol{p}_{1n}^{(1)}\},\tag{14.2}$$

$$\boldsymbol{B}=(\mathrm{i}k/\omega)\sum_{n=1}^{\infty}\frac{(2n+1)(-\mathrm{i})^{n+1}}{2n(n+1)}\{\boldsymbol{p}_{1n}^{(1)}+\boldsymbol{p}_{-1,n}^{(1)}+\boldsymbol{q}_{-1,n}^{(1)}-\boldsymbol{q}_{1n}^{(1)}\}.\tag{14.3}$$

8.15 The dipole expansion

Another important representation is that for an arbitrarily oriented electric dipole placed at $\boldsymbol{x}_0$ or (R_0, θ_0, ϕ_0) in spherical polar coordinates. The field of such a dipole satisfies

$$\operatorname{curl}\boldsymbol{E}+\mathrm{i}\omega\mu\boldsymbol{H}=\boldsymbol{0},\qquad \operatorname{curl}\boldsymbol{H}-\mathrm{i}\omega\varepsilon\boldsymbol{E}=\mathrm{i}\omega A\boldsymbol{p}_0\delta(\boldsymbol{x}-\boldsymbol{x}_0)$$

where A and $\boldsymbol{p}_0$ are invariable. According to §3.8

$$\boldsymbol{E}=\operatorname{grad}\operatorname{div}\boldsymbol{\Pi}+k^2\boldsymbol{\Pi},\qquad \boldsymbol{H}=\mathrm{i}\omega\varepsilon\operatorname{curl}\boldsymbol{\Pi}$$

where $\boldsymbol{\Pi}=A\boldsymbol{p}_0\exp(-\mathrm{i}k|\boldsymbol{x}-\boldsymbol{x}_0|)/4\pi\varepsilon|\boldsymbol{x}-\boldsymbol{x}_0|$. The vector $\boldsymbol{\Pi}$ involves $\boldsymbol{x}$ and $\boldsymbol{x}_0$ only in the combination $\boldsymbol{x}-\boldsymbol{x}_0$ and so

$$\boldsymbol{H}=-\mathrm{i}\omega\varepsilon\operatorname{curl}_0\boldsymbol{\Pi}_0$$

where the suffix 0 on the curve indicates derivatives with respect to $\boldsymbol{x}_0$. It follows from (F.7.6) that

$$\boldsymbol{H}\cdot\boldsymbol{p}_{-m,n}^{(j)}=-\mathrm{i}\omega\varepsilon\operatorname{div}_0(\boldsymbol{\Pi}\wedge\boldsymbol{p}_{-m,n}^{(j)})$$

since $\boldsymbol{p}_{-m,n}$ is independent of $\boldsymbol{x}_0$.

Insert this result in (13.18) and then

$$\begin{aligned}a_{mn}&=\frac{-kA}{4\pi\varepsilon P_{mn}^{(j)}}\operatorname{div}_0\left\{\boldsymbol{p}_0\wedge\int_0^{2\pi}\int_0^{\pi}\boldsymbol{p}_{-m,n}^{(j)}\frac{\exp(-\mathrm{i}k|\boldsymbol{x}-\boldsymbol{x}_0|)}{|\boldsymbol{x}-\boldsymbol{x}_0|}\sin\theta\,\mathrm{d}\theta\,\mathrm{d}\phi\right.\\&=\frac{\mathrm{i}k^2A\mathfrak{z}_n^{(j)}(kR)}{4\pi\varepsilon P_{mn}^{(j)}}\operatorname{div}_0\left[\boldsymbol{p}_0\wedge\int_0^{2\pi}\int_0^{\pi}\right.\\&\quad\times\left\{\left(-\mathrm{i}m\cos\theta\cos\phi P_n^{|m|}+\sin\theta\sin\phi\frac{\mathrm{d}P_n^{|m|}}{\mathrm{d}\theta}\right)\boldsymbol{i}\right.\\&\quad-\left.\left(\mathrm{i}m\cos\theta\sin\phi P_n^{|m|}+\cos\phi\sin\theta\frac{\mathrm{d}P_n^{|m|}}{\mathrm{d}\theta}\right)\boldsymbol{j}+\mathrm{i}m\sin\theta P_n^{|m|}\boldsymbol{k}\right\}\\&\quad\times\mathrm{e}^{-\mathrm{i}m\phi}\mathrm{h}_0^{(2)}(k|\boldsymbol{x}-\boldsymbol{x}_0|)\,\mathrm{d}\theta\,\mathrm{d}\phi,\end{aligned}$$

where P_β^α denotes $\mathrm{P}_\beta^\alpha(\cos\theta)$, from (13.9) and (11.11). When $m\geqslant 0$ the derivatives of the Legendre functions can be eliminated by combining (B.3.15) and (B.3.17) on the one hand, and (B.3.16) and (B.3.18) on the other. Hence

$$a_{mn} = \frac{ik^2 A \mathfrak{z}_n^{(j)}(kR)}{4\pi\varepsilon P_{mn}^{(j)}} \operatorname{div}_0 \left[\boldsymbol{p}_0 \wedge \int_0^{2\pi} \int_0^{\pi} \{\tfrac{1}{2}(\mathrm{i}\boldsymbol{i} - \boldsymbol{j})e^{-\mathrm{i}\phi} P_n^{m+1} \right.$$
$$+ \tfrac{1}{2}(\mathrm{i}\boldsymbol{i} + \boldsymbol{j})e^{\mathrm{i}\phi}(n+m)(n-m+1)P_n^{m-1} + \mathrm{i}mP_n^m \boldsymbol{k}\}$$
$$\left. \times e^{-\mathrm{i}m\phi} h_0^{(2)}(k\,|\boldsymbol{x} - \boldsymbol{x}_0|) \sin\theta \, d\theta \, d\phi \right].$$

It is now evident from (11.13) that, if $\boldsymbol{p}_{mn(0)}^{(j)}$ is $\boldsymbol{p}_{mn}^{(j)}$ when R, θ, ϕ are replaced by R_0, θ_0, ϕ_0 respectively,

$$a_{mn} = \frac{ik^2 A}{\varepsilon P_{mn}^{(j)}} \mathfrak{z}_n^{(j)}(kR) \operatorname{div}_0 \{\boldsymbol{p}_0 \wedge \boldsymbol{p}_{-m,n(0)}^{(4)} \mathrm{j}_n(kR)\} \quad (R_0 > R).$$

Obviously, it is necessary to select $j = 1$ and then, from (13.16) and (13.8),

$$a_{mn} = \frac{(n-|m|)!(2n+1)(-\mathrm{i}k^3 A)}{(n+|m|)!4\pi\varepsilon n(n+1)} \boldsymbol{p}_0 \cdot \boldsymbol{q}_{-m,n(0)}^{(4)} \quad (R_0 > R). \tag{15.1}$$

For $R_0 < R$ the only difference to (15.1) is that $\boldsymbol{q}_{-m,n(0)}^{(4)}$ is replaced by $q_{-m,n(0)}^{(1)}$. It may be confirmed that the same formula holds when $m < 0$.

It may be shown in a similar way that b_{mn} is also given by (15.1) provided that $\boldsymbol{p}_{-m,n(0)}^{(4)}$ is substituted for $\boldsymbol{q}_{-m,n(0)}^{(4)}$. Employing the dyadic notation of §F.6 we have

$$\boldsymbol{E} = \sum_{n=1}^{\infty} \sum_{m=-n}^{n} -(\mathrm{i}k^3 A/\varepsilon) d_{mn} \boldsymbol{p}_0 \cdot \{\boldsymbol{q}_{-m,n(0)}^{(4)} \boldsymbol{q}_{mn}^{(1)} + \boldsymbol{p}_{-m,n(0)}^{(4)} \boldsymbol{p}_{mn}^{(1)}\} \quad (R_0 > R), \tag{15.2}$$

$$\boldsymbol{B} = \sum_{n=1}^{\infty} \sum_{m=-n}^{n} (k^4 A/\omega\varepsilon) d_{mn} \boldsymbol{p}_0 \cdot \{\boldsymbol{q}_{-m,n(0)}^{(4)} \boldsymbol{p}_{mn}^{(1)} + \boldsymbol{p}_{-m,n(0)}^{(4)} \boldsymbol{q}_{mn}^{(1)}\} \quad (R_0 > R) \tag{15.3}$$

where

$$d_{mn} = \frac{(n-|m|)!(2n+1)}{(n+|m|)!4\pi n(n+1)}. \tag{15.4}$$

For $R_0 < R$ the superscripts (4) and (1) are interchanged.

When $\boldsymbol{p}_0 = \boldsymbol{i}$, $\theta_0 = \pi$ and $\phi_0 = 0$ the field, when multiplied by $4\pi\varepsilon R_0 e^{\mathrm{i}kR_0}/Ak^2$, goes over to the plane wave of §8.14 as $R_0 \to \infty$. That the expansions of (15.2), (15.3) become (14.2), (14.3) can be seen from (A.4.13) and (B.3.13).

A comparison of (15.2) and (15.3) with the formulae of §1.33 reveals that the Green's tensor $\boldsymbol{\Gamma}(\boldsymbol{x}, \boldsymbol{x}_0)$ which satisfies

$$\operatorname{curl}\operatorname{curl}\boldsymbol{\Gamma} - k^2\boldsymbol{\Gamma} = -\boldsymbol{I}\delta(\boldsymbol{x} - \boldsymbol{x}_0)$$

everywhere, $\boldsymbol{I}$ being the unit tensor, is given by

$$\boldsymbol{\Gamma}(\boldsymbol{x}_0, \boldsymbol{x}) = \mathrm{i}k \sum_{n=1}^{\infty} \sum_{m=-n}^{n} d_{mn}\{\boldsymbol{q}^{(4)}_{-m,n(0)}\boldsymbol{q}^{(1)}_{mn} + \boldsymbol{p}^{(4)}_{-m,n(0)}\boldsymbol{p}^{(1)}_{mn}\} \quad (R_0 > R). \quad (15.5)$$

The superscripts (4) and (1) are interchanged when $R_0 < R$.

8.16 Electromagnetic scattering by a sphere

The field scattered by a sphere subject to the radiation from a current distribution can be calculated by means of § 1.33 as soon as an appropriate Green's tensor has been determined. The determination will now be undertaken when the sphere is homogeneous with permittivity ε_2 and permeability μ_2. The surrounding medium is assumed to be homogeneous with permittivity ε_1 and permeability μ_1. Let the radius of the sphere be a and select its centre as origin. Only the case when the singularity of the Green's tensor is outside the sphere will be discussed.

The Green's tensor is required to satisfy

$$\text{curl curl}\, \boldsymbol{\Gamma}(\boldsymbol{x}, \boldsymbol{y}) - k_1^2 \boldsymbol{\Gamma}(\boldsymbol{x}, \boldsymbol{y}) = -\boldsymbol{I}\delta(\boldsymbol{x} - \boldsymbol{y}) \quad (R > a), \quad (16.1)$$

$$\text{curl curl}\, \boldsymbol{\Gamma}(\boldsymbol{x}, \boldsymbol{y}) - k_2^2 \boldsymbol{\Gamma}(\boldsymbol{x}, \boldsymbol{y}) = \boldsymbol{0} \quad (16.2)$$

with $|\boldsymbol{y}| > a$. The singularity can be dealt with by taking advantage of (15.5). To indicate the various variables write $\boldsymbol{p}^{(j)}_{mn}(k_1, \boldsymbol{x})$ to signify (13.9) expressed in terms of k_1 and $\boldsymbol{x}$ with a similar notation for $\boldsymbol{q}^{(j)}_{mn}$. Then

$$\boldsymbol{\Gamma}^{\mathrm{i}}(\boldsymbol{x}, \boldsymbol{y}) = \mathrm{i}k_1 \sum_{n=1}^{\infty} \sum_{m=-n}^{n} d_{mn}\{\boldsymbol{q}^{(1)}_{-m,n}(k_1, \boldsymbol{x})\boldsymbol{q}^{(4)}_{mn}(k_1, \boldsymbol{y}) + \boldsymbol{p}^{(1)}_{-m,n}(k_1, \boldsymbol{x})\boldsymbol{p}^{(4)}_{mn}(k_1, \boldsymbol{y})\} \quad (16.3)$$

accounts for the right-hand side of (16.1) when $|\boldsymbol{x}| < |\boldsymbol{y}|$. However, it does not satisfy (16.2) and the question of boundary conditions has also to be settled.

Let the sources of the current density $\boldsymbol{J}$ of the electromagnetic field occupy the volume T_0 outside the sphere. According to §1.34 and (33.7) of §1.33, for $|\boldsymbol{x}| > a$,

$$\boldsymbol{E}(\boldsymbol{x}) = \mathrm{i}\omega\mu_1 \int_{T_0} \boldsymbol{J}(\boldsymbol{y}) \cdot \boldsymbol{\Gamma}(\boldsymbol{y}, \boldsymbol{x})\, \mathrm{d}\boldsymbol{y} - \int_S \{(\boldsymbol{n}_y \wedge \text{curl}_y \boldsymbol{E}) \cdot \boldsymbol{\Gamma}(\boldsymbol{y}, \boldsymbol{x}) + (\boldsymbol{n}_y \wedge \boldsymbol{E}) \cdot \text{curl}_y \boldsymbol{\Gamma}(\boldsymbol{y}, \boldsymbol{x})\}\, \mathrm{d}S_y$$

where S is the surface of the sphere and $\boldsymbol{n}$ is the unit normal to S out of the sphere. Across the surface of the sphere $\boldsymbol{n} \wedge \boldsymbol{E}$ and $\boldsymbol{n} \wedge \boldsymbol{H}$ are continuous. Therefore, if $\boldsymbol{n}_y \wedge \mu \boldsymbol{\Gamma}(\boldsymbol{y}, \boldsymbol{x})$ and $\boldsymbol{n}_y \wedge \text{curl}_y \boldsymbol{\Gamma}(\boldsymbol{y}, \boldsymbol{x})$ are continuous as $\boldsymbol{y}$ moves through S, the surface integral can be moved inside the sphere and will vanish via the divergence theorem since $|\boldsymbol{x}|>a$. Hence, if $\boldsymbol{\Gamma}_{\mathrm{e}}(\boldsymbol{x}, \mathbf{y})$ satisfies the boundary conditions

$$\mu_1[\boldsymbol{n}\wedge\boldsymbol{\Gamma}_{\mathrm{e}}(\boldsymbol{x}, \mathbf{y})]_{R=a+0}=\mu_2[\boldsymbol{n}\wedge\boldsymbol{\Gamma}_{\mathrm{e}}(\boldsymbol{x}, \mathbf{y})]_{R=a-0}, \tag{16.4}$$

$$[\boldsymbol{n}\wedge\operatorname{curl}\boldsymbol{\Gamma}_{\mathrm{e}}(\boldsymbol{x}, \mathbf{y})]_{R=a+0}=[\boldsymbol{n}\wedge\operatorname{curl}\boldsymbol{\Gamma}_{\mathrm{e}}(\boldsymbol{x}, \mathbf{y})]_{R=a-0} \tag{16.5}$$

the electric intensity outside the sphere is given by

$$\boldsymbol{E}(\boldsymbol{x})=\mathrm{i}\omega\mu_1\int_{T_0}\boldsymbol{J}(\mathbf{y})\cdot\boldsymbol{\Gamma}_{\mathrm{e}}(\mathbf{y}, \boldsymbol{x})\,\mathrm{d}\mathbf{y}. \tag{16.6}$$

Similarly, if $\boldsymbol{\Gamma}_{\mathrm{h}}(\boldsymbol{x}, \mathbf{y})$ satisfies the boundary conditions

$$\varepsilon_1[\boldsymbol{n}\wedge\boldsymbol{\Gamma}_{\mathrm{h}}(\boldsymbol{x}, \mathbf{y})]_{R=a+0}=\varepsilon_2[\boldsymbol{n}\wedge\boldsymbol{\Gamma}_{\mathrm{h}}(\boldsymbol{x}, \mathbf{y})]_{R=a-0}, \tag{16.7}$$

$$[\boldsymbol{n}\wedge\operatorname{curl}\boldsymbol{\Gamma}_{\mathrm{h}}(\boldsymbol{x}, \mathbf{y})]_{R=a+0}=[\boldsymbol{n}\wedge\operatorname{curl}\boldsymbol{\Gamma}_{\mathrm{h}}(\boldsymbol{x}, \mathbf{y})]_{R=a-0} \tag{16.8}$$

the magnetic intensity outside the sphere is

$$\boldsymbol{H}(\boldsymbol{x})=-\int_{T_0}\{\operatorname{curl}_y\boldsymbol{J}(\mathbf{y})\}\cdot\boldsymbol{\Gamma}_{\mathrm{h}}(\mathbf{y}, \boldsymbol{x})\,\mathrm{d}\mathbf{y}. \tag{16.9}$$

The two Green's tensors differ only in their boundary conditions involving either the constants μ_1, μ_2 or ε_1, ε_2. It is therefore sufficient to find one of them; the other can be deduced by exchanging μ_1/μ_2 for $\varepsilon_1/\varepsilon_2$.

To find $\boldsymbol{\Gamma}_{\mathrm{e}}(\boldsymbol{x}, \mathbf{y})$ put

$$\boldsymbol{\Gamma}_{\mathrm{e}}(\boldsymbol{x}, \mathbf{y})=\begin{cases}\boldsymbol{\Gamma}^{\mathrm{i}}(\boldsymbol{x}, \mathbf{y})+\boldsymbol{\Gamma}_{\mathrm{s}}(\boldsymbol{x}, \mathbf{y}) & (R>a),\\ \boldsymbol{\Gamma}_{\mathrm{t}}(\boldsymbol{x}, \mathbf{y}) & (R<a).\end{cases}$$

In order that the scattered field is outgoing at infinity and complies with (16.1) (with zero right-hand side) assume

$$\boldsymbol{\Gamma}_{\mathrm{s}}(\boldsymbol{x}, \mathbf{y})=\mathrm{i}k_1\sum_{n=1}^{\infty}\sum_{m=-n}^{n}\{\alpha_{mn}\boldsymbol{q}^{(4)}_{-m,n}(k_1, \boldsymbol{x})\boldsymbol{q}^{(4)}_{mn}(k_1, \mathbf{y})$$
$$+\beta_{mn}\boldsymbol{p}^{(4)}_{-m,n}(k_1, \boldsymbol{x})\boldsymbol{p}^{(4)}_{mn}(k_1, \mathbf{y})\}.$$

The field inside the sphere satisfies (16.2) and is finite, so write

$$\boldsymbol{\Gamma}_{\mathrm{t}}(\boldsymbol{x}, \mathbf{y})=\mathrm{i}k_1\sum_{n=1}^{\infty}\sum_{m=-n}^{n}\{\lambda_{mn}\boldsymbol{q}^{(1)}_{-m,n}(k_2, \boldsymbol{x})\boldsymbol{q}^{(4)}_{mn}(k_1, \mathbf{y})$$
$$+\mu_{mn}\boldsymbol{p}^{(1)}_{-m,n}(k_2, \boldsymbol{x})\boldsymbol{p}^{(4)}_{mn}(k_1, \mathbf{y})\}.$$

The boundary condition (16.4) requires

$$\mu_1 d_{mn}\mathrm{j}_n(k_1a)+\mu_1\beta_{mn}\mathrm{h}_n^{(2)}(k_1a)=\mu_2\mu_{mn}\mathrm{j}_n(k_2a),$$

$$\mu_1 d_{mn}\frac{\mathrm{d}}{\mathrm{d}a}\{a\mathrm{j}_n(k_1a)\}+\mu_1\alpha_{mn}\frac{\mathrm{d}}{\mathrm{d}a}\{a\mathrm{h}_n^{(2)}(k_1a)\}=\frac{k_1\mu_2}{k_2}\lambda_{mn}\frac{\mathrm{d}}{\mathrm{d}a}\{a\mathrm{j}_n(k_2a)\}$$

from (13.9) and (13.10). The condition (16.5) enforces

$$d_{mn}\frac{\mathrm{d}}{\mathrm{d}a}\{a\mathrm{j}_n(k_1a)\}+\beta_{mn}\frac{\mathrm{d}}{\mathrm{d}a}\{a\mathrm{h}_n^{(2)}(k_1a)\}=\mu_{mn}\frac{\mathrm{d}}{\mathrm{d}a}\{a\mathrm{j}_n(k_2a)\},$$

$$k_1d_{mn}\mathrm{j}_n(k_1a)+k_1\alpha_{mn}\mathrm{h}_n^{(2)}(k_1a)=k_2\lambda_{mn}\mathrm{j}_n(k_2a).$$

Denote $\rho\mathrm{j}_n(\rho)$ and $\rho\mathrm{h}_n^{(2)}(\rho)$ by $\varphi_n^{(1)}(\rho)$ and $\varphi_n^{(4)}(\rho)$ respectively. Then

$$\begin{aligned}
\alpha_{mn}&=\{\kappa\varphi^{(1)\prime}(k_1a)\varphi^{(1)}(k_2a)-\varphi^{(1)}(k_1a)\varphi^{(1)\prime}(k_2a)\}d_{mn}/\alpha_n,\\
\lambda_{mn}&=\mathrm{i}\kappa d_{mn}/\alpha_n,\qquad \alpha_n=\varphi^{(1)\prime}(k_2a)\varphi^{(4)}(k_1a)-\kappa\varphi^{(4)\prime}(k_1a)\varphi^{(1)}(k_2a),\\
\beta_{mn}&=\{\kappa\varphi^{(1)}(k_1a)\varphi^{(1)\prime}(k_2a)-\varphi^{(1)\prime}(k_1a)\varphi^{(1)}(k_2a)\}d_{mn}/\beta_n,\\
\mu_{mn}&=-\mathrm{i}\kappa d_{mn}/\beta_n,\qquad \beta_n=\varphi^{(4)\prime}(k_1a)\varphi^{(1)}(k_2a)-\kappa\varphi^{(4)}(k_1a)\varphi^{(1)\prime}(k_2a)
\end{aligned}$$

where $\kappa=\mu_1k_2/\mu_2k_1$ and the primes indicate derivatives with respect to the argument.

The formula for $\boldsymbol{\Gamma}_\mathrm{h}$ can be expressed in the same way except that κ must be identified as $\varepsilon_1k_2/\varepsilon_2k_1$.

It will be observed that the dependence of α_{mn}, β_{mn}, λ_{mn}, and μ_{mn} on m is the same as that of d_{mn}. This observation permits the carrying out of the summation with respect to m. From (13.7)

$$\boldsymbol{p}_{mn}^{(j)}=-\boldsymbol{x}\wedge\mathrm{grad}\ p_{mn}^{(j)}$$

and hence

$$\boldsymbol{p}_{mn}^{(j)}(k_2,\boldsymbol{x})\boldsymbol{p}_{mn}^{(k)}(k_1,\mathbf{y})=(\boldsymbol{x}\wedge\mathrm{grad})(\mathbf{y}\wedge\mathrm{grad}_y)p_{mn}^{(j)}(k_2,\boldsymbol{x})p_{mn}^{(k)}(k_1,\mathbf{y})$$

the gradient of a tensor being defined in §F.7. Hence

$$\begin{aligned}
&\sum_{m=-n}^{n} d_{mn}\boldsymbol{p}_{-m,n}^{(j)}(k_2,\boldsymbol{x})\boldsymbol{p}_{mn}^{(k)}(k_1,\mathbf{y})\\
&=(\boldsymbol{x}\wedge\mathrm{grad})(\mathbf{y}\wedge\mathrm{grad}_y)\sum_{m=-n}^{n} d_{mn}p_{-m,n}^{(j)}(k_2,\boldsymbol{x})p_{mn}^{(k)}(k_1,\mathbf{y})\\
&=(\boldsymbol{x}\wedge\mathrm{grad})(\mathbf{y}\wedge\mathrm{grad}_y)\frac{2n+1}{4\pi n(n+1)}\mathfrak{z}_n^{(j)}(k_2\,|\boldsymbol{x}|)\mathfrak{z}_n^{(k)}(k_1\,|\mathbf{y}|)\mathrm{P}_n(\cos\gamma)
\end{aligned}$$

by (B.3.24), γ being the angle between $\boldsymbol{x}$ and $\mathbf{y}$. There is a similar result for the expansion involving $\boldsymbol{q}_{-m,n}$, $\boldsymbol{q}_{mn}$ stemming from

$$\boldsymbol{q}_{mn}^{(j)}(k_1,\boldsymbol{x})=k_1\boldsymbol{x}p_{mn}^{(j)}(k_1,\boldsymbol{x})+\mathrm{grad}\frac{\partial}{\partial R}(Rp_{mn}^{(j)})/k_1.$$

In particular, if $|\mathbf{y}| \geq R \geq a$,

$$\boldsymbol{\Gamma}_e(\boldsymbol{x}, \mathbf{y}) = (\mathrm{i}/k_1)\left\{k_1^2\hat{\boldsymbol{x}} + \mathrm{grad}\,\frac{\partial}{\partial R}\right\}\left\{k_1^2\hat{\mathbf{y}} + \mathrm{grad}_y\,\frac{\partial}{\partial |\mathbf{y}|}\right\}R\,|\mathbf{y}|\,V_1(\boldsymbol{x}, \mathbf{y})$$
$$+\mathrm{i}k_1(\boldsymbol{x} \wedge \mathrm{grad})(\mathbf{y} \wedge \mathrm{grad}_y)V_2(\boldsymbol{x}, \mathbf{y}) \qquad (16.10)$$

where $\hat{\boldsymbol{x}}$ is a unit vector in the direction of $\boldsymbol{x}$ and

$$V_1(\boldsymbol{x}, \mathbf{y}) = \sum_{n=1}^{\infty} \frac{2n+1}{4\pi n(n+1)}\{\mathrm{j}_n(k_1R) + A_n\mathrm{h}_n^{(2)}(k_1R)\}\mathrm{h}_n^{(2)}(k_1\,|\mathbf{y}|)\mathrm{P}_n(\cos\gamma), \qquad (16.11)$$

$$V_2(\boldsymbol{x}, \mathbf{y}) = \sum_{n=1}^{\infty} \frac{2n+1}{4\pi n(n+1)}\{\mathrm{j}_n(k_1R) + B_n\mathrm{h}_n^{(2)}(k_1R)\}\mathrm{h}_n^{(2)}(k_1\,|\mathbf{y}|)\mathrm{P}_n(\cos\gamma), \qquad (16.12)$$

$$A_n = \alpha_{mn}/d_{mn}, \qquad B_n = \beta_{mn}/d_{mn}.$$

For $R \geq |\mathbf{y}| > a$ it is only necessary to replace the factor $\mathrm{j}_n(k_1R)\mathrm{h}_n^{(2)}(k_1\,|\mathbf{y}|)$ by $\mathrm{h}_n^{(2)}(k_1R)\mathrm{j}_n(k_1\,|\mathbf{y}|)$.

The electric intensity due to an electric dipole at $\boldsymbol{x}_0$ can now be deduced from

$$\boldsymbol{E}(\boldsymbol{x}) = -\omega^2\mu_1A\boldsymbol{p}_0 \cdot \boldsymbol{\Gamma}_e(\boldsymbol{x}_0, \boldsymbol{x}) = -\omega^2\mu_1A\boldsymbol{\Gamma}_e(\boldsymbol{x}, \boldsymbol{x}_0) \cdot \boldsymbol{p}_0, \qquad (16.13)$$

γ being the angle between $\boldsymbol{x}$ and $\boldsymbol{x}_0$. When the dipole is at a large distance from the sphere the asymptotic formula (A.4.13) supplies

$$\left\{k_1^2\hat{\boldsymbol{x}}_0 + \mathrm{grad}_0\,\frac{\partial}{\partial R_0}\right\}R_0V_1(\boldsymbol{x}, \boldsymbol{x}_0)$$
$$\sim \sum_{n=1}^{\infty} \frac{2n+1}{4\pi n(n+1)}\{\mathrm{j}_n(k_1R) + A_n\mathrm{h}_n^{(2)}(k_1R)\}$$
$$\times\frac{\mathrm{e}^{-\mathrm{i}k_1R_0}}{R_0}\,\mathrm{i}^n\left\{\boldsymbol{i}_{20}\,\frac{\partial}{\partial\theta_0} + \frac{\boldsymbol{i}_{30}}{\sin\theta_0}\frac{\partial}{\partial\phi_0}\right\}\mathrm{P}_n(\cos\gamma)$$

where $\boldsymbol{i}_{20}$, $\boldsymbol{i}_{30}$ are unit vectors in the directions of θ_0 increasing, ϕ_0 increasing respectively. If $\boldsymbol{p}_0$ is parallel to the x-axis, $\boldsymbol{p}_0 \cdot \boldsymbol{i}_{20} = \cos\theta_0 \cos\phi_0$, $\boldsymbol{p}_0 \cdot \boldsymbol{i}_{30} = -\sin\phi_0$. Hence, when $\theta_0 = \pi$ and $\phi_0 = 0$, (B.3.1) furnishes

$$\boldsymbol{p}_0 \cdot \left\{k_1^2\hat{\boldsymbol{x}}_0 + \mathrm{grad}_0\,\frac{\partial}{\partial R_0}\right\}R_0V_1(\boldsymbol{x}, \boldsymbol{x}_0)$$
$$\sim -\sum_{n=1}^{\infty} \frac{2n+1}{4\pi n(n+1)}\{\mathrm{j}_n(k_1R) + A_n\mathrm{h}_n^{(2)}(k_1R)\}\frac{\mathrm{e}^{-\mathrm{i}k_1R_0}}{R_0}\,\mathrm{i}^n \cos\phi\mathrm{P}_n^1(-\cos\theta).$$

Multiplication by $4\pi\varepsilon_1R_0\mathrm{e}^{\mathrm{i}k_1R_0}/Ak_1^2$ followed by letting $R_0 \to \infty$ produces a primary field which is a plane wave propagating along the z-axis, namely (14.2). A similar calculation can be carried out for $\bar{V}_2$. The total electric intensity is then, from (16.13),

$$\begin{aligned}\boldsymbol{E}(\boldsymbol{x}) = &\left\{k_1^2\hat{\boldsymbol{x}} + \text{grad}\frac{\partial}{\partial R}\right\}R\sum_{n=1}^{\infty}\frac{(-\mathrm{i})^{n+1}(2n+1)}{k_1 n(n+1)} \\ &\times\{\mathrm{j}_n(k_1R) + A_n\mathrm{h}_n^{(2)}(k_1R)\}\cos\phi\,\mathrm{P}_n^1(\cos\theta) \\ &-(\boldsymbol{x}\wedge\text{grad})\sum_{n=1}^{\infty}\frac{(-\mathrm{i})^{n+2}(2n+1)}{n(n+1)} \\ &\times\{\mathrm{j}_n(k_1R) + B_n\mathrm{h}_n^{(2)}(k_1R)\}\sin\phi\,\mathrm{P}_n^1(\cos\theta) \\ = &\left(k_1^2\hat{\boldsymbol{x}} + \text{grad}\frac{\partial}{\partial R}\right)RW_1 - (\boldsymbol{x}\wedge\text{grad})W_2 \qquad (16.14)\end{aligned}$$

say. This formula can also be derived directly from the expansion (14.2) for the plane wave by assuming appropriate expansions for the scattered field and wave inside the sphere.

If the sphere is perfectly conducting a Green's tensor $\boldsymbol{\Gamma}_1(\boldsymbol{x}, \boldsymbol{y})$ is required such that $\boldsymbol{n}\wedge\boldsymbol{\Gamma}_1(\boldsymbol{x}, \boldsymbol{y})$ vanishes when $\boldsymbol{x}$ is on the sphere. It can be confirmed that (16.10) can be used for $\boldsymbol{\Gamma}_1$ provided that

$$A_n = -\varphi_n^{(1)\prime}(k_1a)/\varphi_n^{(4)\prime}(k_1a), \qquad B_n = -\mathrm{j}_n(k_1a)/\mathrm{h}_n^{(2)}(k_1a). \qquad (6.15)$$

8.17 General discussion of the scattered field

Whatever the primary wave the behaviour of the scattered field is dictated essentially by the coefficients b_{mn}/a_{mn} of (12.2) in the acoustic case and A_n, B_n for electromagnetic waves. These coefficients are the amplitudes of modes excited in the sphere by the primary wave. These are not the natural oscillations, which can occur only at the eigenvalues of the sphere, but forced vibrations with the same frequency as that of the incident field. A natural query is to ask whether resonance can be achieved, i.e. whether the forced frequency can be made equal to that of a natural oscillation. This happens for sound waves if the frequency can be arranged so that b_{mn}/a_{mn} is infinite for some value of n. Now, under the assumption that k_1, k_2, and ζ are all real, this is possible only if $\mathrm{h}_n^{(2)\prime}(k_1b)/\mathrm{h}_n^{(2)}(k_1b)$ is real. But that requires that the imaginary part be zero or $\mathrm{j}_n\mathrm{y}_n' - \mathrm{j}_n'\mathrm{y}_n = 0$. However, this is impossible since the left-hand side is $1/(k_1b)^2$ which cannot be zero for finite k_1b. Hence there is no frequency at which resonance in the strict sense occurs. There may, of course, be frequencies at which the denominator is small although not quite zero. The amplitude of the scattered wave would then be large and a sort of resonance might be said to have been accomplished. This matter will now be investigated a little more fully when the wavelength is large.

At sufficiently low frequencies the approximations $k_1b \ll 1$ and $k_2b \ll 1$

can be introduced. Then, from (A.4.14)–(A.4.21),

$$\frac{b_{m0}}{a_{m0}} \approx \frac{\frac{1}{3}ik_1b^3(k_1^2-k_2^2\zeta)}{1-\frac{1}{3}\zeta(k_2b)^2(1-ik_1b)+\frac{1}{2}b^2(k_1^2-\frac{1}{3}k_2^2)}, \tag{17.1}$$

$$b_{m1}/a_{m1} \approx \tfrac{1}{3}i(k_1b)^3(\zeta-1)/(\zeta+2) \tag{17.2}$$

and generally b_{mn}/a_{mn} is $O\{(k_1b)^{2n+1}\}$ for $n \geqslant 1$. It is clear that in most circumstances the scattered wave is $O\{(k_1b)^3\}$. However, the situation changes dramatically if $k_2b=(3/\zeta)^{\frac{1}{2}}$ because then $b_{m0}/a_{m0} \approx -1$ so that the scattered wave is $O(1)$. This can occur when ζ is large which corresponds to the sphere being almost soft. Thus *sound can be scattered more strongly in a liquid by a small gas bubble than by a hard sphere of the same size;* this is in accordance with experimental observation. Consequently, although there is no resonance in the strict sense, the scattering may be substantially larger at some frequencies than at others.

For electromagnetic waves there is also no strict resonance because A_n and B_n cannot be infinite unless $\varphi_n^{(4)\prime}(k_1a)/\varphi_n^{(4)}(k_1a)$ is real and this is impossible for the same reason as in the acoustic case. At long wavelengths A_n and B_n are $O\{(k_1a)^{2n+1}\}$ and, in particular,

$$A_1 \approx -\tfrac{2}{3}i(k_1a)^3[\mu_1N^2-\mu_2+\tfrac{1}{10}(k_1a)^2\{\mu_2(1+2\kappa^2)-\mu_1N^2(2+\kappa^2)\}]/C_1, \tag{17.3}$$

$$C_1 = 2\mu_2+\mu_1N^2-\tfrac{1}{10}(k_1a)^2 \times\{N^2(4\mu_2+\mu_1N^2)-5(2\mu_2-\mu_1N^2)\}-\tfrac{2}{3}i(k_1a)^3(\mu_2-\mu_1N^2),$$

$$A_2 \approx -\tfrac{1}{15}i(k_1a)^5(\mu_1N^2-\mu_2)/(3\mu_2+2\mu_1N^2), \tag{17.4}$$

$$B_1 \approx -\tfrac{2}{3}i(k_1a)^3[\mu_2-\mu_1+\tfrac{1}{10}(k_1a)^3\{\mu_1(1+2N^2)-\mu_2(2+N^2)\}]/D_1, \tag{17.5}$$

$$D_1 = 2\mu_1+\mu_2-\tfrac{1}{10}(k_1a)^2\{N^2(4\mu_1+\mu_2)-5(2\mu_1-\mu_2)\}-\tfrac{2}{3}i(k_1a)^3(\mu_1-\mu_2),$$

$$B_2 \approx -\tfrac{1}{15}i(k_1a)^5(\mu_2-\mu_1)/(3\mu_1+2\mu_2) \tag{17.6}$$

where $N=k_2/k_1$ is the refractive index of the sphere.

If $\mu_1=\mu_2$ these expressions reduce to

$$A_1 \approx -\tfrac{2}{3}i(k_1a)^3\frac{N^2-1}{N^2+2}\left\{1+\tfrac{3}{5}(k_1a)^2\frac{N^2-2}{N^2+2}\right\}, \tag{17.7}$$

$$A_2 \approx -\tfrac{1}{15}i(k_1a)^5(N^2-1)/(3+2N^2), \tag{17.8}$$

$$B_1 \approx -\tfrac{1}{15}i(k_1a)^5\frac{N^2-1}{3-\frac{1}{2}(k_1a)^2(N^2-1)} \approx -\tfrac{1}{45}i(k_1a)^5(N^2-1), \tag{17.9}$$

$$B_2 \approx 0.$$

In general B_{n+1}, which corresponds to a magnetic oscillation, is of the same order as the coefficient A_n of an electric mode of oscillation.

At first sight it appears that the coefficients may exhibit the same behaviour at certain frequencies as occurs for sound waves. For example, the denominator of B_1 in (17.9) vanishes if $6=(k_1a)^2(N^2-1)=(k_2^2-k_1^2)a^2$. However, it has been assumed in the derivation of (17.9) that both k_1a and k_2a are small so that $(k_2^2-k_1^2)a^2$ can never be as great as 6. It will also be discovered that the denominators in (17.3)–(17.6) can never be small under the conditions that have been imposed in their derivation. Thus electromagnetic scattering does not display the same enhancement at certain frequencies that is a feature of acoustics.

The incident plane wave of the preceding section is linearly polarized with the electric vector parallel to the x-axis. At great distances from the sphere the scattered field is transverse to the direction of propagation. The two components E_θ and E_ϕ are perpendicular to one another and differ in phase in general. Consequently, the scattered wave will be elliptically polarized. There are two exceptional directions: $\phi=0$ where $E_\phi=0$ and $\phi=\frac{1}{2}\pi$ where $E_\theta=0$. Thus, when the scattered radiation is viewed along either the x- or y-axis, it is linearly polarized. Of course, for an unpolarized primary wave, the scattered field will be partially polarized depending upon the direction of observation.

8.18 The effect of conductivity

The presence of dissipation in the sphere does not affect the formulae of §8.12 and §8.16. Some of the parameters may now be complex, e.g. in the electromagnetic case $k_2^2=\omega^2\mu_2(\varepsilon_2-\mathrm{i}\sigma_2/\omega)$ where σ_2 is the conductivity of the sphere. The low frequency expansions will continue to be valid so long as the assumptions on which they are based are still met. They will cease to hold when the dissipation is large because then k_2 may not be small enough to employ the approximations for the Bessel functions. In particular, this will happen when the sphere is perfectly conducting.

When the sphere is perfectly conducting the formulae of (16.15) are relevant. Their close relationship to the soft and hard boundary conditions in (12.4) will be noted. For long wavelengths

$$A_1\approx-\tfrac{2}{3}\mathrm{i}(k_1a)^3\{1+\tfrac{3}{10}(k_1a)^2\},\qquad A_2\approx-\tfrac{1}{30}\mathrm{i}(k_1a)^5, \tag{18.1}$$

$$B_1\approx\tfrac{1}{3}\mathrm{i}(k_1a)^3\{1-\tfrac{3}{5}(k_1a)^2\},\qquad B_2\approx\tfrac{1}{30}\mathrm{i}(k_1a)^5, \tag{18.2}$$

$$(B_{m0}/a_{m0})_\mathrm{S}\approx\mathrm{i}k_1b\{1+\mathrm{i}k_1b-\tfrac{2}{3}(k_1b)^2\},\qquad (B_{m1}/a_{m1})_\mathrm{S}\approx\tfrac{1}{3}\mathrm{i}(k_1b)^3, \tag{18.3}$$

$$(B_{m0}/a_{m0})_\mathrm{H}\approx\tfrac{1}{3}\mathrm{i}(k_1b)^3\{1-\tfrac{3}{5}(k_1b)^2\},\qquad (B_{m1}/a_{m1})_\mathrm{H}\approx-\tfrac{1}{6}\mathrm{i}(k_1b)^3. \tag{18.4}$$

In general, A_n and B_n are of the same order of magnitude.

Resonance can occur for B_n when $\mathrm{h}_n^{(2)}(k_1a)=0$. Roots of this equation are complex (see Table 8.5). Since k_1a is real the most that can be achieved is to make k_1a equal to the real part of a root; even then the imaginary part is so large that there is effectively no resonance phenome-

Table 8.5 Complex zeros of $\mathrm{h}_n^{(2)}$, $\mathrm{h}_n^{(2)\prime}$, and $\{z\mathrm{h}_n^{(2)}(z)\}'$

	$\mathrm{h}_n^{(2)}(z)=0$	$\mathrm{h}_n^{(2)\prime}(z)=0$	$\{z\mathrm{h}_n^{(2)}(z)\}'=0$
$n=0$	—	i	—
$n=1$	i	±1+i	±0.86+0.50i
$n=2$	±0.86+1.50i	1.78i, 1.11i±1.95	1.60i, ±1.81+0.70i
$n=3$	2.26i, ±1.75+1.75i		±0.87+2.17i, ±2.77+0.83i

non. The same conclusion applies to the soft boundary condition in acoustics. For the hard boundary condition the pertinent equation is $\mathrm{h}_n^{(2)\prime}(k_1b)=0$ and again Table 8.5 shows that there is no resonance. Moreover, the possibility of A_n being infinite which requires $\{k_1a\mathrm{h}_n^{(2)}(k_1a)\}'=0$ is also excluded. Thus the phenomenon of resonance does not arise in these cases. Nevertheless, the roots where resonance might have occurred are important in the *singularity expansion method* which will be discussed in §10.3.

Despite the absence of strict resonance in the limiting case when the sphere is impenetrable it is worth while re-examining the behaviour of A_n and B_n when $k_1a\ll 1$ but k_2a is not necessarily small. The denominator of A_n vanishes if

$$k_1a\varphi_n^{(1)\prime}(k_2a)+n\kappa\varphi_n^{(1)}(k_2a)=0 \tag{18.5}$$

while that of B_n disappears if

$$n\varphi_n^{(1)}(k_2a)=-k_1a\kappa\varphi_n^{(1)\prime}(k_2a). \tag{18.6}$$

If $\mu_2=\mu_1$ and $y=k_2a$, (18.6) becomes

$$y\,\mathrm{j}_{n-1}(y)=0 \tag{18.7}$$

on account of (A.4.5) whereas (18.5) goes over to

$$y^2\mathrm{j}_n(y)=-(k_1a)^2\{y\,\mathrm{j}_n(y)\}'/n. \tag{18.8}$$

Thus, if $\mathrm{j}_n(y_{nm})=0$, B_n is much larger than usual when $k_2a=y_{n-1,m}$ so long as $k_1a\ll 1$. An approximate solution of (18.8) is

$$k_2a=y_{nm}-(k_1a)^2/ny_{nm}.$$

These formulae form the basis of solutions to (18.5) and (18.6) when $\mu_1\approx\mu_2$. The expressions do not locate the positions of resonance exactly—for that more terms of α_n and β_n have to be retained—but they should not be in error by more than $O\{(k_1a)^2\}$. Notwithstanding, they cause the magnitudes of A_n and B_n to alter radically. Therefore sharp changes in the radiation can be expected for frequencies near resonance, which may also occur for complex values of k_2.

8.19 The scattering coefficient

The scattering coefficient for an incident plane wave is still defined by (2.1). From (12.5), (B.3.20), (B.3.24), and (B.3.10) it follows that in the acoustic case

$$\sigma_c^A = \frac{4}{(k_1 b)^2} \sum_{n=0}^{\infty} (2n+1) \left| \frac{b_{mn}}{a_{mn}} \right|^2 \tag{19.1}$$

where b_{mn} is given by (12.2). The corresponding formula for the electromagnetic case is

$$\sigma_c^E = \frac{2}{(k_1 a)^2} \sum_{n=1}^{\infty} (2n+1)(|A_n|^2 + |B_n|^2). \tag{19.2}$$

At low frequencies

$$\sigma_c^A \approx \frac{4}{9} \frac{|k_1^2 - k_2^2 \zeta|^2 \, b^4}{\{1 - \frac{1}{3}\zeta (k_2 b)^2\}^2 + \frac{1}{9} k_1^2 k_2^4 b^6} + \frac{4}{3} \left| \frac{\zeta - 1}{\zeta + 2} \right|^2 (k_1 b)^4 \tag{19.3}$$

from (17.1) and (17.2). Likewise

$$\sigma_c^E \approx \tfrac{8}{3}(k_1 a)^4 \left\{ \left(\frac{\mu_1 N^2 - \mu_2}{2\mu_2 + \mu_1 N^2} \right)^2 + \left(\frac{\mu_2 - \mu_1}{2\mu_1 + \mu_2} \right)^2 \right\} \tag{19.4}$$

in general while for perfect conductivity

$$\sigma_c^E \approx \tfrac{10}{3}(k_1 a)^4 \{1 + \tfrac{6}{25}(k_1 a)^2\}. \tag{19.5}$$

The scattering coefficient has been calculated for a wide variety of parameters. Some values for a perfectly conducting sphere are displayed in Table 8.6.

Table 8.6 The scattering coefficient of a perfectly conducting sphere

$k_1 a$	σ_c^E	$k_1 a$	σ_c^E	$k_1 a$	σ_c^E
0.1	0.00034	1.4	2.204	25	2.027
0.2	0.0054	1.5	2.155	30	2.023
0.3	0.028	1.6	2.115	35	2.020
0.4	0.086	1.8	2.136	55	2.013
0.5	0.218	2.0	2.209	60	2.012
0.6	0.466	2.5	2.171	70	2.011
0.7	0.795	3.0	2.172	75	2.010
0.8	1.257	3.5	2.136	80	2.010
0.9	1.696	4.0	2.140	85	2.009
1.0	2.036	5.0	2.116	90	2.009
1.1	2.230	10.0	2.061		
1.2	2.280	15	2.043		
1.3	2.267	20	2.033		

Alternative expressions for the scattering coefficient can be derived along the lines of §8.2. The derivation is legitimate for any finite obstacle not just a sphere. Sound waves will be discussed first.

Let the incident plane wave be $p^{\mathrm{i}} = \mathrm{e}^{-\mathrm{i}k_1 z}$. The average energy flow through a surface S is a real multiple of $\operatorname{Im}\int_S p(\partial p^*/\partial n)\,\mathrm{d}S$. As regards the energy incident on the obstacle this will give $k_1 S_0$ where S_0 is the area of the projection of the obstacle on a plane perpendicular to the direction of propagation of the plane wave.

If the total field is $p^{\mathrm{i}} + p_{\mathrm{s}}$ the energy scattered is $\operatorname{Im}\int_\Omega p_{\mathrm{s}}(\partial p_{\mathrm{s}}^*/\partial n)\,\mathrm{d}\Omega$ where Ω is the surface of a sphere of large radius R. Furthermore

$$\operatorname{Im}\int_\Omega (p^{\mathrm{i}} + p_{\mathrm{s}})\{\partial(p^{\mathrm{i}} + p_{\mathrm{s}})^*/\partial n\}\,\mathrm{d}\Omega = 0$$

when the obstacle is lossless. Hence

$$\sigma_{\mathrm{c}}^{\mathrm{A}} = -(1/k_1 S_0)\operatorname{Im}\int_\Omega \left(p^{\mathrm{i}}\frac{\partial p^{\mathrm{i}*}}{\partial n} + p^{\mathrm{i}}\frac{\partial p_{\mathrm{s}}^*}{\partial n} + p_{\mathrm{s}}\frac{\partial p^{\mathrm{i}*}}{\partial n}\right)\mathrm{d}\Omega.$$

The first term is the responsibility of the incident plane wave and so is zero. With regard to the other two terms let

$$p_{\mathrm{s}} \sim A(\theta, \phi)\mathrm{e}^{-\mathrm{i}k_1 R}/R \tag{19.6}$$

as $R \to \infty$. Then

$$\sigma_{\mathrm{c}}^{\mathrm{A}} = -(1/k_1 S_0)\operatorname{Im}\lim_{R\to\infty}\left(\mathrm{i}k_1 R\int_0^{2\pi}\int_0^{\pi}\{A^*(\theta, \phi)\mathrm{e}^{\mathrm{i}k_1 R(1-\cos\theta)}\right.$$
$$\left. + A(\theta, \phi)\cos\theta\,\mathrm{e}^{\mathrm{i}k_1 R(\cos\theta - 1)}\}\sin\theta\,\mathrm{d}\theta\,\mathrm{d}\phi\right)$$
$$= -(1/k_1 S_0)\operatorname{Im}\lim_{r\to\infty}\left(\int_0^{2\pi}\{A^*(\pi, \phi)\mathrm{e}^{2\mathrm{i}k_1 R} - A^*(0, \phi)\right.$$
$$+ A(\pi, \phi)\mathrm{e}^{-2\mathrm{i}k_1 R} + A(0, \phi)\}\,\mathrm{d}\phi$$
$$- \int_0^{2\pi}\int_0^{\pi}\left[\mathrm{e}^{\mathrm{i}k_1 R(1-\cos\theta)}\frac{\partial}{\partial\theta}A^*(\theta, \phi)\right.$$
$$\left.\left. - \mathrm{e}^{\mathrm{i}k_1 R(\cos\theta - 1)}\frac{\partial}{\partial\theta}\{A(\theta, \phi)\cos\theta\}\right]\mathrm{d}\theta\,\mathrm{d}\phi\right)$$

after an integration by parts with respect to θ. Evaluation of the second integral by the method of stationary phase shows that its contribution vanishes in the limit. Also $A(0, \phi)$ does not, in fact, involve ϕ. Hence

$$\sigma_{\mathrm{c}}^{\mathrm{A}} = -4\pi\operatorname{Im}A(0, \phi)/k_1 S_0.$$

This result can be adapted to the plane wave at arbitrary incidence. If the plane wave $\exp(-\mathrm{i}k_1 \boldsymbol{n}^{\mathrm{i}} \cdot \boldsymbol{x})$ produces a scattered wave $A(\boldsymbol{\nu})\mathrm{e}^{-\mathrm{i}k_1 R}/R$ in the direction of the unit vector $\boldsymbol{\nu}$ then

$$\sigma_c^A = -4\pi \operatorname{Im} A(\boldsymbol{n}^i)/k_1 S_0. \tag{19.7}$$

The same technique may be employed for the electromagnetic field starting from $\int_S \boldsymbol{E} \wedge \boldsymbol{H}^* \cdot \boldsymbol{n} \, dS$. Let the incident plane wave be

$$\boldsymbol{E}^i = \boldsymbol{e}^i \exp(-ik_1 \boldsymbol{n}^i \cdot \boldsymbol{x}), \qquad \boldsymbol{H}^i = (\varepsilon_1/\mu_1)^{\frac{1}{2}} \boldsymbol{n}^i \wedge \boldsymbol{E}^i$$

where $\boldsymbol{e}^i \cdot \boldsymbol{n}^i = 0$. Then, if the scattered wave is such that $\boldsymbol{E}_s \sim \mathbf{A}(\boldsymbol{\nu}) e^{-ik_1 R}/R$,

$$\sigma_c^E = -4\pi \operatorname{Im} \mathbf{A}(\boldsymbol{n}^i) \cdot \boldsymbol{e}^i / k_1 S_0. \tag{19.8}$$

A dissipative obstacle only affects (19.7) and (19.8) by bringing in an absorption coefficient, i.e.

$$\sigma_c^A + \sigma_a^A = -4\pi \operatorname{Im} A(\boldsymbol{n}^i)/k_1 S_0, \tag{19.9}$$

$$\sigma_c^E + \sigma_a^E = -4\pi \operatorname{Im} \mathbf{A}(\boldsymbol{n}^i) \cdot \boldsymbol{e}^i / k_1 S_0. \tag{19.10}$$

The application of (19.9) and (19.10) to the sphere leads to

$$\sigma_c^A + \sigma_a^A = -\frac{4}{(k_1 b)^2} \operatorname{Re} \sum_{n=0}^{\infty} (2n+1)(b_{mn}/a_{mn}), \tag{19.11}$$

$$\sigma_c^E + \sigma_a^E = -\{2/(k_1 a)^2\} \operatorname{Re} \sum_{n=0}^{\infty} (2n+1)(A_n + B_n). \tag{19.12}$$

These formulae should agree with (19.1) and (19.2) respectively when the sphere is non-absorbing. There will be consistency provided that $|b_{mn}/a_{mn}|^2 = -\operatorname{Re} b_{mn}/a_{mn}$, $|A_n|^2 = -\operatorname{Re} A_n$, $|B_n|^2 = -\operatorname{Re} B_n$; confirmation is straightforward. For the dissipative sphere, the difference will account for the absorption coefficient; for example, if $\mu_2 = \mu_1$ and the frequency is low, the inference from (19.4) and (19.12) is that

$$\sigma_a^E \approx -4k_1 a \operatorname{Im}\{(N^2 - 1)/(N^2 + 2)\}. \tag{19.13}$$

8.20 Rayleigh scattering

The assumption of this section is that the wavelength is long enough for powers of $k_1 b$, $k_2 b$ higher than the third to be neglected. This is termed *Rayleigh scattering*. Subject to plane wave incidence (12.5), (17.1), and (17.2) supply

$$p_s = \tfrac{1}{3} i k_1 b^3 (k_1^2 - k_2^2 \zeta) h_0^{(2)}(k_1 R) + (k_1 b)^3 \frac{\zeta - 1}{\zeta + 2} P_1(\cos \Theta) h_1^{(2)}(k_1 R) \tag{20.1}$$

so long as ζ is not too large. Thus the scattered field can be regarded as that *radiated by a simple source and a double source (of the same order in b) placed at the centre of the sphere.* For larger values of ζ the first term of

(20.1) has to be modified as in §8.17 and the predominant effect is caused by a simple source at the centre of the sphere.

Let the direction of propagation be the z-axis. At large distances

$$p_s \sim -\frac{k_1^2 b^3}{3R} e^{-ik_1R}\left(1 - \frac{k_2^2}{k_1^2}\zeta + 3\frac{\zeta - 1}{\zeta + 2}\cos\theta\right)$$

and the intensity is

$$\frac{k_1}{2\omega\rho_0}\frac{k_1^4 b^6}{9R^2}\left|1 - \frac{k_2^2}{k_1^2}\zeta + 3\frac{\zeta - 1}{\zeta + 2}\cos\theta\right|^2.$$

In general the radiation pattern varies with θ. The modification for large values of ζ is that the radiation is uniform in all directions.

The plane wave of §8.14 produces, when $\mu_1 = \mu_2$, a scattered wave

$$\boldsymbol{E}_s = (\text{grad div} + k_1^2)\boldsymbol{p}e^{-ik_1R}/R, \qquad \boldsymbol{H}_s = i\omega\varepsilon_1 \text{ curl } \boldsymbol{p}e^{-ik_1R}/R$$

where

$$\boldsymbol{p} = 4\pi\varepsilon_1 a^3 \boldsymbol{i}(N^2 - 1)/(N^2 + 2).$$

In other words the scattered field is that *radiated by an electric dipole, at the centre of the sphere, oriented parallel to the electric vector of the incident wave and of moment* $4\pi\varepsilon_1 a^3(N^2-1)/(N^2+2)$.

It follows from §3.8 that, at large distances, the electric intensity of the radiation is perpendicular to the radius vector while lying in the plane containing $\boldsymbol{p}$ and the radius vector (Fig. 8.12). The amplitude of the electric intensity is

$$-\frac{N^2-1}{N^2+2}\frac{k_1^2 a^3}{R}e^{-ik_1R}\sin\theta_s$$

where θ_s is depicted in Fig. 8.12. The intensity I of the field is given by

$$I = \frac{1}{2}\left(\frac{\varepsilon_1}{\mu_1}\right)^{\frac{1}{2}}\left|\frac{N^2-1}{N^2+2}\right|^2\frac{k_1^4 a^6}{R^2}\sin^2\theta_s. \qquad (20.2)$$

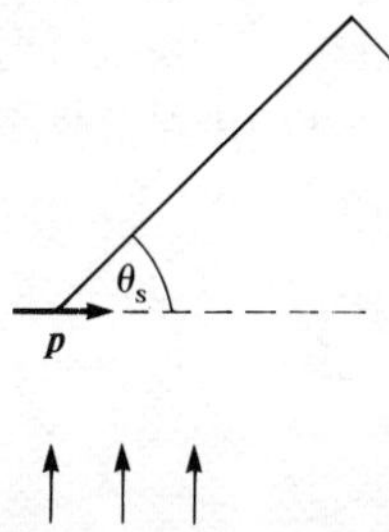

Fig. 8.12 The equivalent electric dipole

Two particular cases are of interest. Firstly, the intensity is a maximum when $\theta_s = \frac{1}{2}\pi$, i.e. in the plane perpendicular to the electric intensity of the incident wave. Secondly, when the point of observation is in the plane containing $\boldsymbol{p}$ and the direction of propagation (the z-axis), $\theta_s = \frac{1}{2}\pi - \theta$ where θ is the angle between the radius vector and the direction of propagation of the incident wave. A convenient way of describing the field can be constructed from these two results. Let P be the plane containing the direction of propagation of the incident wave and the radius vector to the point of observation. The electric intensity of the incident field can be split into two components, E_p perpendicular to P and E_ℓ lying in P. The former produces a field parallel to itself of

$$\frac{N^2-1}{N^2+2} k_1^2 a^3 \frac{e^{-ik_1R}}{R} E_p$$

and the latter one in P of amplitude

$$\frac{N^2-1}{N^2+2} k_1^2 a^3 \frac{e^{-ik_1R}}{R} E_\ell \cos\theta.$$

These fields are perpendicular to one another and so the intensity created is

$$\frac{1}{2}\left(\frac{\varepsilon_1}{\mu_1}\right)^{\frac{1}{2}} \left|\frac{N^2-1}{N^2+2}\right|^2 \frac{k_1^4 a^6}{R^2} (|E_p|^2 + |E_\ell|^2 \cos^2\theta).$$

Natural light is characterized by being unpolarized. Instead, it consists of a succession of plane waves of very short duration, all polarizations occurring equally often. Measured intensities therefore refer to the superposition of an extremely large number of plane waves with independent phases. Each plane wave may have its electric vector resolved as above and, since all polarizations arise equally frequently,

$$|E_p|^2 = |E_\ell|^2 = (\mu_1/\varepsilon_1)^{\frac{1}{2}} I^i$$

where I^i is the incident intensity. Hence the scattered intensity is

$$\frac{1}{2}\left|\frac{N^2-1}{N^2+2}\right|^2 \frac{k_1^4 a^6}{R^2} (1+\cos^2\theta) I^i.$$

The consequent radiation pattern is displayed in Fig. 8.13. There is no

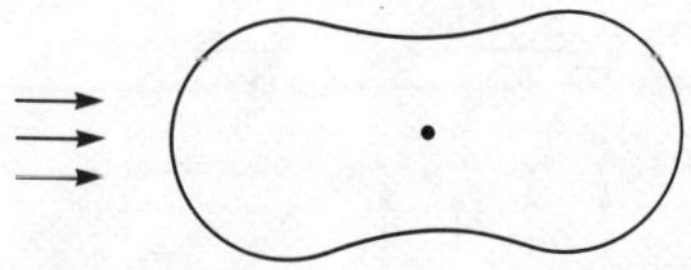

Fig. 8.13 The polar diagram of intensity in Rayleigh scattering of natural light

contribution from E_ℓ when $\theta = \frac{1}{2}\pi$, i.e. the light scattered in any direction perpendicular to the direction of the incoming wave is linearly polarized.

In this Rayleigh scattering the magnitude of the scattered radiation varies as the square of the frequency, and the intensity as the fourth power. The light which arrives from the sky has been scattered by the air molecules in the atmosphere and possibly by water droplets. The radius of these is less than 20 nm which is very small compared with the wavelength, about 600 nm, of visible light. The law of Rayleigh scattering is consequently applicable and the shorter (blue) wavelengths are scattered more than the longer (red) wavelengths in sunlight. Therefore, when looking at the sky away from the sun, one sees a blue colour. The opposite effect is observed when looking directly at the sun for the intensity of the blue, having suffered the greatest scattering, is weakened in the direct beam relative to the red which predominates. The effect is stronger at sunset because the sun's rays then have to traverse a longer path through the atmosphere to reach the observer.

The formulae are unaltered if the spherical particles are absorbent. However, that does not necessarily mean that the scattered intensity is proportional to k_1^4 if the particles are metal because the refractive index of most metals varies considerably with wavelength. The colours of colloidal suspensions of metallic particles usually differ from the blue of the sky as a consequence.

Various graduations in the colour of the sky can be accounted for by the presence of *aerosols* whether in the form of dust or other material particles in the air (see Minnaert (1954)).

As the size of the particle increases the selective character of Rayleigh scattering disappears. The intensity becomes independent of the wavelength; there is a diffuse reflected wave and a sharp diffracted beam. White light is scattered as white light. A simple illustration is the smoke from a cigarette; this is blue when it leaves the cigarette but white when it is expelled from the mouth. The smoke particles are enlarged by a coating of water while in the mouth.

Rayleigh scattering from a perfectly conducting sphere differs from the dielectric case. For perfect conductivity (18.1) and (18.2) are pertinent. Since A_1 and B_1 are of the same order *the scattered wave is caused by an electric dipole of moment* $4\pi\varepsilon_1 a^3\boldsymbol{i}$ *at the centre of the sphere together with a magnetic dipole of moment* $\boldsymbol{m}_s = -2\pi(\mu_1\varepsilon_1)^{\frac{1}{2}}a^3\boldsymbol{j}$ *and Hertz vector*

$$\frac{\boldsymbol{m}_s \wedge \boldsymbol{x}}{4\pi(\mu_1\varepsilon_1)^{\frac{1}{2}}R^2}(1 - \mathrm{i}/k_1R)\mathrm{e}^{-\mathrm{i}k_1R}.$$

The distant electric intensity is perpendicular to the plane containing $\boldsymbol{m}_s$ and the point of observation; it has the value $-\frac{1}{2}k_1^2a^3 \sin\theta_s \mathrm{e}^{-\mathrm{i}k_1R}/R$ where now θ_s is the angle between $\boldsymbol{m}_s$ and $\boldsymbol{x}$. For the plane P introduced earlier, E_p produces a field $k_1^2a^3(1-\frac{1}{2}\cos\theta)\mathrm{e}^{-\mathrm{i}k_1R}E_p/R$ parallel to itself whereas

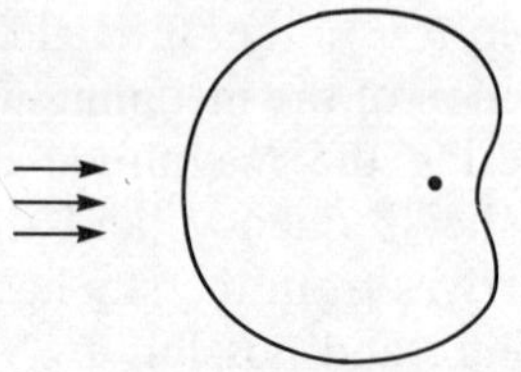

Fig. 8.14 Scattering of natural light by a perfectly conducting sphere

E_ℓ generates $k_1^2 a^3(\cos\theta-\frac{1}{2})e^{-ik_1R}E_\ell/R$. The consequent intensity is

$$\frac{1}{2}\left(\frac{\varepsilon_1}{\mu_1}\right)^{\frac{1}{2}}\frac{k_1^4a^6}{R^2}\{(1-\tfrac{1}{2}\cos\theta)^2|E_p|^2+(\cos\theta-\tfrac{1}{2})^2|E_\ell|^2\}$$

which, for natural light, becomes

$$\frac{5}{8}\frac{k_1^4a^6}{R^2}I^i(1-\tfrac{8}{5}\cos\theta+\cos^2\theta).$$

The energy is strongly directed back towards the source (Fig. 8.14) instead of being distributed nearly uniformly as when the sphere is constructed of dielectric material.

The *back scattering* or *radar cross-section* σ_b is defined as $4\pi R^2$ times the ratio of the backward scattered intensity and the incident intensity, i.e.

$$\sigma_b=4\pi R^2[I]_{\theta=\pi}/I^i. \tag{20.3}$$

Thus, for the dielectric sphere

$$\sigma_b=4\pi\left|\frac{N^2-1}{N^2+2}\right|^2k_1^4a^6$$

and, for the perfectly conducting sphere,

$$\sigma_b=9\pi k_1^4a^6.$$

The corresponding formulae for the acoustic sphere can be deduced without difficulty.

8.21 Rayleigh–Gans scattering by a diaphanous sphere

Rayleigh scattering asks that the radius of the sphere be small compared with the wavelength both outside and inside the sphere but places no restriction on the refractive index. The other extreme is to require the refractive index to be near unity but to dispense with the conditions on the radius. A sphere for which $k_2\approx k_1$ is called *diaphanous*.

For a diaphanous sphere the Rayleigh–Gans approximation of §6.14 can be used. This approximation provides a scattered wave in the acoustic case when $\zeta = 1$ of

$$p_s = \int_T (k_2^2 - k_1^2) p^i(\mathbf{y}) \psi(\mathbf{x}, \mathbf{y})\, d\mathbf{y} \tag{21.1}$$

where $\psi(\mathbf{x}, \mathbf{y}) = \exp(-ik_1 |\mathbf{x} - \mathbf{y}|)/4\pi |\mathbf{x} - \mathbf{y}|$ and p^i is the incident field. At large distances from the obstacle in the direction of the unit vector $\hat{\mathbf{x}}$

$$p_s \sim \frac{k_1^2 e^{-ik_1 R}}{4\pi R} \int_T (N^2 - 1) p^i(\mathbf{y}) e^{ik_1 \hat{\mathbf{x}} \cdot \mathbf{y}}\, d\mathbf{y}$$

where $N = k_2/k_1$. In particular, when $p^i(\mathbf{x}) = e^{-ik\mathbf{n}^i \cdot \mathbf{x}}$,

$$p_s \sim \frac{k_1^2 e^{-ik_1 R}}{4\pi R} \int_T (N^2 - 1) e^{ik_1 \mathbf{y} \cdot (\hat{\mathbf{x}} - \mathbf{n}^i)}\, d\mathbf{y}.$$

When N is the same at all points of the sphere the factor $N^2 - 1$ may be moved outside the integral. To evaluate the integral choose temporarily the z-axis parallel to $\mathbf{n}^i - \hat{\mathbf{x}}$. Then the integral is

$$\int_0^{2\pi} \int_0^{\pi} \int_0^b e^{-2ik_1 R \cos\theta \sin\frac{1}{2}\alpha} R^2 \sin\theta\, dR\, d\theta\, d\phi$$

where α is the angle between $\mathbf{n}^i$ and $\hat{\mathbf{x}}$. From (11.5) the integral is

$$2\pi \sum_{n=0}^{\infty} \int_0^{\pi} \int_0^b (2n+1)(-i)^n j_n(2k_1 R \sin\tfrac{1}{2}\alpha) P_n(\cos\theta) R^2 \sin\theta\, dR\, d\theta$$

$$= 4\pi \int_0^b j_0(2k_1 R \sin\tfrac{1}{2}\alpha) R^2\, dR$$

from (11.6) (with $k = 0$) or (B.3.19). Invoking (A.4.7) we obtain

$$p_s \sim \frac{k_1 b^2 (N^2 - 1)}{2R \sin\frac{1}{2}\alpha} e^{-ik_1 R} j_1(2k_1 b \sin\tfrac{1}{2}\alpha). \tag{21.2}$$

Small values of $k_1 b$ result in a uniform distribution of radiation. For large $k_1 b$ the scattered field is concentrated about $\alpha = 0$, i.e. in the direction in which the incident plane wave is travelling.

The scattering coefficient may be obtained via (21.2) by integration over a large sphere. It is given by

$$\sigma_c^A = \tfrac{1}{2}(k_1 b\, |N^2 - 1|)^2 \int_0^{\pi} \frac{j_1^2(2k_1 b \sin\frac{1}{2}\alpha)}{\sin^2\frac{1}{2}\alpha} \sin\alpha\, d\alpha.$$

Take account of (A.4.14) and substitute $v = 2k_1 b \sin\frac{1}{2}\alpha$; then

$$\sigma_c^A = 2(k_1 b\, |N^2 - 1|)^2 \int_0^{2k_1 b} \left(\frac{\sin v}{v} - \cos v\right)^2 \frac{dv}{v^3}.$$

Complete the square and convert the products of trigonometric functions to have argument $2v$. Then integration by parts leads to

$$\sigma_c^A = 2(|N^2-1|/4k_1b)^2\{4(k_1b)^4-(k_1b)^2+\tfrac{1}{2}k_1b\sin 4k_1b-\tfrac{1}{8}(1-\cos 4k_1b)\}. \tag{21.3}$$

For small values of k_1b

$$\sigma_c^A \approx \tfrac{4}{9}|N^2-1|^2(k_1b)^4$$

consistent with (19.3) when $\zeta=1$. Large values of k_1b give

$$\sigma_c^A \sim \tfrac{1}{2}|N^2-1|^2(k_1b)^2 \tag{21.4}$$

According to (19.9) and (21.2)

$$\sigma_c^A+\sigma_a^A = -\tfrac{4}{3}k_1b\,\mathrm{Im}(N^2-1).$$

In view of the smallness of N^2-1 and (21.3)

$$\sigma_a^A = -\tfrac{4}{3}k_1b\,\mathrm{Im}(N^2-1).$$

The accuracy of the approximation can be estimated by regarding it as the first stage in an iteration in which

$$p_n = \int_T k_1^2(N^2-1)p_{n-1}\psi(\boldsymbol{x},\boldsymbol{y})\,\mathrm{d}\boldsymbol{y}.$$

If p_{n-1} is bounded by M_{n-1}

$$|p_n| \leqslant k_1^2|N_1^2-1|\,M_{n-1}\int_T \frac{\mathrm{d}\boldsymbol{y}}{4\pi|\boldsymbol{x}-\boldsymbol{y}|}.$$

The integral is the potential at an interior point of a uniform distribution of charge. Such a potential is easily calculated for a sphere and is found to be $\tfrac{1}{2}(b^2-\tfrac{1}{3}R^2)$ which never exceeds $\tfrac{1}{2}b^2$. Hence

$$M_n \leqslant \tfrac{1}{2}(k_1b)^2|N^2-1|\,M_{n-1}. \tag{21.5}$$

The iteration scheme, therefore, certainly converges if $\tfrac{1}{2}(k_1b)^2|N^2-1|<1$. However, the retention of the first term alone will be a good approximation only if all the remaining ones are much smaller. Thus

$$(k_1b)^2|N^2-1| \ll 1 \tag{21.6}$$

is the condition of validity of Rayleigh–Gans theory. It will be noted that (21.6) ensures the smallness of σ_c^A in (21.4) despite the largeness of k_1b.

All the theory can be carried over to the electromagnetic case. If $\mu_1=\mu_2$ the approximation for the scattered electric intensity is

$$\boldsymbol{E}_s = (\mathrm{grad\,div}+k_1^2)\int_T (N^2-1)\boldsymbol{E}^i\psi(\boldsymbol{x},\boldsymbol{y})\,\mathrm{d}\boldsymbol{y}.$$

The incident plane wave $\boldsymbol{E}^{\mathrm{i}}=\boldsymbol{e}^{\mathrm{i}}\mathrm{e}^{-\mathrm{i}k_1\boldsymbol{n}^{\mathrm{i}}\cdot\boldsymbol{x}}$ entails the same integral as p_{s} and

$$\boldsymbol{E}_{\mathrm{s}}\sim\frac{k_1a^2(N^2-1)}{2R\sin\frac{1}{2}\alpha}\{\boldsymbol{e}^{\mathrm{i}}-(\boldsymbol{e}^{\mathrm{i}}\cdot\hat{\boldsymbol{x}})\hat{\boldsymbol{x}}\}\mathrm{e}^{-\mathrm{i}k_1R}\mathrm{j}_1(2k_1a\sin\tfrac{1}{2}\alpha). \tag{21.7}$$

The use of the plane P of the preceding section shows that the scattered fields due to E_{p} and E_{ℓ} are

$$\frac{k_1a^2(N^2-1)}{2R\sin\frac{1}{2}\alpha}\mathrm{e}^{-\mathrm{i}k_1R}\mathrm{j}_1(2k_1a\sin\tfrac{1}{2}\alpha)E_{\mathrm{p}}$$

and

$$\frac{k_1a^2(N^2-1)}{2R\sin\frac{1}{2}\alpha}\mathrm{e}^{-\mathrm{i}k_1R}\mathrm{j}_1(2k_1a\sin\tfrac{1}{2}\alpha)E_{\ell}\cos\alpha$$

respectively. Hence the intensity with incident natural light is

$$\frac{k_1^2a^4\,|N^2-1|^2}{8R^2\sin^2\frac{1}{2}\alpha}\mathrm{j}_1^2(2k_1a\sin\tfrac{1}{2}\alpha)I^{\mathrm{i}}(1+\cos^2\alpha). \tag{21.8}$$

For small values of k_1a the radiation pattern is almost uniform but, for large k_1a, the concentration in the forward direction $\alpha=0$ is strongly pronounced.

The scattering coefficient obtained by integrating (21.8) over a sphere is

$$\begin{aligned}\sigma_{\mathrm{c}}^{\mathrm{E}}&=\tfrac{1}{4}k_1^2a^2\,|N^2-1|^2\int_0^{\pi}\frac{\mathrm{j}_1^2(2k_1a\sin\frac{1}{2}\alpha)}{\sin^2\frac{1}{2}\alpha}(1+\cos^2\alpha)\sin\alpha\,\mathrm{d}\alpha\\&=\tfrac{1}{4}|N^2-1|^2\left[\tfrac{5}{2}-\frac{\sin 4k_1a}{4k_1a}+\frac{7}{16k_1^2a^2}(\cos 4k_1a-1)+2k_1^2a^2\right.\\&\quad\left.+\left(\frac{1}{2k_1^2a^2}-2\right)\{\gamma+\ln 4k_1a-\mathrm{Ci}(4k_1a)\}\right]\end{aligned} \tag{21.9}$$

where

$$\mathrm{Ci}(x)=-\int_x^{\infty}\cos v\frac{\mathrm{d}v}{v}.$$

For small k_1a

$$\sigma_{\mathrm{c}}^{\mathrm{E}}\approx\tfrac{8}{27}|N^2-1|^2\,(k_1a)^4$$

consistent with (19.4) and, for large values of k_1a,

$$\sigma_{\mathrm{c}}^{\mathrm{E}}\sim\tfrac{1}{2}|N^2-1|^2\,(k_1a)^2. \tag{21.10}$$

Furthermore

$$\sigma_{\mathrm{a}}^{\mathrm{E}}=-\tfrac{4}{3}k_1a\,\mathrm{Im}(N^2-1).$$

The extra grad div alters (21.6) to

$$\{(k_1a)^2+1\}\,|N^2-1|\ll 1 \tag{21.11}$$

which is still sufficient to keep the right-hand side of (21.10) small.

8.22 High frequency scattering by a sphere

The calculation of the field distribution when the radius is large and N^2 not near unity is a much more difficult task. To limit the ramifications only the electromagnetic problem will be discussed in detail. Broadly speaking k_1a terms of the series must be taken to achieve some kind of accuracy. The same sort of behaviour occurred for the circular cylinder and the method of transforming the series to an integral in §8.3 may be adapted to the present problem. It will therefore be sufficient to bring out the main differences caused by working in three dimensions rather than two. For simplicity, the incident wave will be taken as plane.

An incident plane wave requires the calculation of the series for W_1 and W_2 in (16.14). It will suffice to discuss W_1 since A_n can be changed to B_n by substituting $1/\kappa$ for κ. First, replace P_n^1 from (B.3.14) with $m=-1$ and then write

$$\bar{W}_1\cos\phi = k_1W_1+\mathrm{i}\{\mathrm{j}_0(k_1R)+A_0\mathrm{h}_0^{(2)}(k_1R)\}\mathrm{P}_0^{-1}(\cos\theta)\cos\phi$$

so that

$$\bar{W}_1=-\sum_{n=0}^{\infty}(-\mathrm{i})^{n+1}(2n+1)\{\mathrm{j}_n(k_1R)+A_n\mathrm{h}_n^{(2)}(k_1R)\}\mathrm{P}_n^{-1}(\cos\theta)$$

$$=\tfrac{1}{2}\int_{C_1}(2\nu+1)\{\mathrm{j}_\nu(k_1R)+A_\nu\mathrm{h}_\nu^{(2)}(k_1R)\}\mathrm{P}_\nu^{-1}(\cos\theta)\mathrm{e}^{\frac{1}{2}\nu\pi\mathrm{i}}\frac{\mathrm{d}\nu}{\sin\nu\pi}$$

where C_1 is the contour in the complex ν-plane shown in Fig. 8.15. The integrand can be rewritten through

$$\mathrm{j}_\nu(k_1R)+A_\nu\mathrm{h}_\nu^{(2)}(k_1R)=\tfrac{1}{2}\{\mathrm{h}_\nu^{(1)}(k_1R)+r_\nu\mathrm{h}_\nu^{(2)}(k_1R)\}$$

$$r_\nu=\frac{\kappa\varphi_\nu^{(1)}(k_2a)\varphi_\nu^{(3)\prime}(k_1a)-\varphi_\nu^{(1)\prime}(k_2a)\varphi_\nu^{(3)}(k_1a)}{\varphi_\nu^{(1)\prime}(k_2a)\varphi_\nu^{(4)}(k_1a)-\kappa\varphi_\nu^{(1)}(k_2a)\varphi_\nu^{(4)\prime}(k_1a)}.$$

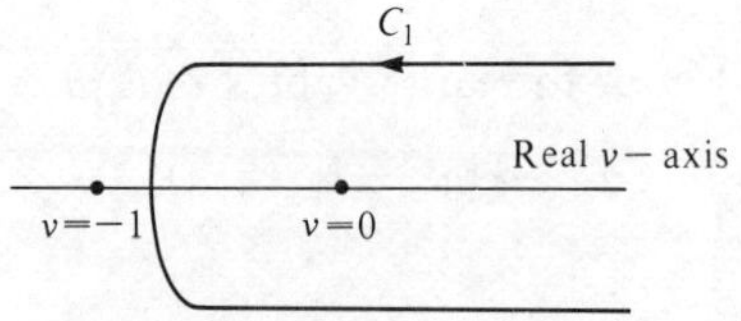

Fig. 8.15 Contour for $\bar{W}_1$

Let

$$r_{21}=\frac{\kappa\varphi_\nu^{(4)\prime}(k_1a)\varphi_\nu^{(4)}(k_2a)-\varphi_\nu^{(4)}(k_1a)\varphi_\nu^{(4)\prime}(k_2a)}{\kappa\varphi_\nu^{(4)\prime}(k_1a)\varphi_\nu^{(3)}(k_2a)-\varphi_\nu^{(4)}(k_1a)\varphi_\nu^{(3)\prime}(k_2a)}\cdot\frac{\varphi_\nu^{(3)}(k_2a)}{\varphi_\nu^{(4)}(k_2a)}.$$

Deform the contour into a path which starts from infinity, goes slightly above the curve on which $-r_{21}\varphi_\nu^{(4)}(k_2a)/\varphi_\nu^{(3)}(k_2a)$ takes the value unity to the origin and then along the positive imaginary axis. It can be verified that

$$\begin{aligned}r_\nu&=r_{12}-\frac{T_{12}T_{21}}{1+r_{21}\varphi_\nu^{(4)}(k_2a)/\varphi_\nu^{(3)}(k_2a)}\cdot\frac{\varphi_\nu^{(3)}(k_1a)\varphi_\nu^{(4)}(k_2a)}{\varphi_\nu^{(4)}(k_1a)\varphi_\nu^{(3)}(k_2a)}\\&=r_{12}-T_{12}T_{21}\frac{\varphi_\nu^{(3)}(k_1a)\varphi_\nu^{(4)}(k_2a)}{\varphi_\nu^{(4)}(k_1a)\varphi_\nu^{(3)}(k_2a)}\sum_{n=0}^{\infty}\left\{-r_{21}\frac{\varphi_\nu^{(4)}(k_2a)}{\varphi_\nu^{(3)}(k_2a)}\right\}^n\end{aligned}$$

where

$$r_{12}=\frac{\kappa\varphi_\nu^{(3)\prime}(k_1a)\varphi_\nu^{(3)}(k_2a)-\varphi_\nu^{(3)}(k_1a)\varphi_\nu^{(3)\prime}(k_2a)}{\kappa\varphi_\nu^{(4)\prime}(k_1a)\varphi_\nu^{(3)}(k_2a)-\varphi_\nu^{(4)}(k_1a)\varphi_\nu^{(3)\prime}(k_2a)}\cdot\frac{\varphi_\nu^{(4)}(k_1a)}{\varphi_\nu^{(3)}(k_1a)},$$

$$T_{12}=1-r_{12},\qquad T_{21}=1-r_{21}.$$

When the series is substituted in the integral it is discovered that the term $\mathrm{h}_\nu^{(1)}(k_1R)+r_{12}\mathrm{h}_\nu^{(2)}(k_1R)$ contributes the primary and reflected waves whereas the other terms account for the field transmitted through the sphere, n corresponding to the ray which has undergone internal reflection $n+1$ times.

The theory is somewhat simpler when the sphere is perfectly conducting because of the absence of the complicated field transmitted through the sphere. In this case

$$A_\nu=-\varphi_\nu^{(1)\prime}(k_1a)/\varphi_\nu^{(4)\prime}(k_1a),\qquad r_\nu=-\varphi_\nu^{(3)\prime}(k_1a)/\varphi_\nu^{(4)\prime}(k_1a).$$

Let C_1 cross the real axis at $\nu=-\frac{1}{2}$ and change the variable of integration in the upper half-plane from ν to $-\nu-1$. Then, from (B.1.8), (A.1.16), $r_{-\nu-1}=-\mathrm{e}^{2\nu\pi\mathrm{i}}r_\nu$ and

$$\bar{W}_1=\tfrac{1}{4}\int_D(2\nu+1)\{\mathrm{h}_\nu^{(1)}(k_1R)+r_\nu\mathrm{h}_\nu^{(2)}(k_1R)\}\mathrm{P}_\nu^{-1}(\cos\theta)\frac{\mathrm{e}^{\frac{1}{2}\nu\pi\mathrm{i}}}{\sin\nu\pi}\,\mathrm{d}\nu$$

where D is the contour of Fig. 8.3. The poles of r_ν can be obtained from those of (3.3) by putting $b=a$, $\eta=1/2a$ and replacing ν by $\nu+\frac{1}{2}$. Since η/k_1 is not large the poles near k_1a are given by (3.8), i.e.

$$\nu_s+\tfrac{1}{2}=k_1a+(\tfrac{1}{2}k_1a)^{\frac{1}{3}}\beta_s\mathrm{e}^{-\frac{1}{3}\pi\mathrm{i}}.$$

Clearly the distribution of poles is much as in Fig. 8.4. On account of (B.2.3) the behaviour of the integrand is very similar to that of (3.3). Thus, in the shadow there will be a residue series of exponentially damped terms. However, there is one important difference from the cylinder. The formula (B.2.3) fails when $\theta \approx 0$ and in this neighbourhood the field is no longer exponentially damped. In fact, it is substantially larger than elsewhere in the shadow. Consequently, there is a tendency for a *bright spot* to appear at $\theta = 0$ on the back of the sphere as $k_1 a$ increases. The phenomenon does not display itself to anything like the same extent on a cylinder.

In order to obtain values in the illuminated region introduce the function

$$\mathrm{R}_\nu^\mu(\cos\theta) = \mathrm{P}_\nu^\mu(\cos\theta) + 2\mathrm{i}\mathrm{Q}_\nu^\mu(\cos\theta)/\pi.$$

Then, from (B.1.24),

$$\mathrm{P}_\nu^{-1}(\cos\theta) = -\mathrm{e}^{-\nu\pi\mathrm{i}}\mathrm{P}_\nu^{-1}(-\cos\theta) - \mathrm{i}\mathrm{R}_\nu^{-1}(-\cos\theta)\sin\nu\pi.$$

The first term gives rise to a residue series while the second supplies the primary and reflected waves. To verify this note that, from (B.2.1) and (B.2.3), if $\varepsilon \leqslant \theta \leqslant \pi - \varepsilon$

$$\mathrm{R}_\nu^\mu(\cos\theta) \sim \frac{(\nu+\mu)!}{(\nu+\frac{1}{2})!}\left(\frac{2}{\pi\sin\theta}\right)^{\frac{1}{2}} \exp[-\mathrm{i}\{(\nu+\tfrac{1}{2})\theta - \tfrac{1}{4}\pi + \tfrac{1}{2}\mu\pi\}] \quad (22.1)$$

as $|\nu| \to \infty$, with the understanding that $-\cos\theta$ is to mean $\cos(\pi-\theta)$.

The functions in the integrand are replaced by their asymptotic approximations and the integral is then evaluated by the method of steepest descents. For $|\nu+\frac{1}{2}| < k_1 a$ there is a saddle-point satisfying

$$\cos^{-1}\{(\nu+\tfrac{1}{2})/k_1 R\} = \tfrac{1}{2}\pi - \theta + 2\cos^{-1}\{(\nu+\tfrac{1}{2})/k_1 a\}.$$

This supplies the reflected ray, $(\nu+\frac{1}{2})/k_1$ being the perpendicular distance y_1 from the centre of the sphere to the incident ray which initiates the reflected one. (Fig. 8.16). There is also a saddle-point where

$$\cos^{-1}\{(\nu+\tfrac{1}{2})/k_1 R\} = \theta - \tfrac{1}{2}\pi$$

for $|\nu+\frac{1}{2}| < k_1 a$ and one of the two possibilities

$$\cos^{-1}\{(\nu+\tfrac{1}{2})/k_1 R\} = \pm(\tfrac{1}{2}\pi - \theta)$$

also exists if $k_1 a < |\nu+\frac{1}{2}| < k_1 R$. These saddle-points are responsible for the primary wave. There results

$$\bar{W}_1 \sim \frac{\mathrm{e}^{-\mathrm{i}k_1 R\cos\theta}}{k_1 R\sin\theta} + R_\mathrm{F} D \frac{\mathrm{e}^{-\mathrm{i}k_1(R_1 - z_1)}}{k_1 R_1 y_1} \quad (22.2)$$

where $z_1 = (a^2 - y_1^2)^{\frac{1}{2}}$,

$$R_\mathrm{F} = -\frac{1+2\mathrm{i}k_1 a z_1}{1-2\mathrm{i}k_1 a z_1},$$

$$D = \left\{\frac{y_1 z_1 R_1^2}{(2R_1 + z_1)R\sin\theta}\right\}^{\frac{1}{2}}$$

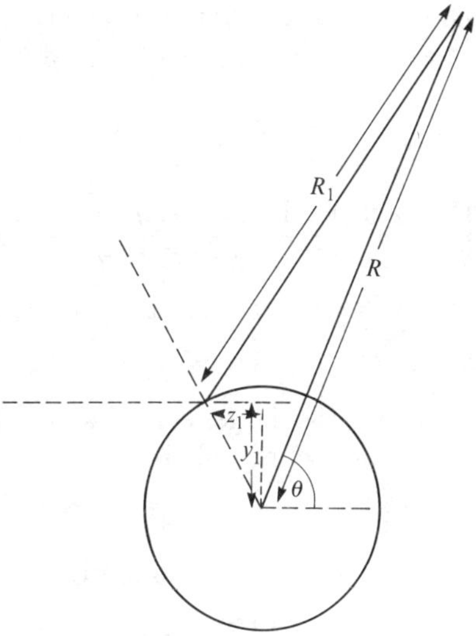

Fig. 8.16 Ray diagram for the perfectly conducting sphere

and R_1 is displayed in Fig. 8.16. R_F is the usual Fresnel reflection coefficient relevant to perfect conductivity. D is the divergence factor allowing for the fact that the reflected rays are not parallel to one another.

Formula (22.2) is not valid when θ is near 0 or π because then (22.1) is vitiated. Now it is appropriate to employ (B.2.4) when θ is small. An analysis similar to that for the cylinder leads to

$$E_\theta \sim \frac{\mathrm{i}e^{-\mathrm{i}k_1R}}{k_1R}\cos\phi\left[\left(\frac{1}{\theta^2}-\frac{1}{\theta\sin\theta}\right)\{1-\mathrm{J}_0(k_1a\theta)\}-\frac{k_1a}{\theta}\mathrm{J}_1(k_1a\theta)\right.$$
$$\left.-(\tfrac{1}{2}k_1a\mathrm{e}^{-\pi\mathrm{i}})^{\frac{1}{3}}\left\{h_1'(\theta)+h_2'(\theta)+\frac{l_1(\theta)+l_2(\theta)}{\sin\theta}\right\}\right],$$

$$E_\phi \sim \frac{\mathrm{i}e^{-\mathrm{i}k_1R}}{k_1R}\sin\phi\left[\left(\frac{1}{\theta\sin\theta}-\frac{1}{\theta^2}\right)\{1-\mathrm{J}_0(k_1a\theta)\}+\frac{k_1a}{\theta}\mathrm{J}_1(k_1a\theta)\right.$$
$$\left.+(\tfrac{1}{2}k_1a\mathrm{e}^{-\pi\mathrm{i}})^{\frac{1}{3}}\left\{\frac{h_1(\theta)+h_2(\theta)}{\sin\theta}+l_1'(\theta)+l_2'(\theta)\right\}\right]$$

at large distances for small θ. Here

$$h_1(\theta)=\int_0^\infty \frac{\mathrm{Ai}(x)-(4k_1a)^{\frac{2}{3}}\,\mathrm{Ai}'(x)}{(4k_1a)^{\frac{2}{3}}\mathrm{e}^{\frac{1}{3}\pi\mathrm{i}}\,\mathrm{Ai}'(x\mathrm{e}^{-\frac{2}{3}\pi\mathrm{i}})+\mathrm{Ai}(x\mathrm{e}^{-\frac{2}{3}\pi\mathrm{i}})}$$
$$\times\frac{\mathrm{J}_1[\{k_1a+(\frac{1}{2}k_1a)^{\frac{1}{3}}x\}\theta]}{1+2^{-\frac{1}{3}}(k_1a)^{-\frac{2}{3}}x}\,\mathrm{d}x,$$

$$h_2(\theta)=\int_0^\infty \frac{(4k_1a)^{\frac{2}{3}}\mathrm{e}^{-\frac{1}{3}\pi\mathrm{i}}\mathrm{Ai}'(x)+\mathrm{Ai}(x)}{(4k_1a)^{\frac{2}{3}}\mathrm{e}^{\frac{1}{3}\pi\mathrm{i}}\,\mathrm{Ai}'(x\mathrm{e}^{\frac{2}{3}\pi\mathrm{i}})+\mathrm{Ai}(x\mathrm{e}^{\frac{2}{3}\pi\mathrm{i}})}\times\frac{\mathrm{J}_1[\{k_1a-(\frac{1}{2}k_1a\mathrm{e}^{\pi\mathrm{i}})^{\frac{1}{3}}x\}\theta]}{1-\mathrm{e}^{\frac{1}{3}\pi\mathrm{i}}(k_1a)^{-\frac{2}{3}}2^{-\frac{1}{3}}x}\,\mathrm{d}x,$$

$$l_1(\theta)=\int_0^\infty \frac{\mathrm{Ai}(x)}{\mathrm{Ai}(x\mathrm{e}^{-\frac{2}{3}\pi\mathrm{i}})}\frac{\mathrm{J}_1[\{k_1a+(\frac{1}{2}k_1a)^{\frac{1}{3}}x\}\theta]}{1+x/2^{\frac{1}{3}}(k_1a)^{\frac{2}{3}}}\,\mathrm{d}x,$$

$$l_2(\theta)=\int_0^\infty \frac{\mathrm{Ai}(x)}{\mathrm{Ai}(x\mathrm{e}^{\frac{2}{3}\pi\mathrm{i}})}\frac{\mathrm{J}_1[\{k_1a-(\frac{1}{2}k_1a\mathrm{e}^{\pi\mathrm{i}})^{\frac{1}{3}}x\}\theta]}{1-\mathrm{e}^{\frac{1}{3}\pi\mathrm{i}}x/2^{\frac{1}{3}}(k_1a)^{\frac{2}{3}}}\,\mathrm{d}x.$$

It follows, from (19.8) and the largeness of k_1a, that the scattering coefficient of the perfectly conducting sphere is

$$\sigma_c^E\sim 2+(\tfrac{1}{2}k_1a)^{-\frac{2}{3}}\,\mathrm{Im}\{\mathrm{f_S}(0)+\mathrm{f_H}(0)\}$$

where $\mathrm{f_S}$ and $\mathrm{f_H}$ are as in Table 8.4. Consequently

$$\sigma_c^E\sim 2+0.1320(k_1a)^{-\frac{2}{3}}.$$

The next term in the series has been calculated and

$$\sigma_c^E\sim 2+0.1320(k_1a)^{-\frac{2}{3}}+0.9706(k_1a)^{-\frac{4}{3}}. \qquad (22.3)$$

The corresponding formulae for the soft and hard spheres are

$$\sigma_c^S\sim 2+1.9924(k_1a)^{-\frac{2}{3}}-0.715(k_1a)^{-\frac{4}{3}}, \qquad (22.4)$$

$$\sigma_c^H\sim 2-1.7284(k_1a)^{-\frac{2}{3}}-2.0104(k_1a)^{-\frac{4}{3}}. \qquad (22.5)$$

It will be observed that, up to the terms in $(k_1a)^{-\frac{2}{3}}$, (22.3) is the mean of (22.4) and (22.5).

The fact that $\sigma_c\to 2$ as the wavelength steadily diminishes calls for comment. It might be expected that at very high frequencies only the energy incident on the sphere would be scattered, implying a scattering coefficient of 1. However, there is a shadow zone behind the sphere in which the scattered field annuls the incident wave. This carries the same energy as the reflected field and so $\sigma_c\to 2$.

8.23 Propagation near a spherical earth

Consider a point sound source placed at (R_0, θ_0, ϕ_0) near a spherical earth, the wavelength being much shorter than the radius b. Let the boundary condition on p at the surface of the earth be

$$\partial p/\partial R=\mathrm{i}\nu_0 p. \qquad (23.1)$$

Then, from (12.4), for $R<R_0$

$$p=(k_1/4\pi\mathrm{i})\sum_{n=0}^{\infty}(2n+1)\{\mathrm{j}_n(k_1R)+C_n\mathrm{h}_n^{(2)}(k_1R)\}\mathrm{h}_n^{(2)}(k_1R_0)\mathrm{P}_n(\cos\Theta)$$

where

$$\cos\Theta = \sin\theta\,\sin\theta_0\cos(\phi-\phi_0)+\cos\theta\,\cos\theta_0,$$

$$C_n = -\frac{k_1 \mathrm{j}_n'(k_1 b) - \mathrm{i}\nu_0 \mathrm{j}_n(k_1 b)}{k_1 \mathrm{h}_n^{(2)\prime}(k_1 b) - \mathrm{i}\nu_0 \mathrm{h}_n^{(2)}(k_1 b)}.$$

An argument similar to that of the preceding section reveals that

$$p = -\frac{k_1}{8\pi}\int_D (2\nu+1)\{\mathrm{j}_\nu(k_1 R) + C_\nu \mathrm{h}_\nu^{(2)}(k_1 R)\}\mathrm{h}_\nu^{(2)}(k_1 R_0)\frac{\mathrm{P}_\nu(-\cos\Theta)}{\sin\nu\pi}\,\mathrm{d}\nu. \tag{23.2}$$

Exclude from the discussion the possibilities that $\Theta \approx 0$ (when the point of observation is virtually coincident with the source) and that $\Theta \approx \pi$ (when the point of observation is in the region of the bright spot). Then (B.2.3) can be employed. Furthermore, it is evident that in the preceding analysis the dominant part of the field at high frequencies can be obtained by assuming in the trigonometric functions that ν has a large negative imaginary part. Making the same assumption here we have

$$p \sim \frac{k_1}{4\pi\mathrm{i}}\frac{\mathrm{e}^{\mathrm{i}(\frac{1}{4}\pi - \frac{1}{2}\Theta)}}{(\frac{1}{2}\pi\sin\Theta)^{\frac{1}{2}}}\int_D \frac{\nu!}{(\nu-\frac{1}{2})!}\{\mathrm{j}_\nu(k_1 R) + C_\nu \mathrm{h}_\nu^{(2)}(k_1 R)\}\mathrm{h}_\nu^{(2)}(k_1 R_0)\mathrm{e}^{-\mathrm{i}\nu\Theta}\,\mathrm{d}\nu. \tag{23.3}$$

When R and R_0 are both nearly equal to b the main contribution to p comes from a neighbourhood of $\nu = k_1 b$. Put $R = b+z$, $R_0 = b+z_0$. Introduce the approximations valid for Bessel functions in the neighbourhood and use Stirling's formula for the factorials. Then, on putting $\nu+\frac{1}{2} = k_1 b(1+\frac{1}{2}\mu)$ and neglecting higher powers of μ, we obtain

$$p \sim \frac{k_1}{4\pi\mathrm{i}}\frac{\mathrm{e}^{\mathrm{i}(\frac{1}{4}\pi - k_1 b\Theta)}}{(\frac{1}{2}k_1 b)^{\frac{1}{6}}}\left(\frac{\pi}{\sin\Theta}\right)^{\frac{1}{2}}\int_{-\infty}^{\infty} \mathrm{e}^{-\frac{1}{2}\mathrm{i}k_1 b\mu\Theta} f\left(\frac{2z_0}{b}-\mu-1\right)$$
$$\times\left\{g\left(\frac{2z}{b}-\mu-1\right) - R(-1-\mu)f\left(\frac{2z}{b}-\mu-1\right)\right\}\mathrm{d}\mu \tag{23.4}$$

where

$$R(\mu) = \frac{g'(\mu) - \frac{1}{2}\mathrm{i}\nu_0 b g(\mu)}{f'(\mu) - \frac{1}{2}\mathrm{i}\nu_0 b f(\mu)}$$

and

$$f(x) = \mathrm{e}^{\frac{1}{3}\pi\mathrm{i}}\,\mathrm{Ai}\{(\tfrac{1}{2}k_1 b)^{\frac{2}{3}}(1+x)\mathrm{e}^{\frac{1}{3}\pi\mathrm{i}}\},$$
$$g(x) = \mathrm{Ai}\{-(\tfrac{1}{2}k_1 b)^{\frac{2}{3}}(1+x)\}.$$

There is a close resemblance between (23.4) and (6.29.9). Now, in the deductions from (6.29.9) it was a neighbourhood of $\lambda = 1$ which played a significant role. If $k_1 r$ is large the Bessel function can be replaced by its asymptotic expression. Then on putting $\lambda = 1 + \frac{1}{2}\mu$, the agreement between (23.4) and (6.29.9) becomes complete, with the identification

$$r = b\Theta, \qquad q = 2/b, \qquad \zeta = 2z/b, \qquad \zeta_0 = 2z_0/b, \qquad \eta = (\tfrac{1}{2}k_1 b)^{\frac{2}{3}}$$

provided that (6.29.9) is multiplied by $k_1(\Theta/\sin\Theta)^{\frac{1}{2}}$. Consequently, *near a spherical earth p behaves as it would above a plane earth with a quasi-homogeneous standard atmosphere horizontally stratified with a refractive index of* $1/b$. Some of the relations joining the parameters for the plane and spherical earths are displayed in Fig. 8.17.

The identification of (23.4) and (6.29.9) makes available all the results of §§6.29 and 6.30. Thus, from (6.29.19) the field beyond the horizon is given by

$$p \sim k_1(2k_1^3 b^3 \pi \sin\Theta)^{-\frac{1}{2}} e^{-\frac{1}{4}\pi i} \sum_{s=1}^{\infty} \frac{f(2z_0/b - \lambda_s^2) f(2z/b - \lambda_s^2) \exp(-i\lambda_s k_1 b\Theta)}{\lambda_s^{\frac{1}{2}}(1 - \lambda_s^2 - \nu_0^2/k_1^2) f^2(-\lambda_s)}.$$

The height-gain factor is $1 + i\nu_0 z$ for $(\frac{1}{2}k_1 b)^{\frac{2}{3}} z \ll \frac{1}{2}b$. The horizon is specified by, according to (6.30.11),

$$\Theta = (2z/b)^{\frac{1}{2}} + (2z_0/b)^{\frac{1}{2}}.$$

Well above the horizon, ray theory is valid, i.e.

$$p \sim \frac{e^{ik_1 R_s}}{4\pi R_s} + D\frac{k_1\cos\theta_2 - \nu_0}{k_1\cos\theta_2 + \nu_0}\frac{\exp\{-ik_1(R_1+R_2)\}}{4\pi(R_1+R_2)}$$

where the divergence factor D is given by

$$D = b(R_1+R_2)\left\{\frac{\sin\theta_2\cos\theta_2}{RR_0\sin\Theta\,(R_1 R\cos\theta_3 + R_2 R_0\cos\theta_1)}\right\}^{\frac{1}{2}}$$

and the various quantities are defined in Fig. 8.18.

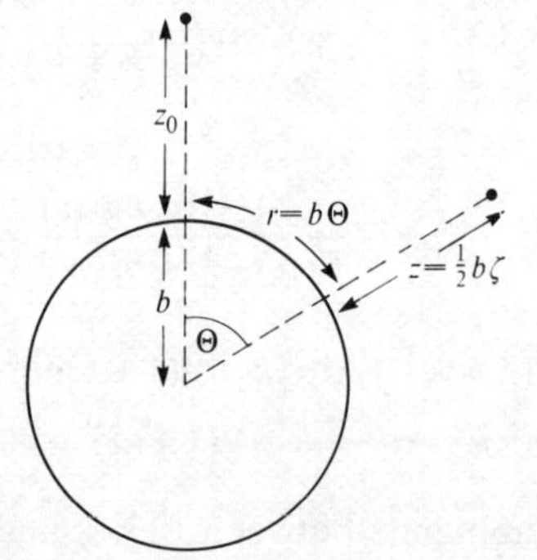

Fig. 8.17 Connection between the parameters for a spherical earth and those for a stratified atmosphere over a plane earth

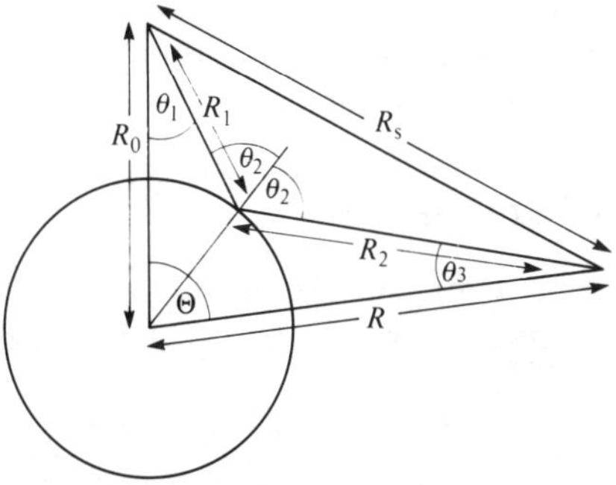

Fig. 8.18 Parameters for ray theory

For the transition region encompassing the horizon (6.30.13) is relevant. In particular, if $(\frac{1}{2}k_1 b)^{\frac{1}{3}}|\nu_0| \gg k_1$, (6.30.15) can be quoted with the result

$$p \sim \frac{k_1 e^{\frac{1}{4}\pi i - ik_1 L_h}}{8\pi(k_1^2 z z_0)^{\frac{1}{4}}(\pi \sin \Theta)^{\frac{1}{2}}}$$
$$\times \left[\frac{1}{(k_1 L_t)^{\frac{1}{2}}} \mathrm{F}\left\{\frac{1}{2}\left(\frac{k_1}{L_t}\right)^{\frac{1}{2}} L_b\right\} + \tfrac{1}{2}\mathrm{i}(\tfrac{1}{2}k_1 b)^{-\frac{2}{3}} \mathrm{f_S}\left\{-\tfrac{1}{2}(\tfrac{1}{2}k_1 b)^{\frac{1}{3}} L_b\right\}\right]$$

where

$$L_h = \tfrac{1}{3}b\{3\Theta + (2z/b)^{\frac{3}{2}} + (2z_0/b)^{\frac{3}{2}}\},$$
$$L_t = \tfrac{1}{2}b\{(2z_0/b)^{-\frac{1}{2}} + (2z/b)^{-\frac{1}{2}} - \Theta\},$$
$$L_b = b\Theta - (2zb)^{\frac{1}{2}} - (2z_0 b)^{\frac{1}{2}}$$

and $\mathrm{f_S}$ is given in Table 8.4.

The electromagnetic field due to an electric dipole oriented along the radius vector from the centre of the sphere can be deduced from the foregoing. Suppose the field is subject to the impedance boundary condition on $R = a$ of

$$\boldsymbol{E}_t = N_{12}(\mu_1/\varepsilon_1)^{\frac{1}{2}} \boldsymbol{i}_1 \wedge \boldsymbol{H}_t$$

where the suffix t signifies a tangential component. The total field can be represented from (16.13) in terms of a Hertz vector Π by

$$\boldsymbol{E} = \mathrm{curl\,curl}(\boldsymbol{x}\Pi/\varepsilon)$$

where, if the primary wave part of Π is $e^{-ik_1 R_s}/4\pi R_s$,

$$\Pi = (k_1/4\pi \mathrm{i}) \sum_{n=0}^{\infty} (2n+1)\{\mathrm{j}_n(k_1 R) + A_n \mathrm{h}_n^{(2)}(k_1 R)\}\mathrm{h}_n^{(2)}(k_1 R_0)\mathrm{P}_n(\cos \Theta),$$

$$A_n = \frac{\varphi_n^{(1)\prime}(k_1 a) - \mathrm{i}N_{12}\varphi_n^{(1)}(k_1 a)}{\mathrm{i}N_{12}\varphi_n^{(4)}(k_1 a) - \varphi_n^{(4)\prime}(k_1 a)}.$$

It is now clear that Π can be expressed in the form (23.4) provided that $\nu_0 = N_{12}k_1$. Consequently, all the information about p can be taken over for Π with this small modification.

8.24 The effect of refraction

Atmospheric effects near the surface of the Earth can be incorporated by allowing the wave number k to vary with the radial distance and a convenient model is

$$k^2 = k_1^2(1-\eta+\eta b^2/R^2)$$

where k_1 and η are constants with η small (of the order of 0.1 in the electromagnetic context). The principal change, because of the smallness of η, is that the governing equation for p or Π has the form

$$\nabla^2 p + k^2 p = 0.$$

The only variation of k is with R so that the angular functions in the separation of variables are unaltered. The radial functions are affected but they remain Bessel functions. Formula (23.2) continues to hold provided that a typical spherical Bessel function $\mathfrak{z}_\nu(k_1R)$ is replaced by $\mathfrak{z}_{\nu^1}\{k_1R(1-\eta)^{\frac{1}{2}}\}$ where $\nu^1+\frac{1}{2} = \{(\nu+\frac{1}{2})^2 - k_1^2b^2\eta\}^{\frac{1}{2}}$. The subsequent investigation of (23.2) was concerned only with large values of ν (near k_1b) and in this region $\nu^1 \approx (\nu - k_1b\eta)(1-\eta)^{-\frac{1}{2}}$ or $\nu = k_1b\eta + (1-\eta)^{\frac{1}{2}}\nu^1$.

Take ν^1 as a new variable of integration. Then (23.3) will be obtained (with ν^1 in place of ν) after the following modifications are made: inside the integral Θ and k_1 are replaced by $\Theta(1-\eta)^{\frac{1}{2}}$ and $k_1(1-\eta)^{\frac{1}{2}}$ respectively; outside the integral there is an additional multiplying factor of $(1-\eta)^{\frac{3}{4}}e^{-ik_1b\eta\Theta}$. The behaviour can now be deduced from the results of the preceding section. If it is agreed to write $(\sin\Theta)^{\frac{1}{2}} = d/(b\Theta)^{\frac{1}{2}}$ where d is interpreted as the actual distance along the Earth's surface between source and recipient the formula is the same as when $\eta = 0$ but in place of b there is $b_e = b/(1-\eta)$. In other words, *propagation with refraction occurs as if the atmosphere were homogeneous but the radius of the Earth b_e*. Thus refraction can be accounted for by giving the earth an *effective radius b_e*.

Meteorological data suggest that, in a dry atmosphere, $a_e = 1.2a = 7640$ km whereas, for a saturated atmosphere, $a_e = 9850$ km. Often an intermediate value of $a_e = 4a/3 = 8470$ km is adopted.

Ducts may be dealt with by proceeding as in §6.32 so long as the effective radius is used.

8.25 The prolate spheroid

Let X, Y, ϕ be prolate spheroidal coordinates such that

$$x = l(X^2-1)^{\frac{1}{2}}(1-Y^2)^{\frac{1}{2}}\cos\phi, \qquad y = x\tan\phi, \qquad z = lXY.$$

Let $X=X^1$ be the equation of the spheroid so that its major axis a and minor axis b are given by

$$a=lX^1, \qquad b=l(X^{12}-1)^{\frac{1}{2}}.$$

A point source at $(X_0, 1, 0)$ on the major axis will produce a primary field

$$\begin{aligned} p^{\mathrm{i}} &= \mathrm{e}^{-\mathrm{i}k_1R_s}/4\pi R_s \\ &= (-\mathrm{i}k_1/4\pi)\sum_{n=0}^{\infty}(2n+1)\mathrm{S}_n^{0(4)}(X_0,\hbar)\mathrm{S}_n^{0(1)}(X,\hbar)\mathrm{ps}_n^0(Y,\hbar^2)\mathrm{A}_n^0(\hbar^2) \end{aligned} \quad (25.1)$$

for $X<X_0$, where $\hbar=k_1l$ and (E.2.1) has been used with $\mathrm{ps}_n^m(1,\hbar^2)=\delta_{m0}\mathrm{A}_n^0(\hbar^2)$.

Suppose that the spheroid is soft. Then the scattered field is given by

$$\begin{aligned} p_s &= (\mathrm{i}k_1/4\pi)\sum_{n=0}^{\infty}(2n+1)\mathrm{S}_n^{0(4)}(X_0,\hbar)\mathrm{S}_n^{0(1)}(X^1,\hbar)\mathrm{S}_n^{0(4)}(X,\hbar) \\ &\quad \times \mathrm{ps}_n^0(Y,\hbar^2)\mathrm{A}_n^0(\hbar^2)/\mathrm{S}_n^{0(4)}(X^1,\hbar) \end{aligned} \quad (25.2)$$

At large distances from the spheroid

$$\begin{aligned} p_s &\sim -(\mathrm{e}^{-\mathrm{i}k_1R}/4\pi R)\sum_{n=0}^{\infty}(2n+1)\mathrm{S}_n^{0(4)}(X_0,\hbar)\mathrm{S}_n^{0(1)}(X^1,\hbar)\mathrm{e}^{\frac{1}{2}n\pi\mathrm{i}} \\ &\quad \times \mathrm{ps}_n^0(Y,\hbar^2)\mathrm{A}_n^0(\hbar^2)/\mathrm{S}_n^{0(4)}(X^1,\hbar) \end{aligned} \quad (25.3)$$

from (E.1.19) since, as $X\to\infty$, lX and Y tend to R and $\cos\theta$ respectively. For low frequencies $\hbar$ is small and

$$p_s \sim \frac{\mathrm{e}^{-\mathrm{i}k_1R}}{8\pi R}\ln\frac{X_0+1}{X_0-1}\bigg/\ln\frac{X^1+1}{X^1-1}. \quad (25.4)$$

The impedance boundary condition can be handled in a similar way as can other locations of the source. The penetrable spheroid is a tougher nut to crack because $\hbar$ has different values inside and outside the spheroid so that the angular functions no longer match on the surface. Expansions of the above type cannot therefore be made to satisfy the boundary conditions term by term.

The analogous electromagnetic problem is much more difficult to discuss even when the spheroid is perfectly conducting. Some progress can be made when the source is an electric dipole located on the axis of the spheroid oriented along this axis. The incident electromagnetic field is then derived from a Hertz vector which has the single component Π_z with $\varepsilon_1\Pi_z$ given by the series in (25.1). Then the sole component of the primary magnetic intensity is

$$H_\phi^{\mathrm{i}} = \frac{\mathrm{i}\omega\varepsilon_1(X^2-1)^{\frac{1}{2}}(1-Y^2)^{\frac{1}{2}}}{l^2(X^2-Y^2)}\left(X\frac{\partial\Pi_z}{\partial X}-Y\frac{\partial\Pi_z}{\partial Y}\right). \quad (25.5)$$

A more helpful formula stems from the perception that $H^{\mathrm{i}}_x = -H^{\mathrm{i}}_\phi \sin\phi$. Since H^{i}_x satisfies Helmholtz's equation and its dependence on ϕ is only to the extent of $\sin\phi$ it can be expanded in spheroidal functions in which $m=1$. Hence, for $X>X_0$,

$$H^{\mathrm{i}}_\phi = \sum_{n=0}^{\infty} c_n \mathrm{S}_n^{1(4)}(X, \boldsymbol{\hbar})\mathrm{ps}_n^1(Y, \boldsymbol{\hbar}^2)$$

where c_n is independent of X and Y. Let $X\to\infty$. Then, on account of (25.1) and (E.1.19),

$$\omega k_1(1-Y^2)^{\frac{1}{2}}\mathrm{e}^{\mathrm{i}\boldsymbol{\hbar}X_0Y} = 4\pi \sum_{n=1}^{\infty} c_n \mathrm{e}^{\frac{1}{2}\pi\mathrm{i}(n+1)}\mathrm{ps}_n^1(Y, \boldsymbol{\hbar}^2).$$

Multiply both sides by $\mathrm{ps}_n^1(Y, \boldsymbol{\hbar}^2)$ and integrate with respect to Y from -1 to 1. The orthogonal properties of ps, together with (E.2.3), supply

$$\begin{aligned}4\pi c_n \mathrm{e}^{\frac{1}{2}\pi\mathrm{i}(n+1)} &= -\frac{(2n+1)\omega k_1}{2n(n+1)}\int_{-1}^{1} \mathrm{e}^{\mathrm{i}\boldsymbol{\hbar}X_0Y}(1-Y^2)^{\frac{1}{2}}\mathrm{ps}_n^1(Y, \boldsymbol{\hbar}^2)\,\mathrm{d}Y \\ &= -(2n+1)\omega k_1 \mathrm{A}_n^{-1}(\boldsymbol{\hbar}^2)\mathrm{S}_n^{1(1)}(X_0, \boldsymbol{\hbar})\mathrm{e}^{\frac{1}{2}\pi\mathrm{i}(n+1)}/\boldsymbol{\hbar}(X_0^2-1)^{\frac{1}{2}}\end{aligned}$$

by virtue of (E.2.4). Hence

$$H^{\mathrm{i}}_\phi = -\omega k_1 \sum_{n=0}^{\infty} (2n+1)\mathrm{A}_n^{-1}(\boldsymbol{\hbar}^2)\mathrm{S}_n^{1(1)}(X_0, \boldsymbol{\hbar})\mathrm{S}_n^{1(4)}(X, \boldsymbol{\hbar})\mathrm{ps}_n^1(Y, \boldsymbol{\hbar}^2)/4\pi\boldsymbol{\hbar}(X_0^2-1)^{\frac{1}{2}}$$

for $X>X_0$. When $X<X_0$ interchange X and X_0 in the spheroidal functions.

The presence of the spheroid is accommodated by introducing a scattered field whose magnetic intensity has just the component H^{s}_ϕ. The vanishing of E_Y on the spheroid can be expressed in terms of H_ϕ by means of

$$\mathrm{i}\omega\varepsilon_1 E_Y = -\frac{1}{(X^2-Y^2)^{\frac{1}{2}}}\frac{\partial}{\partial X}\{(X^2-1)^{\frac{1}{2}}H_\phi\}.$$

Consequently,

$$H^{\mathrm{s}}_\phi = \omega k \sum_{n=0}^{\infty} (2n+1)\mathrm{A}_n^{-1}(\boldsymbol{\hbar}^2)a_n \mathrm{S}_n^{1(4)}(X, \boldsymbol{\hbar})\mathrm{S}_n^{1(4)}(X_0, \boldsymbol{\hbar})\mathrm{ps}_n^1(Y, \boldsymbol{\hbar}^2)/4\pi\boldsymbol{\hbar}(X_0^2-1)^{\frac{1}{2}}$$

where

$$a_n = \left[\frac{\partial}{\partial X}\{(X^2-1)^{\frac{1}{2}}\mathrm{S}_n^{1(1)}(X, \boldsymbol{\hbar})\}\Big/\frac{\partial}{\partial X}\{(X^2-1)^{\frac{1}{2}}\mathrm{S}_n^{1(4)}(X, \boldsymbol{\hbar})\}\right]_{X=X_1}.$$

At large distances from the spheroid

$$H^{\mathrm{s}}_\phi \sim \frac{\omega \mathrm{e}^{-\mathrm{i}k_1R}}{4\pi\boldsymbol{\hbar}R}\sum_{n=0}^{\infty} (2n+1)\mathrm{A}_n^{-1}(\boldsymbol{\hbar}^2)a_n \mathrm{e}^{\frac{1}{2}\pi\mathrm{i}(n+1)}\mathrm{S}_n^{1(4)}(X_0, \boldsymbol{\hbar})\mathrm{ps}_n^1(Y, \boldsymbol{\hbar}^2)(X_0^2-1)^{-\frac{1}{2}}.$$

The approximation for large wavelengths is

$$H^s_\phi \sim \frac{-\omega k X^1(1-Y^2)^{\frac{1}{2}}/2\pi(X_0^2-1)^{\frac{1}{2}}}{X^1 \ln\{(X^1+1)/(X^1-1)\}-2}\left\{\tfrac{1}{2}\ln\frac{X_0+1}{X_0-1}-\frac{X_0}{X_0^2-1}\right\}\frac{e^{-ik_1R}}{R}.$$

Curves of the radiation pattern for various sizes of spheroid have been given by Belkina (1958) when the dipole is placed on the boundary ($X_0 = X^1$). As the ratio of the axes a/b increases for fixed $k_1 l$ the maximum radiated field increases and the pattern steadily deviates from that for a sphere. For fixed a/b, as $k_1 l$ increases so does the number of lobes and often there is a strongly directive pattern in the forward direction ($z < 0$). An equivalent standing wave or travelling wave antenna produces nearly the same pattern in the forward direction but not elsewhere.

8.26 The oblate spheroid

It is possible to go from the prolate spheroidal coordinates X, Y, ϕ to the oblate spheroidal coordinates ξ, η, ϕ defined by

$$x = L(\xi^2+1)^{\frac{1}{2}}(1-\eta^2)^{\frac{1}{2}}\cos\phi, \qquad y = L(\xi^2+1)^{\frac{1}{2}}(1-\eta^2)^{\frac{1}{2}}\sin\phi, \qquad z = L\xi\eta$$

by means of the transformation $X \to -i\xi$, $Y \to \eta$, $l \to iL$. Thus the distant field scattered by the spheroid $\xi = \xi^1$ irradiated by an electric dipole at $(\xi_0, 1, 0)$ is

$$\frac{\omega e^{-ik_1R}}{4\pi\mathcal{H}R}\sum_{n=0}^{\infty}(2n+1)A_n^{-1}(-\mathcal{H}^2)a_n' e^{\frac{1}{2}\pi i(n+1)}S_n^{1(4)}(-i\xi_0, i\mathcal{H})\mathrm{ps}_n^1(\eta, -\mathcal{H}^2)(1+\xi_0^2)^{-\frac{1}{2}} \tag{26.1}$$

where

$$a_n' = \left[\frac{\partial}{\partial\xi}\{(\xi^2+1)^{\frac{1}{2}}S_n^{1(1)}(-i\xi, i\mathcal{H})\}\bigg/\frac{\partial}{\partial\xi}\{(\xi^2+1)^{\frac{1}{2}}S_n^{1(4)}(-i\xi, i\mathcal{H})\}\right]_{\xi=\xi^1}$$

and $\mathcal{H} = k_1 L$. This reduces when $\mathcal{H}$ is small to

$$-\frac{\omega\mathcal{H}\xi^1(1-\eta^2)^{\frac{1}{2}}/2\pi(1+\xi^{12})^{\frac{1}{2}}}{1-\xi^1\cot\xi^1}\left\{\cot^{-1}\xi^1-\frac{\xi^1}{1+\xi^{12}}\right\}\frac{e^{-ik_1R}}{R},$$

the arc cotangent lying between 0 and π, for the particular case of the dipole on the surface of the spheroid.

Arbitrary curved obstacles

As is evident from what has been described in this chapter, considerable analysis has to be drawn on before an evaluation of the scattering from even the simplest obstacles is possible. Moreover, the method of separation of variables is available for very few coordinate systems. Attempts to determine the exact response of an arbitrary obstacle are therefore bound to fail with current mathematical techniques. Recourse to approximate methods cannot be evaded therefore; some devices for approximation will be delineated in the following sections.

8.27 Rayleigh scattering in acoustics

A typical obstacle has a surface S of finite extent which separates the interior T_- from the rest of space T_+ (see Fig. 8.19). It will always be assumed that T_+ is homogeneous and isotropic with wave number k_1.

Consider now p_0 specified by

$$p_0(x) = \int_S f(\mathbf{y})\psi(\boldsymbol{x}, \mathbf{y})\,\mathrm{d}S_y$$

where $\psi(\boldsymbol{x}, \mathbf{y}) = \exp\{-ik_1|\boldsymbol{x}-\mathbf{y}|\}/4\pi|\boldsymbol{x}-\mathbf{y}|$. Evidently, p_0 satisfies

$$\nabla^2 p_0 + k_1^2 p_0 = 0$$

in T_+ and T_-. Also p_0 is continuous as $\boldsymbol{x}$ crosses S. It is not so transparent what happens to the derivatives because of the singularity of ψ on S.

Divide S into two portions, S_1 and S_2, of which S_2 is a small region surrounding the point p of S and S_1 is the remainder of S (Fig. 8.20). So long as S is reasonably smooth near p it may be supposed that S_2 is effectively a small circular disc of radius δ with centre p. The z-axis is chosen to be along the outward normal to S at p. Interest centres on the behaviour of p_0 as the point of observation P approaches p.

The integral over S_1 is continuous as $P \to p$ and has no unusual characteristics. In the limit as $\delta \to 0$ it leads to a principal value.

As far as S_2 is concerned, place P on the z-axis a small distance z from p. Then the phase of ψ can be ignored and

$$\int_{S_2} f(\mathbf{y})\psi(\boldsymbol{x}, \mathbf{y})\,\mathrm{d}S_y \approx \frac{f(p)}{4\pi}\int_0^{2\pi}\int_0^{\delta}\frac{t\,\mathrm{d}t\,\mathrm{d}\varphi}{(t^2+z^2)^{\frac{1}{2}}} = \tfrac{1}{2}f(p)\{(\delta^2+z^2)^{\frac{1}{2}} - |z|\}. \quad (27.1)$$

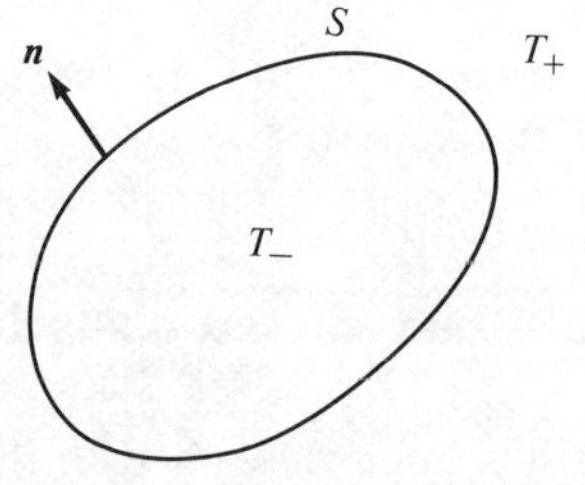

Fig. 8.19 The typical obstacle

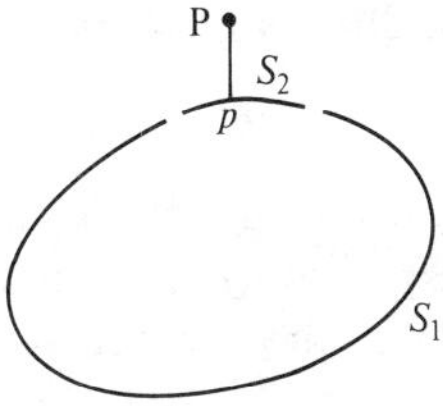

Fig. 8.20 Determination of the discontinuities of surface integrals

For positive z this may be approximated, as $z \to 0$, by $\frac{1}{2}f(p)(\delta - z + \frac{1}{2}z^2/\delta)$. P may now be moved a small distance r off the z-axis so long as a solution of Laplace's equation is retained. Thus, for general positions of P near S_2 with $z > 0$,

$$\int_{S_2} f(\mathbf{y})\psi(\mathbf{x}, \mathbf{y})\, \mathrm{d}S_y \approx \tfrac{1}{2}f(p)\{\delta - z + \tfrac{1}{2}(z^2 - \tfrac{1}{2}r^2)/\delta\}. \tag{27.2}$$

Transparently, a tangential derivative will give no contribution as P approaches p whereas a derivative with respect to z along the normal will be nonzero. Hence

$$\lim_{P \to p+} \operatorname{grad}_P \int_S f(\mathbf{y})\psi(\mathbf{x}, \mathbf{y})\, \mathrm{d}S_y = -\tfrac{1}{2}f(p)\boldsymbol{n}_p + \fint_S f(\mathbf{y}) \operatorname{grad}_x \psi(\mathbf{x}, \mathbf{y})\, \mathrm{d}S_y \tag{27.3}$$

where the bar on the integral sign signifies a principal value and the symbol $p+$ indicates that P tends to the point p of S from T_+.

If P starts in T_- the only difference to (27.2) according to (27.1) is that the sign of z is reversed. Consequently

$$\lim_{P \to p-} \operatorname{grad}_P \int_S f(\mathbf{y})\psi(\mathbf{x}, \mathbf{y})\, \mathrm{d}S_y = \tfrac{1}{2}f(p)\boldsymbol{n}_p + \fint_S f(\mathbf{y}) \operatorname{grad}_x \psi(\mathbf{x}, \mathbf{y})\, \mathrm{d}S_y. \tag{27.4}$$

Moreover, $\operatorname{grad}_x \psi = -\operatorname{grad}_y \psi$ so that

$$\lim_{P \to p+} \int_S f(\mathbf{y}) \operatorname{grad}_y \psi(\mathbf{x}, \mathbf{y})\, \mathrm{d}S_y = \tfrac{1}{2}f(p)\boldsymbol{n}_p + \fint_S f(\mathbf{y}) \operatorname{grad}_y \psi(\mathbf{x}, \mathbf{y})\, \mathrm{d}S_y, \tag{27.5}$$

$$\lim_{P \to p-} \int_S f(\mathbf{y}) \operatorname{grad}_y \psi(\mathbf{x}, \mathbf{y})\, \mathrm{d}S_y = -\tfrac{1}{2}f(p)\boldsymbol{n}_p + \fint_S f(\mathbf{y}) \operatorname{grad}_y \psi(\mathbf{x}, \mathbf{y})\, \mathrm{d}S_y. \tag{27.6}$$

On the basis of this information it is feasible to formulate a representation of the field when p^i impinges on S from T_+ and the total p satisfies

$$(\nabla^2 + k_1^2)p = 0 \quad (\text{in } T_+), \qquad (\nabla^2 + k_2^2)p = 0 \quad (\text{in } T_-) \tag{27.7}$$

subject to the boundary conditions on S of

$$p_+ = p_-, \qquad (\partial p/\partial n)_+ = \zeta(\partial p/\partial n)_- \tag{27.8}$$

where the suffix '+' denotes a value on S as it is approached from T_+ whereas the suffix '−' corresponds to an approach from T_-. The constant ζ is non-negative. Of course, the scattered field must be outgoing at infinity.

A field complying with these conditions is

$$p(\boldsymbol{x}) = p^{\mathrm{i}}(\boldsymbol{x}) + \int_{T_-} (k_2^2 - k_1^2) p(\mathbf{y}) \psi(\boldsymbol{x}, \mathbf{y}) \, \mathrm{d}\mathbf{y}$$
$$+ (1-\zeta) \int_S \left(\frac{\partial p}{\partial n}\right)_- \psi(\boldsymbol{x}, \mathbf{y}) \, \mathrm{d}S_y. \tag{27.9}$$

For, it meets (27.7) and the continuity of both the volume and surface integrals ensures that $p_+ = p_-$. On writing the normal derivative condition as

$$(\partial p/\partial n)_+ - (\partial p/\partial n)_- = (\zeta - 1)(\partial p/\partial n)_-$$

we perceive that, since the derivative of a volume integral is continuous, (27.8) is satisfied by virtue of (27.3) and (27.4). In addition, the volume and surface integrals provide outgoing waves at infinity.

The last statement can be verified from the fact that

$$\psi(\boldsymbol{x}, \mathbf{y}) \sim \exp(-\mathrm{i}k_1 R + \mathrm{i}k_1 \hat{\boldsymbol{x}} \cdot \mathbf{y})/4\pi R$$

as $|\boldsymbol{x}| = R \to \infty$, $\hat{\boldsymbol{x}}$ being a unit vector along $\boldsymbol{x}$. Hence, at a large distance from S

$$p(\boldsymbol{x}) \sim p^{\mathrm{i}}(\boldsymbol{x}) + \frac{\mathrm{e}^{-\mathrm{i}k_1 R}}{4\pi R} \left\{ \int_{T_-} (k_2^2 - k_1^2) p(\mathbf{y}) \mathrm{e}^{\mathrm{i}k_1 \hat{\boldsymbol{x}} \cdot \mathbf{y}} \, \mathrm{d}\mathbf{y} \right.$$
$$\left. + (1-\zeta) \int_S \left(\frac{\partial p}{\partial n}\right)_- \mathrm{e}^{\mathrm{i}k_1 \hat{\boldsymbol{x}} \cdot \mathbf{y}} \mathrm{d}S_y \right\}. \tag{27.10}$$

The surface integral can be converted to one over T_- by the divergence theorem so that, on account of (27.7), the alternative form

$$p(\boldsymbol{x}) \sim p^{\mathrm{i}}(\boldsymbol{x}) + \frac{\mathrm{e}^{-\mathrm{i}k_1 R}}{4\pi R} \left\{ \int_{T_-} (k_2^2 \zeta - k_1^2) p(\mathbf{y}) \mathrm{e}^{\mathrm{i}k_1 \hat{\boldsymbol{x}} \cdot \mathbf{y}} \, \mathrm{d}\mathbf{y} \right.$$
$$\left. + \mathrm{i}k_1 (1-\zeta) \hat{\boldsymbol{x}} \cdot \int_{T_-} \mathrm{e}^{\mathrm{i}k_1 \hat{\boldsymbol{x}} \cdot \mathbf{y}} \operatorname{grad}_y p \, \mathrm{d}\mathbf{y} \right\} \tag{27.11}$$

is obtained.

The imposition of (27.9) when $\boldsymbol{x}$ is in T_- supplies an integral equation to determine p in T_-. Once p is known in T_-, p can be found in T_+ by substituting the known values in the right-hand side of (27.9). In general, the solution of the integral equation will call for approximate or numerical methods or both.

Rayleigh scattering is characterized by $k_1 l \ll 1$ where l is a typical dimension of the body. This suggests that in T_- an expansion in powers of k_1 is appropriate, i.e.

$$p = p_0 - \mathrm{i}k_1 p_1 - k_1^2 p_2 + \dots$$

where $p_0, p_1, \dots$ are independent of k_1. If a similar expansion is made for p^{i}, equating the coefficients of powers of k_1 on the two sides of (27.9) gives, when k_2^2/k_1^2 is treated as independent of k_1,

$$p_0(\boldsymbol{x}) = p_0^{\mathrm{i}}(\boldsymbol{x}) + (1-\zeta)\int_S \left(\frac{\partial p_0}{\partial n}\right)_- \psi_0(\boldsymbol{x}, \mathbf{y})\, \mathrm{d}S_y, \tag{27.12}$$

$$p_1(\boldsymbol{x}) = p_1^{\mathrm{i}}(\boldsymbol{x}) + (1-\zeta)\int_S \left\{\left(\frac{\partial p_1}{\partial n}\right)_- \psi_0(\boldsymbol{x}, \mathbf{y}) + \frac{1}{4\pi}\left(\frac{\partial p_0}{\partial n}\right)_-\right\} \mathrm{d}S_y \tag{27.13}$$

where $\psi_0(\boldsymbol{x}, \mathbf{y}) = 1/4\pi\,|\boldsymbol{x} - \mathbf{y}|$.

It will now be supposed that the incident wave is plane with

$$p^{\mathrm{i}}(\boldsymbol{x}) = \exp(-\mathrm{i}k_1 \boldsymbol{n}^{\mathrm{i}} \cdot \boldsymbol{x}) \tag{27.14}$$

where $\boldsymbol{n}^{\mathrm{i}}$ is a fixed vector. No real loss of generality is incurred by this assumption because any incident field can be approximated by a plane wave in the vicinity of S when the frequency is low, unless the source is very close to the boundary. Thus $p_0^{\mathrm{i}} = 1$ and $p_1^{\mathrm{i}} = \boldsymbol{n}^{\mathrm{i}} \cdot \boldsymbol{x}$.

If now p_0 and p_1 are specified in T_+ by the right-hand side of (27.12) and (27.13) respectively it is evident that both p_0 and p_1 are solutions of Laplace's equation and, from (27.3), (27.4), both satisfy the boundary conditions (27.8).

A possible solution of (27.12) with $p_0^{\mathrm{i}} = 1$ is $p_0 = 1$ everywhere. Indeed, there is no other solution; for, if v is the difference between two solutions, v satisfies Laplace's equation everywhere, complies with (27.8) and has behaviour at infinity such that $|v| = O(1/R)$, $|\mathrm{grad}\, v| = O(1/R^2)$. Hence

$$0 = \int_S v\left\{\zeta\left(\frac{\partial v}{\partial n}\right)_- - \left(\frac{\partial v}{\partial n}\right)_+\right\} \mathrm{d}S = \zeta \int_{T_-} \mathrm{grad}^2 v\, \mathrm{d}\boldsymbol{x} + \int_{T_+} \mathrm{grad}^2 v\, \mathrm{d}\boldsymbol{x}.$$

Consequently, v must be a constant which is forced to be zero because of what happens at infinity.

Since $p_0 = 1$, there is no term of $O(k_1)$ in the scattered field of (27.11) (we are assuming that ζ is finite) and the first term is $O(k_1^2)$. Furthermore, p_0 drops out of (27.13) so that p_1 is a solution of Laplace's equation, satisfying (27.8) and behaving at infinity like $\boldsymbol{n}^{\mathrm{i}} \cdot \boldsymbol{x}$ and a field like v. Suppose now that p_{1j} enjoys the same properties except that its dominant behaviour at infinity is that of x_j. Then $p_1 = n_1^i p_{11} + n_2^i p_{12} + n_3^i p_{13}$ where n_1^i, n_2^i, and n_3^i are the components of $\boldsymbol{n}^i$. Therefore

$$\int_{T_-} \operatorname{grad}_y p_1 \, d\mathbf{y} = \int_S p_1 \boldsymbol{n} \, dS$$

$$= \int_S \{\boldsymbol{n}^i \cdot \boldsymbol{x} + (1-\zeta)(n_1^i t_1 + n_2^i t_2 + n_3^i t_3)\} \boldsymbol{n} \, dS$$

on putting

$$p_{1j} = x_j + (1-\zeta) t_j.$$

Hence, to terms in $O(k_1^2)$, we infer from (27.11) that

$$p(\boldsymbol{x}) \sim p^i(\boldsymbol{x}) + A(\hat{\boldsymbol{x}}) e^{-ikR}/R \tag{27.15}$$

where

$$4\pi A(\hat{\boldsymbol{x}}) = k_1^2 \left[\int_{T_-} (k_2^2 \zeta / k_1^2) \, d\boldsymbol{x} - T_- + (1-\zeta)\{T_- \hat{\boldsymbol{x}} \cdot \boldsymbol{n}^i + (1-\zeta) \hat{\boldsymbol{x}} \cdot \boldsymbol{C} \cdot \boldsymbol{n}^i\} \right], \tag{27.16}$$

T_- is the volume inside S, and the tensor $\boldsymbol{C}$ has the components C_{ij} given by

$$C_{ij} = \int_S n_i t_j \, dS. \tag{27.17}$$

The scattered field is thereby known to $O(k_1^2)$ as soon as the solution t_j of Laplace's equation has been determined.

It may be remarked that the terms in (27.16) can be interpreted as caused by the linear superposition of a simple source and a double source. The internal material of the body contributes only to the simple source (if k_2^2/k_1^2 and ζ are regarded as unconnected). This is as it should be because the simple source arises from the obstacle preventing the fluctuations of pressure that would otherwise take place in the space it occupies. On the other hand, the double source originates from the force on the body which would make it move to and fro were it not fixed.

By proceeding to the next order in the governing integral equation it can be shown that the inclusion of p_2 in the calculation alters the scattered field by changing $A(\hat{\boldsymbol{x}})$ to $A(\hat{\boldsymbol{x}}) + ik_1^3 B(\hat{\boldsymbol{x}})$ where

$$B(\hat{\boldsymbol{x}}) = (\zeta - 1)\hat{\boldsymbol{x}} \cdot \int_S \left\{ \sum_{j=1}^{3} p_{1j} \, \hat{\mathbf{y}}_j \boldsymbol{n}^i \cdot \mathbf{y} \boldsymbol{n}^i \cdot \mathbf{n} - \hat{\mathbf{x}} \cdot \mathbf{y} p_1 \mathbf{n} \right\} dS_y$$

$$+ \zeta \int_{T_-} (1 - k_2^2/k_1^2) \left(p_1 - \hat{\boldsymbol{x}} \cdot \sum_{j=1}^{3} p_{1j} \hat{\mathbf{y}}_j \right) d\mathbf{y}. \tag{27.18}$$

Thus, it is not actually necessary to solve the integral equation for p_2 in order to find the scattered field of $O(k_1^3)$. Note that there is now a quadrupole contribution as well as simple and double sources.

8.28 The content matrix

The matrix with components C_{ij} plays a key role in the scattered field, being the only unknown quantity in (27.16); it is known as the *content matrix* and its trace $C_{11}+C_{22}+C_{33}$ is called the *content.* It is defined by (27.17) where t_j is a solution of Laplace's equation, with behaviour at infinity such that $|t_j|=O(1/R)$, $|\text{grad } t_j|=O(1/R^2)$ and satisfies the boundary conditions

$$(t_j)_+=(t_j)_-, \qquad n_j+(\partial t_j/\partial n)_+=\zeta(\partial t_j/\partial n)_-. \tag{28.1}$$

It will be shown firstly that $[C_{ij}]$ *is symmetric*.. From (27.17) and (28.1)

$$\begin{aligned} C_{ij} &= \int_S t_j\left\{\zeta\left(\frac{\partial t_i}{\partial n}\right)_- - \left(\frac{\partial t_i}{\partial n}\right)_+\right\}\mathrm{d}S \\ &= \zeta\int_{T_-} \text{grad } t_j \cdot \text{grad } t_i \,\mathrm{d}\mathbf{x} + \int_{T_+} \text{grad } t_j \cdot \text{grad } t_i \,\mathrm{d}\mathbf{x} \end{aligned} \tag{28.2}$$

by virtue of the properties of t_j at infinity. The symmetry is now obvious. In addition, if z_j is any complex number, (28.2) reveals that

$$\sum_{i=1}^{3}\sum_{j=1}^{3} z_i^* C_{ij} z_j = \zeta\int_{T_-}\left|\sum_{i=1}^{3} z_i \text{ grad } t_i\right|^2 \mathrm{d}\mathbf{x} + \int_{T_+}\left|\sum_{i=1}^{3} z_i \text{ grad } t_i\right|^2 \mathrm{d}x$$

so that $[C_{ij}]$ *is positive-definite when ζ is real and positive.* From now on, it will be assumed that ζ is real though some of the subsequent results are valid for complex ζ.

Next, an application of Schwarz's inequality to (28.2) gives

$$C_{ij}^2 \leqslant C_{ii}C_{jj}. \tag{28.3}$$

The dependence of C_{jj} on ζ will now be discussed, the notation $C_{jj}(\zeta)$ being employed to denote this. Let $\partial t_j/\partial\zeta=\bar{t}_j$. Then $\bar{t}_j$ has the same properties as t_j except that

$$(\bar{t}_j)_+=(\bar{t}_j)_-, \qquad (\partial t_j/\partial n)_+=\zeta(\partial\bar{t}_j/\partial n)_-+(\partial t_j/\partial n)_-. \tag{28.4}$$

Now

$$\partial C_{jj}(\zeta)/\partial\zeta = \int_S n_j\bar{t}_j \,\mathrm{d}S = \int_S \{\zeta(\partial t_j/\partial n)_- - (\partial t_j/\partial n)_+\}\bar{t}_j \,\mathrm{d}S$$

on account of (28.1). The divergence theorem and (28.4) give

$$\begin{aligned} \partial C_{jj}(\zeta)/\partial\zeta &= \int_S \{\zeta(\partial\bar{t}_j/\partial n)_- - (\partial\bar{t}_j/\partial n)_+\}t_j \,\mathrm{d}S \\ &= -\int_S t_j(\partial t_j/\partial n)_- \,\mathrm{d}S = -\int_{T_-} \text{grad}^2 t_j \,\mathrm{d}\mathbf{x} \end{aligned} \tag{28.5}$$

on account of (28.4) and the divergence theorem. Thus, C_{ij} *decreases monotonically as* ζ *increases.*

The combination of (28.2) and (28.5) supplies

$$\partial C_{ij}(\zeta)/\partial\zeta \geqslant -C_{ij}(\zeta)/\zeta.$$

Accordingly, if $\zeta \geqslant \zeta_1$,

$$C_{ij}(\zeta) \geqslant C_{ij}(\zeta_1)\zeta_1/\zeta. \tag{28.6}$$

An upper bound can be derived by observing that

$$C_{ij}(\zeta) = \int_{T_-} (\partial t_j/\partial x_j)\,\mathrm{d}\mathbf{x}$$

so that Schwarz's inequality and (28.5) give

$$\{C_{ij}(\zeta)\}^2 \leqslant T_-\,\partial C_{ij}(\zeta)/\partial\zeta.$$

Hence, *for* $\zeta \geqslant \zeta_1$, *the content matrix satisfies*

$$C_{ij}(\zeta) \leqslant \frac{T_-}{\zeta - \zeta_1 + T_-/C_{ij}(\zeta_1)}. \tag{28.7}$$

The upper bound (28.7) is the best possible in the sense that equality holds when S is a sphere or a spheroid. It is clear from (28.6) and (28.7) that C_{ij} varies relatively slowly with ζ and that its decay for large ζ is essentially that of $1/\zeta$.

The bounds require the solution of a potential problem for some ζ_1. However, this process can be escaped if ζ_1 is chosen to be unity. There is then an explicit formula available because, when $\zeta = 1$,

$$t_j = \int_S n_j \psi_0(\mathbf{x}, \mathbf{y})\,\mathrm{d}S_y.$$

This is certainly a continuous solution of Laplace's equation, falling properly at infinity and satisfying (28.1) on account of (27.3) and (27.4). Hence

$$C_{ij}(1) = \int_S n_i(\mathbf{x}) \int_S n_j(\mathbf{y})\psi_0(\mathbf{x}, \mathbf{y})\,\mathrm{d}S_y\,\mathrm{d}S_x. \tag{28.8}$$

A deduction from (28.8) is that

$$C_{11}(1) + C_{22}(1) + C_{33}(1) = T_-. \tag{28.9}$$

Variational expressions can also be obtained. *For all continuous functions u which have piecewise continuous first partial derivatives and whose behaviour at infinity satisfies* $u = O(1/R)$, $|\mathrm{grad}\,u| = O(1/R^2)$

$$C_{ij} \geqslant 2\int_{T_-} (\partial u/\partial x_j)\,\mathrm{d}\mathbf{x} - \zeta\int_{T_-} \mathrm{grad}^2 u\,\mathrm{d}\mathbf{x} - \int_{T_+} \mathrm{grad}^2 u\,\mathrm{d}\mathbf{x}. \tag{28.10}$$

Equality occurs in (28.10) if, and only if, $u = t_j$. While (28.10) provides a lower bound, an upper bound stems from the following result.

Let $\boldsymbol{a}$ *have piecewise continuous first partial derivatives such that* div $\boldsymbol{a} = 0$ *and be* $O(1/R^2)$ *as* $R \to \infty$. *Let* $\boldsymbol{a}$ *be continuous except possibly across S where*

$$\boldsymbol{a}_+ \cdot \boldsymbol{n} + n_j = \zeta \boldsymbol{a}_- \cdot \boldsymbol{n}.$$

Then

$$C_{jj} \leqslant \zeta \int_{T_-} |\boldsymbol{a}|^2 \, \mathrm{d}\boldsymbol{x} + \int_{T_+} |\boldsymbol{a}|^2 \, \mathrm{d}\boldsymbol{x}. \tag{28.11}$$

The variational formulae enable one to take advantage of Galerkin's method as in §5.15. The various bounds constitute checks on the numerical accuracy and the errors in any analytical approximation. It is therefore feasible to obtain error bounds for the scattered field.

8.29 Impenetrable obstacles

A particular choice of ζ in the preceding analysis is $\zeta = 0$. Then (27.8) entails $(\partial p/\partial n)_+ = 0$ and the field in T_+ corresponds to scattering by a hard object. Thus the foregoing results are immediately applicable for the hard boundary; there is some simplification of the formulae on placing $\zeta = 0$ but explicit details will not be given.

The other extreme is the soft obstacle for which ζ is infinite. Here it is more convenient to reformulate the problem beginning with the representation

$$p(\boldsymbol{x}) = p^{\mathrm{i}}(\boldsymbol{x}) - \int_S (\partial p/\partial n)_+ \psi(\boldsymbol{x}, \boldsymbol{y}) \, \mathrm{d}S_y \tag{29.1}$$

for p in T_+. The resultant representations for the coefficients in the expansions in powers of k_1 are

$$p_0(\boldsymbol{x}) = 1 - \int_S (\partial p_0/\partial n)_+ \psi_0(\boldsymbol{x}, \boldsymbol{y}) \, \mathrm{d}S_y, \tag{29.2}$$

$$p_1(\boldsymbol{x}) = \boldsymbol{x} \cdot \boldsymbol{n}^{\mathrm{i}} - \int_S \{(\partial p_1/\partial n)_+ \psi_0(\boldsymbol{x}, \boldsymbol{y}) + (\partial p_0/\partial n)_+/4\pi\} \, \mathrm{d}S_y \tag{29.3}$$

for an incident plane wave. The consequent scattered wave has

$$4\pi A(\hat{\boldsymbol{x}}) = \int_S \left\{ ik_1 \left(\frac{\partial p_1}{\partial n}\right)_+ - (1 + ik_1 \hat{\boldsymbol{x}} \cdot \boldsymbol{y}) \left(\frac{\partial p_0}{\partial n}\right)_+ \right\} \mathrm{d}S_y. \tag{29.4}$$

When $\boldsymbol{x}$ is on S, p_0 and p_1 vanish. Then (29.2) becomes the equation for the surface density when the surface S is held at unit potential. Hence

$$\int_S (\partial p_0/\partial n)_+ \, \mathrm{d}S_y = 4\pi C \tag{29.5}$$

where C is the capacity of the body.

Let $p_0 = 1 - q_0$. Then (29.3) (with $\boldsymbol{x}$ on S) and (29.5) require

$$p_1 = \boldsymbol{x} \cdot \boldsymbol{n}^{\mathrm{i}} - C - q_1 + Cq_0$$

where q_1 is the potential function which equals $\boldsymbol{n}^{\mathrm{i}} \cdot \boldsymbol{x}$ on S and vanishes at infinity. Hence

$$\int_S \left(\frac{\partial p_1}{\partial n}\right)_+ \mathrm{d}S = \int_S \left(C\frac{\partial q_0}{\partial n} - \frac{\partial q_1}{\partial n}\right)_+ \mathrm{d}S = -4\pi C^2 - \int_S \left(\frac{\partial q_1}{\partial n}\right)_+ q_0 \,\mathrm{d}S$$

$$= -4\pi C^2 - \int_S q_1 \left(\frac{\partial q_0}{\partial n}\right)_+ \mathrm{d}S$$

by the divergence theorem since q_1 and q_0 decay sufficiently rapidly at infinity. Inserting the known value of q_1 on S we deduce from (29.4) that

$$A(\hat{\boldsymbol{x}}) = -4\pi C(1 + \mathrm{i}k_1 C) + \mathrm{i}k_1(\boldsymbol{n}^{\mathrm{i}} - \hat{\boldsymbol{x}}) \cdot \int_S \mathbf{y}(\partial p_0/\partial n)_+ \,\mathrm{d}S_y \tag{29.6}$$

for the soft boundary.

Evidently, a soft obstacle scatters long wavelengths much more strongly than either a hard or penetrable object. A similar phenomenon was discovered in §8.17 for the sphere.

Only p_0 is needed in order to determine A from (29.6). This involves solving a problem in potential theory. Unfortunately, there are not too many cases where this can be undertaken analytically (some examples occur later) but, as already indicated, numerical procedures and error bounds are available.

8.30 Rayleigh scattering in electromagnetism

Let the material parameters of T_+ be ε_1, μ_1 and of T_- be ε_2, μ_2. Then a suitable representation of the electromagnetic field which ensures that the tangential components of $\boldsymbol{E}$ and $\boldsymbol{H}$ are continuous across S is

$$\boldsymbol{E}(\mathbf{x}) = \boldsymbol{E}^{\mathrm{i}}(\mathbf{x}) - (\text{grad div} + k_1^2)\int_{T_-} (1 - \varepsilon_2/\varepsilon_1)\boldsymbol{E}(\mathbf{y})\psi(\mathbf{x}, \mathbf{y})\,\mathrm{d}\mathbf{y}$$

$$-i\omega \,\text{curl} \int_{T_-} (\mu_2 - \mu_1)\boldsymbol{H}(\mathbf{y})\psi(\mathbf{x}, \mathbf{y})\,\mathrm{d}\mathbf{y}, \tag{30.1}$$

$$\boldsymbol{H}(\mathbf{x}) = \boldsymbol{H}^{\mathrm{i}}(\mathbf{x}) - (\text{grad div} + k_1^2)\int_{T_-} (1 - \mu_2/\mu_1)\boldsymbol{H}(\mathbf{y})\psi(\mathbf{x}, \mathbf{y})\,\mathrm{d}\mathbf{y}$$

$$+\mathrm{i}\omega \,\text{curl} \int_{T_-} (\varepsilon_2 - \varepsilon_1)\boldsymbol{E}(\mathbf{y})\psi(\mathbf{x}, \mathbf{y})\,\mathrm{d}\mathbf{y}. \tag{30.2}$$

The distant radiation is given by

$$\boldsymbol{E}(\mathbf{x}) \sim \boldsymbol{E}^{\mathrm{i}}(\mathbf{x}) + \mathbf{A}(\hat{\mathbf{x}})\mathrm{e}^{-\mathrm{i}k_1 R}/R$$

where

$$4\pi\mathbf{A}(\hat{\boldsymbol{x}}) = k_1^2(\hat{\boldsymbol{x}}\hat{\boldsymbol{x}} - \boldsymbol{I}) \cdot \int_{T_-} (1 - \varepsilon_2/\varepsilon_1)\boldsymbol{E}(\mathbf{y})\exp(\mathrm{i}k_1\hat{\boldsymbol{x}} \cdot \mathbf{y})\,\mathrm{d}\mathbf{y}$$

$$- k_1\omega\hat{\boldsymbol{x}} \wedge \int_{T_-} (\mu_2 - \mu_1)\boldsymbol{H}(\mathbf{y})\exp(\mathrm{i}k_1\hat{\boldsymbol{x}} \cdot \mathbf{y})\,\mathrm{d}\mathbf{y}, \qquad (30.3)$$

$\boldsymbol{I}$ being the unit dyadic.

At long wavelengths the lowest order expressions which spring from (30.1) and (30.2) are

$$\boldsymbol{E}_0(\boldsymbol{x}) = \boldsymbol{E}_0^{\mathrm{i}}(\boldsymbol{x}) - \operatorname{grad}\operatorname{div}\int_{T_-} (1 - \varepsilon_2/\varepsilon_1)\boldsymbol{E}_0(\mathbf{y})\psi_0(\boldsymbol{x}, \mathbf{y})\,\mathrm{d}\mathbf{y}, \qquad (30.4)$$

$$\boldsymbol{H}_0(\boldsymbol{x}) = \boldsymbol{H}_0^{\mathrm{i}}(\boldsymbol{x}) - \operatorname{grad}\operatorname{div}\int_{T_-} (1 - \mu_2/\mu_1)\boldsymbol{H}_0(\mathbf{y})\psi_0(\boldsymbol{x}, \mathbf{y})\,\mathrm{d}\mathbf{y}. \qquad (30.5)$$

These equations are decoupled in that each involves either the electric intensity or magnetic field but not both. To this level of approximation

$$4\pi\mathbf{A}(\hat{\boldsymbol{x}}) = k_1^2\{\hat{\boldsymbol{x}}(\hat{\boldsymbol{x}} \cdot \boldsymbol{p}) - \boldsymbol{p} + \hat{\boldsymbol{x}} \wedge \boldsymbol{q}\} \qquad (30.6)$$

where

$$\boldsymbol{p} = \int_{T_-} (1 - \varepsilon_2/\varepsilon_1)\boldsymbol{E}_0(\mathbf{y})\,\mathrm{d}\mathbf{y}, \qquad \boldsymbol{q} = \int_{T_-} (1 - \mu_2/\mu_1)(\mu_1/\varepsilon_1)^{\frac{1}{2}}\boldsymbol{H}_0(\mathbf{y})\,\mathrm{d}\mathbf{y}. \qquad (30.7)$$

Therefore, *the radiation pattern of the obstacle is the same as that of an electric dipole of moment* $-\varepsilon_1\boldsymbol{p}$ *together with that of a magnetic dipole emanating from* $\boldsymbol{q}$. If $\mu_2 = \mu_1$ the magnetic dipole is absent.

The only difference between (30.4) and (30.5) lies in the exchange of μ_2/μ_1 for $\varepsilon_2/\varepsilon_1$. It is therefore sufficient to discuss (30.4). Let the incident wave be plane with

$$\boldsymbol{E}^{\mathrm{i}}(\boldsymbol{x}) = \boldsymbol{e}^{\mathrm{i}}\exp(-\mathrm{i}k_1\boldsymbol{n}^{\mathrm{i}} \cdot \boldsymbol{x})$$

where $\boldsymbol{e}^{\mathrm{i}}$ is a constant vector such that $\boldsymbol{e}^{\mathrm{i}} \cdot \boldsymbol{n}^{\mathrm{i}} = 0$. Then

$$\boldsymbol{E}_0^{\mathrm{i}}(\boldsymbol{x}) = \boldsymbol{e}^{\mathrm{i}}, \qquad \boldsymbol{H}_0^{\mathrm{i}}(\boldsymbol{x}) = (\varepsilon_1/\mu_1)^{\frac{1}{2}}\boldsymbol{n}^{\mathrm{i}} \wedge \boldsymbol{e}^{\mathrm{i}}.$$

According to (30.4) when ε_2 is constant as will be assumed hereafter, $\operatorname{div}\boldsymbol{E}_0 = 0$ and $\boldsymbol{E}_0(\boldsymbol{x}) = \operatorname{grad} V(\boldsymbol{x})$ where

$$V(\boldsymbol{x}) = \boldsymbol{e}^{\mathrm{i}} \cdot \boldsymbol{x} - \operatorname{div}\int_{T_-} (1 - \varepsilon_2/\varepsilon_1)\boldsymbol{E}_0(\mathbf{y})\psi_0(\boldsymbol{x}, \mathbf{y})\,\mathrm{d}\mathbf{y}.$$

Now $\nabla^2 V = 0$ and so

$$\operatorname{div}\int_{T_-} \psi_0(\boldsymbol{x}, \mathbf{y})\operatorname{grad}_y V(\mathbf{y})\,\mathrm{d}\mathbf{y} = -\int_{T_-} \operatorname{grad}_y \psi_0(\boldsymbol{x}, \mathbf{y}) \cdot \operatorname{grad}_y V(\mathbf{y})\,\mathrm{d}\mathbf{y}$$

$$= -\int_S \left(\frac{\partial V}{\partial n}\right)_- \psi_0(\boldsymbol{x}, \mathbf{y})\,\mathrm{d}S_y.$$

Hence

$$V(\boldsymbol{x}) = \boldsymbol{e}^{\mathrm{i}} \cdot \boldsymbol{x} + (1 - \varepsilon_2/\varepsilon_1) \int_S \left(\frac{\partial V}{\partial n}\right)_- \psi_0(\boldsymbol{x}, \mathbf{y})\, \mathrm{d}S_y. \tag{30.8}$$

It will be instantly perceived that (30.8) has the same structure as the equation for p_1 originating from (27.13). Hence all the results of §§8.27 and 8.28 can be taken over immediately for V. In particular,

$$\boldsymbol{p} = (1 - \varepsilon_2/\varepsilon_1)\{T_- \boldsymbol{e}^{\mathrm{i}} + (1 - \varepsilon_2/\varepsilon_1) \mathbf{C}(\varepsilon_2/\varepsilon_1) \cdot \boldsymbol{e}^{\mathrm{i}}\}. \tag{30.9}$$

It can also be demonstrated that the next order in the scattered field can be expressed explicitly in terms of $\boldsymbol{E}_0$ and $\boldsymbol{H}_0$.

For an isotropic ellipsoid of semi-axes a, b, c it is found that, with the coordinate axes along the principal axes, $\boldsymbol{p} = 4\pi abc\boldsymbol{P}/3$ where the components P_1, P_2, P_3 of $\boldsymbol{P}$ are obtained from the components of $\boldsymbol{e}^{\mathrm{i}}$ via

$$\{I_j - (1 - \varepsilon_2/\varepsilon_1)^{-1}\} P_j = e_j^{\mathrm{i}}.$$

Here $I_1 = \frac{1}{2}abc \int_0^\infty (s + a^2)^{-\frac{3}{2}}(s + b^2)^{-\frac{1}{2}}(s + c^2)^{-\frac{1}{2}}\, \mathrm{d}s$ and I_2, I_3 are supplied by cyclic interchange of a, b, and c. Note that $I_1 + I_2 + I_3 = 1$.

For the sphere $I_1 = I_2 = I_3 = \frac{1}{3}$. In the particular instance of a spheroid with $a > b = c$

$$I_1 = \frac{b^2}{2a^2e^3}\left(\ln\frac{1+e}{1-e} - 2e\right)$$

where $e^2 = 1 - (b/a)^2$; if $a < b = c$

$$I_1 = b^2(e' - \tan^{-1} e')/a^2 e'^3$$

where $e'^2 = (b/a)^2 - 1$.

For a coated isotropic sphere of radius a in which ε_2 is replaced by ε_3 in $0 \leqslant R < qa\,(q \leqslant 1)$

$$\boldsymbol{p} = 4\pi a^3 \frac{(\varepsilon_2 - \varepsilon_1)(\varepsilon_3 + 2\varepsilon_2) + q^3(2\varepsilon_2 + \varepsilon_1)(\varepsilon_3 - \varepsilon_2)}{(\varepsilon_2 + 2\varepsilon_1)(\varepsilon_3 + 2\varepsilon_2) + 2q^3(\varepsilon_2 - \varepsilon_1)(\varepsilon_3 - \varepsilon_2)} \boldsymbol{e}^{\mathrm{i}}.$$

When S is perfectly conducting a convenient starting point is the representation

$$\boldsymbol{E}(\boldsymbol{x}) = \boldsymbol{E}^{\mathrm{i}}(\boldsymbol{x}) + (1/\mathrm{i}\omega\varepsilon_1)(\text{grad div} + k_1^2) \int_S \boldsymbol{n} \wedge \boldsymbol{H}(\mathbf{y}) \psi(\boldsymbol{x}, \mathbf{y})\, \mathrm{d}S_y, \tag{30.10}$$

$$\boldsymbol{H}(\boldsymbol{x}) = \boldsymbol{H}^{\mathrm{i}}(\boldsymbol{x}) + \text{curl} \int_S \boldsymbol{n} \wedge \boldsymbol{H}(\mathbf{y}) \psi(\boldsymbol{x}, \mathbf{y})\, \mathrm{d}S_y \tag{30.11}$$

based on (1.29.9) and (1.29.10). Then

$$4\pi \mathbf{A}(\hat{\boldsymbol{x}}) = k_1^2(\boldsymbol{I} - \hat{\boldsymbol{x}}\hat{\boldsymbol{x}}) \cdot \int_S \boldsymbol{n} \wedge \boldsymbol{H}(\mathbf{y}) \exp(\mathrm{i}k_1 \hat{\boldsymbol{x}} \cdot \mathbf{y})\, \mathrm{d}S_y / \mathrm{i}\omega\varepsilon_1. \tag{30.12}$$

An alternative to (30.12) can be derived from

$$\int_S \boldsymbol{n} \wedge \boldsymbol{a}\, \mathrm{d}S_y = \int_S \mathbf{y}(\boldsymbol{n} \cdot \mathrm{curl}_y\, \boldsymbol{a})\, \mathrm{d}S_y.$$

Thus

$$\int_S \boldsymbol{n} \wedge \boldsymbol{H}(\mathbf{y}) \exp(\mathrm{i}k_1 \hat{\boldsymbol{x}} \cdot \mathbf{y})\, \mathrm{d}S_y$$

$$= \mathrm{i}\omega\varepsilon_1 \int_S [\boldsymbol{n} \cdot \{(\mu_1/\varepsilon_1)^{\frac{1}{2}} \hat{\boldsymbol{x}} \wedge \boldsymbol{H} + \boldsymbol{E}\}]\mathbf{y} \exp(\mathrm{i}k_1 \hat{\boldsymbol{x}} \cdot \mathbf{y})\, \mathrm{d}S_y \quad (30.13)$$

on taking advantage of Maxwell's equations.

The employment of (30.13) requires a knowledge of $\boldsymbol{E}_0$ and $\boldsymbol{H}_0$. These are most easily obtained from the representation ((1.29.7), (1.29.8))

$$\boldsymbol{E}(\boldsymbol{x}) = \boldsymbol{E}^{\mathrm{i}}(\boldsymbol{x}) - \mathrm{grad} \int_S \boldsymbol{n} \cdot \boldsymbol{E}\psi(\boldsymbol{x}, \mathbf{y})\, \mathrm{d}S_y - \mathrm{i}\omega\mu_1 \int_S \boldsymbol{n} \wedge \boldsymbol{H}\psi(\boldsymbol{x}, \mathbf{y})\, \mathrm{d}S_y,$$

while $\boldsymbol{H}$ is still given by (30.11) because $\boldsymbol{n} \cdot \boldsymbol{H} = 0$ on S. Hence

$$\boldsymbol{E}_0(\boldsymbol{x}) = \boldsymbol{e}^{\mathrm{i}} - \mathrm{grad} \int_S \boldsymbol{n} \cdot \boldsymbol{E}_0 \psi_0(\boldsymbol{x}, \mathbf{y})\, \mathrm{d}S_y, \quad (30.14)$$

$$\boldsymbol{H}_0(\boldsymbol{x}) = (\varepsilon_1/\mu_1)^{\frac{1}{2}} \boldsymbol{n}^{\mathrm{i}} \wedge \boldsymbol{e}^{\mathrm{i}} + \mathrm{curl} \int_S \boldsymbol{n} \wedge \boldsymbol{H}_0 \psi_0(\boldsymbol{x}, \mathbf{y})\, \mathrm{d}S_y. \quad (30.15)$$

The form (30.14) is of the type already handled and may be dealt with by a potential V with the structure of (30.8), subject to $V = 0$ on S. On the other hand, (30.15) shows that $\boldsymbol{H}_0$ is the gradient of a potential U which must be such that $\partial U/\partial n = 0$ in order that the normal component of $\boldsymbol{H}_0$ vanish on S. It follows that $\int_S \boldsymbol{n} \wedge \boldsymbol{H}_0\, \mathrm{d}S = \mathbf{0}$, as is also necessary for the consistency of the lowest orders in (30.10).

The combination of these remarks with (30.12) and (30.13) furnishes (30.6) with

$$\boldsymbol{p} = \int_S \boldsymbol{n} \cdot \boldsymbol{E}_0 \mathbf{y}\, \mathrm{d}S_y, \qquad \boldsymbol{q} = (\mu_1/\varepsilon_1)^{\frac{1}{2}} \int_S \mathbf{y} \wedge (\boldsymbol{n} \wedge \boldsymbol{H}_0)\, \mathrm{d}S_y.$$

For the perfectly conducting ellipsoid

$$I_1 p_1 = 4\pi abc e_1^{\mathrm{i}}/3,$$
$$(I_1 - 1) q_1 = 4\pi abc (\boldsymbol{n}^{\mathrm{i}} \wedge \boldsymbol{e}^{\mathrm{i}})_1/3$$

and other components are given by cyclic interchange of a, b, and c. That this is in agreement with the earlier result for a plane wave impinging on a sphere is easily verified.

8.31 Rayleigh–Gans scattering by diaphanous bodies

The general theory for diaphanous bodies has already been expounded in §8.21; among its applications are scattering by opal glass ($N=0.95$) and X-ray scattering. The main point to be made here is that N is not obliged to be a constant. Thus, in a sphere with spherically symmetric refractive index, the derivation of (21.2) can be followed up to the last stage with N kept under the integral sign. Hence

$$\boldsymbol{E}_s \sim k_1^2(e^{-ik_1R}/R)\{\boldsymbol{e}^i-(\boldsymbol{e}^i\cdot\hat{\boldsymbol{x}})\hat{\boldsymbol{x}}\}\int_0^a \{N^2(t)-1\}j_0(2k_1t\sin\tfrac{1}{2}\alpha)t^2\,dt. \tag{31.1}$$

One instance in which integration in (31.1) can be carried out is when the body extends to infinity and $N^2(t)=1+Ae^{-t^2/d^2}$. Since

$$\int_0^\infty J_{\frac{1}{2}}(at)t^{\frac{3}{2}}e^{-t^2/d^2}\,dt = a^{\frac{1}{2}}(\tfrac{1}{2}d^2)^{\frac{3}{2}}e^{-\frac{1}{4}a^2d^2},$$

$$\boldsymbol{E}_s \sim \tfrac{1}{4}(A/R)\pi^{\frac{1}{2}}d^3\{\boldsymbol{e}^i-(\boldsymbol{e}^i\cdot\hat{\boldsymbol{x}})\hat{\boldsymbol{x}}\}\exp(-ik_1R-k_1^2d^2\sin^2\tfrac{1}{2}\alpha).$$

Another example is that in which $N^2(t)=A+dt^2/a^2$. For

$$\int_0^a t^{\frac{7}{2}}J_{\frac{1}{2}}(zt/a)\,dt = a^{\frac{9}{2}}\int_0^{\frac{1}{2}\pi}\sin^{\frac{3}{2}}\theta\, J_{\frac{3}{2}}(z\sin\theta)(1-\sin^2\theta)\cos\theta\,d\theta$$
$$=(a^{\frac{9}{2}}/z^2)\{zJ_{\frac{3}{2}}(z)-2J_{\frac{5}{2}}(z)\}$$

by Sonine's first finite integral (A.1.30). Hence

$$\int_0^a \{N^2(t)-1\}j_0(2k_1t\sin\tfrac{1}{2}\alpha)t^2\,dt$$
$$=(A+d-1)a^3\frac{j_1(2k_1a\sin\frac{1}{2}\alpha)}{2k_1a\sin\frac{1}{2}\alpha}-\frac{da^3j_2(2k_1a\sin\frac{1}{2}\alpha)}{2(k_1a\sin\frac{1}{2}\alpha)^2}.$$

In particular, this gives a formula for the *Luneberg lens* in which $A=2$, $d=-1$ although such a sphere could scarcely be called diaphanous.

General shapes usually defy analytical evaluation unless N is constant. Then some progress can be made by introducing Cartesian coordinates, X, Y, Z with the Z-axis parallel to $\hat{\boldsymbol{x}}-\boldsymbol{n}^i$ and integrating with respect to X and Y first. In this way, the scattered electric intensity due to a diaphanous homogeneous ellipsoid is found to be

$$\boldsymbol{E}_s \sim \frac{k_1abc(N^2-1)}{2l_1R\sin\frac{1}{2}\alpha}e^{-ik_1R}\{\boldsymbol{e}^i-(\boldsymbol{e}^i\cdot\hat{\boldsymbol{x}})\hat{\boldsymbol{x}}\}j_1(2k_1l_1\sin\tfrac{1}{2}\alpha)$$

where $l_1=(l^2a^2+m^2b^2+n^2c^2)^{\frac{1}{2}}$, $(l,\ m,\ n)$ being the direction cosines of $\hat{\boldsymbol{x}}-\boldsymbol{n}^i$ with respect to the principal axes of the ellipsoid.

8.32 High frequency scattering by a diaphanous target

As the frequency increases Rayleigh–Gauss theory eventually breaks down though its practical validity may extend beyond limits which are sufficient to ensure that it is theoretically applicable. Nevertheless, so long as the scatterer is diaphanous, it is possible to consider a high frequency approximation. For simplicity, attention is restricted to the incident plane wave

$$\boldsymbol{E}^{\mathrm{i}} = \boldsymbol{i}\mathrm{e}^{-\mathrm{i}k_1 z}, \qquad \boldsymbol{H}^{\mathrm{i}} = (\varepsilon_1/\mu_1)^{\frac{1}{2}}\boldsymbol{j}\mathrm{e}^{-\mathrm{i}k_1 z}.$$

When the frequency is high a reasonable first approximation is suggested by ray theory, neglecting any contributions from points of glancing incidence. The direction and amplitude on a ray will not be affected appreciably by passage through a diaphanous object but the phase will be altered. Draw a plane P_1 perpendicular to the z-axis and on the opposite side of the obstacle to the source of the incident wave. Provided that P_1 is not too far from the body the rays on it will be the same as in the incident field apart from the change in phase. Let the line passing through $(x, y, 0)$ parallel to the z-axis intercept a length $d(x, y)$ of the obstacle (Fig. 8.21). Then the field on P_1 is $\boldsymbol{E}^{\mathrm{i}}\mathrm{e}^{-\mathrm{i}k_1(N-1)d}$, $\boldsymbol{H}^{\mathrm{i}}\mathrm{e}^{-\mathrm{i}k_1(N-1)d}$ when the body is homogeneous; of course, $d = 0$ if the incident ray does not intersect the body. Then, from (1.29.7), these fields may be used in the representation

$$\boldsymbol{E} = \int_{P_1} \{(\boldsymbol{k} \wedge \boldsymbol{E}) \wedge \operatorname{grad}_y \psi + (\boldsymbol{k} \cdot \boldsymbol{E})\operatorname{grad}_y \psi - \mathrm{i}\omega\mu_1(\boldsymbol{k} \wedge \boldsymbol{H})\psi\}\,\mathrm{d}y_1\,\mathrm{d}y_2 \tag{32.1}$$

for $z > l$ where $z = l$ is the equation of P_1. When there is no obstacle $d = 0$ everywhere on P_1 and the integral must reproduce the incident field $\boldsymbol{E}^{\mathrm{i}}$. Hence, if the integral with $d = 0$ on P_1 is subtracted from the right-hand side of (32.1), $\boldsymbol{E}$ becomes the sum of $\boldsymbol{E}^{\mathrm{i}}$ and an integral over

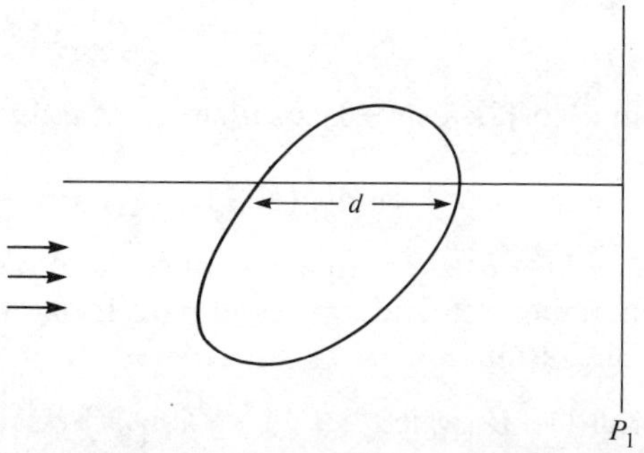

Fig. 8.21 High frequency scattering by a diaphanous body

S_0, the projection of the target on P_1. Thus the far field is given by

$$\boldsymbol{E}_s \sim \frac{ik_1}{4\pi R}\exp(-ik_1 l - ik_1 R + ik_1 l\boldsymbol{k}\cdot\hat{\boldsymbol{x}})(\boldsymbol{j}\wedge\hat{\boldsymbol{x}}+\boldsymbol{i})$$

$$\times \int_{S_0} e^{ik_1\hat{\boldsymbol{x}}\cdot\boldsymbol{y}}\{e^{-ik_1(N-1)d}-1\}\,dy_1\,dy_2. \tag{32.2}$$

The scattering coefficient can be deduced from (19.10) and is

$$\sigma_c^E + \sigma_a^E = -(2/S_0)\mathrm{Im}\left[i\int_{S_0}\{e^{-ik_1(N-1)d}-1\}\,dy_1\,dy_2\right].$$

In order that the field shall not depend on l it is necessary that $\hat{\boldsymbol{x}}$ should not be very different from $\boldsymbol{k}$, i.e. *the approximation can be expected to produce a useful prediction only of forward scattering.* If this restriction be acknowledged the additional approximations may be inserted in (32.2); namely, $\boldsymbol{k}\cdot\hat{\boldsymbol{x}}=1$ and $\hat{\boldsymbol{x}}\cdot\boldsymbol{y}=(y_1\cos\phi+y_2\sin\phi)\theta$ where θ and ϕ are the spherical polar angles of $\hat{\boldsymbol{x}}$.

As an illustration suppose that the incoming wave falls on an ellipsoid, its principal axes of lengths $2a$, $2b$, $2c$ coincident with the coordinate axes that along the z-axis being of length $2c$. Then S_0 is the ellipse $(y_1/a)^2+(y_2/b)^2=1$ and

$$d(y_1, y_2)=2c\{1-(y_1/a)^2-(y_2/b)^2\}^{\frac{1}{2}}.$$

Make the change of variable $y_1=a\sin u\cos v$, $y_2=b\sin u\sin v$ so that the domain of integration is $0\leqslant u\leqslant\frac{1}{2}\pi$, $0\leqslant v\leqslant 2\pi$. Then the integral in (32.2) becomes, if $\delta=(a^2\cos^2\phi+b^2\sin^2\phi)^{\frac{1}{2}}\theta$ and $a\tan\beta=b\tan\phi$,

$$ab\int_0^{2\pi}\int_0^{\frac{1}{2}\pi} e^{ik_1\delta\sin u\cos(v-\beta)}\{e^{-2ik_1c(N-1)\cos u}-1\}\sin u\cos u\,du\,dv$$

$$=2\pi ab\int_0^{\frac{1}{2}\pi} J_0(k_1\delta\sin u)\{e^{-2ik_1c(N-1)\cos u}-1\}\sin u\cos u\,du$$

from (A.1.24). It follows from Sonine's first finite integral (A.1.30) that the integral is

$$2\pi ab\int_0^{\frac{1}{2}\pi} J_0(k_1\delta\sin u)\cos\{2k_1c(N-1)\cos u\}\sin u\cos u\,du$$

$$-4\pi i abc(N-1)j_1(k_1w)/w - 2\pi ab J_1(k_1\delta)/k_1\delta$$

where $w=\{4c^2(N-1)^2+\delta^2\}^{\frac{1}{2}}$. There does not seem to be a simple expression for the remaining integral unless $\theta=0$. If $\theta=0$ the integral is elementary and the result is

$$2\pi ab\{(1+ik_1w_0)e^{-ik_1w_0}-1-\tfrac{1}{2}(k_1w_0)^2\}/(k_1w_0)^2$$

where $w_0=2c(N-1)$.

Consequently the distant scattered field in the direction $\boldsymbol{k}$ is

$$\boldsymbol{E}_s \sim \frac{\mathrm{i}ab\boldsymbol{i}}{k_1 w_0^2 R}\{(1+\mathrm{i}k_1 w_0)\mathrm{e}^{-\mathrm{i}k_1 w_0} - 1 - \tfrac{1}{2}(k_1 w_0)^2\}\mathrm{e}^{-\mathrm{i}k_1 R} \qquad (32.3)$$

whence the scattering coefficient is given by

$$\sigma_c^E + \sigma_a^E = 2 + 4\,\mathrm{Im}\{\mathrm{i} + (k_1 w_0 - \mathrm{i})\mathrm{e}^{-\mathrm{i}k_1 w_0}\}/k_1^2 w_0^2. \qquad (32.4)$$

If N is real (32.4) reduces to

$$\sigma_c^E + \sigma_a^E = 2 + 4(1 - \cos k_1 w_0 - k_1 w_0 \sin k_1 w_0)/k_1^2 w_0^2. \qquad (32.5)$$

Results for the spheroid may be deduced easily. In general, one would expect the approximation to be more accurate the shorter d is and the smaller the area on which incident rays are near to glancing incidence. This seems to be borne out in practice and, indeed, the approximation appears to be tolerably reliable for $k_1 l$ (l maximum dimension of body) in the range 10–20 provided that $|N-1| < 0.2$.

8.33 High frequency scattering

The obvious approach at short wavelengths is via ray theory. However, the investigation of the general obstacle is highly complicated because of the multiple internal reflections of rays. Even the calculation of the scattering coefficient is not straightforward. For instance, if the scattering coefficient of a sphere is evaluated from a consideration of the central refracted ray alone

$$\sigma_c^E + \sigma_a^E = 2 - \mathrm{Re}\frac{8N^2 \mathrm{i}\mathrm{e}^{2\mathrm{i}k_1 a(N-1)}}{(N+1)(N^2-1)k_1 a}.$$

But van de Hulst (1957) found that this deviated markedly from numerical computations of the exact solution and offered various correction terms.

Perfectly conducting obstacles (hard or soft in acoustics) are more amenable because there are no internal rays to muddy the water. Either a ray is reflected according to Snell's laws or it strikes a portion of the surface near the penumbral curve, thereby requiring special attention. It has already been pointed out in §8.7 that locally a parabolic cylinder behaves as a plane in reflecting rays except near a point of glancing incidence where it has the characteristics of a circular cylinder with radius equal to the radius of curvature of the parabola, the relevant surface distributions being set out in §8.3. This observation serves as a basis for predictions for general objects (for another point of view see Jones (1979)).

As regards creeping waves in two dimensions the fundamental picture is Fig. 8.22 which might be called the ABC of the geometrical theory of

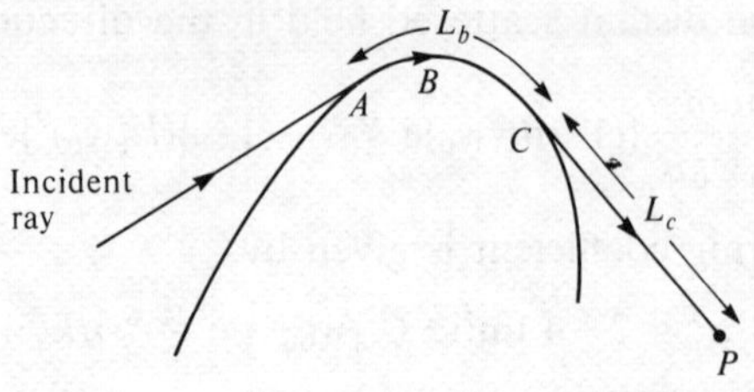

Fig. 8.22 Creeping waves in two dimensions

diffraction. On the basis of the investigation of the circular and parabolic cylinders an incident ray tangential to the boundary at A sets off a creeping ray from A to B to C where it leaves tangentially to the point of observation P. In fact, there is a multiplicity of creeping rays, corresponding to the zeros of the Airy function or its derivative, but only a typical one will be considered here, any necessary summation being left to the reader. The creeping wave field $u(P)$ at P, when $k_1L_b \gg 1$ and $k_1L_c \gg 1$, is postulated to be

$$u(P) \sim u^i(A)d_t(A)d_b(A)d_t(C)d_b(C)$$

$$\times \exp\left\{-ik_1L_b - \int_A^C \alpha(s)\,ds\right\} e^{-ik_1L_c - \frac{1}{4}\pi i}/(8\pi k_1 L_c)^{\frac{1}{2}} \quad (33.1)$$

where s is arc length on the boundary. The function α measures the loss of energy as rays are shed tangentially. It is of the form $K/\rho_s^{\frac{2}{3}}$ where ρ_s is the radius of curvature at s and K is a constant which depends on the boundary conditions. For instance, if $\partial u/\partial n = 0$ on the boundary, $K = e^{\frac{1}{6}\pi i}(\frac{1}{2}k_1)^{\frac{1}{3}}\beta_p$ from (3.8).

The *diffraction coefficients* d_t and d_b are coefficients associated with tangential rays and boundary rays respectively. They are dependent on the boundary conditions. For example, if $\partial u/\partial n - ik_1Zu = 0$,

$$d_t = 2^{\frac{2}{3}} e^{\frac{1}{6}\pi i}(k_1\rho)^{\frac{1}{3}}/\mathrm{Ai}(-\beta_p), \qquad d_b = e^{-\frac{1}{4}\pi i}/Z(k_1\rho)^{\frac{1}{2}} \quad (33.2)$$

when Z is neither very large or very small. When Z is very small

$$d_b = e^{-\frac{1}{12}\pi i}/2^{\frac{1}{3}}\beta_p^{\frac{1}{2}}(k_1\rho)^{\frac{1}{6}} \quad (33.3)$$

and for large Z

$$d_t = 2^{\frac{1}{3}} e^{\frac{1}{3}\pi i} Z(k\rho)^{\frac{2}{3}}/\mathrm{Ai}'(-\alpha_p). \quad (33.4)$$

In $d_t(A)$ and $d_b(A)$, ρ is evaluated at A.

For the field on the boundary at B

$$u(B) = u^i(A)d_t(A)d_b(A)d_b(B)\exp\left\{-ik_1L_b(B) - \int_A^B \alpha(s)\,ds\right\} \quad (33.5)$$

where $L_b(B)$ is the arc length from A to B; it is assumed that $k_1L_b \gg 1$. The values of $\partial u/\partial n$ on the boundary can be obtained by multiplying (33.5) by ik_1Z when the boundary condition is $\partial u/\partial n = ik_1Zu$. If the source is moved to A (not permitted for the boundary condition $u=0$) the terms $u^i(A)d_t(A)$ are deleted from (33.5).

In three dimensions the only essential change to (33.1) is to replace $e^{-ik_1L_c-\frac{1}{4}\pi i}/(8\pi k_1L_c)^{\frac{1}{2}}$ by $e^{-ik_1L_c-\frac{1}{2}\pi i}/8\pi k_1L_c$ for scalar fields (see also Exercise 12, 13, 37–40) but for electromagnetic waves the vector character of the field has to be taken into account.

The neighbourhood of a point of the penumbral curve will act as if it were a cylinder whose axis is perpendicular to the incident ray and whose radius is the radius of curvature ρ of the body in a plane perpendicular to the tangent. The field produced by a perfectly conducting circular cylinder in a plane wave travelling obliquely to the axis of the cylinder can be deduced from §8.10. The contributions from each of the cylinders are additive in the scattering coefficient and hence

$$\sigma_c^E \sim 2 + (1/k_1^{\frac{2}{3}}S_0)\int\{b_2 + n_0^2(b_1 - b_2)\}\rho^{\frac{1}{3}}\sin\beta\, ds \tag{33.6}$$

where β and $\cos^{-1}n_0$ are the angles made by the tangent to the penumbral curve with the direction of propagation and electric vector of the incident wave respectively. Also S_0 is the area of the incident field intercepted by the target and $b_1 = 0.9962$, $b_2 = -0.8642$ being the coefficients which occur in the scattering by a soft and hard cylinder respectively. Confirmation that (33.6) is in harmony with the results of §8.22 for a sphere is almost immediate.

When the target is a solid of revolution and the incident wave travels along the axis of symmetry

$$\sigma_c^E \sim 2 + (b_1 + b_2)\rho^{\frac{1}{3}}/k_1^{\frac{2}{3}}b$$

where b is the radius of the cross-section at the shadow boundary. In this case, *the electromagnetic scattering coefficient is the average of the hard and soft scattering coefficients.*

If the obstacle is a prolate spheroid the integral in (33.6) can be expressed in terms of a hypergeometric function. In particular, for a wave incident along the minor axis,

$$\sigma_c^E \sim 2 + 0.132d/(k_1b)^{\frac{2}{3}}$$

where the value of d is listed in Table 8.7. The column head d_{11} is for $\boldsymbol{E}^i$ parallel to the major axis and the other column is for $\boldsymbol{E}^i$ perpendicular to the major axis. Quite small departures from perfect sphericity create substantial changes in d.

Table 8.7 The coefficient for a prolate spheroid

b/a	d_{11}	$d_{\perp}$
1	1	1
0.9	2.1	−0.21
0.8	3.09	−1.41
0.6	5.08	−3.66
0.4	6.68	−5.47
0.2	8.11	−6.93

8.34 Physical optics

Another route to the determination of the high frequency reflection by an impenetrable obstacle rests on an integral representation for the field supplemented by a physical approximation. It has the advantage of being relatively simple to implement and it also yields information about the passage of rays through caustics. From one aspect, therefore, it serves as a canonical problem elucidating what happens where standard ray theory would fail to supply a finite field.

For sound waves assume that the target is hard so that, from (1.29.6),

$$p(\boldsymbol{x}) = p^{\mathrm{i}}(\boldsymbol{x}) + \int_S p(\mathbf{y}) \frac{\partial}{\partial n_y} \psi(\boldsymbol{x}, \mathbf{y}) \, \mathrm{d}S_y \tag{34.1}$$

for $\boldsymbol{x}$ in T_+. Allow $\boldsymbol{x}$ to tend to S. Then, from (27.5)

$$p(\boldsymbol{x}) = 2p^{\mathrm{i}}(\boldsymbol{x}) + 2 \fint_S p(\mathbf{y}) \frac{\partial}{\partial n_y} \psi(\boldsymbol{x}, \mathbf{y}) \, \mathrm{d}S_y \tag{34.2}$$

for $\boldsymbol{x}$ on S. The integral equation (34.2) must be satisfied by the values of p on S. Once these values have been determined the field in T_+ can be obtained from (34.1) and, in particular, the distant behaviour is given by

$$p(\boldsymbol{x}) \sim p^{\mathrm{i}}(\boldsymbol{x}) + (\mathrm{i}k_1/4\pi R)\mathrm{e}^{-\mathrm{i}k_1 R} \int_S p(\mathbf{y}) \boldsymbol{n}_y \cdot \hat{\boldsymbol{x}} \mathrm{e}^{\mathrm{i}k_1 \hat{\boldsymbol{x}} \cdot \mathbf{y}} \, \mathrm{d}S_y. \tag{34.3}$$

One mode of attack is to try to solve (34.2) numerically although this can be quite laborious at high frequencies. An analytical approach is to remark that, at short wavelengths, ray theory suggests that each small portion of the surface behaves as if it were plane. Each small patch will be subject to the incident radiation and the field produced by the remainder of the surface. But, to a first approximation, the secondary field can be disregarded in comparison with the incident wave. Then the distribution over the illuminated part S_1 of S is determined on the assumption that at every point the incident wave is reflected as if it were impinging on the infinite tangent plane. The distribution on the part of S in shadow is taken to be zero. Thus

$$p = 2p^{\mathrm{i}} \quad (\text{on } S_1), \qquad p = 0 \quad (\text{on } S - S_1) \tag{34.4}$$

This approximation is known as *physical optics.*

Physical optics is likely to be valid only when the radii of curvature of the surface are large compared with the wavelength as well as the radii of curvature of the incident wavefront. The second condition can be met provided that the source of the primary wave is some wavelengths from the point under consideration.

The approximation of physical optics does not always agree well with measured values, especially where electromagnetic crosspolarization is concerned. It can be improved by regarding it as the first stage in an iterative mechanism of solving (34.2). At first sight the presence of the leading term $2p^{\mathrm{i}}$ in the shadow appears to preclude physical optics being a good starting point. However, it will be seen subsequently that the insertion of (34.4) in (34.2) does make p very small on most of the shadow zone on S. Generally, the integral in (34.2) has to be evaluated numerically and our attention will be confined to the predictions of physical optics.

Assume that the source at $\boldsymbol{x}_0$ is such that

$$p^{\mathrm{i}}(\boldsymbol{x}) = \mathrm{e}^{-\mathrm{i}k_1 R_s}/4\pi R_s$$

where $R_s = |\boldsymbol{x} - \boldsymbol{x}_0|$. Then, from (34.1) and (34.4), the physical optics approximation is

$$p(\boldsymbol{x}) = p^{\mathrm{i}}(\boldsymbol{x}) + \frac{1}{8\pi^2}\int_{S_1} \frac{\boldsymbol{n}_y \cdot (\boldsymbol{x} - \boldsymbol{y})}{R_s\,|\boldsymbol{x} - \boldsymbol{y}|^3}(1 + \mathrm{i}k_1\,|\boldsymbol{x} - \boldsymbol{y}|)\exp\{-\mathrm{i}k_1(R_s + |\boldsymbol{x} - \boldsymbol{y}|\}\,\mathrm{d}S_y. \tag{34.5}$$

Before the evaluation of the integral in (34.5) is considered the analogous electromagnetic problem for a perfectly conducting solid will be formulated. The initial point is the representation

$$\boldsymbol{E}(\boldsymbol{x}) = \boldsymbol{E}^{\mathrm{i}}(\boldsymbol{x}) + (1/\mathrm{i}\omega\varepsilon_1)(\operatorname{grad}\operatorname{div} + k_1^2)\int_S \boldsymbol{n}_y \wedge \boldsymbol{H}\psi(\boldsymbol{x}, \boldsymbol{y})\,\mathrm{d}S_y.$$

Following the argument of physical optics we pose

$$\boldsymbol{n} \wedge \boldsymbol{H} = 2\boldsymbol{n} \wedge \boldsymbol{H}^{\mathrm{i}} \quad (\text{on } S_1), \qquad \boldsymbol{n} \wedge \boldsymbol{H} = \boldsymbol{0} \quad (\text{on } S - S_1). \tag{34.6}$$

An electric dipole $\boldsymbol{p}_0$ at $\boldsymbol{x}_0$ produces

$$\boldsymbol{E}^{\mathrm{i}} = (\operatorname{grad}\operatorname{div} + k_1^2)\boldsymbol{p}_0 \exp(-\mathrm{i}k_1 R_s)/4\pi\varepsilon_1 R_s,$$
$$\boldsymbol{H}^{\mathrm{i}} = \omega(\boldsymbol{x} - \boldsymbol{x}_0) \wedge \boldsymbol{p}_0(k_1 - \mathrm{i}/R_s)\exp(-\mathrm{i}k_1 R_s)/4\pi R_s^2.$$

Hence, in physical optics,

$$\mathbf{E}(\mathbf{x}) = \mathbf{E}^{\mathrm{i}} - (1/8\pi^2\varepsilon_1)(\text{grad div} + k_1^2)\int_{S_1} (\mathrm{i}k_1 + 1/R_s)\mathbf{n}_y \wedge \{(\mathbf{y} - \mathbf{x}_0) \wedge \mathbf{p}_0\}$$
$$\times \exp\{-\mathrm{i}k_1(R_s + |\mathbf{x} - \mathbf{y}|)\}\, \mathrm{d}S_y / R_s^2\, |\mathbf{x} - \mathbf{y}| \qquad (34.7)$$

where now $R_s = |\mathbf{y} - \mathbf{x}_0|$.

The integrals in (34.5) and (34.7) have similar structures in view of the shortness of the wavelength. Both are susceptible to asymptotic evaluation by the method of stationary phase and the main contribution will come from those points $\mathbf{y}$ of S_1 where $R_s + |\mathbf{x} - \mathbf{y}|$ is stationary. The close connection with Fermat's principle can be perceived at once.

Let $\mathbf{x}_0$ be designated O. Split space into three regions D_1, D_2, and D_3. D_2 consists of the exterior of the tangent cone from O to S (Fig. 8.23) whereas D_1 is that part of the interior of the tangent cone which lies between O and S_1. The interior which is on the opposite side of S_1 to O constitutes D_3. If S has the shape shown in Fig. 8.23 and the point of observation P is in $D_1 \cup D_2$ the stationary point Q on S is the point of contact of a prolate spheroid, having O and P as foci, which touches S. If the point of observation P' is in D_3 the stationary point Q is the point of intersection of OP' and S_1.

It is now convenient to select Q as origin with coordinates ξ, η, z relative to it, the z-axis being along the outward normal to S_1. The η-axis is selected in the plane of incidence (Fig. 8.24). Let θ be the angle of incidence so that O is the point $(0, -R_0 \sin\theta, R_0 \cos\theta)$ with respect to the axes at Q.

Let the point of observation be in D_3 so that it will be at $(0, R_1 \sin\theta, -R_1 \cos\theta)$ marked as P' on the broken line of Fig. 8.24.

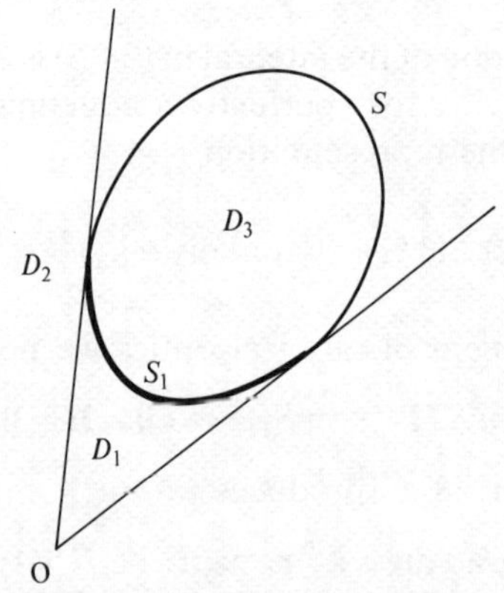

Fig. 8.23 The illuminated and shadow regions of an obstacle

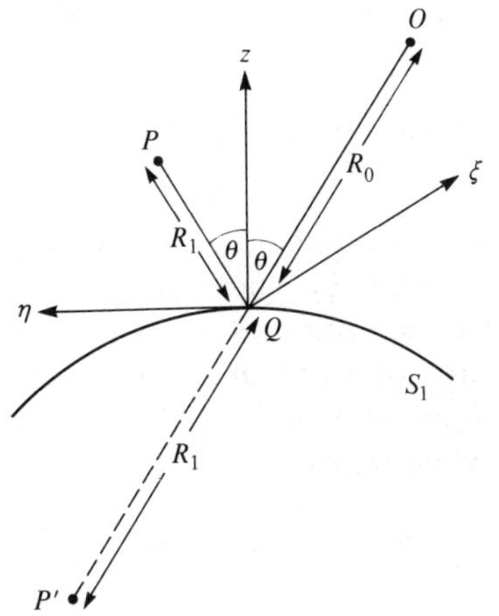

Fig. 8.24 Axes with the stationary point as origin

When the variable of integration (ξ, η) is near Q

$$R_s + |\mathbf{x} - \mathbf{y}| \approx R_0 + R_1 + \tfrac{1}{2}\left(\frac{1}{R_0} + \frac{1}{R_1}\right)(\xi^2 + \eta^2 \cos^2\theta).$$

Hence

$$\begin{aligned}\int &\exp\{-ik_1(R_s + |\mathbf{x} - \mathbf{y}|)\} \, dS_y \\ &\sim \exp\{-ik_1(R_0 + R_1)\} \int_{-\infty}^{\infty} \int_{-\infty}^{\infty} \exp\left\{-\tfrac{1}{2}ik_1\left(\frac{1}{R_0} + \frac{1}{R_1}\right)\right. \\ &\quad \times \left.(\xi^2 + \eta^2 \cos^2\theta)\right\} d\xi \, d\eta \\ &\sim 2\pi R_0 R_1 \exp\{-ik_1(R_0 + R_1)\}/ik_1(R_0 + R_1)\cos\theta.\end{aligned}$$

In this case $\boldsymbol{n}_y \cdot (\mathbf{x} - \mathbf{y}) = -R_1 \cos\theta$ so that (34.5) gives at P'

$$p(\boldsymbol{x}) \sim p^{\mathrm{i}}(\boldsymbol{x}) - \frac{1}{4\pi(R_0 + R_1)}(1 + 1/ik_1R_1)\exp\{-ik_1(R_0 + R_1)\}.$$

Since $R_0 + R_1$ can be identified with R_s of the incident wave at P' we see that for large k_1R_1, p essentially disappears, i.e. p can be regarded as zero in the shadow at points not too close to S_1. This verifies the statement made earlier about the iterative scheme for finding p on S.

Likewise in the electromagnetic case

$$\boldsymbol{E} \sim \boldsymbol{E}^{\mathrm{i}} - (\operatorname{grad} \operatorname{div} + k_1^2)\{\boldsymbol{p}_0 - \boldsymbol{n}_Q \cdot \boldsymbol{p}_0 \hat{\boldsymbol{R}}_0\} \frac{\exp\{-ik_1(R_0 + R_1)\}}{4\pi\varepsilon_1(R_0 + R_1)}$$

and $\boldsymbol{E}$ effectively vanishes in the shadow.

When P is in $D_1 \cup D_2$ a more elaborate analysis is necessary. Let the equation of S_1 near Q be

$$z \approx -\tfrac{1}{2}(A\xi^2 + 2H\xi\eta + B\eta^2) - \tfrac{1}{6}(C\xi^3 + 3D\xi^2\eta + 3E\xi\eta^2 + F\eta^3). \tag{34.8}$$

The coefficients A, B, H are related to the curvature of the surface at Q and, in fact, $AB - H^2 = 1/\rho_1\rho_2$ where ρ_1, ρ_2 are the principal radii of curvature at Q. Here, a radius of curvature is positive or negative according as the centre of curvature is in $z<0$ or $z>0$ so both are positive for the diagram of Fig. 8.24. If terms of the third order are neglected in (34.8), when (ξ, η) is near Q.

$$R_s + |\mathbf{x} - \mathbf{y}| \approx R_0 + R_1 + \tfrac{1}{2}(a\xi^2 + 2h\xi\eta + b\eta^2)$$

where

$$a = 2A \cos\theta + (R_0 + R_1)/R_0R_1, \qquad h = 2H\cos\theta,$$
$$b = 2B\cos\theta + (R_0 + R_1)\cos^2\theta/R_0R_1.$$

Assume that the eigenvalues λ_1, λ_2 of the quadratic form are not small. Then the axes can be rotated to (X, Y) say where the quadratic form is $\lambda_1X^2 + \lambda_2Y^2$. Hence

$$\int \exp\{-\mathrm{i}k_1(R_s + |\mathbf{x} - \mathbf{y}|)\}\,\mathrm{d}S_y$$
$$\sim \exp\{-\mathrm{i}k_1(R_0 + R_1)\}\int_{-\infty}^{\infty}\int_{-\infty}^{\infty} \exp\{-\tfrac{1}{2}\mathrm{i}k_1(\lambda_1X^2 + \lambda_2Y^2)\}\,\mathrm{d}X\,\mathrm{d}Y$$
$$\sim 2\pi\exp\{-\mathrm{i}k_1(R_0 + R_1) - \mathrm{i}\,\delta\}/k_1\,|\lambda_1\lambda_2|^{\frac{1}{2}}$$

where $\delta = 0, \tfrac{1}{2}\pi$, or $-\tfrac{1}{2}\pi$ according as λ_1 and λ_2 have opposite signs, are both positive or are both negative. Now

$$\lambda_1\lambda_2 = ab - h^2 = (\kappa_1 + 1/R_1)(\kappa_2 + 1/R_1)\cos^2\theta$$

where

$$\kappa_1 = A\sec\theta + B\cos\theta + \{(A\sec\theta - B\cos\theta)^2 + 4H^2\}^{\frac{1}{2}} + 1/R_0,$$
$$\kappa_2 = A\sec\theta + B\cos\theta - \{(A\sec\theta - B\cos\theta)^2 + 4H^2\}^{\frac{1}{2}} + 1/R_0.$$

The quantities κ_1 and κ_2 can be identified as the principal curvatures of the reflected wavefront due to a pencil of rays emanating from the vicinity of Q. With the convention that

$$(\kappa + 1/R_1)^{\frac{1}{2}} = |\kappa + 1/R_1|^{\frac{1}{2}}\mathrm{e}^{-\frac{1}{2}\pi\mathrm{i}} \tag{34.9}$$

when $\kappa R_1 + 1 < 0$ the phase change δ is automatically taken care of.

If both k_1R_0 and k_1R_1 are large the result is that

$$p(\boldsymbol{x}) \sim p^{\mathrm{i}}(\boldsymbol{x}) + Dp^{\mathrm{i}}_Q \mathrm{e}^{-\mathrm{i}k_1R_1}/R_1, \tag{34.10}$$

$$\boldsymbol{E}(\boldsymbol{x}) \sim \boldsymbol{E}^{\mathrm{i}}(\boldsymbol{x}) + D\{2(\boldsymbol{E}^{\mathrm{i}} \cdot \boldsymbol{n})\boldsymbol{n} - \boldsymbol{E}^{\mathrm{i}}\}_Q \mathrm{e}^{-\mathrm{i}k_1R_1}/R_1, \tag{34.11}$$

$$D = (\kappa_1 + 1/R_1)^{-\frac{1}{2}}(\kappa_2 + 1/R_1)^{-\frac{1}{2}},$$

the suffix Q signifying the value at Q.

If both κ_1 and κ_2 are both positive, as is true in Fig. 8.23, there is no problem with D becoming infinite. However, if the obstacle is concave instead of convex towards the region under consideration, one or both of κ_1 and κ_2 may be negative. Since $\kappa + 1/R_1$ is certainly positive when R_1 is small, it will decrease, pass through zero and become negative as R_1 increases. The zero occurs when the point of observation is on a caustic. After passage through the caustic, the convention of (34.9) indicates that the phase of D will be augmented by $\frac{1}{2}\pi$, i.e. *the phase of an astigmatic beam advances by* $\frac{1}{2}\pi$ *on crossing a caustic.* At a *focus*, where two caustics coincide, the contribution of both caustics must be allowed for and so *the phase advance on passage through a focus is* π.

The formulae (34.10) and (34.11) are applicable whenever the point of observation is not near a caustic. They are the same as would be predicted by ray theory apart from the phase changes already mentioned. Therefore it may be concluded that *the formulae of ray theory may be used (except in the vicinity of a caustic) when focussing takes place provided that when a caustic of an astigmatic beam is crossed in the direction of propagation the phase is advanced by* $\frac{1}{2}\pi$.

There are two instances when (34.10) and (34.11) break down. The first is when $\theta \approx \frac{1}{2}\pi$. In that event Q is near the penumbral curve and physical optics does not supply a satisfactory surface distribution. The appropriate transitional distribution between light and shadow must then be based on the investigation of earlier sections.

The second case is when Q is well away from the penumbral curve but the point of observation is near a caustic. One at least of the eigenvalues λ_1, λ_2 must then be small and it is no longer legitimate to ignore the cubic terms in the phase of the integral. Suppose that λ_2 is small but λ_1 is not; then the cubic term in Y must be retained so that

$$R_s + |\mathbf{x} - \mathbf{y}| \approx R_0 + R_1 + \tfrac{1}{2}(\lambda_1 X^2 + \lambda_2 Y^2) + \tfrac{1}{6}cY^3.$$

Now

$$\int \exp\{-\mathrm{i}k_1(R_s + |\mathbf{x} - \mathbf{y}|)\}\,\mathrm{d}S_y$$

$$\sim \left(\frac{2\pi}{k_1\lambda_1}\right)^{\frac{1}{2}} \exp\{-\mathrm{i}k_1(R_0 + R_1) - \tfrac{1}{4}\pi\mathrm{i}\} \int_{-\infty}^{\infty} \exp\{\mathrm{i}k_1(\tfrac{1}{2}\lambda_2 Y^2 + \tfrac{1}{6}cY^3)\}\,\mathrm{d}Y$$

$$\sim \frac{(2\pi)^{\frac{3}{2}}2^{\frac{1}{3}}}{k_1^{\frac{5}{6}}\lambda_1^{\frac{1}{2}}|c|^{\frac{1}{3}}} \exp\{-\mathrm{i}k_1(R_0 + R_1) - \tfrac{1}{4}\pi\mathrm{i} - \tfrac{1}{3}\mathrm{i}k_1\lambda_2^3/c^2\}\mathrm{Ai}\left(-\frac{k_1^{\frac{2}{3}}\lambda_2^2}{2^{\frac{2}{3}}|c|^{\frac{4}{3}}}\right) \tag{34.12}$$

by (A.2.11). In (34.12)

$$c = 2\cos\theta\{-C\sin^3\alpha + 3D\sin^2\alpha\cos\alpha - 3E\sin\alpha\cos^2\alpha + F\cos^3\alpha\}$$
$$-(R_0^{-1} - R_1^{-1})3\sin 2\theta\cos\alpha(A\sin^2\alpha - 2H\sin\alpha\cos\alpha + B\cos^2\alpha)$$
$$-3(R_0^{-2} - R_1^{-2})\sin\theta\cos\alpha(\sin^2\alpha + \cos^2\alpha\cos^2\theta)$$

where $\tan 2\alpha = 2h/(a-b)$.

At the caustic $\lambda_2 = 0$ and the Airy function becomes independent of k_1. Therefore, as far as k_1 is concerned, (34.12) is larger than the previous surface integral by a factor of $k^{\frac{1}{6}}$. It follows that *the field at a caustic exceeds that well away from a caustic by a factor of* $k_1^{\frac{1}{6}}$.

The behaviour at the other caustic may be obtained in a similar manner by interchanging the roles of λ_1 and λ_2. At a focus, the two caustics intersect and λ_1, λ_2 vanish simultaneously. In the neighbourhood of a focus it is no longer permissible to discard the cubic terms in X. Thus a product of Airy functions arises and *at the focus the field is larger by the factor* $k_1^{\frac{1}{3}}$ *than that well away from the caustics.* Note that, at a focus $H = 0$ and $A = B\cos^2\theta$.

It may happen that at a caustic c vanishes with λ_2, even when λ_1 is nonzero. Then (34.12) is not valid any more and the term in Y^4 must be kept in $R_s + |\boldsymbol{x} - \mathbf{y}|$. The effect is that at the caustic the field is greater by a factor of $k_1^{\frac{1}{4}}$ than its value some distance away from a caustic. Obviously, one can consider further coefficients vanishing at the same point so that the factor of increase is $k_1^{\frac{1}{2}-1/n}$ if Y^n has the first nonzero coefficient at the caustic, assuming that λ_1 is nonzero.

The substantial rise in the field near a caustic explains why conventional ray theory has to be abandoned when a ray is about to encounter a caustic and other forms of expansion adopted.

The preceding theory has assumed that there is only one stationary point Q which contributes to the scattering. There may, of course, be several such stationary points and then the total scattered field is the sum of their separate contributions. In addition, there may be points $Q, Q_1, \ldots, Q_n$ such that the path $OQ + QQ_1 + \ldots + Q_nP$ is stationary. Now the field radiated from near Q influences the surface distribution at Q_1 as well as the incident field and the calculation becomes much more complicated.

Although the technique has been based on the approximation of physical optics it is, naturally, equally applicable to the radiation from any surface or wavefront on which a suitable surface distribution is specified.

Assemblages of particles

8.35 Independent scattering

The theory of this chapter has so far been wholly concerned with the scattering by a single object but there are many circumstances in which groups of particles play an important role, e.g. scattering by the atmosphere, by colloids, and by biological specimens. The investigation of the effect of several obstacles will be restricted to the more difficult case of electromagnetism since the analysis may be adapted in a straightforward way to acoustics.

The incident wave will be assumed to be plane. In the first case it will be supposed that each particle is unaffected by the radiation from other particles, i.e. *each particle scatters as if it were subject to the incident plane wave alone*; such scattering will be called *independent.*

Let there be M obstacles present and select some easily identified point O_m, such as the centre of mass, in the mth scatter. Choose axes so that in the incident plane wave $\boldsymbol{E}^{\mathrm{i}} = \boldsymbol{i}\mathrm{e}^{-\mathrm{i}k_1 z}$. Suppose that when O_m is placed at the origin and there are no other obstacles, the distant scattered field has electric intensity $\boldsymbol{G}_m(\hat{\boldsymbol{x}})\mathrm{e}^{-\mathrm{i}k_1 R}/R$ in the direction of the unit vector $\hat{\boldsymbol{x}}$, R being the distance from the origin. Then, when O_m is at the point $\boldsymbol{R}_m$ with respect to the origin, it produces a scattered field $\boldsymbol{G}_m(\hat{\boldsymbol{x}})\mathrm{e}^{\mathrm{i}k_1\boldsymbol{R}_m\cdot(\hat{\boldsymbol{x}}-\boldsymbol{k})}\mathrm{e}^{-\mathrm{i}k_1 R}/R$. Hence the total electric intensity radiated by independent scatterers is

$$\left\{\sum_{m=1}^{M} \boldsymbol{G}_m(\hat{\boldsymbol{x}})\mathrm{e}^{\mathrm{i}k_1\boldsymbol{R}_m\cdot(\hat{\boldsymbol{x}}-\boldsymbol{k})}\right\}\mathrm{e}^{-\mathrm{i}k_1 R}/R.$$

It follows from (19.10) that the sum of the average rates at which energy is scattered and absorbed by the assemblage is

$$-2\pi\omega\varepsilon_1 \,\mathrm{Im} \sum_{m=1}^{M} \boldsymbol{G}_m(\boldsymbol{k})\cdot\boldsymbol{i}/k_1^2, \tag{35.1}$$

i.e. *the contributions of the individual particles can be added together separately*. The intensity in the direction $\hat{\boldsymbol{x}}$ is

$$\tfrac{1}{2}(\varepsilon_1/\mu_1)^{\frac{1}{2}} \sum_{m=1}^{M}\sum_{q=1}^{M} \boldsymbol{G}_m(\hat{\boldsymbol{x}})\cdot\boldsymbol{G}_q^*(\hat{\boldsymbol{x}})\exp\{\mathrm{i}k_1(\boldsymbol{R}_m-\boldsymbol{R}_q)\cdot(\hat{\boldsymbol{x}}-\boldsymbol{k})\}/R^2.$$

On occasion the particles occupy a slab perpendicular to the z-axis and there are many of them (Fig. 8.25). If the particles are similarly shaped, all having the same orientation, the function $\boldsymbol{G}_m$ does not differ from particle to particle; the only variation comes from any appreciable differences in the directions of the particles relative to the point of observation. In view of the large number of particles, summation may be replaced by integration and the scattered field may be written

$$\int \{N_0\boldsymbol{G}_1(\hat{\boldsymbol{x}})\mathrm{e}^{-\mathrm{i}k_1(z+R_\mathrm{p})}/R_\mathrm{p}\}\,\mathrm{d}\boldsymbol{z}$$

where the integration is over the volume of the slab, N_0 is the number of particles per unit volume and R_p is the distance between P and a particle at the point of integration.

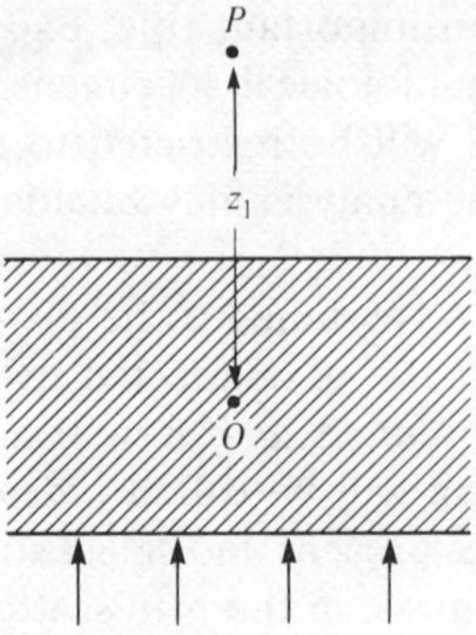

Fig. 8.25 Scattering by a slab

When P is some distance from the slab the main contribution will stem from the points of stationary phase. Carrying out the integration over a plane perpendicular to the z-axis first, we find that such a point occurs at the point of intersection of OP with the plane. After integration with respect to z over the thickness l of the slab, the scattered field is found to be

$$2\pi N_0 l \boldsymbol{G}_1(\boldsymbol{k}) e^{-ik_1 z_1}/ik_1$$

where $z_1 = OP$. If the factor of $e^{-ik_1 z_1}$ is small the $\boldsymbol{i}$-component of the scattered electric intensity can be expressed as

$$e^{-ik_1 z_1}\{e^{-ik_1 l(N-1)} - 1\}$$

where

$$N = 1 + 2\pi N_0 \boldsymbol{G}_1(\boldsymbol{k}) \cdot \boldsymbol{i}/k_1^2.$$

The interpretation of this result is that the transmitted wave (but not the reflected) behaves exactly as if it had passed through a medium of refractive index N. It undergoes attenuation, dictated by $e^{-k_1 l \operatorname{Im} N}$, which is closely related to the rate at which energy is removed from the incident beam as given by (35.1). Some idea of the energy loss in atmospheric haze can be obtained from this formula.

The *optical depth* τ in the direction of the incident wave is such that the intensity of the beam is reduced to $e^{-\tau}$ of its original value by passage through a specimen. Thus, the optical depth of the slab under consideration is $-4\pi N_0 l \operatorname{Im} \boldsymbol{G}_1 \cdot \boldsymbol{i}/k_1$. If it is discovered experimentally that τ is small (<0.1, say) it may be surmised that the hypothesis of independent scattering is fulfilled.

In numerous applications the motion of the particles has to be borne in mind. For example, in atmospheric haze, fog, storm clouds, air supporting dust grains, etc. the scattering is produced by many solid or liquid particles. The particles move relative to one another as they fall because of local turbulence. There is a resultant fluctuation in field strength. So long as the modulation in time is at a rate which is slow compared with $2\pi/\omega$ the scattering can be calculated at any instant as if the particles were motionless. The haphazard movements of the particles makes it plausible to regard the signals of any two as uncorrelated in time. More precisely, if $\langle\cdots\rangle$ denotes a time average, it is assumed that

$$\langle \boldsymbol{G}_m(\hat{\boldsymbol{x}}) \cdot \boldsymbol{G}_q^*(\hat{\boldsymbol{x}}) \exp\{ik_1(\boldsymbol{R}_m - \boldsymbol{R}_q) \cdot (\hat{\boldsymbol{x}} - \boldsymbol{k})\}\rangle = 0 \quad (m \neq q). \tag{35.2}$$

A collection of particles for which (35.2) holds will be called *an assemblage of independent random scatters.* In addition to the examples already quoted, the movement of vegetation in the wind and the motion of *window* or *chaff* (a falling cloud of resonant strips of metallic foil) provide illustrations of independent random scatterers.

One consequence of (35.2) is that the intensity I satisfies

$$\langle I \rangle = \sum_{m=1}^{M} \langle I_m \rangle$$

where I_m is the intensity due to the mth particle, i.e. *the average intensity of an assemblage of independent random scatterers is the sum of the average intensities of the separate particles.* Clearly, this result is also true for the total scattered energy and, moreover, for the total absorbed energy on account of (35.1). A criterion for verifying the hypothesis of independent random scattering is to check that the intensity is doubled when the number of particles is doubled.

One particular case in which (35.2) can be confirmed arises as follows. Let p_m be the probability of receiving $\boldsymbol{G}_m \exp\{ik_1\boldsymbol{R}_m \cdot (\hat{\boldsymbol{x}} - \boldsymbol{k})\}$ from the mth particle whereas p_{mq} is the probability of receiving $\boldsymbol{G}_m \exp\{ik\boldsymbol{R}_m \cdot (\hat{\boldsymbol{x}} - \boldsymbol{k})\}$ and $\boldsymbol{G}_q \exp\{ik_1\boldsymbol{R}_q \cdot (\hat{\boldsymbol{x}} - \boldsymbol{k})\}$ from the mth and qth particles respectively. Suppose that $p_{mq} = p_m p_q$, i.e. the probability distributions are statistically independent. Then the extra condition

$$\langle \boldsymbol{G}_m \exp\{ik_1\boldsymbol{R}_m \cdot (\hat{\boldsymbol{x}} - \boldsymbol{k})\}\rangle = 0$$

ensures validity of (35.2).

For a large number of particles the Central Limit theorem permits the assertion that the probability that

$$\boldsymbol{G} = \sum_{m=1}^{M} \boldsymbol{G}_m \exp\{ik_1\boldsymbol{R}_m \cdot (\hat{\boldsymbol{x}} - \boldsymbol{k})\}$$

lies between $\boldsymbol{G}$ and $\boldsymbol{G} + \mathrm{d}\boldsymbol{G}$ is

$$P(\boldsymbol{G})\,\mathrm{d}\boldsymbol{G} = \exp\{-(X^2 + Y^2)/G_0\}\,\mathrm{d}X\,\mathrm{d}Y/\pi G_0$$

where $G_0 = \langle \boldsymbol{G} \cdot \boldsymbol{G}^* \rangle = 2R^2 \langle I \rangle (\mu_1/\varepsilon_1)^{\frac{1}{2}}$ and X, Y are the components of $\boldsymbol{G}$ parallel to any two perpendicular coordinates in the plane transverse to $\hat{\boldsymbol{x}}$.

This is a bivariate Gaussian distribution with mean $X=0$, $Y=0$. Thus *the mean value of* $\mathbf{G}$ *is zero.*

The distribution of $G=(\mathbf{G}\cdot\mathbf{G}^*)^{\frac{1}{2}}$ is $(2/G_0)e^{-G^2/G_0}G\,dG$ so that the mean value of G is $(\frac{1}{2}\pi G_0)^{\frac{1}{2}}$. The mean value of G^2 is G_0 and its standard deviation is G_0. The fact that the standard deviation is equal to the mean implies that the fluctuation in field strength will be large. The number of particles is irrelevant provided that there are many of them (often ten is sufficient in practice). The disadvantage of this theory is that it gives no information about the rate at which the fluctuation takes place being limited to the magnitude of the variation. To discuss the rate at which signal strength varies it is necessary to derive the probability that G has a particular value at a given time but that problem will not be considered here.

If the particles are confined to a finite domain and the point of observation is a great distance away, variations in $\hat{\mathbf{x}}$ and R can be totally ignored. If, further, the particles are spherical (or with similar shape and orientation) the intensity radiated by any one will be the same as that by any other of the same size. Summation may also be replaced by integration when there are many particles per unit volume. Therefore, if $I_a(\hat{\mathbf{x}})$ is the intensity radiated by a sphere of radius a, the total scattered intensity is

$$T\int_0^\infty m(a)I_a(\hat{\mathbf{x}})\,da$$

where $m(a)$ is the number of particles per unit volume of radius a and T is the volume of the cloud.

The formula is valuable in considering the effects of rain and fog on radio propagation. The attenuation produced may be estimated by taking $\hat{\mathbf{x}}=\mathbf{k}$ and the back scatter by putting $\hat{\mathbf{x}}=-\mathbf{k}$. The magnitude of the back scatter determines the extent to which clouds and storms can be detected by radar. It is also of some moment in radar meteorology and in the observation of *bright bands* generated by the ice particles melting at the bottom of a cloud before dropping as rain. Now, when Rayleigh scattering theory is valid

$$I_a(-\mathbf{k})=\frac{1}{2}\left|\frac{N^2-1}{N^2+2}\right|^2\frac{k_1^4a^6}{R^2}I_0$$

according to §8.20, where I_0 is the incident intensity. The refractive index N of water is complex and varies with frequency. For example, at 18 °C, $N=8.9\text{–}0.69i$ when $\lambda=10$ cm, $N=8.30\text{–}1.90i$ for $\lambda=3.2$ cm, $N=6.4\text{–}2.8i$ at $\lambda=1.24$ cm and $N=3.41\text{–}1.94i$ when $\lambda=3$ mm. The refractive index also depends on temperature; the real part does not change very much unless $\lambda<3$ cm but the imaginary part decreases rapidly, if $\lambda>3$ cm, as the temperature increases. Magnetic resonance, as described in

§8.17, takes place and the exact back scattering deviates from the Rayleigh approximation by a factor of 2 when $k_1 a$ is about $\frac{1}{2}$. Nevertheless, a factor of 2 is insignificant compared with other uncertainties so that the Rayleigh theory may be used for fog ($a = 5$–20×10^{-3} mm) and rain ($a = 0.2$ mm–2 mm in a thunderstorm). The intensity, consequently, involves the sixth power of the radius so that very large drops, even though few in number, account for the major part of the scattering. It is therefore necessary to know the distribution of large drops with some accuracy if a reliable estimate of the scattering is desired.

8.36 General theory for loosely packed objects

The general theory of scattering by an assemblage outside the scope of the preceding section is a matter of some complexity. As a simplification, it will be here assumed that the particles are several wavelengths apart, i.e. they are loosely packed. In addition, the obstacles will be assumed to be composed of the same material, to have the same shape and to be similarly oriented. These assumptions simplify the presentation but the principles can be adapted to more general circumstances.

Consider first a single particle in isolation. Let a representative point of it be O. Suppose that it is irradiated by a plane wave travelling in the direction of the unit vector $\boldsymbol{n}_1$. Let the electric vector of the plane wave at O be $\boldsymbol{E}(\boldsymbol{n}_1)$ and be such that $\boldsymbol{E} = \hat{\mathbf{E}}$ there, $\hat{\mathbf{E}}$ being a unit vector parallel to $\mathbf{E}$. At a large distance R from O the scattered electric intensity in the direction of $\hat{\boldsymbol{x}}$ will be a function of $\hat{\boldsymbol{x}}$ as well as the polarization and direction of propagation of the exciting wave; write it as $\mathbf{A}(\boldsymbol{n}_1, \hat{\mathbf{E}}; \hat{\boldsymbol{x}})\mathrm{e}^{-\mathrm{i}k_1 R}/R$. For small variations in the point of observation only the phase changes and the field will appear as a plane wave of amplitude $\mathbf{A}(\boldsymbol{n}_1, \hat{\mathbf{E}}; \hat{\boldsymbol{x}})\mathrm{e}^{-\mathrm{i}k_1 R}/R$ travelling in the direction $\hat{\boldsymbol{x}}$. Thus, in such a neighbourhood, the scattered field can be regarded as a plane wave with polarization $\hat{\mathbf{A}}$ going parallel to $\hat{\boldsymbol{x}}$.

Now suppose that there are M similarly oriented particles present, their identifying points being $O_1, O_2, \ldots, O_M$. Let the electric intensity of the scattered field due to the jth particle at a large distance R_j from O_j be $\boldsymbol{F}_j(\hat{\boldsymbol{x}})\mathrm{e}^{-\mathrm{i}k_1 R_j}/R_j$ in the direction of $\hat{\boldsymbol{x}}$. Then the electric intensity at the mth particle due to the jth is $\boldsymbol{F}_j(\boldsymbol{n}_{mj})\mathrm{e}^{-\mathrm{i}k_1 R_{mj}}/R_{mj}$ where R_{mj} is the distance between O_j and O_m while $\boldsymbol{n}_{mj}$ is a unit vector from O_j to O_m. As in the preceding paragraph this behaves as a plane wave near O_m and causes a scattered field

$$\mathbf{A}(\boldsymbol{n}_{mj}, \hat{\boldsymbol{F}}_j(\boldsymbol{n}_{mj}); \hat{\boldsymbol{x}})\boldsymbol{F}_j \cdot \hat{\boldsymbol{F}}_j \exp\{-\mathrm{i}k_1(R_m + R_{mj})\}/R_m R_{mj}.$$

By summation over j the field scattered by the mth particle due to its irradiation by the other particles is determined. To this must be added any scattering caused by the plane wave which is impinging on the whole collection. Let the incident plane wave be $\mathbf{E}^{\mathrm{i}}(\mathbf{n}^{\mathrm{i}})$, being $\mathbf{E}^{\mathrm{i}}_m(\mathbf{n}^{\mathrm{i}})$ at the mth particle. Then the total field scattered by the mth particle is

$$\mathrm{e}^{-\mathrm{i}k_1R_m}\{\mathbf{A}(\mathbf{n}^{\mathrm{i}}, \hat{\mathbf{E}}^{\mathrm{i}}_m; \hat{\mathbf{x}})\mathbf{E}^{\mathrm{i}}_m \cdot \hat{\mathbf{E}}^{\mathrm{i}}_m + \sum_{j=1}^{M}{}' \mathbf{A}(\mathbf{n}_{mj}, \hat{\mathbf{F}}_j; \hat{\mathbf{x}})\mathbf{F}_j \cdot \hat{\mathbf{F}}_j \mathrm{e}^{-\mathrm{i}k_1R_{mj}}/R_{mj}\}/R_m$$

where $\sum'$ means omit $j = m$. By definition, this field is $\mathbf{F}_m(\hat{\mathbf{x}})\mathrm{e}^{-\mathrm{i}k_1R_m}/R_m$. Hence

$$\mathbf{F}_m(\hat{\mathbf{x}}) = \mathbf{A}(\mathbf{n}^{\mathrm{i}}, \hat{\mathbf{E}}^{\mathrm{i}}_m; \hat{\mathbf{x}})\mathbf{E}^{\mathrm{i}}_m \cdot \hat{\mathbf{E}}^{\mathrm{i}}_m + \sum_{j=1}^{M}{}' \mathbf{A}(\mathbf{n}_{mj}, \hat{\mathbf{F}}_j; \hat{\mathbf{x}})\mathbf{F}_j \cdot \hat{\mathbf{F}}_j \mathrm{e}^{-\mathrm{i}k_1R_{mj}}/R_{mj} \tag{36.1}$$

for all $\hat{\mathbf{x}}$ and $m = 1, \ldots, M$. By choosing $\hat{\mathbf{x}} = \mathbf{n}_{mq}$ we can obtain from (36.1) $M(M-1)$ equations for the constants $\mathbf{F}_j(\mathbf{n}_{mj}) \cdot \hat{\mathbf{F}}_j(\mathbf{n}_{mj})$. Once these have been solved $\mathbf{F}_m$ is known for all $\hat{\mathbf{x}}$ from (36.1). Since $\mathbf{A}$ is derived from the scattering by a single particle in a plane wave the behaviour of a collection of particles can be determined from that for one particle.

Clearly, the labour of finding the constants grows very rapidly with M and so the number of particles which can be handled is limited unless, perchance, the particles are arranged in a conveniently symmetrical disposition susceptible to analytical treatment.

8.37 The grating

Some easing of the analysis occurs for two-dimensional fields. The field can then be expressed in terms of a single scalar, related to the electric or magnetic intensity according to the polarization under consideration. (It is assumed that the boundary conditions do not alter the polarization.) Let O be a suitably identified point of the cross-section of an obstacle. When the obstacle is subjected to a plane wave of unit amplitude and zero phase at O, travelling in the direction of $\mathbf{n}_1$ the scattered field in the direction of $\hat{\mathbf{x}}$ may be written as $A(\mathbf{n}_1; \hat{\mathbf{x}})\mathrm{e}^{-\mathrm{i}k_1r}/r^{\frac{1}{2}}$. An argument along the lines of §8.36 shows that, if $F_j(\hat{\mathbf{x}})\mathrm{e}^{-\mathrm{i}kr_j}/r^{\frac{1}{2}}$ is the scattered wave due to the jth obstacle,

$$F_m(\hat{\mathbf{x}}) = A(\mathbf{n}^{\mathrm{i}}; \hat{\mathbf{x}})F_0 + \sum_{j=1}^{M}{}' A(\mathbf{n}_{mj}; \hat{\mathbf{x}})F_j(\mathbf{n}_{mj})\mathrm{e}^{-\mathrm{i}k_1r_{mj}}/r_{mj}^{\frac{1}{2}}$$

where F_0 accounts for the incident plane wave. Hence

$$F_m(\mathbf{n}_{pm}) = A(\mathbf{n}^{\mathrm{i}}; \mathbf{n}_{pm})F_0 + \sum_{j=1}^{M}{}' A(\mathbf{n}_{mj}; \mathbf{n}_{pm})F_j(\mathbf{n}_{mj})\mathrm{e}^{-\mathrm{i}k_1r_{mj}}/r_{mj}^{\frac{1}{2}}$$

for $m = 1, \ldots, M$ and $p = 1, \ldots, M$ with $m \neq p$. This is a set of equations to determine the constants $F_j(\mathbf{n}_{mj})$.

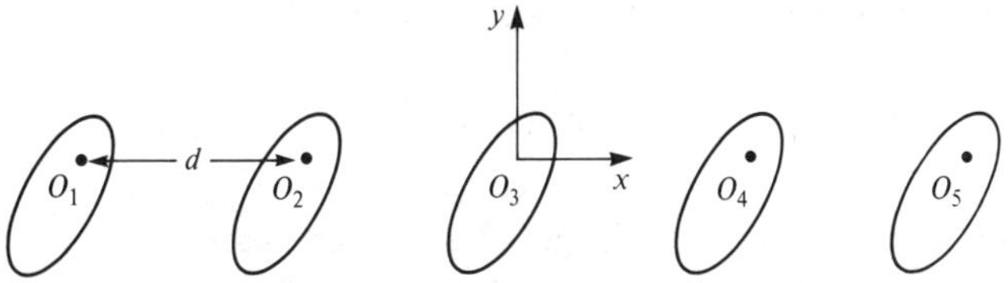

Fig. 8.26 The diffraction grating

One of the applications of this theory is to the *diffraction grating*, which consists of a number of cylinders placed along a row at equal distances (Fig. 8.26). The number of cylinders will be taken to be odd and M replaced by $2M+1$. Choose O_{M+1} as origin and pick axes so that $O_1, O_2, \ldots, O_{2M+1}$ are the points $(-Md, 0)$, $(-(M-1)d, 0), \ldots, (Md, 0)$ respectively (see Fig. 8.26 for $M=2$). Then $\boldsymbol{n}_{mj}$ makes the angle 0 with the positive x-axis if $m>j$ and the angle π if $m<j$. Let the incident plane wave be $\exp\{-\mathrm{i}k_1 r\cos(\phi-\phi^{\mathrm{i}})\}$. Then the wave scattered by the mth cylinder at an angle ϕ to the positive x-axis is given by

$$F_m(\phi)=A(\phi^{\mathrm{i}};\phi)\exp(-\mathrm{i}k_1 md\cos\phi^{\mathrm{i}}) + \sum_{j=-M}^{m-1} A(0;\phi)F_j(0)\mathrm{e}^{-\mathrm{i}k_1 r_{mj}}/r_{mj}^{\frac{1}{2}} + \sum_{j=m+1}^{M} A(\pi;\phi)F_j(\pi)\mathrm{e}^{-\mathrm{i}k_1 r_{mj}}/r_{mj}^{\frac{1}{2}} \tag{37.1}$$

where $r_{mj}=|m-j|\,d$ and $m=-M, -M+1, \ldots, M$. By putting $\phi=0, \pi$ in (37.1) the following equations for $F_j(0)$ and $F_j(\pi)$ are obtained:

$$F_m(0)=A(\phi^{\mathrm{i}};0)\exp(-\mathrm{i}k_1 md\cos\phi^{\mathrm{i}})+A(0;0)\sum_{j=-M}^{m-1} F_j(0)\mathrm{e}^{-\mathrm{i}k_1 r_{mj}}/r_{mj}^{\frac{1}{2}} + A(\pi;0)\sum_{j=m+1}^{M} F_j(\pi)\mathrm{e}^{-\mathrm{i}k_1 r_{mj}}/r_{mj}^{\frac{1}{2}}, \tag{37.2}$$

$$F_m(\pi)=A(\phi^{\mathrm{i}};\pi)\exp(-\mathrm{i}k_1 md\cos\phi^{\mathrm{i}})+A(0;\pi)\sum_{j=-M}^{m-1} F_j(0)\mathrm{e}^{-\mathrm{i}k_1 r_{mj}}/r_{mj}^{\frac{1}{2}} + A(\pi;\pi)\sum_{j=m+1}^{M} F_j(\pi)\mathrm{e}^{-\mathrm{i}k_1 r_{mj}}/r_{mj}^{\frac{1}{2}}. \tag{37.3}$$

If the scattering is independent

$$F_m(\phi)=A(\phi^{\mathrm{i}};\phi)\exp(-\mathrm{i}k_1 md\cos\phi^{\mathrm{i}})$$

and the radiation pattern becomes similar to that of an array of antennas (see Chapter 3) because the total scattered field is

$$A(\phi^{\mathrm{i}}; \phi) \sum_{m=-M}^{M} \exp(-\mathrm{i}k_1 md \cos \phi^{\mathrm{i}} - \mathrm{i}k_1 r_m)/r_m^{\frac{1}{2}}.$$

In particular, if M is not too large, this is approximately

$$A(\phi^{\mathrm{i}}; \phi)\mathrm{e}^{-\mathrm{i}k_1 r} \sin(M+\tfrac{1}{2})t/r^{\frac{1}{2}} \sin \tfrac{1}{2}t$$

where r is the distance from the origin and $t = k_1 d(\cos \phi - \cos \phi^{\mathrm{i}})$. The factor A depends only on the type of cylinder whereas the terms involving t supply information from the structure of the grating. The maxima of the grating radiation pattern occur at $t = 2q\pi$ $(q = 0, \pm 1, \pm 2, \ldots)$ and are all equal to $2M+1$. *They correspond to the observed spectra at the specified angles.* Now, take a reference plane through the origin parallel to the incident wave front. An incident ray has phase given by $\exp(-\mathrm{i}k_1 md \cos \phi^{\mathrm{i}})$ at the mth cylinder and this becomes $\mathrm{e}^{\mathrm{i}mt - \mathrm{i}k_1 r}$ at the point of observation. At the specified angles mt becomes $2mq\pi$, i.e. *at the angles of the maxima the path length from the reference plane along a ray to the point of observation via a scatterer is an integral number of wavelengths.*

When the interaction between the cylinders cannot be omitted (37.2) and (37.3) have to be solved. Except for the lower values of M this task will have to be undertaken numerically but some analytical progress can be achieved when M is infinite. There is then an infinite set of equations to solve, but they can be effectively reduced to two. Change m to $m+1$ in (37.2) and multiply by $\exp(\mathrm{i}k_1 d \cos \phi^{\mathrm{i}})$. Then, since $r_{mj} = r_{m+1,j+1}$,

$$\begin{aligned} F_{m+1}(0)\exp(\mathrm{i}k_1 d \cos \phi^{\mathrm{i}}) = {} & A(\phi^{\mathrm{i}}; 0)\exp(-\mathrm{i}k_1 md \cos \phi^{\mathrm{i}}) \\ & + \Big\{ A(0; 0) \sum_{j=-\infty}^{m-1} F_{j+1}(0)\mathrm{e}^{-\mathrm{i}k_1 r_{mj}}/r_{mj}^{\frac{1}{2}} \\ & + A(\pi; 0) \sum_{j=m+1}^{\infty} F_{j+1}(\pi)\mathrm{e}^{-\mathrm{i}k_1 r_{mj}}/r_{mj}^{\frac{1}{2}} \Big\} \exp(\mathrm{i}k_1 d \cos \phi^{\mathrm{i}}). \end{aligned}$$

Subtract this equation from (37.2) and then

$$G_m(0) = A(0; 0) \sum_{j=-\infty}^{m-1} G_j(0)\mathrm{e}^{-\mathrm{i}k_1 r_{mj}}/r_{mj}^{\frac{1}{2}} + A(\pi; 0) \sum_{j=m+1}^{\infty} G_j(\pi)\mathrm{e}^{-\mathrm{i}k_1 r_{mj}}/r_{mj}^{\frac{1}{2}}$$

where

$$G_m(\phi) = F_m(\phi) - F_{m+1}(\phi)\exp(\mathrm{i}k_1 d \cos \phi^{\mathrm{i}}).$$

A similar set of equations is obtained from (37.3). Comparison with (37.2) and (37.3) reveals that the equations for G_m are of the same form with the incident field missing. The expectation that the scattered field disappears with the incident wave or, more precisely, taking the solution of (37.2) and (37.3) to be unique, leads to

$$G_j(0) = 0, \qquad G_j(\pi) = 0 \quad (\text{all } j)$$

whence

$$F_m(0) = F_{m+1}(0)\exp(ik_1 d \cos\phi^{\mathrm{i}}),$$

$$F_m(\pi) = F_{m+1}(\pi)\exp(ik_1 d \cos\phi^{\mathrm{i}}) \quad (\text{all } m).$$

It follows from (37.1) that

$$F_m(\phi) = F_0(\phi)\exp(-ik_1 md \cos\phi^{\mathrm{i}}).$$

The infinite set (37.2) and (37.3) is thereby reduced to the two equations

$$F_0(0) = A(\phi^{\mathrm{i}}; 0) + A(0; 0)F_0(0)K(\phi^{\mathrm{i}}) + A(\pi; 0)F_0(\pi)K(\pi - \phi^{\mathrm{i}}),$$
$$F_0(\pi) = A(\phi^{\mathrm{i}}; \pi) + A(0; \pi)F_0(0)K(\phi^{\mathrm{i}}) + A(\pi; \pi)F_0(\pi)K(\pi - \phi^{\mathrm{i}})$$

where

$$K(\phi) = \sum_{j=1}^{\infty} (jd)^{-\frac{1}{2}} \exp\{-ik_1 d_j(1 - \cos\phi)\}.$$

Hence $F_0(0)$ and $F_0(\pi)$ are given by

$$\begin{aligned}
&\{1 - A(0; 0)K(\phi^{\mathrm{i}})\}\{1 - A(\pi; \pi)K(\pi - \phi^{\mathrm{i}})\} \\
&\quad - A(0; \pi)A(\pi; 0)K(\phi^{\mathrm{i}})K(\pi - \phi^{\mathrm{i}}) \\
&\quad = [A(\phi^{\mathrm{i}}; 0)\{1 - A(\pi; \pi)K(\pi - \phi^{\mathrm{i}})\} \\
&\qquad + A(\phi^{\mathrm{i}}; \pi)A(\pi; 0)K(\pi - \phi^{\mathrm{i}})]/F_0(0) \qquad (37.4) \\
&\quad = [A(\phi^{\mathrm{i}}; \pi)\{1 - A(0; 0)K(\phi^{\mathrm{i}})\} + A(\phi^{\mathrm{i}}; 0)A(0; \pi)K(\phi^{\mathrm{i}})]/F_0(\pi). \qquad (37.5)
\end{aligned}$$

The field scattered from the grating can now be calculated via (37.1) from

$$\sum_{m=-\infty}^{\infty} F_m(\phi)\mathrm{e}^{-ik_1 r_m}/r_m^{\frac{1}{2}}.$$

An error will be committed if it is not recognized that, in F_m, ϕ is strictly the angle made with the positive x-axis by the radius vector from the mth cylinder to the point of observation and this angle can vary appreciably over an infinite structure. Accordingly, at the point (x, y), the scattered field is

$$\sum_{m=-\infty}^{\infty} F_0\left(\tan^{-1}\frac{y}{x - md}\right)\exp(-ik_1 md \cos\phi^{\mathrm{i}} - ik_1 r_m)/r_m^{\frac{1}{2}}$$

where $r_m = \{(x - md)^2 + y^2\}^{\frac{1}{2}}$.

The series can be converted into an alternative form by means of the

Poisson summation formula (§6.32)

$$\sum_{m=-\infty}^{\infty} g(m) = \sum_{m=-\infty}^{\infty} G(2\pi m)$$

where $G(u) = \int_{-\infty}^{\infty} g(v)\mathrm{e}^{-iuv}\,\mathrm{d}v$. Thus the series becomes

$$\sum_{m=-\infty}^{\infty} \int_{-\infty}^{\infty} F_0\left(\tan^{-1}\frac{y}{x-vd}\right)\exp(-ik_1 dv\cos\phi^{\mathrm{i}} - ik_1 r_v - 2\pi i m v)\frac{\mathrm{d}v}{r_v^{\frac{1}{2}}}$$

or, after the substitution $x - vd = y \cot u$,

$$\sum_{m=-\infty}^{\infty} \exp\{-\mathrm{i}x(k_1\cos\phi^{\mathrm{i}} + 2\pi m/d)\}(y^{\frac{1}{2}}/d)\int_0^{\pi} F_0(u)\operatorname{cosec}^{\frac{3}{2}} u$$

$$\times \exp[-ik_1 y\{\operatorname{cosec} u - (\cos\phi^{\mathrm{i}} + 2\pi m/k_1 d)\cot u\}]\,\mathrm{d}u.$$

For large values of y the integral can be evaluated by the method of stationary phase. The stationary point occurs at

$$u = \phi_m = \cos^{-1}(\cos\phi^{\mathrm{i}} + 2\pi m/k_1 d)$$

and the resulting scattered field is

$$\sum_{m=-\infty}^{\infty} (2\pi k_1)^{\frac{1}{2}} \exp\{-\mathrm{i}x(k_1\cos\phi^{\mathrm{i}} + 2\pi m/d) - \mathrm{i}\kappa_m y - \tfrac{1}{4}\pi\mathrm{i}\}F_0(\phi_m)/\kappa_m d \tag{37.6}$$

where $\kappa_m^2 = k_1^2 - (k_1\cos\phi^{\mathrm{i}} + 2\pi m/d)^2$. The formula (37.6) represents the wave scattered by the infinite grating as a spectrum of plane waves. The plane waves are inhomogeneous or not according as κ_m is negative imaginary or real. For values of m such that κ_m is real *the plane waves are unattenuated and correspond to various observable spectra.* The plane waves associated with other values of m are exponentially damped as they move along.

The problem of the infinite grating is therefore completely resolved as soon as $A(\phi^{\mathrm{i}}; 0)$, $A(0; \phi)$ and $A(\pi; \phi)$ are known. The determination of these quantities requires the solution for the diffraction of a plane wave by a single cylinder. Problems of this type have occupied earlier parts of this chapter and will not be re-examined here.

The derivation of (37.6) rests on the assumption that y is large. A similar plane wave expansion can be obtained for the grating constructed by periodic corrugations on an otherwise plane surface. The *Rayleigh hypothesis* postulates that such a plane wave spectrum exists without the necessity for y to be large and even in the interstices between periodic elements. The hypothesis has been the subject of much investigation in recent years with the conclusion that such a spectral representation is not valid at all points outside the scattering boundaries except in special circumstances.

Another feature of (37.6) merits closer inspection. The series depends upon the solution of (37.4) and (37.5). Now, $K(\phi^{\mathrm{i}})$ tends to infinity as $k_1 d(1-\cos\phi^{\mathrm{i}})$ approaches any integral multiple of 2π. Spacings such that K is infinite are called *critical spacings* and are responsible for *Wood's anomalies.* It is our aim to show that the solution already obtained does, in fact, remain bounded as the spacing becomes critical.

Suppose that $k_1 d(1-\cos\phi^{\mathrm{i}}) = 2\pi n + \delta$ where n is an integer and δ is a small positive quantity which eventually tends to zero. It is convenient to assume firstly that $K(\pi-\phi^{\mathrm{i}})$ stays bounded as $\delta \to 0$. Then, from (37.4) and (37.5),

$$F_0(0)K(\phi^{\mathrm{i}}) \to \{A(\phi^{\mathrm{i}};0) + A_1 K(\pi-\phi^{\mathrm{i}})\}/\{A_3 K(\pi-\phi^{\mathrm{i}}) - A(0;0)\},$$

$$F_0(\pi) \to A_2/\{A_3 K(\pi-\phi^{\mathrm{i}}) - A(0;0)\}$$

where

$$A_1 = A(\phi^{\mathrm{i}};\pi)A(\pi;0) - A(\phi^{\mathrm{i}};0)A(\pi;\pi),$$

$$A_2 = A(\phi^{\mathrm{i}};0)A(0;\pi) - A(\phi^{\mathrm{i}};\pi)A(0;0),$$

$$A_3 = A(0;0)A(\pi;\pi) - A(0;\pi)A(\pi;0).$$

These results may be somewhat simplified by taking advantage of the reciprocity theorem (§1.35) which here takes the form $A(\alpha;\beta) = A(\pi+\beta;\pi+\alpha)$. For example, A_3 is equal to $[A(0;0)]^2 - [A(0;\pi)]^2$.

If $k_1 d$ is so large that $K(\pi-\phi^{\mathrm{i}})$ can be neglected, there results

$$F_0(\phi) = A(\phi^{\mathrm{i}};\phi) - A(0;\phi)A(\phi^{\mathrm{i}};0)/A(0;0). \tag{37.7}$$

Note that $A(0;0)$ cannot be zero because its real part supplies the scattering coefficient of a single cylinder and this cannot vanish.

Each term in (37.6) can now be calculated in a straightforward manner with the exception of that in which $m=n$. In this term both κ_m and ϕ_m tend to zero. Nevertheless, the term has a limit because $\kappa_n = k_1 \sin\phi_n \approx k_1\phi_n$. To see what happens suppose that (37.7) can be used in all terms except $m=n$; it is not legitimate to employ it for $m=n$ because quantities which tend to zero with δ have been excluded from it. Such terms may contribute to $F_0(\phi_n)/\phi_n$ as $\phi_n \to 0$.

Insertion of $\int_0^\infty u^{-\frac{1}{2}}e^{-nu}\,du = (\pi/n)^{\frac{1}{2}}$ in K provides

$$K(\phi^{\mathrm{i}}) = (\pi d)^{-\frac{1}{2}} \int_0^\infty \frac{u^{-\frac{1}{2}}e^{-\mathrm{i}\delta-u}}{1-e^{-\mathrm{i}\delta-u}}\,du.$$

Only the behaviour near the origin is awkward as $\delta \to 0$ and, if ε is a small positive number,

$$K(\phi^{\mathrm{i}}) \approx (\pi d)^{-\frac{1}{2}} \int_0^\varepsilon \frac{u^{-\frac{1}{2}}e^{-u}}{\mathrm{i}\delta+u}\,du + O(1)$$

$$\approx (\pi d\delta)^{-\frac{1}{2}} \int_0^\infty u^{-\frac{1}{2}}(u+\mathrm{i})^{-1}\,du + O(1)$$

$$\approx (\pi/d\delta)^{\frac{1}{2}} e^{\frac{1}{4}\pi\mathrm{i}} + O(1).$$

Hence, if $K(\pi - \phi^{\mathrm{i}})$ is neglected,

$$\frac{F(\phi_n)}{k_1 d\phi_n} \approx \left\{ A(\phi^{\mathrm{i}}; \phi_n) + \frac{A(\phi^{\mathrm{i}}; 0) K(\phi^{\mathrm{i}}) A(0; \phi^{\mathrm{i}})}{1 - A(0; 0) K(\phi^{\mathrm{i}})} \right\} \Big/ k_1 d\phi_n$$

$$\approx \frac{[A(\phi^{\mathrm{i}}; \phi_n) + \{A(\phi^{\mathrm{i}}; 0) A(0; \phi^{\mathrm{i}}) - A(0; 0) A(\phi^{\mathrm{i}}; \phi_n)\} K(\phi^{\mathrm{i}})]}{k_1 d\phi_n \{1 - A(0; 0) K(\phi^{\mathrm{i}})\}}.$$

Since $\phi_n \approx (2\delta/k_1 d)^{\frac{1}{2}}$ we find

$$F(\phi_n)/k_1 d\phi_n \to -A(\phi^{\mathrm{i}}; 0) \mathrm{e}^{-\frac{1}{4}\pi \mathrm{i}} / (2\pi k_1)^{\frac{1}{2}} A(0; 0).$$

A similar formula can be derived when $K(\pi - \phi^{\mathrm{i}})$ becomes infinite while $K(\phi^{\mathrm{i}})$ remains bounded.

When both $k_1 d$ and $k_1 d \cos \phi^{\mathrm{i}}$ are integers $K(\phi^{\mathrm{i}})$ and $K(\pi - \phi^{\mathrm{i}})$ are simultaneously infinite. Suppose that $k_1 d(1 + \cos \phi^{\mathrm{i}})$ approaches $2\pi p$. Then, if $A_3 \neq 0$,

$$F(\phi_n)/k_1 d\phi_n \to A_1 \mathrm{e}^{-\frac{1}{4}\pi \mathrm{i}} / (2\pi k_1)^{\frac{1}{2}} A_3,$$
$$F(\phi_p)/k_1 d\phi_p \to A_2 \mathrm{e}^{-\frac{1}{4}\pi \mathrm{i}} / (2\pi k_1)^{\frac{1}{2}} A_3.$$

If $A_3 = 0$ it can be shown that $A_2 = A_1 = 0$ but the field no longer has a unique limit.

The field in $y < 0$ can be examined in a similar way to the above and analogous phenomena are exhibited. The structure is then behaving as a *reflection grating*. A familiar example of a reflection grating is the long playing record where the spectra produced by the grooves are easily observable.

Exercises on Chapter 8

1. Outside the circular cylinder $r = b$ the field u satisfies (1.1). Between $r = b$ and $r = a$ $(a < b)$ k_1 is replaced by k_2 and the field u_2 vanishes on $r = a$. The field and its normal derivative are continuous across $r = b$. The incident plane wave $u^{\mathrm{i}} = \exp(-\mathrm{i}k_1 r \cos \phi)$ impinges on the cylindrical structure. Show that, in $r > b$,

$$u = u^{\mathrm{i}} + u_{\mathrm{b}} - \frac{2\mathrm{i}}{\pi k_1 b} \sum_{m=-\infty}^{\infty} \frac{\mathrm{H}_m^{(2)}(k_1 r) \mathrm{e}^{\mathrm{i}m(\phi - \frac{1}{2}\pi)}}{A_m \{\mathrm{H}_m^{(2)}(k_1 b)\}^2}$$

where u_{b} is the reflected field when $u = 0$ on $r = b$ and

$$A_m = \frac{\mathrm{H}_m^{(2)\prime}(k_1 b)}{\mathrm{H}_m^{(2)}(k_1 b)} - \frac{k_2}{k_1} \frac{\mathrm{J}_m'(k_2 b) \mathrm{H}_m^{(2)}(k_2 a) - \mathrm{J}_m(k_2 a) \mathrm{H}_m^{(2)\prime}(k_2 b)}{\mathrm{J}_m(k_2 b) \mathrm{H}_m^{(2)}(k_2 a) - \mathrm{J}_m(k_2 a) \mathrm{H}_m^{(2)}(k_2 b)}.$$

The formula may also be written as

$$u = u^{\mathrm{i}} + u_b - \frac{2\mathrm{i}}{\pi k_1 b}\int_{-\infty}^{\infty}\frac{\mathrm{H}_\nu^{(2)}(k_1 r)}{\{\mathrm{H}_\nu^{(2)}(k_1 b)\}^2 \mathrm{A}_\nu}\,\mathrm{e}^{-\mathrm{i}\nu(\phi-\frac{1}{2}\pi)}\,d\nu.$$

Determine the behaviour of the field when $k_1 b \ll 1$ and $k_2 b \ll 1$.

2. Repeat Exercise 1 when $\partial u_2/\partial r = 0$ on $r=a$ and the boundary conditions at $r=b$ are $u=u_2$ and $\partial u/\partial r = \zeta\,\partial u_2/\partial r$.

3. If in Exercise 1 the excitation is due to a line source at $r=d$ find series for the field when (i) $d>b$, (ii) $a<d<b$.

4. The field inside a circular conducting cylinder of radius a varies with ϕ and z according to the factor $\mathrm{e}^{-\mathrm{i}m\phi-\mathrm{i}\kappa z}$. If ε, μ, σ are the parameters of the cylinder show that, at $r=a$, the components of the electromagnetic field satisfy

$$E_z = Z_{zz}H_z + Z_{z\phi}H_\phi, \qquad E_\phi = Z_{\phi z}H_z + Z_{\phi\phi}H_\phi$$

where $Z_{zz} = -\alpha/\beta$, $Z_{z\phi} = 1/\beta$, $Z_{\phi z} = -\gamma - \alpha^2/\beta$, $Z_{\phi\phi} = \alpha/\beta$, $\alpha = -m\kappa/\kappa_2^2 a$, $\beta = -(\sigma+\mathrm{i}\omega\varepsilon)\mathrm{J}_m'(\kappa_2 a)/\kappa_2 \mathrm{J}_m(\kappa_2 a)$, $\gamma = \mathrm{i}\omega\mu\beta/(\sigma+\mathrm{i}\omega\varepsilon)$, $\kappa_2 = (k_2^2 - k^2)^{\frac{1}{2}}$, $k_2^2 = -\mathrm{i}\omega\mu(\sigma+\mathrm{i}\omega\varepsilon)$.

If $|k_2|^2 \gg |\kappa|^2$ and $|k_2 a|^2 \gg m^2$ show that $E_z \approx Z_{\mathrm{i}}H_\phi$ where $Z_{\mathrm{i}} = \mathrm{i}\mu\omega \mathrm{J}_0(k_2 a)/k_2 \mathrm{J}_0'(k_2 a)$.

If $m = \nu a$ and $a \to \infty$ with ν fixed

$$\beta \sim \mathrm{i}(\sigma+\mathrm{i}\omega\varepsilon)(\kappa_2^2-\nu^2)^{\frac{1}{2}}\{1 - \tfrac{1}{2}\mathrm{i}\kappa_2^2/a(\kappa_2^2-\nu^2)^{\frac{3}{2}}\}/\kappa_2^2.$$

The formulae supply the surface impedances and display on the one hand, skin effect, and on the other, the influence of curvature.

5. The cylinder of Exercise 4 is surrounded by free space and is excited by a solenoid on $r=b$ $(b>a)$ occupying $|z|<\frac{1}{2}l$. The solenoid is modelled by requiring E_ϕ to be continuous across it and

$$H_z(b-0) - H_z(b+0) = (1/\pi)\int_{-\infty}^{\infty}\sin(\tfrac{1}{2}vl)\mathrm{e}^{\mathrm{i}(\omega t - \mathrm{i}vz)}\frac{dv}{v}.$$

Find the field in terms of a Hertz vector with a z-component only expressed as an integral of the same type.

6. The plane wave $u^{\mathrm{i}} = \exp\{-\mathrm{i}kr\cos(\phi-\phi^{\mathrm{i}})\}$ is incident on the parabolic cylinder $y^2 = \xi_0^4 - 2\xi_0^2 x$ and the boundary condition is $u=0$. If $k\xi_0^2 \gg 1$ show that well into the shadow

$$u = \left(\frac{2\rho\rho_0}{k}\right)^{\frac{1}{6}}\frac{\mathrm{e}^{-\mathrm{i}kL}}{d^{\frac{1}{2}}}\sum_{s=1}^{\infty}\frac{\exp(-\frac{1}{12}\pi\mathrm{i} - \mathrm{i}\alpha_s\tau\mathrm{e}^{-\frac{1}{3}\pi\mathrm{i}})}{2\pi^{\frac{1}{2}}\{\mathrm{Ai}'(-\alpha_s)\}^2}$$

where $\mathrm{Ai}(-\alpha_s)=0$, d is the length of the tangent from the point of observation to the parabola, ν the inclination of this tangent, ρ the radius of curvature at its point of contact, ρ_0 the radius of curvature at the point of glancing incidence,

$$L = d + \tfrac{1}{2}\xi_0^2\{\ln(\tan\tfrac{1}{2}\nu\cot\tfrac{1}{2}\phi^{\mathrm{i}}) + 4\cos\phi^{\mathrm{i}} - \cos\nu/\sin^2\nu\}$$

and $\tau = (\tfrac{1}{2}k\xi_0^2)^{\frac{1}{3}}\ln(\tan\tfrac{1}{2}\nu\cot\tfrac{1}{2}\phi^{\mathrm{i}})$.

7. Examine the behaviour of (7.7) at low frequencies when $h\xi_0$ is effectively zero and, in particular, verify (7.8).

8. The incident wave $p^{\mathrm{i}} = -\tfrac{1}{4}\mathrm{i}\mathrm{H}_0^{(2)}(k\,|\boldsymbol{x}-\boldsymbol{x}_0|)$ falls on the elliptic cylinder $u = U$ on which the boundary condition is $\partial p/\partial n + (2k^2/\rho)^{\frac{1}{3}}Zp = 0$ where Z is a constant and ρ the radius of curvature. Show that the scattered field has the form of §8.8 with

$$b_m = \frac{\mathrm{Mc}_m^{(1)\prime}(U, h) + Z'\mathrm{Mc}_m^{(1)}(U, h)}{\mathrm{Mc}_m^{(4)\prime}(U, h) + Z'\mathrm{Mc}_m^{(4)}(U, h)}$$

where $Z' = Z(k^2l^2\sinh 2U)^{\frac{1}{3}}$. Find c_m.

Discuss the field at low frequencies when h is small.

9. A plane wave $p^{\mathrm{i}} = \mathrm{e}^{-\mathrm{i}kx}$ is incident on a circular cylinder of radius b in which the wave number $k_2 = k(r/b)^{\frac{3}{2}}$. Show that a possible field dependence in the cylinder is $\mathrm{e}^{\mathrm{i}m\phi}r\mathrm{J}_\nu(2kr^{\frac{5}{2}}/5b^{\frac{3}{2}})$ where $\nu = 2(1+m^2)^{\frac{1}{2}}/5$ and deduce that the reflected wave in $r > b$ is

$$\sum_{m=-\infty}^{\infty} a_m\mathrm{H}_m^{(2)}(kr)\mathrm{e}^{\mathrm{i}m(\phi-\frac{1}{2}\pi)}$$

where

$$a_m = \frac{\mathrm{J}_m(\beta)\mathrm{J}_\nu(\tfrac{2}{5}\beta) + \beta\{\mathrm{J}_m(\beta)\mathrm{J}_\nu'(\tfrac{2}{5}\beta) - \mathrm{J}_m'(\beta)\mathrm{J}_\nu(\tfrac{2}{5}\beta)\}}{\beta\{\mathrm{H}_m^{(2)\prime}(\beta)\mathrm{J}_\nu(\tfrac{2}{5}\beta) - \mathrm{H}_m^{(2)}(\beta)\mathrm{J}_\nu'(\tfrac{2}{5}\beta)\} - \mathrm{H}_m^{(2)}(\beta)\mathrm{J}_\nu(\tfrac{2}{5}\beta)},$$

$\beta = kb$.

10. In the theory of *radiation from a slot* on a perfectly conducting circular the slot consists of an aperture A and, on the surface of the cylinder $r = a$, the tangential electric intensity $E_z = f_1(\phi, z)$, $E_\phi = f_2(\phi, z)$ where f_1 and f_2 vanish outside A. Express the field in terms of E_z and H_z. Use a Fourier transform with respect to z and a Fourier series in ϕ. Show that

$$E_z = (1/2\pi)\int_{-\infty}^{\infty}\sum_{n=-\infty}^{\infty} a_n\mathrm{H}_n^{(2)}(hr)\mathrm{e}^{\mathrm{i}n\phi-\mathrm{i}\alpha z}\,\mathrm{d}\alpha$$

while the formula for H_z has b_n in place of a_n where

$$2\pi a_n\mathrm{H}_n^{(2)}(ha) = \int_A f_1(\phi, z)\mathrm{e}^{-\mathrm{i}n\phi+\mathrm{i}\alpha z}\,\mathrm{d}\phi\,\mathrm{d}z,$$

$$2\pi\{\alpha n a_n\mathrm{H}_n^{(2)}(ha) + \mathrm{i}\omega\mu_0 ab_n\mathrm{H}_n^{(2)\prime}(ha)\} = -ah^2\int_A f_2(\phi, z)\mathrm{e}^{-\mathrm{i}n\phi+\mathrm{i}\alpha z}\,\mathrm{d}\phi\,\mathrm{d}z.$$

Here $h = (k^2 - \alpha^2)^{\frac{1}{2}}$ with $h = k$ when $\alpha = 0$ and the contour of integration in E_z passes below $-k$ and above k.

11. In Exercise 10 show that the field produced inside the cylinder by the slot can be expanded in a similar way with J_n in place of $H_n^{(2)}$. Away from the slot a modal expansion can be obtained from the residues at the simple poles of the integrand.

12. A point source is placed at S outside a cylinder of radius b (Fig. 8.27). The boundary condition on the cylinder is $\partial u/\partial n = ik_1 Zu$. Obtain the exact form of the field and show that, at high frequencies, it verifies the three-dimensional form of (33.1) provided that the radius of curvature is taken as $b \operatorname{cosec}^2\theta$ where θ is the angle made with the z-axis by the incident ray which launches the creeping ray.

13. The field $J_m(j_{mp}r/a)e^{-im\phi}$ satisfies Helmholtz's equation with appropriate wavenumber inside a circular cylinder of radius a and vanishes on the boundary. If m and p are large with $m < j_{mp}$ show, from the asymptotic formula for J_m, that the field is exponentially small for $r < am/j_{mp}$ but oscillatory near the boundary. When j_{mp} is not far from m the dominant field is confined to a thin layer near $r = a$; it may therefore be expected to play an important part in modelling *whispering gallery modes* where the rays are virtually tangential to the boundary, lying between it and a caustic at $r = am/j_{mp}$.

Does any significant alteration occur if the boundary condition is altered from soft to hard?

14. Repeat Exercise 13 for the field inside a sphere.

15. Investigate whispering gallery modes for electromagnetic waves inside a perfectly conducting sphere.

16. In a homogeneous conducting medium with parameters ε, μ, σ show that, if $\omega\varepsilon \ll \sigma$, the vector potential $\mathbf{A}$ satisfies $\nabla^2\mathbf{A} - i\omega\mu\sigma\mathbf{A} = \mathbf{0}$.

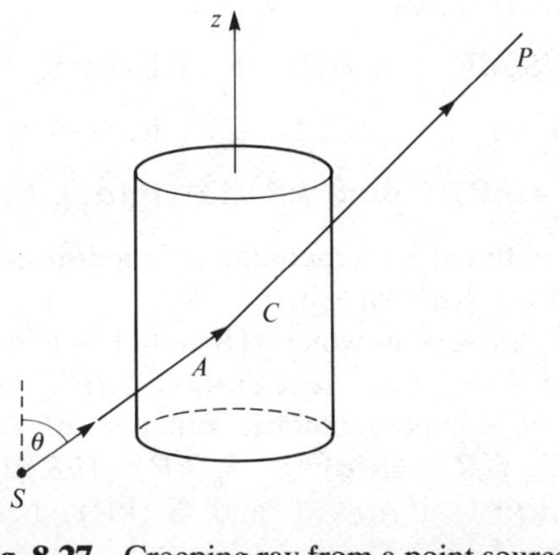

Fig. 8.27 Creeping ray from a point source

A conducting sphere of radius a is placed in free space and subjected to an alternating magnetic field $\boldsymbol{H} = C\boldsymbol{k}e^{i\omega t}$. Show that $\mathbf{A} = (\boldsymbol{j}\cos\phi - \boldsymbol{i}\sin\phi)B$ where

$$B = \begin{cases} \frac{1}{2}C(R + ba^3/R^3)\sin\theta & (R > a) \\ \frac{1}{2}Cb_1(p^{-1}\cosh p - p^{-2}\sinh p)\sin\theta & (R < a) \end{cases}$$

and $p = (i\omega\mu\sigma)^{\frac{1}{2}}R$. Find the constants b, b_1 and show that, in this solution, the current becomes concentrated near the surface of the sphere as the frequency increases.

17. Undertake the summation with respect to m by means of (B.3.24) for G^i, G_1, and G_2 in §8.12. Deduce the effect on (12.5) and (12.6) for an incident plane wave.

18. Show that, in terms of P and M of §8.13,

$$\boldsymbol{E} = \text{curl curl}(P\boldsymbol{x}) - \text{curl}(\boldsymbol{x}\,\partial M/\partial t).$$

19. Show that a solution of $\nabla^2 p + k^2 p = 0$ in spherical polar coordinates is $p = p_0 e^{i\phi}$ where

$$p_0 = \{e^{-ikR\cos\theta} - e^{-ikR} + i(1 - \cos\theta)\sin kR\}/k^2 R\sin\theta.$$

20. In the representation of Exercise 18, $P = p_0\cos\phi$, $M = (k/\omega)p_0\sin\phi$ where p_0 is given in Exercise 19. Prove that the field is the plane wave $\boldsymbol{E} = \boldsymbol{i}e^{-ikz}$, $\boldsymbol{B} = (k/\omega)\boldsymbol{j}e^{-ikz}$.

21. By taking a derivate of $e^{-ikR\cos\theta}$ with respect to θ, then inserting (11.5) and invoking (B.3.15), (B.3.17) show that, after an integration with respect to θ from 0 to θ, p_0 of Exercise 19 is given by

$$p_0 = \sum_{n=1}^{\infty} \frac{(-i)^{n+1}(2n+1)}{n(n+1)k} j_n(kR)P_n^1(\cos\theta).$$

22. The function p satisfies $\nabla^2 p + k^2 f(R)p = 0$. Show that a solution is given by $Rp = S_n(kR)P_n^m(\cos\theta)e^{im\phi}$ where

$$d^2S_n/dR^2 + \{k^2 f(R) - n(n+1)/R^2\}S_n = 0.$$

If $f(R) = 2 - (R/a)^2$ for $R \leqslant a$ (the *Luneberg lens*) show that a solution is

$$S_n(kR) = (kR)^{n+1}\exp(-kR^2/2a)\,{}_1F_1(\alpha; n + \tfrac{3}{2}; kR^2/a)$$

where ${}_1F_1$ is the confluent hypergeometric function (see Exercises 21–27 of Chapter 1) and $\alpha = \frac{1}{2}(n - ka + \frac{3}{2})$.

For the *Maxwell fish-eye* in which $f(R) = 4\{1 - (R/a)^2\}^{-2}$ for $R \leqslant a$, the change of variable $\xi = -(R/a)^2$ reveals that $S_n/\xi^{\frac{1}{2}(n+1)}(\xi - 1)^\nu$ where $\nu = \frac{1}{2}\{1 + (1 + 4k^2a^2)^{\frac{1}{2}}\}$ is a hypergeometric function of ξ. In the *Nomura–Takaku medium* $f(R) = (R/a)^{2q}$; $S_n(kR) = (kR)^{\frac{1}{2}}J_\sigma\{kR^{q+1}/(q+1)a^q\}$ where $\sigma = (n + \frac{1}{2})/(q+1)$ if $q \neq -1$ and $S_n(kR) = (kR)^{\frac{1}{2}}\sin\{(k^2a^2 - n^2 - n - \frac{1}{4})^{\frac{1}{2}}\ln kR\}$ if $q = -1$.

23. The field outside a perfectly conducting sphere of radius a is expressed in terms of P and M. Show that the boundary conditions on the sphere are satisfied by requiring $M=0$, $\partial(RP)/\partial R=0$ on $R=a$.

A plane wave propagating along the z-axis is polarised so that $\boldsymbol{E}=\boldsymbol{i}e^{-ikz}$. Use the results of Exercises 20 and 21 to prove that

$$M=\frac{1}{\omega}\sin\phi\sum_{n=1}^{\infty}\frac{(-i)^{n+1}(2n+1)}{n(n+1)}\left\{j_n(kR)-\frac{j_n(ka)}{h_n^{(2)}(ka)}h_n^{(2)}(kR)\right\}P_n^1(\cos\theta)$$

and obtain the formula for P. Verify that the electric intensity agrees with (16.14).

24. The permittivity and permeability are ε_1, μ_1 outside a sphere of radius a and ε_2, μ_2 inside. The potentials in the two regions are P_1, M_1 and P_2, M_2 respectively. Show that the boundary conditions are met by

$$\varepsilon_1P_1=\varepsilon_2P_2,\qquad M_1=M_2,\qquad \mu_2\,\partial(RM_1)/\partial R=\mu_1\,\partial(RM_2)/\partial R,$$
$$\partial(RP_1)/\partial R=\partial(RP_2)/\partial R$$

on $R=a$.

Solve the problem analogous to Exercise 23.

25. A linearly polarized wave is incident on a small perfectly conducting sphere. Show that the waves scattered in a direction making an angle $\frac{1}{3}\pi$ with the direction of propagation of the incident wave are linearly polarized.

26. A linearly polarized wave is scattered by a small non-conducting sphere with the same permeability as its surroundings. Show that the intensity of the wave scattered in the direction perpendicular to the direction of the incident wave and parallel to the incident electric vector varies as the eighth power of the frequency. Thus incident polarized white light scattered in this direction appears bluer than in Rayleigh scattering. [*Tyndall's residual blue*].

27. A radial electric dipole is placed at a distance $b-a$ from a sphere of radius a. The field of the dipole in the absence of the sphere is given by a radial Hertz vector of magnitude $e^{-ik_1R_s}/R_s$ where R_s is the distance from the dipole. Show that the total field has a radial Hertz vector Π in $R>a$ where

$$\Pi-e^{-ik_1R_s}/R_s=-ik_1\sum_{n=0}^{\infty}(2n+1)a_nh_n^{(2)}(k_1b)h_n^{(2)}(k_1R)P_n(\cos\theta)$$

and

$$a_n=\frac{k_1\varphi_n^{(1)}(k_1a)\varphi_n^{(1)\prime}(k_2a)-k_2\varphi_n^{(1)\prime}(k_1a)\varphi_n^{(1)}(k_2a)}{k_2\varphi_n^{(1)}(k_2a)\varphi_n^{(4)\prime}(k_1a)-k_1\varphi_n^{(1)\prime}(k_2a)\varphi_n^{(4)}(k_1a)}$$

in the notation of §8.16.

28. Calculate the scattering coefficient of a drop of water of radius 1 mm at frequencies from 500 kHz to 2 MHz.

If the conductivity 2×10^{-4} mho m^{-1} of the water is taken into account what is the effect?

29. A linearly polarized plane wave of frequency 500 kHz is scattered by a dielectric sphere in which $\mu_2=\mu_1$, $\varepsilon_2=81\varepsilon_1$. If the radius of the sphere is 1 mm find the moment of the equivalent electric dipole. What is the radar cross-section?

30. A diaphanous sphere of air in which $\varepsilon_2=1.04\varepsilon_1$ is irradiated by a plane wave of wavelength 600 nm. Examine the scattering when the radius of the sphere is (1) 10 nm, (ii) 20 nm, (iii) 100 nm.

31. Draw a graph of the scattering coefficient of a perfectly conducting sphere of radius 1 m in a plane wave as the wavelength varies for frequencies over 500 MHz.

32. In considering the effects of refraction as in §8.24 suppose that (23.1) is altered to $\partial p/\partial R=\mathrm{i}\nu_0(1-\eta)^{-\frac{1}{2}}p$. When $\eta=0$ let $p=f(d, z, z_0, \nu_0)$ where $d=b\Theta\approx b\sin\Theta$. Show that when $\eta\neq 0$

$$p=(1-\eta)^{\frac{2}{3}}f(d(1-\eta)^{\frac{2}{3}}, z(1-\eta)^{\frac{1}{3}}, z_0(1-\eta)^{\frac{1}{3}}, \nu_0(1-\eta)^{\frac{1}{3}}).$$

33. Prove (27.18).

34. Prove that $C_{jj}(\zeta)=T_-/(\zeta+2)$ for a sphere.

35. By means of (28.7) and (28.9) prove that, for all bodies of the same volume and given ζ, the sphere has greatest content when $\zeta\geqslant 1$ and least content when $\zeta\leqslant 1$.

36. Show that increasing the size of a body with fixed ζ decreases $(1-\zeta)T_-+(1-\zeta)^2C_{jj}(\zeta)$ when $\zeta>1$. What happens when $\zeta<1$?

37. The plane wave $p^{\mathrm{i}}=\mathrm{e}^{-\mathrm{i}k(x\cos\theta+z\sin\theta)}$ $(0<\theta<\pi)$ falls on the soft paraboloid $k^2(y^2+z^2)+2ku_0x=u_0^2$. Introduce paraboloidal coordinates ξ, η, ϕ (§1.39(vii)) and put $u=k\xi^2$, $v=k\eta^2$. Use the results of Exercises 21–27 of Chapter 1 to show that on the paraboloid

$$\frac{\partial p}{\partial n}=\frac{\mathrm{i}k/u_0^{\frac{1}{2}}}{4\pi(u_0+v)^{\frac{1}{2}}\sin\theta}\int_{-\mathrm{i}\infty}^{\mathrm{i}\infty}(\tan\tfrac{1}{2}\theta)^{\nu}\mathrm{e}^{-\frac{1}{2}\mathrm{i}(u_0+v)}$$
$$\times\left\{\frac{(-\frac{1}{2}-\frac{1}{4}!\,{}_1\mathrm{F}_1(\frac{1}{2}+\frac{1}{4}\nu;1;\mathrm{i}v)}{\psi(\frac{1}{2}+\frac{1}{4}\nu,1;-\mathrm{i}u_0)}+2\sum_{m=1}^{\infty}\frac{(\frac{1}{2}m-\frac{1}{2}-\frac{1}{4}\nu)!}{m!}\left(\frac{v}{u_0}\right)^{\frac{1}{2}m}\right.$$
$$\left.\times\frac{{}_1F_1(\frac{1}{2}m+\frac{1}{2}+\frac{1}{4}\nu;m+1;\mathrm{i}\nu)\cos m(\frac{1}{2}\pi-\phi)}{\psi(\frac{1}{2}m+\frac{1}{2}+\frac{1}{4}\nu,m+1;-\mathrm{i}u_0)}\right\}\mathrm{d}\nu.$$

Careful consideration indicates that, at high frequencies where $u_0\gg 1$, near the penumbral curve

$$\frac{\partial p}{\partial n}=-\frac{\partial p^{\mathrm{i}}}{\partial n}\exp(-\tfrac{1}{3}\mathrm{i}\zeta^3)\mathrm{J}_{\mathrm{s}}(\zeta)/\zeta$$

where J_{s} is defined in §8.3 and

$$\zeta=\{(u_0v)^{\frac{1}{2}}\sin\theta\sin\phi+u_0\cos\theta\}\{2u_0(u_0+v)\sin^2\theta\}^{-\frac{1}{3}}.$$

This may also be written as $\zeta = (\frac{1}{2}k)^{\frac{1}{3}} s/\rho^{\frac{2}{3}}$ where ρ is the radius of curvature, at a point of the penumbral curve, of the section of the paraboloid by the plane containing the surface normal and the direction of propagation of the incident wave; s is the arc length measured along the surface from the penumbral curve in this plane, being counted positive in the shadow. This may be interpreted as saying that the normal derivative near the penumbral curve behaves as at a point of glancing incidence on a circular cylinder with radius equal to the radius of curvature of the surface in the plane of incidence.

More generally, it has been suggested that, if the exterior medium is not homogeneous, the normal derivative at B in the plane of incidence through the point A of the penumbral curve is given by

$$\frac{\partial p}{\partial n} = \mathrm{i}\left(\frac{2k^2 N_B^2}{\rho_B}\right)^{\frac{1}{3}} \gamma_A^{\mathrm{i}} \mathrm{e}^{-\mathrm{i}kL} \mathrm{J_s}\left\{(\tfrac{1}{2}k)^{\frac{1}{3}} \int_A^B \frac{N_\sigma^{\frac{1}{3}}}{\rho_\sigma^{\frac{2}{3}}} \, \mathrm{d}\sigma\right\}$$

where N_B is the refractive index at B and $1/\rho_B$ is the difference between the curvatures of the tangent ray and boundary at B. The incident field produces $\gamma_A^{\mathrm{i}} \mathrm{e}^{-\mathrm{i}kL_A}$ at A in the absence of the obstacle.

38. Using the physical idea in the last sentence of the second paragraph of Exercise 37 show that the high frequency scattering coefficient of a convex body in a plane wave p^{i} is

$$2 + (b_0/k^{\frac{2}{3}} S_0) \int_D \rho^{\frac{1}{3}} \sin\beta \, \mathrm{d}s$$

where S_0 is the projected area of the body on a plane normal to the direction of propagation of the incident wave, D is the penumbral curve on the obstacle, s its arc length and β is the angle between the tangent to the penumbral curve and the incident direction of propagation. The coefficient is defined so that $2 + b_0(kb)^{-\frac{2}{3}}$ is the scattering coefficient of a circular cylinder of radius b with the same boundary condition.

Deduce that for a soft sphere the scattering coefficient is $2 + 1.9924(kb)^{-\frac{2}{3}}$.

[It is assumed that the radius of curvature of the body is not small or infinite and that β is not near zero.]

39. If a plane wave p^{i} impinges on a prolate spheroid with semi-axes a and b show that the high frequency scattering coefficient is $2 + 2b_0(a^2/k^2b^4)^{\frac{1}{3}}$ when the incident wave travels along the axis of revolution and $2 + 2b_0(kb)^{-\frac{2}{3}}{}_2\mathrm{F}_1(-\frac{2}{3}, \frac{1}{2}; 1; 1 - b^2/a^2)$ for incidence along the minor axis. [Use the theory of Exercise 38].

40. The application of the physical picture of the previous three questions leads to the inference that the current induced by a plane electromagnetic wave on the surface of a perfectly conducting convex object has two components near the penumbral curve at high frequencies.

The components $\boldsymbol{n}\wedge\boldsymbol{H}^{(1)}$ and $\boldsymbol{n}\wedge\boldsymbol{H}^{(2)}$ are given by

$$\boldsymbol{n}\wedge\boldsymbol{H}^{(1)}=-\boldsymbol{n}\wedge\boldsymbol{H}^{\mathrm{i}(1)}\exp(-\tfrac{1}{3}\mathrm{i}\zeta^3)\mathrm{J}_{\mathrm{s}}(\zeta)/\zeta,$$
$$\boldsymbol{n}\wedge\boldsymbol{H}^{(2)}=\boldsymbol{n}\wedge\boldsymbol{H}^{\mathrm{i}(2)}\exp(-\tfrac{1}{3}\mathrm{i}\zeta^3)\mathrm{J}_{\mathrm{h}}(\zeta)$$

where ζ has the same significance as before, $\boldsymbol{H}^{\mathrm{i}(1)}$ is the incident magnetic vector associated with the component of the incident electric intensity tangent to the surface at the point of contact of the incident ray and $\boldsymbol{H}^{\mathrm{i}(2)}$ is the magnetic intensity attached to the component of the incident electric vector normal to the surface.

Deduce that the scattering coefficient is

$$2\mp(1/k^{\frac{2}{3}}S_0)\int_D(b_{\mathrm{s}}\cos^2\gamma_0+b_{\mathrm{h}}\sin^2\gamma_0)\rho^{\frac{1}{3}}\sin\beta\,\mathrm{d}s$$

where γ_0 is the angle between $\boldsymbol{E}^{\mathrm{i}}$ and the tangent to the penumbral curve, b_{s} and b_{h} are the values of b_0 for the soft and hard boundary conditions respectively and other quantities are defined in Exercise 38.

41. Use the theory of Exercise 40 to show that the high frequency scattering coefficient of a perfectly conducting solid of revolution is the mean of the soft and hard scattering coefficients when the incident direction of propagation is along the axis of revolution.

42. A plane electromagnetic wave is propagating along the minor axis of a prolate spheroid with semi-axes a and b. The incident electric vector is parallel to the major axis. Show that the scattering coefficient is, from Exercise 40,

$$2+2(kb)^{-\frac{2}{3}}\{b_{\mathrm{h}}\,{}_2\mathrm{F}_1(-\tfrac{2}{3},\tfrac{1}{2};1;1-b^2/a^2)+\tfrac{1}{2}(b_{\mathrm{s}}-b_{\mathrm{h}})\,{}_2\mathrm{F}_1(\tfrac{1}{3},\tfrac{1}{2};2;1-b^2/a^2)\}.$$

Deduce that the scattering coefficient for a sphere of radius a is $2+0.132(ka)^{-\frac{2}{3}}$.

43. A magnetic dipole is placed, on the axis of revolution of a prolate spheroid of semi-axes a and b, at a distance d ($\gg a$) from the centre. The media outside and inside the spheroid have the same permeability but their permittivites ε_1, ε_2 are related by $\varepsilon_2=2\varepsilon_1$. Discuss the field on the basis of Rayleigh scattering when $k_1a\ll 1$.

What would be the difference if the spheroid were perfectly conducting?

44. A plane wave falls on a diaphanous sphere of radius a in which $N^2=1+\alpha+\beta R/a$. Determine the scattered intensity at low frequencies.

45. A plane wave is incident on a diaphanous spheroid in which N^2 is constant. Determine the low frequency scattered field.

46. Try Exercises 44 and 45 at high frequencies, giving the answers as integrals if necessary.

47. The portion $R=a$, $0\leqslant\theta\leqslant\frac{1}{4}\pi$ of a spherical surface is irradiated by a source. Calculate the scattered wave according to physical optics (§8.34)

when the primary wave is due to a source at (i) $R=d$, $\theta=0$, (ii) $R=d$, $\theta=\pi$ assuming that $d \gg a$.

48. Would you expect many differences in the field of Exercise 47 if the surface were spheroidal instead of spherical?

49. Use the theory of §8.35 to estimate the attenuation in the atmosphere in the presence of (i) fog, (ii) light rain, (iii) heavy rain. [In fog there are about 10^7 drop m^{-3} of 5×10^{-3} mm radius; in light rain about 4000 drop m^{-3} of 0.2 mm radius; in heavy rain about 500 drop m^{-3} of 1 mm radius.]

50. A diffraction grating of perfectly conducting circular cylinders of radius a and separation distance d is illuminated by a plane wave. If the elements are widely spaced and $k_1 a \ll 1$ determine the field produced if the grating contains (i) 5, (ii) 11 cylinders.

If the grating is infinitely long and interactions between the elements cannot be ignored what is the scattered field?

51. If the cylinders in Exercise 50 are of elliptical cross-section with semi-axes a and b $(a > b)$ find the scattered field when $k_1 a \ll 1$.

The following exercises are intended for those who desire some practice in the asymptotic evaluation of integrals. (see Appendix G)

52. Show that, for $|\text{ph}\, s| < \frac{1}{2}\pi$.

$$\int_0^\infty \frac{e^{-sx}}{1+x^2}\,dx \sim \sum_{n=0}^{\infty} \frac{(2n)!(-)^n}{s^{2n+1}}$$

as $|s| \to \infty$.

53. Show that, for $|\text{ph}\, s| < \frac{1}{2}\pi$,

$$\int_0^\infty \frac{e^{-sx^2}}{1+x} \ln x\, dx \sim \sum_{m=1}^{\infty} (\tfrac{1}{2}m-1)!(-)^{m-1}\{\psi(\tfrac{1}{2}m-1) - \ln s\}/4s^{\frac{1}{2}m}$$

as $|s| \to \infty$; $\psi(z) = z!'/z!$.

54. From (A.1.26) show that, as $\tau \to \infty$ with n fixed,

$$J_n(\tau) \sim \left(\frac{2}{\pi\tau}\right)^{\frac{1}{2}} \cos(\tau - \tfrac{1}{2}n\pi - \tfrac{1}{4}\pi).$$

55. Find asymptotic expansions for

$$\int_0^1 x^{\frac{1}{2}} e^{-sx} \frac{dx}{1+x}$$

as $|s| \to \infty$ when (i) $|\text{ph}\, s| < \frac{1}{2}\pi$, (ii) $\frac{1}{2}\pi < |\text{ph}\, s| < \pi$, (iii) $\text{ph}\, s = \frac{1}{2}\pi$.

56. Show that, as $s \to \infty$,

$$\int_{-\infty}^{\infty} e^{-sx^2} \ln(1+x+x^2)\, dx \sim \sum_{m=1}^{\infty} (m-\tfrac{1}{2})!\, d_m/s^{m+\frac{1}{2}}$$

where $d_1 = \frac{1}{2}$, $d_2 = \frac{1}{4}$.

57. Show that, as $s \to \infty$,

$$K_0(s) = \int_1^\infty e^{-sx}(x^2-1)^{-\frac{1}{2}}\,dx \sim (\pi/2s)^{\frac{1}{2}}e^{-s}(1-1/8s+\ldots).$$

58. Obtain the asymptotic expansion as $\sigma \to \infty$ of

$$\int_{c-i\infty}^{c+i\infty} e^{\frac{1}{2}\sigma z^2}\frac{z^{1-\sigma}}{1+z^2}\,dz \quad (c>0).$$

59. Determine the saddle point of

$$\int_{-i\infty}^{i\infty} (z^2-1)^{\frac{1}{2}}\exp[\sigma\{z\cos\phi + i(z^2-1)^{\frac{1}{2}}\sin\phi\}]\,dz$$

and hence find the first term in the asymptotic expansion. Here $0<\phi<\pi$ and $(z^2-1)^{\frac{1}{2}}$ is regular in $|z|<1$, taking the value i at $z=0$.

60. Prove that

$$\int_{C_1} \frac{e^{i\tau\cos(z-\alpha)}}{z-\beta}\,dz \sim -2\pi i e^{i\tau\cos(\beta-\alpha)}f(\beta)$$

$$+\frac{e^{i(\tau-\frac{1}{4}\pi)}}{\alpha-\beta}\left(\frac{2\pi}{\tau}\right)^{\frac{1}{2}}\left[1-\frac{i}{\tau}\left\{\frac{1}{(\alpha-\beta)^2}+\tfrac{1}{8}\right\}+\cdots\right]$$

as $\tau \to \infty$ when $0 \leqslant \alpha < \frac{1}{2}\pi$ and $\alpha \neq \beta$. Here, $f(\beta)$ takes one of the values 1, 0, or -1 according to the position of β. The contour C_1 consists of the straight line from $\frac{1}{2}\pi - i\infty$ to $\frac{1}{2}\pi$, the real axis from $\frac{1}{2}\pi$ to $-\frac{1}{2}\pi$ and the straight line from $-\frac{1}{2}\pi$ to $-\frac{1}{2}\pi + i\infty$.

61. Evaluate asymptotically

$$H_{s\nu}^{(1)}(s) = -\frac{1}{\pi}\int_{C_1} \exp[is\{\cos z + (z-\tfrac{1}{2}\pi)\sin\nu\}]\,dz$$

where C_1 is the contour of Exercise 60, and verify (A.2.29).

9. Diffraction by edges

THE subject of Chapter 8 was the scattering produced by a smooth object, i.e. one on which the direction of the normal changes continuously. Much of the theory developed is applicable to bodies of arbitrary shape but the presence of an edge, where the direction of the surface normal changes discontinuously, introduces some distinctive features. For example, the radius of curvature is zero at an edge so that, no matter how high the frequency, it does not scatter waves like a smooth boundary. Much effort has been devoted to the study of this problem in recent years and only a small part of it can be encompassed in this chapter. Only waves varying harmonically in time will be discussed.

General results

9.1 Uniqueness

Perhaps the most striking characteristic of an edge is that the conditions so far imposed on fields are not sufficient to ensure a unique solution to the diffraction problem. Although this fact was first discovered by Rayleigh in 1897 it is only in the last two decades or so that what must be stipulated further has been elucidated. It is therefore apposite to start with a discussion of uniqueness in general terms.

To commence with the behaviour of solutions of

$$\nabla^2 p + k^2 p = 0 \tag{1.1}$$

outside a closed surface S will be examined. Let the surface Ω of a sphere of radius R totally enclose S (Fig. 9.1). Then, from the representation (29.6) of Chapter 1,

$$p(\mathbf{y}) = \int_\Omega \left\{ \frac{\partial p}{\partial n} \psi(\mathbf{y}, \mathbf{x}) - p \frac{\partial}{\partial n} \psi(\mathbf{y}, \mathbf{x}) \right\} \mathrm{d}\Omega - \int_S \left\{ \frac{\partial p}{\partial n} \psi(\mathbf{y}, \mathbf{x}) - p \frac{\partial}{\partial n} \psi(\mathbf{y}, \mathbf{x}) \right\} \mathrm{d}S_x \tag{1.2}$$

where

$$\psi(\mathbf{x}, \mathbf{y}) = \mathrm{e}^{-\mathrm{i}k|\mathbf{x}-\mathbf{y}|}/4\pi\,|\mathbf{x} - \mathbf{y}|$$

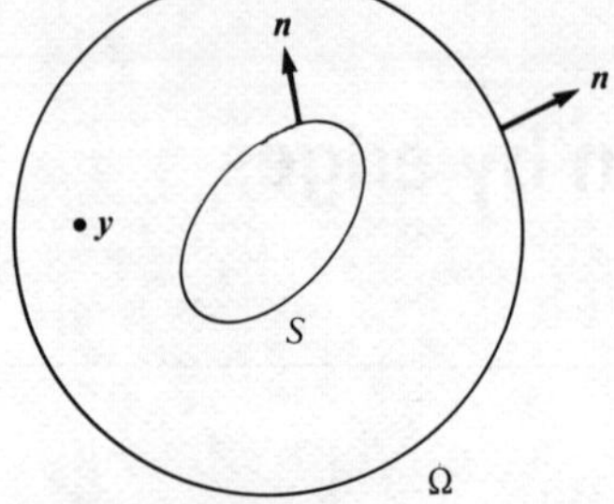

Fig. 9.1 Domain for the representation of the field outside S

and **y** lies between S and Ω. It will now be supposed that k is real and that p satisfies the radiation condition at infinity, i.e.

$$R\left\{\frac{\partial}{\partial R}\,p(\mathbf{x})+ikp\right\}\to 0 \tag{1.3}$$

as $R\to\infty$, R being the length of $\mathbf{x}$.

Let a star signify a complex conjugate. Then

$$\int_\Omega\left(p\frac{\partial p^*}{\partial n}-p^*\frac{\partial p}{\partial n}\right)\mathrm{d}\Omega=\int_S\left(p\frac{\partial p^*}{\partial n}-p^*\frac{\partial p}{\partial n}\right)\mathrm{d}S \tag{1.4}$$

because k is real. It follows that the left-hand side of (1.4), although apparently a function of R, is a finite constant independent of R. Now

$$|\partial p/\partial R|^2+k^2|p|^2-|ikp+\partial p/\partial R|^2=ik(p^*\,\partial p/\partial R-p\,\partial p^*/\partial R).$$

Therefore, by (1.4),

$$\int_\Omega\{|\partial p/\partial R|^2+k^2|p|^2-|ikp+\partial p/\partial R|^2\}\,\mathrm{d}\Omega$$

is a constant independent of R. On account of (1.3) the contribution of the last term in the integrand tends to zero. Consequently

$$\int_\Omega|\partial p/\partial R|^2\,\mathrm{d}\Omega=O(1),\qquad \int_\Omega|p|^2\,\mathrm{d}\Omega=O(1) \tag{1.5}$$

as $R\to\infty$. By virtue of (1.5)

$$\int_\Omega|p|\,\mathrm{d}\Omega\leqslant\left\{4\pi R^2\int_\Omega|p|^2\,\mathrm{d}\Omega\right)^{\frac{1}{2}}=O(R), \tag{1.6}$$

$$\int_\Omega|\partial p/\partial R|\,\mathrm{d}\Omega=O(R) \tag{1.7}$$

In the integral over Ω in (1.2) $R>|\mathbf{y}|$ and so ψ can be expanded in a uniformly and absolutely convergent series of powers of $1/R$, say

$$\psi = e^{-ikR} \sum_{n=0}^{\infty} \psi_n(\theta, \phi)/R^{n+1}. \tag{1.8}$$

Any number of derivatives can be taken without destroying the convergence properties so that

$$\partial\psi/\partial R = -ike^{-ikR}\psi_0/R + O(1/R^2). \tag{1.9}$$

Hence the integral over Ω can be expressed as

$$\int_\Omega \left\{\left(\frac{\partial p}{\partial R} + ikp\right)\psi_0(\theta, \phi)\frac{e^{-ikR}}{R} + \left(\frac{\partial p}{\partial R} + p\right)O\left(\frac{1}{R^2}\right)\right\} d\Omega.$$

As $R \to \infty$ the first term tends to zero by invoking (1.3). The remaining terms disappear because of (1.6) and (1.7).

Accordingly, it has been shown that, *when k is real, a solution of* (1.1) *satisfying* (1.3) *has the representation*

$$p(\mathbf{y}) = \int_S \left\{p\frac{\partial}{\partial n}\psi(\mathbf{y}, \mathbf{x}) - \frac{\partial p}{\partial n}\psi(\mathbf{y}, \mathbf{x})\right\} dS_x \tag{1.10}$$

for $\mathbf{y}$ *outside S.* It will be noticed that the derivation of this representation has assumed less about the behaviour of p at infinity than in §1.31.

Choose a particular value of R, say R_0, such that S is entirely within Ω. Then, when $|\mathbf{y}| > R_0$, (1.8) can be applied to (1.10). The consequence is that

$$p(\mathbf{x}) = e^{-ikR} \sum_{n=0}^{\infty} p_n(\theta, \phi)/R^{n+1} \tag{1.11}$$

for $R > R_0$. On the other hand, (1.11) must satisfy (1.1). Therefore, by expressing (1.1) in spherical polar coordinates and equating to zero the coefficients of the various powers of $1/R$, we arrive at

$$2ik(n+1)p_{n+1} + \frac{1}{\sin\theta}\frac{\partial}{\partial\theta}\left(\sin\theta\frac{\partial p_n}{\partial\theta}\right) + \frac{1}{\sin^2\theta}\frac{\partial^2 p_n}{\partial\phi^2} + n(n+1)p_n = 0$$

$$(n = 0, 1, 2, \ldots). \tag{1.12}$$

Then, *when p complies with* (1.3), *it can be expanded in $R > R_0$ in the series* (1.11) *where the coefficients satisfy* (1.12).

It will be observed that, if $p_0 \equiv 0$, (1.12) implies that $p_n \equiv 0$ $(n \geq 1)$ on the assumption that $k \neq 0$. Consequently, $p_0 \equiv 0$ forces $p \equiv 0$ in $R > R_0$. But the representation (1.10) shows that p is analytic in $\mathbf{y}$ outside S. Therefore, if p is identically zero in $R > R_0$ it is also identically zero everywhere outside S. Hence, we reach the important conclusion that, *if p satisfies* (1.3) *and $p_0 \equiv 0$, p vanishes identically in the exterior of S.*

This result enables us to prove the following:

ACOUSTIC UNIQUENESS THEOREM *If p satisfies* (1.1) *with k positive, complies with* (1.3) *and is such that*

$$\int_{\Omega} (p\,\partial p^*/\partial R - p^*\,\partial p/\partial R)\,\mathrm{d}\Omega \to 0 \tag{1.13}$$

as $R \to \infty$ *then* $p \equiv 0$ *outside S.*

As in the derivation of (1.5), (1.13) forces

$$\int_{\Omega} \{|\partial p/\partial R|^2 + k^2\,|p|^2\}\,\mathrm{d}\Omega \to 0.$$

This requires, on account of (1.11),

$$\int_{\Omega} |p_0|^2 \frac{\mathrm{d}\Omega}{R^2} \to 0$$

which is possible only if $p_0 \equiv 0$. By the preceding paragraph $p \equiv 0$ outside S and the theorem has been demonstrated.

Note that the uniqueness theorem is limited to objects of finite dimensions; it does not include, for example, the semi-infinite plane. Uniqueness theory for infinite obstacles is difficult but it is possible to show that boundaries which are conically shaped at infinity (this is true of the semi-infinite plane and wedge) possess a uniqueness theorem under appropriate boundary conditions (Jones (1953)). For other boundary conditions the theorem is no longer valid; there can be eigenfunction solutions, i.e. solutions which exist even when there is no incident wave.

If k is complex with a negative imaginary part, as when the exterior medium is dissipative, it is no longer necessary to impose (1.3). It is sufficient to ask that the field be exponentially damped at infinity. The representations (1.10) and (1.11) are still valid but the uniqueness theorem has to be modified because (1.13) is satisfied automatically. If $\mathrm{Re}\, k > 0$ we demand

$$\mathrm{Im} \int_S p\,\partial p^*/\partial n\,\mathrm{d}S \leqslant 0. \tag{1.14}$$

Then

$$\mathrm{Im} \int_{\Omega} p\,\partial p^*/\partial R\,\mathrm{d}\Omega = \mathrm{Re}\, k\ \mathrm{Im}\, k \int |p|^2\,\mathrm{d}\mathbf{x} + \mathrm{Im} \int_S p\,\partial p^*/\partial n\,\mathrm{d}S.$$

The integral on the left disappears as $R \to \infty$ and then (1.14) entails $p \equiv 0$.

If Re $k=0$, the equality in (1.14) is dropped and the same conclusion about uniqueness is reached.

There are analogous theorems for the electromagnetic field. The formulae corresponding to (1.2) are, from (29.9) and (29.10) of Chapter 1,

$$\boldsymbol{E}(\mathbf{y}) = -(1/\mathrm{i}\omega\varepsilon)(\operatorname{grad}\operatorname{div} + k^2)\int_{\Omega-S} \boldsymbol{n}\wedge\boldsymbol{H}\psi(\mathbf{y},\boldsymbol{x})\,\mathrm{d}S_x$$
$$-\operatorname{curl}\int_{\Omega-S} \boldsymbol{n}\wedge\boldsymbol{E}\psi(\mathbf{y},\boldsymbol{x})\,\mathrm{d}S_x, \tag{1.15}$$

$$\boldsymbol{H}(\mathbf{y}) = (1/\mathrm{i}\omega\mu)(\operatorname{grad}\operatorname{div} + k^2)\int_{\Omega-S} \boldsymbol{n}\wedge\boldsymbol{E}\psi(\mathbf{y},\boldsymbol{x})\,\mathrm{d}S_x$$
$$-\operatorname{curl}\int_{\Omega-S} \boldsymbol{n}\wedge\boldsymbol{H}\psi(\mathbf{y},\boldsymbol{x})\,\mathrm{d}S_x \tag{1.16}$$

where now $k^2=\omega^2\mu\varepsilon$ and the vector operators apply to $\mathbf{y}$. The relevant radiation conditions are

$$R\{\boldsymbol{E}+(\mu/\varepsilon)^{\frac{1}{2}}\hat{\boldsymbol{x}}\wedge\boldsymbol{H}\}\to\mathbf{0}, \qquad R\{(\mu/\varepsilon)^{\frac{1}{2}}\boldsymbol{H}-\hat{\boldsymbol{x}}\wedge\boldsymbol{E}\}\to\mathbf{0}. \tag{1.17}$$

In place of (1.4) we have

$$\int_{\Omega}(\boldsymbol{E}\wedge\boldsymbol{H}^*+\boldsymbol{E}^*\wedge\boldsymbol{H})\cdot\boldsymbol{n}\,\mathrm{d}S = \int_{S}(\boldsymbol{E}\wedge\boldsymbol{H}^*+\boldsymbol{E}^*\wedge\boldsymbol{H})\cdot\boldsymbol{n}\,\mathrm{d}S. \tag{1.18}$$

Also

$$|\boldsymbol{E}+(\mu/\varepsilon)^{\frac{1}{2}}\hat{\boldsymbol{x}}\wedge\boldsymbol{H}|^2+|(\mu/\varepsilon)^{\frac{1}{2}}\boldsymbol{H}-\hat{\boldsymbol{x}}\wedge\boldsymbol{E}|^2-2\,|\boldsymbol{E}|^2-2(\mu/\varepsilon)\,|\boldsymbol{H}|^2$$
$$+|\hat{\boldsymbol{x}}\cdot\boldsymbol{E}|^2+(\mu/\varepsilon)\,|\hat{\boldsymbol{x}}\cdot\boldsymbol{H}|^2=-2(\mu/\varepsilon)(\boldsymbol{E}\wedge\boldsymbol{H}^*+\boldsymbol{E}^*\wedge\boldsymbol{H})\cdot\hat{\boldsymbol{x}}.$$

The scalar product of (1.17) with $\hat{\boldsymbol{x}}$ shows that $R\hat{\boldsymbol{x}}\cdot\boldsymbol{E}\to 0$, $R\hat{\boldsymbol{x}}\cdot\boldsymbol{H}\to 0$ and therefore, as for acoustic waves, it may be deduced that

$$\int_{\Omega}|\boldsymbol{E}|^2\,\mathrm{d}\Omega=O(1), \qquad \int_{\Omega}|\boldsymbol{H}|^2\,\mathrm{d}\Omega=O(1), \qquad \int_{\Omega}|\boldsymbol{E}|\,\mathrm{d}\Omega=O(R),$$
$$\int_{\Omega}|\boldsymbol{H}|\,\mathrm{d}\Omega=O(R). \tag{1.19}$$

On Ω the derivatives in (1.15) may be taken inside the integrals and converted into ones with respect to $\boldsymbol{x}$ because ψ depends only on $|\boldsymbol{x}-\mathbf{y}|$. Hence, by means of (1.9), (1.17) and (1.19), we perceive that the contribution from Ω vanishes as $R\to\infty$. Thus, *when μ and ε are real, a solution of Maxwell's equations without sources which satisfies* (1.17) *has* *the representation*

$$\mathbf{E}(\mathbf{y}) = (1/\mathrm{i}\omega\varepsilon)(\text{grad div} + k^2) \int_S \boldsymbol{n} \wedge \boldsymbol{H}\psi(\mathbf{y}, \boldsymbol{x})\, \mathrm{d}S_x$$

$$+ \text{curl} \int_S \boldsymbol{n} \wedge \boldsymbol{E}\psi(\mathbf{y}, \boldsymbol{x})\, \mathrm{d}S_x, \tag{1.20}$$

$$\boldsymbol{H}(\mathbf{y}) = \text{curl} \int_S \boldsymbol{n} \wedge \boldsymbol{H}\psi(\mathbf{y}, \boldsymbol{x})\, \mathrm{d}S_x$$

$$-(1/\mathrm{i}\omega\mu)(\text{grad div} + k^2) \int_S (\boldsymbol{n} \wedge \mathbf{E})\psi(\mathbf{y}, \boldsymbol{x})\, \mathrm{d}S_x \tag{1.21}$$

where $\mathbf{y}$ *is outside S and the derivatives are with respect to* $\mathbf{y}$.

The formula (1.20) reveals that

$$\boldsymbol{E} = (1/\mathrm{i}\omega\varepsilon)(\text{grad div} + k^2)\boldsymbol{C} + \text{curl}\, \boldsymbol{D}$$

where, for $\mathbf{y}$ not on S,

$$\nabla^2\boldsymbol{C} + k^2\boldsymbol{C} = \mathbf{0}, \qquad \nabla^2\boldsymbol{D} + k^2\boldsymbol{D} = \mathbf{0}. \tag{1.22}$$

On account of (1.22) it can also be written

$$\boldsymbol{E} = \text{curl curl}\, \boldsymbol{C}/\mathrm{i}\omega\varepsilon + \text{curl}\, \boldsymbol{D}. \tag{1.23}$$

Similarly

$$\boldsymbol{H} = \text{curl}\, \boldsymbol{C} - \text{curl curl}\, \boldsymbol{D}/\mathrm{i}\omega\mu$$

$$= -(1/\mathrm{i}\omega\mu)\text{curl}\{\text{curl}\, \boldsymbol{D} + \text{curl curl}\, \boldsymbol{C}/\mathrm{i}\omega\varepsilon\}. \tag{1.24}$$

By virtue of (1.8)

$$\text{curl}\, \boldsymbol{D} + \text{curl curl}\, \boldsymbol{C}/\mathrm{i}\omega\varepsilon = \mathrm{e}^{-\mathrm{i}kR} \sum_{n=0}^{\infty} \boldsymbol{c}_n(\theta, \phi)/R^{n+1}, \tag{1.25}$$

for $R > R_0$. The coefficients $\boldsymbol{c}_n$ satisfy the vector equations (in Cartesians) obtained by replacing p_n by $\boldsymbol{c}_n$ in (1.12). In addition, the divergence of curl $\boldsymbol{C}$ must vanish; this requires

$$\boldsymbol{c}_0 \cdot \boldsymbol{i}_1 = 0, \tag{1.26}$$

$$-\mathrm{i}k\boldsymbol{c}_{n+1} \cdot \boldsymbol{i}_1 + \frac{1}{\sin\theta}\frac{\partial}{\partial\theta}(\boldsymbol{c}_n \cdot \boldsymbol{i}_2 \sin\theta) + \frac{1}{\sin\theta}\frac{\partial}{\partial\phi}(\boldsymbol{c}_n \cdot \boldsymbol{i}_3) = 0 \quad (n \geqslant 0) \tag{1.27}$$

where $\boldsymbol{i}_1$, $\boldsymbol{i}_2$, $\boldsymbol{i}_3$ are unit vectors along the directions R, θ, ϕ increasing respectively. Thus, *subject to* (1.17), *the electromagnetic field has the representation* (1.23), (1.24) *outside S and, for* $R > R_0$, *the absolutely and uniformly convergent expansion* (1.25) *is valid.* It is now straightforward to infer that, *if* $\boldsymbol{c}_0 = \mathbf{0}$, *the electromagnetic field is identically zero in the exterior of S.*

Observe that, as $R \to \infty$,

$$\boldsymbol{E} = \boldsymbol{c}_0 \mathrm{e}^{-\mathrm{i}kR}/R + O(1/R^2), \tag{1.28}$$

$$\boldsymbol{H} = (\varepsilon/\mu)^{\frac{1}{2}} \hat{\boldsymbol{x}} \wedge \boldsymbol{c}_0 \mathrm{e}^{-\mathrm{i}kR}/R + O(1/R^2). \tag{1.29}$$

One conclusion is that, if

$$\int_\Omega \{|\boldsymbol{E}|^2 + (\mu/\varepsilon)\, |\boldsymbol{H}|^2\}\, \mathrm{d}\Omega \to 0$$

as $R \to \infty$, $\boldsymbol{c}_0 = \boldsymbol{0}$, and $\boldsymbol{E} \equiv \boldsymbol{0}$, $\boldsymbol{H} \equiv \boldsymbol{0}$. This demonstrates the following theorem.

ELECTROMAGNETIC UNIQUENESS THEOREM *A solution of Maxwell's source-free equations with ω, μ, ε positive which satisfies* (1.17) *and is such that*

$$\int_\Omega (\boldsymbol{E} \wedge \boldsymbol{H}^* + \boldsymbol{E}^* \wedge \boldsymbol{H}) \cdot \boldsymbol{n}\, \mathrm{d}S \to 0$$

as $R \to \infty$ is identically zero, i.e. $\boldsymbol{E} \equiv \boldsymbol{0}$, $\boldsymbol{H} \equiv \boldsymbol{0}$ *outside S.*

If the external medium possesses conductivity σ (>0) ε is replaced by $\varepsilon - \mathrm{i}\sigma/\omega$ and k is given a negative imaginary part. The radiation conditions (1.17) are dropped in favour of the requirement that the field is exponentially damped at infinity. It follows from

$$\int_\Omega (\boldsymbol{E} \wedge \boldsymbol{H}^* + \boldsymbol{E}^* \wedge \boldsymbol{H}) \cdot \boldsymbol{n}\, \mathrm{d}\Omega$$

$$= -2\sigma \int \boldsymbol{E} \cdot \boldsymbol{E}^*\, \mathrm{d}\boldsymbol{x} + \int_S (\boldsymbol{E} \wedge \boldsymbol{H}^* + \boldsymbol{E}^* \wedge \boldsymbol{H}) \cdot \boldsymbol{n}\, \mathrm{d}S$$

that, if

$$\int_S (\boldsymbol{E} \wedge \boldsymbol{H}^* + \boldsymbol{E}^* \wedge \boldsymbol{H}) \cdot \boldsymbol{n}\, \mathrm{d}S \leqslant 0, \tag{1.30}$$

$\boldsymbol{E} \equiv \boldsymbol{0}$ and $\boldsymbol{H} \equiv \boldsymbol{0}$ in the exterior of S. A somewhat similar proof can be constructed for the demonstration of uniqueness when ω and μ have certain complex values.

9.2 Edge conditions

The perception that there is a multiplicity of solutions when edges are present is easily realised. Consider the soft semi-infinite plane which occupies the part of $y = 0$ in $x \leqslant 0$. Take cylindrical polar coordinates with $x = r \cos \phi$, $y = r \sin \phi$. Let p_0 be any two-dimensional field which satisfies the boundary and radiation conditions. Then

$$p = p_0 + CH^{(2)}_{\frac{1}{2}}(kr) \sin \tfrac{1}{2}(\phi - \pi) \tag{2.1}$$

satisfies Helmholtz's equation for any value C, complies with the radiation conditions and has the same boundary behaviour as p_0 on the screen $\phi = \pm\pi$. Moreover, p has no singularities off the semi-infinite plane. Therefore, it is a possible solution to the problem. An infinite set of solutions can be generated by giving C different values. Still more can be produced by adding multiples of $\mathrm{H}^{(2)}_{\frac{1}{2}n}(kr)\sin\frac{1}{2}n(\phi-\pi)$ where n is a positive integer.

The phenomenon is not peculiar to the soft boundary condition. For a hard screen multiples of $\mathrm{H}^{(2)}_{\frac{1}{2}}(kr)\cos\frac{1}{2}(\phi-\pi)$ can be added. Nor is the non-uniqueness caused by the infinite length of the edge. Two different solutions have been found for the diffraction by a circular disc.

Clearly, additional restrictions must be applied if uniqueness is to be achieved. One route to discovering suitable constraints is to consider the behaviour on the screen. One would expect the pressure to remain bounded there and the momentum on a finite portion of the boundary to be finite even if the edge is included. The realization of such behaviour can be accomplished by asking that p be finite on the boundary and $\partial p/\partial n$ be integrable in the direction perpendicular to the edge. If, now, p_0 satisfies these conditions then p of (2.1) will not, unless $C = 0$. For, although p is bounded on the screen $\partial p/\partial y$ is not integrable with respect to r since $\mathrm{H}^{(2)}_{\frac{1}{2}}$ has singularity like $1/r^{\frac{1}{2}}$ as $r \to 0$. Similarly, the other possible additions to the solution for the soft screen are excluded. For the hard boundary additional terms are prevented by the finiteness of p. The restrictions on p and $\partial p/\partial n$ near an edge are known as *edge conditions*.

The sort of field dictated by the edge conditions can be determined from the conduct near a wedge with faces $\phi = \pm\delta$ (Fig. 9.2). Very close to the edge it is plausible to approximate Helmholtz's equation by Laplace's equation. Then a possible form for the pressure near the edge is

$$p = (A + B\ln r)(C\phi + D) + r^{\mu}\{E\cos\mu(\phi-\delta) + F\sin\mu(\phi-\delta)\}$$

where $\mu \neq 0$. In order that p be bounded on the wedge it is necessary that $B = 0$ and $\mu > 0$. The other edge condition demands that $\partial p/r\partial\phi$ be integrable with respect to r on $\phi = \pm\delta$ with the upshot $C = 0$. If the

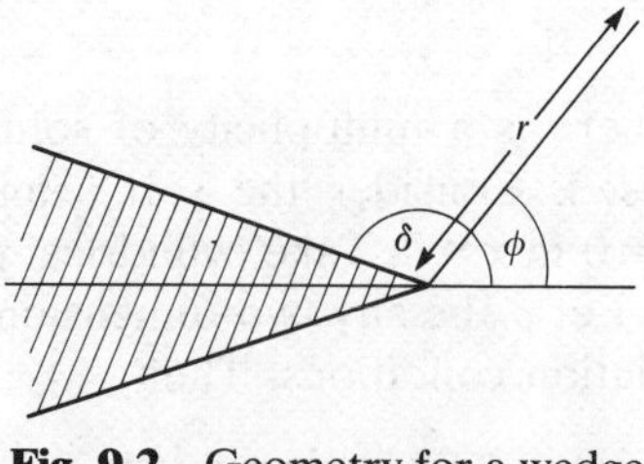

Fig. 9.2 Geometry for a wedge

wedge is soft, we must have $A=0$, $E=0$ and $\sin 2\mu\delta=0$. The smallest value of μ permitted is $\pi/2\delta$ and so *the dominant behaviour of p near a soft wedge is given by*

$$p \approx Fr^{\pi/2\delta} \sin\{\pi(\phi-\delta)/2\delta\} \tag{2.2}$$

A similar argument leads to the assertion that *near a hard wedge the essential form of p is*

$$p \approx A + Fr^{\pi/2\delta} \cos\{\pi(\phi-\delta)/2\delta\}. \tag{2.3}$$

For a semi-infinite plane $\delta=\pi$ and the power of r in both (2.2) and (2.3) is $r^{\frac{1}{2}}$.

It must not be imagined that the behaviour displayed in (2.2) and (2.3) is obligatory for every incident wave. Suppose that p_1^{i} produces F_1 in (2.2) and p_2^{i} creates F_2. Then $F_2p_1^{\mathrm{i}}-F_1p_2^{\mathrm{i}}$ will be an incident field which generates a total field which tends to zero at the edge faster than (2.2). All that (2.2) and (2.3) indicate is the strongest behaviour that is possible, in so far as Laplace's equation is concerned. When $\delta<\frac{1}{4}\pi$ solutions of Helmholtz's equation can have a term in r^2 which is more important than the second term of (2.3).

In three dimensions the same kind of structure can be predicted with r being measured perpendicular to the edge in the vicinity under consideration.

The most singular part of the acoustic energy density for (2.2) or (2.3) is $|\text{grad } p|^2$ which is of the order of $r^{-2+\pi/\delta}$. This is integrable over an area. Conversely, if the acoustic energy density is integrable over an area, (2.2) and (2.3) follow. Therefore, another form of the edge conditions is that the energy in any finite region (including the edge) should be bounded.

Another way of stating this is that $\lim \int_C p\boldsymbol{n} \cdot \text{grad } p \, \mathrm{d}C = 0$ where C is a small circuit enclosing the edge which contracts to the edge in the limit. Thus *the edge conditions prevent an edge from radiating energy on its own account.*

It will now be proved that the edge conditions ensure a unique solution to the scattering by a body of finite size. If the body has edges, surround them by a small surface Σ. The integral on the right of (1.4) becomes one over Σ and the remainder of the obstacle. The latter surface provides no contribution for a hard or soft body and the former can be written

$$\iint_{-\delta}^{\delta} (p \, \partial p^*/\partial r - p^* \, \partial p/\partial r) r \, \mathrm{d}\phi \, \mathrm{d}s$$

where s is tangential to the edge. In view of (2.2) and (2.3) this integral is $O(r^{\pi/2\delta})$ which tends to zero as $r \to 0$, i.e. Σ contracts to the edge. Thus, the integral on the left of (1.4) is zero and the acoustic uniqueness theorem can be invoked. In other words, *for a hard or soft obstacle of finite size, the field satisfying the radiation and edge conditions is identically zero.*

Clearly, the theory can be extended to any other boundary conditions which guarantee that the left-hand side of (1.4) is zero but the details will be left to the reader.

For the electromagnetic field the appropriate edge conditions are that the current and charge on the surface should be finite. If z is parallel to the edge these effectively mean that E_z acts as if the boundary were soft and H_z as if it were hard when the surface is perfectly conducting. Thus, near the edge,

$$E_x \propto r^{\pi/2\delta}, \qquad H_z \propto A, \qquad E_r, E_\phi, H_r, H_\phi \propto r^{(\pi/2\delta)-1} \tag{2.4}$$

ignoring angular variations and understanding that E_r, E_ϕ, H_r, H_ϕ may be proportional to r when $\delta<\frac{1}{4}\pi$. *The field components parallel to the edge are bounded. The components perpendicular to the edge are singular if $\delta>\frac{1}{2}\pi$ and, in any case, are always more singular than the parallel components.* The component of the electric vector parallel to the edge vanishes at the edge. The current density perpendicular to the edge is proportional to the discontinuity in H_z so that it is always bounded; on a semi-infinite plane ($\delta=\pi$) it tends to zero at the edge. On the other hand, (2.4) predicts that the current density parallel to the edge will be infinite there, though it is integrable in harmony with the edge conditions.

With regard to uniqueness (2.4) implies that

$$\int_\Sigma \boldsymbol{E}\wedge\boldsymbol{H}^*\cdot\boldsymbol{n}\,\mathrm{d}S = O(r^{\pi/2\delta})$$

and so *the radiation and edge conditions warrant that the scattering by a finite, perfectly conducting body possesses a unique solution.*

Again the theory can be extended to other boundary conditions and dielectrics but consideration of these extensions will be omitted. It should, however, be emphasized that, if the tangent to the edge changes direction discontinuously at some point, the above theory fails in the neighbourhood of such a corner. Nevertheless, it can be shown that the edge conditions hold arbitrarily close to the corner and that one does not have to worry about accumulation of charge at the point in the electromagnetic case.

9.3 Babinet's principle

In optics there is a well known theorem, named after Babinet, which states that the diffraction patterns produced by complementary screens are the same except for the central spot. It was originally established by an approximate theory due to Kirchhoff but nowadays an exact analysis is available.

Let there be an infinite plane rigid screen from which an aperture A has been removed. Denote by S the remaining portion of the plane. At the same time consider the complementary screen in which A is a soft planar region and S is the aperture (Fig. 9.3). Let the screen occupy the plane $z=0$. The dependence of the field on the coordinate z will be indicated explicitly but the variations with respect to x and y will not be displayed.

Assume that the sources of the incident wave $p^{\mathrm{i}}(z)$ lie in $z<0$ and let $p(z)$ be the resulting total field. The images in $z=0$ of the sources are in $z>0$ and are responsible for an incident field $p^{\mathrm{i}}(-z)$. When the image field is the incident wave the total field is $p(-z)$. The united effect of both sources and images is a total field of $p(z)+p(-z)$. But the combined incident field due to a source and its image automatically satisfies the boundary condition on the screen as well as the edge conditions so that the screen could be eliminated. To put it another way, the resultant field caused by the sources and images is precisely the same as if there were no screen. Therefore

$$p(z)+p(-z)=p^{\mathrm{i}}(z)+p^{\mathrm{i}}(-z). \tag{3.1}$$

The equation continues to hold as $z \to 0$. Now p is continuous across A, i.e. $p(+0)=p(-0)=p(0)$ in A. Hence

$$p(0)=p^{\mathrm{i}}(0) \quad (\text{in } A). \tag{3.2}$$

The pressure in the aperture of a plane rigid screen is exactly the same as the pressure of the incident wave.

On S, where the pressure is not continuous, (3.1) gives

$$p(+0)+p(-0)=2p^{\mathrm{i}}(0) \quad (\text{on } S). \tag{3.3}$$

Construct the pressure field $\tilde{p}$, defined by

$$\tilde{p}(z)=\begin{cases} p(z)-p^{\mathrm{i}}(-z) & (z<0) \\ p^{\mathrm{i}}(z)-p(z) & (z>0) \end{cases}. \tag{3.4}$$

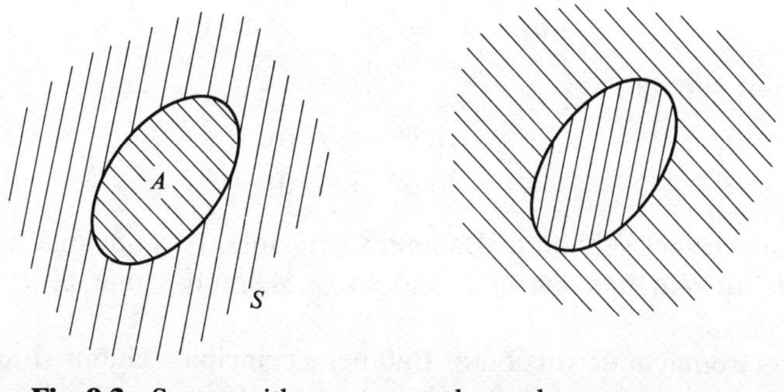

Fig. 9.3 Screen with aperture and complementary screen

Let $z \to +0$; then, from (3.2)

$$\tilde{p}(+0) = \begin{cases} 0 & (\text{in } A) \\ p^{\text{i}}(0) - p(+0) & (\text{on } S) \end{cases}.$$

Allowing $z \to -0$, we have

$$\tilde{p}(-0) = \begin{cases} 0 & (\text{in } A) \\ p(-0) - p^{\text{i}}(0) & (\text{on } S). \end{cases}$$

On account of (3.3), $\tilde{p}$ is continuous across S and vanishes in A. Further, $\partial p/\partial z$ is zero on S and therefore $\partial \tilde{p}/\partial z$ is continuous through S. Accordingly, $\tilde{p}$ satisfies the boundary conditions for the complementary screen. By the edge conditions p is bounded near an edge and therefore so is $\tilde{p}$, i.e. $\tilde{p}$ satisfies the edge conditions.

The sources of $\tilde{p}$ are the same as those of $p^{\text{i}}(z)$ in $z<0$; there are none in $z>0$. In addition, the relevant radiation conditions are complied with at infinity. Consequently, $\tilde{p}$ is the total field in the diffraction of the incident wave $p^{\text{i}}(z)$ by the complementary screen. It will be the one and only one solution if a uniqueness theorem for infinite plane screens is assumed.

The information permits the assertion of

BABINET'S ACOUSTIC PRINCIPLE *Let p be the resultant field in $z>0$ due to p^{i} incident from $z<0$ and let $\tilde{p}$ be the total field when the same incident wave falls on the complementary screen. Then, in $z>0$,*

$$p + \tilde{p} = p^{\text{i}}.$$

In fact, (3.4) tells us what happens in $z<0$. It is, however, advantageous to cast it into a different form. Suppose that S is finite while A stretches to infinity. Then $p = p^{\text{i}} + p^{\text{s}}$ where p^{s} is a scattered field satisfying the usual radiation conditions at infinity. In the complementary problem there will be a reflected wave in $z<0$; in addition, there is a diffracted field $\tilde{p}^{\text{d}}$ which is radiating at infinity in $z>0$ and $z<0$, i.e.

$$\tilde{p} = \begin{cases} p^{\text{i}}(z) - p^{\text{i}}(-z) + \tilde{p}^{\text{d}} & (z<0) \\ \tilde{p}^{\text{d}} & (z>0) \end{cases}.$$

Hence (3.4) supplies

$$p^{\text{s}} = \begin{cases} \tilde{p}^{\text{d}} & (z<0) \\ -\tilde{p}^{\text{d}} & (z>0) \end{cases} \tag{3.5}$$

which is a succinct version of Babinet's principle. It is obvious from (3.1) that *p^{s} is an odd function of z* and so (3.5) implies that *$\tilde{p}^{\text{d}}$ is an even function of z*.

The electromagnetic version of Babinet's principle exhibits some differences. Firstly, in the original problem A is an aperture in the perfectly conducting S. The *complementary screen* has A perfectly conducting and S an aperture. The sources are in $z<0$ and create an incident field $\mathbf{E}^{\mathrm{i}}(z)$, $\mathbf{H}^{\mathrm{i}}(z)$ while the total field is $\mathbf{E}(z)$, $\mathbf{H}(z)$. Denote by $\mathbf{E}_{\mathrm{t}}$ the component of $\mathbf{E}$ transverse to the z-axis and E_{p} the component parallel to the z-axis. Then $\mathbf{E}=\mathbf{E}_{\mathrm{t}}+E_{\mathrm{p}}\mathbf{k}$ where $\mathbf{k}$ is a unit vector along the z-axis; a similar notation will be employed for the magnetic intensity.

The images of the sources produce an incident wave

$$E^{\mathrm{i}}_{\mathrm{p}}(-z)\mathbf{k}-\mathbf{E}^{\mathrm{i}}_{\mathrm{t}}(-z), \qquad \mathbf{H}^{\mathrm{i}}_{\mathrm{t}}(-z)-H^{\mathrm{i}}_{\mathrm{p}}(-z)\mathbf{k}.$$

As before the combination of images and sources is such that the screen could be removed and so

$$\mathbf{E}(z)+E_{\mathrm{p}}(-z)\mathbf{k}-\mathbf{E}_{\mathrm{t}}(-z)=\mathbf{E}^{\mathrm{i}}(z)+E^{\mathrm{i}}_{\mathrm{p}}(-z)\mathbf{k}-\mathbf{E}^{\mathrm{i}}_{\mathrm{t}}(-z), \tag{3.6}$$
$$\mathbf{H}(z)+\mathbf{H}_{\mathrm{t}}(-z)-H_{\mathrm{p}}(-z)\mathbf{k}=\mathbf{H}^{\mathrm{i}}(z)+\mathbf{H}^{\mathrm{i}}_{\mathrm{t}}(-z)-H^{\mathrm{i}}_{\mathrm{p}}(-z)\mathbf{k}. \tag{3.7}$$

Across A, $\mathbf{E}$ and $\mathbf{H}$ are continuous, i.e. $\mathbf{E}(+0)=\mathbf{E}(-0)$, $\mathbf{H}(+0)=\mathbf{H}(-0)$. Therefore, letting $z\to 0$ in (3.6) and (3.7), we have

$$E_{\mathrm{p}}(0)=\mathrm{E}^{\mathrm{i}}_{\mathrm{p}}(0), \qquad \mathbf{H}_{\mathrm{t}}(0)=\mathbf{H}^{\mathrm{i}}_{\mathrm{t}}(0) \quad (\text{in } A) \tag{3.8}$$

The normal component of the electric intensity and the tangential components of the magnetic intensity in the aperture of a perfectly conducting plane screen are precisely the same as these components of the incident wave.

On S, the tangential components of $\mathbf{E}$ and the normal component of $\mathbf{H}$ vanish but other components are not continuous. therefore

$$E_{\mathrm{p}}(+0)+E_{\mathrm{p}}(-0)=2E^{\mathrm{i}}_{\mathrm{p}}(0) \quad (\text{on } S), \tag{3.9}$$
$$\mathbf{H}_{\mathrm{t}}(+0)+\mathbf{H}_{\mathrm{t}}(-0)=2\mathbf{H}^{\mathrm{i}}_{\mathrm{t}}(0) \quad (\text{on } S). \tag{3.10}$$

Let the constants of the medium be ε, μ. Then the electromagnetic field $\tilde{\mathbf{E}}$, $\tilde{\mathbf{H}}$ can be fashioned by

$$\tilde{\mathbf{E}}(z)=\begin{cases}(\mu/\varepsilon)^{\frac{1}{2}}\{\mathbf{H}(z)-\mathbf{H}^{\mathrm{i}}_{\mathrm{t}}(-z)+H^{\mathrm{i}}_{\mathrm{p}}(-z)\mathbf{k}\} & (z<0)\\ (\mu/\varepsilon)^{\frac{1}{2}}\{\mathbf{H}^{\mathrm{i}}(z)-\mathbf{H}(z)\} & (z>0)\end{cases}, \tag{3.11}$$

$$\tilde{\mathbf{H}}(z)=\begin{cases}-(\varepsilon/\mu)^{\frac{1}{2}}\{\mathbf{E}(z)-E^{\mathrm{i}}_{\mathrm{p}}(-z)\mathbf{k}+\mathbf{E}^{\mathrm{i}}_{\mathrm{t}}(-z)\} & (z<0)\\ -(\varepsilon/\mu)^{\frac{1}{2}}\{\mathbf{E}^{\mathrm{i}}(z)-\mathbf{E}(z)\} & (z>0)\end{cases}. \tag{3.12}$$

Let $z\to +0$ and note (3.8); then

$$\begin{gathered}\tilde{\mathbf{E}}_{\mathrm{t}}(+0)=\mathbf{0}, \qquad \tilde{E}_{\mathrm{p}}(+0)=(\mu/\varepsilon)^{\frac{1}{2}}\{H^{\mathrm{i}}_{\mathrm{p}}(0)-H_{\mathrm{p}}(+0)\},\\ \tilde{\mathbf{H}}_{\mathrm{t}}(+0)=-(\varepsilon/\mu)^{\frac{1}{2}}\{\mathbf{E}^{\mathrm{i}}_{\mathrm{t}}(0)-\mathbf{E}_{\mathrm{t}}(+0)\}, \qquad \tilde{H}_{\mathrm{p}}(+0)=0 \quad (\text{in } A),\end{gathered} \tag{3.13}$$

$$\begin{gathered}\tilde{\mathbf{E}}_{\mathrm{t}}(+0)=(\mu/\varepsilon)^{\frac{1}{2}}\{\mathbf{H}^{\mathrm{i}}_{\mathrm{t}}(0)-\mathbf{H}_{\mathrm{t}}(+0)\}, \qquad \tilde{E}_{\mathrm{p}}(+0)=(\mu/\varepsilon)^{\frac{1}{2}}H^{\mathrm{i}}_{\mathrm{p}}(0),\\ \tilde{\mathbf{H}}_{\mathrm{t}}(+0)=-(\varepsilon/\mu)^{\frac{1}{2}}\mathbf{E}^{\mathrm{i}}_{\mathrm{t}}(0), \qquad \tilde{H}_{\mathrm{p}}(+0)=-(\varepsilon/\mu)^{\frac{1}{2}}\{E^{\mathrm{i}}_{\mathrm{p}}(0)-E_{\mathrm{p}}(+0)\}, \quad (\text{on } S).\end{gathered} \tag{3.14}$$

Likewise, from $z \to -0$,

$$\tilde{\boldsymbol{E}}_t(-0) = \mathbf{0}, \qquad \tilde{E}_p = (\mu/\varepsilon)^{\frac{1}{2}}\{H_p(-0) + H_p^i(0)\},$$
$$\tilde{\boldsymbol{H}}_t(-0) = -(\varepsilon/\mu)^{\frac{1}{2}}\{\boldsymbol{E}_t(-0) + \boldsymbol{E}_t^i(0)\}, \qquad \tilde{H}_p(-0) = 0 \quad (\text{in } A), \tag{3.15}$$
$$\tilde{\boldsymbol{E}}_t(-0) = (\mu/\varepsilon)^{\frac{1}{2}}\{\boldsymbol{H}_t(-0) - \boldsymbol{H}_t^i(0)\}, \qquad \tilde{E}_p(-0) = (\mu/\varepsilon)^{\frac{1}{2}}H_p^i(0),$$
$$\tilde{\boldsymbol{H}}_t(-0) = -(\varepsilon/\mu)^{\frac{1}{2}}\boldsymbol{E}_t^i(0), \qquad \tilde{H}_p(-0) = -(\varepsilon/\mu)^{\frac{1}{2}}\{E_p(-0) - E_p^i(0)\} \quad (\text{on } S). \tag{3.16}$$

From (3.14), (3.16), (3.9), (3.10) it follows that $\tilde{\boldsymbol{E}}$ and $\tilde{\boldsymbol{H}}$ are continuous across S. In contrast, (3.14) and (3.16) show that $\tilde{\boldsymbol{E}}$ and $\tilde{\boldsymbol{H}}$ are discontinuous across A but $\tilde{\boldsymbol{E}}_t = \mathbf{0}$ in A. Therefore, $\tilde{\boldsymbol{E}}$, $\tilde{\boldsymbol{H}}$ is an electromagnetic field which satisfies the boundary conditions for the complementary screen. Also the charge density on the complementary screen is proportional to the discontinuity in $\tilde{E}_p$ which, from (3.15) and (3.13), is proportional to $H_p(+0) + H_p(-0)$. It is therefore integrable near an edge because H_p is. The current density is proportional to the discontinuity in $\tilde{\boldsymbol{H}}_t$ and is integrable because $\boldsymbol{E}_t$ is. Therefore, the edge conditions on the complementary screen are satisfied by the constructed field.

The field $\tilde{\boldsymbol{E}}$, $\tilde{\boldsymbol{H}}$ has no sources in $z > 0$ and, in $z < 0$, its sources are those of $(\mu/\varepsilon)^{\frac{1}{2}}\boldsymbol{H}^i(z)$, $-(\varepsilon/\mu)^{\frac{1}{2}}\boldsymbol{E}^i(z)$. Therefore $\tilde{\boldsymbol{E}}$, $\tilde{\boldsymbol{H}}$ is the solution to the problem of the diffraction of the incident wave $(\mu/\varepsilon)^{\frac{1}{2}}\boldsymbol{H}^i(z)$, $-(\varepsilon/\mu)^{\frac{1}{2}}\boldsymbol{E}^i(z)$ by the complementary screen. Thus we can affirm the following result.

BABINET'S ELECTROMAGNETIC PRINCIPLE *Let* $\boldsymbol{E}$, $\boldsymbol{H}$ *be the resultant field in* $z > 0$ *due to* $\boldsymbol{E}^i$, $\boldsymbol{H}^i$ *incident from* $z < 0$ *and let* $\tilde{\boldsymbol{E}}$, $\tilde{\boldsymbol{H}}$ *be the total field when the incident wave* $(\mu/\varepsilon)^{\frac{1}{2}}\boldsymbol{H}^i$, $-(\varepsilon/\mu)^{\frac{1}{2}}\boldsymbol{E}^i$ *impinges on the complementary screen from* $z < 0$. *Then, in* $z > 0$,

$$\boldsymbol{E} - (\mu/\varepsilon)^{\frac{1}{2}}\tilde{\boldsymbol{H}} = \boldsymbol{E}^i,$$
$$\boldsymbol{H} + (\varepsilon/\mu)^{\frac{1}{2}}\tilde{\boldsymbol{E}} = \boldsymbol{H}^i.$$

When S is finite in size, $\boldsymbol{E} = \boldsymbol{E}^i + \boldsymbol{E}^s$, $\boldsymbol{H} = \boldsymbol{H}^i + \boldsymbol{H}^s$ where $\boldsymbol{E}^s$, $\boldsymbol{H}^s$ is an outgoing wave at infinity. In the complementary problem there will be a diffracted field $\tilde{\boldsymbol{E}}^d$, $\tilde{\boldsymbol{H}}^d$ together with a reflected wave, i.e.

$$\left.\begin{aligned}\tilde{\boldsymbol{E}} &= (\mu/\varepsilon)^{\frac{1}{2}}\{\boldsymbol{H}^i(z) - \boldsymbol{H}_t^i(-z) + H_p^i(-z)\boldsymbol{k}\} + \tilde{\boldsymbol{E}}^d \\ \tilde{\boldsymbol{H}} &= -(\varepsilon/\mu)^{\frac{1}{2}}\{\boldsymbol{E}^i(z) - E_p^i(-z)\boldsymbol{k} + \boldsymbol{E}_t^i(-z)\} + \tilde{\boldsymbol{H}}^d\end{aligned}\right\} (z < 0),$$
$$\tilde{\boldsymbol{E}} = \tilde{\boldsymbol{E}}^d, \qquad \tilde{\boldsymbol{H}} = \tilde{\boldsymbol{H}}^d \quad (z > 0).$$

Hence (3.11) and (3.12) provide

$$\boldsymbol{E}^s = (\mu/\varepsilon)^{\frac{1}{2}}\tilde{\boldsymbol{H}}^d, \qquad \boldsymbol{H}^s = -(\varepsilon/\mu)^{\frac{1}{2}}\tilde{\boldsymbol{E}}^d \quad (z > 0), \tag{3.17}$$
$$\boldsymbol{E}^s = -(\mu/\varepsilon)^{\frac{1}{2}}\tilde{\boldsymbol{H}}^d, \qquad \boldsymbol{H}^s = (\varepsilon/\mu)^{\frac{1}{2}}\tilde{\boldsymbol{E}}^d \quad (z < 0). \tag{3.18}$$

It is transparent from (3.6) that $\boldsymbol{E}_t^s$ is an even function of z whereas E_p^s

is an odd function. It follows that $\tilde{\boldsymbol{H}}_t^d$ is odd and $\tilde{H}_p^d$ even in z. Similarly, H_p^s and $\boldsymbol{E}_t^d$ are even functions of z while $\boldsymbol{H}_t^s$ and E_p^d are odd functions.

Observe that, if $\boldsymbol{E}^i$, $\boldsymbol{H}^i$ is a plane wave, the transformation to $(\mu/\varepsilon)^{\frac{1}{2}}\boldsymbol{H}^i$, $-(\varepsilon/\mu)^{\frac{1}{2}}\boldsymbol{E}^i$ is equivalent to rotating the plane of polarization through a right angle clockwise looking in the direction of propagation.

9.4 The scattering and transmission coefficients

The presence of an edge in no way affects the derivation of the scattering coefficient in §8.19 so that the formula is immediately applicable. Suppose that S is finite in size and that, when $p^i = \exp(-ik\boldsymbol{n}^i \cdot \boldsymbol{x})$,

$$p^s \sim A(\hat{\boldsymbol{x}})e^{-ikR}/R.$$

Then, for the scattering coefficient,

$$\sigma = (4\pi/kS_0)\text{Im } A^*(\boldsymbol{n}^i) \tag{4.1}$$

where S_0 is the area of the projection of S on a plane perpendicular to the incident direction of propagation.

The scattering coefficient is not a suitable quantity for dealing with a finite aperture in a plane screen. What is now of interest is the amount of energy passing through the aperture. To measure this define the *transmission coefficient* τ by

$$\tau = \frac{\text{energy transmitted through the aperture}}{\text{energy incident on the aperture}} \tag{4.2}$$

In the complementary problem the transmitted energy is proportional to $\text{Im} \int_{\Omega'} \tilde{p}^d(\partial \tilde{p}^{d*}/\partial n)\, d\Omega$ where Ω' is a large hemisphere in $z > 0$. By (3.5) this is the same as $\text{Im} \int_{\Omega'} p^s(\partial p^{s*}/\partial n)\, d\Omega$. Since p^s is an odd function of z, this integral is one half of the integral over the full sphere. Hence

$$\tau = \tfrac{1}{2}\sigma \tag{4.3}$$

where σ is the scattering coefficient of the complementary screen. An alternative statement is that, if the plane wave $\exp(-ik\boldsymbol{n}^i \cdot \boldsymbol{x})$ falls on a finite aperture in a plane screen and the distant transmitted field is $A(\hat{\boldsymbol{x}})e^{-ikR}/R$,

$$\tau = (2\pi/kS_0)\text{Im } A(\boldsymbol{n}^i). \tag{4.4}$$

The formula (4.3) also holds for two-dimensional fields but (4.4) has to be modified to the appropriate form.

The transmission coefficient for the electromagnetic field is defined by (4.2) and, in view of (3.17) and (3.18), (4.3) is also valid. Allowing for the change of incident field in the complementary problem we see that, if a plane wave in which $\boldsymbol{E}^i = \boldsymbol{e}^i \exp(-ik\boldsymbol{n}^i \cdot \boldsymbol{x})$ strikes a finite aperture in a plane screen and in the distant transmitted field $\mathbf{E} \sim \mathbf{A}(\hat{\boldsymbol{x}})e^{-ikR}/R$,

$$\tau = (2\pi/kS_0)\mathrm{Im}\,\mathbf{A}(\boldsymbol{n}^{\mathrm{i}}) \cdot \boldsymbol{e}^{\mathrm{i}}. \tag{4.5}$$

Transform techniques

9.5 Generalities

The *Laplace transform* $F(s)$ of a complex function f of the real variable t is defined by

$$F(s) = \int_0^\infty f(t)e^{-st}\,dt$$

where s is the complex variable $\sigma + i\tau$. The general properties of such transforms are available in standard texts and only those which are relevant to our purpose will be summarized here.

A function of s is said to be *analytic* at a point when it is *single-valued* (i.e. has a uniquely specified value) and differentiable at that point. A function which is analytic at every point of a domain D is said to be *regular in D*. With this understanding it can be shown that, *if $F(s)$ exists for $s = s_0$, $F(s)$ is a regular function of s in $\sigma > \sigma_0$.*

In most applications the existence of F is dictated by the behaviour of f at infinity. Inspection of the form of f at infinity will therefore reveal the nature of the *singularities* of F, i.e. the points where F ceases to be analytic. For example, if $f(t) \sim At^\beta e^{s_0 t}$ as $t \to \infty$ where A is a complex constant and $\beta > -1$, F differs from a bounded quantity by $\int_M^\infty At^\beta e^{-(s-s_0)t}\,dt$ with $M \gg 1$, assuming that f is integrable over a finite interval. In fact, M may be replaced by zero since the difference is a bounded quantity and hence, as $s \to s_0$ with $|\mathrm{ph}(s - s_0)| < \frac{1}{2}\pi$,

$$F(s) \approx \beta! A(s - s_0)^{-1-\beta} + \text{regular function}. \tag{5.1}$$

Systematic integration by parts enables one to cope with $\mathrm{Re}\,\beta \leqslant -1$ and (5.1) is unaltered if β is not an integer. However, if $\beta = -m$ where m is a positive integer

$$F(s) \approx \{(m-1)!\}^{-1} A(-)^m (s - s_0)^{m-1} \ln(s - s_0) \tag{5.2}$$

as $s \to s_0$. Similarly, when $f(t) \sim At^\beta e^{s_0 t} \ln t$ as $t \to \infty$,

$$F(s) \approx \beta! A(s - s_0)^{-1-\beta}\{\psi(\beta) - \ln(s - s_0)\} \tag{5.3}$$

where $\psi(z)$, defined in Exercise 53 of Chapter 8, is the logarithmic derivative of the factorial function $z!$.

Broadly speaking, converse statements hold, namely that, if $F(s)$ behaves near a singularity in one of the ways (5.1)–(5.3), $f(t)$ conducts itself correspondingly at infinity.

There are also connections between the behaviour of f near the origin and the performance of F at infinity. For instance, if $f(t) \approx At^\beta$ as $t \to 0$ then

$$F(s) \sim \beta! A/s^{1+\beta} \tag{5.4}$$

as $|s| \to \infty$ with $|\text{ph } s| < \frac{1}{2}\pi$; if $f(t) \approx At^\beta \ln t$, $F(s) \sim \beta! A\{\psi(\beta) - \ln s\}/s^{1+\beta}$. In particular, if $f(t) = 0$ for $0 \leqslant t \leqslant a$ then $F(s) = O(e^{-a\sigma})$.

To circumvent singularities the procedure of *analytic continuation* is employed. As an illustration consider the series

$$F(s) = 1 - s + s^2 - s^3 + \ldots$$

which converges only for $|s| < 1$. Nevertheless, inside this circle, $F(s) = (1+s)^{-1}$ and so the function, originally specified for $|s| < 1$, may be attributed a meaning everywhere except $s = -1$ by identifying it with $(1+s)^{-1}$; this is the process of analytic continuation. It permits the by-passing of the singularity at $s = -1$. In general, the description that $F(s)$ is an analytic function will be taken to include its analytic continuations.

If β is a positive integer in (5.1) the singularity at s_0 is called a *pole of order* $1 + \beta$ and if $\beta = 0$ a *simple pole.* In the neighbourhood of a pole F is unique and all the values obtained by analytic continuation are the same. However, if β is not an integer or the form (5.2) is relevant different values can be obtained by analytic continuation and F is said to be *multiple-valued.* The point s_0 is then termed to be a *branch point.* F can be rendered single-valued by the introduction of certain curves or lines in the complex plane which it is not permissible to cross. Such curves are known as *branch lines* or *cuts.* A good deal of freedom is available in choosing what curves are branch lines and it is not uncommon to make them different in different contexts. Whatever the resulting values they constitute a *branch* of the multiple-valued function.

Diffraction theory often encounters branch points of the type $(s - s_0)^{\frac{1}{2}}$ so that it is worth while considering them in more detail. Take

$$(s - s_0)^{\frac{1}{2}} = |s - s_0|^{\frac{1}{2}} \exp\{\tfrac{1}{2}\text{i ph}(s - s_0)\}$$

where the square root is positive. There are two possible choices for $\text{ph}(s - s_0)$ with the branch line drawn as shown dashed in Fig. 9.4. These are $\delta - 2\pi < \text{ph}(s - s_0) \leqslant \delta$ and $\delta < \text{ph}(s - s_0) \leqslant \delta + 2\pi$. Any other selection of an interval for $\text{ph}(s - s_0)$ merely repeats the values of $(s - s_0)^{\frac{1}{2}}$ in one of these two. Each of the intervals corresponds to a different branch of the function, the two branches ascribing opposite signs to $(s - s_0)^{\frac{1}{2}}$. Crossing the branch line causes one to transfer from one branch to the other.

In the first interval $\text{ph}(s - s_0) = -\pi + \phi_0$ when $s = 0$ and

$$(s - s_0)^{\frac{1}{2}} = |s_0|^{\frac{1}{2}} \exp\{-\tfrac{1}{2}\text{i}(\pi - \phi_0)\} = -\text{i}s_0^{\frac{1}{2}}$$

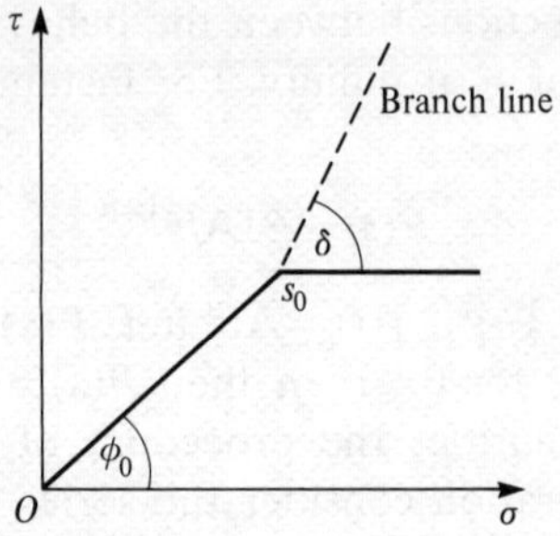

Fig. 9.4 A branch line for $(s-s_0)^{\frac{1}{2}}$

where $s_0^{\frac{1}{2}}$ is defined to have its *principal value*, i.e. $-\pi < \text{ph}\, s_0 \leqslant \pi$ with the branch line of $s_0^{\frac{1}{2}}$ along the negative real axis. If s_0 is real and positive this definition makes $s_0^{\frac{1}{2}}$ positive, fitting in with common usage. When $|s| \to \infty$ with $s = |s|\,e^{i\phi}$

$$(s-s_0)^{\frac{1}{2}} \sim |s|^{\frac{1}{2}} e^{\frac{1}{2}i\phi}$$

for $\delta - 2\pi < \phi \leqslant \delta$. This may also be written as $s^{\frac{1}{2}}$ if, in addition, $-\pi < \phi \leqslant \pi$ but, if $-3\pi < \phi < -\pi$ or $\pi < \phi < 3\pi$, must be changed to $-s^{\frac{1}{2}}$ because of the transition across the branch line of $s^{\frac{1}{2}}$.

If $(s-s_0)^{\frac{1}{2}}$ is specified by the second branch $(\delta < \text{ph}(s-s_0) \leqslant \delta + 2\pi)$ then, when $s = 0$,

$$(s-s_0)^{\frac{1}{2}} = i s_0^{\frac{1}{2}}$$

and, as $|s| \to \infty$, $(s-s_0)^{\frac{1}{2}} \sim -|s|^{\frac{1}{2}}\, e^{\frac{1}{2}i\phi}$ when $\delta - 2\pi < \phi \leqslant \delta$.

The two branches of $(s+s_0)^{\frac{1}{2}}$ may be dealt with in a similar way. If the branch line is selected as in Fig. 9.5 the phase of $s+s_0$ may be taken in either of the intervals $(-3\pi+\delta, -\pi+\delta)$ and $(-\pi+\delta, \pi+\delta)$. If $-\pi+\delta < \text{ph}(s+s_0) \leqslant \pi+\delta$ on the chosen branch $(s+s_0)^{\frac{1}{2}} = s_0^{\frac{1}{2}}$ when $s = 0$.

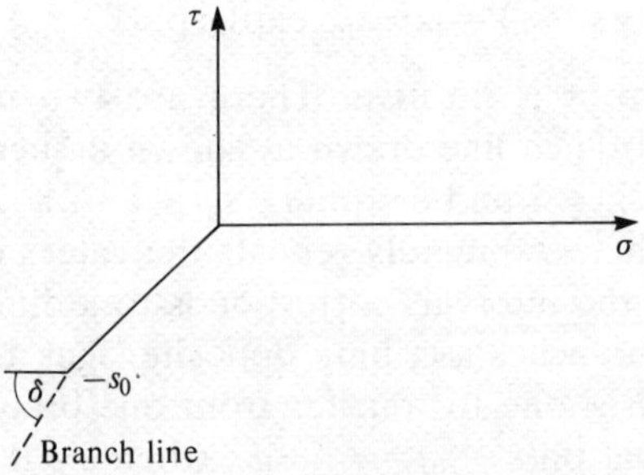

Fig. 9.5 A branch line for $(s+s_0)^{\frac{1}{2}}$

A function such as $(s^2-s_0^2)^{\frac{1}{2}}$ can be handled by treating it as the product of $(s-s_0)^{\frac{1}{2}}$ and $(s+s_0)^{\frac{1}{2}}$. With the branch lines of Figs. 9.4 and 9.5 choose $\delta-2\pi<\mathrm{ph}(s-s_0)\leqslant\pi$ and $-\pi+\delta<\mathrm{ph}(s+s_0)\leqslant\pi+\delta$. Then

$$(s^2-s_0^2)^{\frac{1}{2}}=-\mathrm{i}s_0$$

when $s=0$ and, as $|s|\to\infty$,

$$(s^2-s_0^2)^{\frac{1}{2}}\sim|s|\,\mathrm{e}^{\mathrm{i}\phi}$$

provided that $\delta-2\pi<\phi\leqslant\delta$ and $-\pi+\delta<\phi\leqslant\pi+\delta$. When ϕ has a value which is not common to these two ranges the asymptotic behaviour can be calculated from the rules already given.

A function f can be recovered from its Laplace transform by means of the *Mellin inversion formula*

$$\frac{1}{2\pi\mathrm{i}}\int_{c-\mathrm{i}\infty}^{c+\mathrm{i}\infty}F(s)\mathrm{e}^{st}\,\mathrm{d}s=f(t)\mathrm{H}(t) \tag{5.5}$$

where $\mathrm{H}(t)$ is the step function which is 1 for $t>0$ and 0 for $t<0$, and c is a real constant such that all the singularities of $F(s)$ lie in $\sigma<c$.

The Laplace transform deals exclusively with functions on the positive real axis. A transform on the negative real axis can be defined by $\int_{-\infty}^{0}f(t)\mathrm{e}^{-st}\,\mathrm{d}t$. However, if t is changed to $-t$, the integral converts to a Laplace transform from the positive real axis but with s replaced by $-s$. With this alteration the properties can be deduced from those of the standard Laplace transforms. In particular, a transform from the negative real axis will be regular for $\sigma<\sigma_1$, σ_1 being some appropriate constant.

In some cases it will be pertinent to employ both transforms. Let

$$F_{\mathrm{R}}(s)=\int_0^\infty f(t)\mathrm{e}^{-st}\,\mathrm{d}t \tag{5.6}$$

the suffix R indicating that F_{R} is regular in some right-half plane, say $\sigma>\sigma_{\mathrm{R}}$. Similarly, let

$$F_{\mathrm{L}}(s)=\int_{-\infty}^{0} f(t)\mathrm{e}^{-st}\,\mathrm{d}t, \tag{5.7}$$

F_{L} being regular in the left half plane $\sigma<\sigma_{\mathrm{L}}$. The asymptotic behaviour of F_{R} (F_{L}) is determined by $f(t)$ as $t\to+0$ (-0); the singularities of F_{R} (F_{L}) stem from $f(t)$ as $t\to+\infty$ $(-\infty)$.

Choose $c_{\mathrm{R}}>\sigma_{\mathrm{R}}$ and $c_{\mathrm{L}}<\sigma_{\mathrm{L}}$. Then (5.5) tells us that

$$\frac{1}{2\pi\mathrm{i}}\int_{c_{\mathrm{R}}-\mathrm{i}\infty}^{c_{\mathrm{R}}+\mathrm{i}\infty}F_{\mathrm{R}}(s)\mathrm{e}^{st}\,\mathrm{d}s+\frac{1}{2\pi\mathrm{i}}\int_{c_{\mathrm{L}}-\mathrm{i}\infty}^{c_{\mathrm{L}}+\mathrm{i}\infty}F_{\mathrm{L}}(s)\mathrm{e}^{st}\,\mathrm{d}s=f(t). \tag{5.8}$$

It may happen that $\sigma_{\mathrm{L}}>\sigma_{\mathrm{R}}$. In such an eventuality both F_{R} and F_{L} are well-defined in the common strip $\sigma_{\mathrm{R}}<\sigma<\sigma_{\mathrm{L}}$. Then F_{R} and F_{L} may be combined by addition in the common strip and, from (5.6) and (5.7),

$$F(s)=F_R(s)+F_L(s)=\int_{-\infty}^{\infty} f(t)e^{-st}\,dt$$

for $\sigma_R<\sigma<\sigma_L$. $F(s)$ is known as the *bilateral Laplace transform* of $f(t)$. Furthermore, it is now feasible to select $c_R=c_L=c$ so that (5.8) goes over to

$$\frac{1}{2\pi i}\int_{c-i\infty}^{c+i\infty} F(s)e^{st}\,ds=f(t)$$

for $\sigma_R<c<\sigma_L$.

Related to the bilateral Laplace transform is the *Fourier transform* $\mathscr{F}(\alpha)$ of $f(t)$. Its complex form is obtained by substitution $s=i\alpha$ but it is frequently quoted with α real as

$$\mathscr{F}(\alpha)=\int_{-\infty}^{\infty} f(t)e^{-i\alpha t}\,dt, \qquad f(t)=\frac{1}{2\pi}\int_{-\infty}^{\infty} \mathscr{F}(\alpha)e^{i\alpha t}\,d\alpha$$

though the contour in the inversion integral may need proper indentation if $\mathscr{F}$ has singularities on the real α-axis.

9.6 The line and point source

This section is concerned with the application of transforms to the fields of sources. Suppose firstly that the field is independent of z and that there is a line source at (x_0, y_0). Then the partial differential equation to be solved is

$$\frac{\partial^2 p}{\partial x^2}+\frac{\partial^2 p}{\partial y^2}+k^2p=-\delta(x-x_0)\delta(y-y_0). \tag{6.1}$$

The solution sought is, in fact, already known to be $-\frac{1}{4}iH_0^{(2)}(k\,|\boldsymbol{x}-\boldsymbol{x}_0|)$.

Analytical convenience is acquired by assuming that the medium has a slight amount of dissipation so that $k=k_r-ik_i$ where both k_r and k_i are positive. At the end of the analysis it is the intention to put $k_i=0$ unless the medium actually is dissipative. Strictly, it is then necessary to verify that the expressions so obtained do, indeed, constitute a solution when k is real but the laborious detail of the verification will be omitted here.

The radiation condition demands that p should behave like $e^{-ikr}/r^{\frac{1}{2}}$ as $r=|\boldsymbol{x}|\to\infty$. The field at infinity is therefore exponentially damped by the dissipation by the factor $e^{-k_i r}$. Accordingly, the bilateral Laplace transform $P(s, y)$ can be defined by

$$P(s, y)=\int_{-\infty}^{\infty} p(x, y)e^{-sx}\,dx \tag{6.2}$$

for $-k_i<\sigma<k_i$.

Multiplication of (6.1) by e^{-sx} and integration with respect to x from $-\infty$ to ∞ gives

$$\frac{d^2P}{dy^2}+\kappa^2P=-\delta(y-y_0)e^{-sx_0} \tag{6.3}$$

where $\kappa^2=s^2+k^2$.

The function κ has branch points at $\pm ik$. Pick branch lines as in Figs. 9.4 and 9.5 but with $\delta=\phi_0$, i.e. the branch lines are continuations of the line joining $-ik$ to ik. Select that branch which reduces to k when $s=0$ (this is, in fact, the same as that made for $(s^2-s_0^2)^{\frac{1}{2}}$ in §9.5 with $s_0=ik$). It is then easy to check that κ has a negative imaginary part throughout the strip $-k_i<\sigma<k_i$.

The function p tends to zero exponentially as $|y|\to\infty$. Consequently, P approaches zero as $|y|\to\infty$. On account of the branch chosen for κ the appropriate solution of (6.3) is

$$P(s,y)=\begin{cases}Ae^{-i\kappa(y-y_0)} & (y>y_0)\\ Be^{i\kappa(y-y_0)} & (y<y_0)\end{cases}$$

where A and B depend only on s. Integration of (6.3) with respect to y just less than y_0 to just greater than y_0 supplies

$$[dP/dy]_{y_0-0}^{y_0+0}=-e^{-sx_0}$$

and P is continuous at $y=y_0$. These two conditions force $A=B=e^{-sx_0}/2i\kappa$ and

$$P(s,y)=(1/2i\kappa)e^{-sx_0-i\kappa|y-y_0|}. \tag{6.4}$$

In view of the known solution to our problem we conclude that

$$\int_{-\infty}^{\infty} H_0^{(2)}[k\{(x-x_0)^2+(y-y_0)^2\}^{\frac{1}{2}}]e^{-sx}\,dx=(2/\kappa)\exp(-sx_0-i\kappa\,|y-y_0|).$$

By virtue of the inversion formula in §9.5

$$\frac{1}{\pi i}\int_{c-i\infty}^{c+i\infty}(1/\kappa)\exp\{s(x-x_0)-i\kappa\,|y-y_0|\}\,ds=H_0^{(2)}[k\{(x-x_0)^2+(y-y_0)^2\}^{\frac{1}{2}}] \tag{6.6}$$

where $-k_i<c<k_i$. The formula (6.6) remains valid when $k_i=0$ and $c=0$ provided that the contour of integration is taken as in Fig. 9.6.

There are alternative versions of (6.6) which are useful in various contexts. Suppose that x_0 and y_0 are both zero. Then (6.6) becomes

$$\frac{1}{\pi i}\int_{c-i\infty}^{c+i\infty}(1/\kappa)\exp(sx-i\kappa\,|y|)\,ds=H_0^{(2)}\{k(x^2+y^2)^{\frac{1}{2}}\}. \tag{6.7}$$

Another form for (6.7) is obtained by putting $x=r\cos\phi$, $y=r\sin\phi$

Fig. 9.6 Contour for the inversion integral

$(0<\phi<\pi)$ and deforming the contour of integration into the path $s = -ik\cos(\phi+it)$ where t is real and runs from $-\infty$ to ∞. The new path is a hyperbola, shown in Fig. 9.7 when k is real, and on it $\kappa = k\sin(\phi+it)$. Now (6.7) implies that

$$\int_{-\infty}^{\infty} \exp(-ikr\cosh t)\,dt = -\pi i H_0^{(2)}(kr). \tag{6.8}$$

In the more general case let $x = r\cos\phi$, $y = r\sin\phi$, $x_0 = r_0\cos\phi_0$, $y_0 = r_0\sin\phi_0$ and suppose that $y<y_0$. Make the substitution $s = ik\cos\theta$ in (6.6). The path of Fig. 9.6 is traced in the s-plane if θ follows the contour C from $-i\infty$ to 0 to π to $\pi+i\infty$ in the complex θ-plane (Fig. 9.8).

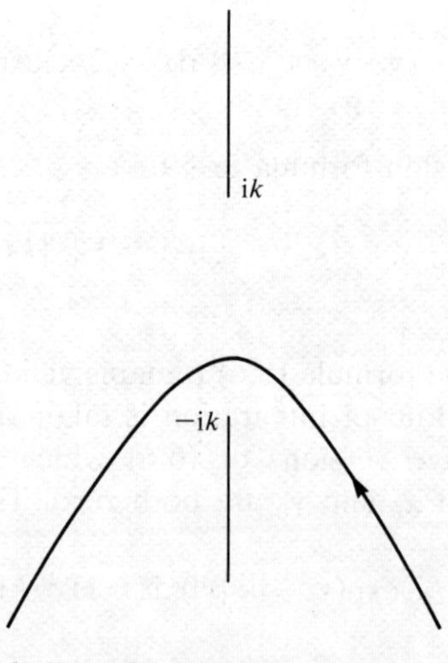

Fig. 9.7 Hyperbolic path after the mapping $s = -ik\cos(\phi+it)$

With this choice of s, $(s^2+k^2)^{\frac{1}{2}} = k \sin\theta$ and

$$\mathrm{H}_0^{(2)}(kr_s) = \frac{1}{\pi}\int_C \exp\{ikr\cos(\phi-\theta) - ikr_0\cos(\phi_0-\theta)\}\,\mathrm{d}\theta \quad (y<y_0) \tag{6.9}$$

where

$$r_s^2 = r^2 + r_0^2 - 2rr_0\cos(\phi-\phi_0). \tag{6.10}$$

The representation of (6.9) expresses the field of a line source as a *spectrum of plane waves*. The separate plane waves corresponding to the portion of the real axis in C are homogeneous. They radiate towards $y<0$, their directions of propagations making angles between 0 and $-\pi$ with the positive x-axis, all possible directions in this range being covered. The individual plane waves on the parts of C on which $\theta=-it$ and on which $\theta=\pi+it$ with $t\geqslant 0$ are inhomogeneous. For one set the direction of phase propagation is along the negative x-axis and the waves are exponentially attenuated as y decreases. In the other set the direction of phase propagation is parallel to the positive x-axis and the waves also decay exponentially with diminishing y. These evanescent waves furnish information about the field which is on a finer scale than a wavelength.

For a point source the solution of

$$\frac{\partial^2 p}{\partial x^2}+\frac{\partial^2 p}{\partial y^2}+\frac{\partial^2 p}{\partial z^2}+k^2 p = -\delta(x-x_0)\delta(y-y_0)\delta(z-z_0) \tag{6.11}$$

is desired. The known answer is $\mathrm{e}^{-ikR_s}/4\pi R_s$ where $R_s = |\boldsymbol{x}-\boldsymbol{x}_0|$. If now $P(x,y,s) = \int_{-\infty}^{\infty} p\mathrm{e}^{-sz}\,\mathrm{d}z$

$$\frac{\partial^2 P}{\partial x^2}+\frac{\partial^2 P}{\partial y^2}+\kappa^2 P = -\delta(x-x_0)\delta(y-y_0)\mathrm{e}^{-sz_0}.$$

This is a return to (6.1) but with κ in place of k. Hence

$$\frac{\mathrm{e}^{-ikR_s}}{4\pi R_s} = -\frac{1}{8\pi}\int_{c-i\infty}^{c+i\infty} \mathrm{H}_0^{(2)}(\kappa r_s)\mathrm{e}^{s(z-z_0)}\,\mathrm{d}s \tag{6.12}$$

with c as in (6.7). Since k and κ both have negative imaginary parts (6.8) may be quoted so that

$$\frac{\mathrm{e}^{-ikR_s}}{4\pi R_s} = \frac{1}{8\pi^2 i}\int_{-\infty}^{\infty}\int_{c-i\infty}^{c+i\infty} \mathrm{e}^{-i\kappa r_s\cosh t + s(z-z_0)}\,\mathrm{d}s\,\mathrm{d}t. \tag{6.13}$$

Now, a derivative of (6.7) with respect to y gives

$$\int_{c-i\infty}^{c+i\infty} \exp(sx - i\kappa y)\,\mathrm{d}s = \frac{\pi k y}{(x^2+y^2)^{\frac{1}{2}}}\mathrm{H}_1^{(2)}\{k(x^2+y^2)^{\frac{1}{2}}\}$$

when $y>0$. Taking advantage of this result with $y = r_s\cosh t$, $x = z - z_0$ in

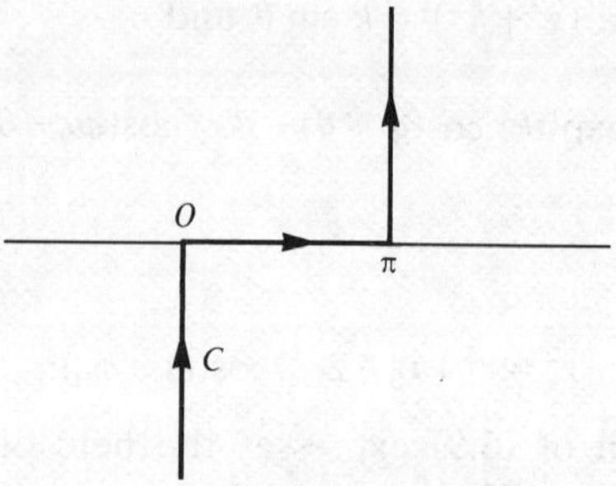

Fig. 9.8 The path of integration in the complex θ-plane

(6.13) we obtain

$$\frac{e^{-ikR_s}}{4\pi R_s} = -\frac{ik}{8\pi}\int_{-\infty}^{\infty} \frac{r_s \cosh t}{\{(z-z_0)^2 + r_s^2 \cosh^2 t\}^{\frac{1}{2}}} H_1^{(2)}[k\{(z-z_0)^2 + r_s^2 \cosh^2 t\}^{\frac{1}{2}}]\, dt$$

$$= -\frac{ik}{8\pi}\int_{-\infty}^{\infty} H_1^{(2)}(kR_s \cosh u)\, du \tag{6.14}$$

after the change of variable $r_s \sinh t = R_s \sinh u$.

A representation as a plane wave spectrum can be obtained via (6.9) and (6.12). It is

$$\frac{e^{-ikR_s}}{4\pi R_s} = \frac{ik}{8\pi^2}\int_C \int_C \sin\psi \exp[ik(z-z_0)\cos\psi + ik\sin\psi\{(x-x_0)\cos\theta + (y-y_0)\sin\theta\}]\, d\theta\, d\psi, \tag{6.15}$$

C being the contour of Fig. 9.8.

9.7 Two-dimensional diffraction of a plane wave by a semi-infinite plane

The simplest diffraction problem which can be solved by transforms and the Wiener–Hopf technique is a two-dimensional plane wave falling on a semi-infinite plane. This is one of the few rigorous solutions available to serve as a bench-mark for checking approximate and numerical methods. Choose axes so that the edge of the plane is the z-axis and the screen occupies $y = 0$, $x \leqslant 0$. The boundary condition on the screen will be taken to be either sound-soft or sound-hard. As in Chapter 8 solutions for these boundary conditions also provide answers for the electromagnetic diffraction by a perfectly conducting screen when the electric vector or magnetic intensity respectively is parallel to the edge. Other boundary conditions can be treated by the method but will not be considered here.

Let the pressure in the incident plane wave be given by

$$p^{\mathrm{i}} = \exp\{-ikr\cos(\phi - \phi^{\mathrm{i}})\} \tag{7.1}$$

where ϕ^{i} is a real angle satisfying $0 < \phi^{\mathrm{i}} < \pi$ (Fig. 9.9). A scattered field p^{s} is produced such that p^{s} complies with (6.1) (with zero right-hand side), the radiation conditions, the boundary condition and $p^{\mathrm{i}} + p^{\mathrm{s}}$ satisfies the edge conditions. For the soft screen the boundary condition is

$$p^{\mathrm{s}} = -\exp(ikr\cos\phi^{\mathrm{i}}) \quad (\phi = \pm\pi). \tag{7.2}$$

A little care is necessary with the radiation conditions because the scatterer extends to infinity. To see what happens imagine that the screen covered the whole of $y = 0$. Then the solution for the soft boundary is

$$p^{\mathrm{s}} = -\exp\{-ikr\cos(\phi + \phi^{\mathrm{i}})\}.$$

Therefore, it must be expected that the semi-infinite plane will generate a reflected plane wave $-\mathrm{e}^{-ikr\cos(\phi+\phi^{\mathrm{i}})}$ in $-\pi < \phi < -\phi^{\mathrm{i}}$. Also the screen will blot out the incident wave in $\phi > \phi^{\mathrm{i}}$ and so must provide a plane wave $-\mathrm{e}^{-ikr\cos(\phi-\phi^{\mathrm{i}})}$ in $\phi^{\mathrm{i}} < \phi < \pi$. Neither of these plane waves created by the semi-infinite plane satisfies the customary radiation conditions. It is the parts of p^{s} after they have been separated off that behave like $\mathrm{e}^{-ikr}/r^{\frac{1}{2}}$ at infinity.

Define $P(s, y)$ by

$$P(s, y) = \int_{-\infty}^{\infty} p^{\mathrm{s}}(x, y)\mathrm{e}^{-sx}\,\mathrm{d}x$$

so that

$$\mathrm{d}^2P/\mathrm{d}y^2 + \kappa^2 P = 0. \tag{7.3}$$

From the above discussion $p^{\mathrm{s}} = O(\mathrm{e}^{-k_{\mathrm{i}}x}/x^{\frac{1}{2}})$ as $x \to \infty$ and $p^{\mathrm{s}} = O(\mathrm{e}^{k_{\mathrm{i}}|x|\cos\phi^{\mathrm{i}}})$

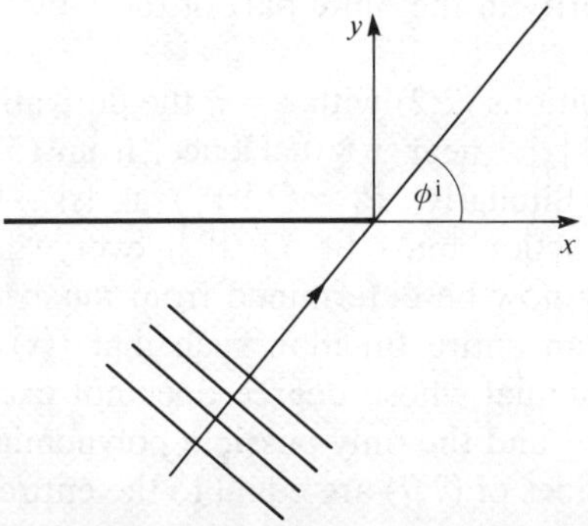

Fig. 9.9 Plane wave incident on a semi-infinite screen

as $x \to -\infty$. Hence P is regular in the strip $-k_i < \sigma < -k_i \cos \phi^i$ and so (7.3) holds in this strip. In order that P should correspond to an outgoing wave the appropriate solution of (7.3) is

$$P(s, y) = \begin{cases} A_1 e^{-i\kappa y} & (y>0) \\ A_2 e^{i\kappa y} & (y<0) \end{cases}.$$

Since p^s is continuous as y crosses $y=0$ so is P and hence $A_1 = A_2 = A$ (say). Therefore $P(s, y) = A e^{-i\kappa |y|}$.

Write $P(s, y) = P_R(s, y) + P_L(s, y)$ where

$$P_R(s, y) = \int_0^\infty p^s(x, y) e^{-sx}\, dx, \qquad P_L(s, y) = \int_{-\infty}^0 p^s(x, y) e^{-sx}\, dx.$$

Then

$$P_R(s, 0) + P_L(s, 0) = A, \tag{7.4}$$

$$P_R'(s, +0) + P_L'(s, +0) = -i\kappa A, \tag{7.5}$$

$$P_R'(s, -0) + P_L'(s, -0) = i\kappa A \tag{7.6}$$

where the prime indicates a derivative with respect to y.

Since $\partial p^s/\partial y$ is continuous in crossing $y=0$, $x>0$

$$P_R'(s, +0) = P_R'(s, -0)$$

and hence, by the addition of (7.5) and (7.6),

$$2P_R'(s, 0) = -P_L'(s, +0) - P_L'(s, -0). \tag{7.7}$$

An equation such as (7.7) is typical of the Wiener–Hopf technique. The left-hand side is regular in $\sigma > -k_i$. The right-hand side is equal to it in $-k_i < \sigma < -k_i \cos \phi^i$ and is regular in $\sigma < -k_i \cos \phi^i$. The right-hand side can be regarded as the analytic continuation of $2P_R'$ into the left half-plane. However, there is no singularity in $\sigma > -k_i$ nor in the analytic continuation in $\sigma < -k_i \cos \phi^i$. Consequently, the function represented can have no singularity in the finite part of the s-plane. It is therefore an *entire function.*

By the edge conditions (2.2) with $\delta = \pi$ the derivative $\partial p^s/\partial y$ cannot be more singular than $|x|^{-\frac{1}{2}}$ near $x=0$. Hence, from (5.4), $P_R' = O(|s|^{-\frac{1}{2}})$ as $|s| \to \infty$ in $\sigma > -k_i$. Similarly, $P_L' = O(|s|^{-\frac{1}{2}})$ as $|s| \to \infty$ in $\sigma < -k_i \cos \phi^i$. Thus the entire function must be $O(|s|^{-\frac{1}{2}})$ everywhere at infinity. The entire function may now be determined from *the extension of Liouville's theorem:* if $f(s)$ is an entire function such that $f(s) = O(|s|^\alpha)$ as $|s| \to \infty$ then $f(s)$ is a polynomial whose degree does not exceed α.

In our case $\alpha = -\frac{1}{2}$ and the only possible polynomial of lower degree is zero. Hence, both sides of (7.7) are equal to the entire function zero, i.e.

$$P_R'(s, 0) = 0, \qquad P_L'(s, +0) + P_L'(s, -0) = 0.$$

The first of these equations states that $\partial p_s/\partial y = 0$ on $y = 0$, $x > 0$, i.e. the normal derivative of the total field is the same as that of the incident wave in the aperture in harmony with the general prediction of Babinet's principle (§9.3).

Insertion of these results and the elimination of A from (7.4) and (7.5) leads to

$$\begin{aligned} P_L'(s, +0) &= -i\kappa\{P_R(s, 0) + P_L(s, 0\} \\ &= -i\kappa\{P_R(s, 0) + 1/(s + ik \cos \phi^i)\} \end{aligned} \tag{7.8}$$

on account of the boundary condition (7.2).

The equation (7.8) is typical of the Wiener–Hopf technique and the aim is to convert it to have the same structure as (7.7), i.e. a function regular in the right half-plane on one side and one regular in the left half-plane on the other side. Now $\kappa = (s + ik)^{\frac{1}{2}}(s - ik)^{\frac{1}{2}}$ and $(s + ik)^{\frac{1}{2}}$ is regular in $\sigma > -k_i$ whereas $(s - ik)^{\frac{1}{2}}$ is regular in $\sigma < k_i$. Division of (7.8) by $(s - ik)^{\frac{1}{2}}$ would then supply the structure of (7.7) except for the term $(s + ik)^{\frac{1}{2}}/(s + ik \cos \phi^i)$ which has singularities in both half-planes. These singularities can be separated, however, by writing

$$\frac{(s + ik)^{\frac{1}{2}}}{s + ik \cos \phi^i} = \frac{(s + ik)^{\frac{1}{2}} - (ik - ik \cos \phi^i)^{\frac{1}{2}}}{s + ik \cos \phi^i} + \frac{(ik - ik \cos \phi^i)^{\frac{1}{2}}}{s + ik \cos \phi^i}.$$

By means of this device the desired form of (7.8) can be achieved, namely

$$iP_L'(s, +0)(s - ik)^{-\frac{1}{2}} - \frac{(ik - ik \cos \phi^i)^{\frac{1}{2}}}{s + ik \cos \phi^i}$$
$$= (s + ik)^{\frac{1}{2}} P_R(s, 0) + \frac{(s + ik)^{\frac{1}{2}} - (ik - ik \cos \phi^i)^{\frac{1}{2}}}{s + ik \cos \phi^i}.$$

The argument following (7.7) can now be repeated and both sides are equal to an entire function. The right-hand side shows that the entire function must be bounded as $|s| \to \infty$ with $\sigma < -k_i \cos \phi^i$. From the extension of Liouville's theorem the entire function can only be a constant, which must be zero because of the behaviour of the left-hand side. Hence

$$P_L'(s, +0) = -i(ik - ik \cos \phi^i)^{\frac{1}{2}}(s - ik)^{\frac{1}{2}}/(s + ik \cos \phi^i).$$

From (7.5)

$$A = (ik - ik \cos \phi^i)^{\frac{1}{2}}/(s + ik)^{\frac{1}{2}}(s + ik \cos \phi^i)$$

and therefore

$$p^s = \frac{1}{2\pi i}\int_{c-i\infty}^{c+i\infty} \frac{(ik - ik \cos \phi^i)^{\frac{1}{2}}}{(s + ik)^{\frac{1}{2}}(s + ik \cos \phi^i)} e^{sx - i\kappa|y|}\, ds \tag{7.9}$$

with $-k_i < c < -k_i \cos \phi^i$.

Several other forms for the field can be deduced from (7.9). Let $x = r\cos\phi$, $y = r\sin\phi$ and, for the moment, impose the limitation that $0 < \phi < \pi$. Map the contour into the hyperbolic path $s = -ik\cos(\phi + it)$ of Fig. 9.7. In the deformation the pole at $s = -ik\cos\phi^{\mathrm{i}}$ will be captured if $\phi > \phi^{\mathrm{i}}$. Accordingly

$$p^{\mathrm{s}} = \frac{1}{\pi}\int_{-\infty}^{\infty} \frac{\sin\frac{1}{2}\phi^{\mathrm{i}}\cos\frac{1}{2}(\phi + it)}{\cos\phi^{\mathrm{i}} - \cos(\phi + it)}\,\mathrm{e}^{-ikr\cosh t}\,\mathrm{d}t - \mathrm{e}^{-ikr\cos(\phi - \phi^{\mathrm{i}})}\mathrm{H}(\phi - \phi^{\mathrm{i}})$$

where H is the Heaviside step function. The term containing the integral can be rewritten as

$$\frac{1}{4\pi}\int_{-\infty}^{\infty}\{\operatorname{cosec}\tfrac{1}{2}(\phi - \phi^{\mathrm{i}} + it) - \operatorname{cosec}\tfrac{1}{2}(\phi + \phi^{\mathrm{i}} + it)\}\mathrm{e}^{-ikr\cosh t}\,\mathrm{d}t.$$

Define $J(\zeta)$ by

$$J(\zeta) = \tfrac{1}{4}\int_{-\infty}^{\infty}\frac{\mathrm{e}^{-ikr\cosh t}}{\sin\frac{1}{2}(\zeta + it)}\,\mathrm{d}t = \int_{0}^{\infty}\frac{\sin\frac{1}{2}\zeta\cosh\frac{1}{2}t}{\cosh t - \cos\zeta}\,\mathrm{e}^{-ikr\cosh t}\,\mathrm{d}t. \qquad (7.10)$$

Then

$$\begin{aligned}\frac{\mathrm{d}}{\mathrm{d}r}\,\mathrm{e}^{ikr\cos\zeta}J(\zeta) &= -ik\sin\tfrac{1}{2}\zeta\int_{0}^{\infty}\cosh\tfrac{1}{2}t\,\mathrm{e}^{ikr(\cos\zeta - \cosh t)}\,\mathrm{d}t\\ &= (\pi k/2r)^{\frac{1}{2}}\sin\tfrac{1}{2}\zeta\exp(-\tfrac{3}{4}\pi\mathrm{i} - 2ikr\sin^{2}\tfrac{1}{2}\zeta).\end{aligned}$$

Integrate this formula to infinity, bearing in mind that $J \to 0$ as $r \to \infty$. There results

$$J(\zeta) = -\pi^{\frac{1}{2}}\mathrm{e}^{-\frac{3}{4}\pi\mathrm{i} - ikr}\mathrm{F}\{(2kr)^{\frac{1}{2}}\sin\tfrac{1}{2}\zeta\} - \pi\mathrm{e}^{-ikr\cos\zeta}\mathrm{H}(-\sin\tfrac{1}{2}\zeta) \qquad (7.11)$$

where

$$\mathrm{F}(z) = \mathrm{e}^{\mathrm{i}z^{2}}\int_{z}^{\infty}\mathrm{e}^{-\mathrm{i}t^{2}}\,\mathrm{d}t. \qquad (7.12)$$

Use of (7.11), coupled with the constraints on ϕ and ϕ^{i}, gives

$$\begin{aligned}p^{\mathrm{s}} = (1/\pi^{\frac{1}{2}})\mathrm{e}^{\frac{1}{4}\pi\mathrm{i} - ikr}[&\mathrm{F}\{(2kr)^{\frac{1}{2}}\sin\tfrac{1}{2}(\phi - \phi^{\mathrm{i}})\}\\ &- \mathrm{F}\{(2kr)^{\frac{1}{2}}\sin\tfrac{1}{2}(\phi + \phi^{\mathrm{i}})\}] - \mathrm{e}^{-ikr\cos(\phi - \phi^{\mathrm{i}})}.\end{aligned}$$

Thus the total field p_{S} for the soft boundary condition, i.e. $p_{\mathrm{S}} = p^{\mathrm{i}} + p^{\mathrm{s}}$ can be expressed as

$$p_{\mathrm{S}} = (1/\pi^{\frac{1}{2}})\mathrm{e}^{\frac{1}{4}\pi\mathrm{i} - ikr}[\mathrm{F}\{(2kr)^{\frac{1}{2}}\sin\tfrac{1}{2}(\phi - \phi^{\mathrm{i}})\} - \mathrm{F}\{(2kr)^{\frac{1}{2}}\sin\tfrac{1}{2}(\phi + \phi^{\mathrm{i}})\}]. \qquad (7.13)$$

Formula (7.13) has been derived subject to $0 < \phi < \pi$. However, it also holds when ϕ is negative. This can be verified either by repetition of the foregoing analysis with ϕ essentially replaced by $-\phi$ or by the observation that each of the terms satisfies the governing partial differential equation on its own and (7.13) complies with the boundary condition on $\phi=-\pi$.

The function $F(z)$ of (7.12) is related to the well-known *Fresnel's integral* via

$$F(z)=\tfrac{1}{2}\pi^{\frac{1}{2}}e^{iz^2}\left[e^{-\frac{1}{4}\pi i}-(2\pi)^{\frac{1}{2}}\left\{C\left(\frac{2^{\frac{1}{2}}z}{\pi^{\frac{1}{2}}}\right)-iS\left(\frac{2^{\frac{1}{2}}z}{\pi^{\frac{1}{2}}}\right)\right\}\right] \tag{7.14}$$

where

$$C(z)-iS(z)=\int_0^z e^{-\frac{1}{2}\pi i t^2}\,dt. \tag{7.15}$$

Another useful property is that

$$F(-z)=\pi^{\frac{1}{2}}e^{iz^2-\frac{1}{4}\pi i}-F(z). \tag{7.16}$$

The behaviour for large $|z|$ can be deduced by integration by parts when $\frac{1}{2}\pi>\mathrm{ph}\,z>-\pi$ and is

$$F(z)=-\frac{i}{2z}+\frac{1}{4z^3}+O\left(\frac{1}{|z|^5}\right). \tag{7.17}$$

Other ranges of ph z can be covered by combining (7.16) and (7.17).

On the other hand, when $|z|$ is small, it is pertinent to write

$$F(z)=e^{iz^2}\left\{\int_0^\infty e^{-it^2}\,dt-\int_0^z e^{-it^2}\,dt\right\}$$

with the deduction that

$$F(z)=\tfrac{1}{2}\pi^{\frac{1}{2}}e^{-\frac{1}{4}\pi i}-z+O(|z|^2) \tag{7.18}$$

if $|z|\ll 1$.

By means of (7.16) and (7.17) it can be seen that (7.13) provides, as $r\to\infty$, a plane wave travelling in the direction ϕ^i for $\phi<\phi^i$ which is expunged when $\phi>\phi^i$. In addition, there is a reflected plane wave in $\phi<-\phi^i$ which is removed in $\phi>-\phi^i$. The corrections to the plane waves are all radiating at infinity. The construction of the solution ensured that the edge conditions are satisfied and this can be confirmed directly from (7.13) by virtue of (7.18). In fact, near the edge

$$p_S\approx 2(2kr/\pi)^{\frac{1}{2}}e^{\frac{1}{4}\pi i}\cos\tfrac{1}{2}\phi\,\sin\tfrac{1}{2}\phi^i. \tag{7.19}$$

Although p_S is finite and continuous at $r=0$, its first derivatives will, in general, diverge like $1/r^{\frac{1}{2}}$. While it might be expected that such infinities should be excluded on physical grounds, they have to be permitted in an idealized model consisting of an infinitesimally thin plane with a sharp edge.

Evidently the field has features which can be attributed to ray theory. In ray theory the disturbance travels in straight lines. So it predicts that

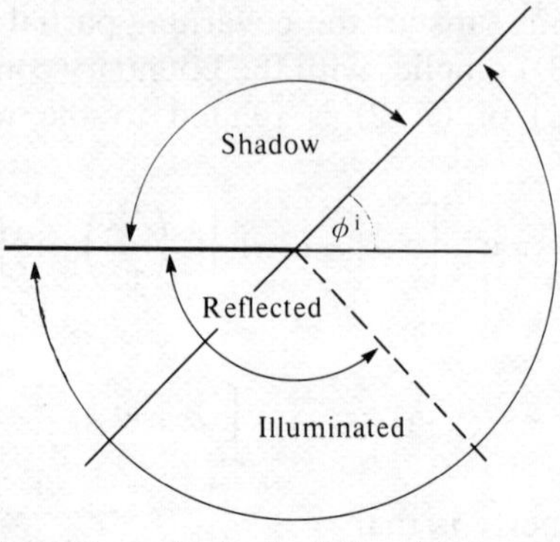

Fig. 9.10 The regions of ray theory in diffraction by a semi-infinite plane

there is a *shadow zone* behind the screen where there is no field (see Fig. 9.10), an *illuminated region* where the incident plane wave exists and a *reflection sector* where the screen produces a reflected plane wave. The field of ray theory is discontinuous across the lines $\phi=\pm\phi^{i}$ and the more recondite expression (7.13) is necessary to accomplish a smooth transition.

The field that is left after the plane waves of ray theory have been extracted is known as the *diffracted field*. From (7.13) the diffracted field when $kr \gg 1$ is given by

$$p_{\mathrm{S}}^{\mathrm{d}} \sim \frac{\cos\frac{1}{2}\phi\,\sin\frac{1}{2}\phi^{i}}{\cos\phi^{i}-\cos\phi}\left(\frac{2}{\pi kr}\right)^{\frac{1}{2}} \mathrm{e}^{-ikr-\frac{1}{4}\pi i}. \tag{7.20}$$

The implication of (7.20) is that the field due to diffraction looks as though it originated from a line source placed along the diffracting edge but with an amplitude factor which depends upon the angle of observation. Experimentally, this can be observed in measurements in the shadow zone where it is found that the disturbance does appear to emanate from the edge.

The asymptotic expression (7.20) is not valid when ϕ is near ϕ^{i} or $-\phi^{i}$. This is because the argument of F may then be small even if kr is large so that (7.17) cannot be invoked. Some idea of the regions of failure can be obtained by drawing the curves $2kr\sin^{2}\frac{1}{2}(\phi\pm\phi^{i})=10$. Each curve is a parabola with focus at the origin and the two axes are $\phi=\pm\phi^{i}$ (Fig. 9.11). The larger kr is the narrower is a parabola. The formula (7.20) should not be used in the regions shown shaded in Fig. 9.11. In these regions the full expression (7.13) must be employed; there is no simpler approximation to it. According to (7.13), when $\phi=\phi^{i}$

$$p_{\mathrm{S}}=\tfrac{1}{2}\mathrm{e}^{-ikr}+O\left\{\frac{1}{(kr)^{\frac{1}{2}}}\right\}$$

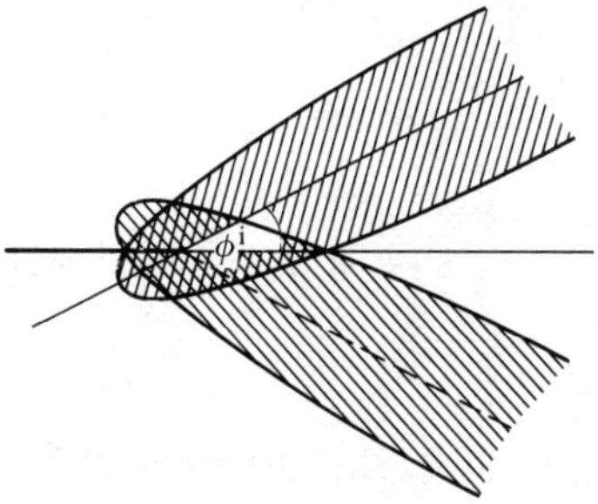

Fig. 9.11 The regions where the asymptotic formula (7.20) does not hold are shaded

so that the diffracted field is of the same order of magnitude as the incident wave. A similar statement is true when $\phi = -\phi^{\mathrm{i}}$ and explains in another fashion why it is not legitimate to use (7.20) in the shaded regions.

Interference will occur between the diffracted field and incident wave in regions where they are of comparable magnitude. An illustration of interference is shown in Fig. 9.12 where the field on a line three wavelengths behind the screen is plotted, i.e. on $y = 6\pi/k$, with an incident plane wave at normal incidence ($\phi^{\mathrm{i}} = \frac{1}{2}\pi$). The diffraction fringes are clearly visible as well as the steady decay which takes place as one moves into the shadow.

As regards the field induced on the semi-infinite plane note that (7.13) implies that the discontinuity in the derivative with respect to y is given

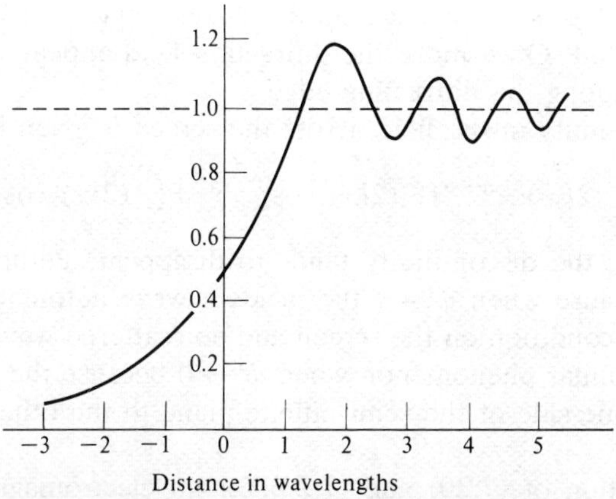

Fig. 9.12 Interference fringes for a plane wave at normal incidence

by

$$\left[\frac{\partial p_S}{\partial y}\right]_{y=-0}^{y=+0} = -\frac{2ik}{\pi^{\frac{1}{2}}} e^{-ikr+\frac{1}{4}\pi i}\left[\sin\phi^i F\{(2kr)^{\frac{1}{2}}\cos\tfrac{1}{2}\phi^i\}\right.$$

$$\left. -\sin\phi^i F\{-(2kr)^{\frac{1}{2}}\cos\tfrac{1}{2}\phi^i\} + i\left(\frac{2}{kr}\right)^{\frac{1}{2}}\sin\tfrac{1}{2}\phi^i\right]. \quad (7.21)$$

When $\phi^i \to \pi$ this tends to

$$[\partial p_S/\partial y]_{y=-0}^{y=+0} = (8k/\pi r)^{\frac{1}{2}} e^{\frac{1}{4}\pi i - ikr}. \quad (7.22)$$

The analogous problem for the hard screen has (7.2) replaced by

$$\partial p^s/\partial\phi = -ikr\sin\phi^i\exp(ikr\cos\phi^i) \quad (\phi = \pm\pi). \quad (7.23)$$

It may be treated by an analysis similar to that for the soft boundary. An alternative approach is to remark that, since both terms in (7.13) satisfy the governing equation, the boundary condition can be met by a reversal of sign. Thus the total field for the hard boundary is

$$p_H = (1/\pi^{\frac{1}{2}}) e^{\frac{1}{4}\pi i - ikr}[F\{(2kr)^{\frac{1}{2}}\sin\tfrac{1}{2}(\phi-\phi^i)\} + F\{(2kr)^{\frac{1}{2}}\sin\tfrac{1}{2}(\phi+\phi^i)\}]. \quad (7.24)$$

Near the edge, where $kr \ll 1$,

$$p_H \approx 1 - 2(2kr/\pi)^{\frac{1}{2}} e^{\frac{1}{4}\pi i}\sin\tfrac{1}{2}\phi\cos\tfrac{1}{2}\phi^i. \quad (7.25)$$

The derivatives are still singular at the edge and the principal difference from the soft case is that the field no longer vanishes at the edge.

The general structure of the ray picture is the same as in Fig. 9.10. For the diffracted field, when $kr \gg 1$,

$$p_H^d \sim \frac{\sin\frac{1}{2}\phi\cos\frac{1}{2}\phi^i}{\cos\phi^i - \cos\phi}\left(\frac{2}{\pi kr}\right)^{\frac{1}{2}} e^{-ikr-\frac{1}{4}\pi i} \quad (7.26)$$

provided that ϕ is not near $\pm\phi^i$, i.e. the shaded regions of Fig. 9.11 must again be omitted. Once more the diffraction field appears to issue from a line source along the diffracting edge.

The discontinuity in the field across the screen is given by

$$[p_H]_{y=-0}^{y=+0} = (2/\pi^{\frac{1}{2}}) e^{\frac{1}{4}\pi i - ikr}[F\{(2kr)^{\frac{1}{2}}\cos\tfrac{1}{2}\phi^i\} - F\{-(2kr)^{\frac{1}{2}}\cos\tfrac{1}{2}\phi^i\}]. \quad (7.27)$$

When $\phi^i \to \pi$, the discontinuity tends to disappear. Perhaps this is not surprising because when $\phi^i = \pi$ the incident wave automatically satisfies the boundary condition on the screen and no scattered wave is produced. There is no similar phenomenon when $\phi^i \to 0$ because the incident wave moves from one side of the semi-infinite plane to the other as ϕ^i passes through zero.

An implication of (7.20) and (7.26) for an electromagnetic perfectly conducting screen is that, well into the shadow, the ratio of the field strengths is

$$E\text{-polarization}/H\text{-polarization} \sim \cot \tfrac{1}{2}\phi \tan \tfrac{1}{2}\phi^{\mathrm{i}}. \qquad (7.28)$$

Thus an unpolarized incident wave will become partially polarized on diffraction, in agreement with what is observed experimentally.

9.8 The line source

Let now the incident excitation be created by a line source at the point L with polar coordinates (r_0, ϕ_0) so that

$$p^{\mathrm{i}} = -\tfrac{1}{4}\mathrm{i}\mathrm{H}_0^{(2)}(k\,|\boldsymbol{x} - \boldsymbol{x}_0|).$$

One method is to draw benefit from (6.9), the various parameters being depicted in Fig. 9.13. Thus the field corresponding to the incident plane wave $\mathrm{e}^{ikr\cos(\phi-\theta)}$ is discovered, multiplied by $\mathrm{e}^{-ikr_0\cos(\phi_0-\theta)}$ and then integrated with respect to θ over C. The representation (6.9) is appropriate since it holds for $y < y_0$ and therefore in the vicinity of the diffracting screen. A seeming difficulty is that angle θ is complex with possible consequences for the plane wave solution. Notwithstanding, it is straightforward to verify that (7.13) and (7.14) remain solutions of the boundary value problem even when ϕ^{i} is complex. So, the integration over C is the only remaining task.

Another way is to mimic the structure of (7.13). In this, the integrals if they were not truncated would produce the incident wave and its image. The argument responsible for the truncation of the first F corresponds to the difference in phase between an incident ray direct to the point of observation and one via the edge; in the second F the same function is performed for the image field. To emulate this, note that (6.8) gives after

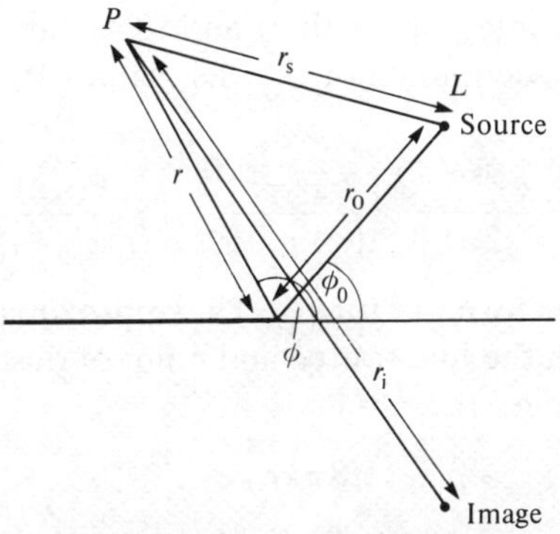

Fig. 9.13 A line source in the presence of a semi-infinite plane

the substitution $u=(2kr)^{\frac{1}{2}}\sinh\frac{1}{2}t$

$$-\tfrac{1}{4}\mathrm{i}H_0^{(2)}(kr_s)=\frac{1}{2\pi}\mathrm{e}^{-ikr_s}\int_{-\infty}^{\infty}\frac{\mathrm{e}^{-\mathrm{i}u^2}}{(u^2+2kr_s)^{\frac{1}{2}}}\,\mathrm{d}u. \tag{8.1}$$

Truncation with the relevant phase difference is achieved by taking the upper limit to be $\{k(r+r_0-r_s)\}^{\frac{1}{2}}$. It can be verified that the right-hand side of (8.1) so changed continues to be a solution of Helmholtz's equation. This suggests

$$p_S=\frac{1}{2\pi}\mathrm{e}^{-ikr_s}\int_{-\infty}^{b}\frac{\mathrm{e}^{-\mathrm{i}u^2}}{(u^2+2kr_s)^{\frac{1}{2}}}\,\mathrm{d}u-\frac{1}{2\pi}\mathrm{e}^{-ikr_i}\int_{-\infty}^{c}\frac{\mathrm{e}^{-\mathrm{i}u^2}}{(u^2+2kr_i)^{\frac{1}{2}}}\,\mathrm{d}u \tag{8.2}$$

where r_i is the distance to the point of observation from the image of the source in $y=0$, i.e.

$$r_i^2=r^2+r_0^2-2rr_0\cos(\phi+\phi_0). \tag{8.3}$$

In order to create the shadow zone and reflected region choose

$$b=\begin{cases}\{k(r+r_0-r_s)\}^{\frac{1}{2}} & \cos\frac{1}{2}(\phi-\phi_0)>0\\ -\{k(r+r_0-r_s)\}^{\frac{1}{2}} & \cos\frac{1}{2}(\phi-\phi_0)<0\end{cases}$$

$$c=\begin{cases}\{k(r+r_0-r_i)\}^{\frac{1}{2}} & \cos\frac{1}{2}(\phi+\phi_0)<0\\ -\{k(r+r_0-r_i)\}^{\frac{1}{2}} & \cos\frac{1}{2}(\phi+\phi_0)>0.\end{cases}$$

On the screen $r_s=r_i$ and $b=c$ so that the soft boundary condition is satisfied. Near the source $r_s\approx 0$ and $b>0$ so that the correct singularity is reproduced. There is no singularity at the image because, although $r_i\to 0$, when $\phi\approx-\phi_0$, $c<0$. Hence the construction has led to the desired solution.

The corresponding solution for the hard boundary condition is

$$p_H=\frac{1}{2\pi}\mathrm{e}^{-ikr_s}\int_{-\infty}^{b}\frac{\mathrm{e}^{-\mathrm{i}u^2}}{(u^2+2kr_s)^{\frac{1}{2}}}\,\mathrm{d}u+\frac{1}{2\pi}\mathrm{e}^{-ikr_i}\int_{-\infty}^{c}\frac{\mathrm{e}^{-\mathrm{i}u^2}}{(u^2+2kr_i)^{\frac{1}{2}}}\,\mathrm{d}u. \tag{8.4}$$

If $k(r+r_0)\gg 1$ the integrals in (8.2) and (8.4) can be approximated by replacing u in the denominator by its value at the upper limit of integration. Then

$$p_S\sim\frac{\mathrm{e}^{-ik(r+r_0)}}{2\pi k^{\frac{1}{2}}}\left\{\frac{F(-b)}{(r+r_0+r_s)^{\frac{1}{2}}}-\frac{F(-c)}{(r+r_0+r_i)^{\frac{1}{2}}}\right\} \tag{8.5}$$

and there is a similar formula for p_H. The approximation (8.5) should be adequate unless both the line source and point of observation are within a wavelength of the edge. If $kr_s\gg 1$,

$$p^{\mathrm{i}}\sim(1/8\pi kr_s)^{\frac{1}{2}}\mathrm{e}^{-ikr_s-\frac{1}{4}\pi\mathrm{i}}. \tag{8.6}$$

Therefore, multiplication by $(8\pi kr_0)^{\frac{1}{2}}\mathrm{e}^{ikr_0+\frac{1}{4}\pi\mathrm{i}}$ and letting $r_0\to\infty$ produces an incident plane wave with $\phi_0=\phi'-\pi$. The same operations applied to

(8.5) recover (7.13) so that the plane wave solution may be deemed to be included in (8.5).

If $|b| \gg 1$ and $|c| \gg 1$ the asymptotic formulae (7.16) and (7.17) may be quoted. The net result is that the diffracted field is given by

$$p_S^d \sim \frac{ie^{-ik(r+r_0)}}{2\pi k(rr_0)^{\frac{1}{2}}} \frac{\cos\frac{1}{2}\phi \cos\frac{1}{2}\phi_0}{\cos\phi + \cos\phi_0}, \tag{8.7}$$

$$p_H^d \sim \frac{-ie^{-ik(r+r_0)}}{2\pi k(rr_0)^{\frac{1}{2}}} \frac{\sin\frac{1}{2}\phi \sin\frac{1}{2}\phi_0}{\cos\phi + \cos\phi_0}. \tag{8.8}$$

The resemblance between (8.7) and (7.20) will be remarked. In fact, if (8.7) is multiplied by $(8\pi kr_0)^{\frac{1}{2}}e^{ikr_0+\frac{1}{4}\pi i}$ and ϕ_0 is replaced by $\phi^i - \pi$, (7.20) is replicated. There is a similar relationship between (8.8) and (7.26).

The expressions (8.7) and (8.8) are reasonable approximations for the diffracted field so long as neither b nor c is small. Roughly speaking, they will be acceptable for points of observation outside the hyperbola $k(r+r_0-r_s)=10$, whose axis coincides with the shadow boundary, and outside the hyperbola $k(r+r_0-r_i)=10$ whose axis is along the reflected wave. Each hyperbola is the analogue of the corresponding parabola which arises for the incident plane wave.

At points near the edge both b and c will be small in magnitude. It then follows from (8.2) and (8.4) that

$$p_S \approx \frac{1}{\pi} e^{-ikr_0}\left(\frac{r}{r_0}\right)^{\frac{1}{2}} \cos\tfrac{1}{2}\phi \cos\tfrac{1}{2}\phi_0, \tag{8.9}$$

$$p_H \approx -\tfrac{1}{4}iH_0^{(2)}(kr_0) + \frac{1}{\pi} e^{-ikr_0}\left(\frac{r}{r_0}\right)^{\frac{1}{2}} \sin\tfrac{1}{2}\phi \sin\tfrac{1}{2}\phi_0. \tag{8.10}$$

The field induced on the screen, deduced from (8.2), is

$$\left[\frac{\partial p_S}{\partial y}\right]_{y=-0}^{y=+0} = \frac{iy_0}{\pi r_s^2} e^{-ikr_s} \int_{-b}^{b} \frac{u^2+kr_s}{(u^2+2kr_s)^{\frac{1}{2}}} e^{-iu^2}\, du + \left\{\frac{y_0(r+r_0-r_s)^{\frac{1}{2}}}{r_s(r+r_0+r_s)^{\frac{1}{2}}} + \left(\frac{r_0}{r}\right)^{\frac{1}{2}} \cos\tfrac{1}{2}\phi_0\right\} \frac{e^{-ik(r+r_0)}}{\pi r_s} \tag{8.11}$$

with b and $(r+r_0-r_s)^{\frac{1}{2}}$ designated as positive. If the source is in the plane of the screen $\phi_0=0$, $r+r_0=r_s$ and (8.4) reduces to

$$\left[\frac{\partial p_S}{\partial y}\right]_{y=-0}^{y=+0} = \frac{r_0^{\frac{1}{2}}e^{-ik(r+r_0)}}{\pi r^{\frac{1}{2}}(r+r_0)}. \tag{8.12}$$

The corresponding quantity for the hard screen is

$$[p_H]_{y=-0}^{y=+0} = \frac{e^{-ikr_s}}{\pi} \int_{-b}^{b} \frac{e^{-iu^2}}{(u^2+2kr_s)^{\frac{1}{2}}}\, du \tag{8.13}$$

with the same conventions as (8.11). As the source approaches the plane of the screen this disappears and there is no discontinuity in p_H. Allied to this phenomenon is the field scattered when the source produces $\frac{\partial}{\partial y_0}\{-\frac{1}{4}\mathrm{i}H_0^{(2)}(kr_s)\}$. This is easily inferred from (8.4) and is

$$\frac{\partial p_H}{\partial y_0}=\frac{-y_0+y}{2\pi r_s}\mathrm{e}^{-ikr_s}\int_{-\infty}^{b}\left\{\frac{ik}{(u^2+2kr_s)^{\frac{1}{2}}}+\frac{k}{(u^2+2kr_s)^{\frac{3}{2}}}\right\}\mathrm{e}^{-iu^2}\,du$$
$$-\frac{y+y_0}{2\pi r_i}\mathrm{e}^{-ikr_i}\int_{-\infty}^{c}\left\{\frac{ik}{(u^2+2kr_i)^{\frac{1}{2}}}\right.$$
$$\left.+\frac{k}{(u^2+2kr_i)^{\frac{3}{2}}}\right\}\mathrm{e}^{-iu^2}\,du$$
$$+\left[\left\{y_0\left(\frac{1}{r_0}-\frac{1}{r_s}\right)+\frac{y}{r_s}\right\}\{(r+r_0)^2-r_s^2\}^{-\frac{1}{2}}\operatorname{sgn} b\right.$$
$$\left.+\left\{y_0\left(\frac{1}{r_0}-\frac{1}{r_i}\right)-\frac{y}{r_i}\right\}\{(r+r_0)^2-r_i^2\}^{-\frac{1}{2}}\operatorname{sgn} c\right]\frac{\mathrm{e}^{-ik(r+r_0)}}{4\pi}. \tag{8.14}$$

In particular, when $\phi_0=0$, (8.14) supplies

$$\left[\frac{\partial p_H}{\partial y_0}\right]_{y=-0}^{y=+0}=\frac{r^{\frac{1}{2}}\mathrm{e}^{-ik(r+r_0)}}{\pi r_0^{\frac{1}{2}}(r+r_0)}. \tag{8.15}$$

It will be noticed that, if r and r_0 are interchanged as well as ϕ and ϕ_0, the formulae (8.2) and (8.4) are unaltered. This illustrates the reciprocity discussed in §1.35 because it indicates that the field produced at P due to a line source at L is the same as the field created by L by a line source at P. There is also reciprocity between line source and plane wave excitations. For suppose that a line source at (r,ϕ) generates the field $A(r,\phi,\phi_0)\mathrm{e}^{-ikr_0}/r_0^{\frac{1}{2}}$ at the distant point (r_0,ϕ_0). Then, after multiplication by $(8\pi kr_0)^{\frac{1}{2}}\mathrm{e}^{ikr_0+\frac{1}{4}\pi i}$, $(8\pi k)^{\frac{1}{2}}A(r,\phi,\phi_0)\mathrm{e}^{\frac{1}{4}\pi i}$ is the field caused at (r,ϕ) by a plane wave of unit amplitude approaching from the direction $\phi=\phi_0$.

9.9 Diffraction of a three-dimensional acoustic plane wave

The direction of propagation of the plane wave in §9.7 was in a plane perpendicular to the diffracting edge. Nevertheless the solution derived there has a part to play when the restriction on the direction of incidence is lifted.

Keep the z-axis along the diffracting edge and let the incident plane wave be

$$p^{(3)\mathrm{i}}=\exp\{-ik(x\sin\theta^{\mathrm{i}}\cos\phi^{\mathrm{i}}+y\sin\theta^{\mathrm{i}}\sin\phi^{\mathrm{i}}+z\cos\theta^{\mathrm{i}})\},$$

the affix (3) distinguishing three-dimensional formulae. The direction of propagation makes an angle θ^i with the edge and $\theta^i = \frac{1}{2}\pi$ returns to the case examined in §9.7.

Take the z-dependence of all fields to be $e^{-ikz\cos\theta^i}$. Then Helmholtz's equation in three dimensions yields

$$\partial^2 p/\partial x^2 + \partial^2 p/\partial y^2 + pk^2 \sin^2\theta^i = 0$$

which is the two-dimensional form considered previously with $k \sin\theta^i$ standing in for k. Since the boundary conditions are independent of z the present problem can be solved by putting $k \sin\theta^i$ for k and then multiplying by $e^{-ikz\cos\theta^i}$. For instance,

$$p_S^{(3)} = \frac{1}{\pi^{\frac{1}{2}}} e^{-ikr\sin\theta^i - ikz\cos\theta^i + \frac{1}{4}\pi i}[F\{(2kr\sin\theta^i)^{\frac{1}{2}} \sin\tfrac{1}{2}(\phi-\phi^i)\} - F\{(2kr\sin\theta^i)^{\frac{1}{2}} \sin\tfrac{1}{2}(\phi+\phi^i)\}] \quad (9.1)$$

where r and ϕ are still polar coordinates in the (x, y)-plane. A simple change of sign in (9.1) furnishes the result for the hard boundary condition.

The diffracted field away from the shadow boundary and boundary of the reflected wave can be deduced from (7.20) and (7.26) provided that, in addition, θ^i is not near 0 or π, i.e. the incident wave is not travelling almost parallel to the edge. Taking $kr \gg 1$, so that points near the edge are precluded

$$p_S^{(3)d} \sim \frac{\cos\frac{1}{2}\phi \sin\frac{1}{2}\phi^i}{\cos\phi^i - \cos\phi} \left(\frac{2}{\pi kr \sin\theta^i}\right)^{\frac{1}{2}} e^{-\frac{1}{4}\pi i - ikr\sin\theta^i - ikz\cos\theta^i}, \quad (9.2)$$

$$p_H^{(3)d} \sim \frac{\sin\frac{1}{2}\phi \cos\frac{1}{2}\phi^i}{\cos\phi^i - \cos\phi} \left(\frac{2}{\pi kr \sin\theta^i}\right)^{\frac{1}{2}} e^{-\frac{1}{4}\pi i - ikr\sin\theta^i - ikz\cos\theta^i}. \quad (9.3)$$

The diffracted wave in both (9.2) and (9.3) at the point $r = R\sin\theta^i$,

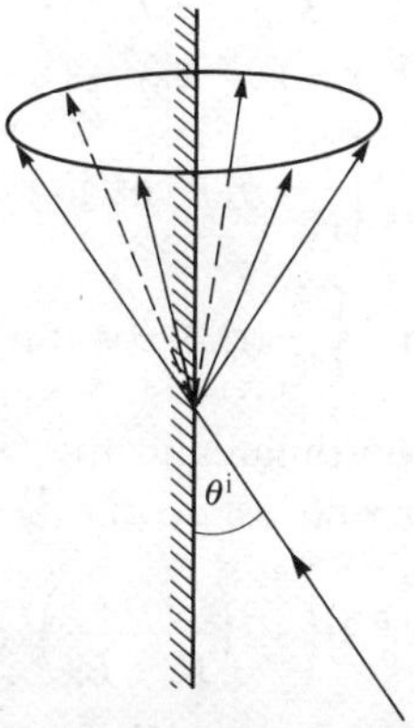

Fig. 9.14 The cone of diffracted rays produced by an edge

$z = R\cos\theta^{\mathrm{i}}$ has $\mathrm{e}^{-\mathrm{i}kR}/R^{\frac{1}{2}}$ as its dependence on R which is the distance from the origin to the point of observation. Thus the fields appears as a progressive wave emanating from the origin. If a cone of semi-angle θ^{i} and apex at the origin be drawn with its axis along the edge, the field on the cone travels along a generator, being constant on any given generator apart from variations in R (Fig. 9.14). Since an alteration in z merely translates the field, the diffracted disturbance at any point appears to be moving along a cone of semi-angle θ^{i} with its apex on the edge. The existence of the cone has been confirmed experimentally.

9.10 Diffraction of an acoustic spherical wave

The remaining case to be discussed for acoustic diffraction by a semi-infinite plane is that in which the incident radiation is caused by a point source. If the source is located at $\boldsymbol{x}_0$ the equation to be solved is

$$\nabla^2 p^{(3)} + k^2 p^{(3)} = -\delta(\boldsymbol{x} - \boldsymbol{x}_0)$$

and the incident wave is

$$p^{(3)\mathrm{i}} = \mathrm{e}^{-\mathrm{i}kR_s}/4\pi R_s$$

where R_s is the distance from $\boldsymbol{x}_0$ to the point of observation (Fig. 9.15). Invoke (6.14) and carry out a construction similar to that of §9.8. Then

$$p_S^{(3)} = -\frac{\mathrm{i}k}{8\pi}\int_{-b_1}^{\infty} \mathrm{H}_1^{(2)}(kR_s\cosh u)\,\mathrm{d}u + \frac{\mathrm{i}k}{8\pi}\int_{-c_1}^{\infty} \mathrm{H}_1^{(2)}(kR_i\cosh u)\,\mathrm{d}u, \tag{10.1}$$

$$p_H^{(3)} = -\frac{\mathrm{i}k}{8\pi}\int_{-b_1}^{\infty} \mathrm{H}_1^{(2)}(kR_s\cosh u)\,\mathrm{d}u - \frac{\mathrm{i}k}{8\pi}\int_{-c_1}^{\infty} \mathrm{H}_1^{(2)}(kR_i\cosh u)\,\mathrm{d}u \tag{10.2}$$

where R_i is the distance from the image of S in the plane $y=0$ and

$$R_s^2 = (x-x_0)^2 + (y-y_0)^2 + (z-z_0)^2,$$
$$R_i^2 = (x-x_0)^2 + (y+y_0)^2 + (z-z_0)^2,$$
$$b_1 = \sinh^{-1}\left\{\frac{2}{R_s}(rr_0)^{\frac{1}{2}}\cos\tfrac{1}{2}(\phi-\phi_0)\right\},$$
$$c_1 = -\sinh^{-1}\left\{\frac{2}{R_i}(rr_0)^{\frac{1}{2}}\cos\tfrac{1}{2}(\phi+\phi_0)\right\}$$

with (r,ϕ), (r_0,ϕ_0) polar coordinates in the (x,y)-plane as before.

If $kR' \gg 1$ where $R' = \{(r+r_0)^2 + (z-z_0)^2\}^{\frac{1}{2}}$, (10.1) simplifies to

$$p_S^{(3)} \sim \frac{\mathrm{e}^{\frac{1}{4}\pi\mathrm{i}}}{4\pi^{\frac{3}{2}}}\left\{\frac{2}{R'(R_s+R')}\right\}^{\frac{1}{2}}\mathrm{e}^{-\mathrm{i}kR'}\mathrm{F}\left\{-2\left(\frac{krr_0}{R_s+R'}\right)^{\frac{1}{2}}\cos\tfrac{1}{2}(\phi-\phi_0)\right\}$$
$$-\frac{\mathrm{e}^{\frac{1}{4}\pi\mathrm{i}}}{4\pi^{\frac{3}{2}}}\left\{\frac{2}{R'(R_i+R')}\right\}^{\frac{1}{2}}\mathrm{e}^{-\mathrm{i}kR'}\mathrm{F}\left\{2\left(\frac{krr_0}{R_i+R'}\right)^{\frac{1}{2}}\cos\tfrac{1}{2}(\phi+\phi_0)\right\}. \tag{10.3}$$

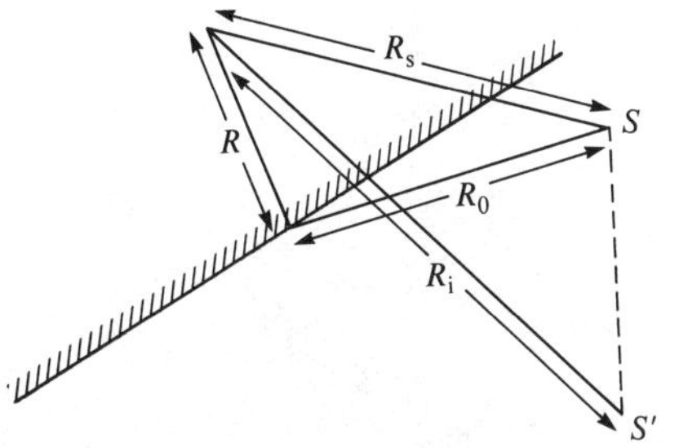

Fig. 9.15 The configuration for a point source

Well away from the shadow and reflection boundaries the diffraction field is

$$p_S^{(3)d} \sim \frac{2^{\frac{1}{2}} e^{\frac{3}{4}\pi i - ikR'}}{4\pi^{\frac{3}{2}}(krr_0R')^{\frac{1}{2}}} \frac{\cos\frac{1}{2}\phi \cos\frac{1}{2}\phi_0}{\cos\phi + \cos\phi_0}, \tag{10.4}$$

$$p_H^{(3)d} \sim -\frac{2^{\frac{1}{2}} e^{\frac{3}{4}\pi i - ikR'}}{4\pi^{\frac{3}{2}}(krr_0R')^{\frac{1}{2}}} \frac{\sin\frac{1}{2}\phi \sin\frac{1}{2}\phi_0}{\cos\phi + \cos\phi_0} \tag{10.5}$$

for $krr_0 \gg R_s + R'$. The close family likeness of (10.4) and (10.5) with (8.7) and (8.8) will be observed.

Near the edge

$$p_S^{(3)} \approx -ik \frac{H_1^{(2)}[k\{r_0^2+(z-z_0)^2\}^{\frac{1}{2}}]}{2\pi\{r_0^2+(z-z_0)^2\}^{\frac{1}{2}}} (rr_0)^{\frac{1}{2}} \cos\tfrac{1}{2}\phi \cos\tfrac{1}{2}\phi_0, \tag{10.6}$$

$$p_H^{(3)} \approx \frac{\exp[-ik\{r_0^2+(z-z_0)^2\}^{\frac{1}{2}}]}{4\pi\{r_0^2+(z-z_0)^2\}^{\frac{1}{2}}} - \frac{ikH_1^{(2)}[k\{r_0^2+(z-z_0)^2\}^{\frac{1}{2}}]}{2\pi\{r_0^2+(z-z_0)^2\}^{\frac{1}{2}}} (rr_0)^{\frac{1}{2}} \sin\tfrac{1}{2}\phi \sin\tfrac{1}{2}\phi_0 \tag{10.7}$$

on account of (6.14).

Note that reciprocity is available for both (10.1) and (10.2).

9.11 The diffracted sound wave

On several occasions it has been noted how the diffracted fields for different excitations resemble one another. It is now convenient to combine them all into a single formula which copes with all the cases discussed so far. In the derivation it is advantageous to make a slight change of notation and measure angles from the screen (Fig. 9.16). In the parameters of the diagram

$$R' = \{(r+r_0)^2 + d^2\}^{\frac{1}{2}} \tag{11.1}$$

Let β_0 be the angle such that

$$d = r\tan\beta = R'\cos\beta_0, \qquad r + r_0 = R'\sin\beta_0. \tag{11.2}$$

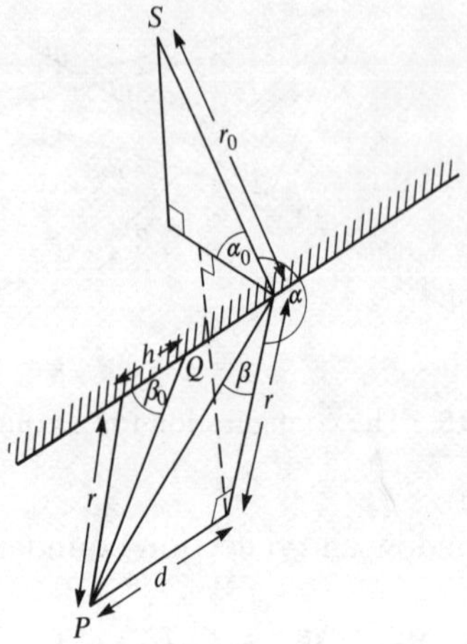

Fig. 9.16 Geometry for diffraction by a semi-infinite plane

Then (10.4) and (10.5) can be expressed as

$$p_S^d = D_S\left(\frac{r_0}{rR'}\right)^{\frac{1}{2}} p_Q^i e^{-ikr\,\mathrm{cosec}\,\beta_0}, \tag{11.3}$$

$$p_H^d = D_H\left(\frac{r_0}{rR'}\right)^{\frac{1}{2}} p_Q^i e^{-ikr\,\mathrm{cosec}\,\beta_0} \tag{11.4}$$

where

$$D_S = \left(\frac{2}{\pi k}\right)^{\frac{1}{2}} e^{-\frac{1}{4}\pi i}\frac{\sin\frac{1}{2}\alpha_0 \sin\frac{1}{2}\alpha}{\cos\alpha_0 + \cos\alpha}, \tag{11.5}$$

$$D_H = -\left(\frac{2}{\pi k}\right)^{\frac{1}{2}} e^{-\frac{1}{4}\pi i}\frac{\cos\frac{1}{2}\alpha_0 \cos\frac{1}{2}\alpha}{\cos\alpha_0 + \cos\alpha}. \tag{11.6}$$

The quantity p_Q^i is the incident field at the point Q of the edge which is at a distance $h = r\cot\beta_0$ along the edge towards the source from the projection of P, the point of observation, on the edge. It is straightforward to check that (11.3) and (11.4) also reproduce (9.2) and (9.3) when R' is replaced by $r_0\,\mathrm{cosec}\,\beta_0$ and $\beta_0 = \theta^i$. Furthermore, (8.7) and (8.8) are supplied as well if r_0/R' is taken as unity and $\beta_0 = \frac{1}{2}\pi$ on the understanding that (8.6) is used for p^i.

Consequently, (11.3) and (11.4) can be regarded as universal formulae for the diffracted field. They can be expected to hold for angular deviations from the shadow or reflection boundaries in excess of 10° if $kr_0 = 200$.

9.12 Three-dimensional diffraction of an electromagnetic plane wave

Let the electric intensity in the primary wave be

$$\boldsymbol{E}^{\mathrm{i}} = (l^{\mathrm{i}}\boldsymbol{i} + m^{\mathrm{i}}\boldsymbol{j} + n^{\mathrm{i}}\boldsymbol{k})\exp\{-ik(x \sin\theta^{\mathrm{i}} \cos\phi^{\mathrm{i}} + y \sin\theta^{\mathrm{i}} \sin\phi^{\mathrm{i}} + z\cos\theta^{\mathrm{i}})\}$$

where $\boldsymbol{i}$, $\boldsymbol{j}$, $\boldsymbol{k}$ are unit vectors parallel to the Cartesian axes and

$$(l^{\mathrm{i}}\cos\phi^{\mathrm{i}} + m^{\mathrm{i}}\sin\phi^{\mathrm{i}})\sin\theta^{\mathrm{i}} + n^{\mathrm{i}}\cos\theta^{\mathrm{i}} = 0. \tag{12.1}$$

As in §9.9 assume that the z-dependence of all fields is $e^{-ikz\cos\theta^{\mathrm{i}}}$. It is then found that the governing equations are those of two dimensions except that $k\sin\theta^{\mathrm{i}}$ is substituted for k. The deduction is that, if $\boldsymbol{E}^{(3)}$ denotes the total field,

$$\begin{aligned}\boldsymbol{E}^{(3)} &= (n^{\mathrm{i}}/k)\operatorname{cosec}^2\theta^{\mathrm{i}}\exp(-ikz\cos\theta^{\mathrm{i}})(\boldsymbol{k}k\sin^2\theta^{\mathrm{i}} - \mathrm{i}\cos\theta^{\mathrm{i}}\,\mathrm{grad})p_{\mathrm{S}}^{(3)}\\ &\quad - (\mathrm{i}/k)\operatorname{cosec}\theta^{\mathrm{i}}\exp(-ikz\cos\theta^{\mathrm{i}})(l^{\mathrm{i}}\sin\phi^{\mathrm{i}} - m^{\mathrm{i}}\cos\phi^{\mathrm{i}})\boldsymbol{k}\wedge\mathrm{grad}\,p_{\mathrm{H}}^{(3)},\end{aligned} \tag{12.2}$$

$$\begin{aligned}\boldsymbol{H}^{(3)}(\mu_0/\varepsilon_0)^{\frac{1}{2}} &= -\mathrm{i}(n^{\mathrm{i}}/k)\operatorname{cosec}^2\theta^{\mathrm{i}}\exp(-ikz\cos\theta^{\mathrm{i}})\boldsymbol{k}\wedge\mathrm{grad}\,p_{\mathrm{S}}^{(3)}\\ &\quad -(1/k\sin\theta^{\mathrm{i}})\exp(-ikz\cos\theta^{\mathrm{i}})(l^{\mathrm{i}}\sin\phi^{\mathrm{i}} - m^{\mathrm{i}}\cos\phi^{\mathrm{i}})\\ &\quad \times(\boldsymbol{k}k\sin^2\theta^{\mathrm{i}} - \mathrm{i}\cos\theta^{\mathrm{i}}\,\mathrm{grad})p_{\mathrm{H}}^{(3)}\end{aligned} \tag{12.3}$$

where $p_{\mathrm{S}}^{(3)}$ is given by (9.1) and $p_{\mathrm{H}}^{(3)}$ by the same formula with the sign of the second term reversed.

The diffracted field when $kr\sin\theta^{\mathrm{i}}$ is large is given in cylindrical polar coordinates by

$$E_r^{(3)\mathrm{d}} \sim -n^{\mathrm{i}}\cot\theta^{\mathrm{i}}p_{\mathrm{S}}^{(3)\mathrm{d}}, \qquad E_\phi^{(3)\mathrm{d}} \sim -(l^{\mathrm{i}}\sin\phi^{\mathrm{i}} - m^{\mathrm{i}}\cos\phi^{\mathrm{i}})p_{\mathrm{H}}^{(3)\mathrm{d}},$$

$$E_z^{(3)\mathrm{d}} \sim n^{\mathrm{i}}p_{\mathrm{S}}^{(3)\mathrm{d}}, \tag{12.4}$$

$$(\mu_0/\varepsilon_0)^{\frac{1}{2}}\boldsymbol{H}^{(3)\mathrm{d}} \sim (\hat{\boldsymbol{r}}\sin\theta^{\mathrm{i}} + \boldsymbol{k}\cos\theta^{\mathrm{i}})\wedge\boldsymbol{E}^{(3)\mathrm{d}} \tag{12.5}$$

where $\hat{\boldsymbol{r}}$ is a unit vector along $\boldsymbol{r}$ and $p_{\mathrm{S}}^{(3)\mathrm{d}}$, $p_{\mathrm{H}}^{(3)\mathrm{d}}$ are specified by (9.2), (9.3) respectively.

It is clear from (12.4) and (12.5) that the electric and magnetic intensities are perpendicular to each other and both are orthogonal to $\hat{\boldsymbol{r}}\sin\theta^{\mathrm{i}} + \boldsymbol{k}\cos\theta^{\mathrm{i}}$. Thus, locally, the diffracted field has the appearance of a plane wave moving parallel to $\hat{\boldsymbol{r}}\sin\theta^{\mathrm{i}} + \boldsymbol{k}\cos\theta^{\mathrm{i}}$ and so is propagating along the generators of the cone of Fig. 9.14. If $\theta^{\mathrm{i}} = \frac{1}{2}\pi$, the cone becomes a plane perpendicular to the edge and the generators are radial lines from the edge, consistent with the two-dimensional behaviour already described.

The expressions (12.4) and (12.5) can be cast into a form related to (11.3) by introducing the coordinates ξ, η on a generator so that both are perpendicular to the generator and form a right-handed system with the direction of propagation. Choose ξ to lie in the plane containing the generator and edge. Thus η will be in the opposite direction to the unit ϕ vector and $E_\xi^{\mathrm{e}} = E_z^{(3)\mathrm{d}}\sin\theta^{\mathrm{i}} - E_r^{(3)\mathrm{d}}\cos\theta^{\mathrm{i}}$. For the incoming wave define ξ

and η in like manner relative to an incident ray. Then

$$\begin{pmatrix} E^{\mathrm{e}}_{\xi} \\ E^{\mathrm{e}}_{\eta} \end{pmatrix} = \begin{pmatrix} D_{\mathrm{S}} & 0 \\ 0 & D_{\mathrm{H}} \end{pmatrix} \begin{pmatrix} E^{\mathrm{i}}_{\xi Q} \\ E^{\mathrm{i}}_{\eta Q} \end{pmatrix} \frac{\exp(-\mathrm{i}kr \operatorname{cosec} \theta^{\mathrm{i}})}{(r \operatorname{cosec} \theta^{\mathrm{i}})^{\frac{1}{2}}} \tag{12.6}$$

where D_{S}, D_{H} are the same as in (11.5), (11.6). The magnetic intensity follows from (12.5).

9.13 Excitation by a dipole

The elementary point source of an electromagnetic wave is a dipole. Only the electric dipole need be considered in detail since the diffraction pattern of a semi-infinite plane irradiated from a magnetic dipole can be deduced from that for an electric dipole by means of Babinet's principle.

Let the direction of the electric dipole be the unit vector $\boldsymbol{p}_0$. Then the primary radiation can be derived from the Hertz vector $\boldsymbol{\Pi}^{\mathrm{i}} = \boldsymbol{p}_0 \mathrm{e}^{-\mathrm{i}kR_{\mathrm{s}}}/4\pi\varepsilon R_{\mathrm{s}}$. A promising line of attack therefore is to suppose that the total field can be expressed in terms of an electric Hertz vector via

$$\boldsymbol{E} = (\operatorname{grad} \operatorname{div} + k^2)\boldsymbol{\Pi}, \qquad \boldsymbol{H} = \mathrm{i}\omega\varepsilon \operatorname{curl} \boldsymbol{\Pi}$$

and invoke the results of §9.10.

If $\boldsymbol{p}_0 = \boldsymbol{k}$ so that the dipole is parallel to the edge the choice to try is $\boldsymbol{\Pi} = p^{(3)}_{\mathrm{S}}\boldsymbol{k}/\varepsilon$ where $p^{(3)}_{\mathrm{S}}$ is given by (10.1). It can be verified that the boundary, edge, and radiation conditions are satisfied. The field clearly has the correct singularity at the dipole and so the desired solution to the diffraction problem has been found.

When $\boldsymbol{p}_0 = \boldsymbol{i}$ the corresponding choice for $\boldsymbol{\Pi}$ would seem to be $\boldsymbol{\Pi} = p^{(3)}_{\mathrm{S}}\boldsymbol{i}/\varepsilon$. However, although this meets the boundary and radiation conditions it is too singular at the edge. Therefore, it is necessary to add solutions of Maxwell's equations (with no source present) which satisfy the radiation and boundary conditions while adjusting the behaviour of the field near the edge. According to §1.13 such a field can be represented by means of an electric Hertz vector $P\boldsymbol{k}$ and a magnetic Hertz vector $M\boldsymbol{k}$; we take the representation for the magnetic vector to be such that the contribution to the field is specified by $\boldsymbol{E} = \operatorname{curl} M\boldsymbol{k}$, $-\mathrm{i}\omega\mu\boldsymbol{H} = (\operatorname{grad} \operatorname{div} + k^2)M\boldsymbol{k}$. The near-field pattern of (10.6) suggests that we should try $M = -q(r, z)\sin\frac{1}{2}\phi$. Since M satisfies Helmholtz's equation it is necessary, if $q = h/r^{\frac{1}{2}}$, that

$$\partial^2 h/\partial r^2 + \partial^2 h/\partial z^2 + k^2 h = 0.$$

As this is Helmholtz's equation (with r, z treated as Cartesian coordinates) solutions such as $\mathrm{H}^{(2)}_0(kR')$ are easily constructed.

The procedure is eased by isolating the edge behaviour through

$$\partial p_S^{(3)}/\partial x + \partial p_S^{(3)}/\partial x_0 = \zeta \cos \tfrac{1}{2}\phi \cos \tfrac{1}{2}\phi_0,$$
$$\partial p_H^{(3)}/\partial x + \partial p_H^{(3)}/\partial x_0 = -\zeta \sin \tfrac{1}{2}\phi \sin \tfrac{1}{2}\phi_0,$$
$$\partial p_S^{(3)}/\partial y + \partial p_H^{(3)}/\partial y_0 = \zeta \sin \tfrac{1}{2}\phi \cos \tfrac{1}{2}\phi_0,$$
$$\partial p_S^{(3)}/\partial y_0 + \partial p_H^{(3)}/\partial y = \zeta \cos \tfrac{1}{2}\phi \sin \tfrac{1}{2}\phi_0,$$
$$\partial p_S^{(3)}/\partial z + \partial p_S^{(3)}/\partial z_0 = 0$$
$$\partial p_H^{(3)}/\partial z + \partial p_H^{(3)}/\partial z_0 = 0$$

where

$$\zeta = \frac{-ik(r+r_0)}{4\pi R'(rr_0)^{\frac{1}{2}}} H_1^{(2)}(kR').$$

Consideration of H_z leads one to pose

$$P = (1/k^2 r^{\frac{1}{2}})(\partial h/\partial z)\cos \tfrac{1}{2}\phi, \qquad M = -(h/r^{\frac{1}{2}})\sin \tfrac{1}{2}\phi$$

where

$$\frac{\partial h}{\partial r} = \frac{ik^2}{4\pi r_0^{\frac{1}{2}}} H_0^{(2)}(kR')\cos \tfrac{1}{2}\phi_0.$$

It is then discovered that all the edge conditions are met and

$$\varepsilon E_x = -\frac{\partial^2 p_S^{(3)}}{\partial x\, \partial x_0} + k^2 p_S^{(3)} - \frac{ik^2}{4\pi(rr_0)^{\frac{1}{2}}} H_0^{(2)}(kR')\cos \tfrac{1}{2}\phi \cos \tfrac{1}{2}\phi_0,$$

$$\varepsilon E_y = -\frac{\partial^2 p_S^{(3)}}{\partial y\, \partial x_0} - \frac{ik^2}{4\pi(rr_0)^{\frac{1}{2}}} H_0^{(2)}(kR')\sin \tfrac{1}{2}\phi \cos \tfrac{1}{2}\phi_0,$$

$$\varepsilon E_z = -\partial^2 p_S^{(3)}/\partial z\, \partial x_0,$$

$$H_x/\omega\varepsilon = \frac{z-z_0}{4\pi R'(rr_0)^{\frac{1}{2}}} H_1^{(2)}(kR')\sin \tfrac{1}{2}\phi \cos \tfrac{1}{2}\phi_0,$$

$$H_y/\omega\varepsilon = -i\frac{\partial p_S^{(3)}}{\partial z_0} - \frac{z-z_0}{4\pi R'(rr_0)^{\frac{1}{2}}} H_1^{(2)}(kR')\cos \tfrac{1}{2}\phi \cos \tfrac{1}{2}\phi_0,$$

$$H_z/\omega\varepsilon = i\,\partial p_H^{(3)}/\partial y_0.$$

It is interesting to note that the screen introduces a component of H_x although this is absent from the incident wave. This is one of the reasons why the two additional Hertz vectors have to be incorporated.

For $\mathbf{p}_0 = \mathbf{j}$ the same phenomenon occurs when one tries $\boldsymbol{\Pi} = p_H^{(3)}\mathbf{j}/\varepsilon$. The necessary extra fields are the same as before except that, in h, $\cos \tfrac{1}{2}\phi_0$ is replaced by $\sin\frac{1}{2}\phi_0$. The solution is therefore

$$\varepsilon E_x = -\frac{\partial^2 p_S^{(3)}}{\partial x\, \partial y_0} - \frac{ik^2}{4\pi(rr_0)^{\frac{1}{2}}} H_0^{(2)}(kR')\cos\tfrac{1}{2}\phi \sin\tfrac{1}{2}\phi_0,$$

$$\varepsilon E_y = -\frac{\partial^2 p_S^{(3)}}{\partial y\, \partial y_0} + k^2 p_H^{(3)} - \frac{ik^2}{4\pi(rr_0)^{\frac{1}{2}}} H_0^{(2)}(kR')\sin\tfrac{1}{2}\phi \sin\tfrac{1}{2}\phi_0,$$

$$\varepsilon E_z = -\partial^2 p_S^{(3)}/\partial z\, \partial y_0,$$

$$H_x/\omega\varepsilon = i\frac{\partial p_H^{(3)}}{\partial z_0} + \frac{z - z_0}{4\pi R'(rr_0)^{\frac{1}{2}}} H_1^{(2)}(kR')\sin\tfrac{1}{2}\phi \sin\tfrac{1}{2}\phi_0,$$

$$H_y/\omega\varepsilon = -\frac{z - z_0}{4\pi R'(rr_0)^{\frac{1}{2}}} H_1^{(2)}(kR')\cos\tfrac{1}{2}\phi \sin\tfrac{1}{2}\phi_0,$$

$$H_z/\omega\varepsilon = -i\, \partial p_H^{(3)}/\partial x_0.$$

As regards the diffracted field it can be expressed in an analogous manner to (12.6) as

$$\boldsymbol{E}^e = \boldsymbol{D}\left(\frac{r_0}{rR'}\right)^{\frac{1}{2}} e^{-ikr \operatorname{cosec} \beta_0} \boldsymbol{E}_Q^i$$

in the notation of §§9.11, 9.12 with $\boldsymbol{D}$ the matrix of diffraction coefficients.

9.14 Radiation from a semi-infinite pipe

Another problem amenable to the Wiener–Hopf technique is the radiation from a semi-infinite cylinder. The connected problem of the scattering of a plane wave by a semi-infinite pipe can also be solved. In fact, the two can be dealt with simultaneously because the reciprocity theorem of §1.35 reveals that there is a relation between the field produced inside the cylinder by an external source and that radiated externally from internal excitation. In particular, when the external source moves off to infinity the reciprocity when the primary field is a plane wave can be deduced.

Draw a large closed surface part of which penetrates the interior of the cylinder and part of which is outside (a cross-section is shown dashed in Fig. 9.17). Let p_1 and p_2 be two sound waves which have the same boundary condition on the pipe and which have no sources within the dashed surface of Fig. 9.17. Then

$$\int \left(p_1 \frac{\partial p_2}{\partial n} - \frac{\partial p_1}{\partial n} p_2\right) dS = 0,$$

the integral being taken over the dashed surface. Since either p or $\partial p/\partial n$ vanishes on the walls of the cylinder

$$\int_{L \cup \Omega} \left(p_1 \frac{\partial p_2}{\partial n} - \frac{\partial p_1}{\partial n} p_2\right) dS = 0 \qquad (14.1)$$

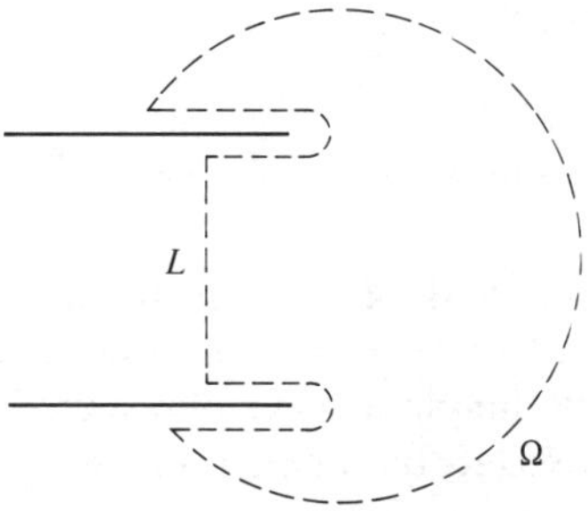

Fig. 9.17 The surface of integration for the reciprocity theorem

where L is the portion of the plane $z=-l$ inside the pipe and Ω is the part of the sphere of radius $\tilde{\omega}$ outside the cylinder.

Choose p_1 to be the total field caused by a plane wave travelling in the direction $\boldsymbol{n}^{\text{i}}$ where $\boldsymbol{n}^{\text{i}} \neq \boldsymbol{k}$ so that the incident wave is not initially moving along the pipe towards the mouth. Write

$$p_1 = \text{e}^{-\text{i}k\boldsymbol{x}\cdot\boldsymbol{n}^{\text{i}}} + p^{\text{s}}.$$

Inside the tube p_1 can be represented in terms of modes. Only those propagating to the left will, however, be present. Taking the cylinder to be rigid (there is a similar analysis when the boundary is soft) we see from §5.1 that

$$p_1 = \sum_{m=0}^{\infty} A_m \text{e}^{\text{i}\kappa_m z} \psi_m.$$

Suppose that p_2 is generated by some modes in the pipe moving from left to right. Inside the tube

$$p_2 = \sum_{m=0}^{\infty} (B_m \text{e}^{-\text{i}\kappa_m z} + C_m \text{e}^{\text{i}\kappa_m z})\psi_m.$$

The orthogonal properties of the eigenfunctions ψ_m now lead to

$$\int_L \left(p_1 \frac{\partial p_2}{\partial n} - \frac{\partial p_1}{\partial n} p_2 \right) \text{d}S = 2\text{i} \sum_{m=0}^{\infty} \kappa_m A_m B_m. \tag{14.2}$$

As regards the integral over Ω, p_2 will consist of an outgoing wave and

$$p_2 \sim A(\hat{\boldsymbol{x}})\text{e}^{-\text{i}kR}/R$$

where $R=|\boldsymbol{x}|$ and $\hat{\boldsymbol{x}}$ is a unit vector in the direction of $\boldsymbol{x}$. Then, as $\tilde{\omega} \to \infty$, the principle of stationary phase gives

$$\int_\Omega \left(p_1 \frac{\partial p_2}{\partial n} - \frac{\partial p_1}{\partial n} p_2 \right) \text{d}S \sim -\text{i}k\tilde{\omega} \int_\Omega (1 - \hat{\boldsymbol{x}}\cdot\boldsymbol{n}^{\text{i}}) A(\hat{\boldsymbol{x}}) \text{e}^{-\text{i}k\boldsymbol{x}\cdot\boldsymbol{n}^{\text{i}} - \text{i}k\tilde{\omega}}\, \text{d}S$$

$$\to -4\pi A(-\boldsymbol{n}^{\text{i}}). \tag{14.3}$$

Combining (14.1)–(14.3) we have

$$2\pi i A(-\boldsymbol{n}^{i}) = \sum_{m=0}^{\infty} \kappa_m A_m B_m. \tag{14.4}$$

The left-hand side concerns the field of p_2 radiated in the direction of the source of p_1 due to incident modes in the tube whereas the modes produced in the pipe by the plane wave play a role on the right-hand side. In particular, if p_2 is excited by a single mode so that $B_m = 0$ $(m \neq M)$

$$2\pi i A(-\boldsymbol{n}^{i}) = \kappa_M A_M B_M \tag{14.5}$$

which connects the radiation of the single mode with the amplitude of the same mode created by the incident plane wave.

Having established how the solution of one problem is interrelated to the other we discuss the matter of radiation only. Let the cross-section of the cylinder, which was previously unrestricted, be circular and of radius a. Assume that there is only one incident mode and that it possesses circular symmetry, though the method works equally well for asymmetric modes. Then

$$p^{i} = \mathrm{J}_0(j_n' r/a)\mathrm{e}^{-i\kappa_n z}$$

where $\mathrm{J}_0'(j_n') = -\mathrm{J}_1(j_n') = 0$ and $\kappa_n^2 = k^2 - j_n'^2/a^2$. The mode takes a particularly simple form for $n = 0$ since $j_0' = 0$.

Let the total field in $r \leqslant a$ be $p^{i} + p^{s}$ and write it as p^{s} in $r \geqslant a$. Then p^{s} must be a solution of Helmholtz's equation such that its normal derivative vanishes on the pipe while the total field and its derivatives are continuous across $r = a$, $z > 0$. Define the bilateral Laplace transform by

$$P(r) = \int_{-\infty}^{\infty} p^{s} \mathrm{e}^{-sz}\, \mathrm{d}z$$

where only the dependence on r is explicitly displayed since p^{s} is axially symmetric. As before, take $k = k_r - ik_i$; it follows without difficulty that $\mathrm{Im}(\kappa_m) \leqslant -k_i$, $\mathrm{Re}(\kappa_m) \leqslant k_r$ so that the Laplace transform exists in the strip $-k_i < \sigma < k_i$.

The equation satisfied by P is

$$\frac{1}{r}\frac{\mathrm{d}}{\mathrm{d}r}\left(r\frac{\mathrm{d}P}{\mathrm{d}r}\right) + \kappa^2 P = 0.$$

The solutions are Bessel functions. The appropriate ones are dictated by the field being finite at $r = 0$ and outgoing as $r \to \infty$. Thus

$$P(r) = \begin{cases} A\mathrm{J}_0(\kappa r) & (r < a) \\ B\mathrm{H}_0^{(2)}(\kappa r) & (r > a) \end{cases}.$$

Since $P'(=\mathrm{d}P/\mathrm{d}r)$ is continuous across $r=a$

$$AJ_0'(\kappa a)=BH_0^{(2)\prime}(\kappa a). \tag{14.6}$$

Define

$$P_R(r)=\int_0^\infty p^s e^{-sz}\,dz, \qquad P_L(r)=\int_{-\infty}^0 p^s e^{-sz}\,dz.$$

Then

$$P_L'(a)=0, \qquad P_R'(a+0)=P_R'(a-0), \tag{14.7}$$

$$P_R(a+0)=P_R(a-0)+J_0(j_n')/(s+i\kappa_n). \tag{14.8}$$

The implication of (14.6)–(14.8) is that

$$P_L(a+0)-P_L(a-0)=\frac{2iP_R'(a)}{\pi\kappa^2 aJ_1(\kappa a)H_1^{(2)}(\kappa a)}-\frac{J_0(j_n')}{s+i\kappa_n} \tag{14.9}$$

on taking advantage of the Wronskian relation (A.1.21).

To tackle (14.9) by the Wiener–Hopf technique let

$$K(s)=-\pi iJ_1(\kappa a)H_1^{(2)}(\kappa a). \tag{14.10}$$

It is shown in the next section that $K(s)=K_+(s)/K_-(s)$ where K_+ is regular and nonzero in $\sigma>-k_i$ whereas K_- is regular and nonzero in $\sigma<k_i$. Further $K_+(s)K_-(-s)=1$ and $K_+(s)\sim(as)^{-\frac{1}{2}}$ as $|s|\to\infty$ in the right half-plane.

Rewrite (14.9) as

$$\frac{(s-ik)}{K_-(s)}\{P_L(a+0)-P_L(a-0)\}+\frac{J_0(j_n')}{s+i\kappa_n}\left\{\frac{s-ik}{K_-(s)}+\frac{i(k+\kappa_n)}{K_-(-i\kappa_n)}\right\}$$

$$=\frac{2P_R'(a)}{a(s+ik)K_+(s)}+\frac{i(k+\kappa_n)J_0(j_n')}{K_-(-i\kappa_n)(s+i\kappa_n)}.$$

By an argument similar to that adopted earlier both sides are equal to an entire function. On account of the edge conditions the left-hand side is $O(|s|^{-\frac{1}{2}})$ as $|s|\to\infty$ in the negative half-plane; on the other hand, the right-hand side is $O(1/|s|)$ as $|s|\to\infty$ in the positive half-plane. Hence the entire function is, in fact, zero and

$$P_R'(a)=-ai(k+\kappa_n)J_0(j_n')(s+ik)K_+(s)/2K_-(-i\kappa_n)(s+i\kappa_n).$$

Since $A=-P_R'(a)/\kappa J_1(\kappa a)$ the field in $r<a$ is given by

$$p^s=\frac{a(k+\kappa_n)J_0(j_n')}{4\pi K_-(-i\kappa_n)}\int_{c-i\infty}^{c+i\infty}\frac{(s+ik)K_+(s)J_0(\kappa r)}{\kappa(s+i\kappa_n)J_1(\kappa a)}e^{sz}\,ds. \tag{14.10}$$

The only singularities of the integrand in the right half plane are simple poles at $s=i\kappa_m$, the zeros of $J_1(\kappa a)$. Therefore, when $z<0$, the contour of integration can be deformed to the right with the result

$$p^{s} = \sum_{m=0}^{\infty} R_{nm} J_0(j'_n) J_0(j'_m r/a) e^{i\kappa_m z} / J_0(j'_m) \tag{14.11}$$

where

$$R_{nm} = -\tfrac{1}{2}(k+\kappa_n)(k+\kappa_m)K_+(i\kappa_n)K_+(i\kappa_m)/\kappa_m(\kappa_m+\kappa_n). \tag{14.12}$$

In the expression for R_{nm} the relation $J'_1(j'_m) = J_0(j'_m)$ $(m \neq 0)$ has been used $(J'_1(j'_0) = \frac{1}{2}J_0(j'_0))$. The quantities R_{nm} can be regarded as reflection coefficients indicating the amplitude of the mth reflected mode when the nth mode is incident on the open end. At large distances from the mouth the most important reflection coefficients are those of the propagating modes. Some values of these are depicted graphically in Fig. 9.18.

The reflection coefficient has the property that

$$\kappa_m R_{nm} = \kappa_n R_{mn} \tag{14.13}$$

in view of (14.12). For the power flow the reflection of energy is governed by

$$r_{nm} = \kappa_m |R_{nm}|^2 / \kappa_n$$

as can be seen from §5.5. On account of (14.13) $r_{nm} = r_{mn}$ displaying the symmetric nature of the conversion of energy from one mode to another.

The field radiated from the mouth is given in $r > a$ by

$$p^{s} = \frac{a(k+\kappa_n)J_0(j'_n)}{4\pi K_-(-i\kappa_n)} \int_{c-i\infty}^{c+i\infty} \frac{(s+ik)K_+(s)H_0^{(2)}(\kappa r)}{\kappa(s+i\kappa_n)H_1^{(2)}(\kappa a)} e^{sz}\, ds. \tag{14.14}$$

The distant pattern can be estimated by the method of steepest descent (Appendix G). Let $z = R\cos\theta$, $r = R\sin\theta$ where $0 < \theta < \pi$. The saddle-point of $sz - i\kappa r$ occurs at $s = -ik\cos\theta$ and the curve of steepest descent does not pass near $s = \pm ik$ when θ is not near 0 or π. Hence, as $kR \to \infty$, $H_0^{(2)}(\kappa r)e^{i\kappa r}$ may be replaced on this path by $(2/\pi\kappa r)^{\frac{1}{2}} e^{\frac{1}{4}\pi i}$.

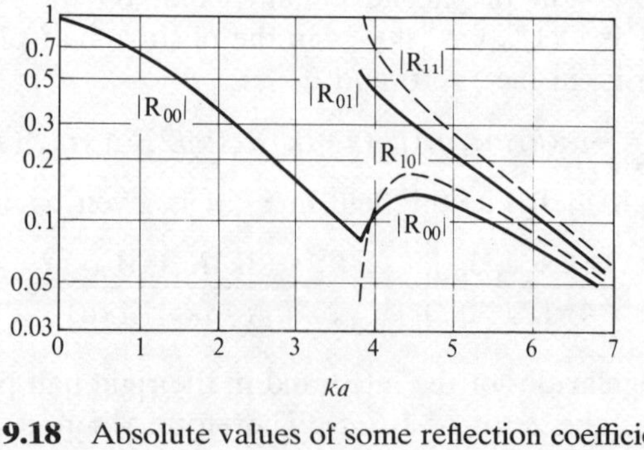

Fig. 9.18 Absolute values of some reflection coefficients

The question of whether any poles are captured in deforming the contour of integration into the path of steepest descent must now be faced. It is known that $H_1^{(2)}(z)$ has no zeros for $\frac{1}{2}\pi \geqslant \text{ph}\, z \geqslant -\frac{3}{2}\pi$. Now the curve of steepest descent always lies in a region where $\frac{1}{2}\pi \geqslant \text{ph}\, \kappa \geqslant -\frac{3}{2}\pi$ since it crosses from one sheet of the Riemann surface of κ to another when the contrary would arise. Since the contour of integration can be distorted into the curve of steepest descent without leaving a region in which $\frac{1}{2}\pi \geqslant \text{ph}\kappa \geqslant -\frac{3}{2}\pi$ there is no contribution from the zeros of $H_1^{(2)}(\kappa a)$.

Hence, if θ is not near 0 or π,

$$p^s \sim -\frac{a(k+\kappa_n)(1-\cos\theta)J_0(j_n')K_+(i\kappa_n)K_+(-ik\cos\theta)}{2\pi(\kappa_n - k\cos\theta)\sin\theta H_1^{(2)}(ka\sin\theta)}\frac{e^{-ikR}}{R}. \quad (14.15)$$

The polar diagram has been compared by Weinstein (1969) with the approximation derived from Kirchhoff's formulation (§9.26) with the conclusion that the Kirchhoff approximation is tolerably accurate in the forward direction (θ not too far from 0) but is gravely inadequate in the backward direction ($\theta \approx \pi$). In particular, the Kirchhoff formula is incapable of reproducing the directional effect in which the perpendicular radiation ($\theta = \frac{1}{2}\pi$) is always less than that backwards.

The combination of (14.5) and (14.15) shows that an incident plane wave $p^i = \exp(-ikr\sin\theta^i - ikz\cos\theta^i)$ falling on the open-ended pipe produces modes inside the tube of which the nth is

$$-\frac{ai(k+\kappa_n)(1+\cos\theta^i)J_0(j_n')K_+(i\kappa_n)K_+(ik\cos\theta^i)}{\kappa_n(\kappa_n + k\cos\theta^i)\sin\theta^i H_1^{(2)}(ka\sin\theta^i)} J_0\left(\frac{j_n' r}{a}\right)e^{i\kappa_n z}.$$

9.15 The 'split' functions

The matter of finding functions such that $K(s) = K_+(s)/K_-(s)$ may be addressed by looking for a decomposition such that

$$\ln K(s) = \ln K_+(s) - \ln K_-(s).$$

Construct a rectangle with vertices $\pm c \pm iM$ where $0 < c < k_i$ (Fig. 9.19). Since $J_1(\kappa a)H_1^{(2)}(\kappa a)$ has no zeros in the rectangle Cauchy's theorem asserts that

$$\ln K(s) = \frac{1}{\pi i}\int \frac{\ln K(w)}{w-s}\,dw$$

where the integral is taken round the contour in the positive sense and s is inside the rectangle. The growth of $K(w)$ at infinity ensures that the contributions from the sides $\text{Im}\, w = \pm M$ tend to zero as $M \to \infty$. Hence

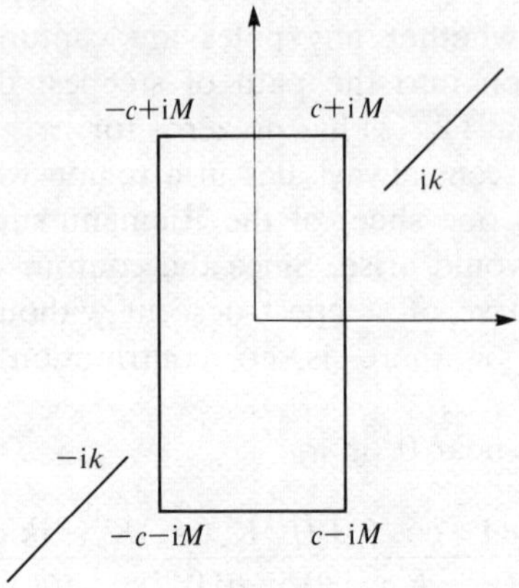

Fig. 9.19 The rectangular contour for the split functions

we can write

$$\ln K_+(s) = -\frac{1}{2\pi i}\mathscr{P}\int_{-c-i\infty}^{-c+i\infty}\frac{\ln K(w)}{w-s}\,dw \tag{15.1}$$

where $\mathrm{Re}(s) > -c$ and the symbol $\mathscr{P}$ signifies that the integral is to be calculated as the limit when both ends of the straight line go off to infinity at the same time. In fact, (15.1) specifies a function regular in $\mathrm{Re}(s) > -k_i$ for any c in $-k_i < c < k_i$ so long as $\mathrm{Re}(s) > -c$. Similarly

$$\ln K_-(s) = -\frac{1}{2\pi i}\mathscr{P}\int_{c-i\infty}^{c+i\infty}\frac{\ln K(w)}{w-s}\,dw \tag{15.2}$$

provided that $\mathrm{Re}(s) < c$. Since s is not on the counter of integration the integrals are always bounded and neither $K_+(s)$ nor $K_-(s)$ has a zero in its domain of regularity.

Change the variable of integration in (15.1) from w to $-w$. Then, because $K(w)$ is an even function of w,

$$\ln K_+(s) = \frac{1}{2\pi i}\mathscr{P}\int_{c-i\infty}^{c+i\infty}\frac{\ln K(w)}{w+s}\,dw. \tag{15.3}$$

It follows from (15.2) and (15.3) that

$$\ln K_+(s) + \ln K_-(-s) = 0, \tag{15.4}$$

i.e.

$$K_+(s)K_-(-s) = 1. \tag{15.5}$$

Given the flexibility in the choice of c, it may be replaced in (15.3) by $-c$ and then the addition of (15.1) gives

$$\ln K_+(s) = -\frac{s}{2\pi i}\int_{-c-i\infty}^{-c+i\infty}\frac{\ln K(w)}{w^2-s^2}\,dw.$$

Table 9.1 Values of the split functions

ka	$K_+(\mathrm{i}k)$	$K_+(\mathrm{i}\kappa_1)$	$K_+(\mathrm{i}\kappa_2)$
0	1	0.4533	0.3594
1.0	0.7206−0.4235i	0.4836−0.0147i	0.3721−0.0061i
2.0	0.3942−0.4391i	0.5395−0.0996i	0.3946−0.0371i
3.0	0.2348−0.3163i	0.5028−0.3026i	0.3954−0.1004i
4.0	0.2867−0.1903i	0.3440−0.1148i	0.3145−0.0976i
4.5	0.2790−0.2391i	0.3553−0.2531i	0.3488−0.0902i
5.0	0.2439−0.2532i	0.2880−0.2851i	0.3748−0.1137i
5.5	0.2078−0.2464i	0.2286−0.2742i	0.3878−0.1608i
6.0	0.1789−0.2260i	0.1870−0.2437i	0.3705−0.2294i
6.5	0.1612−0.1954i	0.1644−0.2026i	0.2842−0.2997i
7.0	0.1705−0.1388i	0.1749−0.1328i	0.0167−0.1197i
7.25	0.2068−0.1640i	0.2202−0.1614i	0.2805−0.1131i
7.5	0.2054−0.1801i	0.2194−0.1809i	0.2921−0.1750i
8.0	0.1884−0.1957i	0.1997−0.1998i	0.2507−0.2256i
8.5	0.1668−0.1970i	0.1748−0.2015i	0.2029−0.2295i
9.0	0.1473−0.1882i	0.1527−0.1915i	0.1666−0.2119i
9.5	0.1334−0.1717i	0.1372−0.1730i	0.1443−0.1832i
10.0	0.1305−0.1452i	0.1338−0.1437i	0.1398−0.1426i

Hence, for large values of $|s|$,

$$\ln K_+(s) = -\frac{s}{\pi \mathrm{i}}\int_{\mathrm{i}N}^{\mathrm{i}\infty} \frac{\ln K(w)}{w^2 - s^2}\,\mathrm{d}w + O(|s|^{-1})$$

where N is a fixed large positive number. Deploy the asymptotic formulae for Bessel functions in the integral and then

$$\ln K_+(s) \sim \frac{s}{\pi \mathrm{i}}\int_{\mathrm{i}N}^{\mathrm{i}\infty} \frac{\ln |w|\, a}{w^2 - s^2}\,\mathrm{d}w + O(|s|^{-1}\ln|sa|) \sim -\tfrac{1}{2}\ln(as) + O(|s|^{-1}\ln|sa|).$$

Consequently, $K_+(s) \sim (as)^{-\frac{1}{2}}$.

Properties of $K_-(s)$ follow immediately from (15.4) and (15.5). Some values of K_+ have been computed by Jones (1955) and Matsui (1960); an extract from these is displayed in Table 9.1.

9.16 The circular waveguide

The emission of an electromagnetic wave from a perfectly conducting circular waveguide can also be tackled by the Wiener–Hopf technique. It presents a new feature in that simultaneous Wiener–Hopf equations are involved. No general practical method has so far been found for dealing with a system but the particular case of a circular waveguide can be handled (for other possibilities see Exercises 24, 25).

It is known from Chapter 5 that the field inside a circular waveguide can be expressed as the sum of TE- and TM-modes, of which the TE_{11}-mode is dominant. First, the analogue of the reciprocity theorem established for sound waves will be derived.

Let $\boldsymbol{E}_1$, $\boldsymbol{H}_1$ and $\boldsymbol{E}_2$, $\boldsymbol{H}_2$ be any two electromagnetic fields which satisfy the boundary condition of perfect conductivity on the guide. Then, as for (14.1),

$$\int_{L\cup\Omega} (\boldsymbol{E}_1\wedge\boldsymbol{H}_2-\boldsymbol{E}_2\wedge\boldsymbol{H}_1)\cdot\boldsymbol{n}\,dS=0. \tag{16.1}$$

Let $\boldsymbol{E}_1$ be due to a plane wave propagating in the direction of $\boldsymbol{n}^{\mathrm{i}}$ ($\neq \boldsymbol{k}$) and write

$$\boldsymbol{E}_1=\boldsymbol{e}^{\mathrm{i}}\mathrm{e}^{-\mathrm{i}k\boldsymbol{x}\cdot\boldsymbol{n}^{\mathrm{i}}}+\boldsymbol{E}_1^{\mathrm{s}}, \qquad \boldsymbol{H}_1=\boldsymbol{h}^{\mathrm{i}}\mathrm{e}^{-\mathrm{i}k\boldsymbol{x}\cdot\boldsymbol{n}^{\mathrm{i}}}+\boldsymbol{H}_1^{\mathrm{s}}$$

where $\boldsymbol{e}^{\mathrm{i}}\cdot\boldsymbol{n}^{\mathrm{i}}=0$ and $(\mu/\varepsilon)^{\frac{1}{2}}\boldsymbol{h}^{\mathrm{i}}=\boldsymbol{n}^{\mathrm{i}}\wedge\boldsymbol{e}^{\mathrm{i}}$. Inside the guide use the lumped circuit representation of §5.4 so that the transverse components are given by

$$(\boldsymbol{E}_1)_{\mathrm{t}}=\sum_{m=1}^{\infty}\{C_{1m}\mathrm{e}^{\mathrm{i}\kappa_m z}(\boldsymbol{e}_m^{\mathrm{E}})_{\mathrm{t}}+A_{1m}\mathrm{e}^{\mathrm{i}\lambda_m z}(\boldsymbol{e}_m^{\mathrm{M}})_{\mathrm{t}}\},$$

$$(\boldsymbol{H}_1)_{\mathrm{t}}=-(1/\omega\mu)\sum_{m=1}^{\infty}\{\kappa_m C_{1m}\mathrm{e}^{\mathrm{i}\kappa_m z}(\boldsymbol{h}_m^{\mathrm{E}})_{\mathrm{t}}+k^2A_{1m}\mathrm{e}^{\mathrm{i}\lambda_m z}(\boldsymbol{h}_m^{\mathrm{M}})_{\mathrm{t}}/\lambda_m\}.$$

Assume that $\boldsymbol{E}_2$, $\boldsymbol{H}_2$ originates from some modes in the waveguide travelling from left to right so that

$$(\boldsymbol{E}_2)_{\mathrm{t}}=\sum_{m=1}^{\infty}\{(C_{2m}\mathrm{e}^{\mathrm{i}\kappa_m z}+D_{2m}\mathrm{e}^{-\mathrm{i}\kappa_m z})(\boldsymbol{e}_m^{\mathrm{E}})_{\mathrm{t}}+(A_{2m}\mathrm{e}^{\mathrm{i}\lambda_m z}-B_{2m}\mathrm{e}^{-\mathrm{i}\lambda_m z})(\boldsymbol{e}_m^{\mathrm{M}})_{\mathrm{t}}\},$$

$$(\boldsymbol{H}_2)_{\mathrm{t}}=-(1/\omega\mu)\sum_{m=1}^{\infty}\{\kappa_m(C_{2m}\mathrm{e}^{\mathrm{i}\kappa_m z}-D_{2m}\mathrm{e}^{-\mathrm{i}\kappa_m z})(\boldsymbol{h}_m^{\mathrm{E}})_{\mathrm{t}}$$
$$+k^2(A_{2m}\mathrm{e}^{\mathrm{i}\lambda_m z}+B_{2m}\mathrm{e}^{-\mathrm{i}\lambda_m z})(\boldsymbol{h}_m^{\mathrm{M}})_{\mathrm{t}}/\lambda_m\}.$$

The orthogonal properties of the $(\boldsymbol{e}_m)_{\mathrm{t}}$ entail

$$\int_{L}(\boldsymbol{E}_1\wedge\boldsymbol{H}_2-\boldsymbol{E}_2\wedge\boldsymbol{H}_1)\cdot\boldsymbol{n}\,\mathrm{d}S=-(2/\omega\mu)\sum_{m=1}^{\infty}(\kappa_m C_{1m}D_{2m}-k^2A_{1m}B_{2m}/\lambda_m).$$

Also, if $\boldsymbol{E}_2\sim\mathbf{A}(\hat{\boldsymbol{x}})\mathrm{e}^{-\mathrm{i}kR}/R$ as $R\to\infty$,

$$\int_{\Omega}(\boldsymbol{E}_1\wedge\boldsymbol{H}_2-\boldsymbol{E}_2\wedge\boldsymbol{H}_1)\cdot\boldsymbol{n}\,\mathrm{d}S\sim\left(\frac{\varepsilon}{\mu}\right)^{\frac{1}{2}}\frac{\mathrm{e}^{-\mathrm{i}kR}}{R}\int_{\Omega}\{\boldsymbol{e}^{\mathrm{i}}\cdot\mathbf{A}(\hat{\boldsymbol{x}})-\mathbf{A}(\hat{\boldsymbol{x}})$$
$$\wedge(\boldsymbol{n}^{\mathrm{i}}\wedge\boldsymbol{e}^{\mathrm{i}})\cdot\hat{\boldsymbol{x}}\}\mathrm{e}^{-\mathrm{i}kR\hat{\boldsymbol{x}}\cdot\boldsymbol{n}^{\mathrm{i}}}\,\mathrm{d}S\to-(4\pi\mathrm{i}/\omega\mu)\boldsymbol{e}^{\mathrm{i}}\cdot\mathbf{A}(-\boldsymbol{n}^{\mathrm{i}})$$

by the principle of stationary phase. Hence

$$\boldsymbol{e}^{\mathrm{i}}\cdot\mathbf{A}(-\boldsymbol{n}^{\mathrm{i}})=-(1/2\pi\mathrm{i})\sum_{m=1}^{\infty}(\kappa_m C_{1m}D_{2m}-k^2A_{1m}B_{2m}/\lambda_m) \tag{16.2}$$

is the desired reciprocity result; it holds irrespective of the cross-section of the cylinder.

Now suppose that the circularly symmetric TE_{on}-mode is propagating along the positive z-axis. Its nonzero components are

$$H_{nz} = \mathrm{J}_0(j_n' r/a)\mathrm{e}^{-\mathrm{i}\kappa_n z}, \qquad H_{nr} = (\mathrm{i}\kappa_n/j_n' a)\mathrm{J}_1(j_n' r/a)\mathrm{e}^{-\mathrm{i}\kappa_n z},$$
$$E_{n\phi} = -(\omega\mu/\kappa_n)H_{nr}.$$

Let the total field in $r \leqslant a$ be $\boldsymbol{E}_n + \boldsymbol{e}$, $\boldsymbol{H}_n + \boldsymbol{h}$ and, in $r \geqslant a$, be $\boldsymbol{e}$, $\boldsymbol{h}$. Introduce the bilateral Laplace transforms $\mathscr{E}(r)$, $\mathscr{H}(r)$ where

$$\mathscr{E}(r) = \int_{-\infty}^{\infty} \boldsymbol{e}\mathrm{e}^{-sz}\,\mathrm{d}z, \qquad \mathscr{H}(r) = \int_{-\infty}^{\infty} \boldsymbol{h}\mathrm{e}^{-sz}\,\mathrm{d}z.$$

Then

$$\mathscr{E}_z = a_0\mathrm{J}_0(\kappa r), \qquad \mathscr{H}_z = b_0\mathrm{J}_0(\kappa r), \quad (r < a)$$
$$\mathscr{E}_z = A_0\mathrm{H}_0^{(2)}(\kappa r), \qquad \mathscr{H}_z = B_0\mathrm{H}_0^{(2)}(\kappa r) \quad (r > a).$$

The continuity of e_z (and hence of $\mathscr{E}_z$) across $r = a$ requires

$$a_0\mathrm{J}_0(\kappa a) = A_0\mathrm{H}_0^{(2)}(\kappa a). \tag{16.3}$$

From Maxwell's equations

$$\mathrm{i}\omega\varepsilon\mathscr{E}_\phi = s\mathscr{H}_r - \partial\mathscr{H}_z/\partial r, \qquad \mathrm{i}\omega\mu\mathscr{H}_r = s\mathscr{E}_\phi$$

whence

$$\kappa^2\mathscr{E}_\phi = \mathrm{i}\omega\mu\,\partial\mathscr{H}_z/\partial r. \tag{16.4}$$

Therefore $\mathscr{E}_\phi$ is continuous through $r = a$ provided that

$$b_0\mathrm{J}_0'(\kappa a) = B_0\mathrm{H}_0^{(2)\prime}(\kappa a). \tag{16.5}$$

Let

$$\mathscr{E}_\mathrm{R}(r) = \int_0^{\infty} \boldsymbol{e}\mathrm{e}^{-sz}\,\mathrm{d}z, \qquad \mathscr{E}_\mathrm{L}(r) = \int_{-\infty}^{0} \boldsymbol{e}\mathrm{e}^{-sz}\,\mathrm{d}z.$$

Because

$$\mathscr{H}_{z\mathrm{R}}(a+0) = \mathscr{H}_{z\mathrm{R}}(a-0) + \mathrm{J}_0(j_n')/(s + \mathrm{i}\kappa_n)$$

we obtain from (16.5)

$$J_\mathrm{L} = \mathscr{H}_{z\mathrm{L}}(a+0) - \mathscr{H}_{z\mathrm{L}}(a-0) = \{2\mathrm{i}B_0/\pi\kappa a\mathrm{J}_0'(\kappa a)\} - \mathrm{J}_0(j_n')/(s + \mathrm{i}\kappa_n)$$

on taking advantage of the Wronskian relation (A.1.21).

On the other hand, $\mathscr{E}_{\phi\mathrm{L}}(a) = 0$ and so (16.4) gives

$$\kappa\mathscr{E}_{\phi\mathrm{R}}(a) = -\mathrm{i}\omega\mu B_0\mathrm{H}_1^{(2)}(\kappa a).$$

Consequently,

$$J_\mathrm{L} = \frac{-2\mathrm{i}\mathscr{E}_{\phi\mathrm{R}}(a)}{\omega\mu a K(s)} - \frac{\mathrm{J}_0(j_n')}{s + \mathrm{i}\kappa_n}. \tag{16.6}$$

Next, note that Maxwell's equations imply that $\kappa^2\mathscr{H}_\phi = -\mathrm{i}\omega\varepsilon\, \partial\mathscr{E}_z/\partial r$. Since $\mathscr{H}_{\phi R}$ is continuous

$$Z_L = \mathscr{H}_{\phi L}(a+0) - \mathscr{H}_{\phi L}(a-0) = -2\omega\varepsilon A_0/\pi\kappa^2 a J_0(\kappa a)$$

from (16.3). Hence, since $\mathscr{E}_{zL}(a) = 0$,

$$Z_L = 2\mathrm{i}\omega\varepsilon\mathscr{E}_{zR}(a)/\kappa^2 a L(s) \tag{16.7}$$

where $L(s) = -\pi \mathrm{i} J_0(\kappa a) H_0^{(2)}(\kappa a)$.

The function $L(s)$ can be split as $L_+(s)/L_-(s)$ where $L_+(s)L_-(-s) = 1$ and $L_+(s) \sim (as)^{-\frac{1}{2}}$ as $|s| \to \infty$ in the positive half-plane. Write (16.7) as

$$(s - \mathrm{i}k)Z_L/L_-(s) = 2\mathrm{i}\omega\varepsilon\mathscr{E}_{zR}(a)/a(s+\mathrm{i}k)L_+(s).$$

By the standard argument both sides are equal to the same entire function. By virtue of the edge conditions the right-hand side is $O(1/|s|)$ as $|s| \to \infty$ in the right half-plane whereas the left-hand side is $O(1/|s|)$ as $|s| \to \infty$ in the left half-plane. Thus the entire function is zero. Hence $\mathscr{E}_{zR}(a) = 0$ whence $a_0 = A_0 = 0$ so that $\mathscr{E}_z$, $\mathscr{H}_\phi$ and $\mathscr{E}_r$ vanish as might have been expected.

Equation (16.6) may be dealt with in the same way with the result

$$J_L = (s + \mathrm{i}\kappa_n)^{-1}\{K_-(s)K_+(\mathrm{i}\kappa_n) - 1\}J_0(j'_n).$$

Hence

$$B_0 = \tfrac{1}{2}\pi \mathrm{i}\kappa a J_1(\kappa a)K_-(s)K_+(\mathrm{i}\kappa_n)(s+\mathrm{i}\kappa_n)^{-1}J_0(j'_n) \tag{16.6}$$

and, in $r < a$,

$$h_z = -\frac{aK_+(\mathrm{i}k_n)}{4\pi \mathrm{i}} J_0(j'_n) \int_{c-\mathrm{i}\infty}^{c+\mathrm{i}\infty} \frac{\kappa K_+(s)J_0(\kappa r)}{(s+\mathrm{i}\kappa_n)J_1(\kappa a)} \mathrm{e}^{sz}\, \mathrm{d}s.$$

Deformation of the contour to the right supplies

$$h_z = \sum_{m=1}^{\infty} R'_{nm} J_0(j'_n) J_0(j'_m r/a) \mathrm{e}^{\mathrm{i}\kappa_m z}/J_0(j'_m) \tag{16.7}$$

where

$$R'_{nm} = -\tfrac{1}{2}K_+(\mathrm{i}\kappa_n)K_+(\mathrm{i}\kappa_m) j'^2_m/\kappa_m(\kappa_n + \kappa_m)a^2.$$

The reflection coefficients R_{nm} have the property that

$$\kappa_m R'_{nm}/j'^2_m = \kappa_n R'_{mn}/j'^2_n$$

Some values of the reflection coefficient are tabulated in Table 9.2. It will be noticed that $|R_{12}|$ is appreciably larger than $|R_{11}|$ when the first two modes can propagate.

A similar analysis can be carried out for TM-modes. The behaviour of the reflection coefficients is qualitatively similar but they are much greater

Table 9.2 Reflection coefficients $R'_{nm} = |R'_{nm}| \exp i(\pi + \theta_{nm})$

ka	$\lvert R'_{11}\rvert \times 10^2$	θ_{11} (degrees)	$\lvert R'_{12}\rvert \times 10^2$	θ_{12} (degrees)	$\lvert R'_{22}\rvert \times 10^2$	θ_{22} (degrees)
4	36.6	57				
4.5	12.6	71				
5	5.84	90				
5.5	3.01	100				
6	1.62	105				
6.5	0.906	102				
7	0.516	74				
7.25	0.722	72	1.23	58	33.7	54
7.5	0.714	79	1.21	70	20.3	62
8	0.594	90	0.915	87	9.47	84
8.5	0.454	98	0.638	98	5.01	97
9	0.332	103	0.432	103	2.81	104
9.5	0.237	103	0.288	103	1.63	103
10	0.166	94	0.191	93	0.966	91

than those for TE-modes. In other words, *incident TE-modes radiate more energy outside the guide than incident TM-modes.*

The radiation pattern for incident TE-modes can be calculated from (16.4) and (16.6). Thus, for $r > a$,

$$e_\phi = K_+(i\kappa_n)J_0(j'_n)\frac{\omega\mu a}{4\pi}\int_{c-i\infty}^{c+i\infty}\frac{K_+(s)H_1^{(2)}(\kappa r)}{(s+i\kappa_n)H_1^{(2)}(\kappa a)}e^{sz}\,ds. \tag{16.8}$$

As in §9.14 we deduce that

$$e_\phi \sim -\frac{\omega\mu a K_+(i\kappa_n)J_0(j'_n)K_+(-ik\cos\theta)}{2\pi(\kappa_n - k\cos\theta)H_1^{(2)}(ka\sin\theta)}\frac{e^{-ikR}}{R}. \tag{16.9}$$

The modes induced by a plane wave falling on the open end of the waveguide can be inferred from (16.2) and (16.9).

9.17 The strip

The Wiener–Hopf method is an extremely powerful weapon for attacking certain types of diffraction problem. Examples have already been given illustrating its ability to supply exact solutions but it is also capable of providing approximate answers if the circumstances are right. To indicate how this is achieved the diffraction of a plane wave by a soft strip will be discussed although other excitations and different boundary conditions can be coped with; indeed the technique can be applied to structures other than the strip.

Let the strip occupy the region $-l \leqslant x \leqslant 0$, $y = 0$, $-\infty < z < \infty$. The irradiating field is the plane wave $p^i = \exp\{-ik(x\cos\phi^i + y\sin\phi^i)\}$ where, without loss of generality, the restriction $0<\phi^{\mathrm{i}}\leqslant\frac{1}{2}\pi$ can be imposed. Write the total field as $p^{\mathrm{i}}+p^{\mathrm{s}}$. If

$$P(s, y)=\int_{-\infty}^{\infty} p^{\mathrm{s}}\mathrm{e}^{-sx}\,\mathrm{d}x$$

it is evident from the discussion of §9.7 that

$$P(s, y)=A\mathrm{e}^{-\mathrm{i}\kappa|y|}$$

since p^{s} is continuous across $y=0$.

Put

$$P_{\mathrm{R}}(y)=\int_{0}^{\infty} p^{\mathrm{s}}\mathrm{e}^{-sx}\,\mathrm{d}x,\qquad P_{\mathrm{L}}(y)=\int_{-\infty}^{-l} p^{\mathrm{s}}\mathrm{e}^{-sx}\,\mathrm{d}x,\qquad P_{\mathrm{E}}(y)=\int_{-l}^{0} p^{\mathrm{s}}\mathrm{e}^{-sx}\,\mathrm{d}x$$

so that

$$P_{\mathrm{R}}(0)+P_{\mathrm{L}}(0)+P_{\mathrm{E}}(0)=A. \tag{17.1}$$

The additional function which has been defined, $P_{\mathrm{E}}(y)$, is an entire function. As $|s|\to\infty$ in the right half-plane the dominant behaviour of P_{E} comes from that of p^{s} near $x=-l$ and involves the factor e^{sl}. In the negative half-plane as $|s|\to\infty$, P_{E} will be $O(1)$ but P_{L} will contain e^{sl}. These general remarks are confirmed by the explicit form of P_{E} which is

$$P_{\mathrm{E}}(0)=(s+\mathrm{i}k\cos\phi^{\mathrm{i}})^{-1}[1-\exp\{(s+\mathrm{i}k\cos\phi^{\mathrm{i}})l\}] \tag{17.2}$$

since $p^{\mathrm{s}}=-p^{\mathrm{i}}$ on the strip.

Derivatives with respect to y are continuous away from the strip. Therefore

$$P_{\mathrm{E}}'(0+0)-P_{\mathrm{E}}'(0-0)=-2\mathrm{i}\kappa A. \tag{17.3}$$

Substitute for A from (17.1) and rewrite the equation as

$$-2\mathrm{i}(s+\mathrm{i}k)^{\frac{1}{2}}\{P_{\mathrm{R}}(0)+P_{\mathrm{L}}(0)+P_{\mathrm{E}}(0)\}=(s-\mathrm{i}k)^{-\frac{1}{2}}\{P_{\mathrm{E}}'(0+0)-P_{\mathrm{E}}'(0-0)\}. \tag{17.4}$$

The function $(s+\mathrm{i}k)^{\frac{1}{2}}\{P_{\mathrm{L}}(0)+P_{\mathrm{E}}(0)\}$ can be split into the sum of two functions regular in the right and left half-planes respectively by means of the rectangular contour of Fig. 9.19. Thus

$$(s+\mathrm{i}k)^{\frac{1}{2}}\{P_{\mathrm{L}}(0)+P_{\mathrm{E}}(0)\}=X_{\mathrm{R}}(s)+X_{\mathrm{L}}(s)$$

where

$$X_{\mathrm{R}}(s)=-\frac{1}{2\pi\mathrm{i}}\int_{-c-\mathrm{i}\infty}^{-c+\mathrm{i}\infty}\frac{(w+\mathrm{i}k)^{\frac{1}{2}}}{w-s}\{P_{\mathrm{L}}(w,0)+P_{\mathrm{E}}(w,0)\}\,\mathrm{d}w \quad (\mathrm{Re}\, s>-c),$$

$$X_{\mathrm{L}}(s)=\frac{1}{2\pi\mathrm{i}}\int_{c-\mathrm{i}\infty}^{c+\mathrm{i}\infty}\frac{(w+\mathrm{i}k)^{\frac{1}{2}}}{w-s}\{P_{\mathrm{L}}(w,0)+P_{\mathrm{E}}(w,0)\}\,\mathrm{d}w \quad (\mathrm{Re}\, s<c).$$

Equation (17.4) can then be rearranged as

$$-2\mathrm{i}X_{\mathrm{L}}(s)-(s-\mathrm{i}k)^{-\frac{1}{2}}\{P_{\mathrm{E}}'(0+0)-P_{\mathrm{E}}'(0-0)\}=2\mathrm{i}(s+\mathrm{i}k)^{\frac{1}{2}}P_{\mathrm{R}}(0)+2\mathrm{i}X_{\mathrm{R}}(s).$$

By the standard Wiener–Hopf argument both sides are equal to the entire function which is zero. Hence

$$P_{\mathrm{R}}(s,0)=-(s+ik)^{-\frac{1}{2}}X_{\mathrm{R}}(s). \tag{17.5}$$

Instead of deriving (17.4) from (17.3) by multiplying by $(s-ik)^{-\frac{1}{2}}$ the choice of multiplication by $(s+ik)^{-\frac{1}{2}}$ is open. However, it is better to elect for $\mathrm{e}^{-sl}(s+ik)^{-\frac{1}{2}}$ because of the exponential growth of P'_{E} at infinity in the positive half-plane. Thus, (17.4) is rewritten as

$$-2\mathrm{i}(s-ik)^{\frac{1}{2}}\mathrm{e}^{-sl}\{P_{\mathrm{R}}(0)+P_{\mathrm{L}}(0)+P_{\mathrm{E}}(0)\}=(s+ik)^{-\frac{1}{2}}\mathrm{e}^{-sl}\{P'_{\mathrm{E}}(0+0)-P'_{\mathrm{E}}(0-0)\}.$$

Following the same route as before we obtain

$$\mathrm{e}^{-sl}P_{\mathrm{L}}(s,0)=-(s-ik)^{-\frac{1}{2}}Y_{\mathrm{L}}(s) \tag{17.6}$$

where

$$Y_{\mathrm{L}}(s)=\frac{1}{2\pi\mathrm{i}}\int_{c-\mathrm{i}\infty}^{c+\mathrm{i}\infty}\frac{(w-ik)^{\frac{1}{2}}}{w-s}\{P_{\mathrm{R}}(w,0)+P_{\mathrm{E}}(w,0)\}\mathrm{e}^{-wl}\,\mathrm{d}w \quad (\mathrm{Re}\, s<c).$$

The Wiener–Hopf technique has not furnished an exact solution but has expressed P_{R} as an integral involving P_{L} via (17.5) and P_{L} in terms of P_{R} through (17.6). More symmetrical relations can be achieved by changing s to $-s$ in (17.6) so that

$$\mathrm{e}^{sl}P_{\mathrm{L}}(-s,0)=-\mathrm{i}(s+ik)^{-\frac{1}{2}}Y_{\mathrm{L}}(-s). \tag{17.7}$$

Switch the variable of integration in X_{R} from w to $-w$; then

$$X_{\mathrm{R}}(s)=\frac{1}{2\pi}\int_{c-\mathrm{i}\infty}^{c+\mathrm{i}\infty}\frac{(w-ik)^{\frac{1}{2}}}{w+s}\{P_{\mathrm{L}}(-w,0)+P_{\mathrm{E}}(-w,0)\}\,\mathrm{d}w \quad (\mathrm{Re}\, s>-c)$$

which has the same range of validity as the integral for $Y_{\mathrm{L}}(-s)$. It is reasonable to denote $P_{\mathrm{R}}(s)+\mathrm{e}^{sl}P_{\mathrm{L}}(-s)$ by $I_{\mathrm{R}}(s)$ because $\mathrm{e}^{sl}P_{\mathrm{L}}(-s)$ is regular in the right half-plane. Likewise $P_{\mathrm{R}}(s)-\mathrm{e}^{sl}P_{\mathrm{L}}(-s)$ can be called $H_{\mathrm{R}}(s)$. The addition and subtraction of (17.5) and (17.7) then provide

$$(s+ik)^{\frac{1}{2}}I_{\mathrm{R}}(s)=Q_1(s)-\frac{1}{2\pi}\int_{c-\mathrm{i}\infty}^{c+\mathrm{i}\infty}\frac{(w-ik)^{\frac{1}{2}}}{w+s}I_{\mathrm{R}}(w)\mathrm{e}^{-wl}\,\mathrm{d}w, \tag{17.8}$$

$$(s+ik)^{\frac{1}{2}}H_{\mathrm{R}}(s)=Q_2(s)+\frac{1}{2\pi}\int_{c-\mathrm{i}\infty}^{c+\mathrm{i}\infty}\frac{(w-ik)^{\frac{1}{2}}}{w+s}H_{\mathrm{R}}(w)\mathrm{e}^{-wl}\,\mathrm{d}w \tag{17.9}$$

where

$$Q_1(s)=-\frac{1}{2\pi}\int_{c-\mathrm{i}\infty}^{c+\mathrm{i}\infty}\frac{(w-ik)^{\frac{1}{2}}}{w+s}\{P_{\mathrm{E}}(w,0)\mathrm{e}^{-wl}+P_{\mathrm{E}}(-w,0)\}\,\mathrm{d}w,$$

$$Q_2(s)=\frac{1}{2\pi}\int_{c-\mathrm{i}\infty}^{c+\mathrm{i}\infty}\frac{(w-ik)^{\frac{1}{2}}}{w+s}\{P_{\mathrm{E}}(w,0)\mathrm{e}^{-wl}-P_{\mathrm{E}}(-w,0)\}\,\mathrm{d}w$$

and P_{E} is given by (17.2).

The two integral equations (17.8), (17.9) are of Fredholm type and of the second kind. They both have essentially the same kernel and any method for dealing with one will cope with the other. Even if they cannot be solved exactly they are susceptible to known methods of approximation.

Choose $k_i \cos\phi^i < c < k_i$ and put

$$J_R(s) = I_R(s) + (s + ik\cos\phi^i)^{-1} + (s - ik\cos\phi^i)\exp(ikl\cos\phi^i),$$
$$G_R(s) = H_R(s) + (s + ik\cos\phi^i)^{-1} - (s - ik\cos\phi^i)\exp(ikl\cos\phi^i).$$

The integrals from which J_R and G_R are absent can be evaluated by deforming the contour to the left. Hence

$$(s+ik)^{\frac{1}{2}}J_R(s) = S_1(s) - \frac{1}{2\pi}\int_{c-i\infty}^{c+i\infty}\frac{(w-ik)^{\frac{1}{2}}}{w+s}J_R(w)e^{-wl}\,dw, \quad (17.10)$$

$$(s+ik)^{\frac{1}{2}}G_R(s) = S_2(s) + \frac{1}{2\pi}\int_{c-i\infty}^{c+i\infty}\frac{(w-ik)^{\frac{1}{2}}}{w+s}G_R(w)e^{-wl}\,dw \quad (17.11)$$

where

$$S_1(s) + S_2(s) = 2(2k)^{\frac{1}{2}}e^{\frac{1}{4}\pi i}(s + ik\cos\phi^i)^{-1}\sin\tfrac{1}{2}\phi^i, \quad (17.12)$$
$$S_1(s) - S_2(s) = 2(2k)^{\frac{1}{2}}e^{\frac{1}{4}\pi i}(s - ik\cos\phi^i)^{-1}e^{ikl\cos\phi^i}\cos\tfrac{1}{2}\phi^i. \quad (17.13)$$

For the normally incident plane wave $\phi^i = \frac{1}{2}\pi$ and $S_2(s) = 0$. Then (17.11) implies that $G_R = 0$ in conformity with what is dictated by symmetry. Also (17.10) reduces to

$$(s+ik)^{\frac{1}{2}}J_R(s) = \frac{2k^{\frac{1}{2}}}{s}e^{\frac{1}{4}\pi i} - \frac{1}{2\pi}\int_{c-i\infty}^{c+i\infty}\frac{(w-ik)^{\frac{1}{2}}}{w+s}J_R(w)e^{-wl}\,dw. \quad (17.14)$$

The total field in $y > 0$ is

$$p = p^i + \frac{1}{2\pi i}\int_{c-i\infty}^{c+i\infty}P(s,0)e^{sx - i\kappa y}\,ds.$$

When $x = r\cos\phi$ and $y = r\sin\phi$ the integral can be estimated asymptotically by the method of steepest descents for large kr. The saddle-point occurs at $s = -ik\cos\phi$ and

$$p \sim p^i + P(-ik\cos\phi, 0)\left(\frac{k}{2\pi r}\right)^{\frac{1}{2}}e^{-ikr + \frac{1}{4}\pi i}\sin\phi \quad (17.15)$$

since $y > 0$. This can be expressed in terms of the solutions of (17.10) and (17.11) because

$$2P(s,0) = J_R(s) + G_R(s) + \{J_R(-s) - G_R(-s)\}e^{sl}. \quad (17.16)$$

One implication of (17.15) is that the scattering coefficient of a soft strip is determined from §8.2 as

$$\sigma_c = -(2/l)\mathrm{Re}\,P(-ik\cos\phi^i, 0). \quad (17.17)$$

Iteration suggests itself as a course of action for solving (17.10) and (17.11). No attempt will be made here to demonstrate the convergence of such a process. Instead it will merely be observed that when kl is large the integrals should be small so that one or two terms of the iteration should be a suitable approximation.

The first approximation is obviously

$$(s+ik)^{\frac{1}{2}}J_R(s)=S_1(s), \qquad (s+ik)^{\frac{1}{2}}G_R(s)=S_2(s). \qquad (17.18)$$

The physical interpretation of this is that the field on $y=0$, $x>0$ is the same as if there were a semi-infinite soft screen going from $x=0$ to $x=-\infty$ whereas the field on $y=0$, $x<-l$ is the same as if there were a semi-infinite plane extending from $x=-l$ to $x=\infty$. The approximation is therefore physically plausible.

A better approximation should be obtained by inserting (17.18) in the integrals of (17.10) and (17.11). To assist in the evaluation of the integrals consider

$$I=\frac{1}{2\pi}\int_{c-i\infty}^{c+i\infty}\frac{(w-ik)^{-\frac{1}{2}}}{w+s}\,e^{-wb}\,dw \quad (\mathrm{Re}\, s>-c).$$

Then

$$\begin{aligned} d(e^{-sb}I)/db &= -(1/2\pi)\int_{c-i\infty}^{c+i\infty}(w-ik)^{-\frac{1}{2}}e^{-b(w+s)}\,dw \\ &= (\pi b)^{-\frac{1}{2}}e^{-b(s+ik)} \end{aligned}$$

by deforming the contour to the right and wrapping it round the branch-line from ik to infinity. When $b=0$, the contour in I can be deformed to the left and then only the pole at $w=-s$ contributes resulting in the value $-(s+ik)^{-\frac{1}{2}}$. Hence

$$e^{-sb}I=\int_0^b(\pi t)^{-\frac{1}{2}}e^{-t(s+ik)}\,dt-(s+ik)^{-\frac{1}{2}}$$

whence

$$I=-\{2/\pi^{\frac{1}{2}}(s+ik)^{\frac{1}{2}}\}e^{\frac{1}{4}\pi i-ikb}\mathrm{F}\{(s+ik)^{\frac{1}{2}}b^{\frac{1}{2}}e^{-\frac{1}{4}\pi i}\}.$$

Therefore

$$\begin{aligned} &\frac{1}{2\pi}\int_{c-i\infty}^{c+i\infty}\frac{(w-ik)^{\frac{1}{2}}}{w+s}\,e^{-wl}\,dw \\ &\qquad =-(\pi l)^{-\frac{1}{2}}e^{-ikl}+\frac{2}{\pi^{\frac{1}{2}}}(s+ik)^{\frac{1}{2}}e^{\frac{1}{4}\pi i-ikl}\mathrm{F}\{(s+ik)^{\frac{1}{2}}l^{\frac{1}{2}}e^{-\frac{1}{4}\pi i}\}. \quad (17.19) \end{aligned}$$

Let $\mathscr{J}(s)=(s+ik)^{\frac{1}{2}}J_R(s)-S_1(s)$ so that, from (17.10),

$$\mathscr{J}(s)=-\frac{1}{2\pi}\int_{c-i\infty}^{c+i\infty}\frac{(w-ik)^{\frac{1}{2}}}{w+s}\,J_R(w)e^{-wl}\,dw.$$

Substitute (17.18) in the integral. The occurrence of $(s+\mathrm{i}k)^{\frac{1}{2}}$ in the denominator is a complicating factor. In essence, though, (17.19) stems from bending the contour round the branch line from $s=\mathrm{i}k$. The main contribution comes from a neighbourhood of $s=\mathrm{i}k$ when kl is large. Therefore, it is legitimate to put $(s+\mathrm{i}k)^{\frac{1}{2}}$ equal to $(2\mathrm{i}k)^{\frac{1}{2}}$ when $kl \gg 1$, the relative error being $O(1/kl)$. With this gloss

$$\begin{aligned}\mathscr{J}(s) &= \frac{2\pi^{-\frac{1}{2}}\mathrm{e}^{-\mathrm{i}kl}\sin\frac{1}{2}\phi^{\mathrm{i}}}{s-\mathrm{i}k\cos\phi^{\mathrm{i}}}[\mathrm{e}^{\frac{1}{4}\pi\mathrm{i}}(s+\mathrm{i}k)^{\frac{1}{2}}\mathrm{F}\{(s+\mathrm{i}k)^{\frac{1}{2}}l^{\frac{1}{2}}\mathrm{e}^{-\frac{1}{4}\pi\mathrm{i}}\} \\ &\quad -(2k)^{\frac{1}{2}}\mathrm{i}\cos\tfrac{1}{2}\phi^{\mathrm{i}}\mathrm{F}\{(2kl)^{\frac{1}{2}}\cos\tfrac{1}{2}\phi^{\mathrm{i}}\}] \\ &\quad +\frac{2\pi^{-\frac{1}{2}}\mathrm{e}^{\mathrm{i}kl(\cos\phi^{\mathrm{i}}-1)}}{s+\mathrm{i}k\cos\phi^{\mathrm{i}}}\cos\tfrac{1}{2}\phi^{\mathrm{i}}[\mathrm{e}^{\frac{1}{4}\pi\mathrm{i}}(s+\mathrm{i}k)^{\frac{1}{2}}\mathrm{F}\{(s+\mathrm{i}k)^{\frac{1}{2}}l^{\frac{1}{2}}\mathrm{e}^{-\frac{1}{4}\pi\mathrm{i}}\} \\ &\quad -(2k)^{\frac{1}{2}}\mathrm{i}\sin\tfrac{1}{2}\phi^{\mathrm{i}}\mathrm{F}\{(2kl)^{\frac{1}{2}}\sin\tfrac{1}{2}\phi^{\mathrm{i}}\}]\end{aligned} \tag{17.20}$$

when $kl \gg 1$. Except when s is near $-\mathrm{i}k$ or when ϕ^{i} is close to zero the arguments of F are large in magnitude and the asymptotic formula (7.17) can be invoked, with the consequence

$$\mathscr{J}(s)=\frac{-\mathrm{i}\mathrm{e}^{-\mathrm{i}kl}}{4\pi^{\frac{1}{2}}kl^{\frac{3}{2}}(s+\mathrm{i}k)}\left\{\frac{\sin\frac{1}{2}\phi^{\mathrm{i}}}{\cos^2\frac{1}{2}\phi^{\mathrm{i}}}+\frac{\cos\frac{1}{2}\phi^{\mathrm{i}}}{\sin^2\frac{1}{2}\phi^{\mathrm{i}}}\exp(\mathrm{i}kl\cos\phi^{\mathrm{i}})\right\}. \tag{17.21}$$

If $\mathscr{G}(s)=(s+\mathrm{i}k)^{\frac{1}{2}}G_{\mathrm{R}}(s)-S_2(s)$, the same process leads to

$$\mathscr{G}(s)=\frac{\mathrm{i}\mathrm{e}^{-\mathrm{i}kl}}{4\pi^{\frac{1}{2}}kl^{\frac{3}{2}}(s+\mathrm{i}k)}\left\{\frac{\sin\frac{1}{2}\phi^{\mathrm{i}}}{\cos^2\frac{1}{2}\phi^{\mathrm{i}}}-\frac{\cos\frac{1}{2}\phi^{\mathrm{i}}}{\sin^2\frac{1}{2}\phi^{\mathrm{i}}}\exp(\mathrm{i}kl\cos\phi^{\mathrm{i}})\right\} \tag{17.22}$$

to the same degree of approximation.

Hence, from (17.16),

$$\begin{aligned}P(-\mathrm{i}k\cos\phi,0) &= \frac{1/\mathrm{i}k}{\cos\phi^{\mathrm{i}}-\cos\phi}\left[\frac{\sin\frac{1}{2}\phi^{\mathrm{i}}}{\sin\frac{1}{2}\phi}-\frac{\cos\frac{1}{2}\phi^{\mathrm{i}}}{\cos\frac{1}{2}\phi}\exp\{\mathrm{i}kl(\cos\phi^{\mathrm{i}}-\cos\phi)\}\right] \\ &\quad -\frac{\mathrm{e}^{-\mathrm{i}kl-\frac{1}{4}\pi\mathrm{i}}}{4k\pi^{\frac{1}{2}}(2kl)^{\frac{3}{2}}}\left[\frac{\cos\frac{1}{2}\varphi^{\mathrm{i}}\exp(\mathrm{i}kl\cos\varphi^{\mathrm{i}})}{\sin^3\frac{1}{2}\varphi\sin^2\frac{1}{2}\varphi^{\mathrm{i}}}\right. \\ &\quad \left.+\frac{\sin\frac{1}{2}\phi^{\mathrm{i}}\exp(-\mathrm{i}kl\cos\phi)}{\cos^3\frac{1}{2}\phi\cos^2\frac{1}{2}\phi^{\mathrm{i}}}\right]\end{aligned}$$

which leads to the far field via (17.15). It follows from (17.17) that

$$\begin{aligned}\sigma_{\mathrm{c}} &= 2+\frac{1}{\pi^{\frac{1}{2}}(2kl)^{\frac{5}{2}}}\left[\frac{\cos\frac{1}{2}\phi^{\mathrm{i}}}{\sin^5\frac{1}{2}\phi^{\mathrm{i}}}\cos\{kl(\cos\phi^{\mathrm{i}}-1)-\tfrac{1}{4}\pi\}\right. \\ &\quad \left.+\frac{\sin\frac{1}{2}\phi^{\mathrm{i}}}{\cos^5\frac{1}{2}\phi^{\mathrm{i}}}\cos\{kl(\cos\phi^{\mathrm{i}}+1)+\tfrac{1}{4}\pi\}\right]\end{aligned} \tag{17.23}$$

so long as ϕ^{i} is not near zero.

The correction due to the presence of $\mathscr{J}(s)$ in the integrand, which has so far been ignored, can be estimated by the device which took care of $(s+ik)^{\frac{1}{2}}$ in the denominator. It adds

$$-\{\mathcal{J}(ik)e^{-ikl}/(2\pi k)^{\frac{1}{2}}e^{\frac{1}{4}\pi i}\}[-l^{-\frac{1}{2}}+2(s+ik)^{\frac{1}{2}}e^{\frac{1}{4}\pi i}F\{(s+ik)^{\frac{1}{2}}l^{\frac{1}{2}}e^{-\frac{1}{4}\pi i}\}]$$

to the right-hand side of (17.20). This affects (17.21) by the insertion of $\mathcal{J}(ik)(2k)^{\frac{1}{2}}e^{\frac{1}{4}\pi i}$ inside the brace. Placing $s=ik$ then reveals that the term $\mathcal{J}(ik)$ can be safely neglected provided that ϕ^{i} is not near zero.

When ϕ^{i} is near zero (17.21) is no longer valid because one of the arguments of F can be small. Nor is it legitimate any more to omit $\mathcal{J}(ik)$. Actually the substitution $s=ik$ in (17.20), with $\mathcal{J}(ik)$ included, gives

$$\mathcal{J}(ik)=\frac{-e^{-ikl}}{8\pi^{\frac{1}{2}}k^{2}l^{\frac{3}{2}}}\frac{\sin\frac{1}{2}\phi^{i}}{\cos^{2}\frac{1}{2}\phi^{i}}+\frac{\exp\{ikl(\cos\phi^{i}-1)\}}{2ik(\pi l)^{\frac{1}{2}}\cos\frac{1}{2}\phi^{i}}$$
$$\times[1-(8kl)^{\frac{1}{2}}i\sin\tfrac{1}{2}\phi^{i}F\{(2kl)^{\frac{1}{2}}\sin\tfrac{1}{2}\phi^{i}\}-1/4ikl].$$

The evaluation of the scattering coefficient from (17.20) must also be undertaken carefully because, in it, s approaches $-ik$ as $\phi^{i}\to 0$. In fact, the scattering cross-section $\sigma_{c}l\sin\phi^{i}$ tends to the limit $2(l/\pi k)^{\frac{1}{2}}(1-1/8kl)$.

The penetration through a *wide slit* in a hard screen can be deduced from the foregoing by means of Babinet's principle. As pointed out in §9.4 the transmission coefficient τ_{c} can be obtained from $\tau_{c}=\frac{1}{2}\sigma_{c}$ where σ_{c} is given by (17.23) when ϕ^{i} is not near zero.

9.18 The Kontorovich–Lebedev transform

The customary form of the Kontorovich–Lebedev transform has the reciprocal formulae

$$g(y)=\int_{0}^{\infty}f(x)K_{ix}(y)\,dx,$$

$$f(x)=2\pi^{-2}\sinh(\pi x)\int_{0}^{\infty}g(y)K_{ix}(y)y^{-1}\,dy.$$

For harmonic waves it is less convenient than another version, involving Hankel functions, namely: if

$$g(\nu)=\int_{0}^{\infty}f(y)H_{\nu}^{(2)}(y)\,dy\quad(\mathrm{Re}\,\nu=0),\tag{18.1}$$

then

$$xf(x)=\lim_{\varepsilon\to+0}-\tfrac{1}{2}\int_{-i\infty}^{i\infty}e^{\varepsilon\nu^{2}}\nu J_{\nu}(x)g(\nu)\,d\nu.\tag{18.2}$$

Note that (18.1) and (18.2) are not reciprocal, in the sense that $g(\nu)$ must be constructed first from f via (18.1). The conditions of validity are (Jones 1980) that (a) $\int_{0}^{1}|f(y)\ln y|\,dy<\infty$, (b) $\int_{c}^{\infty}f(y)y^{-\frac{1}{2}}e^{iy}\,dy$ is finite for any positive c. If f is discontinuous but of bounded variation $f(x)$ is replaced in (18.2) by $\frac{1}{2}\{f(x+0)+f(x-0)\}$.

9.19 Application to the wedge

The faces of a soft wedge occupy $\phi=\pm\delta$ as in Fig. 9.2 and the wedge is irradiated by a line source at (r_0, ϕ_0). The primary field is then $p^{\mathrm{i}}=-\frac{1}{4}\mathrm{i}\mathrm{H}_0^{(2)}(k\,|\boldsymbol{x}-\boldsymbol{x}_0|)$ and the total field satisfies, in polar coordinates,

$$\frac{1}{r}\frac{\partial}{\partial r}\left(r\frac{\partial p}{\partial r}\right)+\frac{1}{r^2}\frac{\partial^2 p}{\partial \phi^2}+k^2p=-\frac{1}{r_0}\delta(r-r_0)\delta(\phi-\phi_0). \tag{19.1}$$

A solution is required such that $p=0$ on the sides of the wedge and p is an outgoing wave at infinity.

Define a Kontorovich–Lebedev transform with respect to r by

$$P=\int_0^\infty pr^{-1}\mathrm{H}_\nu^{(2)}(kr)\,\mathrm{d}r$$

with Re $\nu=0$. The integral exists because $p\sim \mathrm{e}^{-\mathrm{i}kr}/r^{\frac{1}{2}}$ as $r\to\infty$ and, by the edge conditions $p\approx r^{\pi/2\delta}$ as $r\to 0$; indeed this behaviour enables one to verify that (18.2) will also hold.

Multiply (19.1) by $r\mathrm{H}_\nu^{(2)}(kr)$ and integrate from 0 to ∞. After integration by parts the following differential equation is obtained for P

$$\mathrm{d}^2P/\mathrm{d}\phi^2+\nu^2P=-\mathrm{H}_\nu^{(2)}(kr_0)\delta(\phi-\phi_0).$$

In order that P vanishes on $\phi=\pm\delta$ the appropriate solution is

$$-P\nu\sin 2\nu\delta/\mathrm{H}_\nu^{(2)}(kr_0)=\begin{cases}\sin\nu(\phi-\delta)\sin\nu(\phi_0+\delta) & (\phi\geqslant\phi_0)\\ \sin\nu(\phi+\delta)\sin\nu(\phi_0-\delta) & (\phi\leqslant\phi_0).\end{cases}$$

Quoting (18.2) we can now derive p_{S} from

$$p_{\mathrm{S}}=\lim_{\varepsilon\to+0}-\tfrac{1}{2}\int_{-\mathrm{i}\infty}^{\mathrm{i}\infty}\mathrm{e}^{\varepsilon\nu^2}\nu\mathrm{J}_\nu(kr)P\,\mathrm{d}\nu. \tag{19.2}$$

Various alternative expressions for p can be obtained. When $r<r_0$, the contour in (19.2) can be deformed to the right so that it goes off to infinity along straight lines making angles $\pm\theta$ with the positive ν-axis, θ being slightly greater than $\frac{1}{4}\pi$. On the new contour the integral converges absolutely even when $\varepsilon=0$ and so ε can be replaced by zero. The contour can now be deformed completely to the right capturing the simple poles at the zeros of $\sin 2\nu\delta$ in the process. Hence, when $r<r_0$,

$$p_{\mathrm{S}}=-(\pi\mathrm{i}/2\delta)\sum_{n=1}^{\infty}\mathrm{J}_{\nu_n}(kr)\mathrm{H}_{\nu_n}^{(2)}(kr_0)\sin\nu_n(\phi+\delta)\sin\nu_n(\phi_0+\delta) \tag{19.3}$$

where $\nu_n=n\pi/2\delta$, the same result being obtained whether ϕ be greater or less than ϕ_0. The formula (19.3) confirms (2.2) as $r \to 0$ and is a fuller expansion of the field near a soft edge. The series (19.3) is also furnished by tackling (19.1) by the method of separation of variables.

If $r > r_0$, the aforementioned deformation cannot be carried out because the integral fails to converge when $\varepsilon = 0$ although we know that (19.3) holds with r and r_0 interchanged by the reciprocity theory of §1.35. A comparable result can be achieved for the integral representation by the observation that $P(\nu) = Q(\nu)\mathrm{H}_\nu^{(2)}(kr_0)$ where $Q(-\nu) = Q(\nu)$. Therefore, the substitution $\mathrm{H}_\nu^{(2)}(kr_0) = \{e^{\nu\pi i}\mathrm{J}_\nu(kr_0) - \mathrm{J}_{-\nu}(kr_0)\}/\mathrm{i}\sin\nu\pi$ followed by a change of sign of the variable of integration in the term involving $\mathrm{J}_{-\nu}(kr_0)$, leads to

$$p_\mathrm{S} = \lim_{\varepsilon \to +0} -\tfrac{1}{2}\int_{-\mathrm{i}\infty}^{\mathrm{i}\infty} e^{\varepsilon\nu^2}\nu \mathrm{J}_\nu(kr_0)\mathrm{H}_\nu^{(2)}(kr)Q(\nu)\,\mathrm{d}\nu. \tag{19.4}$$

The difficulty over convergence referred to earlier now disappears when $r > r_0$ (it arises when $r < r_0$) and the contour can be deformed to the right as above so that the desired expansion is recovered.

A further integral formula originates from substituting (A.3.21), followed by (A.3.19), in (19.3) when $r < r_0$. The summation of the series is then immediate. Taking advantage of (A.1.37) we have

$$p_\mathrm{S} = w(\phi - \phi_0) - w(\phi + \phi_0 + 2\delta) \tag{19.5}$$

where

$$w(\phi) = -\frac{1}{8\pi}\int_{\infty - \pi\mathrm{i}\nu_1}^{\infty + \pi\mathrm{i}\nu_1} \frac{\mathrm{H}_0^{(2)}[k\{r^2 + r_0^2 - 2rr_0\cosh(w/\nu_1)\}^{\frac{1}{2}}]}{\cosh w - \cos\nu_1\phi}\sinh w\,\mathrm{d}w, \tag{19.6}$$

the contour of integration passing to the left of the point where $2rr_0\cosh(w/\nu_1) = r^2 + r_0^2$ and to the right of the imaginary axis (Fig. 9.20).

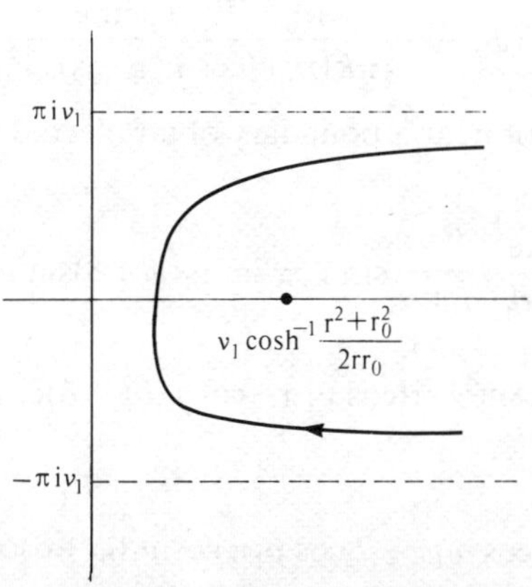

Fig. 9.20 The contour of integration for the wedge

For the hard wedge a solution p_H of (19.1) is required such that $\partial p_H/\partial\phi = 0$ on the surface of the wedge. The Kontorovich–Lebedev transform cannot be directly applied because p_H need not vanish at the edge. Notwithstanding, p_H is bounded there so this awkward feature can be circumvented by considering the transform of $p_H - [p_H]_{r=0} e^{-ikr}$.

It is, however, simpler to observe that $\partial p_H/\partial\phi$ vanishes at the edge as well as on the faces of the wedge and satisfies (19.1) with $\delta(\phi-\phi_0)$ replaced by $\delta'(\phi-\phi_0)$. Hence $\partial p_H/\partial\phi = -\partial p_S/\partial\phi_0$. Accordingly, (19.5) implies that $p_H = w(\phi-\phi_0) + w(\phi+\phi_0+2\delta) + f(r)$ where f is a function of r only. Now w is bounded at the origin; in fact, (19.6) shows that $w(\phi) \to -\frac{1}{4}i\nu_1 H_0^{(2)}(kr_0)$ as $r \to 0$. Hence f is an outgoing solution of Helmholtz's equation which is bounded; it must therefore be zero. Consequently,

$$p_H = w(\phi-\phi_0) + w(\phi+\phi_0+2\delta). \tag{19.7}$$

The distant field can be determined from (19.6) by moving the contour into the straight lines joining $\infty - \nu_1\pi i$, $-\nu_1\pi i$, $\nu_1\pi i$, $\infty + \nu_1\pi i$ with indentations round any poles encountered. The contributions of the poles represent the field of geometrical ray theory, each corresponding to one of the various images of the source in the faces of the wedge. The remainder of the integral along the imaginary axis disappears because the integrand is odd. The integrals over the other two straight lines may be combined to give

$$w(\phi) = w_p(\phi) - \frac{1}{8\pi}\int_{-\infty}^{\infty} \frac{\sinh(\nu_1\pi i + u)}{\cosh(\nu_1\pi i + u) - \cos\nu_1\phi} H_0^{(2)}[k\{r^2 + r_0^2 + 2rr_0\cosh(u/\nu_1)\}^{\frac{1}{2}}]\,du \tag{19.8}$$

where w_p covers any subvention from poles. When $kr \gg 1$ the Hankel function can be replaced by its asymptotic approximation. There is then a point of stationary phase at $u = 0$ and

$$w^d(\phi) \sim -\frac{i\nu_1 e^{-ik(r+r_0)}\sin\nu_1\pi}{4\pi k(rr_0)^{\frac{1}{2}}(\cos\nu_1\pi - \cos\nu_1\phi)} \tag{19.9}$$

provided that ϕ is not near a boundary of a reflected wave or the shadow. Consequently,

$$p_S^d \sim -\frac{i\nu_1 e^{-ik(r+r_0)}}{2\pi k(rr_0)^{\frac{1}{2}}\alpha}\sin\nu_1\pi\sin\nu_1(\phi+\delta)\sin\nu_1(\phi_0+\delta), \tag{19.10}$$

$$p_H^d \sim -\frac{i\nu_1 e^{-ik(r+r_0)}}{2\pi k(rr_0)^{\frac{1}{2}}\alpha}\sin\nu_1\pi\{\cos\nu_1\pi - \cos\nu_1(\phi+\delta)\cos\nu_1(\phi_0+\delta)\} \tag{19.11}$$

where

$$\alpha = \cos^2\nu_1\pi - 2\cos\nu_1\pi\cos\nu_1(\phi+\delta)\cos\nu_1(\phi_0+\delta) + \tfrac{1}{2}\cos 2\nu_1(\phi+\delta) + \tfrac{1}{2}\cos 2\nu_1(\phi_0+\delta).$$

Expressions (19.10) and (19.11) reduce to (8.7) and (8.8) respectively when $\delta=\pi$ and $\nu_1=\frac{1}{2}$.

The approximation (19.9) ceases to hold when ϕ approaches a value which causes the denominator to become zero. It is then necessary to return to (19.8) and re-examine the behaviour of the integrand near $u=0$. Noting that

$$\frac{2\mathrm{i}\sinh(\mathrm{i}\nu_1\pi+u)\cos\frac{1}{2}\nu_1\phi}{\cosh(\mathrm{i}\nu_1\pi+u)-\cos\nu_1\phi}$$
$$=\cos\tfrac{1}{2}(\nu_1\pi-\mathrm{i}u)\{\operatorname{cosec}\tfrac{1}{2}(\nu_1\pi-\mathrm{i}u+\nu_1\phi)+\operatorname{cosec}\tfrac{1}{2}(\nu_1\pi-\mathrm{i}u-\nu_1\phi)\}$$

we have

$$w(\phi)-w_{\mathrm{p}}(\phi)\sim\frac{\mathrm{i}}{16\pi}\left(\frac{2}{\pi kr}\right)^{\frac{1}{2}}\sec\tfrac{1}{2}\nu_1\phi\,\mathrm{e}^{\frac{1}{4}\pi\mathrm{i}-\mathrm{i}kr}\{h(\phi)+h(-\phi)\}$$

where

$$h(\phi)=\int_{-\infty}^{\infty}\frac{\cos\frac{1}{2}(\nu_1\pi-\mathrm{i}u)}{\sin\frac{1}{2}(\nu_1\pi-\mathrm{i}u+\nu_1\phi)}\exp\{-\mathrm{i}kr_0\cosh(u/\nu_1)\}\,\mathrm{d}u.$$

The main contribution comes from the point of stationary phase $u=0$ so that $\cos\frac{1}{2}(\nu_1\pi-\mathrm{i}u)$ may be placed equal to $\cos\frac{1}{2}\nu_1\pi$. Also, since the exponent is expanded to the second power of u, there is no loss of accuracy in replacing $\cosh(u/\nu_1)$ by $1+(\cosh u-1)/\nu_1^2$. Hence

$$h(\phi)\sim\exp\{-\mathrm{i}kr_0(1-1/\nu_1^2)\}\cos\tfrac{1}{2}\nu_1\pi\int_{-\infty}^{\infty}\frac{\exp\{-\mathrm{i}k(r_0/\nu_1^2)\cosh u\}}{\sin\frac{1}{2}(\nu_1\pi-\mathrm{i}u+\nu_1\phi)}\,\mathrm{d}u$$
$$\sim-4\mathrm{e}^{-\mathrm{i}kr_0}\cos\tfrac{1}{2}\nu_1\pi[\pi^{\frac{1}{2}}\mathrm{e}^{-\frac{3}{4}\pi\mathrm{i}}\mathrm{F}\{(2kr_0/\nu_1^2)^{\frac{1}{2}}\sin\tfrac{1}{2}\nu_1(\pi+\phi)\}$$
$$+\pi\mathrm{H}\{-\sin\tfrac{1}{2}\nu_1(\pi+\phi)\}\exp\{2\mathrm{i}k(r_0/\nu_1^2)\sin^2\tfrac{1}{2}\nu_1(\pi+\phi)\}] \qquad (19.12)$$

from (7.11). The formula (19.12) is uniformly valid in ϕ and includes that already found because F can be replaced by its asymptotic expression when ϕ is not near a value which makes its argument small.

For excitation due to a point source, the field may be deduced by means of a Fourier transform with respect to z, the coordinate parallel to the edge. There is no loss of generality in taking $z_0=0$ since other positions merely correspond to a translation of the origin parallel to the edge. With this understanding

$$p_{\mathrm{S}}^{(3)}=W(\phi-\phi_0)-W(\phi+\phi_0+2\delta), \qquad (19.13)$$
$$p_{\mathrm{H}}^{(3)}=W(\phi-\phi_0)+W(\phi+\phi_0+2\delta) \qquad (19.14)$$

where

$$W(\phi)=-\frac{\mathrm{i}}{8\pi^2}\int_{\infty-\nu_1\pi\mathrm{i}}^{\infty+\nu_1\pi\mathrm{i}}\frac{\exp[-\mathrm{i}k\{r^2+r_0^2-2rr_0\cosh(w/\nu_1)+z^2\}^{\frac{1}{2}}]\sinh w}{\{r^2+r_0^2-2rr_0\cosh(w/\nu_1)+z^2\}^{\frac{1}{2}}(\cosh w-\cos\nu_1\phi)}\,\mathrm{d}w. \qquad (19.15)$$

For points near the edge

$$W(\phi)\approx\frac{k\nu_1}{4\pi i}\,h_0^{(2)}\{k(r_0^2+z^2)^{\frac{1}{2}}\}$$
$$+\frac{k(\frac{1}{2}krr_0)^{\nu_1}}{(\nu_1-1)!2\pi i}\,\frac{h_{\nu_1}^{(2)}\{k(r_0^2+z^2)^{\frac{1}{2}}\}}{(r_0^2+z^2)^{\frac{1}{2}\nu_1}}\cos\nu_1\phi \tag{19.16}$$

which enables the performance of $p_S^{(3)}$ and $p_H^{(3)}$ around the edge to be determined.

As regards the diffracted field when $krr_0 \gg R_s+R'$

$$W^d(\phi)\sim\frac{\nu_1}{8\pi^2}\left(\frac{2\pi}{rr_0R'}\right)^{\frac{1}{2}}e^{-\frac{1}{4}\pi i-ikR'}\frac{\sin\nu_1\pi}{\cos\nu_1\pi-\cos\nu_1\phi} \tag{19.17}$$

in the notation of §9.10. It may be confirmed that the resulting $p_S^{(3)d}$ and $p_H^{(3)d}$ turn into (10.4) and (10.5) when $\delta=\pi$.

Formulae for an incident plane wave can be obtained without difficulty by allowing the source to go off to infinity in a suitable direction. These have the interpretation that a ray incident on the edge is diffracted onto a right circular cone with axis the edge, and of which one generator is the continuation of the incident ray (c.f. Fig. 9.14).

9.20 The cone

Another obstacle which can be treated by the Kontorovich–Lebedev transform is the cone illuminated by a point source.

Take spherical polar coordinates R, θ, ϕ with the tip of the cone as origin and such that the surface of the cone is $\theta=\beta$. Then a solution is required in $0\leqslant\theta\leqslant\beta$ of

$$\frac{1}{R^2}\frac{\partial}{\partial R}\left(R^2\frac{\partial p}{\partial R}\right)+\frac{1}{R^2\sin\theta}\frac{\partial}{\partial\theta}\left(\sin\theta\frac{\partial p}{\partial\theta}\right)+\frac{1}{R^2\sin^2\theta}\frac{\partial^2 p}{\partial\phi^2}+k^2p$$
$$=-\delta(R-R_0)\delta(\theta-\theta_0)\delta(\phi-\phi_0)R^{-2}\operatorname{cosec}\theta_0. \tag{20.1}$$

The soft boundary condition will be discussed.

Let

$$P=\int_0^\infty pR^{-\frac{1}{2}}H_\nu^{(2)}(kR)\,dR.$$

Multiply (20.1) by $R^{\frac{3}{2}}H_\nu^{(2)}(kR)$ and integrate with respect to R from 0 to ∞. Since p is bounded at the origin

$$\frac{1}{\sin\theta}\frac{\partial}{\partial\theta}\left(\sin\theta\frac{\partial P}{\partial\theta}\right)+\frac{1}{\sin^2\theta}\frac{\partial^2P}{\partial\phi^2}+(\nu^2-\tfrac{1}{4})P$$
$$=-\delta(\theta-\theta_0)\delta(\phi-\phi_0)H_\nu^{(2)}(kR_0)R_0^{-\frac{1}{2}}\operatorname{cosec}\theta_0.$$

Write $P=\sum_{p=-\infty}^{\infty} a_p e^{ip\phi}$ and then the equation

$$\frac{1}{\sin\theta}\frac{d}{d\theta}\left(\sin\theta\frac{da_p}{d\theta}\right)+\left(\nu^2-\tfrac{1}{4}-\frac{p^2}{\sin^2\theta}\right)a_p=-\frac{\delta(\theta-\theta_0)}{2\pi R^{\frac{1}{2}}\sin\theta_0}e^{-ip\phi_0}H_\nu^{(2)}(kR_0)$$

has to be solved for a_p.

The differential operator on the left-hand side occurs in Legendre's associated equation. There a_p is a linear combination of $P^p_{\nu-\frac{1}{2}}(\cos\theta)$ and $Q^p_{\nu-\frac{1}{2}}(\cos\theta)$. It is found that, when the Wronskian relation (B.1.18) is deployed, in order to make $a_p=0$ on $\theta=\beta$

$$a_p=-\frac{(\nu-p-\frac{1}{2})!e^{-ip\phi_0}H_\nu^{(2)}(kR_0)}{(\nu+p-\frac{1}{2})!2\pi R_0^{\frac{1}{2}}P^p_{\nu-\frac{1}{2}}(\cos\beta)}P^p_{\nu-\frac{1}{2}}(\cos\theta)$$
$$\times\{Q^p_{\nu-\frac{1}{2}}(\cos\beta)P^p_{\nu-\frac{1}{2}}(\cos\theta)-Q^p_{\nu-\frac{1}{2}}(\cos\theta)P^p_{\nu-\frac{1}{2}}(\cos\beta)\}$$

when $\theta>\theta_0$; for $\theta<\theta_0$ interchange θ and θ_0.

The inverse of the Kontorovich–Lebedev transform gives

$$p_S^{(3)}=\lim_{\varepsilon\to+0}-\tfrac{1}{2}R^{-\frac{1}{2}}\int_{-i\infty}^{i\infty}e^{\varepsilon\nu^2}\nu J_\nu(kR)\sum_{p=-\infty}^{\infty}a_p e^{ip\phi}\,d\nu.$$

As in the preceding section the contour can be deformed to the right when $R<R_0$. The only singularities are simple poles at $\nu=\nu_q+\frac{1}{2}$ where the real positive ν_q satisfies $P^p_{\nu_q}(\cos\beta)=0$. The positive poles of $(\nu-p-\frac{1}{2})!$ play no role because $P_{\nu-\frac{1}{2}}$ vanishes identically at them. Hence, for $R<R_0$,

$$p_S^{(3)}=\frac{ik}{\pi}\sum_{p=-\infty}^{\infty}\sum_{q=1}^{\infty}\frac{j_{\nu_q}(kR)h^{(2)}_{\nu_q}(kR_0)P^p_{\nu_q}(\cos\theta)P^p_{\nu_q}(\cos\theta_0)}{\sin^2\beta P^{p\prime}_{\nu_q}(\cos\beta)[\partial P^p_\nu(\cos\beta)/\partial\nu]_{\nu=\nu_q}}(\nu_q+\tfrac{1}{2})e^{ip(\phi-\phi_0)}\tag{20.2}$$

where Q_ν has been eliminated by implementation of (B.1.18). For $R>R_0$, interchange R and R_0 in (20.2), as can be seen either from reciprocity or from exploiting the device of the last section.

The special case when $\beta\approx\pi$ is of some interest because it conveys an idea of the scattering by a slender object although this notion must not be pushed too far since, even if the cone has a small non-zero angle, the cross-section must eventually become large at a great enough distance from the vertex. Let $\beta=\pi-\delta$ where δ is a small positive quantity. For simplicity, put $\theta_0=0$ so that the excitation is axially symmetric. Then all terms in (20.2) are removed except those in which $p=0$. As $\theta\to\pi$, P_ν becomes logarithmically infinite unless ν is an integer or zero. Hence ν_q differs by little from $q-1$. The difference can be calculated from (B.3.12) and

$$\nu_q=q-1-\tfrac{1}{2}(\ln\tfrac{1}{2}\delta)^{-1}.$$

It then follows from (B.3.12) that

$$P'_{\nu_q}(\cos\beta) \approx (-)^q(\delta^2 \ln \tfrac{1}{2}\delta)^{-1},$$
$$[\partial P_\nu(\cos\beta)/\partial\nu]_{\nu=\nu_q} \sim (-)^{q-1} 2 \ln \tfrac{1}{2}\delta.$$

Hence

$$p_S^{(3)} = -\frac{ik}{2\pi}\sum_{q=0}^{\infty}(q+\tfrac{1}{2})\mathrm{j}_q(kR)\mathrm{h}_q^{(2)}(kR_0)\mathrm{P}_q(\cos\theta) = -ik\mathrm{h}_0^{(2)}(kR_s)/4\pi$$

from (§8.11.11) so that the field is essentially the primary wave, the scattered field being $O(1/\ln \tfrac{1}{2}\delta)$.

It can also be shown that the Green's tensor $\boldsymbol{\Gamma}_1$ (§1.33) for the cone is given by

$$\boldsymbol{\Gamma}_1(\boldsymbol{x}_0, \boldsymbol{x}) = \frac{\mathrm{i}}{2\pi k}\sum_{p=-\infty}^{\infty}\sum_{q=1}^{\infty}\{c_{pq}k^2 \mathbf{g}^{(4)}_{-p,q}(\boldsymbol{x}_0)\mathbf{g}^{(1)}_{p,q}(\boldsymbol{x})$$
$$+ d_{pq}\boldsymbol{f}^{(4)}_{p,q}(\boldsymbol{x}_0)\boldsymbol{f}^{(1)}_{p,q}(\boldsymbol{x})\} \qquad (20.3)$$

where

$$\boldsymbol{f}^{(1)}_{p,q}(\boldsymbol{x}) = \mathrm{curl}\ \mathrm{curl}\{\mathrm{j}_{\nu_q}(kR)\mathrm{P}^{p}_{\nu_q}(\cos\theta)\mathrm{e}^{\mathrm{i}p\phi}\boldsymbol{x}\}$$

and $\boldsymbol{f}^{(4)}_{p,q}$ is the same with $\mathrm{h}^{(2)}_{\nu_q}$ in place of j_{ν_q}. The vectors $\mathbf{g}$ are defined in a similar way but there is a single curl instead of a double one and for ν_q is substituted μ_q where $\partial \mathrm{P}^{p}_{\mu_q}(\cos\beta)/\partial\beta = 0$. If there were no difference between μ_q and ν_q then $\boldsymbol{f}_{pq} = \mathrm{curl}\ \mathbf{g}_{pq}$ would be valid. The coefficient c_{pq} is specified by

$$c_{pq} = \mu_q(\mu_q+1)\int_0^\beta \{\mathrm{P}^{p}_{\mu_q}(\cos\theta)\}^2 \sin\theta\, \mathrm{d}\theta;$$

for d_{pq} exchange μ_q for ν_q.

Separation of variables

The method of separation of variables, already used in §8.1, can also be applied here. Two illustrations are given—the first comparatively straightforward and the second involving both the vector nature of the electromagnetic field and the edge conditions in a significant fashion.

9.21 The strip

The scattering by an elliptic cylinder has been investigated by means of separation of variables in §8.8. The surface of the elliptic cylinder was taken to be $u = U$ in the elliptic cylinder coordinates $x = l\cosh u \cos v$, $y = l \sinh u \sin v$. The cylinder degenerates to the strip $-l \leqslant x \leqslant l$, $y = 0$ when $U = 0$.

According to (8.8.6) the scattering coefficient of the strip in a plane wave is

$$\sigma_c = \frac{(4/kl)}{\sin\phi^i}\left[\sum_{m=0}^{\infty}|b_m|^2\{\mathrm{ce}_m(\phi^i, h)\}^2 + \sum_{m=1}^{\infty}|c_m|^2\{\mathrm{se}_m(\phi^i, h)\}^2\right] \quad (21.1)$$

where $h = \frac{1}{4}k^2l^2$ and, for the soft boundary, $b_m = \mathrm{Mc}_m^{(1)}(0, h)/\mathrm{Mc}_m^{(4)}(0, h)$, $c_m = \mathrm{Ms}_m^{(1)}(0, h)/\mathrm{Ms}_m^{(4)}(0, h)$; the coefficients for the hard boundary can be deduced from (8.8.3).

For the *general elliptic cylinder* the approximations of §C.4 can be inserted at low frequencies when h is small, provided that U is not too large. The result is

$$(\cosh^2 U - \cos^2\phi^i)^{\frac{1}{2}}\sigma_c^S \approx (2\pi^2/klC)(1 - \tfrac{1}{4}k^2l^2\cos 2\phi^i + \tfrac{3}{256}k^4l^4\cos 4\phi^i), \quad (21.2)$$

$$(\cosh^2 U - \cos^2\phi^i)^{\frac{1}{2}}\sigma_c^H \approx \tfrac{1}{32}\pi^2k^3l^3\{\sinh^2 2U + e^{2U}(\cosh^2 U - \cos^2\phi^i)\} \quad (21.3)$$

where

$$C = \pi^2 + \{2Q + \tfrac{1}{4}k^2l^2\sinh 2U - \tfrac{1}{16}k^4l^4(\tfrac{1}{4} - \tfrac{5}{32}\sinh 4U)\}^2 + \pi^2k^3l^3e^{2U}(\cos^2\phi^i + \sinh^2 U)/16,$$

$$Q = \gamma + \ln(\tfrac{1}{4}kle^U).$$

Particularizing these to the strip $U = 0$ at normal incidence ($\phi^i = \frac{1}{2}\pi$) we obtain

$$\sigma_c^S \approx 2\pi^2(1 + \tfrac{1}{4}k^2l^2 + \tfrac{3}{256}k^4l^4)/kl(\pi^2 + 4Q_0^2 - Q_0k^4l^4/16), \quad (21.4)$$

$$\sigma_c^H \approx \pi^2k^3l^3/16 \quad (21.5)$$

where $Q_0 = \gamma + \ln\frac{1}{4}kl$. The expressions exemplify the awkwardness of trying to find a parameter for a low frequency power series expansion based on a combination of kl and Q_0. Indeed, although the term $m = 0$ originates a series in powers of k^4l^4/Q_0, terms with greater values of m involve a parameter of the form $k^nl^nQ_0$ because Q_0 is an integral part of the definition of the Bessel function Y and therefore $\mathrm{Mc}^{(2)}$ but it dominates all other terms only in $\mathrm{Mc}_0^{(2)}$.

The series (21.1) has been computed by Skavlem (1951) and some values are displayed in Table 9.3. From them it can be seen that if natural light is normally incident on a narrow strip the diffracted light is essentially polarized parallel to the edge provided that $kl < 0.8$. For scattering by a narrow slit Babinet's principle indicates that the transmission coefficients are $\frac{1}{2}\sigma_c^H$ and $\frac{1}{2}\sigma_c^S$ respectively; now the diffracted light is mainly polarized perpendicular to the edge. Equations (21.2) and (21.3) reveal that these assertions are true for arbitrary angles of incidence.

Table 9.3 The soft and hard scattering coefficients of a strip at normal incidence

kl	$\frac{1}{2}\sigma_c^S$	$\frac{1}{2}\sigma_c^H$
0		0
0.2		0.0026
0.24	1.3965	
0.4		0.0239
0.48	1.1216	
0.6		0.0948
0.8	1.0143	0.2606
1.0	0.9908	0.5454
1.4	0.9813	1.1172
2.0	0.9948	1.1843
3.0		0.9720
4	0.9990	0.9424
5		1.0499
6		0.9956
7		0.9717
8	0.9998	1.0233
9		1.0020
10		0.9822

9.22 The circular disc

The scattering by a circular disc will be studied only for electromagnetic waves in order that the full complications of the analysis can be brought out. One result can be deduced at once from the formulae of §8.26 for the oblate spheroid. The spheroid becomes a perfectly conducting circular disk of radius L on putting $\xi^1 = 0$ and then (8.26.1) supplies the scattered wave due to a certain vertical electric dipole. Belkina (1958) computed the series for the dipole at the centre of the disc and discovered that for $kL \leqslant 1$ the dipole radiates as if it were in free space but that a strong shadow zone developes for $kL > 3$.

More general orientations of the dipole run into trouble because the vector wave equation is not separable in spheroidal coordinates. Nevertheless, some progress can be made and the particular case of a *horizontal magnetic dipole* will now be discussed.

It is convenient to replace L by a, the radius of the disc, and take oblate spheroidal coordinates ξ, η, ϕ where

$$x = a(\xi^2+1)^{\frac{1}{2}}(1-\eta^2)^{\frac{1}{2}} \cos \phi, \qquad y = x \tan \phi, \qquad z = a\xi\eta.$$

The surface of the perfectly conducting disc is then given by $\xi = 0$. Let the horizontal magnetic dipole be at $\xi = \xi_0$, $\eta = -1$, i.e. at $(0, 0, -a\xi_0)$ in the

Cartesian system. The primary wave can be described by a magnetic Hertz vector $\boldsymbol{M}^{\mathrm{i}}$ which has the sole component

$$M_x^{\mathrm{i}} = m^{\mathrm{i}} \mathrm{e}^{-\mathrm{i}kR_s}/R_s;$$

the intensities are obtained from $\boldsymbol{E}^{\mathrm{i}} = -\mathrm{i}\omega\mu \,\mathrm{curl}\, \boldsymbol{M}^{\mathrm{i}}$, $\boldsymbol{H}^{\mathrm{i}} = (\mathrm{grad}\ \mathrm{div} + k^2)\boldsymbol{M}^{\mathrm{i}}$. The tangential components of $\boldsymbol{E}^{\mathrm{i}}$ involve only the derivative of M_x^{i} with respect to z. Therefore the tangential components of the total electric intensity can be made to vanish on the disc by adding a scattered field, derived from a magnetic Hertz vector $\boldsymbol{M}^{\mathrm{s}}$ with only the component M_x^{s}, whose normal derivative cancels that of the incident field. From the expansion for M_x^{i} in oblate spheroidal coordinates (Appendix E) it is clear that the appropriate choice is

$$M_x^{\mathrm{s}} = \mathrm{i}m^{\mathrm{i}}k \sum_{n=0}^{\infty} (2n+1) a_n \mathrm{S}_n^{0(4)}(-\mathrm{i}\xi, \mathrm{i}\alpha)\mathrm{ps}_n^0(\eta, -\alpha^2)(-)^n \mathrm{A}_n^0(-\alpha^2)$$

where $\alpha = ka$ and

$$a_n = \mathrm{S}_n^{0(4)}(-\mathrm{i}\xi_0, \mathrm{i}\alpha)\mathrm{S}_n^{0(1)\prime}(-\mathrm{i}0, \mathrm{i}\alpha)/\mathrm{S}_n^{0(4)\prime}(-\mathrm{i}0, \mathrm{i}\alpha).$$

Before $\boldsymbol{M}^{\mathrm{i}} + \boldsymbol{M}^{\mathrm{s}}$ can be accepted as the solution it is necessary to check the behaviour near the edge. Let r, ϕ, z be cylindrical polar coordinates. Then the ϕ-component of the magnetic intensity is

$$-\{r^{-1}\, \partial(M_x^{\mathrm{i}} + M_x^{\mathrm{s}})/\partial r + k^2(M_x^{\mathrm{i}} + M_x^{\mathrm{s}})\}\sin\phi.$$

The discontinuity of this quantity across the disc is proportional to the radial component of the current density and must therefore vanish at the edge. The required discontinuity is the difference between the values at $\pm\eta$ on $\xi = 0$ as $\eta \to 0$. On $\xi = 0$, $\partial/\partial r = -(r/a^2\eta)\,\partial/\partial\eta$ so that the desired vanishing of the radial current does not take place since the discontinuity is singular like $1/\eta$. It is therefore necessary to find an additional field which will correct the edge behaviour without violating the boundary conditions on the surface of the disc.

In terms of a general magnetic Hertz vector $\boldsymbol{M}$

$$H_\phi = \frac{1}{r}\frac{\partial}{\partial\phi}\left\{\frac{1}{r}\frac{\partial}{\partial r}(rM_x \cos\phi) - \frac{1}{r}\frac{\partial}{\partial\phi}(M_x \sin\phi) + \frac{\partial M_z}{\partial z}\right\} - k^2 M_x \sin\phi$$

when $M_y = 0$. Evidently, H_ϕ will depend on ϕ only to the extent of the factor $\sin\phi$ provided that M_x is chosen to be independent of ϕ and M_z depends on ϕ only through the factor $\cos\phi$. Therefore an additional magnetic Hertz vector $\boldsymbol{M}^{\mathrm{a}}$ is sought such that M_x^{a} is independent of ϕ, $M_y^{\mathrm{a}} = 0$ and $M_z^{\mathrm{a}} = W\cos\phi$ where W is independent of ϕ. In order that the tangential components of the electric intensity of this additional field vanish on the disc

$$\partial M_x^{\mathrm{a}}/\partial z - W/r = 0, \qquad \partial M_x^{\mathrm{a}}/\partial z - \partial W/\partial r = 0$$

so that $\partial W/\partial r = W/r$. Hence, on the disc,

$$W = Cr, \qquad \partial M_x^a/\partial z = C \tag{22.1}$$

where C is a constant to be determined.

Both M_z^a and M_x^a satisfy Helmholtz's equation and can therefore be expanded in spheroidal functions, the former involving those in which $m = 1$ because of its dependence on ϕ whereas $m = 0$ is pertinent for the latter. Assume therefore that

$$M_x^a = \sum_{n=0}^{\infty} b_n S_n^{0(4)}(-i\xi, i\alpha)ps_n^0(\eta, -\alpha^2),$$

$$W = \sum_{n=0}^{\infty} c_n S_n^{1(4)}(-i\xi, i\alpha)ps_n^1(\eta, -\alpha^2).$$

The boundary conditions (22.1) are satisfied provided that

$$\sum_{n=0}^{\infty} c_n S_n^{1(4)}(-i0, i\alpha)ps_n^1(\eta, -a^2) = Ca(1-\eta^2)^{\frac{1}{2}}, \tag{22.2}$$

$$\sum_{n=0}^{\infty} b_n S_n^{0(4)\prime}(-i0, i\alpha)ps_n^0(\eta, -\alpha^2) = iCa\eta. \tag{22.3}$$

Multiply both sides of (22.2) by $ps_n^1(\eta, -\alpha^2)$ and use (E.2.3), (E.2.4). Then

$$c_n = (2n+1)e^{\frac{1}{2}\pi i(n+1)}CaA_n^{-1}(-\alpha^2)S_n^{1(1)}(-i0, i\alpha)/\alpha S_n^{1(4)}(-i0, i\alpha). \tag{22.4}$$

Similarly (22.3), coupled with a derivative of (E.2.4) with respect to ξ, gives

$$b_n = -i(2n+1)e^{\frac{1}{2}\pi in}CaA_n^0(-\alpha^2)S_n^{0(1)\prime}(-i0, i\alpha)/\alpha S_n^{0(4)\prime}(-i0, i\alpha). \tag{22.5}$$

Since $S_n^{0(1)\prime}(0, i\alpha) = 0$ for n even it follows that $b_n = 0$ if n is even. Also $S_n^{1(1)}(0, i\alpha) = 0$ for even n so that $c_n = 0$ if n is even. Note, however, that although $c_0 \neq 0$ there is no term involving it because ps_0^1 is identically zero.

The condition for the radial current density to vanish at the rim is that, as $d \to +0$,

$$\left(\frac{\partial M_x}{\partial \eta}\right)_{\eta=d} + \left(\frac{\partial M_x}{\partial \eta}\right)_{\eta=-d} - \frac{1}{(1-d^2)^{\frac{1}{2}}}\left\{\left(\frac{\partial W}{\partial \xi}\right)_{\eta=d} + \left(\frac{\partial W}{\partial \xi}\right)_{\eta=-d}\right\} = O(d^2) \tag{22.6}$$

where $M_x = M_x^s + M_x^a$ and $\xi = 0$. Both ps_n^1 and $ps_n^{0\prime}$ have expansions in powers of η^2 in a neighbourhood of $\eta = 0$ and both are odd or even about $\eta = 0$ according as n is even or odd. Since only odd values of n arise the edge condition can be met if there is no constant term on the left-hand side of (22.6), i.e. if

$$m^{\mathrm{i}}k\sum_{n=0}^{\infty}(2n+1)\mathrm{S}_n^{0(4)}(-\mathrm{i}0,\mathrm{i}\alpha)\mathrm{ps}_n^{0\prime}(0,-\alpha^2)\{(-)^n\mathrm{A}_n^0(-\alpha^2)a_n$$

$$+b_n/\mathrm{i}m^{\mathrm{i}}k(2n+1)\}+\sum_{n=0}^{\infty}c_n\mathrm{S}_n^{1(4)\prime}(-\mathrm{i}0,\mathrm{i}\alpha)\mathrm{ps}_n^1(0,-\alpha^2)=0.$$

Hence $C=P/Q$ where

$$P=\mathrm{i}m^{\mathrm{i}}k\sum_{m=0}^{\infty}(4m+3)\mathrm{S}_{2m+1}^{0(4)}(-\mathrm{i}0,\mathrm{i}\alpha)\mathrm{ps}_{2m+1}^{0\prime}(0,-\alpha^2)\mathrm{A}_{2m+1}^{0}(-\alpha^2)a_{2m+1},\tag{22.7}$$

$$Q=\frac{-\mathrm{i}}{k}\sum_{m=0}^{\infty}(4m+3)(-)^m\left\{\frac{\mathrm{i}\mathrm{S}_{2m+1}^{0(4)}(-\mathrm{i}0,\mathrm{i}\alpha)}{\mathrm{S}_{2m+1}^{0(4)\prime}(-\mathrm{i}0,\mathrm{i}\alpha)}\right.$$

$$\times\mathrm{S}_{2m+1}^{0(1)\prime}(-\mathrm{i}0,\mathrm{i}\alpha)\mathrm{A}_{2m+1}^{0}(-\alpha^2)\mathrm{ps}_{2m+1}^{0\prime}(0,-\alpha^2)$$

$$\left.+\frac{\mathrm{S}_{2m+1}^{1(4)\prime}(-\mathrm{i}0,\mathrm{i}\alpha)}{\mathrm{S}_{2m+1}^{1(4)}(-\mathrm{i}0,\mathrm{i}\alpha)}\mathrm{S}_{2m+1}^{1(1)}(-\mathrm{i}0,\mathrm{i}\alpha)\mathrm{A}_{2m+1}^{-1}(-\alpha^2)\mathrm{ps}_{2m+1}^{1}(0,-\alpha^2)\right\}.\tag{22.8}$$

It is a straightforward matter to verify that the other edge conditions on the current and charge densities are complied with. The complete field is now fully determined.

For an incident plane wave multiply the solution by $-a\xi_0\mathrm{e}^{\mathrm{i}ka\xi_0}/m^{\mathrm{i}}\omega k\mu$ and let $\xi_0\to\infty$. The incident wave goes over into the plane wave, normally incident on the disc, with nonzero components $E_y^{\mathrm{i}}=\mathrm{e}^{-\mathrm{i}kz}$, $H_x^{\mathrm{i}}=-(\varepsilon/\mu)^{\frac{1}{2}}E_y^{\mathrm{i}}$. The scattered field $\boldsymbol{E}^{\mathrm{s}}$ is expressed in terms of a magnetic Hertz vector $\boldsymbol{M}$ which has components

$$M_x=\frac{-\mathrm{i}}{\omega\mu k}\sum_{n=0}^{\infty}(2n+1)\mathrm{e}^{\frac{1}{2}\pi\mathrm{i}(n+1)}\mathrm{S}_n^{0(4)}(-\mathrm{i}\xi,\mathrm{i}\alpha)\mathrm{ps}_n^0(\eta,-\alpha^2)(-)^n\frac{S_n^{0(1)\prime}}{S_n^{0(4)\prime}}$$

$$-(\mathrm{i}/\omega\mu)\sum_{n=0}^{\infty}b_n\mathrm{S}_n^{0(4)}(-\mathrm{i}\xi,\mathrm{i}\alpha)\mathrm{ps}_n^0(\eta,-\alpha^2),$$

$$M_z=-(\mathrm{i}/\omega\mu)\cos\phi\sum_{n=0}^{\infty}c_n\mathrm{S}_n^{1(4)}(-\mathrm{i}\xi,\mathrm{i}\alpha)\mathrm{ps}_n^1(\eta,-\alpha^2)$$

where b_n, c_n are given by (22.5), (22.4) but C is replaced by $C_{\mathrm{p}}=P_{\mathrm{p}}/Q$ with

$$P_{\mathrm{p}}=\sum_{m=0}^{\infty}(4m+3)S_{2m+1}^{0(4)}\mathrm{ps}_{2m+1}^{0\prime}(0,-\alpha^2)\mathrm{A}_{2m+1}^{0}(-\alpha^2)\mathrm{e}^{\pi\mathrm{i}(m+1)}S_{2m+1}^{0(1)\prime}/kS_{2m+1}^{0(4)\prime}\tag{22.9}$$

and $S_r^{p(q)}$ denotes $\mathrm{S}_r^{p(q)}(-\mathrm{i}0,\mathrm{i}\alpha)$, with a similar convention for the derivatives.

At a large distance R from the origin, the spheroidal coordinates approach spherical polars for $a\xi \to R$, $\eta \to \cos\theta$. Hence

$$E^s_\theta \sim -\omega\mu k M_x \sin\phi$$

$$\sim -\mathrm{i}\frac{\mathrm{e}^{-\mathrm{i}kR}}{kR}\left\{\sum_{n=0}^{\infty}(2n+1)\mathrm{ps}^0_n(\cos\theta,-\alpha^2)\mathrm{A}^0_n(-\alpha^2)S^{0(1)\prime}_n/S^{0(4)\prime}_n\right.$$

$$\left.-k\sum_{n=0}^{\infty} b_n \mathrm{e}^{\frac{1}{2}\pi\mathrm{i}(n+1)}\mathrm{ps}^0_n(\cos\theta,-\alpha^2)\right\}\sin\phi,$$

$$E^s_\phi \sim -\omega\mu k(M_x\cos\phi\cos\theta - M_z\sin\theta)$$

$$\sim E^s_\theta \cot\phi\cos\theta - \mathrm{i}\cos\phi\sin\theta\frac{\mathrm{e}^{-\mathrm{i}kR}}{R}\sum_{n=0}^{\infty} c_n \mathrm{e}^{\frac{1}{2}\pi\mathrm{i}(n+1)}\mathrm{ps}^1_n(\cos\theta,-\alpha^2).$$

In particular, §9.4 shows that the scattering coefficient is given by

$$\sigma = -\frac{4}{ka^2}\,\mathrm{Im}\left[-\frac{\mathrm{i}}{k}\sum_{n=0}^{\infty}(2n+1)\{\mathrm{A}^0_n(-\alpha^2)\}^2 S^{0(1)\prime}_n/S^{0(4)\prime}_n\right.$$

$$\left.+\mathrm{i}\sum_{n=0}^{\infty} b_n \mathrm{e}^{\frac{1}{2}\pi\mathrm{i}(n+1)}\mathrm{A}^0_n(-\alpha^2)\right]$$

$$= -\frac{4}{k^2a^2}\,\mathrm{Re}\left[\sum_{n=0}^{\infty}(2n+1)\{1-(-)^n C_\mathrm{p}\}\{\mathrm{A}^0_n(-\alpha^2)\}^2 S^{0(1)\prime}_n/S^{0(4)\prime}_n\right].$$

For low frequencies the formulae can be approximated by the power series expansions of §E.2 since α is small. Some computation is saved by observing that the Wronskian relation (E.1.18) permits the assertion in (22.8) and (22.9) of

$$S^{0(4)}_{2m+1}S^{0(1)\prime}_{2m+1} = -1/\alpha, \qquad S^{1(4)\prime}_{2m+1}S^{1(1)}_{2m+1} = 1/\alpha$$

because $S^{0(1)}_{2m+1}$ and $S^{1(1)\prime}_{2m+1}$ vanish. The resulting approximations are

$$C_\mathrm{p} = 1 + \tfrac{2}{3}\alpha^2 + \ldots\ldots - \frac{4\mathrm{i}\alpha^3}{3\pi}(1 + \tfrac{34}{45}\alpha^2 + \ldots),$$

$$E^s_\theta \sim \frac{4\alpha^3}{3\pi k}\frac{\mathrm{e}^{-\mathrm{i}kR}}{R}\cos\theta\sin\phi\left[1 + \tfrac{13}{30}(1 + \tfrac{3}{13}\cos^2\theta)\alpha^2 + \ldots\right.$$

$$\left.-\frac{8\mathrm{i}\alpha^3}{9\pi}\{1 + \tfrac{39}{50}(1 + \tfrac{5}{39}\cos^2\theta)\alpha^2 + \ldots\}\right],$$

$$E^s_\phi \sim \frac{4\alpha^3}{3\pi k}\frac{\mathrm{e}^{-\mathrm{i}kR}}{R}\cos\phi\left[1 + \tfrac{11}{30}(1 + \tfrac{5}{11}\cos^2\theta)\alpha^2 + \ldots\right.$$

$$\left.-\frac{8\mathrm{i}\alpha^3}{9\pi}\{1 + \tfrac{17}{25}(1 + \tfrac{5}{17}\cos^2\theta)\alpha^2 + \ldots\}\right],$$

$$\sigma \approx \frac{128\alpha^4}{27\pi^2}(1 + \tfrac{22}{25}\alpha^2 + \tfrac{7312}{18375}\alpha^6 + \ldots). \qquad (22.10)$$

From these formulae expressions for the field and transmission coefficient of a circular aperture in a perfectly conducting plane screen can be deduced by means of Babinet's principle.

Approximate methods

Most of the methods of approximation described in Chapter 8 can be applied to the problems encountered in this chapter, but there are some differences caused by the presence of edges. Low and high frequencies, which represent opposite ends of the scales, have to be treated by quite different techniques. Yet both may be related by integral equations, one of the most fruitful approaches to diffraction problems.

9.23 Integral equations

The field radiated from a body with surface S can be deduced from §1.29. Suppose now that the body is confined to the plane $z=0$. Then S can be collapsed into a portion of $z=0$, one side being at $z=+0$ and the other at $z=-0$. The surface integral over S can be written as one over one side only provided that the difference between the integrands is employed. Let the variable of integration $\mathbf{y}$ on S have components ξ, η, ζ. Then $\mathscr{S}$ is written for S in $y=+0$ and $[p]$ for $(p)_{\zeta=+0}-(p)_{\zeta=-0}$. With these conventions

$$p^{\mathrm{s}}(\boldsymbol{x})=\int_{\mathscr{S}}\left\{[p^{\mathrm{s}}]\frac{\partial}{\partial\zeta}\psi(\boldsymbol{x},\mathbf{y})-\left[\frac{\partial p^{\mathrm{s}}}{\partial\zeta}\right]\psi(\boldsymbol{x},\mathbf{y})\right\}\mathrm{d}\xi\,\mathrm{d}\eta$$

where $\psi(\boldsymbol{x},\mathbf{y})=\mathrm{e}^{-\mathrm{i}k|\boldsymbol{x}-\mathbf{y}|}/4\pi\,|\boldsymbol{x}-\mathbf{y}|$ and $\zeta=0$. The total field is obtained by adding $p^{\mathrm{i}}(\boldsymbol{x})$.

If the screen is soft, $[p^{\mathrm{s}}]$ disappears and the integral equation

$$0=p^{\mathrm{i}}(\boldsymbol{x})-\int_{\mathscr{S}}\left[\frac{\partial p^{\mathrm{s}}}{\partial\zeta}\right]\psi(\boldsymbol{x},\mathbf{y})\,\mathrm{d}\xi\,\mathrm{d}\eta \tag{23.1}$$

holds for $\boldsymbol{x}$ on $\mathscr{S}$. On the other hand, when the screen is hard, the integral equation is

$$0=\frac{\partial p^{\mathrm{i}}}{\partial z}+\left(\frac{\partial^2}{\partial x^2}+\frac{\partial^2}{\partial y^2}+k^2\right)\int_{\mathscr{S}}[p^{\mathrm{s}}]\psi(\boldsymbol{x},\mathbf{y})\,\mathrm{d}\xi\,\mathrm{d}\eta \tag{23.2}$$

for $\boldsymbol{x}$ on $\mathscr{S}$.

The analogous electromagnetic integral equation for a perfectly conducting obstacle is

$$\mathbf{0}=\boldsymbol{k}\wedge\boldsymbol{E}^{\mathrm{i}}4\pi\varepsilon\mathrm{i}\omega+\boldsymbol{k}\wedge(\operatorname{grad}\operatorname{div}+k^2)\int_{\mathscr{S}}\boldsymbol{k}\wedge[\boldsymbol{H}]\psi\,\mathrm{d}\xi\,\mathrm{d}\eta. \tag{23.3}$$

The three integral equations are also valid for two-dimensional disturbances provided that ψ is interpreted as $-\frac{1}{4}\mathrm{i}\mathrm{H}_0^{(2)}(k|\boldsymbol{x}-\mathbf{y}|)$ and the surface integrals replaced by line integrals.

9.24 Small objects

The first problem to be considered is the scattering of a plane wave by a small rigid screen occupying a finite portion of $z=0$. As in §8.27 all quantities are expanded in powers of k in order to aim for a solution of (23.2). Thus, we write,

$$[p^{\mathrm{s}}]=p_0-\mathrm{i}kp_1-\frac{1}{2!}k^2p_2+\frac{1}{3!}\mathrm{i}k^3p_3\ldots,$$

$$p^{\mathrm{i}}=\mathrm{e}^{-\mathrm{i}k\boldsymbol{n}^{\mathrm{i}}.\boldsymbol{x}}=1-\mathrm{i}k\boldsymbol{n}^{\mathrm{i}}\cdot\boldsymbol{x}-\frac{1}{2!}k^2(\boldsymbol{n}^{\mathrm{i}}\cdot\boldsymbol{x})^2+\ldots.$$

Then (23.2) breaks up into

$$0=\left(\frac{\partial^2}{\partial x^2}+\frac{\partial^2}{\partial y^2}\right)\int_{\mathcal{S}}p_0\psi_0(\boldsymbol{x},\mathbf{y})\,\mathrm{d}\xi\,\mathrm{d}\eta, \tag{24.1}$$

$$0=\cos\theta^{\mathrm{i}}+\left(\frac{\partial^2}{\partial x^2}+\frac{\partial^2}{\partial y^2}\right)\int_{\mathcal{S}}p_1\psi_0(\boldsymbol{x},\mathbf{y})\,\mathrm{d}\xi\,\mathrm{d}\eta, \tag{24.2}$$

$$\begin{aligned}0={}&\sin 2\theta^{\mathrm{i}}(x\cos\phi^{\mathrm{i}}+y\sin\phi^{\mathrm{i}})-2\int_{\mathcal{S}}p_0\psi_0(\boldsymbol{x},\mathbf{y})\,\mathrm{d}\xi\,\mathrm{d}\eta\\&+\left(\frac{\partial^2}{\partial x^2}+\frac{\partial^2}{\partial y^2}\right)\int_{\mathcal{S}}\left\{p_2\psi_0(\boldsymbol{x},\mathbf{y})+\frac{p_0}{4\pi}|\boldsymbol{x}-\mathbf{y}|\right\}\mathrm{d}\xi\,\mathrm{d}\eta,\end{aligned} \tag{24.3}$$

$$\begin{aligned}0={}&3\cos\theta^{\mathrm{i}}\sin^2\theta^{\mathrm{i}}(x\cos\phi^{\mathrm{i}}+y\sin\phi^{\mathrm{i}})^2-6\int_{\mathcal{S}}(p_1\psi_0+p_0/4\pi)\,\mathrm{d}\xi\,\mathrm{d}\eta\\&+\left(\frac{\partial^2}{\partial x^2}+\frac{\partial^2}{\partial y^2}\right)\int_{\mathcal{S}}\left\{p_3\psi_0+\frac{1}{4\pi}(3p_1+p_0\,|\boldsymbol{x}-\mathbf{y}|)\,|\boldsymbol{x}-\mathbf{y}|\right\}\mathrm{d}\xi\,\mathrm{d}\eta\end{aligned} \tag{24.4}$$

when $\psi_0(\boldsymbol{x},\mathbf{y})=1/4\pi\,|\boldsymbol{x}-\mathbf{y}|$. Equations (24.1)–(24.4) constitute four static problems to be solved successively for p_0, p_1, p_2, and p_3. Once their solutions have been found the scattered far field is determined from

$$\begin{aligned}p^{\mathrm{s}}(\boldsymbol{x})\sim\frac{\mathrm{i}k\cos\theta}{4\pi R}\mathrm{e}^{-\mathrm{i}kR}\int_{\mathcal{S}}\Big[&p_0+\mathrm{i}k(p_0\hat{\boldsymbol{x}}\cdot\mathbf{y}-p_1)\\&-\frac{k^2}{2!}\{p_2-2p_1\hat{\boldsymbol{x}}\cdot\mathbf{y}+p_0(\hat{\boldsymbol{x}}\cdot\mathbf{y})^2\}\\&+\frac{1}{3!}\mathrm{i}k^3\{p_3-3p_2\hat{\boldsymbol{x}}\cdot\mathbf{y}+3p_1(\hat{\boldsymbol{x}}\cdot\mathbf{y})^2-p_0(\hat{\boldsymbol{x}}\cdot\mathbf{y})^3\}\Big]\,\mathrm{d}\xi\,\mathrm{d}\eta.\end{aligned} \tag{24.5}$$

The static problem in (24.1) consists of finding the discontinuity in a solution of Laplace's equation which decays at infinity and has zero normal derivative on $\mathscr{S}$. Such a static field is clearly 0 and therefore the solution of (24.1) is $p_0 \equiv 0$.

For (24.2) a solution of Laplace's equation which vanishes at infinity and whose normal derivative is $-\cos\theta^{\text{i}}$ on the screen is needed. Let

$$V(\boldsymbol{x}) = \int_{\mathscr{S}} \chi \frac{\partial}{\partial \zeta} \psi_0(\boldsymbol{x}, \boldsymbol{y})\, \mathrm{d}\xi\, \mathrm{d}\eta$$

be such that $\partial V/\partial z = -1$ on $\mathscr{S}$. Then $p_1 = \chi \cos\theta^{\text{i}}$.

Similarly, if V_1 and V_2, specified in the same way in terms of χ_1 and χ_2, are such that $\partial V_1/\partial z = -x$ and $\partial V_2/\partial z = -y$ on $\mathscr{S}$, (24.3) implies that

$$p_2 = 2\cos\theta^{\text{i}} \sin\theta^{\text{i}}(\chi_1 \cos\phi^{\text{i}} + \chi_2 \sin\phi^{\text{i}}).$$

One of the integrals involving p_2 in (24.5) can be expressed in terms of χ (van Bladel, 1967, 1968) via the static reciprocity theorem

$$\int_{\mathscr{S}} [\phi_1] \frac{\partial \phi_2}{\partial \zeta}\, \mathrm{d}\xi\, \mathrm{d}\eta = \int_{\mathscr{S}} [\phi_2] \frac{\partial \phi_1}{\partial \zeta}\, \mathrm{d}\xi\, \mathrm{d}\eta.$$

Thus, with $[\phi_1] = p_2$ and $\phi_2 = V$,

$$\int_{\mathscr{S}} p_2\, \mathrm{d}\xi\, \mathrm{d}\eta = (M_1 \cos\phi^{\text{i}} + M_2 \sin\phi^{\text{i}}) \sin 2\theta^{\text{i}}$$

where

$$M_1 = \int_{\mathscr{S}} \chi\xi\, \mathrm{d}\xi\, \mathrm{d}\eta, \qquad M_2 = \int_{\mathscr{S}} \chi\eta\, \mathrm{d}\xi\, \mathrm{d}\eta.$$

Similarly, from (24.4)

$$\int_{\mathscr{S}} p_3\, \mathrm{d}\xi\, \mathrm{d}\eta = 3\cos\theta^{\text{i}} \sin^2\theta^{\text{i}}(M_{11}\cos^2\phi^{\text{i}} + 2M_{12}\cos\phi^{\text{i}}\sin\phi^{\text{i}} + M_{22}\sin^2\phi^{\text{i}}) - 3M_0\cos\theta^{\text{i}}$$

where

$$M_{11} = \int_{\mathscr{S}} \chi\xi^2\, \mathrm{d}\xi\, \mathrm{d}\eta, \qquad M_{12} = \int_{\mathscr{S}} \chi\xi\eta\, \mathrm{d}\xi\, \mathrm{d}\eta, \qquad M_{22} = \int_{\mathscr{S}} \chi\eta^2\, \mathrm{d}\xi\, \mathrm{d}\eta,$$

$$M_0 = (1/4\pi) \int_{\mathscr{S}} \chi(x, y) \int_{\mathscr{S}} \chi(\xi, \eta)\, |\boldsymbol{x} - \boldsymbol{y}|^{-1}\, \mathrm{d}\xi\, \mathrm{d}\eta\, \mathrm{d}x\, \mathrm{d}y.$$

Accordingly

$$\begin{aligned} p^{\text{s}}(\boldsymbol{x}) \sim \frac{k^2 \mathrm{e}^{-\mathrm{i}kR}}{4\pi R} \cos\theta \cos\theta^{\text{i}} \Big[& \int_{\mathscr{S}} \chi\, \mathrm{d}\xi\, \mathrm{d}\eta - \mathrm{i}k(M_1\cos\phi^{\text{i}} + M_2\sin\phi^{\text{i}})\sin\theta^{\text{i}} \\ & + \mathrm{i}k(M_1\cos\phi + M_2\sin\phi)\sin\theta + \tfrac{1}{2}M_0 k^2 \\ & - \tfrac{1}{2}k^2 \sin^2\theta^{\text{i}}(M_{11}\cos^2\phi^{\text{i}} + 2M_{12}\cos\phi^{\text{i}}\sin\phi^{\text{i}} + M_{22}\sin^2\phi^{\text{i}}) \\ & - \tfrac{1}{2}k^2 \sin^2\theta(M_{11}\cos^2\phi + 2M_{12}\cos\phi\sin\phi + M_{22}\sin^2\phi) \\ & + k^2 \sin\theta^{\text{i}} \sin\theta \int_{\mathscr{S}} (\xi\cos\phi + \eta\sin\phi)(\chi_1\cos\phi^{\text{i}} + \chi_2\sin\phi^{\text{i}})\, \mathrm{d}\xi\, \mathrm{d}\eta \Big]. \end{aligned} \tag{24.6}$$

The scattering coefficient can be found by integrating the energy of (24.6) over a sphere (the theory of §9.4 supplies zero). One obtains

$$\sigma_S = \frac{k^4}{2\pi\mathscr{S}} \int_{\mathscr{S}} \chi \, d\xi \, d\eta \left[\int_{\mathscr{S}} \chi \, d\xi \, d\eta + M_0 k^2 - \tfrac{1}{5} k^2 (M_{11} + M_{22}) \right.$$
$$- k^2 \sin^2\theta^i (M_{11} \cos^2\phi^i + 2M_{12} \cos\phi^i \sin\phi^i + M_{22} \sin^2\phi^i)$$
$$\left. + k^2 \{(M_1 \cos\phi^i + M_2 \sin\phi^i)^2 \sin^2\theta^i + \tfrac{1}{5}(M_1^2 + M_2^2)\} \Big/ \int_{\mathscr{S}} \chi \, d\xi \, d\eta \right]. \tag{24.7}$$

The determination of χ alone is therefore sufficient for the calculation of the scattering coefficient to this order. For χ only the solution of a static problem is required.

The transmission coefficient for a small opening in a soft screen can be deduced immediately from (24.7) and §9.4.

The scattering of a sound wave by a small soft screen and of an electromagnetic wave by a small perfectly conducting object may be dealt with in the same way. For example, the scattering coefficient of a perfectly conducting circular disc is

$$\sigma_E = \frac{128(ka)^4}{27\pi^2 \cos\theta^i} \{1 + (\tfrac{1}{4} \sin^2\phi^i - \cos^2\phi^i) \sin^2\theta^i\} \tag{24.8}$$

which agrees with (22.10) at normal incidence.

9.25 Wide apertures

For an aperture many wavelengths across the approximation of §9.24 is a complete failure and a different technique must be adopted. It will be enough to confine attention to the two-dimensional wide slit in a plane screen but the subject will be returned to in more generality later (§9.29).

Let the gap occupy the interval $-l \leqslant x \leqslant 0$ of the x-axis and let the screen on the remaining part of the axis be soft (Fig. 9.21). The basic idea is that, because the aperture is many wavelengths across, the two soft semi-infinite planes scatter independently of each other to a first approximation. The method will be illustrated for the incident plane wave $e^{-ikr\cos(\phi-\phi^i)}$ though it is applicable to fields from other sources.

The total field produced by the plane wave falling on a single soft semi-infinite plane filling $x \leqslant 0$ is, according to (7.13),

$$\pi^{-\frac{1}{2}} e^{\frac{1}{4}\pi i - ikr} [F\{(2kr)^{\frac{1}{2}} \sin \tfrac{1}{2}(\phi - \phi^i)\} - F\{(2kr)^{\frac{1}{2}} \sin \tfrac{1}{2}(\phi + \phi^i)\}].$$

Denote this field by $e^{-ikr\cos(\phi-\phi^i)} + p^s(r, \phi, \phi^i)$ so that p^s is the wave scattered by a soft screen in $x \leqslant 0$. For the gap problem the left screen extends only over $x \leqslant -l$ so that the field scattered by it is $p^s(r_1, \phi_1, \phi^i)e^{ikl\cos\phi^i}$ where r_1 and ϕ_1 are measured from the edge at $x = -l$

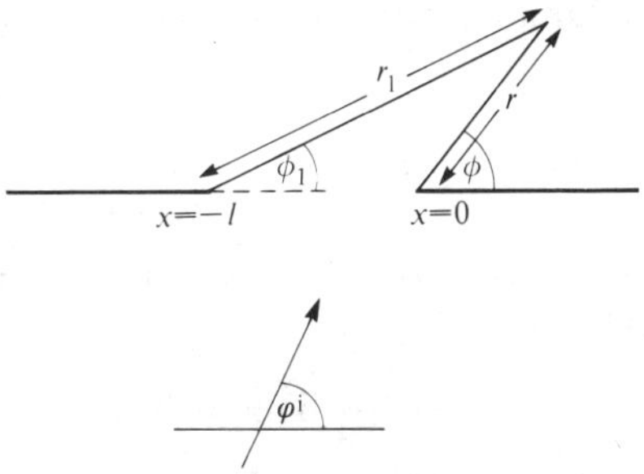

Fig. 9.21 The wide aperture

(Fig. 9.21). This scattered field will satisfy the boundary condition on $x \leqslant -l$ but fail to meet that on the screen in $x \geqslant 0$. However, because of the distance over the gap, $p^s(r_1, \phi_1, \phi^i)$ can be expected to be small on the right screen. Therefore, to a first approximation, the screen on the right is subject only to the incident plane wave and is responsible for the scattered field $p^s(r, \pi - \phi, \pi - \phi^i)$. Thus the first approximation to the total field is

$$e^{-ikr\cos(\phi-\phi^i)} + e^{ikl\cos\phi^i} p^s(r_1, \phi_1, \phi^i) + p^s(r, \pi-\phi, \pi-\phi^i).$$

This can be expressed entirely in terms of r and ϕ since $r_1 \cos \phi_1 = r \cos \phi + l$ and $r_1 \sin \phi_1 = r \sin \phi$.

At large distances ($kr \gg 1$) asymptotic expansion via (7.17) and (7.16) leads to

$$\frac{e^{-ikr-\frac{1}{4}\pi i}}{(8\pi kr)^{\frac{1}{2}}}\left[\frac{\exp\{-ikl(\cos\phi - \cos\phi^i)\} - 1}{\sin\frac{1}{2}(\phi - \phi^i)} - \frac{\exp\{-ikl(\cos\phi - \cos\phi^i)\} + 1}{\sin\frac{1}{2}(\phi + \phi^i)}\right] \quad (25.1)$$

when $0 \leqslant \phi \leqslant \pi$ and ϕ is not near ϕ^i. If ϕ is near ϕ^i the first term in the bracket is replaced by

$$4ikr^{\frac{1}{2}}\{r_1 \sin\tfrac{1}{2}(\phi_1 - \phi^i) - r \sin\tfrac{1}{2}(\phi - \phi^i)\}$$

which supplies a scattered field of

$$\frac{e^{\frac{1}{4}\pi i - ikr}}{(2\pi kr)^{\frac{1}{2}}}\left\{kl \sin\phi^i + \frac{i}{\sin\phi^i}\right\}$$

when $\phi = \phi^i$. It follows from the two-dimensional form of (4.4) that, to this order of approximation, the transmission coefficient is 1, in line with the prediction of geometrical acoustics.

A more accurate field can be achieved by noting that our first approximation furnishes $p^s(r, 0, \pi - \phi^i)$ on the left screen in $x \leqslant -l$ instead of zero. Therefore an additional field is required to cancel this. Now, since $kl \gg 1$,

$$p^{s}(r, 0, \pi-\phi^{i}) \sim -\frac{e^{-ikr-\frac{1}{4}\pi i}}{(2\pi kr)^{\frac{1}{2}} \cos \frac{1}{2}\phi^{i}}$$

on $r \geqslant l$. In view of the asymptotic behaviour of the Hankel function this can be expressed as

$$p^{s}(r, 0, \pi-\phi^{i}) \sim \tfrac{1}{2}i \sec \tfrac{1}{2}\phi^{i} H_{0}^{(2)}(kr)$$

to the same degree of approximation. In other words, the excitation is effectively due to a line source of strength $-2 \sec \frac{1}{2}\phi^{i}$ placed at $x=0$. This problem has been discussed in §9.8. The far field of the additional scattered wave is, from (8.7) (with due allowance for the difference in notation),

$$-\frac{ie^{-ik(r_1+l)} \cos \frac{1}{2}\phi_1}{\pi k(r_1 l)^{\frac{1}{2}}(1+\cos \phi_1)\cos \frac{1}{2}\phi^{i}}.$$

Similarly, the additional distant field from the wave counteracting $p^{s}(r_1, 0, \phi^{i})e^{ikl\cos\phi^{i}}$ on $x \geqslant 0$ is

$$\frac{-i \exp\{ikl \cos \phi^{i} - ik(r+l)\}}{2\pi(rl)^{\frac{1}{2}} \sin \frac{1}{2}\phi^{i} \sin \frac{1}{2}\phi}.$$

Hence the total extra field to be added to the first approximation (25.1) is

$$\frac{-ie^{-ik(r+l)}}{2\pi k(rl)^{\frac{1}{2}}}\left\{\frac{e^{-ikl\cos\phi}}{\cos \frac{1}{2}\phi^{i} \cos \frac{1}{2}\phi}+\frac{e^{ikl\cos\phi^{i}}}{\sin \frac{1}{2}\phi^{i} \sin \frac{1}{2}\phi}\right\}. \tag{25.2}$$

Placing $\phi=\phi^{i}$ in (25.2) is straightforward and so the transmission coefficient of a wide aperture in a soft screen is

$$\tau_S = 1-\frac{(2/\pi)^{\frac{1}{2}}}{(kl)^{\frac{3}{2}} \sin \phi^{i}}\left[\frac{\cos\{kl(1+\cos \phi^{i})-\frac{1}{4}\pi\}}{1+\cos \phi^{i}} + \frac{\cos\{kl(1-\cos \phi^{i})-\frac{1}{4}\pi\}}{1-\cos \phi^{i}}\right]. \tag{25.3}$$

At normal incidence $\phi^{i}=\frac{1}{2}\pi$ and (25.3) reduces to

$$\tau_S = 1-\left(\frac{2}{kl}\right)^{\frac{3}{2}} \pi^{-\frac{1}{2}} \cos(kl-\tfrac{1}{4}\pi). \tag{25.4}$$

By Babinet's principle expressions (25.3) and (25.4) are one-half of the scattering coefficient of the wide hard strip.

Although the second approximation is more accurate than the first it is still in error in respect of the boundary conditions. Nevertheless, these can be made yet smaller by the same technique and the field found to any desired accuracy so long as kl is large enough for higher powers of $1/kl$ to be negligible.

For a wide slit in a hard screen a similar procedure can be adopted, but there are some differences in detail. The first approximation to the scattered field is

$$\left(\frac{2}{\pi kr}\right)^{\frac{1}{2}}\frac{e^{-ikr-\frac{1}{4}\pi i}}{\cos\phi^{i}-\cos\phi}\{\sin\tfrac{1}{2}\phi\cos\tfrac{1}{2}\phi^{i}e^{-ikl(\cos\phi-\cos\phi^{i})}-\cos\tfrac{1}{2}\phi\sin\tfrac{1}{2}\phi^{i}\}. \tag{25.5}$$

The transmission coefficient is again unity to this level of approximation.

The first approximation vanishes on $x \leqslant -l$ but, since the boundary condition is now on the normal derivative, correction is still necessary in order to attain a second approximation. The normal derivative in the direction of positive y in the first approximation is

$$-\frac{e^{-ikr-\frac{1}{4}\pi i}\sin\tfrac{1}{2}\phi^{i}}{(8\pi kr^{3})^{\frac{1}{2}}\cos^{2}\tfrac{1}{2}\phi^{i}}.$$

Now suppose that there is a simple line source at $(0, y_0)$. A double line source can be generated by applying the operator $\partial/\partial y_0$ so that the incident wave is $(\partial/\partial y_0)(-\frac{1}{4}iH_0^{(2)})$. Put $y_0=0$; then the normal derivative of the incident field is $ik(1/8\pi kr^3)^{\frac{1}{2}}e^{-ikr-\frac{1}{4}\pi i}$ at a large distance. Thus the field to be corrected on the left plane can be regarded as originating from a double line source of strength $(i/k)\sin\frac{1}{2}\phi^{i}\sec^{2}\frac{1}{2}\phi^{i}$ at $x=0$ with its moment parallel to the y-axis. The scattering that results is available from (8.14). In particular, the scattered wave for $kr_1 \gg 1$ is

$$\frac{e^{-ik(l+r_1)}\sin\tfrac{1}{2}\phi^{i}\sin\tfrac{1}{2}\phi_1}{8\pi k^{2}r_1^{\frac{1}{2}}l^{\frac{3}{2}}\cos^{2}\tfrac{1}{2}\phi_1\cos^{2}\tfrac{1}{2}\phi^{i}}.$$

Similar considerations apply to the corrective field for the right screen and the field to be added to (25.5) to give the second approximation is

$$\left(\frac{2}{\pi kr}\right)^{\frac{1}{2}}\frac{e^{-ik(r+l)}}{4\pi^{\frac{1}{2}}(2kl)^{\frac{3}{2}}}\left\{\frac{\cos\frac{1}{2}\phi\cos\frac{1}{2}\phi^{i}}{\sin^{2}\frac{1}{2}\phi\sin^{2}\frac{1}{2}\phi^{i}}e^{ikl\cos\phi^{i}}+\frac{\sin\frac{1}{2}\phi\sin\frac{1}{2}\phi^{i}}{\cos^{2}\frac{1}{2}\phi\cos^{2}\frac{1}{2}\phi^{i}}e^{-ikl\cos\phi}\right\}. \tag{25.6}$$

The transmission coefficient for a wide slit in a hard screen is therefore

$$\tau_H = 1+\frac{1}{2(2kl)^{\frac{3}{2}}\pi^{\frac{1}{2}}}\left[\frac{\cos\frac{1}{2}\phi^{i}}{\sin^{5}\frac{1}{2}\phi^{i}}\cos\{kl(\cos\phi^{i}-1)-\tfrac{1}{4}\pi\}+\frac{\sin\frac{1}{2}\phi^{i}}{\cos^{5}\frac{1}{2}\phi^{i}}\cos\{kl(1+\cos\phi^{i})+\tfrac{1}{4}\pi\}\right] \tag{25.7}$$

which reduces at normal incidence to

$$\tau_H = 1+\frac{1}{(2\pi)^{\frac{1}{2}}(kl)^{\frac{3}{2}}}\cos(kl+\tfrac{1}{4}\pi). \tag{25.8}$$

The formulae (25.7) and (25.8) also provide one-half the scattering coefficient of a wide soft strip.

In general, the deviation from geometrical acoustics tends to be larger for the aperture in the soft screen than that in a hard boundary. However, the oscillatory nature of (25.4) and (25.8) prevents any simple comparison between the total energies transmitted in the two cases.

This section has concentrated on the transmission through the aperture but there is no difficulty in employing the same technique to determine the back scattering from a gap.

9.26 Kirchhoff's approximation

When $\boldsymbol{x}$ is in the interior of a closed surface S which contains no sources the formulae of §1.29 give

$$p(\boldsymbol{x}) = \int_S \left\{\frac{\partial p}{\partial n_y}\psi(\boldsymbol{x}, \mathbf{y}) - p\frac{\partial}{\partial n_y}\psi(\boldsymbol{x}, \mathbf{y})\right\} \mathrm{d}S_y. \tag{26.1}$$

If $\boldsymbol{x}$ is outside S the left-hand side of (26.1) is replaced by zero. The relation (26.1) is the embodiment of the Huygens–Fresnel principle that every point of a wavefront may be viewed as the centre of a secondary disturbance, which gives rise to secondary wavelets, and the wavefront at any later instant is the envelope of the wavelets. Unfortunately, the construction can be undertaken only when both p and $\partial p/\partial n$ are known on the wavefront, a circumstance of relatively rare occurrence. Despite this drawback, Kirchhoff was able to develop an approximate theory which escaped this defect and which is valuable in some diffraction problems.

Let the wave from a source S_0 propagate through the opening in an infinite plane screen and let P be the point of observation (Fig. 9.22). Take S to be a large hemisphere Ω (in the space not containing S_0) closed by a diametral plane coinciding with the plane of the aperture on its non-illuminated side. On Ω, p is an outgoing wave satisfying the radiation conditions and so the contribution of the integral over Ω tends to zero as the radius of Ω tends to infinity. Choose axes so that the screen is $z = 0$ with S_0 in $z < 0$. Then, if $\mathbf{y} \equiv (\xi, \eta, \zeta)$, at P

$$p(\boldsymbol{x}) = \int_{\mathscr{A}\cup\mathscr{S}} \left\{p\frac{\partial}{\partial\zeta}\psi(\boldsymbol{x}, \mathbf{y}) - \frac{\partial p}{\partial\zeta}\psi(\boldsymbol{x}, \mathbf{y})\right\} \mathrm{d}S_y \tag{26.2}$$

with $\zeta = 0$, the integration being over the whole plane $z = 0$ and $\mathscr{S}$ denoting the back of the screen.

The fundamental idea of Kirchhoff is that, if the aperture is many wavelengths across, the situation cannot be much different from that of geometrical acoustics, i.e. there is a shadow zone behind the screen extending from the back to the incident rays through the boundary of the

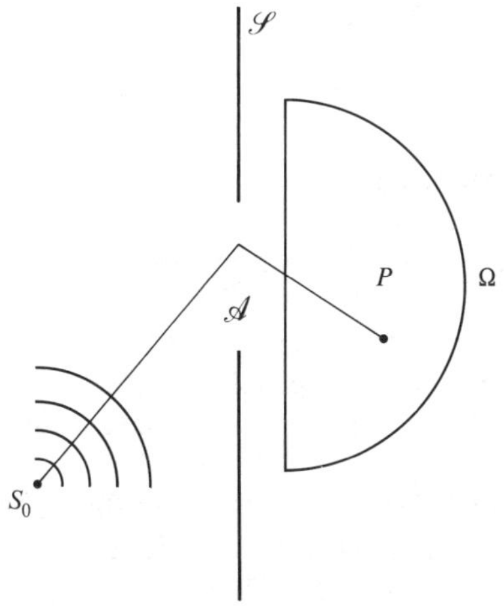

Fig. 9.22 Diffraction through an aperture

aperture and elsewhere the primary wave is undisturbed. Kirchhoff therefore assumed that p and $\partial p/\partial n$ are zero on the back of the screen whereas they have the same values as in the incident field on the aperture $\mathscr{A}$. All quantities in the integral of (26.2) are thereby known and the transmitted field can be calculated from an integration over the aperture alone.

It is straightforward to check that the predictions of Kirchhoff are consistent with Babinet's principle, which is a point in its favour. Regrettably, the mathematical objections to the scheme are serious. The first one is that (26.2) specifies p to be analytic so that the prescription that p and $\partial p/\partial n$ vanish on a portion of surface of nonzero area forces p to be zero everywhere. Thus Kirchhoff's approximate integrals cannot reproduce the p assumed on the screen and aperture. Another awkward feature is that the material of the screen is irrelevant—the same transmitted wave is obtained for an opening of given shape no matter what blanking sheet it is excised from.

In spite of these imperfections the basic concept is valuable if handled with care. This is because only a small part of the assumed field plays a significant role and because it is possible to reformulate the argument so as to take some account of the substance of the screen.

Let P' be the point $\boldsymbol{x}' \equiv (x, y, -z)$ which is the image of P in $z=0$ (Fig. 9.22). Then, since $\boldsymbol{x}'$ is not inside S,

$$0 = \int_{\mathscr{A}\cup\mathscr{S}} \left\{ p \frac{\partial}{\partial \zeta} \psi(\boldsymbol{x}', \mathbf{y}) - \frac{\partial p}{\partial \zeta} \psi(\boldsymbol{x}', \mathbf{y}) \right\} \mathrm{d}S_y. \tag{26.3}$$

Now, since $\zeta = 0$,

$$|\boldsymbol{x}' - \mathbf{y}| = |\boldsymbol{x} - \mathbf{y}|, \qquad \frac{\partial}{\partial \zeta}|\boldsymbol{x}' - \mathbf{y}| = -\frac{\partial}{\partial \zeta}|\boldsymbol{x} - \mathbf{y}|.$$

Hence, if (26.2) and (26.3) are added,

$$p(\boldsymbol{x}) = -2\int_{\mathscr{A}\cup\mathscr{S}} \frac{\partial p}{\partial \zeta}\, \psi(\boldsymbol{x}, \mathbf{y})\, \mathrm{d}S_y \tag{26.4}$$

whereas, if they are subtracted,

$$p(\boldsymbol{x}) = 2\int_{\mathscr{A}\cup\mathscr{S}} p \frac{\partial}{\partial \zeta}\, \psi(\boldsymbol{x}, \mathbf{y})\, \mathrm{d}S_y. \tag{26.5}$$

Both (26.4) and (26.5) are exact. They have the advantage that an assumption need be made about p only in (26.5) or its normal derivative in (26.4), thereby avoiding some of the difficulties in Kirchhoff's original formulation.

If the screen is rigid $\partial p/\partial\zeta = 0$ on $\mathscr{S}$. Insert the approximation in (26.4) that $\partial p/\partial\zeta$ is the same on $\mathscr{A}$ as in the incident wave on the grounds that there are many wavelengths in $\mathscr{A}$. Then

$$p_{\mathrm{KH}}(\boldsymbol{x}) = -2\int_{\mathscr{A}} \frac{\partial p^{\mathrm{i}}}{\partial \zeta}\, \psi(\boldsymbol{x}, \mathbf{y})\, \mathrm{d}S_y. \tag{26.6}$$

An alternative argument can be pursued. According to (3.2) p agrees with p^{i} in $\mathscr{A}$. Therefore, if p is assumed to be negligibly small on $\mathscr{S}$, (26.5) furnishes

$$p_{\mathrm{KS}}(\boldsymbol{x}) = 2\int_{\mathscr{A}} p^{\mathrm{i}} \frac{\partial}{\partial \zeta}\, \psi(\boldsymbol{x}, \mathbf{y})\, \mathrm{d}S_y. \tag{26.7}$$

For the soft screen $p = 0$ on $\mathscr{S}$ and the approximation $p = p^{\mathrm{i}}$ in $\mathscr{A}$ leads to (26.7). Similarly, (26.6) stems from the alternative argument. Thus, for either screen, (26.6) and (26.7) are the approximations offered by the modification to Kirchhoff's method.

Both (26.6) and (26.7) suffer from one deficiency. In one p has to change discontinuously across the edge of the aperture. Such behaviour conflicts with the continuity of p demanded by the edge conditions of §9.2. In the other $\partial p/\partial\zeta$ alters by a finite discontinuity in passage through the edge and this does not conform to the singularity that (2.2) and (2.3) entail. At first sight these errors may not seem to be of great significance but they are magnified in the evaluation of the integrals because it is discovered that the character of the diffraction is highly dependent upon the structure of the field near the edge. (For the asymptotics of multiple integrals see Jones (1982).)

There are corresponding approximations when the shield is of finite extent but the aperture goes off to infinity. They are easily deduced to be

$$p_{\mathrm{KH}}(\boldsymbol{x}) = p^{\mathrm{i}}(\boldsymbol{x}) + 2\int_{\mathscr{S}} \frac{\partial p^{\mathrm{i}}}{\partial \zeta}\, \psi(\boldsymbol{x}, \mathbf{y})\, \mathrm{d}S_y, \tag{26.8}$$

$$p_{\mathrm{KS}}(\boldsymbol{x}) = p^{\mathrm{i}}(\boldsymbol{x}) - 2\int_{\mathscr{S}} p^{\mathrm{i}} \frac{\partial}{\partial \zeta}\, \psi(\boldsymbol{x}, \mathbf{y})\, \mathrm{d}S_y. \tag{26.9}$$

For two-dimensional problems all the foregoing carries over provided that $\psi(\boldsymbol{x}, \mathbf{y})$ is taken to mean $-\frac{1}{4}\mathrm{i}\mathrm{H}_0^{(2)}(k\,|\boldsymbol{x}-\mathbf{y}|)$.

The diffraction of electromagnetic waves by a gap in a plane screen can be tackled in the same way by starting from Helmholtz's representation (§1.29). One analogue of the exact formulae (26.4) and (26.5) is

$$\boldsymbol{E}(\boldsymbol{x}) = (2/\mathrm{i}\omega\varepsilon)(\mathrm{grad\,div} + k^2)\int_{\mathcal{A}\cup\mathcal{S}} \boldsymbol{k}\wedge\boldsymbol{H}\psi(\boldsymbol{x}, \mathbf{y})\,\mathrm{d}S_y, \tag{26.10}$$

$$\boldsymbol{H}(\boldsymbol{x}) = 2\,\mathrm{curl}\int_{\mathcal{A}\cup\mathcal{S}} \boldsymbol{k}\wedge\boldsymbol{H}\psi(\boldsymbol{x}, \mathbf{y})\,\mathrm{d}S_y \tag{26.11}$$

where $\boldsymbol{k}$ is a unit vector along the positive z-axis. This exact representation expresses the transmitted electromagnetic field solely in terms of the tangential components of the magnetic intensity on $z=0$. Exact expressions involving the tangential components of the electric intensity alone are

$$\boldsymbol{E}(\boldsymbol{x}) = 2\,\mathrm{curl}\int_{\mathcal{A}\cup\mathcal{S}} \boldsymbol{k}\wedge\boldsymbol{E}\psi(\boldsymbol{x}, \mathbf{y})\,\mathrm{d}S_y, \tag{26.12}$$

$$\boldsymbol{H}(\boldsymbol{x}) = -(2/\mathrm{i}\omega\mu)(\mathrm{grad\,div} + k^2)\int_{\mathcal{A}\cup\mathcal{S}} \boldsymbol{k}\wedge\boldsymbol{E}\psi(\boldsymbol{x}, \mathbf{y})\,\mathrm{d}S_y. \tag{26.13}$$

Assume that the screen is perfectly conducting. The integrals over $\mathcal{S}$ in (26.12) and (26.13) disappear. With the approximation $\boldsymbol{k}\wedge\boldsymbol{E} = \boldsymbol{k}\wedge\boldsymbol{E}^{\mathrm{i}}$ in $\mathcal{A}$

$$\boldsymbol{E}_{\mathrm{K1}}(\boldsymbol{x}) = 2\,\mathrm{curl}\int_{\mathcal{A}} \boldsymbol{k}\wedge\boldsymbol{E}^{\mathrm{i}}\psi(\boldsymbol{x}, \mathbf{y})\,\mathrm{d}S_y, \tag{26.14}$$

$$\boldsymbol{H}_{\mathrm{K1}}(\boldsymbol{x}) = -(2/\mathrm{i}\omega\mu)(\mathrm{grad\,div} + k^2)\int_{\mathcal{A}} \boldsymbol{k}\wedge\boldsymbol{E}^{\mathrm{i}}\psi(\boldsymbol{x}, \mathbf{y})\,\mathrm{d}S_y. \tag{26.15}$$

These formulae satisfy the boundary conditions on the screen exactly but give an approximate aperture field.

In contrast, §9.3 demonstrates that $\boldsymbol{k}\wedge\boldsymbol{H} = \boldsymbol{k}\wedge\boldsymbol{H}^{\mathrm{i}}$ in $\mathcal{A}$. Therefore, if the current on the shadow side is assumed to vanish, (26.10) and (26.11) supply

$$\boldsymbol{E}_{\mathrm{K2}}(\boldsymbol{x}) = (2/\mathrm{i}\omega\varepsilon)(\mathrm{grad\,div} + k^2)\int_{\mathcal{A}} \boldsymbol{k}\wedge\boldsymbol{H}^{\mathrm{i}}\psi(\boldsymbol{x}, \mathbf{y})\,\mathrm{d}S_y, \tag{26.16}$$

$$\boldsymbol{H}_{\mathrm{K2}}(\boldsymbol{x}) = 2\,\mathrm{curl}\int_{\mathcal{A}} \boldsymbol{k}\wedge\boldsymbol{H}^{\mathrm{i}}\psi(\boldsymbol{x}, \mathbf{y})\,\mathrm{d}S_y. \tag{26.17}$$

In this approximation the tangential aperture field is reproduced correctly but the boundary conditions on the perfect conductor are in error.

If $\mathscr{S}$ is of finite size but $\mathscr{A}$ is infinite

$$\boldsymbol{E}_{K1}(\boldsymbol{x}) = \boldsymbol{E}^{i}(\boldsymbol{x}) - 2\,\mathrm{curl}\int_{\mathscr{S}} \boldsymbol{k}\wedge\boldsymbol{E}^{i}\psi(\boldsymbol{x},\boldsymbol{y})\,\mathrm{d}S_y, \tag{26.18}$$

$$\boldsymbol{E}_{K2}(\boldsymbol{x}) = \boldsymbol{E}^{i}(\boldsymbol{x}) - (2/\mathrm{i}\omega\varepsilon)(\mathrm{grad\,div} + k^2)\int_{\mathscr{S}} \boldsymbol{k}\wedge\boldsymbol{E}^{i}\psi(\boldsymbol{x},\boldsymbol{y})\,\mathrm{d}S_y \tag{26.19}$$

with appropriate expressions for the magnetic intensity.

9.27 Kirchhoff's approximations for a semi-infinite plane

A test of the adequacy of Kirchhoff's approximations is to compare them with the exact solutions for a semi-infinite plane. For simplicity, consider a plane wave at normal incidence, i.e. $p^{i} = \mathrm{e}^{-\mathrm{i}kz}$. Then, with the positive x-axis in the aperture and the y-axis along the edge,

$$p_{KH}(\boldsymbol{x}) = \tfrac{1}{2}k\int_0^\infty \mathrm{H}_0^{(2)}[k\{(x-t)^2 + z^2\}^{\frac{1}{2}}]\,\mathrm{d}t.$$

The Hankel function can be replaced by its asymptotic formula when $kz \gg 1$ and

$$p_{KH} \sim \left(\frac{k}{2\pi}\right)^{\frac{1}{2}} \mathrm{e}^{\frac{1}{4}\pi\mathrm{i}} \int_0^\infty \frac{\exp[-\mathrm{i}k\{(x-t)^2 + z^2\}^{\frac{1}{2}}]}{\{(x-t)^2 + z^2\}^{\frac{1}{4}}}\,\mathrm{d}t.$$

In the shadow $x < 0$ and the exponent changes monotonically with t. Therefore, the principal contribution comes from a neighbourhood of $t = 0$. Expand the integrand about $t = 0$ and evaluate the result asymptotically. There results

$$p_{KH} \sim \mathrm{e}^{-\mathrm{i}kr - \frac{1}{4}\pi\mathrm{i}}/(2\pi kr)^{\frac{1}{2}} \cos\phi$$

where $x = r\cos\phi$, $z = r\sin\phi$. Comparison with (7.26) reveals that

$$p_{KH}/p_H^{d} = 2^{-\frac{1}{2}}\,\mathrm{cosec}\,\tfrac{1}{2}\phi.$$

Similarly

$$p_{KS}/p_S^{d} = 2^{\frac{1}{2}}\sin\tfrac{1}{2}\phi.$$

It is perhaps not surprising that these ratios are not exactly unity since the expressions for p_{KH} and p_{KS} have been dictated by the behaviour near $t = 0$ and it has already been observed that the approximations involved in Kirchhoff's method are inaccurate near the edge. One curiosity is that $p_{KH}p_{KS} = p_H^{d}p_S^{d}$ so that the geometric mean of the Kirchhoff approximations agrees precisely with the geometric mean of the exact solutions.

Generally speaking, the discrepancies in the Kirchhoff results are serious. The ratios are never close to unity unless ϕ is near $\frac{1}{2}\pi$, i.e. for small angles of deviation from the incident rays to the edge.

The same qualitative conclusions are reached if the incidence of the plane wave is oblique or if the primary wave is due to a line source. If the line source is not far from the screen one of the above ratios can be very small and the other very large. Moreover, the Kirchhoff approximation does not reproduce the necessary reciprocity between source and point of observation. Summing up then, Kirchhoff's method supplies reliable answers only for small angular displacements of the observer into the shadow. Both approximations are of about equal accuracy in this region and one criterion for adopting Kirchhoff predictions is to require that they should agree within acceptable limits. The approximations should be treated with suspicion in the deep shadow—only the dependence on wavelength is likely to be trustworthy.

9.28 Fraunhofer and Fresnel diffraction

The Kirchhoff theory will now be applied to the diffraction of a plane wave normally incident on an aperture of finite extent. Other angles of incidence can be handled—the details are modified but the qualitative considerations are unaltered.

The integral formula to be discussed is

$$p_{\mathrm{KH}}(\boldsymbol{x}) = 2\mathrm{i}k \int_{\mathscr{A}} \psi(\boldsymbol{x}, \mathbf{y})\,\mathrm{d}S_y. \tag{28.1}$$

If $\boldsymbol{E}^{\mathrm{i}} = \boldsymbol{i}\mathrm{e}^{-\mathrm{i}kz}$ the same integral occurs in the electromagnetic field so that it will suffice to examine (28.1).

Since $\boldsymbol{x} \equiv (x, y, z)$ and $\mathbf{y} \equiv (\xi, \eta, 0)$

$$|\boldsymbol{x} - \mathbf{y}| = R - \frac{1}{R}(x\xi + y\eta) + \frac{1}{2R}(\xi^2 + \eta^2) - \frac{1}{2R^3}(x\xi + y\eta)^2$$

correct to the second order when $R = (x^2 + y^2 + z^2)^{\frac{1}{2}}$ is large. Hence (28.1) simplifies for large R to

$$p_{\mathrm{KH}}(\boldsymbol{x}) = \frac{\mathrm{i}k\mathrm{e}^{-\mathrm{i}kR}}{2\pi R} \int_{\mathscr{A}} \mathrm{e}^{-\mathrm{i}kf(\xi,\eta)}\,\mathrm{d}\xi\,\mathrm{d}\eta \tag{28.2}$$

where

$$f(\xi, \eta) = -\frac{1}{R}(x\xi + y\eta) + \frac{1}{2R}(\xi^2 + \eta^2) - \frac{1}{2R^3}(x\xi + y\eta)^2 \tag{28.3}$$

if terms of $O(R^{-2})$ are neglected in the phase.

When the quadratic terms in ξ and η in (28.3) must be retained the point of observation is said to be in a region of *Fresnel diffraction.* In a few cases the integral can be evaluated exactly. For example, if $\mathscr{A}$ is the circle $x^2+y^2 \leqslant a^2$ and the point of observation is on the z-axis

$$(\boldsymbol{E}_{K1})_x = (\mu/\varepsilon)^{\frac{1}{2}}(\boldsymbol{H}_{K2})_y = e^{-ikz} - zde^{-ikd},$$

$$(\boldsymbol{E}_{K2})_x = (\mu/\varepsilon)^{\frac{1}{2}}(\boldsymbol{H}_{K1})_y = e^{-ikz} - \tfrac{1}{2}\left(1 + \frac{z^2}{d^2} + \frac{ia^2}{kd^3}\right)e^{-ikd}$$

where $d^2 = a^2 + z^2$. Neither formula seems preferable to the other since both are in tolerable agreement with experiment.

If the quadratic terms in ξ and η in (28.3) can be ignored *Fraunhofer diffraction* occurs. Then

$$p_{KH}(\boldsymbol{x}) = \frac{ike^{-ikR}}{2\pi R}\, q\left(\frac{x}{R}, \frac{y}{R}\right),$$

$$\boldsymbol{E}_{K1} \sim \frac{-ike^{-ikR}}{2\pi R}\, \hat{\boldsymbol{x}} \wedge \boldsymbol{j} q\left(\frac{x}{R}, \frac{y}{R}\right),$$

$$\boldsymbol{E}_{K2} \sim -\frac{ike^{-ikR}}{2\pi R}\, \hat{\boldsymbol{x}} \wedge (\hat{\boldsymbol{x}} \wedge \boldsymbol{i}) q\left(\frac{x}{R}, \frac{y}{R}\right),$$

$$\boldsymbol{H}_{K1} \sim \hat{\boldsymbol{x}} \wedge \boldsymbol{E}_{K1}(\mu/\varepsilon)^{\frac{1}{2}}, \quad \boldsymbol{H}_{K2} \sim \hat{\boldsymbol{x}} \wedge \boldsymbol{E}_{K2}(\mu/\varepsilon)^{\frac{1}{2}}$$

where

$$q(x, y) = \int_{\mathscr{A}} e^{ik(x\xi + y\eta)}\, d\xi\, d\eta. \tag{28.4}$$

Let $\chi(\xi, \eta)$ be defined to be 1 if (ξ, η) is in $\mathscr{A}$ and 0 if (ξ, η) is not in $\mathscr{A}$. The function χ is then called the *characteristic function* of the aperture. Equation (28.4) may then be interpreted as the Fourier transform of the characteristic function, i.e.

$$q(x, y) = \int_{-\infty}^{\infty}\int_{-\infty}^{\infty} \chi(\xi, \eta) e^{ik(x\xi + y\eta)}\, d\xi\, d\eta. \tag{28.5}$$

Obviously

$$\chi(\xi, \eta) = \frac{k^2}{4\pi^2}\int_{-\infty}^{\infty}\int_{-\infty}^{\infty} q(x, y) e^{-ik(x\xi + y\eta)}\, dx\, dy. \tag{28.6}$$

The intensity perpendicular to the plane z = constant is proportional to

$$\frac{k^2 z}{R^3}\left|q\left(\frac{x}{R}, \frac{y}{R}\right)\right|^2.$$

If the intensity can be regarded as negligibly small unless x and y are both near zero, z/R can be taken as unity. The energy crossing z = constant is then proportional to

$$k^2\int_{-\infty}^{\infty}\int_{-\infty}^{\infty} |q(x, y)|^2\, dx\, dy.$$

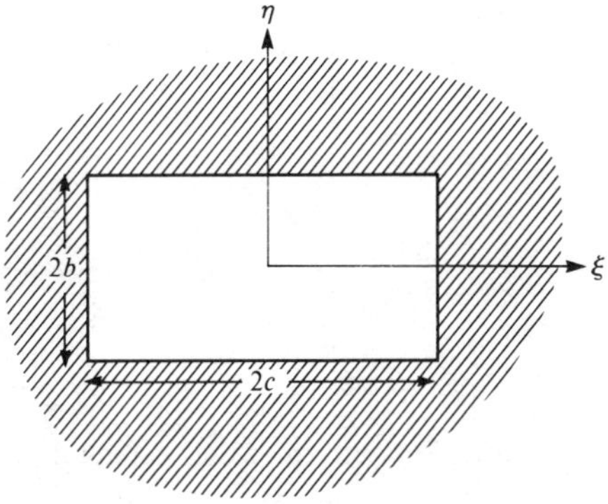

Fig. 9.23 The rectangular aperture

From (28.5)

$$\int_{-\infty}^{\infty}\int_{-\infty}^{\infty} |q(x, y)|^2 \, dx \, dy = \frac{4\pi^2}{k^2}\int_{-\infty}^{\infty}\int_{-\infty}^{\infty} |\chi(\xi, \eta)|^2 \, d\xi \, d\eta = \frac{4\pi^2 \mathscr{A}}{k^2}$$

where $\mathscr{A}$ is the area of the aperture. The energy and intensity are now immediately available. In particular, the transmission coefficient (either acoustic or electromagnetic) is unity to this order of accuracy.

An example of an aperture is a rectangle with sides of lengths $2c$ and $2b$ parallel to the coordinate axes (Fig. 9.23). The formula (28.4) gives

$$q(x, y) = (4/k^2xy)\sin kxc \sin kyb,$$

so that the intensity involves the function $(\sin^2 u)/u^2$. As u increases from zero $(\sin^2 u)/u^2$ starts from a maximum of 1 and then steadily decreases to zero at $u = \pi$. Thereafter, it increases to a maximum of about 0.05 at $u = 1.43\pi$ and then diminishes to zero at $u = 2\pi$. The next maximum is 0.02 at $u = 2.46\pi$ and the next zero at $u = 3\pi$. The pattern carries on with succeeding maxima becoming progressively smaller.

The intensity disappears on the lines $x/R = m\pi/kc$ and $y/R = n\pi kb$ where m and n are any non-zero integers. These lines form a pattern of rectangles whose longer sides are parallel to the shorter sides of the aperture (Fig. 9.24). Within each rectangle framed by pairs of consecutive lines of no intensity, the intensity rises to a maximum. These maxima are, however, only a fraction of that in the central rectangle; those on the axes are at least 1/20 of (or 13 dB below) the central maximum and, off the axes, they are 1/400 (or 26 dB below). The maxima decrease rapidly with increasing distance from the centre. Therefore, there is a tendency for only the central rectangle to be observed and possibly one or two of the side rectangles of the cross of Fig. 9.24. Notice that the central rectangle is smaller, the larger the size of the aperture.

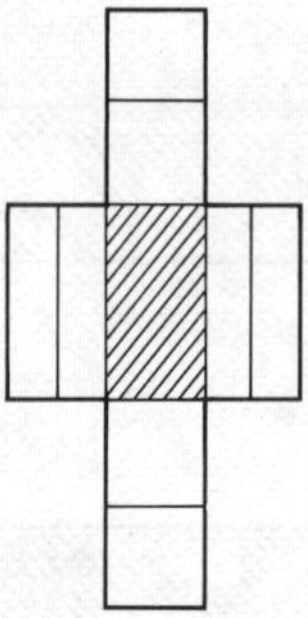

Fig. 9.24 Diffraction pattern of the rectangular aperture

For a circular aperture of radius b

$$q(x, y) = 2\pi b \mathrm{J}_1\{kb(x^2+y^2)^{\frac{1}{2}}\}/k(x^2+y^2)^{\frac{1}{2}}.$$

The intensity now vanishes on circles corresponding to the zeros of the Bessel function. If $z = R\cos\theta$, the central maximum is 1 at $\theta = 0$ and the first subsidiary maximum is 0.02 at $kb\sin\theta = 1.64\pi$. The pattern consists of a strong central disc of radius $0.61\lambda/b$ surrounded by rings of which the strongest is some 17 dB down from the central zone. In fact, the central disc contains over 80% of the energy and more than 90% is within the circle whose perimeter is the second curve of no intensity. The electromagnetic transmission coefficient is

$$\tau_{\mathrm{KE}} = 1 - (1/2kb)\int_0^{2kb} \mathrm{J}_0(u)\,du,$$

a tabulated function. For $kb \gg 1$

$$\tau_{\mathrm{KE}} \sim 1 - (1/2kb) - \{1/2\pi^{\frac{1}{2}}(kb)^{\frac{3}{2}}\}\cos(2kb - \tfrac{3}{4}\pi). \tag{28.7}$$

One question of importance, especially in connection with telescopes, is the extent to which observations of the diffraction pattern can separate out the influences of two distinct sources. In other words, the displacement of a pattern necessary to distinguish it from another is desired. This topic is that of the *resolving power* of an aperture.

In part the answer depends upon the method of observation. If the pattern is made visible the differences hang on what the eye can detect and the eye has a limit of resolution of about 1 minute of arc. With photographic and image processing techniques it may be possible to enhance the contrast and so decrease the limit of resolution. Therefore, any criterion which is independent of the means of observation must be viewed as rough and ready.

If the aperture is a circle, a plane wave with a direction of propagation at a slight angle to the z-axis may be expected to produce mainly a circle of radius $0.61\lambda/b$ but with its centre not on the z-axis. It is reasonable to suppose that this circle could be distinguished from that at normal incidence if its circumference passed through the z-axis but perhaps not if its centre were closer to the z-axis. Assuming that $0.61\lambda/b$ is small this means that two plane waves (or the positions of two distant sources) can be just resolved if the angle between their directions of propagation is $0.61\lambda/b$.

The resolving power can be improved by blanking out the centre of the aperture by a circular screen, thereby creating an annular aperture. For instance, if the blanking circle is of radius $\frac{1}{2}b$, the resolving power is $0.5\lambda/b$. However, the overall intensity of the pattern is reduced and the subsidiary maxima are intensified so that the contrast is less pronounced.

9.29 Keller's edge rays

Another approximate method for dealing with high frequencies has been devised by Keller (1957). It has been generalized by him and his colleagues to cover many applications; the general approach is often known as the *geometrical theory of diffraction.* Keller's theory has now been checked in a variety of theoretical and experimental circumstances (Bechtel and Ross, 1966) and can be reckoned to be fully substantiated as a good first approximation.

Keller's theory commences with the observation that for various straight edges the radiation pattern can be visualized as lying on cones which have an incident ray as generator (see §§9.12, 9.19 and Fig. 9.14). He now introduces the Ansatz that the rays behave in the same way even if the edge is curved. The incident ray to a point O of the edge is imagined to initiate a set of *edge rays* all lying on a right circular cone with vertex O, axis the tangent to the edge at O and with one generator the continuation of the incident ray. If the incident ray happens to strike the edge at right angles, the edge rays spread out all over the plane perpendicular to the edge.

An edge ray may itself encounter an edge. It will then stimulate its own cone of edge rays, often called *doubly-diffracted rays* to indicate that two edges have been involved. There is nothing new in principle. Considerable complication of detail may result particularly when trying to decide which rays pass through a given point.

Suppose that the sound field on an edge ray is $\gamma_e' e^{-ikL'}$ at the point Q obeying the formulae of §6.15 well away from the edge. Let γ_e be the amplitude on the edge ray at a point a distance s further from the edge than Q. Then the field at this point is $\gamma_e e^{-ik(L'+s)}$ and, according to §6.16,

$$\gamma_e = \left\{\frac{\rho_1\rho_2}{(\rho_1+s)(\rho_2+s)}\right\}^{\frac{1}{2}} \gamma_e' \qquad (29.1)$$

where ρ_1 and ρ_2 are the principal radii of curvature of the wavefront at Q.

One would like to choose Q as a point O of the edge so as to relate γ_e' to the incident ray. Unhappily, all the rays on the cone intersect at the vertex O so that the edge is a caustic of the diffracted edge rays. Hence, if Q coincides with O, ρ_2 is zero and γ_e' would need to be infinite to supply a finite nonzero value for γ_e from (29.1).

Another way of looking at (29.1) is that it makes $\rho_2^{\frac{1}{2}}\gamma_e'$ remain finite as Q approaches O. Let its limiting value be A_0. Then

$$p = A_0\left\{\frac{\rho_1}{(\rho_1+s)s}\right\}^{\frac{1}{2}} e^{-ik(L_0+s)} \tag{29.2}$$

where L_0 is the value of L' at O and s is measured along the edge ray from the point O on the edge. The radius of curvature ρ_1 is the distance of O from the second caustic of the rays. Its evaluation is straightforward. Let θ^i be the angle between the incident ray and the tangent OT to the edge whereas δ_0 is the angle between the edge ray OP and the principal normal ON which lies in the osculating plane (Fig. 9.25). Then, if ρ_0 is the radius of curvature of the edge at O

$$\rho_1 = -\rho_0 \sin^2\theta^i \bigg/ \left(\rho_0 \frac{d\theta^i}{d\sigma_0} \sin\theta^i + \cos\delta_0\right) \tag{29.3}$$

where σ_0 is the arc-length of the edge and the derivative is evaluated at O. Note that $\cos\delta_0 = \hat{\mathbf{s}}^e \cdot \boldsymbol{\nu}$ where $\hat{\mathbf{s}}^e$ and $\boldsymbol{\nu}$ are unit vectors along OP and ON respectively. Consequently, only A_0 remains to be determined in (29.2).

The suggestion is that, if γ^i is the amplitude on the incident ray, A_0 and γ^i should be related in a similar manner to that in which they are when the edge is straight, i.e. $A_0 = D\gamma^i$ where the diffraction coefficient D is obtained by solving a straight edge problem. Finding D entails the exact solution of the *canonical problem* with straight edge and appropriate

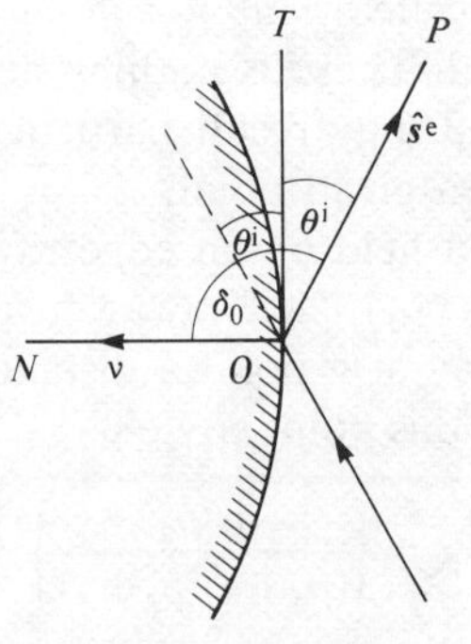

Fig. 9.25 Parameters for the calculation of ρ_1

boundary conditions. Earlier in this chapter such solutions have been derived for the edge on a semi-infinite plane and on a wedge. Therefore D can be determined when the boundary has similar local properties near the edge.

The method will be demonstrated for an edge on an infinitesimally thin surface. For points not too far from the edge a locally plane wave approximation can be utilized. Therefore the primary field can be treated as a plane wave. Moreover, s will be negligible compared with ρ_1 and, in fact, ρ_1 is infinite for a straight edge subject to an incident plane wave. Hence, in this neighbourhood, the diffracted field will have the form

$$(A_0/s^{\frac{1}{2}})e^{-ik(L^i+s)}. \tag{29.4}$$

The theory of a straight edge illuminated by a three-dimensional plane wave has been delineated in §9.9. In the notation of that section an edge ray passing through (x, y, z) is excited from the point on the edge with coordinates $(0, 0, z-r\cot\theta^i)$. Therefore $s=r\operatorname{cosec}\theta^i$ and $L^i=(z-r\cot\theta^i)\cos\theta^i$. Consequently

$$L^i+s=r\sin\theta^i+z\cos\theta^i$$

so that (29.4) reproduces correctly the exponential terms of (9.2) and (9.3). For the soft boundary condition the remainder of (9.2) will be obtained if A_0 is chosen as

$$A_0=\frac{\cos\frac{1}{2}\phi\sin\frac{1}{2}\phi^i}{\cos\phi^i-\cos\phi}\left(\frac{2}{\pi k\sin^2\theta^i}\right)^{\frac{1}{2}}e^{-\frac{1}{4}\pi i}. \tag{29.5}$$

In this case $\gamma^i=1$ and D^S may therefore be taken as the right-hand side of (29.5).

Support for this selection of D^S is forthcoming from scrutiny of the excitation of a straight edge by a line source. Now ρ_1 is infinite and $s=r$. From (8.6)

$$\gamma^i=(1/8\pi kr_0)^{\frac{1}{2}}e^{-\frac{1}{4}\pi i}$$

at the edge and $L^i=r_0$. Hence edge ray theory predicts the diffracted field

$$\{D^S/(8\pi krr_0)^{\frac{1}{2}}\}e^{-\frac{1}{4}\pi i-ik(r+r_0)}$$

with $\phi^i=\phi_0-\pi$, $\theta^i=\frac{1}{2}\pi$ in (29.5). There is complete agreement with (8.7).

As a further confirmation let the primary wave be caused by the point source of §9.10 with $z_0=0$. At a point distant h along the edge from the source (Fig. 9.26) $\cos\theta^i=h(r_0^2+h^2)^{-\frac{1}{2}}$ and, in (29.3), σ_0 may be identified with h. Since ρ_0 is infinite it follows that $\rho_1=\tilde{\omega}_1$ where $\tilde{\omega}_1$ is the distance of the point of the edge from the source P_0. In order that the edge ray pass through P it is necessary that $h=r_0z/(r+r_0)$; hence

$$\tilde{\omega}_1=r_0R'/(r+r_0), \qquad \tilde{\omega}_2=rR'/(r+r_0).$$

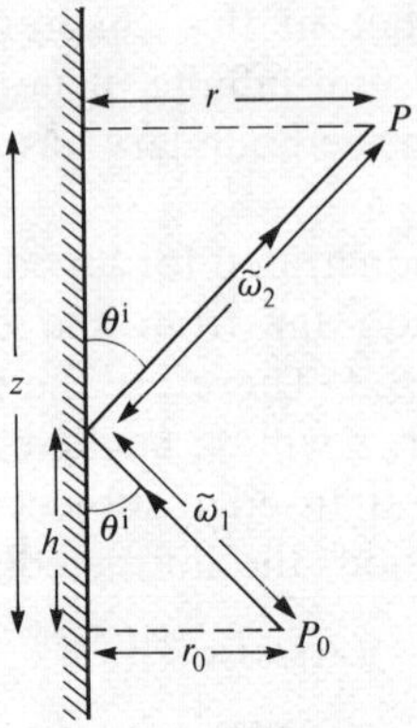

Fig. 9.26 Geometry for the edge ray from a point source

Now $L^i = \tilde{\omega}_1$ and $\gamma^i = 1/4\pi\tilde{\omega}_1$ at the edge. Hence the theory of edge rays predicts

$$\frac{D^S}{4\pi}\left\{\frac{1}{\tilde{\omega}_1\tilde{\omega}_2(\tilde{\omega}_1+\tilde{\omega}_2)}\right\}^{\frac{1}{2}} e^{-ik(\tilde{\omega}_1+\tilde{\omega}_2)}$$

with $\phi^i = \phi_0 - \pi$ and $\sin\theta^i = r_0/\tilde{\omega}_1$ in (29.5). Substitution of the values of $\tilde{\omega}_1$ and $\tilde{\omega}_2$ recovers (10.4).

The analysis can be repeated for the hard boundary condition and again complete agreement is secured.

In view of this concord, the field on an edge ray is taken as

$$p^e = D'\left\{\frac{\rho_1}{(\rho_1+s)s}\right\}^{\frac{1}{2}} \gamma^i e^{-ik(L^i+s)} \tag{29.6}$$

where γ^i has its value at the point on the edge from which the edge ray is emitted and the coefficient D' is drawn from the canonical problem (via (29.5) for the soft boundary). If ϕ^i and ϕ are the angles made with the extension of the principal normal NO by the projections of the continuation of the incident ray P_0O and of OP respectively on the normal plane at O (Fig. 9.27), (29.5) supplies

$$D'_S = \frac{\cos\frac{1}{2}\phi\,\sin\frac{1}{2}\phi^i}{\cos\phi^i - \cos\phi}\left(\frac{2}{\pi k\sin^2\theta^i}\right)^{\frac{1}{2}} e^{-\frac{1}{4}\pi i}. \tag{29.7}$$

Correspondingly, for the hard boundary condition, (9.3) gives

$$D'_H = \frac{\sin\frac{1}{2}\phi\,\cos\frac{1}{2}\phi^i}{\cos\phi^i - \cos\phi}\left(\frac{2}{\pi k\sin^2\theta^i}\right)^{\frac{1}{2}} e^{-\frac{1}{4}\pi i}. \tag{29.8}$$

For electromagnetic waves striking a perfectly conducting screen there

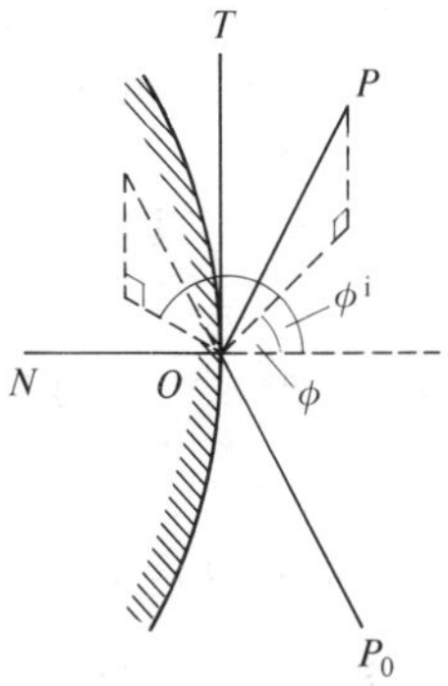

Fig. 9.27 Parameters for the diffraction coefficient

is an analogous formula stemming from (12.6). It has the form

$$\boldsymbol{E}^{\mathrm{e}} = \left\{\frac{\rho_1}{(\rho_1 + s)s}\right\}^{\frac{1}{2}} \mathrm{e}^{-\mathrm{i}ks} \boldsymbol{D}\boldsymbol{E}^{\mathrm{i}}(0) \tag{29.9}$$

where $\boldsymbol{D}$ is the apposite matrix and $\boldsymbol{E}^{\mathrm{i}}(0)$ is the incident electric intensity at O.

Keller's theory of edge rays is superior to Kirchhoff's approximation in every respect. It is not limited to plane screens. The rays are easy to imagine and the calculation of the field on them is usually simpler than the evaluation of integrals, though complications can arise when many rays are involved. Furthermore, the field is predicted as accurately as a first asymptotic approximation to an exact solution when one exists. It is, therefore, not prone to the large errors which can occur with Kirchhoff's method as mentioned in §9.27. In addition, it is capable of coping with inhomogeneous media which would be a severe embarrassment to Kirchhoff's approach.

9.30 Uniformly valid approximations

The formulae (29.7)–(29.9) are nugatory at $\phi = \pm\phi^{\mathrm{i}}$, which correspond to the shadow and reflection boundaries. Indeed, they would predict an infinite field there. They are, therefore, *non-uniform* in the sense that certain regions are excluded from their domains of validity. Transition formulae are consequently needed which change smoothly through the shadow and reflection boundaries while retaining the character of (29.7)–(29.9) away from troublesome spots. Such approximations which are also consistent with edge conditions are known as *uniformly valid.*

The derivation of a uniformly valid expansion will be restricted to acoustic waves since the electromagnetic case is treated fully in Jones (1979). The analysis will be limited to a finite aperture in a plane screen.

A point of the edge will be identified by the arc length σ_0 and then other points specified by the parameters s, ϕ of an edge ray as in the preceding section.

Assume that the incident wave has an asymptotic expansion in accordance with §6.15, namely

$$p^{\mathrm{i}} = \mathrm{e}^{-\mathrm{i}kL^{\mathrm{i}}} \sum_{m=0}^{\infty} p_m^{\mathrm{i}}/k^m \tag{30.1}$$

where

$$p_0^{\mathrm{i}}\nabla^2 L^{\mathrm{i}} + 2\,\mathrm{grad}\,L^{\mathrm{i}}\cdot\mathrm{grad}\,p_0^{\mathrm{i}} = 0,$$
$$p_j^{\mathrm{i}}\nabla^2 L^{\mathrm{i}} + 2\,\mathrm{grad}\,L^{\mathrm{i}}\cdot\mathrm{grad}\,p_j^{\mathrm{i}} = -\mathrm{i}\nabla^2 p_{j-1}^{\mathrm{i}} \quad (j=1,2,\ldots)$$

and the eikonal L^{i} is such that $\mathrm{grad}^2 L^{\mathrm{i}} = 1$.

Let $L^{\mathrm{e}} = L^{\mathrm{i}}(\sigma_0) + s$ so that L^{e} is the eikonal of an edge ray. Introduce the quantity ψ defined by

$$\psi^2 = L^{\mathrm{e}} - L^{\mathrm{i}}.$$

The distance along an incident ray to the point of observation is never more than that traversed in reaching the point via an edge ray (Fermat's principle) and so $L^{\mathrm{e}} \geqslant L^{\mathrm{i}}$. Further, $\sin\frac{1}{2}(\phi-\phi^{\mathrm{i}})$ is zero on the shadow boundary; indeed, this is true whenever $\phi-\phi^{\mathrm{i}} = 2n\pi$ for some integer n. Accordingly, *choose* ψ *to have the same sign as* $\sin\frac{1}{2}(\phi-\phi^{\mathrm{i}})$. Then, as ϕ varies, ψ will repeat itself at intervals of 4π but not of 2π; ψ is double-valued in the physical space.

Next, observe that the smooth transition over shadow and reflection boundaries in exact solutions comes from the presence of the function F. So, a field is sought which complies with the Ansatz (Ahluwalia, Lewis, and Boersma 1968)

$$p(\phi) = \pi^{-\frac{1}{2}} \exp(-\mathrm{i}kL^{\mathrm{e}} + \tfrac{1}{4}\pi\mathrm{i}) \left\{ \mathrm{F}(k^{\frac{1}{2}}\psi) \sum_{m=0}^{\infty} (p_m^{\mathrm{i}}/k^m) + k^{-\frac{1}{2}} \sum_{m=0}^{\infty} (q_m/k^m) \right\} \tag{30.2}$$

which has a similar structure to exact solutions for straight edges near a shadow boundary. To check whether (30.2) satisfies the partial differential equation for p we invoke the transport equations. Also $\mathrm{d}F(w)/\mathrm{d}w = 2\mathrm{i}wF(w) - 1$ and the partial differential equation is satisfied if

$$2\,\mathrm{grad}\,L^{\mathrm{e}}\cdot\mathrm{grad}\,q_0 + q_0\nabla^2 L^{\mathrm{e}} = \mathrm{i}p_0^{\mathrm{i}}\nabla^2\psi + 2\mathrm{i}\,\mathrm{grad}\,p_0^{\mathrm{i}}\cdot\mathrm{grad}\,\psi, \tag{30.3}$$

$$2\,\mathrm{grad}\,L^{\mathrm{e}}\cdot\mathrm{grad}\,q_m + q_m\nabla^2 L^{\mathrm{e}} = \mathrm{i}p_m^{\mathrm{i}}\nabla^2\psi + 2\mathrm{i}\,\mathrm{grad}\,p_m^{\mathrm{i}}\cdot\mathrm{grad}\,\psi - \mathrm{i}\nabla^2 q_{m-1} \quad (m\geqslant 1). \tag{30.4}$$

The coordinates of a point $\boldsymbol{x}$ are expressed in terms of s, σ_0, ϕ. Denote these temporarily by σ_1, σ_2, σ_3 and form the Jacobian

$$J = \partial(x_1, x_2, x_3)/\partial(\sigma_1, \sigma_2, \sigma_3).$$

Then

$$\frac{\partial J}{\partial \sigma_1} = \sum_{j=1}^{3} \sum_{i=1}^{3} \frac{\partial}{\partial \sigma_1}\left(\frac{\partial x_i}{\partial \sigma_j}\right) \operatorname{cof} \frac{\partial x_i}{\partial \sigma_j}$$

where cof a_{ij} stands for the co-factor of a_{ij} in J. Hence

$$\frac{\partial J}{\partial \sigma_1} = \sum_{j=1}^{3} \sum_{i=1}^{3} \frac{\partial}{\partial \sigma_j}\left(\frac{\partial L^{\mathrm{e}}}{\partial x_i}\right) \operatorname{cof} \frac{\partial x_i}{\partial \sigma_j}.$$

Now $\operatorname{cof}(\partial x_i/\partial \sigma_j) = J\, \partial \sigma_j/\partial x_i$ and so

$$\frac{\partial J}{\partial \sigma_1} = J \sum_{j=1}^{3} \sum_{i=1}^{3} \frac{\partial}{\partial \sigma_j}\left(\frac{\partial L^{\mathrm{e}}}{\partial x_i}\right) \frac{\partial \sigma_j}{\partial x_i} = J \sum_{i=1}^{3} \frac{\partial^2 L^{\mathrm{e}}}{\partial x_i^2} = J \nabla^2 L^{\mathrm{e}}. \tag{30.5}$$

Accordingly, if the right-hand side of (30.4) be written as f_m,

$$2\, \mathrm{d}(J^{\frac{1}{2}} q_m)/\mathrm{d}s = J^{\frac{1}{2}} f_m.$$

Integration along a ray now leads to

$$J^{\frac{1}{2}}(s) q_m(s) = \tfrac{1}{2} \int_0^s J^{\frac{1}{2}}(\sigma) f_m(\sigma)\, \mathrm{d}\sigma. \tag{30.6}$$

Equation (30.6) takes cognizance of the edge conditions forcing q_m to be finite as $s \to 0$ whereas $J(s)$ vanishes like s. The integrand in (30.6) is finite so that q_m does not have any singularities except possibly at $s = 0$. But here the boundedness of f_m ensures that $q_m \to 0$ as $s \to 0$.

Since f_0 is known, q_0 can be found from (30.6). Then f_1 is known and so q_1 can be determined. Thus all of q_m can be obtained in a recursive manner from (30.6).

With $p(\phi)$ now completely specified consider the field

$$p_{\mathrm{S}} = p(\phi) - p(2\pi - \phi) \tag{30.7}$$

as a candidate for solving the diffraction problem in $-\pi \leqslant \phi \leqslant \pi$ subject to the soft boundary condition.

Clearly, p_{S} vanishes on $\phi = \pi$. It is also zero on $\phi = -\pi$ because of the built-in 4π-periodicity of p. Therefore, the boundary conditions are fulfilled. Next, by (7.16) and (7.17),

$$\mathrm{F}(k^{\frac{1}{2}}\psi) \sim \pi^{\frac{1}{2}} \exp(\mathrm{i}k\psi^2 - \tfrac{1}{4}\pi\mathrm{i})\mathrm{H}(-\psi) - \frac{\mathrm{i}}{2k^{\frac{1}{2}}\psi} + \frac{1}{4k^{\frac{3}{2}}\psi^3}$$

when ψ is not near zero. Thus the incident wave is properly reproduced in $\phi \leqslant \phi^{\mathrm{i}}$, while the reflected wave arises in $\phi \leqslant -\phi^{\mathrm{i}}$. Hence (30.7) can be regarded as a satisfactory formula.

The corresponding solution for the hard boundary condition is

$$p_{\mathrm{H}} = p(\phi) + p(2\pi - \phi) \tag{30.8}$$

as may be confirmed without difficulty.

The non-uniform diffracted wave part of p is

$$p^{\mathrm{d}}(\phi)=\pi^{-\frac{1}{2}}\exp(-\mathrm{i}kL^{\mathrm{e}}+\tfrac{1}{4}\pi\mathrm{i})\left\{k^{-\frac{1}{2}}\left(q_0-\frac{\mathrm{i}p_0^{\mathrm{i}}}{2\psi}\right)+k^{-\frac{3}{2}}\left(q_1+\frac{p_0^{\mathrm{i}}}{4\psi^3}-\frac{\mathrm{i}p_1^{\mathrm{i}}}{2\psi}\right)\right\} \tag{30.9}$$

and more terms can be added by carrying the asymptotic expansion of F further. Now, it may be checked from (30.2), (30.5), and the definition of ψ that

$$\frac{\mathrm{d}}{\mathrm{d}s}\left(\mathrm{i}\frac{J^{\frac{1}{2}}p_0^{\mathrm{i}}}{\psi}\right)=J^{\frac{1}{2}}f_0.$$

Hence $J^{\frac{1}{2}}(q_0-\mathrm{i}p_0^{\mathrm{i}}/2\psi)$ does not vary on an edge ray.

To discover the behaviour of ψ near $s=0$ introduce local Cartesian axes $\boldsymbol{t}_1$, $\boldsymbol{t}_2$, $\boldsymbol{t}_3$ at a point of the edge so that $\boldsymbol{t}_1$ and $\boldsymbol{t}_3$ coincide with $-\boldsymbol{\nu}$ and σ_0 is previous notation (Fig. 9.28). Then, on the ray near the origin,

$$L^{\mathrm{i}}=L^{\mathrm{i}}(0)+s\frac{\partial}{\partial s}L^{\mathrm{i}}(0)+\tfrac{1}{2}Bs^2+\ldots$$

where

$$B=\sum_{j=1}^{3}\sum_{k=1}^{3}\left[\frac{\partial^2 L^{\mathrm{i}}}{\partial t_j\,\partial t_k}\right]_0\chi_j\chi_k,$$

$$\chi_1=\sin\theta^{\mathrm{i}}\cos\phi,\qquad \chi_2=\sin\theta^{\mathrm{i}}\sin\phi,\qquad \chi_3=\cos\theta^{\mathrm{i}}.$$

Since $\partial L^{\mathrm{i}}(0)/\partial s=\sin^2\theta^{\mathrm{i}}\cos(\phi-\phi^{\mathrm{i}})+\cos^2\theta^{\mathrm{i}}$,

$$\psi=(2s)^{\frac{1}{2}}\sin\theta^{\mathrm{i}}\sin\tfrac{1}{2}(\phi-\phi^{\mathrm{i}})\left\{1-\frac{Bs}{8\sin^2\theta^{\mathrm{i}}\sin^2\frac{1}{2}(\phi-\phi^{\mathrm{i}})}\right\} \tag{30.10}$$

for small s.

Also

$$J(s)=s(1+s/\rho_1)\sin^2\theta^{\mathrm{i}}$$

so that

$$J^{\frac{1}{2}}(q_0-\mathrm{i}p_0^{\mathrm{i}}/2\psi)=-\mathrm{i}2^{-\frac{3}{2}}(p_0^{\mathrm{i}})_{s=0}\operatorname{cosec}\tfrac{1}{2}(\phi-\phi^{\mathrm{i}}).$$

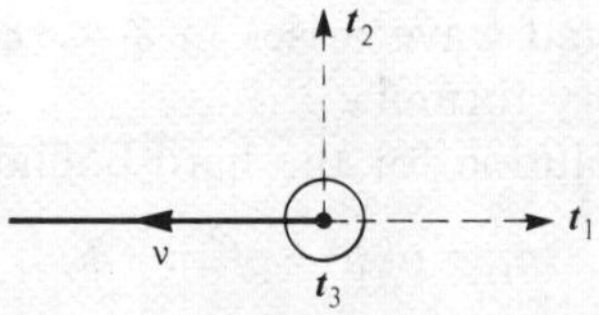

Fig. 9.28 Local Cartesian axes at a point of the edge

Hence the term in $k^{-\frac{1}{2}}$ in (30.9) contributes

$$\frac{\pi^{-\frac{1}{2}}(p_0^{\mathrm{i}})_{s=0}\exp(-\mathrm{i}kL^{\mathrm{e}}-\frac{1}{4}\pi\mathrm{i})}{2^{\frac{3}{2}}s^{\frac{1}{2}}(1+s/\rho_1)^{\frac{1}{2}}\sin\theta^{\mathrm{i}}\sin\frac{1}{2}(\phi-\phi^{\mathrm{i}})}$$

to $p^{\mathrm{d}}(\phi)$. When the corresponding term for $p^{\mathrm{d}}(2\pi-\phi)$ is included there is total agreement with Keller's estimates of $p_{\mathrm{S}}^{\mathrm{d}}$ and $p_{\mathrm{H}}^{\mathrm{d}}$. This is very satisfactory but there is additional bonus that (30.9) provides higher order terms which are not available from Keller's theory.

The higher order terms may be significant. Consider *grazing incidence* on a hard screen with $\phi^{\mathrm{i}}=\pi$, i.e. the incident ray travels towards the edge in the plane of the aperture. Then the leading term, just evaluated, is zero and it is necessary to proceed to the next order in (30.9). Now

$$\frac{\mathrm{d}}{\mathrm{d}s}\left\{J^{\frac{1}{2}}\left(q_1+\frac{p_0^{\mathrm{i}}}{4\psi^3}-\frac{\mathrm{i}p_1^{\mathrm{i}}}{2\psi}\right)\right\}=-\tfrac{1}{2}\mathrm{i}J^{\frac{1}{2}}\nabla^2\left(q_0-\frac{\mathrm{i}p_0^{\mathrm{i}}}{2\psi}\right) \tag{30.11}$$

and, if (30.10) is written as $\psi=s^{\frac{1}{2}}\beta(1+\alpha s)$,

$$J^{\frac{1}{2}}\left(q_1+\frac{p_0^{\mathrm{i}}}{4\psi^3}-\frac{\mathrm{i}p_1^{\mathrm{i}}}{2\psi}\right)\approx\frac{\sin\theta^{\mathrm{i}}}{2\beta}\left\{-\mathrm{i}p_1^{\mathrm{i}}+\frac{p_0^{\mathrm{i}}}{2\beta^2}\left(\frac{1}{s}-3\alpha+\frac{1}{2\rho_1}\right)+\frac{1}{2\beta^2}\frac{\partial p_0^{\mathrm{i}}}{\partial s}\right\} \tag{30.12}$$

for s near zero, where

$$\frac{\partial p_0^{\mathrm{i}}}{\partial s}=\sin\theta^{\mathrm{i}}\cos\phi\frac{\partial p_0^{\mathrm{i}}}{\partial t_1}+\sin\theta^{\mathrm{i}}\sin\phi\frac{\partial p_0^{\mathrm{i}}}{\partial t_2}+\cos\theta^{\mathrm{i}}\frac{\partial p_0^{\mathrm{i}}}{\partial t_3}.$$

In the total field the sum of $q_0-\mathrm{i}p_0^{\mathrm{i}}/2\psi$ for ϕ and $2\pi-\phi$ cancel by what has already been said. Therefore, from (30.11), the sum of the two types of

$$J^{\frac{1}{2}}\left(q_1+\frac{p_0^{\mathrm{i}}}{4\psi^3}-\frac{\mathrm{i}p_1^{\mathrm{i}}}{2\psi}\right)$$

is invariable on a ray and the constant can be determined from (30.12). In going from ϕ to $2\pi-\phi$, β changes sign but α and ρ_1 are unaffected as can be seen from (29.3). Only the term in $\partial p_0^{\mathrm{i}}/\partial t_2$ alters sign in $\partial p/\partial s$ and consequently the contribution to the field on the ray is

$$p_{\mathrm{H}}^{\mathrm{d}}=-\frac{\sin\phi\,\partial p_0^{\mathrm{i}}/\partial t_2}{k^{\frac{3}{2}}\pi^{\frac{1}{2}}2^{\frac{3}{2}}\sin^2\theta^{\mathrm{i}}\cos^2\frac{1}{2}\phi\,s^{\frac{1}{2}}(1+s/\rho_1)^{\frac{1}{2}}}\exp(-\mathrm{i}kL^{\mathrm{e}}+\tfrac{1}{4}\pi\mathrm{i}). \tag{30.13}$$

It will be remarked that the coefficient of $\partial p_0^{\mathrm{i}}/\partial t_2$ is closely related to $\partial/\partial\phi^{\mathrm{i}}$ of the corresponding factor for $k^{-\frac{1}{2}}$ when ϕ^{i} is placed equal to π in the derivative.

The occurrence of the derivative of p_0^{i} in (30.13) explains why it is difficult to find the non-uniform expansion when the uniformly valid approximation is unknown.

Finally, it should be pointed out for grazing incidence from the screen ($\phi^{\mathrm{i}}=0$) when the boundary is soft

$$p_{\mathrm{s}}^{\mathrm{d}}=\frac{\cos\frac{1}{2}\phi\ \partial p_0^{\mathrm{i}}/\partial t_2}{k^{\frac{3}{2}}\pi^{\frac{1}{2}}2^{\frac{3}{2}}\sin^2\theta^{\mathrm{i}}\sin^2\frac{1}{2}\phi\ s^{\frac{1}{2}}(1+s/\rho_1)^{\frac{1}{2}}}\exp(-\mathrm{i}kL^{\mathrm{e}}+\tfrac{1}{4}\pi\mathrm{i}). \qquad (30.14)$$

Exercises on Chapter 9

1. A two-dimensional field p outside the closed curve C satisfies the radiation condition

$$r^{\frac{1}{2}}\left\{\frac{\partial p}{\partial r}+\mathrm{i}kp\right\}\to 0$$

as $r\to\infty$, k being positive. Obtain the analogues of the representation (1.10) and the expansions (1.11), (1.12). Show that, if C_0 is a circle of radius r and

$$\int_{C_0}(p\ \partial p^*/\partial r-p^*\ \partial p/\partial r)\,\mathrm{d}s\to 0$$

as $r\to\infty$, $p\equiv 0$ outside C.

Indicate the appropriate modifications to achieve uniqueness if $\operatorname{Im} k<0$, $\operatorname{Re} k\geqslant 0$.

2. The two-dimensional field p satisfies $\nabla^2 p+k^2 p=0$ with k positive in the connected domain T possessing a soft boundary, which in $r\geqslant r_0>0$ consists of the straight lines $\phi=0$, $\phi=\Phi$ $(0<\Phi\leqslant 2\pi)$ but is otherwise arbitrary in $r<r_0$. There are no sources in T and the part of T in $r\geqslant r_0$ lies in $0\leqslant\phi\leqslant\Phi$. If C_1 is the portion of the circumference of a circle of radius r which is common to T and

$$\lim_{r\to\infty}\int_{C_1}|\partial p/\partial r+\mathrm{i}kp|^2\,\mathrm{d}s=0,$$

prove that $p\equiv 0$ in T.

Obtain the same result for a hard boundary.

3. Are the results of Exercise 2 still valid if the boundary condition in $r<r_0$ is replaced by $\partial p/\partial n+\sigma p=0$?

4. An electromagnetic field occupies a domain of which the part in $R\geqslant R_0>0$ is the interior of the cone $\theta=\Theta$. The boundary is perfectly conducting and there are no sources present. If the radiation conditions are satisfied show that $\boldsymbol{E}\equiv\mathbf{0}$, $\boldsymbol{H}\equiv\mathbf{0}$. [The representation of §8.13 may be helpful.]

5. Modify the proof of the Electromagnetic Uniqueness Theorem so that

$$\lim_{R\to\infty}\int_{\Omega}|(\mu/\varepsilon)^{\frac{1}{2}}\boldsymbol{n}\wedge\boldsymbol{H}+\boldsymbol{E}|^2\,\mathrm{d}\Omega=0$$

is sufficient to ensure uniqueness instead of (1.17).

6. Two plane screens $z=0$ and $z=-h$ contain apertures. The surface S_R consists of the cylinder $x^2+y^2=R^2$, $0\geqslant z\geqslant -h$, the hemisphere of radius R with centre $(0,0,0)$ in $z\geqslant 0$ and the hemisphere with centre $(0,0,-h)$ of radius R in $z\leqslant -h$. Show that

$$\lim_{R\to\infty}\int_{S_R}\left|\frac{\partial p}{\partial n}+ikp\right|^2 dS=0$$

ensures a unique solution whether a screen be hard or soft provided that the edge conditions are imposed.

7. Is there an analogue of Exercise 7 for an electromagnetic field when the two screens are perfectly conducting?

8. The domain $0<\phi\leqslant\phi_1$ consists of dielectric of permittivity ε and permeability μ; the parameters in $\phi_1<\phi<\phi_2<2\pi$ are ε_1 and μ_1. The boundaries $\phi=0$ and $\phi=\phi_2$ are perfectly conducting. Show that, to satisfy the edge conditions, the components of the field parallel to the edge are bounded. If r^ν is the lowest power of r which occurs show that $(\varepsilon-\varepsilon_1)\sin(\nu+1)(2\phi_1-\phi_2)=(\varepsilon+\varepsilon_1)\sin(\nu+1)\phi_2$ if $E_z=0$; if $H_z=0$ replace ε and ε_1 in the equation by μ_1 and μ respectively.

Deduce that there are no infinite fields if $\phi_2\leqslant\frac{1}{2}\pi$ and that, if $\phi_1=\frac{1}{2}\phi_2$, $\nu+1=\pi/\phi_2$.

9. If, in Exercise 8, $\phi_2=2\pi$ and the perfectly conducting boundaries are omitted show that similar equations for ν are obtained but with π substituted for ϕ_2.

10. Devise uniqueness theorems on the lines of Exercises 4 and 6 which will be suitable for the proof of Babinet's principle.

11. The plane wave $p^{\mathrm{i}}=\exp(-ikx\cos\phi^{\mathrm{i}}-iky\sin\phi^{\mathrm{i}})$ falls on the infinite set of soft plates $y=na$, $x\geqslant nb$ $(n=0,\pm1,\pm2,\ldots)$. The discontinuity in $\partial p/\partial y$ across the plate $y=na$ is denoted by $[p']_{y=na}$ and $[p']_{y=na}=[p']_{y=0}e^{-ikn\eta}$ where $\eta=b\cos\phi^{\mathrm{i}}+a\sin\phi^{\mathrm{i}}$. If P is the Laplace transform of the scattered field and $P=A_n\cos\kappa(y-na)+B_n\sin\kappa(y-na)$ in $na\leqslant y\leqslant(n+1)a$ show that

$$2A_0=(A_n+iB_n+A_{-n}-iB_{-n})e^{in\kappa a}-(i/\kappa)P'\sum_{r=1-n}^{n}e^{-ikr\eta-srb+i\kappa|r|a}$$

where P' is the Laplace transform of $[p']_{y=0}$. If the first term on the right-hand side can be neglected for sufficiently large n show that the equation which arises in the Wiener–Hopf method is

$$P_L(0)-\frac{1}{s+ik\cos\phi^{\mathrm{i}}}=-\frac{iP'\sin\kappa a}{2k\{\cos\kappa a-\cos(k\eta-isb)\}}.$$

12. The function u satisfies $(\nabla^2-q^2)u=0$ and the boundary conditions $u=1$ on $\phi=0$, $u=0$ on $\phi=2\pi-\beta$ $(0\leqslant\beta<2\pi)$. Show that the representation

$$u=\int_{-\infty}^{\infty}e^{-iqr\sinh t}[f\{t+i(\phi-2\pi+\beta)\}-f\{t-i(\phi-2\pi+\beta)\}]\,dt$$

satisfies the governing equation and complies with the boundary conditions if

$$f\{t-\mathrm{i}(2\pi-\beta)\}-f\{t+\mathrm{i}(2\pi-\beta)\}=\delta(t).$$

The Fourier transform of this equation with respect to t leads to $F(\alpha)=\frac{1}{2}\operatorname{cosech}(2\pi-\beta)\alpha$ where $F(\alpha)$ is the Fourier transform of $f(t)$. Hence $f(t)=(\mathrm{i}\tau/2\pi)\tanh \tau t$ where $\tau=\pi/2(2\pi-\beta)$.

Find the corresponding solution if $\partial u/\partial\phi=r$ on $\phi=0$ and $\partial u/\partial\phi=0$ on $\phi=2\pi-\beta$.

13. In Exercise 12 the boundary conditions are changed to $u=1$ on $\phi=0$ and $\partial u/\partial\phi+\mathrm{i}qrZu=0$ on $\phi=2\pi-\beta$. The starting representation is

$$u=\int_{-\infty}^{\infty}\mathrm{e}^{-\mathrm{i}qr\sinh t}[(\cosh t-Z)f\{t+\mathrm{i}(\phi-2\pi+\beta)\}$$
$$+(\cosh t+Z)f\{t-\mathrm{i}(\phi-2\pi+\beta)\}]\,\mathrm{d}t$$

and the equation for f is

$$(\cosh t-Z)f\{t-\mathrm{i}(2\pi-\beta)\}+(\cosh t+Z)f\{t+\mathrm{i}(2\pi-\beta)\}=\delta(t).\quad\text{(A)}$$

The solution is $\mathrm{i}\tau g(t)\tanh \tau t/2\pi g\{\mathrm{i}(\beta-2\pi)\}(1-Z)$ where g is a solution of (A) when the right-hand side is zero.

14. In Exercises 12 and 13 replace t by $t\pm\frac{1}{2}\pi\mathrm{i}$ and then deform both contours into the real axis, thereby obtaining a formula for u when $q=\mathrm{i}k$, k real.

15. Determine the three-dimensional Green's functions and tensors for a semi-infinite plane.

16. Verify the formula

$$\partial p_S^{(3)}/\partial x+\partial p_S^{(3)}/\partial x_0=\zeta\cos\tfrac{1}{2}\phi\cos\tfrac{1}{2}\phi_0$$

of §9.13 and its associates.

17. A *unidirectionally conducting surface* is one on which the current is constrained to flow in a given direction but is otherwise perfectly conducting. If $\boldsymbol{i}_1$ is a unit vector in the given direction and $\boldsymbol{i}_2$ a tangential unit vector perpendicular to $\boldsymbol{i}_1$ the boundary conditions are $\boldsymbol{E}\cdot\boldsymbol{i}_1=0$, $[\boldsymbol{H}]\cdot\boldsymbol{i}_1=0$, $[\boldsymbol{E}]\cdot\boldsymbol{i}_2=0$, [ ] signifying the discontinuity across the surface. At an edge $[\boldsymbol{H}]\cdot\boldsymbol{i}_2=0$.

The semi-infinite screen $x\geqslant 0$, $y=0$ is unidirectionally conducting with

$$\boldsymbol{i}_1=\boldsymbol{i}\cos\beta-\boldsymbol{k}\sin\beta\quad(0<\beta<\tfrac{1}{2}\pi)$$

and there is an electric dipole at (x_0, y_0, z_0) $(y_0>0)$ with Hertz vector $\Pi^{\mathrm{i}}\boldsymbol{j}$ where $\Pi^{\mathrm{i}}=\mathrm{e}^{-\mathrm{i}kR_s}/4\pi\varepsilon R_s$. If E_1 and E_2 are the components of $\boldsymbol{E}$ in the directions of $\boldsymbol{i}_1$ and $\boldsymbol{i}_2$ show that

$$\varepsilon E_1=-\frac{\partial^2 p_S^{(3)}}{\partial\xi\,\partial y_0}+\Psi_1,$$

$$\varepsilon E_y = -\frac{\partial^2 p_S^{(3)}}{\partial y\, \partial y_0} + k^2 \Pi^{\mathrm{i}} + e^{-ik\xi} \int_{-\infty}^{\xi} e^{ikt} \left[\frac{\partial \Psi_1}{\partial y}\right]_{\xi=t} dt,$$

$$\varepsilon E_2 = -\frac{\partial^2 p_S^{(3)}}{\partial \eta\, \partial y_0} + e^{-ik\xi} \int_{-\infty}^{\xi} e^{ikt} \left[\frac{\partial \Psi_1}{\partial \eta}\right]_{\xi=t} dt$$

where $p_S^{(3)}$ is in §9.10, ξ, η are parallel to $\boldsymbol{i}_1$, $\boldsymbol{i}_2$ and

$$\Psi_1 = \frac{\pi i k}{r^{\frac{1}{2}}} \cos\beta \sin\tfrac{1}{2}\phi \sin\tfrac{1}{2}\phi_0 e^{-ik\xi_0} \frac{\partial}{\partial y_0} \int_{-\infty}^{\xi_0} \frac{r + r_0'}{r_0'^{\frac{1}{2}}} \frac{H_1^{(2)}(k\rho')}{\rho'} e^{ik\xi'} d\xi_0'.$$

18. The plane wave $\exp\{-ik(x\cos\phi^{\mathrm{i}} + y\sin\phi^{\mathrm{i}})\}$ is incident on the two soft semi-infinite planes $y = \pm d$, $x \leqslant 0$. Find two equations for the Laplace transforms of the discontinuity in normal derivative across a plane. Show that addition and subtraction enables them to be solved by the Wiener–Hopf technique.

19. Repeat Exercise 18 for hard semi-infinite planes.

20. Derive a reciprocity theorem which permits the results of Exercises 18 and 19 to be used for finding the radiation pattern due to a mode travelling to the right between the planes.

21. If, in §9.14, $R_{00} = -|R_{00}|\, e^{-2ik\alpha}$, α is known as the *end correction* of the tube. Use Table 9.1 to find the end correction and check that, when ka is small, α is about $0.6a$.

22. Consider the theory of §9.14 when the exciting field is the asymmetric mode

$$p^{\mathrm{i}} = J_m(j_{mn} r/a) e^{-i\kappa_{mn} z} \cos m\phi.$$

23. An electromagnetic plane wave moving in a direction making an angle θ^{i} with the positive z-axis has

$$\boldsymbol{E}^{\mathrm{i}} = -(\cos\theta^{\mathrm{i}} \sin\phi\, \boldsymbol{i}_1 - \cos\theta^{\mathrm{i}} \cos\phi\, \boldsymbol{i}_2 + \sin\theta^{\mathrm{i}} \boldsymbol{k}) e^{-ik(r\sin\theta^{\mathrm{i}}\sin\phi + z\cos\theta^{\mathrm{i}})}$$

where $\boldsymbol{i}_1$, $\boldsymbol{i}_2$, $\boldsymbol{k}$ are in the directions of cylindrical polars. The semi-infinite circular tube $r = a$, $z \geqslant 0$ is perfectly conducting. Let the total field be $\boldsymbol{E}^{\mathrm{i}} + \boldsymbol{E}^{\mathrm{s}}$ with $\boldsymbol{E}^{\mathrm{s}} = \sum_n \boldsymbol{E}_n e^{in\phi}$. If $\boldsymbol{e}^{(n)} = \int_{-\infty}^{\infty} e^{-sz} \boldsymbol{E}_n\, dz$ show that

$$\Theta^{(n)} = h_z^{(n)}(a+0) - h_z^{(n)}(a-0) = \frac{2[e_\phi^{(n)}\kappa^2 - ins e_z^{(n)}/a]_{r=a}}{\pi\kappa^2 a\mu\omega J_n'(\kappa a) H_n^{(2)\prime}(\kappa a)}$$

and, if $Z^{(n)} = h_\phi^{(n)}(a+0) - h_\phi^{(n)}(a-0)$,

$$\frac{ins}{a}\Theta^{(n)} - \kappa^2 Z^{(n)} = \frac{-2\varepsilon\omega[e_z^{(n)}]_{r=a}}{\pi a J_n(\kappa a) H_n^{(2)}(\kappa a)}.$$

The first equation can be solved for $\Theta_R^{(n)}$ by the Wiener–Hopf method and then $Z_R^{(n)}$ can be found from the second.

24. In the matrix $\begin{pmatrix} f - ge_3 & ge_1 \\ ge_2 & f + ge_3 \end{pmatrix}$ e_1, e_2, and e_3 are entire functions but f

and g need not be. Verify that the matrix can be written as $\boldsymbol{C}_+\boldsymbol{C}_-$ where

$$\boldsymbol{C}_+ = r_+^{\frac{1}{2}}\begin{pmatrix}\cosh(G_{\mathrm{R}}t)-(e_3/t)\sinh(G_{\mathrm{R}}t) & (e_1/t)\sinh(G_{\mathrm{R}}t)\\ (e_2/t)\sinh(G_{\mathrm{R}}t) & \cosh(G_{\mathrm{R}}t)+(e_3/t)\sinh(G_{\mathrm{R}}t)\end{pmatrix},$$

$(G_{\mathrm{R}}+G_{\mathrm{L}})t=\frac{1}{2}\ln p$, $p=(f+q)/(f-q)$, $r=f^2-q^2$, $q=gt$, $t=(e_1e_2+e_3^2)^{\frac{1}{2}}$. For $\boldsymbol{C}_-$ replace r_+ by r_- and G_{R} by G_{L}.

Confirm that $\boldsymbol{C}_+\boldsymbol{C}_-=\boldsymbol{C}_-\boldsymbol{C}_+$. [The splitting of the $n\times n$-matrix $\sum_{m=1}^{n} a_m\boldsymbol{E}^m$ where $\boldsymbol{E}$ is an entire matrix such that the trace of $\boldsymbol{E}^r$ is zero for $r=1,\ldots,n-1$ and $\boldsymbol{E}^n$ is a scalar multiple of the unit matrix can also be carried out explicitly.]

25. Check whether the splitting of Exercise 24 is equivalent to the methods in Exercises 18, 19, and 23.

26. Use the Kontorovich–Lebedev transform to find $p_{\mathrm{H}}^{(3)}$ for a cone.

27. Find the electromagnetic field produced by a given current on a perfectly conducting (a) wedge, (b) cone.

28. Obtain a solution of $\partial^2u/\partial x^2+\partial^2u/\partial y^2=\partial^2u/c^2\,\partial t^2$ in the form

$$u=\mathrm{e}^{\mathrm{i}(\omega t-ky)}\int_0^{(r+y)^{\frac{1}{2}}}\mathrm{e}^{\mathrm{i}kv^2}\,\mathrm{d}v=\mathrm{e}^{\mathrm{i}\omega t}f(x,y).$$

Show that $\mathrm{e}^{\mathrm{i}\omega t}[\cos ky+(k/\pi)^{\frac{1}{2}}\mathrm{e}^{-\frac{1}{4}\pi\mathrm{i}}\{f(x,y)+f(x,-y)\}]$ solves a half-plane problem.

29. The medium $y>0$ has constants ε_1, μ_1 and that in $y<0$ has constants ε_2, μ_2. There is a perfectly conducting plane on $y=0$ with a slit of width $2l$ in it. A plane wave is incident on the slit from $y>0$. If $\varepsilon_1\mu_1=\varepsilon_2\mu_2$ show that the transmitted wave multiplied by $\mu_2/(\mu_1+\mu_2)$ is the same as if the medium were homogeneous.

30. Is there a result comparable to that of Exercise 29 for a circular aperture in (a) acoustic waves, (b) electromagnetic waves?

31. A narrow slit of width l in a hard plane screen is irradiated by the plane wave $\mathrm{e}^{-\mathrm{i}kz}$ at normal incidence. Show that in the transmitted far field

$$p_{\mathrm{H}}\sim(\gamma+\ln\tfrac{1}{4}kl+\tfrac{1}{2}\pi\mathrm{i})^{-1}(\pi/2kr)^{\frac{1}{2}}\mathrm{e}^{-\mathrm{i}kr+\frac{3}{4}\pi\mathrm{i}}$$

where $\gamma=0.5772\ldots$ is Euler's constant. Deduce the transmission coefficient.

32. If the screen in Exercise 31 is soft show that

$$p_{\mathrm{S}}\sim\tfrac{1}{4}(kl)^2(2\pi/kr)^{\frac{1}{2}}\mathrm{e}^{-\mathrm{i}kr+\frac{3}{4}\pi\mathrm{i}}\cos\phi^1$$

where ϕ^1 is the angular deviation from the normal to the slit. Deduce that $\tau_{\mathrm{S}}\approx\pi^2(kl)^3/32$.

33. One model of a perfectly absorbing semi-infinite plane assumes that there is no reflected wave so that the effect of the image source is removed. Thus the solution is asserted to be the first term of (7.13) or (7.24). Deduce the diffracted field.

Generalize the result to a perfectly absorbing wedge.

34. A point source at $\boldsymbol{x}_0$ produces the primary wave $p^{\mathrm{i}} = 4\pi\psi(\boldsymbol{x}, \boldsymbol{x}_0)$ on an aperture in a plane screen. If Kirchhoff's formula (26.2) is adopted show that to a first approximation

$$p = \frac{ik}{4\pi} \int_{\mathscr{A}} |\boldsymbol{x} - \mathbf{y}|^{-1} |\boldsymbol{x}_0 - \mathbf{y}|^{-1} \left(\frac{\boldsymbol{x} - \mathbf{y}}{|\boldsymbol{x} - \mathbf{y}|} - \frac{\boldsymbol{x}_0 - \mathbf{y}}{|\boldsymbol{x}_0 - \mathbf{y}|} \right) \cdot \boldsymbol{k} \exp\{-ik(|\boldsymbol{x} - \mathbf{y}| + |\boldsymbol{x}_0 - \mathbf{y}|)\} \, \mathrm{d}S_y.$$

If both $|\boldsymbol{x}|$ and $|\boldsymbol{x}_0|$ are large compared with the dimensions of the aperture show that p takes the form

$$p = \frac{ik}{4\pi R R_0} \left(\frac{\boldsymbol{x}}{R} - \frac{\boldsymbol{x}_0}{R_0} \right) \cdot \boldsymbol{k} \mathrm{e}^{-ik(R+R_0)} \int_{\mathscr{A}} \mathrm{e}^{-ikf} \, \mathrm{d}S_y$$

where $R = |\boldsymbol{x}|$, $R_0 = |\boldsymbol{x}_0|$.

35. Show that for a circular aperture of radius a in a plane wave $f = x\xi/R$ in Exercise 34. Deduce that the intensity is proportional to the square of $\mathrm{J}_1(ka \sin\theta)/k \sin\theta$.

36. If both $\boldsymbol{x}_0$ and $\boldsymbol{x}$ in Exercise 34 are on the axis of a circular aperture of radius a show that the intensity is $4 \sin^2 \frac{1}{4}ka^2(R^{-1} + R_0^{-1})/(R + R_0)^2$.

37. Verify the statement made at the end of §9.28 about the improvement in resolving power resulting from the introduction of a blanking circle.

38. Show that there is a $\boldsymbol{W}(\boldsymbol{x}, \mathbf{y}) = \boldsymbol{W}(\boldsymbol{x} - \mathbf{y}, \mathbf{0})$ such that

$$\operatorname{curl} \boldsymbol{W}(\boldsymbol{x}, \mathbf{0}) = \psi(\boldsymbol{x}, \mathbf{0}) \operatorname{grad} p^{\mathrm{i}}(\boldsymbol{x}) - p^{\mathrm{i}}(\boldsymbol{x}) \operatorname{grad} \psi(\boldsymbol{x}, \mathbf{0}).$$

Assuming that $\boldsymbol{W}(\boldsymbol{x}, \mathbf{0}) = \boldsymbol{x} \wedge \mathbf{A}(\boldsymbol{x}, \mathbf{0})$ use (F.7.5) to prove that

$$\frac{\partial}{\partial R}(R\boldsymbol{W}) = -\frac{\mathrm{e}^{-ikR}}{4\pi R} \boldsymbol{x} \wedge \operatorname{grad} p^{\mathrm{i}}$$

where $R = |\boldsymbol{x}|$. Thus

$$R_2 \boldsymbol{W}(\boldsymbol{x}_2, \mathbf{0}) - R_1 \boldsymbol{W}(\boldsymbol{x}_1, 0) = -\frac{1}{4\pi} \int_{R_1}^{R_2} \frac{\boldsymbol{x}}{R} \wedge \mathrm{e}^{-ikR} \operatorname{grad} p^{\mathrm{i}} \, \mathrm{d}R.$$

Verify that the correct form of curl $\boldsymbol{W}$ is obtained.

If the radiation condition at infinity is satisfied find $\boldsymbol{W}(\boldsymbol{x}, \mathbf{0})$.

39. If p^{i} is as in Exercise 34 show that $\boldsymbol{W}$ of Exercise 38 is given by

$$\boldsymbol{W}(\boldsymbol{x}, \mathbf{y}) = -\frac{(\boldsymbol{x} - \mathbf{y}) \wedge (\boldsymbol{x}_0 - \mathbf{y}) \exp\{-ik(|\boldsymbol{x} - \mathbf{y}| + |\boldsymbol{x}_0 - \mathbf{y}|)\}}{4\pi |\boldsymbol{x} - \mathbf{y}| \, |\boldsymbol{x}_0 - \mathbf{y}| \, \{|\boldsymbol{x} - \mathbf{y}| \, |\boldsymbol{x}_0 - \mathbf{y}| + (\boldsymbol{x} - \mathbf{y}) \cdot (\boldsymbol{x}_0 - \mathbf{y})\}}.$$

The integral in Kirchhoff's theory can be written $\int_C \boldsymbol{W}(\mathbf{y}, \boldsymbol{x}) \, \mathrm{d}s_y$ where C in the edge of $\mathscr{A}$.

40. Generalize the results of Exercises 38 and 39 to the electromagnetic field by utilizing the fact that each Cartesian component satisfies

Helmholtz's equation. Can the edge integral be expressed in terms of currents and charges?

41. Use edge rays to determine the scattering by a wide slit.

42. It is surmised that the transmission coefficient of a large circular aperture of radius a is given by the following formulae

$$\tau_S = 1 - \frac{2}{\pi^{\frac{1}{2}}(ka)^{\frac{3}{2}}}\sin(2ka - \tfrac{1}{4}\pi) - \frac{1}{(ka)^2}(\tfrac{1}{4} + \cos 4ka),$$

$$\tau_H = 1 - \frac{1}{4(ka)^2} + \frac{1}{4\pi^{\frac{1}{2}}(ka)^{\frac{5}{2}}}\cos(2ka - \tfrac{1}{4}\pi),$$

$$\tau_E = 1 - \frac{1}{\pi^{\frac{1}{2}}(ka)^{\frac{3}{2}}}\sin(2ka - \tfrac{1}{4}\pi)$$

at normal incidence. By means of edge ray theory decide whether or not you agree with these conjectures.

43. Verify (30.14).

10. Transient waves

THE discussion of the last two chapters has concentrated on the characteristics of disturbances which vary harmonically with time. Harmonic waves are of great practical importance but there are also significant fields whose time variation is not harmonic. The solution for any time dependence can be written down from the harmonic wave by means of a Laplace or Fourier transform. The interpretation of this solution may not be easy for all values of the time but it is usually possible to estimate asymptotically the behaviour for small and large time by the method of stationary phase and allied techniques. Some examples of this approach are given.

Another line of attack is to eschew the harmonic problem and work directly in the time domain. The procedures then are quite different as will be seen from the illustrations.

Time transforms

10.1 Reflection of a pulse by a plane interface

Let the plane $z=0$ separate two homogeneous isotropic dielectric media and suppose that there is a line source of current parallel to the y-axis at $(0, 0, h)$. The whole electromagnetic field is then independent of y and may be expressed in terms of E_y. The time variation of the line source will be assumed to be a δ-function; formulae for arbitrary variations in time can then be deduced immediately by integration. With these assumptions the equation satisfied by E_y in $z>0$ is

$$\nabla^2 E_y - \frac{1}{c_1^2}\frac{\partial^2 E_y}{\partial t^2} = -\delta(x)\delta(z-h)\delta(t) \tag{1.1}$$

where $c_1 = 1/(\mu_1\varepsilon_1)^{\frac{1}{2}}$ is the speed of light in the upper medium. In the lower medium c_1 is replaced in (1.1) by $c_2 = 1/(\mu_2\varepsilon_2)^{\frac{1}{2}}$ and the right-hand side is, of course, zero.

Define $\mathscr{E}$, the Fourier transform in time of E_y, by

$$\mathscr{E} = \int_{-\infty}^{\infty} E_y \mathrm{e}^{-\mathrm{i}\omega t}\,\mathrm{d}t.$$

Then $\mathscr{E}$ satisfies in $z>0$

$$\frac{\partial^2 \mathscr{E}}{\partial x^2}+\frac{\partial^2 \mathscr{E}}{\partial z^2}+k_1^2\mathscr{E}=-\delta(x)\delta(z-h) \tag{1.2}$$

where $k_1=\omega/c_1$. Now introduce the Laplace transform $\tilde{\mathscr{E}}$ of $\mathscr{E}$ by the definition

$$\tilde{\mathscr{E}}=\int_{-\infty}^{\infty}\mathscr{E}\mathrm{e}^{-sx}\,\mathrm{d}x.$$

Then the equations governing $\tilde{\mathscr{E}}$ are

$$\mathrm{d}^2\tilde{\mathscr{E}}/\mathrm{d}z^2+(s^2+k_1^2)\tilde{\mathscr{E}}=-\delta(z-h)\quad(z>0), \tag{1.3}$$

$$\mathrm{d}^2\tilde{\mathscr{E}}/\mathrm{d}z^2+(s^2+k_2^2)\tilde{\mathscr{E}}=0\quad(z<0) \tag{1.4}$$

where $k_2=\omega/c_2$.

The first objective is to discover the primary wave; i.e. (1.1), and therefore (1.3), is supposed for the moment to hold for all z. The solution of (1.3) which is outgoing at infinity is

$$\tilde{\mathscr{E}}^{\mathrm{i}}=\frac{\exp\{-\mathrm{i}\,|z-h|\,(s^2+k_1^2)^{\frac{1}{2}}\}}{2\mathrm{i}(s^2+k_1^2)^{\frac{1}{2}}}$$

where the branch lines of $(s^2+k_1^2)^{\frac{1}{2}}$ are defined as in §9.5. Therefore

$$\mathscr{E}^{\mathrm{i}}=-\frac{1}{4\pi}\int_{-\mathrm{i}\infty}^{\mathrm{i}\infty}\frac{\exp\{sx-\mathrm{i}\,|z-h|\,(s^2+k_1^2)^{\frac{1}{2}}\}}{(s^2+k_1^2)^{\frac{1}{2}}}\,\mathrm{d}s,$$

the path of integration being to the left of $s=\mathrm{i}k_1$ and to the right of $s=-\mathrm{i}k_1$. Put $x=r_{\mathrm{s}}\cos\phi$ and $z-h=r_{\mathrm{s}}\sin\phi$ and make the transformation $s=-\mathrm{i}k_1\cos(\phi+\mathrm{i}u)$ as in §9.6; then

$$\mathscr{E}^{\mathrm{i}}=\frac{1}{4\pi}\int_{-\infty}^{\infty}\exp(-\mathrm{i}k_1r_{\mathrm{s}}\cosh u)\,\mathrm{d}u.$$

Hence

$$E_y^{\mathrm{i}}=\frac{1}{2\pi}\int_{-\infty}^{\infty}\mathscr{E}^{\mathrm{i}}\mathrm{e}^{\mathrm{i}\omega t}\,\mathrm{d}\omega=\frac{1}{4\pi}\int_{-\infty}^{\infty}\delta\left(t-\frac{r_{\mathrm{s}}}{c_1}\cosh u\right)\mathrm{d}u$$

on interchanging the order of integration. The substitution $r_{\mathrm{s}}\cosh u=c_1v$ now gives

$$E_y^{\mathrm{i}}=\frac{1}{2\pi}\int_{r_{\mathrm{s}}/c_1}^{\infty}\delta(t-v)\frac{\mathrm{d}v}{(v^2-r_{\mathrm{s}}^2/c_1^2)^{\frac{1}{2}}}=\frac{\mathrm{H}(t-r_{\mathrm{s}}/c_1)}{2\pi(t^2-r_{\mathrm{s}}^2/c_1^2)^{\frac{1}{2}}} \tag{1.5}$$

where $\mathrm{H}(x)$ is the Heaviside step function which is equal to 1 if $x>0$ and zero if $x<0$. It can easily be confirmed that (1.5) is a solution of (1.1) when $t>r_{\mathrm{s}}/c_1$ either directly or from the observation that the mapping

$c_1 t = \mathrm{i}y$ converts (1.1) into Laplace's equation with solution $\{x^2+y^2+(z-h)^2\}^{-\frac{1}{2}}$. The initial impulse of the source lasts an infinitesimally short time yet (1.5) demonstrates that *the incident wave has no sharp cut-off behind its wavefront.* This special feature of cylindrical waves is due to the fact that some disturbance is always arriving from the distant parts of the line source (regarded as a distribution of point sources).

Return now to the matter of solving (1.3) and (1.4). A solution is required that guarantees the continuity of E_y and H_x across $z=0$. It is sufficient to ask that $\tilde{\mathscr{E}}$ and $\mathrm{d}\tilde{\mathscr{E}}/\mu\,\mathrm{d}z$ be continuous there. Consequently

$$\tilde{\mathscr{E}} = \frac{1}{2\mathrm{i}(s^2+k_1^2)^{\frac{1}{2}}}\left[\exp\{-\mathrm{i}\,|z-h|\,(s^2+k_1^2)^{\frac{1}{2}}\}\right.$$
$$\left. + \frac{\mu_2(s^2+k_1^2)^{\frac{1}{2}} - \mu_1(s^2+k_2^2)^{\frac{1}{2}}}{\mu_2(s^2+k_1^2)^{\frac{1}{2}} + \mu_1(s^2+k_2^2)^{\frac{1}{2}}}\exp\{-\mathrm{i}(z+h)(s^2+k_1^2)^{\frac{1}{2}}\}\right]$$

in $z>0$ and

$$\tilde{\mathscr{E}} = \frac{-\mathrm{i}\mu_2\exp\{-\mathrm{i}h(s^2+k_1^2)^{\frac{1}{2}} + \mathrm{i}z(s^2+k_2^2)^{\frac{1}{2}}\}}{\mu_2(s^2+k_1^2)^{\frac{1}{2}} + \mu_1(s^2+k_2^2)^{\frac{1}{2}}}$$

in $z<0$.

Let $x = r_\mathrm{i}\cos\phi_\mathrm{i}$, $z+h = r_\mathrm{i}\sin\phi_\mathrm{i}$ and suppose that $c_1>c_2$. Map the contour into the hyperbolic path $s = -\mathrm{i}k_1\cos(\phi_\mathrm{i}+\mathrm{i}u)$. Then, on account of (1.5),

$$E_y = E_y^\mathrm{i} + \frac{1}{8\pi^2}\int_{-\infty}^{\infty}\mathrm{e}^{\mathrm{i}\omega t}$$
$$\times\int_{-\infty}^{\infty}\frac{\mu_2 c_2\sin(\phi_\mathrm{i}+\mathrm{i}u) - \mu_1\{c_1^2 - c_2^2\cos^2(\phi_\mathrm{i}+\mathrm{i}u)\}^{\frac{1}{2}}}{\mu_2 c_2\sin(\phi_\mathrm{i}+\mathrm{i}u) + \mu_1\{c_1^2 - c_2^2\cos^2(\phi_\mathrm{i}+\mathrm{i}u)\}^{\frac{1}{2}}}\,\mathrm{e}^{-\mathrm{i}k_1 r_\mathrm{i}\cosh u}\,\mathrm{d}u\,\mathrm{d}\omega$$

in $z>0$. An integration with respect to ω followed by one with respect to u leads to

$$E_y = E_y^\mathrm{i} + \mathrm{Re}\,\frac{\mu_2 c_2\sin(\phi_\mathrm{i}+\mathrm{i}u_\mathrm{i}) - \mu_1\{c_1^2 - c_2^2\cos^2(\phi_\mathrm{i}+\mathrm{i}u_\mathrm{i})\}^{\frac{1}{2}}}{\mu_2 c_2\sin(\phi_\mathrm{i}+\mathrm{i}u_\mathrm{i}) + \mu_1\{c_1^2 - c_2^2\cos^2(\phi_\mathrm{i}+\mathrm{i}u_\mathrm{i})\}^{\frac{1}{2}}}$$
$$\times\frac{\mathrm{H}(t - r_\mathrm{i}/c_1)}{2\pi(t^2 - r_\mathrm{i}^2/c_1^2)^{\frac{1}{2}}} \tag{1.6}$$

in $z>0$, u_i being such that $\cosh u_\mathrm{i} = c_1 t/r_\mathrm{i}$.

While the first term of (1.6) is the primary wave the second term is the wave reflected by the interface; it appears to emanate from the image of the source in $z=0$ (Fig. 10.1). At the reflected wavefront $c_1 t = r_\mathrm{i}$ and $u_\mathrm{i}=0$. The first factor in the second term then becomes the same as the Fresnel reflection coefficient for a plane wave incident at an angle $\frac{1}{2}\pi - \phi_\mathrm{i}$ with the electric vector perpendicular to the plane of incidence (§6.6(b)), i.e. *near the reflected wavefront predictions based on geometrical optics are valid.*

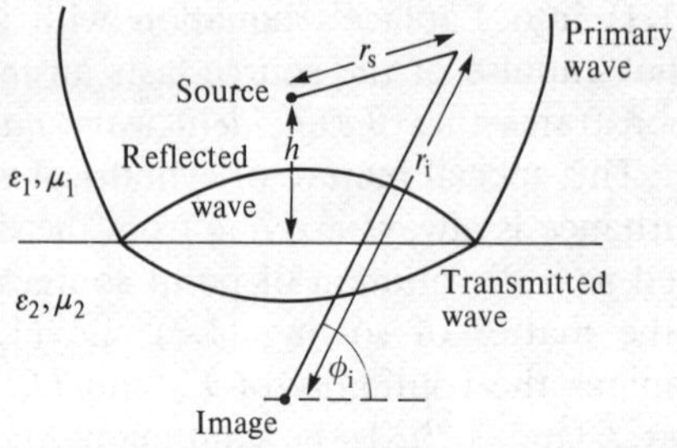

Fig. 10.1 Reflection of cylindrical wave when $c_1 > c_2$

Next suppose that the speed of light in the lower medium is greater than in the upper so that $c_2 > c_1$. If $|\cos\phi_1| < c_1/c_2$ deformation of the contour into the hyperbolic path is still feasible and (1.6) continues to hold. When $|\cos\phi_1| > c_1/c_2$ the hyperbola cannot be reached without intercepting one of the branch lines from $\pm ik_2$. So, for $\cos\phi_i > c_1/c_2$ the original contour is switched to the hyperbolic path $s = -ik_1\cos(\phi_i + iu)$ together with a suitable loop round the branch line encountered. A typical case is illustrated in Fig. 10.2. The contribution of the hyperbolic path proves to be the same as in (1.6). As for the branch-line loop it can be evaluated by the change of variables $s = -ik_1\cos v$ and provides

$$\frac{1}{4\pi^2}\int_{-\infty}^{\infty} e^{i\omega t}\int_{\phi_i}^{\cos^{-1}c_1/c_2} \exp\{-ik_1 r_i \cos(v-\phi_i)\}$$

$$\times \operatorname{Im}\frac{\mu_2 c_2 \sin v + \mu_1 i(c_2^2\cos^2 v - c_1^2)^{\frac{1}{2}}}{\mu_2 c_2 \sin v - \mu_1 i(c_2^2\cos^2 v - c_1^2)^{\frac{1}{2}}}\,dv\,d\omega$$

$$= \frac{H(r_i - c_1 t)H\{c_1c_2t - c_1x - (c_2^2 - c_1^2)^{\frac{1}{2}}(z+h)\}}{2\pi\{(r_i/c_1)^2 - t^2\}^{\frac{1}{2}}}$$

$$\times \operatorname{Im}\frac{\mu_2 c_2 \sin v_i + \mu_1 i(c_2^2\cos^2 v_i - c_1^2)^{\frac{1}{2}}}{\mu_2 c_2 \sin v_i - \mu_1 i(c_2^2\cos^2 v_i - c_1^2)^{\frac{1}{2}}} \qquad (1.7)$$

where v_i satisfies $r_i\cos(v_i - \phi_i) = c_1 t$.

If $c_1 t > r_i$ the field of (1.7) disappears and the electric intensity is given by (1.6). For $c_1 t < r_i$, the second term of (1.6) is removed leaving only the primary wave but if, in addition, $c_1c_2t > c_1x + (c_2^2 - c_1^2)^{\frac{1}{2}}(z+h)$ there is also

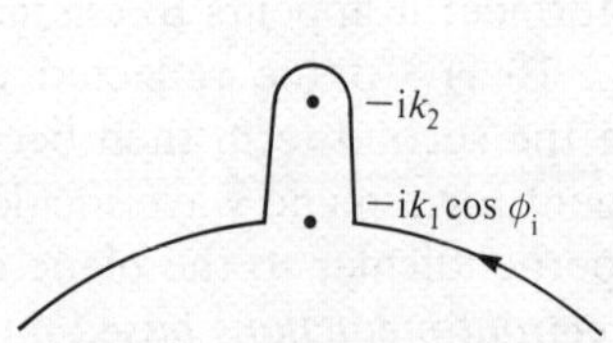

Fig. 10.2 The deformed contour when $\cos\phi_i > c_1/c_2$

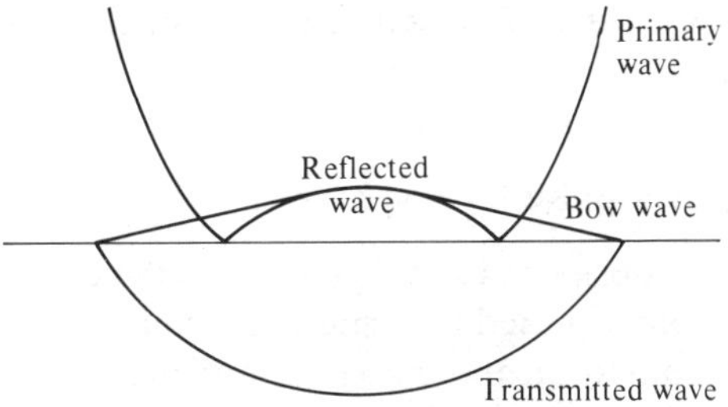

Fig. 10.3 Bow wave produced when $c_2 > c_1$

the field of (1.7) present. The *secondary wave extends beyond the cylindrical reflected wavefront centred on the image* (Fig. 10.3). The upper limit to the additional region in $x > 0$ is the plane whose equation at time t is $c_1c_2t = c_1x + (c_2^2 - c_1^2)^{\frac{1}{2}}(z+h)$. The plane intersects the interface $z=0$ in $c_2t = x + (c_2^2 - c_1^2)^{\frac{1}{2}}h/c_1$ and the line of intersection therefore moves along the interface at speed c_2, i.e. at the speed of waves in the lower medium. The plane also touches the reflected wave, the line of contact lying always in the plane $x(c_2^2 - c_1^2)^{\frac{1}{2}} = c_1(z+h)$.

What happens physically is that the arrival of the primary wave at the surface separating the media initiates a disturbance in the lower medium travelling at speed c_2. At first this disturbance takes longer to move along the interface than for the primary wave to arrive and Fig. 10.1 is pertinent. Eventually, energy in the incident wave comes in on a ray in the direction for *total reflection* (§6.7); this occurs on $z=0$ at $x = c_1h(c_2^2 - c_1^2)^{-\frac{1}{2}}$. Thereafter the disturbance in the lower medium travels faster along the interface than the incoming radiation. It has already been shown in §3.4(b) that a source moving faster than the speed of light in a medium generates a Mach cone of semi-angle $\sin^{-1} c_1/c_2$. Since cylindrical waves are involved in the current phenomenon the wavefront will be plane and make an angle $\sin^{-1} c_1/c_2$ with the direction of motion of the source, i.e. along the interface. In this context it is known as a *bow wave.* The strength of the bow wave increases with distance from the interface, being zero there where $\cos v_i = c_1/c_2$. The existence of the bow wave has been demonstrated experimentally. Notwithstanding, the major part of the energy is concentrated near the reflected wavefront and may be predicted by geometrical optics.

Should the time dependence of the source be $\delta(t-w)$ instead of $\delta(t)$ it is only necessary to replace t throughout by $t-w$. Consequently, if the source has time dependence $f(t)$ we can, by writing $f(t) = \int_{-\infty}^{\infty} f(w)\delta(t-w)\,\mathrm{d}w$, deduce that the primary wave is

$$\int_{-\infty}^{\infty} \frac{\mathrm{H}(t-w-r_s/c_1)f(w)}{2\pi\{(t-w)^2 - r_s^2/c_1^2\}^{\frac{1}{2}}}\,\mathrm{d}w$$

and the secondary field can be expressed at once as an integral of the preceding solution.

10.2 Scattering of a pulse by a sphere

The problem to be considered in this section is the effect of a rigid sphere on the radiation from a pulsed acoustic point source. Let the sphere be of radius b and the source on the z-axis at a distance R_0 from the centre. Then a solution of

$$\nabla^2 p - \frac{1}{a^2}\frac{\partial^2 p}{\partial t^2} = -\delta(\boldsymbol{x}-\boldsymbol{x}_0)\delta(t) \tag{2.1}$$

is required such that the normal derivative vanishes on $R=b$. Define the Laplace transform in time P by

$$P = \int_{-\infty}^{\infty} p\mathrm{e}^{-st}\,\mathrm{d}t.$$

Then

$$\nabla^2 P - (s^2/a^2)P = -\delta(\boldsymbol{x}-\boldsymbol{x}_0). \tag{2.2}$$

It is assumed that $\mathrm{Re}\, s > 0$.

The primary wave solution of (2.2) is

$$P^{\mathrm{i}} = \exp(-sR_{\mathrm{s}}/a)/4\pi R_{\mathrm{s}} \tag{2.3}$$

where R_{s} is, as usual, the distance from the source to the point of observation. Hence

$$p^{\mathrm{i}} = \mathrm{H}(t - R_{\mathrm{s}}/a)/4\pi R_{\mathrm{s}}. \tag{2.4}$$

The incident field (2.3) can be expressed in terms of spherical Bessel functions (see §8.11) by

$$P^{\mathrm{i}} = -(s/4\pi)\sum_{n=0}^{\infty}(2n+1)\mathrm{h}_n^{(2)}(-\mathrm{i}sR_0/a)\mathrm{j}_n(-\mathrm{i}sR/a)\mathrm{P}_n(\cos\theta)$$

for $R<R_0$. Therefore, assume a scattered wave

$$P^{\mathrm{s}} = -(s/4\pi)\sum_{n=0}^{\infty}(2n+1)A_n\mathrm{h}_n^{(2)}(-\mathrm{i}sR/a)\mathrm{P}_n(\cos\theta).$$

The boundary condition is met by the choice

$$A_n = -\mathrm{j}_n'(-\mathrm{i}sb/a)\mathrm{h}_n^{(2)}(-\mathrm{i}sR_0/a)/\mathrm{h}_n^{(2)\prime}(-\mathrm{i}sb/a).$$

It follows that on the sphere

$$[P]_{R=b} = -\frac{a^2\mathrm{i}}{4\pi b^2 s}\sum_{n=0}^{\infty}(2n+1)\frac{\mathrm{h}_n^{(2)}(-\mathrm{i}sR_0/a)}{\mathrm{h}_n^{(2)\prime}(-\mathrm{i}sb/a)}\mathrm{P}_n(\cos\theta) \tag{2.5}$$

when the Wronskian relation (A.4.11) is invoked. Consequently

$$[p]_{R=b} = -\frac{a^2}{8\pi^2 b^2}\int_{c-i\infty}^{c+i\infty}\sum_{n=0}^{\infty}(2n+1)\frac{h_n^{(2)}(-isR_0/a)}{s\,h_n^{(2)\prime}(-isb/a)}P_n(\cos\theta)e^{st}\,ds. \quad (2.6)$$

The singularities of the integrand of (2.6) will now be examined. At first sight one might suggest that there is a pole at $s=0$ but the behaviour of the spherical Hankel functions near $s=0$ displayed in §A.4 indicates that this suggestion is false and there is no singularity at $s=0$. There remains the possibility of zeros of $h_n^{(2)\prime}$. Some of these are set out in Table 8.5 of §8.18 and show that when $n=0$ there is a simple pole at $s=-a/b$, when $n=1$ poles at $bs/a=-1\pm i$ and, when $n=2$, three poles at $bs/a=-1.78$, $-1.11\pm1.95i$. In these three cases the poles are either on the negative real axis or occur as complex conjugate pairs in $\mathrm{Re}\,s<0$. This statement can be shown to be correct for every positive integer n.

All the singularities are in $\mathrm{Re}\,s<0$. Therefore, when the contour can be deformed to the right there is no disturbance on the sphere. From the exponential dependence of the spherical Hankel functions this will be true if $at<R_0-b$, i.e. before the primary wavefront strikes the sphere.

When $at>R_0-b$ the contour has to be pushed to the left and then there is a field on the sphere. Let $s=s_m$ be a typical pole; then

$$[p]_{R=b} = -\frac{a^3}{4\pi b}\sum_{n=0}^{\infty}(2n+1)P_n(\cos\theta)$$

$$\times\sum_m \frac{s_m h_n^{(2)}(-is_m R_0/a)e^{s_m t}}{\{n(n+1)a^2+b^2 s_m^2\}h_n^{(2)}(-is_m b/a)}. \quad (2.7)$$

The summation with respect to m depends, naturally, on the value of n. Since the s_m are real or occur in complex conjugate pairs the field specified by (2.7) is indeed real.

Two aspects of (2.7) will be noticed. Firstly, $\mathrm{Re}\,s_m<0$ so that each term decays exponentially in time. Secondly, s_m is independent of the incident wave; its value is fixed purely by the size of the sphere and the boundary condition. Thus each exponential time factor is unaffected by an alteration of the source though the multiplying coefficient will be influenced.

10.3 Singularity Expansion Method

The structure of (2.7) with respect to the time factors invites the conjecture that it might carry over to bodies other than the sphere. It is found that this is indeed so and such series are a mark of the *singularity expansion method* or *SEM* for short.

SEM will be exemplified by a soft body with surface S surrounded by a medium in which the speed of sound $a=1$. An appropriate representation for P is

$$P(\boldsymbol{x})=P^{\rm i}(\boldsymbol{x})+\int_S v\frac{\partial}{\partial n_y}\frac{{\rm e}^{-s\,|\boldsymbol{x}-\mathbf{y}|}}{4\pi\,|\boldsymbol{x}-\mathbf{y}|}\,{\rm d}S_y.$$

Allowing $\boldsymbol{x}$ to approach S we can comply with the boundary condition $P=0$ by imposing the integral equation

$$0=P^{\rm i}(\boldsymbol{x})+\tfrac{1}{2}v(\boldsymbol{x})+\int_S v\frac{\partial}{\partial n_y}\frac{{\rm e}^{-s\,|\boldsymbol{x}-\mathbf{y}|}}{4\pi\,|\boldsymbol{x}-\mathbf{y}|}\,{\rm d}S_y \tag{3.1}$$

for v. Here $\boldsymbol{x}$ is on S and Re $s>0$.

It can be shown that (3.1) possesses a unique solution and that there is a resolvent $\Gamma(\boldsymbol{x}, \mathbf{y}, s)$. Thus

$$v(\boldsymbol{x})=-2P^{\rm i}(\boldsymbol{x})-2\int_S \Gamma(\boldsymbol{x}, \mathbf{y}, s)P^{\rm i}(\mathbf{y})\,{\rm d}S_y. \tag{3.2}$$

The properties of the resolvent are more difficult to come by. They are certainly not affected by the incident field but are intrinsic to S and the boundary condition on it. Our experience with the sphere inclines us to speculate that Γ has only simple poles which either lie on the negative real axis or form complex conjugate pairs in Re $s<0$ (the difficult part to prove of this conjecture is that the poles are simple). If this be accepted the residue of Γ at a typical pole $s=s_m$ can be taken as $P_m(\boldsymbol{x})Q_m(\mathbf{y})$. Another idea is that as Re $s\to\infty$

$$\Gamma\sim{\rm e}^{-s\,|\boldsymbol{x}-\mathbf{y}|}$$

and that there are contours going off to infinity in the direction of Re $s\to-\infty$ on which this behaviour is reproduced.

Suppose that the incident wave is plane with $P^{\rm i}={\rm e}^{-s\boldsymbol{n}^{\rm i}\cdot\boldsymbol{x}}$. Make an inversion of (3.2) back into the time domain. Then, with the assumptions on Γ a typical pole contributes a constant multiple of

$${\rm e}^{s_m t}{\rm H}(t-\boldsymbol{n}^{\rm i}\cdot\boldsymbol{x})P_m(\boldsymbol{x})\int_S Q_m(\mathbf{y}){\rm e}^{-s_m\boldsymbol{n}^{\rm i}\cdot\mathbf{y}}\,{\rm d}S_y$$

when $t\geqslant\boldsymbol{n}^{\rm i}\cdot\boldsymbol{x}$. Again the exponential time factor is settled by the body and not by the primary wave. The incident field affects only a *coupling coefficient* involving integration over S.

The basic simplicity of the structure of SEM has stimulated a good deal of research recently into numerical methods to implement it but details will be left on one side here.

10.4 Distant radiation

A general way of tackling problems that offers itself is to take Fourier transforms with respect to time and all space variables, deploying generalized functions to advantage. On inversion a formula of the type

$$u=\int_{-\infty}^{\infty}\int_{-\infty}^{\infty}\int_{-\infty}^{\infty}\int_{-\infty}^{\infty}\frac{\exp(\mathrm{i}\omega t-\mathrm{i}\alpha_1x_1-\mathrm{i}\alpha_2x_2-\mathrm{i}\alpha_3x_3)}{F(\omega,\alpha_1,\alpha_2,\alpha_3)}\,\mathrm{d}\alpha_1\,\mathrm{d}\alpha_2\,\mathrm{d}\alpha_3\,\mathrm{d}\omega \tag{4.1}$$

$$=\int_{-\infty}^{\infty} v\mathrm{e}^{\mathrm{i}\omega t}\,\mathrm{d}\omega \tag{4.2}$$

is obtained for the sought solution u. It is rarely feasible to evaluate (4.1) exactly and the question of approximation at large distances justifies study. A general treatment of asymptotics of integrals is in Jones (1982).

We begin by discussing v as obtained from (4.1) and (4.2), i.e. the integration with respect to ω will be ignored. Actually ω will be assumed to have a small negative imaginary part whereas α_1, α_2, and α_3 are real. Another simplification is that F is a polynomial in α_1, α_2, and α_3 because this is a not unusual occurrence in practice.

Regard α_1, α_2, and α_3 as Cartesian axes and make a rotation to new Cartesian axes α, β, γ in which the γ-axis coincides with the direction of (x_1, x_2, x_3). If $R=(x_1^2+x_2^2+x_3^2)^{\frac{1}{2}}$

$$v=\int_{-\infty}^{\infty}\int_{-\infty}^{\infty}\int_{-\infty}^{\infty}\frac{\mathrm{e}^{-\mathrm{i}\gamma R}}{G(\omega,\alpha,\beta,\gamma)}\,\mathrm{d}\alpha\,\mathrm{d}\beta\,\mathrm{d}\gamma \tag{4.3}$$

where G is what F becomes after the mapping. Obviously, G is a polynomial in α, β, and γ.

The integration with respect to γ can be accomplished by pushing the path of integration down to infinity over the zeros of G in the lower half-plane. Each of these poles contributes $-2\pi\mathrm{i}$ (residue). Normally, the presence of the complex ω in G prevents any pole occurring for real γ. If there is a real simple pole $\gamma=\gamma_{\mathrm{p}}$ arising from the factor $(\gamma-\gamma_{\mathrm{p}})^{-1}$ in the integrand then we adopt the convention that the factor is a generalized function so that the principal value of the integral is required. Such a pole will contribute $-\pi\mathrm{i}$ (residue); the factor 2 is the only difference between the real and non-real poles so that our convention does permit the discussion of real poles should that be necessary.

The topic merits spending a little more time on it. Consider the equation $\partial^2u/\partial x_1^2-\partial^2u/\partial t^2=0$. For this the integral for v is

$$v=\int_{-\infty}^{\infty}\frac{\mathrm{e}^{-\mathrm{i}\alpha_1x_1}}{\omega^2-\alpha_1^2}\,\mathrm{d}\alpha_1.$$

There are simple poles at $\alpha_1=\pm\omega$. If $\operatorname{Im}\omega<0$, only $\alpha_1=\omega$ is relevant and the value of the integral is $(\pi\mathrm{i}/\omega)\exp(-\mathrm{i}\omega x_1)$ when $x_1>0$. When ω is real both poles must be taken into account and, according to our convention, the integral is $\frac{1}{2}(\pi\mathrm{i}/\omega)\{\exp(-\mathrm{i}\omega x_1)-\exp(\mathrm{i}\omega x_1)\}$. Thus, although any particular pole which exists for both real ω and when $\operatorname{Im}\omega<0$ gives a contribution differing only by a factor 2, the net results when *all poles* are considered are distinct. Indeed, in the first case only a radiating wave is present whereas in the second there is a standing wave involving both incoming and outgoing waves. Generally, only solutions which produce outgoing waves are of interest. This is one reason for forcing ω to have a small negative imaginary part in the analysis.

Let a typical pole be $\gamma = \gamma_0(\alpha, \beta)$, the notation indicating that its position will depend on α and β in general. Assume that the pole has a negative imaginary part and is simple; there is no difficulty in extending the theory to poles of higher order but no formulae will be given. The contribution to v is

$$-2\pi i \int_{-\infty}^{\infty} \int_{-\infty}^{\infty} \frac{e^{-i\gamma_0 R}}{[\partial G/\partial\gamma]_{\gamma=\gamma_0}} \, d\alpha \, d\beta.$$

So far the theory is exact. It will now be assumed that R is large so that the integral with respect to α and β can be estimated asymptotically by the method of stationary phase. (Compare §8.34 and see also Jones (1982).)

All points (α, β) such that $\partial\gamma_0/\partial\alpha = 0$ and $\partial\gamma_0/\partial\beta = 0$ must be considered. Let a typical one be (α_0, β_0). Then, in its neighbourhood,

$$\gamma_0 \approx \gamma_0(\alpha_0, \beta_0) + \tfrac{1}{2}a(\alpha-\alpha_0)^2 + h(\alpha-\alpha_0)(\beta-\beta_0) + \tfrac{1}{2}b(\beta-\beta_0)^2. \quad (4.4)$$

Suppose that $ab \neq h^2$. The contribution of this point to the asymptotic expansion is

$$\frac{-2\pi i \exp\{-i\gamma_0(\alpha_0, \beta_0)R\}}{[\partial G/\partial\gamma]_{\alpha=\alpha_0, \beta=\beta_0, \gamma=\gamma_0}} \int_{-\infty}^{\infty} \int_{-\infty}^{\infty} \exp\{-\tfrac{1}{2}iR(a\alpha^2 + 2h\alpha\beta + b\beta^2)\} \, d\alpha \, d\beta. \quad (4.5)$$

For later developments it is convenient to avail oneself of matrix notation. Let $\boldsymbol{\alpha}^T$, the transpose of $\boldsymbol{\alpha}$, be $(\alpha \quad \beta)$ and let $\boldsymbol{A} = \begin{pmatrix} a & h \\ h & b \end{pmatrix}$. Then the quadratic form $a\alpha^2 + 2h\alpha\beta + b\beta^2$ can be expressed as $\boldsymbol{\alpha}^T\boldsymbol{A}\boldsymbol{\alpha}$. A rotation of axes can be written as $\boldsymbol{x} = \boldsymbol{L}\boldsymbol{\alpha}$ where $\boldsymbol{x}^T = (x \quad y)$ and the orthogonal matrix $\boldsymbol{L}$ satisfies $\boldsymbol{L}^T\boldsymbol{L} = \boldsymbol{L}\boldsymbol{L}^T = \boldsymbol{I}$; the determinant of $\boldsymbol{L}$ or $\det \boldsymbol{L}$ is unity. After the rotation the quadratic form becomes $\boldsymbol{x}^T\boldsymbol{L}^T\boldsymbol{A}\boldsymbol{L}\boldsymbol{x}$. It is known that $\boldsymbol{L}$ can be chosen so that the quadratic form is $(x^2/\rho_1) + (y^2/\rho_2)$ whence

$$\boldsymbol{L}^T\boldsymbol{A}\boldsymbol{L} = \begin{pmatrix} 1/\rho_1 & 0 \\ 0 & 1/\rho_2 \end{pmatrix}.$$

Therefore

$$\det \boldsymbol{A} = \det \boldsymbol{L}^T \det \boldsymbol{A} \det \boldsymbol{L} = \det \boldsymbol{L}^T\boldsymbol{A}\boldsymbol{L} = 1/\rho_1\rho_2$$

so that $ab - h^2 = 1/\rho_1\rho_2$.

Assuming that $0 \leqslant \text{ph}\,\rho_1 \leqslant \pi$, $0 \leqslant \text{ph}\,\rho_2 \leqslant \pi$, the double integral becomes after the rotation

$$\int_{-\infty}^{\infty}\int_{-\infty}^{\infty} \exp[-\tfrac{1}{2}\mathrm{i}R\{(x^2/\rho_1)+(y^2/\rho_2)\}\,\mathrm{d}x\,\mathrm{d}y = -2\pi\mathrm{i}(\rho_1\rho_2)^{\frac{1}{2}}/R.$$

The quantity $(\rho_1\rho_2)^{\frac{1}{2}}$ is such that, when both ρ_1 and ρ_2 are real,

$$-\mathrm{i}(\rho_1\rho_2)^{\frac{1}{2}} = |\rho_1\rho_2|^{\frac{1}{2}}\,\mathrm{e}^{-\frac{1}{4}\pi\mathrm{i}\sigma} = |\det \mathbf{A}|^{-\frac{1}{2}}\,\mathrm{e}^{-\frac{1}{4}\pi\mathrm{i}\sigma}$$

where σ is the number of positive ρs less the number of negative. Thus $\sigma = 2$ if both ρ_1 and ρ_2 are positive but $\sigma = -2$ if they are both negative; if ρ_1 and ρ_2 have opposite signs $\sigma = 0$. The entity σ is called the *signature* of $\mathbf{A}$.

The contribution to v of the typical point of stationary phase associated with γ_0 can thus be written as

$$-4\pi^2(\rho_1\rho_2)^{\frac{1}{2}}\exp(-\mathrm{i}\gamma_0 R)/R\;\partial G/\partial\gamma_0. \tag{4.6}$$

In terms of the original variables this is

$$-4\pi^2(\rho_1\rho_2)^{\frac{1}{2}}\exp(-\mathrm{i}\boldsymbol{k}_0\cdot\boldsymbol{x})/\boldsymbol{x}\cdot\text{grad}\,F \tag{4.7}$$

where $\boldsymbol{k}_0$ is the vector from the origin to the point $\alpha = \alpha_0$, $\beta = \beta_0$, $\gamma = \gamma_0$ and the scalar product in the denominator is evaluated at this point.

The approximation for v is completed by adding the contributions from all points of stationary phase attached to a pole and then summing over all the relevant poles. Since γ_0 has a negative imaginary part the field is exponentially damped as $R \to \infty$ and often all poles except those which cause the smallest decay can be discarded.

It is transparent from the foregoing remarks that any poles which are real when ω is real tend to be dominant in v. Suppose that when ω is real F is a real polynomial possessing at least one real pole which has a negative imaginary part when ω is not real. Draw the surface $F = 0$ when ω has no imaginary part. This surface, which is the same as $G = 0$, determines the location of the real poles. Since γ_0 in (4.6) is stationary with respect to α and β the points $(\alpha_0, \beta_0, \gamma_0)$ correspond to the stationary points on $F = 0$ in the direction $\hat{\boldsymbol{x}}$, the unit vector along $\boldsymbol{x}$. Such a point is P of Fig. 10.4. Then $\boldsymbol{k}_0$ in (4.7) is the vector joining the origin to P and $\boldsymbol{x}\cdot\text{grad}\,F$ is evaluated at P. Also (4.4) represents the equation of the surface $F = 0$ in the neighbourhood of P so that ρ_1 and ρ_2 are the principal radii of curvature of the surface at P. The signature depends upon whether the surface is synclastic or anticlastic at P.

This geometrical interpretation often allows a rapid reckoning of the important terms. The exponential factor in (4.7) corresponds to a plane wave travelling in the direction $\boldsymbol{k}_0$. The planes of constant phase are therefore perpendicular to $\boldsymbol{k}_0$ but not to $\hat{\boldsymbol{x}}$ in general. Outgoing waves are obtained provided that the angle between $\boldsymbol{k}_0$ and $\hat{\boldsymbol{x}}$ is acute. All stationary

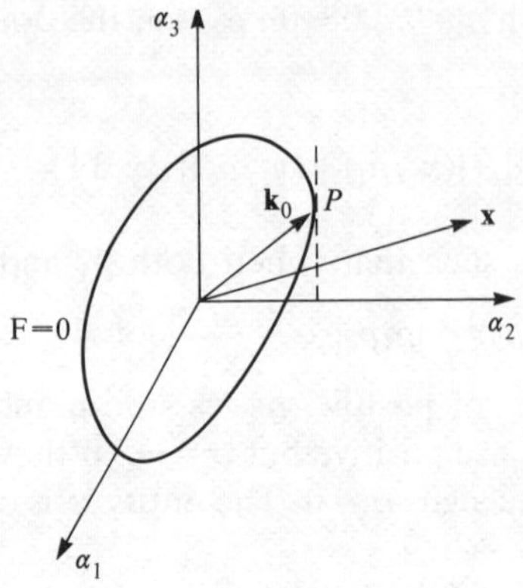

Fig. 10.4 Geometrical interpretation when ω is real

points at which the angle between $\boldsymbol{k}_0$ and $\hat{\boldsymbol{x}}$ is obtuse can therefore be ignored. It is also permissible to replace $\hat{\boldsymbol{x}} \cdot \text{grad}\, F$ by one of $\pm|\text{grad}\, F|$ depending on the direction of $\hat{\boldsymbol{x}}$ since at a stationary point the two vectors are parallel.

The picture can be inverted; i.e. instead of imagining that $\boldsymbol{k}_0$ is determined by $\hat{\boldsymbol{x}}$, a given $\boldsymbol{k}_0$ can be visualized as fixing $\hat{\boldsymbol{x}}$. From this point of view the number of intersections of $\boldsymbol{k}_0$ with the surface $F=0$ tells us how many waves there are with phase front normals parallel to $\boldsymbol{k}_0$ and the normals to $F=0$ show the directions of energy flow (see also §10.10).

As an example of the theory consider

$$a^2\frac{\partial^2 u}{\partial x_1^2}+\frac{\partial^2 u}{\partial x_2^2}+\frac{\partial^2 u}{\partial x_3^2}-\frac{1}{c^2}\frac{\partial^2 u}{\partial t^2}=0.$$

Here $F\equiv(\omega/c)^2-a^2\alpha_1^2-\alpha_2^2-\alpha_3^2$. The spheroid $a^2\alpha_1^2+\alpha_2^2+\alpha_3^2=(\omega/c)^2$ is therefore drawn and those points where the tangent plane is perpendicular to $\hat{\boldsymbol{x}}$ are determined. There are two, of which the one nearer the point of observation is responsible for an outgoing wave. Thus $\boldsymbol{k}_0$ is known and the product of the two principal radii of curvature can be derived without great difficulty.

The extension of the theory to

$$\int_{-\infty}^{\infty}\ldots\int_{-\infty}^{\infty}\frac{\exp\{-\mathrm{i}(\alpha_1 x_1+\ldots+\alpha_M x_M)\}}{F(\alpha_1,\ldots,\alpha_M)}\,\mathrm{d}\alpha_1\ldots\mathrm{d}\alpha_M$$

will be described briefly. The same method may be employed, with the difference that (4.5) now involves a quadratic form of $M-1$ variables instead of 2. The quadratic form can still be represented as $\boldsymbol{\alpha}^{\mathrm{T}}\boldsymbol{A}\boldsymbol{\alpha}$ on the understanding that $\boldsymbol{\alpha}$ is an $(M-1)\times 1$ column matrix and $\boldsymbol{A}$ is a $(M-1)\times(M-1)$ square matrix of which the element in the ith row and jth column is $\partial^2\gamma_0/\partial\alpha_i\,\partial\alpha_j$. By rotation, the quadratic form is reduced to $(x_1^2/\rho_1)+\ldots+(x_{M-1}^2/\rho_{M-1})$ where $1/\rho_1\rho_2\ldots\rho_{M-1}=\det\mathbf{A}$. Hence

$$\int_{-\infty}^{\infty}\cdots\int_{-\infty}^{\infty}\frac{\exp\{-\mathrm{i}(\alpha_1x_1+\ldots+\alpha_Mx_M)\}}{F(\alpha_1,\ldots,\alpha_M)}\,\mathrm{d}\alpha_1\ldots\mathrm{d}\alpha_M$$

$$\sim\sum-(2\pi)^{\frac{1}{2}(M+1)}\frac{\mathrm{i}(\rho_1\rho_2\ldots\rho_{M-1})^{\frac{1}{2}}}{R^{\frac{1}{2}(M-1)}\hat{\boldsymbol{x}}\cdot\operatorname{grad}F}\exp\{-\mathrm{i}\boldsymbol{k}_0\cdot\boldsymbol{x}-\tfrac{1}{4}\pi\mathrm{i}(M-1)\}. \quad (4.8)$$

If $\rho_1,\ldots,\rho_{M-1}$ are all real $(\rho_1\ldots\rho_{M-1})^{\frac{1}{2}}\mathrm{e}^{-\frac{1}{4}\pi\mathrm{i}(M-1)}$ has the alternative form $|\rho_1\ldots\rho_{M-1}|^{\frac{1}{2}}\,\mathrm{e}^{-\frac{1}{4}\pi\mathrm{i}\sigma}$ or $|\det\mathbf{A}|^{-\frac{1}{2}}\,\mathrm{e}^{-\frac{1}{4}\pi\mathrm{i}\sigma}$ where σ is the signature of $\mathbf{A}$, i.e. the number of ρ which are positive less the number which are negative. The geometrical interpretation is unchanged except that $F=0$ is a hypersurface in M-space and $\rho_1\ldots\rho_{M-1}$ its principal radii of curvature.

10.5 The time dependence and group velocity

Once v has been calculated by the process of the preceding section the time dependence of u follows from integration with respect to ω. This involves an integral of the type

$$\int_{-\infty}^{\infty}f(\omega)\mathrm{e}^{\mathrm{i}(\omega t-\boldsymbol{k}_0\cdot\boldsymbol{x})}\,\mathrm{d}\omega.$$

Of course, explicit evaluation of such an integral is usually impossible unless f and $\boldsymbol{k}_0$ are particularly simple functions of ω. However, it has already been assumed in the derivation of v that R is large so that, for finite speeds of propagation, t will also be large before any disturbance is observed. Therefore put $t=R/v$ where, for fixed R, v will vary from something of the order of speed of propagation when the disturbance arrives to very small values much later. The integral becomes

$$\int_{-\infty}^{\infty}f(\omega)\exp[\mathrm{i}R\{(\omega/v)-\boldsymbol{k}_0\cdot\hat{\boldsymbol{x}}\}]\,\mathrm{d}\omega$$

which may be tackled by the method of steepest descent on account of the largeness of R. (If t were known to be large rather than R the appropriate substitution would be $R=vt$.)

A typical saddle-point $\omega=\omega_v$ satisfies

$$1/v=\hat{\boldsymbol{x}}\cdot\mathrm{d}\boldsymbol{k}_0/\mathrm{d}\omega \quad (5.1)$$

and contributes to the integral

$$(2\pi)^{\frac{1}{2}}f(\omega_v)\frac{\exp\{\mathrm{i}(\omega_vt-\boldsymbol{k}_v\cdot\boldsymbol{x})-\frac{1}{4}\pi\mathrm{i}\}}{(\mathrm{d}^2\boldsymbol{k}_v\cdot\boldsymbol{x}/\mathrm{d}\omega_v^2)^{\frac{1}{2}}} \quad (5.2)$$

where

$$0\geqslant\operatorname{ph}\boldsymbol{x}\cdot\mathrm{d}^2\boldsymbol{k}_v/\mathrm{d}\omega_v^2\geqslant-\pi$$

and $\boldsymbol{k}_v$ is $\boldsymbol{k}_0$ when $\omega = \omega_v$. A replacement of v by R/t supplies ω_v and $\boldsymbol{k}_v$ as functions of R/t and (5.2) offers the dependence of the field on R and t at large distances from the source of the disturbance.

Another way of looking at (5.2) is profitable. Treat v as a constant; i.e. examine those points of observation and those times for which R/t has the constant value v. An observer moving with constant speed v in the direction $\boldsymbol{x}$ would find such points and times noteworthy. When v is constant so are ω_v and $\boldsymbol{k}_v$; the moving observer sees, according to (5.2), a plane wave of circular frequency ω_v travelling in the direction $\boldsymbol{k}_v$ apart from the decay due to the factor $R^{\frac{1}{2}}$ in the denominator (the plane wave is inhomogeneous if ω_v and $\boldsymbol{k}_v$ are not real). Furthermore, a slight alteration in ω_v, with consequent modification of $\boldsymbol{k}_v$, does not affect $\omega_0 t - \boldsymbol{k}_v \cdot \boldsymbol{x}$ because of (5.1). Hence *waves with frequency near $\omega_v/2\pi$, associated with the direction $\boldsymbol{k}_v$, group together to form a packet travelling at speed v.* For this reason $(\hat{\boldsymbol{x}} \cdot \mathrm{d}\boldsymbol{k}_0/\mathrm{d}\omega)^{-1}$ is known as the *group speed* of waves of frequency $\omega/2\pi$ in the direction $\hat{\boldsymbol{x}}$. Note that in general the group speed is distinct from the phase speed ω/k_0 even when $\boldsymbol{k}_0$ and $\hat{\boldsymbol{x}}$ are parallel.

If $\boldsymbol{k}_0$ is parallel to $\hat{\boldsymbol{x}}$, the group speed becomes $\mathrm{d}\omega/\mathrm{d}k_0$. A medium in which $\mathrm{d}k_0/\mathrm{d}\omega$ is not independent of ω is called *dispersive* (c.f. §5.5). The dispersion is said to be *normal* if an increase in wavelength results in a greater phase speed; if a decrease in phase speed occurs the dispersion is known as *anomalous.* The group speed in a medium with normal dispersion is *less* than the phase speed but is *greater* than the phase speed when the dispersion is anomalous. If the speed of energy transport is the same as the group speed energy can be transmitted with a speed greater than light in a medium with anomalous dispersion. Since this would be contrary to the theory of relativity the subject will be studied in more detail in the next section.

Finally, the reader should be warned that, in taking (5.2) to be the pertinent asymptotic representation, the singularities of $f(\omega)$ have been entirely neglected. If $f(\omega)$ has a pole which is captured in shifting the contour of integration to the path of steepest descent its contribution could be more significant than that of (5.2). Should the pole be near the saddle-point further investigation is necessary; see §G.3.

10.6 Dispersive media

The basic phenomena that are prevalent in dispersive media take place for a single space dimension so that it will be sufficient to limit the discussion to this case. Suppose that after a Fourier transform with respect to t the governing equation is

$$\mathrm{d}^2 U/\mathrm{d}x^2 + k^2 U = 0. \tag{6.1}$$

Let the excitation be due to a source at the origin suddenly switched on at

$t=0$ and forcing $u=e^{i\omega_0 t}$ at $x=0$ subsequently, ω_0 being positive. Then a solution of (6.1) is desired such that $U=-i(\omega-\omega_0)^{-1}$ at $x=0$. As before, a negative imaginary part will be assigned to ω.

Some assumption has to be made about the dependence of k on ω. A standard form is

$$k^2=\frac{\omega^2}{c^2}\left\{1-\frac{\omega_c^2}{\omega(\omega-i\nu)-\Omega^2}\right\}. \tag{6.2}$$

It arises in dispersion by molecules. With $\Omega=0$ it is appropriate for dispersion in a metal or the ionosphere (§6.34). For propagation in tubes and waveguides the case $\Omega=0$ and $\nu=0$ is relevant. However, it is simpler to begin with

$$k^2=(\omega/c)^2-1/d^2 \tag{6.3}$$

with d a positive constant and mention later the modifications that (6.2) entails.

Outgoing waves are ensured by

$$U=-i(\omega-\omega_0)^{-1}e^{-ik|x|}$$

as the solution of (6.1), the branch lines of k being chosen so that $k\sim\omega/c$ as $|\omega|\to\infty$.

From now on x is positive and so

$$u=\frac{1}{2\pi i}\int_{-\infty}^{\infty}\frac{e^{i(\omega t-kx)}}{\omega-\omega_0}\,d\omega.$$

In order that there shall be no signal for $t<0$ the path of integration must be placed below the pole $\omega=\omega_0$ and the branch points $\omega=\pm c/d$. In fact, there is then no disturbance for $t<x/c$.

Put $t=x/v$; only $v\leqslant c$ will be of interest in view of the last paragraph. Now

$$u=\frac{1}{2\pi i}\int_{-\infty}^{\infty}\frac{\exp[ix\{(\omega/v)-k\}]}{\omega-\omega_0}\,d\omega.$$

There are saddle-points at $\omega=\pm\beta c/d$ where $\beta=(1-v^2/c^2)^{-\frac{1}{2}}$. At the saddle-points the second derivative of the factor of x is $\pm icd/v^3\beta^3$. The paths of steepest descent are shown in Fig. 10.5. They cross the branch line of k at $\omega=\pm c/d\beta$ and swing over to the second sheet of the Riemann surface. On deforming the contour into the paths of steepest descent we obtain, as $x\to\infty$,

$$u\sim\exp\{i(\omega_0 t-k_0 x)\}H(\omega_0-\beta c/d)$$
$$+\left(\frac{v^3\beta^3 d}{2\pi x c}\right)^{\frac{1}{2}}\left\{\frac{e^{-\frac{1}{4}\pi i+icx/dv\beta}}{\beta c-\omega_0 d}+\frac{e^{\frac{1}{4}\pi i-icx/dv\beta}}{\beta c+\omega_0 d}\right\}H(ct-x) \tag{6.4}$$

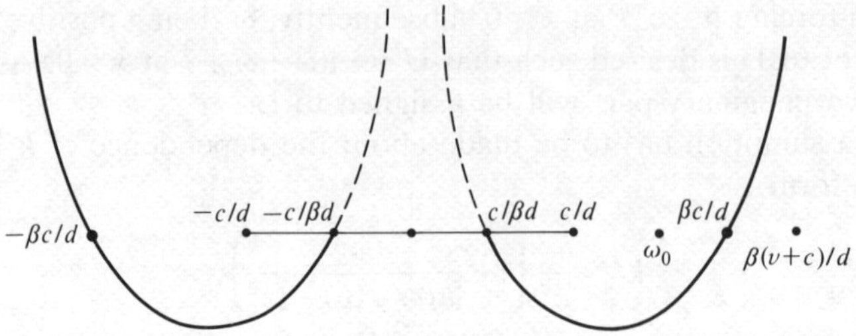

Fig. 10.5 The paths of steepest descent

provided that βc is not near $\omega_0 d$. Here k_0 is the value of k when $\omega = \omega_0$. The form of the first term of (6.4) is appropriate to $\omega_0 > c/d$. If $\omega_0 < c/d$ there is an exponentially damped term from the pole when $\omega_0 < c/d\beta$.

The formula (6.4) can be expressed in terms of x and t instead of v with the result

$$u \sim \exp\{i(\omega_0 t - k_0 x)\}\mathrm{H}\{ct(\omega_0^2 - c^2/d^2)^{\frac{1}{2}} - \omega_0 x\}$$
$$+\frac{xcd^{\frac{1}{2}}}{(2\pi)^{\frac{1}{2}}(c^2t^2-x^2)^{\frac{3}{4}}}\left[\frac{\exp\{-\frac{1}{4}\pi i + i(c^2t^2-x^2)^{\frac{1}{2}}/d\}}{\beta c - \omega_0 d}\right.$$
$$\left.+\frac{\exp\{\frac{1}{4}\pi i - i(c^2t^2-x^2)^{\frac{1}{2}}/d\}}{\beta c + \omega_0 d}\right]\mathrm{H}(ct-x) \qquad (6.5)$$

where now $\beta = (1 - x^2/c^2t^2)^{-\frac{1}{2}}$.

Although (6.5) was achieved by letting $x \to \infty$, precisely the same result is obtained if x is fixed and $t \to \infty$. We infer that at any given point the disturbance will eventually consist of the first term of (6.5) alone. This term is therefore called the *main signal*. The remaining term, which may be designated as a *transient*, oscillates with diminishing amplitude and decreasing frequency (the minimum being $c/2\pi d$) as t increases.

The representation (6.5) fails when $ct \approx x$, i.e. during the arrival of the first disturbance. The saddle-points are then at infinity and the theory breaks down.

Nor is (6.5) available when $\beta c \approx \omega_0 d$ for then the saddle-point is near the pole. In this region the theory of §G.3 must be employed and from (G.3.3),

$$u \sim \exp\{i(\omega_0 t - k_0 x)\}[\mathrm{H}(\beta c - \omega_0 d) - \exp(Zx^{\frac{1}{2}})]$$
$$+\left(\frac{v^3\beta^3 d}{2\pi xc}\right)^{\frac{1}{2}}\frac{\exp(\frac{1}{4}\pi ic - ix/dv\beta)}{\beta c + \omega_0 d} \qquad (6.6)$$

where $Z = (\omega_0 d - \beta c)(c/2dv^3\beta^3)^{\frac{1}{2}}e^{\frac{1}{4}\pi i}$.

The solution corresponding to an initiating signal of $\cos\omega_0 t$ follows by taking the real part of u. The envelope of the signal given by (6.5) and

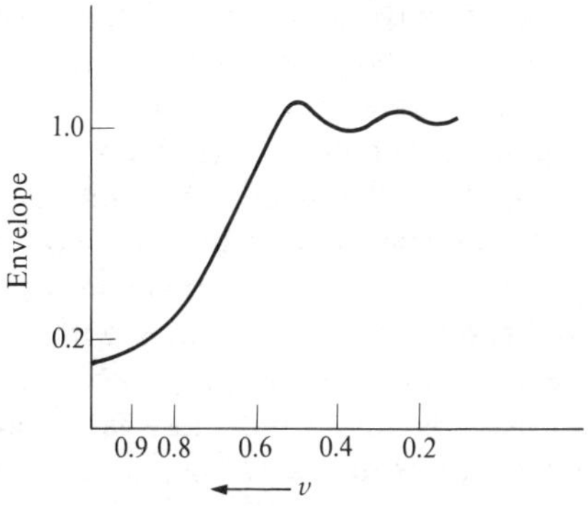

Fig. 10.6 Transition to main signal

(6.6) has been plotted in Fig. 10.6 for a fixed time as x varies. The actual values selected were $ct = 500\pi$, $d = 0.1$ cm, $\omega_0 = 4000$ Hz. It can be seen that the envelope has reached about half its final value when the saddle-point coalesces with the pole ($v = 0.56$). For this reason the value of v when the saddle-point and pole coincide is known as the *signal speed.* The signal speed equals the group speed of waves of frequency $\omega_0/2\pi$ as might be expected since the medium exhibits normal dispersion. It should be emphasized that the signal speed is not necessarily the speed that would be determined by measuring the time from $t = 0$ until the first detection of a disturbance by receiving apparatus. Such a speed would depend wholly on the sensitivity of the detector. If the apparatus were sufficiently insensitive no signal would be detected and if it were highly sensitive a speed near c would be measured.

The main features of propagation when k is given by (6.3) may now be summarized in: *A disturbance travelling with speed c reaches the point of observation first pursued by transients which are eventually overwhelmed by the main signal.*

Next the modifications to cope with (6.2) will be delineated. Suppose firstly that $\nu = 0$ and $\omega_c^2 = c^2/d^2$; then

$$k^2 = \frac{\omega^2}{c^2} - \frac{\omega^2}{(\omega^2 - \Omega^2)d^2}.$$

There are now four branch points of k at $\omega = \pm\Omega$ and $\omega = \pm(\Omega^2 + c^2/d^2)^{\frac{1}{2}}$. Two saddle-points outside the interval $\omega^2 \leqslant \Omega^2 + c^2/d^2$ provide transients qualitatively similar to those above. If $\omega_0 > (\Omega^2 + c^2/d^2)^{\frac{1}{2}}$ the bulk of the main signal arrives as the saddle-point passes the pole and the signal speed can be defined as before. There are, however, two additional saddle-points when $v < cd\Omega(c^2 + d^2\Omega^2)^{-\frac{1}{2}}$. These lie in the range $\omega^2 \leqslant \Omega^2$ and move from the origin to $\omega = \pm\Omega$ as v drops to zero. The transients due to these saddle points have a period which is large at first, decreases as time increases and finally becomes equal to the natural period $2\pi/\Omega$. Whether these transients arrive at the observer before or after the main signal depends on the magnitude of ω_0 when $\omega_0 > (\Omega^2 + c^2/d^2)^{\frac{1}{2}}$. If $\omega_0 < \Omega$ the propagating waves are still possible (the only cut-off region is $\Omega^2 < \omega_0^2 < \Omega^2 + c^2/d^2$) and the transients from the additional saddle-points always reach the receiver before the main signal which is initiated when one of the additional saddle-points coincides with the pole. Once again the signal speed agreed with the group speed—the dispersion is normal.

When $\nu \neq 0$ it is normally small and there is no substantial change in behaviour. The branch points now have negative imaginary parts and all signals are damped to some extent. The main signal travels with much the same speed as before unless ω_0 is near Ω. In this event absorption effects are pronounced and a much more elaborate analysis has to be undertaken.

Anomalous dispersion is a feature of

$$k^2 = \frac{\omega^2}{c^2} + \frac{1}{d^2}. \tag{6.7}$$

The contour of integration must pass below the branch point $\omega = -\mathrm{i}c/d$ in order that there shall be no disturbance for $t < 0$. There continue to be saddle-points at $\omega = \pm \mathrm{i}\beta c/d$ but now the path of steepest descent from the lower saddle-point terminates at the upper one (being the path of steepest ascent there). On account of there always being a path of integration in the negative imaginary half-plane the solution grows exponentially with t. Indeed, for large values of t the exponential factor is e^{ct}/d as is also evident by deforming the contour upwards round the branch line. This exponential term dominates all others, even that stemming from the pole, and is virtually independent of ω_0. The concept of a main signal is therefore valueless here; nor are the notions of signal speed and group speed of any use. It cannot be said that the energy travels with the group speed; actually since the energy augments steadily, it cannot be asserted that the energy has any particular speed. The process is akin to diffusion.

The integral for u has been estimated above by asymptotic techniques but another approach is on the cards when k is given by (6.3) or (6.7). A derivative of (A.3.22) with respect to x gives

$$\frac{1}{2\pi\mathrm{i}}\int_{-\infty}^{\infty} \frac{\mathrm{e}^{\mathrm{i}(\omega t - kx)}}{\omega - \omega_0}\,\mathrm{d}\omega = -\frac{\partial}{\partial x}\frac{1}{2\pi\mathrm{i}}\int_{-\infty}^{\infty} \frac{\mathrm{e}^{\mathrm{i}\omega t}}{\omega - \omega_0} \times \int_x^{\infty} \mathrm{e}^{-\mathrm{i}\omega u/c} \mathrm{I}_0\{(u^2 - x^2)^{\frac{1}{2}}/d\}\,\mathrm{d}u\,\mathrm{d}\omega$$

when k is specified by (6.7). An inversion of the order of integration leads to

$$u = -\frac{\partial}{\partial x}\int_x^{ct} e^{i\omega_0(t-w/c)} I_0\{(w^2-x^2)^{\frac{1}{2}}/d\}\,dw. \tag{6.8}$$

The formula may also be applied to (6.3) by making d purely imaginary.

One immediate consequence of (6.8) is that *near the wavefront* where $ct \approx x$, $u \approx \exp\{i\omega_0(t-x/c)\}H(ct-x)$ displaying the discontinuous change which occurs with the first arrival of the disturbance.

If d is purely imaginary the integral in (6.8) can be split as $\int_x^\infty - \int_{ct}^\infty$ which, via (A.3.22), separates the field into the main signal and transient when t is large. This decomposition is not permissible when d is real because the integral to infinity does not converge. The exponential growth of the field is, however, manifest.

The disadvantage of (6.8) is that it does not demonstrate the physical significance of the saddle-points and pole; moreover, it is difficult to generalize the ideas to the more general form of k in (6.2).

10.7 Time and frequency domains

The relations between the responses of an object in the time and frequency domains are of interest. Define the time transform $F(\omega)$ of $f(t)$ by

$$F(\omega) = \int_{-\infty}^{\infty} f(t)e^{-i\omega t}\,dt.$$

For real f this implies that

$$\{F(\omega)\}^* = F(-\omega), \tag{7.1}$$

the star indicating a complex conjugate.

The connection (7.1) shows that it is sufficient to determine F for positive ω; values for negative ω can be derived from (7.1) so long as F is the transform of a real signal. Further, it permits one to write

$$f(t) = \frac{1}{\pi}\,\mathrm{Re}\int_0^\infty F(\omega)e^{i\omega t}\,d\omega. \tag{7.2}$$

Suppose that a target is illuminated in the time domain by the plane sound wave

$$p^i = f(t - \boldsymbol{n}^i \cdot \boldsymbol{x}/a).$$

The excitation in the frequency domain is then the plane wave $P^i = F(\omega)\exp(-ik\boldsymbol{n}^i \cdot \boldsymbol{x})$. Assume that the scattered field when $F(\omega) \equiv 1$ has a distant behaviour of $(A/|\boldsymbol{x}|)\exp(-ik\,|\boldsymbol{x}|)$. Then the far scattered field produced by P^{i} for positive ω is

$$P^{\mathrm{s}}(\boldsymbol{x}) \sim F(\omega)A(\boldsymbol{n}^{\mathrm{i}}, \hat{\boldsymbol{x}}, \omega)\exp(-ik|\boldsymbol{x}|)/|\boldsymbol{x}|$$

the dependence of A on the directions of the plane wave, of the point of observation and on ω being explicitly displayed. The formula for P^{s} can be extended to negative values of ω by virtue of (7.1).

An inverse transform gives for the distant scattered field in the time domain

$$p^{\mathrm{s}}(\boldsymbol{x}, t) \sim p(\boldsymbol{n}^{\mathrm{i}}, \hat{\boldsymbol{x}}, t-|\boldsymbol{x}|/a)/|\boldsymbol{x}|$$

where

$$p(\boldsymbol{n}^{\mathrm{i}}, \hat{\boldsymbol{x}}, t) = \frac{1}{2\pi}\int_{-\infty}^{\infty} F(\omega)A(\boldsymbol{n}^{\mathrm{i}}, \hat{\boldsymbol{x}}, \omega)\mathrm{e}^{\mathrm{i}\omega t}\,\mathrm{d}\omega \tag{7.3}$$

or, on account of (7.2),

$$p(\boldsymbol{n}^{\mathrm{i}}, \hat{\boldsymbol{x}}, t) = \frac{1}{\pi}\operatorname{Re}\int_{0}^{\infty} F(\omega)A(\boldsymbol{n}^{\mathrm{i}}, \hat{\boldsymbol{x}}, \omega)\mathrm{e}^{\mathrm{i}\omega t}\,\mathrm{d}\omega. \tag{7.4}$$

If

$$A(\boldsymbol{n}^{\mathrm{i}}, \hat{\boldsymbol{x}}, \omega) = \int_{-\infty}^{\infty} a(\boldsymbol{n}^{\mathrm{i}}, \hat{\boldsymbol{x}}, t)\mathrm{e}^{-\mathrm{i}\omega t}\,\mathrm{d}t$$

(7.3) can be expressed in the convolution form

$$p(\boldsymbol{n}^{\mathrm{i}}, \hat{\boldsymbol{x}}, t) = \int_{-\infty}^{\infty} f(t-\tau)a(\boldsymbol{n}^{\mathrm{i}}, \hat{\boldsymbol{x}}, \tau)\,\mathrm{d}\tau. \tag{7.5}$$

Equations (7.3)–(7.5) are equivalent forms of an exact relation between scattering in the time and frequency domains respectively. In particular, $f(t)=\delta(t)$ implies that $F(\omega)=1$ and that $p(\boldsymbol{n}^{\mathrm{i}}, \hat{\boldsymbol{x}}, t)=a(\boldsymbol{n}^{\mathrm{i}}, \hat{\boldsymbol{x}}, t)$. It follows that, in order to accommodate the finite rate of spreading of the disturbance caused by the impulsive source, we must have $a(\boldsymbol{n}^{\mathrm{i}}, \hat{\boldsymbol{x}}, t)=0$ for $t<0$.

In general, one conclusion from (7.5) is that *the scattered wave in the time domain is the convolution of the incident signal waveform f and the scattered waveform for an impulsive source.*

There are analogous results for electromagnetic waves. If the harmonic electric intensity $\boldsymbol{e}_0\exp(-ik\boldsymbol{n}^{\mathrm{i}}\cdot\boldsymbol{x})$ creates a far scattered electric intensity $\mathbf{A}(\boldsymbol{n}^{\mathrm{i}}, \hat{\boldsymbol{x}}, \omega)\exp(-ik|\boldsymbol{x}|)/|\boldsymbol{x}|$ then the incident wave

$$\boldsymbol{e}^{\mathrm{i}} = \boldsymbol{e}_0 f(t-\boldsymbol{n}^{\mathrm{i}}\cdot\boldsymbol{x}/c)$$

generates the distant electric intensity

$$\boldsymbol{e}^{\mathrm{s}}(\boldsymbol{x}, t) = \boldsymbol{e}(\boldsymbol{n}^{\mathrm{i}}, \hat{\boldsymbol{x}}, t-|\boldsymbol{x}|/c)/|\boldsymbol{x}|$$

where

$$\boldsymbol{e}(\boldsymbol{n}^{\mathrm{i}}, \hat{\boldsymbol{x}}, t) = \frac{1}{2\pi}\int_{-\infty}^{\infty} F(\omega)\mathbf{A}(\boldsymbol{n}^{\mathrm{i}}, \hat{\boldsymbol{x}}, \omega)\mathrm{e}^{\mathrm{i}\omega t}\,\mathrm{d}\omega \tag{7.6}$$

$$= \frac{1}{\pi}\,\mathrm{Re}\int_{0}^{\infty} F(\omega)\mathbf{A}(\boldsymbol{n}^{\mathrm{i}}, \hat{\boldsymbol{x}}, \omega)\mathrm{e}^{\mathrm{i}\omega t}\,\mathrm{d}\omega \tag{7.7}$$

$$= \int_{-\infty}^{\infty} f(t-\tau)\boldsymbol{a}(\boldsymbol{n}^{\mathrm{i}}, \hat{\boldsymbol{x}}, \tau)\,\mathrm{d}\tau \tag{7.8}$$

with

$$\mathbf{A}(\boldsymbol{n}^{\mathrm{i}}, \hat{\boldsymbol{x}}, \omega) = \int_{-\infty}^{\infty} \boldsymbol{a}(\boldsymbol{n}^{\mathrm{i}}, \hat{\boldsymbol{x}}, t)\mathrm{e}^{-\mathrm{i}\omega t}\,\mathrm{d}t.$$

The same interpretation of (7.8) is available and $\boldsymbol{a}(\boldsymbol{n}^{\mathrm{i}}, \hat{\boldsymbol{x}}, t) = \mathbf{0}$ for $t<0$.

10.8 Moving scatterers

The formulae of the preceding section refer to a target which is stationary with respect to the source of the irradiating field. Modification is necessary if the source or scatterer is moving. Since the source, if it is mobile, can always be reduced to rest by a simple transformation it will be enough to consider a target in motion. In fact, an obstacle travelling with constant velocity while the source is fixed will be discussed. The analysis when different parts of the target possess different velocities is much more complicated and details will be omitted.

Let $\boldsymbol{x}$, t be the coordinate system in which the sources are at rest and $\boldsymbol{x}'$, t' a system in which the scatterer is stationary. The two coordinate systems are related by a Lorentz transformation (§2.2). It will be assumed that the spatial axes of the two systems are aligned with the origins coinciding at $t=0=t'$. In addition, the target will be assumed to move along the positive x_1-axis with speed v. However, the consequent formulae will be expressed in terms of the velocity $\boldsymbol{v}$ of the obstacle so that the special choice of axes can be eliminated.

The formula for the Lorentz transformation is, in the acoustic case,

$$x_1' = \beta(x_1 - vt), \qquad x_2' = x_2, \qquad x_3' = x_3, \qquad t' = \beta(t - vx_1/a^2), \tag{8.1}$$

$$x_1 = \beta(x_1' + vt'), \qquad x_2 = x_2', \qquad x_3 = x_3', \qquad t = \beta(t' + vx_1'/a^2) \tag{8.2}$$

where $\beta = (1 - v^2/a^2)^{-\frac{1}{2}}$.

Let the incident wave in the $\boldsymbol{x}$, t system be

$$p^{\mathrm{i}} = f(t - \boldsymbol{n}^{\mathrm{i}}\cdot\boldsymbol{x}/a).$$

On substitution from (8.2)

$$f(t - \boldsymbol{n}^{\mathrm{i}}\cdot\boldsymbol{x}/a) = f\{\gamma(t' - \boldsymbol{n}'\cdot\boldsymbol{x}'/a)\} = g(t' - \boldsymbol{n}'\cdot\boldsymbol{x}'/a)$$

where

$$\gamma = \beta(1 - n_1^{\mathrm{i}} v/a) = \beta(1 - \boldsymbol{n}^{\mathrm{i}} \cdot \boldsymbol{v}/a), \tag{8.3}$$

$$n_1' = \beta(n_1^{\mathrm{i}} - v/a)/\gamma, \qquad n_2' = n_2^{\mathrm{i}}/\gamma, \qquad n_3' = n_3^{\mathrm{i}}/\gamma. \tag{8.4}$$

Since p^{i} is a scalar and the wave equation is invariant under a Lorentz transformation the incident wave in the $\boldsymbol{x}'$, t' system is

$$p^{\mathrm{i}\prime} = g(t' - \boldsymbol{n}' \cdot \boldsymbol{x}'/a). \tag{8.5}$$

Thus, if the incident wave has angular frequency ω in the $\boldsymbol{x}$, t space it appears to have angular frequency $\gamma\omega$ in the system in which the scatterer is at rest. This change in frequency represents the *Doppler shift* due to motion. In addition, the apparent direction of propagation is $\boldsymbol{n}'$ instead of $\boldsymbol{n}^{\mathrm{i}}$.

The next stage is to apply (7.3) in the $\boldsymbol{x}'$, t system where the obstacle is motionless. Hence, if

$$G(\omega) = \int_{-\infty}^{\infty} g(t) \mathrm{e}^{-\mathrm{i}\omega t} \, \mathrm{d}t,$$

$$p^{\mathrm{s}\prime}(\boldsymbol{x}', t') \sim p'(\boldsymbol{n}', \hat{\boldsymbol{x}}', t' - |\boldsymbol{x}'|/a)/|\boldsymbol{x}'|$$

where

$$p'(\boldsymbol{n}', \hat{\boldsymbol{x}}', t') = \frac{1}{2\pi} \int_{-\infty}^{\infty} G(\omega) A(\boldsymbol{n}', \hat{\boldsymbol{x}}', \omega) \mathrm{e}^{\mathrm{i}\omega t'} \, \mathrm{d}\omega.$$

Now $\gamma G(\omega) = F(\omega/\gamma)$ and, from (8.1),

$$t' - |\boldsymbol{x}'|/a = t' - \boldsymbol{x}' \cdot \hat{\boldsymbol{x}}'/a = (t - \boldsymbol{x} \cdot \hat{\boldsymbol{x}}/a)/\gamma_1$$

where $\gamma_1 = \beta(1 - \hat{\boldsymbol{x}} \cdot \boldsymbol{v}/c)$,

$$\gamma_1 \hat{x}_1' = \beta(\hat{x}_1 - v/a), \qquad \gamma_1 \hat{x}_2' = \hat{x}_2, \qquad \gamma_1 \hat{x}_3' = \hat{x}_3 \tag{8.6}$$

in analogy with (8.4). Hence

$$p'(\boldsymbol{n}', \hat{\boldsymbol{x}}', t' - |\boldsymbol{x}'|/a) = p_1(\boldsymbol{n}', \hat{\boldsymbol{x}}', t - |\boldsymbol{x}|/a)$$

with

$$p_1(\boldsymbol{n}', \hat{\boldsymbol{x}}', t) = \frac{\gamma_1}{2\pi\gamma} \int_{-\infty}^{\infty} F\left(\frac{\gamma_1 \omega}{\gamma}\right) A(\boldsymbol{n}', \hat{\boldsymbol{x}}', \gamma_1 \omega) \mathrm{e}^{\mathrm{i}\omega t} \, \mathrm{d}\omega. \tag{8.7}$$

The distant field can now be expressed in terms of $\boldsymbol{x}$, t by means of (8.3)–(8.6).

Comparison of (8.7) with (7.3) reveals that (8.7) reduces to (7.3) when the target is immobile so that $v = 0$, $\beta = 1 = \gamma = \gamma_1$, $\hat{\boldsymbol{x}} = \hat{\boldsymbol{x}}'$, $\boldsymbol{n}^{\mathrm{i}} = \boldsymbol{n}'$. Also (8.7) conveys evidence of the effect of motion in three places. There is a frequency shift in the incident signal waveform due to the factor γ_1/γ. The scattering amplitude A is also shifted in frequency because of the presence of $\gamma_1\omega$ and suffers an angular distortion as well due to $\boldsymbol{n}'$, $\hat{\boldsymbol{x}}'$ replacing $\boldsymbol{n}^{\mathrm{i}}$, $\boldsymbol{x}$.

If desired, the integral in (8.7) can be rewritten in forms corresponding to (7.4) and (7.5).

Other phenomena are exhibited by electromagnetic waves. Let the incident wave in the $\boldsymbol{x}$, t system be

$$\boldsymbol{e}^{\mathrm{i}} = \boldsymbol{e}_0 f(t - \boldsymbol{n}^{\mathrm{i}} \cdot \boldsymbol{x}/c), \qquad \boldsymbol{h}^{\mathrm{i}} = (\varepsilon/\mu)^{\frac{1}{2}} \boldsymbol{n}^{\mathrm{i}} \wedge \boldsymbol{e}^{\mathrm{i}}.$$

Because the electric and magnetic fields are coupled by the Lorentz transformation, the incident electric intensity in the $\boldsymbol{x}'$, t' system is

$$\boldsymbol{e}^{\mathrm{i}'} = \boldsymbol{e}_0' g(t' - \boldsymbol{n}' \cdot \boldsymbol{x}/c)$$

where

$$\boldsymbol{e}_0' = \gamma \boldsymbol{e}_0 + (\boldsymbol{e}_0 \cdot \boldsymbol{v})\{(1-\beta)\boldsymbol{v} + \beta v^2 \boldsymbol{n}^{\mathrm{i}}/c\}/v^2. \tag{8.8}$$

Thus, in addition to the effects already described for the incident sound wave, the polarization is also altered (unless $\boldsymbol{e}_0$ is perpendicular to $\boldsymbol{v}$); in part this is caused by the change in the direction of propagation and, in part, by the interdependence of the electric and magnetic fields.

The change of polarization forces one to show explicitly the dependence of $\mathbf{A}$ in (7.6) on the polarization of the incident wave by writing $\mathbf{A}(\boldsymbol{e}_0, \boldsymbol{n}^{\mathrm{i}}, \hat{\boldsymbol{x}}, \omega)$. Then the distant field in the $\boldsymbol{x}'$, t' system is

$$\boldsymbol{e}^{\mathrm{s}'}(\boldsymbol{x}', t') = \boldsymbol{e}_1(\boldsymbol{n}', \hat{\boldsymbol{x}}', t)/|\boldsymbol{x}'|$$

where

$$\boldsymbol{e}_1(\boldsymbol{n}', \hat{\boldsymbol{x}}', t) = \frac{\gamma_1}{2\pi\gamma} \int_{-\infty}^{\infty} F\left(\frac{\gamma_1}{\gamma}\omega\right) \mathbf{A}(\boldsymbol{e}_0', \boldsymbol{n}', \hat{\boldsymbol{x}}', \gamma_1\omega) \mathrm{e}^{\mathrm{i}\omega t}\, \mathrm{d}\omega. \tag{8.9}$$

The far field in the $\boldsymbol{x}$, t system is now given by

$$\boldsymbol{e}^{\mathrm{s}}(\boldsymbol{x}, t) = \beta \mathbf{e}^{\mathrm{s}'} + (1-\beta)(\boldsymbol{e}^{\mathrm{s}'} \cdot \boldsymbol{v})(\boldsymbol{v}/v^2) - \beta \boldsymbol{v} \wedge (\boldsymbol{x}' \wedge \boldsymbol{e}^{\mathrm{s}'})/c. \tag{8.10}$$

The new features are the polarization warping of $\mathbf{A}$ in (8.9) because $\boldsymbol{e}_0'$ replaces $\boldsymbol{e}_0$. Moreover, there is a further polarization twist in the conversion back to source coordinates in (8.10).

Time domain methods

10.9 Integral equations

A fertile route for attacking problems in the time domain directly is via integral equations. Only the scattering by a perfectly conducting body will be mentioned; the analysis for acoustic waves is similar and simpler.

Let S be the surface of the body subjected to the incident wave $\mathbf{E}^{\mathrm{i}}(\boldsymbol{x}, t)$,

$\boldsymbol{H}^{\mathrm{i}}(\boldsymbol{x}, t)$ from the exterior. Then the total magnetic intensity can be expressed from §1.19, as

$$\boldsymbol{H}(\boldsymbol{x}, t) = \boldsymbol{H}^{\mathrm{i}}(\boldsymbol{x}, t) - \int_S \left\{ \boldsymbol{j}(\mathbf{y}, T) + \frac{\partial}{\partial t}\boldsymbol{j}(\mathbf{y}, T)\frac{1}{c\Psi} \right\} \wedge \operatorname{grad}_y \Psi(\boldsymbol{x}, \mathbf{y})\, \mathrm{d}S_y \quad (9.1)$$

where $T = t - |\boldsymbol{x} - \mathbf{y}|/c$, $\Psi(\boldsymbol{x}, \mathbf{y}) = 1/4\pi\,|\boldsymbol{x} - \mathbf{y}|$ and $\boldsymbol{j}$ is the surface current on the body. The surface current satisfies the integral equation

$$-\tfrac{1}{2}\boldsymbol{j}(\boldsymbol{x}, t) + \boldsymbol{n} \wedge ⨍_S \left\{ \boldsymbol{j}(\mathbf{y}, T) + \frac{\partial}{\partial t}\boldsymbol{j}(\mathbf{y}, T)\frac{1}{c\Psi} \right\} \wedge \operatorname{grad}_y \Psi\, \mathrm{d}S_y = \boldsymbol{n} \wedge \boldsymbol{H}^{\mathrm{i}}(\boldsymbol{x}, t) \quad (9.2)$$

for $\boldsymbol{x}$ on S, the bar on the integral sign signifying a principal value.

Equation (9.2) is a time dependent integral equation for $\boldsymbol{j}$ in which the values of $\boldsymbol{j}$ in the integrand correspond to time T strictly less than t except when $\boldsymbol{x} = \mathbf{y}$. This observation is the basis of a convenient method of numerical solution. The surface of the body is divided into a set of patches and the time variable is discretized into a sequence $t_0, t_1, t_2, \ldots$. On the assumption that the incident wave is pulsed t_0 can be selected so that all fields and currents in (9.2) are zero for $t \leqslant t_0$.

Suppose that for some r the solution is known for all $t_i \leqslant t_r$ and that $\boldsymbol{j}(\boldsymbol{x}, t_{r+1})$ is wanted. The integral over the patch containing $\boldsymbol{x}$ can be neglected if the patch sizes are small enough because it is a principal value. On the other patches $\boldsymbol{j}(\mathbf{y}, T)$ is known if the time increments are not too large and so $\boldsymbol{j}(\boldsymbol{x}, t_{r+1})$ can be calculated from the integral in (9.2). Thus the procedure is to start at t_0 and march forward in time, evaluating the solution at chosen values of $\boldsymbol{x}$ at each time step by a numerical integration over the previous values of $\boldsymbol{j}$.

The method works only if the values of $\boldsymbol{j}$ in the integration of (9.2) are drawn from previous time steps; i.e. $t_{r+1} - |\boldsymbol{x} - \mathbf{y}|/c \leqslant t_r$ must be imposed for all $\mathbf{y}$ employed in the method of quadrature.

Once $\boldsymbol{j}$ has been found the total field is available from (9.1). The simpler formula

$$\boldsymbol{H}^{\mathrm{s}}(\boldsymbol{x}, t) \sim -\frac{1}{4\pi c R} \int_S \frac{\partial}{\partial t}\boldsymbol{j}(\mathbf{y}, T) \wedge \hat{\boldsymbol{x}}\, \mathrm{d}S_y \quad (9.3)$$

can be invoked for the distant field.

A mixture of marching in time and transforms can be rewarding in some circumstances. Consider

$$\nabla^2 u - \frac{1}{c^2}\frac{\partial^2 u}{\partial t^2} = v \quad (9.4)$$

where it is supposed that there is no signal before $t = 0$. Let

$$U = \int_{-\infty}^{\infty}\int_{-\infty}^{\infty}\int_{-\infty}^{\infty} u \mathrm{e}^{-\mathrm{i}(\alpha_1 x_1 + \alpha_2 x_2 + \alpha_3 x_3)}\, \mathrm{d}x_1\, \mathrm{d}x_2\, \mathrm{d}x_3.$$

Then

$$\frac{d^2U}{dt^2}+\alpha^2c^2U=-c^2V \tag{9.5}$$

where $\alpha^2=\alpha_1^2+\alpha_2^2+\alpha_3^2$. It has been assumed in (9.5) that there are no surfaces of discontinuity otherwise extra terms from the boundary conditions would have to be added to the right-hand side of (9.5). The solution of (9.5) such that there is no disturbance in $t<0$ is

$$U(t)=-\frac{c}{\alpha}\int_0^t V(\tau)\sin\alpha c(t-\tau)\,d\tau. \tag{9.6}$$

The vanishing of the sinusoidal factor at $\tau=t$ ensures that only earlier values of V are needed in the calculation of U. Once U is known a fast Fourier transform supplies u. If boundary terms were added to V, (9.6) would be an integral equation to be solved by marching in time. Notice that with discretization in time the whole process of finding u involves only the summation of Fourier series.

A related problem is the propagation of harmonic waves in an inhomogeneous medium where

$$\nabla^2u+\frac{\omega^2}{a^2(\boldsymbol{x})}u=0 \tag{9.7}$$

with $a(\boldsymbol{x})\to c$ as $|\boldsymbol{x}|\to\infty$. Let the incident field be u^{i}, satisfying $\nabla^2u^{\mathrm{i}}+\omega^2u^{\mathrm{i}}/c^2=0$, and put $u=u^{\mathrm{i}}+u^{\mathrm{s}}$. Write the equation as

$$\nabla^2u^{\mathrm{s}}+\frac{\omega^2}{c^2}u^{\mathrm{s}}=\omega^2\left(\frac{1}{c^2}-\frac{1}{a^2}\right)(u^{\mathrm{i}}+u^{\mathrm{s}}). \tag{9.8}$$

Then, if $\tilde{u}=\frac{1}{2\pi}\int_{-\infty}^{\infty}u^{\mathrm{s}}e^{i\omega t}\,d\omega$,

$$\nabla^2\tilde{u}-\frac{1}{c^2}\frac{\partial^2\tilde{u}}{\partial t^2}=\left(\frac{1}{a^2}-\frac{1}{c^2}\right)\frac{\partial^2}{\partial t^2}(\tilde{u}+\tilde{u}^{\mathrm{i}}).$$

Treating this as (9.4) we obtain

$$\tilde{U}(t)=\alpha c\int_0^b w(\tau)\sin\alpha c(t-\tau)\,d\tau-w(t) \tag{9.9}$$

where

$$w=\int_{-\infty}^{\infty}\int_{-\infty}^{\infty}\int_{-\infty}^{\infty}\left(\frac{c^2}{a^2}-1\right)(\tilde{u}+\tilde{u}^{\mathrm{i}})e^{-i(\alpha_1x_1+\alpha_2x_2+\alpha_3x_3)}\,dx_1\,dx_2\,dx_3. \tag{9.10}$$

Again boundary terms have been omitted though they can be incorporated easily. The simultaneous equations (9.9) and (9.10) can be handled by marching in time and fast Fourier transforms in a numerically efficient way. There is no conceptual difficulty in extending the theory to dispersive media or Maxwell's equations though the detailed implementation is more awkward.

10.10 Characteristics

Wavefronts and their accompanying rays have an important part to play in harmonic waves. Their analogues in waves of general time dependence are therefore of interest. Before commencing a general discussion it is convenient to deal with a simpler problem. Let u satisfy the linear partial differential equation

$$a\frac{\partial u}{\partial x}+b\frac{\partial u}{\partial y}+cu+d=0 \tag{10.1}$$

where a, b, c, d are functions of x and y. Postulating that u is known to be equal to f on the curve $C(x, y)=0$ we enquire whether u can be found at points off the curve.

Introduce new coordinates ξ and η where $\xi=C(x, y)$. The choice of η is fairly free, the only restriction being that a curve $\eta=$ constant intersects any curve $\xi=$ constant just once in the domain under consideration. On $\xi=0$, u is a given function which can be written as $f(\eta)$. Clearly $\partial u/\partial\eta=f'(\eta)$ on $\xi=0$. In terms of ξ and η equation (10.1) becomes

$$\left(a\frac{\partial C}{\partial x}+b\frac{\partial C}{\partial y}\right)\frac{\partial u}{\partial \xi}=-\left(a\frac{\partial \eta}{\partial x}+b\frac{\partial \eta}{\partial y}\right)f'-cf-d \tag{10.2}$$

on $\xi=0$. The right-hand side of (10.2) is known and so $\partial u/\partial\xi$ can be determined on $\xi=0$ provided that

$$a\frac{\partial C}{\partial x}+b\frac{\partial C}{\partial y}\neq 0.$$

Partial derivatives of (10.1) with respect to ξ and η reveal that all derivatives of u can be found on $\xi=0$ with the same proviso (the coefficients are assumed to have sufficient differentiability). Hence, if u can be expanded in a Taylor series, it can be determined near $C=0$.

However, if

$$a\frac{\partial C}{\partial x}+b\frac{\partial C}{\partial y}=0, \tag{10.3}$$

$\partial u/\partial\xi$ disappears from (10.2) and is indeterminate on $\xi=0$. Thus, u cannot be found off $C=0$. In fact, u cannot now be ascribed arbitrary values on $C=0$ because f must satisfy the differential equation

$$\left(a\frac{\partial \eta}{\partial x}+b\frac{\partial \eta}{\partial y}\right)f'+cf+d=0. \tag{10.4}$$

The curves $C=0$ which are solutions of (10.3) are known as *characteristics*.

Another point of view is helpful. Let u be continuous across $C=0$; then so is $\partial u/\partial\eta$. Apply (10.2) at $\xi=+0$ and at $\xi=-0$; subtraction of the two resulting equations leads to

$$\left(a\frac{\partial C}{\partial x}+b\frac{\partial C}{\partial y}\right)\left[\frac{\partial u}{\partial \xi}\right]=0$$

where [ ] denotes the difference between values at $\xi=+0$ and at $\xi=-0$. If $[\partial u/\partial\xi]\neq 0$ then C is forced to obey (10.3). Hence, *discontinuities in the derivatives of u occur only on characteristics.*

The generalization of these ideas to other partial differential equations is straightforward. For example, suppose that u satisfies the wave equation

$$\frac{\partial^2 u}{\partial x^2}+\frac{\partial^2 u}{\partial y^2}+\frac{\partial^2 u}{\partial z^2}-\frac{1}{c^2}\frac{\partial^2 u}{\partial t^2}=0$$

and we ask whether u is available off $C(x, y, z, t)=0$ given u and its first derivatives on the curve. It transpires that this is impossible if C is such that

$$a^2\,\mathrm{grad}^2 C-(\partial C/\partial t)^2=0 \tag{10.5}$$

which is therefore the equation for the characteristics. If $C=0$ is solved for t and written as $t=L(\boldsymbol{x})$, (10.5) changes to

$$\mathrm{grad}^2 L=1/a^2 \tag{10.6}$$

which is the *eikonal equation* (§6.15). The theory of characteristics is therefore closely connected to ray theory.

For Maxwell's equations

$$\mathrm{curl}\,\boldsymbol{E}+\mu\,\partial\boldsymbol{H}/\partial t=\boldsymbol{0}, \qquad \mathrm{curl}\,\boldsymbol{H}-\varepsilon\,\partial\boldsymbol{E}/\partial t=\boldsymbol{0} \tag{10.7}$$

with $\boldsymbol{E}$ and $\boldsymbol{H}$ given on $t=L(\boldsymbol{x})$ the equation for the characteristics turns out to be

$$\mathrm{grad}^2 L=\mu\varepsilon; \tag{10.8}$$

again the eikonal equation arises.

Reverting to §10.4 and Fig. 10.4 we remark that, if F is a homogeneous polynomial in α_1, α_2, α_3, ω such as originates from a partial differential equation in which all the derivatives are of the same order, the equation for the characteristics $t=L(\boldsymbol{x})$ is

$$F(\omega\,\partial L/\partial x_1, \omega\,\partial L/\partial x_2, \omega\,\partial L/\partial x_3)=0.$$

So, to construct $F=0$, draw all the characteristics through the origin and then trace there a normal vector to each of length $\omega\,|\text{grad}\,L|$. The ends of these vectors describe the surface $F=0$.

10.11 Transport equations

It has already been explained that special relations exist on characteristics (see, for example, (10.4)) and that discontinuities occur on them. The purpose of this section is to investigate in more detail the discontinuities on characteristics.

For sound waves

$$a^2\nabla^2 p - \partial^2 p/\partial t^2 = 0 \tag{11.1}$$

the discontinuities on $t=L(\boldsymbol{x})$ are considered. Since (11.1) is satisfied on both sides of $t=L(\boldsymbol{x})$, subtraction gives for the discontinuity [ ]

$$a^2[\nabla^2 p]-[\partial^2 p/\partial t^2]=0,$$

a being taken to be continuous. Now

$$\text{grad}[p]=[\text{grad}\,p]+[\partial p/\partial t]\text{grad}\,L$$

because the variables in $[p]$ are related by $t=L(\boldsymbol{x})$. Also

$$\text{div}[\boldsymbol{v}]=[\text{div}\,\boldsymbol{v}]+[\partial\boldsymbol{v}/\partial t]\cdot\text{grad}\,L.$$

Hence, on account of (10.6),

$$2\,\text{grad}\,L\cdot\text{grad}[\partial p/\partial t]+[\partial p/\partial t]\nabla^2 L=\nabla^2[p]. \tag{11.2}$$

Derivatives with respect to time of (11.1) show that time derivatives of p also satisfy (11.1). Hence (11.2) remains valid if p is replaced by any of its time derivatives. Denote $[\partial^m p/\partial t^m]$ by $p_m(\boldsymbol{x})$. Then

$$2\,\text{grad}\,L\cdot\text{grad}\,p_m+p_m\nabla^2 L=\nabla^2 p_{m-1}. \tag{11.3}$$

Equations (11.3) constitute a set of ordinary differential equations for the p_m along the rays normal to the wavefront $L=\text{constant}$. They are called *transport equations*.

When $p_{m-1}=0$, (11.3) can be written as

$$\text{div}(p_m^2\,\text{grad}\,L)=0.$$

Integrate over a ray tube beginning at the small area $\mathrm{d}S_1$ on $L=t_1$ and proceeding to the small area $\mathrm{d}S_2$ on $L=t_2$. There is no contribution from the sides of the tube and the unit normals on $\mathrm{d}S_1$ and $\mathrm{d}S_2$ are $\pm\text{grad}\,L/|\text{grad}\,L|$. Consequently, bearing in mind (10.6), we have

$$(p_m^2/a)_1\,\mathrm{d}S_1=(p_m^2/a)_2\,\mathrm{d}S_2. \tag{11.4}$$

Thus *p_m^2/a satisfies the intensity law of geometrical acoustics.* Fermat's principle and other aspects can be developed as in Chapter 6.

The same procedure may be adopted for the electromagnetic equations (10.7). Analogous to (11.2) are

$$\operatorname{grad} L \wedge \boldsymbol{E}_m - \mu \boldsymbol{H}_m = \operatorname{curl} \boldsymbol{E}_{m-1}, \tag{11.5}$$

$$\operatorname{grad} L \wedge \boldsymbol{H}_m + \varepsilon \boldsymbol{E}_m = \operatorname{curl} \boldsymbol{H}_{m-1} \tag{11.6}$$

with $\boldsymbol{E}_m$, $\boldsymbol{H}_m$ the discontinuities $[\partial^m \boldsymbol{E}/\partial t^m]$, $[\partial^m \boldsymbol{H}/\partial t^m]$.

For $\boldsymbol{E}_{m-1} = \mathbf{0}$ and $\boldsymbol{H}_{m-1} = \mathbf{0}$, (11.5) and (11.6) furnish

$$\boldsymbol{E}_m \cdot \operatorname{grad} L = 0, \qquad \boldsymbol{H}_m \cdot \operatorname{grad} L = 0 \tag{11.7}$$

i.e. $\boldsymbol{E}_m$ and $\boldsymbol{H}_m$ *are transverse to the direction of propagation.* Furthermore, the elimination of $\boldsymbol{H}_m$ leaves

$$(\mu\varepsilon - \operatorname{grad}^2 L)\boldsymbol{E}_m = \mathbf{0}$$

which makes another demonstration that the discontinuities in the first and higher derivatives of $\boldsymbol{E}$ and $\boldsymbol{H}$ propagate along characteristics.

A certain amount of manipulation of (11.5) and (11.6) fabricates transport equations for $\boldsymbol{E}_m$ and $\boldsymbol{H}_m$ separately, namely

$$2(\operatorname{grad} L \cdot \operatorname{grad})\boldsymbol{E}_m + \mu \boldsymbol{E}_m \operatorname{div}\{(1/\mu)\operatorname{grad} L\} + (1/\mu\varepsilon)\{\boldsymbol{E}_m \cdot \operatorname{grad}(\mu\varepsilon)\}\operatorname{grad} L$$
$$= \operatorname{grad}\{(1/\varepsilon)\operatorname{div} \varepsilon \boldsymbol{E}_{m-1}\} - \mu \operatorname{curl}\{(1/\mu)\operatorname{curl} \boldsymbol{E}_{m-1}\}, \tag{11.8}$$

$$2(\operatorname{grad} L \cdot \operatorname{grad})\boldsymbol{H}_m + \varepsilon \boldsymbol{H}_m \operatorname{div}\{(1/\varepsilon)\operatorname{grad} L\} + (1/\mu\varepsilon)\{\boldsymbol{H}_m \cdot \operatorname{grad}(\mu\varepsilon)\}\operatorname{grad} L$$
$$= \operatorname{grad}\{(1/\mu)\operatorname{div} \mu \boldsymbol{H}_{m-1}\} - \varepsilon \operatorname{curl}\{(1/\varepsilon)\operatorname{curl} \boldsymbol{H}_{m-1}\}. \tag{11.9}$$

From these it can be shown that, when $\boldsymbol{E}_{m-1} = \mathbf{0}$ and $\boldsymbol{H}_{m-1} = \mathbf{0}$, $(\varepsilon/\mu)^{\frac{1}{2}} E_m^2$ *satisfies the intensity law of geometrical optics.*

Discontinuities in $\boldsymbol{E}$ and $\boldsymbol{H}$ are not covered by the foregoing. For these introduce the tensor $G^{\alpha\beta}$ defined in (9.5) of §2.9 and the coordinates x^1 $(=x)$, x^2 $(=y)$, x^3 $(=z)$, x^4 $(=ct)$. Then two of Maxwell's equations can be expressed as

$$\partial G^{\alpha\beta}/\partial x^\beta = 0. \tag{11.10}$$

Integrate (11.10) over a four-dimensional region bounded by the closed hypersurface Σ. The tensor form of the divergence theorem (F.8.1) gives

$$\int_\Sigma n_\beta G^{\alpha\beta} \, \mathrm{d}S = 0. \tag{11.11}$$

Equation (11.11), valid for all Σ, may be regarded as the integral form of (11.10) and is more general than it. For, in any domain where $G^{\alpha\beta}$ and its first derivatives exist (11.11) implies (11.10). In contrast, for a discontinuous $G^{\alpha\beta}$ (11.10) may have no meaning while (11.11) does. Therefore (11.11) replaces (11.10). Now choose Σ to be small and to cross $t = L$. By allowing the normal distance between $t = L$ and Σ to shrink to zero but not dS (cf. §1.22) we see that $n_\beta G^{\alpha\beta}$ is continuous through $t=L$. Since

$$n_1:n_2:n_3:n_4=\frac{\partial L}{\partial x}:\frac{\partial L}{\partial y}:\frac{\partial L}{\partial z}:-\frac{1}{c}$$

the substitution of the components of $G^{\alpha\beta}$ shows that $\operatorname{grad} L\wedge\boldsymbol{H}+\boldsymbol{D}$ and $(\operatorname{grad} L)\cdot\boldsymbol{D}$ are continuous across $t=L$. In other words

$$\operatorname{grad} L\wedge[\boldsymbol{H}]+\varepsilon[\boldsymbol{E}]=\mathbf{0}, \tag{11.12}$$

$$[\boldsymbol{E}]\cdot\operatorname{grad} L=0 \tag{11.13}$$

for ε continuous. Indeed, (11.13) follows from (11.12) by taking a scalar product with grad L. Similarly, it can be proved that

$$\operatorname{grad} L\wedge[\boldsymbol{E}]-\mu[\boldsymbol{H}]=\mathbf{0}. \tag{11.14}$$

Equations (11.12) and (11.14) are the same as (11.5) and (11.6) with $m=0$ provided that $\boldsymbol{E}_{-1}$ and $\boldsymbol{H}_{-1}$ are understood to be zero. All deductions made about $\boldsymbol{E}_m$ and $\boldsymbol{H}_m$ are therefore applicable to $[\boldsymbol{E}]$ and $[\boldsymbol{H}]$. In particular, *discontinuities in* **E** *and* **H** *propagate along characteristics.* Also they satisfy the transport equations with $m=0$ so that the right-hand sides of (11.8) and (11.9) are zero. As a consequence *the intensity law of geometrical optics is always valid for discontinuities in* **E** *and* **H**.

A similar approach may be followed for discontinuities in (11.1).

10.12 Uniqueness

The question of whether information at $t=0$ determines uniquely the subsequent behaviour of the field is clearly of some substance. Electromagnetic waves will be discussed here but the argument can be transferred to acoustics without difficulty.

In view of the linearity of Maxwell's equations a sufficient demonstration of uniqueness will be to prove that, if $\boldsymbol{E}=\mathbf{0}$ and $\boldsymbol{H}=\mathbf{0}$ at $t=0$, they vanish for all later times.

Introduce the vector S_4^α defined in (2.12.4) as

$$S_4^\alpha=(\boldsymbol{E}\wedge\boldsymbol{H})_\alpha/c \quad (\alpha\neq 4), \qquad S_4^4=\tfrac{1}{2}(\boldsymbol{B}\cdot\boldsymbol{H}+\boldsymbol{D}\cdot\boldsymbol{E})=\tfrac{1}{2}(\mu\boldsymbol{H}^2+\varepsilon\boldsymbol{E}^2).$$

It follows from (2.12.3) that, in the absence of currents,

$$\partial S_4^\alpha/\partial x^\alpha=0.$$

The integral of this equation over a four-dimensional domain surrounded by the closed hypersurface Σ is

$$\int_\Sigma n_\alpha S_4^\alpha\,dS=0 \tag{12.1}$$

after invoking the tensor divergence theorem (F.8.1).

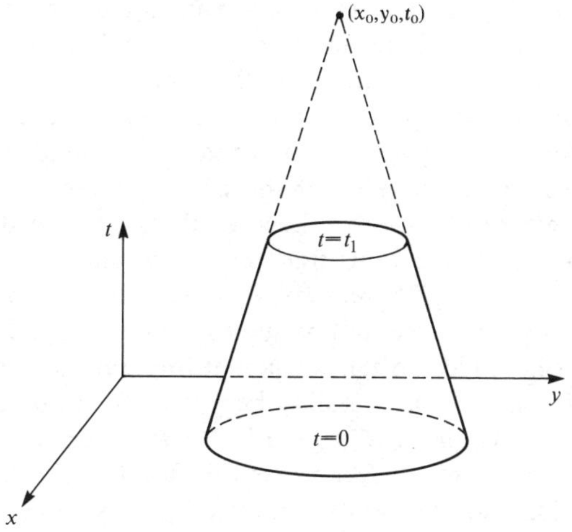

Fig. 10.7 Domain of integration for the uniqueness theorem

Choose for Σ the hypersurface generated by the rays through (x_0, y_0, z_0, t_0) in the region $t \leqslant t_0$, limited by the hyperplanes $t=0$ and $t=t_1$ with $0 \leqslant t_1 < t_0$. See Fig. 10.7 where one space dimension has been omitted for ease of pictorial representation. The conoid produced by the rays will be characteristic and have an equation of the form $t = t_0 - L$. On $t=0$, $n_\alpha = -\delta_{\alpha 4}$ and on $t = t_1$, $n_\alpha = \delta_{\alpha 4}$. Hence

$$\int_{t=t_1} \tfrac{1}{2}(\mu \boldsymbol{H}^2 + \varepsilon \boldsymbol{E}^2)\,\mathrm{d}x\,\mathrm{d}y\,\mathrm{d}z - \int_{t=0} \tfrac{1}{2}(\mu \boldsymbol{H}^2 + \varepsilon \boldsymbol{E}^2)\,\mathrm{d}x\,\mathrm{d}y\,\mathrm{d}z = -\int n_\alpha S_4^\alpha\,\mathrm{d}S \tag{12.2}$$

the integral on the right being over the conoidal surface between $t=0$ and $t=t_1$. On the conoid $n_4 = (c^2\mu\varepsilon + 1)^{-\frac{1}{2}}$ and $n_\alpha = (\text{grad } L)_\alpha/(\mu\varepsilon + 1/c^2)^{\frac{1}{2}}$. Hence, on the conoid

$$\begin{aligned} n_\alpha S_4^\alpha &= \tfrac{1}{2}(c^2\mu\varepsilon+1)^{-\frac{1}{2}}\{\mu \boldsymbol{H}^2 + \varepsilon \boldsymbol{E}^2 + 2\boldsymbol{E} \cdot \boldsymbol{H} \wedge \text{grad } L\} \\ &= \tfrac{1}{2}\varepsilon(c^2\mu\varepsilon+1)^{-\frac{1}{2}}[\{\boldsymbol{E} + (1/\varepsilon)\boldsymbol{H} \wedge \text{grad } L\}^2 \\ &\quad + (\mu/\varepsilon)\boldsymbol{H}^2 - (\boldsymbol{H} \wedge \text{grad } L)^2/\varepsilon^2] \\ &= \tfrac{1}{2}\varepsilon(c^2\mu\varepsilon+1)^{-\frac{1}{2}}[\{\boldsymbol{E} + (1/\varepsilon)\boldsymbol{H} \wedge \text{grad } L\}^2 + (\boldsymbol{H} \cdot \text{grad } L)^2/\varepsilon^2] \\ &\geqslant 0 \end{aligned}$$

since L satisfies the eikonal equation. Applying this inequality to the right-hand side of (12.2) we obtain

$$\int_{t=t_1} \tfrac{1}{2}(\mu \boldsymbol{H}^2 + \varepsilon \boldsymbol{E}^2)\,\mathrm{d}x\,\mathrm{d}y\,\mathrm{d}z \leqslant \int_{t=0} \tfrac{1}{2}(\mu \boldsymbol{H}^2 + \varepsilon \boldsymbol{E}^2)\,\mathrm{d}x\,\mathrm{d}y\,\mathrm{d}z.$$

The vanishing of $\boldsymbol{E}$ and $\boldsymbol{H}$ at $t=0$ now entails them being zero at $t=t_1$. Since t_1 can be any time between 0 and t_0, $\boldsymbol{E}$, and $\boldsymbol{H}$ are zero within the conoid from (x_0, y_0, z_0, t_0). No restriction was placed on the choice of (x_0, y_0, z_0, t_0) so that the proof of the uniqueness theorem is complete.

Another inference from the proof is that the dependence domain of a point P_0 consists of the interior of the characteristic conoid, with P_0 as vertex, drawn towards $t=-\infty$. Likewise, the influence domain is composed of the inside of the characteristic conoid aimed towards $t=+\infty$.

The above proof fails if there are fixed obstacles present and the conoid intersects the (four-dimensional) boundary of the body. Then part of the boundary must be added to the surface of integration to make Σ closed. Let the equation of such a boundary be $f(x, y, z)=0$; on it $n_\alpha \propto (\operatorname{grad} f)_\alpha$ $(\alpha \neq 4)$ and $n_4=0$ so that $n_\alpha S_4^\alpha \propto \operatorname{grad} f \cdot \boldsymbol{E} \wedge \boldsymbol{H}$. Therefore, if the tangential component of $\boldsymbol{E}$ or of $\boldsymbol{H}$ vanishes on the obstacle the integral from the body provides no contribution and uniqueness follows as before. If the tangential components of $\boldsymbol{E}$ and $\boldsymbol{H}$ are continuous across the surface of the scatterer the integral may be continued inside the scatterer and then (12.1) permits the closure of the gaps in the conoidal surface so that (12.2) and the succeeding argument remain unaltered.

Broadly speaking it may be asserted that *in the situations likely to arise in practice the electromagnetic field is determined uniquely by its values at any earlier time.*

10.13 Nonlinear waves

An area where a setting in the time domain is essential is that of nonlinear phenomena. Extensive research has been carried out in this area in recent years but space precludes anything but a very constricted description. An excellent account of all kinds of nonlinear effects is in the book of Whitham (1974).

The governing equations that will be discussed are

$$\operatorname{curl} \boldsymbol{E} + \partial \boldsymbol{B}/\partial t = \boldsymbol{0}, \qquad \operatorname{curl} \boldsymbol{H} - \partial \boldsymbol{D}/\partial t = \boldsymbol{0}, \tag{13.1}$$

$$\operatorname{div} \boldsymbol{D} = 0, \qquad \operatorname{div} \boldsymbol{B} = 0 \tag{13.2}$$

where the medium is such that the constitutive equations do not relate $\boldsymbol{D}$ to $\boldsymbol{E}$ or $\boldsymbol{B}$ to $\boldsymbol{H}$ linearly. As a specific assumption the constitutive relations will be taken as isotropic and such that

$$\partial \boldsymbol{D}/\partial \eta = \varepsilon(E)\, \partial \boldsymbol{E}/\partial \eta, \qquad \partial \boldsymbol{B}/\partial \eta = \mu(H)\, \partial \boldsymbol{H}/\partial \eta \tag{13.3}$$

where $E=(\boldsymbol{E} \cdot \boldsymbol{E})^{\frac{1}{2}}$, $H=(\boldsymbol{H} \cdot \boldsymbol{H})^{\frac{1}{2}}$ and η is any of the space or time coordinates.

Plane waves play a significant role in linear media so that it is natural to seek something similar for the nonlinear case. Therefore solutions are searched for whose only dependence on the space and time coordinates is through ξ where

$$\xi = t - \mathbf{s} \cdot \boldsymbol{x}. \tag{13.4}$$

Such solutions, when they exist, are called *simple waves*. The vector $\mathbf{s}$, which has the dimensions of the reciprocal of speed, is known as the *slowness vector* and $s = (\mathbf{s} \cdot \mathbf{s})^{\frac{1}{2}}$ as the *slowness*.

Subject to the dependence (13.4)

$$\text{grad}\, u = -\mathbf{s}\, \partial u/\partial \xi, \qquad \partial u/\partial t = \partial u/\partial \xi$$

so that (13.1) and (13.2) become

$$-\mathbf{s} \wedge \partial \mathbf{E}/\partial \xi + \mu(H)\, \partial \mathbf{H}/\partial \xi = \mathbf{0}, \qquad \mathbf{s} \wedge \partial \mathbf{H}/\partial \xi + \varepsilon(E)\, \partial \mathbf{E}/\partial \xi = 0, \tag{13.5}$$

$$\mathbf{s} \cdot \partial \mathbf{E}/\partial \xi = 0, \qquad \mathbf{s} \cdot \partial \mathbf{H}/\partial \xi = 0 \tag{13.6}$$

provided that neither $\varepsilon(E)$ nor $\mu(H)$ vanishes. Actually, (13.6) can be deduced from (13.5) by taking a scalar product with $\mathbf{s}$.

Elimination of $\partial \mathbf{E}/\partial \xi$ from (13.5) leads to

$$\{s^2 - \mu(H)\varepsilon(E)\}\, \partial \mathbf{H}/\partial \xi = \mathbf{0}.$$

Hence the slowness is given by

$$s = \{\mu(H)\varepsilon(E)\}^{\frac{1}{2}}. \tag{13.7}$$

Energy flow is governed by the Poynting vector $\boldsymbol{S} = \mathbf{E} \wedge \boldsymbol{H}$. From (13.5)

$$\frac{\partial \boldsymbol{S}}{\partial \xi} = \{1/\varepsilon(E)\mu(H)\}\left[\left\{\mu(H)\boldsymbol{H} \cdot \frac{\partial \boldsymbol{H}}{\partial \xi} + \varepsilon(E)\boldsymbol{E} \cdot \frac{\partial \mathbf{E}}{\partial \xi}\right\}\mathbf{s} - \mu(H)(\boldsymbol{H} \cdot \mathbf{s})\frac{\partial \boldsymbol{H}}{\partial \xi} - \varepsilon(E)(\mathbf{E} \cdot \mathbf{s})\frac{\partial \mathbf{E}}{\partial \xi}\right]. \tag{13.8}$$

The first term of (13.8) is similar to that for a linear medium representing flow parallel to the slowness vector $\mathbf{s}$ due to changes in the densities of electric and magnetic energy. The remaining two terms, which are transverse to $\mathbf{s}$ by virtue of (13.6) are caused by the nonlinearities and the wave velocity depending on E and H.

It may happen that the slowness vector is constant in direction, i.e. $\hat{\mathbf{s}}$ is a constant vector. Then division of (13.6) by s and integration shows that

$$\mathbf{s} \cdot \boldsymbol{E} = 0, \qquad \mathbf{s} \cdot \boldsymbol{H} = 0 \tag{13.9}$$

as in the linear case. The transverse flow is then removed from (13.8) and the formula is very like that for linear waves.

A *constant-amplitude wave* is one in which E and H are constant. For particular values of E and H the customary constitutive relations follow from (13.3). Thus *a constant-amplitude wave behaves as if it were in a linear medium.* The effect of a change of amplitude is to alter the properties of the medium in which the wave is propagating.

Reflection at a plane interface indicates some of the problems to be faced in the presence of nonlinearities. Let the medium in $z<0$ be linear with constants ε_1, μ_1. Suppose that in $z>0$ there is a nonlinear medium of the type just described. Impose on the linear medium an incident wave which depends only on $\xi^{\mathrm{i}}=t-\mathbf{s}^{\mathrm{i}}\cdot\boldsymbol{x}$ and has only the single component E_x^{i} in the electric intensity. It is necessary to emphasize that although a general plane wave can be split into two waves of independent polarizations the reflection problem cannot be solved by superposing the two solutions because of the nonlinearity of the second medium. The simplicity of approach possible when both media are linear is therefore lost right at the very beginning.

Assume that there is a reflected wave of same character as the incident, with E_x^{r} depending only on $\xi^{\mathrm{r}}=t-\mathbf{s}^{\mathrm{r}}\cdot\boldsymbol{x}$. Take for the transmitted disturbance a simple wave of the type described above and with E_x the only nonzero component of $\boldsymbol{E}$. This is a definite assumption, the feasibility of which is open to question. It is not by any means obvious that the exciting wave will create a simple wave with such polarization. But, even if it be accepted, there would seem to be nothing to prevent the simple wave forming a shock wave somewhere and so the region in which the assumed solution is valid is a fuzzy area whose precise limitations are unknown at the moment.

Let the component of $\mathbf{s}$ perpendicular to the z-axis be denoted by $\mathbf{s}_{\mathrm{T}}$. Then the continuity of the tangential component of $\boldsymbol{E}$ across $z=0$ can be met for all points if

$$\mathbf{s}_{\mathrm{T}}^{\mathrm{i}}=\mathbf{s}_{\mathrm{T}}^{\mathrm{r}}=\mathbf{s}_{\mathrm{T}}. \tag{13.10}$$

One implication of (13.10) is that $\mathbf{s}^{\mathrm{i}}\cdot\boldsymbol{k}=-\mathbf{s}^{\mathrm{r}}\cdot\boldsymbol{k}$ so that the angles of incidence and reflection are equal. Another inference is that the directions of propagation of the three waves are coplanar at $z=0$. In these respects the situation is analogous to that for linear media. A further point is that $\mathbf{s}_{\mathrm{T}}$ is independent of $\boldsymbol{E}$ and $\boldsymbol{H}$ so that only $\mathbf{s}\cdot\boldsymbol{k}$ exhibits dependence; the actual dependence is available from (13.7).

Further progress requires the solution of (13.5). Now

$$H_y=\int_0^{\xi}(\mathbf{s}\cdot\boldsymbol{k}/\mu)(\partial E_x/\partial\zeta)\,\mathrm{d}\zeta, \tag{13.11}$$

$$H_z=-\int_0^{\xi}(\mathbf{s}\cdot\boldsymbol{j}/\mu)(\partial E_x/\partial\zeta)\,\mathrm{d}\zeta \tag{13.12}$$

on the assumption that H_y and H_z are zero for $\xi=0$. In essence H, which is now $(H_y^2+H_z^2)^{\frac{1}{2}}$, has to be determined in terms of E_x by solving the implicit equations (13.11) and (13.12).

The boundary conditions on $z=0$ are

$$E_x^i + E_x^r = E_x, \tag{13.13}$$

$$(\varepsilon_1/\mu_1)^{\frac{1}{2}}\mathbf{s}^i \cdot \boldsymbol{k}(E_x^i - E_x^r) = \int_0^{\xi} (\mathbf{s}\cdot\boldsymbol{k}/\mu)(\partial E_x/\partial\zeta)\,\mathrm{d}\zeta. \tag{13.14}$$

Regarding H as a known function of E_x, we combine (13.13) and (13.14) into the integral equation

$$(\varepsilon_1/\mu_1)^{\frac{1}{2}}\mathbf{s}^i \cdot \boldsymbol{k}(2E_x^i - E_x) = \int_0^{E_x} (\mathbf{s}\cdot\boldsymbol{k}/\mu)\,\mathrm{d}E_x \tag{13.15}$$

for E_x on $z=0$.

The simultaneous solution of (13.11), (13.12), and (13.15) is not practicable in general other than by numerical means. Accordingly, we shall not take the matter beyond this point.

Exercises on Chapter 10

1. In the problem of the plane interface of §10.1 the excitation is magnetic rather than electric so that the sole component of the magnetic intensity is H_y, satisfying

$$\frac{\partial^2 H_y}{\partial x^2} + \frac{\partial^2 H_y}{\partial z^2} - \frac{1}{c_1^2}\frac{\partial^2 H_y}{\partial t^2} = -\delta(x)\delta(z-h)\delta(t)$$

in $z>0$. If $c_1 > c_2$ and r_i is nearly equal to $c_1 t$ show that, in $z>0$,

$$H_y = \frac{\mathrm{H}(t - r_s/c_1)}{2\pi(t^2 - r_s^2/c_1^2)^{\frac{1}{2}}} + \mathrm{Re}\, R_i(\tfrac{1}{2}\pi - \phi_i)\frac{\mathrm{H}(t - r_i/c_1)}{2\pi(t^2 - r_i^2/c_1^2)^{\frac{1}{2}}}$$

where

$$R_i(\phi) = \frac{\mu_1 c_1^2 \cos\phi - \mu_2 c_2 (c_1^2 - c_2^2 \sin^2\phi)^{\frac{1}{2}}}{\mu_1 c_1^2 \cos\phi + \mu_2 c_2 (c_1^2 - c_2^2 \sin^2\phi)^{\frac{1}{2}}}.$$

2. In Exercise 1 show, by making the substitution $s = -(\mathrm{i}\omega/c_1)\sin\theta$, that

$$\begin{aligned}\mathscr{H} &= \int_{-\infty}^{\infty} H_y \mathrm{e}^{-\mathrm{i}\omega t}\,\mathrm{d}t \\ &= \mathscr{H}^i + \frac{1}{4\pi\mathrm{i}}\int_{-\frac{1}{2}\pi - \mathrm{i}\infty}^{\frac{1}{2}\pi + \mathrm{i}\infty} R_i(\theta)\exp\{-\mathrm{i}k_1 r_i \sin(\theta + \phi_i)\}\,\mathrm{d}\theta\end{aligned}$$

where $k_1 = \omega/c_1$ and the integration is along the straight lines from $-\frac{1}{2}\pi - \mathrm{i}\infty$ to $-\frac{1}{2}\pi$, from $-\frac{1}{2}\pi$ to $\frac{1}{2}\pi$ and from $\frac{1}{2}\pi$ to $\frac{1}{2}\pi + \mathrm{i}\infty$.

If $\mu_1 = \mu_2$ and $k_1 r_i \gg 1$ show by the method of steepest descent that

$$\mathcal{H} \sim \mathcal{H}^i + \frac{R_i(\frac{1}{2}\pi - \phi_i)}{2i(2\pi k_1 r_i)^{\frac{1}{2}}} \exp\{-i(k_1 r_i - \tfrac{1}{4}\pi)\}$$

for $\cos \phi_i < \min(1, c_1/c_2)$.

3. If, in the last part of Exercise 2, $c_1 \gg c_2$ and $\phi_i \approx 0$, show, by replacing $R_i(\theta)$ by $1 + R_i(\theta) - 1$, that

$$\mathcal{H} \sim \mathcal{H}^i + \frac{\exp\{-i(k_1 r_i - \frac{1}{4}\pi)\}}{2i(2\pi k_1 r_i)^{\frac{1}{2}}} \{1 - 2i\gamma F(\gamma)\}$$

where $\gamma = (k_1 r_i)^{\frac{1}{2}}\{1 + \sin(\phi_i - \theta_B)\}^{\frac{1}{2}}$, $\tan \theta_B = c_1/c_2$ and F is defined in §9.7.

4. If, in the last part of Exercise 2, $c_1 < c_2$ and $\cos \phi_i > c_1/c_2$ show that an extra term

$$\frac{\exp\{-ik_1 r_i \cos(\phi_i - \delta_0) + \frac{1}{4}\pi i\}}{(2\pi \sin \delta_0)^{\frac{1}{2}}\{k_1 c_1 r_i \sin(\delta_0 - \phi_i)/c_2\}^{\frac{3}{2}}}$$

corresponding to the bow wave must be added to the asymptotic expression for $\mathcal{H}$. Here $\cos \delta_0 = c_1/c_2$.

The phase can be written as $-ik_1 L$ where L is the optical path of the ray which goes from the source to the boundary at the angle of *total reflection*, then along the boundary and finally leaving at the angle of total reflection to travel to the point of observation. [c.f. Exercise 40 of Chapter 6.]

5. The formula of Exercise 4 is not valid if $\phi_i \approx \delta_0$, i.e. the point of observation is near the join of the bow and reflected waves. In this region

$$\mathcal{H} \sim \mathcal{H}^i + \frac{\exp\{-i(k_1 r_i + \frac{1}{4}\pi)\}}{(8\pi k_1 r_i)^{\frac{1}{2}}}$$
$$\times \left[1 - \frac{c_2 D_{\frac{1}{2}}(\zeta)\exp(-\frac{1}{8}\pi i - \frac{1}{4}\zeta^2)}{c_1(\cos \phi_i \sin \delta_0)^{\frac{1}{2}}(k_1 r_i)^{\frac{1}{4}}\{\frac{1}{2}\cos \frac{1}{2}(\delta_0 - \phi_i)\}^{\frac{3}{2}}}\right]$$

where $\zeta = (k_1 r_i)^{\frac{1}{2}} 2e^{-\frac{1}{4}\pi i} \sin \frac{1}{2}(\delta_0 - \phi_i)$ and $D_{\frac{1}{2}}(z)$ is the parabolic cylinder function defined by

$$D_{\frac{1}{2}}(z) = \left(\frac{2}{\pi}\right)^{\frac{1}{2}} e^{\frac{1}{4}z^2} \int_0^\infty e^{-\frac{1}{2}t^2} t^{\frac{1}{2}} \cos(zt - \tfrac{1}{4}\pi)\, dt.$$

6. A vertical electric dipole is suddenly switched on at $t = 0$ and thereafter has constant moment. Prove that H_ϕ is proportional to $R_s^{-2}\delta(ct - R_s)\sin \theta$.

Such a dipole produces a vertical Hertz vector $\Pi = H(t - R/c_1)/R$ above the horizontal plane $z = 0$ separating two media which possess the same permeability. Show that the Hertz vector on $z=0$ tends to

$$\frac{N}{N^2-1}\left[\frac{N}{r}\mathrm{H}(c_1t-r)-\frac{1}{Nr}\mathrm{H}(c_2t-r)-\frac{\mathrm{H}(c_1t-r)\mathrm{H}(r-c_2t)}{\{(1+N^2)c_2^2t^2-r^2\}^{\frac{1}{2}}}\right]$$

eventually where $N=c_1/c_2>1$.

7. Repeat the analysis of §10.2 for a soft sphere.

8. Consider a plane electromagnetic pulse striking a perfectly conducting sphere by the method of §10.2.

9. A radiating solution of

$$\nabla^2 G-\frac{1}{a^2}\frac{\partial^2 G}{\partial t^2}=-\delta(\boldsymbol{x}-\boldsymbol{x}_0)\delta(t)$$

is required in the presence of a hard semi-infinite plane along $\phi=0$. Show that, if $at<r+r_0$, $G=0$ for $\pi+\phi_0<\phi<2\pi$, $G=a\mathrm{H}(at-r_s)/2\pi(a^2t^2-r_s^2)^{\frac{1}{2}}$ for $\pi-\phi_0<\phi<\pi<\phi_0$ and

$$G=\frac{a\mathrm{H}(at-r_s)}{2\pi(a^2t^2-r_s^2)^{\frac{1}{2}}}+\frac{a\mathrm{H}(at-r_i)}{2\pi(a^2t^2-r_i^2)^{\frac{1}{2}}}$$

for $0<\phi<\pi-\phi_0$ where $r_s^2=r^2+r_0^2-2rr_0\cos(\phi-\phi_0)$, $r_i^2=r^2+r_0^2-2rr_0\cos(\phi+\phi_0)$. If $at>r+r_0$

$$G=(a/4\pi)\{(a^2t^2-r_s^2)^{-\frac{1}{2}}+(a^2t^2-r_i^2)^{-\frac{1}{2}}\}.$$

Prove that changing the boundary condition from hard to soft merely alters the sign of the terms involving r_i.

10. The incident pulse $p^i=\mathrm{H}\{at+r\cos(\phi-\phi^i)\}$ where $0<\phi^i<\pi$ falls on the hard semi-infinite plane $\phi=0$. Show that the total p is zero in $\pi+\phi^i<\phi<2\pi$ for $at<r$ and in $0<\phi<\pi+\phi^i$ for $at<-r\cos(\phi-\phi^i)$. Show also that in $0<\phi<\pi-\phi^i$, $p=1$ for $-r\cos(\phi-\phi^i)<at<-r\cos(\phi+\phi^i)$ and $p=2$ for $-r\cos(\phi+\phi^i)<at<r$ while, in $\pi-\phi^i<\phi<\pi+\phi^i$, $p=2$ for $-r\cos(\phi+\phi^i)<at<r$.

If $at>r$ prove that in $0<\phi<\pi-\phi^i$

$$p=2-\frac{1}{\pi}\tan^{-1}\left\{\left(\frac{at-r}{2r}\right)^{\frac{1}{2}}\sec\tfrac{1}{2}(\phi-\phi^i)\right\}-\frac{1}{\pi}\tan^{-1}\left\{\left(\frac{at-r}{2r}\right)^{\frac{1}{2}}\sec\tfrac{1}{2}(\phi+\phi^i)\right\}$$

and find p for the other ranges of ϕ.

11. The pulse of Exercise 10 is incident in $0<\phi<\beta$ on a hard wedge with faces $\phi=0$ and $\phi=\beta$. Find the consequent field.

12. Find the solution of the telegraph equation

$$c^2\frac{\partial^2 V}{\partial z^2}=\frac{\partial^2 V}{\partial t^2}+2a\frac{\partial V}{\partial t}+a^2V\quad(c>0,\ a>0)$$

in $z>0$ such that $V=V_0$ at $z=0$, $t>0$ and initially V and I are zero.

$$\left[\int_0^\infty\frac{tc^2}{a^2+t^2c^2}\mathrm{e}^{\mathrm{i}tx}\,\mathrm{d}t=\tfrac{1}{2}\pi(c+\mathrm{i}\,\mathrm{sgn}\,x)\mathrm{e}^{-a|x|/c}\right].$$

13. A Heaviside non-distorting transmission line ($R/L = G/C$) of length l is initially at zero potential. From time $t = 0$ the end $z = 0$ is open circuit and $z = l$ is maintained at the constant potential difference V_0. Prove that the potential at $z = 0$ at time t is

$$V = V_0 \operatorname{sech}(a/b) - 2V_0 b \mathrm{e}^{-at} \sum_{n=1}^{\infty} (-)^n [a \sin\{(n-\tfrac{1}{2})\pi bt\}$$
$$+ (n-\tfrac{1}{2})\pi b \cos\{(n-\tfrac{1}{2})\pi bt\}]/[a^2 + \{(n-\tfrac{1}{2})\pi b\}^2]$$

where $a = R/L$ and $b = 1/l(LC)^{\frac{1}{2}}$.

14. Initially the line in Exercise 13 is at unit potential and the end $z = l$ is insulated. At $t = 0$ the end $z = 0$ is short-circuited. Show that the subsequent potential difference is

$$\frac{4}{\pi} \mathrm{e}^{-at} \sum_{n=0}^{\infty} \frac{1}{2n+1} \sin\left\{(2n+1)\frac{\pi x}{2l}\right\} \cos\{\tfrac{1}{2}(2n+1)\pi bt\}.$$

15. The method of §10.4 is applied to

$$\frac{\partial^4 u}{\partial t^4} - (a_0^2 + A^2)\frac{\partial^2}{\partial t^2}\nabla^2 u + a_0^2 A^2 \frac{\partial^2}{\partial z^2}\nabla^2 u = \delta(\boldsymbol{x})\mathrm{e}^{\mathrm{i}\omega t}$$

where ω has a small negative imaginary part. Show that if $\hat{\boldsymbol{x}}$ and $\boldsymbol{k}_0$ make angles θ and θ_1 respectively with the z-axis

$$c' = c \tan(\theta - \theta_1), \qquad c^4 - (a_0^2 + A^2)c^2 + a_0^2 A^2 \cos^2\theta_1 = 0$$

where $c = \omega/k_0$, $c' = \mathrm{d}c/\mathrm{d}\theta_1$. Hence determine the distant field.

16. In Exercise 15 $\mathrm{e}^{\mathrm{i}\omega t}$ is replaced by $\delta(t)$. Find the distant field.

17. Try an analysis similar to that of Exercises 15 and 16 for the equation with left-hand side

$$(1 + \mu\varepsilon A^2)\frac{\partial^4 u}{\partial t^4} - (a_0^2 + A^2)\frac{\partial^2}{\partial t^2}\nabla^2 u - \mu\varepsilon a_0^2 \frac{\partial^4 u}{\partial z^2\,\partial t^2} + a_0^2 A^2 \frac{\partial^2}{\partial z^2}\nabla^2 u.$$

18. What are the approximations (if any) involved when it is said that a signal travels down a waveguide with group speed?

19. Show that in Exercise 15 the group speed of waves of frequency $\omega/2\pi$ in the direction making an angle θ with the z-axis is $(c^2 + c'^2)^{\frac{1}{2}}$.

20. The electric intensity $\boldsymbol{E}$ in a crystal satisfies

$$\mu_0 \boldsymbol{\varepsilon} \cdot \partial^2 \boldsymbol{E}/\partial t^2 = \nabla^2 \boldsymbol{E} - \operatorname{grad} \operatorname{div} \boldsymbol{E}$$

and the axes are chosen so that $\boldsymbol{\varepsilon}$ is diagonal with elements ε_1, ε_2, ε_3. Show that in §10.4, $F = 0$ is

$$\omega^4 - \omega^2 \phi(\alpha) + \alpha^2 \psi(\alpha) = 0$$

where

$$\phi(\alpha) = v_1^2(\alpha^2 - \alpha_1^2) + v_2^2(\alpha^2 - \alpha_2^2) + v_3^2(\alpha^2 - \alpha_3^2),$$
$$\psi(\alpha) = \alpha_1^2 v_2^2 v_3^2 + \alpha_2^2 v_3^2 v_1^2 + \alpha_3^2 v_1^2 v_2^2,$$

$v_j^2 = 1/\mu_0\varepsilon_j$, $\alpha^2 = \alpha_1^2 + \alpha_2^2 + \alpha_3^2$. Deduce that $\sum_{j=1}^3 v_j^2\hat{x}_j^2/\{v_j^2 - \omega^2/(\boldsymbol{k}_0 \cdot \hat{\boldsymbol{x}})^2\} = 0$,

$$\frac{\hat{x}_j}{v_j^2 - \omega^2/(\boldsymbol{k}_0 \cdot \hat{\boldsymbol{x}})^2} = \frac{\alpha_j(\boldsymbol{k}_0 \cdot \hat{\boldsymbol{x}})}{\alpha^2 v_j^2 - \omega^2},$$

α_j and $\hat{x}_j$ being the components of $\boldsymbol{k}_0$ and $\hat{\boldsymbol{x}}$ respectively.

21. The function $f(t)$ in §10.7 is $\delta'(t)$. Find the connection between the time and frequency domains.

22. A perfectly reflecting moving target is illuminated by a beam from a fixed source and the Doppler shift of the reflected wave is measured at the source. If the object is large enough for the phase of the incident beam to vary over it a maximum of the Doppler spectrum is selected. Show that such a maximum is a characteristic of the phase variation as much as the speed of the target so that it is an unreliable measure of the speed. [Assume that the distance from the source to target is much larger than vT, T being the relevant time of illumination.]

23. Prove that

$$\frac{\partial u}{\partial x} = (y-1)\left(u - \frac{\partial u}{\partial y}\right)$$

has a solution $u = e^y f\{(y-1)e^{-x}\}$.

24. Find the solution of

$$\frac{\partial u}{\partial x} - (x-y-1)\frac{\partial u}{\partial y} = x(y-x)$$

such that $u = y^2$ when $x = 0$.

25. Prove that

$$\frac{\partial^2 u}{\partial x^2} - \frac{1}{x}\frac{\partial u}{\partial x} - 4x^2\frac{\partial^2 u}{\partial y^2} = 0$$

has a solution $u = f(x^2 - y) + g(x^2 + y)$.

26. Solve

$$(u+x)\frac{\partial u}{\partial y} - (u+y)\frac{\partial u}{\partial x} = y - x.$$

27. An electromagnetic field is continuous across $t = L$ in a medium in which ε and μ are continuous. Prove that $\text{grad}^2 L = \mu\varepsilon$, $[\boldsymbol{Q}] \cdot \text{grad}\, L = 0$ where $\boldsymbol{Q} = \boldsymbol{B} + i(\mu\varepsilon)^{\frac{1}{2}}\boldsymbol{E}$.

Show that a possible form for an electromagnetic field in a homogeneous isotropic medium is

$$\boldsymbol{H} = -\boldsymbol{k} \wedge \text{grad}\frac{\partial f}{\partial t}, \qquad \varepsilon\boldsymbol{E} = (\boldsymbol{k} \cdot \text{grad})\text{grad}\, f - \frac{1}{c^2}\frac{\partial^2 f}{\partial t^2}\boldsymbol{k}$$

for $R < ct$, $\boldsymbol{k}$ being a constant unit vector and

$$f = (A_1/R)e^{a(R-ct)} \cos \omega(R - ct + A_2)$$

whereas, for $R > ct$,

$$\boldsymbol{H} = \boldsymbol{0}, \qquad \varepsilon \boldsymbol{E} = \mu \operatorname{grad}(\boldsymbol{k} \cdot \operatorname{grad}) R^{-1}.$$

28. Prove that the characteristics $t = L$ of

$$\operatorname{div}(b \operatorname{grad} u) - \frac{b}{a^2} \frac{\partial^2 u}{\partial t^2} = 0$$

satisfy $a^2 \operatorname{grad}^2 L = 1$.

By substituting $u = u_1(t-L)^{-p} + u_2$ $(0 < p < 1)$ where u_1 and u_2 are regular and by considering dominant terms as $t - L \to 0$ show that the surfaces on which u has an algebraic singularity are characteristics. The same result is true if $(t-L)^{-p}$ is replaced by $\ln |t - L|$.

Are there corresponding conclusions for singular solutions of Maxwell's equations?

29. Show that the transport equations of

$$\operatorname{div}(b \operatorname{grad} p) - \frac{b}{a^2} \frac{\partial^2 p}{\partial t^2} = 0$$

are

$$2b \operatorname{grad} L \cdot \operatorname{grad} p_m + p_m \operatorname{div}(b \operatorname{grad} L) = \operatorname{div}(b \operatorname{grad} p_{m-1})$$

30. Investigate the behaviour of p_0 for the reflection of a plane pulse by a perfectly reflecting cylindrical obstacle.

31. Formulate the problem of the interface in §10.13 when the incident wave has only the component H_x^{i} in the magnetic intensity and depends only on ξ^{i}.

32. Prove that, as $R \to \infty$ and $0 \leqslant z_0 < \frac{1}{2}\pi$,

$$\int_{-\frac{1}{2}\pi - \mathrm{i}\infty}^{\frac{1}{2}\pi + \mathrm{i}\infty} \frac{\mathrm{e}^{-\mathrm{i}R\cos(z - z_0)}}{z - z_{\mathrm{p}}} \, \mathrm{d}z \sim \pi \mathrm{i} \mathrm{e}^{-\mathrm{i}R\cos(z_{\mathrm{p}} - z_{\mathrm{s}})} \operatorname{erfc}\{(\tfrac{1}{2}R)^{\frac{1}{2}}(z_0 - z_{\mathrm{p}})\mathrm{e}^{\frac{1}{4}\pi\mathrm{i}}\}$$

for $\operatorname{Im}(z_{\mathrm{p}} - z_0)\mathrm{e}^{-\frac{1}{4}\pi\mathrm{i}} > 0$, the contour of integration being that of Exercise 2. [§G.3]

33. If a saddle-point is near a pole which is not simple then, in §G.3, $f(z) = h(z)/(z - z_{\mathrm{p}})^m$ where $h(z)$ has no singularity. By taking a derivative $m-1$ times with respect to z_{p} of the result in §G.3 obtain the appropriate asymptotic formula.

Appendix A. Bessel functions

A.1 Bessel's equation

THE differential equation known as *Bessel's equation* is

$$z^2y'' + zy' + (z^2 - \nu^2)y = 0 \tag{1.1}$$

where both z and ν may be complex and the primes indicate derivatives with respect to z. When it is desired to specialize the constant ν to an integer it will be denoted by n. The *Bessel function of the first kind and νth order $J_\nu(z)$* is the solution of (1.1) defined by

$$J_\nu(z) = \sum_{m=0}^{\infty} (-)^m (\tfrac{1}{2}z)^{\nu+2m}/m!(\nu+m)! \tag{1.2}$$

where $\xi!$ is the factorial function such that $\xi! = (\xi-1)!\xi$. For Re $\xi > 0$

$$\xi! = \int_0^\infty e^{-t} t^\xi \, dt$$

and formulae for other complex values of ξ can be obtained by analytic continuation.

The series for $J_\nu(z)/z^\nu$ converges uniformly and absolutely for all finite values of z. Indeed, the only singularity of $J_\nu(z)$ is a branch point at the origin. A particular branch is selected by specifying $z^\nu = e^{\nu \ln z}$ where $\ln z = \ln|z| + i \,\text{ph}\, z$ and ph z has its principal value, i.e. $-\pi < \text{ph}\, z \leqslant \pi$. To retain the property that $J_\nu(z)/z^\nu$ is single-valued

$$J_\nu(ze^{m\pi i}) = e^{m\nu\pi i} J_\nu(z) \tag{1.3}$$

where m is any integer and $\text{ph}(ze^{m\pi i}) = m\pi + \text{ph}\, z$ with ph z having its principal value.

Bessel functions of different orders are related by

$$z\{J_{\nu-1}(z) + J_{\nu+1}(z)\} = 2\nu J_\nu(z), \tag{1.4}$$

$$J_{\nu-1}(z) - J_{\nu+1}(z) = 2J'_\nu(z), \tag{1.5}$$

$$zJ'_\nu(z) + \nu J_\nu(z) = zJ_{\nu-1}(z), \tag{1.6}$$

$$zJ'_\nu(z) - \nu J_\nu(z) = -zJ_{\nu+1}(z). \tag{1.7}$$

When ν is not an integer a second independent solution of (1.1) is $J_{-\nu}(z)$. Independence fails when ν is the integer n because

$$J_{-n}(z)=(-)^n J_n(z). \tag{1.8}$$

Therefore it is usual to pick as a second solution of Bessel's equation a function that will also serve when ν is an integer. An appropriate definition is

$$Y_\nu(z)=\{J_\nu(z)\cos\nu\pi - J_{-\nu}(z)\}/\sin\nu\pi, \qquad Y_n(z)=\lim_{\nu\to n} Y_\nu(z). \tag{1.9}$$

The general solution of (1.1) can then be written as

$$y = AJ_\nu(z)+BY_\nu(z)$$

for any value of ν.

$Y_\nu(z)$ is called the *Bessel function of the second kind and νth order.* It satisfies the same relations as (1.4)–(1.7) with J_ν replaced by Y_ν. Series expansions for Y_ν follow from (1.2) and, for n a non-negative integer,

$$Y_n(z)=\frac{2}{\pi}\{\gamma+\ln(\tfrac{1}{2}z)\}J_n(z)-(\tfrac{1}{2}z)^{-n}\sum_{m=0}^{n-1}\frac{(n-m-1)!}{m!\pi}(\tfrac{1}{2}z)^{2m}$$
$$-\sum_{m=0}^{\infty}\frac{(-)^m(\tfrac{1}{2}z)^{n+2m}}{m!(n+m)!\pi}\left\{1+\frac{1}{2}+\frac{1}{3}+\ldots+\frac{1}{m}+1+\frac{1}{2}+\frac{1}{3}+\ldots+\frac{1}{n+m}\right\} \tag{1.10}$$

where γ is Euler's constant 0.5772... and $1+\frac{1}{2}+\ldots.+1/m$ is taken to be zero when $m=0$. It is clear that Y_ν is always singular at the origin; $J_\nu(z)$ is finite if $\text{Re}\,\nu\geqslant 0$.

It is found convenient in order to achieve certain behaviour as $|z|\to\infty$ to construct as solutions of (1.1) the *Bessel functions of the third kind and νth order* defined by

$$H_\nu^{(1)}(z)=J_\nu(z)+iY_\nu(z), \qquad H_\nu^{(2)}(z)=J_\nu(z)-iY_\nu(z). \tag{1.11}$$

They are more frequently known as the *Hankel functions of the first and second kinds.* Both Hankel functions satisfy (1.4)–(1.7) and both are singular at the origin.

A unifying notation that is sometimes employed is to write $\mathscr{Z}_\nu^{(1)}$, $\mathscr{Z}_\nu^{(2)}$, $\mathscr{Z}_\nu^{(3)}$, and $\mathscr{Z}_\nu^{(4)}$ for J_ν, Y_ν, $H_\nu^{(1)}$, and $H_\nu^{(2)}$ respectively.

Series for the Hankel functions can be deduced from what has gone before. The following approximations are valid for small z

$$J_\nu(z)\approx z^\nu/\nu!2^\nu - z^{\nu+2}/(\nu+1)!2^{\nu+2}, \tag{1.12}$$

$$H_0^{(2)}(z)\approx[1-(2i/\pi)\{\gamma+\ln(\tfrac{1}{2}z)\}](1-z^2/4)-iz^2/2\pi, \tag{1.13}$$

$$H_1^{(2)}(z)\approx 2i/\pi z+\tfrac{1}{2}z[1-(2i/\pi)\{\gamma-\tfrac{1}{2}+\ln(\tfrac{1}{2}z)\}], \tag{1.14}$$

$$H_n^{(2)}(z)\approx(2^n i/\pi z^n)\{n-1)!+(n-2)!z^2/4\} \quad (n>1). \tag{1.15}$$

Note also the following

$$\mathrm{H}_{-\nu}^{(1)}(z)=\mathrm{e}^{\nu\pi\mathrm{i}}\mathrm{H}_{\nu}^{(1)}(z), \qquad \mathrm{H}_{-\nu}^{(2)}(z)=\mathrm{e}^{-\nu\pi\mathrm{i}}\mathrm{H}_{\nu}^{(2)}(z), \tag{1.16}$$

$$2\mathrm{J}_{-\nu}(z)=\mathrm{e}^{\nu\pi\mathrm{i}}\mathrm{H}_{\nu}^{(1)}(z)+\mathrm{e}^{-\nu\pi\mathrm{i}}\mathrm{H}_{\nu}^{(2)}(z), \tag{1.17}$$

$$2\mathrm{i}\mathrm{Y}_{-\nu}(z)=\mathrm{e}^{\nu\pi\mathrm{i}}\mathrm{H}_{\nu}^{(1)}(z)-\mathrm{e}^{-\nu\pi\mathrm{i}}\mathrm{H}_{\nu}^{(2)}(z), \tag{1.18}$$

$$\mathrm{H}_{\nu}^{(1)}(z\mathrm{e}^{m\pi\mathrm{i}})\sin\nu\pi=\mathrm{H}_{\nu}^{(1)}(z)\sin(1-m)\nu\pi-\mathrm{e}^{-\nu\pi\mathrm{i}}\mathrm{H}_{\nu}^{(2)}(z)\sin m\nu\pi, \tag{1.19}$$

$$\mathrm{H}_{\nu}^{(2)}(z\mathrm{e}^{m\pi\mathrm{i}})\sin\nu\pi=\mathrm{H}_{\nu}^{(2)}(z)\sin(1+m)\nu\pi+\mathrm{e}^{\nu\pi\mathrm{i}}\mathrm{H}_{\nu}^{(1)}(z)\sin m\nu\pi. \tag{1.20}$$

It may be deduced from Bessel's equation that $z(\mathrm{H}_{\nu}'\mathrm{J}_{\nu}-\mathrm{H}_{\nu}\mathrm{J}_{\nu}')$ is a constant, which may be evaluated by putting $z=0$; consequently

$$\mathrm{H}_{\nu}^{(2)\prime}(z)\mathrm{J}_{\nu}(z)-\mathrm{H}_{\nu}^{(2)}(z)\mathrm{J}_{\nu}'(z)=-2\mathrm{i}/\pi z. \tag{1.21}$$

An expansion in powers of t gives

$$\mathrm{e}^{\frac{1}{2}z(t-1/t)}=\mathrm{J}_0(z)+\sum_{n=1}^{\infty}\{t^n+(-)^n t^{-n}\}\mathrm{J}_n(z). \tag{1.22}$$

The substitution $t=\mathrm{i}\mathrm{e}^{\mathrm{i}\phi}$ supplies

$$\mathrm{e}^{\mathrm{i}z\cos\phi}=\mathrm{J}_0(z)+2\sum_{n=1}^{\infty}\mathrm{e}^{\frac{1}{2}\mathrm{i}n\pi}\mathrm{J}_n(z)\cos n\phi. \tag{1.23}$$

On the other hand, putting $t=\mathrm{e}^{-\mathrm{i}\phi}$ in (1.22) and integrating from α to $2\pi+\alpha$, where α is any real angle, provides

$$\mathrm{J}_n(z)=(1/2\pi)\int_{\alpha}^{2\pi+\alpha}\mathrm{e}^{\mathrm{i}(n\phi-z\sin\phi)}\,\mathrm{d}\phi \tag{1.24}$$

$$=(\mathrm{e}^{\frac{1}{2}n\pi\mathrm{i}}/2\pi)\int_0^{2\pi}\mathrm{e}^{-\mathrm{i}(n\phi+z\cos\phi)}\,\mathrm{d}\phi \tag{1.25}$$

$$=(\mathrm{e}^{\frac{1}{2}n\pi\mathrm{i}}/\pi)\int_0^{\pi}\mathrm{e}^{-\mathrm{i}z\cos\phi}\cos n\phi\,\mathrm{d}\phi. \tag{1.26}$$

From the series for J_{ν} follows

$$\mathrm{J}_{\nu}(z)=\frac{(\frac{1}{2}z)^{\nu}}{(\nu-\frac{1}{2})!(-\frac{1}{2})!}\int_0^{\pi}\cos(z\cos\phi)\sin^{2\nu}\phi\,\mathrm{d}\phi \quad (\mathrm{Re}\,\nu>-\tfrac{1}{2}) \tag{1.27}$$

$$=\frac{(\frac{1}{2}z)^{\nu}}{(\nu-\frac{1}{2})!(-\frac{1}{2})!}\int_0^{\pi}\mathrm{e}^{-\mathrm{i}z\cos\phi}\sin^{2\nu}\phi\,\mathrm{d}\phi, \tag{1.28}$$

$$\int_0^{\infty}\mathrm{J}_{\nu}(at)t^{\nu+1}\mathrm{e}^{-b^2t^2}\,\mathrm{d}t=a^{\nu}\mathrm{e}^{-a^2/4b^2}/(2b^2)^{\nu+1} \quad (\mathrm{Re}\,\nu>-1) \tag{1.29}$$

where $|\mathrm{ph}\,b|<\frac{1}{4}\pi$ and a is any complex number. Similarly, *Sonine's first finite integral*

$$\mathrm{J}_{\mu+\nu+1}(z)=\frac{z^{\nu+1}}{\nu!2^{\nu}}\int_0^{\frac{1}{2}\pi}\mathrm{J}_{\mu}(z\sin\phi)\sin^{\mu+1}\phi\cos^{2\nu+1}\phi\,\mathrm{d}\phi \tag{1.30}$$

can be obtained. It is valid for Re $\mu>-1$, Re $\nu>-1$.

An integral representation is

$$\mathrm{J}_{\nu}(z)=\frac{(\frac{1}{2}z)^{\nu}}{2\pi\mathrm{i}}\int_{-\infty}^{(0+)}t^{-\nu-1}\exp\left(t-\frac{z^2}{4t}\right)\mathrm{d}t \tag{1.31}$$

where the contour starts at $-\infty$ and returns to it, after going round the origin in the positive sense. We infer that

$$\mathrm{J}_{\nu}(z)=\frac{1}{2\pi\mathrm{i}}\int_{-\infty}^{(0+)}u^{-\nu-1}\exp\{\tfrac{1}{2}z(u-1/u)\}\,\mathrm{d}u\quad(|\mathrm{ph}\,z|<\tfrac{1}{2}\pi) \tag{1.32}$$

$$=(1/2\pi\mathrm{i})\int_{\infty-\pi\mathrm{i}}^{\infty+\pi\mathrm{i}}\mathrm{e}^{z\sinh w-\nu w}\,\mathrm{d}w \tag{1.33}$$

where the contour consists of the three sides of the rectangle with vertices $\infty-\pi\mathrm{i}$, $-\pi\mathrm{i}$, $\pi\mathrm{i}$, and $\infty+\pi\mathrm{i}$.

It may also be remarked that

$$\int_0^{\pi}\mathscr{Z}_0^{(j)}\{z^2+y^2-2zy\cos\phi)^{\frac{1}{2}}\}\,\mathrm{d}\phi=\pi\mathscr{Z}_0^{(j)}(y)\mathrm{J}_0(z)\quad(y\geqslant z) \tag{1.34}$$

which is a special case of

$$\int_0^{\pi}\frac{\mathscr{Z}_{\nu}^{(j)}\{(z^2+y^2-2zy\cos\phi)^{\frac{1}{2}}\}}{(z^2+y^2-2zy\cos\phi)^{\frac{1}{2}\nu}}\sin^{2\nu}\phi\,\mathrm{d}\phi=(\nu-\tfrac{1}{2})!(-\tfrac{1}{2})!2^{\nu}\frac{\mathscr{Z}_{\nu}^{(j)}(y)}{y^{\nu}}\frac{\mathrm{J}_{\nu}(z)}{z^{\nu}} \tag{1.35}$$

valid for Re $\nu>-\frac{1}{2}$.

Integral representations for the Hankel function are

$$\mathrm{H}_{\nu}^{(2)}(z)=-(1/\pi\mathrm{i})\int_{-\infty}^{\infty-\pi\mathrm{i}}\mathrm{e}^{z\sinh w-\nu w}\,\mathrm{d}w \tag{1.36}$$

$$=-(1/\pi\mathrm{i})\int_0^{\infty\exp(-\pi\mathrm{i})}u^{-\nu-1}\exp\{\tfrac{1}{2}z(u-1/u)\}\,\mathrm{d}u\quad(|\mathrm{ph}\,z|<\tfrac{1}{2}\pi). \tag{1.37}$$

Other variants are

$$\int_1^{\infty}\frac{\mathrm{e}^{-\mathrm{i}xt}}{(t^2-1)^{\frac{1}{2}}}\mathrm{d}t=-\tfrac{1}{2}\pi\mathrm{i}\mathrm{H}_0^{(2)}(x)\quad(x>0), \tag{1.38}$$

$$\int_1^{\infty}\frac{\mathrm{e}^{\mathrm{i}xt}}{(t^2-1)^{\frac{1}{2}}}\mathrm{d}t=\tfrac{1}{2}\pi\mathrm{i}\mathrm{H}_0^{(1)}(x)\quad(x>0), \tag{1.39}$$

$$\int_{-\infty}^{\infty}\frac{\mathrm{e}^{-\mathrm{i}k(a^2+x^2)^{\frac{1}{2}}+\mathrm{i}tx}}{(a^2+x^2)^{\frac{1}{2}}}\mathrm{d}x=-\pi\mathrm{i}\mathrm{H}_0^{(2)}\{a(k^2-t^2)^{\frac{1}{2}}\} \tag{1.40}$$

where a, k, t are real and $(k^2-t^2)^{\frac{1}{2}}$ means $-\mathrm{i}(t^2-k^2)^{\frac{1}{2}}$ when $t>k$. Another useful result is that, when $\operatorname{Re} s>0$,

$$\int_0^\infty e^{-sx}H_\nu^{(2)}(x)\,dx=\frac{i(t^\nu-e^{i\nu\pi}t^{-\nu})}{(s^2+1)^{\frac{1}{2}}\sin\nu\pi}\quad(|\operatorname{Re}\nu|<1)\tag{1.41}$$

where $t=s+(s^2+1)^{\frac{1}{2}}$.

A further integral of value is

$$\int^z t\mathscr{Z}_\nu^{(m)}(kt)\mathscr{Z}_\nu^{(n)}(lt)\,dt=z\{k\mathscr{Z}_{\nu+1}^{(m)}(kz)\mathscr{Z}_\nu^{(n)}(lz)$$

$$-l\mathscr{Z}_\nu^{(m)}(kz)\mathscr{Z}_{\nu+1}^{(n)}(lz)\}/(k^2-l^2)\quad(k\neq l)\tag{1.42}$$

$$=\tfrac{1}{2}z^2\{\mathscr{Z}_\nu^{(m)\prime}(kz)\mathscr{Z}_\nu^{(n)\prime}(kz)$$

$$+\mathscr{Z}_\nu^{(m)}(kz)\mathscr{Z}_\nu^{(n)}(kz)(1-\nu^2/k^2z^2)\}\quad(k=l).\tag{1.43}$$

If $j_{\nu m}$ is the mth zero of $J_\nu(z)$ it can be deduced from (1.42) and (1.43) that

$$\int_0^1 tJ_\nu(j_{\nu m}t)J_\nu(j_{\nu n}t)\,dt=\tfrac{1}{2}\delta_{mn}J_{\nu+1}^2(j_{\nu n})\tag{1.44}$$

where δ_{mn} is zero if $m\neq n$ and 1 if $m=n$. Some values of $j_{\nu n}$ are in Table A.1. More generally, if $\lambda_{\nu m}$ is a positive zero of $z^{-\nu}\{zJ_\nu'(z)+CJ_\nu(z)\}$, C being a constant,

$$\int_0^1 tJ_\nu(\lambda_{\nu m}t)J_\nu(\lambda_{\nu n}t)\,dt=\delta_{mn}\{(\lambda_{\nu m}^2-\nu^2)J_\nu^2(\lambda_{\nu m})+\lambda_{\nu m}^2J_\nu'^2(\lambda_{\nu m})\}/2\lambda_{\nu m}^2.\tag{1.45}$$

Two special formulae of interest are

$$H_{\frac{1}{2}}^{(2)}(z)=iH_{-\frac{1}{2}}^{(2)}(z)=i(2/\pi z)^{\frac{1}{2}}e^{-iz},\tag{1.46}$$

$$H_{\frac{1}{2}}^{(1)}(z)=-iH_{-\frac{1}{2}}^{(1)}(z)=-i(2/\pi z)^{\frac{1}{2}}e^{iz}.\tag{1.47}$$

A.2 Asymptotic behaviour

It may be shown from (1.28) that, when $|z|\to\infty$ with $|\mathrm{ph}\,z|<\pi$ and ν fixed,

$$J_\nu(z)=(2/\pi z)^{\frac{1}{2}}[\cos(z-\tfrac{1}{2}\nu\pi-\tfrac{1}{4}\pi)\{1-(4\nu^2-1)(4\nu^2-9)/32z^2+O(z^{-4})\}$$

$$-\sin(z-\tfrac{1}{2}\nu\pi-\tfrac{1}{4}\pi)\{(4\nu^2-1)/2z+O(z^{-3})\}].\tag{2.1}$$

An immediate deduction is that

$$H_\nu^{(1)}(z)=(2/\pi z)^{\frac{1}{2}}e^{i(z-\frac{1}{2}\nu\pi-\frac{1}{4}\pi)}$$

$$\times\{1-(4\nu^2-1)/8iz+O(z^{-2})\}\quad(-\pi<\mathrm{ph}\,z<2\pi),\tag{2.2}$$

$$H_\nu^{(2)}(z)=(2/\pi z)^{\frac{1}{2}}e^{-i(z-\frac{1}{2}\nu\pi-\frac{1}{4}\pi)}$$

$$\times\{1+(4\nu^2-1)/8iz+O(z^{-2})\}\quad(-2\pi<\mathrm{ph}\,z<\pi).\tag{2.3}$$

Formulae for other ranges of ph z can be derived from (1.3), (1.19), and (1.20).

For fixed z, as $|\nu|\to\infty$, an obvious conclusion from (1.2) is

$$J_\nu(z)=\frac{(\frac{1}{2}z)^\nu}{\nu!}\{1+O(\nu^{-1})\}. \tag{2.4}$$

The factorial can be approximated by *Stirling's formula*

$$\nu!\sim(2\pi)^{\frac{1}{2}}e^{(\nu+\frac{1}{2})\ln\nu-\nu}\quad(|\text{ph}\,\nu|<\pi) \tag{2.5}$$

as $|\nu|\to\infty$, the logarithm having its principal value. If ν is real and negative

$$\nu!\sim-(\tfrac{1}{2}\pi)^{\frac{1}{2}}\operatorname{cosec}\nu\pi e^{(\nu+\frac{1}{2})\ln\nu-\nu}. \tag{2.6}$$

When both z and ν become large none of (2.1)–(2.4) is satisfactory. If z and ν are not of the same order of magnitude one may take advantage of *Meissel's formula*

$$J_\nu(\nu x)\sim\left\{\frac{2}{\pi\nu(x^2-1)^{\frac{1}{2}}}\right\}^{\frac{1}{2}}\cos[\nu\{\sec^{-1}x-(x^2-1)^{\frac{1}{2}}\}+\tfrac{1}{4}\pi]\quad(x>1) \tag{2.7}$$

or *Carlini's formula*

$$J_\nu(\nu x)\sim\{2\pi\nu(1-x^2)^{\frac{1}{2}}\}^{-\frac{1}{2}}\exp[\nu\{(1-x^2)^{\frac{1}{2}}-\operatorname{sech}^{-1}x\}]\quad(0<x<1). \tag{2.8}$$

More elaborate approximations are needed when ν and z are of the same order of magnitude. An important factor in them is the *Airy function* Ai defined by

$$\text{Ai}(\eta e^{\frac{1}{3}\pi i})=-\tfrac{1}{2}i(\tfrac{1}{3}\eta)^{\frac{1}{2}}H^{(2)}_{\frac{1}{3}}(\tfrac{2}{3}\eta^{\frac{3}{2}}). \tag{2.9}$$

An integral representation is

$$\text{Ai}(\zeta)=-(i/2\pi)\int_{\infty\exp(-\frac{1}{3}\pi i)}^{\infty\exp(\frac{1}{3}\pi i)}\exp(\tfrac{1}{3}z^3-\zeta z)\,dz, \tag{2.10}$$

the contour lying to the right of the origin. For real x

$$\text{Ai}(x)=(1/2\pi)\int_{-\infty}^{\infty}\exp\{-i(\tfrac{1}{3}z^3+xz)\}\,dz. \tag{2.11}$$

It is straightforward to check that $v=\text{Ai}(\zeta)$ satisfies the differential equation

$$d^2v/d\zeta^2-\zeta v=0. \tag{2.12}$$

From (1.20)

$$\text{Ai}(\eta e^{-\frac{1}{3}\pi i})=\tfrac{1}{2}i(\tfrac{1}{3}\eta)^{\frac{1}{2}}H^{(1)}_{\frac{1}{3}}(\tfrac{2}{3}\eta^{\frac{3}{2}}) \tag{2.13}$$

and

$$\mathrm{Ai}(\eta e^{\pm\pi i}) = e^{\frac{1}{3}\pi i}\,\mathrm{Ai}(\eta e^{\frac{1}{3}\pi i}) + e^{-\frac{1}{3}\pi i}\,\mathrm{Ai}(\eta e^{-\frac{1}{3}\pi i}). \tag{2.14}$$

Note that

$$\mathrm{Ai}(0) = (-\tfrac{2}{3})!/2\pi 3^{\frac{1}{6}} = 3^{-\frac{2}{3}}/(-\tfrac{1}{3})!, \tag{2.15}$$

$$\mathrm{Ai}'(0) = (-\tfrac{1}{3})!3^{\frac{1}{6}}/2\pi = -3^{-\frac{1}{3}}/(-\tfrac{2}{3})!. \tag{2.16}$$

As $|\xi| \to \infty$

$$\mathrm{Ai}(\xi) = \tfrac{1}{2}\pi^{-\frac{1}{2}}\xi^{-\frac{1}{4}}\exp(-\tfrac{2}{3}\xi^{\frac{3}{2}})\{1 + O(1/|\xi|)\} \quad (|\mathrm{ph}\,\xi| < \pi), \tag{2.17}$$

$$\mathrm{Ai}(\xi) = \tfrac{1}{2}\mathrm{i}\pi^{-\frac{1}{2}}\xi^{-\frac{1}{4}}\exp(\tfrac{2}{3}\xi^{\frac{3}{2}})\{1 + O(1/|\xi|)\} \quad (\pi < \mathrm{ph}\,\xi < 3\pi), \tag{2.18}$$

$$\mathrm{Ai}(\xi) = \pi^{-\frac{1}{2}}\xi^{-\frac{1}{4}}e^{\frac{1}{4}\pi i} \times \{\cos(\tfrac{2}{3}\mathrm{i}\xi^{\frac{3}{2}} - \tfrac{1}{4}\pi) + \exp(|\mathrm{Re}\,\xi^{\frac{3}{2}}|)O(|\xi|^{-1})\} \quad (\tfrac{1}{3}\pi < \mathrm{ph}\,\xi < \tfrac{5}{3}\pi), \tag{2.19}$$

$$\mathrm{Ai}(\xi) = \pi^{-\frac{1}{2}}\xi^{-\frac{1}{4}}e^{-\frac{1}{4}\pi i} \times \{\cos(\tfrac{2}{3}\mathrm{i}\xi^{\frac{3}{2}} + \tfrac{1}{4}\pi) + \exp(|\mathrm{Re}\,\xi^{\frac{3}{2}}|)O(|\xi|^{-1})\} \quad (-\tfrac{5}{3}\pi < \mathrm{ph}\,\xi < -\tfrac{1}{3}\pi), \tag{2.20}$$

$$\mathrm{Ai}'(\xi) = -\tfrac{1}{2}\pi^{-\frac{1}{2}}\xi^{\frac{1}{4}}\exp(-\tfrac{2}{3}\xi^{\frac{3}{2}})\{1 + O(1/|\xi|)\} \quad (|\mathrm{ph}\,\xi| < \pi), \tag{2.21}$$

$$\mathrm{Ai}'(\xi) = \pi^{-\frac{1}{2}}\xi^{\frac{1}{4}}e^{-\frac{1}{4}\pi i} \times \{\sin(\tfrac{2}{3}\mathrm{i}\xi^{\frac{3}{2}} - \tfrac{1}{4}\pi) + \exp(|\mathrm{Re}\,\xi^{\frac{3}{2}}|)O(|\xi|^{-1})\} \quad (\tfrac{1}{3}\pi < \mathrm{ph}\,\xi < \tfrac{5}{3}\pi). \tag{2.22}$$

Some zeros of the Airy function and its derivative will be found in Tables A.5 and A.6 at the end of this Appendix.

As regards the Bessel function of the first kind

$$J_\nu(\nu z) \sim 2^{\frac{1}{2}}\nu^{-\frac{1}{3}}\{\xi/(z^2 - 1)\}^{\frac{1}{4}}\,\mathrm{Ai}(\nu^{\frac{2}{3}}\xi e^{\pi i}) \tag{2.23}$$

where

$$\tfrac{2}{3}\xi^{\frac{3}{2}} = (z^2 - 1)^{\frac{1}{2}} - \sec^{-1} z. \tag{2.24}$$

First find ξ for $z > 1$ and then determine its value for complex z by continuous variation. Near $z = 1$, $\xi = 2^{\frac{1}{3}}(z - 1)$ and for $z < 1 (\mathrm{ph}\, z = 0)$, $\xi = |\xi|e^{i\pi}$; in fact

$$\tfrac{2}{3}\xi^{\frac{3}{2}} = e^{\frac{3}{2}\pi i}\{\mathrm{sech}^{-1} z - (1 - z^2)^{\frac{1}{2}}\} \tag{2.25}$$

in this interval. As $|z| \to \infty$

$$\tfrac{2}{3}\xi^{\frac{3}{2}} \sim z - \tfrac{1}{2}\pi \tag{2.26}$$

and, as $z \to 0$,

$$\tfrac{2}{3}\xi^{\frac{3}{2}} \approx -e^{\frac{3}{2}\pi i}(1 + \ln \tfrac{1}{2}z). \tag{2.27}$$

With this definition of ξ, (2.23) and subsequent formulae are valid for $|\mathrm{ph}\, z| < \pi - \delta$ $(\delta > 0)$ subject to ν being real and for complex ν with $|\mathrm{ph}\, \nu| \leqslant \frac{1}{2}\pi$ provided that the z-plane is cut in a certain way (for precise details see Olver (1974)). Asymptotic expressions for other ranges of ph z

and ph ν can be inferred by means of the continuation formulae (1.16) and (1.20).

For the other Bessel functions

$$\mathrm{Y}_\nu(\nu z) \sim -2^{\frac{1}{2}}\nu^{-\frac{1}{3}}\{\xi/(z^2-1)\}^{\frac{1}{4}}\{\mathrm{e}^{\frac{1}{6}\pi \mathrm{i}}\,\mathrm{Ai}(\nu^{\frac{2}{3}}\xi \mathrm{e}^{-\frac{1}{3}\pi \mathrm{i}})+\mathrm{e}^{-\frac{1}{6}\pi \mathrm{i}}\,\mathrm{Ai}(\nu^{\frac{2}{3}}\xi \mathrm{e}^{\frac{1}{3}\pi \mathrm{i}})\}, \tag{2.28}$$

$$\mathrm{H}_\nu^{(1)}(\nu z) \sim 2^{\frac{3}{2}}\nu^{-\frac{1}{3}}\mathrm{e}^{-\frac{1}{3}\pi \mathrm{i}}\{\xi/(z^2-1)\}^{\frac{1}{4}}\,\mathrm{Ai}(\nu^{\frac{2}{3}}\xi \mathrm{e}^{-\frac{1}{3}\pi \mathrm{i}}), \tag{2.29}$$

$$\mathrm{H}_\nu^{(2)}(\nu z) \sim 2^{\frac{3}{2}}\nu^{-\frac{1}{3}}\mathrm{e}^{\frac{1}{3}\pi \mathrm{i}}\{\xi/(z^2-1)\}^{\frac{1}{4}}\,\mathrm{Ai}(\nu^{\frac{2}{3}}\xi \mathrm{e}^{\frac{1}{3}\pi \mathrm{i}}), \tag{2.30}$$

$$\mathrm{H}_\nu^{(1)\prime}(\nu z) \sim 2^{\frac{3}{2}}\nu^{-\frac{2}{3}}z^{-1}\mathrm{e}^{\frac{4}{3}\pi \mathrm{i}}\{(z^2-1)/\xi\}^{\frac{1}{4}}\,\mathrm{Ai}'(\nu^{\frac{2}{3}}\xi \mathrm{e}^{-\frac{1}{3}\pi \mathrm{i}}), \tag{2.31}$$

$$\mathrm{H}_\nu^{(2)\prime}(\nu z) \sim 2^{\frac{3}{2}}\nu^{-\frac{2}{3}}z^{-1}\mathrm{e}^{\frac{2}{3}\pi \mathrm{i}}\{(z^2-1)/\xi\}^{\frac{1}{4}}\,\mathrm{Ai}'(\nu^{\frac{2}{3}}\xi \mathrm{e}^{\frac{1}{3}\pi \mathrm{i}}). \tag{2.32}$$

If z is close enough to ν for $\xi = 2^{\frac{1}{3}}(z-1)$ to be an acceptable approximation

$$\mathrm{H}_\nu^{(2)}(\nu z) \sim 2^{\frac{1}{2}}3^{-\frac{1}{2}}\mathrm{e}^{-\frac{1}{6}\pi \mathrm{i}}(z-1)^{\frac{1}{2}}\mathrm{H}_{\frac{1}{3}}^{(2)}\{2^{\frac{3}{2}}\nu(z-1)^{\frac{3}{2}}/3\}, \tag{2.33}$$

$$\mathrm{H}_\nu^{(2)\prime}(\nu z) \sim 2\mathrm{i}3^{-\frac{1}{2}}\mathrm{e}^{\frac{2}{3}\pi \mathrm{i}}(z-1)\mathrm{H}_{\frac{2}{3}}^{(2)}\{2^{\frac{3}{2}}\nu(z-1)^{\frac{3}{2}}/3\}. \tag{2.34}$$

One conclusion from (2.28)–(2.32) and (2.15), (2.16) is that

$$\mathrm{J}_\nu(\nu) \sim 2^{\frac{1}{3}}3^{-\frac{2}{3}}\nu^{-\frac{1}{3}}/(-\tfrac{1}{3})!, \qquad \mathrm{Y}_\nu(\nu) \sim -2^{\frac{1}{3}}3^{-\frac{1}{6}}\nu^{-\frac{1}{3}}/(-\tfrac{1}{3})!, \tag{2.35}$$

$$\mathrm{J}_\nu'(\nu) \sim 2^{\frac{2}{3}}3^{-\frac{1}{3}}\nu^{-\frac{2}{3}}/(-\tfrac{2}{3})!, \qquad \mathrm{Y}_\nu'(\nu) \sim -2^{\frac{2}{3}}3^{\frac{1}{6}}\nu^{-\frac{2}{3}}/(-\tfrac{2}{3})!, \tag{2.36}$$

$$\mathrm{H}_\nu^{(2)}(\nu) \sim 2^{\frac{4}{3}}\mathrm{e}^{\frac{1}{3}\pi \mathrm{i}}/(-\tfrac{1}{3})!3^{\frac{2}{3}}\nu^{\frac{1}{3}}, \qquad \mathrm{H}_\nu^{(1)}(\nu) \sim 2^{\frac{4}{3}}\mathrm{e}^{-\frac{1}{3}\pi \mathrm{i}}/(-\tfrac{1}{3})!3^{\frac{2}{3}}\nu^{\frac{1}{3}}, \tag{2.37}$$

$$\mathrm{H}_\nu^{(2)\prime}(\nu) \sim -2^{\frac{5}{3}}\mathrm{e}^{\frac{2}{3}\pi \mathrm{i}}/(-\tfrac{2}{3})!3^{\frac{1}{3}}\nu^{\frac{2}{3}}, \qquad \mathrm{H}_\nu^{(1)\prime}(\nu) \sim -2^{\frac{5}{3}}\mathrm{e}^{-\frac{2}{3}\pi \mathrm{i}}/(-\tfrac{2}{3})!3^{\frac{1}{3}}\nu^{\frac{2}{3}}. \tag{2.38}$$

The Wronskian relation

$$\mathrm{e}^{\frac{1}{3}\pi \mathrm{i}}\,\mathrm{Ai}'(x\mathrm{e}^{\frac{1}{3}\pi \mathrm{i}})\mathrm{Ai}(-x)+\mathrm{Ai}(x\mathrm{e}^{\frac{1}{3}\pi \mathrm{i}})\mathrm{Ai}'(-x) = -\mathrm{i}\mathrm{e}^{-\frac{1}{3}\pi \mathrm{i}}/2\pi \tag{2.39}$$

can be confirmed by taking a derivative of the left-hand side and then applying (2.15), (2.16).

A.3 Modified Bessel functions

The substitution of $\mathrm{i}z$ for z in (1.1) leads to the differential equation

$$z^2y'' + zy' - (z^2+\nu^2)y = 0 \tag{3.1}$$

which is known as *Bessel's modified equation.* A solution is

$$\mathrm{I}_\nu(z) = \sum_{m=0}^{\infty} (\tfrac{1}{2}z)^{\nu+2m}/m!(\nu+m)!. \tag{3.2}$$

It is related to the Bessel function of the first kind by

$$\mathrm{I}_\nu(z) = \mathrm{e}^{-\frac{1}{2}\nu\pi \mathrm{i}}\mathrm{J}_\nu(z\mathrm{e}^{\frac{1}{2}\pi \mathrm{i}}) \quad (-\pi < \mathrm{ph}\, z \leqslant \tfrac{1}{2}\pi), \tag{3.3}$$

$$\mathrm{I}_\nu(z) = \mathrm{e}^{\frac{3}{2}\nu\pi \mathrm{i}}\mathrm{J}_\nu(z\mathrm{e}^{-\frac{3}{2}\pi \mathrm{i}}) \quad (\tfrac{1}{2}\pi < \mathrm{ph}\, z \leqslant \pi). \tag{3.4}$$

A second solution of (3.1) is $K_\nu(z)$ defined by

$$K_\nu(z)=\tfrac{1}{2}\pi\{I_{-\nu}(z)-I_{\nu}(z)\}/\sin\nu\pi, \qquad K_n(z)=\lim_{\nu\to n}K_\nu(z). \tag{3.5}$$

I_ν and K_ν are called *modified Bessel functions*. It is evident that

$$K_{-\nu}(z)=K_\nu(z), \qquad I_{-n}(z)=I_n(z). \tag{3.6}$$

From (3.3), (3.5), (1.9), and (1.11) it can be deduced that

$$K_\nu(z)=\tfrac{1}{2}\pi i e^{\frac{1}{2}\nu\pi i}H_\nu^{(1)}(ze^{\frac{1}{2}\pi i}), \tag{3.7}$$

$$K_\nu(ze^{\frac{1}{2}\pi i})=-\tfrac{1}{2}\pi i e^{-\frac{1}{2}\nu\pi i}H_\nu^{(2)}(z). \tag{3.8}$$

It follows from (3.8) and (2.13) that

$$\mathrm{Ai}(\xi)=\{(\tfrac{1}{3}\xi)^{\frac{1}{2}}/\pi\}K_{\frac{1}{3}}(\tfrac{2}{3}\xi^{\frac{3}{2}}), \qquad \mathrm{Ai}'(\xi)=-(\xi/\pi 3^{\frac{1}{2}})K_{\frac{2}{3}}(\tfrac{2}{3}\xi^{\frac{3}{2}}). \tag{3.9}$$

By means of (3.3) and (3.7) we can derive from the analogous relations for Bessel functions

$$I_\nu(ze^{m\pi i})=e^{m\nu\pi i}I_\nu(z), \tag{3.10}$$

$$K_\nu(ze^{m\pi i})=e^{-m\nu\pi i}K_\nu(z)-\pi i\sin m\nu\pi I_\nu(z)/\sin\nu\pi, \tag{3.11}$$

$$z\{I_{\nu-1}(z)-I_{\nu+1}(z)\}=2\nu I_\nu(z), \qquad z\{K_{\nu-1}(z)-K_{\nu+1}(z)=-2\nu K_\nu(z), \tag{3.12}$$

$$I_{\nu-1}(z)+I_{\nu+1}(z)=2I_\nu'(z), \qquad K_{\nu-1}(z)+K_{\nu+1}(z)=-2K_\nu'(z), \tag{3.13}$$

$$zI_\nu'(z)+\nu I_\nu(z)=zI_{\nu-1}(z), \qquad zK_\nu'(z)+\nu K_\nu(z)=-zK_{\nu-1}(z), \tag{3.14}$$

$$zI_\nu'(z)-\nu I_\nu(z)=zI_{\nu+1}(z), \qquad zK_\nu'(z)-\nu K_\nu(z)=-zK_{\nu+1}(z). \tag{3.15}$$

Evidently, when z is small and $n\geqslant 1$,

$$K_n(z)\approx(n-1)!2^{n-1}/z^n. \tag{3.16}$$

Corresponding to (1.28) is

$$I_\nu(z)=\frac{(\frac{1}{2}z)^\nu}{(\nu-\frac{1}{2})!(-\frac{1}{2})!}\int_0^\pi e^{\pm z\cos\phi}\sin^{2\nu}\phi\,d\phi \quad (\mathrm{Re}\,\nu>-\tfrac{1}{2}) \tag{3.17}$$

and to (1.31) is

$$I_\nu(z)=\frac{(\frac{1}{2}z)^\nu}{2\pi i}\int_{-\infty}^{(0+)}t^{-\nu-1}\exp(t+z^2/4t)\,dt \tag{3.18}$$

$$=(1/2\pi i)\int_{\infty-\pi i}^{\infty+\pi i}e^{z\cosh w-\nu w}\,dw \quad (|\mathrm{ph}\,z|<\tfrac{1}{2}\pi). \tag{3.19}$$

It follows from (3.19) that

$$K_\nu(z)=\int_0^\infty e^{-z\cosh t}\cosh\nu t\,dt \quad (|\mathrm{ph}\,z|<\tfrac{1}{2}\pi). \tag{3.20}$$

From (1.35) with $m=4$ it may be concluded that

$$J_\nu(z)H_\nu^{(2)}(Z)=-(1/\pi i)\int_0^{\infty\exp(-\pi i)} e^{\frac{1}{2}\{t-(Z^2+z^2)/t\}}I_\nu(Zz/t)\frac{dt}{t} \tag{3.21}$$

for $|z|<|Z|$ and $\operatorname{Re}\nu>-1$. A further result is

$$\int_x^\infty e^{-i\omega u}I_0\{a(u^2-x^2)^{\frac{1}{2}}\}\,du=-i(\omega^2+a^2)^{-\frac{1}{2}}e^{-ix(\omega^2+a^2)^{\frac{1}{2}}} \tag{3.22}$$

for $\operatorname{Im}\omega>|a|$.

Additional information on Bessel functions can be found in Watson (1944) and Erdélyi (1953).

A.4 Spherical Bessel functions

The differential equation

$$z^2y''+2zy'+\{z^2-\nu(\nu+1)\}y=0 \tag{4.1}$$

can be converted to Bessel's equation of order $\nu+\frac{1}{2}$ by the transformation $y=w/z^{\frac{1}{2}}$. Therefore a solution of (4.1) is $j_\nu(z)$ defined by

$$j_\nu(z)=(\pi/2z)^{\frac{1}{2}}J_{\nu+\frac{1}{2}}(z). \tag{4.2}$$

Other solutions are $y_\nu(z)$, $h_\nu^{(1)}(z)$, and $h_\nu^{(2)}(z)$ defined in a similar way. These four solutions of (4.1) are called *spherical Bessel functions*. It is convenient to use $\mathfrak{z}_\nu^{(m)}(z)$, specified in terms of $\mathscr{Z}_\nu^{(m)}(z)$ by the rule of (4.2), to identify a typical member of the set of four.

Results for spherical Bessel functions can be deduced from the preceding sections and only a selection will be reproduced here. There are the recurrence relations

$$z\{\mathfrak{z}_{\nu-1}^{(m)}(z)+\mathfrak{z}_{\nu+1}^{(m)}(z)\}=(2\nu+1)\mathfrak{z}_\nu^{(m)}(z), \tag{4.3}$$

$$z\{\mathfrak{z}_{\nu-1}^{(m)}(z)-\mathfrak{z}_{\nu+1}^{(m)}(z)\}=2z\mathfrak{z}_\nu^{(m)\prime}(z)+\mathfrak{z}_\nu^{(m)}(z), \tag{4.4}$$

$$z\mathfrak{z}_\nu^{(m)\prime}(z)+(\nu+1)\mathfrak{z}_\nu^{(m)}(z)=z\mathfrak{z}_{\nu-1}^{(m)}(z), \tag{4.5}$$

$$z\mathfrak{z}_\nu^{(m)\prime}(z)-\nu\mathfrak{z}_\nu^{(m)}(z)=-z\mathfrak{z}_{\nu+1}^{(m)}(z). \tag{4.6}$$

A consequence of (4.5) and (4.6) is that

$$(d/z\,dz)^r\{z^{\nu+1}\mathfrak{z}_\nu^{(m)}(z)\}=z^{\nu-r+1}\mathfrak{z}_{\nu-r}^{(m)}(z), \tag{4.7}$$

$$(d/z\,dz)^r\{z^{-\nu}\mathfrak{z}_\nu^{(m)}(z)\}=(-)^r z^{-\nu-r}\mathfrak{z}_{\nu+r}^{(m)}(z). \tag{4.8}$$

It follows from (4.8) via a Taylor expansion that

$$h_0^{(2)}\{(\rho+\sigma)^{\frac{1}{2}}\}=\sum_{m=0}^{\infty}(-\tfrac{1}{2}\sigma)^m h_m^{(2)}(\rho^{\frac{1}{2}})/m!\rho^{\frac{1}{2}m} \tag{4.9}$$

if $|\sigma|<|\rho|$.

Other particular results are

$$y_n(z) = (-)^{n-1} j_{-n-1}(z), \tag{4.10}$$

$$j_\nu(z) h_\nu^{(2)\prime}(z) - j_\nu'(z) h_\nu^{(2)}(z) = -i/z^2, \tag{4.11}$$

$$[d^n j_n(z)/dz^n]_{z=0} = (n!)^2 2^n/(2n+1)!. \tag{4.12}$$

Also, as $|z| \to \infty$ with $-2\pi < \text{ph } z < \pi$,

$$h_n^{(2)}(z) \sim e^{\frac{1}{2}(n+1)\pi i - iz}/z. \tag{4.13}$$

The following special formulae are worthy of note

$$j_0(z) = (\sin z)/z, \qquad j_1(z) = \{-\cos z + (\sin z)/z\}/z, \tag{4.14}$$

$$y_0(z) = -(\cos z)/z, \qquad y_1(z) = -\{\sin z + (\cos z)/z\}/z, \tag{4.15}$$

$$h_0^{(1)}(z) = e^{iz}/iz, \qquad h_1^{(1)}(z) = -(1 + i/z)e^{iz}/z, \tag{4.16}$$

$$h_0^{(2)}(z) = -e^{-iz}/iz, \qquad h_1^{(2)}(z) = -(1 - i/z)e^{-iz}/z. \tag{4.17}$$

For small $|z|$ and non-negative integral n

$$j_n(z) \approx (2z)^n \left\{ \frac{n!}{(2n+1)!} - \frac{(n+1)!}{(2n+3)!} z^2 \right\}, \tag{4.18}$$

$$h_n^{(2)}(z) \approx \frac{2i}{(2z)^{n+1}} \left\{ \frac{(2n)!}{n!} + \frac{(2n-2)!}{(n-1)!} z^2 \right\} + j_n(z) \quad (n \geqslant 1), \tag{4.19}$$

$$j_n'(z) + j_n(z)/z \approx 2^n z^{n-1} \left\{ \frac{(n+1)!}{(2n+1)!} - \frac{(n+1)!(n+3)}{(2n+3)!} z^2 \right\}, \tag{4.20}$$

$$h_n^{(2)\prime}(z) + \frac{h_n^{(2)}(z)}{z} \approx \frac{-i}{2^n z^{n+2}} \left\{ \frac{(2n)!}{(n-1)!} + \frac{(2n-2)!(n-3)}{(n-1)!} z^2 \right\}. \tag{4.21}$$

A.5 Tables

Some useful zeros of Bessel functions and Airy functions are listed in the following tables.

Table A.1 Value of j_{mp} such that $J_m(j_{mp}) = 0$

	m							
p	0	1	2	3	4	5	6	7
1	2.405	3.832	5.136	6.380	7.588	8.771	9.936	11.086
2	5.520	7.016	8.147	9.761	11.065	12.339	13.589	14.821
3	8.654	10.173	11.620	13.015	14.372			
4	11.792	13.323	14.796					

Table A.2 Value of positive j'_{mp} such that $J'_m(j'_{mp})=0$

	m							
p	0	1	2	3	4	5	6	7
1	3.832	1.841	3.054	4.201	5.317	6.416	7.501	8.578
2	7.016	5.331	6.706	8.015	9.282	10.520	11.735	12.932
3	10.173	8.536	9.969	11.346	12.682	13.987		
4	13.323	11.706	13.170					

Table A.3 Value of $(d-1)\chi_{mp}$ where
$J_m(d\chi_{mp})Y_m(\chi_{mp})-Y_m(d\chi_{mp})J_m(\chi_{mp})=0$

	$m=0$			$m=1$		$m=2$		$m=3$
d	$p=1$	$p=2$	$p=3$	$p=1$	$p=2$	$p=1$	$p=2$	$p=1$
1	3.142	6.283	9.425	3.142	6.283	3.142	6.283	3.142
1.4	3.137	6.281	9.423	3.155	6.290	3.208	6.317	3.294
1.8	3.128	6.276	9.420	3.182	6.304	3.36	6.387	3.6
2.5	3.110	6.266	9.413	3.235	6.335			
3.5	3.085	6.250	9.402	3.305	6.381			

Table A.4 Determination of χ'_{mp} such that
$J'_m(d\chi'_{mp})Y'_m(\chi'_{mp})-Y'_m(d\chi'_{mp})J'_m(\chi'_{mp})=0.$

	$(d+1)\chi'_{m1}$			$(d-1)\chi'_{mp}$	
d	$m=1$	$m=2$	$m=3$	$m=1, p=2$	$m=2, p=2$
1	2.000	4.000	6.000	3.142	3.142
1.4	2.009	4.015	6.017	3.174	3.229
1.8	2.024	4.026	5.986	3.241	3.4
2.5	2.048	3.980	5.751	3.396	
3.5	2.057	3.834	5.382	3.636	

Note that $\chi'_{0p}=\chi_{1p}$ where χ_{1p} is given in Table A.3.

Table A.5 Value of α_s such that

$$\mathrm{Ai}(-\alpha_s)=0.$$

s	α_s
1	2.338
2	4.088
3	5.520
4	6.787
5	7.944
6	9.023

Table A.6 Value of β_s where

$$\mathrm{Ai}'(-\beta_s)=0.$$

s	β_s
1	1.019
2	3.248
3	4.820
4	6.163
5	7.372
6	8.488

Appendix B. Legendre functions

B.1 Legendre's equation

LEGENDRE'S *associated equation* is

$$(1-z^2)y''-2zy'+\{\nu(\nu+1)-\mu^2/(1-z^2)\}y=0 \tag{1.1}$$

though it is commonly referred to as *Legendre's equation* when $\mu=0$. It goes over to

$$\sin\theta\frac{d}{d\theta}\left(\sin\theta\frac{dy}{d\theta}\right)+\{\nu(\nu+1)\sin^2\theta-\mu^2\}y=0 \tag{1.2}$$

on making the substitution $z=\cos\theta$.

In general, the solutions of (1.1) are hypergeometric functions. Two are selected as the *associated Legendre functions* P_ν^μ and Q_ν^μ, namely

$$P_\nu^\mu(z)=\frac{1}{(-\mu)!}\left(\frac{z+1}{z-1}\right)^{\frac{1}{2}\mu}F(-\nu,\nu+1;1-\mu;\tfrac{1}{2}-\tfrac{1}{2}z), \tag{1.3}$$

$$Q_\nu^\mu(z)=\frac{(\nu+\mu)!\pi^{\frac{1}{2}}e^{i\mu\pi}(z^2-1)^{\frac{1}{2}\mu}}{(\nu+\frac{1}{2})!2^{\nu+1}z^{\nu+\mu+1}}F\left(\tfrac{1}{2}\nu+\tfrac{1}{2}\mu+1,\tfrac{1}{2}\nu+\tfrac{1}{2}\mu+\tfrac{1}{2};\nu+\tfrac{3}{2};\frac{1}{z^2}\right) \tag{1.4}$$

where

$$F(a,b;c;z)=1+\frac{ab}{1!c}z+\frac{a(a+1)b(b+1)}{2!c(c+1)}z^2+\dots.$$

The meaning to be attached to quantities such as $(z^2-1)^{\frac{1}{2}\mu}$ is determined by the rule

$$|\mathrm{ph}(z+1)|<\pi,\qquad |\mathrm{ph}(z-1)|<\pi,\qquad |\mathrm{ph}\,z|<\pi$$

so that $z+1=-e^{\pm\pi i}(z+1)$, $z-1=e^{\pm\pi i}(1-z)$, $z^2-1=e^{\pm\pi i}(1-z^2)$, the upper or lower sign being taken according as $\mathrm{Im}\,z\gtrless 0$. The series in (1.3) and (1.4) converge only for certain domains of z but, in other regions, there are other convergent expansions via the many relations which connect hypergeometric functions. As a result the associated Legendre functions are single-valued and regular in the z-plane cut from 1 to $-\infty$.

The existence of the cut means that the value of P_ν^μ as $z\to x+i0$ is different from that as $z \to x - \mathrm{i}0$ when $x < 1$. Solutions are, however, required of (1.1) when $|x| < 1$ especially when it has the form (1.2) and so particular combinations of (1.3) and (1.4) are employed in this interval. They are denoted by $\mathrm{P}_\nu^\mu(x)$ and $\mathrm{Q}_\nu^\mu(x)$ to distinguish from, for example, the limits $\mathrm{P}_\nu^\mu(x+\mathrm{i}0)$ and $\mathrm{P}_\nu^\mu(x-\mathrm{i}0)$ of (1.3) as $z \to x \pm \mathrm{i}0$. The definitions are

$$\mathrm{P}_\nu^\mu(x) = \mathrm{e}^{\frac{1}{2}\mathrm{i}\mu\pi}\mathrm{P}_\nu^\mu(x+\mathrm{i}0), \tag{1.5}$$

$$\mathrm{Q}_\nu^\mu(x) = \tfrac{1}{2}\mathrm{e}^{-\mathrm{i}\mu\pi}\{\mathrm{e}^{-\frac{1}{2}\mathrm{i}\mu\pi}\mathrm{Q}_\nu^\mu(x+\mathrm{i}0) + \mathrm{e}^{\frac{1}{2}\mathrm{i}\mu\pi}\mathrm{Q}_\nu^\mu(x-\mathrm{i}0)\}. \tag{1.6}$$

The consequential series for $\mathrm{P}_\nu^\mu(x)$ is

$$\mathrm{P}_\nu^\mu(x) = \frac{1}{(-\mu)!}\left(\frac{1+x}{1-x}\right)^{\frac{1}{2}\mu}\mathrm{F}(-\nu, \nu+1; 1-\mu; \tfrac{1}{2}-\tfrac{1}{2}x). \tag{1.7}$$

It is immediately evident from (1.3) and (1.7) that

$$\mathrm{P}_{-\nu-1}^\mu(z) = \mathrm{P}_\nu^\mu(z), \qquad \mathrm{P}_{-\nu-1}^\mu(x) = \mathrm{P}_\nu^\mu(x). \tag{1.8}$$

The transformations available for hypergeometric functions enable one to show that

$$\mathrm{P}_\nu^\mu(+\mathrm{i}0) = 2^\mu\pi^{\frac{1}{2}}\mathrm{e}^{-\frac{1}{2}\mu\pi\mathrm{i}}/(-\tfrac{1}{2}\nu-\tfrac{1}{2}\mu-\tfrac{1}{2})!(\tfrac{1}{2}\nu-\tfrac{1}{2}\mu)!, \tag{1.9}$$

$$\mathrm{P}_\nu^{\mu\prime}(+\mathrm{i}0) = -2^{\mu+1}\pi^{\frac{1}{2}}\mathrm{e}^{-\frac{1}{2}\mu\pi\mathrm{i}}/(-\tfrac{1}{2}+\tfrac{1}{2}\nu-\tfrac{1}{2}\mu)!(-\tfrac{1}{2}\nu-\tfrac{1}{2}\mu-1)!. \tag{1.10}$$

Values at $-\mathrm{i}0$ are obtained by altering the sign of i on the right-hand sides of (1.9) and (1.10). Combining this information with (1.5) we deduce that

$$\mathrm{P}_\nu^\mu(0) = 2^\mu\pi^{\frac{1}{2}}/(-\tfrac{1}{2}\nu-\tfrac{1}{2}\mu-\tfrac{1}{2})!(\tfrac{1}{2}\nu-\tfrac{1}{2}\mu)!, \tag{1.11}$$

$$\mathrm{P}_\nu^{\mu\prime}(0) = -2^{\mu+1}\pi^{\frac{1}{2}}/(-\tfrac{1}{2}+\tfrac{1}{2}\nu-\tfrac{1}{2}\mu)!(-\tfrac{1}{2}\nu-\tfrac{1}{2}\mu-1)!. \tag{1.12}$$

Furthermore

$$\mathrm{Q}_\nu^\mu(+\mathrm{i}0) = (\tfrac{1}{2}\nu+\tfrac{1}{2}\mu-\tfrac{1}{2})!2^{\mu-1}\pi^{\frac{1}{2}}\mathrm{e}^{\frac{1}{2}\pi\mathrm{i}(2\mu-\nu-1)}/(\tfrac{1}{2}\nu-\tfrac{1}{2}\mu)!, \tag{1.13}$$

$$\mathrm{Q}_\nu^{\mu\prime}(+\mathrm{i}0) = (\tfrac{1}{2}\nu+\tfrac{1}{2}\mu)!2^\mu\pi^{\frac{1}{2}}\mathrm{e}^{\frac{1}{2}\pi\mathrm{i}(2\mu-\nu)}/(\tfrac{1}{2}\nu-\tfrac{1}{2}\mu-\tfrac{1}{2})!. \tag{1.14}$$

For values at $-\mathrm{i}0$ replace $\mathrm{e}^{-\frac{1}{2}\pi\mathrm{i}(\nu+1)}$ in (1.13) by $\mathrm{e}^{\frac{1}{2}\pi\mathrm{i}(\nu+1)}$, and $\mathrm{e}^{-\frac{1}{2}\pi\mathrm{i}\nu}$ in (1.14) by $\mathrm{e}^{\frac{1}{2}\pi\mathrm{i}\nu}$. These imply that

$$\mathrm{Q}_\nu^\mu(0) = -(\tfrac{1}{2}\nu+\tfrac{1}{2}\mu-\tfrac{1}{2})!2^{\mu-1}\pi^{\frac{1}{2}}\sin\tfrac{1}{2}\pi(\nu+\mu)/(\tfrac{1}{2}\nu-\tfrac{1}{2}\mu)!, \tag{1.15}$$

$$\mathrm{Q}_\nu^{\mu\prime}(0) = (\tfrac{1}{2}\nu+\tfrac{1}{2}\mu)!2^\mu\pi^{\frac{1}{2}}\cos\tfrac{1}{2}\pi(\nu+\mu)/(\tfrac{1}{2}\nu-\tfrac{1}{2}\mu-\tfrac{1}{2})!. \tag{1.16}$$

If the differential equation for P_ν^μ is multiplied by Q_ν^μ and subtracted from the corresponding equation with P_ν^μ and Q_ν^μ interchanged, it is found from (1.9), (1.10), (1.13), and (1.14) that

$$P_\nu^\mu(z)\mathrm{Q}_\nu^{\mu\prime}(z) - \mathrm{P}_\nu^{\mu\prime}(z)\mathrm{Q}_\nu^\mu(z) = \frac{(\mu+\nu)!\mathrm{e}^{\mathrm{i}\mu\pi}}{(\nu-\mu)!(1-z^2)}. \tag{1.17}$$

Similarly

$$\mathrm{P}_\nu^\mu(x)\mathrm{Q}_\nu^{\mu\prime}(x)-\mathrm{P}_\nu^{\mu\prime}(x)\mathrm{Q}_\nu^\mu(x)=\frac{(\mu+\nu)!}{(\nu-\mu)!(1-x^2)}. \tag{1.18}$$

The following relations between associated Legendre functions can be substantiated:

$$\mathrm{P}_\nu^{-\mu}(z)=\frac{(\nu-\mu)!}{(\nu+\mu)!}\left\{\mathrm{P}_\nu^\mu(z)-\frac{2}{\pi}\mathrm{e}^{-\mathrm{i}\mu\pi}\mathrm{Q}_\nu^\mu(z)\sin\mu\pi\right\}, \tag{1.19}$$

$$\mathrm{Q}_\nu^{-\mu}(z)=(\nu-\mu)!\mathrm{e}^{-2\mathrm{i}\mu\pi}\mathrm{Q}_\nu^\mu(z)/(\nu+\mu)!, \tag{1.20}$$

$$\mathrm{Q}_{-\nu-1}^\mu(z)\sin(\nu-\mu)\pi=\mathrm{Q}_\nu^\mu(z)\sin(\mu+\nu)\pi-\pi\mathrm{e}^{\mathrm{i}\mu\pi}\mathrm{P}_\nu^\mu(z)\cos\nu\pi, \tag{1.21}$$

$$\mathrm{P}_\nu^\mu(-z)=\mathrm{e}^{\mp\nu\pi\mathrm{i}}\mathrm{P}_\nu^\mu(z)-(2/\pi)\mathrm{e}^{-\mathrm{i}\mu\pi}\mathrm{Q}_\nu^\mu(z)\sin(\mu+\nu)\pi, \tag{1.22}$$

$$\mathrm{Q}_\nu^\mu(-z)=-\mathrm{e}^{\pm\nu\pi\mathrm{i}}\mathrm{Q}_\nu^\mu(z) \tag{1.23}$$

the upper or lower sign being taken in (1.22) and (1.23) according as $\mathrm{Im}\, z \gtrless 0$. An immediate deduction from (1.5) and (1.6) is that

$$\mathrm{P}_\nu^\mu(-x)=\mathrm{P}_\nu^\mu(x)\cos(\mu+\nu)\pi-(2/\pi)\mathrm{Q}_\nu^\mu(x)\sin(\mu+\nu)\pi, \tag{1.24}$$

$$\mathrm{Q}_\nu^\mu(-x)=-\mathrm{Q}_\nu^\mu(x)\cos(\mu+\nu)\pi-\tfrac{1}{2}\pi\mathrm{P}_\nu^\mu(x)\sin(\mu+\nu)\pi. \tag{1.25}$$

B.2 Asymptotic behaviour

Manipulation of the hypergeometric functions reveals that, for large ν with fixed μ,

$$\mathrm{Q}_\nu^\mu(\cos\theta)=\frac{(\nu+\mu)!}{(\nu+\frac{1}{2})!}\left(\frac{\pi}{2\sin\theta}\right)^{\frac{1}{2}}[\cos\{(\nu+\tfrac{1}{2})\theta+\tfrac{1}{4}\pi+\tfrac{1}{2}\mu\pi\}+O(\nu^{-1})] \tag{2.1}$$

subject to $0<\theta<\pi$. Since

$$\mathrm{i}\pi\mathrm{e}^{\mathrm{i}\mu\pi}\mathrm{P}_\nu^\mu(x)=\mathrm{e}^{-\frac{1}{2}\mu\pi\mathrm{i}}\mathrm{Q}_\nu^\mu(x+\mathrm{i}0)-\mathrm{e}^{\frac{1}{2}\mu\pi\mathrm{i}}\mathrm{Q}_\nu^\mu(x-\mathrm{i}0) \tag{2.2}$$

it may be inferred that

$$\mathrm{P}_\nu^\mu(\cos\theta)=\frac{(\nu+\mu)!}{(\nu+\frac{1}{2})!}\left(\frac{2}{\pi\sin\theta}\right)^{\frac{1}{2}}[\cos\{(\nu+\tfrac{1}{2})\theta-\tfrac{1}{4}\pi+\tfrac{1}{2}\mu\pi\}+O(\nu^{-1})] \tag{2.3}$$

subject to $0<\theta<\pi$. When θ is small (2.1) and (2.3) are not satisfactory. Now (1.7), with $(\nu+n)!/(\nu-n)!$ expanded in powers of $\nu+\frac{1}{2}$ leads to

$$\mathrm{P}_\nu^{-\mu}(\cos\theta)=\frac{\mathrm{J}_\mu\{(2\nu+1)\sin\frac{1}{2}\theta\}}{\{(\nu+\frac{1}{2})\cos\frac{1}{2}\theta\}^\mu}\{1+O(\sin^2\tfrac{1}{2}\theta)\}. \tag{2.4}$$

B.3 Legendre polynomials

When ν is the non-negative integer n the infinite series in (1.3) and (1.7) terminate and P_n^μ is $\{(1+x)/(1-x)\}^{\frac{1}{2}\mu}$ times a polynomial of degree n

unless μ is a non-negative integer m. In the latter case P_n^m is $(1-x^2)^{\frac{1}{2}m}$ times a polynomial of degree $n-m$ which vanishes identically when $m>n$. The functions in which $m=0$ are written P_n and called *Legendre polynomials.*

Cases in which n and m are negative can be handled via (1.8), (1.19) and (1.20) so they will be taken as non-negative integers from now on.

Derivatives of (1.7) give

$$P_\nu^m(x)=(-)^m(1-x^2)^{\frac{1}{2}m}\,d^mP_\nu(x)/dx^m \tag{3.1}$$

and the same relation holds if P is replaced by Q. It may also be shown that

$$P_n(x)=(1/n!2^n)d^n(x^2-1)^n/dx^n \tag{3.2}$$

which is *Rodrigues' formula.* Examples of some of the functions of integer order are

$$P_0(x)=1,\qquad P_1(x)=x,\qquad P_2(x)=\tfrac{1}{2}(3x^2-1),\qquad P_3(x)=\tfrac{1}{2}(5x^3-3x), \tag{3.3}$$

$$P_4(x)=(35x^4-30x^2+3)/8,\qquad P_5(x)=(63x^5-70x^3+15x)/8, \tag{3.4}$$

$$P_1^1(x)=-(1-x^2)^{\frac{1}{2}},\qquad P_2^1(x)=-3x(1-x^2)^{\frac{1}{2}},\qquad P_2^2(x)=3(1-x^2), \tag{3.5}$$

$$Q_0(x)=\tfrac{1}{2}\ln\frac{1+x}{1-x},\qquad Q_1(x)=\tfrac{1}{2}x\ln\frac{1+x}{1-x}-1, \tag{3.6}$$

$$Q_2(x)=P_2(x)Q_0(x)-\tfrac{3}{2}x,\qquad Q_3(x)=P_3(x)Q_0(x)-\tfrac{5}{2}x^2+\tfrac{2}{3}, \tag{3.7}$$

$$Q_1^1(x)=-(1-x^2)^{\frac{1}{2}}\left\{Q_0(x)+\frac{x}{1-x^2}\right\},\qquad Q_2^1(x)=\tfrac{3}{2}x(1-x^2)^{\frac{1}{2}}Q_0(x)+\frac{3x^2-2}{(1-x^2)^{\frac{1}{2}}}. \tag{3.8}$$

From (1.24) and (1.25)

$$P_n^m(-x)=(-)^{m+n}P_n^m(x),\qquad Q_n^m(-x)=(-)^{m+n+1}Q_n^m(x). \tag{3.9}$$

When $x\approx 1$, (1.7) shows that

$$P_\nu^m(x)\approx(\nu+m)!(-)^m(1-x)^{\frac{1}{2}m}/(\nu-m)!m!2^{\frac{1}{2}m} \tag{3.10}$$

whereas

$$Q_\nu(x)\approx-\gamma-\psi(\nu)-\tfrac{1}{2}\ln(\tfrac{1}{2}-\tfrac{1}{2}x)\quad(\nu\neq-1,-2,\ldots) \tag{3.11}$$

where $\psi(z)=z!'/z!$ and γ is Euler's constant.

When $x\approx-1$, relevant formulae are

$$P_\nu(x)\approx\{\ln(\tfrac{1}{2}+\tfrac{1}{2}x)+\gamma+2\psi(\nu)+\pi\cot\nu\pi\}\sin\nu\pi/\pi, \tag{3.12}$$

$$P_n^m(x)\approx(n+m)!(-)^n(1+x)^{\frac{1}{2}m}/(n-m)!m!2^{\frac{1}{2}m}. \tag{3.13}$$

The following recurrence relations can be established:

$$\mathrm{P}_n^{m+2}(x)+2(m+1)x(1-x^2)^{-\frac{1}{2}}\mathrm{P}_n^{m+1}(x)+(n-m)(n+m+1)\mathrm{P}_n^m(x)=0, \tag{3.14}$$

$$(1-x^2)\,\mathrm{dP}_n^m(x)/\mathrm{d}x=(m+n)\mathrm{P}_{n-1}^m(x)-nx\mathrm{P}_n^m(x) \tag{3.15}$$

$$=(n+1)x\mathrm{P}_n^m(x)-(n-m+1)\mathrm{P}_{n+1}^m(x), \tag{3.16}$$

$$\mathrm{P}_{n-1}^m(x)=x\mathrm{P}_n^m(x)+(n-m+1)(1-x^2)^{\frac{1}{2}}\mathrm{P}_n^{m-1}(x), \tag{3.17}$$

$$(1-x^2)^{\frac{1}{2}}\mathrm{P}_n^{m+1}(x)=(n-m+1)\mathrm{P}_{n+1}^m(x)-(n+m+1)x\mathrm{P}_n^m(x). \tag{3.18}$$

If the equation for P_n is multiplied by P_p and subtracted from the equation with n and p exchanged it is found from (3.2), (3.10), and (3.13) that

$$\int_{-1}^{1}\mathrm{P}_n(x)\mathrm{P}_p(x)\,\mathrm{d}x=2\delta_{np}/(2n+1). \tag{3.19}$$

Similarly

$$\int_{-1}^{1}\mathrm{P}_n^m(x)\mathrm{P}_p^m(x)\,\mathrm{d}x=(n+m)!2\delta_{np}/(n-m)!(2n+1), \tag{3.20}$$

$$\int_{-1}^{1}\left\{\frac{m^2}{1-x^2}\mathrm{P}_n^m(x)\mathrm{P}_p^m(x)+(1-x^2)\frac{\mathrm{dP}_n^m}{\mathrm{d}x}\frac{\mathrm{dP}_p^m}{\mathrm{d}x}\right\}\mathrm{d}x=\frac{(n+m)!2n(n+1)\delta_{np}}{(n-m)!(2n+1)} \tag{3.21}$$

The application of Rodrigues' formula and integration by parts gives

$$\int_{-1}^{1}x^n\mathrm{P}_n(x)\,\mathrm{d}x=(n!)^2 2^{n+1}/(2n+1)!. \tag{3.22}$$

The formula

$$(1-2hz+h^2)^{-\frac{1}{2}}=\sum_{n=0}^{\infty}h^n\mathrm{P}_n(z) \qquad \{|h|<\min|z\pm(z^2-1)^{\frac{1}{2}}|\} \tag{3.23}$$

can be verified by expanding the left-hand side in powers of h and applying (3.2). Another valuable result is

$$\mathrm{P}_n(\cos\Theta)=\mathrm{P}_n(\cos\theta)\mathrm{P}_n(\cos\theta_0)+2\sum_{m=1}^{n}\frac{(n-m)!}{(n+m)!}\mathrm{P}_n^m(\cos\theta)\mathrm{P}_n^m(\cos\theta_0)\cos m\phi \tag{3.24}$$

where $\cos\Theta=\sin\theta\sin\theta_0\cos\phi+\cos\theta\cos\theta_0$ and θ, θ_0, ϕ are any complex numbers.

Further information about Legendre functions can be found in Erdélyi (1953) and Hobson (1931).

Appendix C. Mathieu functions

C.1 Mathieu's equation

THE differential equation

$$V'' + (\lambda - 2h \cos 2v) V = 0 \tag{1.1}$$

where the primes indicate derivatives with respect to v, is called *Mathieu's equation*. The constant h is regarded as given but the parameter λ may have to be adjusted to achieve a certain type of solution.

The substitution $z = \sin^2 v$ converts Mathieu's equation to algebraic form; in this form it has regular singularities at $z = 0$ and $z = 1$, both with exponents 0 and $\frac{1}{2}$, and an irregular singularity at infinity. The irregular singularity makes the differential equation difficult to handle.

Mathieu's equation has solutions for all values of λ but only for certain eigenvalues (which depend on h) are the solutions periodic in v; i.e. they are unaltered when v is changed by 2π. Periodic solutions can be classified according as they are even or odd functions of v. If the eigenvalue $\lambda = \mathrm{a}_n(h)$ the even periodic solution is

$$\mathrm{ce}_n(v, h) = \sum_r{}' \mathrm{A}_r^n(h) \cos rv \quad (n = 0, 1, \ldots) \tag{1.2}$$

where the prime signifies that the summation is over $r = 0, 2, 4, \ldots$ when n is even and over $r = 1, 3, 5, \ldots$ when n is odd. The coefficients A_r^n have to satisfy certain recurrence relations in order to ensure that ce_n is a solution of Mathieu's equation and are chosen so that

$$\int_0^{2\pi} \mathrm{ce}_n^2(v, h)\, \mathrm{d}v = \pi. \tag{1.3}$$

The odd functions occur for $\lambda = \mathrm{b}_n(h)$ and are given by

$$\mathrm{se}_n(v, h) = \sum_r{}' \mathrm{B}_r^n(h) \sin rv \quad (n = 1, 2, \ldots). \tag{1.4}$$

They are normalized so that

$$\int_0^{2\pi} \mathrm{se}_n^2(v, h)\, \mathrm{d}v = \pi. \tag{1.5}$$

The functions ce_n and se_n are known as *Mathieu functions of the first kind of order n.*

In the same way that (B.3.19) was derived may be obtained

$$\int_0^{2\pi} \mathrm{ce}_m(v, h)\mathrm{ce}_n(v, h)\,\mathrm{d}v = \pi\delta_{mn}, \tag{1.6}$$

$$\int_0^{2\pi} \mathrm{se}_m(v, h)\mathrm{se}_n(v, h)\,\mathrm{d}v = \pi\delta_{mn}, \tag{1.7}$$

$$\int_0^{2\pi} \mathrm{ce}_m(v, h)\mathrm{se}_n(v, h)\,\mathrm{d}v = 0. \tag{1.8}$$

For a general solution of Mathieu's equation a solution independent of the function of the first kind is needed. This leads to the *Mathieu functions of the second kind* defined by

$$\mathrm{fe}_n(v, h) = \mathrm{C}_n(h)\{v\ \mathrm{ce}_n(v, h) + \sum_r{}' \mathrm{f}_r^n(h)\sin rv\}, \tag{1.9}$$

$$\mathrm{ge}_n(v, h) = \mathrm{S}_n(h)\{v\ \mathrm{se}_n(v, h) + \sum_r{}' \mathrm{g}_r^n(h)\cos rv\}. \tag{1.10}$$

The coefficients f_r^n and g_r^n are related to A_r^n and B_r^n by recurrence relations whereas C_n and S_n are normalizing constants. Note that the functions of the second kind are not periodic.

C.2 Modified Mathieu functions

The change of variable $v = \mathrm{i}u$ transforms (1.1) to the *modified Mathieu equation*

$$\mathrm{d}^2V/\mathrm{d}u^2 - (\lambda - 2h\cosh 2u)V = 0. \tag{2.1}$$

Clearly this has solutions

$$\mathrm{Ce}_n(u, h) = \mathrm{ce}_n(\mathrm{i}u, h), \qquad \mathrm{Se}_n(u, h) = -\mathrm{i}\,\mathrm{se}_n(\mathrm{i}u, h), \tag{2.2}$$

$$\mathrm{Fe}_n(u, h) = -\mathrm{i}\,\mathrm{fe}_n(\mathrm{i}u, h), \qquad \mathrm{Ge}_n(u, h) = \mathrm{ge}_n(\mathrm{i}u, h). \tag{2.3}$$

Ce_n and Se_n are *modified Mathieu functions of the first kind*; Fe_n and Ge_n are those of the second kind.

For some purposes different forms of solution are desirable. It may be checked that solutions of (2.1) are provided by

$$\mathrm{Mc}_n^{(j)}(u, h) = \{1/\mathrm{ce}_n(0, h)\} \sum_r{}' (-)^{\frac{1}{2}n - \frac{1}{2}r} \mathrm{A}_r^n(h)\mathscr{Z}_r^{(j)}(2h^{\frac{1}{2}}\cosh u) \qquad (n = 0, 1, \ldots), \tag{2.4}$$

$$\mathrm{Ms}_n^{(j)}(u, h) = \{1/\mathrm{se}_n'(0, h)\}\tanh u \sum_r{}' (-)^{\frac{1}{2}n - \frac{1}{2}r} r\mathrm{B}_r^n(h)\mathscr{Z}_r^{(j)}(2h^{\frac{1}{2}}\cosh u) \qquad (n = 1, 2, \ldots) \tag{2.5}$$

where $j=1$, 2, 3, 4 and $\mathscr{Z}_r^{(j)}$ is the notation for Bessel functions of Appendix A.1. When $j=1$ these solutions are multiples of Ce_n and Se_n; in fact

$$\mathrm{Mc}_n^{(1)}(u,h)=\mathrm{Ce}_n(u,h)/\mathrm{p}_n(h), \qquad \mathrm{Ms}_n^{(1)}(u,h)=\mathrm{Se}_n(u,h)/\mathrm{s}_n(h) \quad (2.6)$$

where p_n, s_n are defined in (3.3)–(3.6) below.

The other values of j supply various ways of expressing a second solution of the modified Mathieu equation. The series converge for $|\cosh u|>1$ and $\mathrm{Re}\,u>0$; for other values of u alternative series are required. (For further details see Meixner and Schäfke (1954).) It is obvious from (A.1.11) that

$$\mathrm{Mc}_n^{(3)}=\mathrm{Mc}_n^{(1)}+\mathrm{i}\,\mathrm{Mc}_n^{(2)}, \qquad \mathrm{Mc}_n^{(4)}=\mathrm{Mc}_n^{(1)}-\mathrm{i}\,\mathrm{Mc}_n^{(2)} \quad (2.7)$$

and there are similar relations for Ms_n. The Wronskian relation

$$\mathrm{Mc}_n^{(1)}(u,h)\mathrm{Mc}_n^{(2)\prime}(u,h)-\mathrm{Mc}_n^{(2)}(u,h)\mathrm{Mc}_n^{(1)\prime}(u,h)=2/\pi \quad (2.8)$$

should also be noted; it is valid if Mc_n is replaced by Ms_n throughout.

C.3 Asymptotic behaviour

When $\mathrm{Re}\,u\to\infty$ with $-\pi<\frac{1}{2}\mathrm{ph}\,h+\mathrm{Im}\,u<\pi$

$$\mathrm{Ce}_n(u,h)\sim(2/\pi h^{\frac{1}{2}})^{\frac{1}{2}}\mathrm{p}_n(h)\mathrm{e}^{-\frac{1}{2}u}\cos(h^{\frac{1}{2}}\mathrm{e}^u-\tfrac{1}{4}\pi-\tfrac{1}{2}n\pi), \quad (3.1)$$

$$\mathrm{Se}_n(u,h)\sim(2/\pi h^{\frac{1}{2}})^{\frac{1}{2}}\mathrm{S}_n(h)\mathrm{e}^{-\frac{1}{2}u}\cos(h^{\frac{1}{2}}\mathrm{e}^u-\tfrac{1}{4}\pi-\tfrac{1}{2}n\pi) \quad (3.2)$$

where

$$\mathrm{p}_{2n}(h)=(-)^n\mathrm{ce}_{2n}(0,h)\mathrm{ce}_{2n}(\tfrac{1}{2}\pi,h)/\mathrm{A}_0^{2n}(h), \quad (3.3)$$

$$\mathrm{p}_{2n+1}(h)=(-)^{n+1}\mathrm{ce}_{2n+1}(0,h)\mathrm{ce}'_{2n+1}(\tfrac{1}{2}\pi,h)/h^{\frac{1}{2}}\mathrm{A}_1^{2n+1}(h), \quad (3.4)$$

$$\mathrm{s}_{2n}(h)=(-)^n\mathrm{se}'_{2n}(0,h)\mathrm{se}'_{2n}(\tfrac{1}{2}\pi,h)/h\mathrm{B}_2^{2n}(h), \quad (3.5)$$

$$\mathrm{s}_{2n+1}(h)=(-)^n\mathrm{se}'_{2n+1}(0,h)\mathrm{se}_{2n+1}(\tfrac{1}{2}\pi,h)/h^{\frac{1}{2}}\mathrm{B}_1^{2n+1}(h). \quad (3.6)$$

Also

$$\mathrm{Mc}_n^{(3)}(u,h)\sim(2/\pi h^{\frac{1}{2}})^{\frac{1}{2}}\exp\{-\tfrac{1}{2}u+\mathrm{i}(h^{\frac{1}{2}}\mathrm{e}^u-\tfrac{1}{4}\pi-\tfrac{1}{2}n\pi)\}, \quad (3.7)$$

$$\mathrm{Mc}_n^{(4)}(u,h)\sim(2/\pi h^{\frac{1}{2}})^{\frac{1}{2}}\exp\{-\tfrac{1}{2}u-\mathrm{i}(h^{\frac{1}{2}}\mathrm{e}^u-\tfrac{1}{4}\pi-\tfrac{1}{2}n\pi)\}, \quad (3.8)$$

$$\mathrm{Ms}_n^{(3)}(u,h)\sim\mathrm{Mc}_n^{(3)}(u,h), \qquad \mathrm{Ms}_n^{(4)}(u,h)\sim\mathrm{Mc}_n^{(4)}(u,h). \quad (3.9)$$

C.4 Some expansions

For small values of h

$$\mathrm{A}_r^n(h)=O(h^{\frac{1}{2}|r-n|}), \qquad \mathrm{B}_r^n(h)=O(h^{\frac{1}{2}|r-n|}). \quad (4.1)$$

Consequently, as $h\to0$ with u not too large,

$$\mathrm{ce}_0(v,h)=2^{-\frac{1}{2}}+O(h), \qquad \mathrm{se}_n(v,h)=\sin nv+O(h), \quad (4.2)$$

$$\mathrm{ce}_n(v,h)=\cos nv+O(h) \quad (n\neq0). \quad (4.3)$$

Furthermore

$$\mathrm{Mc}_0^{(1)}(u, h) = \{2^{\frac{1}{2}} + O(h)\}\mathrm{A}_0^0(h) \tag{4.4}$$

$$\mathrm{Mc}_n^{(1)}(u, h) = \{\cosh nu + O(h)\}\mathrm{A}_0^n(h) \quad (n \text{ even but nonzero}), \tag{4.5}$$

$$\mathrm{Mc}_n^{(1)}(u, h) = \{(1/n)\cosh nu + O(h)\}h^{\frac{1}{2}}\mathrm{A}_1^n(h) \quad (n \text{ odd}), \tag{4.6}$$

$$\mathrm{Mc}_n^{(2)}(u, h) = O(h^{-\frac{1}{2}n}) \quad (n \neq 0), \tag{4.7}$$

$$\mathrm{Ms}_n^{(1)}(u, h) = \{(1/n^2)\sinh nu + O(h)\}h\mathrm{B}_2^n(h) \quad (n \text{ even}), \tag{4.8}$$

$$\mathrm{Ms}_n^{(1)}(u, h) = \{(1/n)\sinh nu + O(h)\}h^{\frac{1}{2}}\mathrm{B}_1^n(h) \quad (n \text{ odd}), \tag{4.9}$$

$$\mathrm{Ms}_n^{(2)}(u, h) = O(h^{-\frac{1}{2}n}). \tag{4.10}$$

More extensive expansions for one or two values of n are

$$\mathrm{ce}_0(v, h) = 2^{-\frac{1}{2}}(1 - \tfrac{1}{2}h\cos 2v - \tfrac{1}{16}h^2 + \tfrac{1}{32}h^2\cos 4v) + O(h^3), \tag{4.11}$$

$$\mathrm{Mc}_0^{(1)}(u, h) = 1 - \tfrac{1}{2}h\cosh 2u + (6 + \cosh 4u)h^2/32 + O(h^3), \tag{4.12}$$

$$\begin{aligned}\mathrm{Mc}_0^{(2)}(u, h) = {} & (2/\pi)\{\gamma + \ln(\tfrac{1}{2}h^{\frac{1}{2}}e^u)\}\mathrm{Mc}_0^{(1)}(u, h) \\ & + (1/\pi)\{h\sinh 2u - h^2(\tfrac{1}{4} + \tfrac{3}{32}\sinh 4u)\} + O(h^3),\end{aligned} \tag{4.13}$$

$$\mathrm{Mc}_1^{(1)}(u, h) = h^{\frac{1}{2}}\cosh u + O(h^{\frac{3}{2}}), \tag{4.14}$$

$$\mathrm{Ms}_1^{(1)}(u, h) = h^{\frac{1}{2}}\sinh u + O(h^{\frac{3}{2}}), \tag{4.15}$$

$$\mathrm{Mc}_1^{(2)}(u, h) \approx \mathrm{Ms}_1^{(2)}(u, h) \approx -(2/\pi h^{\frac{1}{2}})e^{-u}. \tag{4.16}$$

Two important expansions that merit recording are

$$\begin{aligned}&\exp\{-2ih^{\frac{1}{2}}(\cosh u\cos v\cos\alpha + \sinh u\sin v\sin\alpha)\} \\ &\quad = 2\sum_{m=0}^{\infty}(-i)^m\mathrm{ce}_m(\alpha, h)\mathrm{ce}_m(v, h)\mathrm{Mc}_m^{(1)}(u, h) \\ &\qquad + 2\sum_{m=1}^{\infty}(-i)^m\mathrm{se}_m(\alpha, h)\mathrm{se}_m(v, h)\mathrm{Ms}_m^{(1)}(u, h)\end{aligned} \tag{4.17}$$

and, if $x = l\cosh u\cos v$, $y = l\sinh u\sin v$,

$$\begin{aligned}&\mathrm{H}_0^{(2)}[k\{(x - x_1)^2 + (y - y_1)^2\}^{\frac{1}{2}}] \\ &\quad = 2\sum_{m=0}^{\infty}\mathrm{ce}_m(v, h)\mathrm{ce}_m(v_1, h)\mathrm{Mc}_m^{(1)}(u, h)\mathrm{Mc}_m^{(4)}(u_1, h) \\ &\qquad + 2\sum_{m=1}^{\infty}\mathrm{se}_m(v, h)\mathrm{se}_m(v_1, h)\mathrm{Ms}_m^{(1)}(u, h)\mathrm{Ms}_m^{(4)}(u_1, h)\end{aligned} \tag{4.18}$$

where $h = \frac{1}{4}k^2l^2$ and $u_1 > u$. If $u_1 < u$ interchange u and u_1 on the right-hand side of (4.18).

Appendix D. Parabolic cylinder functions

D.1 Weber's equation

Weber's equation is

$$y''+(\nu+\tfrac{1}{2}-\tfrac{1}{4}z^2)y=0. \tag{1.1}$$

It can be transformed to a number of different forms. For instance $z=Xe^{-\frac{1}{4}\pi i}$ makes it

$$d^2y/dX^2+\{\tfrac{1}{4}X^2-i(\nu+\tfrac{1}{2})\}y=0 \tag{1.2}$$

and then, if $i(\nu+\frac{1}{2})=(\mu^2+k^2)^{-\frac{1}{2}}h/2$, the substitution $X=\{4(\mu^2+k^2)\}^{\frac{1}{4}}\xi$ produces

$$d^2y/d\xi^2+\{(\mu^2+k^2)\xi^2-h\}y=0. \tag{1.3}$$

Solutions of (1.1) are known as *parabolic cylinder functions* or *Weber–Hermite functions.* One is

$$D_\nu(z)=2^{\frac{1}{2}\nu}e^{-\frac{1}{4}z^2}\left\{\frac{(-\frac{3}{2})!z}{(-1-\frac{1}{2}\nu)!2^{\frac{1}{2}}}\,{}_1F_1(\tfrac{1}{2}-\tfrac{1}{2}\nu;\tfrac{3}{2};\tfrac{1}{2}z^2)\right.$$
$$\left.+\frac{(-\frac{1}{2})!}{(-\frac{1}{2}-\frac{1}{2}\nu)!}\,{}_1F_1(-\tfrac{1}{2}\nu;\tfrac{1}{2};\tfrac{1}{2}z^2)\right\} \tag{1.4}$$

where

$${}_1F_1(a;c;z)=1+\frac{a}{c}\frac{z}{1!}+\frac{a(a+1)}{c(c+1)}\frac{z^2}{2!}+\dots. \tag{1.5}$$

Changing z to $-z$ leaves Weber's equation unaltered. So $D_\nu(-z)$ is also a solution. Moreover, the simultaneous replacement of ν and z by $-\nu-1$ and $\pm iz$ has no effect on (1.1). Therefore $D_{-\nu-1}(iz)$ and $D_{-\nu-1}(-iz)$ are also solutions. They are connected by

$$D_\nu(z)=\nu!\{e^{\frac{1}{2}\nu\pi i}D_{-\nu-1}(iz)+e^{-\frac{1}{2}\nu\pi i}D_{-\nu-1}(-iz)\}/(2\pi)^{\frac{1}{2}} \tag{1.6}$$
$$=e^{-\nu\pi i}D_\nu(-z)+(2\pi)^{\frac{1}{2}}e^{-\frac{1}{2}(\nu+1)\pi i}D_{-\nu-1}(iz)/(-\nu-1)! \tag{1.7}$$
$$=e^{\nu\pi i}D_\nu(-z)+(2\pi)^{\frac{1}{2}}e^{\frac{1}{2}(\nu+1)\pi i}D_{-\nu-1}(-iz)/(-\nu-1)!. \tag{1.8}$$

When ν is the non-negative integer n

$$D_n(z)=(-)^n e^{\frac{1}{4}z^2}\frac{d^n}{dz^n}(e^{-\frac{1}{2}z^2})=2^{-\frac{1}{2}n}e^{-\frac{1}{4}z^2}H_n(2^{-\frac{1}{2}}z) \tag{1.9}$$

where $H_n(z) = (-)^n e^{z^2} d^n (e^{-z^2})/dz^n$ is the *Hermite polynomial.* Where ν is the negative integer $-m-1$

$$D_{-m-1}(z) = \frac{(-)^m}{m!}(\tfrac{1}{2}\pi e^{-\frac{1}{2}z^2})^{\frac{1}{2}} \frac{d^m}{dz^m}\{e^{\frac{1}{2}z^2}\,\mathrm{erfc}(2^{-\frac{1}{2}}z)\} \tag{1.10}$$

where $\mathrm{erfc}\ z = (2/\pi^{\frac{1}{2}})\int_z^\infty e^{-t^2}\,dt$.

The Wronskian relations connecting the solutions are

$$D'_\nu(z)D_\nu(-z) + D_\nu(z)D'_\nu(-z) = -(2\pi)^{\frac{1}{2}}/(-1-\nu)! = \nu!(2/\pi)^{\frac{1}{2}} \sin \nu\pi, \tag{1.11}$$

$$iD_\nu(z)D'_{-\nu-1}(iz) - D_{-\nu-1}(iz)D'_\nu(z) = -ie^{-\frac{1}{2}\nu\pi i}. \tag{1.12}$$

For $|z| \to \infty$ with ν fixed

$$D_\nu(z) \sim z^\nu e^{-\frac{1}{4}z^2} \tag{1.13}$$

when $|\mathrm{ph}\ z| < \frac{3}{4}\pi$. The asymptotic behaviour for $-\frac{5}{4}\pi < \mathrm{ph}\ z < -\frac{1}{4}\pi$ can be derived from (1.13) and (1.7) whereas that for $\frac{1}{4}\pi < \mathrm{ph}\ z < \frac{5}{4}\pi$ stems from (1.13) and (1.8).

A uniformly valid asymptotic expression as $|\nu| \to \infty$ is

$$D_{\nu-\frac{1}{2}}(x) \sim (2\pi)^{\frac{1}{2}} \exp\{\tfrac{1}{2}(\nu - \tfrac{1}{6})\ln \nu - \tfrac{1}{2}\nu\}\left(\frac{\xi}{1 - x^2/4\nu}\right)^{\frac{1}{4}} \mathrm{Ai}(\nu^{\frac{2}{3}}\xi e^{\pi i}) \tag{1.14}$$

where

$$\tfrac{2}{3}\xi^{\frac{3}{2}} = \cos^{-1}(x/2\nu^{\frac{1}{2}}) - (x/2\nu^{\frac{1}{2}})(1 - x^2/4\nu)^{\frac{1}{2}} \tag{1.15}$$

and Ai is the Airy function of Appendix A.

Appendix E. Spheroidal functions

E.1 Basic properties

THE *spheroidal functions* are solutions of the differential equation

$$\frac{d}{dz}\left\{(1-z^2)\frac{dy}{dz}\right\}+\left\{\lambda+\hbar^2(1-z^2)-\frac{\mu^2}{1-z^2}\right\}y=0. \qquad (1.1)$$

This equation includes several others as limiting cases. If $\hbar=0$ and $\lambda=\nu(\nu+1)$, (1.1) becomes Legendre's associated equation of Appendix B. If $\mu=\frac{1}{2}$ and $z=\cos v$, $(\sin v)^{-\frac{1}{2}}y$ satisfies Mathieu's equation of Appendix C. Putting $\lambda=\nu(\nu+1)$ and letting $\hbar\to 0$, after the change of variable $z=\xi/\hbar$, leads to the equation for spherical Bessel functions (Appendix A) in terms of ξ.

The differential equation (1.1) has regular singularities at $z=\pm 1$, with exponents $\pm\frac{1}{2}\mu$, and an irregular singularity at infinity. The behaviour at infinity is such that there are solutions which are z^{ν} or $z^{-\nu-1}$ times a single-valued function, the exponent ν being a certain function of λ, $\hbar$ and μ. However, it turns out to be more convenient to regard λ as a function of ν, $\hbar$, μ and write $\lambda=\lambda_{\nu}^{\mu}(\hbar^2)$.

It can be verified that, if the coefficients $a_{\nu,2r}^{\mu}(\hbar^2)$ satisfy the appropriate recurrence relations, a solution of (1.1) in terms of spherical Bessel functions is

$$S_{\nu}^{\mu(j)}(z,\hbar)=\{z^{\mu}(z^2-1)^{-\frac{1}{2}\mu}/A_{\nu}^{\mu}(\hbar^2)\}\sum_{r=-\infty}^{\infty}a_{\nu,2r}^{\mu}(\hbar^2)\mathfrak{z}_{\nu+2r}^{(j)}(\hbar z). \qquad (1.2)$$

The notation for the spherical Bessel functions is that of Appendix A.4. The coefficient A_{ν}^{μ} is defined by

$$A_{\nu}^{\mu}(\hbar^2)=\sum_{r=-\infty}^{\infty}(-)^r a_{\nu,2r}^{\mu}(\hbar^2). \qquad (1.3)$$

The spheroidal functions, as specified by (1.2), are connected by many relations because of the interlocking of Bessel functions. In particular

$$S_{\nu}^{\mu(3)}=S_{\nu}^{\mu(1)}+iS_{\nu}^{\mu(2)}, \qquad S_{\nu}^{\mu(4)}=S_{\nu}^{\mu(1)}-iS_{\nu}^{\mu(2)}, \qquad (1.4)$$

$$S_{\nu}^{-\mu(j)}=S_{\nu}^{\mu(j)}, \qquad S_{\nu}^{\mu(2)}\cos\nu\pi=-S_{-\nu-1}^{\mu(1)}-S_{\nu}^{\mu(1)}\sin\nu\pi. \qquad (1.5)$$

The solutions of (1.2) are most helpful when z is not small. For lower values of z it is convenient to have available expansions in Legendre functions, namely

$$\mathrm{Ps}_\nu^\mu(z, \hbar^2) = \sum_{r=-\infty}^{\infty} (-)^r \mathrm{a}_{\nu,2r}^\mu(\hbar^2)\mathrm{P}_{\nu+2r}^\mu(z), \tag{1.6}$$

$$\mathrm{Qs}_\nu^\mu(z, \hbar^2) = \sum_{r=-\infty}^{\infty} (-)^r \mathrm{a}_{\nu,2r}^\mu(\hbar^2)\mathrm{Q}_{\nu+2r}^\mu(z) \tag{1.7}$$

where the coefficients are the same as in (1.2). It is permissible to replace z by x throughout (1.6), (1.7) (in which case Ps and Qs are written as ps and qs) under the same convention on z and x as for Legendre functions in Appendix B. There are numerous interconnections between these spheroidal functions because of those between Legendre functions. For example,

$$(n+m)!\mathrm{ps}_n^{-m}(x, \hbar^2) = (n-m)!(-)^m \mathrm{ps}_n^m(x, \hbar^2) \tag{1.8}$$

and there are equations analogous to (B.1.8) and (B.1.19)–(B.1.23).

The definition of $\mathrm{S}_\nu^{\mu(j)}$ fails when $\nu+\frac{1}{2}$ is an integer and that for Qs_ν^μ when $\mu+\nu$ is an integer. This difficulty can be overcome (see Meixner and Schäfke (1954) for information). The failure for Qs_ν^μ is more apparent than real because it is due to $\mathrm{Q}_{\nu+2r}^\mu$ being infinite when $\nu+\mu+2r$ is a negative integer. However, $\mathrm{a}_{\nu,2r}^\mu$ vanishes at the same time and the product tends to a limit which can be calculated from (B.1.21). Thus, if $n+m$ is even,

$$\mathrm{Qs}_n^m(z) = \sum_{r=-\ (n+m)} (-)^r \mathrm{a}_{n,2r}^m \mathrm{Q}_{n+2r}^m(z) + \sum_{r=\frac{1}{2}(n+m)+1} (-)^r b_{n,2r}^m \mathrm{P}_{2r-n-1}^m(z) \tag{1.9}$$

where

$$b_{n,2r}^m = \lim_{\varepsilon\to 0} a_{n-\varepsilon,2r}^m/\varepsilon.$$

If $n+m$ is odd, the two series in (1.9) start at $r=-\frac{1}{2}(n+m-1)$ and $r=\frac{1}{2}(n+m+1)$ respectively.

The connection between $\mathrm{S}_\nu^{\mu(1)}$ and Qs is expressed by

$$\mathrm{S}_\nu^{\mu(1)}(z, \hbar) = \{\mathrm{K}_\nu^\mu(\hbar)/\pi\}\mathrm{e}^{-(\nu+\mu+1)\pi\mathrm{i}}\mathrm{Qs}_{-\nu-1}^\mu(z, \hbar^2)\sin(\nu-\mu)\pi. \tag{1.10}$$

By equating powers of z on both sides we obtain

$$\begin{aligned} &\mathrm{A}_\nu^{-\mu}(\hbar^2)\mathrm{K}_\nu^\mu(\hbar) \sum_{r=-\infty}^{\infty} (-)^r \mathrm{a}_{\nu,2r}^\mu/(r-s)!(-\nu-r-s-\tfrac{1}{2})! \\ &\quad = (\nu+2s-\mu)!\tfrac{1}{2}\mathrm{e}^{\mathrm{i}(\nu+s)\pi}(\tfrac{1}{4}\hbar)^{\nu+2s} \sum_{r=-\infty}^{\infty} (-)^r a_{\nu,2r}^{-\mu}/(s-r)!(\nu+s+r+\tfrac{1}{2})! \end{aligned} \tag{1.11}$$

for any integer s for which the sums do not vanish. In particular, when

$\nu=n$ and $\mu=m$ with $n-m$ an even integer, put $s=-\frac{1}{2}(n-m)$. Since $a^m_{n,2r}$ is zero for positive $r\geqslant\frac{1}{2}(m-n)$ when $n-m$ is a negative integer we see from (B.1.9) that, if $n\geqslant|m|\geqslant 0$,

$$\mathrm{K}^m_n(\hbar)=\tfrac{1}{2}\pi^{\frac{1}{2}}\mathrm{e}^{\frac{1}{2}m\pi\mathrm{i}}(\tfrac{1}{2}\hbar)^m\frac{a^{-m}_{n,m-n}(\hbar^2)}{(m+\frac{1}{2})!\mathrm{A}^{-m}_n(\hbar^2)\mathrm{Ps}^m_n(+\mathrm{i}0,\hbar^2)}. \tag{1.12}$$

Similarly, when $n-m$ is an odd integer and $n\geqslant|m|\geqslant 0$,

$$\mathrm{K}^m_n(\hbar)=\tfrac{1}{2}\pi^{\frac{1}{2}}\mathrm{e}^{\frac{1}{2}m\pi\mathrm{i}}(\tfrac{1}{2}\hbar)^{m+1}\frac{a^{-m}_{n,m-n+1}(\hbar^2)}{(m+\frac{3}{2})!\mathrm{A}^{-m}_n(\hbar^2)\mathrm{Ps}^{m\prime}_n(+\mathrm{i}0,\hbar^2)}. \tag{1.13}$$

For $m>n\geqslant 0$ replace m by $-m$ throughout both sides of (1.12) and (1.13). The invocation of (1.5) and (B.1.20) enables the assertion

$$\mathrm{K}^\mu_\nu=(\nu-\mu)!\mathrm{K}^{-\mu}_\nu/(\nu+\mu)!. \tag{1.14}$$

Another method for finding K^m_n is to put $z=0$ in (1.10) or its derivative.

These results are especially valuable because, from (B.1.19), (B.1.21), and (1.10),

$$\mathrm{S}^{m(1)}_n(z,\hbar)=\mathrm{K}^m_n(\hbar)\mathrm{Ps}^m_n(z,\hbar^2)\quad(n\geqslant|m|\geqslant 0), \tag{1.15}$$

$$\mathrm{S}^{m(1)}_n(z,\hbar)=\mathrm{K}^{-m}_n(\hbar)\mathrm{Ps}^{-m}_n(z,\hbar^2)\quad(m>n\geqslant 0) \tag{1.16}$$

and, from (1.9),

$$\mathrm{S}^{m(2)}_n(z,\hbar)=(-)^{m+1}\mathrm{Qs}^m_n(z,\hbar^2)/\hbar\mathrm{K}^{-m}_n(\hbar)\mathrm{A}^m_n(\hbar^2)\mathrm{A}^{-m}_n(\hbar^2)\quad(n\geqslant 0,\ m\geqslant 0). \tag{1.17}$$

A useful Wronskian is

$$\mathrm{S}^{m(1)}_n(z,\hbar)\mathrm{S}^{m(4)\prime}_n(z,\hbar)-\mathrm{S}^{m(1)\prime}_n(z,\hbar)\mathrm{S}^{m(4)}_n(z,\hbar)=-\mathrm{i}/\hbar(z^2-1). \tag{1.18}$$

Asymptotic behaviour for large z, other parameters being fixed, can be deduced from

$$\mathrm{S}^{\mu(j)}_\nu(z,\hbar^2)\sim\mathfrak{z}^{(j)}_\nu(\hbar z) \tag{1.19}$$

and the foregoing interrelations.

E.2 Various results

Let $x=l(X^2-1)^{\frac{1}{2}}(1-Y^2)^{\frac{1}{2}}\cos\phi$, $y=x\tan\phi$, $z=lXY$ and define $\boldsymbol{x}_1$ similarly. Then

$$\frac{\mathrm{e}^{-ik|\boldsymbol{x}-\boldsymbol{x}_1|}}{-ik\,|\boldsymbol{x}-\boldsymbol{x}_1|}=\sum_{n=0}^{\infty}(2n+1)\sum_{m=-n}^{n}(-)^m$$
$$\times\,\mathrm{S}^{m(4)}_n(X,\hbar)\mathrm{S}^{m(1)}_n(X_1,\hbar)\mathrm{ps}^m_n(Y,\hbar^2)\mathrm{ps}^{-m}_n(Y_1,\hbar^2)\mathrm{e}^{\mathrm{i}m(\phi-\phi_1)} \tag{2.1}$$

where $\hbar=kl$ and $X>X_1$. For $X<X_1$ interchange X and X_1 on the right-hand side of (2.1). The formula for $\mathrm{e}^{ik|\boldsymbol{x}-\boldsymbol{x}_1|}/ik\,|\boldsymbol{x}-\boldsymbol{x}_1|$ is obtained by

replacing $S_n^{(m)(4)}$ with $S_n^{m(3)}$. In oblate spheroidal coordinates in which $x = L(1+\xi^2)^{\frac{1}{2}}(1-\eta^2)^{\frac{1}{2}}\cos\phi$, $y = x\tan\phi$, $z = L\xi\eta$ replace X by $-i\xi$, Y by η, and $\hbar$ by ikL on the right-hand side of (2.1).

Another representation is

$$\exp[i\hbar\{(X^2-1)^{\frac{1}{2}}(1-Y^2)^{\frac{1}{2}}\sin\theta\cos\phi + XY\cos\theta\}]$$
$$= \sum_{n=0}^{\infty}\sum_{m=-n}^{n}(2n+1)i^{n+2m}S_n^{m(1)}(X,\hbar)ps_n^m(Y,\hbar^2)ps_n^{-m}(\cos\theta,\hbar^2)e^{im\phi}. \quad (2.2)$$

The functions ps_n^m form an orthogonal set and

$$\int_{-1}^{1}\{ps_n^m(x,\hbar^2)\}^2\,dx = (n+m)!2/(n-m)!(2n+1). \quad (2.3)$$

One further result of interest is

$$A_n^{-m}(\hbar^2)S_n^{m(1)}(X,\hbar)$$
$$= \frac{(n-m)!}{(n+m)!}\tfrac{1}{2}e^{-\frac{1}{2}\pi i(m+n)}\hbar^m\int_{-1}^{1}e^{i\hbar Xx}(X^2-1)^{\frac{1}{2}m}(1-x^2)^{\frac{1}{2}m}ps_n^m(x,\hbar^2)\,dx \quad (2.4)$$

For $\hbar \approx 0$ and other quantities not too large

$$a_{\nu,2r}^{\mu}(\hbar^2) \approx \frac{(\nu+2r-\mu)!(\nu-\frac{1}{2})!(\nu+\frac{1}{2})!2^{-4r}}{(\nu-\mu)!(\nu+2r-\frac{1}{2})!(\nu+r+\frac{1}{2})!r!}\hbar^{2r} \quad (r \geqslant 0), \quad (2.5)$$

$$a_{\nu,2r}^{\mu}(\hbar^2) \approx \frac{(\nu+\mu)!(\nu+2r+\frac{1}{2})!(\nu+r-\frac{1}{2})!2^{4r}}{(\nu+2r+\mu)!(\nu+\frac{1}{2})!(-r)!(\nu-\frac{1}{2})!}(-)^r(\hbar)^{-2r} \quad (r \leqslant 0). \quad (2.6)$$

Thus

$$S_n^{m(1)}(z,\hbar) = O(\hbar^n), \qquad S_n^{m(2)}(z,\hbar) = O(\hbar^{-n-1}) \quad (2.7)$$

as $\hbar \to 0$. In fact

$$S_n^{m(1)}(z,\hbar) \approx K_n^m(\hbar)P_n^m(z) \quad (n \geqslant |m| \geqslant 0), \quad (2.8)$$

$$S_n^{m(2)}(z,\hbar) \approx (-)^{m+1}Q_n^m(z)/\hbar K_n^{-m}(\hbar) \quad (n \geqslant 0,\ m \geqslant 0) \quad (2.9)$$

where

$$K_n^m(\hbar) \approx (n-m)!(-)^{n+m+1}\tfrac{1}{2}\pi(\tfrac{1}{4}\hbar)^n/(n+\tfrac{1}{2})!(n-\tfrac{1}{2})! \quad (n \geqslant |m| \geqslant 0), \quad (2.10)$$

$$K_n^{-m}(\hbar) \approx (n+m)!(-)^{n+1-m}\tfrac{1}{2}\pi(\tfrac{1}{4}\hbar)^n/(n+\tfrac{1}{2})!(n-\tfrac{1}{2})! \quad (m > n \geqslant 0). \quad (2.11)$$

Some expansions with additional terms as $\hbar \to 0$ will now be listed:

$$a_{n,0}^m(\hbar^2) \approx 1 - \frac{\{(n+2)^2-m^2\}\{(n+1)^2-m^2\}}{8(2n+1)(2n+3)^4(2n+5)}\hbar^4 - \frac{(n^2-m^2)\{(n-1)^2-m^2\}}{8(2n-3)(2n-1)^4(2n+1)}\hbar^4, \quad (2.12)$$

$$a^m_{n,2}(\hbar^2) \approx \frac{(n-m+2)(n-m+1)}{2(2n+1)(2n+3)^2}\hbar^2\left\{1+\frac{2(1-4m^2)\hbar^2}{(2n-1)(2n+3)^2(2n+7)}\right\}, \tag{2.13}$$

$$a^m_{n,-2}(\hbar^2) \approx -\frac{(n+m)(n+m-1)}{2(2n+1)(2n-1)^2}\hbar^2\left\{1+\frac{2(1-4m^2)\hbar^2}{(2n-5)(2n-1)^2(2n+3)}\right\}, \tag{2.14}$$

$$\mathrm{ps}^0_0(x,\hbar^2) \approx 1-\frac{\hbar^2}{18}(3x^2-1)+\frac{\hbar^4}{60}\left(\frac{x^4}{20}-\frac{x^2}{9}-\frac{37}{270}\right), \tag{2.15}$$

$$\mathrm{ps}^1_1(x,\hbar^2) \approx -(1-x^2)^{\frac{1}{2}}\{1-\hbar^2(5x^2-1)/50\}, \tag{2.16}$$

$$\mathrm{ps}^0_1(x,\hbar^2) \approx x-\hbar^2(5x^3-3x)/50, \tag{2.17}$$

$$\mathrm{S}^{m(1)}_n(-\mathrm{i}0,\hbar)=0 \quad \text{if } n-m \text{ is odd}, \tag{2.18}$$

$$\mathrm{S}^{1(1)}_1(-\mathrm{i}0,\hbar) \approx -\tfrac{1}{3}\mathrm{i}\hbar(1+2\hbar^2/25),\ \mathrm{S}^{1(1)}_3(-\mathrm{i}0,\hbar) \approx \mathrm{i}\hbar^3/525, \tag{2.19}$$

$$\mathrm{S}^{m(1)\prime}_n(-\mathrm{i}0,\hbar)=0 \quad \text{if } n-m \text{ is even}, \tag{2.20}$$

$$\mathrm{S}^{0(1)\prime}_1(-\mathrm{i}0,\hbar) \approx \tfrac{1}{3}\hbar(1+\hbar^2/25), \qquad \mathrm{S}^{0(1)\prime}_3(-\mathrm{i}0,\hbar) \approx -\hbar^3/175, \tag{2.21}$$

$$\mathrm{S}^{0(2)}_1(-\mathrm{i}0,\hbar) \approx 3(1-\hbar^2/25)/\hbar^2, \qquad \mathrm{S}^{1(2)}_3(-\mathrm{i}0,\hbar) \approx -1575\pi/32\hbar^4, \tag{2.22}$$

$$S^{1(2)}_1(-\mathrm{i}0,\hbar) \approx 3\pi(1-7\hbar^2/25)/4\hbar^2, \tag{2.23}$$

$$\mathrm{S}^{0(2)\prime}_1(-\mathrm{i}0,\hbar) \approx -3\pi\mathrm{i}(1-4\hbar^2/25)/2\hbar^2, \qquad \mathrm{S}^{0(2)\prime}_3(-\mathrm{i}0,\hbar) \approx 1575\pi\mathrm{i}/8\hbar^4, \tag{2.24}$$

$$S^{1(2)\prime}_1(-\mathrm{i}0,\hbar) \approx -3\mathrm{i}(1-2\hbar^2/25)/\hbar^2. \tag{2.25}$$

Appendix F. Tensor calculus

F.1 Coordinate transformations

THE development of a tensor calculus is facilitated by working in a space of n dimensions, i.e. every point of the space can be identified by giving appropriate values to the n coordinates $x^1, x^2, \ldots, x^n$. The upper position must not be regarded as signifying a power; it is used to indicate the number of the coordinate but does not occupy the lower position for a reason that will become clear shortly. The coordinates are not limited to any particular type; for instance, when $n=3$, they might be Cartesian or spherical polar or elliptic cylinder. The space is *abstract* in the sense that no special relationship is assumed with the physical world. A new set of coordinates $(x'^1, x'^2, \ldots, x'^n)$ can be obtained by giving $x'^1, \ldots, x'^n$ as functions of $x^1, \ldots, x^n$. A small change $\mathrm{d}x'^1, \ldots, \mathrm{d}x'^n$ in the primed coordinates is connected to a small change in the unprimed coordinates by

$$\mathrm{d}x'^1 = \sum_{\beta=1}^{n} \frac{\partial x'^1}{\partial x^\beta}\,\mathrm{d}x^\beta,$$

$$\mathrm{d}x'^2 = \sum_{\beta=1}^{n} \frac{\partial x'^2}{\partial x^\beta}\,\mathrm{d}x^\beta,$$

$$\cdots\cdots\cdots\cdots$$

$$\mathrm{d}x'^n = \sum_{\beta=1}^{n} \frac{\partial x'^n}{\partial x^\beta}\,\mathrm{d}x^\beta.$$

All these equations can be combined into

$$\mathrm{d}x'^\alpha = \sum_{\beta=1}^{n} a^\alpha_\beta\,\mathrm{d}x^\beta \quad (\alpha = 1, 2, \ldots, n) \tag{1.1}$$

where $a^\alpha_\beta = \partial x'^\alpha/\partial x^\beta$.

The appearance of (1.1) can be simplified by adopting the *summation convention* according to which a product in which a Greek letter occurs twice is to be summed for the values $1, 2, \ldots, n$ of the letter. Thus, under the summation convention, $x_\beta x^\beta$ means $x_1x^1 + x_2x^2 + \ldots + x_nx^n$, and $x_{\alpha\beta}x^\beta$ means $x_{\alpha 1}x^1 + x_{\alpha 2}x^2 + \ldots + x_{\alpha n}x^n$, but $x_{\alpha\alpha}x^\alpha$ is outside the convention because the letter α has more than two occurrences. The notation is chosen so that a repeated letter appears once in the lower position and once in the higher spot. The repeated letter is often called a *dummy suffix* because it can be replaced by any other Greek letter (not already in use) without affecting the result, e.g. $x_\beta x^\beta = x_\gamma x^\gamma$. With the summation convention (1.1) can be expressed as

$$dx'^\alpha = a^\alpha_\beta \, dx^\beta. \tag{1.2}$$

The displacement $dx^1, dx^2, \ldots, dx^n$ constitutes what is customarily regarded as a vector and this suggests that a vector could be defined as something which satisfies (1.2). Thus, if the n quantities $(A^1, A^2, \ldots, A^n)$ become $(A'^1, A'^2, \ldots, A'^n)$ after the coordinate transformation and

$$A'^\alpha = a^\alpha_\beta A^\beta, \tag{1.3}$$

$(A^1, A^2, \ldots, A^n)$ is said to be a *contravariant vector.* A brief notation for the vector $(A^1, A^2, , \ldots, A^n)$ is A^α. Since a^α_β is a function of the coordinates, the transformation law has a definite meaning only with reference to the particular point with which the vector A^α is connected.

The adjective contravariant has been attached because in the tensor calculus it is necessary to distinguish between two kinds of vectors. If ϕ is an *invariant scalar,* i.e. its value at a point is independent of the coordinate system, grad ϕ is expected to behave like a vector. Now

$$\frac{\partial \phi}{\partial x'^\alpha} = \frac{\partial \phi}{\partial x^\beta} \frac{\partial x^\beta}{\partial x'^\alpha}$$

which, in general, is not the same law as that exemplified by (1.3). To cover this possibility the notion of a *covariant vector* is introduced. A covariant vector is a set of n quantities $(A_1, A_2, \ldots, A_n)$, denoted briefly by A_α, with the transformation law

$$A'_\alpha = b^\beta_\alpha A_\beta \tag{1.4}$$

where $b^\beta_\alpha = \partial x^\beta / \partial x'^\alpha$.

The quantities a^β_α and b^β_α are related because

$$a^\alpha_\beta b^\beta_\gamma = \frac{\partial x'^\alpha}{\partial x^\beta} \frac{\partial x^\beta}{\partial x'^\gamma} = \frac{dx'^\alpha}{dx'^\gamma}.$$

The coordinates x'^α and x'^γ belong to the same system and so are independent unless $\alpha = \gamma$. Hence the right-hand side is zero except when x'^α and x'^γ are the same coordinate in which case it is 1. Therefore

$$a^\alpha_\beta b^\beta_\gamma = \delta^\alpha_\gamma \tag{1.5}$$

where

$$\delta^\alpha_\gamma = \begin{cases} 0 & (\alpha \neq \gamma) \\ 1 & (\alpha = \gamma) \end{cases}.$$

The *Kronecker symbol* is the name used to describe δ^α_γ. It may be proved in a similar way that

$$a^\beta_\gamma b^\alpha_\beta = \delta^\alpha_\gamma. \tag{1.6}$$

It follows from (1.5) that, if $A(\gamma)$ is any quantity involving the affix γ,

$$a^\alpha_\beta b^\beta_\gamma A(\gamma) = \delta^\alpha_\gamma A(\gamma) = A(\alpha).$$

On account of this property $a^\alpha_\beta b^\beta_\gamma$ is sometimes known as a *substitution operator*—it substitutes α for γ.

F.2 Tensors

The notion of a tensor arises from consideration of the transformation law when two vectors are multiplied together. Let A^α and B^β be contravariant vectors. Then as α and β run through their values there are n^2 possibilities for $A^\alpha B^\beta$. They transform according to

$$A'^\alpha B'^\beta = a^\alpha_\mu A^\mu a^\beta_\nu B^\nu = a^\alpha_\mu a^\beta_\nu A^\mu B^\nu$$

on account of (1.3). This is a transformation law with two suffixes. Accordingly, the n^2 quantities $T^{\mu\nu}$ are called a *contravariant tensor of rank 2* if they are such that

$$T'^{\alpha\beta} = a^\alpha_\mu a^\beta_\nu T^{\mu\nu}. \tag{2.1}$$

In this nomenclature a vector is sometimes referred to as a tensor of rank 1 while a scalar is a tensor of rank 0.

Similarly, a *covariant tensor* $T_{\alpha\beta}$ satisfies

$$T'_{\alpha\beta} = b^\mu_\alpha b^\nu_\beta T_{\mu\nu}$$

and a *mixed tensor* T^α_β is such that

$$T'^\alpha_\beta = a^\alpha_\mu b^\nu_\beta T^\mu_\nu.$$

Although the product of two vectors is a tensor of rank 2 it is not necessary for a tensor to be the product of two vectors.

It is obvious that the sum of two tensors of the same character (contravariant, covariant, or mixed) is a tensor of the same type. Thus a law expressed by the vanishing of the sum of a number of tensors is independent of the coordinate system. In particular, if a tensor vanishes in one coordinate system it is zero in any transformed system also.

Tensors of higher rank may be defined by further generalization of the transformation law. For example, a contravariant tensor of rank m possesses m upper suffixes and satisfies a transformation containing m factors of the type a^α_β. It can be shown that the product of two tensors is a tensor whose rank is the sum of the ranks of the two. The reader may confirm, for example, that $T^{\alpha\beta}T^\mu_\nu$ satisfies a transformation law for a tensor of rank 4.

F.3 Contraction

If A_α is a covariant vector and B^β is a contravariant vector $A_\alpha B^\alpha$ is called the *inner product* of the vectors. In Cartesian coordinates it coincides with the scalar product of vector analysis. The inner product is a scalar; for

$$\begin{aligned} A'_\alpha B'^\alpha &= b^\mu_\alpha A_\mu a^\alpha_\nu B^\nu \\ &= \delta^\mu_\nu A_\mu B^\nu = A_\mu B^\mu \end{aligned}$$

from (1.4), (1.3) and (1.6).

The inner product can be viewed as being formed from the tensor $A_\mu B^\nu$ by replacing μ by ν. This process is called *contraction* because it produces a tensor two ranks lower. The device is applicable to tensors of higher rank; for example, contraction of the tensor of rank 4 $T^{\mu\nu\rho}_\sigma$ gives $T^{\mu\nu\sigma}_\sigma$ a tensor of rank 2 because

$$T'^{\alpha\beta\gamma}_\gamma = a^\alpha_\mu a^\beta_\nu a^\gamma_\rho b^\sigma_\gamma T^{\mu\nu\rho}_\sigma = a^\alpha_\mu a^\beta_\nu T^{\mu\nu\rho}_\rho$$

on account of (1.6). It is only when one upper and one lower suffix are the same that contraction occurs; $T_{\mu\nu\nu}$ is not an acceptable candidate.

It may be confirmed easily that T^μ_μ is a scalar, $T^{\mu\nu}A_\nu$ a contravariant vector, $T^\mu_\nu A_\mu$ a covariant vector and that $T^{\mu\nu}_{\mu\nu}$, $T^{\mu\nu}A_\mu B_\nu$, $T^\mu_\nu A_\mu B^\nu$ are scalars. The last three are illustrations of double contraction producing a tensor four ranks lower.

These results lead to an important method of determining whether a quantity is a tensor or not, namely: *if $T(\mu\nu)B^\nu$ is a covariant vector for every choice of the contravariant vector B^ν then $T(\mu\nu)$ is a covariant tensor of rank 2*. For, the fact that $T(\mu\nu)B^\nu$ is a covariant vector means

$$T'(\alpha\beta)B'^\beta = b^\mu_\alpha T(\mu\nu)B^\nu.$$

Now $B'^\beta = a^\beta_\mu B^\mu$ so that $b^\nu_\beta B'^\beta = B^\nu$ from (1.6) and hence

$$T'(\alpha\beta)B'^\beta = b^\mu_\alpha b^\nu_\beta T(\mu\nu)B'^\beta.$$

Since B'^β can be chosen arbitrarily this is possible only if

$$T'(\alpha\beta) = b^\mu_\alpha b^\nu_\beta T(\mu\nu)$$

and the result is proved.

Clearly, the theorem could be stated in the more general form: *if the inner product of a quantity and every contravariant (covariant) vector is a tensor, then the quantity is a tensor.*

Another theorem stems from this one: *if $T(\mu\nu)A^\mu B^\nu$ is a scalar for every choice of the contravariant vectors A^μ and B^ν then $T(\mu\nu)$ is a tensor of rank 2*. For, since the product is a scalar for every A^μ the preceding theorem implies that $T(\mu\nu)B^\nu$ is a vector for every B^ν. A further application of the preceding theorem shows that $T(\mu\nu)$ is a tensor.

The reader should be careful to observe that, in this theorem, A^μ and B^ν are different vectors. In general, it is *not* true that $T(\mu\nu)$ is a tensor if $T(\mu\nu)A^\mu A^\nu$ is a scalar for every A^μ. All that can be proved in this case is that

$$T'(\alpha\beta)+T'(\beta\alpha)=b^\mu_\alpha b^\nu_\beta\{T(\mu\nu)+T(\nu\mu)\}$$

which is insufficient to demonstrate that T is a tensor. However, if $T(\mu\nu)=T(\nu\mu)$ (in which case T is said to be *symmetric*) and $T'(\alpha\beta)=T'(\beta\alpha)$ the tensor law follows. Hence: *if $T(\mu\nu)A^\mu A^\nu$ is a scalar for every A^μ and T is symmetric, then $T(\mu\nu)$ is a symmetric tensor of rank* 2.

Note that a tensor of rank 2 such that $T^{\mu\nu}=-T^{\nu\mu}$ is described as *antisymmetric*.

F.4 Metric tensors

Suppose that in the coordinate system $(x^1,\ldots,x^n)$ the distance ds between two nearby points is given by

$$ds^2=g_{\mu\nu}\,dx^\mu\,dx^\nu$$

where $g_{\mu\nu}=g_{\nu\mu}$. Typical examples can be found in §1.39. For example, in three-dimensional Cartesian space $g_{\mu\nu}=0$ $(\mu\neq\nu)$, $=1$ $(\mu=\nu)$; in spherical polar coordinates R, θ, ϕ we have $g_{\mu\nu}=0$ $(\mu\neq\nu)$, $g_{11}=1$, $g_{22}=R^2$, $g_{33}=R^2\sin^2\theta$. Nevertheless, the reader should observe that, since an abstract space is being dealt with, it cannot be assumed that ds^2 is always positive.

Since ds^2 is a scalar, dx^μ an arbitrary contravariant vector and $g_{\mu\nu}$ is symmetric, the theorem at the end of the preceding section tells us that $g_{\mu\nu}$ is a covariant tensor; it will be called the *metric tensor.*

If A^ν is a contravariant vector, $g_{\mu\nu}A^\nu$ is a covariant vector which may be written B_μ, i.e.

$$g_{\mu\nu}A^\nu=B_\mu.$$

Let $G^{\mu\nu}$ be the co-factor of $g_{\mu\nu}$ in the determinant

$$g=\begin{vmatrix} g_{11} & g_{12} & \cdots & g_{1n}\\ g_{21} & & \cdots & g_{2n}\\ & & \cdots & \\ g_{n1} & & \cdots & g_{nn}\end{vmatrix}$$

and define $g^{\mu\nu}=G^{\mu\nu}/g$. Then, by standard algebra,

$$A^\mu=g^{\nu\mu}B_\nu. \qquad (4.1)$$

Evidently, $g^{\mu\nu}$ is a (symmetric) contravariant tensor by the theory of the previous section.

Furthermore,

$$g^{\nu\mu}g_{\nu\rho} = \delta^{\mu}_{\rho} = g_{\rho\nu}g^{\mu\nu} \tag{4.2}$$

shows that a mixed tensor can be defined whose components in one system are δ^{α}_{β}; its components in any other system are the same.

Given a covariant vector A_ν an associated contravariant vector A^μ can be defined by

$$A^\mu = g^{\mu\nu}A_\nu. \tag{4.3}$$

From (4.2)

$$A_\nu = g_{\mu\nu}A^\mu. \tag{4.4}$$

Conversely, starting from A^μ a covariant vector can be defined by (4.4) and then (4.3) holds. Thus A^μ and A_μ can be regarded as the contravariant and covariant components of a vector.

In terms of these components the length l of a vector is defined by

$$l^2 = A_\mu A^\mu.$$

Clearly, this is consistent with the customary definition in Cartesian coordinates where $g^{\mu\nu} = 0$ $(\mu \neq \nu)$, $= 1$ $(\mu = \nu)$ and $A^\mu = A_\mu$.

The angle θ between two vectors may also be defined in terms of contravariant and covariant components by

$$\cos\theta = A_\mu B^\mu / (A_\alpha A^\alpha B_\beta B^\beta)^{\frac{1}{2}}$$

because $A_\mu B^\mu = g_{\nu\mu}A^\nu B^\mu = A^\nu B_\nu$ on account of the symmetry of the metric tensor. Consequently, two vectors of nonzero length are perpendicular when $A_\mu B^\mu = 0$.

The operations in (4.3) and (4.4) are called *raising* and *lowering the suffixes* respectively. The processes of raising and lowering suffixes may also be applied to tensors. Thus, the contravariant and covariant components of a tensor are connected by

$$T^{\alpha\beta} = g^{\alpha\mu}g^{\beta\nu}T_{\mu\nu},$$
$$T_{\alpha\beta} = g_{\mu\alpha}g_{\nu\beta}T^{\mu\nu}.$$

Two types of mixed components can be defined via

$$T_{\alpha\cdot}^{\cdot\beta} = g^{\beta\mu}T_{\alpha\mu}, \qquad T^{\beta}_{\cdot\alpha} = g^{\beta\mu}T_{\mu\alpha}.$$

For a symmetric tensor $T_{\alpha\cdot}^{\cdot\beta}$ is equal to $T^{\beta}_{\cdot\alpha}$ but, in general, they are different.

F.5 Derivatives

It has been explained in §F.1 that, if ϕ is a scalar, the components of grad ϕ, given by

$$(\text{grad}\,\phi)_\alpha = \frac{\partial\phi}{\partial x^\alpha}, \tag{5.1}$$

form a covariant vector. It might be conjectured that the derivatives of a vector would supply a tensor but this surmise is false. Instead a special derivative, called the *covariant derivative*, has to be introduced in general tensor calculus (see Exercises on Chapter 2). However, the covariant derivative is not necessary for our purposes; it will be sufficient to note that

$$T_{\mu\nu} = \frac{\partial A_\nu}{\partial x^\mu} - \frac{\partial A_\mu}{\partial x^\nu}$$

is a covariant tensor which is obviously related to the curl of a covariant vector. The reader will find it instructive to prove that it is a covariant tensor bearing in mind the identity $b^\mu_\alpha \, \partial b^\nu_\beta / \partial x^\mu = b^\mu_\beta \, \partial b^\nu_\alpha / \partial x^\mu$.

The divergence of a contravariant vector is defined by

$$\operatorname{div} A = \frac{1}{|g|^{\frac{1}{2}}} \frac{\partial}{\partial x^\mu} (A^\mu \, |g|^{\frac{1}{2}}) \tag{5.2}$$

and the divergence of an *antisymmetric* contravariant tensor of rank 2 by

$$(\operatorname{div} T)^\mu = \frac{1}{|g|^{\frac{1}{2}}} \frac{\partial}{\partial x^\nu} (T^{\mu\nu} \, |g|^{\frac{1}{2}}). \tag{5.3}$$

It can be checked that the divergence of (5.2) is a scalar and that the one in (5.3) is a contravariant vector.

The reader should be aware that vectors are not treated in the same way in vector analysis and tensor calculus. Both kinds of vector have in common an addition law by which a component of the sum of two vectors is the sum of the corresponding components of the vectors but the vectors of tensor calculus must also comply with the transformation law. The vector product of vector analysis does not satisfy the transformation law and so the vector product is not a vector in the tensor calculus (it is, in fact, a pseudo-tensor). Another difference, even for the vectors of vector analysis which obey the transformation law, is that the components of a vector are not specified by the same rule. The components in the tensor calculus are computed from the transformation law whereas those in vector analysis are obtained by resolving in appropriate directions. To see the significance of this, suppose that x^μ is a Cartesian system so that there is no difference between contravariant and covariant components. Given a vector $\mathbf{A}$ in vector analysis it will have components parallel to the coordinate axes of A_1, A_2, A_3 say. Now take A_1, A_2, A_3 to be the components of a covariant vector in the tensor calculus. Make a change of coordinates to the curvilinear system x'^μ which, for simplicity, will be assumed to be orthogonal. According to the tensor calculus the component of the vector to be associated with x'^1 is A'_1, as determined by the transformation law for covariant vectors. In vector analysis the corresponding component is $\mathbf{A} \cdot \boldsymbol{i}_1$ where $\boldsymbol{i}_1$ is a unit vector tangent to the

x'^1-curve. If dx^μ/ds are the direction cosines of $\boldsymbol{i}_1$ the required component is $A_\mu\, dx^\mu/ds$ which, as a scalar, may also be written as $A'_\mu\, dx'^\mu/ds$, i.e. $A'_1/(g'_{11})^{\frac{1}{2}}$. Thus the components of the tensor calculus and vector analysis differ by the factor $(g'_{11})^{\frac{1}{2}}$ when **A** is regarded as covariant. If **A** had been expressed in terms of contravariant components the corresponding quantities would have been A'^1 and $A'^1(g'_{11})^{\frac{1}{2}}$.

When these differences are allowed for the formulae (5.1) and (5.2) lead at once to the expressions in Chapter 1 ((38.1) and (38.3)) for the calculation of grad and div in orthogonal curvilinear coordinates.

F.6 Cartesian tensors

For coordinate changes which transform one Cartesian system into another there is no difference between contravariant and covariant components; it does not matter then whether the suffix is in the upper or lower position. To fix ideas the lower side will be selected. Then, $x^\alpha x_\alpha = x_\alpha x_\alpha$ and, since this quantity is a scalar,

$$x_\alpha x_\alpha = x'_\alpha x'_\alpha = a^\alpha_\beta a^\alpha_\gamma x_\beta x_\gamma.$$

From the arbitrariness of x_α is inferred

$$a^\alpha_\beta a^\alpha_\gamma = \delta_{\beta\gamma} \tag{6.1}$$

where $\delta_{\beta\gamma} = 0$ $(\beta \neq \gamma)$, $= 1$ $(\beta = \gamma)$. Similarly

$$b^\beta_\alpha b^\gamma_\alpha = \delta_{\beta\gamma}. \tag{6.2}$$

However, $a^\alpha_\beta b^\beta_\gamma = \delta_{\alpha\gamma}$ and therefore

$$b^\mu_\gamma a^\alpha_\beta b^\beta_\gamma = b^\mu_\gamma\, \delta_{\alpha\gamma}$$

or, from (6.2), $a^\alpha_\mu = b^\mu_\alpha$. If a and b are determinants derived from a^α_β and b^α_β in the same way as g from $g_{\mu\nu}$ we see that $ab = 1$ and $a = b$ so that

$$a = b = \pm 1.$$

The coordinates x'_α may be reached from x_α by means of a *rotation* about the origin. In a continous rotation a cannot change sign and so $a = 1$ because that is the value obtained in the identical transformation $x'_\alpha = x_\alpha$. On the other hand, when x'_α requires not only a rotation from x_α but also a *reflection*, i.e. the direction of one of the axes is reversed, then $a = -1$.

It will now be assumed that the coordinate changes are restricted to rotations so that $a = 1$. The analysis of tensors when Cartesian axes are converted to Cartesian axes subject to $a = 1$, i.e. only rotations are permitted, is known as the *theory of Cartesian tensors*. In the theory of Cartesian tensors the vectors of the tensor calculus and of vector analysis have the same properties. From this point onwards only Cartesian tensors in three dimensions will be considered.

Let $\boldsymbol{i}_1$, $\boldsymbol{i}_2$, $\boldsymbol{i}_3$ be unit vectors along the coordinate axes. Then

$$\mathbf{A} = A_\alpha \boldsymbol{i}_\alpha \tag{6.3}$$

and the components in any other Cartesian system obtained by rotation can be derived from the transformation law for tensors. Instead of (6.3) we could write $(\mathbf{A})_\alpha = A_\alpha$ and this suggests that there should be associated with the tensor $T_{\alpha\beta}$ the quantity $\boldsymbol{T}$ defined by

$$\boldsymbol{T} = T_{\alpha\beta}\boldsymbol{i}_\alpha \boldsymbol{i}_\beta. \tag{6.4}$$

The order of the unit vectors in (6.4) must be preserved. Another notation for (6.4) is $(\boldsymbol{T})_{\alpha\beta} = T_{\alpha\beta}$. The quantity $\boldsymbol{T}$ defined by (6.4) is often called a *dyadic*.

A formal procedure leads to definitions for the inner products $\boldsymbol{T}\cdot\mathbf{A}$, $\mathbf{A}\cdot\boldsymbol{T}$. For example

$$\boldsymbol{T}\cdot\mathbf{A} = T_{\alpha\beta}\boldsymbol{i}_\alpha\boldsymbol{i}_\beta \cdot A_\gamma \boldsymbol{i}_\gamma = T_{\alpha\beta}\boldsymbol{i}_\alpha \delta_{\beta\gamma} A_\gamma = T_{\alpha\beta}A_\beta \boldsymbol{i}_\alpha$$

since $\boldsymbol{i}_\beta \cdot \boldsymbol{i}_\gamma = \delta_{\beta\gamma}$. Therefore, we define

$$(\boldsymbol{T}\cdot\mathbf{A})_\alpha = T_{\alpha\beta}A_\beta, \tag{6.5}$$

$$(\mathbf{A}\cdot\boldsymbol{T})_\alpha = A_\beta T_{\beta\alpha}, \tag{6.6}$$

the two inner products not being equal in general. Similarly, the inner product of two tensors $\boldsymbol{T}^{(1)}$, $\boldsymbol{T}^{(2)}$ is given by

$$(\boldsymbol{T}^{(1)}\cdot\boldsymbol{T}^{(2)})_{\alpha\beta} = T^{(1)}_{\alpha\mu}T^{(2)}_{\mu\beta}. \tag{6.7}$$

By contraction, we arrive at the *double inner product*, namely

$$\boldsymbol{T}^{(1)}:\boldsymbol{T}^{(2)} = T^{(1)}_{\alpha\mu}T^{(2)}_{\mu\alpha}. \tag{6.8}$$

In particular, since $\mathbf{AA}$ is a tensor,

$$\boldsymbol{T}:\mathbf{AA} = T_{\alpha\beta}A_\alpha A_\beta.$$

Consider now the tensor $\boldsymbol{I}$ in which $I_{\alpha\beta} = \delta_{\alpha\beta}$, i.e.

$$\boldsymbol{I} = \boldsymbol{i}_\alpha\boldsymbol{i}_\alpha.$$

Then

$$I'_{\alpha\beta} = a^\mu_\alpha a^\nu_\beta \delta_{\mu\nu} = a^\mu_\alpha a^\mu_\beta = \delta_{\alpha\beta}$$

from (6.1). Consequently,

$$\boldsymbol{I}\cdot\mathbf{A} = \mathbf{A} = \mathbf{A}\cdot\boldsymbol{I}$$

in any system obtained by rotation. For this reason, $\boldsymbol{I}$ is called the *unit tensor.*

The vector product can be defined by means of the *Levi–Civita* symbol $\varepsilon_{\alpha\beta\gamma}$. This symbol is zero unless all three α, β, γ are different. If α, β, γ are all different, $\varepsilon_{\alpha\beta\gamma} = 1$ when α, β, γ form a cyclic permutation of 1, 2,

3, i.e. $\varepsilon_{123} = \varepsilon_{231} = \varepsilon_{312} = 1$; in contrast, when the permutation is anticyclic, $\varepsilon_{\alpha\beta\gamma} = -1$, i.e. $\varepsilon_{132} = \varepsilon_{321} = \varepsilon_{213} = -1$. It is now straightforward to confirm

$$(\mathbf{A} \wedge \boldsymbol{B})_\alpha = \varepsilon_{\alpha\beta\gamma} A_\beta B_\gamma. \tag{6.9}$$

A valuable formula for $\varepsilon_{\alpha\beta\gamma}\varepsilon_{\alpha\mu\nu}$ originates from the product $\varepsilon_{\alpha\beta\gamma}\varepsilon_{\alpha\mu\nu}T_{\mu\nu}$. The product is zero if β is the same as γ. If β is different from γ the only nonzero contribution can come from α being equal to neither; then, however, $\varepsilon_{\alpha\mu\nu}$ is zero unless $\mu = \beta$, $\nu = \gamma$ or $\mu = \gamma$, $\nu = \beta$. In the first case the product of εs is 1 and in the second is -1. Thus

$$\varepsilon_{\alpha\beta\gamma}\varepsilon_{\alpha\mu\nu}T_{\mu\nu} = T_{\beta\gamma} - T_{\gamma\beta}.$$

Since any number of suffixes could be added to T without affecting this result we deduce that

$$\varepsilon_{\alpha\beta\gamma}\varepsilon_{\alpha\mu\nu} = I_{\beta\mu}I_{\gamma\nu} - I_{\beta\nu}I_{\gamma\mu}. \tag{6.10}$$

Use of (6.9) and (6.10) gives

$$\begin{aligned}((\mathbf{A} \wedge \boldsymbol{B}) \wedge \boldsymbol{C})_\alpha &= \varepsilon_{\alpha\beta\gamma}(\mathbf{A} \wedge \boldsymbol{B})_\beta C_\gamma \\ &= \varepsilon_{\alpha\beta\gamma}\varepsilon_{\beta\mu\nu}A_\mu B_\nu C_\gamma \\ &= A_\gamma B_\alpha C_\gamma - A_\alpha B_\gamma C_\gamma \\ &= ((\mathbf{A} \cdot \boldsymbol{C})\boldsymbol{B} - (\boldsymbol{B} \cdot \boldsymbol{C})\mathbf{A})_\alpha,\end{aligned}$$

a well-known formula.

F.7 Derivatives of Cartesian tensors

If ϕ is a scalar, grad ϕ is a vector with components

$$(\text{grad } \phi)_\alpha = \partial\phi/\partial x_\alpha$$

according to (5.1). The gradient of a vector $\mathbf{A}$ may be defined by

$$(\text{grad } \mathbf{A})_{\alpha\beta} = \partial A_\beta/\partial x_\alpha.$$

That this is a tensor may be checked directly by verifying the transformation law or by noting that $\mathrm{d}\boldsymbol{x} \cdot \text{grad } \mathbf{A}$ is the vector $\mathrm{d}\mathbf{A}$ and invoking the theory of §F.3. Similarly, the gradient of a tensor $\boldsymbol{T}$ of rank m can be defined by

$$(\text{grad } \boldsymbol{T})_{\alpha\beta\ldots} = \partial T_{\beta\ldots}/\partial x_\alpha;$$

it is a tensor of rank $m+1$.

Contraction applied to the gradient leads to a tensor two ranks lower, which will be called the divergence. Thus

$$\text{div } \mathbf{A} = (\text{grad } \mathbf{A})_{\alpha\alpha} = \partial A_\alpha/\partial x_\alpha,$$

$$(\text{div } \boldsymbol{T})_{\alpha\alpha\ldots} = \partial T_{\alpha\ldots}/\partial x_\alpha.$$

In particular, the divergence of a tensor of rank 2 is a vector with components $\partial T_{\beta\alpha}/\partial x_\beta$.

The curl of a vector can be written as

$$(\text{curl}\,\mathbf{A})_\alpha = \varepsilon_{\alpha\beta\gamma}\,\partial A_\gamma/\partial x_\beta \tag{7.1}$$

and this suggests the corresponding formula for a tensor, namely

$$(\text{curl}\,\boldsymbol{T})_{\alpha\beta} = \varepsilon_{\alpha\mu\nu}\,\partial T_{\nu\beta}/\partial x_\mu.$$

Many formulae in vector analysis can be proved by means of these definitions. It must be recognized, however, that when Cartesian tensors are employed in their proof, their validity is assured only when Cartesian components are taken of each side. As an illustration of the method consider

$$\begin{aligned}(\text{curl curl}\,\mathbf{A})_\alpha &= \varepsilon_{\alpha\beta\gamma}\,\partial(\text{curl}\,\mathbf{A})_\gamma/\partial x_\beta\\ &= \varepsilon_{\alpha\beta\gamma}\varepsilon_{\gamma\mu\nu}\frac{\partial^2 A_\nu}{\partial x_\mu\,\partial x_\beta}\end{aligned}$$

since $\varepsilon_{\gamma\mu\nu}$ is a constant. An application of (6.10) converts the right-hand side to

$$\frac{\partial^2 A_\beta}{\partial x_\beta\,\partial x_\alpha} - \frac{\partial^2 A_\alpha}{\partial x_\beta\,\partial x_\beta} = (\text{grad div}\,\mathbf{A} - \nabla^2\mathbf{A})_\alpha.$$

It follows that

$$\text{curl curl}\,\mathbf{A} = \text{grad div}\,\mathbf{A} - \nabla^2\mathbf{A}. \tag{7.2}$$

It may be proved in a similar manner that

$$\text{div curl}\,\mathbf{A} = 0,$$

$$\text{curl grad}\,\mathbf{A} = \mathbf{0},$$

$$\text{div}(\phi\mathbf{A}) = (\text{grad}\,\phi)\cdot\mathbf{A} + \phi\,\text{div}\,\mathbf{A}, \tag{7.3}$$

$$\text{curl}(\phi\mathbf{A}) = (\text{grad}\,\phi)\wedge\mathbf{A} + \phi\,\text{curl}\,\mathbf{A}, \tag{7.4}$$

$$\begin{aligned}\text{grad}(\mathbf{A}\cdot\boldsymbol{B}) &= (\text{grad}\,\mathbf{A})\cdot\boldsymbol{B} + (\text{grad}\,\boldsymbol{B})\cdot\mathbf{A}\\ &= (\mathbf{A}\cdot\text{grad})\boldsymbol{B} + (\boldsymbol{B}\cdot\text{grad})\mathbf{A} + \mathbf{A}\wedge\text{curl}\,\boldsymbol{B} + \boldsymbol{B}\wedge\text{curl}\,\mathbf{A},\end{aligned} \tag{7.5}$$

$$\text{div}(\mathbf{A}\wedge\boldsymbol{B}) = \boldsymbol{B}\cdot\text{curl}\,\mathbf{A} - \mathbf{A}\cdot\text{curl}\,\boldsymbol{B}, \tag{7.6}$$

$$\text{curl}(\mathbf{A}\wedge\boldsymbol{B}) = \mathbf{A}\,\text{div}\,\boldsymbol{B} - \boldsymbol{B}\,\text{div}\,\mathbf{A} + (\boldsymbol{B}\cdot\text{grad})\mathbf{A} - (\mathbf{A}\cdot\text{grad})\boldsymbol{B}. \tag{7.7}$$

In (7.5) and (7.7)

$$((\mathbf{A}\cdot\text{grad})\boldsymbol{B})_\alpha = A_\beta\,\partial B_\alpha/\partial x_\beta; \tag{7.8}$$

it is related to grad $\boldsymbol{B}$ by

$$(\text{grad}\,\boldsymbol{B})\cdot\mathbf{A} = (\mathbf{A}\cdot\text{grad})\boldsymbol{B} + \mathbf{A}\wedge\text{curl}\,\boldsymbol{B}. \tag{7.9}$$

F.8 The divergence and Stokes's theorem

The divergence theorem of vector analysis can be generalized so as to handle tensors. Let S be a closed surface and T its interior. Let $\boldsymbol{n}$ be the unit outward normal to S and $T_{\alpha\beta\ldots}$ the components of a Cartesian tensor defined at all points of T and S. Then

$$\int_S n_\lambda T_{\alpha\beta\ldots}\,\mathrm{d}S = \int_T \frac{\partial}{\partial x_\lambda} T_{\alpha\beta\ldots}\,\mathrm{d}\boldsymbol{x} \tag{8.1}$$

which is the generalization of the divergence theorem.

For the proof of (8.1) assume firstly that λ is different from all the suffixes $\alpha, \beta, \ldots$ so that there is no summation with respect to it. Draw a cylinder, of small cross-section, with generators parallel to the direction of $\boldsymbol{i}_\lambda$ and let it intersect S in the surface elements A and B (Fig. F.1). Then, if x_μ and x_ν are the other coordinates,

$$\int_T \frac{\partial}{\partial x_\lambda} T_{\alpha\beta\ldots}\,\mathrm{d}\boldsymbol{x} = \iint \left(\int_A^B \frac{\partial}{\partial x_\lambda} T_{\alpha\beta\ldots}\,\mathrm{d}x_\lambda\right) \mathrm{d}x_\mu\,\mathrm{d}x_\nu$$

where no summation is implied by the repetition of λ. The right-hand side is

$$\iint [T_{\alpha\beta\ldots}]_A^B\,\mathrm{d}x_\mu\,\mathrm{d}x_\nu = \int_S n_\lambda T_{\alpha\beta\ldots}\,\mathrm{d}S$$

since $n_\lambda\,\mathrm{d}S = \mathrm{d}x_\mu\,\mathrm{d}x_\nu$ at B and $n_\lambda\,\mathrm{d}S = -\mathrm{d}x_\mu\,\mathrm{d}x_\nu$ at A. The theorem is thereby proved.

Since the theorem holds as a tensor equation it remains true after contraction. Hence

$$\int_S n_\alpha T_{\alpha\beta\ldots}\,\mathrm{d}S = \int_T \frac{\partial}{\partial x_\alpha} T_{\alpha\beta\ldots}\,\mathrm{d}\boldsymbol{x}. \tag{8.2}$$

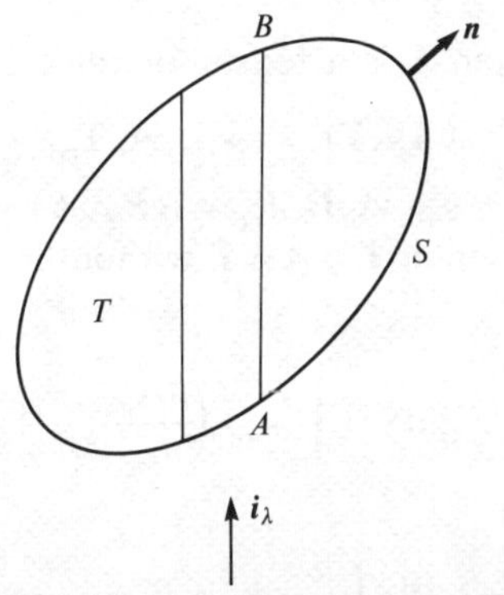

Fig. F.1 The generalized divergence theorem

Several special cases of (8.1) and (8.2) are worthy of attention.

(a) Let $T_{\alpha\beta\ldots}=A_\alpha$ in (8.2); i.e. the components of the tensor are actually components of a vector. Then

$$\int_S n_\alpha A_\alpha \,\mathrm{d}S=\int_T \frac{\partial A_\alpha}{\partial x_\alpha}\,\mathrm{d}\boldsymbol{x},$$

i.e.

$$\int_S \boldsymbol{n}\cdot\mathbf{A}\,\mathrm{d}S=\int_T \operatorname{div}\mathbf{A}\,\mathrm{d}\boldsymbol{x} \tag{8.3}$$

which is the standard divergence theorem.

(b) In (8.1) take the tensor T to be the scalar ϕ. Then

$$\int_S n_\lambda\phi\,\mathrm{d}S=\int_T \frac{\partial\phi}{\partial x_\lambda}\,\mathrm{d}\boldsymbol{x},$$

i.e.

$$\int_S \phi\boldsymbol{n}\,\mathrm{d}S=\int_T \operatorname{grad}\phi\,\mathrm{d}\boldsymbol{x}. \tag{8.4}$$

(c) A formula involving curl is obtained by putting $T_{\alpha\beta\ldots}=\varepsilon_{\alpha\gamma\beta}A_\gamma$. Then (8.2) gives

$$\int_S n_\alpha\varepsilon_{\alpha\gamma\beta}A_\gamma\,\mathrm{d}S=\int_T \varepsilon_{\alpha\gamma\beta}\frac{\partial A_\gamma}{\partial x_\alpha}\,\mathrm{d}\boldsymbol{x},$$

i.e.

$$\int_S \boldsymbol{n}\wedge\mathbf{A}\,\mathrm{d}S=\int_T \operatorname{curl}\mathbf{A}\,\mathrm{d}\boldsymbol{x} \tag{8.5}$$

from (6.9) and (7.1).

(d) The combination of (8.3) and (7.6) supplies

$$\int_S \boldsymbol{n}\cdot\mathbf{A}\wedge\operatorname{curl}\boldsymbol{B}\,\mathrm{d}S=\int_T (\operatorname{curl}\boldsymbol{B}\cdot\operatorname{curl}\mathbf{A}-\mathbf{A}\cdot\operatorname{curl}\operatorname{curl}\boldsymbol{B})\,\mathrm{d}\boldsymbol{x}. \tag{8.6}$$

(e) Define $\mathbf{A}\wedge\boldsymbol{T}$ when $\boldsymbol{T}$ is a tensor of rank 2 by

$$(\mathbf{A}\wedge\boldsymbol{T})_{\alpha\beta}=\varepsilon_{\alpha\mu\nu}A_\mu T_{\nu\beta}. \tag{8.7}$$

Then $(\boldsymbol{B}\cdot(\mathbf{A}\wedge\boldsymbol{T}))_\beta=\varepsilon_{\alpha\mu\nu}A_\mu B_\alpha T_{\nu\beta}=((\boldsymbol{B}\wedge\mathbf{A})\cdot\boldsymbol{T})_\beta$, showing that $\boldsymbol{B}\cdot(\mathbf{A}\wedge\boldsymbol{T})$ can be written as $\boldsymbol{B}\cdot\mathbf{A}\wedge\boldsymbol{T}$ without loss of clarity. Now, if the tensor in (8.2) is $\mathbf{A}\wedge\boldsymbol{T}$,

$$\int_S n_\alpha\varepsilon_{\alpha\mu\nu}A_\mu T_{\nu\beta}\,\mathrm{d}S=\int_T \varepsilon_{\alpha\mu\nu}\left(\frac{\partial A_\mu}{\partial x_\alpha}T_{\nu\beta}+A_\mu\frac{\partial T_{\nu\beta}}{\partial x_\alpha}\right)\mathrm{d}\boldsymbol{x}$$

whence

$$\int_S \boldsymbol{n}\cdot\mathbf{A}\wedge\boldsymbol{T}\,\mathrm{d}S=\int_T \{(\operatorname{curl}\mathbf{A})\cdot\boldsymbol{T}-\mathbf{A}\cdot\operatorname{curl}\boldsymbol{T}\}\,\mathrm{d}\boldsymbol{x}. \tag{8.8}$$

Another theorem which can be generalized to tensors is that of Stokes. In this case the surface S is not closed but has the closed curve C as rim. Then

$$\int_C T_{\alpha\beta\ldots}\,\mathrm{d}x_\nu = \int_S n_\lambda \varepsilon_{\lambda\mu\nu} \frac{\partial}{\partial x_\mu} T_{\alpha\beta\ldots}\,\mathrm{d}S \tag{8.9}$$

where the direction of the unit normal $\boldsymbol{n}$ to S is related to the sense of integration round C by the same rule as in Stokes's theorem (see §38 of Chapter 1).

For a plausible demonstration of (8.9) it is convenient to designate the Cartesian coordinates by x, y, z. Suppose that the equation of S can be written as $z = f(x, y)$. Then it must be remembered that, in the derivatives on the right-hand side of (8.9), the operations $\partial/\partial x$, $\partial/\partial y$ and $\partial/\partial z$ are performed *before* z is put equal to $f(x, y)$.

Consider what happens when $\nu = 3$, i.e. x_ν coincides with z. The direction of the normal to S is given by

$$n_1/n_3 = -\partial f/\partial x, \qquad n_2/n_3 = -\partial f/\partial y.$$

Hence the right-hand side of (8.9) is

$$\int_{S_0} \left(-\frac{\partial f}{\partial x}\frac{\partial T_{\alpha\beta}}{\partial y} + \frac{\partial f}{\partial y}\frac{\partial T_{\alpha\beta}}{\partial x}\right) \mathrm{d}x\,\mathrm{d}y$$

where S_0 is the projection of S on $z = 0$. Let

$$T^0_{\alpha\beta}(x, y) = T_{\alpha\beta}(x, y, f(x, y)).$$

Then

$$\frac{\partial T^0_{\alpha\beta}}{\partial x} = \frac{\partial T_{\alpha\beta}}{\partial x} + \frac{\partial T_{\alpha\beta}}{\partial z}\frac{\partial f}{\partial x}, \qquad \frac{\partial T^0_{\alpha\beta}}{\partial y} = \frac{\partial T_{\alpha\beta}}{\partial y} + \frac{\partial T_{\alpha\beta}}{\partial z}\frac{\partial f}{\partial y}$$

and so the integral can be expressed as

$$\int_{S_0} \left\{-\frac{\partial}{\partial y}\left(T^0_{\alpha\beta}\frac{\partial f}{\partial x}\right) + \frac{\partial}{\partial x}\left(T^0_{\alpha\beta}\frac{\partial f}{\partial y}\right)\right\} \mathrm{d}x\,\mathrm{d}y$$

$$= \int_{C_0} \left(T^0_{\alpha\beta}\frac{\partial f}{\partial x}\,\mathrm{d}x + T^0_{\alpha\beta}\frac{\partial f}{\partial y}\,\mathrm{d}y\right) = \int_C T_{\alpha\beta}\,\mathrm{d}z$$

where C_0 is the projection of C on $z = 0$. This verifies (8.9) when $\nu = 3$ and there are similar arguments when $\nu = 1, 2$.

Since (8.9) is a tensor equation contraction in which $\nu = \alpha$ does not destroy its validity. Hence

$$\int_C T_{\alpha\beta\ldots}\,\mathrm{d}x_\alpha = \int_S n_\lambda \varepsilon_{\lambda\mu\alpha} \frac{\partial T_{\alpha\beta\ldots}}{\partial x_\mu}\,\mathrm{d}S. \tag{8.10}$$

Some particular cases of the theorems now follow.

(i) In (8.10) put $T_{\alpha\beta\ldots}=A_\alpha$. Then

$$\int_C A_\alpha \,\mathrm{d}x_\alpha = \int_S n_\lambda \varepsilon_{\lambda\mu\alpha} \frac{\partial A_\alpha}{\partial x_\mu}\,\mathrm{d}S,$$

i.e.

$$\int_C \mathbf{A}\cdot\mathrm{d}\mathbf{s} = \int_S \boldsymbol{n}\cdot\operatorname{curl}\mathbf{A}\,\mathrm{d}S \tag{8.11}$$

which is the standard form of Stokes's theorem

(ii) If the choice $T_{\alpha\beta\ldots}=\phi$ is made in (8.9)

$$\int_C \phi \,\mathrm{d}x_\nu = \int_S n_\lambda \varepsilon_{\lambda\mu\nu} \frac{\partial \phi}{\partial x_\mu}\,\mathrm{d}S,$$

i.e.

$$\int_C \phi\,\mathrm{d}\mathbf{s} = \int_S \boldsymbol{n}\wedge\operatorname{grad}\phi\,\mathrm{d}S. \tag{8.12}$$

(iii) When $T_{\alpha\beta\ldots}=\varepsilon_{\alpha\gamma\beta}A_\gamma$, (8.10) gives

$$\int_C \varepsilon_{\alpha\gamma\beta}A_\gamma\,\mathrm{d}x_\alpha = \int_S n_\lambda \varepsilon_{\lambda\mu\alpha}\varepsilon_{\alpha\gamma\beta}\frac{\partial A_\gamma}{\partial x_\mu}\,\mathrm{d}S$$
$$= \int_S \left(n_\gamma \frac{\partial A_\gamma}{\partial x_\beta} - n_\beta \frac{\partial A_\gamma}{\partial x_\gamma}\right)\mathrm{d}S$$

from (6.10). Hence

$$-\int_C \mathbf{A}\wedge\mathrm{d}\mathbf{s} = \int_S \{(\operatorname{grad}\mathbf{A})\cdot\boldsymbol{n} - \boldsymbol{n}\operatorname{div}\mathbf{A}\}\,\mathrm{d}S$$
$$= \int_S \{(\boldsymbol{n}\cdot\operatorname{grad})\mathbf{A} + \boldsymbol{n}\wedge\operatorname{curl}\mathbf{A} - \boldsymbol{n}\operatorname{div}\mathbf{A}\}\,\mathrm{d}S \tag{8.13}$$

on account of (7.9).

If S is closed C disappears and we conclude that

$$\int_S \{(\boldsymbol{n}\cdot\operatorname{grad})\mathbf{A} + \boldsymbol{n}\wedge\operatorname{curl}\mathbf{A} - \boldsymbol{n}\operatorname{div}\mathbf{A}\}\,\mathrm{d}S = \mathbf{0} \tag{8.14}$$

for closed S.

Appendix G. Asymptotic evaluation of integrals

G.1 The method of stationary phase

KELVIN'S method of stationary phase is concerned with estimating

$$I = \int_a^b f(x)\mathrm{e}^{\mathrm{i}\tau g(x)}\,\mathrm{d}x \tag{1.1}$$

as $\tau \to \infty$ when g is a real function of the real variable x. (For multiple integrals see Jones (1982).) Suppose that an integration by parts is made in (1.1). Then

$$I = [\mathrm{e}^{\mathrm{i}\tau g} f/\mathrm{i}\tau g']_a^b - (1/\mathrm{i}\tau)\int_a^b \mathrm{e}^{\mathrm{i}\tau g}(f/g')'\,\mathrm{d}x. \tag{1.2}$$

The first term is $O(1/\tau)$ and the second, after another integration by parts, should be $O(1/\tau^2)$. Therefore our estimate for the integral is

$$I \sim [\mathrm{e}^{\mathrm{i}\tau g} f/\mathrm{i}\tau g']_a^b. \tag{1.3}$$

Such a procedure is legitimate provided that g′ does not vanish in the interval of integration. If g′ does have a zero the argument fails because the integral in (1.2) is not finite. To remedy the deficiency consider the case where g′ has a single zero x_0 in the interval, i.e. $g'(x_0) = 0$. Select a very small number η and cut out the segment from $x_0 - \eta$ to $x_0 + \eta$. In the stretch from a to $x_0 - \eta$, g′ is nonzero and (1.3) can be applied; likewise (1.3) is applicable for $x_0 + \eta$ to b. The remaining question is how the integral from $x_0 - \eta$ to $x_0 + \eta$ compares with these two contributions.

Because of the smallness of η the integrand can be expanded in a power series about x_0, going to the second power in g on account of the largeness of τ. Recalling that $g'(x_0) = 0$ and putting $x - x_0 = t$ we see that our estimate is

$$f(x_0)\int_{-\eta}^{\eta} \exp[\mathrm{i}\tau\{g(x_0) + \tfrac{1}{2}g''(x_0)t^2\}]\,\mathrm{d}t.$$

It is plausible that (1.3) should be applicable to $(-\infty, -\eta)$ and (η, ∞) (though quite difficult to prove rigorously) and so making the limits

$(-\infty, \infty)$ does not alter the order of the terms set on one side. Since $\int_{-\infty}^{\infty} e^{iax^2}\, dx = \pi^{\frac{1}{2}} e^{\frac{1}{4}\pi i}/a^{\frac{1}{2}}$ when $a > 0$ we see that the contribution of the small interval is going to dominate and

$$I \sim \{2\pi/\tau g''(x_0)\}^{\frac{1}{2}} f(x_0) \exp\{i\tau g(x_0) + \tfrac{1}{4}\pi i\} \tag{1.4}$$

when $g''(x_0) > 0$. If $g''(x_0) < 0$, the factor $|g''(x_0)| e^{\pi i}$ is put instead of $g''(x_0)$.

If $g''(x_0) = 0$ the procedure breaks down and additional terms must be taken in the exponent. Reconsideration is also necessary if $f(x)$ possesses a singularity near x_0 for then $f(x_0)$ may not be a reasonable approximation to $f(x)$ in the small interval.

If $g''(x_0)$ is small but not precisely zero the situation is rather more complicated. There is now another point of stationary phase nearby, namely at $x = x_0 - 2g''(x_0)/g'''(x_0)$ if g''' is well away from zero as will be assumed. When the two points are close enough it is not permissible to treat them separately. So g should be expanded about their mid-point x_1 which is recognised by $g''(x_1) = 0$. Therefore, in this case,

$$I \sim f(x_1) \int_{-\infty}^{\infty} \exp[i\tau\{g(x_1) + g'(x_1)t + \tfrac{1}{6}g'''(x_1)t^3\}]\, dx.$$

Quoting (A.2.11) we have

$$I \sim 2\pi f(x_1) e^{i\tau g(x_1)} \left\{\frac{2}{\tau g'''(x_1)}\right\}^{\frac{1}{3}} \mathrm{Ai}\left[\tau^{\frac{2}{3}} g'(x_1) \left\{\frac{2}{g'''(x_1)}\right\}^{\frac{1}{3}}\right] \tag{1.5}$$

when $g'''(x_1) > 0$. If $g'''(x_1) < 0$ it is only necessary to reverse the signs of g' and g''' in (1.5).

For sufficient separation between the two points of stationary phase the argument of the Airy function is big enough to employ (A.2.19) (g' and g''' have opposite signs). It is then found that (1.5) splits into two individual contributions, each of the type (1.4), making each point of stationary phase a distinctive entity. At the other extreme, placing $g'(x_1) = 0$ in (1.5) and using (A.2.15) supplies the asymptotic behaviour when the two points of stationary phase coalesce at a point where $g' = g'' = 0$ and $g''' \neq 0$.

G.2 The method of steepest descent

A more general integral is

$$I_1 = \int_C f(z) e^{\tau g(z)}\, dz$$

where C is a contour in the complex z-plane. By writing $\tau g = (\tau e^{i\theta}) g e^{-i\theta}$ we realize that we can cope with complex parameters just as well as the positive τ. Sometimes I_1 can be converted easily into the type discussed in §G.1 and there is nothing further to be said.

It again looks as if the dominating contribution will come from a neighbourhood of a zero of g'. Let z_0 be such a zero; it is called a *saddle-point* or *col.*

Set $z = x + iy$ and $g(z) = u(x, y) + iv(x, y)$ where u and v are real. Denote $\partial u/\partial x$ by u_x with a similar notation for the y-derivative. Then u and v are connected by the *Cauchy–Riemann relations* $u_x = v_y$, $u_y = -v_x$. One consequence of these relations is that u and v are solutions of Laplace's equation. Another is that $g' = u_x - iu_y$. Thus, at a saddle-point $u_x = 0$ and $u_y = 0$. It follows from the Cauchy–Riemann relations that $v_x = 0$ and $v_y = 0$. Hence, at a saddle-point, all four partial derivatives vanish.

Near a saddle-point

$$g(z) - g(z_0) = \tfrac{1}{2}g''(z_0)(z - z_0)^2 = \tfrac{1}{2}Ar^2 e^{i(\alpha + 2\theta)}$$

if $z - z_0 = re^{i\theta}$ and $g''(z_0) = Ae^{i\alpha}$ where A, α are real with A positive. If $g(z_0) = u_0 + iv_0$

$$u - u_0 = \tfrac{1}{2}Ar^2 \cos(2\theta + \alpha), \qquad v - v_0 = \tfrac{1}{2}Ar^2 \sin(2\theta + \alpha).$$

Now $u - u_0$ is negative if θ is such that the cosine is negative and drops fastest with r if $\theta = \frac{1}{2}(\pi - \alpha)$ or $\theta = \frac{1}{2}(3\pi - \alpha)$. These are desirable directions because they force exponential decay on the integrand as one moves away from the saddle-point. They are called *paths of steepest descent.* Note that on a path of steepest descent $v = v_0$.

Having seen that $v = v_0$ is a good route to start on from a saddle-point let us see what happens if we stay on it. Suppose we reach a point z_1 and that $z_1 + \rho e^{i\phi}$ is a nearby point on $v = v_0$. Since v does not change in the move from one point to the other ϕ must satisfy $v_x \cos\phi + v_y \sin\phi = 0$. By means of the Cauchy–Riemann relations this may be expressed as

$$-u_y \cos\phi + u_x \sin\phi = 0. \tag{2.1}$$

The change of u in the move is $(u_x \cos\phi + u_y \sin\phi)\rho$. This quantity is known to be negative at z_0. It will therefore remain negative until a point z_1 is arrived at where it is zero. But, on account of (2.1), this is impossible unless $u_x = 0$, $u_y = 0$, i.e. z_1 is a saddle-point. Hence, *on a path of steepest descent, u decreases steadily from a saddle-point until another saddle-point is reached.* Should $g'(z)$ have a singularity on the path, that can upset the apple-cart too.

It will now be supposed that the path of steepest descent goes off to infinity without encountering another saddle-point or singularity of $g'(x)$. Consider I_1 taken along the path of steepest descent that begins from z_0 along $\theta = \frac{1}{2}(\pi - \alpha)$. Convergence of the integral at infinity is assumed because of the property of u demonstrated above. Indeed, that property guarantees that the main contribution to the integral comes from a neighbourhood of the saddle-point. In this vicinity $g(z) - g(z_0) = -\frac{1}{2}Ar^2$

and $z-z_0=re^{\frac{1}{2}i(\pi-\alpha)}$. Hence the integral is essentially

$$f(z_0)\int_0^\infty \exp[\tau\{g(z_0)-\tfrac{1}{2}Ar^2\}]e^{\frac{1}{2}i(\pi-\alpha)}\,dr,$$

i.e.

$$\int_{z_0}^\infty f(z)e^{\tau g(z)}\,dz \sim \left(\frac{\pi}{2\tau A e^{i\alpha}}\right)^{\frac{1}{2}} f(z_0)e^{\tau g(z_0)+\frac{1}{2}\pi i}. \tag{2.2}$$

Going from z_0 to infinity via the path of steepest descent on which $\theta=\frac{1}{2}(3\pi-\alpha)$ merely reverses the sign of the right-hand side of (2.2).

The strategy for dealing with I_1 therefore is to deform the contour C as far as possible into a path or paths of steepest descent. Then the contribution of each saddle-point is calculated by means of (2.2). However, it must not be forgotten that in the deformation of C poles or other singularities of the integrand may be captured; their contribution may be as significant as that from the saddle-point.

If $g''(z_0)=0$ there will be more than two paths of steepest descent from the saddle-point but on each one the argument about u is unaffected. The mode of calculation is still effective though the asymptotic formula will differ from (2.2).

As an illustration *Stirling's formula* for the factorial function will be obtained. From Euler's integral $s!=\int_0^\infty e^{-z}z^s\,dz$ we have

$$s!/s^{s+1}=\int_0^\infty e^{s(\ln z-z)}\,dz.$$

This is, in fact, valid for $s>-1$ but s will now be limited to being real and positive. In this case $g(z)=\ln z-z$ so that there is a saddle-point at $z=1$ and $g''(1)=-1$. Hence $A=1$ and $\alpha=\pi$. One path of steepest descent is from 1 to $+\infty$ along the real axis; the other goes from 1 down to the singularity at the origin. No deformation of the contour is necessary but note that one path of steepest descent is in a direction *towards* the saddle-point whereas (2.2) refers to the opposite direction. The result is

$$s!\sim(2\pi)^{\frac{1}{2}}e^{(s+\frac{1}{2})\ln s-s} \tag{2.3}$$

as $|s|\to\infty$. Although proved by this technique only for positive s it is actually true for $|\text{ph}\, s|<\pi$ with the interpretation $\ln s=\ln|s|+i\,\text{ph}\, s$. When s is real and negative

$$s!\sim-(\tfrac{1}{2}\pi)^{\frac{1}{2}}e^{(s+\frac{1}{2})\ln|s|-s}\operatorname{cosec}\pi s.$$

G.3 Saddle-point near a pole

If the contour of integration consists of the path of steepest descent on which $\theta=\frac{1}{2}(3\pi-\alpha)$, but traversed from infinity to z_0, followed by the path of steepest descent on which $\theta=\frac{1}{2}(\pi-\alpha)$ from z_0 to infinity the foregoing theory predicts

$$\int_P f(z)e^{\tau g(z)}\,dz \sim \left(\frac{2\pi}{\tau A e^{i\alpha}}\right)^{\frac{1}{2}} f(z_0)e^{\tau g(z_0)+\frac{1}{2}\pi i} \tag{3.1}$$

where P stands for the stated path. It may happen that $f(z)$ has a pole at $z=z_p$ which approaches z_0 as some parameter varies. The formula (3.1) then needs modification because the right-hand side will become unduly large.

Suppose that $f(z)=h(z)/(z-z_p)$ where $h(z)$ is regular. It is assumed that z_p does not lie on P except when it coincides with z_0. Make the same change of variable as before, namely $g(z)-g(z_0)=-\frac{1}{2}Ar^2$. Then

$$\int_P f(z)e^{\tau g(z)}\,dz = e^{\tau g(z_0)}\int_{-\infty}^{\infty} f(z)\frac{dz}{dr}e^{-\frac{1}{2}A\tau r^2}\,dr.$$

The principal contribution will again come from a neighbourhood of $r=0$. In this vicinity $z-z_0=re^{\frac{1}{2}i(\pi-\alpha)}$. The pole lies at $r=r_p$ where $g(z_p)-g(z_0)=-\frac{1}{2}Ar_p^2$, that value of r_p being chosen which satisfies $z_p-z_0=r_pe^{\frac{1}{2}i(\pi-\alpha)}$ when z_p is close to z_0. Now, near the pole,

$$f(z)\frac{dz}{dr}=\frac{B}{r-r_p}+C+k(r)$$

where $k(r)$ is a regular function of r such that $k(0)=0$. The constants B and C are given by

$$\begin{aligned} B=h(z_p), \qquad C&=f(z_0)(dz/dr)_{r=0}+B/r_p \\ &=\frac{h(z_0)}{z_0-z_p}e^{\frac{1}{2}i(\pi-\alpha)}+\frac{h(z_p)}{r_p}. \end{aligned} \tag{3.2}$$

Clearly, $k(r)$ will make a smaller contribution than the other terms and it will be ignored from now on.

If $\operatorname{Im} r_p<0$

$$\begin{aligned}\int_{-\infty}^{\infty}\frac{e^{-\frac{1}{2}A\tau r^2}}{r-r_p}\,dr &= -i\int_{-\infty}^{\infty}e^{-\frac{1}{2}A\tau r^2}\int_0^{\infty}e^{iy(r-r_p)}\,dy\,dr \\ &= -i(2\pi/A\tau)^{\frac{1}{2}}\int_0^{\infty}\exp(-iyr_p-y^2/2A\tau)\,dy \\ &= -\pi i e^{-\frac{1}{2}A\tau r_p^2}\operatorname{erfc}\{ir_p(\tfrac{1}{2}A\tau)^{\frac{1}{2}}\}\end{aligned}$$

where

$$\operatorname{erfc} z=\frac{2}{\pi^{\frac{1}{2}}}\int_z^{\infty}e^{-t^2}\,dt.$$

If Im $r_p > 0$, it is necessary only to change i to $-$i in this result. Hence, as $\tau \to \infty$,

$$\int_P f(z)e^{\tau g(z)}\,dz \sim \pi i B \exp\{\tau g(z_p)\}[2H(\text{Im } r_p)$$

$$-\operatorname{erfc}\{ir_p(\tfrac{1}{2}A\tau)^{\frac{1}{2}}\}] + (2\pi/A\tau)^{\frac{1}{2}}Ce^{\tau g(z_0)} \quad (3.3)$$

where $H(x)$ is the Heaviside unit function which is 1 for $x > 0$ and zero for $x < 0$.

When $|z_p - z_0| \geq \delta > 0$ and $\tau^{\frac{1}{2}}\delta \gg 1$ the asymptotic formula

$$\operatorname{erfc} z \sim e^{-z^2}/\pi^{\frac{1}{2}}z \quad (|\text{ph } z| < \tfrac{1}{2}\pi)$$

may be employed in (3.3) with the consequence that (3.1) is recovered. On the other hand, as $z_p \to z_0$, B and C remain finite while the quantity in [ ] tends to 1. Thus (3.3) stays bounded and provides a uniformly valid asymptotic expression for the integral which makes a smooth transition as z_p passes through z_0.

References

Ahluwalia, D. S., Lewis, R. M., and Boersma, J. (1968). *SIAM J. Appl. Math.* **16,** 783.

Baños, A. (1966). *Dipole Radiation in the Presence of a Conducting Half-Space.* Pergamon, Oxford.

Bechtel, M. E. and Ross, R. A. (1966). *Cornell Aero. Lab. Rep.* ER/RIS-10.

Belkina, M. G. (1958). *Diffraction of Electromagnetic Waves by Certain Bodies of Revolution* (Russian).

Bowman, J. J., Senior, T. B. A., and Uslenghi, P. L. E. (1969). *Electromagnetic and Acoustic Scattering by Simple Shapes.* North-Holland, Amsterdam.

Brillouin, L. (1938). *Elect. Comm.* **16,** 350.

—— (1938). *Bull. Soc. Fr. Elect.* **8,** 899.

Chu, L. J. (1938). *J. appl. Phys.* **9,** 583.

Eichenwald, A. (1903). *Ann. Phys.* (4), **11,** 1 and 421.

Erdélyi, A. (1953). *Higher Transcendental Functions.* McGraw-Hill, New York.

Felsen, L. B. (1976). *Transient Electromagnetic Fields.* Springer, Berlin.

Gear, J. F. (1975). *J. Fluid Mech.* **67,** 817,

Hobson, E. W. (1931). *The Theory of Spherical and Ellipsoidal Harmonics.* Cambridge University Press, Cambridge.

Jones, D. S. (1952). *Proc. Camb. Phil. Soc.* **48,** 118.

—— (1953). *Proc. Camb. Phil. Soc.* **49,** 668.

—— (1955). *Phil. Trans.* **247A,** 499.

—— (1979). *Methods in Electromagnetic Wave Propagation.* Oxford University Press, Oxford.

—— (1980). *J. Inst. Maths. Applics.* **26,** 133.

—— (1981). *I.E.E. Proc.* **128,** 114.

—— (1982). *The Theory of Generalised Functions,* Cambridge University Press, Cambridge.

Karp, S. N. and Williams, W. E. (1951). *Proc. Camb. Phil Soc.* **53,** 683.

Keller, J. B. (1957). *J. appl. Phys.* **28,** 426.

Kellogg, O. D. (1929). *Foundations of Potential Theory.* Springer, Berlin.

Matsui, E. (1960). *Nat. Bur. Stand. Rep.* 7038.

Meixner, J. and Schäfke, F. W. (1954). *Mathieusche Funktionen und Sphäroidfunktionen.* Springer, Berlin.

—— Schäfke, F. W. and Wolf, G. (1980). *Mathieu Functions and Spheroidal Functions and Their Mathematical Foundations.* Springer, Berlin.

Minnaert, M. (1954). *Light and Colour.* Dover, New York.

Mittra, R. (1973). *Computer Techniques for Electromagnetics.* Pergamon, Oxford.

Murray, F. H. (1931). *Amer. J. Math.* **53,** 275 and 873.

—— (1933). *Proc. I.R.E.* **21,** 154.

Olver, F. W. J. (1974). *Asymptotics and Special Functions.* Academic Press, New York.

Pidduck, F. B. (1946). *Currents in Aerials and High-Frequency Networks*. Oxford University Press, Oxford.
Röntgen, W. C. (1888). *Ann. Phys.* (3), **35,** 264.
—— (1890). *Ann. Phys.* (3), **40,** 93.
Rowland, H. A. (1876). *Sitzber, Akad. Wiss. Berlin* 211.
Rozzi, T. E. (1973). *Int. J. circuit Theory Appl.* **1,** 161.
Skavlem, S. (1951). *Arch. Math. Naturvid.* **51,** 61.
Skwirzynski, S. K. (1981). *Theoretical Method for Determining the Interaction of Electromagnetic Waves with Structures.* Sitjhoff and Noordhoff, Alpen aan den Rijn, The Netherlands.
Stratton, J. A. (1941). *Electromagnetic Theory.* McGraw-Hill, New York.
Uslenghi, P. L. E. (1978). *Electromagnetic Scattering.* Academic Press, New York.
Van Bladel, J. (1967). *J. Sound Vib.* **6,** 386.
—— (1968). *J. Sound Vib.* **8**(2), 186.
Van de Hulst, H. C. (1957). *Light Scattering by Small Particles.* Wiley, New York.
Wait, J. R. (1959). *Electromagnetic Radiation from Cylindrical Structures.* Pergamon, Oxford.
Watson, G. N. (1944). *The Theory of Bessel Functions* 2nd ed., Cambridge University Press, Cambridge.
Weinstein, L. A. (1969). *The Theory of Diffraction and the Factorization Method.* Golem Press, Boulder, Colorado.
Whitham, G. B. (1974). *Linear and Nonlinear Waves.* Wiley, New York.
Whittaker, E. T. and Watson, G. N. (1940). *Modern Analysis,* 4th edn. Cambridge University Press. Cambridge.
Wilson, H. A. (1905). *Phil. Trans.* **204,** 121.
—— and Wilson, M. (1913). *Proc. Roy. Soc.* **89,** 99.

Index